THE CULTURAL LANDSCAPE

An Introduction to HUMAN GEOGRAPHY

TENTH EDITION

James M. Rubenstein

Miami University, Oxford, Ohio

Prentice Hall

Boston Columbus Indianapolis New York San Francisco Upper Saddle River
Amsterdam Cape Town Dubai London Madrid Milan Munich Paris Montréal Toronto
Delhi Mexico City São Paulo Sydney Hong Kong Seoul Singapore Taipei Tokyo

Library of Congress Cataloging-in-Publication Data

Rubenstein, James M.
An Introduction to human geography : the cultural landscape / James M. Rubenstein.—10th ed.
p. cm.
Includes bibliographical references and index.
ISBN-13: 978-0-321-67735-8
ISBN-10: 0-321-67735-8
1. Human geography. I. Title. II. Title: Cultural landscape.
GF41.R82 2011
304.2—dc22

2009047535

Geography Editor: Christian Botting
Editor in Chief, Geosciences and Chemistry: Nicole Folchetti
Marketing Manager: Maureen McLaughlin
Project Manager, Editorial: Tim Flem
Assistant Editor: Jennifer Aranda
Editorial Assistant: Christina Ferraro
Marketing Assistant: Nicola Houston
Managing Editor, Geosciences and Chemistry: Gina M. Cheselka
Project Manager, Production: Shari Toron
Copy Editor: Pamela Rockwell
Proofreader: Jeff Georgeson
Senior Manufacturing and Operations Manager: Nick Sklitsis
Operations Specialist: Maura Zaldivar
Art Director: Mark Ong
Interior and Cover Designer: Emily Friel
Senior Technical Art Specialist: Connie Long
Spatial Graphics: Kevin Lear
Photo Research Manager: Elaine Soares
Photo Researcher: Regalle Jaramillo
Senior Media Producer: Angela Bernhardt
Media Producer: Victoria Prather
Senior Media Production Supervisor: Liz Winer
Media Production Coordinator: Shannon Kong
Full-Service/Composition: GGS Higher Education Resources
Full-Service Project Management: Cindy Miller/Saraswathi Muralidhar/GGS Higher Education Resources
Cover Photo Credit: epa/Corbis

Pearson Prentice Hall
Pearson Education, Inc.
Upper Saddle River, NJ 07458

Printed in the United States of America
10 9 8 7 6 5 4 3 2 1

ISBN: 0-321-67735-8 / 978-0-321-67735-8 [Student Edition]
ISBN: 0-321-69520-8 / 978-0-321-69520-8 [Books á la Carte]

Prentice Hall
is an imprint of

www.pearsonhighered.com

Brief Contents

About the Author

Dr. James M. Rubenstein received his Ph.D. from Johns Hopkins University in 1975. His dissertation on French urban planning was later developed into a book entitled *The French New Towns* (Johns Hopkins University Press). In 1976 he joined the faculty at Miami University, where he is currently Professor of Geography. Besides teaching courses on urban and human geography and writing textbooks, Dr. Rubenstein also conducts research in the automotive industry and has published three books on the subject, *The Changing U.S. Auto Industry: A Geographical Analysis* (Routledge); *Making and Selling Cars: Innovation and Change in the U.S. Auto Industry* (The Johns Hopkins University Press); and *Who Really Made Your Car? Restructuring and Geographic Change in the Auto Industry* (W.E. Upjohn Institute, with Thomas Klier). Originally from Baltimore, he is an avid Orioles fan. Winston, a lab mix, takes Dr. Rubenstein for a long walk in the woods every day.

This book is dedicated to the memory of Bernard Rubenstein, Dr. Rubenstein's father, the smartest man he knew. Dr. Rubenstein also gratefully thanks his wife Bernadette Unger and the rest of his family for their love and support.

Contents

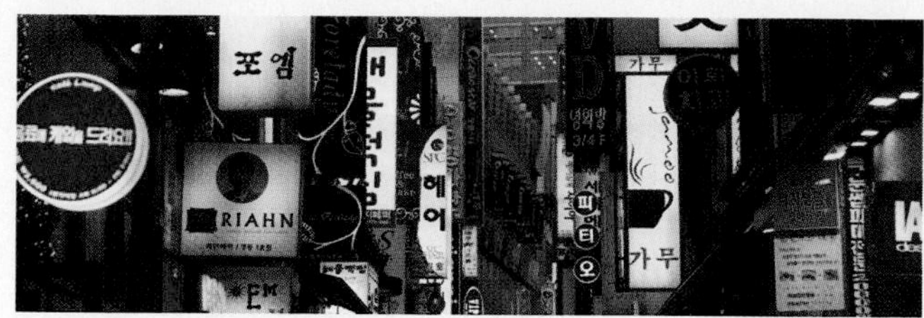

5 Language 134

6 Religion 168

lessons from
Rwanda
Genocide
Rwanda

9 Development 272

10 Agriculture 306

Wai Yan Dispensary

Preface

What is geography? Geography is the study of where things are located on Earth's surface and the reasons for the location. The word *geography,* invented by the ancient Greek scholar Eratosthenes, is based on two Greek words. *Geo* means "Earth," and *graphy* means "to write." Geographers ask two simple questions: where and why. Where are people and activities located across Earth's surface? Why are they located in particular places?

New to this Edition

This edition was prepared as the world faced its most severe economic challenges since the Great Depression of the 1930s. The effects of the recession will linger well into the second decade of the twenty-first century. The world's economy as a whole, as well as individual countries and regions, are being restructured.

- Economic challenges in the wake of the recession are discussed in several chapters. The geographic implications of the bankruptcy and U.S. government rescue of Chrysler and General Motors are addressed in Chapter 11 ("Industry"). Also discussed is evidence concerning distinctions between "American" and "foreign" cars, and future prospects for alternate fuel vehicles (see Chapter 14, "Resource Issues"). The author of this book is also the author of three books on the U.S. auto industry.
- The severe recession has affected many other topics in human geography. For example, migration within the United States has declined as a result of poor economic conditions (see Chapter 3, "Migration"). Chapter 12 ("Services") examines not only the impact of the recession on the service sector of the economy but also the role that the service sector played in bringing on the recession. Similarly, Chapter 13 ("Urban Patterns") discusses both the role and the impact of the housing market in the recession.
- Economic changes in the world extend beyond the effects of the severe recession. Most notable has been the economic growth of regions outside of North America and Europe, notably the so-called BRIC countries (Brazil, Russia, India, and China). The increasingly important role played by the BRIC countries is addressed in Chapter 11. Fresh information is provided on the origin of agriculture in Chapter 10 ("Agriculture"), on outsourcing to back offices in Chapter 12 ("Services"), and on millennium development goals in Chapter 9 ("Development").
- The world is also changing culturally. The map of regions of the world is being redrawn, with the combination of Western Europe and Eastern Europe into one region and the emergence of a distinctive region in Central Asia (see Chapter 9, "Development"). Conflicts in Sudan, Somalia, and Sri Lanka are addressed in Chapter 7 ("Ethnicity"), and in Afghanistan, Iran, Iraq, and Pakistan in Chapter 8 ("Political Geography").
- Chapter 2 ("Population") addresses the new realities of population change. Population growth has slowed in much of the world, but how does this slowing relate to changes in food supply and other resources? What fresh perspectives do geographers bring to popular culture, such as the diffusion of Facebook, and regional preferences for beverages (see Chapter 4, "Folk and Popular Culture")?
- Finally, online **Multimedia and Assessment** for *A Cultural Landscape* can be found at *www.mygeoscienceplace.com*, including a fully-integrated Pearson eText, numerous geography video clips, interactive maps, activities, assessments, flash cards, RSS feeds, and additional resources.

Geography as a Social Science

The main purpose of this book is to introduce students to the study of geography as a social science by emphasizing the relevance of geographic concepts to human problems. It is intended for use in college-level introductory human or cultural geography courses, as well as the equivalent advanced placement course in high school.

A central theme in this book is a tension between two important themes—globalization and cultural diversity. In many respects, we are living in a more unified world economically, culturally, and environmentally. The actions of a particular corporation or country affect people around the world.

For example, geographers examine the prospects for an energy crisis by relating the distributions of energy production and consumption. Geographers find that the users of energy are located in places with different social, economic, and political institutions than the producers of energy. The United States and Japan consume far more energy than they produce, whereas Russia and Saudi Arabia produce far more energy than they consume.

This book argues that after a period when globalization of the economy and culture has been a paramount concern in geographic analysis, local diversity now demands equal time. People are taking deliberate steps to retain distinctive cultural identities. They are preserving little-used languages, fighting fiercely to protect their religions, and carving out distinctive economic roles.

Local diversity even extends to addressing issues, such as the energy crisis, that at first glance are considered global. For example, Israel is working with the French carmaker Renault and the Silicon Valley company Project Better Place to encourage electric vehicles by installing tens of thousands of recharging stations. Brazil has passed laws to require more use of biofuels, produced from crops grown in Brazil and processed in factories there.

Meanwhile, the United Arab Emirates has invested in a subway system as an alternative to motor vehicles, even though the country is one of the world's leading producers of petroleum.

Divisions Within Geography

Because geography is a broad subject, some specialization is inevitable. At the same time, one of geography's strengths is its diversity of approaches. Rather than being forced to adhere rigorously to established disciplinary laws, geographers can combine a variety of methods and approaches. This tradition stimulates innovative thinking, although students who are looking for a series of ironclad laws to memorize may be disappointed.

Human Versus Physical Geography

Geography is both a physical and a social science. When geography concentrates on the distribution of physical features, such as climate, soil, and vegetation, it is a natural science. When it studies cultural features, such as language, industries, and cities, geography is a social science. This division is reflected in some colleges, where physical geography courses may carry natural science credit while human and cultural geography courses carry social science credit.

While this book is concerned with geography from a social science perspective, one of the distinctive features of geography is its use of natural science concepts to help understand human behavior. The distinction between physical and human geography reflects differences in emphasis, not an absolute separation.

Topical Versus Regional Approach

Geographers face a choice between a topical and a regional approach. The topical approach, which is used in this book, starts by identifying a set of important cultural issues to be studied, such as population growth, political disputes, and economic restructuring. Geographers using the topical approach examine the location of different aspects of the topic, the reasons for the observed pattern, and the significance of the distribution.

The alternative approach is regional. Regional geographers select a portion of Earth and studying the environment, people, and activities within the area. The regional geography approach is used in courses on Europe, Africa, Asia, and other areas of the world. Although this book is organized by topics, geography students should be aware of the location of places in the world. A separate index section lists the book's maps by location. One indispensable aid in the study of regions is an atlas, which can also be used to find unfamiliar places that may pop up in the news.

Descriptive Versus Systematic Method

Whether using a topical or a regional approach, geographers can select either a descriptive or a systematic method. Again, the distinction is one of emphasis, not an absolute separation. The descriptive method emphasizes the collection of a variety of details about a particular location. This method has been used primarily by regional geographers to illustrate the uniqueness of a particular location on Earth's surface. The systematic method emphasizes the identification of several basic theories or techniques developed by geographers to explain the distribution of activities.

This book uses both the descriptive and systematic methods because total dependence on either approach is unsatisfactory. An entirely descriptive book would contain a large collection of individual examples not organized into a unified structure. A completely systematic approach suffers because some of the theories and techniques are so abstract that they lack meaning for the student. Geographers who depend only on the systematic approach may have difficulty explaining important contemporary issues.

Outline

The book discusses the following main topics:

- What basic concepts do geographers use? Chapter 1 provides an introduction to ways that geographers think about the world. Geographers employ several concepts to describe the distribution of people and activities across Earth, to explain reasons underlying the observed distribution, and to understand the significance of the arrangements.
- Where are people located in the world? Chapters 2 and 3 examine the distribution and growth of the world's population, as well as the movement of people from one place to another. Why do some places on Earth contain large numbers of people or attract newcomers while other places are sparsely inhabited?
- How are different cultural groups distributed? Chapters 4 through 8 analyze the distribution of different cultural traits and beliefs and the problems that result from those spatial patterns. Important cultural traits discussed in Chapter 4 include food, clothing, shelter, and leisure activities. Chapters 5 through 7 examine three main elements of cultural identity: language, religion, and ethnicity. Chapter 8 looks at political problems that arise from cultural diversity. Geographers look for similarities and differences in the cultural features at different places, the reasons for their distribution, and the importance of these differences for world peace.
- How do people earn a living in different parts of the world? Human survival depends on acquiring an adequate food supply. One of the most significant distinctions in the world is whether people produce their food directly from the land or buy it with money earned by performing other types of work. Chapters 9 through 12 look at the three main ways of earning a living: agriculture, manufacturing, and services. Chapter 13 discusses cities, the centers for economic as well as cultural activities.
- What issues result from using Earth's resources? The final chapter is devoted to a study of issues related to the use of Earth's natural resources. Geographers recognize that cultural problems result from the depletion, destruction, and inefficient use of the world's natural resources.

A Cultural Landscape
A STRUCTURED LEARNING PATH

Key Issues in Human Geography

Each chapter is organized around a set of three or four key issues. All issues include one of the two key geographic concerns: where or why? These questions reappear as major headings within the chapter and are revisited in the end-of-chapter Summary.

KEY ISSUES

1 Why Do People Migrate?
2 Where Are Migrants Distributed?
3 Why Do Migrants Face Obstacles?
4 Why Do People Migrate Within a Country?

KEY ISSUE 1
Why Do People Migrate?

- Reasons for Migrating
- Distance of Migration
- Characteristics of Migrants

Geography has no comprehensive theory of migration, although a nineteenth-century outline of 11 migration "laws" written by E. G. Ravenstein is the basis for contemporary geographic migration studies. To understand where and why migration occurs, Ravenstein's "laws" can be organized into three groups: the reasons why migrants move, the distance they typically move, and their characteristics. Each of these elements is addressed in this section of the chapter. ■

SUMMARY

Migration plays a major role in determining population change. For countries in stage 4 of the demographic transition, net in-migration is the only source of population growth. On the other hand, for countries in stage 2 of the demographic transition, net out-migration reduces the stress of a high NIR.

For example, Spain, in stage 4 of the demographic transition, has around 400,000 births and 400,000 deaths annually, and therefore a natural increase rate of around 0. But the country has had net in-migration of around 400,000 per year, thereby generating a total population growth of around 1 percent.

Cape Verde, described in Chapter 2 as a country in stage 2 of the demographic transition, has around 10,000 births and 2,500 deaths per year, producing a natural increase of 7,500. But Cape Verde's total annual population growth is actually only 2,500 because of net-outmigration of 5,000. Similarly, Jamaica has around 58,000 births and 18,000 deaths, but total annual growth is only 20,000 rather than 40,000, again because of net out-migration.

Here again are the key issues for Chapter 3:

1. **Why Do People Migrate?** We can group the reasons into push and pull factors. People feel compelled (pushed) to emigrate from a location for political, economic, and environmental reasons. Similarly, people are induced (pulled) to immigrate because of the political, economic, or environmental attractiveness of a new location. We can also distinguish between international and internal migration.
2. **Where Are Migrants Distributed?** On a global scale, the largest flows of migrants are from Asia to Europe and from Asia and Latin America to the United States. The United States receives by far the largest number of migrants.
3. **Why Do Migrants Face Obstacles?** Migrants have difficulty getting permission to enter other countries, and they face hostility from local citizens once they arrive. Immigration laws restrict the number who can legally enter the United States. In Europe and the Middle East, guest workers migrate temporarily to perform menial jobs.
4. **Why Do People Migrate Within a Country?** We can distinguish between interregional and intraregional migration within an individual country. Historically, interregional migration was especially important in settling the frontier of large countries such as the United States, Russia, and Brazil. The most important intraregional migration trends are from rural to urban areas within LDCs and from cities to suburbs within MDCs.

Case Studies

Each chapter opens with a case study that illustrates some of the key concepts presented in the text. The case studies are generally drawn from news events or issues that will interest students. Each case study is revisited at the end of the chapter, where additional related information may be used to reinforce some of the main points.

CASE STUDY / Population Growth in India

The Phatak family lives in a village of 600 inhabitants in India. At age 40, Indira Phatak has been pregnant five times. Four of her children have survived; they are aged 5 to 18.

When the two Phatak daughters marry a few years from now, how many children will each of them bear? The Indian government hop
than their moth
year in India, an
lion annually. U
in the next few y
would reach 1.8

Three-fourths
fewer than 5,00
dren are an eco
on the farm an
their old age. T

before they reach working age also encourages large families. One out of every 18 infants in India dies within 1 year of birth, and 1 out of every 14 pregnant women dies annually during pregnancy and childbirth.

In recent years, India has made significant progress in dif-

CASE STUDY REVISITED / India Versus China

The world's two most populous countries, China and India, will heavily influence future prospects for global overpopulation (Figure 2-34). These two countries—together encompassing more than one-third of the world's population—have adopted different family-planning programs. As a result of less effective policies, India adds 12 million more people each year than does China. Current projections show that India would surpass China as the world's most populous country around 2030.

India's Population Policies

India, like most countries in Africa, Asia, and Latin America, remained in stage 1 of the demographic transition until the late 1940s. During the first half of the twentieth century, population increased modestly—less than 1 percent per year—and even decreased in some years because of malaria, famines, plagues, and cholera epidemics.

Immediately following independence from England in 1947, India's death rate declined sharply (to 20 per 1,000 in 1951), whereas the CBR remained relatively high (about 40). Consequently, the NIR jumped to 2 percent per year. In response to this rapid growth, India became the first country to embark on a national family-planning program, in 1952. The government has established clinics and has provided information about alternative methods of birth control. Birth-control devices have been distributed free or at subsidized prices. Abortions, legalized in 1972, have been performed at a rate of several million per year. All together, the government spends several hundred million dollars annually on various family-planning programs.

FIGURE 2-34 Parents and their only child in Beijing, China.

(Continued)

Key Terms & Thinking Geographically

The key terms in each chapter are indicated in bold type when they are introduced. These terms are also defined at the end of the chapter and again at the end of the book.

The Thinking Geographically end-of-chapter section offers five questions based on concepts and themes developed in the chapter. The questions help students apply geographic concepts to explore issues more intensively.

Global Forces, Local Impacts

Each chapter has a one-page box that explores in depth an issue related to the interplay between the cultural and economic forces that push toward greater global unity while at the same time preserving a diverse local landscape.

GLOBAL FORCES, LOCAL IMPACTS
Hurricane Katrina

Because geographers are trained in both social and physical sciences, they are particularly well equipped to understand interactions between people and their environment. An example is the devastation in the southern United States after Hurricane Katrina in 2005.

Physical geography concepts explain the process by which hurricanes, such as Katrina, form in the Atlantic Ocean during the late summer and autumn and gather strength over the warms waters of the Gulf of Mexico. When it passes over land, a hurricane can generate a powerful storm surge that floods low-lying areas (Figure 1-18, left).

It is here that physical and human geography intersect. Katrina caused massive damage, in part because it made landfall near heavily populated areas, including the cities of Biloxi and Gulfport, Mississippi; Mobile, Alabama; and New Orleans, Louisiana. In an effort to protect these low-lying cities from flooding, government agencies constructed a complex system of levees, dikes, seawalls, canals, and pumps. The experience of Katrina proved that humans are not able to control and tame all of the forces of nature.

Human geographers are especially concerned with the uneven impact of destruction. Hurricane Katrina's victims were primarily poor, African American, and older individuals (Figure 1-18, center). They lived in the lowest-lying areas, most vulnerable to flooding, and many lacked transportation, money, and information that would have enabled them to evacuate in advance of the storm.

The wealthy portions of New Orleans, such as tourist attractions like the French Quarter, were spared the worst because they were located on slightly higher ground. The slow and incompetent response to the destruction by local, state, and federal emergency teams was attributed by many analysts to the victims' lack of a voice in the political, economic, and social life of New Orleans and other impacted communities.

Inequalities persist several years after the hurricane (Figure 1-18, right). In 2009, four years after the hurricane, more than 90 percent of householders in predominantly white areas were back in their homes, as measured by whethe contras househ African receivi

FIGURE 1-18 Cultural ecology: New Orleans after Hurricane Katrina. From a physical geography perspective, 80 percen city's flood-protection levees broke (left). The 20 percent that was not flooded was land at slightly higher elevations, inclu Vieux Carre (French Quarter). From a social science perspective, at the time of the hurricane two-thirds of the populat (middle). However, the population in the area that was not flooded was less than one-fourth African American. The percen reoccupied since the hurricane is lower in the areas that had relatively large African American populations (right).

Contemporary Geographic Tools

Each chapter also contains a box that examines how geographic tools, such as geographic information systems, aerial photography, and remotely-sensed images, have been used to resolve—or at least understand—cultural, political, and economic issues discussed in the chapter.

CONTEMPORARY GEOGRAPHIC TOOLS
English on the Internet

English was the dominant language of the Internet during the 1990s. In 1998, 71 percent of people online were using English (Figure 5-26). The early dominance of English on the Internet was partly a reflection of the fact that the most populous English-speaking country, the United States, had a head start on the rest of the world in making the Internet available to most of its citizens (refer to Figure 4-15).

English continued as the leading Internet language in the first years of the twenty-first century, but it was far less dominant. The percentage of English-language online users declined from 71 percent in 1998 to 29 percent in 2008. Chinese (Mandarin) language online users increased from 2 percent of the world total in 1998 to 20 percent in 2008. English may be less dominant as the language of the Internet in the twenty-first century. But the United States remains the Internet leader in key respects—and with it the English language.

The United States created the English-language nomenclature for the Internet that the rest of the world has followed. The designation "www," which English speakers recognize as an abbreviation of "World Wide Web," is awkward in other languages, most of which do not have an equivalent sound to the English "w." In French, for example, "w" is pronounced "doo-blah-vay."

The U.S.-based Internet Corporation for Assigned Names and Numbers (ICANN) has been responsible for assigning domain names and for the suffixes following the dot, such as "com" and "edu." Domain names in the rest of the world include a two-letter suffix for the country, such as "fr" for France and "jp" for Japan, whereas U.S.-based domain names don't need the suffix.

U.S.-based companies provide the principal search engines for Internet users everyone. In 2007, 67 percent of all searches worldwide used Google. Second place was another U.S.-based company, Yahoo, with 15 percent. These companies offer search engines in languages other than English. Google was heavily criticized when its Mandarin-language Google.cn was designed to block web sites deemed unsuitable by China's government. ■

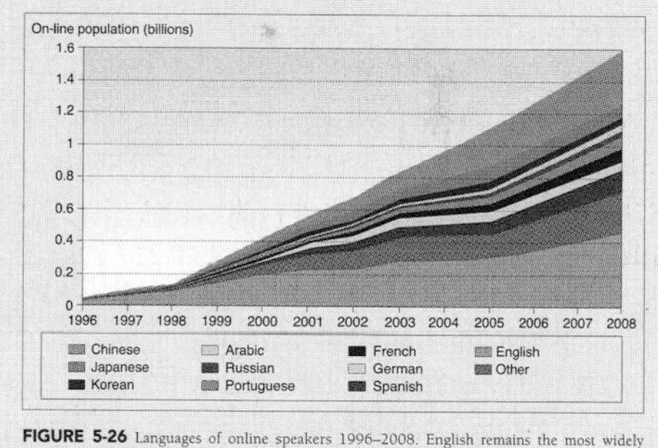

FIGURE 5-26 Languages of online speakers 1996–2008. English remains the most widely used language on the Internet, but Chinese is growing more rapidly.

Resources

Key Internet sites related to the topic are provided at the end of each chapter, as well as a list of books and articles for students who wish to study the subject further.

Premium Web Site

Includes an eText version of the textbook with linked/integrated multimedia, assignable geography videos with associated assessments, Interactive Maps, Flashcards, RSS Feeds, weblinks, annotated resources for further exploration, and Class Manager & GradeTracker functionality for instructors **www.mygeoscienceplace.com.**

The Teaching and Learning Package

In addition to the text itself, the author and publisher have been pleased to work with a number of talented people to produce an excellent instructional package.

Instructor Resources

- ***Instructor Manual*** (0-321-68217-3) Each chapter of the *Instructor Manual* opens with a specific Introduction highlighting core learning objectives presented in the specific chapter, Icebreakers to start classroom discussion, Challenges to Comprehension, assignments of Review/Reflection Questions and Demographic Data Collection and Analysis, and Additional Resources to examine during classroom sessions or to assign to students.
- ***TestGen®/Test Bank*** (0-321-68216-5) TestGen® is a computerized test generator that lets instructors view and edit *Test Bank* questions, transfer questions to tests, and print the test in a variety of customized formats. This *Test Bank* includes approximately 1,000 multiple-choice, true/false, and short answer/essay questions. Questions correlate to the U.S. National Geography Standards and Bloom's Taxonomy to help instructors better map the assessments against both broad and specific teaching and learning objectives. The *Test Bank* is also available in Microsoft Word© and is importable into Blackboard and WebCT.
- ***Instructor Resource Center on DVD*** (0-321-66740-9) Everything instructors need where they want it. The Pearson Prentice Hall Instructor Resource Center helps make instructors more effective by saving them time and effort. All digital resources can be found in one, well-organized, easy-to-access place. The IRC on DVD includes:
    - All textbook images as JPEGs, PDFs, and PowerPoint™ presentations
    - Preauthored Lecture Outline PowerPoint™ presentations, which outline the concepts of each chapter with embedded art and can be customized to fit instructors' lecture requirements
    - CRS "Clicker" Questions in PowerPoint™ format, mapped against the U.S. National Geography Standards and Bloom's Taxonomy
    - The TestGen software, *Test Bank* questions, and answers for both MACs and PCs
    - Electronic files of the *Instructor Manual* and *Test Bank*
    - Over 120 Geography Video Clips

This Instructor Resource content is also available completely online via the Instructor Resources section of **www.mygeoscienceplace.com** and **www.pearsonhighered.com/irc**.

- ***Online Course Management Systems*** Pearson Prentice Hall offers content specific to *The Cultural Landscape* in the BlackBoard and CourseCompass course management system platforms. Each of these platforms lets the instructor easily post his or her syllabus, communicate with students online or off-line, administer quizzes, and record student results and track their progress. Please call your local Pearson Prentice Hall representative for details. **http://www.pearsonhighered.com/elearning/**.
- **Television for the Environment *Earth Report* Geography Videos on DVD** (0-321-66298-9) This three-DVD set is designed to help students visualize how human decisions and behavior have affected the environment and how individuals are taking steps toward recovery. With topics ranging from the poor land management promoting the devastation of river systems in Central America, to the struggles for electricity in China and Africa, these 13 videos from Television for the Environment's global *Earth Report* series recognize the efforts of individuals around the world to unite and protect the planet.
- **Television for the Environment *Life* Human Geography Videos on DVD** (0-132-41656-5) This three-DVD set is designed to enhance any human geography course. These DVDs include 14 full-length video programs from Television for the Environment's global *Life* series, covering a wide array of issues affecting people and places in the contemporary world, including the serious health risks of pregnant women in Bangladesh, the social inequalities of the "untouchables" in the Hindu caste system, and Ghana's struggle to compete in a global market.
- **Television for the Environment *Life* World Regional Geography Videos on DVD** (0-131-59348-X) From Television for the Environment's global *Life* series this two-DVD set brings globalization and the developing world to the attention of any world regional geography course. These 10 full-length video programs highlight matters such as the growing number of homeless children in Russia, the lives of immigrants living in the United States trying to aid family still living in their native countries, and the European conflict between commercial interests and environmental concerns.
- ***Aspiring Academics: A Resource Book for Graduate Students and Early Career Faculty*** (0-136-04891-9) Drawing on several years of research, this set of essays is designed to help graduate students and early career faculty start their careers in geography and related social and environmental sciences. This teaching aid stresses the interdependence of teaching, research, and service—and the importance of achieving a healthy balance in professional and personal life—in faculty work, and does not view it as a collection of unrelated tasks. Each chapter provides accessible, forward-looking advice on topics that often cause the most stress in the first years of a college or university appointment.
- ***Teaching College Geography: A Practical Guide for Graduate Students and Early Career Faculty*** (0-136-05447-1) Provides a starting point for becoming an effective geography teacher from the very first day of class. Divided in two parts, the first set of chapters addresses "nuts-and-bolts" teaching issues in the context of the new technologies, student demographics, and institutional expectations that are the hallmarks of higher education in the twenty-first century. The second part explores other

important issues: effective teaching in the field; supporting critical thinking with GIS and mapping technologies; engaging learners in large geography classes; and promoting awareness of international perspectives and geographic issues.

- ***AAG Community Portal for Aspiring Academics and Teaching College Geography*** This web site is intended to support community-based professional development in geography and related disciplines. Here you will find activities providing extended treatment of the topics covered in both books. The activities can be used in workshops, graduate seminars, brown bags, and mentoring programs offered on campus or within an academic department. You can also use the discussion boards and contributions tool to share advice and materials with others. **http://pearsonhighered.com/aag/**.

Student Resources

- ***Premium Web Site*** A dedicated Premium Web Site with eText offers a variety of resources for students and professors, including an eText version of the textbook with linked/integrated multimedia, geography videos with associated assessments, Interactive Maps, Flashcards, RSS Feeds, weblinks, annotated resources for further exploration, and Class Manager & GradeTracker Gradebook functionality for instructors. Available at **www.mygeoscienceplace.com**.
- ***Study Guide*** (0-321-68173-8) Used as a framework to outline the chapters to encourage a full grasp of concepts and as a means to integrate the material with course notes. The basic design goal is to provide a uniformly distilled version of the text to help students recall classroom discussions and lectures in preparation for exams. Features include chapter introduction, a detailed breakdown of key issues, and a review of key terms. For assessment of material, different types of questions will test comprehension beyond memorization.
- ***Goode's World Atlas 22nd Edition*** (0-321-65200-2) Goode's World Atlas has been the world's premier educational atlas since 1923, and for good reason. It features over 250 pages of maps, from definitive physical and political maps to important thematic maps that illustrate the spatial aspects of many important topics. The 22nd edition includes 160 pages of new, digitally produced reference maps, as well as new thematic maps on global climate change, sea level rise, CO_2 emissions, polar ice fluctuations, deforestation, extreme weather events, infectious diseases, water resources, and energy production.
- ***Encounter Human Geography*** **Workbook and Premium Web Site** (0-321-68220-3) *Encounter Human Geography* provides rich, interactive explorations of human geography concepts through Google Earth™ explorations. All chapter explorations are available in print format as well as in online quizzes, accommodating different classroom needs. All worksheets are accompanied by corresponding Google Earth™ media files, available for download from **www.mygeoscienceplace.com**.
- ***Encounter Earth: Interactive Geoscience Explorations*** **Workbook and Premium Web Site** (0-321-58129-6) Ideal for professors who want to integrate Google Earth™ in their classrooms, *Encounter Earth* gives students a way to visualize key topics in their introductory geoscience courses. Each exploration consists of a worksheet, available in the workbook and as a PDF file, and a Google Earth™ KMZ file, containing placemarks, overlays, and annotations referred to in the worksheets. The accompanying *Encounter Earth* Premium WebSite is located at **www.mygeoscienceplace.com**.
- ***Dire Predictions*** (978-0-136-04435-2) Periodic reports from the Intergovernmental Panel on Climate Change (IPCC) evaluate the risk of climate change brought on by humans. But the sheer volume of scientific data remains inscrutable to the general public, particularly to those who may still question the validity of climate change. In just over 200 pages, this practical text presents and expands upon the essential findings in a visually stunning and undeniably powerful way to the lay reader. Scientific findings that provide validity to the implications of climate change are presented in clear-cut graphic elements, striking images, and understandable analogies.

Suggestions for Use

This book can be used in an introductory human or cultural geography course that extends over one semester, one quarter, or two quarters. An instructor in a one-semester course could devote one week to each of the chapters, leaving time for examinations. In a one-quarter course, the instructor might need to omit some of the book's material.

A course with more of a cultural orientation could use Chapters 1 through 8, plus Chapter 14. If the course has more of an economic orientation, then the appropriate chapters would be 1 through 3 and 8 through 14.

A two-quarter course could be organized around the culturally oriented Chapters 1 through 8 during the first quarter and the more economically oriented Chapters 9 through 14 during the second quarter. Topics of particular interest to the instructor or students could be discussed for more than one week.

Acknowledgments

A major reason for the long-term success of this book has been the quality and stability of leadership in geography at Pearson Education. Two individuals have served as geography editors for most of the past three decades. Paul F. Corey, who guided development of the third, fourth, and fifth editions of this book, is now President of Science at Pearson. Dan Kaveney, who guided development of the sixth, seventh, eighth, and ninth editions, is now Chemistry Publisher at Pearson. Because Pearson is the dominant publisher of college geography textbooks, the person in charge of geography wields considerable influence in shaping what is taught in the nation's geography curriculum. I will always value the sound judgment, outstanding vision, and friendship of both Paul and Dan.

Now it is time to welcome a new team to Pearson geography. Christian Botting, Acquisitions Editor for Geography and Atmospheric Sciences, has helped to reinvigorate Pearson's geography program with fresh, creative perspectives from his post in Boston, which non-geographers, of course, regard as the "Hub" of the nation. Tim Flem, Geography Project Manager, has managed three projects with me now. Tim can solve any problem.

In this age of outsourcing, Pearson works with many independent companies to create books. This edition has been the beneficiary of a top-notch team. Cindy Miller, Publishing Services Director at GGS Higher Education Resources, coordinated the flow of production work to the author. Cindy was especially helpful in sorting out the correct order to undertake various tasks. Saraswathi Muralidhar, GGS Production Editor, handled the copyediting work with sensitivity. Regalle Jaramillo found the neat photos. Kevin Lear and his team at Spatial Graphics produced outstanding maps for this book. Shari Toron, Project Manager for Science at Pearson, coordinated the production work at the Pearson end. I am also grateful for the great work done on a variety of ancillaries by Owen Dwyer, Indiana University; Tim Scharks, Green River Community College; Gillian Acheson, Southern Illinois University, Edwardsville and Eike Reichardt, Lehigh Carbon Community College.

10th Edition Reviewers

I would like to extend a special thanks to all of my colleagues who have, over the years, offered a good deal of feedback and constructive criticism. Colleagues who served as reviewers as we prepared the 10th edition are Andrew Baker, Purdue University: Indianapolis; Don Ziegler, Old Dominion University; Mark Goodman, Grossmont College; Patricia Trusty, Tulsa Community College; Roger M. Selya, University of Cincinnati; and Thomas M. Tharp, Purdue University.

About Our Sustainability Initiatives

This book is carefully crafted to minimize environmental impact. The materials used to manufacture this book originated from sources committed to responsible forestry practices. The interior page is FSC certified. The printing, binding, cover, and paper come from facilities that minimize waste, energy consumption, and the use of harmful chemicals.

Pearson closes the loop by recycling every out-of-date text returned to our warehouse. We pulp the books, and the pulp is used to produce items such as paper coffee cups and shopping bags. In addition, Pearson aims to become the first climate neutral educational publishing company.

The future holds great promise for reducing our impact on Earth's environment, and Pearson is proud to be leading the way. We strive to publish the best books with the most up-to-date and accurate content, and to do so in ways that minimize our impact on Earth.

©
FSC
Mixed Sources
Product group from well-managed forests, controlled sources and recycled wood or fibre
Cert no. SW-COC-002985
www.fsc.org
© 1996 Forest Stewardship Council

Prentice Hall
is an imprint of

CHAPTER 1

Basic Concepts

What do you expect from this geography course? You may think that geography involves memorizing lists of countries and capitals or exports and imports. Perhaps you associate geography with photographic essays of exotic places in popular magazines. Contemporary geography is the scientific study of the location of people and activities across Earth and the reasons for their distribution

Geographers ask "where" things are and "why" they are there. Historians organize material by time, because they understand that action at one point in time can result from past actions and can affect future ones. Geographers organize material by place, because they understand that something happening at one place can result from

KEY ISSUES

1 How Do Geographers Describe Where Things Are?

2 Why Is Each Point on Earth Unique?

3 Why Are Different Places Similar?

something that happened elsewhere and can affect conditions at other places. Historians study the logical sequence of human activities in time, whereas geographers study the logical arrangement of human activities in space.

As in all sciences, the study of geography requires understanding some basic concepts. For example, the definition of geography in the first paragraph included the words *location* and *distribution*. We use these words commonly in daily speech, but geographers give them precise meanings. This first chapter introduces how human geographers think about the world.

Geographers observe that people are being pulled in opposite directions by two factors—***globalization*** and ***local diversity***. Modern communications and technology have fostered globalization, pulling people into greater cultural and economic interaction with others. At the same time, people are searching for more ways to express their unique cultural traditions and economic practices. Tensions between the simultaneous geographic trends of globalization and local diversity underlie many of the world's problems that geographers study, such as political conflicts, economic uncertainty, and environmental management.

Traffic, Kolkata (Calcutta), India.

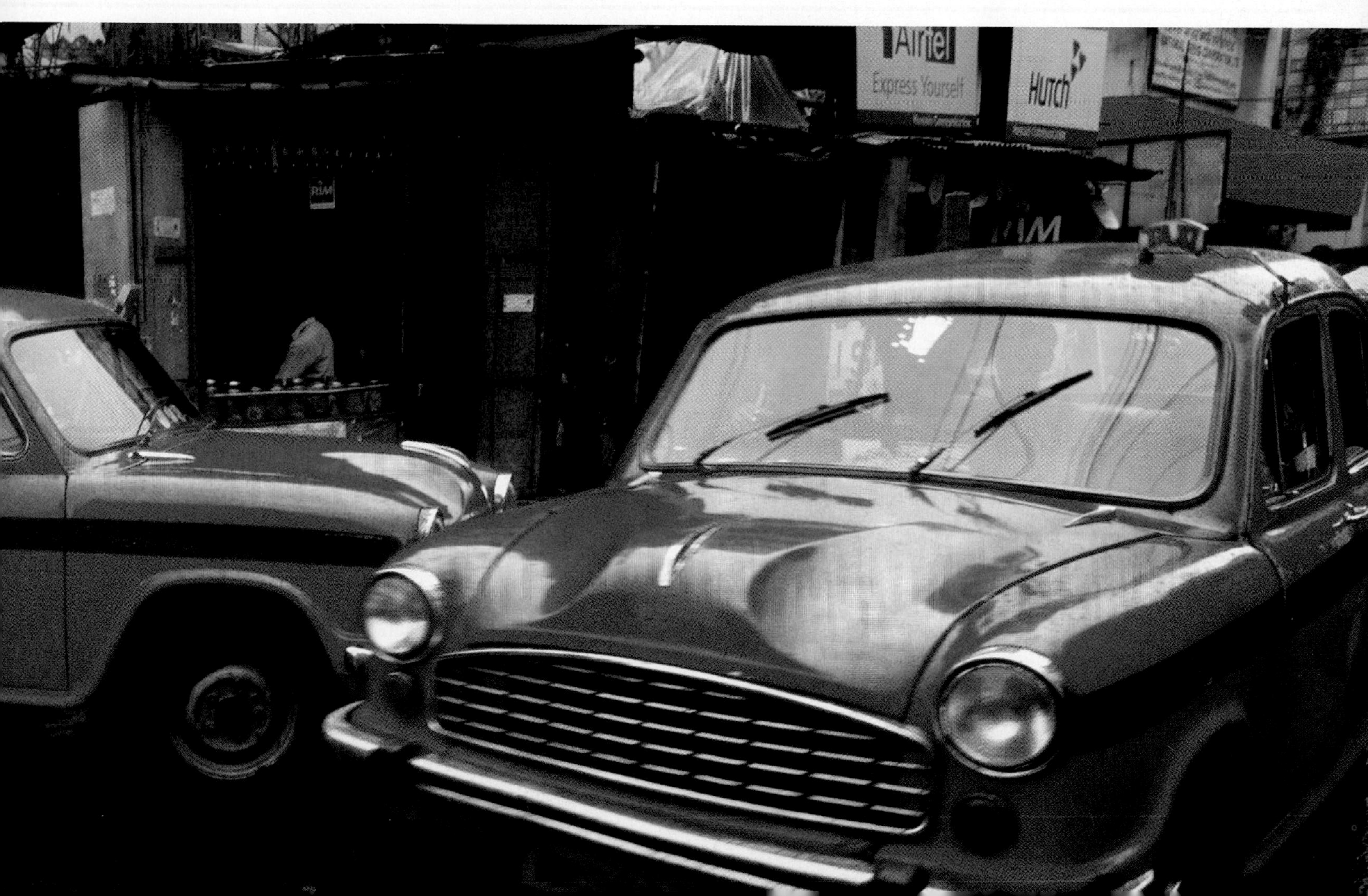

CASE STUDY / Big Mac Attack

If when driving across the vast expanses of the United States on an interstate highway you are hit by hunger pangs, you are unlikely to be thinking about geography. At the next interchange you scan the horizon for fast-food restaurant signs, again in vain.

Now, very hungry, you are again disappointed at the second interchange. Finally, as you approach the third interchange, you spot a familiar image atop a very large pole—McDonald's "golden arches." When you drive up the ramp from the highway to the local road, you are confronted with a choice of a half-dozen fast-food restaurants. Very annoyed, you wonder why none of these establishments were located at the two previous interchanges.

Why cluster a half dozen at a single interchange instead of dispersing one or two at each interchange? Now you are asking questions about geography. Geographers ask where things are located and why.

Geographers are interested in the location of McDonald's restaurants around the world, not just around a U.S. interstate exit. The spread of McDonald's from a single establishment in Des Plaines, Illinois, in 1955, to 32,000 establishments worldwide reflects what for many human geographers was the defining trend of the late twentieth century—globalization of economy and culture. Human geographers are interested in understanding the economic and cultural conditions that permitted, and even encouraged, companies such as McDonald's to spread around the world during that time. Especially significant for some human geographers is the prominent role played by corporations such as Coca-Cola, Toyota, and Microsoft in the creation of a global economy and culture.

In the twenty-first century, human geographers also recognize that global forces have not eliminated local diversity in economic conditions and cultural preferences. McDonald's success has been built on many individual decisions concerning the local economy and culture. The company encourages local operators to tailor menu items to local tastes—such as India's Maharaja Mac (with lamb patties) and Uruguay's McHuevo (hamburger with poached egg)—and it avoids countries where few people can afford its meals.

Human geography is an especially exciting subject in the twenty-first century because of the constant interplay between the common and the exotic, between global forces and local distinctiveness. Every McDonald's—every place on Earth—is in some way tied to a global economy and culture, yet at the same time reflects certain characteristics that are unlike anywhere else. ■

The word *geography*, invented by the ancient Greek scholar Eratosthenes, is based on two Greek words. *Geo* means "Earth," and *graphy* means "to write."

Thinking geographically is one of the oldest human activities. Perhaps the first geographer was a prehistoric human who crossed a river or climbed a hill, observed what was on the other side, returned home to tell about it, and scratched the route in the dirt. Perhaps the second geographer was a friend or relative who followed the dirt map to reach the other side. Today, geographers are still trying to reach the other side, to understand more about the world in which we live. Geography is the study of where things are found on Earth's surface and the reasons for the location. Human geographers ask two simple questions: Where are people and activities found on Earth? Why are they found there?

Geography is divided broadly into two categories—*human* geography and *physical* geography—and within each category, slightly different "where" and "why" questions are addressed.

- **Human geography** is the study of where and why human activities are located where they are—for example, religions, businesses, and cities.
- **Physical geography** studies where and why natural forces occur as they do—for example, climates, landforms, and types of vegetation.

This book focuses on human geography, but it never forgets Earth's atmosphere, land, water, vegetation, and other living creatures. Relationships such as these between humans and nature will be examined throughout. The final chapter of the book will explicitly tie human activities to the physical environment.

To introduce human geography, we concentrate on two main features of human behavior—culture and economy. The first half of the book explains why the most important cultural features, such as major languages, religions, and ethnicities, are arranged as they are across Earth. The second half of the book looks at the locations of the most important economic activities, including agriculture, manufacturing, and services.

This first chapter introduces basic concepts that geographers employ to address their "where" and "why" questions. Many of these concepts are words commonly employed in English but given particular meaning by geographers. The first key issue in this chapter looks at geography's most important tool—mapping. A **map** is a two-dimensional or flat-scale model of Earth's surface, or a portion of it. Geography is immediately distinguished from other disciplines by its reliance on maps to display and analyze information (Figure 1-1).

The second and third sections of this chapter look at basic concepts geographers use to ask two principal "why" questions. First, geographers want to know why each place on Earth is in some ways unique. For example, why do people living close to each other speak different languages and employ different methods of agriculture? Geographers use two basic concepts to explain why every place is unique—place and region.

FIGURE 1-1 Satellite image of the world. The composite image was assembled by the Geosphere Project of Santa Monica, California. Thousands of images were recorded over a ten-month period by satellites of the National Oceanographic and Atmospheric Administration. The images were then electronically assembled, much like a jigsaw puzzle.

- A **place** is a specific point on Earth distinguished by a particular characteristic. Every place occupies a unique location, or position, on Earth's surface, and geographers have many ways to identify location.
- A **region** is an area of Earth distinguished by a distinctive combination of cultural and physical features. Human geographers are especially concerned with the cultural features of a group of people in a region—their body of beliefs and traditions, as well as their political and economic practices.

The third key issue in this chapter looks at geography's other main "why" question. Geographers want to know why different places on Earth have similar features. For example, why do people living far apart from each other practice the same religion and earn a living in similar ways?

Three basic concepts—scale, space, and connections—help geographers explain why these similarities do not result from coincidence.

- **Scale** is the relationship between the portion of Earth being studied and Earth as a whole. Although geographers study every scale from the individual to the entire Earth, increasingly they are concerned with global-scale patterns and processes.
- **Space** refers to the physical gap or interval between two objects. Geographers observe that many objects are distributed across space in a regular manner, for discernable reasons.
- **Connections** are relationships among people and objects across the barrier of space. Geographers are concerned with the various means by which connections occur.

KEY ISSUE 1

How Do Geographers Describe Where Things Are?

- Maps
- Contemporary Tools

Geography's most important tool for thinking spatially about the distribution of features across Earth is a map: "[B]efore travel began a map existed first" (Zbigniew Herbert, "Home," in *Still Life with a Bridle*).

As you turn the pages of this book, the first thing you may notice is the large number of maps—more than 200. These maps range in size from small boxes covering part of a city (Figure 2-31) to two-page spreads of the entire world (Figures 5-16 and 6-2). Some are highly detailed, with complex colors, lines, points, and shadings, whereas others seem highly generalized and unrealistic. For centuries, geographers have worked to perfect the science of mapmaking, called **cartography**. Contemporary cartographers are assisted by computers and satellite imagery. ■

Maps

A map is a scale model of the real world, made small enough to work with on a desk or computer. It can be a hasty here's-how-to-get-to-the-party sketch, an elaborate work of art, or a precise computer-generated product. A map serves two purposes: It is a tool for storing reference material and a tool for communicating geographic information.

- **As a reference tool.** A map helps us to find the shortest route between two places and to avoid getting lost along the way. We consult maps to learn where in the world something is found, especially in relation to a place we know, such as a town, body of water, or highway. The maps in an atlas or a road map are especially useful for this purpose.
- **As a communications tool.** A map is often the best means for depicting the distribution of human activities or physical features, as well as for thinking about reasons underlying a distribution.

A series of maps of the same area over several years can reveal dynamic processes at work, such as human migration or spread of a disease. Patterns on maps may suggest interactions among different features of Earth. Placing information on a map is a principal way that geographers share data or results of scientific analysis.

FIGURE 1-2 The oldest known maps. (top) A seventh-century BC map of a plan for the town of Çatalhöyük, in present-day Turkey. Archaeologists found the map on the wall of a house that was excavated in the 1960s. (middle) A color version of the Çatalhöyük map. A volcano rises above the buildings of the city. (bottom) A world map from the sixth century BC depicts a circular land area surrounded by a ring of water. The ancient city of Babylon is thought to be shown in the center of the land area and other cities are shown as circles. Extending out from the water ring are seven islands that together form a star shape.

Early Mapmaking

From the earliest human occupancy of Earth, people have been creating maps to assist with navigation. The earliest surviving maps were drawn in the Middle East in the seventh or sixth century BC (Figure 1-2). Miletus, a port in present-day Turkey, became a center for geographic thought and mapmaking in the ancient world. Thales (624?–546? BC) applied principles of geometry to measuring land area. His student, Anaximander (610–546? BC), made a world map based on information from sailors, though he portrayed Earth's shape as a cylinder. Hecateus may have produced the first geography book around 500 BC.

Aristotle (384–322 BC) was the first to demonstrate that Earth was spherical. He observed that matter falls together toward a common center, that Earth's shadow on the Moon is circular during an eclipse, and that the visible groups of stars change as one travels north or south.

Eratosthenes (276?–194? BC), the first person of record to use the word *geography*, also accepted that Earth was spherical and calculated its circumference within a remarkable 0.5 percent accuracy. He prepared one of the earliest maps of the known world, correctly dividing Earth into five climatic regions—a torrid zone across the middle, two frigid zones at the extreme north and south, and two temperate bands in between.

Two thousand years ago, the Roman Empire controlled an extensive area of the known world, including much of Europe, northern Africa, and western Asia. Taking advantage of information collected by merchants and soldiers who traveled through the Roman Empire, the Greek Ptolemy (AD 100?–170?) wrote an eight-volume *Guide to Geography*. He codified basic principles of mapmaking and prepared numerous maps, which were not improved upon for more than a thousand years. Ancient Greek and Roman maps were compiled in the *Barrington Atlas of the Greek and Roman World*. "We can't truly understand the Greeks and Romans without good maps that show us their world," explained *Barrington Atlas* editor Richard J. A. Talbert.

After Ptolemy, little progress in mapmaking or geographic thought was made in Europe for several hundred years. Maps became less mathematical and more fanciful, showing Earth as a flat disk surrounded by fierce animals and monsters. Geographic inquiry continued, though, outside of Europe.

- The oldest Chinese geographical writing, from the fifth century BC, describes the economic resources of the country's different provinces. Phei Hsiu (or Fei Xiu), the "father of Chinese cartography," produced an elaborate map of the country in AD 267.
- The Muslim geographer al-Idrisi (1100–1165?) prepared a world map and geography text in 1154, building on Ptolemy's long-neglected work. Ibn-Battutah (1304–1368?) wrote *Rihlah* ("Travels") based on three decades of journeys covering more than 120,000 kilometers (75,000 miles) through the Muslim world of northern Africa, southern Europe, and much of Asia.

A revival of geography and mapmaking occurred during the Age of Exploration and Discovery. Ptolemy's maps were rediscovered, and his writings were translated into European languages. Columbus, Magellan, and other explorers who sailed across the oceans in search of trade routes and resources required accurate maps to reach desired destinations without wrecking their ships. In turn, cartographers such as Gerardus Mercator (1512–1594) and Abraham Ortelius (1527–1598) took information collected by the explorers to create more accurate maps (Figure 1-3).

By the seventeenth century, maps accurately displayed the outline of most continents and the positions of oceans. Bernhardus Varenius (1622–1650) produced *Geographia Generalis*, which stood for more than a century as the standard treatise on systematic geography.

Map Scale

The first decision a cartographer faces is how much of Earth's surface to depict on the map. Is it necessary to show the entire globe, or just one continent, or a country, or a city? To make a scale model of the entire world, many details must be omitted because there simply is not enough space. Conversely, if a map shows only a small portion of Earth's surface, such as a street map of a city, it can provide a wealth of detail about a particular place.

The level of detail and the amount of area covered on a map depend on its scale. When specifically applied to a map, scale refers to the relationship of a feature's size on a map to its actual size on Earth. Map scale is presented in three ways (Figure 1-4).

- **A ratio or fraction** shows the numerical ratio between distances on the map and Earth's surface. A scale of 1:24,000 or 1/24,000 means that 1 unit (inch, centimeter, foot, finger length) on the map represents 24,000 of the same unit (inch, centimeter, foot, finger length) on the ground. The unit chosen for distance can be anything, as long as the units of measure on both the map and the ground are the same. The 1 on the left side of the ratio always refers to a unit of distance *on the map*, and the number on the right always refers to the *same unit* of distance *on Earth's surface.*
- **A written scale** describes this relation between map and Earth distances in words. For example, the statement "1 inch equals 1 mile" on a map means that 1 inch on the map represents 1 mile on Earth's surface. Again, the first number always refers to map distance, and the second to distance on Earth's surface.
- **A graphic scale** usually consists of a bar line marked to show distance on Earth's surface. To use a bar line, first determine with a ruler the distance on the map in inches or centimeters. Then hold the ruler against the bar line and read the number on the bar line opposite the map distance on the ruler. The number on the bar line is the equivalent distance on Earth's surface.

Maps often display scale in more than one of these three ways.

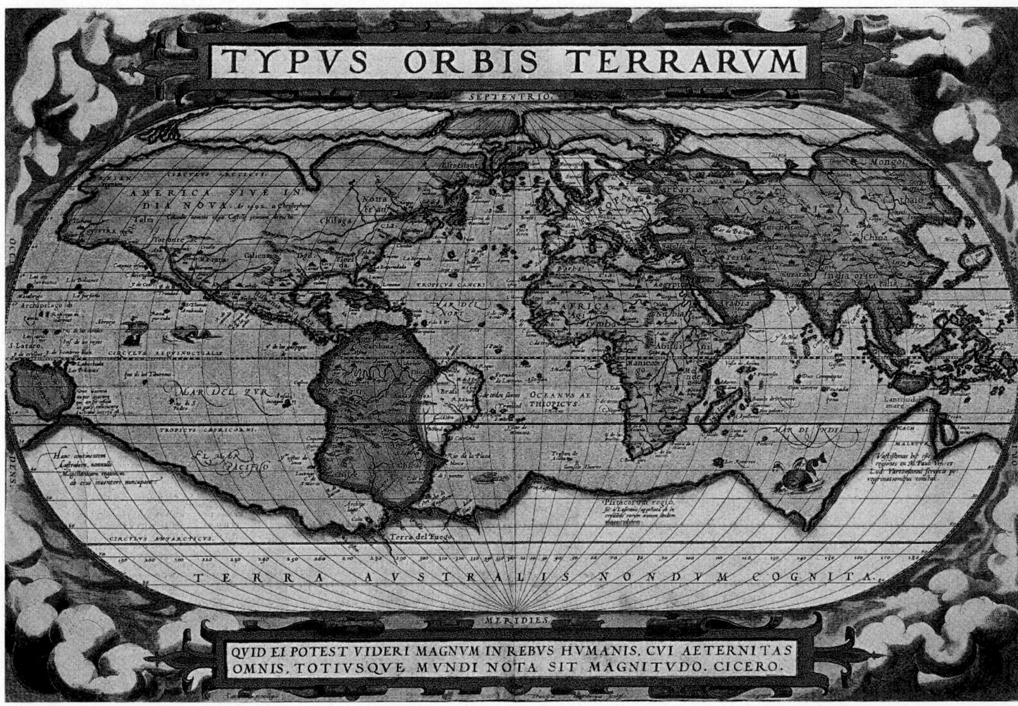

FIGURE 1-3 Map of the world made in 1571 by Flemish cartographer Abraham Ortelius (1527–1598). Compare the accuracy of the coastlines on Ortelius's map with the recent image of the world based on satellite photographs (Figure 1-1).

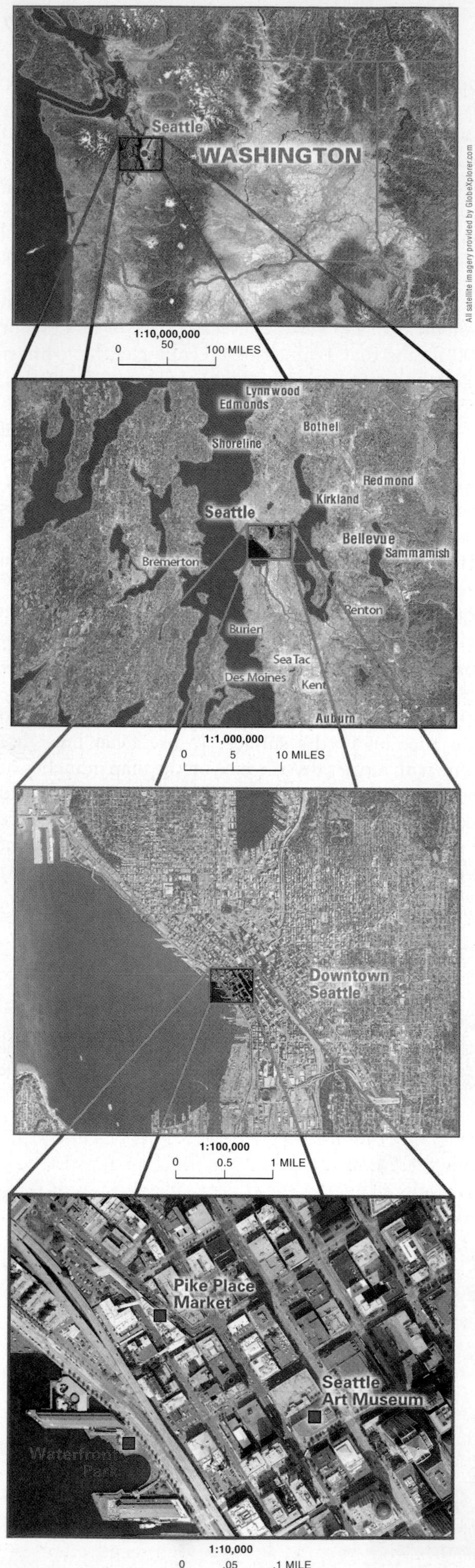

The appropriate scale for a map depends on the information being portrayed. A map of a downtown area, such as Figure 1-4 bottom, has a scale of 1:10,000, whereas the map of Washington State (Figure 1-4 top) has a scale of 1:10,000,000. One inch represents about 1/6 mile on the downtown Seattle map and about 170 miles on the Washington State map.

At the scale of a small portion of Earth's surface, such as a downtown area, a map provides a wealth of details about the place. At the scale of the entire globe, a map must omit many details because of lack of space, but it can effectively communicate processes and trends that affect everyone.

Projection

Earth is very nearly a sphere and therefore accurately represented in the form of a globe. However, a globe is an extremely limited tool with which to communicate information about Earth's surface. A small globe does not have enough space to display detailed information, whereas a large globe is too bulky and cumbersome to use. And a globe is difficult to write on, photocopy, display on a computer screen, or carry in the glove box of a car. Consequently, most maps—including those in this book—are flat. Three-dimensional maps can be made but are expensive and difficult to reproduce.

Earth's spherical shape poses a challenge for cartographers because drawing Earth on a flat piece of paper unavoidably produces some distortion. Cartographers have invented hundreds of clever methods of producing flat maps, but none has produced perfect results. The scientific method of transferring locations on Earth's surface to a flat map is called **projection**.

The problem of distortion is especially severe for maps depicting the entire world. Four types of distortion can result:

1. The *shape* of an area can be distorted, so that it appears more elongated or squat than in reality.
2. The *distance* between two points may become increased or decreased.
3. The *relative size* of different areas may be altered, so that one area may appear larger than another on a map but is in reality smaller.
4. The *direction* from one place to another can be distorted.

Most of the world maps in this book, such as Figure 1-19, are *equal area projections*. The primary benefit of this type of projection is that the relative sizes of the landmasses on the map are the same as in reality. The projection minimizes distortion in the shapes of most landmasses. Areas toward the North and South poles—such as Greenland and Australia—become more distorted, but they are sparsely inhabited, so distorting their shapes usually is not important.

FIGURE 1-4 Map scale. The four images show Washington State (first), western Washington (second), the Seattle region (third), and downtown Seattle (fourth). The map of Washington State has a fractional scale of 1:10,000,000. Expressed as a written statement, 1 inch on the map represents 10 million inches (about 158 miles) on the ground. Look what happens to the scale on the other three maps. As the area covered gets smaller, the maps get more detailed, and 1 inch on the map represents smaller distances.

To largely preserve the size and shape of landmasses, however, the projection in Figure 1-19 forces other distortions:

- The Eastern and Western hemispheres are separated into two pieces, a characteristic known as interruption.
- The meridians (the vertical lines), which in reality converge at the North and South poles, do not converge at all on the map. Also, they do not form right angles with the parallels (the horizontal lines).

Two types of uninterrupted projections display information as shown in Figure 1-13 and 1-23 on pages 18 and 30.

- The Robinson projection, in Figure 1-23, is useful for displaying information across the oceans. Its major disadvantage is that by allocating space to the oceans, the land areas are much smaller than on interrupted maps of the same size.
- The Mercator projection in Figure 1-13 has several advantages: Shape is distorted very little, direction is consistent, and the map is rectangular. Its greatest disadvantage is that area is grossly distorted toward the poles, making high-latitude places look much larger than they actually are.

Compare the sizes of Greenland and South America in the maps shown in Figures 1-13 and 1-19. The map in Figure 1-19 illustrates their size accurately.

U.S. Land Ordinance of 1785

In addition to the global system of latitude and longitude, other mathematical indicators of locations are used in different parts of the world. In the United States, the **Land Ordinance of 1785** divided much of the country into a system of townships and ranges to facilitate the sale of land to settlers in the West. The initial surveying was performed by Thomas Hutchins, who was appointed geographer to the United States in 1781. After Hutchins died in 1789, responsibility for surveying was transferred to the Surveyor General.

In this system, a **township** is a square 6 miles on each side. Some of the north–south lines separating townships are called **principal meridians**, and some east–west lines are designated **base lines** (Figure 1-5, upper left).

Each township has a number corresponding to its distance north or south of a particular base line. Townships in the first row north of a base line are called T1N (Township 1 North), the second row to the north is T2N, the first row to the south is T1S, and so on.

Each township has a second number, known as the range, corresponding to its location east or west of a principal meridian. Townships in the first column east of a principal meridian are designated R1E (Range 1 East). The Tallahatchie River, for example, is in township T23N R1E, north of a base line that runs east–west across Mississippi and east of a principal meridian along 90° west longitude.

A township is divided into 36 **sections**, each of which is 1 mile by 1 mile (Figure 1-5, lower left). Sections are numbered in a consistent order, from 1 in the northeast to 36 in the southeast. Each section is divided into four quarter-sections, designated as the northeast, northwest, southeast, and southwest quarters of a particular section. A quarter-section, which is 0.5 mile by 0.5 mile, or 160 acres, was the amount of land many Western pioneers bought as a homestead. The Tallahatchie River is located in the southeast and southwest quarter-sections of Section 32.

The township and range system remains important in understanding the location of objects across much of the United States. It explains the location of highways across the Midwest, farm fields in Iowa, and major streets in Chicago.

Contemporary Tools

Having largely completed the formidable task of accurately mapping Earth's surface, which required several centuries, geographers have turned to Geographic Information Science (GIScience) to learn more about places. GIScience helps geographers to create more accurate and complex maps and to measure changes over time in the characteristics of places.

Satellite-based Imagery

GIScience is made possible by satellites in orbit above Earth sending information to electronic devices on Earth to record and interpret information. Satellite-based information allows us to know the precise location of something on Earth and data about that place.

GPS. The system that accurately determines the precise position of something on Earth is **GPS (Global Positioning System)**. The GPS system in the United States includes three elements:

- Satellites placed in predetermined orbits by the U.S. military (24 in operation and 3 in reserve)
- Tracking stations to monitor and control the satellites
- A receiver that can locate at least 4 satellites, figure out the distance to each, and use this information to pinpoint its own location.

GPS is most commonly used for navigation. Pilots of aircraft and ships stay on course with GPS. On land, GPS detects a vehicle's current position, the motorist programs the desired destination, and GPS provides instructions on how to reach the destination. GPS can also be used to find the precise location of a vehicle, enabling a motorist to summon help in an emergency or monitoring the progress of a delivery truck or position of a city bus. Geographers find GPS to be particularly useful in coding the precise location of objects collected in fieldwork. That information can later be entered as a layer in a GIS.

GPS devices enable private individuals to contribute to the production of accurate digital maps, through web sites like Google's OpenStreetMap.org. Travelers can enter information about streets, buildings, and bodies of water in their GPS devices, so that digital maps can be improved or in some cases be created for the first time.

REMOTE SENSING. The acquisition of data about Earth's surface from a satellite orbiting Earth or from other long-distance methods is known as **remote sensing**. Remote-sensing satellites scan Earth's surface, much like a television camera scans an image in the thin lines you can see on a TV screen. Images are transmitted in digital form to a receiving station on Earth.

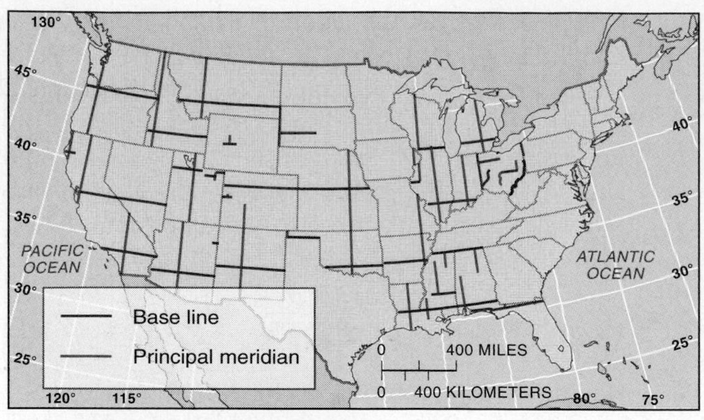

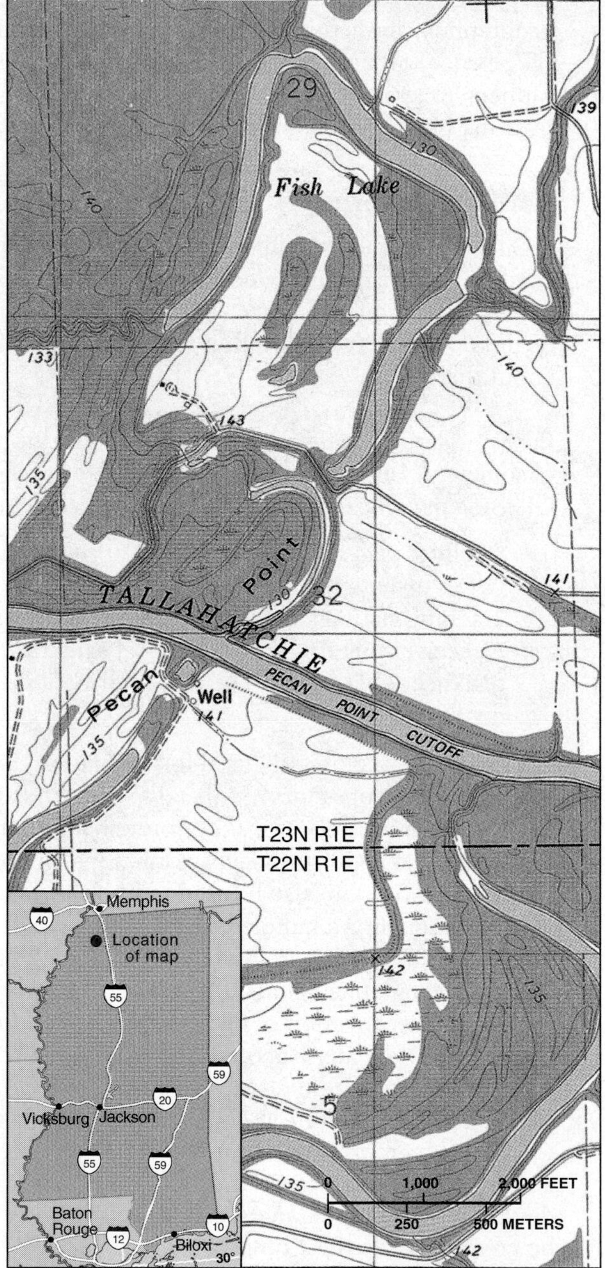

N

T24N R1W

T24N R1E

T23N R1W

6	5	4	3	2	1
7	8	9	10	11	12
18	17	16	15	14	13
19	20	21	22	23	24
30	29	28	27	26	25
31	32	33	34	35	36

Tallahatchie River

T22N R1W

T22N R1E

FIGURE 1-5 Township and range system. To facilitate the numbering of townships, the U.S. Land Ordinance of 1785 designated several north–south lines as principal meridians and several east–west lines as base lines (upper left). As territory farther west was settled, additional lines were delineated. Townships are typically 6 miles by 6 miles, although physical features, such as rivers and mountains, result in some irregularly shaped ones (upper right). The Tallahatchie River, for example, is located in the twenty-third township north of a base line that runs east–west across Mississippi and in the first range east of the principal meridian at 90° west longitude. Townships are divided into 36 sections, each 1 square mile. Sections are divided into four quarter-sections. The Tallahatchie River is located in the southeast and southwest quarter-sections of Section 32, T23N R1E. The topographic map (lower left), published by the U.S. Geological Survey, has a fractional scale of 1:24,000. Expressed as a written statement, 1 inch on the map represents 24,000 inches (2,000 feet) on the ground. The map displays portions of two townships, shown on the above map. The brown lines on the map are contour lines that show the elevation of any location.

CONTEMPORARY GEOGRAPHIC TOOLS

Navigation Devices from Hand-Drawn to Electronic

The earliest maps were simple navigation devices designed to show the traveler how to get from Point A to Point B. For example, Polynesian peoples navigated among South Pacific islands for thousands of years using three-dimensional maps called stick charts, made of strips from palm trees and seashells. The shells represented islands, and the palm strips represented patterns of waves between the islands (Figure 1-6).

After 3,000 years of ever more complex, detailed, and accurate cartography, contemporary maps have reverted to their earliest purpose, as simple navigation devices. But to figure out how to get from one place to another, you no longer have to unfurl an ungainly map filled with hard-to-read information irrelevant to your immediate journey. Instead, you program your desired destination into an electronic navigation device. Because it knows where you are now, the device can tell you the route to take from your current location to your desired location. Electronic navigation devices have been installed in the dashboards of motor vehicles and in handheld devices such as mobile phones, personal digital assistants (PDAs), and personal navigation devices (PNDs). All of these devices depend on GPS receivers to pinpoint your current location.

Most trips involve making a choice from among alternative routes. Navigation devices calculate which route will get you from Point A to Point B in the fastest time. Time is a function of a combination of speed and distance. The shortest route may not always be the quickest, because every road segment has an expected speed depending on its nature—an interstate highway has a higher expected speed than a local road.

The best route is also affected by attributes of the road, such as the presence of crosswalks, traffic lights, and turn restrictions. Current technology does not incorporate every possible attribute, such as construction, weather, and time of day, but presumably, future models will.

Two companies are responsible for supplying most of the information fed into navigation devices: Navteq, short for Navigation Technologies, and Tele Atlas, originally known as Etak. Navteq, based in the United States, and Tele Atlas, based in the Netherlands and Belgium; both were founded in 1985. Navteq and Tele Atlas get their information from what they call "ground truthing." Hundreds of field researchers drive around, building the database. One person drives while the other feeds information into a notebook computer. Hundreds of attributes are recorded, such as crosswalks, turn restrictions, and name changes. Thus, electronic navigation systems ultimately depend on human observation. ■

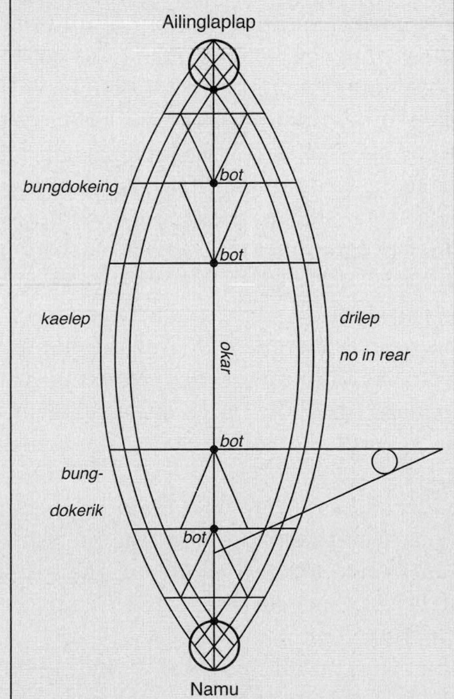

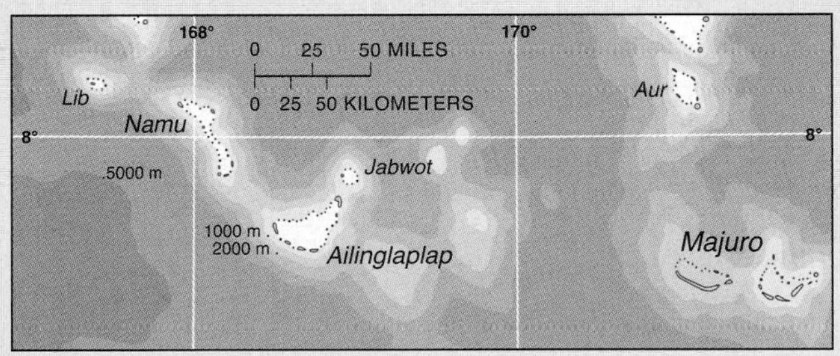

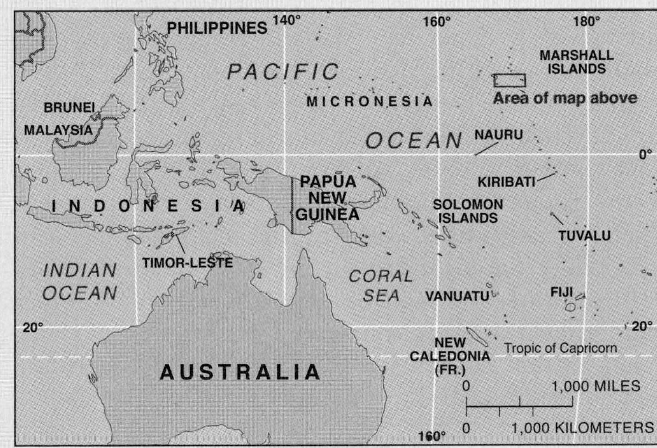

FIGURE 1-6 Polynesian "stick chart," a type of ancient map. Islands were shown with shells, and patterns of swelling of waves were shown with palm strips. Curved strips and straight strips made of palm represented different wave swells. This ancient example depicted the sea route between Ailinglapalap and Namu, two islands in the present-day Marshall Islands, in the South Pacific Ocean. The top of the stick chart faces southeast.

At any moment a satellite sensor records the image of a tiny area called a picture element or pixel. Scanners are detecting the radiation being reflected from that tiny area. A map created by remote sensing is essentially a grid containing many rows of pixels. The resolution of the scanner determines the smallest feature on Earth's surface that can be detected by a sensor. Some can sense objects as small as 1 meter across.

GIS

A computer system that can capture, store, query, analyze, and display geographic data is a **GIS (geographic information system)**. The key to GIS is geocoding: The position of any object on Earth can be measured and recorded with mathematical precision and then stored in a computer.

GIS can be used to produce maps (including those in this book) that are more accurate and attractive than those drawn by hand. A map can be created by asking the computer to retrieve a number of stored objects and combine them to form an image. In the past, when cartographers drew maps with pen and paper, a careless moment could result in an object being placed in the wrong location, and a slip of the hand could ruin hours of work. GIS is more efficient for making a map than pen and ink: Objects can be added or removed, colors brightened or toned down, and mistakes corrected (as long as humans find them!) without having to tear up the paper and start from scratch.

Each type of information can be stored in a layer. For example, separate layers could be created for boundaries of countries, bodies of water, roads, and names of places. A simple map might display only a single layer by itself, but most maps combine several layers (Figure 1-7), and GIS permits construction of much more complex maps than can be drawn by hand.

The value of GIS extends beyond the ability to make complex maps more easily. Layers can be compared to show relationships among different kinds of information. To understand the impact of farming practices on water pollution, a physical geographer may wish to compare a layer of vegetation with a layer of bodies of water. To protect hillsides from development, a human geographer may wish to compare a layer of recently built houses with a layer of steep slopes.

Scottish environmentalist Ian McHarg pioneered a technique of comparing layers of various physical and social features to determine where new roads and houses should be built and where the landscape should be protected from development. When McHarg was developing the technique during the 1960s—before the diffusion of powerful microcomputers and GIS software—he painstakingly created layers by laying hand-drawn plastic transparencies on top of each other. A half-century later, his pioneering technique can be replicated quickly on a desktop computer with GIS software.

GIS enables geographers to calculate whether relationships between objects on a map are significant or merely coincidental. For example, maps showing where cancer rates are relatively high and low (such as those in Figure 1-17) can be combined with layers showing the location of people with various incomes and ethnicities, the location of different types of factories, and the location of mountains and valleys. Desktop computer users have the ability to do their own GIS, because computer mapping services provide access to the application programming interface (API), which is the language that links a database such as an address list with software such as mapping. The API for mapping software, available at such sites as www.google.com/apis/maps, enables a computer programmer to create a mash-up that places data on a map.

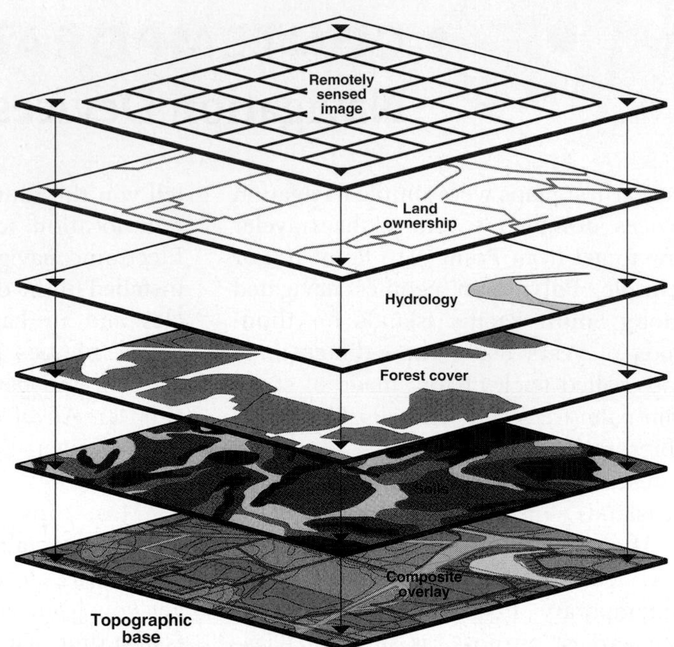

FIGURE 1-7 GIS. Geographic information systems involve storing information about a location in layers. Each layer represents a different piece of human or environmental information. The layers can be viewed individually or in combination.

The term *mash-up* refers to the practice of overlaying data from one source on top of one of the mapping services and comes from the hip-hop practice of mixing two or more songs. Mash-up maps can show the locations of businesses and activities near a particular street or within a neighborhood in a city. The requested information could be all restaurants within ½ mile of an address or, to be even more specific, all pizza parlors. Mapping software can show the precise location of commercial airplanes currently in the air, the gas stations with the cheapest prices, and current traffic tie-ups on highways and bridges (Figure 1-8).

In some cities, mash-ups assist in finding housing. They can pinpoint the location of houses currently for sale and apartments currently for rent. A map showing the prices of recently sold houses in the area can help a potential buyer determine how much to offer. A map showing the locations of crime in the city can help the buyer determine the safety of the

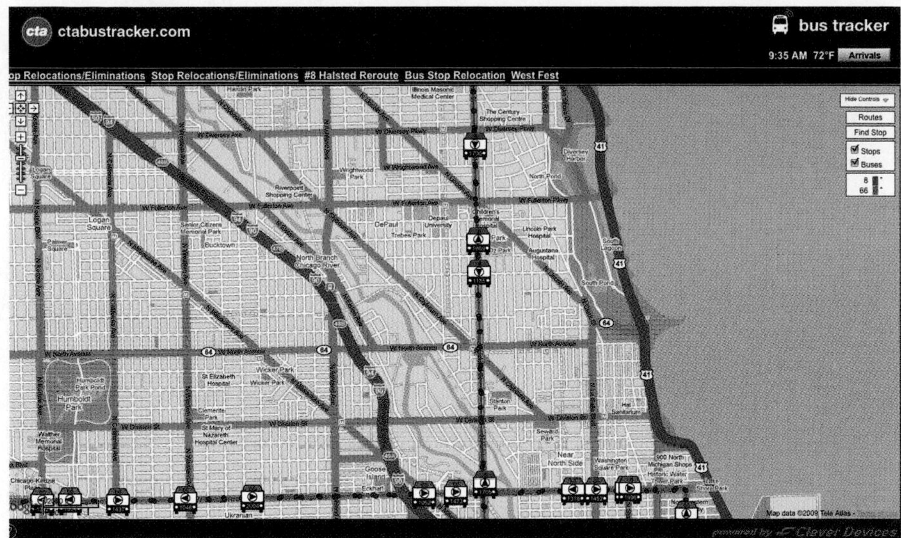

FIGURE 1-8 Mash-up. Chicago Transit Authority mash-up shows location of buses and bus stops along three routes. Rolling the mouse over a bus stop shows when the next three buses are expected.

surrounding area. Bars, hotels, sports facilities, transit stops, and other information about the neighborhood can be mapped.

KEY ISSUE 2

Why Is Each Point on Earth Unique?

- **Place: Unique Location of a Feature**
- **Regions: Areas of Unique Characteristics**
- **Spatial Association**

Each place on Earth is in some respects unique and in other respects similar to other places. The interplay between the uniqueness of each place and the similarities among places lies at the heart of geographic inquiry into why things are found where they are. ■

Two basic concepts help geographers to explain why every point on Earth is in some ways unique—place and region. The difference between the two concepts is partly a matter of scale: A place is a point, whereas a region is an area.

Place: Unique Location of a Feature

Humans possess a strong sense of place—that is, a feeling for the features that contribute to the distinctiveness of a particular spot on Earth, perhaps a hometown, vacation destination, or part of a country. Describing the features of a place or region is an essential building block for geographers to explain similarities, differences, and changes across Earth. Geographers think about where particular places and regions are located and the combination of features that make each place and region on Earth distinct.

Geographers describe a feature's place on Earth by identifying its **location**, the position that something occupies on Earth's surface, and in doing so consider four ways to identify location: place name, site, situation, and mathematical location.

Place Names

Because all inhabited places on Earth's surface—and many uninhabited places—have been named, the most straightforward way to describe a particular location is often by referring to its place name. A **toponym** is the name given to a place on Earth (Figure 1-9).

A place may be named for a person, perhaps its founder or a famous person with no connection to the community. George Washington's name has been selected for one state, counties in 31 other states, and dozens of cities, including the national capital. Places may be named for an obscure person, such as Jenkinjones, West Virginia, named for a mine operator, and Gassaway, West Virginia, named for a U.S. senator.

Some settlers select place names associated with religion, such as St. Louis and St. Paul, whereas other names derive from ancient history, such as Athens, Attica, and Rome. A place name may also indicate the origin of its settlers. Place names commonly have British origins in North America and Australia, Portuguese origins in Brazil, Spanish origins elsewhere in Latin America, and Dutch origins in South Africa.

Pioneers lured to the American West by the prospect of finding gold or silver placed many picturesque names on the landscape. Place names in Nevada selected by successful miners include Eureka, Lucky Boy Pass, Gold Point, and Silver Peak. Unsuccessful Nevada pioneers sadly or bitterly named other places, such as Battle Mountain, Disaster Peak, and Massacre Lake. The name Jackpot was given in 1959 by the Elko, Nevada, county commissioners to a town near the Idaho state border in recognition of the importance of legalized gambling to the local economy.

Some place names derive from features of the physical environment. Trees, valleys, bodies of water, and other natural features appear in the place names of most languages. The capital of the Netherlands, called *'s-Gravenhage* in Dutch (in English, The Hague), means "the prince's forest." Aberystwyth, in Wales, means "mouth of the River Ystwyth," while 22 kilometers (13 miles) upstream lies the tiny village of Cwmystwyth, which means "valley of the Ystwyth." The name of the river, Ystwyth, in turn, is the Welsh word for "meandering," descriptive of a stream that bends like a snake.

FIGURE 1-9 Welsh toponym. The town's name originally was Llanfairpwllgwyngyll, but when the railway was built in the nineteenth century, the townspeople lengthened it. They decided that signs with the 58-letter name in the railway station would attract attention and bring more business and visitors to the town.

Places can change names. The city of Cincinnati was originally named Losantiville. The name was derived as follows: *L* is for Licking River; *os* is Latin for mouth; *anti* is Latin for opposite; *ville* is Latin for town—hence, "town opposite the mouth of the Licking River." The name was changed to Cincinnati in honor of a society of Revolutionary War heroes named after Cincinnatus, an ancient Roman general.

Hot Springs, New Mexico, was renamed Truth or Consequences in 1950 in honor of a long-running radio and television program of that name. The name was changed by an overwhelmingly favorable vote of the residents in order to promote publicity for the economically struggling town.

The Board of Geographical Names, operated by the U.S. Geological Survey, was established in the late nineteenth century to be the final arbiter of names on U.S. maps. In recent years the board has been especially concerned with removing offensive place names, such as those with racial or ethnic connotations.

Names can also change as a result of political upheavals. For example, following World War II, Poland gained control over territory that was formerly part of Germany and changed many of the place names from German to Polish. Among the larger cities, Danzig became Gdánsk, Breslau became Wrocław, and Stettin became Szczecin. After the fall of communism in the early 1990s, names throughout Eastern Europe were changed, in many cases reverting to those used before the Communists had gained power some decades earlier. For example after the demise of communism, Gottwaldov (named for a Communist president of Czechoslovakia) reverted to its former name, Zlín, in the Czech Republic; Leningrad (the second-largest city in the Soviet Union) reverted to St. Petersburg, Russia; and Karl-Marx-Stadt (in East Germany) reverted to Chemnitz in a reunified Germany.

Site

The second way that geographers describe the location of a place is by **site**, which is the physical character of a place. Important site characteristics include climate, water sources, topography, soil, vegetation, latitude, and elevation. The combination of physical features gives each place a distinctive character.

Site factors have always been essential in selecting locations for settlements, although people have disagreed on the attributes of a good site, depending on cultural values. Some have preferred a hilltop site for easy defense from attack. Others located settlements near convenient river-crossing points to facilitate communication with people in other places.

Humans have the ability to modify the characteristics of a site. The southern portion of New York City's Manhattan Island is twice as large today as it was in 1626, when Peter Minuit bought the island from its native inhabitants for the equivalent of $23.75 worth of Dutch gold and silver coins (Figure 1-10). Manhattan's additional land area was created by filling in portions of the East River and the Hudson River. In the eighteenth century, landfills were created by sinking old ships and dumping refuse on top of them. More recently, New York City permitted construction of Battery Park City, a 57-hectare (142-acre) site designed to house more than 20,000 residents and 30,000 office workers. The central areas of Boston and Tokyo have also been expanded through centuries of landfilling in nearby bays, substantially changing these sites.

Situation

Situation is the location of a place relative to other places. Situation is a valuable way to indicate location, for two reasons—finding an unfamiliar place and understanding its importance.

First, situation helps us find an unfamiliar place by comparing its location with a familiar one. We give directions to people by referring to the situation of a place: "It's down past the courthouse, on Locust Street, after the third traffic light, beside the yellow-brick bank." We identify important buildings, streets, and other landmarks to direct people to the desired location.

Second, situation helps us understand the importance of a location. Many locations are important because they are accessible to other places. For example, because of its situation, Singapore has become a center for the trading and

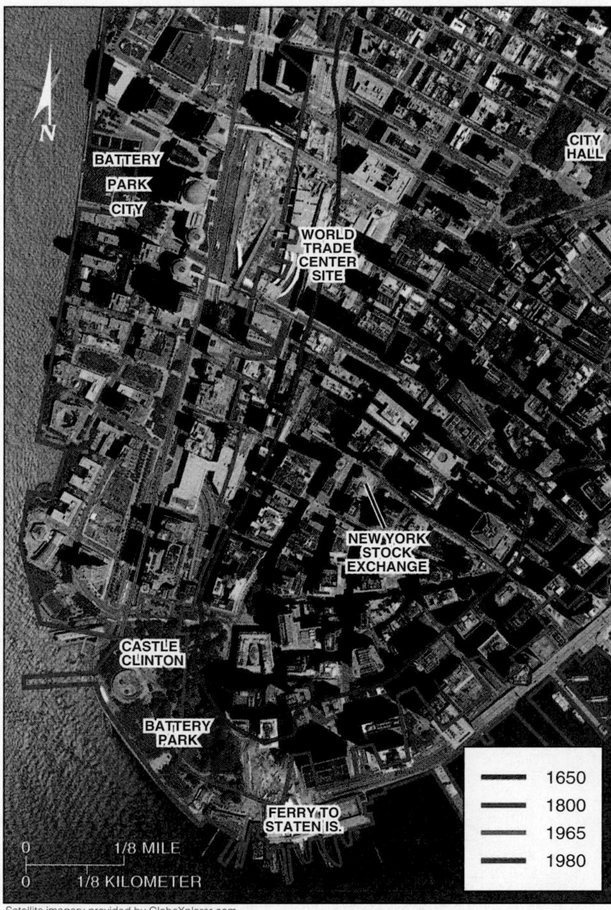

FIGURE 1-10 Changing site of New York City. Much of the southern part of New York City's Manhattan Island was built on landfill. Several times in the past 200 years, the waterfront has been extended into the Hudson and East rivers to provide more land for offices, homes, parks, warehouses, and docks. The World Trade Center was built during the late 1960s and early 1970s partially on landfill in the Hudson River from the colonial era. Battery Park City (at left in the photograph) was built on landfill removed from the World Trade Center construction site. Excavating the landfill from the World Trade Center site unearthed a large number of maritime objects, such as anchors, because the site was underwater in the seventeenth century.

distribution of goods for much of Southeast Asia (Figure 1-11). Singapore is situated near the Strait of Malacca, which is the major passageway for ships traveling between the South China Sea and the Indian Ocean. Some 50,000 vessels, one-fourth of the world's maritime trade, pass through the strait each year.

Mathematical Location

The location of any place on Earth's surface can be described precisely by meridians and parallels, two sets of imaginary arcs drawn in a grid pattern on Earth's surface (Figure 1-12).

- A **meridian** is an arc drawn between the North and South poles. The location of each meridian is identified on Earth's surface according to a numbering system known as **longitude**.

  The meridian that passes through the Royal Observatory at Greenwich, England, is 0° longitude, also called the **prime meridian**. The meridian on the opposite side of the globe from the prime meridian is 180° longitude. All other meridians have numbers between 0° and 180° east or west, depending on whether they are east or west of the prime meridian. For example, New York City is located at 74° west longitude, and Lahore, Pakistan, at 74° east longitude. San Diego is located at 117° west longitude, and Tianjin, China, at 117° east longitude.

- A **parallel** is a circle drawn around the globe parallel to the equator and at right angles to the meridians. The numbering system to indicate the location of a parallel is called **latitude**.

  The equator is 0° latitude, the North Pole 90° north latitude, and the South Pole 90° south latitude. New York City is located at 41° north latitude, and Wellington, New Zealand, at 41° south latitude. San Diego is located at 33° north latitude, and Santiago, Chile, at 33° south latitude.

Latitude and longitude are used together to identify locations. For example, Midland, Texas, is located at 32° north latitude and 102° west longitude.

The mathematical location of a place can be designated more precisely by dividing each degree into 60 minutes (′) and each minute into 60 seconds (″). For example, the official mathematical location of Denver, Colorado, is 39°44′ north latitude and 104°59′ west longitude. The state capitol building in Denver is located at 39°42′52″ north latitude and 104°59′04″ west longitude. GPS systems typically divide degrees into decimal fractions rather than minutes and seconds. Toyota's factory in Georgetown, Kentucky, for example is located at 38.233407° north latitude and 84.550239° west longitude.

Measuring latitude and longitude is a good example of how geography is partly a natural science and partly a study of human behavior. Latitudes are scientifically derived by Earth's

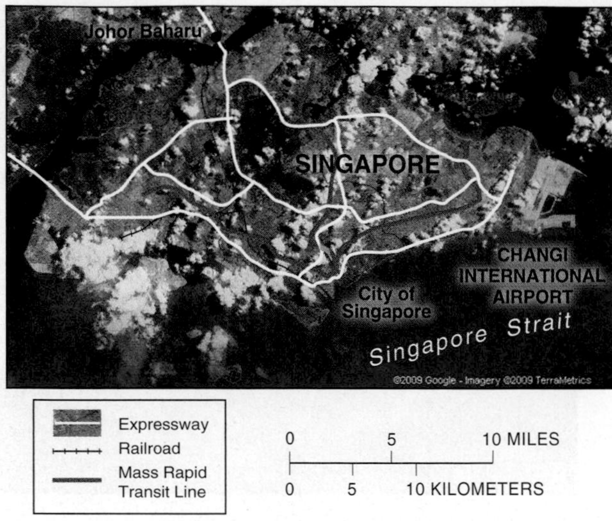

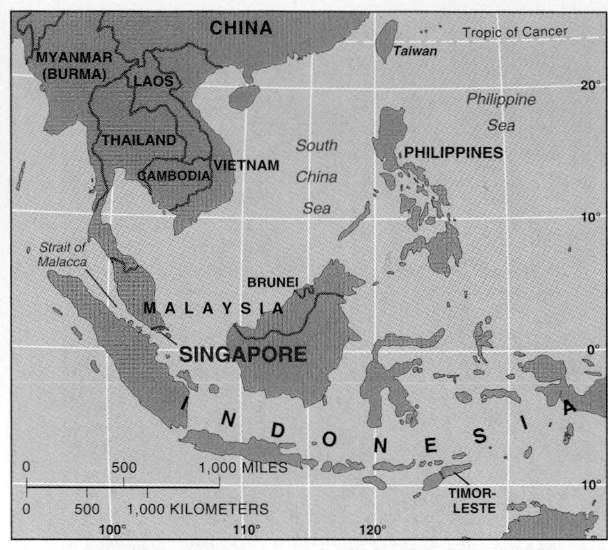

FIGURE 1-11 Situation and site of Singapore. The site of the country of Singapore is a small island approximately 1 kilometer off the southern tip of the Malay Peninsula at the eastern end of the Strait of Malacca. The city of Singapore covers nearly 20 percent of the island. Its situation is the confluence of several straits that serve as major passageways for shipping between the South China Sea and the Indian Ocean. Downtown Singapore is situated near where the Singapore River flows into the Singapore Strait.

shape and its rotation around the Sun. The equator (0° latitude) is the parallel with the largest circumference and is the place where every day has 12 hours of daylight. Even in ancient times, latitude could be accurately measured by the length of daylight and the position of the Sun and stars.

On the other hand, 0° longitude is a human creation. Any meridian could have been selected as 0° longitude, because all have the same length and all run between the poles. The 0° longitude runs through Greenwich because England was the world's most powerful country when longitude was first accurately measured and the international agreement was made. For many centuries, inability to measure longitude was the greatest obstacle to exploration and discovery. Ships ran aground or were lost at sea because no one on board could pinpoint longitude. In 1714, the British Parliament enacted the Longitude Act, which offered a prize equivalent to several million in today's dollars to the person who could first measure longitude accurately.

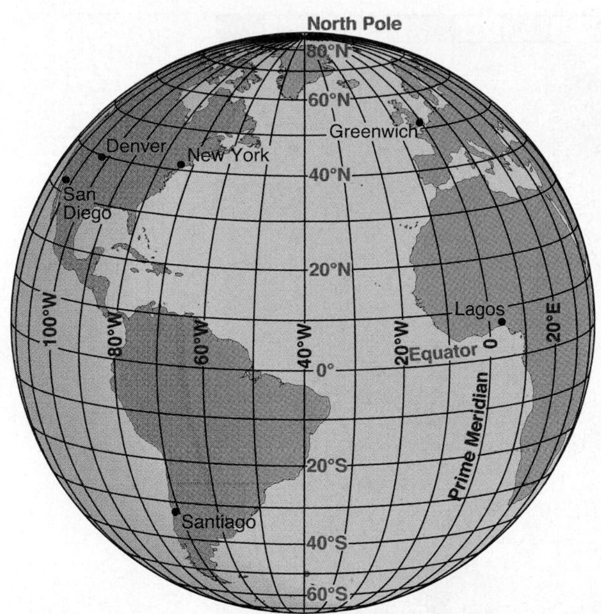

FIGURE 1-12 Geographic grid. Meridians are arcs that connect the North and South poles. The meridian through Greenwich, England, is the prime meridian, or 0° longitude. Parallels are circles drawn around the globe parallel to the equator. The equator is 0° latitude and the North Pole is 90° north latitude.

English clockmaker John Harrison won the prize by inventing the first portable clock that could keep accurate time on a ship—because it did not have a pendulum. When the Sun was directly overhead of the ship—noon local time—Harrison's portable clock set to Greenwich time could say it was 2 PM in Greenwich, for example, so the ship would be at 30° west longitude, because each hour of difference was equivalent to traveling 15° longitude. (Most eighteenth-century scientists were convinced that longitude could be determined only by the position of the stars, so Harrison was not actually awarded the prize until 40 years after his invention.)

Regions: Areas of Unique Characteristics

The "sense of place" that humans possess may apply to a larger area of Earth rather than to a specific point. A person may feel attachment as a native or resident of the Los Angeles area, or the area of attachment could encompass southern California or the U.S. Southwest. An area of Earth defined by one or more distinctive characteristics is a region.

Cultural Landscape

A region derives its unified character through the **cultural landscape**—a combination of cultural features such as language and religion, economic features such as agriculture and industry, and physical features such as climate and vegetation. The Los Angeles region can be distinguished from the New York region, southern California from northern California, the Southwest from the Midwest.

The contemporary cultural landscape approach in geography—sometimes called the **regional studies** approach—was initiated in France by Paul Vidal de la Blache (1845–1918) and Jean Brunhes (1869–1930). It was later adopted by several American geographers, including Carl Sauer (1889–1975) and Robert Platt (1880–1950). Sauer defined cultural landscape as an area fashioned from nature by a cultural group. "Culture is the agent, the natural area the medium, the cultural landscape is the result."

Cultural landscape geographers argued that each region has its own distinctive landscape that results from a unique combination of social relationships and physical processes. People, activities, and environment display similarities and regularities within a region and differ in some way from those of other regions. A region gains uniqueness from possessing not a single human or environmental characteristic, but a combination of them. Not content to merely identify these characteristics, geographers seek relationships among them. Geographers recognize that in the real world, characteristics are integrated.

The fundamental principle underlying the cultural landscape approach is that people are the most important agents of change to Earth's surface. The distinctive character of a particular landscape may derive in part from natural features, such as vegetation and soil. However, the physical environment is not always the most significant factor in human decisions. People can fashion a landscape by superimposing new forms on the physical environment. For example, the critical factor in selecting a site for a cotton textile factory is not proximity to a place where cotton is grown. A more important factor in selecting a suitable location is access to a supply of low-cost labor. Economic systems, political structures, living arrangements, religious practices, and human activities can produce distinctive landscapes that do not stem primarily from distinctive physical features.

The geographer's job is to sort out the associations among various social characteristics, each of which is uniquely distributed across Earth's surface. For example, geographers conclude that political unrest in sub-Saharan Africa, Southwest Asia, and other areas derives in large measure from the fact that the distributions of important features, such as ethnicity and resources, do not match the political boundaries of individual countries.

Types of Regions

The designation of "region" can be applied to any area larger than a point and smaller than the entire planet. Geographers most often apply the concept at one of two scales:

- Several neighboring countries that share important features, such as those in Latin America
- Many localities within a country, such as those in southern California.

A particular place can be included in more than one region depending on how the region is defined.

Geographers identify three types of regions—formal, functional, and vernacular.

FORMAL REGION. A **formal region**, also called a uniform region or a homogeneous region, is an area within which everyone shares in common one or more distinctive characteristics. The shared feature could be a cultural value such as a common

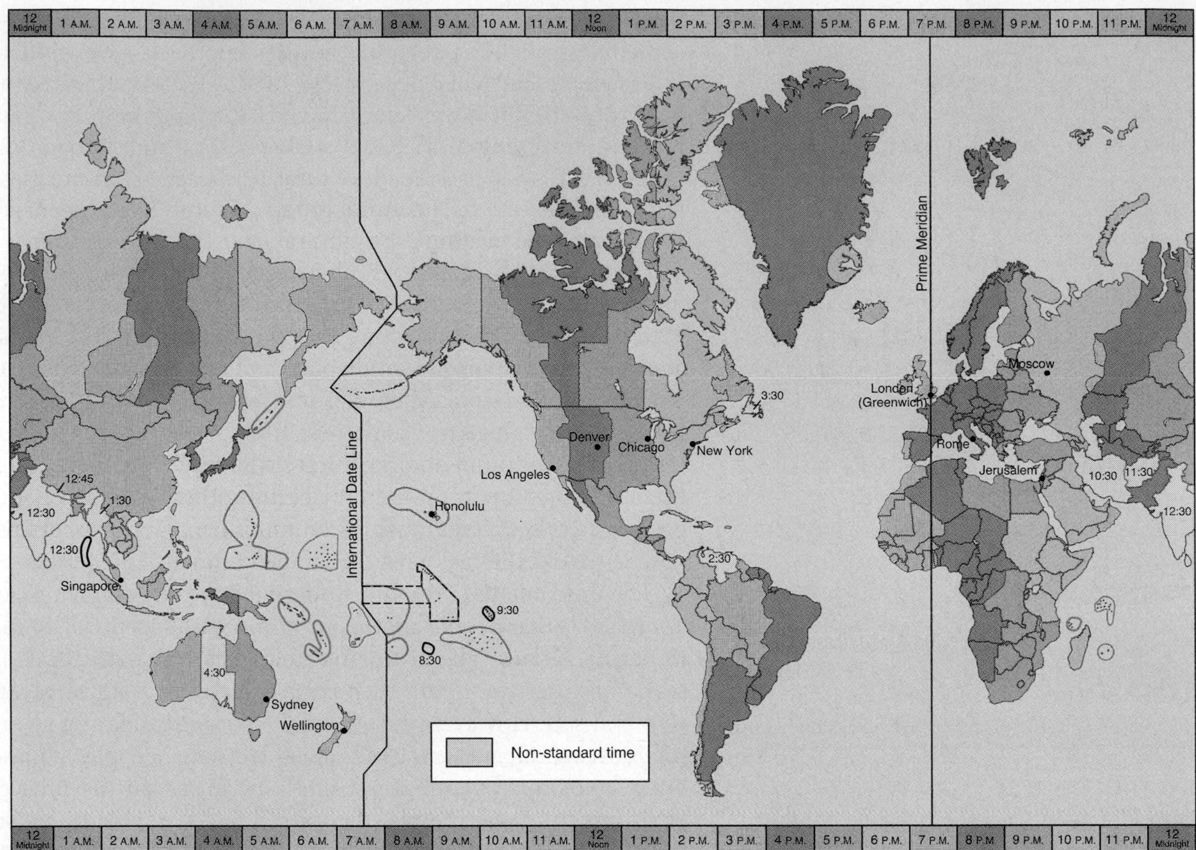

FIGURE 1-13 Time zones. Longitude plays an important role in calculating time. Earth as a sphere is divided into 360° of longitude (the degrees from 0° to 180° west longitude, plus the degrees from 0° to 180° east longitude).

As Earth rotates daily, these 360 imaginary lines of longitude pass beneath the cascading sunshine. If we let every fifteenth degree of longitude represent one time zone, and divide the 360° by 15°, we get 24 time zones, or one for each hour of the day. By international agreement, **Greenwich Mean Time (GMT)** or Universal Time (UT), which is the time at the prime meridian (0° longitude), is the master reference time for all points on Earth.

As Earth rotates eastward, any place to the east of you always passes "under" the Sun earlier. Thus as you travel eastward from the prime meridian, you are "catching up" with the Sun, so you must turn your clock ahead from GMT by 1 hour for each 15°. If you travel westward from the prime meridian, you are "falling behind" the Sun, so you turn your clock back from GMT by 1 hour for each 15°.

The eastern United States, which is near 75° west longitude, is therefore 5 hours earlier than GMT (the 75° difference between the prime meridian and 75° west longitude, divided by 15° per hour, equals 5 hours). Thus when the time is 11 AM GMT, the time in the eastern United States is 5 hours earlier, or 6 AM.

Each 15° band of longitude is assigned to a standard time zone. The 48 contiguous U.S. States and Canada share four standard time zones, known as Eastern, Central, Mountain, and Pacific:

- The Eastern Standard Time Zone is near 75° west longitude, which passes close to Philadelphia, and is 5 hours earlier than GMT.
- The Central Standard Time Zone is near 90° west longitude, which passes through Memphis, Tennessee, and is 6 hours earlier than GMT.
- The Mountain Standard Time Zone is near 105° west longitude, which passes through Denver, Colorado, and is 7 hours earlier than GMT.
- The Pacific Standard Time Zone is near 120° west longitude, which passes through Lake Tahoe in California, and is 8 hours earlier than GMT.

Most of Alaska is in the Alaska Time Zone, which is 9 hours earlier than GMT. Hawaii and some of the Aleutian Islands are in the Hawaii-Aleutian Time Zone, which is 10 hours earlier than GMT.

Eastern Canada is in the Atlantic Time Zone, which is 4 hours earlier than GMT. The residents of Newfoundland assert that their island, which lies between 53° and 59° west longitude, would face dark winter afternoons if it were 4 hours earlier than GMT, like the rest of eastern Canada, and dark winter mornings if it were 3 hours earlier than GMT. Therefore, Newfoundland is 3½ hours earlier than GMT.

Before standard time zones were created, each locality set its own time, usually that kept by a local jeweler. When railroads became the main cross-country transportation during the nineteenth century, each rail company kept its own time, normally that of the largest city it served. Train timetables listed two sets of arrival and departure times, one for local time and one for railroad company time. Railroad stations had one clock for local time and a separate clock for each of the railroad companies using the station.

To reduce the confusion from the multiplicity of local times, the railroads urged adoption of standard time zones. Standard time zones were established in the United States in 1883 and in the rest of the world following the international meridian conference in Washington, D.C., in 1884. At noon on November 18, 1883, time stood still in the United States so that each locality could adjust to the new standard time zones. In New York City, for example, time stopped for 3 minutes and 58 seconds to adjust to the new Eastern Standard Time.

When you cross the **International Date Line**, which, for the most part, follows 180° longitude, you move the clock back 24 hours, or one entire day, if you are heading eastward toward America. You turn the clock ahead 24 hours if you are heading westward toward Asia.

To see the need for the International Date Line, try counting the hours around the world from the time zone in which you live. As you go from west to east, you add 1 hour for each time zone. When you return to your starting point, you will reach the absurd conclusion that it is 24 hours later in your locality than it really is.

Therefore, when the time in New York City is 2 PM, it is 7 PM Sunday in London, 8 PM Sunday in Rome, 9 PM Sunday in Jerusalem, 10 PM Sunday in Moscow, 3 AM Monday in Singapore, and 5 AM Monday in Sydney, Australia. Continuing farther east, it is 7 AM Monday in Wellington, New Zealand—but when you get to Honolulu, it is 9 AM Sunday, because the International Date Line lies between New Zealand and Hawaii.

The International Date Line for the most part follows 180° longitude. However, in 1997, Kiribati, a collection of small islands in the Pacific Ocean, moved the International Date Line 3,000 kilometers (2,000 miles) to its eastern border near 150° west longitude. As a result, Kiribati is the first country to see each day's sunrise. Kiribati hoped that this feature would attract tourists to celebrate the start of the new millennium on January 1, 2000 (or January 1, 2001, when sticklers pointed out the new millennium really began). But it did not.

language, an economic activity such as production of a particular crop, or an environmental property such as climate. In a formal region the selected characteristic is present throughout.

Some formal regions are easy to identify, such as countries or local government units. Montana is an example of a formal region, characterized with equal intensity throughout the state by a government that passes laws, collects taxes, and issues license plates. The formal region of Montana has clearly drawn and legally recognized boundaries, and everyone living within them shares the status of being subject to a common set of laws.

In other kinds of formal regions a characteristic may be predominant rather than universal. For example, the North American wheat belt is a formal region in which wheat is the most commonly grown crop, but other crops are grown there as well. And the wheat belt can be distinguished from the corn belt—a region where corn is the most commonly grown crop.

Similarly, we can distinguish formal regions within the United States characterized by a predominant voting for Republican candidates, although Republicans do not get 100 percent of the votes in these regions—nor in fact do they always win (Figure 1-14, left). However, in a presidential election, the candidate with the largest number of votes receives all of the electoral votes of a state, regardless of the margin of victory. Consequently, a state that usually has Democratic electors can be considered a Democratic state (Figure 1-14, right).

Geographers typically identify formal regions to help explain broad global or national patterns, such as variations in religions and levels of economic development. The characteristic selected to distinguish a formal region often illustrates a general concept rather than a precise mathematical distribution.

A cautionary step in identifying formal regions is the need to recognize the diversity of cultural, economic, and environmental factors, even while making a generalization. Problems may arise because a minority of people in a region speak a language, practice a religion, or possess resources different from those of the majority. People in a region may play distinctive roles in the economy and hold different positions in society based on their gender or ethnicity.

FUNCTIONAL REGION.

A **functional region**, also called a nodal region, is an area organized around a node or focal point. The characteristic chosen to define a functional region dominates at a central focus or node and diminishes in importance outward. The region is tied to the central point by transportation or communications systems or by economic or functional associations.

Geographers often use functional regions to display information about economic areas. The region's node may be a shop or service, and the boundaries of the region mark the limits of the trading area of the activity. People and activities may be attracted to the node, and information may flow from the node to the surrounding area.

An example of a functional region is the reception area of a television station. A television station's signal is strongest at the center of its service area, becomes weaker at the edge, and eventually can no longer be distinguished from snow (Figure 1-15). At some distance from the center, more people are watching a station originating in another city. That place is the boundary between the nodal regions of the two TV market areas.

Other examples of functional regions include the circulation area of a newspaper and the trading area of a department store. A newspaper dominates circulation figures in the city in which it is published. Farther away from the city, fewer people read that newspaper, whereas more people read a newspaper published in a neighboring city. A department store attracts fewer customers from the edge of a trading area, and beyond that edge customers will most likely choose to shop elsewhere.

New technology is breaking down traditional functional regions. Television stations are broadcast to distant places by cable or satellite. Newspapers such as *USA Today, The Wall Street Journal,* and *The New York Times* are composed in one place, transmitted by satellite to printing machines in other places, and delivered to yet other places by airplane, truck, or the Internet. Customers can shop at distant stores by mail or the Internet.

VERNACULAR REGION.

A **vernacular region,** or perceptual region, is a place that people believe exists as part of their cultural identity. Such regions emerge from people's

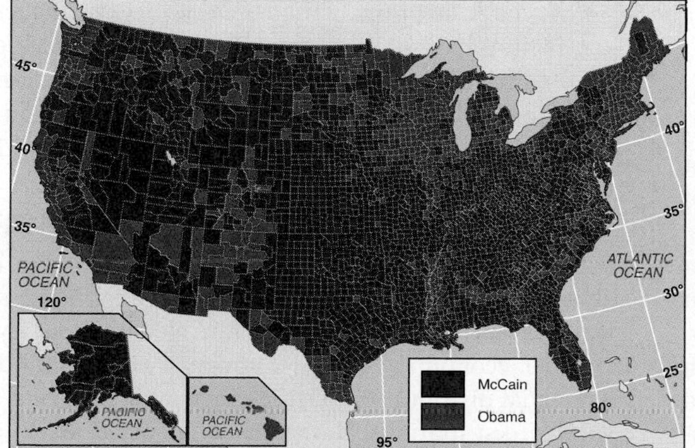

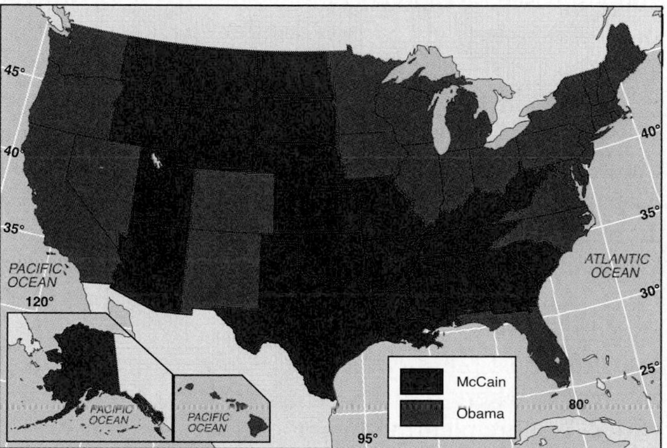

FIGURE 1-14 Formal regions. The two maps show the winner by county (left) and state (right) in the 2008 presidential election. Counties and states are examples of formal regions. The extensive areas of support for Obama (blue) and McCain (red) are also examples of formal regions. McCain carried 82 percent of the counties (left), but most of these were in rural areas, where population was lower than in the urban counties that Obama won. Obama won 28 of 50 states (right), and 53 percent of the overall vote.

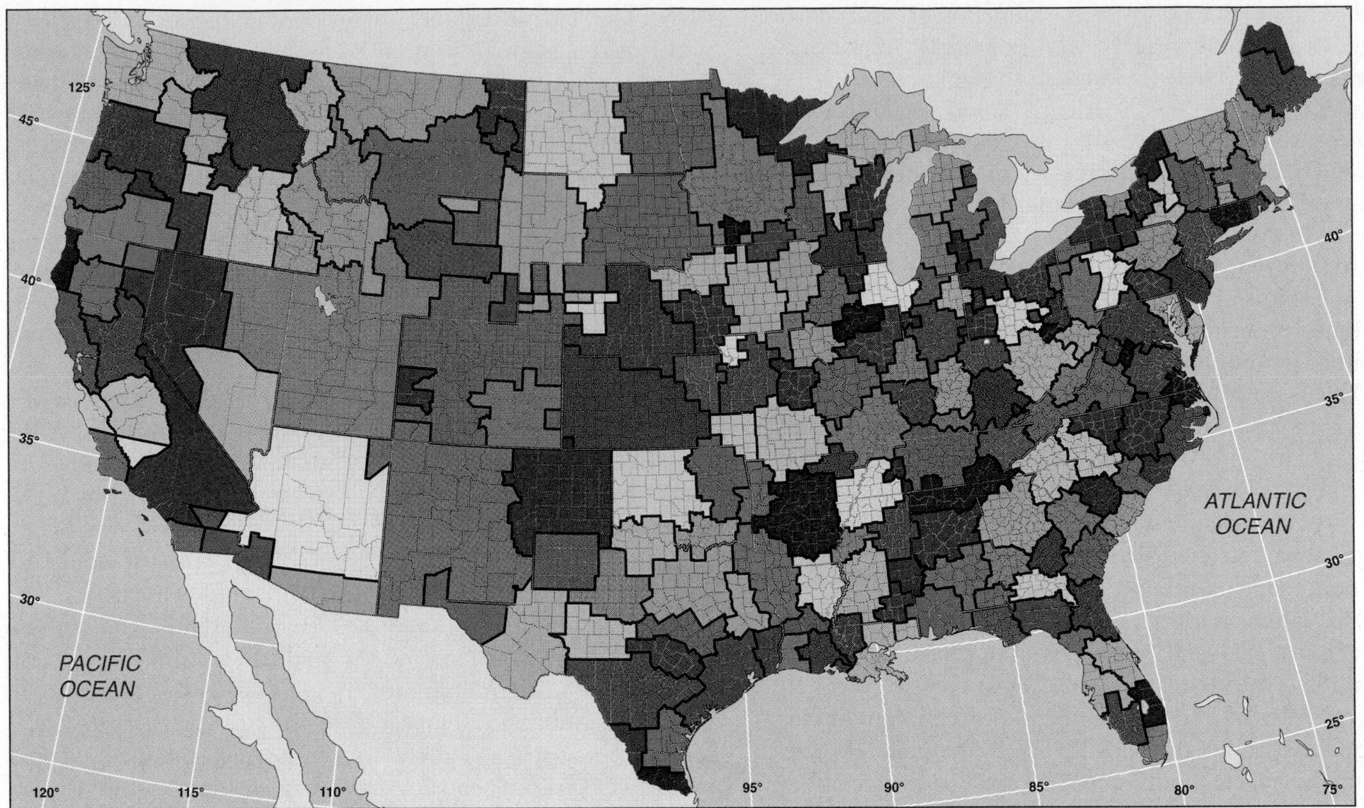

FIGURE 1-15 Functional regions. The United States is divided into functional regions based on television markets, which are groups of counties served by a collection of TV stations. Many of these TV market functional regions cross state lines.

informal sense of place rather than from scientific models developed through geographic thought.

A useful way to identify a perceptual region is to get someone to draw a **mental map**, which is an internal representation of a portion of Earth's surface. A mental map depicts what an individual knows about a place, containing personal impressions of what is in a place and where places are located. A student and a professor are likely to have different mental maps of a college campus, based on differences in where they work, live, and eat, and a senior is likely to have a more detailed and "accurate" map than a first-year student.

As an example of a vernacular region, Americans frequently refer to the South as a place with environmental, cultural, and economic features perceived to be quite distinct from the rest of the United States (Figure 1-16). Many of these features can be measured. Economically, the South is a region of high cotton production and low high school graduation rates. Culturally, the South includes the states that joined the Confederacy during the Civil War and where Baptist is the most prevalent religion. Environmentally, the South is a region where the last winter frost occurs in March, and rainfall is more plentiful in winter than in summer. Southerners and other Americans alike share a strong sense of the American South as a distinctive place that transcends geographic measurement. The perceptual region known as the South is a source of pride to many Americans—and for others it is a place to avoid.

Spatial Association

A region can be constructed to encompass an area of widely varying scale, from a very small portion of Earth to a very large portion. Different conclusions may be reached concerning a region's characteristics depending on its scale. Consider the percentage of Americans who die each year from cancer. Death rates vary widely among scales within the United States (Figure 1-17):

- At the scale of the United States, the Great Lakes and South regions have higher levels of cancer than the West.
- At the scale of the state of Maryland, the eastern region has a higher level of cancer than the western region.
- At the scale of the city of Baltimore, Maryland, lower levels of cancer are found in the northern region.

Maps showing regions of high and low cancer rates do not communicate useful information to someone who knows little about the regions. To explain why regions possess distinctive features, such as a high cancer rate, geographers try to identify cultural, economic, and environmental factors that display similar spatial distributions. Geographers conclude that factors with similar distributions have spatial association. By integrating other spatial information about people, activities, and environments, we can begin to see factors that may be associated with regional differences in cancer.

At the national scale, the Great Lakes region may have higher cancer rates in part because the distribution of cancer is

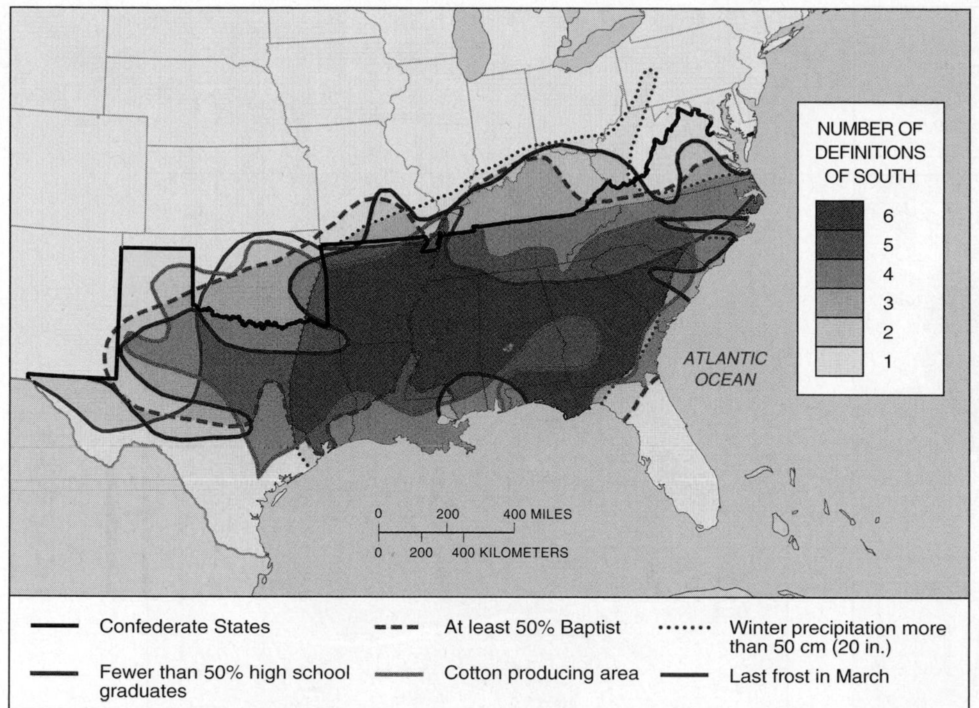

FIGURE 1-16 Vernacular regions. The South is popularly distinguished as a distinct vernacular region within the United States according to a number of factors, such as mild climate, propensity for growing cotton, and importance of the Baptist Church.

spatially associated with the distribution of factories. Residents of the South may have high cancer rates because, with lower levels of education and income, they may be less aware of the risks associated with activities such as smoking and less able to afford medical care to minimize the risk of dying from cancer.

Similarly, at the state scale, variations in regions may be associated with a combination of economic, cultural, and environmental factors. Baltimore City may have higher cancer rates because of a concentration of people with lower levels of income and education. People living in the rural Eastern Shore region may be exposed to runoff of chemicals from farms into the nearby Chesapeake Bay, as well as discharges carried by prevailing winds from factories further west.

At the urban scale, again, a combination of economic, cultural, and environmental factors may form a spatial association with the distribution of cancer. The ZIP codes on the north side of Baltimore City contain a higher percentage of people with high incomes, and are further from the city's factories and port facilities.

Regional Integration of Culture

In thinking about *why* each region on Earth is distinctive, geographers refer to **culture**, which is the body of customary beliefs, material traits, and social forms that together constitute the distinct tradition of a group of people. Geographers distinguish groups of people according to important cultural characteristics, describe where particular cultural groups are distributed, and offer reasons to explain the observed distribution.

In everyday language we think of *culture* as the collection of novels, paintings, symphonies, and other works produced by talented individuals. A person with a taste for these intellectual outputs is said to be "cultured." Intellectually challenging culture is often distinguished from *popular* culture, such as television programs. *Culture* also refers to small living organisms, such as those found under a microscope or in yogurt. *Agriculture* is a term for the growing of living material at a much larger scale than in a test tube.

The origin of the word *culture* is the Latin *cultus,* which means "to care for." Culture is a complex concept because "to care for" something has two very different meanings:

- To care *about*—to adore or worship something, as in the modern word *cult*.
- To take care *of*—to nurse or look after something, as in the modern word *cultivate.*

Geography looks at both of these facets of the concept of culture to see why each region in the world is unique.

When geographers think about culture, they may be referring to either one of the two main meanings of the concept. Some geographers study what people care about (their ideas, beliefs, values, and customs), whereas other geographers emphasize what people take care of (their ways of earning a living and obtaining food, clothing, and shelter).

WHAT PEOPLE CARE ABOUT. Geographers study why the customary ideas, beliefs, and values of a people produce a distinctive culture in a particular place. Especially important cultural values derive from a group's language, religion, and ethnicity. These three cultural traits are both an excellent way of identifying the location of a culture and the principal means by which cultural values become distributed around the world.

Language is a system of signs, sounds, gestures, and marks that have meanings understood within a cultural group. People communicate the cultural values they care about through language, and the words themselves tell something about where different cultural groups are located. The distribution of speakers of different languages and reasons for the distinctive distribution are discussed in Chapter 5.

Religion is an important cultural value because it is the principal system of attitudes, beliefs, and practices through which people worship in a formal, organized way. As discussed in Chapter 6, geographers look at the distribution of religious groups around the world and the different ways that the various groups interact with their environment.

Ethnicity encompasses a group's language, religion, and other cultural values, as well as its physical traits. A group possesses these cultural and physical characteristics as a product of

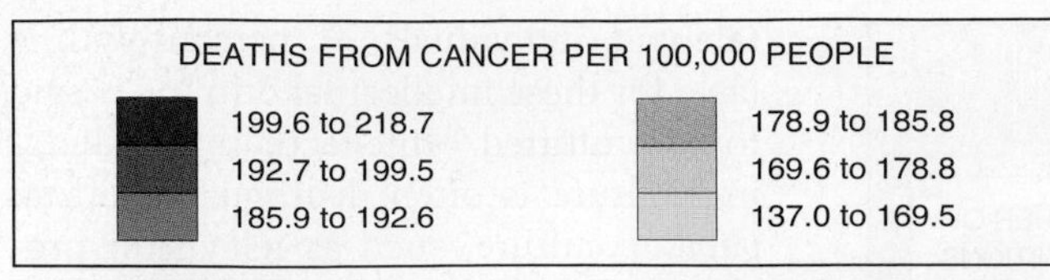

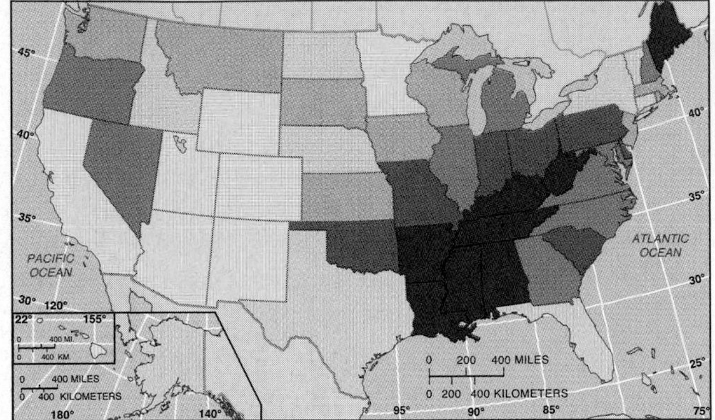

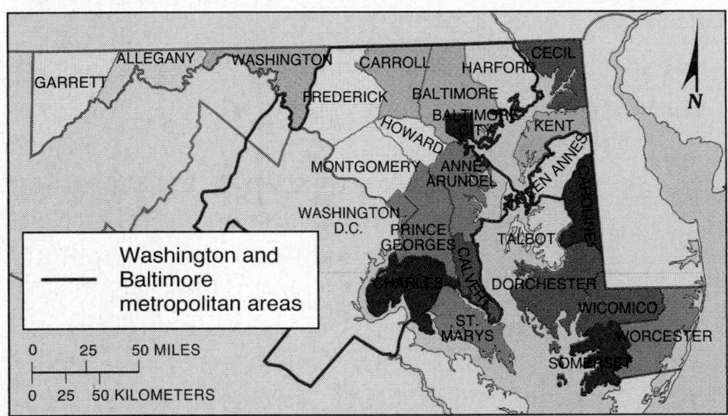

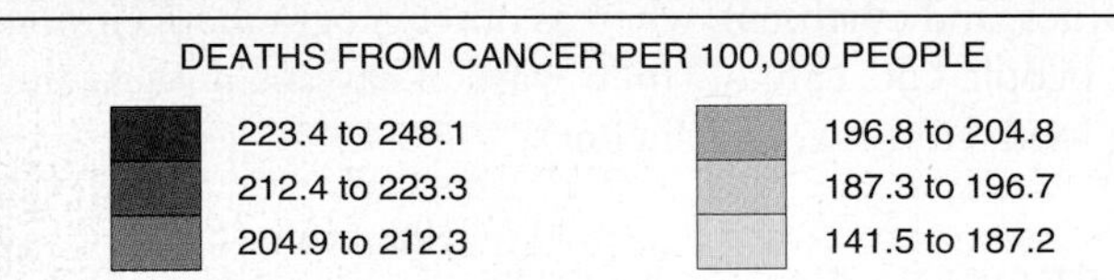

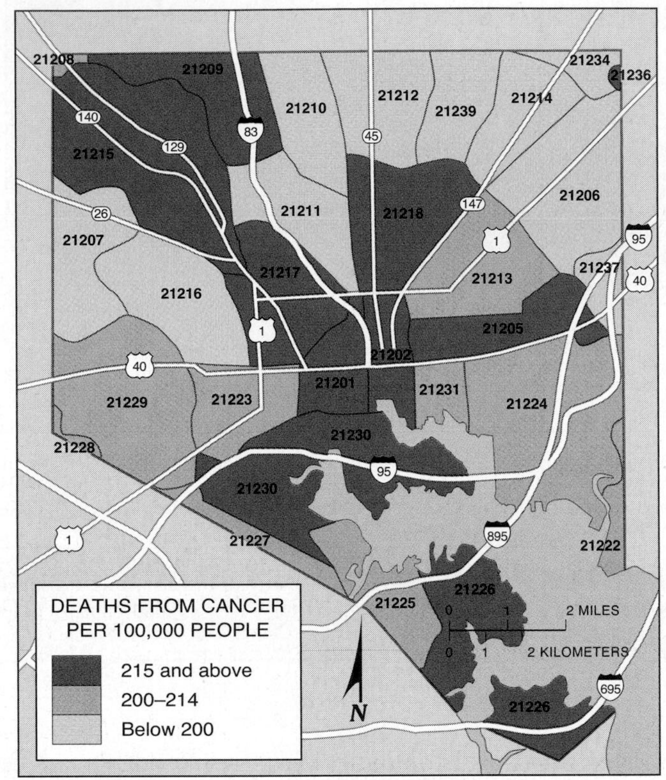

FIGURE 1-17 Spatial association. On the national scale, the Great Lakes and South regions have higher cancer rates than the Western region. On the scale of the state of Maryland, the eastern region has a higher cancer rate than the western region. On the urban scale, southern and northwestern neighborhoods of Baltimore City have higher cancer rates than northeastern ones. Geographers try to understand the reason for these variations.

its common traditions and heredity. As addressed in Chapter 7, geographers find that problems of conflict and inequality tend to occur in places where more than one ethnic group inhabits and seeks to organize the same territory.

WHAT PEOPLE TAKE CARE OF. The second element of culture of interest to geographers is production of material wealth—the food, clothing, and shelter that humans need in order to survive and thrive. All people consume food, wear clothing, build shelter, and create art, but different cultural groups obtain their wealth in different ways.

Geographers divide the world into regions of more (or relatively) developed countries (abbreviated MDCs), and regions of less developed (or developing) countries (abbreviated LDCs). Regions of MDCs include North America, Europe, and Japan, and regions of LDCs include sub-Saharan Africa, the Middle East, East Asia, South Asia, Southeast Asia, and Latin America. Various shared characteristics—such as per capita income, literacy rates, televisions per capita, and hospital beds per capita—distinguish regions of MDCs and regions of LDCs. These differences are reviewed in Chapter 9.

Possession of wealth and material goods is higher in MDCs because of different types of economic activities than those in LDCs. Most people in LDCs are engaged in agriculture, whereas most people in MDCs earn their living through manufacturing products or performing services in exchange for wages. This fundamental economic difference between MDCs and LDCs is discussed in more detail in Chapters 10 through 13.

Geographers are also interested in the political institutions that protect material artifacts, as well as cultural values. The world is organized into a collection of countries, or states, controlled by governments put in place through various representative and unrepresentative means. A major element of a group's cultural identity is its citizenship, the country or countries that it inhabits and in which it pays taxes, votes, and otherwise participates in the administration of space.

GLOBAL FORCES, LOCAL IMPACTS

Hurricane Katrina

Because geographers are trained in both social and physical sciences, they are particularly well equipped to understand interactions between people and their environment. An example is the devastation in the southern United States after Hurricane Katrina in 2005.

Physical geography concepts explain the process by which hurricanes, such as Katrina, form in the Atlantic Ocean during the late summer and autumn and gather strength over the warms waters of the Gulf of Mexico. When it passes over land, a hurricane can generate a powerful storm surge that floods low-lying areas (Figure 1-18, left).

It is here that physical and human geography intersect. Katrina caused massive damage, in part because it made landfall near heavily populated areas, including the cities of Biloxi and Gulfport, Mississippi; Mobile, Alabama; and New Orleans, Louisiana. In an effort to protect these low-lying cities from flooding, government agencies constructed a complex system of levees, dikes, seawalls, canals, and pumps. The experience of Katrina proved that humans are not able to control and tame all of the forces of nature.

Human geographers are especially concerned with the uneven impact of destruction. Hurricane Katrina's victims were primarily poor, African American, and older individuals (Figure 1-18, center). They lived in the lowest-lying areas, most vulnerable to flooding, and many lacked transportation, money, and information that would have enabled them to evacuate in advance of the storm.

The wealthy portions of New Orleans, such as tourist attractions like the French Quarter, were spared the worst because they were located on slightly higher ground. The slow and incompetent response to the destruction by local, state, and federal emergency teams was attributed by many analysts to the victims' lack of a voice in the political, economic, and social life of New Orleans and other impacted communities.

Inequalities persist several years after the hurricane (Figure 1-18, right). In 2009, four years after the hurricane, more than 90 percent of householders in predominantly white areas were back in their homes, as measured by whether they were receiving mail. In contrast, less than two-thirds of the households in several predominantly African American neighborhoods were receiving mail. ■

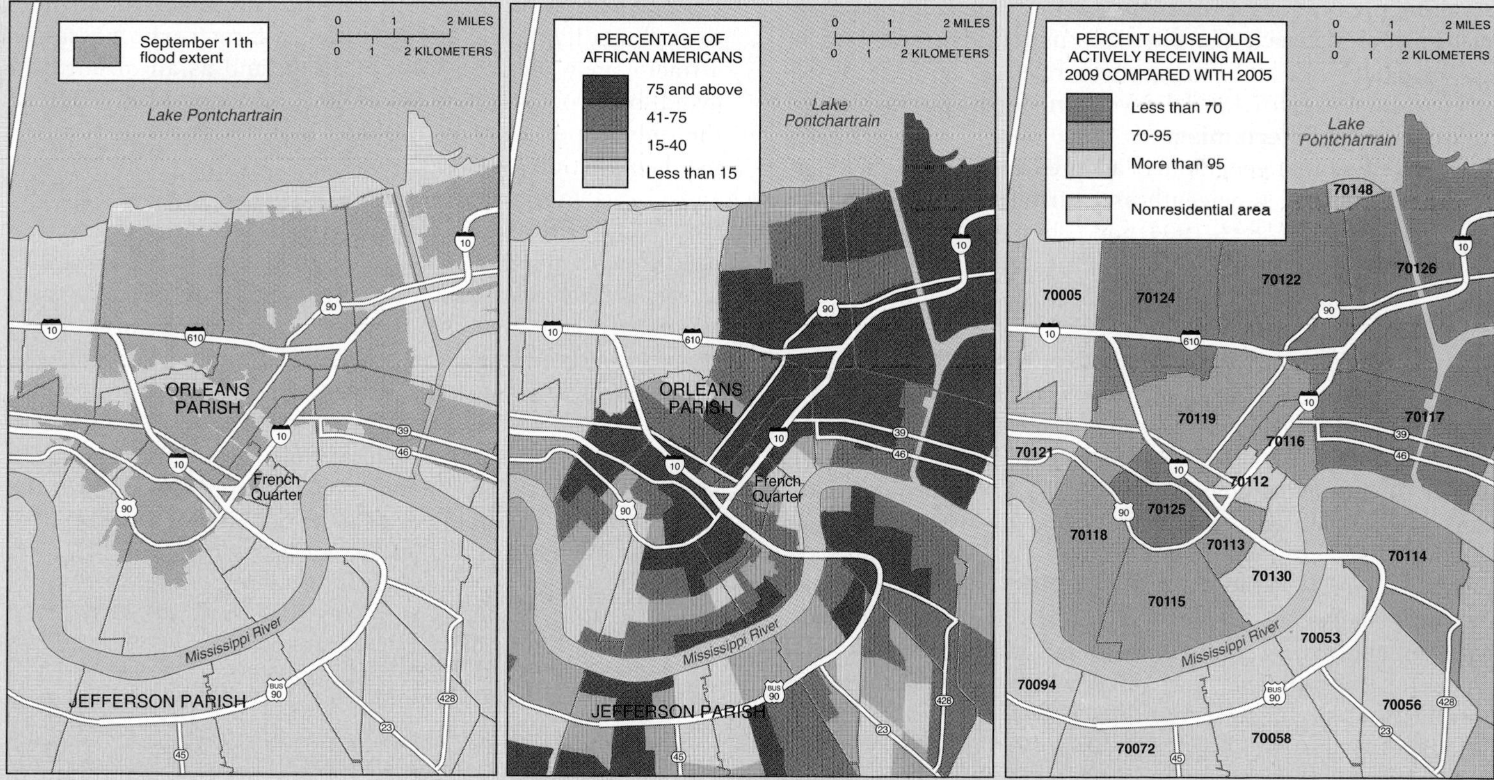

FIGURE 1-18 Cultural ecology: New Orleans after Hurricane Katrina. From a physical geography perspective, 80 percent of New Orleans was underwater after the city's flood-protection levees broke (left). The 20 percent that was not flooded was land at slightly higher elevations, including the leading tourist destinations in the Vieux Carré (French Quarter). From a social science perspective, at the time of the hurricane two-thirds of the population of New Orleans was African American (middle). However, the population in the area that was not flooded was less than one-fourth African American. The percentage of homes that have been fixed up and reoccupied since the hurricane is lower in the areas that had relatively large African American populations (right).

As discussed in Chapter 8, cultural groups in the modern world are increasingly asserting their right to organize their own affairs at the local scale rather than submit to the control of other cultural groups. Political problems are found in places where the area occupied by a cultural group does not coincide with the boundaries of a country.

Cultural Ecology: Integrating Culture and Environment

In constructing regions, geographers consider environmental as well as cultural factors. Distinctive to geography is the importance given to relationships between culture and the natural environment. Different cultural groups modify the natural environment in distinctive ways to produce unique regions. The geographic study of human–environment relationships is known as **cultural ecology**.

Pioneering nineteenth-century German geographers Alexander von Humboldt (1769–1859) and Carl Ritter (1779–1859) urged human geographers to adopt the methods of scientific inquiry used by natural scientists. They argued that the scientific study of social and natural processes is fundamentally the same. Natural scientists have made more progress in formulating general laws than have social scientists, so an important goal of human geographers is to discover general laws.

According to Humboldt and Ritter, human geographers should apply laws from the natural sciences to understanding relationships between the physical environment and human actions. Humboldt and Ritter concentrated on how the physical environment *caused* social development, an approach called **environmental determinism**.

Other influential geographers adopted environmental determinism in the late nineteenth and early twentieth centuries. Friedrich Ratzel (1844–1904) and his American student, Ellen Churchill Semple (1863–1932), claimed that geography was the study of the influences of the natural environment on people.

Another early American geographer, Ellsworth Huntington (1876–1947), argued that climate was a major determinant of civilization. For instance, according to Huntington, the temperate climate of maritime northwestern Europe produced greater human efficiency as measured by better health conditions, lower death rates, and higher standards of living.

HUMAN AND PHYSICAL FACTORS. To explain relationships between human activities and the physical environment in a region, modern geographers reject environmental determinism in favor of possibilism. According to **possibilism**, the physical environment may limit some human actions, but people have the ability to adjust to their environment. People can choose a course of action from many alternatives in the physical environment. Humans endow the physical environment with cultural values by regarding it as a collection of **resources**, which are substances that are useful to people, economically and technologically feasible to access, and socially acceptable to use.

For example, the climate of any location influences human activities, especially food production. From one generation to the next, people learn that different crops thrive in different climates—rice requires plentiful water, whereas wheat survives on limited moisture and actually grows poorly in very wet environments. On the other hand, wheat is more likely than rice to be grown successfully in colder climates. Thus, under possibilism, it is possible for people to choose the crops they grow and to be compatible with their environment.

Human geographers use this cultural ecology, or human–environment, approach to explain many global issues. For example, world population growth is a problem if the number of people exceeds the capacity of the physical environment to produce food. However, people can adjust to the capacity of the physical environment by controlling their numbers, adopting new technology, consuming different foods, migrating to new locations, and taking other actions.

Some human impacts on the environment are casual, and some are based on deep-seated cultural values. Why do we plant our front yard with grass, water it to make it grow, mow it to keep it from growing tall, and impose fines on those who fail to mow often enough? Why not let dandelions grow or pour concrete instead? Why does one group of people consume the fruit from deciduous trees and chop down the conifers for building materials, whereas another group chops down the deciduous trees for furniture while preserving the conifers as religious symbols?

A people's level of wealth can also influence its attitude toward modifying the environment. A farmer who possesses a tractor may regard a hilly piece of land as an obstacle to avoid, but a poor farmer with a hoe may regard hilly land as the only opportunity to produce food for survival through hand cultivation.

PHYSICAL PROCESSES: CLIMATE. Human geographers need some familiarity with global environmental processes to understand the distribution of human activities, such as where people live and how they earn a living. Important physical processes include climate, vegetation, soil, and landforms.

Climate is the long-term average weather condition at a particular location. Geographers frequently classify climates according to a system developed by German climatologist Vladimir Köppen. The modified Köppen system divides the world into five main climate regions that are identified by the letters A through E as well as by names:

- A Tropical Climates
- B Dry Climates
- C Warm Mid-Latitude Climates
- D Cold Mid-Latitude Climates
- E Polar Climates

The modified Köppen system divides the five main climate regions into several subtypes (Figure 1-19). For all but the B climate, the basis for the subdivision is the amount of precipitation and the season in which it falls. For the B climate, subdivision is made on the basis of temperature and precipitation.

Humans have a limited tolerance f[illegible]treme temperature and precipitation levels and thus a[illegible] in places that are too hot, too cold, too wet, o[illegible]mpare the map of global climate to the distri[illegible]pulation (see Figure 2-3). Relatively few pe[illegible]e Dry (B) and Polar (E) climate regions.

The climate of a pa[illegible]on influences human activities, especially pr[illegible]he food needed to survive. People in parts of th[illegible] region, especially southwestern India, Bangladesh, and [illegible]Myanmar (Burma) coast, anxiously await the annual monsoon rain, which is essential for successful agriculture and provides nearly 90 percent of India's water supply (Figure 1-20). For most of the year, the region receives dry, somewhat cool air from the northeast. In June, the wind direction suddenly shifts, bringing moist, warm, southwesterly air, known as the *monsoon*, from the Indian Ocean. The monsoon rain lasts until September. In years when the monsoon rain is delayed or fails to arrive—in recent decades, at least one-fourth of the time—agricultural output falls and famine threatens in the countries of South Asia, where nearly 20 percent of the world's people live. The monsoon rain is so important in India that the words for "year," "rain," and "rainy season" are identical in many local languages.

PHYSICAL PROCESSES: VEGETATION. Plant life covers nearly the entire land surface of Earth. Earth's land vegetation includes four major forms of plant communities, called biomes. Their location and extent are influenced by both climate and human activities. Vegetation and soil, in turn, influence the types of agriculture that people practice in a particular region. The four main biomes are forest, savanna, grassland, and desert.

- **Forest biome.** Trees form a continuous canopy over the ground; grasses and shrubs may grow beneath the cover. The forest biome covers a large percentage of Earth's surface, including much of North America, Europe, and Asia, as well as tropical areas of South America, Africa, and Southeast Asia.
- **Savanna biome.** The trees do not form a continuous canopy, and the resultant lack of shade allows grass to grow. Savanna covers large areas of Africa, South Asia, South America, and Australia.
- **Grassland biome.** Land is covered by grass rather than trees; few trees grow in the region because of low precipitation. Early explorers from northern Europe and eastern North America regarded the American prairies—the

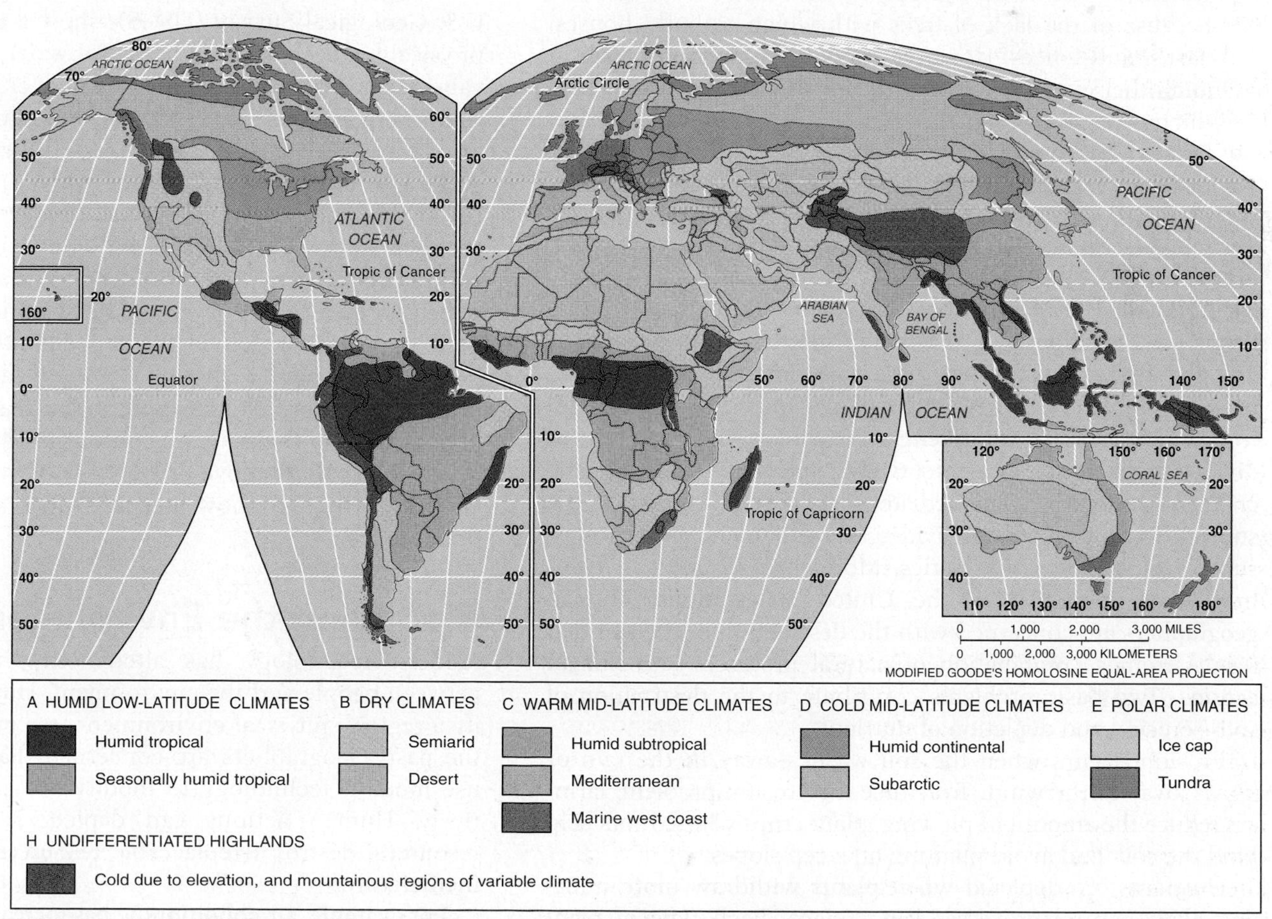

FIGURE 1-19 Climate regions. Geographers frequently classify global climates according to a system developed by Vladimir Köppen. The modified Köppen system divides the world into five main climate regions, represented by the letters A, B, C, D, and E.

FIGURE 1-20 Monsoon in India. Camels graze on a sand dune turned green by summer monsoon rains in Mathania, Rajasthan state, India.

world's most extensive grassland area—to be uninhabitable because of the lack of trees with which to build houses, barns, and fences. However, modern cultivation of wheat and other crops has turned the grasslands into a very productive region.

- **Desert biome.** Although many desert areas have essentially no vegetation, the region contains dispersed patches of plants adapted to dry conditions. Vegetation is often sufficient for the survival of small numbers of animals.

PHYSICAL PROCESSES: SOIL. Soil, the material that forms on Earth's surface, is the thin interface between the air and the rocks. Not merely dirt, soil contains the nutrients necessary for successful growth of plants, including those useful to humans. The U.S. Comprehensive Soil Classification System divides global soil types into twelve *orders*, according to the characteristics of the immediate surface soil layers and the subsoil. The orders are subdivided into suborders, great groups, subgroups, families, and series. More than 12,000 soil types have been identified in the United States alone. Human geographers are concerned with the destruction of the soil that results from a combination of natural processes and human actions. Two basic problems contribute to the destruction of soil—erosion and depletion of nutrients.

Erosion occurs when the soil washes away in the rain or blows away in the wind. To reduce the erosion problem, farmers reduce the amount of plowing, plant crops whose roots help bind the soil, and avoid planting on steep slopes.

Nutrients are depleted when plants withdraw more nutrients than natural processes can replace. Each type of plant withdraws certain nutrients from the soil and restores others. Repeated harvesting of the same type of crop year after year can remove certain nutrients and reduce the soil's productivity. To minimize depletion, farmers in MDCs sometimes plant crops that offer no economic return but restore nutrients to the soil and keep the land productive over a longer term. Farmers also restore nutrients to the soil by adding fertilizers, either natural or synthetic. Farmers in LDCs may face greater problems with depletion of nutrients because they lack knowledge of proper soil management practices and funds to buy fertilizer.

PHYSICAL PROCESSES: LANDFORMS. Earth's surface features, or landforms, vary from relatively flat to mountainous. Geographers find that the study of Earth's landforms—a science known as geomorphology—helps to explain the distribution of people and the choice of economic activities at different locations. People prefer living on flatter land, which generally is better suited for agriculture. Great concentrations of people and activities in hilly areas may require extensive effort to modify the landscape.

Topographic maps, published (for the United States) by the U.S. Geological Survey (USGS), show a remarkable detail of physical features, such as bodies of water, forests, mountains, valleys, and wetlands. They also show cultural features, such as buildings, roads, parks, farms, and dams. "Topos," as they are called, are used by engineers, hikers, hunters, people seeking a homesite, and anyone who really needs to see the lay of the land. Geographers use topographic maps to study the relief and slope of localities. Relief is the difference in elevation between any two points, and it measures the extent to which an area is flat or hilly. The steepness of hills is measured by slope, which is the relief divided by the distance between two points. Figure 1-5 shows a portion of a USGS map for northern Mississippi, at the scale of 1:24,000. The brown lines on the map are contour lines that connect points of equal elevation above or below sea level. Contour lines are closer together to show steeper slopes and farther apart in flatter areas.

Modifying the Environment

Modern technology has altered the historic relationship between people and the environment. Humans now can modify a region's physical environment to a greater extent than in the past. Geographers are concerned that people sometimes use modern technology to modify the environment insensitively. Human actions can deplete scarce environmental resources, destroy irreplaceable resources, and use resources inefficiently.

For example, air-conditioning has increased the attractiveness of living in regions with warmer climates. But the refrigerants in the air conditioners have also increased the amount of chlorofluorocarbons in the atmosphere, damaging the ozone layer that protects living things from UV rays and contributing to global warming. We explore the consequences of such use, abuse, and misuse of the environment in more detail in Chapter 14.

Few regions have been as thoroughly modified by humans as the Netherlands and Florida's Everglades. Because more than half of the Netherlands lies below sea level, most of the country today would be under water if it were not for massive projects to modify the environment by holding back the sea. Meanwhile, the fragile landscape of south Florida has been altered in insensitive ways.

THE NETHERLANDS: SENSITIVE ENVIRONMENTAL MODIFICATION. The Dutch have a saying that "God made Earth, but the Dutch made the Netherlands." The Dutch have modified their environment with two distinctive types of construction projects—polders and dikes.

A **polder** is a piece of land that is created by draining water from an area. Polders, first created in the thirteenth century, were constructed primarily by private developers in the sixteenth and seventeenth centuries and by the government during the past 200 years. All together, the Netherlands has 6,500 square kilometers (2,600 square miles) of polders, comprising 16 percent of the country's land area (Figure 1-21). The Dutch government has reserved most of the polders for agriculture to reduce the country's dependence on imported food. Some of the polders are used for housing, and one contains Schiphol, one of Europe's busiest airports.

The second distinctive modification of the landscape in the Netherlands is the construction of massive dikes to prevent the North Sea, an arm of the Atlantic Ocean, from flooding much of the country. The Dutch have built dikes in two major locations—the Zuider Zee project in the north and the Delta Plan project in the southwest.

The Zuider Zee, an arm of the North Sea, once threatened the heart of the Netherlands with flooding. A dike completed in 1932 caused the Zuider Zee to be converted from a saltwater sea to a freshwater lake. The newly created body of water was named the IJsselmeer, or Lake IJssel, because the IJssel River now flows into it. Some of the lake has been drained to create several polders, encompassing an area of 1,600 square kilometers (620 square miles).

A second ambitious project in the Netherlands is the Delta Plan in the southwestern part of the country. Flowing through the Netherlands are several important rivers, including the Rhine (Europe's busiest river), the Maas (known as the Meuse in France), and the Scheldt (known as the Schelde in Belgium). As these rivers flow into the North Sea, they split into many branches and form a low-lying delta that is vulnerable to flooding. After a devastating flood in January 1953 killed nearly 2,000

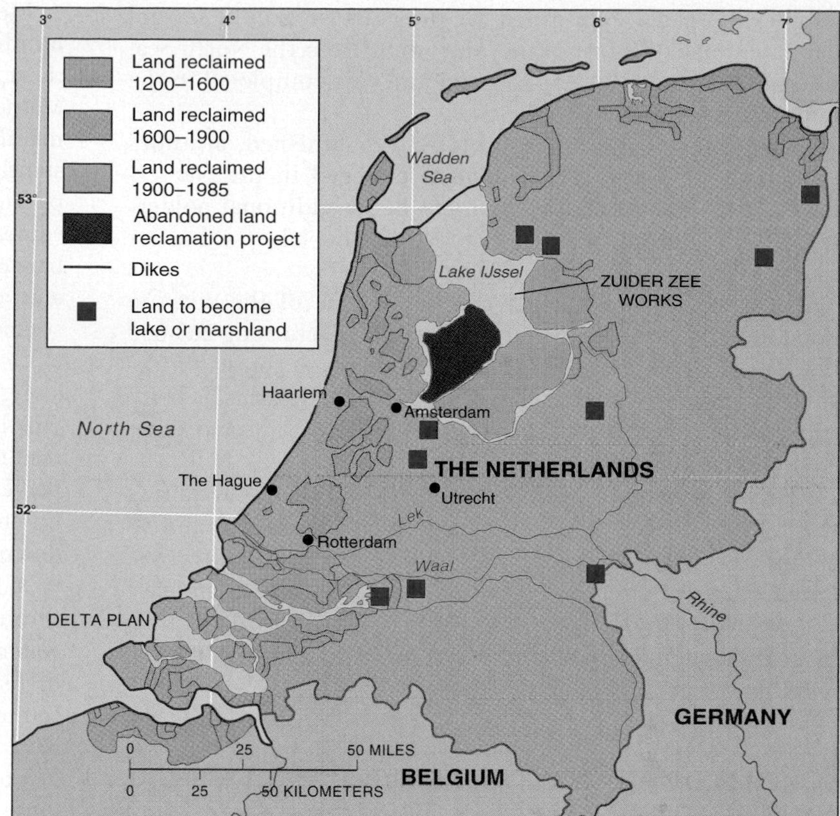

FIGURE 1-21 Environmentally sensitive cultural ecology in the Netherlands. The Dutch people have considerably altered the site of the Netherlands through creation of polders and dikes.

The first step in making a polder is to build a wall encircling the site, which is still underwater. Then the water inside the walled area is pumped from the site into either nearby canals or the remaining portion of the original body of water. Once dry, the site is prepared for human activities.

In the late nineteenth century, Dutch engineer Cornelis Lely proposed an ambitious project to seal off the Zuider Zee permanently from the North Sea, the ultimate source of the floodwaters. In accordance with Lely's plan, a dike was built, 32 kilometers (20 miles) long, across the mouth of the Zuider Zee to block the flow of North Sea water and create Lake IJssel.

After a devastating flood in 1953, the Delta Plan built dikes to close off most of the waterways in the southwestern part of the country. Because Rotterdam, Europe's largest port, is located nearby, some of the waterways were kept open. The Delta Works are barely visible in the background of the photograph of a polder.

people, the Delta Plan called for the construction of several dams to close off most of the waterways from the North Sea. The project took 30 years to build and was completed in the mid-1980s.

Once these two massive projects were finished, attitudes toward modifying the environment changed in the Netherlands. The Dutch scrapped plans to build additional polders in the IJsselmeer in order to preserve the lake's value for recreation.

The Dutch are deliberately breaking some of the dikes to flood fields. A plan adopted in 1990 called for returning 263,000 hectares (650,000 acres) of farms to wetlands or forests. Widespread use of insecticides and fertilizers on Dutch farms has contributed to contaminated drinking water, acid rain, and other environmental problems.

Global warming could threaten the Netherlands by raising the level of the sea around the country by between 20 and 58 centimeters (8 and 23 inches) within the next 100 years. Rather than build new dikes and polders, the Dutch have become world leaders in reducing the causes of global warming by acting to reduce industrial pollution and increase solar and wind power use, among other actions.

SOUTH FLORIDA: NOT-SO-SENSITIVE ENVIRONMENTAL MODIFICATION. Sensitive environmental areas in South Florida include barrier islands along the Atlantic and Gulf coasts, the wetlands between Lake Okeechobee and the Everglades National Park, and the Kissimmee River between Lake Kissimmee and Lake Okeechobee (Figure 1-22). These lowlands have been modified less sensitively than those in the Netherlands.

The Everglades was once a very wide and shallow freshwater river 80 kilometers (50 miles) wide and 15 centimeters (6 inches) deep, slowly flowing south from Lake Okeechobee to the Gulf of Mexico. A sensitive ecosystem of plants and animals once thrived in this distinctive landscape, but much of it has been destroyed by human actions.

The U.S. Army Corps of Engineers built a levee around Lake Okeechobee during the 1930s, drained the northern one-third of the Everglades during the 1940s, diverted the Kissimmee River into canals during the 1950s, and constructed dikes and levees near Miami and Fort Lauderdale during the 1960s. The southern portion of the Everglades became a National Park. These modifications opened up hundreds of thousands of hectares of land for growing sugarcane and protected farmland as well as the land occupied by the growing South Florida population from flooding. But they had unintended consequences for South Florida's environment.

Polluted water mainly from cattle grazing along the banks of the canals flowed into Lake Okeechobee, which is the source of fresh water for half of Florida's population. Fish in the lake began to die from the high levels of mercury, phosphorous, and other contaminants. The polluted water then continued to flow south into the National Park, threatening native vegetation such as sawgrass and endangering rare birds and other animals.

Meanwhile, Florida's barrier islands are home to several hundred thousand people. These barrier islands, as well as those elsewhere along the Atlantic and Gulf coasts between Maine and Texas, are essentially large sandbars that shield the mainland from flooding and storm damage. They are constantly being eroded and shifted from the force of storms and pounding surf, and after a major storm, large sections are sometimes washed away. Despite their fragile condition, the barrier islands are attractive locations for constructing homes and recreational facilities to take advantage of proximity to the seashore. Most of the barrier islands are linked with the mainland by bridge, causeway, or ferry service. To fight erosion along the barrier islands, people build seawalls and jetties, which are structures extending into the sea, but these projects result in more damage than protection. A seawall or jetty can prevent sand from drifting away, but by trapping sand along the up-current side, it causes erosion on the barrier islands on the down-current side.

A 2000 plan called for restoring the historic flow of water through South Florida while improving flood control and water quality. A 2008 plan called for the state to acquire hundreds of thousands of acres of land from sugarcane growers. But to date, few elements of the plans to restore the Everglades have been implemented. One-half of the Everglades has been lost to development. In an ironic reminder of the Dutch saying quoted earlier, Floridians say, "God made the world in six days, and the Army Corps of Engineers has been tinkering with it ever since."

KEY ISSUE 3

Why Are Different Places Similar?

- **Scale: From Local to Global**
- **Space: Distribution of Features**
- **Connections Between Places**

Although accepting that each place or region on Earth is unique, geographers recognize that human activities are rarely confined to one location. Discussed in this section are three basic concepts—scale, space, and connections—that help geographers understand why two places or regions can display similar features. ■

Scale: From Local to Global

Geographers think about scale at many levels, from local to global. At a local scale, such as an urban neighborhood, geographers tend to see unique features. At the global scale, encompassing the entire world, geographers tend to see broad patterns.

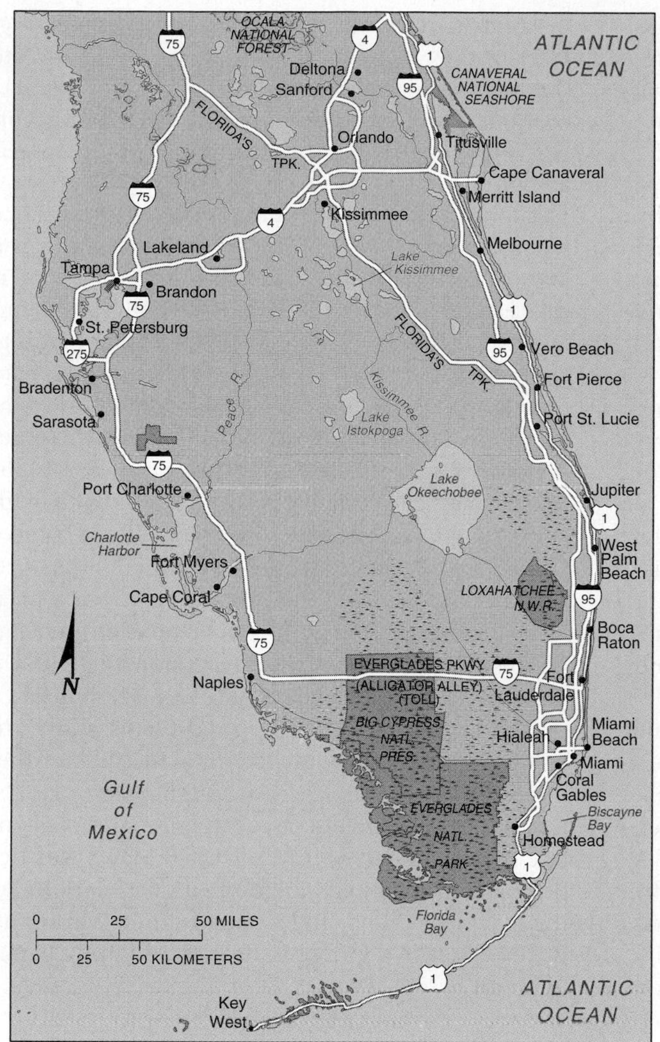

FIGURE 1-22 Environmentally insensitive cultural ecology in Florida. To control flooding in central Florida, the U.S. Army Corps of Engineers straightened the course of the Kissimmee River, which had meandered for 160 kilometers (98 miles) from near Orlando to Lake Okeechobee. The water was rechanneled into a canal 90 meters wide (300 feet) and 9 meters deep (30 feet), running in a straight line for 84 kilometers (52 miles). After the canal, known as C-38, opened in 1971, millions of gallons of polluted water—mainly runoff from cattle grazing—began pouring into Lake Okeechobee, which is the major source of freshwater for about half of Florida's population. The U.S. Army Corps of Engineers has returned the river from the canal (on the right side of the photograph) to its original course (on the left side).

A generation ago, people concerned with environmental quality proclaimed, "Think global, act local." The phrase meant that the environment was being harmed by processes such as global warming that were global in scale, but it could be improved by actions, such as consuming less gasoline, that were local in scale. Contemporary geographers offer a different version of the phrase: "Think and act both global and local." All scales from local to global are important in geography—the appropriate scale depends on the specific subject.

Geography matters in the contemporary world because it can explain human actions at all scales, from local to global. At the national and international scales, geography is concerned with such questions as where the population is growing rapidly, where the followers of different religions live, and where corporations place factories. Geography also studies why these arrangements can cause problems. Why can rapid population growth exceed available food supply? Why are different religious groups unable to live in peace with each other? Why are some places unable to attract or retain industries?

Globalization of Economy

Scale is an increasingly important concept in geography because of **globalization**, which is a force or process that involves the entire world and results in making something worldwide in scope. Globalization means that the scale of the world is shrinking—not literally in size, of course, but in the ability of a person, object, or idea to interact with a person, object, or idea in another place.

People are plugged into a global economy and culture, producing a world that is more uniform, integrated, and interdependent. The world contains only a handful of individuals who lead such isolated and sheltered lives that they have never watched a television set, used a telephone, or been in a motor vehicle. Even extremely isolated and sheltered people are at least aware of the existence of these important means of connection.

A few people living in very remote regions of the world may be able to provide all of their daily necessities. The crop grown or product manufactured in a particular place may be influenced by the distinctive features and assets of the place.

But most economic activities undertaken in one region are influenced by interaction with decision makers located elsewhere. The choice of crop is influenced by demand and prices set in markets elsewhere. The factory is located to facilitate bringing in raw materials and shipping out products to the markets.

Globalization of the economy has been led primarily by transnational corporations, sometimes called multinational corporations (Figure 1-23). A **transnational corporation** conducts research, operates factories, and sells products in many countries, not just where its headquarters and principal shareholders are located.

Historically, people and companies had difficulty moving even small sums of money from one country to another. International transfer of money involved a cumbersome set of procedures, and funds could be frozen for several weeks until all of the paperwork cleared. Most governments prohibited the removal of large sums of money, and in the case of Communist countries, no money could be removed without government approval. Modern technology provides the means to easily move money—as well as materials, products, technology, and other economic assets—around the world. Thanks to the electronic superhighway, companies can now organize economic activities at a global scale.

Every place in the world is part of the global economy, but globalization has led to more specialization at the local level. Each place plays a distinctive role, based on its local assets. A place may be near valuable minerals, or it may be inhabited by especially well-educated workers. Transnational corporations assess the particular economic assets of each place. A locality may be especially suitable for a transnational corporation to conduct research, to develop new engineering systems, to extract raw materials, to produce parts, to store finished products, to sell them, or to manage operations. In a global economy, transnational corporations remain competitive by correctly identifying the optimal location for each of these activities.

Globalization of the economy has heightened economic differences among places. Factories are closed in some locations and opened in others. Some places become centers for technical research, whereas others become centers for low-skilled tasks. Changes in production have led to a spatial division of labor, in which a region's workers specialize in particular tasks. Transnationals decide where to produce things in response to characteristics of the local labor force, such as level of skills, prevailing wage rates, and attitudes toward unions. Transnationals may close factories in locations with high wage rates and strong labor unions.

The deep recession that began in 2008 has been called the first global recession. Past recessions were typically confined to one country or region. For example, financial policies in Thailand triggered a severe recession there and in neighboring countries of Southeast Asia in 1997 but had little impact on the economies of the United States and Europe. In contrast, the global economy declined in 2009 for the first time in more than a half-century. Although every region suffered economic decline, the effects of the global recession varied. The fate of a home buyer in the United States was tied to the fate of a banker in United Kingdom, a sales clerk in Japan, a clothing maker in China, and a construction worker in Nigeria. All were caught in a global-scale web of falling demand and lack of credit.

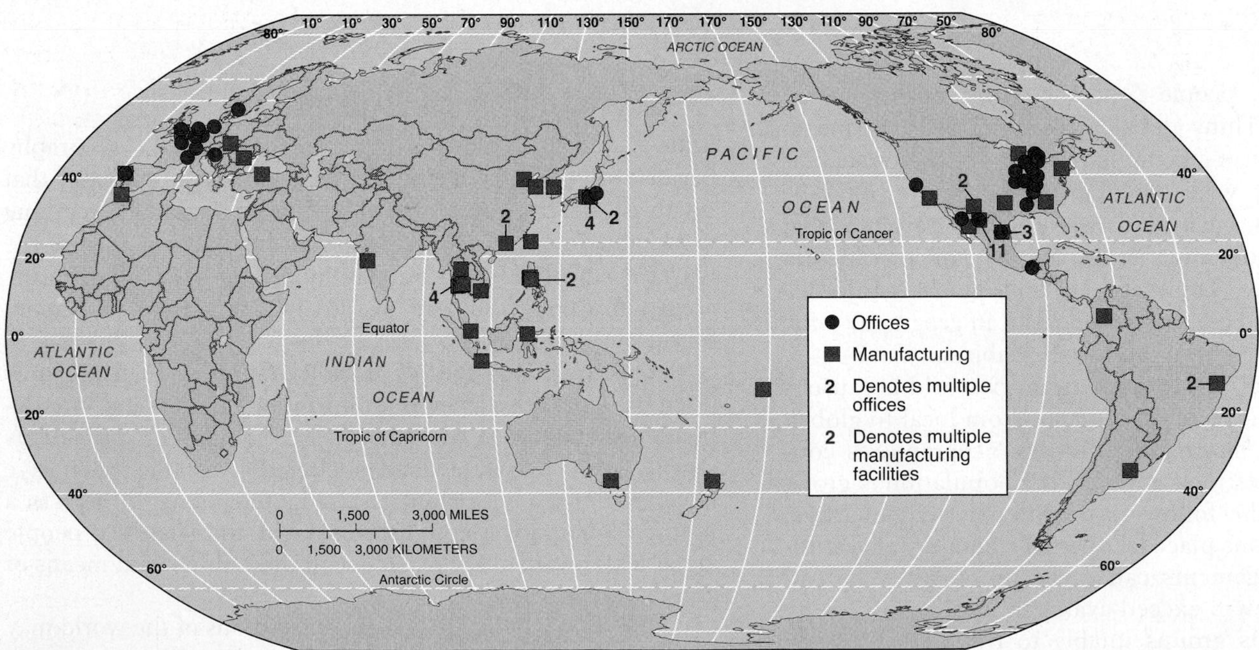

FIGURE 1-23 Globalization of economy. Yazaki, a transnational corporation that makes parts for cars has factories primarily in Asia and Latin America, where labor costs are relatively low, and offices primarily in Europe, North America, and Japan, where most of the customers (carmakers) are located.

Globalization of Culture

Geographers observe that increasingly uniform cultural preferences produce uniform "global" landscapes of material artifacts and of cultural values (Figure 1-24). Houses built on the edge of one urban area will look very much like houses built on the edge of urban areas in other regions. Fast-food restaurants, service stations, and retail chains deliberately create a visual appearance that varies among locations as little as possible. That way, customers know what to expect regardless of where in the world they happen to be.

Regardless of local cultural traditions, people around the world aspire to drive an automobile, watch television, and own a house. The survival of a local culture's distinctive beliefs, forms, and traits may be threatened by interaction with such social customs as wearing jeans and Nike shoes, consuming Coca-Cola and McDonald's hamburgers, and communicating by cell phone and computer. Underlying the uniform cultural landscape is globalization of cultural beliefs and forms, especially religion and language. Africans, in particular, have moved away from traditional religions and have adopted Christianity or Islam, religions shared with hundreds of millions of people throughout the world. Globalization requires a form of common communication, and the English language is increasingly playing that role.

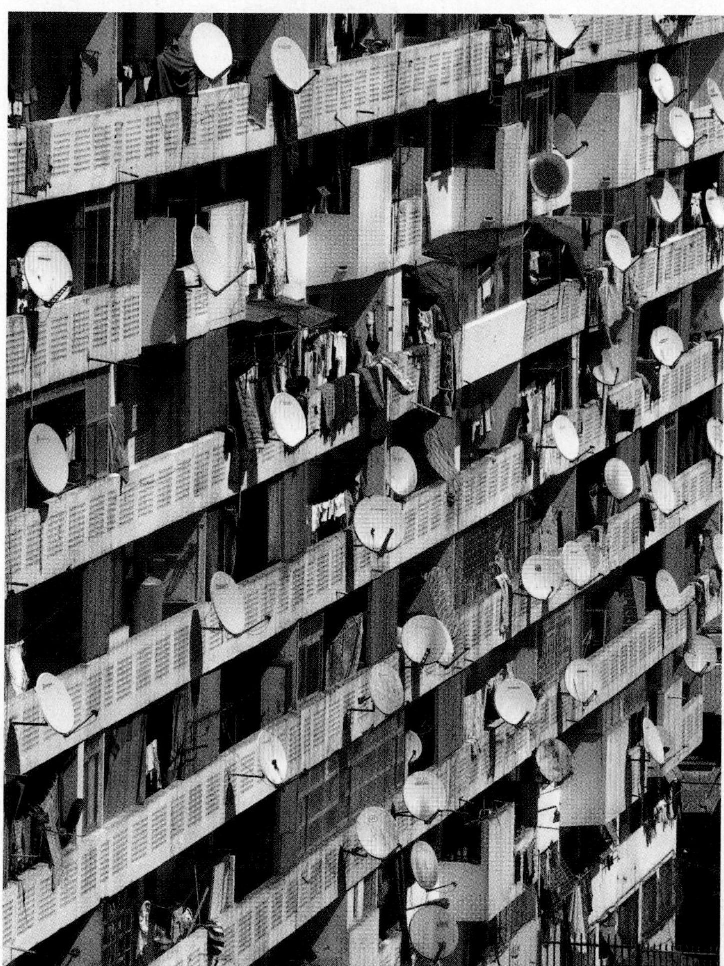

FIGURE 1-24 Globalization of culture. Algiers, the capital of Algeria, has one of the world's highest per capita ownership of satellite dishes.

As more people become aware of elements of global culture and aspire to possess them, local cultural beliefs, forms, and traits are threatened with extinction. Yet despite globalization, cultural differences among places not only persist but actually flourish in many places. Global standardization of products does not mean that everyone wants the same cultural products.

The communications revolution that promotes globalization of culture also permits preservation of cultural diversity. TV, for example, was once limited to a handful of channels displaying one set of cultural values. With the distribution of programming through cable and satellite systems, people now can choose from hundreds of programs in more than one language.

With the globalization of communications, people in two distant places can watch the same television program. At the same time, with the fragmentation of the broadcasting market, two people in the same house can watch different programs. Groups of people on every continent may aspire to wear jeans, but they might live with someone who prefers skirts. In a global culture, companies can target groups of consumers with similar tastes in different parts of the world.

Strong determination on the part of a group to retain its local cultural traditions in the face of globalization of culture can lead to intolerance of people who display other beliefs, social forms, and material traits. Political disputes, unrest, and wars have erupted in places such as Southeast Europe, East Africa, and the Middle East, where different cultural groups have been unable to share the same space peacefully (see Chapter 7).

A much more extreme opposition to globalization led to the attack by al-Qaeda terrorists against the United States on September 11, 2001, with support from the Taliban, then in control of Afghanistan (Chapter 8). Al-Qaeda selected targets—the World Trade Center and the Pentagon—it considered especially visible symbols of U.S. domination of globalization trends in culture, politics, and the economy. Afghanistan's Taliban leaders justified such actions as banning television and restricting women's activities as being consistent with local traditions, and such punishments as public floggings and severing of limbs as being a necessary counterbalance to strong forces of globalization.

Culturally, people residing in different places are displaying fewer differences and more similarities in their cultural preferences. But although consumers in different places express increasingly similar cultural preferences, they do not share the same access to them. And the desire of some people to retain their traditional cultural elements, in the face of increased globalization of cultural preferences, has led to political conflict and market fragmentation in some regions. Globalization has not destroyed the uniqueness of an individual place's culture and economy. Human geographers understand that many contemporary social problems result from a tension between forces promoting global culture and economy on the one hand and preservation of local economic autonomy and cultural traditions on the other hand.

Space: Distribution of Features

Chess and computer games, where pieces are placed on a grid-shaped playing surface, require thinking about space. Pieces are arranged on the game board or screen in order to outmaneuver an opponent or form a geometric pattern. To excel at these games, a player needs spatial skills, the ability to perceive the future arrangement of pieces. Similarly, spatial thinking is the most fundamental skill that geographers possess to understand the arrangement of objects across surfaces considerably larger than a game board. Geographers think about the arrangement of people and activities found in space and try to understand why those people and activities are distributed across space as they are.

In his framework of all scientific knowledge, the German philosopher Immanuel Kant (1724–1804) compared geography's concern for space to history's concern for time. Historians identify the dates of important events and explain why human activities follow one another chronologically. Geographers identify the location of important places and explain why human activities are located beside one another in space. Historians ask when and why. Geographers ask where and why. Historians organize material chronologically because they understand that an action at one point in time can result from past actions that can in turn affect future ones. Geographers organize material spatially because they understand that an action at one point in space can result from something happening at another point, which can consequently affect conditions elsewhere.

History and geography differ in one especially important manner: A historian cannot enter a time machine to study other eras firsthand; however, a geographer can enter an automobile or airplane to study other spaces. This ability to reach other spaces lends excitement to the discipline of geography—and geographic training raises the understanding of other spaces to a level above that of casual sightseeing.

Distribution

Look around the space you currently occupy—perhaps a classroom, residence hall, or room in a house. Tables, chairs, and other large objects are arranged regularly, such as in a row in a classroom or against a wall at home (though books and papers may be strewn about the space randomly). The room is located in a building that occupies an organized space—along a street, a side of a quadrangle, or next to a park. Similarly, the community containing the campus or house is part of a system of communities arranged across the country and around the world.

Each building and community, as well as every other human or natural object, occupies a unique space on Earth, and geographers explain how these features are arranged across Earth. On Earth as a whole, or within an area of Earth, features may be numerous or scarce, close together or far apart. The arrangement of a feature in space is known as its **distribution**. Geographers identify three main properties of distribution across Earth—density, concentration, and pattern (Figure 1-25).

FIGURE 1-25 Distribution. The top plan for a residential area has a lower density than the middle plan (24 houses compared to 32 houses on the same 82-acre piece of land), but both have dispersed concentrations. The middle and lower plans have the same density (32 houses on 82 acres), but the distribution of houses is more clustered in the lower plan. The lower plan has shared open space, whereas the middle plan provides a larger, private yard surrounding each house.

DENSITY. The frequency with which something occurs in space is its **density**. The feature being measured could be people, houses, cars, volcanoes, or anything. The area could be measured in square kilometers, square miles, hectares, acres, or any other unit of area.

Arithmetic density, which is the total number of objects in an area, is commonly used to compare the distribution of population in different countries. The arithmetic density of Belgium, for example, is 345 persons per square kilometer (900 persons per square mile). This density is the country's total population (10.5 million people) divided by its area (30,278 square kilometers, or 11,690 square miles).

Remember that a large *population* does not necessarily lead to a high *density*. Arithmetic density involves two measures—the number of people and the land area. The most populous country in the world, China, with approximately 1.3 billion inhabitants, by no means has the highest density. The arithmetic density of China—139 persons per square kilometer (360 persons per square mile)—is less than half that of Belgium. Although China has 123 times more inhabitants than Belgium, it has more than 300 times more land.

High population density is also unrelated to poverty. The Netherlands, one of the world's wealthiest countries, has an arithmetic density of 398 persons per square kilometer (1,031 persons per square mile). One of the poorest countries, Mali, has an arithmetic density of only 10 persons per square kilometer (26 persons per square mile).

Geographers measure density in other ways, depending on the subject being studied. A high **physiological density**—the number of persons per unit of area suitable for agriculture—may mean that a country has difficulty growing enough food to sustain its population. A high **agricultural density**—the number of farmers per unit area of farmland—may mean that a country has inefficient agriculture (see Chapter 2). A high housing density—the number of dwelling units per unit of area—may mean that people live in overcrowded housing.

CONCENTRATION. The extent of a feature's spread over space is its **concentration**. If the objects in an area are close together, they are *clustered*; if relatively far apart, they are *dispersed*. To compare the level of concentration most clearly, two areas need to have the same number of objects and the same size area.

Geographers use concentration to describe changes in distribution. For example, the distribution of people across the United States is increasingly dispersed. The total number of people living in the United States is growing slowly—less than 1 percent per year—and the land area is essentially unchanged. But the population distribution is changing from *relatively clustered* in the Northeast to more *evenly dispersed* across the country.

Concentration is not the same as density (Figure 1-26). Two neighborhoods could have the same density of housing but different concentrations. In a dispersed neighborhood, each house has a large private yard, whereas in a clustered neighborhood, the houses are close together and the open space is shared as a community park.

We can illustrate the difference between density and concentration at a far larger scale than a neighborhood. Within North America the distribution of major-league baseball teams changed during the second half of the twentieth century after remaining unchanged during the first half of the twentieth century. The major leagues expanded from 16 to 30 teams in North America between 1960 and 1998, thus increasing the density. At the same time, 6 of the 16 original teams moved to other locations. In 1952, every team was clustered in the Northeast United States, but the moves dispersed several teams to the West Coast and Southeast. These moves, as well as the spaces occupied by the expansion teams, resulted in a more dispersed distribution.

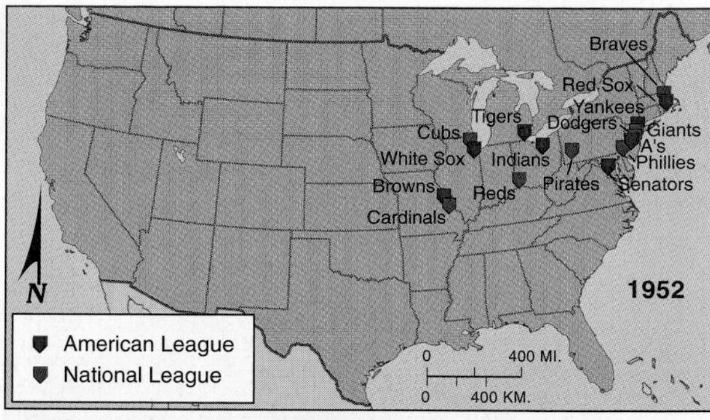

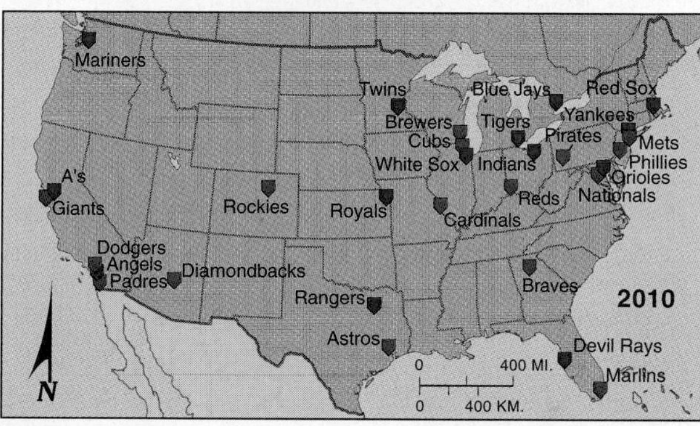

FIGURE 1-26 Distribution of baseball teams. The changing distribution of North American baseball teams illustrates the difference between density and concentration.

These six teams moved to other cities during the 1950s and 1960s:

- Braves—Boston to Milwaukee in 1953, then to Atlanta in 1966
- Browns—St. Louis to Baltimore (Orioles) in 1954
- Athletics—Philadelphia to Kansas City in 1955, then to Oakland in 1968
- Dodgers—Brooklyn to Los Angeles in 1958
- Giants—New York to San Francisco in 1958
- Senators—Washington to Minneapolis (Minnesota Twins) in 1961

These 14 teams were added between the 1960s and 1990s:

- Angels—Los Angeles in 1961, then to Anaheim (California) in 1965
- Senators—Washington in 1961, then to Dallas (Texas Rangers) in 1971
- Mets—New York in 1962
- Astros—Houston (originally Colt .45s) in 1962
- Royals—Kansas City in 1969
- Padres—San Diego in 1969
- Expos—Montreal in 1969, then to Washington (Nationals) in 2005
- Pilots—Seattle in 1969, then to Milwaukee (Brewers) in 1970
- Blue Jays—Toronto in 1977
- Mariners—Seattle in 1977
- Marlins—Miami (Florida) in 1993
- Rockies—Denver (Colorado) in 1993
- Devil Rays—Tampa Bay in 1998
- Diamondbacks—Phoenix (Arizona) in 1998

As a result of these relocations and additions, the density of teams increased, and the distribution became more dispersed.

PATTERN. The third property of distribution is **pattern**, which is the geometric arrangement of objects in space. Some features are organized in a geometric pattern, whereas others are distributed irregularly (Figure 1-27). Geographers observe that many objects form a linear distribution, such as the arrangement of houses along a street or stations along a subway line.

Objects are frequently arranged in a square or rectangular pattern. Many American cities contain a regular pattern of streets, known as a grid pattern, which intersect at right angles at uniform intervals to form square or rectangular blocks. The system of townships, ranges, and sections established by the

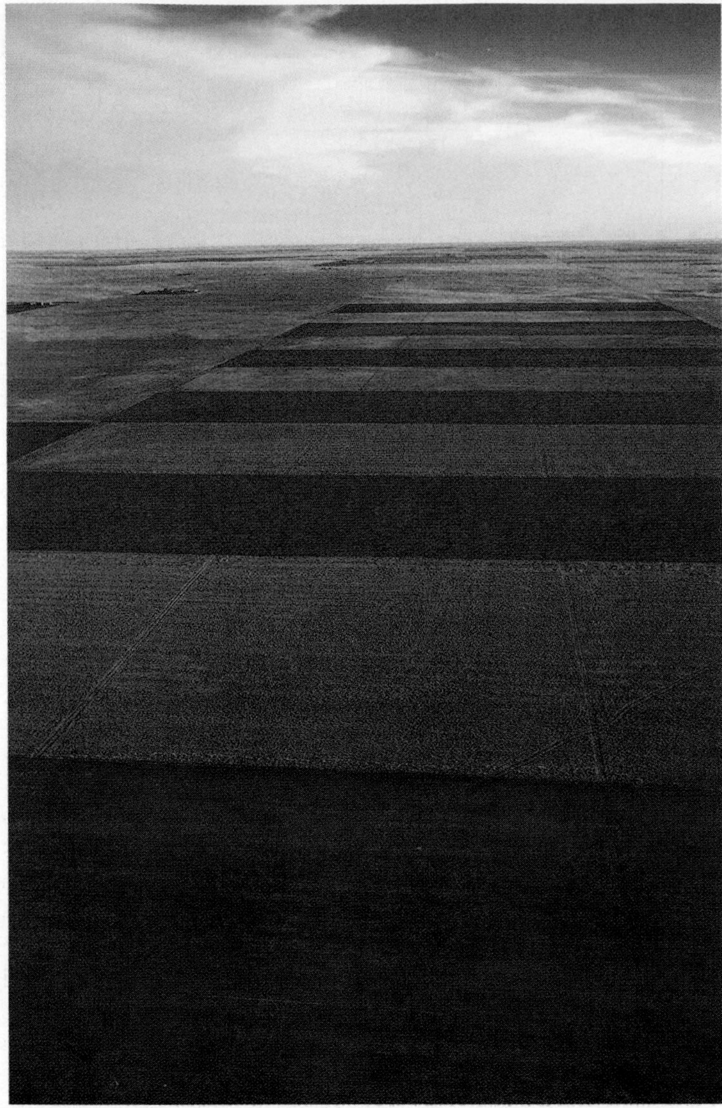

FIGURE 1-27 Pattern. Much of the farmland in the United States, including in Eastern Washington State, forms a checkerboard pattern. The pattern is a result of the township and range system (see Figure 1-5).

Land Ordinance of 1785 is another example of a square or grid pattern.

A sinister pattern of two dozen pipe bombs was placed on the American landscape in 2002 by Lucas Helder, a University of Wisconsin-Stout art student. The bomber confessed that he was trying to create a large "smile" pattern across the U.S. interior. He got as far as creating the two "eyes" by placing bombs in two large circles, one in Nebraska and one in eastern Iowa and western Illinois. Before being caught, he also placed bombs in Colorado and Texas to start the "mouth."

Gender and Ethnic Diversity in Space

Patterns in space vary according to gender and ethnicity. Consider first the daily patterns of an "all-American" family of mother, father, son, and daughter. Leave aside for the moment that this type of family constitutes less than one-fourth of American households.

In the morning Dad gets in his car and drives from home to work, where he parks the car and spends the day; then, in the late afternoon, he collects the car and drives home. The location of the home was selected in part to ease Dad's daily commute to work.

The mother's local-scale travel patterns are likely to be far more complex than the father's. Mom takes the children to school and returns home. She also drives to the supermarket, visits Grandmother, and walks the dog. In between she organizes the several thousand square feet of space that the family calls home. In the afternoon, she picks up the youngsters at school and takes them to Little League or ballet lessons. Later, she brings them home, just in time for her to resume her responsibility for organizing the home.

Most American women are now employed at work outside the home, adding a substantial complication to an already complex pattern of moving across urban space. Where is her job located? The family house was already selected largely for access to Dad's place of employment, so Mom may need to travel across town. Who leaves work early to drive a child to a doctor's office? Who takes a day off work when a child is at home sick?

The importance of gender in space is learned as a child. Which child—the boy or girl—went to Little League and which went to ballet lessons? To which activity is substantially more land allocated in a city—ballfields or dance studios?

If the family described above consisted of persons of color, its connections with space would change. The effects of race on spatial interaction can be seen across America. In downtown Dayton, Ohio, for example, watch the people at the bus stops along the main east–west street, Third Street. In the afternoon, when office workers are heading home, persons of color are waiting on the north side of Third Street for westbound buses, while whites are waiting on the south side for eastbound buses. Why do persons of color head west on Dayton's afternoon buses? Virtually all African Americans in Dayton live on the west side, whereas the east side is home to a virtually all-white population. In most U.S. neighborhoods, the residents are virtually all white people or virtually all persons of color.

Although it is illegal to discriminate against people of color, segregation persists in part because people want to reinforce their cultural identity by living near persons of similar background and in part because persons of color have lower-than-average incomes. But many Americans of European ancestry still practice discrimination because of a deep-seated fear of spatial interaction with a person of color.

Openly homosexual men and lesbian women may be attracted to some locations to reinforce spatial interaction with other gays. San Francisco reinforces its reputation as a sympathetic home for homosexuals and lesbians through inclusive public policies (Figure 1-28). Specific neighborhoods in other cities are known to have large gay populations.

A pet dog doesn't care if you are male or female, black or white, gay or not. As long as you feed it, take care of it, and maintain close spatial interaction with it, your dog will respond with

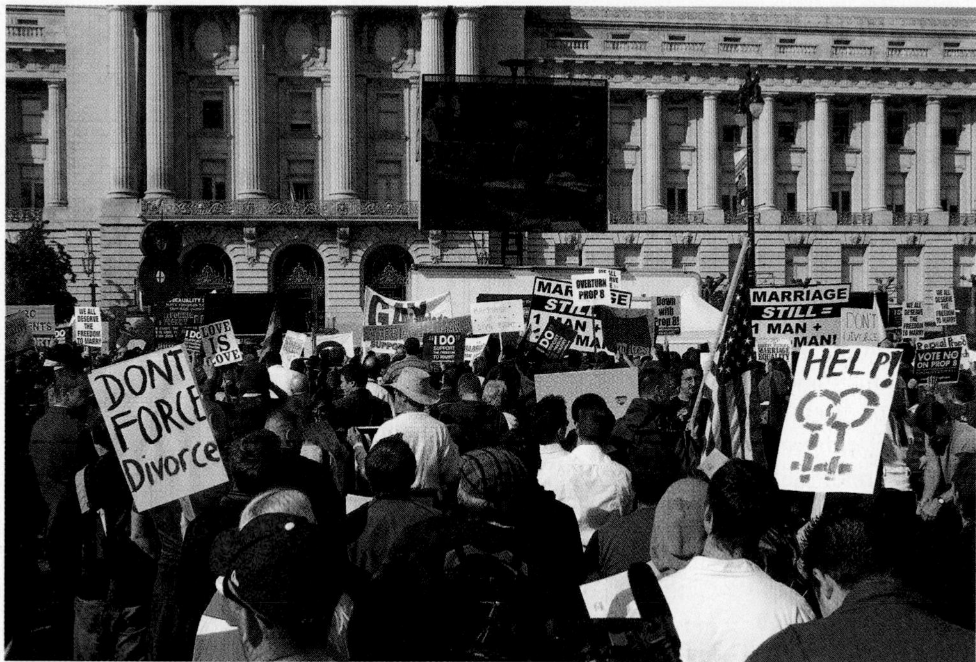

FIGURE 1-28 Diversity in space. Supporters and opponents of same-sex marriage in San Francisco watch a live broadcast of the California Supreme Court in 2009 as it considered the legality of a proposition passed a year earlier banning same-sex marriage in the state.

total, unquestioned devotion. Although dogs don't care about these cultural traits, people do. They are key characteristics to which people refer in order to identify who they are.

Cultural identity is a source of pride to people at the local scale and an inspiration for personal values. Even more important than self-identification, these traits matter to other people. They are the criteria by which other people classify us and choose to interact with us. Whatever biological basis may or may not exist for distinguishing among humans, differences in gender, race, and sexual orientation are first and foremost constructed by the attitudes and actions of others. Geographers consider cultural identity to be important in understanding spatial interaction, because humans repeatedly demonstrate that these factors are important in explaining why they sort themselves out in space and move across the landscape in distinctive ways.

All academic disciplines and workplaces have proclaimed sensitivity to issues of cultural diversity. For geographers, concern for cultural diversity is not merely a politically correct expediency; it lies at the heart of geography's spatial tradition. Nor is geographers' deep respect for the dignity of all cultural groups merely a matter of political correctness; it lies at the heart of geography's explanation of why each place on Earth is unique.

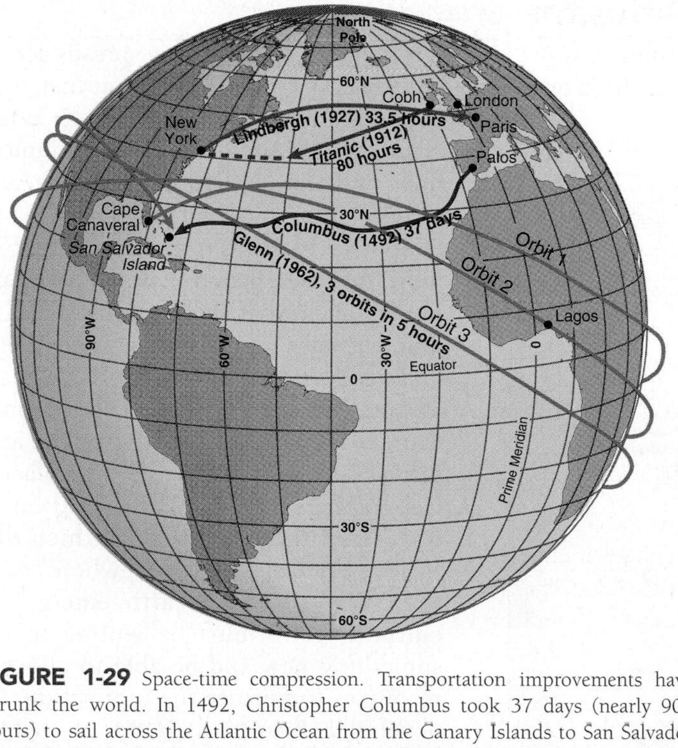

FIGURE 1-29 Space-time compression. Transportation improvements have shrunk the world. In 1492, Christopher Columbus took 37 days (nearly 900 hours) to sail across the Atlantic Ocean from the Canary Islands to San Salvador Island. In 1912, the *Titanic* was scheduled to sail from Queenstown (now Cobh), Ireland, to New York in about 5 days, although two-thirds of the way across, after 80 hours at sea, it hit an iceberg and sank. In 1927, Charles Lindbergh was the first person to fly nonstop across the Atlantic, taking 33.5 hours to go from New York to Paris. In 1962, John Glenn, the first American to orbit in space, crossed above the Atlantic in about a quarter-hour and circled the globe three times in 5 hours.

Connections Between Places

Geographers increasingly think about connections among places and regions. More rapid connections have reduced the distance across space between places, not literally in miles, of course, but in time.

Geographers apply the term **space-time compression** to describe the reduction in the time it takes for something to reach another place. Distant places seem less remote and more accessible to us. We know more about what is happening elsewhere in the world, and we know sooner. Space-time compression promotes rapid change, as the culture and economy of one place reach other places much more quickly than in the past (Figure 1-29). With better connections between places, people in one region are now exposed to a constant barrage of cultural traits and economic initiatives from people in other regions, and they may adopt some of these cultural and economic elements. Geographers explain the process, called *diffusion*, by which connections are made between regions, as well as the mechanism by which connections are maintained through networks.

Spatial Interaction

In the past, most forms of interaction among cultural groups required the physical movement of settlers, explorers, and plunderers from one location to another. As recently as A.D. 1800, people traveled in the same ways and at about the same speeds, as in 1800 B.C.—they were carried by an animal, took a sailboat, or walked.

Today, travel by motor vehicle or airplane is much quicker. But we do not even need to travel to know about another place. We can transmit images and messages from one part of the world to another at the touch of a button. We can communicate instantly with people in distant places through computers and telecommunications, and we can instantly see people in distant places on television. The various forms of communication have made it possible for people in different places to be aware of the same cultural beliefs, forms, and traits. When places are connected to each other through a network, geographers say there is spatial interaction between them. Interaction takes place through networks, which are chains of communication that connect places. A well-known example of a network in the United States is the television network (ABC, CBS, FOX, NBC, PBS). Each comprises a chain of stations around the country simultaneously broadcasting the same program, such as a football game.

Transportation systems also form networks that connect places to each other. Airlines in the United States, for example, have adopted distinctive networks known as "hub-and-spokes" (Figure 1-30). Under the hub-and-spokes system, airlines fly planes from a large number of places into one hub airport within a short period of time and then a short time later send the planes to another set of places. In principle, travelers originating in relatively small towns can reach a wide variety of destinations by changing planes at the hub airport.

Interaction among groups can be retarded by barriers. These can be physical, such as oceans and deserts, or cultural, such as language and traditions. We regard the landscape as part of our inheritance from the past. As a result, we may be reluctant to modify it unless we are under heavy pressure to do so. A major change in the landscape may reflect an upheaval in a people's culture. Typically, the farther away one group is from another, the less likely the two groups are to interact. Contact diminishes with increasing distance and eventually disappears. This trailing-off phenomenon is called **distance decay**. Electronic communications, such as text messaging and e-mail, have removed barriers to interaction between people who are far from each other. The birth of these electronic communications was once viewed as the "death" of geography, because they made it cheap and easy to stay in touch with someone on the other side of the planet. Regardless of its location, a business could maintain instantaneous communications among employees and with customers.

In reality, geography matters even more than before. Internet access depends upon availability of electricity to power the computer and a service provider. Broadband service requires proximity to a digital subscriber line (DSL) or cable line. The Internet has also magnified the importance of geography, because when an individual is online, the specific place in the world where the individual is located is known. This knowledge is valuable information for businesses that target advertisements and products to specific tastes and preferences of particular places (see Chapter 12).

Diffusion

Diffusion is the process by which a characteristic spreads across space from one place to another over time. Today, ideas that originate in one area diffuse rapidly to other areas through sophisticated communications and transportation networks. As a result of diffusion, interaction in the contemporary world is complex. People in more than one region may improve and modify an idea at the same time but in different ways.

The place from which an innovation originates is called a **hearth**. Something originates at a hearth or node and diffuses from there to other places. Geographers document the location of nodes and the processes by which diffusion carries things elsewhere over time. How does a hearth emerge? A cultural group must be willing to try something new and be able to allocate resources to nurture the innovation. To develop a hearth, a group of people must also have the technical ability to achieve the desired idea and the economic structures, such as financial institutions, to facilitate implementation of the innovation.

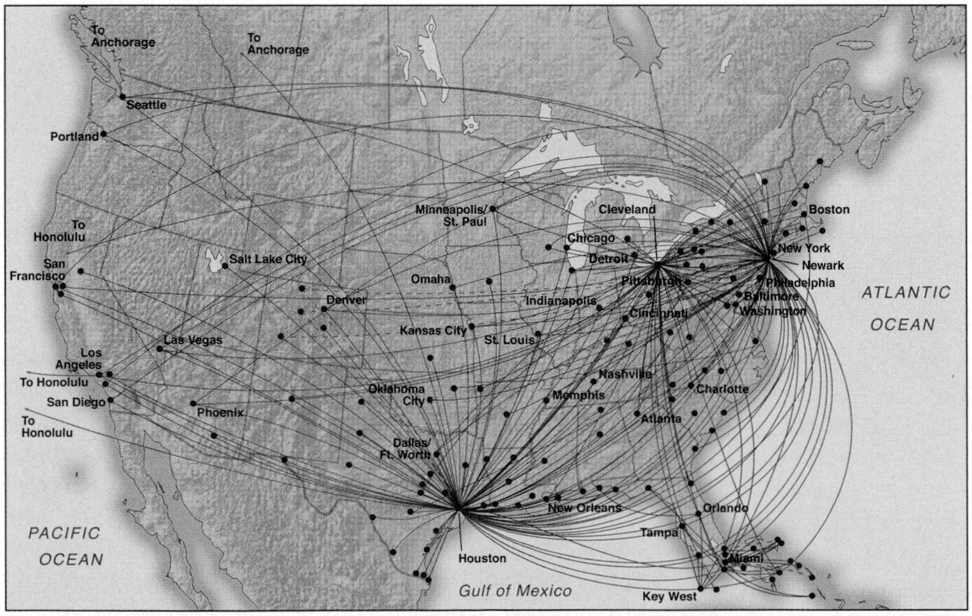

FIGURE 1-30 Continental Airlines' network. Continental, like other major U.S. airlines, has configured its route network in a system known as "hub and spokes." Lines connect each airport to the city to which it sends the most nonstop flights. Most flights originate or end at one of the company's hubs, especially at Houston, Newark, and Cleveland.

As discussed in subsequent chapters, geographers can trace the dominant cultural, political, and economic features of the contemporary United States and Canada primarily to hearths in Europe and the Middle East. Other regions of the world also contain important hearths. In some cases an idea, such as an agricultural practice, may originate independently in more than one hearth. In other cases, hearths may emerge in two regions because two cultural groups modify a shared concept in two different ways.

For a person, object, or idea to have interaction with persons, objects, or ideas in other regions, diffusion must occur. Geographers observe two basic types of diffusion—relocation and expansion.

RELOCATION DIFFUSION. The spread of an idea through physical movement of people from one place to another is termed **relocation diffusion**. We shall see in Chapter 3 that people migrate for a variety of political, economic, and environmental reasons. When they move, they carry with them their culture, including language, religion, and ethnicity.

The most commonly spoken languages in North and South America are Spanish, English, French, and Portuguese, primarily because several hundred years ago Europeans who spoke those languages comprised the largest number of migrants. Thus these languages spread through relocation diffusion. We will examine the diffusion of languages, religions, and ethnicity in Chapters 5 through 7.

Introduction of a common currency, the euro, in 12 Western European countries gave scientists an unusual opportunity to measure relocation diffusion from hearths (Figure 1-31). Although a single set of paper money was issued, each of the 12 countries minted its own coins in proportion to its share of the region's economy. A country's coins were initially distributed only inside its borders, although the coins could also be used in the other 11 countries. Scientists took month-to-month samples in France to monitor the proportion of coins from each of the other 11 countries. The percentage of coins from a particular country is a measure of the level of relocation diffusion to and from France.

Relocation diffusion helps us understand the distribution of acquired immunodeficiency syndrome (AIDS) within the United States. New York, California, and Florida were the nodes of origin for the disease within the United States during the early 1980s (Figure 1-32). Half of the 50 states had no reported cases, whereas New York City, with only 3 percent of the nation's population, contained more than one-fourth of the AIDS cases. New AIDS cases diffused to every state during the 1980s and early 1990s, although California, Florida, and New York remained the hearths. These three states, plus Texas, accounted for half of the nation's new AIDS cases in the peak year of 1993. At a national scale, the

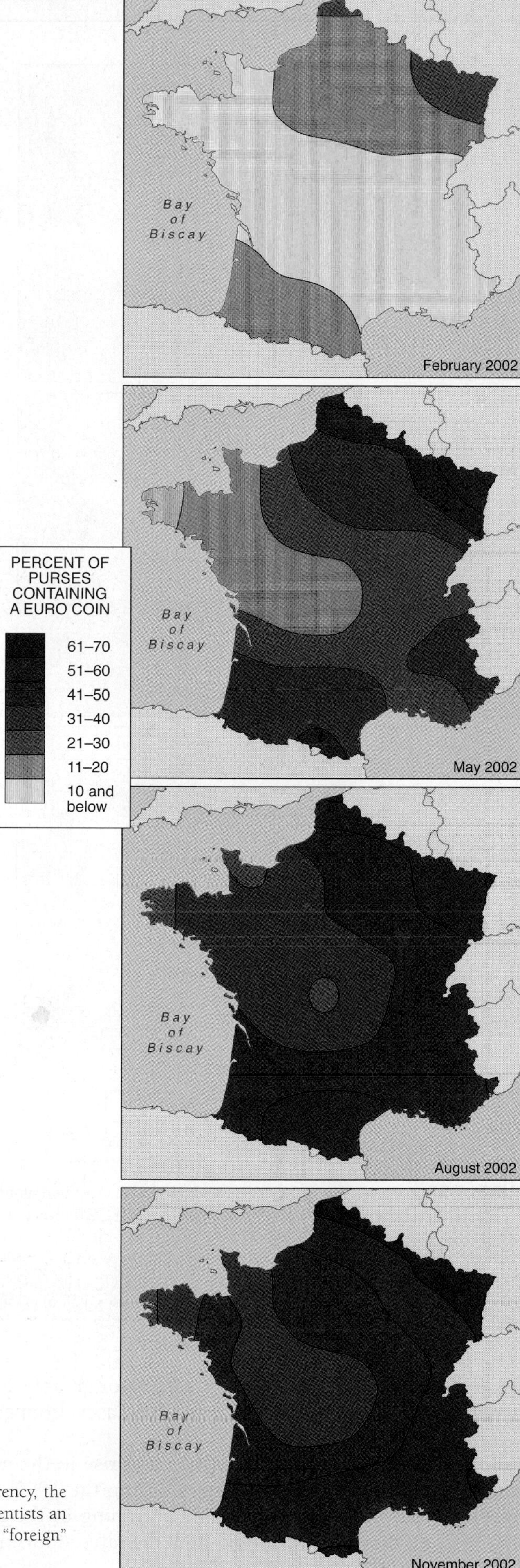

FIGURE 1-31 Relocation diffusion. Introduction of a common currency, the euro, in 12 Western European countries on January 1, 2002, gave scientists an unusual opportunity to measure relocation diffusion. The percentage of "foreign" euro coins is a measure of the level of relocation diffusion into France.

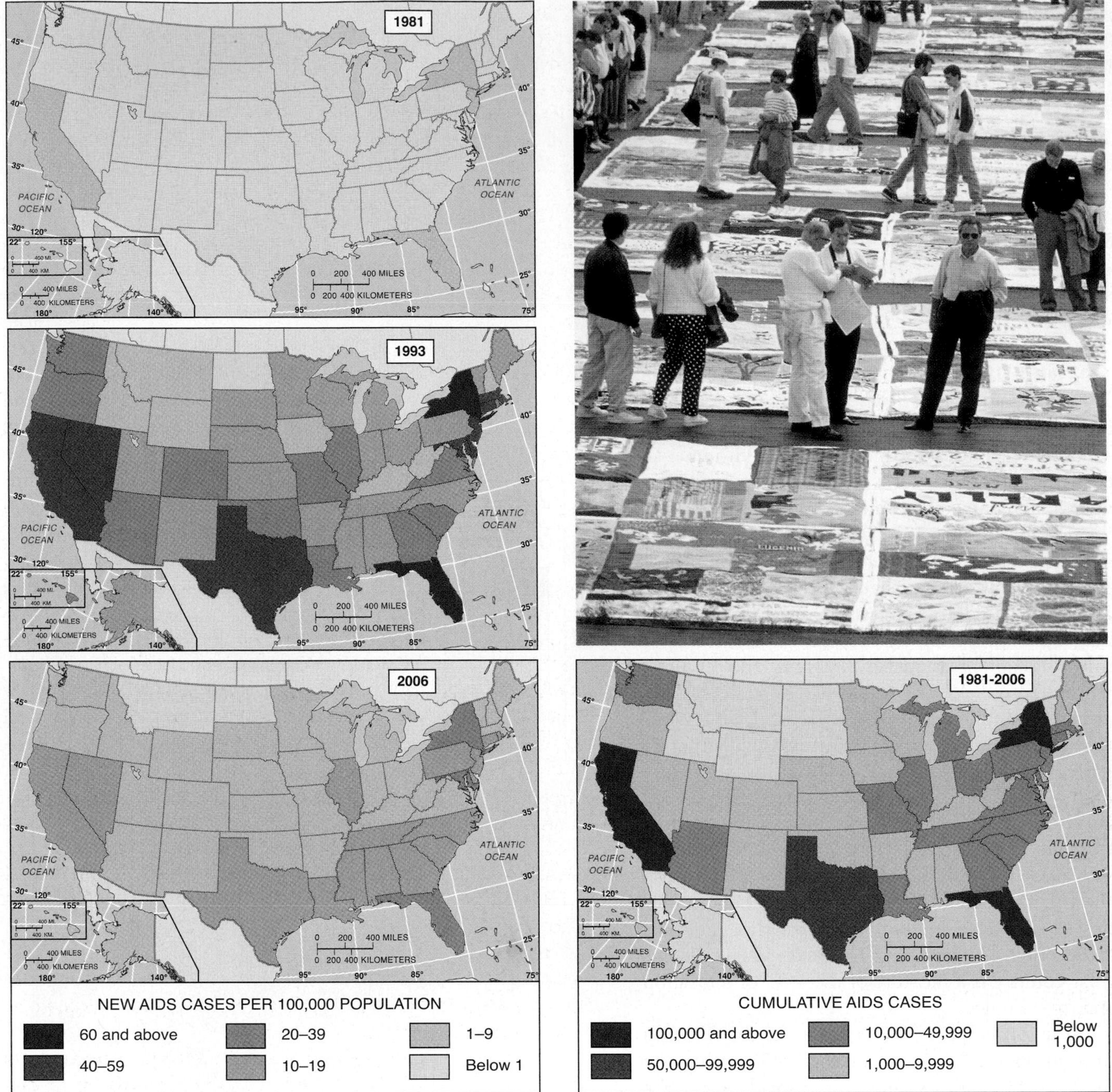

FIGURE 1-32 Diffusion of AIDS in the United States. Acquired immunodeficiency syndrome (AIDS) diffused across the United States from nodes in New York, California, and Florida. In 1981, virtually all people with AIDS were found in these three nodes. During the 1980s, the number of cases increased everywhere, but the incidence remained highest in the three original nodes. The number of cases declined relatively rapidly in the original nodes during the 1990s. The AIDS Memorial Quilt, on display in Washington, DC, was assembled as a memorial to people who died of AIDS.

diffusion of AIDS in the United States through relocation halted after 1993. The number of new AIDS cases dropped by one-fourth in just two years.

Relocation diffusion can explain the rapid rise in the number of AIDS cases in the United States during the 1980s and early 1990s, but not the rapid decline beginning in the mid-1990s. Instead, the decline resulted from the rapid diffusion of preventive methods and medicines such as AZT. The rapid spread of these innovations is an example of expansion diffusion rather than relocation diffusion.

EXPANSION DIFFUSION. The spread of a feature from oneplace to another in a snowballing process is **expansion diffusion**. This expansion may result from one of three processes:

- Hierarchical diffusion
- Contagious diffusion
- Stimulus diffusion.

Hierarchical diffusion is the spread of an idea from persons or nodes of authority or power to other persons or places. Hierarchical diffusion may result from the spread of ideas from political leaders, socially elite people, or other important persons to others in the community.

Innovations may also originate in a particular node or place of power, such as a large urban center, and diffuse later to isolated rural areas. Hip-hop or rap music is an example of an innovation that diffused from low-income African Americans rather than from socially elite people, but it originated in urban areas.

Contagious diffusion is the rapid, widespread diffusion of a characteristic throughout the population. As the term implies, this form of diffusion is analogous to the spread of a contagious disease, such as influenza. Contagious diffusion spreads like a wave among fans in a stadium, without regard for hierarchy and without requiring permanent relocation of people.

The rapid adoption throughout the United States of AIDS prevention methods and new medicines is an example of contagious diffusion. An idea placed on the World Wide Web spreads through contagious diffusion, because Web surfers throughout the world have access to the same material simultaneously—and quickly.

Stimulus diffusion is the spread of an underlying principle, even though a characteristic itself apparently fails to diffuse. For example, early desktop computer sales in the United States were divided about evenly between Macintosh Apple and IBM-compatible DOS systems. By the 1990s, Apple sales had fallen far behind IBM-compatibles in the United States, and the company had limited presence in rapidly expanding overseas markets. But principles pioneered by Apple, notably making selections by pointing a mouse at an icon rather than typing a string of words, diffused through a succession of IBM-compatible Windows systems.

Expansion diffusion occurs much more rapidly in the contemporary world than in the past:

- Hierarchical diffusion is encouraged by modern methods of communications, such as computers, facsimile machines, and electronic mail systems
- Contagious diffusion is encouraged by use of the Internet, especially the World Wide Web.
- Stimulus diffusion is encouraged by all of the new technologies.

Diffusion from one place to another can be instantaneous in time, even if the physical distance between two places—as measured in kilometers or miles—is large.

DIFFUSION OF CULTURE AND ECONOMY. In a global culture and economy, transportation and communications systems rapidly diffuse raw materials, goods, services, and capital from nodes of origin to other regions. Every area of the world plays some role intertwined with the roles played by other regions. Workers and cultural groups that in the past were largely unaffected by events elsewhere in the world now share a single economic and cultural world with other workers and cultural groups. The fate of an autoworker in Detroit is tied to investment decisions made in Mexico City, Seoul, Stuttgart, and Tokyo.

Global culture and economy are increasingly centered on the three core or hearth regions of North America, Western Europe, and Japan. These three regions have a large percentage of the world's advanced technology, capital to invest in new activities, and wealth to purchase goods and services. From "command centers" in the three major world cities of New York, London, and Tokyo, key decision makers employ modern telecommunications to send out orders to factories, shops, and research centers around the world, an example of hierarchical diffusion. Meanwhile, "nonessential" employees of the companies can be relocated to lower-cost offices outside the major financial centers. For example, Fila maintains headquarters in Italy but has moved 90 percent of its production of sportswear to Asian countries. Mitsubishi's corporate offices are in Japan, but all of its VCRs and DVDs are produced in other Asian countries.

Countries in Africa, Asia, and Latin America contain three-fourths of the world's population and nearly all of its population growth, but they find themselves on a periphery, or outer edge, of global investment that arrives through hierarchical diffusion of decisions made by transnational corporations through hierarchical diffusion. People in peripheral regions, who once toiled in isolated farm fields to produce food for their families, now produce crops for sale in core regions or have given up farm life altogether and migrated to cities in search of jobs in factories and offices. As a result, the global economy has produced greater disparities than in the past between the levels of wealth and well-being enjoyed by people in the core and in the periphery. The increasing gap in economic conditions between regions in the core and periphery that results from the globalization of the economy is known as **uneven development**.

Many people take for granted the ability to watch events in distant places through television, speak to others in distant places by telephone, and travel to far-off places by motor vehicle. An increasing number of the world's population regard access to these communications systems as novelties, perhaps recently experienced for the first time. For some people, access to these cultural elements is a distant aspiration. Knowledge of these communications systems is global, but the ability to purchase them is not. Access to television, telephones, motor vehicles, and other means of communicating culture is restricted by an uneven division of wealth in the world. In some regions possession of these objects is widespread, but in other regions few people have enough wealth to buy them. Even within regions, access to cultural elements may be restricted because of uneven distribution of wealth or because of discrimination against women or minority groups.

SUMMARY

Each chapter has a summary that reviews the chapter's most important concepts. The summary is organized around the major headings within the chapter.

In all of the subsequent chapters, these headings will be in the form of questions that are answered in the text. In this first chapter, the principal headings concern thinking about five key concepts in geography (place, region, scale, space, and connections):

1. **How Do Geographers Describe Where Things Are?** Geography is most fundamentally a spatial science. Geographers use maps to display the location of objects and to extract information about places. Early geographers drew maps of Earth's surface based on exploration and observation. GIS and other contemporary tools assist geographers in understanding reasons for observed regularities across Earth.
2. **Why Is Each Point on Earth Unique?** Every place in the world has a unique location or position on Earth's surface. Geographers also identify regions as areas distinguished by distinctive combinations of cultural as well as economic and environmental features. The distributions of features help us to understand why every place and every region is unique.
3. **Why Are Different Places Similar?** Geographers work at all scales, from local to global. The global scale is increasingly important because few places in the contemporary world are totally isolated. Because places are connected to each other, they display similarities. Geographers study the interactions of groups of people and human activities across space, and they identify processes by which people and ideas diffuse from one location to another over time.

CASE STUDY REVISITED / The Geography of a Big Mac Attack

Each chapter in this book concludes by reviewing the opening case study in light of the issues raised in the chapter. This chapter presents five basic concepts— place, region, scale, space, and connections. The opening case study offers a typical everyday geographic concern—a search for a restaurant—to which these five concepts can be applied.

Geography is fundamentally concerned with the organization of space. McDonald's restaurants are not distributed randomly across the landscape; rather, each restaurant has a unique location that can be depicted on a map (Figure 1-33).

FIGURE 1-33 Fast-food nation. Franchised restaurants cluster near each other, such as along Highway 412 in Springdale Arkansas.

Geographers use maps to describe where these establishments are found and explain why they are so arranged. Because "where" and "why" are the questions most fundamental to geographic inquiry, they are used to organize the material presented within all of the other chapters in this book.

Geographers observe from a map that McDonald's restaurants cluster in some regions, whereas other regions have few. A world map of McDonald's restaurants helps us to understand global-scale patterns of investment by a major international corporation. Most McDonald's are located in countries where average incomes are high enough to buy the products.

A world map of McDonald's doesn't help a hungry American driving on an interstate highway. The motorist needs a local-scale map showing the location of McDonald's in relation to specific highway exit ramps. As McDonald's have diffused from the United States to other regions of the world, each McDonald's is connected to all other McDonald's by a communications network through which uniform standards and practices are set.

In subsequent chapters, these five basic concepts will be applied to elements of human geography:

- Chapters 2 and 3 where humans are clustered in the world, why the number of people has increased in some places, and why people have moved to certain places

(Continued)

CASE STUDY REVISITED (*Continued*)

- Chapters 4 through 8 where important cultural traits, including popular and folk customs, language, religion, ethnicity, and political institutions, are distributed and why these cultural features are so distributed and why these distributions can lead to conflict
- Chapters 9 through 14 where different economic activities are found around the world, why people earn a living in different ways in different regions of the world, and why people increasingly earn a living by residing in urban areas. ■

KEY TERMS

Agricultural density (p. 33) The ratio of the number of farmers to the total amount of land suitable for agriculture.

Arithmetic density (p. 32) The total number of people divided by the total land area.

Base line (p. 9) An east–west line designated under the Land Ordinance of 1785 to facilitate the surveying and numbering of townships in the United States.

Cartography (p. 5) The science of making maps.

Concentration (p. 33) The spread of something over a given area.

Connections (p. 5) Relationships among people and objects across the barrier of space.

Contagious diffusion (p. 39) The rapid, widespread diffusion of a feature or trend throughout a population.

Cultural ecology (p. 24) Geographic approach that emphasizes human–environment relationships.

Cultural landscape (p. 17) Fashioning of a natural landscape by a cultural group.

Culture (p. 21) The body of customary beliefs, social forms, and material traits that together constitute a group's distinct tradition.

Density (p. 32) The frequency with which something exists within a given unit of area.

Diffusion (p. 36) The process of spread of a feature or trend from one place to another over time.

Distance decay (p. 36) The diminishing in importance and eventual disappearance of a phenomenon with increasing distance from its origin.

Distribution (p. 32) The arrangement of something across Earth's surface.

Environmental determinism (p. 24) A nineteenth- and early twentieth-century approach to the study of geography which argued that the general laws sought by human geographers could be found in the physical sciences. Geography was therefore the study of how the physical environment caused human activities.

Expansion diffusion (p. 38) The spread of a feature or trend among people from one area to another in a snowballing process.

Formal region (or uniform or homogeneous region) (p. 17) An area in which everyone shares in one or more distinctive characteristics.

Functional region (or nodal region) (p. 19) An area organized around a node or focal point.

Geographic information system (GIS) (p. 12) A computer system that stores, organizes, analyzes, and displays geographic data.

Global Positioning System (GPS) (p. 9) A system that determines the precise position of something on Earth through a series of satellites, tracking stations, and receivers.

Globalization (p. 29) Actions or processes that involve the entire world and result in making something worldwide in scope.

Greenwich Mean Time (GMT) (p. 18) The time in that zone encompassing the prime meridian, or 0° longitude.

Hearth (p. 36) The region from which innovative ideas originate.

Hierarchical diffusion (p. 39) The spread of a feature or trend from one key person or node of authority or power to other persons or places.

International Date Line (p. 18) An arc that for the most part follows 180° longitude, although it deviates in several places to avoid dividing land areas. When you cross the International Date Line heading east (toward America), the clock moves back 24 hours, or one entire day. When you go west (toward Asia), the calendar moves ahead one day.

Land Ordinance of 1785 (p. 9) A law that divided much of the United States into townships to facilitate the sale of land to settlers.

Latitude (p. 15) The numbering system used to indicate the location of parallels drawn on a globe and measuring distance north and south of the equator (0°).

Location (p. 13) The position of anything on Earth's surface.

Longitude (p. 15) The numbering system used to indicate the location of meridians drawn on a globe and measuring distance east and west of the prime meridian (0°).

Map (p. 4) A two-dimensional, or flat, representation of Earth's surface or a portion of it.

Mental map (p. 20) A representation of a portion of Earth's surface based on what an individual knows about a place, containing personal impressions of what is in a place and where places are located.

Meridian (p. 15) An arc drawn on a map between the North and South poles.

Parallel (p. 15) A circle drawn around the globe parallel to the equator and at right angles to the meridians.

Pattern (p. 33) The geometric or regular arrangement of something in a study area.

Physiological density (p. 33) The number of people per unit of area of arable land, which is land suitable for agriculture.

Place (p. 5) A specific point on Earth distinguished by a particular character.

Polder (p. 27) Land created by the Dutch by draining water from an area.

Possibilism (p. 24) The theory that the physical environment may set limits on human actions, but people have the ability to adjust to the physical environment and choose a course of action from many alternatives.

Prime meridian (p. 15) The meridian, designated as 0° longitude, that passes through the Royal Observatory at Greenwich, England.

Principal meridian (p. 9) A north–south line designated in the Land Ordinance of 1785 to facilitate the surveying and numbering of townships in the United States.

Projection (p. 8) The system used to transfer locations from Earth's surface to a flat map.

Region (p. 5) An area distinguished by a unique combination of trends or features.

Regional (or cultural landscape) studies (p. 17) An approach to geography that emphasizes the relationships among social and physical phenomena in a particular study area.

Relocation diffusion (p. 37) The spread of a feature or trend through bodily movement of people from one place to another.

Remote sensing (p. 9) The acquisition of data about Earth's surface from a satellite orbiting the planet or from other long-distance methods.

Resource (p. 24) A substance in the environment that is useful to people, is economically and technologically feasible to access, and is socially acceptable to use.

Scale (p. 5) Generally, the relationship between the portion of Earth being studied and Earth as a whole; specifically, the relationship between the size of an object on a map and the size of the actual feature on Earth's surface.

Section (p. 9) A square normally 1 mile on a side. The Land Ordinance of 1785 divided townships in the United States into 36 sections.

Site (p. 14) The physical character of a place.

Situation (p. 14) The location of a place relative to another place.

Space (p. 5) The physical gap or interval between two objects.

Space-time compression (p. 35) The reduction in the time it takes to diffuse something to a distant place as a result of improved communications and transportation systems.

Stimulus diffusion (p. 39) The spread of an underlying principle, even though a specific characteristic is rejected.

Toponym (p. 13) The name given to a portion of Earth's surface.

Township (p. 9) A square normally 6 miles on a side. The Land Ordinance of 1785 divided much of the United States into a series of townships.

Transnational corporation (p. 30) A company that conducts research, operates factories, and sells products in many countries, not just where its headquarters or shareholders are located.

Uneven development (p. 39) The increasing gap in economic conditions between core and peripheral regions as a result of the globalization of the economy.

Vernacular region (or perceptual region) (p. 19) An area that people believe exists as part of their cultural identity.

THINKING GEOGRAPHICALLY

1. Cartography is not simply a technical exercise in penmanship and coloring, nor are decisions confined to scale and projection. Mapping is a politically sensitive undertaking. Look at how maps in this book distinguish between the territories of Israel and its neighbors and the locations of borders in South Asia, the Arabian Peninsula, and northwest Africa. Are there other logical ways to draw boundaries and distinguish among territories in these regions? What might they be?
2. Imagine that a transportation device (perhaps the one in Star Trek or *Harry Potter*) would enable all humans to travel instantaneously to any location on Earth's surface. What would be the impact of that invention on the distribution of peoples and activities across Earth?
3. When earthquakes, hurricanes, or other environmental disasters strike, humans tend to "blame" nature and see themselves as the innocent victims of a harsh and cruel nature. To what extent do environmental hazards stem from unpredictable nature and to what extent do they originate from human actions? Should victims blame nature, other humans, or themselves for the disaster? Why?
4. The construction of dams is a particularly prominent example of human–environment interaction in regions throughout the world. Turkey built the Ataturk Dam on the Euphrates River, a move opposed by Syria and Iraq, the two downstream countries. Similarly, the Balbina Dam on the Uatruma River, a tributary of the Amazon, generated considerable opposition in Brazil. Some Russians oppose construction of the St. Petersburg Dam in the Gulf of Finland. Egypt, which operates the Aswan Dam on the Nile River, has blocked loans to Ethiopia that could be used to divert the source of the Nile. Why do governments push the construction of dams so forcefully, and why do others oppose their construction so passionately?
5. Geographic concepts are supposed to help explain contemporary issues. Are there any stories in your newspaper to which geographic concepts can be applied to help understand the issues? Discuss.

RESOURCES

Some recent and classic books and articles on human geography:

Anderson, Kay, Mona Domosh, Nigel Thrift, and Steve Pile, eds. *Handbook of Cultural Geography*. Thousand Oaks, CA: Sage, 2003.

Arendt, Randall. *Designing Open Space Subdivisions*. Media, PA: Natural Lands Trust, 1994.

Brunn, Stanley D. "Sunbelt USA." *Focus* 36 (1986): 34–35.

Claval, Paul. "The Region as a Geographical, Economic and Cultural Concept." *International Social Science Journal* 39 (1987): 159–72.

Constandse, A. K. *Planning and Creation of an Environment*. Lelystad, Netherlands: Rijksdienst voor de IJsselmeerpolders, 1976.

Cresswell, Tim. *Place: A Short Introduction*. Malden, MA: Blackwell, 2004.

Cutter, Susan L., Reginald Golledge, and William L. Graf. "The Big Questions in Geography." *Professional Geographer* 54 (2002): 235–55.

Dicken, Peter. "Geographers and 'Globalization': (Yet) Another Missed Boat?" *Transactions of the Institute of British Geographers New Series* 29 (2004): 5–26.

Dodgshon, Robert A. "Human Geography at the End of Time? Some Thoughts on the Notion of Time-Space Compression." *Environment and Planning D: Society and Space* 17 (1999): 607–20.

Domosh, Mona. "Geography and Gender: The Personal and the Political." *Progress in Human Geography* 21 (1997): 81–87.

Geography Education Standards Project. *Geography for Life: National Geography Standards*. Washington, DC: National Geographic Research and Exploration, 1994.

Golledge, Reginald G. "Geographical Theories." *International Social Science Journal* 48 (1996): 461–76.

Gregory, Derek, Ron Johnston, Geraldine Pratt, Michael Watts, and Sarah Whatmore, eds. *The Dictionary of Human Geography*, 5th ed. Malden, MA: Wiley-Blackwell, 2009.

Hägerstrand, Torsten. *Innovation Diffusion as a Spatial Process*. Chicago: University of Chicago Press, 1967.

Hanson, Susan, ed. *Ten Geographic Ideas That Changed the World*. New Brunswick, NJ: Rutgers University Press, 1997.

Hartshorne, Richard. *The Nature of Geography*. Lancaster, PA: Association of American Geographers, 1939.

James, Preston E. *All Possible Worlds: A History of Geographical Ideas*. New York: Bobbs-Merrill, 1972.

Johnston, Ron. "Geography: A Different Sort of Discipline?" *Transactions of the Institute of British Geographers New Series* 28 (2003): 133–41.

Marston, Sallie A. "The Social Construction of Scale." *Progress in Human Geography* 24 (2000): 219–42.

Massey, Doreen B. *Space, Place, and Gender*. Minneapolis: University of Minnesota Press, 1994.

McDowell, Linda, and Joanne P. Sharp, eds. *A Feminist Glossary of Human Geography*. London and New York: Arnold, 1999.

Meinig, D. W. *Cultural Geography: A Critical Introduction*. Malden, MA: Blackwell, 2000.

Mitchell, Don. "Cultural Landscapes: The Dialectical Landscape—Recent Landscape Research in Human Geography." *Progress in Human Geography* 26 (2002): 381–90.

Rice, Bradley R. "Searching for the Sunbelt." *American Demographics* 3 (1981): 22–23.

Sauer, Carl O. "Morphology of Landscape." *University of California Publications in Geography* 2 (1925): 19–54.

Sobel, Dava. *Longitude: The True Story of a Lone Genius Who Solved the Greatest Scientific Problem of His Time*. New York: Walker and Co., 1995.

Talbert, Richard J. A., ed. *Barrington Atlas of the Greek and Roman World*. Princeton, NJ: Princeton University Press, 2000.

Tuan, Yi-Fu. *Topophilia: A Study of Environmental Perception, Attitudes, and Values*. New York: Columbia University Press, 1990.

Wallach, Bret. *Understanding the Cultural Landscape*. New York: Guilford Press, 2005.

Journals featuring human geography:

Annals of the Association of American Geographers, Antipode, Area, Canadian Geographer (Géographe canadien), Focus, Geografiska Annaler Series B Human Geography, Geographical Analysis, Geographical Review, Geography, Journal of Geography, Professional Geographer, Progress in Human Geography, Transactions of the Institute of British Geographers.

Key Internet sites:

www.aag.org The Association of American Geographers web site includes information about careers in geography, as well as its publications.

www.ncge.org National Council for Geographic Education resources include national standards for geographic literacy.

www.amergeog.org The American Geographical Society provides access to its publications.

www.nationalgeographic.com The National Geographic Society offers access to material from its magazine and television programs, as well as on-line mapping.

Log in to www.mygeoscienceplace.com for videos, interactive maps, RSS feeds, casestudies,and self-study quizzes to enhance your study of Basic Geography Concepts.

CHAPTER 2

Population

How many brothers and sisters do you have? How many brothers and sisters did your parents or grandparents have? Did they have more, fewer, or the same number of siblings as yourself? How many children do you have, or intend to have? Is that figure larger, smaller, or the same number as your parents and grandparents had?

The typical family in a more developed country (MDC) today contains fewer people than in the past, and the number of children is declining. In much of North America and Europe a majority of people have

KEY ISSUES

1 Where Is the World's Population Distributed?

2 Where Has the World's Population Increased?

3 Why Is Population Increasing at Different Rates in Different Countries?

4 Why Might the World Face an Overpopulation Problem?

the same number or fewer siblings than their parents and grandparents. And the number of children your generation has, or will have, appears to be fewer on average, although only the future can reveal the actual trend.

In other regions of the world the number of children per household tends to be much higher than in the MDCs. The ability of less developed countries (LDCs) to provide food, clothing, and shelter for their people is severely hampered by the continued rapid growth of their population.

The scientific study of population characteristics is **demography**. Demographers look statistically at how people are distributed spatially and by age, gender, occupation, fertility, health, and so on.

A study of population is the basis for understanding a wide variety of issues in human geography. To study the challenge of increasing the food supply, reducing pollution, and encouraging economic growth, geographers must ask where and why a region's population is distributed as it is. Therefore, our study of human geography begins with a study of population.

Listening to speech in Baltimore by President-elect Barack Obama, January 17, 2009, three days before his inauguration.

CASE STUDY / Population Growth in India

The Phatak family lives in a village of 600 inhabitants in India. At age 40, Indira Phatak has been pregnant five times. Four of her children have survived; they are aged 5 to 18.

When the two Phatak daughters marry a few years from now, how many children will each of them bear? The Indian government hopes that they will choose to have fewer children than their mother. About 27 million babies will be born this year in India, and the country's population is growing by 18 million annually. Unless attitudes and behavior drastically change in the next few years, India's population—currently 1.2 billion—would reach 1.8 billion in 2050.

Three-fourths of Indians live in rural settlements that have fewer than 5,000 inhabitants. For many of these people, children are an economic asset because they help perform chores on the farm and are expected to provide for their parents in their old age. The high percentage of children who will die before they reach working age also encourages large families. One out of every 18 infants in India dies within 1 year of birth, and 1 out of every 14 pregnant women dies annually during pregnancy and childbirth.

In recent years, India has made significant progress in diffusing modern agricultural practices, building new industry, and developing natural resources, all of which have increased national wealth. However, in a country with a rapidly expanding population, much of the newly created wealth must be used to provide food, housing, and other basic services for the additional people. With one-third of the population under the age of 15, the government must build schools, hospitals, and day-care centers. Therefore, the growing wealth is going primarily to provide a reasonable standard of living for an expanding population. Further, will employment be available to these 375 million children when they are old enough to work? ■

The study of population is critically important for three reasons:

- More people are alive at this time—nearly 7 billion—than at any point in Earth's long history.
- The world's population increased at a faster rate during the second half of the twentieth century than ever before in history.
- Virtually all global population growth is concentrated in LDCs.

These facts lend urgency to the task of understanding the diversity of population problems in the world today.

As introduced in Chapter 1, geographers ask "where" and "why" questions. As we begin our study of the major topics in human geography, note the wording of the four key issues that organize the material in this chapter. The first two issues ask "where" questions, the second two ask "why" questions. These four issues rely on the five basic concepts presented in Chapter 1.

Geographers study population problems by first describing *where* people are found across Earth's *space*. The location of Earth's nearly 7 billion people forms a regular distribution. The second key issue looks at another "where" question, this time the *places* where population is growing.

The chapter then turns to explaining *why* population is growing at different rates in different places. From the perspective of *globalization*, geographers argue that the world's so-called **overpopulation** problem is not simply a matter of the total number of people on Earth but also includes the relationship between the number of people and the availability of resources. Problems result when an area's population exceeds the capacity of the environment to support it at an acceptable standard of living.

At a local *scale*, geographers find that overpopulation is a threat in some *regions* of the world but not in others. The capacity of Earth as a whole to support human life may be high, but some regions have a favorable balance between people and available resources, whereas others do not. Further, the regions with the most people are not necessarily the same as the regions with an unfavorable balance between population and resources.

The final key issue explains why geographers consider *local diversity* in growth rates to be important. Some demographers predict that the world may become overburdened with too many people in the future. They ask whether the world's population will exceed the capacity of Earth to provide food, space, and resources for the people. Geographers who specialize in demography cannot offer a simple "yes" or "no" answer, but they recognize the *connections* among regions of high and low population growth, discussed in more detail in Chapter 3.

KEY ISSUE 1

Where Is the World's Population Distributed?

- **Population Concentrations**
- **Sparsely Populated Regions**
- **Population Density**

Human beings are not distributed uniformly across Earth's surface. We can understand how population is distributed by examining two basic properties—concentration and density. Geographers identify regions of Earth's surface where population is clustered and regions where it is sparse. We also construct several density measures that help geographers explain the relationship between the number of people and available resources. ■

Population Concentrations

Two-thirds of the world's inhabitants are clustered in four regions—East Asia, South Asia, Southeast Asia, and Europe. The clustering of the world's population can be displayed on a cartogram, which depicts the size of countries according to population rather than land area, as is the case with most maps (Figure 2-1). The shapes of several large or populous countries, including Brazil, Canada, China, Indonesia, Russia, and the United States, have been exaggerated to show the regions within the countries where most of the population is clustered.

When compared to a more typical equal-area map, such as that shown in Figure 2-2, the population cartogram displays the major population clusters of Europe and East, South, and Southeast Asia as much larger, and Africa and the Western Hemisphere as much smaller. As you look at maps of population growth and other topics in this and subsequent chapters, pay special attention to Asia and Europe, because global patterns are heavily influenced by conditions in these regions, where two-thirds of the world's people live.

The four regions display some similarities. Most of the people in these regions live near an ocean or near a river with easy access to an ocean, rather than in the interior of major landmasses (compare Figure 2-2 with Figure 2-3). In fact, approximately two-thirds of the world's population live within 500 kilometers (300 miles) of an ocean, and four-fifths live within 800 kilometers (500 miles).

The four population clusters occupy generally low-lying areas, with fertile soil and temperate climate. The regions all are located in the Northern Hemisphere between 10° and 55° north latitude, with the exception of part of the Southeast Asia concentration. Despite these similarities, we can see significant differences in the pattern of occupancy of the land in the five concentrations.

East Asia

Nearly one-fourth of the world's people live in East Asia. The region, bordering the Pacific Ocean, includes eastern China, the islands of Japan, the Korean peninsula, and the island of Taiwan.

Five-sixths of the people in this concentration live in the People's Republic of China, the world's most populous country. China is the world's fourth-largest country in land area, but much of its interior is sparsely inhabited mountains and deserts. The Chinese population is clustered near the Pacific Coast and in several fertile river valleys that extend inland, such as the Huang and the Yangtze. Although China has 25 urban areas with more than 2 million inhabitants and 61 with more than 1 million, more than one-half of the people live in rural areas where they work as farmers.

In Japan and South Korea, population is not distributed uniformly either. Forty percent of the people live in three large metropolitan areas—Tokyo and Osaka in Japan, and Seoul in South Korea—that cover less than 3 percent of the two countries' land area. In sharp contrast to China, more than three-fourths of all Japanese and Koreans live in urban areas and work at industrial or service jobs.

South Asia

Nearly one-fourth of the world's people also live in South Asia, which includes India, Pakistan, Bangladesh, and the island of Sri Lanka. India, the world's second most populous country, contains more than three-fourths of the South Asia population concentration.

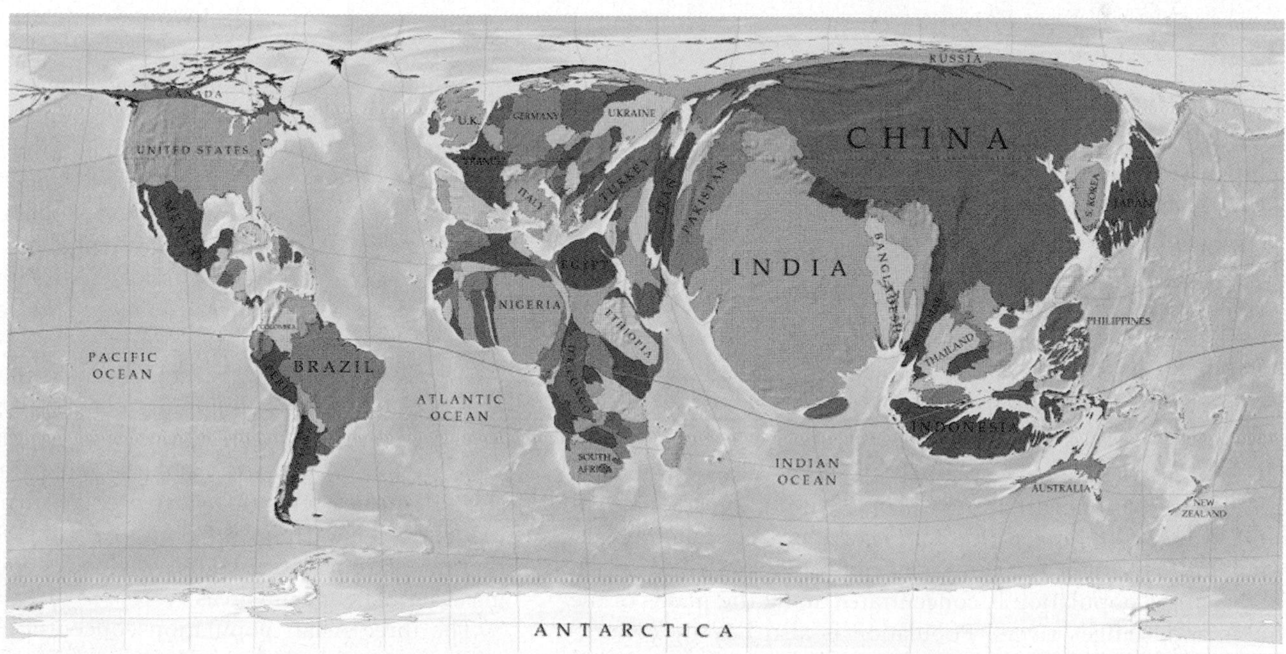

FIGURE 2-1 Population cartogram. Countries are displayed by size of population rather than land area. Countries named on the cartogram have at least 50 million inhabitants.

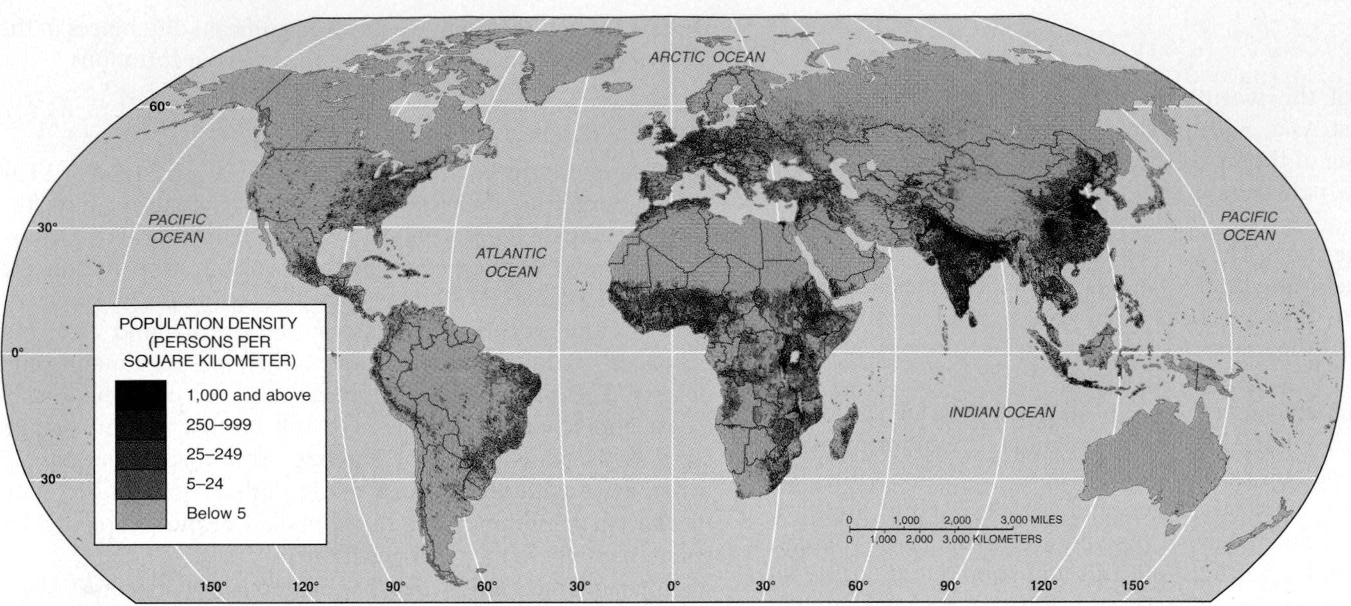

FIGURE 2-2 Population distribution. People are not distributed uniformly across Earth's surface.

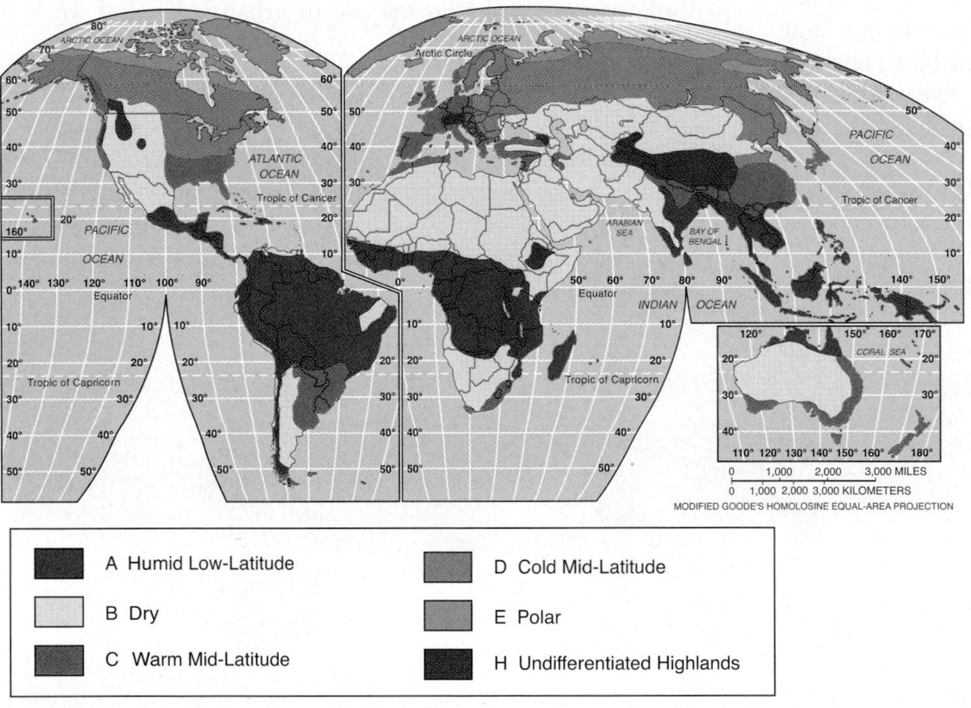

FIGURE 2-3 World climate. Map is simplified version of climate map by Vladimir Köppen (Figure 1-19). Note that most humans live in C climate regions.

The largest concentration of people within South Asia lives along a 1,500-kilometer (900-mile) corridor from Lahore, Pakistan, through India and Bangladesh to the Bay of Bengal. Much of this area's population is concentrated along the plains of the Indus and Ganges rivers. Population is also heavily concentrated near India's two long coastlines—the Arabian Sea to the west and the Bay of Bengal to the east.

Like the Chinese, most people in South Asia are farmers living in rural areas. The region contains 18 urban areas with more than 2 million inhabitants and 46 with more than 1 million, but only one-fourth of the total population lives in an urban area.

Southeast Asia

A third important Asian population cluster, and the world's fourth largest (after Europe, described next), is in Southeast Asia. Around 600 million people live in Southeast Asia, mostly on a series of islands that lie between the Indian and Pacific oceans. These islands include Java, Sumatra, Borneo, Papua New Guinea, and the Philippines.

The largest concentration is on the island of Java, inhabited by more than 100 million people. Indonesia, which consists of 13,677 islands, including Java, is the world's fourth most populous country.

Several islands that belong to the Philippines contain high population concentrations, and population is also clustered along several river valleys and deltas at the southeastern tip of the Asian mainland, known as Indochina. Like China and South Asia, the Southeast Asia concentration is characterized by a high percentage of people working as farmers in rural areas.

The three Asian population concentrations together comprise more than half of the world's total population, but together they live on less than 10 percent of Earth's land area. The same held true 2,000 years ago, when approximately half of the world's population was found in these same regions.

Europe

Europe, including the European portion of Russia, forms the world's third-largest population cluster, one-ninth of the world's people. The region includes four dozen countries, ranging from Monaco, with 1 square kilometer (0.7 square miles) and a population of 33,000, to Russia, the world's largest country in land area when its Asian part is included.

In contrast to the three Asian concentrations, three-fourths of Europe's inhabitants live in cities, and less than 10 percent are farmers. A dense network of road and rail lines links settlements. The highest population concentrations in Europe are near the coalfields of England, Germany, and Belgium, historically the major source of energy for industry.

Although the region's temperate climate permits cultivation of a variety of crops, Europeans do not produce enough food for themselves. Instead, they import food and other resources from elsewhere in the world. The search for additional resources was a major incentive for Europeans to explore and colonize other parts of the world during the previous six centuries. Today, Europeans turn many of these resources into manufactured products.

Other Population Clusters

The largest population concentration in the Western Hemisphere is in the northeastern United States and southeastern Canada. This cluster extends along the Atlantic Coast from Boston to Newport News, Virginia, and westward along the Great Lakes to Chicago. About 2 percent of the world's people live in the area. Like the Europeans, most Americans are urban dwellers; less than 2 percent are farmers.

Another 2 percent of the world's population is clustered in West Africa, especially along the south-facing Atlantic coast. Approximately half of the West Africa concentration is found in Nigeria, the most populous country in Africa, and the other half is divided among several small countries west of Nigeria. As in the three Asian concentrations, most West Africans work in agriculture, although the region has 5 urban areas with more than 2 million inhabitants and 11 with more than 1 million.

Sparsely Populated Regions

Human beings avoid clustering in certain physical environments (Figure 2-3). Relatively few people live in regions that are too dry, too wet, too cold, or too mountainous for activities such as agriculture. The portion of Earth's surface occupied by permanent human settlement is called the **ecumene**.

The areas of Earth that humans consider too harsh for occupancy have diminished over time, whereas the ecumene has increased (Figure 2-4). Seven thousand years ago humans occupied only a small percentage of Earth's land area, primarily in the Middle East, Eastern Europe, and East Asia. Even 500 years ago much of North America and Asia lay outside the ecumene.

Still, approximately three-fourths of the world's population live on only 5 percent of Earth's surface. The balance of Earth's

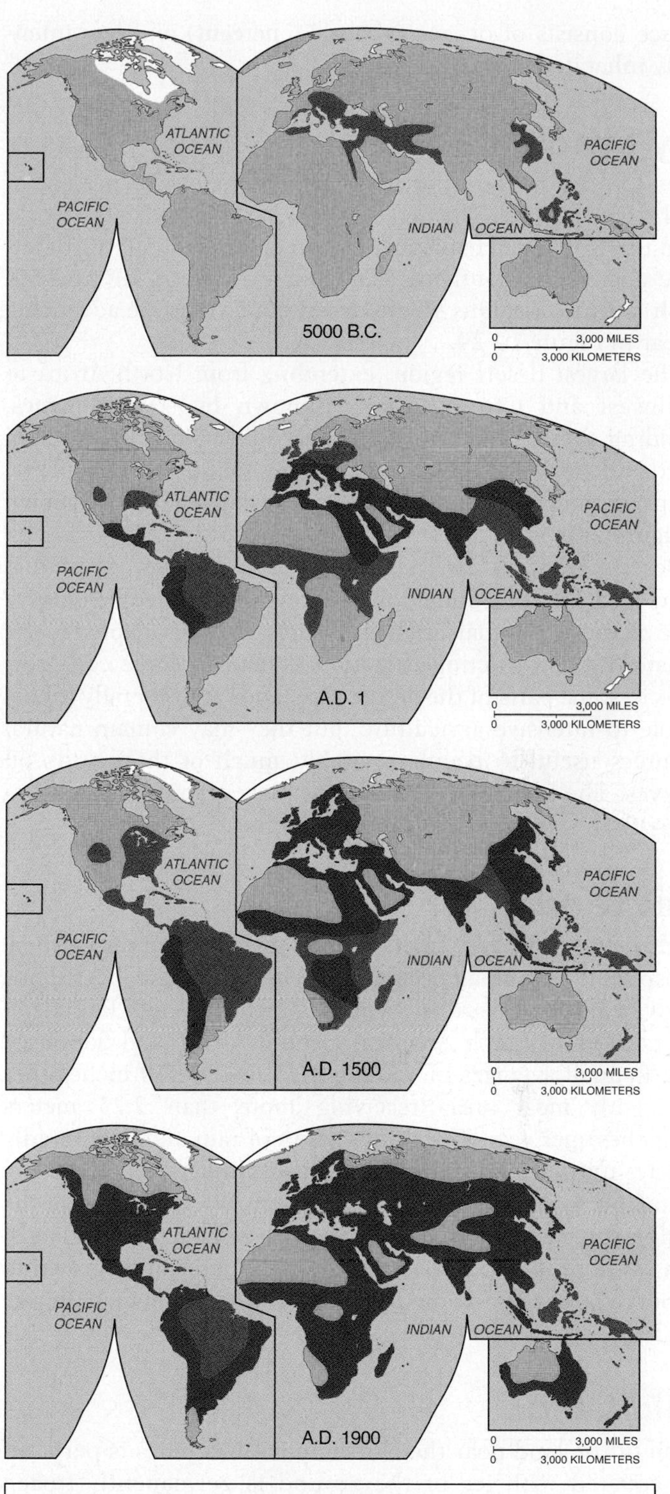

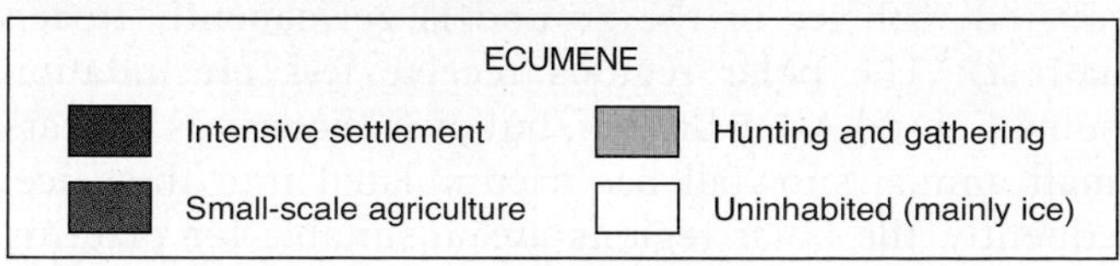

FIGURE 2-4 Ecumene. The portion of Earth occupied by permanent human settlement—the ecumene—has expanded from the Middle East and East Asia to encompass most of the world's land area.

surface consists of oceans (about 71 percent) and less intensively inhabited land.

Dry Lands

Areas too dry for farming cover approximately 20 percent of Earth's land surface. The two largest desert regions in the world lie in the Northern Hemisphere between 15° and 50° north latitude and in the Southern Hemisphere between 20° and 50° south latitude. Regions where desert conditions are advancing appear in Figure 10-29.

The largest desert region, extending from North Africa to Southwest and Central Asia, is known by several names, including the Sahara, Arabian, Thar, Takla Makan, and Gobi deserts. A smaller desert region, in the Southern Hemisphere, comprises much of Australia. Earth's desert regions are shown in Figure 2-3.

Deserts generally lack sufficient water to grow crops that could feed a large population, although some people survive there by raising animals, such as camels, that are adapted to the climate. By constructing irrigation systems, people can grow crops in some parts of the desert. Dry lands are generally inhospitable to intensive agriculture, but they may contain natural resources useful to people—notably, much of the world's oil reserves. The increasing demand for these resources has led to a growth in settlements in or near deserts.

Wet Lands

Lands that receive very high levels of precipitation may also be inhospitable for human occupation. These lands are located primarily near the equator between 20° north and south latitude in the interiors of South America, Central Africa, and Southeast Asia. Rainfall averages more than 1.25 meters (50 inches) per year, with most areas receiving more than 2.25 meters (90 inches) per year. The combination of rain and heat rapidly depletes nutrients from the soil and thus hinders agriculture.

Precipitation may be concentrated into specific times of the year or spread throughout the year. In seasonally wet lands, such as those in Southeast Asia, enough food can be grown to support a large population (see the rice production map, Figure 10-12).

Cold Lands

Much of the land near the North and South poles is perpetually covered with ice or the ground is permanently frozen (permafrost). The polar regions receive less precipitation than some Central Asian deserts, but over thousands of years the small annual snowfall has accumulated into thick ice. Consequently, the polar regions are unsuitable for planting crops; few animals can survive the extreme cold, and few human beings live there.

High Lands

Relatively few people live at high elevations. The highest mountains in the world are steep, snow covered, and sparsely settled. For example, approximately half of Switzerland's land is more than 1,000 meters (3,300 feet) above sea level, and only 5 percent of the country's people live there.

We can find some significant exceptions, especially in Latin America and Africa. People may prefer to occupy higher lands if temperatures and precipitation are uncomfortably high at lower elevations. In fact, Mexico City, one of the world's largest cities, is located at an elevation of 2,243 meters (7,360 feet).

Population Density

Density, defined in Chapter 1 as the number of people occupying an area of land, can be computed in several ways, including arithmetic density, physiological density, and agricultural density. These measures of density help geographers to describe the distribution of people in comparison to available resources.

Arithmetic Density

Geographers most frequently use **arithmetic density**, which was defined in Chapter 1 as the total number of objects in an area. In population geography, arithmetic density refers to the total number of people divided by total land area. Geographers rely on the arithmetic density to compare conditions in different countries because the two pieces of information needed to calculate the measure—total population and total land area—are easy to obtain.

For example, to compute the arithmetic or population density for the United States, we can divide the population (approximately 310 million people) by the land area (approximately 9.6 million square kilometers, or 3.7 million square miles). The result shows that the United States has an arithmetic density of 32 persons per square kilometer (84 persons per square mile).

By comparison, the arithmetic density is much higher in South Asia. In Bangladesh, it is approximately 1,127 persons per square kilometer (2,919 persons per square mile), and in India it is 356 (922). On the other hand, the arithmetic density is only 3 persons per square kilometer (7 persons per square mile) in Australia and Canada (Figure 2-5).

Arithmetic density varies even more within individual countries. In the United States, for example, New York County (Manhattan Island) has a population density of approximately 27,500 persons per square kilometer (71,200 persons per square mile), whereas Loving County, Texas, has a population density of approximately 0.024 persons per square kilometer (0.06 per square mile). In Egypt the arithmetic density is only 79 persons per square kilometer (205 persons per square mile) overall, but it is 2,000 persons per square kilometer (5,400 persons per square mile) in the delta and valley of the Nile River.

Arithmetic density enables geographers to compare the number of people trying to live on a given piece of land in different regions of the world. Thus, arithmetic density answers the "where" question. However, to explain *why* people are not uniformly distributed across Earth's surface, other density measures are more useful (Table 2-1).

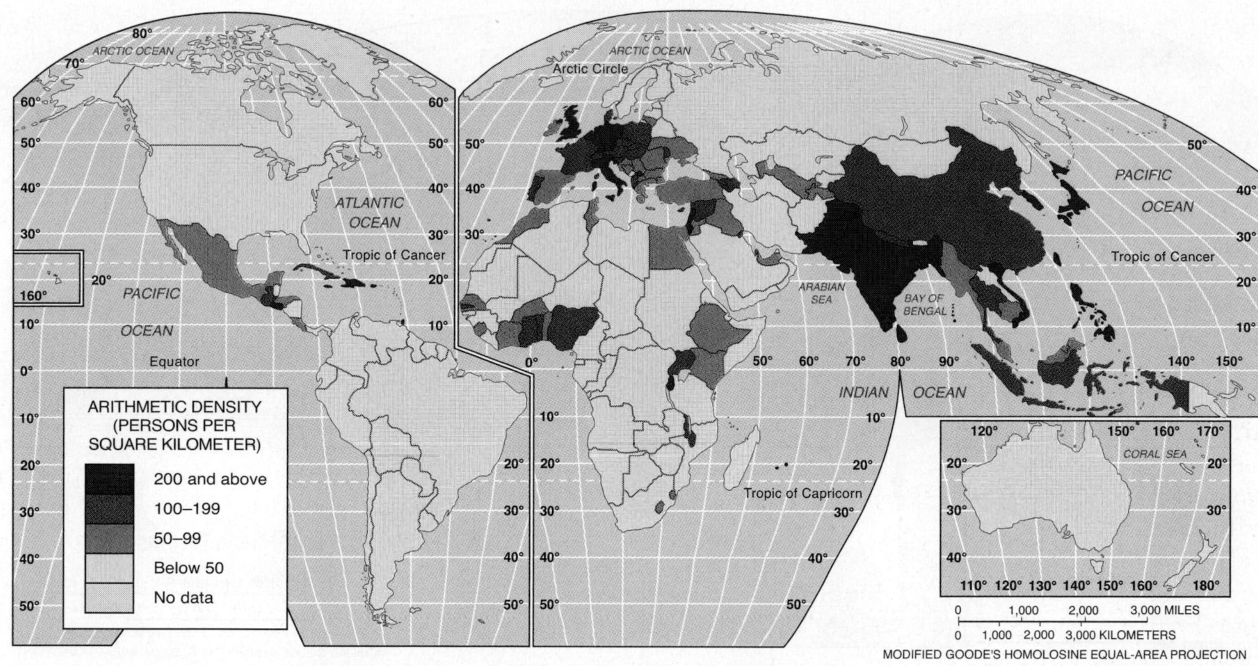

FIGURE 2-5 Arithmetic density. Arithmetic, or population, density is the total number of people divided by the total land area. The highest population densities are found in Asia, Europe, and Central America, whereas the lowest are in North and South America and Australia.

Physiological Density

A more meaningful population measure is afforded by looking at the number of people per area of a certain type of land in a region. Land suited for agriculture is called arable land. In a region, the number of people supported by a unit area of arable land is called the **physiological density** (Figure 2-6).

The United States has a physiological density of 175 persons per square kilometer (453 per square mile) of arable land. This contrasts sharply with Egypt, which has 2,296 persons per square kilometer (5,947 per square mile) of arable land. This large difference in physiological densities demonstrates that crops grown on a hectare of land in Egypt must feed far more people than in the United States. The higher the physiological density, the greater the pressure that people may place on the land to produce enough food. Physiological density provides insights into the relationship between the size of a population and the availability of resources in a region.

Comparing physiological and arithmetic densities helps geographers to understand the capacity of the land to yield enough food for the needs of the people. In Egypt, the large difference between the physiological density (2,296 people per square kilometer of arable land) and arithmetic density (79 persons per square kilometer over the entire country) indicates that most of the country's land is unsuitable for intensive agriculture. In fact, all but 5 percent of the Egyptian people live in the Nile River valley and delta, because it is the only area in the country that receives enough moisture (by irrigation from the river) to allow intensive cultivation of crops.

TABLE 2-1 MEASURES OF DENSITY IN SELECTED COUNTRIES

	ARITHMETIC DENSITY*	PHYSIOLOGICAL DENSITY*	AGRICULTURAL DENSITY*	PERCENT FARMERS	PERCENT ARABLE
Canada	3	65	1	2	5
United States	32	175	2	2	18
Egypt	79	2,296	251	31	3
United Kingdom	255	1,083	9	2	23
Japan	338	2,695	46	3	13
India	356	690	163	58	52
Netherlands	398	1,748	23	3	23
Bangladesh	1,127	1,927	472	52	58

*Population per square kilometer

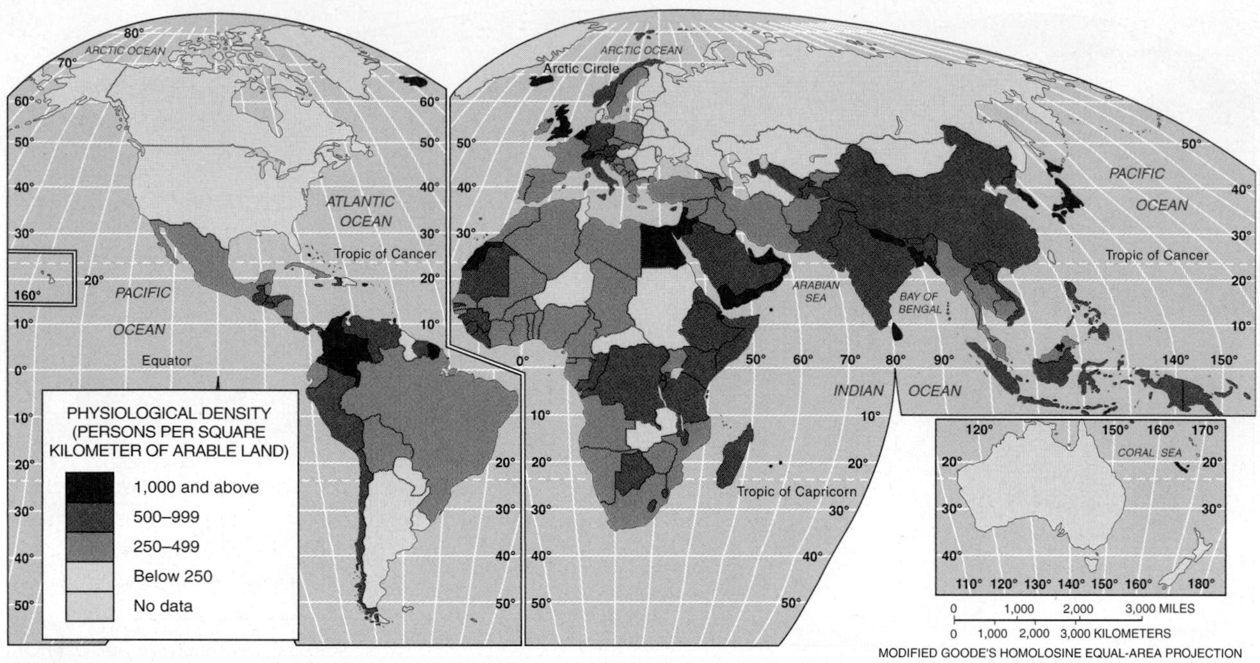

FIGURE 2-6 Physiological density. Physiological density is the number of people per unit area of arable land, which is land suitable for agriculture. Physiological density is a better measure than arithmetic density of the relationship between population and the availability of resources in a society.

Agricultural Density

Two countries can have similar physiological densities, but they may produce significantly different amounts of food because of different economic conditions. **Agricultural density** is the ratio of the number of farmers to the amount of arable land, which is land suitable for agriculture. This density measure helps account for economic differences (Figure 2-7).

The United States has an extremely low agricultural density (1.6 farmers per square kilometer of arable land), whereas Egypt has a very high density (251 farmers per square kilometer of arable land). MDCs have lower agricultural densities because technology and finance allow a few people to farm extensive land areas and feed many people. This frees most of the MDC population to work in factories, offices, or shops rather than in the fields.

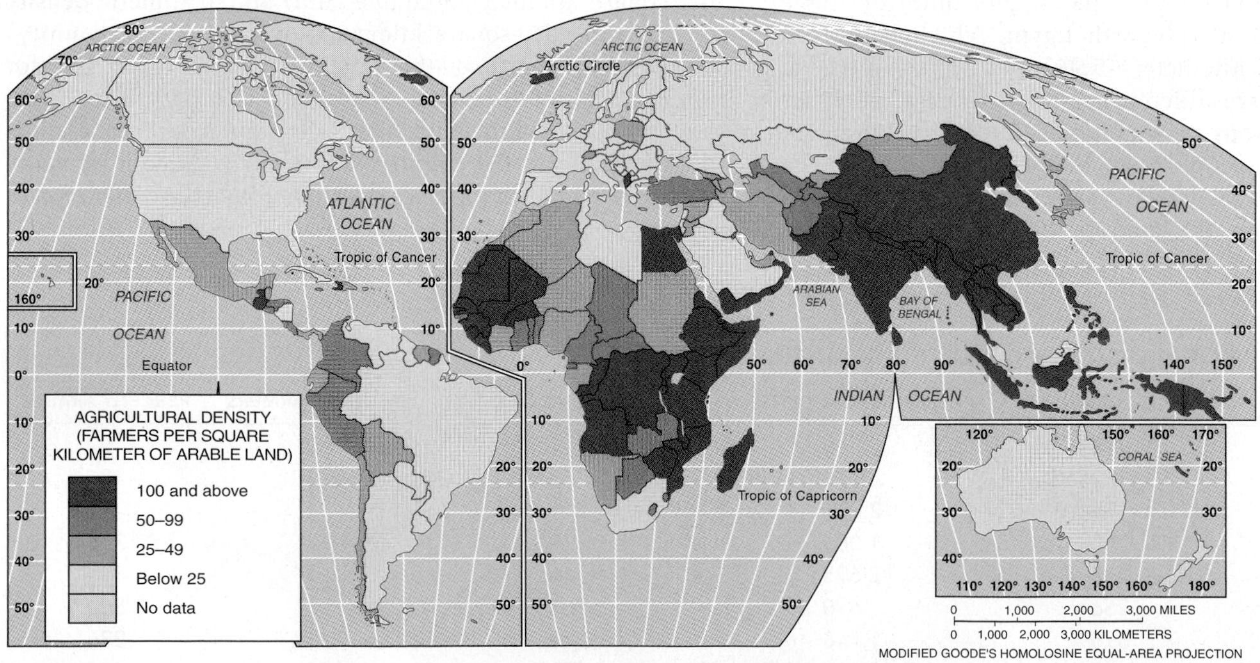

FIGURE 2-7 Agricultural density. Agricultural density is the ratio of the number of farmers to the amount of arable land.

To understand the relationship between population and resources in a country, geographers examine a country's physiological and agricultural densities together. As shown in Table 2-1, the physiological densities of both Bangladesh and the Netherlands are high, but the Dutch have a much lower agricultural density than the Bangladeshi. Geographers conclude that both the Dutch and Bangladeshi put heavy pressure on the land to produce food, but the Dutch agricultural system utilizes fewer farmers than does the Bangladeshi system.

Similarly, the Netherlands has a much higher physiological density than does India but a much lower agricultural density. This difference demonstrates that, compared with India, the Dutch have extremely limited arable land to meet the needs of their population.

Recall from Chapter 1 how the Dutch have built dikes and created polders, areas of land made usable by draining water from them. The highly efficient Dutch farmers can generate a large food supply from a limited resource.

KEY ISSUE 2

Where Has the World's Population Increased?

- Natural Increase
- Fertility
- Mortality

After identifying where people are distributed across Earth's surface, we can describe the locations where the numbers of people are increasing. Population increases rapidly in places where many more people are born than die, increases slowly in places where the number of births exceeds the number of deaths by only a small margin, and declines in places where deaths outnumber births. The population of a place also increases when people move in and decreases when people move out. This element of population change—migration—is discussed in Chapter 3. ■

Natural Increase

Geographers most frequently measure population change in a country or the world as a whole through three measures—crude birth rate, crude death rate, and natural increase rate.

- **Crude birth rate (CBR)** is the total number of live births in a year for every 1,000 people alive in the society. A CBR of 20 means that for every 1,000 people in a country, 20 babies are born over a 1-year period.
- **Crude death rate (CDR)** is the total number of deaths in a year for every 1,000 people alive in the society. Comparable to the CBR, the CDR is expressed as the annual number of deaths per 1,000 population.
- **Natural increase rate (NIR)** is the percentage by which a population grows in a year. It is computed by subtracting CDR from CBR, after first converting the two measures from numbers per 1,000 to percentages (numbers per 100). Thus if the CBR is 20 and the CDR is 5 (both per 1,000), then the NIR is 15 per 1,000, or 1.5 percent. The term natural means that a country's growth rate excludes migration.

The world NIR during the early twenty-first century has been 1.2, meaning that the population of the world had been growing each year by 1.2 percent. The world NIR is lower today than its all-time peak of 2.2 percent in 1963, and it has declined sharply since the 1990s. However, the NIR during the second half of the twentieth century was high by historical standards.

About 80 million people are being added to the population of the world annually (Figure 2-8). That number represents a decline from the historic high of 87 million in 1989. The number of people added each year has dropped much more slowly than the NIR because the population base is much higher now than in the past.

World population increased from 3 to 4 billion in 14 years, from 4 to 5 billion in 13 years, and from 5 to 6 billion in 12 years. As the base continues to grow in the twenty-first century, a change of only one-tenth of 1 percent can produce very large swings in population growth.

The rate of natural increase affects the **doubling time**, which is the number of years needed to double a population, assuming a constant rate of natural increase. At the early twenty-first-century rate of 1.2 percent per year, world population would double in about 54 years. Should the same NIR continue through the twenty-first century, global population

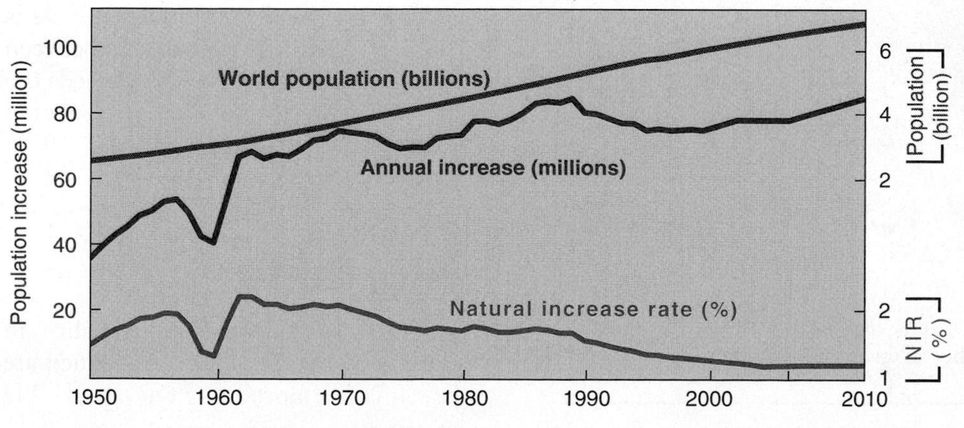

FIGURE 2-8 World population growth, 1950–2010. The percentage by which the population grew (that is, the natural increase rate [NIR]) declined during the late twentieth century from its historic peak in the early 1960s, but the number of people added each year did not decline very much, because with world population increasing from 2.5 billion to nearly 7 billion people during the period, the percentage was applied to an ever larger base.

in the year 2100 would reach 24 billion. When the NIR was 2.2 percent back in 1963, doubling time was 35 years. Had the 2.2 percent rate continued into the twenty-first century, Earth's population in 2010 would be nearly 10 billion instead of nearly 7 billion. A 2.2 percent NIR through the twenty-first century would have produced a total population of more than 50 billion in 2100.

More than 95 percent of the natural increase is clustered in LDCs (Figure 2-9). The NIR exceeds 2.0 percent in most countries of sub-Saharan Africa and the Middle East, whereas it is negative in Europe, meaning that in the absence of immigrants, population actually is declining. About one-third of the world's population growth during the past decade has been in South Asia, one-fourth in sub-Saharan Africa, and the remainder divided about equally among East Asia, Southeast Asia, Latin America, and the Middle East.

Regional differences in NIRs mean that most of the world's additional people live in the countries that are least able to maintain them. To explain these differences in growth rates, geographers point to regional differences in fertility and mortality rates.

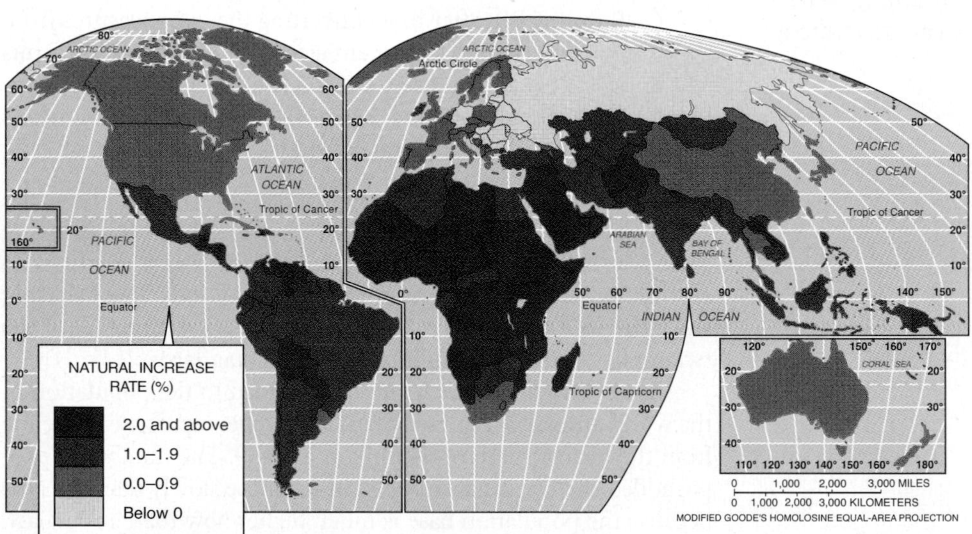

FIGURE 2-9 Natural increase rate (NIR). The natural increase rate is the percentage by which the population of a country grows in a year. The world average is currently about 1.2 percent. The countries with the highest natural increase rates are concentrated in Africa and Southwest Asia.

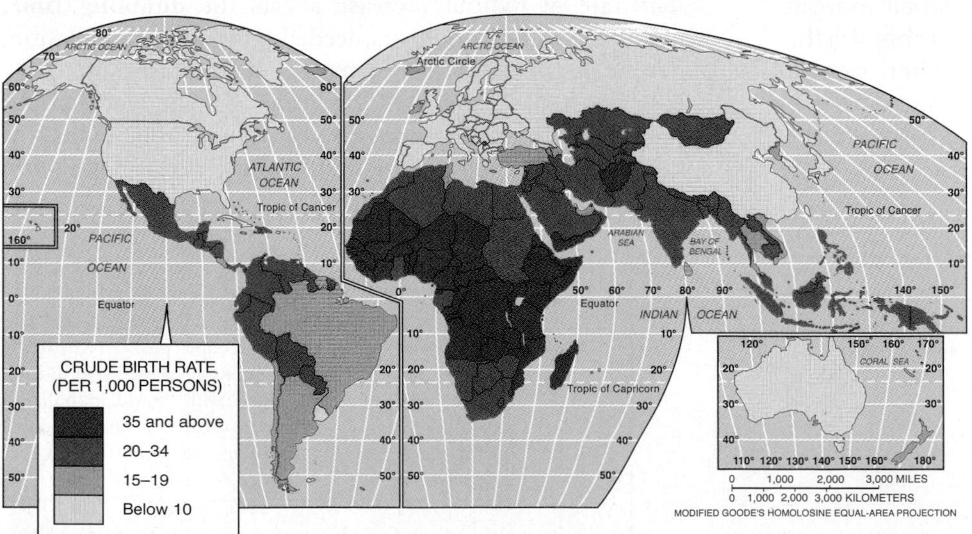

FIGURE 2-10 Crude birth rate (CBR). The crude birth rate is the total number of live births in a year for every 1,000 people alive in the society. The global distribution of crude birth rates parallels that of natural increase rates.

Fertility

The world map of crude birth rates (Figure 2-10) mirrors the distribution of NIRs (compare with Figure 2-9). As was the case with NIRs, the highest CBRs are in sub-Saharan Africa, and the lowest are in Europe. Many sub-Saharan African countries have a CBR over 40, whereas many European countries have a CBR below 10.

The word *crude* in *crude birth rate* and *crude death rate* means that we are concerned with society as a whole rather than a refined look at particular individuals or groups. In communities with an unusually large number of people of a certain age—such as a college town—we may study separate birth rates for women of each age. These numbers are *age-specific birth rates* rather than CBRs.

Geographers also use the **total fertility rate (TFR)** to measure the number of births in a society (Figure 2-11). The TFR is the average number of children a woman will have throughout her childbearing years (roughly ages 15 through 49). To compute the TFR, demographers assume that a woman reaching a particular age in the future will be just as likely to have a child as are women of that age today. Thus, the CBR provides a picture of a society as a whole in a given year, whereas the TFR attempts to predict the future behavior of individual women in a world of rapid cultural change.

The TFR for the world as a whole is 2.6, and, again, the figures vary between MDCs and LDCs. The TFR exceeds 6.0 in many countries of sub-Saharan Africa, compared to less than 1.9 in most European countries.

Mortality

Two useful measures of mortality in addition to the CDR already defined are the infant mortality rate and life expectancy. The **infant mortality rate (IMR)** is the annual number of deaths of infants under 1 year of age, compared with total live births (Figure 2-12). As was the case with the CBR and CDR, the IMR is usually expressed as the number of deaths among infants per 1,000 births rather than as a percentage (per 100).

The global distribution of IMRs follows the pattern that by now has become familiar. The highest rates are in the LDCs of sub-Saharan Africa, whereas the lowest rates are in Europe. The IMR approaches 100 in sub-Saharan Africa, meaning that nearly 10 percent of all babies in the region die before reaching their first birthday. The IMR is less than 5 percent throughout Western Europe. In general, the IMR reflects a country's health-care system. Lower IMRs are found in countries with well-trained doctors and nurses, modern hospitals, and large supplies of medicine.

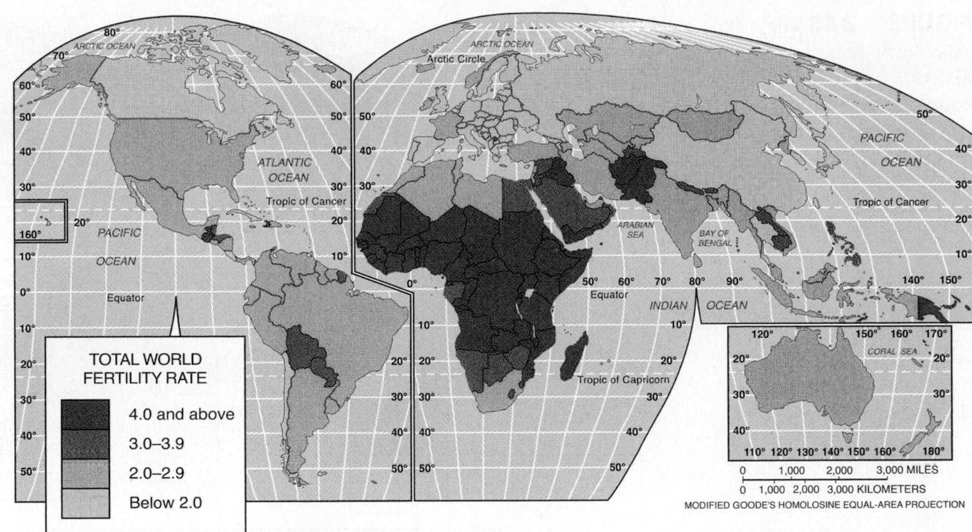

FIGURE 2-11 Total fertility rate (TFR). Total fertility rate is the number of children a woman will have throughout her childbearing years. Again, the highest rates are in sub-Saharan Africa and the Middle East, whereas the lowest are in Europe.

Although the United States is well endowed with medical facilities, it suffers from a higher IMR than Canada and every country in Western Europe. African Americans and other minorities in the United States have IMRs that are twice as high as the national average, comparable to levels in Latin America and Asia. Some health experts attribute this to the fact that many poor people in the United States, especially minorities, cannot afford good health care for their infants.

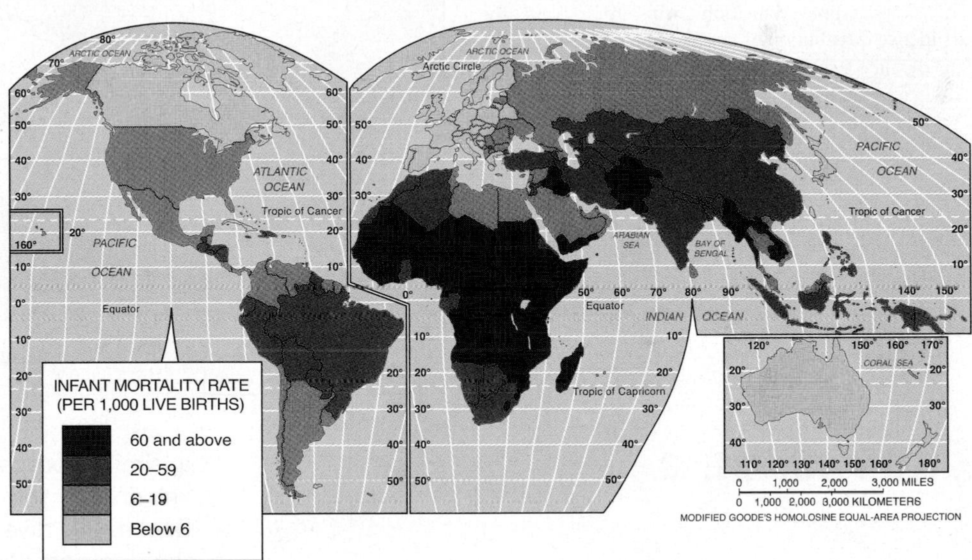

FIGURE 2-12 Infant mortality rate (IMR). The infant mortality rate is the number of deaths of infants under age 1 per 1,000 live births in a year. European and North American countries generally have infant mortality rates of under 10 per 1,000, whereas rates of more than 100 per 1,000 are common in Africa

Life expectancy at birth measures the average number of years a newborn infant can expect to live at current mortality levels (Figure 2-13). Like every other mortality and fertility rate discussed thus far, life expectancy is most favorable in the wealthy countries of Western Europe and least favorable in the poor countries of sub-Saharan Africa. Babies born today can expect to live to around 80 in Western Europe but only to around 50 in sub-Saharan Africa.

Natural increase, crude birth, total fertility, infant mortality, life expectancy—the descriptions have become repetitious because their distributions follow similar patterns. MDCs have lower rates of natural increase, crude birth, total fertility, and infant mortality, and higher average life expectancy. Higher natural increase, crude birth, total fertility rates, and IMRs and lower average life expectancy are found in LDCs.

The final world map of demographic variables—CDR—does not follow the familiar pattern (Figure 2-14). The combined CDR for all LDCs is actually lower than the combined rate for all MDCs. Furthermore, the variation between the world's highest and lowest CDRs is much less extreme than the variation in CBRs. The highest CDR in the world is 23 per 1,000, and the lowest is 1—a difference of 22—whereas CBRs for individual countries range from 7 per 1,000 to 53, a spread of 46.

Why does Denmark, one of the world's wealthiest countries, have a higher CDR than Cape Verde, one of the poorest? Why does the United States, with its extensive system of hospitals and physicians, have a higher CDR than Mexico and nearly every country in Latin America? The answer is that the populations of different countries are at various stages in an important process known as the demographic transition, upon which we focus in the third key issue of this chapter.

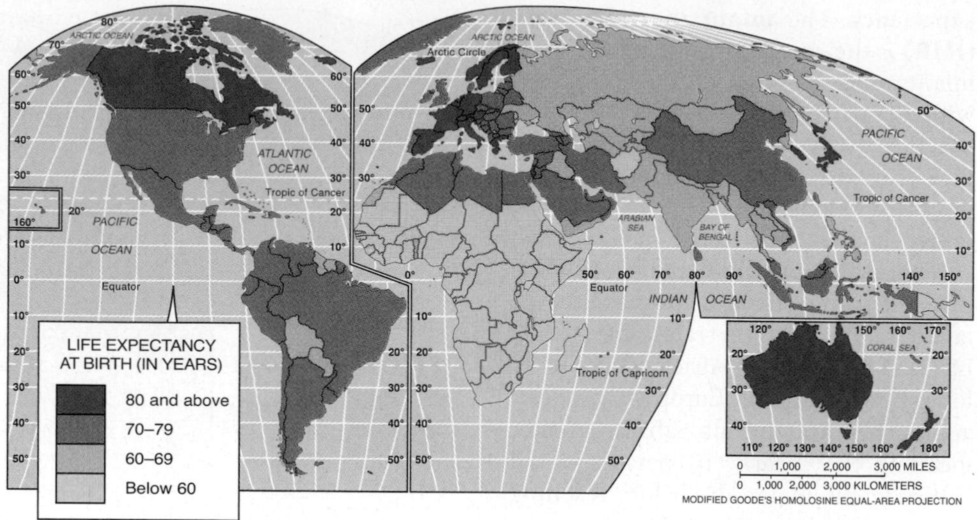

FIGURE 2-13 Life expectancy at birth. Life expectancy at birth is the average number of years a newborn infant can expect to live. Worldwide, babies born this year are expected to live until their mid-sixties. Life expectancy for babies ranges from the 40s in several African countries to the 80s in some MDCs.

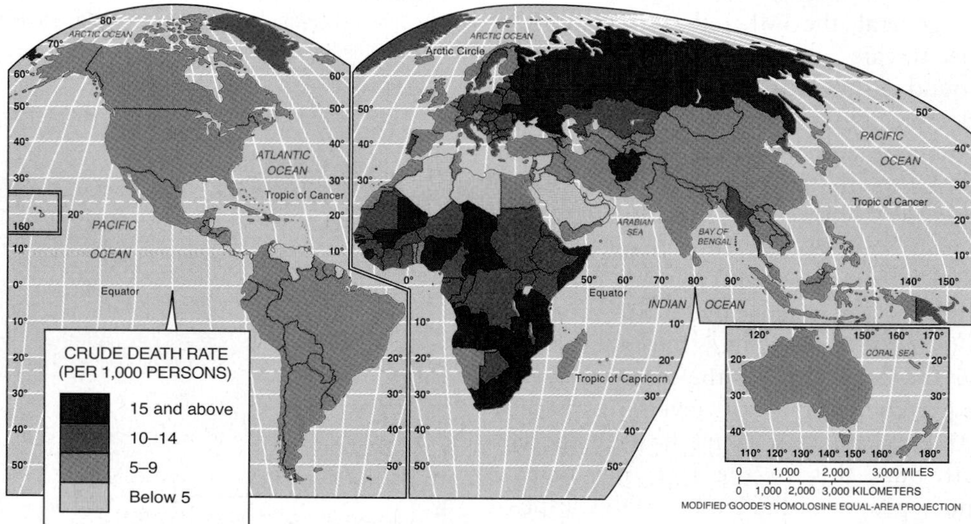

FIGURE 2-14 Crude death rate (CDR). Crude death rate is the total number of deaths in a year for every 1,000 people alive in the society. The global pattern of crude death rates varies from those for the other demographic variables already mapped in this chapter. The demographic transition helps to explain the distinctive distribution of crude death rates.

KEY ISSUE 3

Why Is Population Increasing at Different Rates in Different Countries?

- **The Demographic Transition**
- **Population Pyramids**
- **Countries in Different Stages of Demographic Transition**
- **Demographic Transition and World Population Growth**

All countries have experienced some changes in natural increase, fertility, and mortality rates, but at different times and at different rates. Although rates vary among countries, a similar process of change in a society's population, known as the **demographic transition,** is operating. Because of diverse local cultural and economic conditions, the demographic transition diffuses to individual countries at different rates and produces local variations in natural increase, fertility, and mortality. ■

The Demographic Transition

The demographic transition is a process with several stages, and every country is in one of them. The process has a beginning, middle, and end. Historically, once a country has moved from one stage of the process to the next, it has not reverted to an earlier stage. However, a reversal may be occurring in some African countries because of the AIDS epidemic. The four stages are shown in Figure 2-15.

Stage 1: Low Growth

Most of humanity's several-hundred-thousand-year occupancy of Earth was characterized by stage 1 of the demographic transition. Crude birth and death rates varied considerably from

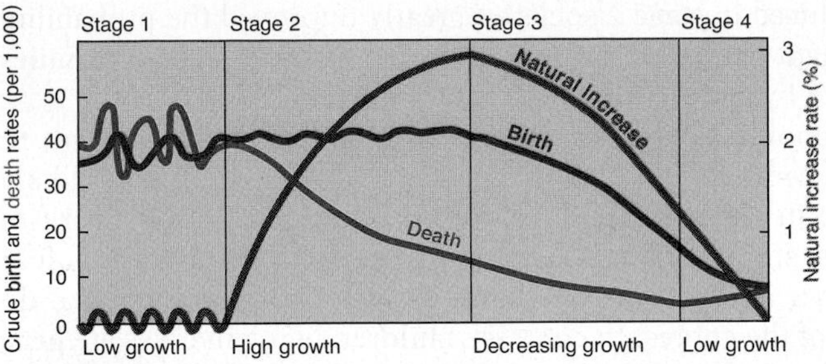

FIGURE 2-15 Demographic transition. The demographic transition consists of four stages:

- **Stage 1:** Very high birth and death rates produce virtually no long-term natural increase.
- **Stage 2:** Rapidly declining death rates combined with very high birth rates produce a very high natural increase.
- **Stage 3:** Birth rates rapidly decline, death rates continue to decline, and natural increase rates begin to moderate.
- **Stage 4:** Very low birth and death rates produce virtually no long-term natural increase, and possibly a decrease.

one year to the next and from one region to another, but over the long term they were roughly comparable, at very high levels. As a result, the NIR was essentially zero, and Earth's population was unchanged, at perhaps a half-million.

During most of this period, people depended on hunting and gathering for food (see Chapter 10). When food was easily obtained, a region's population increased, but it declined when people were unable to locate enough animals or vegetation nearby.

About the year 8000, the world's population began to grow by several thousand per year. Between 8000 BC and AD 1750, Earth's human population increased from approximately 5 million to 800 million (Figure 2-16). The burst of population growth around 8000 BC was caused by the **agricultural revolution**, which was the time when human beings first domesticated plants and animals and no longer relied entirely on hunting and gathering. By growing plants and raising animals, human beings created larger and more stable sources of food, so more people could survive.

Despite the agricultural revolution, the human population remained in stage 1 of the demographic transition because food supplies were still unpredictable. Farmers prospered in regions with abundant harvests, and the population expanded, but when unfavorable climatic conditions resulted in low food production, the CDR would soar. War and disease also took their toll in stage 1 societies.

Most of human history was spent in stage 1 of the demographic transition, but today no such country remains there. Every nation has moved on to at least stage 2 of the demographic transition, and, with that transition, has experienced profound changes in population.

Stage 2: High Growth

For nearly 10,000 years after the agricultural revolution, world population grew at a modest pace. After around 1750, the world's population suddenly began to grow ten times faster than in the past. The average annual increase jumped from about 0.05 percent (one-twentieth of 1 percent) to 0.5 percent (one-half of 1 percent). World population grew by about 5 million in 1800, compared to only about one-half million in 1750.

The sudden burst of population growth occurred in the late eighteenth and early nineteenth centuries because several countries moved on to stage 2 of the demographic transition. In stage 2 of the demographic transition, the CDR suddenly plummets, while the CBR remains roughly the same as in stage 1. Because the difference between the CBR and CDR is suddenly very high, the NIR is also very high, and population grows rapidly. Some demographers divide stage 2 of the demographic transition into two parts. The first part is the period of accelerating population growth. During the second part, the growth rate begins to slow, although the gap between births and deaths remains high.

Countries entered stage 2 of the demographic transition after 1750 as a result of the **Industrial Revolution**, which began in England in the late eighteenth century and spread to the European continent and North America during the nineteenth century. The Industrial Revolution was a conjunction of major improvements in industrial technology (invention of the steam engine, mass production, powered

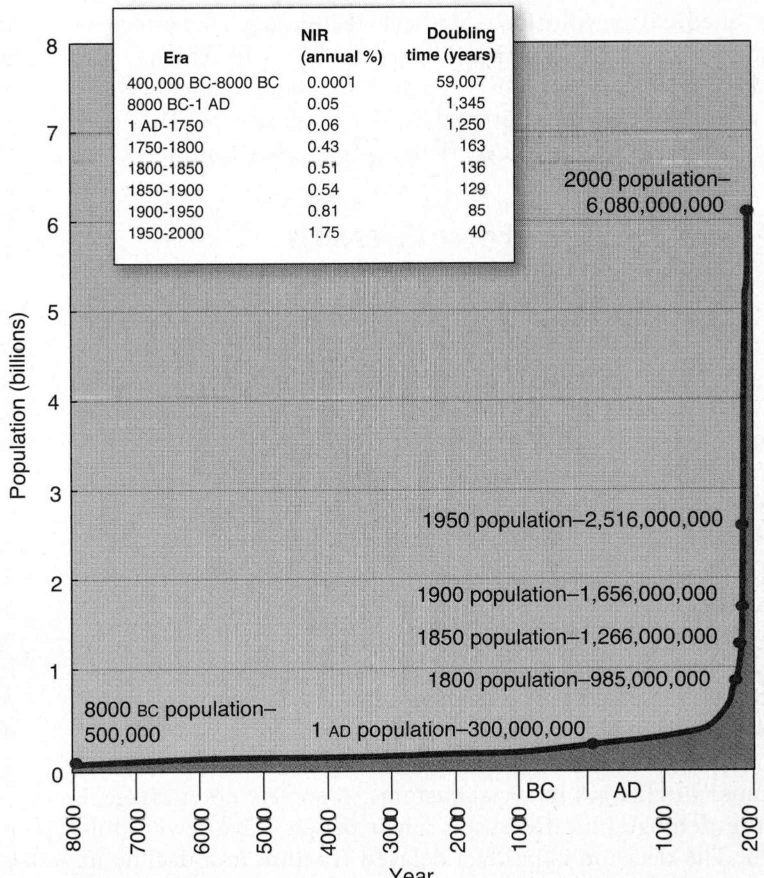

Era	NIR (annual %)	Doubling time (years)
400,000 BC-8000 BC	0.0001	59,007
8000 BC-1 AD	0.05	1,345
1 AD-1750	0.06	1,250
1750-1800	0.43	163
1800-1850	0.51	136
1850-1900	0.54	129
1900-1950	0.81	85
1950-2000	1.75	40

FIGURE 2-16 Population growth through history. Through most of human history population growth was virtually nil. Population increased rapidly beginning in the eighteenth century.

transportation) that transformed the process of manufacturing goods and delivering them to market (see Chapter 11). The result of this transformation was an unprecedented level of wealth, some of which was used to make communities healthier places to live.

New machines helped farmers increase agricultural production and feed the rapidly growing population. More efficient agriculture freed people to work in factories, producing other goods and generating enough food for the industrial workers. The wealth produced by the Industrial Revolution was also used to improve sanitation and personal hygiene. Sewer systems were installed in cities, and food and water supplies were protected against contamination. As a result of these public improvements, people were healthier and therefore lived longer.

Countries in Europe and North America entered stage 2 of the demographic transition about 1800, but stage 2 did not diffuse to most countries in Africa, Asia, and Latin America until around 1950. With the diffusion of stage 2 of the demographic transition, world population grew by 1.7 percent per year during the second half of the twentieth century, compared to 0.5 percent per year during the nineteenth century. The world added about 80 million people in 2000, compared to 8 million in 1900.

Countries in Africa, Asia, and Latin America moved on to stage 2 of the demographic transition during the second half of the twentieth century for a different reason than was the case for Europe and North America 200 years earlier. The late-twentieth-century push of countries into stage 2 was caused by the **medical revolution**. Medical technology invented in Europe and North America diffused to LDCs in Africa, Asia, and Latin America. Improved medical practices suddenly eliminated many of the traditional causes of death in LDCs and enabled more people to experience longer and healthier lives.

Stage 3: Moderate Growth

A country moves from stage 2 to stage 3 of the demographic transition when the CBR begins to drop sharply. The CDR continues to fall in stage 3 but at a much slower rate than in stage 2. The population continues to grow because the CBR is still greater than the CDR. But the rate of natural increase is more modest in countries in stage 3 than in those in stage 2 because the gap between the CBR and the CDR narrows. European and North American countries generally moved from stage 2 to stage 3 of the demographic transition during the first half of the twentieth century. Most countries in Asia and Latin America have moved to stage 3 in recent years, while most African countries remain in stage 2.

The sudden drop in the CBR during stage 3 occurs for different reasons than the rapid decline of the CDR during stage 2. The CDR declined in stage 2 following the introduction of new technology into the society, but the CBR declines in stage 3 because of changes in social customs. A society enters stage 3 of the demographic transition when people have fewer children. The decision is partly a delayed reaction to a decline in mortality, especially the IMR. In stage 1 societies, the survival of any one infant could not be confidently predicted, and families typically had a large number of babies so as to improve the chances of some surviving to adulthood. Medical practices introduced in stage 2 societies greatly improved the probability of infant survival, but many years elapsed before families reacted by conceiving fewer babies.

Economic changes in stage 3 societies also induce people to have fewer offspring. People in stage 3 societies are more likely to live in cities rather than in the countryside and to work in offices, shops, or factories rather than on farms. Farmers often consider a large family to be an asset because children can do some of the chores. In contrast, children living in cities are generally not economic assets to their parents, because they are prohibited from working in most types of urban jobs. In addition, urban homes are relatively small and may not have space to accommodate large families.

Stage 4: Low Growth

A country reaches stage 4 of the demographic transition when the CBR declines to the point where it equals the CDR, and the NIR approaches zero (Figure 2-17). This condition is called **zero population growth (ZPG)**, a term often applied to stage 4 countries.

ZPG may occur when the CBR is still slightly higher than the CDR, because some females die before reaching childbearing years, and the number of females in their childbearing years can vary. To account for these discrepancies, demographers more precisely define ZPG as the TFR that results in a lack of change in the total population over a long term. A TFR of approximately 2.1 produces ZPG, although a country that receives many immigrants may need a lower TFR to achieve ZPG.

Countries in stage 4 of the demographic transition can be identified on the map of total fertility rate (Figure 2-11). Most European countries have reached stage 4 of the demographic transition because they have TFRs well below the ZPG replacement level of 2.1. In the United States, the TFR has hovered around ZPG since 2000.

Social customs again explain the movement from one stage of the demographic transition to the next. Increasingly, women in stage 4 societies enter the labor force rather than remain at home as full-time homemakers. When most families lived on farms, employment and child rearing were conducted at the same place, but in urban societies most parents must leave the home to work in an office, shop, or factory. An employed parent must arrange for someone to take care of their preschool-age children during working hours.

Changes in lifestyle also encourage smaller families. People who have access to a wider variety of birth-control methods are more likely to use some of them. With increased income and leisure time, more people participate in entertainment and recreation activities that may not be suitable for young children, such as attending cultural events, traveling overseas, going to bars, and eating at upscale restaurants.

Several Eastern European countries, most notably Russia, have negative NIRs, meaning that the number of deaths exceeds the number of births (refer to Figure 2-9). Eastern Europe's relatively high death rates and low birth rates are a legacy of a half-century of Communist rule. Higher death rates may be a result of inadequate pollution controls and inaccurate reporting by the Communists. Lower birth rates may stem from

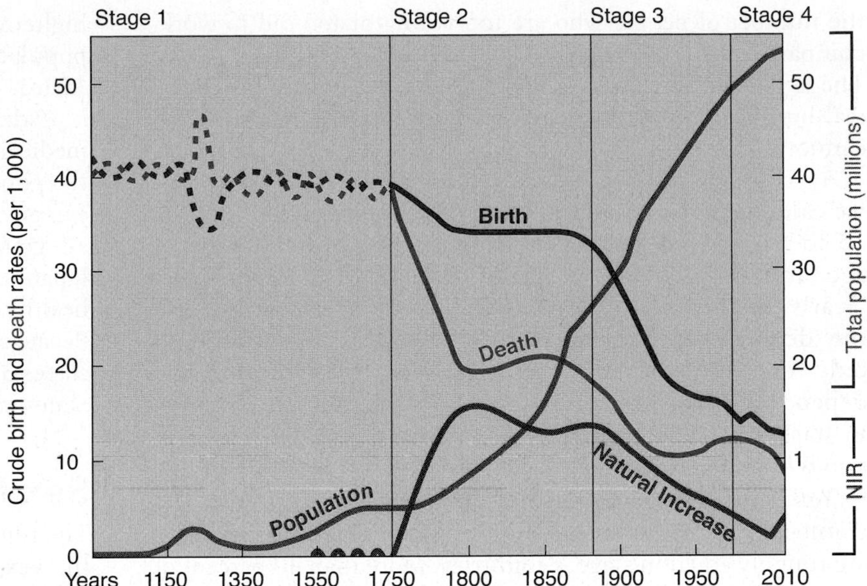

FIGURE 2-17 Demographic transition for England. England provides a good case study of the long-term impact of the demographic transition. It has reached stage 4, and at least fragmentary information on its population is available for the past 1,000 years. Further, unlike the United States and many other countries, England has not changed its boundaries, nor has it been affected by migration of enough people to affect national trends.

- **Stage 1:** Demographers must estimate birth and death rates prior to 1750 because precise records are not available. Church parish records of births, baptisms, marriages, and burials help in making estimates.

  In 1066, when the Normans invaded England, the country's population was approximately 1 million. Seven hundred years later, the population was only 6 million, and the country was still in stage 1 of the demographic transition. During that 700-year period, the population rose in some years and fell in others. For example, England's population declined from 4 million in the year 1250 to 2 million a century later after the Black Death (bubonic plague) and famines swept the country. As recently as the 1740s, the CDR skyrocketed following a series of bad harvests.
- **Stage 2:** In 1750, the CBR and CDR in England were both 40 per 1,000. In 1800, the CBR remained very high at 34, but the CDR had plummeted to 20.

  This 50-year period marked the start of the Industrial Revolution in England. New production techniques increased the nation's food supply and generated money that was spent on improvements in public health. England remained in stage 2 of the demographic transition for about 125 years. During that period the population rose from 6 million to 30 million, an average annual NIR of 1.4 percent.
- **Stage 3:** After 1880 England entered stage 3 of the demographic transition. The CBR declined sharply over the next century, while the CDR continued to fall somewhat. The population increased between 1880 and 1970 from 26 million to 49 million, about 0.7 percent per year.
- **Stage 4:** England has been in stage 4 of the demographic transition since the early 1970s. The CBR has varied between 12 and 14 per 1,000; the CDR has varied between 10 and 12.

  The CBR increases slightly in some years because the number of women in their childbearing years is greater, not because of decisions by women to have more children. The TFR has long been well below the 2.1 needed for replacement. England's population has grown by 3 million since 1970, primarily because of immigration from former colonies.

When England began to progress through the demographic transition around 1750, the country had 6 million people, crude birth and death rates of 40 per 1,000, and a record of little population growth over the previous 700 years. For the past three decades, England has been in another period of little population growth. The difference is that the crude birth and death rates are now around 11 rather than 40, and the country has 53 million inhabitants instead of 6 million.

very strong family-planning programs and deep-seated pessimism about having children in an uncertain world. As memories of the Communist era fade, Russians and other Eastern Europeans may display birth and death rates more comparable to those in Western Europe. Alternatively, demographers in the future may identify a fifth stage of the demographic transition, characterized by higher death rates than birth rates and an irreversible population decline.

A country that has passed through all four stages of the demographic transition has in some ways completed a cycle—from little or no natural increase in stage 1, to little or no natural increase in stage 4. Two crucial demographic differences underlie this process, however. First, at the beginning of the demographic transition, the CBRs and CDRs are high—35 to 40 per 1,000—whereas at the end of the process the rates are very low, approximately 10 per 1,000. Second, the total population of the country is much higher in stage 4 than in stage 1.

Population Pyramids

A country's stage of demographic transition gives it a distinctive population structure. Population in a country is influenced by the demographic transition in two principal ways—the percentage of the population in each age group and the distribution of males and females.

A country's population can be displayed by age and gender groups on a bar graph called a **population pyramid**. A population pyramid normally shows the percentage of the total population in 5-year age groups, with the youngest group (0 to 4 years old) at the base of the pyramid and the oldest group at the top. The length of the bar represents the percentage of the total population contained in that group. By convention, males are usually shown on the left side of the pyramid and females on the right.

The shape of a pyramid is determined primarily by the CBR in the community. A country in stage 2 of the demographic transition, with a high CBR, has a relatively large number of young children, making the base of the population pyramid very broad. On the other hand, a country in stage 4, with a relatively large number of older people, has a graph with a wider top that looks more like a rectangle than a pyramid.

Age Distribution

The age structure of a population is extremely important in understanding similarities and differences among countries. The most important factor is the **dependency ratio**, which is the number of people who are too young or too old to work, compared to the number of people in their productive years. The larger the percentage of dependents, the greater the financial burden on those who are working to support those who cannot.

To compare the dependency ratios of different countries, we can divide the population into three age groups—0 to 14, 15 to 64, and 65 and older. People who are 0–14 years of age and 65-plus are normally classified as dependents. Nearly one-half of all people living in countries in stage 2 of the demographic transition are dependents, compared to only one-third in stage 4 countries. Consequently, the dependency ratio is nearly 1:1 in stage 2 countries, whereas in stage 4 countries the ratio is 1:2 (one dependent for every two workers).

Young dependents outnumber elderly ones by 10:1 in stage 2 countries, but the numbers of young and elderly dependents are roughly equal in stage 4 countries. More than 40 percent of the people are under age 15 in sub-Saharan Africa, compared to 20 percent or less in Europe and North America (Figure 2-18). The large percentage of children in sub-Saharan Africa strains the ability of these relatively poor countries to provide needed services such as schools, hospitals, and day-care centers. When children reach the age of leaving school, jobs must be found for them, but the government must continue to allocate scarce resources to meet the needs of the still growing number of young people.

As countries pass through the stages of the demographic transition, the percentage of elderly people increases. The higher percentage partly reflects the lower percentage of young people produced by declining CBRs. Older people also benefit in stage 4 countries from improved medical care and higher incomes. People over age 65 comprise 16 percent of the population in Europe compared to 3 percent in sub-Saharan Africa.

Older people must receive adequate levels of income and medical care after they retire from their jobs. The "graying" of the population places a burden on European and North American governments to meet these needs. More than one-fourth of all government expenditures in the United States, Canada, Japan, and many European countries go to Social Security, health care, and other programs for the older population. Because of the larger percentage of older people, countries in stages 3 and 4 of the demographic transition, such as the United States and Sweden, have higher CDRs than do stage 2 countries.

Sex Ratio

The number of males per hundred females in the population is the **sex ratio**. It varies among countries, depending on birth and death rates. In general, slightly more males than females are born, but males have higher death rates. The ratio of men to women is about 93:100 (that is, 93 men for each 100 women) in Europe and 97:100 in North America. In LDCs, the ratio is 103:100.

In the United States, males under age 15 exceed females 105:100. Women start outnumbering men at about age 40, and they comprise 58 percent of the population over age 65.

In stage 2 countries, the high mortality rate during childbirth partly explains the lower percentage of women. The difference also relates to the age structure, because stage 2 countries have a larger percentage of young people—where males generally outnumber females—and a lower percentage of older people, where females are much more numerous.

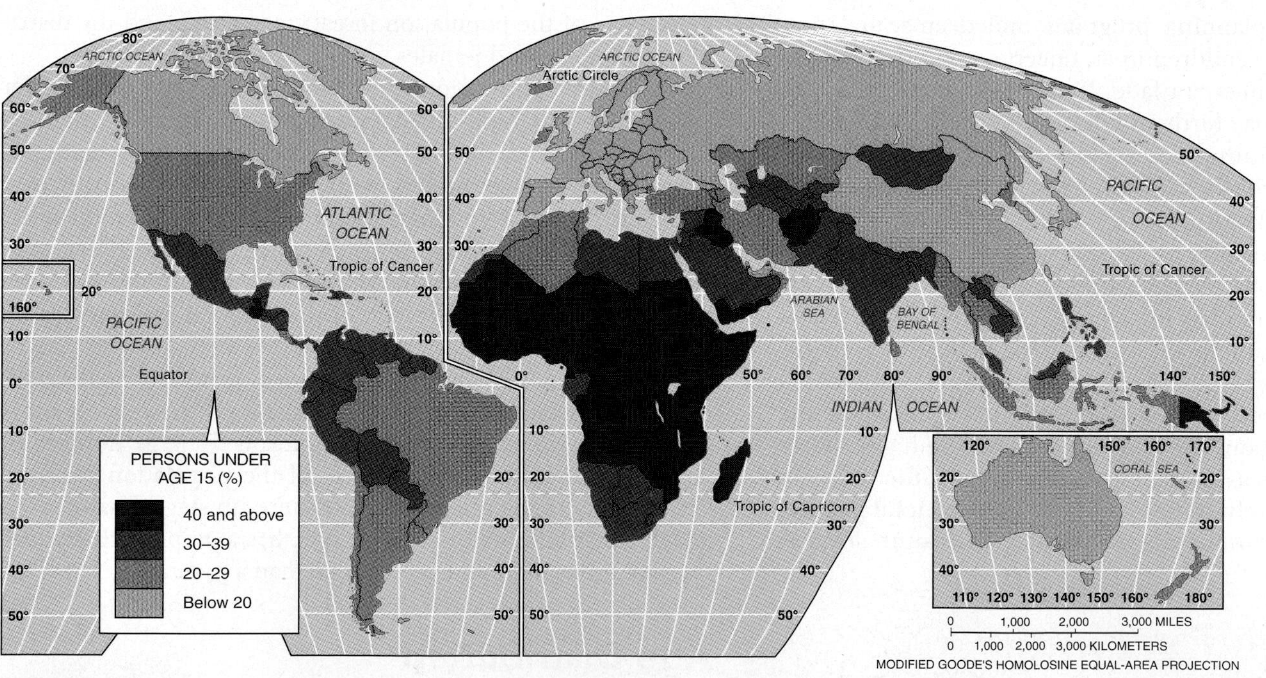

FIGURE 2-18 Percent of population under age 15. A map of the percentage of people over age 65 would show a reverse pattern, with the highest percentages in Europe and the lowest in sub-Saharan Africa and the Middle East.

The shape of a community's population pyramid tells a lot about its distinctive character, especially compared with other places. In Figure 2-19, compare the shapes of the overall U.S. population pyramid with those for Cedar Rapids, Detroit, Honolulu, and Laredo. Cedar Rapids and Honolulu have relatively flat pyramids; Detroit and Laredo have relatively broad-based ones. The different shapes result from differences in the ethnic composition of the four cities. Detroit and Laredo have relatively broad-based pyramids because birth rates are relatively high among African Americans and Hispanic Americans, who form the majority in these two cities. On the other hand, birth rates are relatively low among the Asian American and European-descended communities, the majorities in Honolulu and Cedar Rapids, respectively.

The population pyramid for Naples, where 42 percent of the people are over age 65, resembles an upside-down pyramid. Unalaska, a small town with a military base, has an exceptionally high percentage of males, whereas Naples, with a large percentage of elderly people, has substantially more females than males because females have longer life expectancies. Cities with large universities, such as Lawrence, have an exceptionally high percentage of people in their twenties. See the Contemporary Geographic Tools box for information on collecting statistics about a country's population.

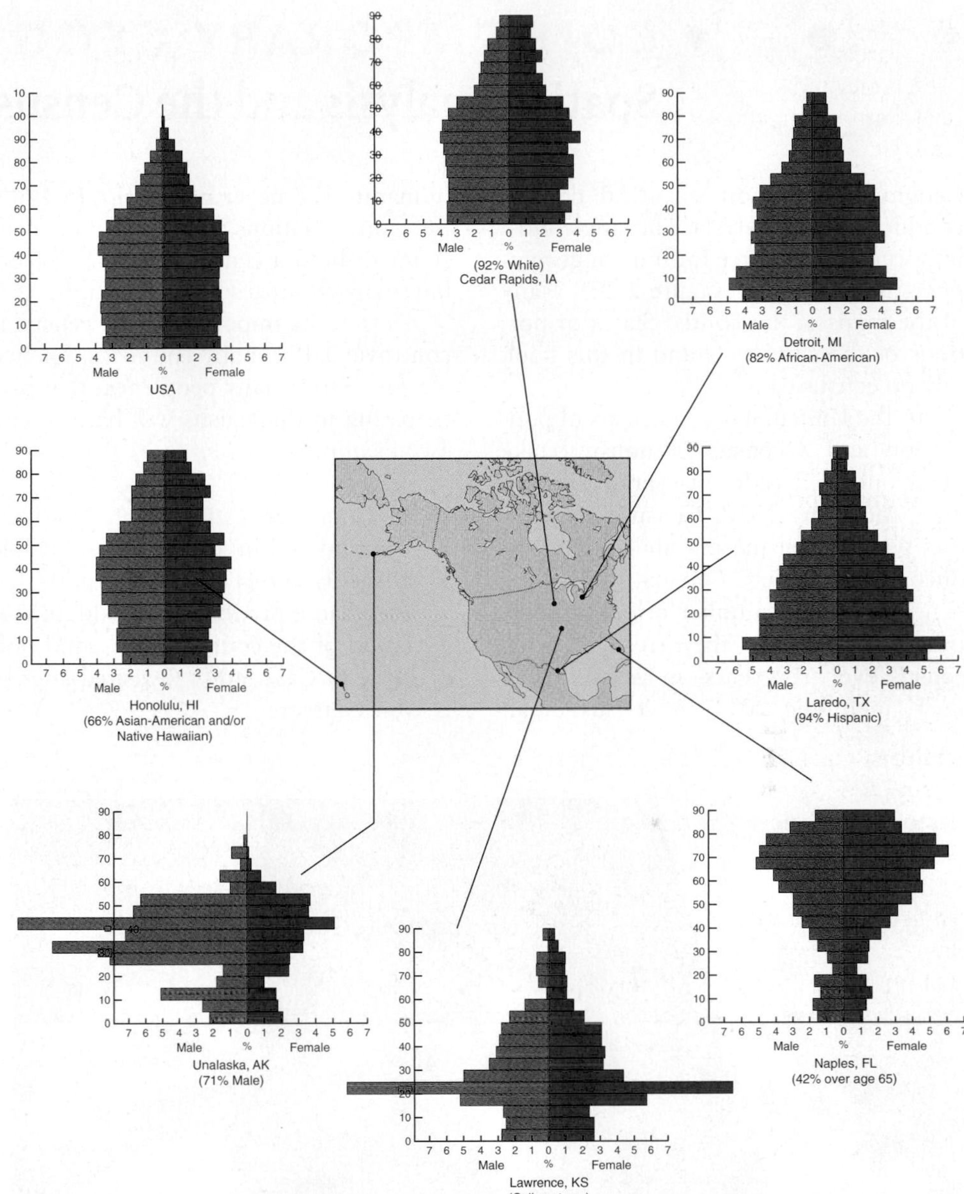

FIGURE 2-19 Population pyramids for the United States and selected communities. Detroit and Laredo have broader pyramids than Cedar Rapids and Honolulu, indicating higher percentages of young people and higher fertility rates. Unalaska has a high percentage of males because it contains an isolated military base. Lawrence has a high percentage of people in their twenties because it is the home of the University of Kansas. Naples has a high percentage of elderly people, especially women, so its pyramid is upside down.

Countries in Different Stages of Demographic Transition

Countries display distinctive population characteristics depending on their stage in the demographic transition. No country today remains in stage 1 of the demographic transition, but it is instructive to compare countries in each of the other three stages. Let us look at three case studies of countries in stages 2, 3, and 4.

Cape Verde: Stage 2 (High Growth)

Cape Verde, a collection of ten small islands in the Atlantic Ocean off the coast of West Africa, moved from stage 1 to stage 2 about 1950. Cape Verde was a colony of Portugal until it became independent in 1975, and the Portuguese administrators left better records of births and deaths than are typical for a colony in stage 1.

During the first half of the twentieth century, Cape Verde's population declined, from 147,000 in 1900 to 137,000 in 1949. CBRs were generally in the forties and CDRs in the twenties (Figure 2-21). The large gap between births and deaths most years produced a high NIR typical of stage 2 of the demographic transition, yet Cape Verde remained in stage 1 until 1950 because several severe famines dramatically disrupted the typical patterns of birth, death, and natural increase. For example, famine made Cape Verde's CDR rocket to 74 per 1,000 in 1941 and 101 in 1942. Because fewer babies were conceived at the height of the famine in 1942, the CBR fell in 1943 to only 22. Population also declined during periods of famine because survivors migrated to other countries. Wide fluctuations in the crude birth and death

CONTEMPORARY GEOGRAPHIC TOOLS

Spatial Analysis and the Census

Geography relies on statistical data to conduct spatial analysis. The single most important data source for human geographers is the **census** (Figure 2-20). Many of the maps of the United States or portions of the country found in this book rely on census data.

In the United States, a census of population and a census of housing take place once a decade, in years ending in zero, including 2010. Censuses of various types of businesses are undertaken once every 5 years. Canada, the United Kingdom, and a number of other countries once ruled by the British take the census every 10 years, in years ending in 1, including 2011. A census of all humanity has never been held. In 1995, the United Nations called for all member states to hold a census within a decade, but many countries did not comply.

Despite its importance, the census is controversial in many countries, for two reasons. First, many people fear that participating in the census will harm them. Two examples:

- In China, an estimated 100 million were missed in 2000 because people allegedly didn't want to report that they had more than one child, in violation of the country's one-child policy (see Case Study Revisited later in this chapter).
- In the United States, homeless people, ethnic minorities, and citizens of other countries without proper immigration documents may be less likely to complete the census form. These individuals may fear that the Bureau of the Census could turn over the forms to another government agency, such as the FBI or the Department of Homeland Security.

The second controversy is how to account for the people who are missed by the census. Geographers, census bureau officials, and others interested in spatial analysis are able to apply statistical sampling techniques to get a more accurate count, as well as to identify detailed characteristics of people, housing, and businesses. Sampling is routinely applied to the British census to improve its accuracy, but it has proved controversial in the United States.

The method of counting matters in the United States, because the district boundaries of the U.S. House of Representatives, as well as of the 50 state legislatures, must be redrawn every decade in accordance with the census. Also, the U.S. and state governments allocate many types of funds to communities on the basis of population. Politicians sympathetic to the needs of the homeless and immigrants have been especially vocal in support of sampling, whereas those from other constituencies are more inclined to oppose it. The U.S. Supreme Court has ruled that Article 1, Section 2 of the U.S. Constitution prevents using statistical techniques to apportion seats. Otherwise, spatial patterns and trends are analyzed using statistically valid samples. ■

FIGURE 2-20 Taking the census. Homeless people sleeping in New York City's Penn Station are counted by a census taker.

rates from one year to the next, depending on economic and environmental conditions, are typical of stage 1 countries.

This long-term pattern of demographic uncertainty suddenly ended in 1950, and Cape Verde quickly moved to stage 2 of the demographic transition. Since entering stage 2 a half-century ago, the population of Cape Verde has surged to 427,000, and natural increase has averaged more than 2.0 percent per year. Cape Verde moved into stage 2 when an antimalarial campaign was launched. The CDR dropped by around 60 percent in just one year, between 1949 and 1950, and another 70 percent in the half-century since then. Meanwhile, as is typical of stage 2 countries, Cape Verde's CBR has remained relatively high and still fluctuates wildly.

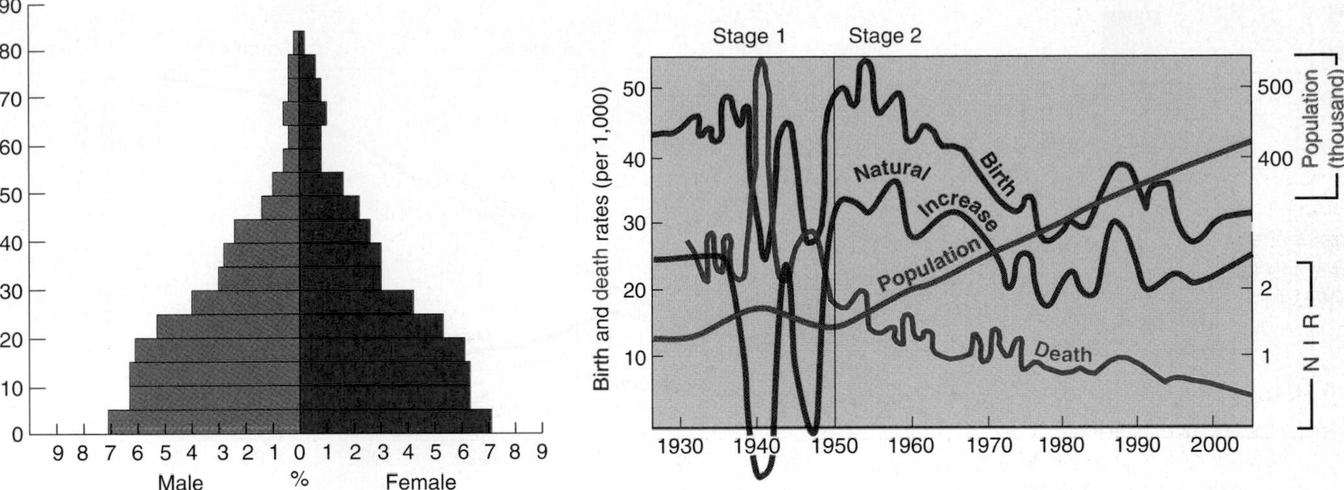

FIGURE 2-21 Demographic transition and population pyramid for Cape Verde. Cape Verde entered stage 2 of the demographic transition in approximately 1950, as indicated by the large gap between birth and death rates since then. As is typical of countries in stage 2 of the demographic transition, Cape Verde has a population pyramid with a very wide base.

The wild fluctuations in Cape Verde's CBR are a legacy of the severe famine during the 1940s. Birth rates were lower during the 1960s because Cape Verde had relatively few women in their twenties, the prime childbearing years. Women in their twenties during the 1960s would have been born during the 1940s, when the famine kept birth rates very low. Similarly, declining birth rates during the 1990s reflected the small number of women in prime childbearing years, who would have been born during the 1960s and 1970s. Conversely, the higher birth rates in the 1950s and 1980s resulted from a larger number of women in their childbearing years.

The population pyramid shows that Cape Verde has a large number of females age 5–14 who will soon start moving into their prime childbearing years. For Cape Verde to enter stage 3 of the demographic transition during the next decade, these females must bear considerably fewer children than did their mothers.

Chile: Stage 3 (Moderate Growth)

Chile provides an example of a country outside Europe and North America that has reached stage 3 of the demographic transition but is likely to take some time before continuing to stage 4 (Figure 2-22). Chile has changed from a predominantly rural society based on agriculture to an urban society in which most people now work in factories, offices, and shops. However, many Chileans still have large families.

Like most countries outside Europe and North America, Chile entered the twentieth century still in stage 1 of the demographic

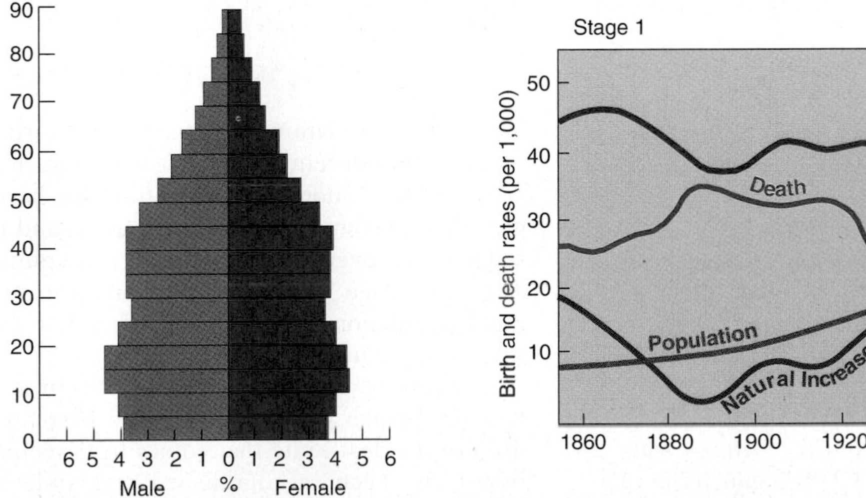

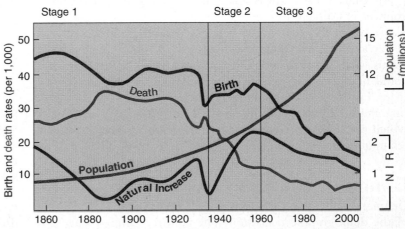

FIGURE 2-22 Demographic transition and population pyramid for Chile. Chile entered stage 2 of the demographic transition in the 1930s, when death rates declined sharply, and stage 3 in the 1960s, when birth rates declined sharply.

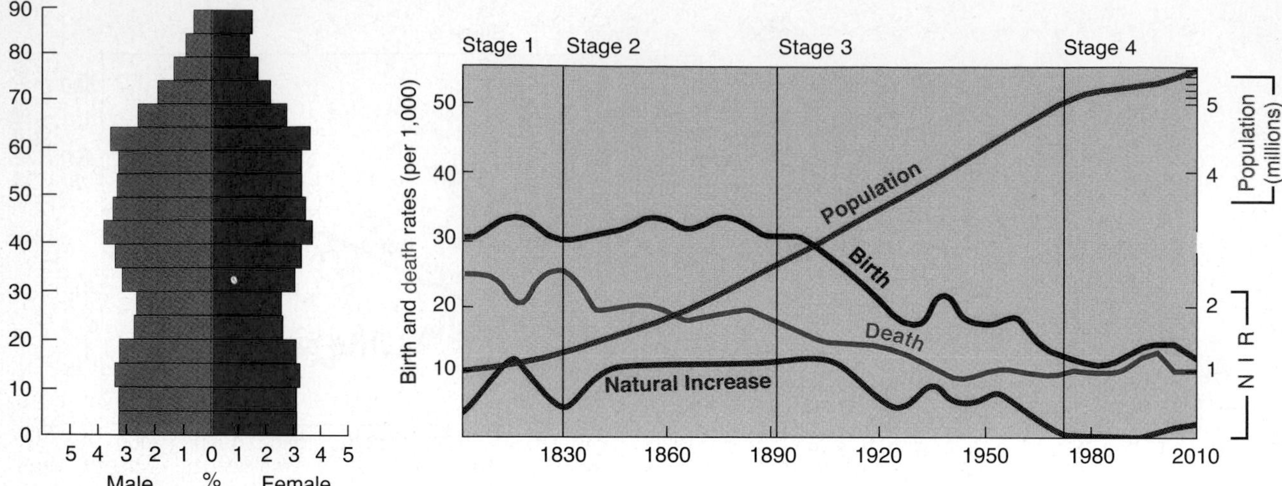

FIGURE 2-23 Demographic transition and population pyramid for Denmark. Denmark has been in stage 4 of the demographic transition and has experienced virtually no change in total population since the 1970s. The population pyramid is much straighter than that of Cape Verde and that of Chile, a reflection of the relatively large percentage of elderly people and small percentage of children

transition. Population had grown modestly during the nineteenth century at a NIR of less than 1 percent per year. However, much of Chile's population growth—as in other countries in the Western Hemisphere—resulted from European immigration.

Chile's CDR declined sharply in the 1930s, moving the country into stage 2 of the demographic transition. As elsewhere in Latin America, Chile's CDR was lowered by the infusion of medical technology from MDCs such as the United States, bringing under control such diseases as smallpox, malaria, and dysentery.

Chile has been in stage 3 of the demographic transition since the 1960s. The CDR continued to decline more modestly than in the past, while the CBR dropped sharply. Chile moved on to stage 3 of the demographic transition primarily because of a vigorous government family-planning policy, initiated in 1966. Reduced income and high unemployment at that time also induced couples to postpone marriage and delay childbearing.

Although Chile's NIR is lower today than in the 1950s, the country is unlikely to move into stage 4 of the demographic transition in the near future. Chile's government reversed its policy and renounced support for family planning during the 1970s. The government policy was that population growth could help promote national security and economic development. Further reduction in the CBR is also hindered by the fact that most Chileans belong to the Roman Catholic Church, which opposes the use of what it calls artificial birth-control techniques.

Denmark: Stage 4 (Low Growth)

Denmark, like most European countries, has reached stage 4 of the demographic transition. The country entered stage 2 in the nineteenth century, when the CDR began its permanent decline. The CBR then dropped in the late nineteenth century, and the country moved on to stage 3 (Figure 2-23). Since the 1970s the CBR and the CDR have been roughly equal, at around 10–12 per year. The country has reached ZPG, and the population is increasing almost entirely because of immigration.

Denmark's population pyramid shows the impact of the demographic transition. Instead of a classic pyramid shape, Denmark has a column, demonstrating that the percentages of young and elderly people are nearly the same. With further medical advances, the number of elderly people may actually exceed the number of young people in a few years.

Denmark's CDR has actually increased somewhat in recent years because of the increasing number of elderly people. The CDR is unlikely to decline unless another medical revolution, such as a cure for cancer, keeps older elderly people alive much longer (see the Global Forces, Local Impacts box).

Demographic Transition and World Population Growth

Worldwide population increased rapidly during the second half of the twentieth century because few countries were in the two stages of the demographic transition that have low population growth—no country remains in stage 1, and few have reached stage 4. The overwhelming majority of countries are in either stage 2 or stage 3 of the demographic transition—stages with rapid population growth—and only a few are likely to reach stage 4 in the near future.

The four-stage demographic transition is characterized by two big breaks with the past. The first break—the sudden drop in the death rate that comes from technological innovation—has been accomplished everywhere. The second break—the sudden drop in the birth rate that comes from changing social customs—has yet to be achieved in many countries.

GLOBAL FORCES, LOCAL IMPACTS
Japan's Population Decline

Japan is the most populous country outside Europe and North America to reach stage 4 of the demographic transition. Japan has a NIR of approximately zero because crude birth and death rates are nearly equal. Its TFR of 1.4 is well below the replacement rate of approximately 2.1.

Japan's population is expected to decline from an all-time peak of 128 million in 2006 to 119 million in 2025 and 95 million in 2050 (Figure 2-24). With the population decline will come an increasing percentage of elderly people. Japan's current percentage of persons over age 65—22 percent—is higher than in any other country, and nearly double the percentage under age 15.

Unlike Japan, the United States is expecting an increase in population during the twenty-first century. A low NIR in the United States will probably be offset by large-scale immigration from countries in Latin America and Asia still in stage 2 or 3 (see Chapter 3). Japan, on the other hand, has discouraged immigration from nearby countries in Asia. Japanese society, having placed a high value on social conformity for thousands of years, does not welcome outsiders from other cultural traditions.

With few immigrants compared to other stage 4 societies, Japan faces a severe shortage of workers. Rather than increasing immigration, Japan is addressing the labor force shortage primarily by encouraging more Japanese people to work, especially older people and women. Programs make it more attractive for older people to continue working, to receive more health-care services at home instead of in hospitals, and to borrow against the value of their homes to pay for health care.

In the long run, more women in the labor force may translate into an even lower birth rate and therefore an even lower NIR in the future. Rather than combine work with child rearing, Japanese women are expected to make a stark choice: either marry and raise children or remain single and work. According to the 2005 census, the majority has chosen the work option: More than half of women in the prime childbearing years of 20 to 34 are not married. ■

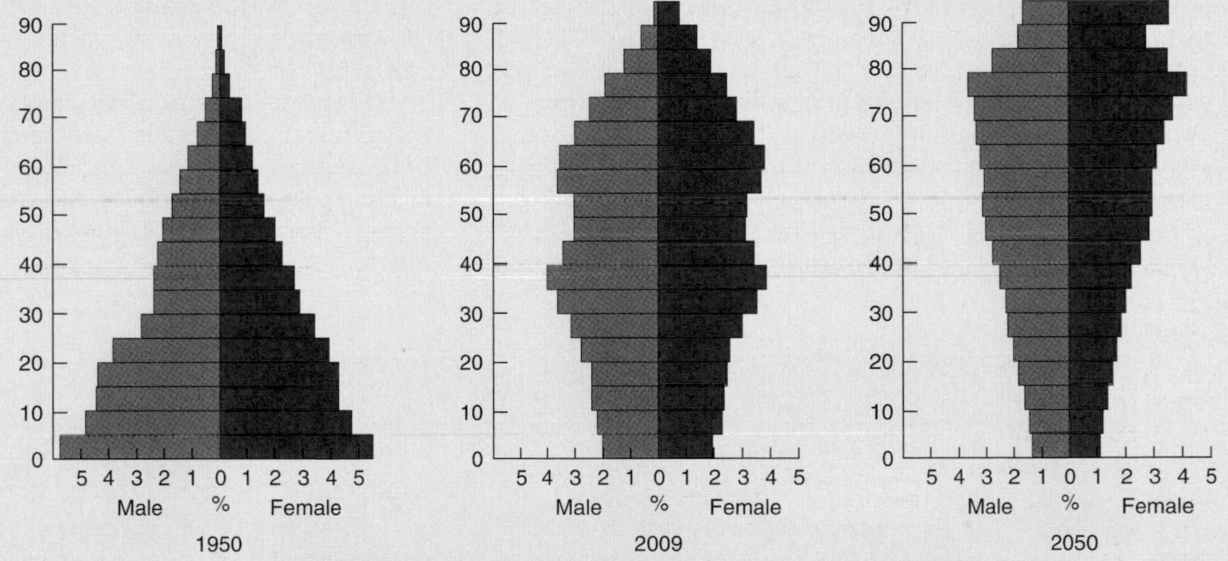

FIGURE 2-24 Japan's changing population pyramid. Japan's population pyramid has shifting from a broad base in 1950 to a rectangular shape. In the future, the bottom of the pyramid is expected to contract and the top to expand.

If most countries in Europe and North America have reached—or at least are approaching—stage 4 of the demographic transition, why aren't countries elsewhere in the world? The answer is that fundamental problems prevent other countries from replicating the experience in Europe and North America.

The first demographic change—the sudden decline in CDR—occurred for different reasons in the past. The nineteenth-century decline in the CDR in Europe and North America took place in conjunction with the Industrial Revolution. The unprecedented level of wealth generated by the Industrial Revolution was used in part to stimulate research by European and North American scientists into the causes and cures for diseases. These studies ultimately led to medical advances, such as pasteurization, X-rays, penicillin, and insecticides.

In contrast, the sudden drop in the CDR in Africa, Asia, and Latin America in the twentieth century was accomplished by different means and with less internal effort by local citizens. For example, the CDR on the island of Sri Lanka (then known as Ceylon) plummeted 43 percent between 1946 and 1947. The most important reason for the sharp drop was the use of the insecticide DDT to control the mosquitoes that spread malaria. European and North American countries invented and manufactured the DDT and trained the experts to supervise its use. The spraying of Sri Lankans' houses and other medical services, which cost only $2 per person per year, were paid for primarily by international organizations.

Thus Sri Lanka's CDR was reduced by nearly one-half in a single year with no change in the country's economy or culture. Medical technology was injected from Europe and North America instead of arising within the country as part of an economic revolution. This pattern has been repeated throughout Latin America, Asia, and Africa.

Having caused the first break with the past through diffusion of medical technology, European and North American countries now urge other countries to complete the second break with the past—the reduction in the birth rate. However, reducing the CBR is more difficult. A decline in the CDR can be induced through introduction of new technology by outsiders, but the CBR will drop only when people decide for themselves to have fewer children. Many LDCs, especially in Asia and Latin America, have moved in recent years to stage 3 of the demographic transition thanks to rapidly declining birth rates. Other countries, especially in Africa, have not yet made this second break with the past and in some cases may be slipping back into stage 1.

In the past, stage 2 of the demographic transition lasted for approximately 100 years in Europe and North America, but today's stage 2 countries are being asked to move through to stage 3 in much less time in order to curtail population growth. When European and North American countries were in stage 2, the global population was increasing by only about 6 million per year, compared to 80 million per year now.

KEY ISSUE 4

Why Might the World Face an Overpopulation Problem?

- **Malthus On Overpopulation**
- **Declining Birth Rates**
- **World Health Threats**

Why does global population growth matter? In view of the current size of Earth's population and the NIR, will there soon be too many of us? Will continued population growth lead to global starvation, war, and a lower quality of life?

Geographers are particularly well suited to address these questions because answers require understanding both human behavior and the physical environment. Further, geographers observe that diverse local cultural and environmental conditions may produce different answers in different places. ■

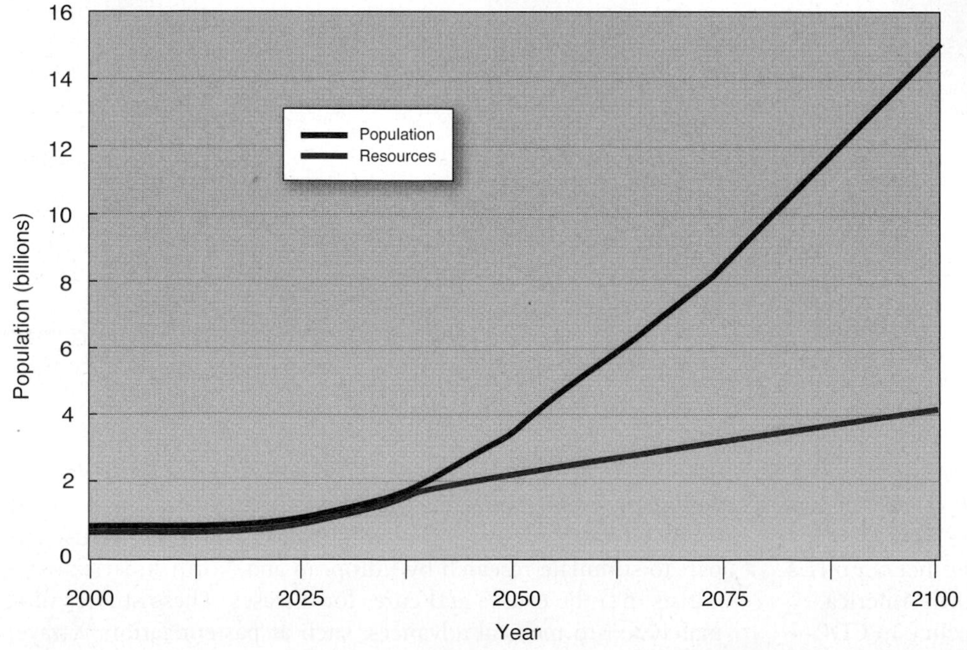

FIGURE 2-25 Malthus's theory. Malthus expected population to grow more rapidly than food production.

Malthus on Overpopulation

English economist Thomas Malthus (1766–1834) was one of the first to argue that the world's rate of population increase was far outrunning the development of food supplies. In *An Essay on the Principle of Population*, published in 1798, Malthus claimed that the population was growing much more rapidly than Earth's food supply because population increased geometrically, whereas food supply increased arithmetically (Figure 2-25).

According to Malthus, these growth rates would produce the following relationships between people and food in the future:

- Today: 1 person, 1 unit of food
- 25 years from now: 2 persons, 2 units of food
- 50 years from now: 4 persons, 3 units of food
- 75 years from now: 8 persons, 4 units of food
- 100 years from now: 16 persons, 5 units of food

Malthus made these conclusions several decades after England had become the first country to enter stage 2 of the demographic transition, in association with the Industrial Revolution. He concluded that population growth would press against available resources in every country, unless "moral restraint" produced lower CBRs or unless disease, famine, war, or other disasters produced higher CDRs.

Contemporary Neo-Malthusians

Malthus's views remain influential today. Contemporary geographers and other analysts are taking another look at Malthus's theory because of Earth's unprecedented rate of natural increase during the late twentieth century.

Neo-Malthusians argue that two characteristics of recent population growth make Malthus's thesis more frightening than when it was first written more than 200 years ago.

1. First, in Malthus's time only a few relatively wealthy countries had entered stage 2 of the demographic transition, characterized by rapid population increase. Malthus failed to anticipate that relatively poor countries would have the most rapid population growth because of transfer of medical technology (but not wealth) from MDCs. As a result, the gap between population growth and resources is wider in some countries than even Malthus anticipated. Many LDCs have expanded their food production significantly in recent years, but they have more poor people than ever before.
2. The second argument made by neo-Malthusians is that world population growth is outstripping a wide variety of resources, not just food production (Figure 2-26). Neo-Malthusians Robert Kaplan and Thomas Fraser Homer-Dixon paint a frightening picture of a world in which billions of people are engaged in a desperate search for food and energy. They assert that wars and civil violence will increase in the coming years because of scarcities of food as well as such resources as clean air, suitable farmland, and fuel.

Malthus's Critics

Malthus's theory has been severely criticized from a variety of perspectives. Criticism has been leveled at both the population growth and resource depletion sides of Malthus's equation.

Many geographers consider Malthusian beliefs unrealistically pessimistic because they are based on a belief that the world's supply of resources is fixed rather than expanding. According to the principles of possibilism discussed in Chapter 1, our well-being is influenced by conditions in the physical environment, but humans have some ability to choose courses of action that can expand the supply of food and other resources. A steady flow of new technology can offset scarcity of minerals and arable land by using existing resources more efficiently and substituting new resources for scarce ones.

Contemporary analysts such as Esther Boserup and Simon Kuznets criticize Malthus's theory that population growth produces problems. To the contrary, a larger population could stimulate economic growth and, therefore, production of more food. Population growth could generate more customers and more ideas for improving technology.

Julian Simon argued that population growth stimulated economic growth. More people means more brains to invent good ideas for improving life. Asked Simon, "Does anyone seriously doubt that Europe is more prosperous with a population of hundreds of millions than it would be with a population of hundreds of thousands?"

Marxists maintain that no cause-and-effect relationship exists between population growth and economic development. Poverty, hunger, and other social welfare problems associated with lack of economic development are a result of unjust social and economic institutions, not population growth.

FIGURE 2-26 Overpopulation in Mali. A region can be sparsely inhabited yet overpopulated if it has rapid population growth and limited resources, as is the case in Mali.

Marxist theorist Friedrich Engels (1820–1895) dismissed Malthus's arithmetic as an artifact of capitalism. Engels argued that the world possessed sufficient resources to eliminate global hunger and poverty, if only these resources were shared equally. Under capitalism, workers do not have enough food because they do not control the production and distribution of food and are not paid sufficient wages to purchase it.

The world is much better off economically with 7 billion people than it was with 1 billion, argue Malthus's critics, because too few people can retard economic development as surely as can too many people. A large population of consumers can generate a greater demand for goods, which results in more jobs.

Some political leaders, especially in Africa, argue that high population growth is good for a country because more people will result in greater power. Population growth is desired in order to increase the supply of young men who could serve in the armed forces. On the other side of the coin, more developed countries are viewed as pushing for lower population growth as a means of preventing further expansion in the percentage of the world's population living in poorer countries.

Malthus's Theory and Reality

On a global scale, conditions during the past half-century have not supported Malthus's theory. Even though the human population has grown at its most rapid rate ever, world food production has consistently grown at a faster rate than the NIR since 1950, according to geographer Vaclav Smil, Distinguished Professor at the University of Manitoba. Smil has shown that Malthus was fairly close to the mark on food production but much too pessimistic on population growth.

Overall food production has increased during the last half-century somewhat more rapidly than Malthus predicted. In India, for example, rice production has followed Malthus's expectations fairly closely, but wheat production has increased twice as fast as Malthus expected (Figure 2-27). Better growing techniques, higher-yielding seeds, and cultivation of more land have contributed to the expansion in food supply (see Chapter 10). Many people in the world cannot afford to buy food or do not have access to sources of food, but these are problems of distribution of wealth rather than insufficient global production of food, as Malthus theorized.

It is on the population side of the equation, though, that Malthus has proved to be inaccurate. His model expected population to quadruple during a half-century, but even in India—a country known for relatively rapid growth (see Case Study Revisited at the end of this chapter)—population has increased more slowly than food supply.

However, neo-Malthusians point out that production of both wheat and rice has slowed in India in recent years, as shown in Figure 2-27. Without new breakthroughs in food production, India will not be able to keep food supply ahead of population growth.

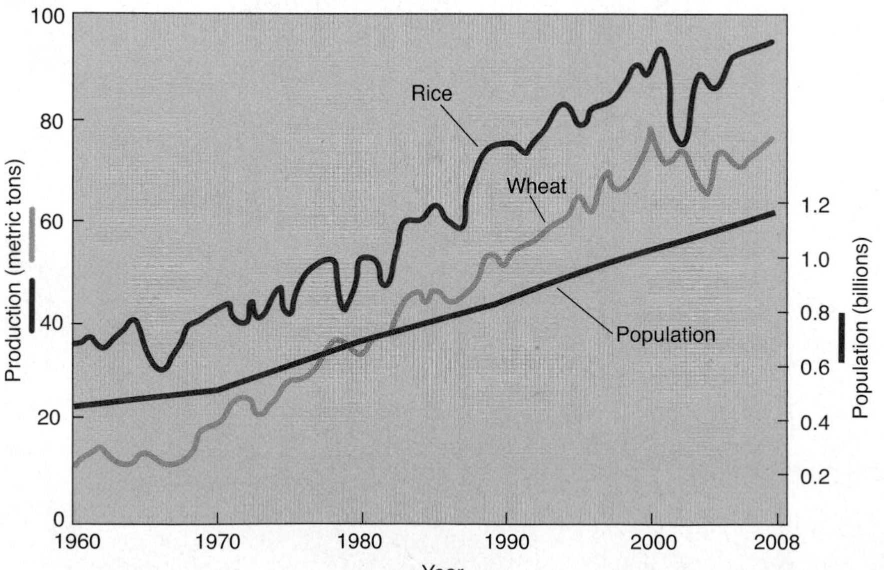

FIGURE 2-27 Population and food production in India. Production of wheat and rice has increased at a more rapid rate than has population.

Declining Birth Rates

The Malthus theory seems unduly pessimistic on a global scale, but geographers recognize the diversity of conditions among regions of the world. Although the world as a whole may not be in danger of "running out" of food, some regions with rapid population growth do face shortages of food.

The NIR can decline for only two reasons—lower birth rates or higher death rates. Few people wish to see the NIR decline because of an increase in death rates. The only demographic alternative is to reduce birth rates. In most countries, the decline in the NIR has occurred because of a lower birth rate, but in some countries of sub-Saharan Africa, the CDR is increasing.

The CBR has declined rapidly since 1990 from 27 to 21 in the world as a whole and from 31 to 23 in LDCs. A substantial decline in the birth rate has been recorded since 1990 by nearly every country in Asia, Latin America, and the Middle East, as well as selected countries in sub-Saharan Africa (Figure 2-28).

Two strategies have been successful in reducing birth rates. One alternative emphasizes reliance on economic development, the other on distribution of contraceptives. Because of varied economic and cultural conditions, the most effective method varies among countries.

Reasons for Declining Birth Rates

One approach to lowering birth rates emphasizes the importance of improving local economic conditions. A wealthier community has more money to spend on education and health-care programs that would promote lower birth rates. According to this approach, if more women are able to attend school and to remain in school longer, they are more likely to learn

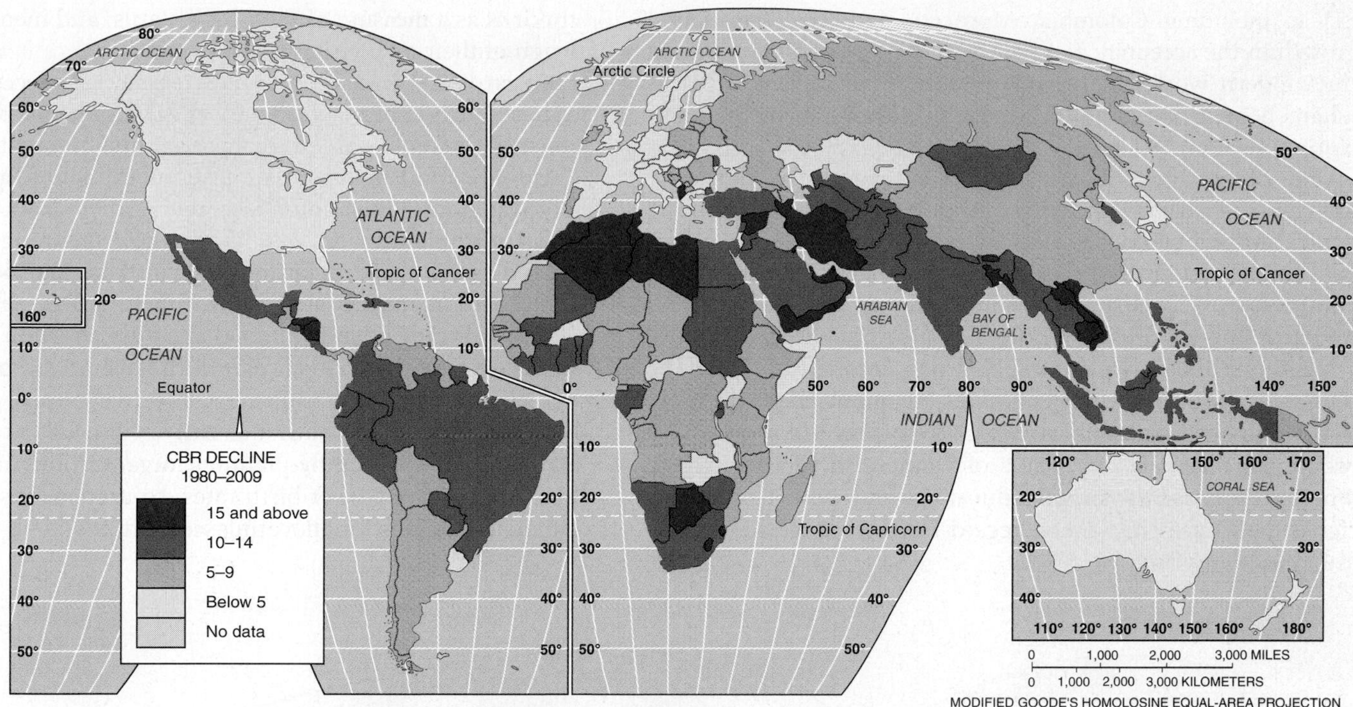

FIGURE 2-28 Crude birth rate change, 1980–2008. The crude birth rate has declined in all but a handful of countries. Declines have been most rapid in the Middle East, Latin America, and South Asia. Still, the number of births in the world increased during the three decades from about 120 million to 140 million per year.

employment skills and gain more economic control over their lives. With better education, women would better understand their reproductive rights, make more informed reproductive choices, and select more effective methods of contraception. With improved health-care programs, IMRs would decline through such programs as improved prenatal care, counseling about sexually transmitted diseases, and child immunization. With the survival of more infants ensured, women would be more likely to choose to make more effective use of contraceptives to limit the number of children.

Reducing Births Through Contraception

The other approach to lowering birth rates emphasizes the importance of rapidly diffusing modern contraceptive methods (Figure 2-29). Economic development may promote lower birth rates in the long run, but the world cannot wait around for that alternative to take effect. Putting resources into family-planning programs can reduce birth rates much more rapidly. In LDCs, demand for contraceptive devices is greater than the available supply. Therefore, the most effective way to increase their use is to distribute more of them, cheaply and quickly. According to this approach, contraceptives are the best method for lowering the birth rate.

Bangladesh is an example of a country that has had little improvement in the wealth and literacy of its people, but 56 percent of the women in the country used contraceptives in 2009 compared to 6 percent three decades earlier. Similar growth in the use of contraceptives has occurred in other

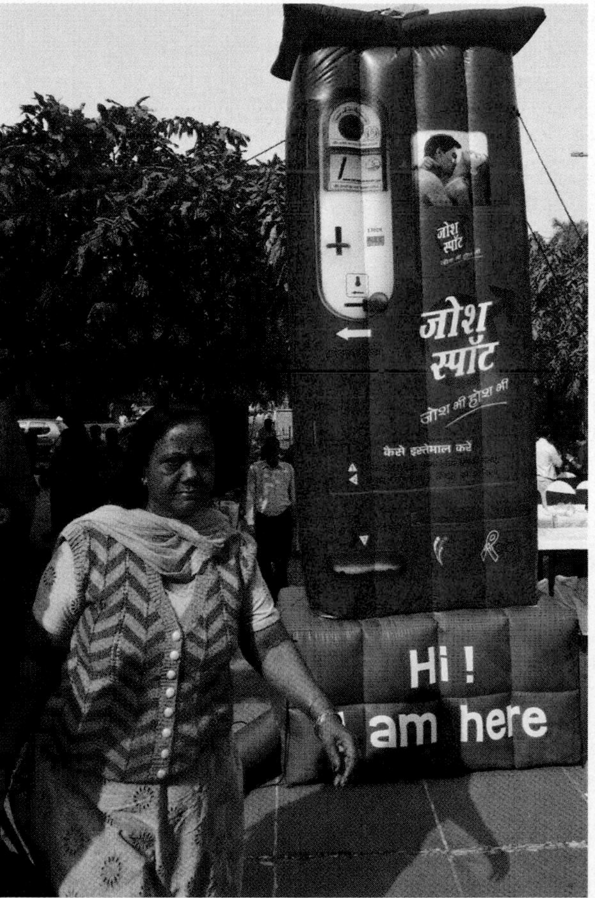

FIGURE 2-29 Promoting contraceptives in India. An inflatable vending machine promotes use of condoms in India.

LDCs, including Colombia, Morocco, and Thailand. Rapid growth in the acceptance of family planning is evidence that in the modern world, ideas can diffuse rapidly, even to places where people have limited access to education and modern communications.

The percentage of women using contraceptives is especially low in sub-Saharan Africa, so the alternative of distributing contraceptives could have an especially strong impact there. Less than one-fourth of women in sub-Saharan Africa employ contraceptives, compared to more than two-thirds in Asia and in Latin America (Figure 2-30).

Methods of family planning also vary among countries. The reason for this is partly economics, religion, and education. Very high birth rates in Africa and southwestern Asia also reflect the relatively low status of women. In societies where women receive less formal education and hold fewer legal rights than do men, women regard having a large number of children as a measure of their high status, and men regard it as a sign of their own virility.

Regardless of which alternative is more successful, many oppose birth-control programs for religious and political reasons. Adherents of several religions, including Roman Catholics, fundamentalist Protestants, Muslims, and Hindus, have religious convictions that prevent them from using some or all birth-control devices. Opposition is strong within the United States to terminating pregnancy by abortion, and the U.S. government has at times withheld aid to countries and family-planning organizations that advise abortion, even when such advice is only a small part of the overall aid program.

Analysts agree that the most effective means of reducing births would employ both alternatives. But LDC governments and international family-planning organizations have limited funds to promote lower birth rates, so they must set priorities and make choices for allocating scarce funds.

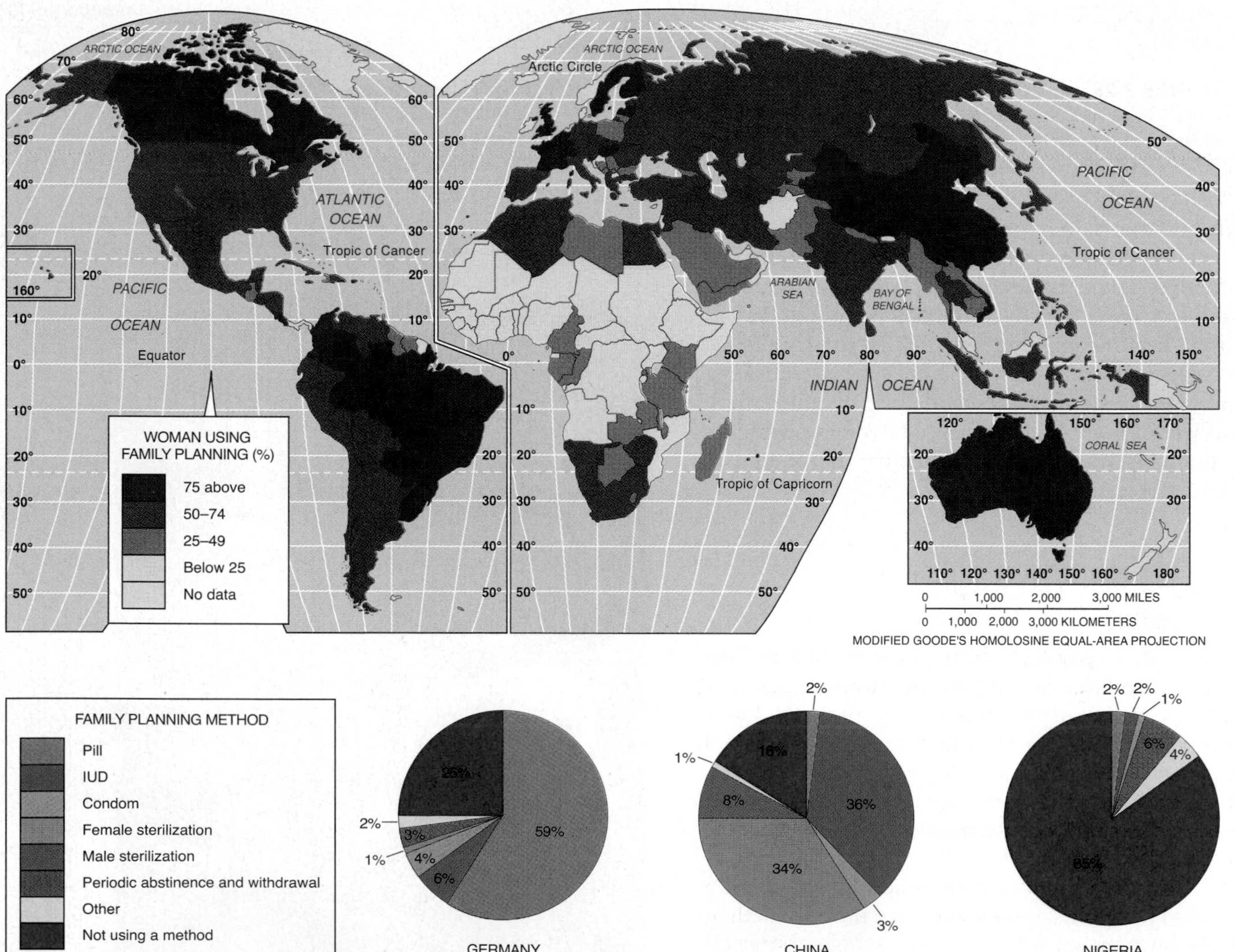

FIGURE 2-30 Family planning. More than two-thirds of couples in MDCs use a family-planning method, primarily condoms or birth-control pills. Family-planning practices vary more widely in other regions. China reports the world's highest rate of family planning, primarily with the use of intrauterine devices (IUDs) and female sterilization. The lowest rates are in countries of sub-Saharan Africa, such as Nigeria.

World Health Threats

Lower CBRs have been responsible for declining NIRs in most countries. However, in some countries of sub-Saharan Africa, lower NIRs have also resulted from higher CDRs, especially through the diffusion of AIDS.

Medical researchers have identified an **epidemiologic transition** that focuses on distinctive causes of death in each stage of the demographic transition. The term *epidemiologic transition* comes from **epidemiology**, which is the branch of medical science concerned with the incidence, distribution, and control of diseases that are prevalent among a population at a special time and are produced by some special causes not generally present in the affected locality. Epidemiologists rely heavily on geographic concepts such as scale and connection because measures to control and prevent an epidemic derive from understanding its distinctive distribution and method of diffusion.

Epidemiologic Transition Stages 1 and 2

Stage 1 of the epidemiologic transition, as originally formulated by epidemiologist Abdel Omran in 1971, has been called the stage of pestilence and famine. Infectious and parasitic diseases were principal causes of human deaths, along with accidents and attacks by animals and other humans. Malthus called these causes of deaths "natural checks" on the growth of the human population in stage 1 of the demographic transition.

BLACK PLAGUE. Well documented is the origin and diffusion of history's most violent stage 1 epidemic—the Black Plague, or bubonic plague, which was probably transmitted to humans by fleas from migrating infected rats. The Black Plague originated in present-day Kyrgyzstan and was brought from there by a Tatar army when it attacked an Italian trading post on the Black Sea in present-day Ukraine. Italians fleeing the trading post then carried the infected rats on ships west to the major coastal cities of southeastern Europe in 1347.

The plague spread from the coast to inland towns and then to rural areas. The plague reached Western Europe in 1348 and northern Europe in 1349. About 25 million Europeans died between 1347 and 1350, at least one-half of the continent's population. Five other epidemics in the late fourteenth century added to the toll in Europe. In China, 13 million died from the plague in 1380.

The plague wiped out entire villages and families, leaving farms with no workers and estates with no heirs. Churches were left without priests and parishioners, schools without teachers and students. Ships drifted aimlessly at sea after entire crews succumbed to the plague.

Stage 2 of the epidemiologic transition has been called the stage of receding pandemics. A **pandemic** is disease that occurs over a wide geographic area and affects a very high proportion of the population. Improved sanitation, nutrition, and medicine during the Industrial Revolution reduced the spread of infectious diseases.

Death rates did not decline immediately and universally during the early years of the Industrial Revolution. Poor people crowded into rapidly growing industrial cities had especially high death rates. Cholera—uncommon in rural areas—became an especially virulent epidemic in urban areas during the Industrial Revolution. A half-million people died of cholera in New York City in 1832, and one-eighth of the population of Cairo in 1831.

Geographic methods played a key role in understanding the cause of cholera during the early nineteenth century. The *Report of Sanitary Condition of the Labouring Population of Great Britain*, written in 1842 by Edwin Chadwick (1800–1890), showed that residents of poorer neighborhoods had a much higher incidence of cholera and other diseases and died at a younger age. Dr. John Snow (1813–1858) mapped the distribution of deaths from cholera in 1854 in the poor London neighborhood of Soho (Figure 2-31).

Many in the nineteenth century believed that epidemic victims were being punished for sinful behavior and that most victims were poor because poverty was considered a sin. Dr. Snow, however, showed that cholera was not distributed uniformly among the poor. Predating GIS by more than a century, he overlaid a map of the distribution of cholera victims with a map of the distribution of water pumps—for poor people the source of water for drinking, cleaning, and cooking.

Dr. Snow found that a large percentage of cholera victims were clustered around one pump, on Broad Street (refer to Figure 2-31). Tests proved that the water at the Broad Street pump was contaminated, and further investigation revealed that contaminated sewage was getting into the water supply near the pump.

Construction of water and sewer systems eradicated cholera by the late nineteenth century. However, cholera reappeared a century later in rapidly growing cities of LDCs as they moved into stage 2 of the demographic transition.

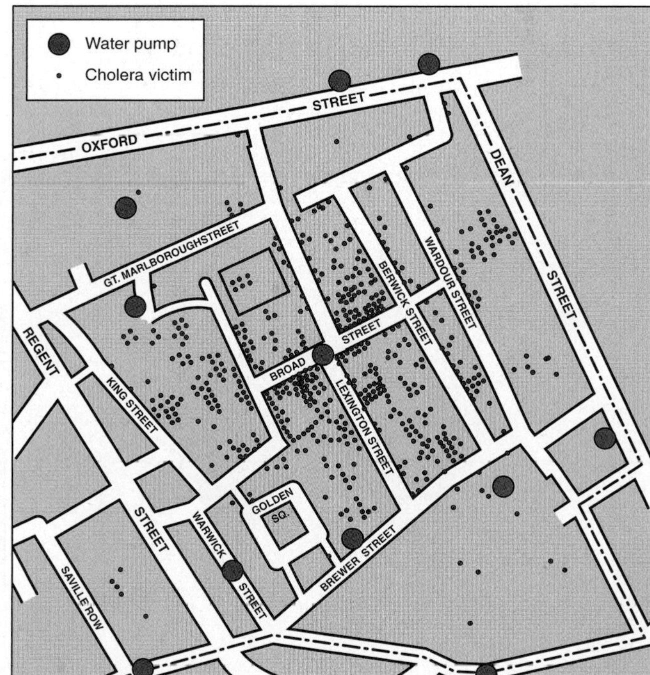

FIGURE 2-31 Cholera in Soho, London, 1854. Dr. John Snow mapped the distribution of cholera victims and water pumps to prove that the cause of the infection was contamination of the pump near the corner of Broad and Lexington streets.

Epidemiologic Transition Stages 3 and 4

Stage 3 of the epidemiologic transition, the stage of degenerative and human-created diseases, is characterized by a decrease in deaths from infectious diseases and an increase in chronic disorders associated with aging. The two especially important chronic disorders in stage 3 are cardiovascular diseases, such as heart attacks, and various forms of cancer.

The decline in infectious diseases has been sharp in stage 3 countries. Cases of polio declined in the United States from 14,000 in 1954 to 167 in 1965, 20 in 1975, and 0 in the entire Western Hemisphere during the 1990s. Worldwide polio cases declined from 39,000 in 1985 to 6,000 in 1994. The number of measles cases per year declined in the United States from 760,000 in 1958 to 2,000 during the 1980s and 1,000 during the 1990s. Fatalities from measles for children under age 15 declined in England from 110 per 100,000 during the nineteenth century to 10 during the 1940s and none during the 1960s. Effective vaccines were responsible for these declines.

As LDCs moved recently from stage 2 to stage 3, infectious diseases also declined. The number of cases of polio, neonatal tetanus, diphtheria, and pertussis declined by more than three-fourths in Southeast Asia between 1988 and 1994. The number of cases of leprosy declined from 483,000 in 1990 to 159,000 in 1993 in Africa.

Omran's epidemiologic transition was extended by S. Jay Olshansky and Brian Ault to stage 4, the stage of delayed degenerative diseases. The major degenerative causes of death—cardiovascular diseases and cancers—linger, but the life expectancy of older people is extended through medical advances. Through medicine, cancers spread more slowly or are removed altogether. Operations such as bypasses repair deficiencies in the cardiovascular system. Also improving health are behavior changes such as better diet, reduced use of tobacco and alcohol, and exercise.

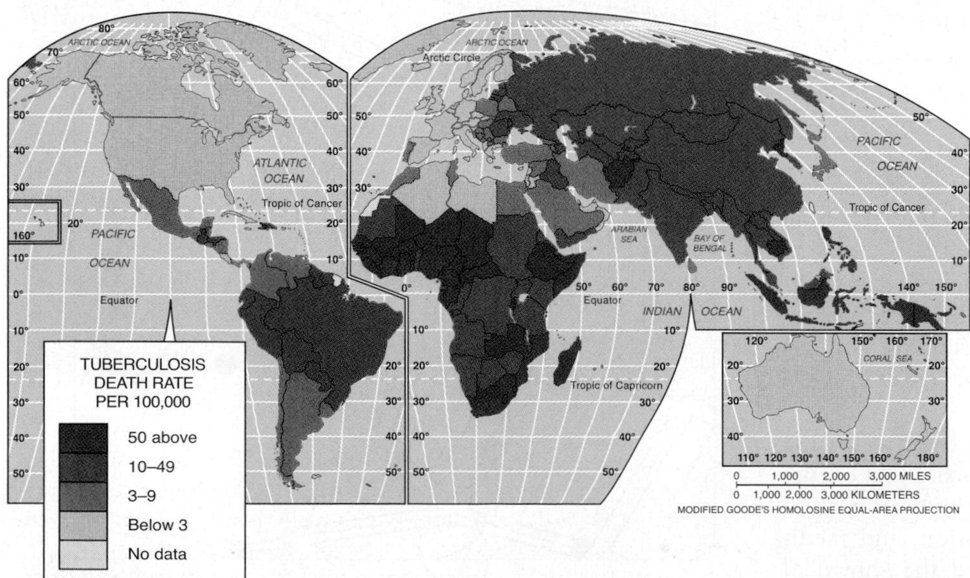

FIGURE 2-32 Tuberculosis (TB) cases, 2009. Death from tuberculosis is a good indicator of a country's ability to invest in health care, because treating the disease is expensive.

Epidemiologic Transition Possible Stage 5

Some medical analysts argue that the world is moving into stage 5 of the epidemiologic transition, the stage of reemergence of infectious and parasitic diseases. Infectious diseases thought to have been eradicated or controlled have returned, and new ones have emerged. A consequence of stage 5 would be higher CDRs. Other epidemiologists dismiss recent trends as a temporary setback in a long process of controlling infectious diseases.

Three reasons help to explain the possible emergence of a stage 5 of the epidemiologic transition:

1. **Evolution.** Infectious disease microbes have continuously evolved and changed in response to environmental pressures by developing resistance to drugs and insecticides. Antibiotics and genetic engineering contribute to the emergence of new strains of viruses and bacteria.

   Malaria was nearly eradicated in the mid-twentieth century by spraying DDT in areas infested with the mosquito that carried the parasite. For example, new malaria cases in Sri Lanka fell from 1 million in 1955 to 18 in 1963. The disease returned after 1963, however, and now causes more than 1 million deaths worldwide annually. A major reason was the evolution of DDT-resistant mosquitoes.
2. **Poverty.** Tuberculosis (TB) is an example of an infectious disease that has been largely controlled in relatively developed countries like the United States but remains a major cause of death in LDCs (Figure 2-32). An airborne disease, TB spreads principally through coughing and sneezing, damaging lungs.

   TB was one of the principal causes of death among the urban poor in the nineteenth century during the Industrial Revolution. The death rate from TB declined in the United States from 200 per 100,000 in 1900 to 60 in 1940 and 4 today. However, in LDCs, the TB rate is more than ten times higher than in MDCs, and nearly 2 million worldwide die from it annually. TB is more prevalent in poor areas because the long, expensive treatment poses a significant economic burden. Patients stop taking the drugs before the treatment cycle is completed.
3. **Improved travel.** A pandemic is a disease that occurs over a wide geographic area and affects an exceptionally high proportion of the population. Motor vehicles allow rural residents to easily reach urban areas and urban residents to reach rural areas. Airplanes allow residents of one country to easily reach another. As they travel, people carry diseases with them and are exposed to the diseases of others.

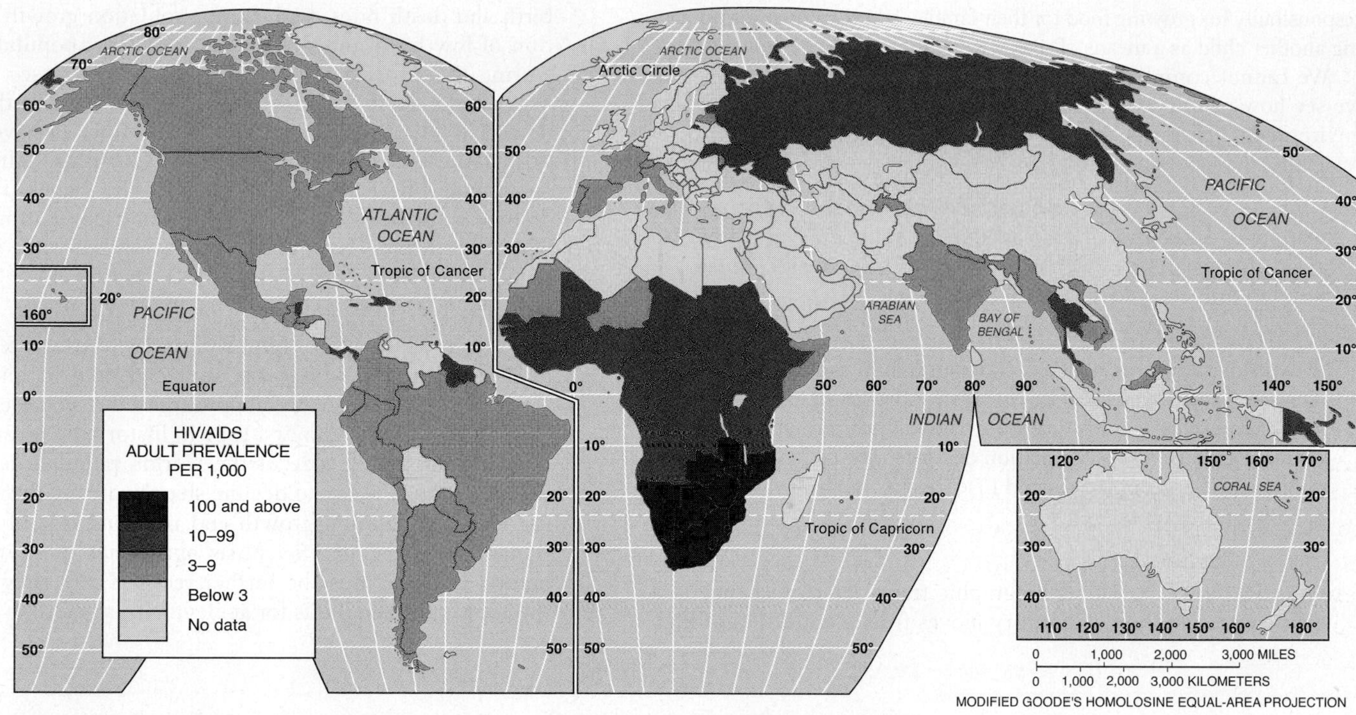

FIGURE 2-33 HIV/AIDS, 2007. The highest rates of HIV infection are in sub-Saharan Africa. India and China have relatively high numbers of HIV-positive adults, but they constitute a lower percentage of the total population.

Several dozen "new" infectious diseases have emerged over the past three decades and have spread through travel. Most prominent currently is H1N1, commonly known as swine flu, which was first identified in Mexico in early 2009 and spread around the world very rapidly.

The Bio.Diaspora Project, based at St. Michael's Hospital in Toronto, matched the global diffusion of H1N1 to airline travel patterns. The number of passengers arriving by air from Mexico was a strong predictor of the number of H1N1 cases in a particular city or country.

AIDS. The most lethal epidemic in recent years has been AIDS (acquired immunodeficiency syndrome). Worldwide, 25 million people died of AIDS as of 2007, and 33 million were living with HIV (human immunodeficiency virus, the cause of AIDS). The distribution of AIDS within the United States was discussed in Chapter 1 (see Figure 1-22), but 90 percent of people living with HIV come from LDCs. There were 22 million people infected with HIV in sub-Saharan Africa in 2007, 5 million in Asia, 2 million each in Eastern Europe and Latin America, and 1 million each in North America and Western Europe (Figure 2-33).

The impact of AIDS has been felt most strongly in sub-Saharan Africa. With one-tenth of the world's population, sub-Saharan Africa had two-thirds of the world's total HIV-positive population and nine-tenths of the world's infected children. South Africa had the most cases, 6 million, and Botswana, Lesotho, and Swaziland had the highest rates of infection—one-fourth of the three countries' adults were HIV-positive.

CDRs in many sub-Saharan Africa countries rose sharply during the 1990s as a result of AIDS, from the mid-teens to the low twenties. The populations of Lesotho and Swaziland are forecast to decline between now and 2050 as a result of AIDS. Life expectancy has declined in these two countries, from the 50s during the 1980s to the 40s currently.

SUMMARY

Overpopulation—too many people for the available resources—has already hit regions of Africa and threatens other countries in Asia and Latin America. The world as a whole does not face overpopulation immediately, but current trends must be reversed to prevent a future crisis.

Geographers caution that the number of people living in a region is not by itself an indication of overpopulation. Some densely populated regions are not overpopulated, whereas some sparsely inhabited areas are. Instead, overpopulation is a relationship between the size of the population and a region's level of resources. The capacity of the land to support life derives partly from characteristics of the natural environment and partly from human actions to modify the environment through agriculture, industry, and exploitation of raw materials.

The track toward overpopulation already may be irreversible in Africa. Rapid population growth has led to the overuse of land. As the land declines in quality, more effort is needed to yield the same amount of crops. This extends the working day of women, who have the primary responsibility for growing food for their families. Women then regard having another child as a means of securing additional help in growing food.

We cannot completely explain the overpopulation problem until we see how people in different regions earn a living and modify the environment. However, we can reach some conclusions by briefly reviewing the key issues raised at the beginning of this chapter.

1. **Where Is the World's Population Distributed?** Global population is concentrated in a few places. Human beings tend to avoid those parts of Earth's surface that they consider to be too wet, too dry, too cold, or too mountainous.
2. **Where Has the World's Population Increased?** Virtually all the world's natural increase is concentrated in the LDCs of Africa, Asia, and Latin America. In contrast, most European and North American countries now have low population growth rates, and some are experiencing population declines. The difference in natural increase between MDCs and LDCs is attributable to differences in CBRs rather than in CDRs.
3. **Why is Population Increasing at Different Rates in Different Countries?** The demographic transition is a change in a country's population. A country moves from a condition of high birth and death rates, with little population growth, to a condition of low birth and death rates, with low population growth. During this process the total population increases enormously, because the death rate declines some years before the birth rate does. The MDCs of Europe and North America have reached or neared the end of the demographic transition. African, Asian, and Latin American countries are at the stages of the demographic transition characterized by rapid population growth, in which death rates have declined sharply, but birth rates remain relatively high.
4. **Why Might the World Face an Over Population Problem?** The rate at which global population grew during the second half of the twentieth century was unprecedented in history. A dramatic decline in the death rate produced the increase. With death rates controlled, for the first time in history the most critical factor determining the size of the world's population is the birth rate. Birth rates began to decline sharply during the 1990s, slowing world population growth and reducing fear of overpopulation in most regions. Scientists agree that the current rate of natural increase must be further reduced, but they disagree on the appropriate methods for achieving this goal.

CASE STUDY REVISITED / India Versus China

The world's two most populous countries, China and India, will heavily influence future prospects for global overpopulation (Figure 2-34). These two countries—together encompassing more than one-third of the world's population—have adopted different family-planning programs. As a result of less effective policies, India adds 12 million more people each year than does China. Current projections show that India would surpass China as the world's most populous country around 2030.

India's Population Policies

India, like most countries in Africa, Asia, and Latin America, remained in stage 1 of the demographic transition until the late 1940s. During the first half of the twentieth century, population increased modestly—less than 1 percent per year—and even decreased in some years because of malaria, famines, plagues, and cholera epidemics.

Immediately following independence from England in 1947, India's death rate declined sharply (to 20 per 1,000 in 1951), whereas the CBR remained relatively high (about 40). Consequently, the NIR jumped to 2 percent per year. In response to this rapid growth, India became the first country to embark on a national family-planning program, in 1952. The government has established clinics and has provided information about alternative methods of birth control. Birth-control devices have been distributed free or at subsidized prices. Abortions, legalized in 1972, have been performed at a rate of several million per year. All together, the government spends several hundred million dollars annually on various family-planning programs.

FIGURE 2-34 Parents and their only child in Beijing, China.

(Continued)

CASE STUDY REVISITED (*Continued*)

India's most controversial family-planning program was the establishment of camps in 1971 to perform sterilizations—surgical procedures by which people were made incapable of reproduction. A sterilized person was entitled to a payment, which has been adjusted several times but generally has been equivalent to the average monthly income in India. At the height of the program, in 1976, 8.3 million sterilizations were performed during a 6-month period, mostly on women.

The birth-control drive declined in India after 1976. Widespread opposition to the sterilization program grew in the country because people feared that they would be forcibly sterilized. The prime minister, Indira Gandhi, was defeated in 1977, and the new government emphasized the voluntary nature of birth-control programs. The term *family planning*, which the Indian people associated with the forced sterilization policy, was replaced by the term *family welfare* to indicate that compulsory birth-control programs had been terminated. Although Mrs. Gandhi served again as prime minister from 1980 until she was assassinated in 1984, she did not emphasize family planning because of the opposition during her previous administration.

Government-sponsored family-planning programs have instead emphasized education, including advertisements on national radio and television networks and information distributed through local health centers. Given the cultural diversity of the Indian people, the national campaign has had only limited success. The dominant form of birth control continues to be sterilization of women, many of whom have already borne several children.

China's Population Policies

In contrast to India, China has made substantial progress in reducing its rate of growth. Since 2000, China has actually had a lower CBR than the United States.

The core of the Chinese government's family-planning program has been the One Child Policy, adopted in 1980. Under the One Child Policy, couples need a permit to have a child. Couples receive financial subsidies, a long maternity leave, better housing, and (in rural areas) more land if they agree to have just one child. The government prohibits marriage for men until they are 22 and women until they are 20. To further discourage births, people receive free contraceptives, abortions, and sterilizations. Rules are enforced by a government agency, the State Family Planning Commission.

As China moves toward a market economy in the twenty-first century and Chinese families become wealthier, the harsh rules in the One Child Policy have been relaxed, especially in urban areas. Clinics provide counseling on a wider range of family-planning options. Instead of fines, Chinese couples wishing a second child pay a "family-planning fee" to cover the cost to the government of supporting the additional person. Fears that relaxing the One Child Policy would produce a large increase in the birth rate have been unfounded. After a quarter-century of intensive educational programs, as well as coercion, the Chinese people have accepted the benefits of family planning. ■

KEY TERMS

Agricultural density (p. 52) The ratio of the number of farmers to the total amount of land suitable for agriculture.

Agricultural revolution (p. 57) The time when human beings first domesticated plants and animals and no longer relied entirely on hunting and gathering.

Arithmetic density (p.50) The total number of people divided by the total land area.

Census (p. 62) A complete enumeration of a population.

Crude birth rate (CBR) (p. 53) The total number of live births in a year for every 1,000 people alive in the society.

Crude death rate (CDR) (p. 53) The total number of deaths in a year for every 1,000 people alive in the society.

Demographic transition (p. 56) The process of change in a society's population from a condition of high crude birth and death rates and low rate of natural increase to a condition of low crude birth and death rates, low rate of natural increase, and a higher total population.

Demography (p. 45) The scientific study of population characteristics.

Dependency ratio (p. 59) The number of people under the age of 15 and over age 64 compared to the number of people active in the labor force.

Doubling time (p. 53) The number of years needed to double a population, assuming a constant rate of natural increase.

Ecumene (p. 49) The portion of Earth's surface occupied by permanent human settlement.

Epidemiologic transition (p. 71) Distinctive causes of death in each stage of the demographic transition.

Epidemiology (p. 71) Branch of medical science concerned with the incidence, distribution, and control of diseases that are prevalent among a population at a special time and are produced by some special causes not generally present in the affected locality.

Industrial Revolution (p. 57) A series of improvements in industrial technology that transformed the process of manufacturing goods.

Infant mortality rate (IMR) (p. 55) The total number of deaths in a year among infants under 1 year old for every 1,000 live births in a society.

Life expectancy (p. 55) The average number of years an individual can be expected to live, given current social, economic, and medical conditions. Life expectancy at birth is the average number of years a newborn infant can expect to live.

Medical revolution (p. 58) Medical technology invented in Europe and North America that is diffused to the poorer countries of Latin America, Asia, and Africa. Improved medical practices have eliminated many of the traditional causes of death in poorer countries and enabled more people to live longer and healthier lives.

Natural increase rate (NIR) (p. 53) The percentage growth of a population in a year, computed as the crude birth rate minus the crude death rate.

Overpopulation (p. 46) The number of people in an area exceeds the capacity of the environment to support life at a decent standard of living.

Pandemic (p. 71) Disease that occurs over a wide geographic area and affects a very high proportion of the population.

Physiological density (p. 51) The number of people per unit of area of arable land, which is land suitable for agriculture.

Population pyramid (p. 59) A bar graph representing the distribution of population by age and sex.

Sex ratio (p. 60) The number of males per 100 females in the population.

Total fertility rate (TFR) (p. 54) The average number of children a woman will have throughout her childbearing years.

Zero population growth (ZPG) (p. 58) A decline of the total fertility rate to the point where the natural increase rate equals zero.

THINKING GEOGRAPHICALLY

1. As discussed in the Contemporary Geographic Tools box, the current method of counting a country's population by requiring every household to complete a census form once every ten years has been severely criticized as inaccurate. The undercounting produces a geographic bias, because people who are missed are more likely to live in inner cities, remote rural areas, or communities that attract a relatively high number of recent immigrants. Given the availability of reliable statistical tests, should the current method of trying to count 100 percent of the population be replaced by a survey of a carefully drawn sample of the population, as is done with political polling and consumer preferences? Why or why not?
2. Scientists disagree about the effects of high density on human behavior. Some laboratory tests have shown that rats display evidence of increased aggressiveness, competition, and violence when very large numbers of them are placed in a box. Is there any evidence that very high density causes humans to behave especially aggressively or violently? Discuss.
3. Paul and Anne Ehrlich argue in *The Population Explosion* (1990) that a baby born in an MDC such as the United States poses a graver threat to global overpopulation than a baby born in an LDC. The reason is that people in MDCs place much higher demands on the world's supply of energy, food, and other limited resources. Do you agree with this view? Why?
4. Members of the baby-boom generation—people born between 1946 and 1964—constitute nearly one-third of the U.S. population. Baby boomers have received more education than their parents, and women are more likely to enter the labor force. They have delayed marriage and parenthood and have fewer children compared to their parents. They are more likely to divorce, to bear children while unmarried, and to cohabit. As they grow older, what impact will baby boomers have on the American population in the years ahead?
5. What policies should governments in MDCs pursue to reduce global population growth? If an MDC provides funds and advice to promote family planning, does it gain the right to tell developing countries how to spend the funds and how to use the expertise? Explain your answer.

RESOURCES

Recent and classic books and articles on population geography:

Bailey, Adrian. *Making Population Geography.* London: Hodder Arnold, 2005.

Haub, Carl, and O. P. Sharma. "India's Population: Reconciling Change and Tradition." *Population Bulletin* 61 (3). Washington, DC: Population Reference Bureau, 2006.

Lamptey, Peter, Jami L. Johnson, and Marya Khan. "The Global Challenge of HIV and AIDS." *Population Bulletin*. Washington, DC: Population Reference Bureau, 2006.

Malthus, Thomas. *An Essay on the Principles of Population.* 1978 (reprint). London: Royal Economic Society, 1926 (first published 1798).

McFalls, Joseph A., Jr. "Population: A Lively Introduction," 5th ed. *Population Bulletin* 62 (1). Washington, DC: Population Reference Bureau, 2007.

Newbold, K. Bruce. *Six Billion Plus: World Population in the Twenty-first Century*, 2nd ed. Lanham, MD: Rowman & Littlefield, 2007.

Riley, Nancy E. "China's Population: New Trends and Challenges." *Population Bulletin* 59 (2). Washington, DC: Population Reference Bureau, 2004.

Sanderson, Warren, and Sergei Scherbov. "Rethinking Age and Aging." *Population Bulletin* 63 (4). Washington, DC: Population Reference Bureau, 2008.

Simon, Julian. *Theory of Population and Economic Growth.* Oxford and New York: Blackwell, 1986.

Smil, Vaclav. "How Many People Can the Earth Feed?" *Population and Development Review* 20 (1994): 255–92.

St. Michael's Hospital. *The Bio.Diaspora Project: An Analysis of Canada's Vulnerability to Emerging Infectious Disease Threats via the Global Airline Transportation Network.* Toronto: St. Michael's Hospital (2009).

Journals featuring population geography:

American Demographics, Demography, Intercom, Population, Population and Development Review, Population Bulletin, Population Studies

Key internet sites:

www.prb.org. The Population Reference Bureau (PRB) provides authoritative demographic information for every country and world region. The PRB also provides electronic access to many of the articles and reports it publishes.

www.census.gov. The U.S. Bureau of the Census provides access to the results of the various censuses at its web site. The census's American Factfinder has the latest information on communities within the United States. *The Statistical Abstract of the United States* publishes many census and other government statistics in book form, available in libraries or in PDF format through the census web site. A chapter of international statistics is included.

Log in to www.mygeoscienceplace.com for videos, interactive maps, RSS feeds, case studies, and self-study quizzes to enhance your study of Population.

CHAPTER 3

Migration

Refer back to Figure 2-4 (ecumene) for a moment. Humans have spread across Earth during the past 7,000 years. This diffusion of human settlement from a small portion of Earth's land area to most of it resulted from migration. To accomplish the spread across Earth, humans have permanently changed their place of residence—where they sleep, store their possessions, and receive legal documents. Geographers document *from where* people migrate and *to where* they migrate. They also study reasons *why* people migrate.

How many times has your family moved? In the United States, the average family moves once every six

KEY ISSUES

1 Why Do People Migrate?
2 Where Are Migrants Distributed?
3 Why Do Migrants Face Obstacles?
4 Why Do People Migrate Within a Country?

years. Was your last move traumatic or exciting? The loss of old friends and familiar settings can hurt, but the experiences awaiting you at a new location can be stimulating. Think about the multitude of Americans—maybe including yourself—who have migrated from other countries. Imagine the feelings of people migrating from another country when they arrive in a new land without a job, friends, or—for many—the ability to speak the local language.

Why would people make a perilous journey across thousands of kilometers of ocean? Why did the pioneers cross the Great Plains, the Rocky Mountains, or the Mojave Desert to reach the American West? Why do people continue to migrate by the millions today? The hazards that many migrants have faced are a measure of the strong lure of new locations and the desperate conditions in their former homelands. Most people migrate in search of three objectives: economic opportunity, cultural freedom, and environmental comfort. This chapter will study the reasons why people migrate.

Migrant workers break for lunch on a construction site in Beijing, China.

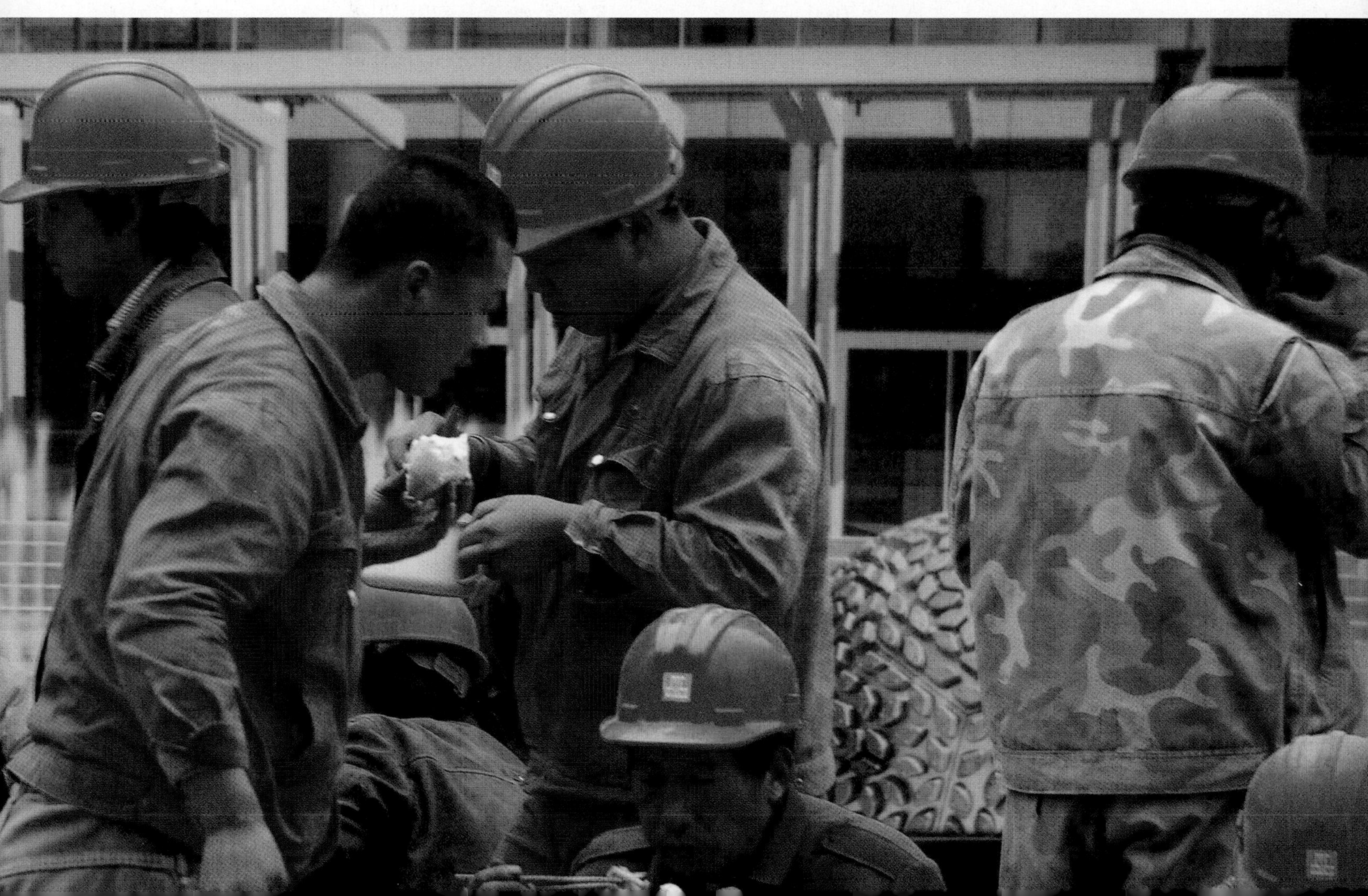

CASE STUDY / Migrating from Uruguay to Russia

Vasily Kilin was born and raised in Uruguay, where his parents and grandparents live. Kilin is migrating to Vladivostok, a city of one-half million in the far east region of Russia. Kilin is migrating from Uruguay to Russia with the encouragement of the Russian government. Programs worth several thousand dollars have paid for Kilin's relocation expenses and assisted in his job search.

In Chapter 2, we saw that Russia is in stage 4 of the demographic transition. Russia's NIR is one of the world's lowest at around −0.3. Contributing to the low NIR is a CBR of 12 per 1,000, one of the world's lowest, and a CDR of 15, one of the world's highest. Poor health practices, such as a high rate of alcoholism, have contributed to the high CDR. Family-planning practices and a deep pessimism about the future have contributed to the low CBR. As a result, Russia's population is expected to decline sharply from 142 million in 2009 to 117 million in 2050. Yet Russia's population decline would be even steeper were it not for immigration. More people have been migrating into Russia than leaving.

Like people on the move elsewhere in the world, Russia is attracting many immigrants for economic reasons. With the fall of communism and the breakup of the Soviet Union in 1991, Russians have more opportunity to start businesses. Although economic conditions in Russia are difficult, they are worse in neighboring countries.

Russia has also attracted immigrants for cultural reasons. When the Soviet Union was disbanded in 1991, many Russians suddenly found themselves living as minorities in newly created countries such as Ukraine and Kazakhstan (see Chapter 7), so they moved to Russia.

To reduce the anticipated population decline, Russia launched a program in 2006 to induce more immigration by people like Vasily Kilin. But in the first two years of the program, only 10,000 people participated. Response has been low in part because most Russians who have migrated to other countries have no interest in migrating back to Russia. Having migrated a generation ago to London or New York in search of economic gain and political freedom, few Russians are interested in returning to Russia. Further, the program is open only to ethnic Russians who speak fluent Russian.

Barely 100 kilometers from Vladivostok is the border between Russia and China. Many Chinese would be willing to move the short distance into Russia in search of better economic prospects, but Russia doesn't want them. ■

Diffusion was defined in Chapter 1 as a process by which a characteristic spreads from one area to another, and relocation diffusion was the spread of a characteristic through the bodily movement of people from one place to another. The subject of this chapter is a specific type of relocation diffusion called **migration**, which is a permanent move to a new location. Geographers document ***where*** people migrate to and from across the ***space*** of Earth.

The flow of migration always involves two-way ***connections***. Given two locations, A and B, some people migrate from A to B, while at the same time others migrate from B to A. **Emigration** is migration *from* a location; **immigration** is migration *to* a location.

The difference between the number of immigrants and the number of emigrants is the **net migration**. If the number of immigrants exceeds the number of emigrants, the net migration is positive, and the region has *net in-migration*. If the number of emigrants exceeds the number of immigrants, the net migration is negative, and the region has *net out-migration*.

Migration is a form of **mobility**, which is a more general term covering all types of movements from one place to another. People display mobility in a variety of ways, such as by journeying every weekday from their homes to places of work or education and once a week to shops, places of worship, or recreation areas. These types of short-term, repetitive, or cyclical movements that recur on a regular basis, such as daily, monthly, or annually, are called **circulation**. College students display another form of mobility—seasonal mobility—by moving to a dormitory each fall and returning home the following spring.

Geographers are especially interested in ***why*** people migrate, even though migration occurs much less frequently than other forms of mobility, because it produces profound changes for individuals and entire cultures. A permanent move to a new location disrupts traditional cultural ties and economic patterns in one ***region***. At the same time, when people migrate, they take with them to their new home their language, religion, ethnicity, and other cultural traits, as well as their methods of farming and other economic practices.

The changing ***scale*** generated by modern transportation systems, especially motor vehicles and airplanes, makes relocation diffusion more feasible than in the past, when people had to rely on walking, animal power, or slow ships. However, thanks to modern communications systems, relocation diffusion is no longer essential for transmittal of ideas from one place to another. Culture and economy can diffuse rapidly around the world through forms of expansion diffusion.

If people can participate in the ***globalization*** of culture and economy regardless of place of residence, why do they still migrate in large numbers? The answer is that ***place*** is still important to an individual's cultural identity and economic prospects. Within a global economy, an individual's ability to earn a living varies by location. Within a global culture, people migrate to escape from domination by other cultural groups or to be reunited with others of similar culture. Migration of people with similar cultural values creates pockets of ***local diversity***.

Although migration is a form of relocation diffusion, reasons for migrating can be gained from expansion diffusion. Someone may migrate and send back a message that gives others the idea of migrating. For example, many Europeans migrated to the United States in the nineteenth century because very favorable reports from early migrants led them to believe that the streets of American cities were paved with gold.

KEY ISSUE 1

Why Do People Migrate?

- Reasons for Migrating
- Distance of Migration
- Characteristics of Migrants

Geography has no comprehensive theory of migration, although a nineteenth-century outline of 11 migration "laws" written by E. G. Ravenstein is the basis for contemporary geographic migration studies. To understand where and why migration occurs, Ravenstein's "laws" can be organized into three groups: the reasons why migrants move, the distance they typically move, and their characteristics. Each of these elements is addressed in this section of the chapter. ■

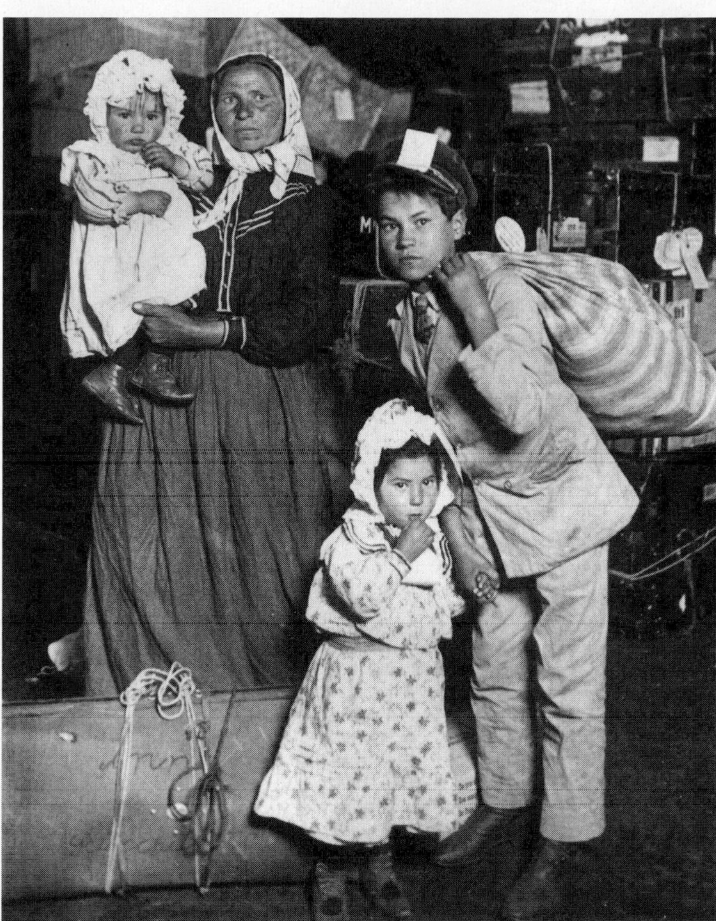

FIGURE 3-1 U.S. immigration. This mother and three children immigrated to the United States from Italy around 1900.

Reasons for Migrating

- Most people migrate for economic reasons.
- Cultural and environmental factors also induce migration, although not as frequently as economic factors.

People decide to migrate because of push factors and pull factors. A **push factor** induces people to move out of their present location, whereas a **pull factor** induces people to move into a new location. As migration for most people is a major step not taken lightly, both push and pull factors typically play a role. To migrate, people view their current place of residence so negatively that they feel pushed away, and they view another place so attractively that they feel pulled toward it.

We can identify three major kinds of push and pull factors: economic, cultural, and environmental. Usually, one of the three factors emerges as most important, although as will be discussed later in this chapter, ranking the relative importance of the three factors can be difficult and even controversial.

Economic Push and Pull Factors

Most people migrate for economic reasons. People think about emigrating from places that have few job opportunities, and they immigrate to places where jobs seem to be available. Because of economic restructuring, job prospects often vary from one country to another and within regions of the same country.

The United States and Canada have been especially prominent destinations for economic migrants (Figure 3-1). Many European immigrants to North America in the nineteenth century truly expected to find streets paved with gold. While not literally so gilded, the United States and Canada did offer Europeans prospects for economic advancement. This same perception of economic plenty now lures people to the United States and Canada from Latin America and Asia.

Cultural Push and Pull Factors

Cultural factors can be especially compelling push factors, forcing people to emigrate from a country. Forced international migration has historically occurred for two main cultural reasons: slavery and political instability.

Millions of people were shipped to other countries as slaves or as prisoners, especially from Africa to the Western Hemisphere, during the eighteenth and early nineteenth centuries (see Chapter 7). Large groups of people are no longer forced to migrate as slaves, but forced international migration persists because of political instability resulting from cultural diversity.

According to the United Nations, **refugees** are people who have been forced to migrate from their homes and cannot

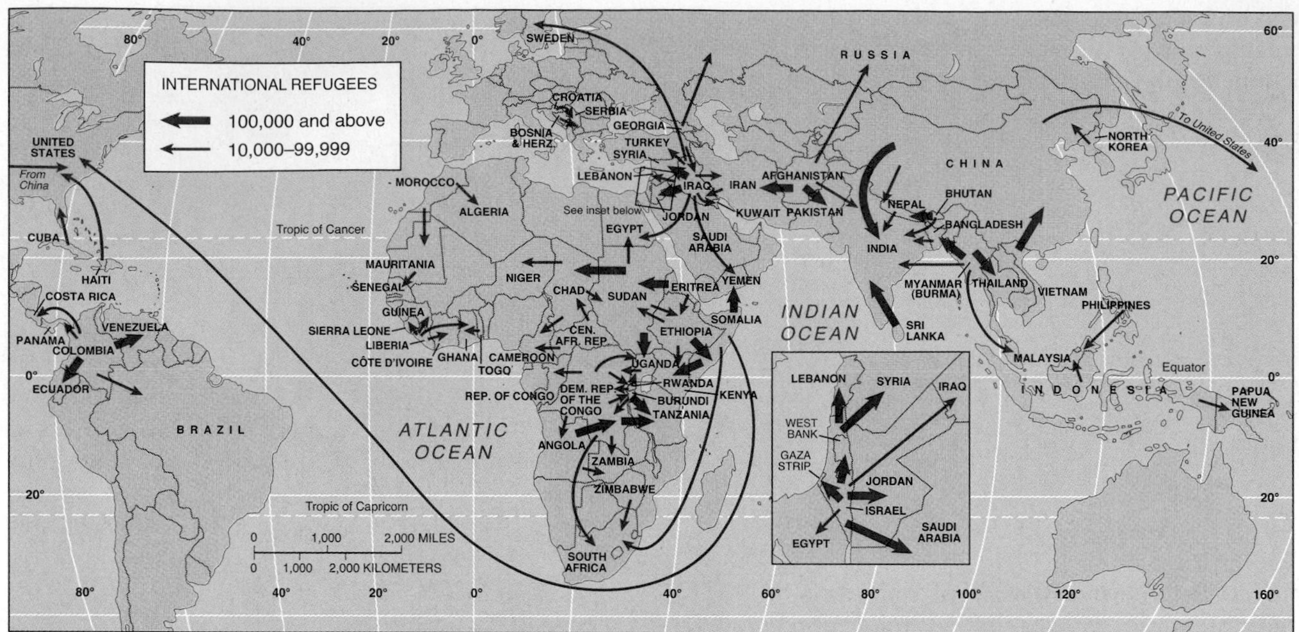

FIGURE 3-2 Major sources and destinations of refugees. A refugee is a person who is forced to migrate from a country, usually for political reasons. The U.S. Committee for Refugees estimates that the three largest groups of refugees are Afghans, Palestinians, and Iraqis. The large number of refugees from Afghanistan has resulted from more than three decades of civil war (see Chapter 8). Palestinians are people who left Israel after the country was created in 1948, or those who left territories captured by Israel in 1967 (see Chapter 6). The number of Iraqi refugees increased rapidly after the United States invaded Iraq in 2003 (see Chapter 8).

return for fear of persecution because of their race, religion, nationality, membership in a social group, or political opinion. The U.S. Committee for Refugees, a nonprofit organization independent of the U.S. government (www.refugees.org), counted 14 million refugees in 2007 (Figure 3-2). Refugees have no home until another country agrees to allow them in, or improving conditions make possible a return to their former home. In the interim, they must camp out in tents, board in shelters, or lie down by the side of a road.

Political conditions can also operate as pull factors. People may be attracted to democratic countries that encourage individual choice in education, career, and place of residence. After Communists gained control of Eastern Europe in the late 1940s, many people in that region were pulled toward the democracies in Western Europe and North America. Communist governments in Eastern Europe clamped down on emigration for fear of losing their most able workers. The most dramatic symbol of restricted emigration was the Berlin Wall, which the Communists built to prevent emigration from Communist-controlled East Berlin into democratic West Berlin.

With the election of democratic governments in Eastern Europe during the 1990s, Western Europe's political pull disappeared as a migration factor. Eastern Europeans now can visit where they wish, although few have the money to pay for travel-related expenses beyond a round-trip bus ticket. However, Western Europe pulls an increasing number of migrants from Eastern Europe for economic reasons, as discussed later in this chapter.

Environmental Push and Pull Factors

People also migrate for environmental reasons, pulled toward physically attractive regions and pushed from hazardous ones. In an age of improved communications and transportation systems, people can live in environmentally attractive areas that are relatively remote and still not feel too isolated from employment, shopping, and entertainment opportunities.

Attractive environments for migrants include mountains, seasides, and warm climates. Proximity to the Rocky Mountains lures Americans to the state of Colorado, and the Alps pull French people to eastern France. Some migrants are shocked to find polluted air and congestion in these areas. The southern coast of England, the Mediterranean coast of France, and the coasts of Florida attract migrants, especially retirees, who enjoy swimming and lying on the beach. Of all elderly people who migrate from one U.S. state to another, one-third select Florida as their destination. Regions with warm winters, such as southern Spain and the southwestern United States, attract migrants from harsher climates.

Migrants are also pushed from their homes by adverse physical conditions. Water—either too much or too little—poses the most common environmental threat (Figure 3-3). Many people are forced to move by water-related disasters because they live in a vulnerable area, such as a floodplain. The **floodplain** of a river is the area subject to flooding during a specific number of years, based on historical trends. People living in the "100-year

FIGURE 3-3 Environmental push factor: Too much water. The widespread flooding in New Orleans and other Gulf Coast communities in 2005 following Hurricane Katrina caused around 1,400 deaths and forced several hundred thousand people from their homes. Americans watching on television were shocked by the plight of residents stranded by the flooding: the squalid conditions in the evacuation centers, the lawlessness in the streets of New Orleans, and above all the unsatisfactory response of emergency management officials.

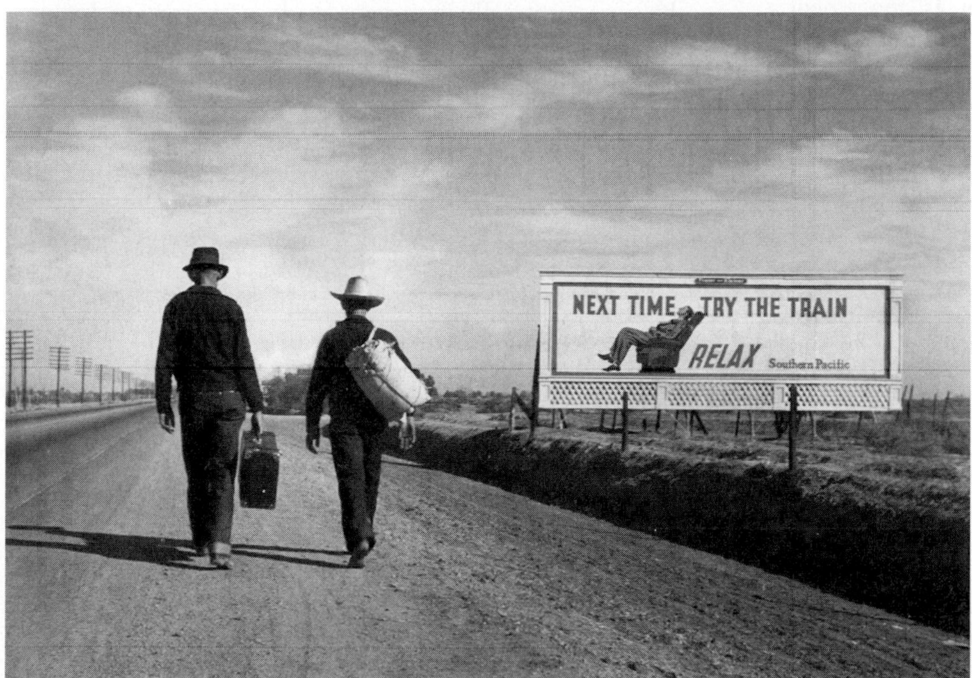

FIGURE 3-4 Environmental push factor: Lack of water. People were pushed from their land in Oklahoma and adjacent states during the 1930s by severe drought, known as the Dust Bowl. Thousands of families, known as Okies, abandoned their farms and migrated 1,000 miles west to California, some on foot.

floodplain," for example, can expect flooding on average once every century. Many people are unaware that they live in a floodplain, and even people who do know often choose to live there anyway.

A lack of water pushes others from their land (Figure 3-4). Hundreds of thousands have been forced to move from the Sahel region of northern Africa because of drought conditions. The people of the Sahel have traditionally been pastoral nomads, a form of agriculture adapted to dry lands but effective only at low population densities (see Chapter 10).

The capacity of the Sahel to sustain human life—never very high—has declined recently because of population growth and several years of unusually low rainfall. Consequently, many of these nomads have been forced to move into cities and rural camps, where they survive on food donated by the government and international relief organizations.

Intervening Obstacles

Where migrants go is not always their desired destination. The reason is that they may be blocked by an **intervening obstacle**, which is an environmental or cultural feature that hinders migration.

In the past, intervening obstacles were primarily environmental. Bodies of water have long been important intervening obstacles. The Atlantic Ocean proved a particularly significant intervening obstacle for most European immigrants to North America. Tens of millions of Europeans spent their life savings for the right to cross the rough and dangerous Atlantic in the hold of a ship shared with hundreds of other immigrants.

Before the invention of modern transportation, such as railroads and motor vehicles, people migrated across landmasses by horse or on foot. Such migration was frequently difficult because of hostile features in the physical environment, such as mountains and deserts. For example, many migrants lured to California during the nineteenth century by the economic pull factor of the Gold Rush failed to reach their destination because they could not cross such intervening obstacles as the Great Plains, the Rocky Mountains, or desert country.

Transportation improvements that have promoted globalization, such as motor vehicles and airplanes, have diminished the importance of environmental features as intervening obstacles. However, today's migrant faces intervening obstacles created by local diversity in government and politics. A migrant needs a passport to legally emigrate from a country and a visa to legally immigrate to a new country.

Distance of Migration

Ravenstein's theories made two main points about the distance that migrants travel to their new homes:

- Most migrants relocate a short distance and remain within the same country.
- Long-distance migrants to other countries head for major centers of economic activity.

Internal Migration

International migration is permanent movement from one country to another, whereas **internal migration** is permanent movement within the same country. Consistent with the distance-decay principle presented in Chapter 1, the farther away a place is located, the less likely that people will migrate to it. Thus, international migrants are much less numerous than internal migrants.

Most people find migration within a country less traumatic than international migration because they find familiar language, foods, broadcasts, literature, music, and other social customs after they move. Moves within a country also generally involve much shorter distances than those in international migration. However, internal migration can involve long-distance moves in large countries, such as in the United States and Russia.

Internal migration can be divided into two types: **Interregional migration** is movement from one region of a country to another; **intraregional migration** is movement within one region. Historically, the main type of interregional migration has been from rural to urban areas in search of jobs. In recent years, some developed countries have seen migration from urban to environmentally attractive rural areas. The main type of intraregional migration has been within urban areas, from older cities to newer suburbs.

International Migration

International migration is further divided into two types: forced and voluntary. **Voluntary migration** implies that the migrant has *chosen* to move for economic improvement, whereas **forced migration** means that the migrant has been *compelled* to move by cultural factors. Economic push and pull factors usually induce voluntary migration, and cultural factors normally compel forced migration. In one sense, migrants may also feel compelled by pressure inside themselves to migrate for economic reasons, such as to search for food or jobs, but they have not been explicitly compelled to migrate by the violent actions of other people.

Geographer Wilbur Zelinsky identified a **migration transition**, which consists of changes in a society comparable to those in the demographic transition. The migration transition is a change in the migration pattern in a society that results from the social and economic changes that also produce the demographic transition. According to the migration transition, international migration is primarily a phenomenon of countries in stage 2 of the demographic transition, whereas internal migration is more important in stages 3 and 4.

- A country in ***stage 1*** of the demographic transition (high CBR and CDR and low NIR) is characterized by high daily or seasonal mobility in search of food rather than permanent migration to a new location.
- A country in ***stage 2*** (high NIR because of rapidly declining CDR) is at the point when international migration becomes especially important, as does interregional migration from one country's rural areas to its cities. Like the sudden decline in the crude death rate, migration patterns in stage 2 societies are a consequence of technological change. Improvement in agricultural practices reduces the number of people needed in rural areas, and jobs in factories attract migrants to the cities in another region of the same country or in another country.
- Countries in ***stages 3*** and ***4*** (moderating NIR because of rapidly declining CBR) are the principal destinations of the international migrants leaving the stage 2 countries in search of economic opportunities (Figure 3-5). The principal form of internal migration within countries in stages 3 and 4 of the demographic transition is intraregional, from cities to surrounding suburbs.

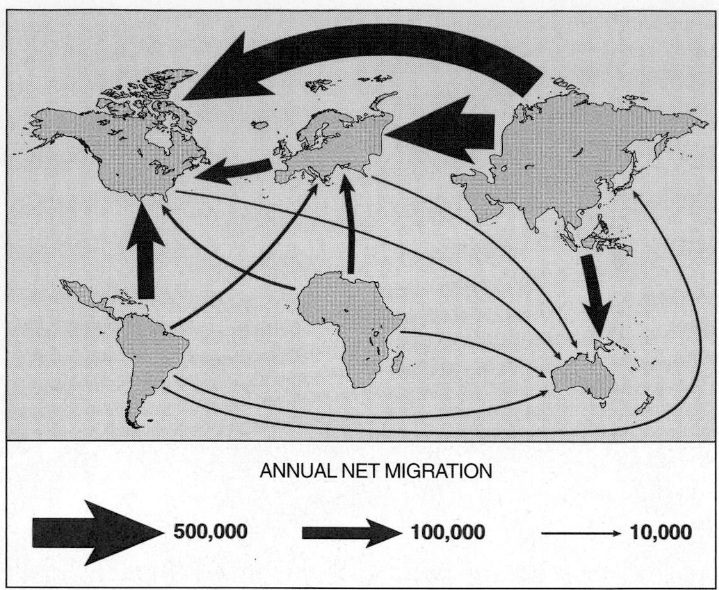

FIGURE 3-5 Global migration patterns. The major flows of international migrants are from LDCs to MDCs, especially from Asia and Latin America to North America and from Asia to Europe.

Characteristics of Migrants

Ravenstein noted distinctive gender and family-status patterns in his migration theories:

- Most long-distance migrants are male.
- Most long-distance migrants are adult individuals rather than families with children.

Gender of Migrants

Ravenstein theorized that males were more likely than females to migrate long distances to other countries because searching for work was the main reason for international migration and males were much more likely than females to be employed. This held true for U.S. immigrants during the nineteenth and much of the twentieth centuries, when about 55 percent were male. But the gender pattern reversed in the 1990s, and in the twenty-first century women constitute about 55 percent of U.S. immigrants (Figure 3-6).

Mexicans who come to the United States without authorized immigration documents—currently the largest group of U.S. immigrants—show similar gender changes. As recently as the late 1980s, males constituted 85 percent of the Mexican migrants arriving in the United States without proper documents, according to U.S. census and immigration service estimates. But since the 1990s, women have accounted for about half of the unauthorized immigrants from Mexico.

The increased female migration to the United States partly reflects the changing role of women in Mexican society. In the past, rural Mexican women were obliged to marry at a young age and to remain in the village to care for children. Now some Mexican women are migrating to the United States to join husbands or brothers already in the United States, but most are seeking jobs. At the same time, women also feel increased pressure to get a job in the United States because of poor economic conditions in Mexico.

Family Status of Migrants

Ravenstein also believed that most long-distance migrants were young adults seeking work, rather than children or elderly people. For the most part, this pattern continues for the United States.

- About 40 percent of immigrants are young adults between the ages of 25 and 39, compared to about 23 percent of the entire U.S. population.
- Immigrants are less likely to be elderly people; only 5 percent of immigrants are over age 65, compared to 12 percent of the entire U.S. population.

FIGURE 3-6 Family status of migrants. A Mexican family crosses into the United States near Ciudad Juárez.

However, an increasing percentage of U.S. immigrants are children—16 percent of immigrants are under age 15, compared to 21 percent for the total U.S. population. With the increase in women migrating to the United States, more children are coming with their mothers.

Recent immigrants to the United States have attended school for fewer years and are less likely to have high school diplomas than are U.S. citizens. The typical unauthorized Mexican immigrant has attended school for four years, less than the average American but a year more than the average Mexican.

KEY ISSUE 2
Where Are Migrants Distributed?

- **Global Migration Patterns**
- **U.S. Immigration Patterns**
- **Impact of Immigration on the United States**

About 9 percent of the world's people are international migrants—that is, they currently live in countries other than the ones in which they were born. The country with by far the largest number of international migrants is the United States. ■

Global Migration Patterns

At a global scale, Asia, Latin America, and Africa have net out-migration, and North America, Europe, and Oceania have net in-migration. The three largest flows of migrants are to Europe from Asia and to North America from Asia and from Latin America. The global pattern reflects the importance of migration from LDCs to MDCs. Migrants from countries with relatively low incomes and high natural increase rates head for relatively wealthy countries, where job prospects are brighter.

The United States has more foreign-born residents than any other country, approximately 40 million as of 2010, and growing annually by around 1 million. Other MDCs have higher rates of net in-migration, including Australia and Canada, which are much less populous than the United States (Figure 3-7). The highest rates can be found in petroleum-exporting countries of the Middle East, which attract immigrants primarily from poorer Middle Eastern countries and from Asia to perform many of the dirty and dangerous functions in the oil fields.

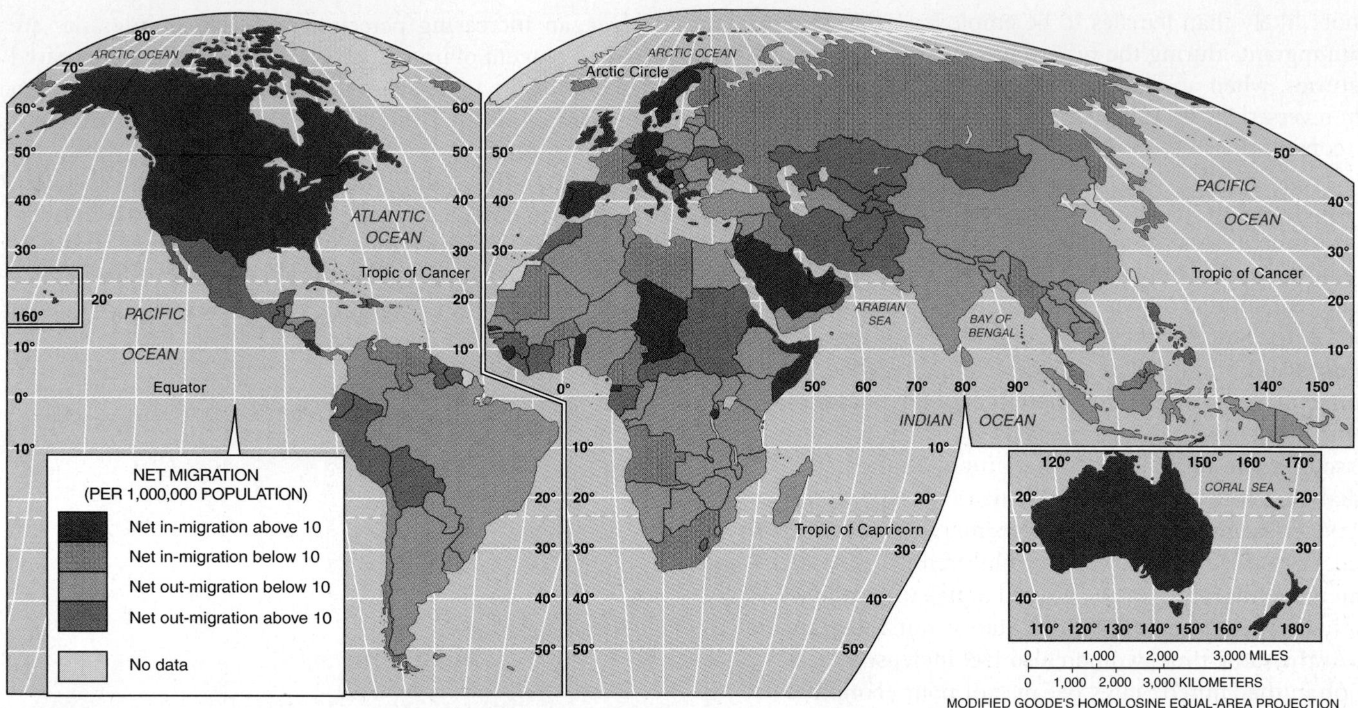

FIGURE 3-7 Net migration by country. High net in-migration is found in oil-rich Middle East countries, as well as in MDCs perceived as having more job opportunities.

U.S. Immigration Patterns

The United States plays a special role in the study of international migration. The world's third most populous country is inhabited overwhelmingly by direct descendants of immigrants. About 75 million people migrated to the United States between 1820 and 2010, including 40 million who were alive in 2010.

The United States has had three main eras of immigration (Figure 3-8). The first era was the initial settlement of colonies. The second era began in the mid-nineteenth century and culminated in the early twentieth century. The third era began in the 1970s and continues today.

The three eras have drawn migrants from different regions. Most immigrants were English or African slaves during the first era, nearly all were European during the second era, and more than three-fourths were from Latin America and Asia during the third era.

Although the origins vary, the reason for migrating has remained essentially the same. Rapid population growth limited prospects for economic advancement at home. Europeans left when their countries entered stage 2 of the demographic transition in the nineteenth century, and Latin Americans and Asians began to leave in large numbers in recent years after their countries entered stage 2. But Europeans arriving in the United States in the nineteenth century found a very different country than Latin Americans and Asians who have recently arrived.

Colonial Immigration from England and Africa

Immigration to the American colonies and the newly independent United States came from two principal sources: Europe and Africa. Most of the Africans were forced to migrate to the United States as slaves, whereas most Europeans were voluntary migrants—although harsh economic conditions and persecution in Europe blurred the distinction between forced and voluntary migration for many Europeans.

About 1 million Europeans migrated to the American colonies prior to independence, and another million from the late 1700s until 1840. From the first permanent English settlers to arrive at the Virginia colony's Jamestown, in 1607, until 1840, a steady stream of Europeans migrated to the American colonies (and after 1776 to the newly independent United States of America). Ninety percent of European immigrants to the United States prior to 1840 came from Great Britain.

Most African Americans are descended from Africans forced to migrate to the Western Hemisphere as slaves. During the eighteenth century, about 400,000 Africans were shipped as slaves to the 13 colonies that later formed the United States, primarily by the British. The importation of Africans as slaves was made illegal in 1808, but another 250,000 Africans were brought to the United States during the next half-century (see Chapter 7).

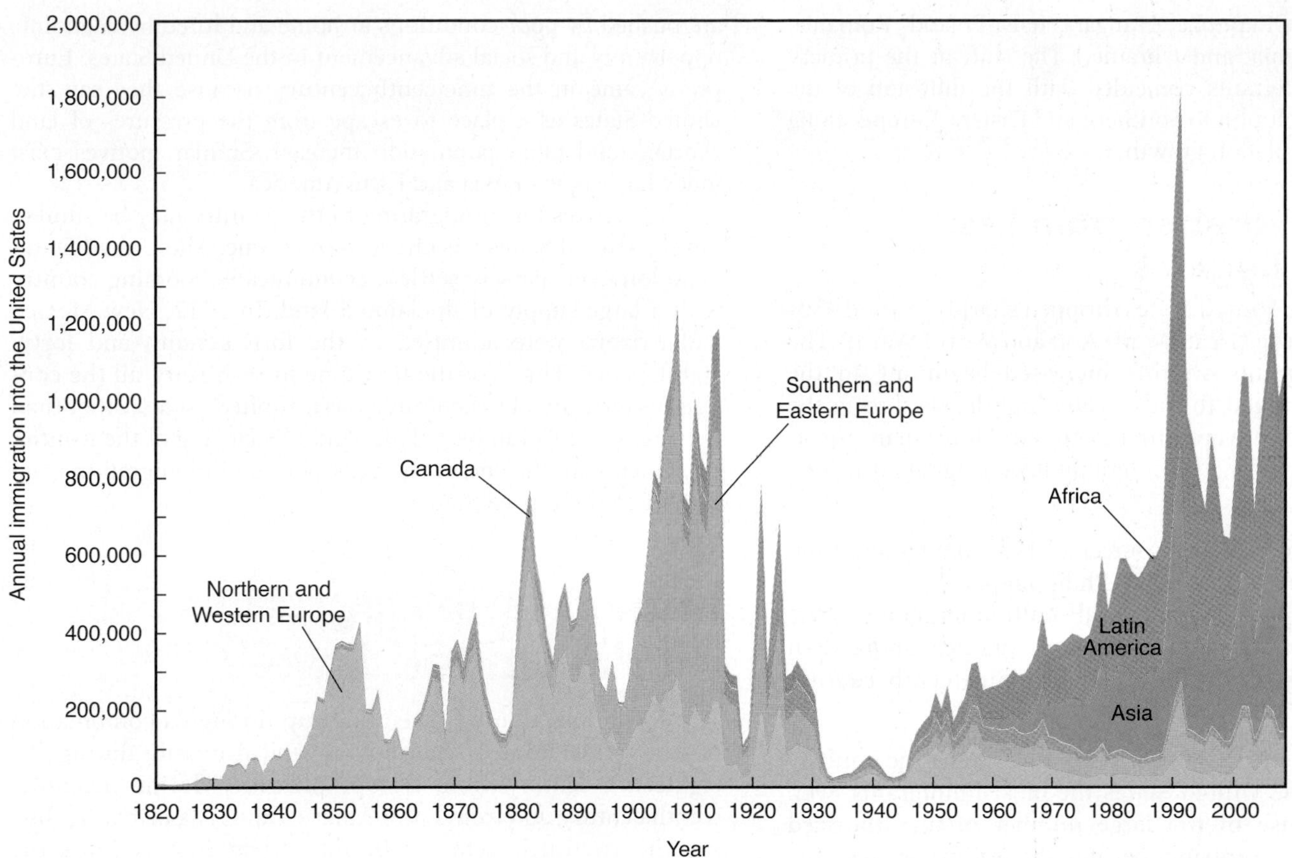

FIGURE 3-8 Migration to the United States by region of origin. Europeans comprised more than 90 percent of immigrants to the United States during the nineteenth century, and even as recently as the 1950s still accounted for more than 50 percent. Latin America and Asia are now the dominant sources of immigrants to the United States.

Nineteenth-Century Immigration from Europe

In the 500-plus years since Christopher Columbus sailed from Spain to the Western Hemisphere, about 65 million Europeans have migrated to other continents. For 40 million of them, the destination was the United States. The remainder went primarily to the temperate climates of Canada, Australia, New Zealand, southern Africa, and southern South America, where farming methods used in Europe could be most easily transplanted. For European migrants, the United States offered the greatest opportunity for economic success. Early migrants extolled the virtues of the country to friends and relatives back in Europe, which encouraged still others to come.

Among European countries, Germany has sent the largest number of immigrants to the United States, 7.2 million. Other major European sources include Italy, 5.4 million; the United Kingdom, 5.3 million; Ireland, 4.8 million; and Russia and the former Soviet Union, 4.1 million. About one-fourth of Americans trace their ancestry to German immigrants, and one-eighth each to Irish and English immigrants.

Note that frequent boundary changes in Europe make precise national counts impossible. For example, most Poles migrated to the United States at a time when Poland did not exist as an independent country. Therefore, most were counted as immigrants from Germany, Russia, or Austria.

Migration from Europe to the United States peaked at several points during the nineteenth century.

- **1840s and 1850s.** Annual immigration jumped from 20,000 to more than 200,000. Three-fourths of all U.S. immigrants during those two decades came from Ireland and Germany. Desperate economic push factors compelled the Irish and Germans to cross the Atlantic. Germans also emigrated to escape from political unrest.
- **1870s.** Emigration from Western Europe resumed following a temporary decline during the U.S. Civil War (1861–1865).
- **1880s.** Immigration increased to one-half million per year. Increasing numbers of Scandinavians, especially Swedes and Norwegians, joined Western Europeans in migrating to the United States. The Industrial Revolution had diffused to Scandinavia, triggering a rapid population increase.
- **1900–1914.** Nearly a million people a year immigrated to the United States. Two-thirds of all immigrants during this period came from Southern and Eastern Europe, especially Italy, Russia, and Austria-Hungary. (Austria-Hungary encompassed portions of present-day Austria, Bosnia-Herzegovina,

Croatia, Czech Republic, Hungary, Italy, Poland, Romania, Slovakia, Slovenia, and Ukraine.) The shift in the primary source of immigrants coincided with the diffusion of the Industrial Revolution to Southern and Eastern Europe, along with rapid population growth.

Recent Immigration from Less Developed Regions

Immigration to the United States dropped sharply in the 1930s and 1940s during the Great Depression and World War II. The number of immigrants steadily increased beginning in the 1950s, and then surged to historically high levels during the first decade of the twenty-first century. More than three-fourths of the recent U.S. immigrants have originated in two regions:

- **Asia.** The three leading sources of U.S. immigrants from Asia are China, India, and the Philippines.
- **Latin America.** Nearly one-half million emigrate to the United States annually from Latin America, more than twice as many as during the entire nineteenth century (Figure 3-9).

Officially, Mexico passed Germany in 2006 as the country that has sent to the United States the most immigrants ever. Unofficially, because of the large number of unauthorized immigrants, Mexico probably became the leading source during the 1980s. In the early 1990s, an unusually large number of immigrants came from Mexico and other Latin American countries as a result of the 1986 Immigration Reform and Control Act, which issued visas to several hundred thousand people who had entered the United States in previous years without legal documents.

Although the pattern of immigration to the United States has changed from predominantly European to Asian and Latin American, the reason for immigration remains the same. People are pushed by poor conditions at home and lured by economic opportunity and social advancement in the United States. Europeans came in the nineteenth century because they saw the United States as a place to escape from the pressures of land shortage and rapid population increase. Similar motives exist today for people in Asia and Latin America.

The motives for immigrating to the country may be similar, but the United States has changed over time. The United States is no longer a sparsely settled, economically booming country with a large supply of unclaimed land. In 1912, New Mexico and Arizona were admitted as the forty-seventh and forty-eighth states. Thus, for the first time in its history, all the contiguous territory of the country was a "united" state (other than the District of Columbia). This symbolic closing of the frontier coincided with the end of the peak period of immigration from Europe to the United States.

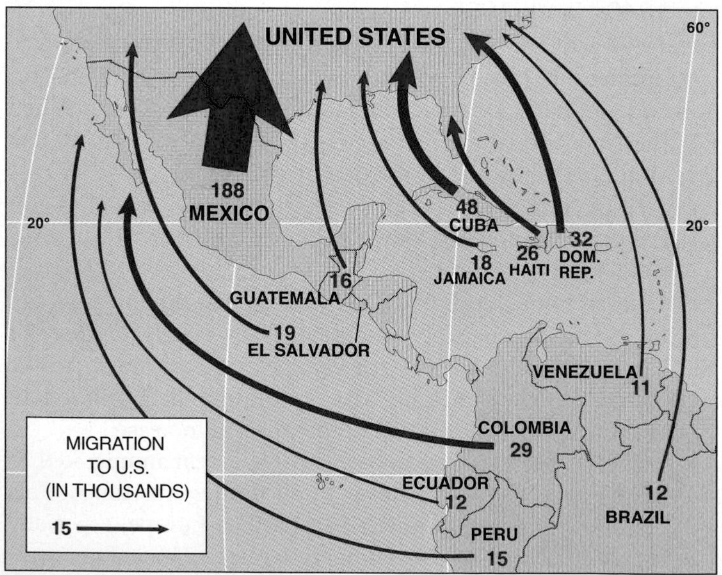

FIGURE 3-9 Migration to the United States from Latin America. Mexico has been the largest source of immigrants to the United States in recent decades.

Impact of Immigration on the United States

The U.S. population has been built up through a combination of emigration from Africa and England primarily during the eighteenth century, from Europe primarily during the nineteenth century, and from Latin America and Asia primarily during the twentieth century. In the twenty-first century, the impact of immigration varies around the country.

Legacy of European Migration

The era of massive European migration to the United States ended with the start of World War I in 1914, because the war involved the most important source countries, such as Austria-Hungary, Germany, and Russia, as well as the United States. The level of European emigration has steadily declined since that time.

EUROPE'S DEMOGRAPHIC TRANSITION. Rapid population growth in Europe fueled emigration, especially during the nineteenth century. Application of new technologies spawned by the Industrial Revolution—in areas such as public health, medicine, and food—produced a rapid decline in the CDR and pushed much of Europe into stage 2 of the demographic transition (high NIR). As the population increased, many Europeans found limited opportunities for economic advancement. Migration to the United States served as a safety valve, draining off some of that increase. People remaining in Europe enjoyed more of the economic and social benefits from the Industrial Revolution.

Most European countries are now in stage 4 of the demographic transition (very low or negative NIR) and have economies capable of meeting the needs of their people. Countries such as Germany, Italy, and Ireland, which once sent several hundred thousand people annually to the United States, now send only a few thousand. The safety valve is no longer needed.

DIFFUSION OF EUROPEAN CULTURE. The emigration of 65 million Europeans has profoundly changed world culture. As do all migrants, Europeans brought their cultural heritage to

CONTEMPORARY GEOGRAPHIC TOOLS

Claiming Ellis Island

Twelve million immigrants to the United States between 1892 and 1954 were processed at Ellis Island, situated in New York Harbor (Figure 3-10). Incorporated as part of the Statue of Liberty National Monument in 1965, Ellis Island was restored and reopened in 1990 as a museum of immigration. Before building the immigration center, the U.S. government used Ellis Island as a fort and powder magazine beginning in 1808.

An 1834 agreement approved by the U.S. Congress gave Ellis Island to New York State and the submerged lands surrounding the island to New Jersey. When the agreement was signed, Ellis Island was only 1.1 hectares (2.75 acres), but beginning in the 1890s, the U.S. government enlarged the island, eventually to 10.6 hectares (27.5 acres).

New Jersey state officials claimed that the 10.6-hectare Ellis Island was part of their state, not New York. The claim was partly a matter of pride on the part of New Jersey officials to stand up to their more glamorous neighbor. After all, Ellis Island was only 1,300 feet from the New Jersey shoreline, yet tourists—like immigrants a century ago—are transported by ferry to Lower Manhattan more than a mile away (Figure 3-11). More practically, the sales tax collected by the Ellis Island museum gift shop was going to New York rather than to New Jersey.

After decades of dispute, New Jersey took the case to the U.S. Supreme Court. In 1998, the Supreme Court ruled 6–3 that New York owned the original island but that New Jersey owned the rest. New York's jurisdiction was set as the low waterline of the original island. Critical evidence in the decision was a series of maps prepared by New Jersey Department of Environmental Protection (NJDEP) officials using GIS. NJDEP officials scanned into an image file an 1857 U.S. coast map that was considered to be the most reliable from that era. The image file of the old map was brought into ArcView, and then the low waterline shown on the 1857 map was edited and depicted by a series of dots. The perimeter of the current island was mapped, using global positioning system (GPS) surveying.

After ruling in favor of New Jersey's claim, the Supreme Court directed the NJDEP to delineate the precise boundary between the two states, again using GIS. Overlaying the 1857 low waterline onto the current map identified New York's territory, and the rest of the current island belonged to New Jersey. ■

FIGURE 3-10 New York Harbor. Ellis Island is connected to New Jersey by bridge. The other two islands in the harbor are Governor's Island (the larger one, formerly a military base, now a park) and Liberty Island (containing the Statue of Liberty). Manhattan is in the upper right, Brooklyn in the lower right, and Jersey City in the upper left.

FIGURE 3-11 Ellis Island. Ellis Island is behind Liberty Island. The large building on Ellis Island behind and to the right of the Statue of Liberty is the Great Hall, now the Immigration Museum, where immigrants were processed. The Great Hall is part of New York, but most of the remainder of the island is part of New Jersey. The buildings behind Ellis Island are in New Jersey. New York is barely visible in the distant background.

their new homes. Because of migration, Indo-European languages now are spoken by half of the world's people (as discussed in Chapter 5), and Europe's most prevalent religion, Christianity, has the world's largest numbers of adherents (see Chapter 6). European art, music, literature, philosophy, and ethics have also diffused throughout the world.

Regions that were sparsely inhabited prior to European immigration, such as North America and Australia, have become closely integrated into Europe's cultural traditions. Distinctive European political structures and economic systems have also diffused to these regions. Europeans also planted the seeds of conflict by migrating to regions with large indigenous populations, especially in Africa and Asia. They frequently imposed political domination on existing populations and injected their cultural values with little regard for local traditions. Economies in Africa and Asia became based on raising crops and extracting resources for export to Europe rather than on growing crops for local consumption and using resources to build local industry. Many of today's conflicts in former European colonies result from past practices by European immigrants, such as drawing arbitrary boundary lines and discriminating among different local ethnic groups.

Unauthorized Immigration to the United States

The number of people allowed to immigrate into the United States is at a historically high level, yet the number who wish to come is even higher. Many who cannot legally enter the United States immigrate illegally. Those who do so are entering without proper documents and thus are called **unauthorized (or undocumented) immigrants**.

The Pew Hispanic Center estimated that there were 11.9 million unauthorized immigrants living in the United States in 2008, and around 500,000 arrived that year without documentation. Around 59 percent of unauthorized immigrants came from Mexico, 22 percent from elsewhere in Latin America, and 12 percent from Asia. The Pew Hispanic Center's 2008 estimate of unauthorized immigrants included 6.3 million adult males, 4.1 million adult females, and 1.5 million children. In addition, 4 million children who were born in the United States—and therefore U.S. citizens—were living in families with an unauthorized immigrant.

People are in the United States without authorization primarily because they wish to work but do not have permission to do so from the government. About 8.3 million of the 11.9 unauthorized immigrants were employed, according to the Pew Hispanic Center's 2008 estimate, accounting for 5.4 percent of the total U.S. civilian labor force. Unauthorized immigrants were much more likely than the average American to be employed in construction and hospitality (food service and lodging) jobs and less likely to be in white-collar jobs such as education, health care, and finance.

Crossing the U.S.–Mexican border illegally has not been difficult. The border is 3,141 kilometers (1,951 miles) long and runs mostly through sparsely inhabited regions. The United States has constructed a barrier covering approximately one-fourth of the border. Guards heavily patrol border crossings in urban areas such as El Paso, Texas, and San Diego, California, or along highways, but rural areas are guarded by only a handful of agents.

Americans are divided concerning whether unauthorized migration helps or hurts the country. Most Americans recognize that unauthorized immigrants take jobs that no one else wants, and a majority would support some type of work-related program to make them legal. At the same time, Americans would like more effective border patrols so that fewer unauthorized immigrants can get into the country.

Destination of Immigrants within the United States

Recent immigrants are not distributed uniformly throughout the United States. One-fifth are in California and one-sixth in the New York metropolitan area. One-fourth of unauthorized immigrants are in California (Figure 3-12).

Individual states attract immigrants from different countries. In 2008, more than 50,000 migrated from Mexico to California. Between 10,000 and 50,000 migrated from China to California, from China to New York, from Colombia to Florida, from Cuba to Florida, from the Dominican Republic to New York, from Haiti to Florida, from India to California, from Mexico to Texas, from the Philippines to California, and from Vietnam to California.

Proximity clearly influences some decisions, such as Mexicans preferring California or Texas and Cubans preferring Florida. But proximity is not a factor in Poles heading for Illinois or Iranians for California. Immigrants cluster in communities where people from the same country previously settled. **Chain migration** is the migration of people to a specific location because relatives or members of the same nationality previously migrated there.

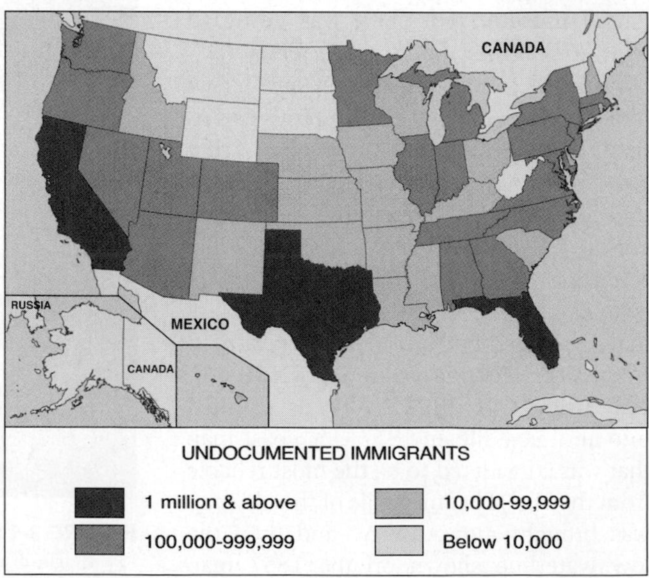

FIGURE 3-12 Destination of unauthorized immigrants by U.S. states. California, Texas, and Florida are the leading destinations for unauthorized immigrants.

GLOBAL FORCES, LOCAL IMPACTS
Unauthorized Immigration Viewed from the Mexican Side

From the United States, the view to the south may seem straightforward. Millions of Mexicans are trying to cross the border by whatever means, legal or otherwise, in search of employment, family reunification, and a better way of life in the United States.

The view from Mexico is more complex. Along its northern border with the United States, Mexico is the source of the unauthorized immigrants. At the same time, along its southern border with Guatemala, Mexico is the destination for unauthorized immigrants. When talking with its neighbor to the north, Mexicans urge understanding and sympathy for the plight of the immigrants. When talking with its neighbor to the south, Mexicans urge stronger security along the border.

Along the Mexican–U.S. border, the contrast in wealth between the two countries is apparent, even in satellite imagery. Small houses packed close together on the Mexican side face parks and open space on the American side (Figure 3-13). Along the Mexican–Guatemalan border, the Suchiate River is sometimes only ankle deep. Immigrants from other Latin American countries, especially El Salvador and Honduras, travel through Guatemala without need of a passport in order to cross into Mexico. Although a passport is needed to cross the border from Guatemala into Mexico, the Mexican government estimates that 2 million a year do so illegally. Some migrate illegally from Guatemala to Mexico for higher-paying jobs in tropical fruit plantations. For most, the ultimate destination is the United States.

Meanwhile, the millions of Mexicans living legally and illegally in the United States have constituted a powerful political and economic force back in Mexico. The Inter-American Development Bank estimated that immigrants in the United States sent $14 billion back to Mexico in 2007. Most of these remittances were used by relatives for food, clothing, and shelter, but government officials have tried to channel some of the money into development projects. ■

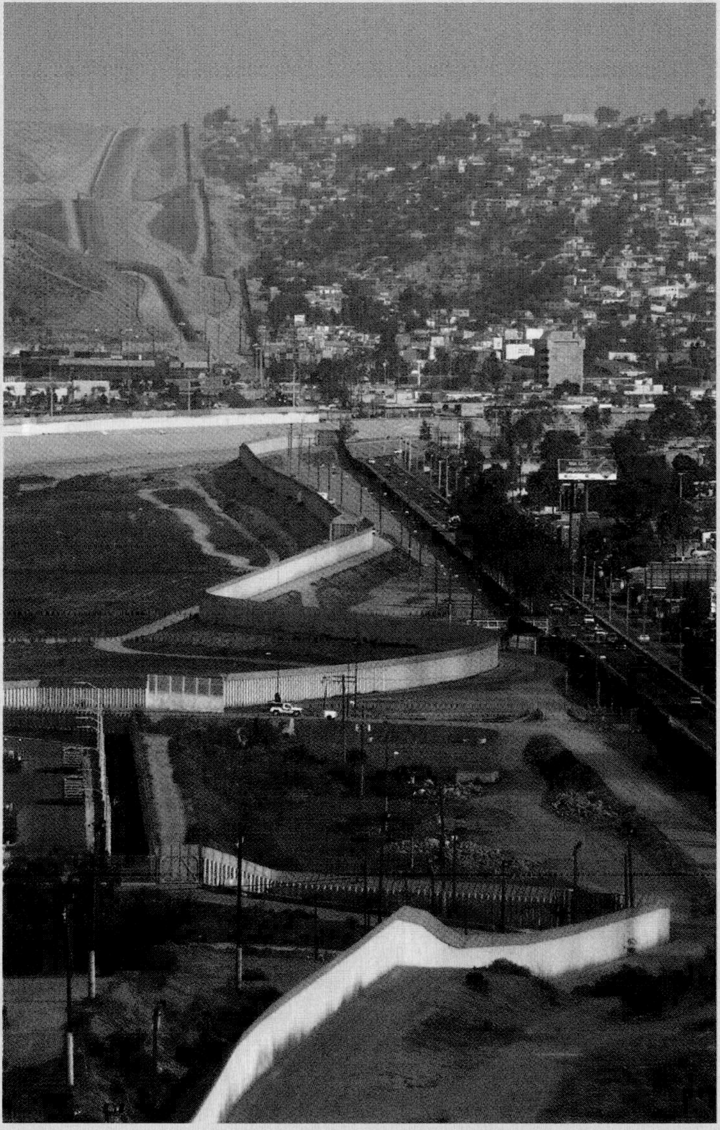

FIGURE 3-13 U.S.–Mexico border at Tijuana. Poorly constructed houses for low-income people are built adjacent to the fence on the Mexican side (right), but the U.S. side (left) is uninhabited. The irregular shape of the fence was designed to create a park for people on one side of the border to meet and talk with people on the other side. But the U.S. Department of Homeland Security claimed that the parks were being used for illegal activity, so it installed additional fences to make face-to-face meetings impossible.

Job prospects affect the states to which immigrants head. The South and West have attracted a large percentage of immigrants because the regions have had more rapid growth in jobs than the Northeast or Midwest (see Key Issue 4). In recent years, though, many immigrants—especially Mexicans—have migrated to the Midwest to take industrial jobs, such as in meatpacking and related food processing.

KEY ISSUE 3

Why Do Migrants Face Obstacles?

- **Immigration Policies of Host Countries**
- **Cultural Challenges Faced While Living in Other Countries**

The principal obstacle traditionally faced by migrants to other countries was environmental: the long, arduous, and expensive passage over land or by sea. Think of the cramped and unsanitary conditions endured by nineteenth-century immigrants to the United States who had to sail across the Atlantic or Pacific Ocean in tiny ships.

Today, the major obstacles faced by most immigrants are cultural. Motor vehicles and airplanes bring most immigrants speedily and reasonably comfortably to the United States and other countries. But once they arrive, immigrants face two major difficulties—gaining permission to enter a new country in the first place and hostile attitudes of citizens once they have entered the new country. ■

Immigration Policies of Host Countries

Countries to which immigrants wish to migrate have adopted two policies to control the arrival of foreigners seeking work. The United States uses a quota system to limit the number of foreign citizens who can migrate permanently to the country and obtain work. Other major recipients of immigrants, especially in Western Europe and the Middle East, permit guest workers to work temporarily but not stay permanently.

U.S. Quota Laws

The era of unrestricted immigration to the United States ended when Congress passed the Quota Act in 1921 and the National Origins Act in 1924. These laws established **quotas**, or maximum limits on the number of people who could immigrate to the United States from each country during a one-year period. According to the quota, for each country that had native-born persons already living in the United States, 2 percent of their number (based on the 1910 census) could immigrate each year. This limited the number of immigrants from the Eastern Hemisphere to 150,000 per year, virtually all of whom had to be from Europe. The system continued with minor modifications until the 1960s.

Quota laws were designed to ensure that most immigrants to the United States continued to be Europeans. Although Asians never accounted for more than 5 percent of immigrants during the late nineteenth and early twentieth centuries, many Americans were, nevertheless, alarmed at the prospect of millions of Asians flooding into the country, especially to states along the Pacific Coast.

Following passage of the Immigration Act of 1965, quotas for individual countries were eliminated in 1968 and replaced with hemisphere quotas. The annual number of U.S. immigrants was restricted to 170,000 from the Eastern Hemisphere and 120,000 from the Western Hemisphere. In 1978, the hemisphere quotas were replaced by a global quota of 290,000, including a maximum of 20,000 per country. The current law has a global quota of 620,000, with no more than 7 percent from one country, but numerous qualifications and exceptions can alter the limit considerably.

Because the number of applicants for admission to the United States far exceeds the quotas, Congress has set preferences. About three-fourths of the immigrants are admitted to reunify families, primarily spouses or unmarried children of people already living in the United States. The typical wait for a spouse to gain entry is currently about five years. Skilled workers and exceptionally talented professionals receive most of the remaining one-fourth of the visas. Others are admitted by lottery under a diversity category for people from countries that historically sent few people to the United States.

The quota does not apply to refugees, who are admitted if they are judged genuine refugees. Also admitted without limit are spouses, children, and parents of U.S. citizens. The number of immigrants can vary sharply from year to year, primarily because numbers in these two groups are unpredictable.

Asians have made especially good use of the priorities set by the U.S. quota laws. Many well-educated Asians enter the United States under the preference for skilled workers. Once admitted, they can bring in relatives under the family-reunification provisions of the quota. Eventually, these immigrants can bring in a wider range of other relatives from Asia, through a process of chain migration.

Some of today's immigrants to the United States and Canada are poor people pushed from their homes by economic desperation, but most are young, well-educated people lured to economically growing countries. Other countries charge that by giving preference to skilled workers, immigration policies in the United States and Europe contribute to a **brain drain**, which is a large-scale emigration by talented people. Scientists, researchers, doctors, and other professionals migrate to countries where they can make better use of their abilities.

Temporary Migration for Work

People unable to migrate permanently to a new country for employment opportunities may be allowed to migrate temporarily. Prominent forms of temporary-work migrants include guest workers in Europe and the Middle East and, historically, time-contract workers in Asia (Figure 3-14).

FIGURE 3-14 Guest workers. Turkish guest workers gather at the railway station in Munich, Germany. Most Turkish guest workers arrive in Germany by rail. The station and nearby businesses are a major gathering place for Turks.

Citizens of poor countries who obtain jobs in Western Europe and the Middle East are known as **guest workers**. In Europe, guest workers are protected by minimum-wage laws, labor union contracts, and other support programs. Foreign-born workers comprise more than one-half of the labor force in Luxembourg; one-sixth in Switzerland; and one-tenth in Austria, Belgium, and Germany. The influx of migrants has reached Western European countries that had previously experienced net out-migration. For example between 1999 and 2008, the foreign-born population rose in Spain from around ¾ million to 5¼ million, and in Ireland from ¼ million to around ½ million.

Guest workers serve a useful role in Western Europe because they take low-status and low-skilled jobs that local residents won't accept. In cities such as Berlin, Brussels, Paris, and Zurich, guest workers provide essential services, such as driving buses, collecting garbage, repairing streets, and washing dishes. Although relatively low paid by European standards, guest workers earn far more than they would at home. The economy of the guest worker's native country also gains from the arrangement. By letting their people work elsewhere, poorer countries reduce their own unemployment problems. Guest workers also help their native countries by sending a large percentage of their earnings back home to their families. The injection of foreign currency then stimulates the local economy.

Most guest workers in Europe come from North Africa, the Middle East, Eastern Europe, and Asia. Distinctive migration routes have emerged among the exporting and importing countries. Millions of Asians migrated in the nineteenth century as time-contract laborers, recruited for a fixed period to work in mines or on plantations. When their contracts expired, many would settle permanently in the new country. For example, Chinese worked on the U.S. West Coast and helped build the first railroad to span the United States, completed in 1869. More than 33 million ethnic Chinese currently live permanently in other countries, for the most part in Asia.

Distinguishing Between Economic Migrants and Refugees

It is sometimes difficult to distinguish between migrants seeking economic opportunities and refugees fleeing from government persecution. The distinction between economic migrants and refugees is important because the United States, Canada, and Western European countries treat the two groups differently. Economic migrants are generally not admitted unless they possess special skills or have a close relative already there, and even then they must compete with similar applicants from other countries. However, refugees receive special priority in admission to other countries.

Cuba, Haiti, and Vietnam have each sent around 25,000 emigrants a year to the United States in recent decades. Distinguishing between economic migrants and refugees has been especially difficult for people trying to get to the United States from these three countries.

- **Emigrants from Cuba.** The U.S. government regarded emigrants from Cuba as political refugees after the 1959 revolution that brought the Communist government of Fidel Castro to power. Under Castro's leadership, the Cuban government took control of privately owned banks, factories, and farms, and political opponents of the government were jailed. The U.S. government closed its embassy and prevented companies from buying and selling in Cuba.

  In the years immediately following the revolution, more than 600,000 Cubans were admitted to the United States. The largest number settled in southern Florida, where they became prominent in the region's economy and politics. A second flood of Cuban emigrants reached the United States in 1980, when Castro suddenly decided to permit political prisoners, criminals, and mental patients to leave the country (Figure 3-15). More than 125,000 Cubans left within a few weeks to seek political asylum in the United States, a migration stream that became known as the "Mariel

FIGURE 3-15 Mariel boatlift. When Fidel Castro suddenly permitted some Cubans to leave the country, more than 125,000 sailed across the Straits of Florida, including this freighter, *Red Diamond*, with 850 refugees on board.

boatlift," named for the port from which the Cubans were allowed to embark. Beginning in 1987, the United States agreed to permit 20,000 Cubans per year to migrate to the United States.

- **Emigrants from Haiti.** Shortly after the 1980 Mariel boatlift from Cuba, several thousand Haitians also sailed in small vessels for the United States. Claiming that they had migrated for economic advancement rather than political asylum, U.S. immigration officials would not let the Haitians aboard the boats stay in the United States.

  Under the dictatorship of Francois (Papa Doc) Duvalier (1957–1971) and his son Jean-Claude (Baby Doc) Duvalier (1971–1986), the Haitian government persecuted its political opponents at least as harshly as did the Cuban government. But the U.S. government drew a distinction between the governments of the two neighboring Caribbean countries because Castro was allied with the Soviet Union and the Duvaliers were not. Haitians brought a lawsuit against the U.S. government, arguing that if the Cubans were admitted, they should be too. The government settled the case by agreeing to admit some Haitians.

  After a 1991 coup that replaced Haiti's elected president, Jean-Bertrand Aristide, with military leaders, thousands of Haitians fled their country in small boats. Although political persecution has subsided, many Haitians still try to migrate to the United States, reinforcing the view that economic factors may always have been important in emigration from the Western Hemisphere's poorest country.

- **Emigrants from Vietnam.** The Vietnam War ended in 1975 when Communist-controlled North Vietnam captured South Vietnam's capital city of Saigon (since renamed Ho Chi Minh City). The United States, which had supported the government of South Vietnam, evacuated from Saigon several thousand people who had been closely identified with the American position during the war and who were, therefore, vulnerable to persecution after the Communist victory. Thousands of other pro-U.S. South Vietnamese who were not politically prominent enough to get space on an American evacuation helicopter tried to leave by boat.

  A second surge of Vietnamese boat people came in the late 1980s. As memories of the Vietnam War faded, officials in other countries no longer considered Vietnamese boat people as refugees. Most of the boat people were now judged economic migrants, so they were placed in detention camps monitored by the United Nations until they could be sent back to Vietnam. Vietnam remains a major source of immigrants to the United States, but the pull of economic opportunity is a greater incentive than the push of political persecution.

Cultural Challenges Faced While Living in Other Countries

For many immigrants, admission to another country does not end the challenges. Citizens of the host country may dislike the newcomers' cultural differences. More significantly, politicians exploit immigrants as scapegoats for local economic problems.

U.S. Attitudes Toward Immigrants

Americans have always regarded new arrivals with suspicion but tempered their dislike during the nineteenth century because immigrants helped to settle the frontier and extend U.S. control across the continent. European immigrants converted the forests and prairies of the vast North American interior into productive farms. By the early twentieth century, most Americans saw the frontier as closed and thought that therefore entry into the country should be closed as well.

Opposition to immigration also intensified into the twentieth century when the majority of immigrants no longer came from Northern and Western Europe. Italians, Russians, Poles, and other Southern and Eastern Europeans who poured into the United States after 1900 faced much more hostility than did British, German, and Irish immigrants a half-century earlier. A government study in 1911 reflected popular attitudes when it concluded that immigrants from Southern and Eastern Europe were racially inferior, "inclined toward violent crime," resisted assimilation, and "drove old-stock citizens out of some lines of work."

More recently, hostile citizens in California and other states have voted to deny unauthorized immigrants access to most public services, such as schools, day-care centers, and health clinics. The laws have been difficult to enforce and of dubious constitutionality, but their enactment reflects the unwillingness on the part of many Americans to help out needy immigrants. Whether children of recent immigrants should be entitled to attend school and receive social services is much debated in the United States.

Attitudes Toward Guest Workers

In Europe, many guest workers suffer from poor social conditions. The guest worker is typically a young man who arrives alone in a city. He has little money for food, housing, or entertainment because his primary objective is to send home as much money as possible. He is likely to use any surplus money for a railway ticket home for the weekend.

Far from his family and friends, the guest worker can lead a lonely life. His isolation may be heightened by unfamiliarity with the host country's language and distinctive cultural activities. Many guest workers pass their leisure time at the local railway station. There they can buy native-language newspapers, mingle with other guest workers, and meet people who have just arrived by train from home.

Both guest workers and their host countries regard the arrangement as temporary. In reality, however, many guest workers remain indefinitely, especially if they are joined by other family members. Some guest workers apply their savings to starting a grocery store, restaurant, or other small shop. These businesses can fill a need in European cities by remaining open on weekends and evenings when most locally owned establishments are closed.

Many Western Europeans dislike the guest workers and oppose government programs to improve their living conditions. Political parties that support restrictions on immigration have gained support in France, Germany, and other European countries, and attacks by local citizens on immigrants have increased. In the Middle East, petroleum-exporting countries fear that the increasing numbers of guest workers will spark political unrest and abandonment of traditional Islamic customs. After the 1991 Gulf War, Kuwaiti officials expelled hundreds of thousands of Palestinian guest workers who had sympathized with Iraq's invasion of Kuwait in 1990. To minimize long-term stays, other host countries in the Middle East force migrants to return home if they wish to marry and prevent them from returning once they have wives and children.

The severe global recession of the early twenty-first century has sharply reduced the number of guest workers and economic migrants. With high unemployment and limited job opportunities in the principal destination countries, potential migrants have much less incentive to risk the uncertainties and expenses of international migration.

KEY ISSUE 4
Why Do People Migrate Within a Country?

- **Migration Between Regions of a Country**
- **Migration Within One Region**

Internal migration for most people is less disruptive than international migration. Two main types of internal migration are interregional (between regions of a country) and intraregional (within a region). ■

Migration Between Regions of a Country

In the past, people migrated from one region of a country to another in search of better farmland. Lack of farmland pushed many people from the more densely settled regions of the country and lured them to the frontier, where land was abundant. Today, the principal type of interregional migration is from rural areas to urban areas. Most jobs, especially in services, are clustered in urban areas (see Chapter 12).

Migration Between Regions Within the United States

An especially prominent example of large-scale internal migration is the opening of the American West. Two hundred years ago, the United States consisted of a collection of settlements concentrated on the Atlantic Coast. Through mass interregional migration, the interior of the continent was settled and developed.

The U.S. Census Bureau computes the country's population center at the time of each census. The changing location of the population center graphically demonstrates the march of the American people across the North American continent over the past 200 years (Figure 3-16).

- **Colonial Settlement.** When the first U.S. census was taken, in 1790, the population center was located in the Chesapeake Bay, near Chestertown, Maryland. This location reflects the fact that virtually all colonial-era settlements were near the Atlantic Coast.

  Few colonists ventured far from coastal locations because they depended on shipping links with Europe to receive products and to export raw materials. Settlement in the interior was also hindered by an intervening obstacle, the Appalachian Mountains. The Appalachians blocked western development because of their steep slopes, thick forests, and few gaps that allowed easy passage. Hostile indigenous residents, commonly called "Indians," also retarded western settlement.

FIGURE 3-16 Changing center of population in the United States. The population center is the average location of everyone in the country, the "center of population gravity." If the United States were a flat plane placed on top of a pin, and each individual weighed the same, the population center would be the point where the population distribution causes the flat plane to balance on the pin. The center has consistently shifted westward, although the rate of movement has varied in different eras. In recent decades, the center has also started to shift southward, a reflection of recent migration to the South.

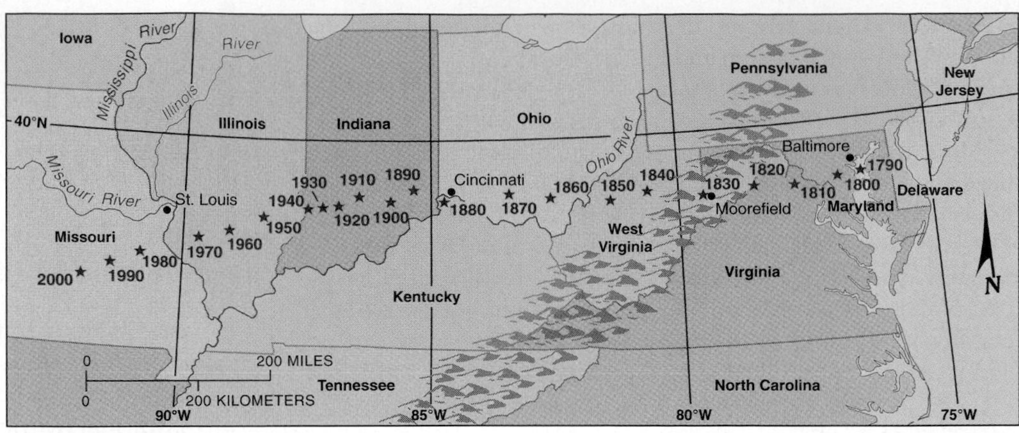

- **Early Settlement in the Interior.** Transportation improvements, especially the building of canals, helped to open the interior in the early 1800s. Most important was the Erie Canal, which enabled people to travel inexpensively by boat between New York City and the Great Lakes. In 1840, the United States had 5,352 kilometers (3,326) miles of canals, and the U.S. center of population had moved to Weston, West Virginia, 250 miles west of Chestertown.

  Encouraged by the opportunity to obtain a large amount of land at a low price, people moved into forested river valleys between the Appalachians and the Mississippi River. They cut down the trees and used the wood to build homes, barns, and fences.
- **Migration to California.** The population center shifted more rapidly during the mid-nineteenth century, reaching Greensburg, Indiana, in 1890, a 400-mile westward movement in 50 years. Rather than continuing to expand agriculture into the next available westward land, mid-nineteenth century pioneers kept going all the way to California.

  The principal pull to California was the Gold Rush beginning in the late 1840s. Mid-nineteenth century pioneers also passed over the Great Plains because of the physical environment. The region's dry climate, lack of trees, and tough grassland sod convinced early explorers such as Zebulon Pike that the region was unfit for farming, and maps at the time labeled the Great Plains as the Great American Desert.
- **Settlement of the Great Plains.** The westward movement of the U.S. population center slowed in the late nineteenth and early twentieth centuries. In 1940, the center of population was still in Indiana, only 150 miles west of its 1890 position.

The rate slowed, in part, because large-scale migration to the East Coast from Europe offset some of the migration from the East Coast to the U.S. West. Also, immigrants began to fill in the area between the 98th meridian and California that earlier generations had bypassed. Advances in agricultural technology enabled people to cultivate the Great Plains (Figure 3-17). Farmers used barbed wire to reduce dependence on wood fencing, the steel plow to cut the thick sod, and windmills and well-drilling equipment to pump more water.

The expansion of the railroads encouraged settlement of the Great Plains. The federal government gave large land grants to the railroad companies, which financed construction of their lines by selling portions to farmers. The extensive rail network then permitted settlers to

FIGURE 3-17 Great Plains settlement. Large-scale migration into the Great Plains began in the 1880s. This family was photographed on their farm in Loup Valley, Nebraska, in 1886.

transport their products to the large concentrations of customers in East Coast cities.

- **Recent Growth of the South.** The population center resumed a more vigorous migration during the late twentieth century, moving 250 miles further west between 1940 and 2000, across Illinois to central Missouri. The population center also moved southward by 75 miles between 1940 and 2000. The population center drifted southward because of net migration into southern states, especially during the last two decades of the twentieth century (Figure 3-18).

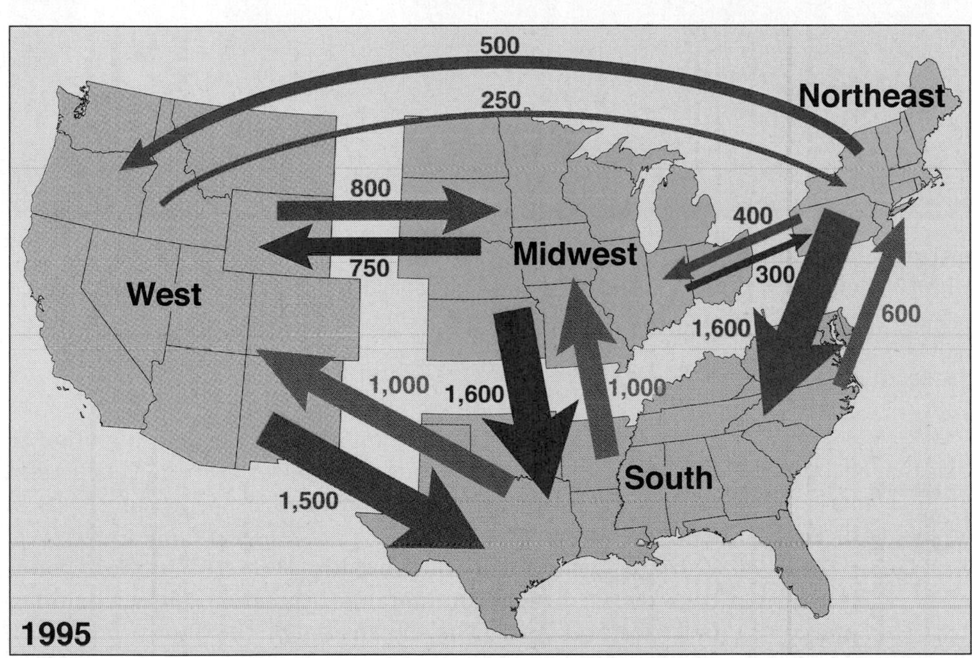

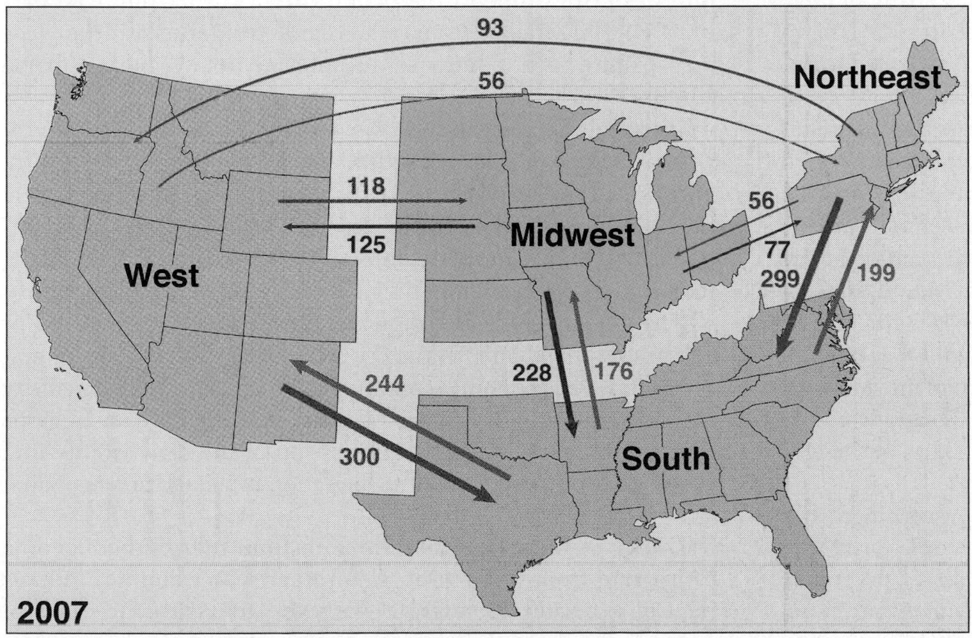

FIGURE 3-18 U.S. interregional migration. Figures show average annual migration (in thousands) in 1995 (left) and 2007 (right). Compared to 2007, the 1995 figures show much higher total interregional migration and migration into the South.

Americans migrated to the South primarily for job opportunities and environmental conditions. Americans commonly refer to the South as the "sunbelt" because of its more temperate climate and the Midwest as the "rustbelt" because of its dependency on declining manufacturing (as well as the ability of the climate to rust out cars relatively quickly).

The rapid growth of population and employment in the South has aggravated interregional antagonism. Some people in the Northeast and Midwest believe that southern states have stolen industries from them. In reality, some industries have relocated from the Northeast and Midwest, but most of the South's industrial growth comes from newly established companies.

Interregional migration has slowed considerably in the United States into the twenty-first century; net migration between each pair of regions is now close to zero. Regional differences in employment prospects have become less dramatic. With most new jobs in the service sector of the economy, jobs are expanding and contracting at similar rates around the country.

Migration Between Regions in Other Countries

As in the United States, long-distance interregional migration has been an important means of opening new regions for economic development in other large countries. Incentives have been used to stimulate migration to other regions.

RUSSIA. Interregional migration was important in developing the former Soviet Union. Soviet policy encouraged factory construction near raw materials rather than near existing population concentrations (see Chapter 11). Not enough workers lived nearby to fill all the jobs at the mines, factories, and construction sites established in these remote, resource-rich regions. To build up an adequate labor force, the Soviet government had to stimulate interregional migration.

Soviet officials were especially eager to develop Russia's Far North, which included much of Siberia, because it is rich in natural resources—fossil fuels, minerals, and forests. The Far North encompassed 45 percent of the Soviet Union's land area but contained less than

2 percent of its people. The Soviet government forced people to migrate to the Far North to construct and operate steel mills, hydroelectric power stations, mines, and other enterprises. In later years, the government encouraged, instead, voluntary migration to the Far North, including higher wages, more paid holidays, and earlier retirement.

The incentives failed to pull as many migrants to the Far North as Soviet officials desired. People were reluctant because of the region's harsh climate and remoteness from population clusters. Each year, as many as half of the people in the Far North migrated back to other regions of the country and had to be replaced by other immigrants, especially young males willing to work in the region for a short period. One method the Soviet government used was to send a brigade of young volunteers, known as *Komsomol*, during school vacations to help construct projects. An example is the Baikal-Amur Railroad, which runs for 3,145 kilometers (1,955 miles) from Taishet to Sovetskaia Gavan.

FIGURE 3-19 Brasília. Brazil's capital was moved here in 1960. Since then, thousands of immigrants have arrived in search of jobs. Many live in poor-quality housing on the edge of the city, a contrast to the carefully planned high-rises in the background.

The collapse of the Soviet Union ended policies that encouraged interregional migration. In the transition to a market-based economy, Russian government officials no longer dictate "optimal" locations for factories.

BRAZIL. Another large country, Brazil, has encouraged interregional migration. Most Brazilians live in a string of large cities near the Atlantic Coast. São Paulo and Rio de Janeiro have become two of the world's largest cities. In contrast, Brazil's tropical interior is very sparsely inhabited.

To increase the attractiveness of the interior, the government moved its capital in 1960 from Rio to a newly built city called Brasília, situated 1,000 kilometers (600 miles) from the Atlantic Coast. From above, Brasília's design resembles an airplane, with government buildings located at the center of the city and housing arranged along the "wings." Thousands of people have migrated to Brasília in search of jobs (Figure 3-19). In a country with rapid population growth, many people will migrate where they think they can find employment. Many of these workers could not afford housing in Brasília and were living instead in hastily erected shacks on the outskirts of the city.

INDONESIA. Since 1969, the Indonesian government has paid for the migration of more than 5 million people, primarily from the island of Java, where nearly two-thirds of its people live, to less populated islands. Under the government program, families receive a one-way air ticket, 2 hectares (5 acres) of land, materials to build a house, seeds and pesticides, and food—a year's worth of rice—to tide them over until the crops are ready.

EUROPE. The principal flow of interregional migration in Europe is from east and south to west and north. This pattern reflects the relatively low incomes and bleak job prospects in eastern and southern Europe.

In the twentieth century, wealthy Western European countries received many immigrants from their former colonies in Africa and Asia. The expansion of the European Union into Eastern Europe in the twenty-first century removed barriers for Bulgarians, Romanians, and residents of other former Communist countries to migrate to Western Europe (Figure 3-20).

Interregional migration flows can also be found within individual European countries. Italians migrate from the south, known as the Mezzogiorno, to the north, and Britons migrate from the north to the south. In both cases, economic conditions are stronger in the regions to which migrants are heading than in the regions where they originated.

The attractiveness of regions within Europe can change. For centuries, Ireland and Scotland were regions with net out-migration. Improved economic conditions in the late twentieth century induced a reversal of historic patterns, and both became regions of net in-migration. The deep recession of the early twenty-first century discouraged further in-migration to Ireland and Scotland.

INDIA. A number of governments limit the ability of people to migrate from one region to another. For example, Indians require a permit to migrate—or even to visit—the State of Assam in the northeastern part of the country. The restrictions, which date from the British colonial era, are designed to protect the ethnic identity of Assamese by limiting the ability of outsiders to compete for jobs and purchase land.

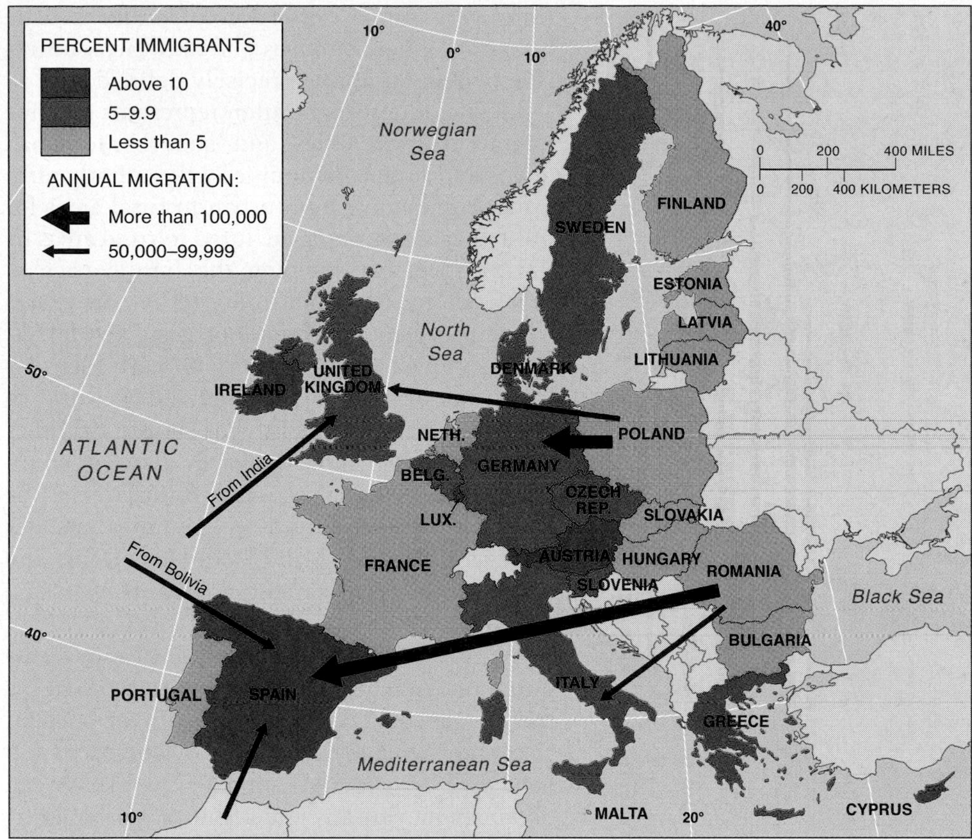

FIGURE 3-20 Migration in Europe. Migrants originate primarily in Eastern Europe and North Africa to work in the wealthier MDCs of Western Europe. Guest workers follow distinctive migration routes. The selected country may be a former colonial ruler, have a similar language, or have an agreement with the exporting country.

Because Assam is situated on the border with Bangladesh, the restrictions also limit international migration.

Migration Within One Region

Interregional migration attracts considerable attention, but far more people move within the same region, which is *intraregional* migration. Worldwide, the most prominent type of intraregional migration is from rural areas to urban areas. In the United States, the principal intraregional migration is from cities to suburbs.

Migration from Rural to Urban Areas

Migration from rural (or nonmetropolitan) areas to urban (or metropolitan) areas began in the 1800s in Europe and North America as part of the Industrial Revolution. The percentage of people living in urban areas in the United States, for example, increased from 5 percent in 1800 to 50 percent in 1920. Today, approximately three-fourths of the people in the United States and other MDCs live in urban areas.

In recent years, urbanization has diffused to LDCs, especially in Asia. The number of Asians living in urban areas increased from ½ billion in 1982 to 1¾ billion in 2007, and the number in rural areas declined from 1¾ billion to 1¼ billion. The percentage of Asians living in urban areas increased during that quarter-century from 23 percent to 42 percent. Worldwide, more than 20 million people are estimated to migrate each year from rural to urban areas.

Like interregional migrants, most people who move from rural to urban areas seek economic advancement. They are pushed from rural areas by declining opportunities in agriculture and are pulled to the cities by the prospect of work in factories or in service industries.

Migration from Urban to Suburban Areas

Most intraregional migration in MDCs is from cities out to surrounding suburbs. The population of most cities in MDCs declined during the second half of the twentieth century, and suburbs grew rapidly. Into the twenty-first century, nearly twice as many Americans migrate from central cities to suburbs each year than migrate from suburbs to central cities (Figure 3-21). Comparable patterns are found in Canada, the United Kingdom, and other Western European countries.

The major reason for the large-scale migration to the suburbs is not related to employment, as is the case with other forms of migration. For most people, migration to suburbs does not coincide with changing jobs. Instead, people are pulled by a suburban lifestyle. Suburbs offer the opportunity to live in a detached house rather than an apartment, surrounded by a private yard where children can play safely. A garage or driveway on the property guarantees space to park automobiles at no charge. Suburban schools tend to be more modern, better equipped, and safer than those in cities. Automobiles and trains enable people to live in suburbs yet have access to jobs, shops, and recreational facilities throughout the urban area (see Chapter 13).

As a result of suburbanization, the territory occupied by urban areas has rapidly expanded. To accommodate suburban growth, farms on the periphery of urban areas are converted to housing developments, where new roads, sewers, and other services must be built.

Migration from Urban to Rural Areas

MDCs witnessed a new migration trend during the late twentieth century. For the first time, more people immigrated into rural areas than emigrated out of them. Net migration from urban to rural areas is called **counterurbanization**. Counterurbanization results in part

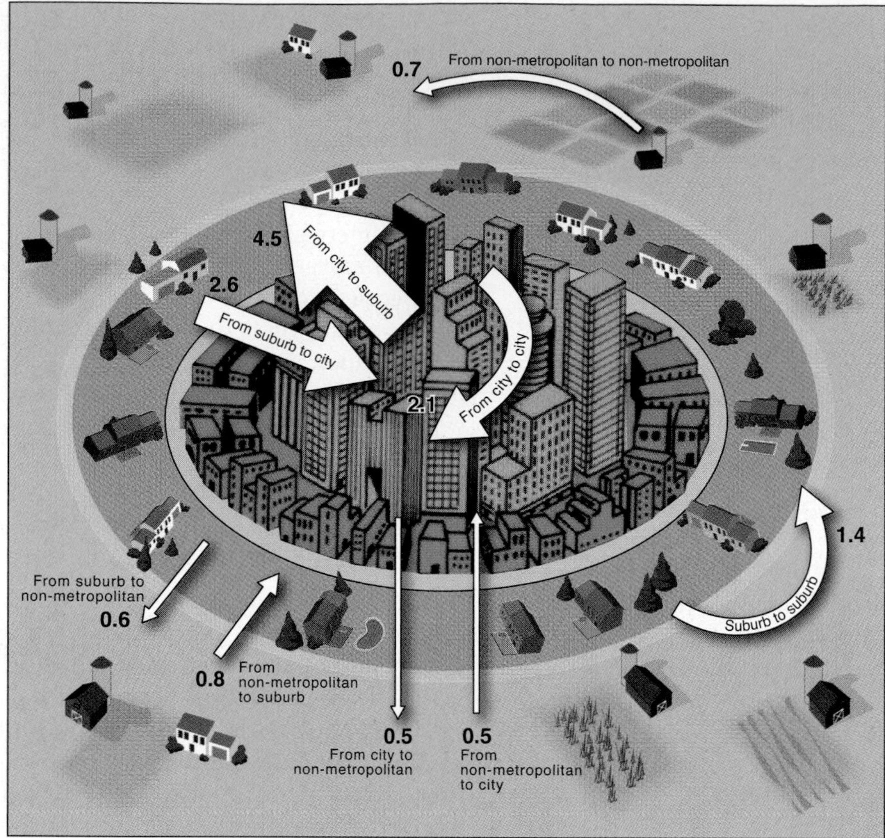

FIGURE 3-21 U.S. intraregional migration. Figures show migration (in millions) in 2007. Excluded are 8.5 million who moved elsewhere in the same city, 8.7 million who moved within the same suburbs, and 3.1 million who moved elsewhere in the same nonmetropolitan county.

from very rapid expansion of suburbs. The boundary where suburbs end and the countryside begins cannot be precisely defined.

Most counterurbanization represents genuine migration from cities and suburbs to small towns and rural communities. Like suburbanization, people move from urban to rural areas for lifestyle reasons. Some are lured to rural areas by the prospect of swapping the frantic pace of urban life for the opportunity to live on a farm where they can own horses or grow vegetables. Others move to farms but do not earn their living from agriculture; instead, they work in nearby factories, small-town shops, or other services. In the United States, evidence of counterurbanization can be seen primarily in the Rocky Mountain states. Rural counties in states such as Colorado, Idaho, Montana, Utah, and Wyoming have experienced net in-migration (Figure 3-22).

With modern communications and transportation systems, no location in an MDC is truly isolated, either economically or socially. Computers enable us to work anywhere and still have access to an international network. We can obtain money at any time from a conveniently located electronic transfer machine rather than by going to a bank building. We can select clothing from a mail-order catalog, place the order by telephone, pay by credit card, and have the desired items delivered within a few days. We can follow the fortunes of our favorite baseball teams on television anywhere in the country, thanks to satellite dishes and computer webcasts.

Roughly the same number of people now migrate from urban to rural areas as from rural to urban areas. Net in-migration into Rocky Mountain states has been offset by out-migration from the Great Plains states, where the economy has been hurt by poor agricultural conditions. Future migration trends in MDCs are unpredictable, because future economic conditions are difficult to forecast. Have these countries reached long-term equilibrium, in which approximately three-fourths of the people live in urban areas and one-fourth in rural areas? Will counterurbanization resume in the future because people prefer to live in rural areas? Is the decline of the rural economy reversible?

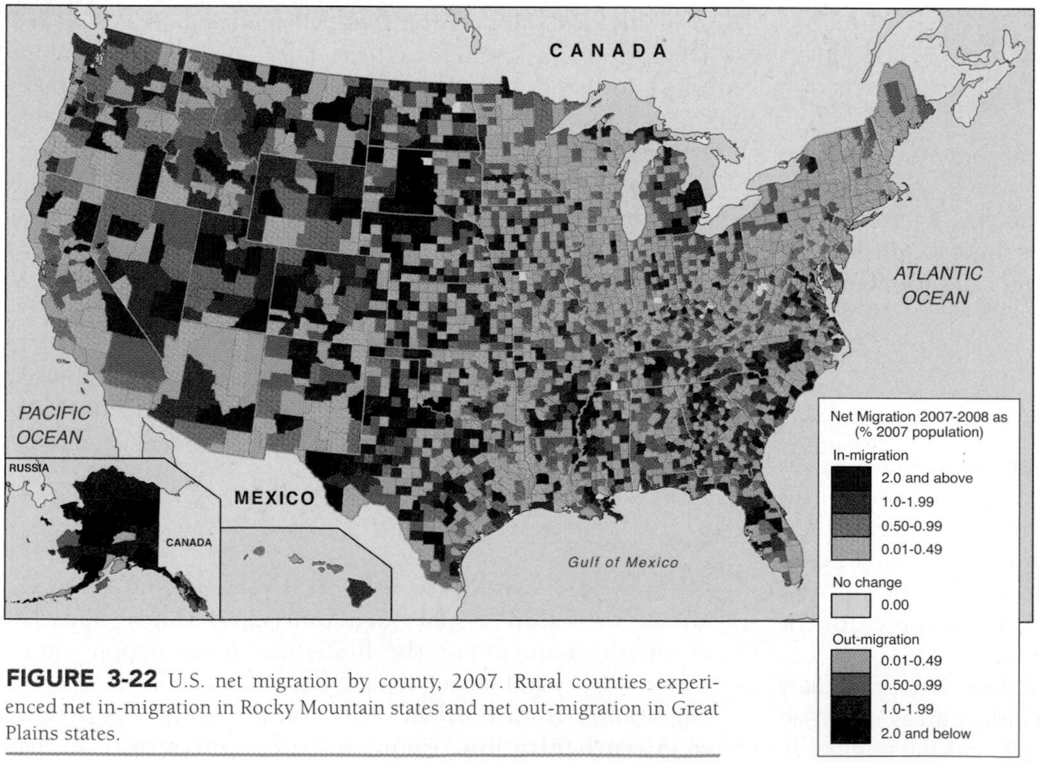

FIGURE 3-22 U.S. net migration by county, 2007. Rural counties experienced net in-migration in Rocky Mountain states and net out-migration in Great Plains states.

SUMMARY

Migration plays a major role in determining population change. For countries in stage 4 of the demographic transition, net in-migration is the only source of population growth. On the other hand, for countries in stage 2 of the demographic transition, net out-migration reduces the stress of a high NIR.

For example, Spain, in stage 4 of the demographic transition, has around 400,000 births and 400,000 deaths annually, and therefore a natural increase rate of around 0. But the country has had net in-migration of around 400,000 per year, thereby generating a total population growth of around 1 percent.

Cape Verde, described in Chapter 2 as a country in stage 2 of the demographic transition, has around 10,000 births and 2,500 deaths per year, producing a natural increase of 7,500. But Cape Verde's total annual population growth is actually only 2,500 because of net-outmigration of 5,000. Similarly, Jamaica has around 58,000 births and 18,000 deaths, but total annual growth is only 20,000 rather than 40,000, again because of net out-migration.

Here again are the key issues for Chapter 3:

1. **Why Do People Migrate?** We can group the reasons into push and pull factors. People feel compelled (pushed) to emigrate from a location for political, economic, and environmental reasons. Similarly, people are induced (pulled) to immigrate because of the political, economic, or environmental attractiveness of a new location. We can also distinguish between international and internal migration.
2. **Where Are Migrants Distributed?** On a global scale, the largest flows of migrants are from Asia to Europe and from Asia and Latin America to the United States. The United States receives by far the largest number of migrants.
3. **Why Do Migrants Face Obstacles?** Migrants have difficulty getting permission to enter other countries, and they face hostility from local citizens once they arrive. Immigration laws restrict the number who can legally enter the United States. In Europe and the Middle East, guest workers migrate temporarily to perform menial jobs.
4. **Why Do People Migrate Within a Country?** We can distinguish between interregional and intraregional migration within an individual country. Historically, interregional migration was especially important in settling the frontier of large countries such as the United States, Russia, and Brazil. The most important intraregional migration trends are from rural to urban areas within LDCs and from cities to suburbs within MDCs.

CASE STUDY REVISITED / Give Me Your Tired, Your Poor?

Most people migrate for economic reasons, involving a combination of push factors and pull factors. Poor economic conditions at home push people to consider migration, and promising job prospects pull them to other locations. For many potential migrants, push factors are as strong as ever in the early twenty-first century. High unemployment levels at home still induce thoughts of migration. The pull factor has changed dramatically. Places offering abundant job opportunities have been hard to find. As the recent recession deepened, migration declined at all scales.

On the international scale, net migration from LDCs to MDCs slowed considerably in the wake of the recession. In the Western Hemisphere, fewer Mexicans migrated to the United States, and more Mexicans returned home from the United States. Similarly, in Europe fewer Romanians migrated to Spain and more returned home from Spain. Faced with double-digit unemployment, European countries have encouraged migrants to return to their place of origin with cash payments and one-way tickets. Back home, jobs are scarce. But in grim economic times, migrants have family and friends at home who could provide food, shelter, and other support for the unemployed.

Within individual countries, the deep recession altered interregional migration. The principal impact in the United States was sharply reduced movement from one state to another (Figure 3-23). One in five Americans moved annually as recently as the 1980s; now the rate is one in nine. In China, many migrants who had moved from one province to another returned home when their jobs disappeared. Provinces that had been destinations for migrant workers reported that one-third of the migrants had left. China had an estimated 120 million interregional migrant workers, 20 million of whom were unemployed in 2009.

Also slowed by the deep recession was intraregional migration in MDCs, especially from cities to suburbs. Intraregional migrants, who move primarily for lifestyle reasons rather than jobs, found that they couldn't get loans to buy new homes nor find buyers for their old homes.

As individual countries pull out of the deep recession, they will once again become destinations for migrants from countries with economic difficulties. And within countries, migration between and within regions will also increase. In fact, an increase in internal migration will be one of the best indicators of economic recovery. ■

FIGURE 3-23 Percent of Americans moving in a year.

KEY TERMS

Brain drain (p. 92) Large-scale emigration by talented people.
Chain migration (p. 90) Migration of people to a specific location because relatives or members of the same nationality previously migrated there.
Circulation (p. 80) Short-term, repetitive, or cyclical movements that recur on a regular basis.
Counterurbanization (p. 99) Net migration from urban to rural areas in more developed countries.
Emigration (p. 80) Migration *from* a location.
Floodplain (p. 82) The area subject to flooding during a given number of years according to historical trends.
Forced migration (p. 84) Permanent movement compelled usually by cultural factors.
Guest workers (p. 93) Workers who migrate to the more developed countries of Northern and Western Europe, usually from Southern and Eastern Europe or from North Africa, in search of higher-paying jobs.
Immigration (p. 80) Migration *to* a new location.
Internal migration (p. 84) Permanent movement within a particular country.
International migration (p. 84) Permanent movement from one country to another.
Interregional migration (p. 84) Permanent movement from one region of a country to another.
Intervening obstacle (p. 83) An environmental or cultural feature of the landscape that hinders migration.
Intraregional migration (p. 84) Permanent movement within one region of a country.
Migration (p. 80) Form of relocation diffusion involving a permanent move to a new location.
Migration transition (p. 84) Change in the migration pattern in a society that results from industrialization, population growth, and other social and economic changes that also produce the demographic transition.
Mobility (p. 80) All types of movement from one location to another.
Net migration (p. 80) The difference between the level of immigration and the level of emigration.
Pull factor (p. 81) Factor that induces people to move to a new location.
Push factor (p. 81) Factor that induces people to leave old residences.
Quotas (p. 92) In reference to migration, laws that place maximum limits on the number of people who can immigrate to a country each year.
Refugees (p. 81) People who are forced to migrate from their home country and cannot return for fear of persecution because of their race, religion, nationality, membership in a social group, or political opinion.
Unauthorized immigrants (p. 90) People who enter a country without proper documents.
Voluntary migration (p. 84) Permanent movement undertaken by choice.

THINKING GEOGRAPHICALLY

1. Should preference for immigrating to the United States and Canada be given to individuals with special job skills, or should priority be given to reunification of family members? Should quotas be raised to meet increasing demand for both types of immigrants? Why or why not?
2. What is the impact of large-scale emigration on the places from which migrants depart? On balance, do these places suffer because of the loss of young, upwardly mobile workers, or do these places benefit from the draining away of surplus labor? In the communities from which migrants depart, is the quality of life improved overall through reduced pressures on local resources, or is it damaged overall through the deterioration of social structures and institutions? Explain.
3. According to the concept of chain migration, current migrants tend to follow the paths of relatives and friends who have moved earlier. Can you find evidence of chain migration in your community? Does chain migration apply primarily to the relocation of people from one community in a less developed country to one community in a more developed country, or is chain migration more applicable to movement within a more developed country? Explain.
4. Which demographic characteristics (such as rates of natural increase, crude birth, and crude death) prevail in the Middle East, which is the world region with the largest numbers of refugees? Is large-scale forced migration alleviating or exacerbating population growth in these regions? Explain.
5. At the same time that some people are migrating from LDCs to MDCs in search of employment, transnational corporations have relocated some low-skilled jobs to LDCs to take advantage of low wage rates. Should LDCs care whether their surplus workers emigrate or remain as employees of foreign companies? Why?

RESOURCES

Some recent and classic books and articles on migration geography:

Bankston, Carl L., III, Danielle Antoinette Hidalgo, and R. Kent Rasmussen, eds. *Immigration in U.S. History.* Pasadena, CA: Salem Press, 2006.

Bartram, David. *International Labor Migration: Foreign Workers and Public Policy.* New York: Palgrave Macmillan, 2005.

Fan, C. Cindy. *China on the Move: Migration, the State, and the Household.* London and New York: Routledge, 2008.

Geddes, Andrew. *The Politics of Migration and Immigration in Europe.* Thousand Oaks, CA: Sage, 2003.

Grigg, D. B. "E. G. Ravenstein and the 'Laws of Migration.'" *Journal of Historical Geography* 3 (1977): 41–54.

Herm, Anne. "Recent Migration Trends: Citizens of EU-27 Member States Become Ever More Mobile while EU Remains Attractive to non-EU Citizens." *Population and Social Conditions Statistics in Focus.* Luxembourg: Office of Official Publications of the European Communities 98, 2008.

Kosinski, Leszek A., and R. Mansell Prothero. *People on the Move.* London: Methuen, 1975.

Lee, Everett. "A Theory of Migration." *Demography* 3 (1966): 47–57.

Manning, Patrick. *Migration in World History*. New York: Routledge, 2005.

Martin, Philip, and Elizabeth Midgley. "Immigration: Shaping and Reshaping America," 2nd ed. *Population Bulletin* 61 (4). Washington, DC: Population Reference Bureau, 2006.

Martin, Philip, and Gottfried Zürcher. "Managing Migration: The Global Challenge." *Population Bulletin* 63 (1). Washington, DC: Population Reference Bureau, 2008.

Powell, John. *Encyclopedia of North American Immigration*. New York: Facts On File, 2005.

Ravenstein, Ernest George. "The Laws of Migration." *Journal of the Royal Statistical Society* 48 (1885): 167–227.

Zelinsky, Wilbur. "The Hypothesis of the Mobility Transition." *Geographical Review* 61 (1971): 219–49.

Journals featuring migration geography:

International Journal of Migration, Health and Social Care, International Migration, International Migration Review, Journal of Ethnic and Migration Studies, Journal of International Migration and Integration

Key Internet sites:

http://www.dhs.gov/ximgtn/statistics/publications/yearbook.shtm The Yearbook of Immigration Statistics is published annually by the Office of Immigration Statistics in the U.S. Department of Homeland Security. Immigration statistics are available by country of origin extending back to 1820.

http://pewhispanic.org The Pew Hispanic Center offers reports and data on unauthorized immigrants to the United States, especially from Latin America.

Log in to www.mygeoscienceplace.com for videos, interactive maps, RSS feeds, case studies, and self-study quizzes to enhance your study of Migration.

CHAPTER 4

Folk and Popular Culture

What did you do today? Presumably, your first activity was to get out of bed—for some of us, the most difficult task of the day. Shortly thereafter, you got dressed. What did you wear? That depended on both the weather (shorts or sweater) and the day's activities (suit or T-shirt).

After work or school, you returned home (house, apartment, or dorm room). You then ate dinner (pizza or salad). After studying or finishing some work, you may now have some free time during the evening for leisure

KEY ISSUES

1 Where Do Folk and Popular Cultures Originate and Diffuse?

2 Why Is Folk Culture Clustered?

3 Why Is Popular Culture Widely Distributed?

4 Why Does Globalization of Popular Culture Cause Problems?

activities (watching television, listening to music, or playing or watching sports).

This narrative may not precisely describe you, but you can recognize the day of a "typical" North American. However, the routine described and the choices mentioned in parentheses do not accurately reflect the practices of many people elsewhere in the world. People living in other locations often have extremely different social customs. Geographers ask why such differences exist and how social customs are related to the cultural landscape.

As you watch television in your single-family dwelling, wearing jeans and munching on a pizza, consider the impact if people from rural Botswana or Papua New Guinea were suddenly placed in the room. Despite striking differences in social customs across the landscape, you might be surprised to find that your visitors are familiar with most of your customs, as Earth becomes more and more a "global village." Your visitors might be inclined within a short period of time to change their customs—or to strongly condemn yours.

Playing the pan flute near Lake Titicaca, Peru.

CASE STUDY / Food Preferences

In the popular culture of twenty-first century America, food preferences seem far removed from folk traditions. Yet even hamburgers, subs, pizzas, "French" fries, and the other staples of a contemporary globalized culture all have place-specific origins.

In folk cultures, certain foods are eaten because their natural properties are perceived to enhance qualities considered desirable by the society. The Abipone Indians of Paraguay eat jaguars, stags, and bulls to make them strong, brave, and swift. The Abipones believe that consuming hens or tortoises will make them cowardly. The Ainu people in Japan avoid eating otters because they are believed to be forgetful animals and consuming them could cause loss of memory. Before becoming pregnant, the Mbum Kpau women of Chad do not eat chicken or goat. Abstaining from consumption of these animals is thought to help escape pain in childbirth and to prevent birth of a child with abnormalities. During pregnancy, they avoid meat from antelopes with twisted horns, which could cause them to bear offspring with deformities. What foods do you avoid?

Food customs are inevitably affected by the availability of products, but people do not simply eat what is available in their particular environment. Food habits are strongly influenced by cultural traditions. What is eaten establishes one's social, religious, and ethnic memberships. The surest way to identify a family's ethnic origins is to look in its kitchen.

According to the nineteenth-century geographer Vidal de la Blache, "Among the connections that tie [people] to a certain environment, one of the most tenacious is food supply; clothing and weapons are more subject to modification than the dietary regime, which experience has shown to be best suited to human needs in a given climate. ■

In Chapter 1, *culture* was shown to combine three things—values, material artifacts, and political institutions. Geographers are interested in all three components of the definition of culture. They search for where these various elements of culture are found in the world and for reasons why the observed distributions occur.

This chapter deals with the material artifacts of culture, the visible objects that a group possesses and leaves behind for the future. Chapters 5, 6, and 7 examine three important components of a group's beliefs and values, including language, religion, and ethnicity. Chapter 8 concludes the emphasis on the cultural elements of human geography by looking at the political institutions that maintain values and protect their artifacts.

Culture follows logically from the discussion of migration in Chapter 3. Two locations have similar cultural beliefs, objects, and institutions because people bring along their culture when they migrate. Differences emerge when two groups have limited interaction. In this chapter, two facets of material culture are examined.

- Material culture deriving from the survival activities of everyone's daily life—food, clothing, and shelter.
- Culture involving leisure activities—the arts and recreation.

Each cultural group provides for the activities of daily life in distinctive ways. And each cultural group has its own definition of meaningful art and stimulating recreation.

Culture can be distinguished from habit and custom.

- A **habit** is a repetitive act that a particular *individual* performs, such as wearing jeans to class every day.
- A **custom** is a repetitive act of a *group*, performed to the extent that it becomes *characteristic* of the group—American university students wear jeans to class every day.

Unlike custom, habit does not imply that the act has been adopted by most of the society's population. A custom is therefore a habit that has been widely adopted by a group of people.

A collection of social customs produces a group's material culture—jeans typically represent American informality and a badge of youth. In this chapter, *custom* may be used to denote a specific element of material culture, such as wearing jeans, whereas *culture* refers to a group's entire collection of customs.

Material culture falls into two basic categories that differ according to scale—folk and popular.

- **Folk culture** is traditionally practiced primarily by small, homogeneous groups living in isolated rural areas and may include a custom such as wearing a sarong (a loose skirt made of a long strip of cloth wrapped around the body) in Malaysia or a sari (a long cloth draped so that one end forms a skirt and the other a head or shoulder covering) in India (Figure 4-1).
- **Popular culture** is found in large, heterogeneous societies that share certain habits (such as wearing jeans) despite differences in other personal characteristics. The ***scale*** of territory covered by a folk culture is typically much smaller than that covered by a popular culture.

Geographers focus on two aspects of ***where*** folk and popular cultures are located in ***space***. First, each cultural activity, like wearing jeans, has a distinctive spatial distribution. Geographers study a particular social custom's origin, its diffusion, and its integration with other social characteristics.

Second, geographers study the relation between material culture and the physical environment. Each cultural group takes particular elements from the environment into its

FIGURE 4-1 Vietnamese folk songs. Singers perform Quan Ho folk songs as part of the annual Lim Festival.

culture and in turn constructs landscapes (what geographers call "built environments") that modify nature in distinctive ways.

Geographers observe that popular culture has a more widespread distribution than folk culture. The reason *why* the distributions are different is interaction, or lack of it. A group develops distinctive customs from experiencing local social and physical conditions in a *place* that is isolated from other groups.

Even groups living in proximity may generate a variety of folk customs in a limited geographic area, because of limited communication. Landscapes dominated by a collection of folk customs change relatively little over time. In contrast, popular culture is based on rapid simultaneous global *connections* through communications systems, transportation networks, and other modern technology. Rapid diffusion facilitates frequent changes in popular customs. Thus, folk culture is more likely to vary from place to place at a given time, whereas popular culture is more likely to vary from time to time at a given place.

In Earth's *globalization*, popular culture is becoming more dominant, threatening the survival of unique folk cultures. These folk customs—along with language, religion, and ethnicity—provide a unique identity to each group of people who occupy a specific *region* of Earth's surface. The disappearance of local folk customs reduces *local diversity* in the world and the intellectual stimulation that arises from differences in backgrounds.

The dominance of popular culture can also threaten the quality of the environment. Folk culture derived from local natural elements may be more sensitive to the protection and enhancement of the environment. Popular culture is less likely to reflect the diversity of local physical conditions and is more likely to modify the environment in accordance with global values.

KEY ISSUE 1
Where Do Folk and Popular Cultures Originate and Diffuse?

- **Origin of Folk and Popular Cultures**
- **Diffusion of Folk and Popular Cultures**

Each social custom has a unique spatial distribution, but in general, distribution is more extensive for popular culture than for folk culture. Two basic factors help explain the spatial differences between popular and folk cultures—the process of origin and the pattern of diffusion. ■

Origin of Folk and Popular Cultures

A social custom originates at a hearth, a center of innovation. Folk customs often have anonymous hearths, originating from anonymous sources, at unknown dates, through unidentified originators. They may also have multiple hearths, originating independently in isolated locations.

In contrast to folk customs, popular culture is most often a product of MDCs, especially in North America, Western Europe, and Japan. Popular music and fast food are good examples. They arise from a combination of advances in industrial technology and increased leisure time. Industrial technology permits the uniform reproduction of objects in large quantities (CDs, T-shirts, pizzas). Many of these objects help people enjoy leisure time, which has increased as a result of the widespread change for the labor force from predominantly agricultural work to predominantly service and manufacturing jobs.

Origin of Folk Music

Music exemplifies the differences in the origins of folk and popular culture. Folk songs tell a story or convey information about daily activities such as farming, life-cycle events (birth, death, and marriage), or mysterious events such as storms and earthquakes.

In Vietnam, where most people are subsistence farmers, information about agricultural technology is conveyed through folk songs. For example, the following folk song provides advice about the difference between seeds planted in summer and seeds planted in winter:

Ma chiêm ba tháng không già
Ma mùa tháng rưỡi ắt la'không non[1]

This song can be translated as follows:

While seedlings for the summer crop are not old when they are three months of age,
Seedlings for the winter crop are certainly not young when they are one-and-a-half months old.

The song hardly sounds lyrical to a Western ear. But when English-language folk songs appear in cold print, similar themes emerge, even if the specific information conveyed about the environment differs.

[1]From John Blacking and Joann W. Kealiinohomoku, eds., The Performing Arts: Music and Dance (The Hague: Mouton, 1979), 144. Reprinted by permission of the publisher.

According to a Chinese legend, music was invented in 2697 BC when the Emperor Huang Ti sent Ling Lun to cut bamboo poles that would produce a sound matching the call of the phoenix bird. In reality, folk songs are usually composed anonymously and transmitted orally. A song may be modified from one generation to the next as conditions change, but the content is most often derived from events in daily life that are familiar to the majority of the people.

Origin of Popular Music

In contrast to folk music, popular music is written by specific individuals for the purpose of being sold to a large number of people. It displays a high degree of technical skill and is frequently capable of being performed only in a studio with electronic equipment.

Popular music as we know it today originated around 1900. At that time, the main popular musical entertainment in the United States and Western Europe was the variety show, called the *music hall* in the United Kingdom and *vaudeville* in the United States. To provide songs for music halls and vaudeville, a music industry was developed in a district of New York that became known as Tin Pan Alley.

Tin Pan Alley was located along 28th Street between Fifth Avenue and Sixth Avenue (now Avenue of the Americas). It later moved uptown to Broadway and 32nd Street and then again along Broadway between 42nd and 50th streets. The district was home to songwriters, music publishers, orchestrators, and arrangers. The name Tin Pan Alley derived from the sound of pianos being furiously pounded by people called song pluggers, who were demonstrating tunes to publishers. Companies in Tin Pan Alley originally tried to sell as many printed songsheets as possible, although sales of recordings ultimately became the most important measure of success. After World War II, Tin Pan Alley disappeared as recorded music became more important than printed songsheets.

The diffusion of American popular music worldwide began in earnest during World War II, when the Armed Forces Radio Network broadcast music to American soldiers and to citizens of countries where American forces were stationed or fighting. English became the international language for popular music. Today, popular musicians in Japan, Poland, Russia, and other countries often write and perform in English, even though few people in their audiences understand the language (Figure 4-2).

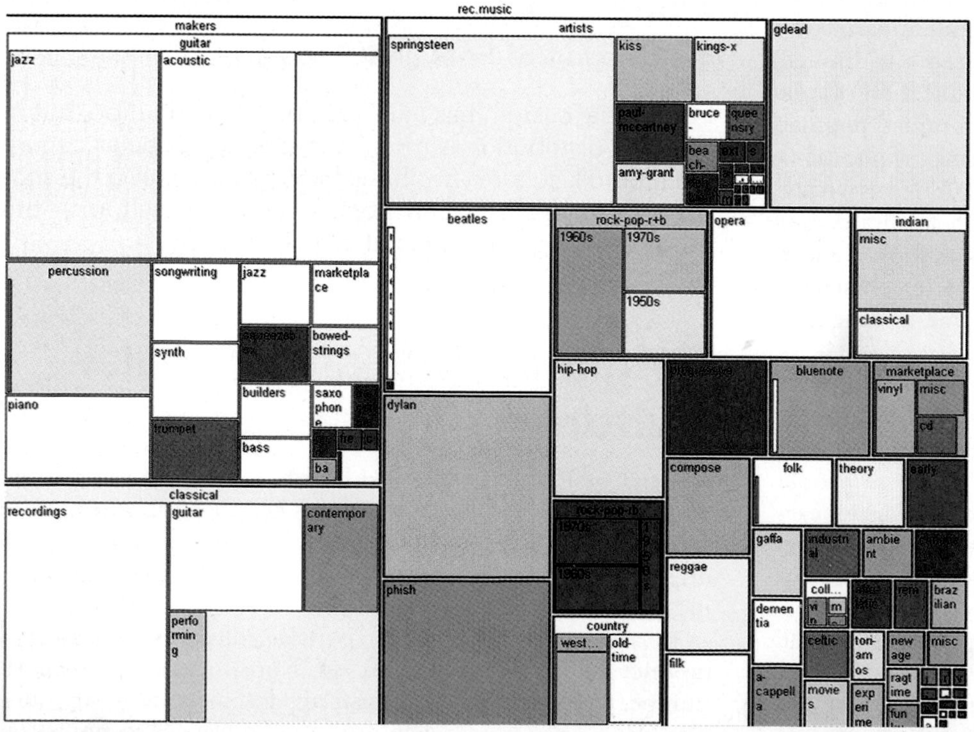

FIGURE 4-2 Popular music "map." This "map," prepared by Marc Smith and Andrew Fiore, shows the hierarchy of popularity of artists and types of music as reflected in the rec.music newsgroup (accessed at http://groups.google.com/group/rec.music.info).

Hip-hop is a more recent form of popular music that also originated in New York (Figure 4-3). Whereas the music industry of Tin Pan Alley originated in Manhattan office buildings, hip-hop originated in the late 1970s in the South Bronx, a neighborhood predominantly

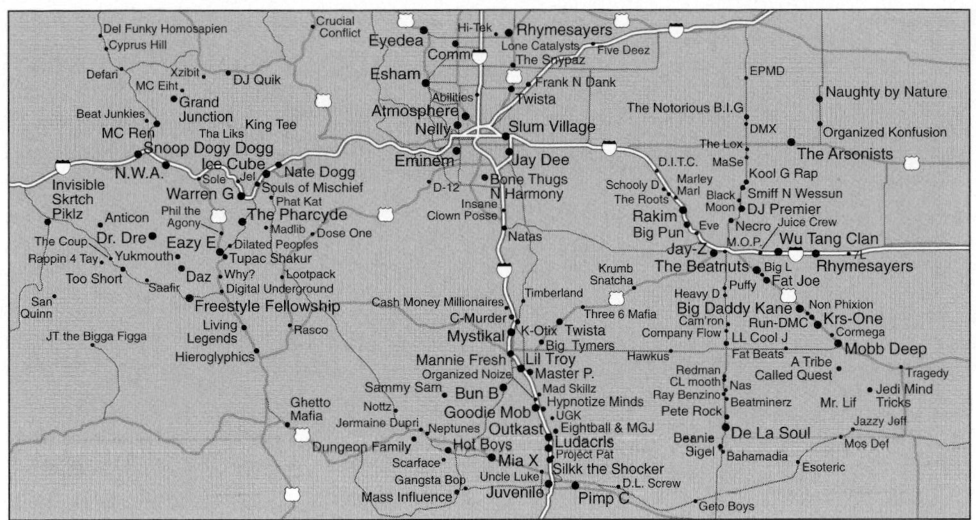

FIGURE 4-3 Hip-hop map. The fictional "map" attempts to place prominent hip-hop performers in proximity to similar performers as well as in the region of the country (Northeast, South, Midwest, West, inner city, suburbs) where they performed or drew inspiration.

populated by low-income African American and Puerto Rican people (a changeover from its predominant population of middle-class white people of European origin). Rappers in other low-income New York City neighborhoods of Queens, Brooklyn, and Harlem adopted the style with local twists—"thug" rap in Queens and clever lines in Brooklyn. Hip-hop remained predominantly a New York phenomenon until the late 1980s, when it spread to Oakland and Atlanta and then to other large cities in the South, Midwest, and West.

Hip-hop demonstrates well the interplay between globalization and local diversity that is a prominent theme of this book. On the one hand, hip-hop is a return to a very local form of music expression rather than a form that is studio manufactured. Lyrics make local references and represent a distinctive hometown scene. The KRS-One song "The Bridge Is Over," for example, was a slam by a South Bronx rapper against Queens (located on the other side of the bridge from the Bronx). At the same time, hip-hop has diffused rapidly around the world through instruments of globalization: The music is broadcast online and sold through Web marketing. Artists are expressing a sense of a specific place across the boundless space of the Internet.

Diffusion of Folk and Popular Cultures

The broadcasting of American popular music on Armed Forces Radio during the 1940s and online today illustrates the difference in diffusion of folk and popular cultures. The spread of popular culture typically follows the process of hierarchical diffusion from hearths or nodes of innovation.

In the United States, prominent nodes of innovation for popular culture include Hollywood, California, for the film industry and Madison Avenue in New York City for advertising agencies. Popular culture diffuses rapidly and extensively through the use of modern communications and transportation.

In contrast, folk culture is transmitted from one location to another more slowly and on a smaller scale, primarily through migration rather than electronic communication. One reason why hip-hop music is classified as popular rather than folk music is that it diffuses primarily through electronics. In contrast, the spread of folk culture occurs through relocation diffusion, the spread of a characteristic through migration.

The Amish: Relocation Diffusion of Folk Culture

Amish customs illustrate how relocation diffusion distributes folk culture. Although the Amish number only about one-quarter million, their folk culture remains visible on the landscape in at least 19 states (Figure 4-4). Shunning mechanical and electrical power, the Amish still travel by horse and buggy and continue to use hand tools for farming. The Amish have distinctive clothing, farming, religious practices, and other customs.

The distribution of Amish folk culture across a major portion of the U.S. landscape is explained by examining the diffusion of their culture through migration. In the 1600s, a Swiss Mennonite bishop named Jakob Ammann gathered a group of followers who became known as the Amish. The Amish originated in Bern, Switzerland; Alsace in northeastern France; and the Palatinate region of southwestern Germany. They migrated to other portions of northwestern Europe in the 1700s, primarily for religious freedom. In Europe, the Amish did not develop distinctive language, clothing, or farming practices and gradually merged with various Mennonite church groups.

Several hundred Amish families migrated to North America in two waves. The first group, primarily from Bern and the Palatinate, settled in Pennsylvania in the early 1700s, enticed by William Penn's offer of low-priced land. Because of lower land prices, the second group, from Alsace, settled in Ohio, Illinois, and Iowa in the United States and Ontario, Canada, in the early 1800s. From these core areas, groups of Amish migrated to other locations where inexpensive land was available.

Living in rural and frontier settlements relatively isolated from other groups, Amish communities retained their traditional customs, even as other European immigrants to the United States adopted new ones. We can observe Amish customs on the landscape in such diverse areas as southeastern Pennsylvania, northeastern Ohio, and east-central Iowa. These communities are relatively isolated from each other but share cultural traditions distinct from those of other Americans.

Amish folk culture continues to diffuse slowly through interregional migration within the United States. In recent years, a number of Amish families have sold their farms in Lancaster County, Pennsylvania—the oldest and at one time largest Amish community in the United States—and migrated to Christian and Todd counties in southwestern Kentucky. According to Amish tradition, every son is given a farm when he is an adult,

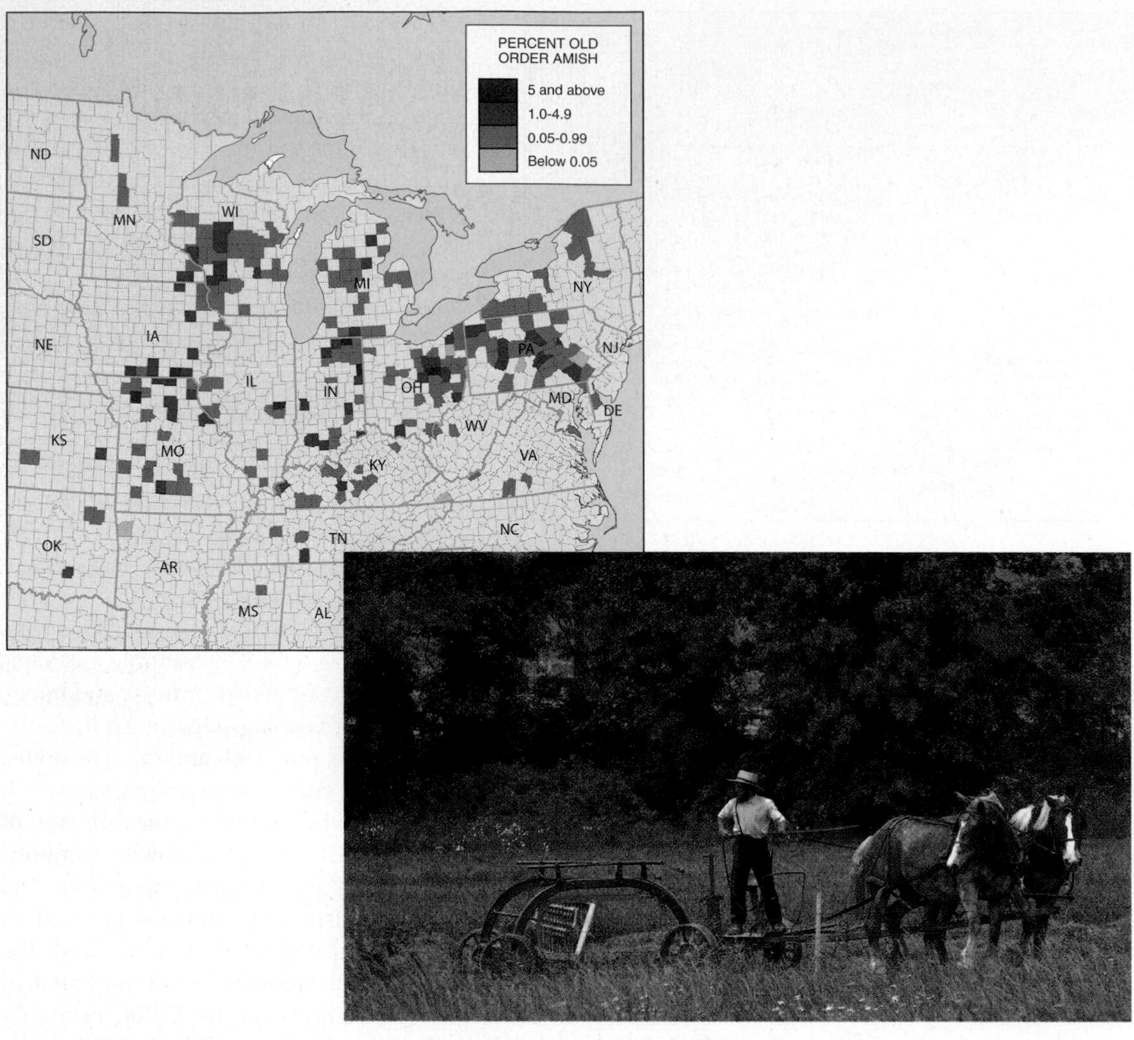

FIGURE 4-4 Distribution of Amish. Amish settlements are distributed throughout the northeastern United States. Amish farmers minimize the use of mechanical devices.

but land suitable for farming is expensive and hard to find in Lancaster County because of its proximity to growing metropolitan areas. With the average price of farmland in southwestern Kentucky less than one-fifth that in Lancaster County, an Amish family can sell its farm in Pennsylvania and acquire enough land in Kentucky to provide adequate farmland for all their sons. Amish families are also migrating from Lancaster County to escape the influx of tourists who come from the nearby metropolitan areas to gawk at the distinctive folk culture.

Sports: Hierarchical Diffusion of Popular Culture

In contrast with the diffusion of folk customs, organized sports provide examples of how popular culture is diffused. Many sports originated as isolated folk customs and were diffused like other folk culture, through the migration of individuals. The contemporary diffusion of organized sports, however, displays the characteristics of popular culture.

FOLK CULTURE ORIGIN OF SOCCER. Soccer (called *football* outside North America) is the world's most popular sport. Its origin is obscure. The earliest documented contest took place in England in the eleventh century. According to football historians, after the Danish invasion of England between 1018 and 1042, workers excavating a building site encountered a Danish soldier's head, which they began to kick. "Kick the Dane's head" was imitated by boys, one of whom got the idea of using an inflated cow bladder.

Early football games resembled mob scenes. A large number of people from two villages would gather to kick the ball. The winning side was the one that kicked the ball into the center of the rival village. In the twelfth century, the game—by then commonly called football—was confined to smaller vacant areas, and the rules became standardized. Because football disrupted village life, King Henry II banned the game from England in the late twelfth century. It was not legalized again until 1603 by King James I. At this point, football was an English folk custom rather than a global popular custom.

GLOBALIZATION OF SOCCER. The transformation of football from an English folk custom to global popular culture began in the 1800s. Football and other recreation clubs were founded in Britain, frequently by churches, to provide factory workers with organized recreation during leisure hours. Sport became a subject that was taught in school.

Increasing leisure time permitted people not only to view sporting events but also to participate in them. With higher incomes, spectators paid to see first-class events. To meet public demand, football clubs began to hire professional players. Several British football clubs formed an association in 1863 to standardize the rules and to organize professional leagues. Organization of the sport into a formal structure in Great Britain marks the transition of football from folk to popular culture.

The word *soccer* originated after 1863, when supporters of the game formed the Football Association. *Association* was shortened to *assoc,* which ultimately became twisted around into the word *soccer.* The terms *soccer* and *association football* also helped to distinguish the game from rugby football, which permits both kicking and carrying of the ball. Rugby originated in 1823, when

FIGURE 4-5 Iroquois lacrosse. Iroquois Nationals reached the finals of the 2007 World Indoor Lacrosse Championships, but lost to Canada in overtime. Canada forced overtime when Gavin Prout, wearing number 9, scored the tying goal with 3 seconds to play.

a football player at Rugby School (in Rugby, England) picked up the ball and ran with it.

Beginning in the late 1800s, the British exported association football around the world, first to continental Europe and then to other countries. Football was first played in continental Europe in the late 1870s by Dutch students who had been in Britain. The game was diffused to other countries through contact with English players. For example, football went to Spain via English engineers working in Bilbao in 1893 and was quickly adopted by local miners. British citizens further diffused the game throughout the worldwide British Empire. In the twentieth century, soccer, like other sports, was further diffused by new communication systems, especially radio and television.

SPORTS IN POPULAR CULTURE. Each country has its own preferred sports. Cricket is popular primarily in Britain and former British colonies. Ice hockey prevails, logically, in colder climates, especially in Canada, Northern Europe, and Russia. The most popular sports in China are martial arts, known as *wushu,* including archery, fencing, wrestling, and boxing. Baseball, once confined to North America, became popular in Japan after it was introduced by American soldiers who occupied the country after World War II.

Lacrosse has fostered cultural identity among the Iroquois Confederation of Six Nations (Cayugas, Mohawks, Oneidas, Onondagas, Senecas, and Tuscaroras) who live in the northeastern United States and southeastern Canada (Figure 4-5). As early as 1636, European explorers observed the Iroquois playing lacrosse, known in their language as *guhchigwaha,* which means "bump hips." European colonists in Canada picked up the game from the Iroquois and diffused it to a handful of U.S. communities, especially in Maryland, upstate New York, and Long Island. The name *lacrosse* derived from the French words *la crosse,* for a bishop's crosier or staff, which has a similar shape to the lacrosse stick.

In recent years, the Federation of International Lacrosse has invited the Iroquois National team to participate in world championships, along with teams from the United States, Canada, and other countries. Although the Iroquois have not won, they have had the satisfaction of hearing their national anthem played and seeing their flag fly alongside those of the other participants.

Despite the diversity in distribution of sports across Earth's surface and the anonymous origin of some games, organized spectator sports today are part of popular culture. The common element in professional sports is the willingness of people throughout the world to pay for the privilege of viewing, in person or on TV, events played by professional athletes.

KEY ISSUE 2
Why Is Folk Culture Clustered?

- **Influence of the Physical Environment**
- **Isolation Promotes Cultural Diversity**

Folk culture typically has unknown or multiple origins among groups living in relative isolation. Folk culture diffuses slowly to other locations through the process of migration. A combination of physical and cultural factors influences the distinctive distributions of folk culture. ■

Influence of the Physical Environment

Recall from Chapter 1 that a century ago environmental determinists theorized how processes in the environment caused social customs. Most contemporary geographers reject environmental determinism. Nonetheless, the physical environment does influence human actions, especially in folk culture.

Folk societies are particularly responsive to the environment because of their limited technology and the prevailing agricultural economy. People living in folk cultures are likely to be farmers growing their own food, using hand tools and animal power.

Customs such as provision of food, clothing, and shelter are clearly influenced by the prevailing climate, soil, and vegetation. With regard to clothing, for example, residents of arctic climates may wear fur-lined boots, which protect against the cold, and snowshoes, with which to walk on soft, deep snow without sinking in. People living in warm and humid climates may not need any footwear if heavy rainfall and time spent in water discourage such use. The custom in the Netherlands of wearing wooden shoes may appear quaint, but it actually derives from environmental conditions. Dutch farmers wear the wooden shoes, which are waterproof, as they work in fields that often are extremely wet because much of the Netherlands is below sea level.

Yet folk culture may ignore the environment. Not all arctic residents wear snowshoes, nor do all people in wet temperate climates wear wooden shoes. Geographers observe that broad differences in folk culture arise in part from physical conditions and that these conditions produce varied customs.

More than clothing, the other two material necessities of daily life—food and shelter—demonstrate the influence of the environment on the development of unique folk culture. Different folk societies prefer different foods and styles of house construction.

Food Preferences and the Environment

Folk food habits are embedded especially strongly in the environment. Humans eat mostly plants and animals—living things that spring from the soil and water of a region. Inhabitants of a region must consider the soil, climate, terrain, vegetation, and other characteristics of the environment in deciding to produce particular foods.

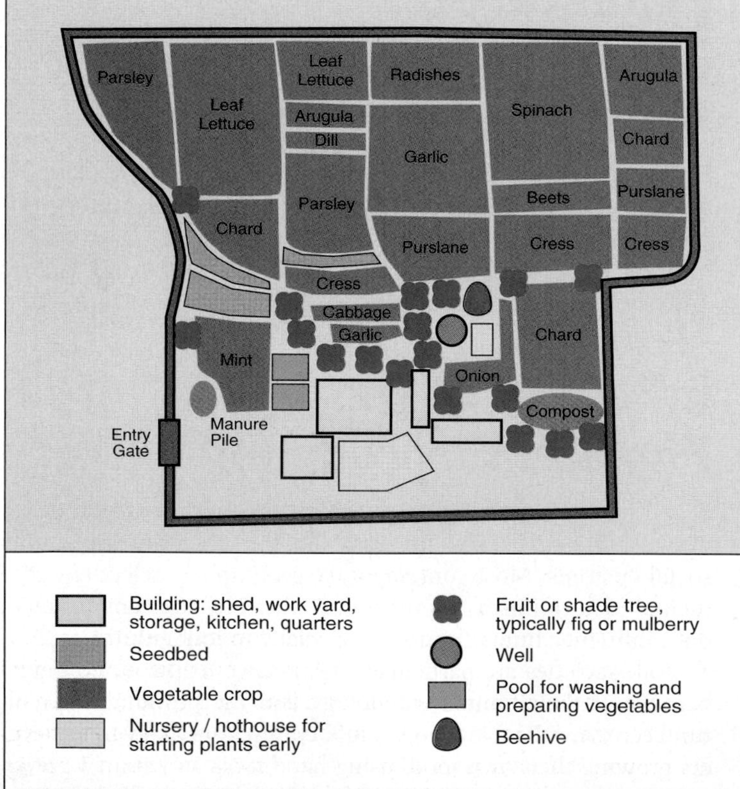

FIGURE 4-6 Istanbul vegetable garden. Geographer Paul Kaldjian sketched a typical bostan, a traditional vegetable garden in the center of Istanbul, Turkey. Bostans provide residents of the large city of Istanbul with a source of fresh vegetables.

Bostans, which are small gardens inside Istanbul, Turkey, have been supplying the city with fresh produce for hundreds of years (Figure 4-6). According to geographer Paul Kaldjian, Istanbul has around 1,000 bostans, run primarily by immigrants from Cide, a rural village in Turkey's Kastamonu province. Bostan farmers are able to maximize yields from their small plots of land (typically 1 hectare) through what Kaldjian calls clever and efficient manipulation of space, season, and resources. Fifteen to twenty different types of vegetables are planted at different times of the year, and the choice is varied from year to year, in order to reduce the risk of damage from poor weather. Most of the work is done by older men, who prepare beds for planting, sow, irrigate, and operate motorized equipment, according to Kaldjian. Women weed, and both men and women harvest.

People adapt their food preferences to conditions in the environment. In Asia, rice is grown in milder, moister regions; wheat thrives in colder, drier regions. In Europe, traditional preferences for quick-frying foods in Italy resulted in part from fuel shortages. In Northern Europe, an abundant wood supply encouraged the slow stewing and roasting of foods over fires, which also provided home heat in the colder climate.

Soybeans, an excellent source of protein, are widely grown in Asia. In the raw state they are toxic and indigestible. Lengthy cooking renders them edible, but fuel is scarce in Asia. Asians have adapted to this environmental challenge by deriving foods from soybeans that do not require extensive cooking. These include bean sprouts (germinated seeds), soy sauce (fermented soybeans), and bean curd (steamed soybeans).

According to many folk customs, everything in nature carries a signature, or distinctive characteristic, based on its appearance and natural properties. Consequently, people may desire or avoid certain foods in response to perceived beneficial or harmful natural traits.

People refuse to eat particular plants or animals that are thought to embody negative forces in the environment. Such a restriction on behavior imposed by social custom is a **taboo**. Other social customs, such as sexual practices, carry prohibitions, but taboos are especially strong in the area of food. Some folk cultures may establish food taboos because of concern for the natural environment. These taboos may help to protect endangered animals or to conserve scarce natural resources. To preserve scarce animal species, only a few high-ranking people in some tropical regions are permitted to hunt, whereas the majority cultivate crops.

Relatively well-known taboos against consumption of certain foods can be found in the Bible. The ancient Hebrews were prohibited from eating a wide variety of foods, including animals that do not chew their cud or that have cloven feet and fish lacking fins or scales (Figure 4-7). These taboos arose partially from concern for the environment by the Hebrews, who lived as pastoral nomads in lands bordering the eastern Mediterranean. The pig, for example, is prohibited in part because it is more suited to sedentary farming than pastoral nomadism and in part because its meat spoils relatively quickly in hot climates, such as the Mediterranean. These biblical taboos were developed through oral tradition and by rabbis into the kosher laws observed today by some Jews.

FIGURE 4-7 Kosher McDonald's. The restaurant is located in Tel Aviv, Israel, to serve Jews who keep the kosher dietary laws. The sign says "McDonald's" in Hebrew.

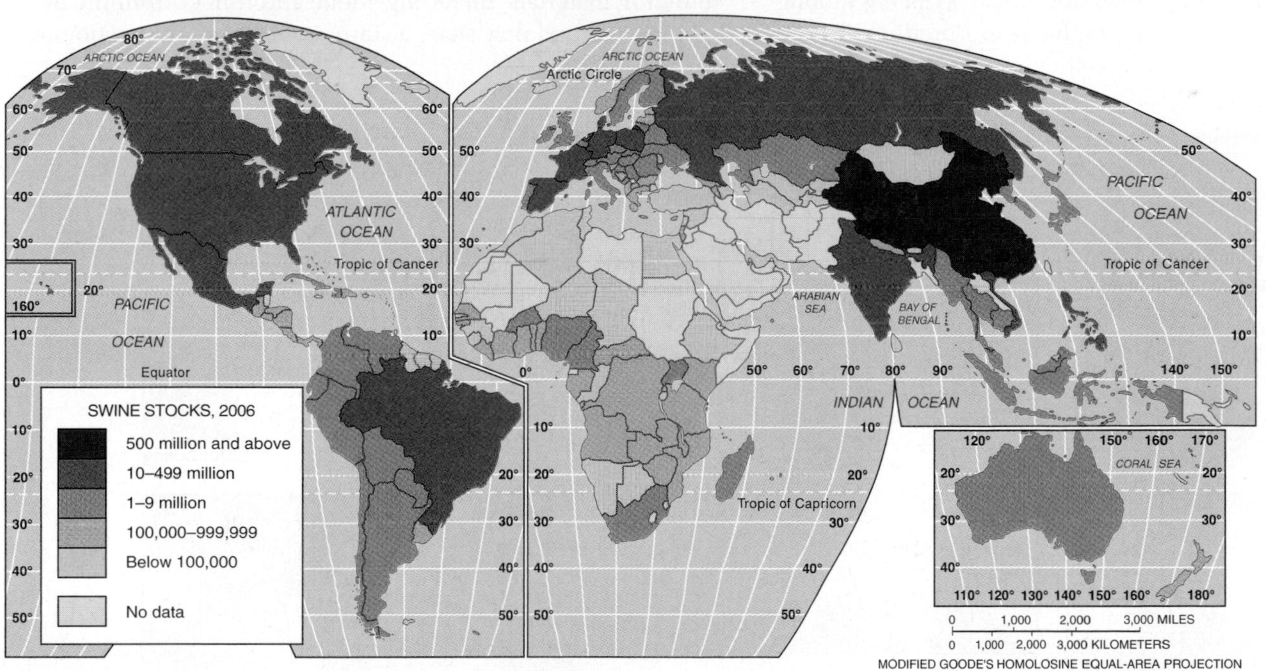

FIGURE 4-8 Swine stock. The number of swine produced in different parts of the world is influenced to a considerable extent by religious taboos against consuming pork. Swine are scarce in predominantly Muslim regions, such as northern Africa and southwestern Asia. China has more than one-half of the world's swine stock.

Similarly, Muslims embrace the taboo against pork, because pigs are unsuited for the dry lands of the Arabian Peninsula (Figure 4-8). Pigs would compete with humans for food and water without offering compensating benefits, such as being able to pull a plow, carry loads, or provide milk and wool. Widespread raising of pigs would be an ecological disaster in Islam's hearth.

Hindu taboos against consuming cows can also be partly explained by environmental reasons. Cows are the source of oxen (castrated male bovine), the traditional choice for pulling plows as well as carts. A large supply of oxen must be maintained in India because every field has to be plowed at approximately the same time—when the monsoon rains arrive. Religious sanctions have kept India's cow population large as a form of insurance against the loss of oxen and increasing population.

But the taboo against consumption of meat among many people, including Muslims, Hindus, and Jews, cannot be explained primarily by environmental factors. Social values must influence the choice of diet, because people in similar climates and with similar levels of income consume different foods. The biblical food taboos were established in part to set the Hebrew people apart from others. That Christians ignore the biblical food injunctions reflects their desire to distinguish themselves from Jews. Furthermore, as a universalizing religion, Christianity was less tied to taboos that originated in the Middle East (see Chapter 6).

The contribution of a location's distinctive physical features to the way food tastes is known by the French term **terroir**. The word comes from the same root as *terre* (French word for land or earth), but terroir does not translate precisely into English; it has a similar meaning to the English expressions "grounded" or "sense of place." Terroir is the sum of the effects of the local environment on a particular food item. The term is frequently used to refer to the combination of soil, climate, and other physical features that contribute to the distinctive taste of a wine.

Folk Housing and the Environment

French geographer Jean Brunhes, a major contributor to the cultural landscape tradition, views the house as being among the essential facts of human geography. It is a product of both cultural tradition and natural conditions. American cultural geographer Fred Kniffen considered the house to be a good reflection of cultural heritage, current fashion, functional needs, and the impact of environment.

The type of building materials used to construct folk houses is influenced partly by the resources available in the environment. The two most common building materials in the world are wood and brick; stone, grass, sod, and skins are also used. If available, wood is generally preferred for house construction because it is easy to build with it. In the past, pioneers who settled in forested regions built log cabins for themselves. In hot, dry climates—such as the U.S. Southwest, Mexico, northern China, and parts of the Middle East—bricks are made by baking wet mud in the sun. Stone is used to build houses in parts of Europe and South America and as decoration on the outside of brick or wood houses in other countries.

Even in areas that share similar climates and available building materials, folk housing can vary because of minor differences in environmental features. For example, R. W. McColl compared house types in four villages situated in the dry lands of northern and western China (Figure 4-9). All use similar building materials, including adobe and timber from the desert poplar tree, and they share a similar objective—protection from

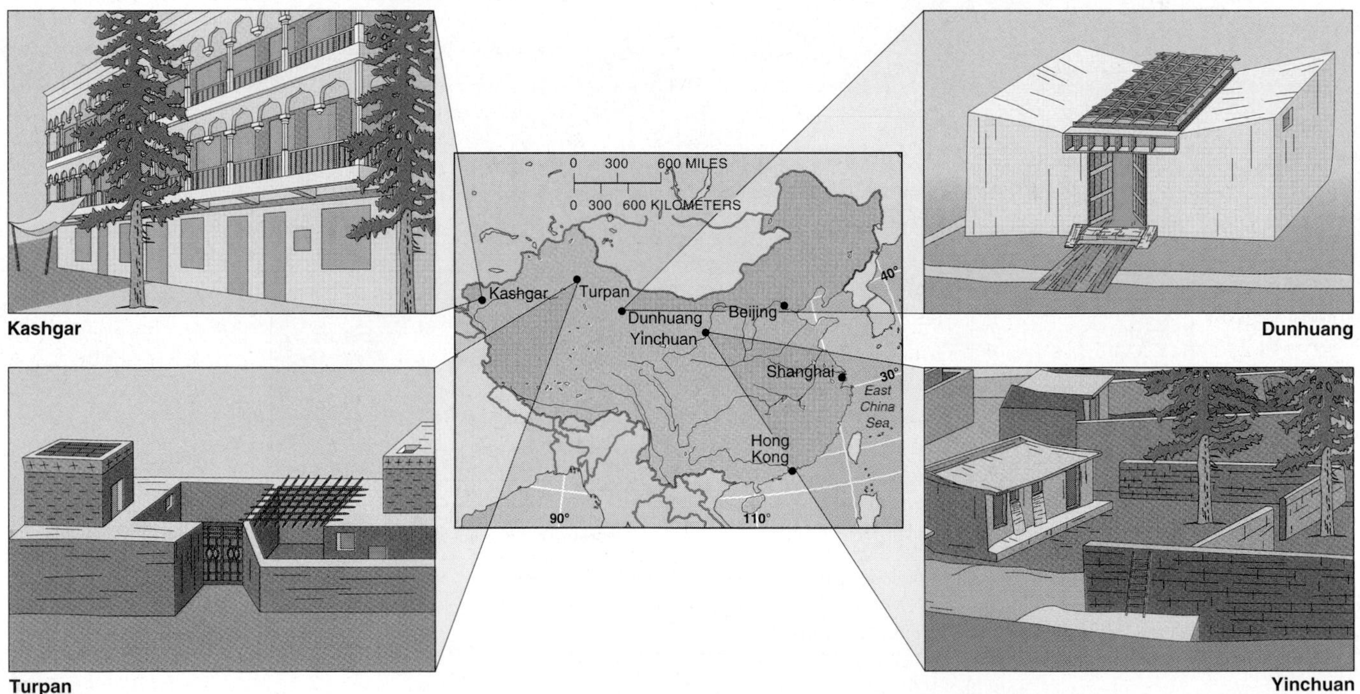

FIGURE 4-9 House types in four communities of western China. (upper left) Kashgar houses have second-floor open-air patios, where the residents can catch evening breezes. Poplar and fruit trees can be planted around the houses because the village has a river that is constantly flowing rather than seasonal, as is the case in much of China's dry lands. These deciduous trees provide shade in the summer and openings for sunlight in the winter. (lower left) Turpan houses have small, open courtyards for social gatherings. Turpan is situated in a deep valley with relatively little open land because much of the space is allocated to drying raisins. Second-story patios, which would use even less land, are avoided because the village is subject to strong winds. (lower right) Yinchuan houses are built around large, open-air courtyards, which contain tall trees to provide shade. Most residents are Muslims, who regard courtyards as private spaces to be screened from outsiders. The adobe bricks are square or cubic rather than rectangular, as is the case in the other villages, though R. W. McColl found no reason for this distinctive custom. (upper right) Dunhuang houses are characterized by walled central courtyards, covered by an open-lattice grape arbor. The cover allows for the free movement of air but provides shade from the especially intense direct summer heat and light. Rather than the flat roofs characteristic of dry lands, houses in Dunhuang have sloped roofs, typical of wetter climates, so that rainfall can run off. The practice is apparently influenced by Dunhuang's relative proximity to the population centers of eastern China, where sloped roofs predominate.

extreme temperatures, from very hot summer days to subfreezing winter nights. Despite their similarities, the houses in these four Chinese villages have individual designs. Houses have second-floor open-air patios in Kashgar, small open courtyards in Turpan, large private courtyards in Yinchuan, and sloped roofs in Dunhuang. McColl attributed the differences to local cultural preferences.

The construction of a pitched roof is important in wet or snowy climates to facilitate runoff and to reduce the weight of accumulated snow. Windows may face south in temperate climates to take advantage of the Sun's heat and light. In hot climates, on the other hand, window openings may be smaller to protect the interior from the full heat of the Sun.

Today, people in MDCs buy lumber that has been cut by machine into the needed shapes. Cut lumber is used to erect a frame, and sheets or strips of wood are attached for the floors, ceilings, and roof. Shingles, stucco, vinyl, aluminum, or other materials may be placed on the exterior for insulation or decoration.

Isolation Promotes Cultural Diversity

A group's unique folk customs develop through centuries of relative isolation from customs practiced by other cultural groups. As a result, folk customs observed at a point in time vary widely from one place to another, even among nearby places.

Himalayan Art

In a study of artistic customs in the Himalaya Mountains, geographers P. Karan and Cotton Mather demonstrated that distinctive views of the physical environment emerge among neighboring cultural groups that are isolated. The study area, a narrow corridor of 2,500 kilometers (1,500 miles) in the Himalaya Mountains of Bhutan, Nepal, northern India, and southern Tibet (China), contains four religious groups: Tibetan Buddhists in the north, Hindus in the south, Muslims in the west, and Southeast Asian animists in the east (Figure 4-10). Despite their spatial proximity, limited interaction among these groups produces distinctive folk customs.

Through their choices of subjects of paintings, each group reveals how their folk culture mirrors their religions and individual views of their environment:

- ***Buddhists*** in the northern region paint idealized divine figures, such as monks and saints. Some of these figures are depicted as bizarre or terrifying, perhaps reflecting the inhospitable environment.
- ***Hindus*** in the southern region create scenes from everyday life and familiar local scenes. Their paintings sometimes portray a deity in a domestic scene and frequently represent the region's violent and extreme climatic conditions.
- ***Muslims*** in the Islamic western portion show the region's beautiful plants and flowers because the Muslim faith prohibits displaying animate objects in art. In contrast with the paintings from the Buddhist and Hindu regions, these paintings do not depict harsh climatic conditions.
- ***Animists*** from Myanmar (Burma) and elsewhere in Southeast Asia, who have migrated to the eastern region of the study area, paint symbols and designs that derive from their religion rather than from the local environment.

The distribution of artistic subjects in the Himalayas shows how folk customs are influenced by cultural institutions like religion and by environmental processes such as climate, landforms, and vegetation. These groups display similar uniqueness in their dance, music, architecture, and crafts.

Beliefs and Folk House Forms

The distinctive form of folk houses may derive primarily from religious values and other customary beliefs rather than from environmental factors. Some compass directions may be more important than other directions.

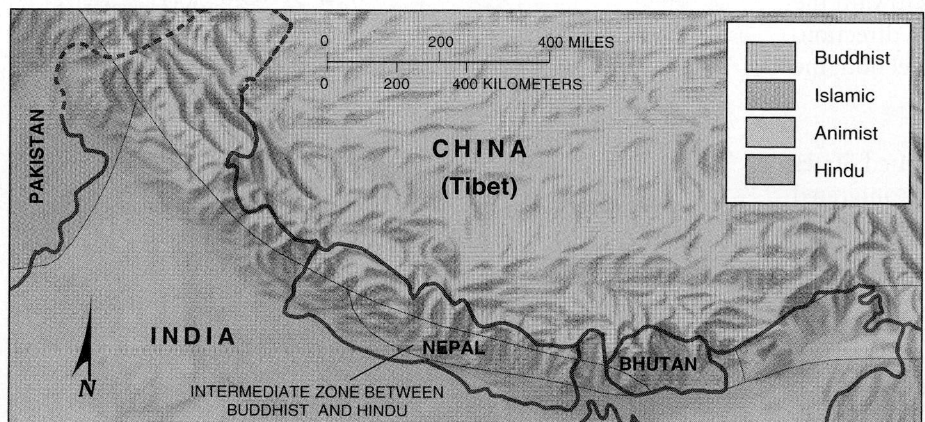

FIGURE 4-10 Cultural diversity in isolated folk regions. Cultural geographers P. Karan and Cotton Mather found four cultural areas in the rugged Himalayan region of Bhutan, Nepal, and northern India. Variations among the four groups were found in painting, dance, and other folk customs.

SACRED SPACES. Houses may have sacred walls or corners. In the south-central part of the island of Java, for example, the front door always faces south, the direction of the South Sea Goddess, who holds the key to Earth. The east wall of a house is considered sacred in Fiji, as is the northwest wall in parts of China. Sacred walls or corners are also noted in parts of the Middle East, India, and Africa.

In Madagascar, the main door is on the west, considered the most important direction, and the northeast corner is the most sacred. The north wall is for honoring ancestors; in addition, important guests enter a room from the north and are seated against the north wall. The bed is placed against the east wall of the house, with the head facing north.

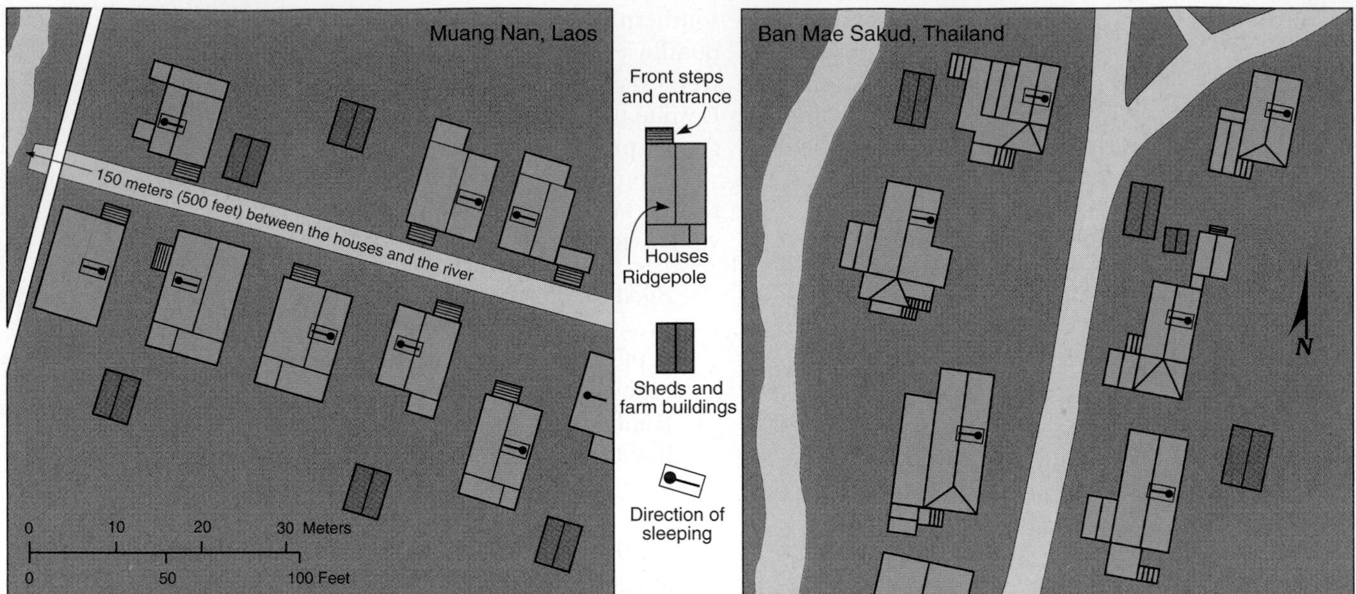

FIGURE 4-11 Sacred housing space. (left) Houses of Lao people in northern Laos. The fronts of Lao houses, such as those in the village of Muang Nan, Laos, face one another across a path and the backs face each other at the rear. Their ridgepoles (the centerline of the roof) are set perpendicular to the path but parallel to a stream if one is nearby. Inside adjacent houses, people sleep in the orientation shown, so neighbors are head-to-head or feet-to-feet. (right) Houses of Yuan and Shan peoples in northern Thailand. In the village of Ban Mae Sakud, Thailand, the houses are not set in a straight line because of a belief that evil spirits move in straight lines. Ridgepoles parallel the path, and the heads of all sleeping persons point eastward.

The Lao people in northern Laos arrange beds perpendicular to the center ridgepole of the house (Figure 4-11, left). Because the head is considered high and noble and the feet low and vulgar, people sleep so that their heads will be opposite their neighbor's heads and their feet opposite their neighbor's feet. The principal exception to this arrangement: A child who builds a house next door to the parents sleeps with his or her head toward the parents' feet as a sign of obeying the customary hierarchy.

Although they speak similar Southeast Asian languages and adhere to Buddhism, the Lao do not orient their houses in the same manner as the Yuan and Shan peoples in nearby northern Thailand (Figure 4-11, right). The Yuan and Shan ignore the position of neighbors and all sleep with their heads toward the east, which Buddhists consider the most auspicious direction. Staircases must not face west, the least auspicious direction, the direction of death and evil spirits.

U.S. FOLK HOUSING. Older houses in the United States display local folk-culture traditions. When families migrated westward in the 1700s and 1800s, they cut trees to clear fields for planting and used the wood to build houses, barns, and fences. The style of pioneer homes reflected whatever upscale style was prevailing at the place on the East Coast from which they migrated.

Geographer Fred Kniffen identified three major hearths or nodes of folk house forms in the United States: New England, Middle Atlantic, and Lower Chesapeake (Figure 4-12).

- *The Lower Chesapeake* or Tidewater style of house typically comprised one story, with a steep roof and chimneys at either

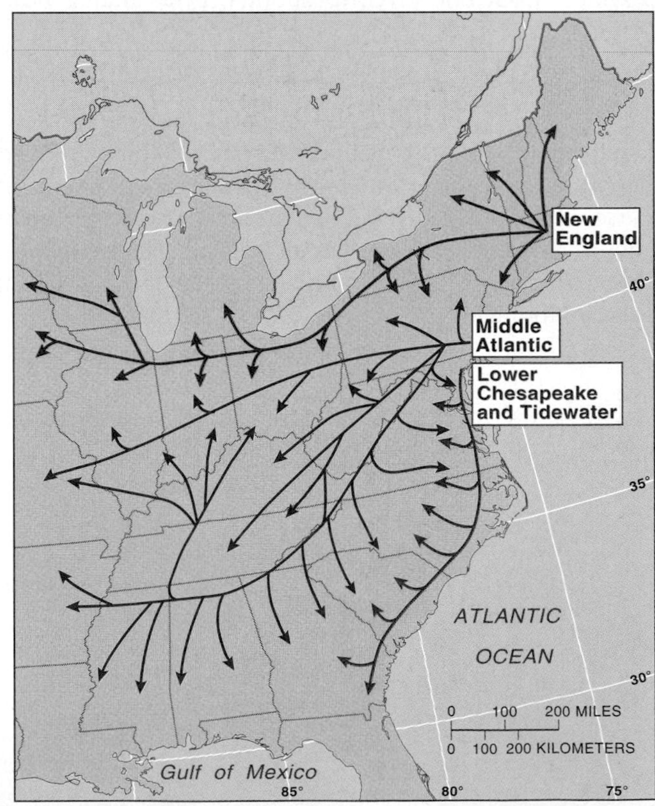

FIGURE 4-12 Hearths of U.S. house types. U.S. house types in the United States originated in three main source areas and diffused westward along different paths. These paths coincided with predominant routes taken by migrants from the East Coast toward the interior of the country.

end. Migrants spread these houses from the Chesapeake Bay–Tidewater, Virginia, area along the southeast coast.

As was the case with the Middle Atlantic "I"-house, the form of housing that evolved along the southeast coast typically was only one room deep. In wet areas, houses in the coastal southeast were often raised on piers or on a brick foundation.

- ***The Middle Atlantic*** region's principal house type was known as the "I"-house, typically two full stories in height, with gables to the sides. The "I"-house resembled the letter "I"—it was only one room deep and at least two rooms wide.

  Middle Atlantic migrants carried their house type westward across the Ohio Valley and southwestward along the Appalachian trails. As a result, the "I"-house became the most extensive style of construction in much of the eastern half of the United States, especially in the Ohio Valley and Appalachia.
- ***New England*** migrants carried house types northward to upper New England and westward across the southern Great Lakes region. The New England house types can be found throughout the Great Lakes region as far west as Wisconsin because this area was settled primarily by migrants from New England.

  Four major house types were popular in New England at various times during the eighteenth and early nineteenth centuries (Figure 4-13). As the house preferred by New Englanders changed over time, the predominant form found on the landscape varies based on the date of initial settlement.

Today, such distinctions are relatively difficult to observe in the United States. Houses built in the United States during the past half-century display popular culture influences. The degree of regional distinctiveness in housing style has diminished because rapid communication and transportation systems provide people throughout the country with knowledge of alternative styles. Furthermore, most people do not build the houses in which they live. Instead, houses are usually mass-produced by construction companies.

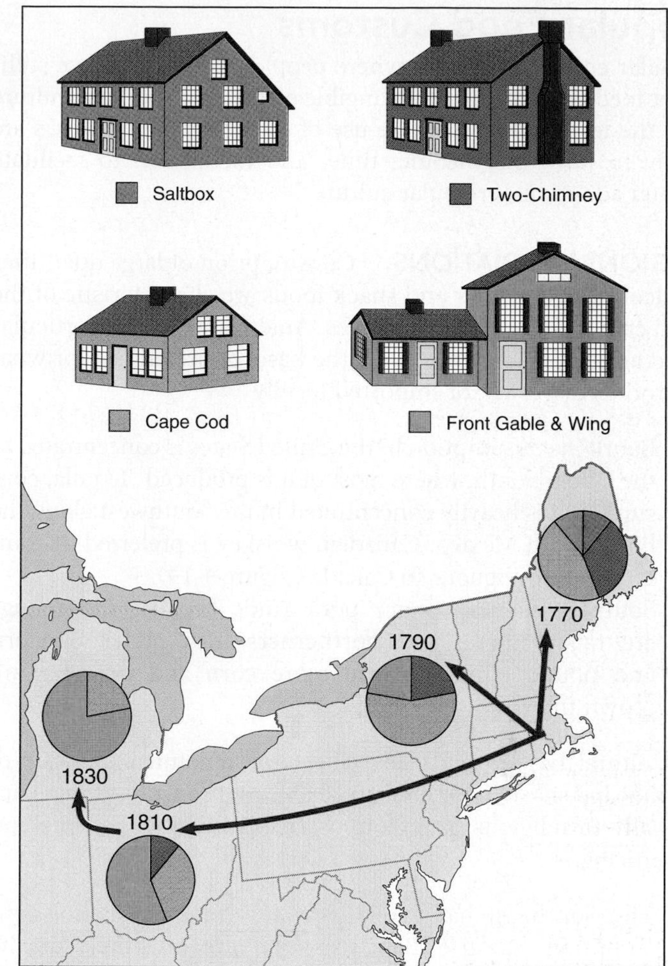

FIGURE 4-13 Diffusion of New England house types. Fred Kniffen suggests that these four major house types were popular in New England at various times during the eighteenth and early nineteenth centuries. As settlers migrated, they carried memories of familiar house types with them and built similar structures on the frontier. Thus New Englanders were most likely to build houses like the Cape Cod (green) when they began to migrate to upstate New York in the 1790s because that was the predominant house type they knew. During the 1800s, when New Englanders began to migrate farther westward to Ohio and Michigan, they built the front gable type of house typical in New England at that time, shown here in yellow.

KEY ISSUE 3

Why Is Popular Culture Widely Distributed?

- **Diffusion of Popular Housing, Clothing, and Food**
- **Electronic Diffusion of Popular Culture**

Popular culture varies more in time than in place. Like folk culture, it may originate in one location, within the context of a particular society and environment. But, in contrast to folk culture, it diffuses rapidly across Earth to locations with a variety of physical conditions. Rapid diffusion depends on a group of people having a sufficiently high level of economic development to acquire the material possessions associated with popular culture. ■

Diffusion of Popular Housing, Clothing, and Food

Some regional differences in food, clothing, and shelter persist in MDCs, but differences are much less than in the past. Go to any recently built neighborhood on the outskirts of an American city from Portland, Maine, to Portland, Oregon: The houses look the same, the people wear jeans, and the same chains deliver pizza.

Popular Food Customs

Popular culture flourishes where people in a society have sufficient income to acquire the tangible elements of popular culture and the leisure time to make use of them. People in MDCs are likely to have the income, time, and inclination to facilitate greater adoption of popular culture.

REGIONAL VARIATIONS. Consumption of large quantities of alcoholic beverages and snack foods are characteristic of the food customs of popular societies. Americans choose particular beverages or snacks in part on the basis of preference for what is produced, grown, or imported locally.

- Bourbon consumption in the United States is concentrated in the Upper South, where most of it is produced. Tequila, consumption is heavily concentrated in the Southwest along the border with Mexico. Canadian whiskey is preferred in communities contiguous to Canada (Figure 4-14).
- Southerners may prefer pork rinds because more hogs are raised there, and northerners may prefer popcorn and potato chips because more corn and potatoes are grown there.

Cultural backgrounds also affect the amount and types of alcohol and snack foods consumed. Alcohol consumption relates partially to religious backgrounds and partially to income and advertising.

- The Southeast has a relatively low rate of alcohol consumption because Baptists—who are clustered in the region—drink less than do adherents of other denominations; Utah also has a low rate because of a concentration of Latter-day Saints. Nevada has a high rate because of the heavy concentration of gambling and other resort activities there.
- Texans may prefer tortilla chips because of the large number of Hispanic Americans there. Westerners may prefer multigrain chips because of greater concern for the nutritional content of snack foods.

Geographers cannot explain all the regional variations in food preferences. Why do urban residents prefer Scotch and New Englanders consume more nuts? Why is per capita consumption of snack food one-third higher in the Midwest than in the West? Why does consumption of gin and vodka show little spatial variation within the United States? In general, consumption of alcohol and snack foods is part of popular culture primarily dependent on two factors—high income and national advertising. Variations within the United States are much less significant than differences between the United States and LDCs in Africa and Asia.

WINE. The spatial distribution of wine production demonstrates that the environment plays a role in the distribution of popular as well as folk food customs (Figure 4-15). The distinctive character of a wine derives from a unique combination of soil, climate, and other physical characteristics at the place where the grapes are grown.

Vineyards are best cultivated in temperate climates of moderately cold, rainy winters and fairly long, hot summers. Hot, sunny weather is necessary in the summer for the fruit to mature properly, whereas winter is the preferred season for rain, because plant diseases that cause the fruit to rot are more active in hot, humid weather. Vineyards are planted on hillsides, if possible, to maximize exposure to sunlight and to facilitate drainage. A site near a lake or river is also desirable because

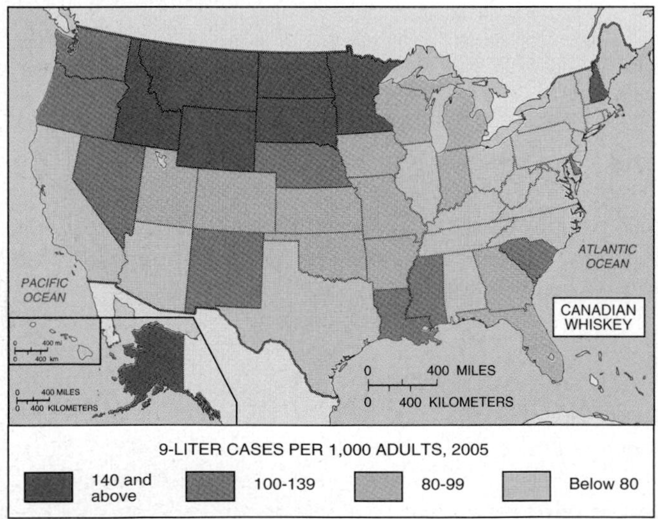

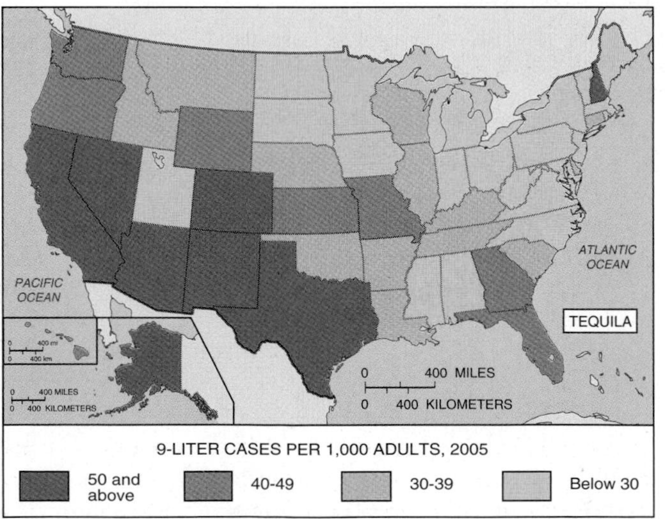

FIGURE 4-14 Consumption of Canadian whiskey (left) and tequila (right). States that have high per capita consumption of Canadian whiskey are located in the north, along the Canadian border. States that have high per capita consumption of tequila are located in the southwest, along the Mexican border. Preference for Canadian whiskey has apparently diffused southward from Canada into the United States, and for tequila northward from Mexico.

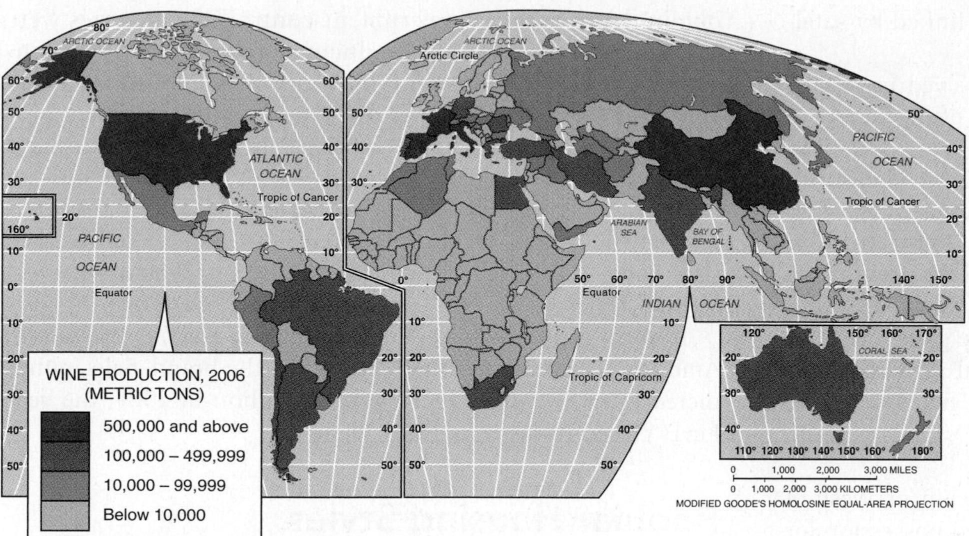

FIGURE 4-15 Wine production. The distribution of wine production is influenced in part by the physical environment and in part by social customs. Most grapes used for wine are grown near the Mediterranean Sea or in areas of similar climate. Income, preferences, and other social customs also influence the distribution of wine consumption, as seen in the lower production levels of predominantly Muslim countries south of the Mediterranean.

water can temper extremes of temperature. Grapes can be grown in a variety of soils, but the best wine tends to be produced from grapes grown in soil that is coarse and well drained—a soil not necessarily fertile for other crops. The soil is generally sandy and gravelly in the Bordeaux wine region, chalky in Champagne country, and of a slate composition in the Moselle Valley. The distinctive character of each region's wine is especially influenced by the unique combination of trace elements, such as boron, manganese, and zinc, in the rock or soil. In large quantities these elements could destroy the plants, but in small quantities they lend a unique taste to the grapes.

Because of the unique product created by the distinctive soil and climate characteristics, the world's finest wines are most frequently identified by their place of origin. Wines may be labeled with the region, town, district, or specific estate. A wine expert can determine the precise origin of a wine just by tasting because of the unique taste imparted to the grapes by the specific soil composition of each estate. The year of the harvest is also indicated on finer wines because specific weather conditions each year affect the quality and quantity of the harvest. Wines may also be identified by the variety of grape used rather than the location of the vineyard. Less expensive wines might contain a blend of grapes from a variety of estates and years.

Although grapes can be grown in a wide variety of locations, wine distribution is based principally on cultural values, both historical and contemporary. The distribution of wine production shows that the diffusion of popular customs depends less on the distinctive environment of a location than on the presence of beliefs, institutions, and material traits conducive to accepting those customs. Wine is made today primarily in locations that have a tradition of excellence in making it and people who like to drink it and can afford to purchase it.

The social custom of wine production in much of France and Italy extends back at least to the Roman Empire. Wine consumption declined after the fall of Rome, and many vineyards were destroyed. Monasteries preserved the wine-making tradition in medieval Europe for both sustenance and ritual. Wine consumption has become extremely popular again in Europe in recent centuries, as well as in the Western Hemisphere, which was colonized by Europeans. Vineyards are now typically owned by private individuals and corporations rather than religious organizations.

Wine production is discouraged in regions of the world dominated by religions other than Christianity. Hindus and Muslims in particular avoid alcoholic beverages. Thus wine production is limited in the Middle East (other than Israel) and southern Asia primarily because of cultural values, especially religion.

Rapid Diffusion of Clothing Styles

Individual clothing habits reveal how popular culture can be distributed across the landscape with little regard for distinctive physical features. Such habits reflect availability of income, as well as social forms such as job characteristics.

In the MDCs of North America and Western Europe, clothing habits generally reflect occupations rather than particular environments. A lawyer or business executive, for example, tends to wear a dark suit, light shirt or blouse, and necktie or scarf, whereas a factory worker wears jeans and a work shirt. A lawyer in California is more likely to dress like a lawyer in New York than like a steelworker in California.

A second influence on clothing in MDCs is higher income. Women's clothes, in particular, change in fashion from one year to the next. The color, shape, and design of dresses change to imitate pieces created by clothing designers. For social purposes, people with sufficient income may update their wardrobe frequently with the latest fashions.

Improved communications have permitted the rapid diffusion of clothing styles from one region of Earth to another. Original designs for women's dresses, created in Paris, Milan, London, or New York, are reproduced in large quantities at factories in Asia and sold for relatively low prices in North American and European chain stores. Speed is essential in manufacturing copies of designer dresses because fashion tastes change quickly. Until recently, a year could elapse from the time an original dress was displayed to the time that inexpensive reproductions were available in the stores. Now the time lag is only a few weeks because of the diffusion of fax machines, computers, and satellites. Sketches, patterns, and specifications are sent instantly from European fashion centers to American corporate headquarters and then on to Asian factories. Buyers from the major retail chains can view the fashions on large, high-definition televisions linked by satellite networks.

The globalization of clothing styles has involved increasing awareness by North Americans and Europeans of the variety of folk costumes around the world. Increased travel and the diffusion of television have exposed people in MDCs to other forms of dress, just as people in other parts of the world have come into contact with Western dress. The poncho from South America, the dashiki of the Yoruba people of Nigeria, and the Aleut parka have been adopted by people elsewhere in the world. The continued use of folk costumes in some parts of the globe may persist not because of distinctive environmental conditions or traditional cultural values, but to preserve past memories or to attract tourists.

JEANS. An important symbol of the diffusion of Western popular culture is jeans, which became a prized possession for young people throughout the world. In the 1960s, jeans acquired an image of youthful independence in the United States as young people adopted a style of clothing previously associated with low-status manual laborers and farmers.

Jeans became an obsession and a status symbol among youth in the former Soviet Union when the Communist government prevented their import. Gangs would attack people to steal their American-made jeans, and authentic jeans would sell for $400 on the black market. Ironically, jeans were brought into the Soviet Union by the elite, including diplomats, bureaucrats, and business executives—essentially those who were permitted to travel to the West. These citizens obtained scarce products in the West and resold them inside the Soviet Union for a considerable profit.

The scarcity of high-quality jeans was just one of many consumer problems that were important motives in the dismantling of Communist governments in Eastern Europe around 1990. Eastern Europeans, who were aware of Western fashions and products—thanks to television—could not obtain them, because government-controlled industries were inefficient and geared to producing tanks rather than consumer-oriented goods. With the end of communism, Levi's and other brands of jeans are freely sold and even produced in the former Soviet Union. But Levi's retain their American image. In Belarus, a former Soviet republic now an independent country, an antigovernment protest in 2006 was termed the denim revolution when protesters were urged to wear jeans at a rally. "Jeans evoke the West," said a protest leader.

Ironically, as access to Levi's increased around the world, American consumers turned away from the brand. Sales plummeted from $7 billion in 1996 to $4 billion in 2004, the year Levi's closed its last U.S. factory.

Popular Housing Styles

Housing built in the United States since the 1940s demonstrates how popular customs vary more in time than in place. In contrast with folk housing characteristic of the early 1800s, newer housing in the United States has been built to reflect rapidly changing fashion concerning the most suitable house form.

Houses show the influence of shapes, materials, detailing, and other features of architectural style in vogue at any one point in time. In the years immediately after World War II, which ended in 1945, most U.S. houses were built in a *modern style.* Since the 1960s, styles that architects call *neo-eclectic* have predominated (Figure 4-16).

MODERN HOUSE STYLES (1945–1960). Specific types of modern-style houses were popular at different times:

- **Minimal traditional:** Dominant in the late 1940s and early 1950s, reminiscent of Tudor-style houses popular in the 1920s and 1930s; usually one story, with a dominant

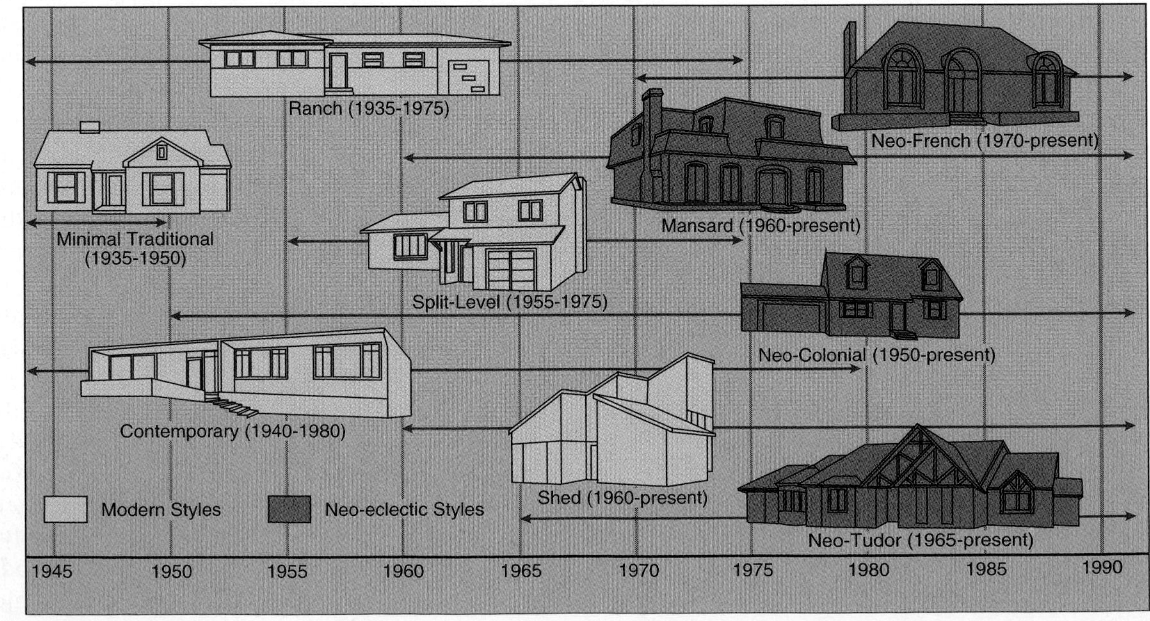

FIGURE 4-16 U.S. house types 1945–present. The dominant type of house construction in the United States was *minimal traditional* during the late 1940s and early 1950s, followed by ranch houses during the late 1950s and 1960s. The split-level was a popular variant of the ranch between the 1950s and 1970s, and the *contemporary* style was popular for architect-designed houses during the same period. The *shed* style was widely built in the late 1960s. Neo-eclectic styles, beginning with the *mansard,* were in vogue during the late 1960s. The *neo-Tudor* was popular in the 1970s and the *neo-French* in the 1980s. The *neo-colonial* style has been widely built since the 1950s but never dominated popular architecture.

CONTEMPORARY GEOGRAPHIC TOOLS

Documenting House Types Through Fieldwork

Fieldwork has been regarded as an important geographic method since the development of geography as a modern science two centuries ago. Geographers head for destinations near and far—to bustling urban areas and to remote rural areas, within their own countries or abroad. Given their concern with regularities in space, geographers need to get out of their classrooms and laboratories to observe the visible elements of other places with their own eyes. Fieldwork has been especially important for understanding the unique character of a place or the collection of features that distinguish one region from another.

Geographers make use of fieldwork in two principal ways. First, collecting information in the field can be the basis for drawing conclusions about expected patterns. Second, observing conditions in the field can be a source of inspiration for thinking about problems to address in future scientific studies. In other words, fieldwork helps some geographers to *answer* questions and helps others to *ask* questions.

Especially well suited to field studies have been visible everyday elements of folk and popular culture, such as house styles. Statistical studies and questionnaires such as the census can help geographers determine the size and date of construction of a house, but not the style that inspired its design. Only by looking at a house can its style of design be classified. Field material can be collected by delineating one or more areas on a map and visiting the sites. Armed with a chart or a spreadsheet, the geographer counts the number of times that something appears in the area, such as a particular type of house.

According to fieldwork by geographers John Jakle, Robert Bastian, and Douglas Meyer, regional differences in the predominant type of house persist to some extent in the United States (Figure 4-17). Differences in housing among U.S. communities derive largely from differences in the time period in which the houses were built. A housing development built in one region will resemble more closely developments built at the same time elsewhere in the country than will developments built in the same region at other points in time. ■

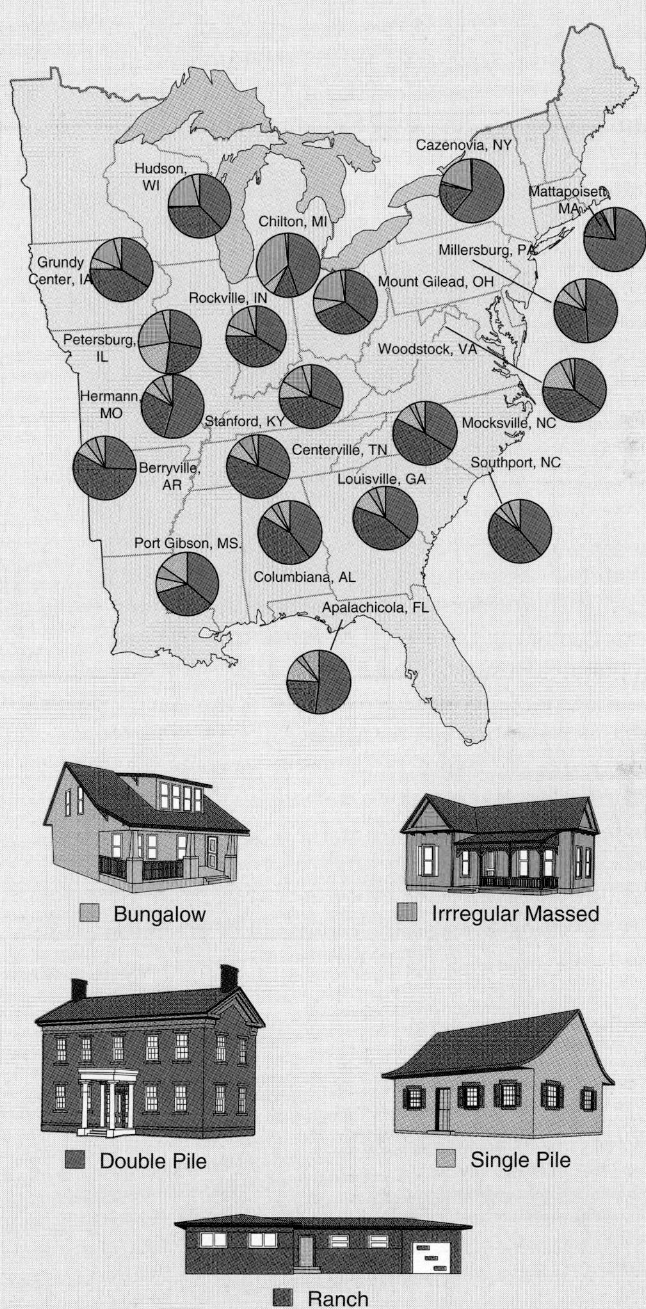

FIGURE 4-17 Regional differences in house types. Jakle, Bastian, and Meyer allocated the single-family housing in 20 small towns in the eastern United States into five groups: *bungalow, double pile, irregularly massed, ranch,* and *single pile.* Ranch houses were more common in the southeastern towns; double-pile houses predominated in northeastern areas.

front gable and few decorative details; small, modest houses designed to house young families and veterans returning from World War II.

- **Ranch house:** Replaced minimal traditional in the 1950s and into the 1960s; one story, with the long side parallel to the street; with all the rooms on one level rather than two or three, it took up a larger lot and encouraged the sprawl of urban areas.
- **Split-level:** A popular variant of the ranch house between the 1950s and 1970s; the lower level contained the garage and the newly invented "family" room, where the television set was placed; the kitchen and formal living and dining rooms were placed on the intermediate level, and the bedrooms on the top level above the family room and garage.
- **Contemporary:** Especially popular between the 1950s and 1970s for architect-designed houses; they frequently had flat or low-pitched roofs.
- **Shed:** Popular in the late 1960s; characterized by high-pitched shed roofs, giving the house the appearance of a series of geometric forms.

NEO-ECLECTIC (SINCE 1960). In the late 1960s, neo-eclectic styles became popular, and by the 1970s had surpassed modern styles in vogue:

- **Mansard:** The first popular neo-eclectic style, in the late 1960s and early 1970s; the shingle-covered second-story walls sloped slightly inward and merged into the roofline.
- **Neo-Tudor:** Popular in the 1970s; characterized by dominant, steep-pitched front-facing gables and half-timbered detailing.
- **Neo-French:** Also appeared in the early 1970s, and by the early 1980 was the most fashionable style for new houses; it featured dormer windows, usually with rounded tops, and high-hipped roofs.
- **Neo-colonial:** An adaptation of English colonial houses, it has been continuously popular since the 1950s but never dominant; inside many neo-colonial houses, a large central "great room" has replaced separate family and living rooms, which were located in different wings or floors of ranch and split-level houses.

Electronic Diffusion of Popular Culture

Watching television has been an especially significant popular custom for two reasons. First, it has been the most popular leisure activity in MDCs throughout the world. Second, television has been the most important mechanism by which knowledge of popular culture, such as professional sports, is rapidly diffused across Earth. In the twenty-first century, other electronic media have become important transmitters of popular culture.

Diffusion of Television

Television technology was developed simultaneously in the United Kingdom, France, Germany, Japan, and the Soviet Union, as well as in the United States, but in the early years of broadcasting the United States held a near monopoly. Through the second half of the twentieth century, television diffused from the United States, first to Europe and other MDCs, then to LDCs (Figure 4-18).

- In 1954, the first year that the United Nations published data on the subject, the United States had 86 percent of the world's 37 million TV sets. The United States had approximately 200 TV sets per 1,000 inhabitants in 1954, and the rest of the world had approximately 2 per 1,000.
- In 1970, the United States still had far more TV sets per capita than any other country except Canada. However, rapid growth of ownership in Europe meant that the share of the world's sets in the United States declined to one-fourth. Still, in 1970, half of the countries in the world, including most of those in Africa and Asia, had little if any TV broadcasting.
- By 2005, international differences in TV ownership had diminished, although not disappeared altogether. Other MDCs had similar rates of ownership as the United States, and ownership rates climbed sharply in many LDCs.

Diffusion of the Internet

The diffusion of Internet service follows the pattern established by television a generation earlier, but at a more rapid pace (Figure 4-19):

- In 1995, there were 40 million Internet users worldwide, including 25 million in the United States, and Internet service had not yet reached most countries.
- In 2000, Internet usage increased rapidly in the United States, from 9 percent to 44 percent of the population. But the worldwide increase was much greater, from 40 million Internet users in 1995 to 361 million in 2000. As Internet usage diffused rapidly, the U.S. percentage share declined rapidly in five years, from 62 to 31 percent.
- In 2008, Internet usage further diffused rapidly. World usage more than quadrupled in 8 years, to 1.6 billion. U.S. usage continued to increase, but at a more modest rate, to 74 percent of the population, and the share of the world's Internet users found in the United States continued to decline to 14 percent in 2008.

Note that all six maps in Figures 4-18 and 4–19 use the same intervals and colors. For example, the highest class in all maps is 300 or more per 1,000. What is different is the time interval. The diffusion of television from the United States to the rest of the world took a half-century, whereas the diffusion of the Internet has taken only a decade. Given the history of television, the Internet is likely to diffuse further in the years ahead at a rapid rate (Figure 4-20).

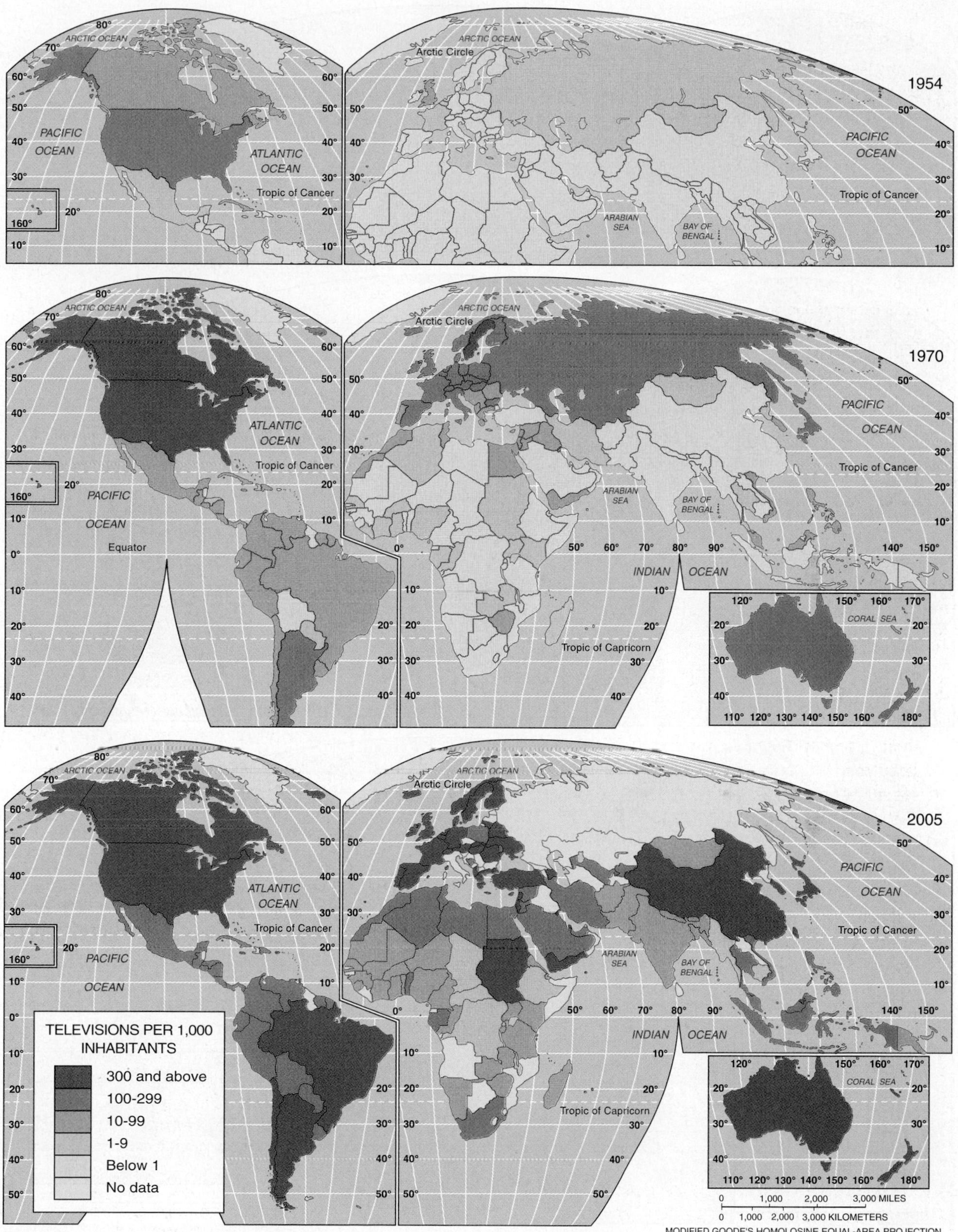

FIGURE 4-18 Diffusion of TV. Televisions per 1,000 inhabitants in 1954 (top), 1970 (middle), and 2005. Television has diffused from North America and Europe to other regions of the world. The United States and Canada had far more TV sets per capita than any other country as recently as the 1970s, but several European countries now have higher rates of ownership.

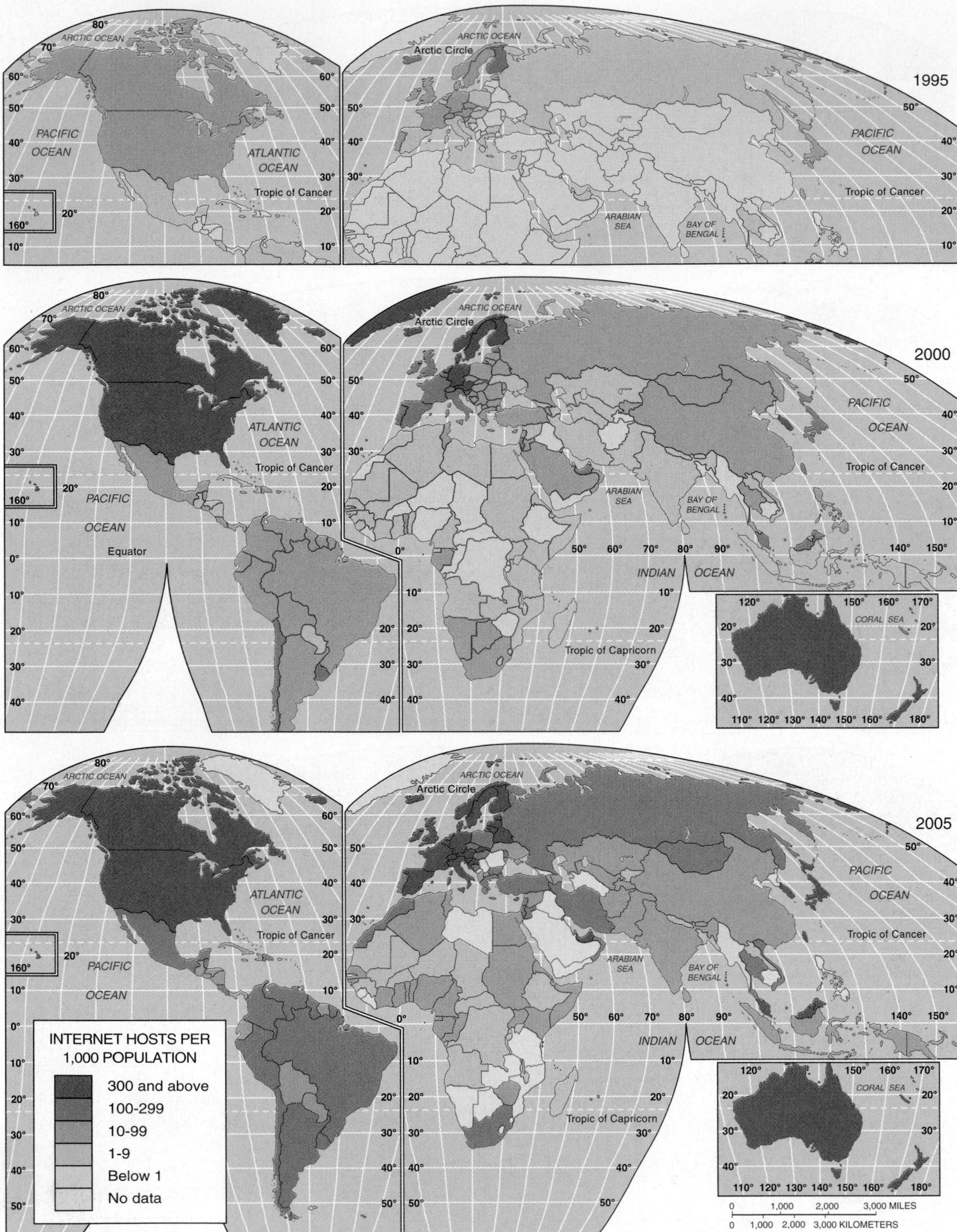

FIGURE 4-19 Diffusion of the Internet. Internet users per 1,000 inhabitants in 1995 (top), 2000 (middle), and 2005 (bottom). Compare to the diffusion of TV (Figure 4-18). The two sets of maps were drawn with the same scale and color scheme. Internet service is following a pattern in the twenty-first century similar to the diffusion of television in the twentieth century. The United States started out with a much higher rate of usage than elsewhere, until other countries caught up. The difference is that the diffusion of television took a half-century and the Internet only a decade.

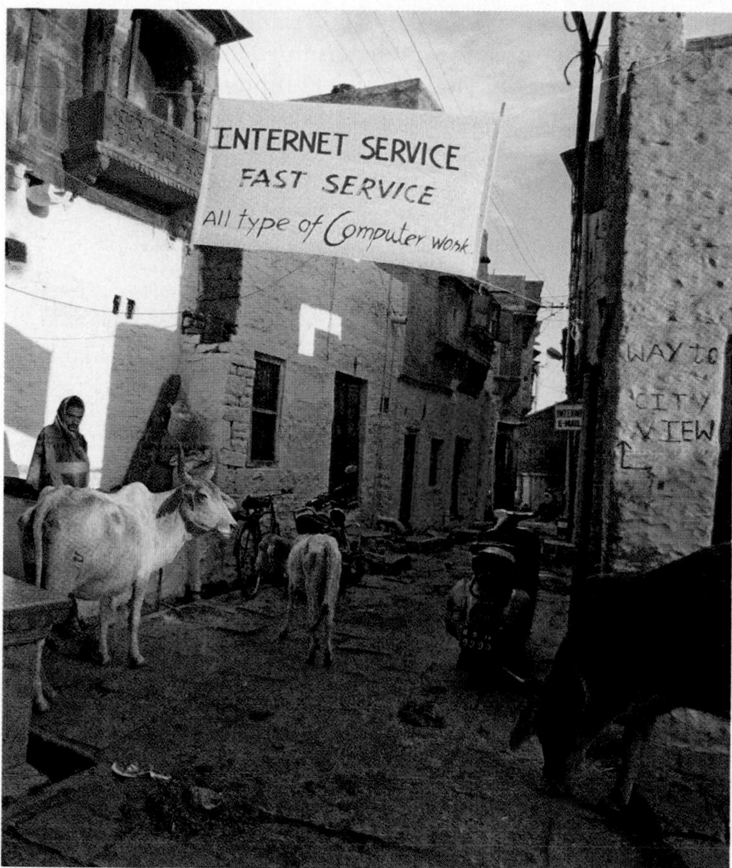

FIGURE 4-20 Diffusion of the Internet to India. Access to the Internet is available in even many rural areas of many LDCs.

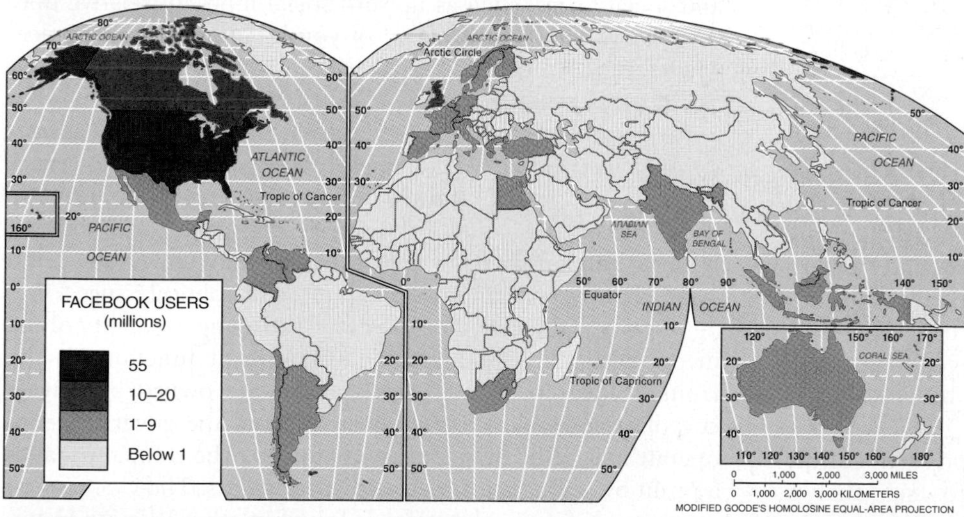

FIGURE 4-21 Diffusion of Facebook. In 2009, most Facebook users were located in the United States. In future years, Facebook may diffuse around the world in a similar pattern to Figures 4-18 and 19, or it may be overtaken by other electronic communications.

Diffusion of Facebook

Facebook, founded in 2004 by Harvard University students, has begun to diffuse rapidly. In 2009, five years after its founding, Facebook had 200 million active users (Figure 4-21). As with the first few years of TV and the Internet, once again the United States had far more Facebook users than any other country. In the years ahead, Facebook is likely to either diffuse to other parts of the world, or it will be overtaken by other electronic social networking programs and be relegated to a footnote in the continuous repeating pattern of diffusing electronic communications.

KEY ISSUE 4
Why Does Globalization of Popular Culture Cause Problems?

- **Threat to Folk Culture**
- **Environmental Impact of Popular Culture**

The international diffusion of popular culture has led to two issues, both of which can be understood from geographic perspectives. First, the diffusion of popular culture may threaten the survival of traditional folk culture in many countries. Second, popular culture may be less responsive to the diversity of local environments and consequently may generate adverse environmental impacts. ■

Threat to Folk Culture

Many fear the loss of folk culture, especially because rising incomes can fuel demand for the possessions typical of popular culture. When people turn from folk to popular culture, they may also turn away from the society's traditional values. And the diffusion of popular culture from MDCs can lead to dominance of Western perspectives.

Loss of Traditional Values

People in folk societies may turn away from traditional material culture, such as food, clothing, and shelter. Exposure to popular culture may stimulate desire to adopt similar practices.

One example of the symbolic importance of folk culture is clothing. In African and Asian countries today, there is a contrast between the clothes of rural farmworkers and of urban business and government leaders. Adoption of clothing from MDCs is part of a process of imitation and replication of foreign symbols of success. Leaders of African and Asian countries have traveled to MDCs and experienced the sense of social status attached to clothes, such as men's business suits. Back home, executives and officials may wear Western business suits as a symbol of authority and leadership.

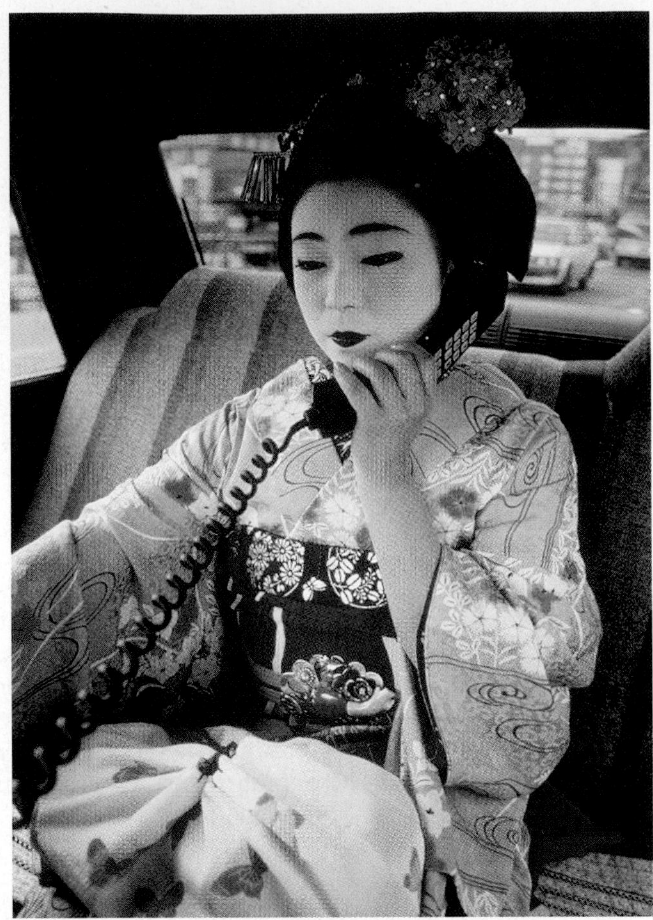

FIGURE 4-22 Role of women. Exposure to modern technology does not necessarily change the traditional role of women in many societies. In Kyoto, Japan, a geisha girl, who is trained to provide entertainment for men, arranges appointments on her way to the restaurant where she entertains her male clients.

Wearing clothes typical of MDCs is controversial in some Middle Eastern countries. Some political leaders in the region choose to wear Western business suits as a sign that they are trying to forge closer links with the United States and Western European countries. Others, such as fundamentalist Muslims, may oppose the widespread adoption of Western clothes, especially by women living in cities. Women are urged to abandon skirts and blouses in favor of the traditional black *chador,* a combination head covering and veil.

Beyond clothing, the global diffusion of popular culture may threaten the subservience of women to men that is embedded in some folk customs (Figure 4-22). Women may have been traditionally relegated to performing household chores, such as cooking and cleaning, and to bearing and raising large numbers of children. Those women who worked outside the home were likely to be obtaining food for the family, either through agricultural work or by trading handicrafts (see Global Forces, Local Impacts box). Contact with popular culture also has brought negative impacts for women in LDCs. For example, prostitution has increased in some LDCs to serve men from MDCs traveling on "sex tours." These tours, primarily from Japan and Northern Europe (especially Norway, Germany, and the Netherlands), include airfare, hotels, and the use of a predetermined number of women. Leading destinations include the Philippines, Thailand, and South Korea. International prostitution is encouraged in these countries as a major source of foreign currency. Through this form of global interaction, popular culture may regard women as essentially equal at home but as objects that money can buy in foreign folk societies.

Threat of Foreign Media Imperialism

Leaders of some LDCs consider the dominance of popular customs by MDCs as a threat to their independence. The threat is posed primarily by the media, especially news-gathering organizations and television.

WESTERN CONTROL OF MEDIA. Three MDCs—the United States, the United Kingdom, and Japan—dominate the television industry in LDCs. The Japanese operate primarily in South Asia and East Asia, selling their electronic equipment. British companies have invested directly in management and programming for television in Africa. U.S. corporations own or provide technical advice to many Latin American stations. These three countries are also the major exporters of programs. Even in Europe, the United States has been the source of two-thirds of the entertainment programs.

Leaders of many LDCs view the spread of television as a new method of economic and cultural imperialism on the part of the MDCs, especially the United States. American television, like other media, presents characteristically American beliefs and social forms, such as upward social mobility, relative freedom for women, glorification of youth, and stylized violence. These attractive themes may conflict with and drive out traditional social customs.

To avoid offending traditional values, many satellite broadcasters in Asia do not carry MTV or else allow governments to censor unacceptable videos. Cartoons featuring Porky Pig may be banned in Muslim countries, where people avoid pork products. Instead, entertainment programs emphasize family values and avoid controversial cultural, economic, and political issues.

LDCs fear the effects of the news-gathering capability of the media even more than their entertainment function. In the United States most television stations are owned by private corporations, which receive licenses from the government to operate at specific frequencies (channels). The company makes a profit by selling air time for advertisements. The U.S. pattern of private commercial stations is found in other Western Hemisphere countries but is rare elsewhere in the world.

The news media in most LDCs are dominated by the government, which typically runs the radio and TV service as well as the domestic news-gathering agency. Newspapers may be owned by the government, a political party, or a private individual, but in any event they are dependent on the government news-gathering organization for information. Veteran travelers and journalists invariably pack a portable shortwave radio when they visit other countries. In many regions of the world, the only reliable and unbiased news accounts come from the

GLOBAL FORCES, LOCAL IMPACTS
India's Marriage Dowries

Global diffusion of popular social customs has had an unintended negative impact for women in India: an increase in demand for dowries. A dowry is a "gift" from the family of a bride to the family of a groom, as a sign of respect. Though illegal in India since 1961, the dowry has regained popularity in recent years.

Traditionally, the local custom in much of India was for the groom to provide a small dowry to the bride's family. Now, the custom has reversed, and the family of a bride is often expected to provide a substantial dowry to the husband's family (Figure 4-23). Dowries have become much larger in modern India and an important source of income for the groom's family. A dowry can take the form of either cash or expensive consumer goods, such as motor vehicles, electronics, and household appliances.

The government has tried to ban dowries because of the adverse impact on women. If the bride's family is unable to pay a promised dowry or installments, the groom's family may cast the bride out on the street, and her family may refuse to take her back. Husbands and in-laws angry over the small size of dowry payments killed 5,000 to 7,000 women during the 1990s and early twenty-first century, according to government statistics.

Because a boy will generate revenue, whereas a girl will impose a significant burden, a fetus is more likely to be aborted if it is found to be a girl. A study of a Mumbai (Bombay) clinic found that 7,999 of 8,000 aborted fetuses were female. In families where food is scarce, girls age 1 to 5 are 43 percent more likely than boys to die of hunger or malnutrition, according to another study.

In a highly publicized case, just before the start of a wedding ceremony in 2003, a groom's family demanded a dowry of $25,000 in cash, in addition to two televisions, two home theater sets, two refrigerators, two air conditioners, and one car that had already been paid. The bride halted the ceremony and called the police on her cell phone. The family was arrested for violation of the 1961 antidowry law. The story appeared in *The Times of India* with the headline "It Takes Guts to Send Your Groom Packing." ■

FIGURE 4-23 India's marriage dowries. A mother mourns the death of her daughter, who drowned in a well after repeated demands for an increased dowry by her husband's family. The husband claimed that she jumped in the well, but police charged the husband with killing her.

BBC World Service shortwave and satellite radio newscasts. Reliance on BBC newscasts is especially strong in war zones.

Sufficient funds are not available to establish a private news service in LDCs. The process of gathering news worldwide is expensive, and most broadcasters and newspapers are unable to afford their own correspondents. Instead, they buy the right to use the dispatches of one or more of the main news organizations. The diffusion of information to newspapers around the world is dominated by the Associated Press (AP) and Reuters, which are owned by American and British companies, respectively. The AP and Reuters also supply most of the world's television news video.

Many African and Asian government officials criticize the Western concept of freedom of the press. They argue that the American news organizations reflect American values and do not provide a balanced, accurate view of other countries. U.S. news-gathering organizations are more interested in covering earthquakes, hurricanes, or other sensational disasters than more meaningful but less visual and dramatic domestic stories, such as birth-control programs, health-care innovations, or construction of new roads.

In the past, many governments viewed television as an important tool for fostering cultural integration; television could extol the exploits of the leaders or the accomplishments of the political system. People turned on their TV sets and watched what the government wanted them to see. Because television signals weaken with distance and are strong up to roughly 100 kilometers (60 miles), few people could receive television broadcasts from other countries.

SATELLITES. George Orwell's novel *1984*, published in 1949, anticipated that television—then in its infancy—would play a major role in the ability of a totalitarian government to control people's daily lives. In recent years, changing technology—especially the diffusion of small satellite dishes—has made television a force for political change rather than stability. Satellite dishes enable people to choose from a wide variety of programs produced in other countries, not just the local government-controlled station.

A number of governments in Asia have tried to prevent consumers from obtaining satellite dishes. The Chinese government banned private ownership of satellite dishes by its citizens, although foreigners and upscale hotels were allowed to keep them. The government of Singapore banned ownership of satellite dishes, yet it encourages satellite services, including MTV and HBO, to locate their Asian headquarters in the country. The government of Saudi Arabia ordered 150,000 satellite dishes dismantled, claiming that they were "un-Islamic."

Governments have had little success in shutting down satellite technology. Despite the threat of heavy fines, several hundred thousand Chinese still own satellite dishes. Consumers can outwit the government because the small size of satellite dishes makes them easy to smuggle into the country and erect out of sight, perhaps behind a brick wall or under a canvas tarpaulin. A dish may be expensive by local standards—twice the annual salary of a typical Chinese, for example—but several neighbors can share the cost and hook up all of their TV sets to it.

Satellite dishes represent only one assault on government control of the flow of information. Fax machines, portable video recorders, the Internet, and cellular telephones have also put chinks in government censorship. TV broadcasting has also migrated to new media, such as computers, cellular telephones, and other handheld devices. Programs can be viewed on demand, sometimes for a fee.

Environmental Impacts of Popular Culture

Popular culture is less likely than folk culture to be distributed with consideration for physical features. The spatial organization of popular culture reflects the distribution of social and economic features. In a global economy and culture, popular culture appears increasingly uniform.

Modifying Nature

Popular culture can significantly modify or control the environment. It may be imposed on the environment rather than spring forth from it, as with many folk customs. For many popular customs the environment is something to be modified to enhance participation in a leisure activity or to promote the sale of a product. Even if the resulting built environment looks "natural," it is actually the deliberate creation of people in pursuit of popular social customs.

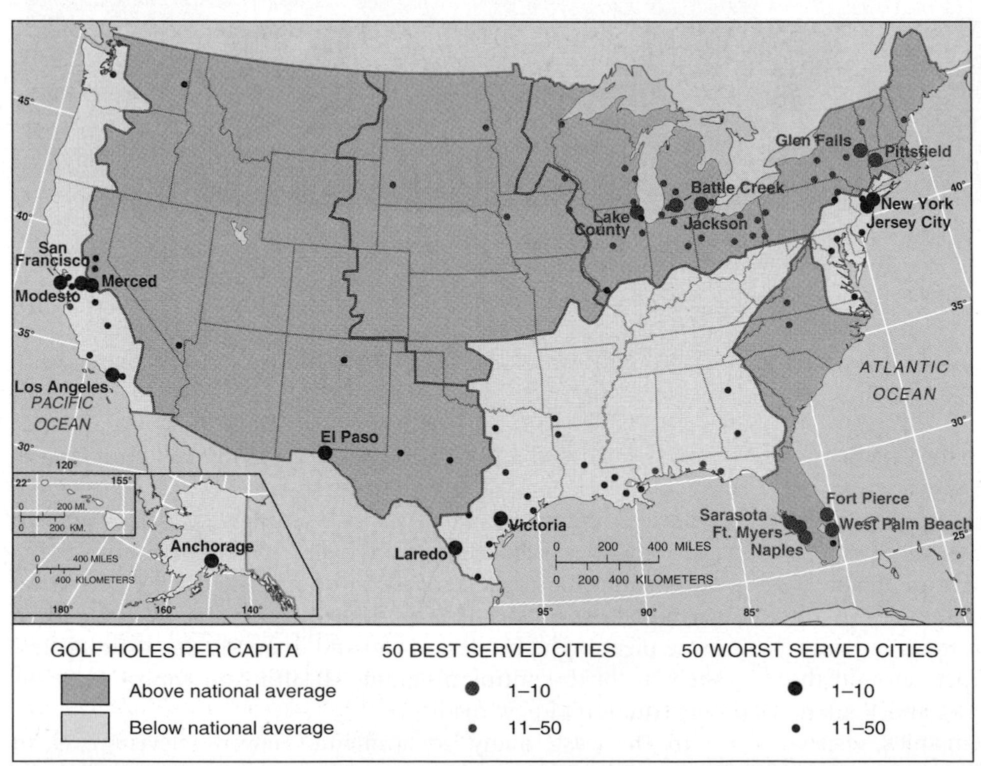

FIGURE 4-24 Golf courses. John Rooney identified the 50 best-served and worst-served metropolitan areas in terms of the number of golf holes per capita.

DISTRIBUTION OF GOLF. Golf courses, because of their large size (80 hectares, or 200 acres), provide a prominent example of imposing popular culture on the environment. A surge in U.S. golf popularity has spawned construction of roughly 200 courses during the past two decades. Geographer John Rooney attributes this to increased income and leisure time, especially among recently retired older people and younger people with flexible working hours.

According to Rooney, the provision of golf courses is not uniform across the United States. Although perceived as a warm-weather sport, the number of golf courses per person is actually greatest in north-central states, from Kansas to North Dakota, as well as the northeastern states abutting the Great Lakes, from Wisconsin to upstate New York (Figure 4-24). People in these regions have a long tradition of playing golf, and social clubs with golf courses are important institutions in the fabric of the regions' popular customs.

In contrast, access to golf courses is more limited in the South, in California, and in the heavily urbanized Middle Atlantic region between New York City and Washington, D.C. Rapid population growth in the South and West and lack of land on which to build in the Middle Atlantic region have reduced the number of courses per capita. Selected southern and western areas, such as coastal South Carolina, southern Florida, and central Arizona, have high concentrations of golf courses as a result of the arrival of large numbers of golf-playing northerners, either as vacationers or as permanent residents.

Golf courses are designed partially in response to local physical conditions. Grass species are selected to thrive in the local climate and still be suitable for the needs of greens, fairways, and roughs. Existing trees and native vegetation are retained if possible (few fairways in Michigan are lined by palms). Yet, like other popular customs, golf courses remake the environment—creating or flattening hills, cutting grass or letting it grow tall, carting in or digging up sand for traps, and draining or expanding bodies of water to create hazards.

Uniform Landscapes

The distribution of popular culture around the world tends to produce more uniform landscapes. The spatial expression of a popular custom in one location will be similar to another. In fact, promoters of popular culture want a uniform appearance to generate "product recognition" and greater consumption (Figure 4-25).

The diffusion of fast-food restaurants is a good example of such uniformity (refer to Figure 4-7). Such restaurants are usually organized as franchises. A franchise is a company's agreement with businesspeople in a local area to market that company's product. The franchise agreement lets the local outlet use the company's name, symbols, trademarks, methods, and architectural styles. To both local residents and travelers, the buildings are immediately recognizable as part of a national or multinational company. A uniform sign is prominently displayed.

Much of the attraction of fast-food restaurants comes from the convenience of the product and the use of the building as a low-cost socializing location for teenagers or families with young children. At the same time, the success of fast-food restaurants depends on large-scale mobility: People who travel or move to another city immediately recognize a familiar place. Newcomers to a particular place know what to expect in the restaurant because the establishment does not reflect strange and unfamiliar local customs that could be uncomfortable.

Fast-food restaurants were originally developed to attract people who arrived by car. The buildings generally were brightly colored, even gaudy, to attract motorists. Recently built fast-food restaurants are more subdued, with brick facades, pseudo-antique fixtures, and other stylistic details. To facilitate reuse of the structure in case the restaurant fails, company signs are often free-standing rather than integrated into the building design.

Uniformity in the appearance of the landscape is promoted by a wide variety of other popular structures in North America, such as gas stations, supermarkets, and motels. These structures are designed so that both local residents and visitors immediately recognize the purpose of the building, even if not the name of the company.

Physical expression of uniformity in popular culture has diffused from North America to other parts of the world. American motels and fast-food chains have opened in other countries. These establishments appeal to North American travelers, yet most customers are local residents who wish to sample American customs they have seen on television.

FIGURE 4-25 Route 66. When it connected Chicago and Los Angeles, Route 66 was a well-known symbol of an especially prominent element of U.S. popular culture—the freedom to drive a car across the country's wide-open spaces. This stretch of Route 66 in New Mexico is cluttered by unattractive strip development and large signs for national gasoline, lodging, and restaurant chains. Most of Route 66 has been replaced by interstate highways.

Negative Environment Impact

The diffusion of some popular customs can adversely impact environmental quality in two ways—depletion of scarce natural resources and pollution of the landscape.

INCREASED DEMAND FOR NATURAL RESOURCES. Diffusion of some popular customs increases demand for raw materials, such as minerals and other substances found beneath Earth's surface. The depletion of resources used to produce energy, especially petroleum, is discussed in Chapter 14.

Popular culture may demand a large supply of certain animals, resulting in depletion or even extinction of some species. For example, some animals are killed for their skins, which can be shaped into fashionable clothing and sold to people living thousands of kilometers from the animals' habitat. The skins of the mink, lynx, jaguar, kangaroo, and whale have been heavily consumed for various articles of clothing, to the point that the survival of these species is endangered. This unbalances ecological systems of which the animals are members. Folk culture may also encourage the use of animal skins, but the demand is usually smaller than for popular culture.

Increased demand for some products can strain the capacity of the environment. An important example is increased meat consumption. This has not caused extinction of cattle and poultry—we simply raise more. But animal consumption is an inefficient way for people to acquire calories—90 percent less efficient than if people simply ate grain directly. To produce 1 kilogram (2.2 pounds) of beef sold in the supermarket, nearly 10 kilograms (22 pounds) of grain are consumed by the animal. For every kilogram of chicken, nearly 3 kilograms (6.6 pounds) of grain are consumed by the fowl. This grain could be fed to people directly, bypassing the inefficient meat step. With a large percentage of the world's population undernourished, some question this inefficient use of grain to feed animals for eventual human consumption.

POLLUTION. Popular culture also can pollute the environment. The environment can accept and assimilate some level of waste from human activities. But popular culture generates a high volume of waste—solids, liquids, and gases—that must be absorbed into the environment. Although waste is discharged in all three forms, the most visible is solid waste—cans, bottles, old cars, paper, and plastics. These products are often discarded rather than recycled. With more people adopting popular customs worldwide, this problem grows.

Folk culture, like popular culture, can also cause environmental damage, especially when natural processes are ignored. A widespread belief exists that indigenous peoples of the Western Hemisphere practiced more "natural," ecologically sensitive agriculture before the arrival of Columbus and other Europeans. Geographers increasingly question this. In reality, pre-Columbian folk customs included burning grasslands for planting and hunting, cutting extensive forests, and overhunting some species. Very high rates of soil erosion have been documented in Central America from the practice of folk culture.

The MDCs that produce endless supplies for popular culture have created the technological capacity both to create large-scale environmental damage and to control it. However, a commitment of time and money must be made to control the damage. The adverse environmental impact of popular culture is further examined in Chapter 14.

SUMMARY

Material culture can be divided into two types—folk and popular. Folk culture most often exists among small, homogeneous groups living in relative isolation at a low level of economic development. Popular culture is characteristic of societies with good communications and transportation, which enable rapid diffusion of uniform concepts. Geographers are concerned with several aspects of folk and popular culture.

Geographers study an array of thousands of social customs with distinctive spatial distributions. Groups display preferences in providing for material needs such as food, clothing, and shelter, and in leisure activities such as performing arts and recreation. Examining where various social customs are practiced helps us to understand the extent of cultural diversity in the world.

Folk culture is especially interesting to geographers because its distribution is relatively clustered and its preservation can be seen as enhancing diversity in the world. Popular culture is important, too, because it derives from the high levels of material wealth characteristic of societies that are economically developed. As societies seek to improve their economic level, they may abandon traditional folk culture and embrace popular culture associated with MDCs.

Underlying the patterns of material culture are differences in the ways people relate to their environment. Material culture contributes to the modification of the environment, and in turn, nature influences the cultural values of an individual or a group.

Geographers, then, classify culture into popular and folk based on differences in the ways the environment is modified and meaning is derived from environmental conditions. Popular culture makes relatively extensive modifications of the environment, given society's greater technological means and inclination to do so. Here again are the key issues concerning folk and popular culture:

1. **Where Do Folk and Popular Cultures Originate and Diffuse?** Because of distinctive processes of origin and diffusion, folk culture has different distribution patterns than does popular culture. Folk culture is more likely to have an anonymous origin and to diffuse slowly through migration, whereas popular culture is more likely to be invented and diffuse rapidly with the use of modern communications.
2. **Why Is Folk Culture Clustered?** Unique regions of folk culture arise because of lack of interaction among groups, even those living nearby. Folk culture is more likely to be influenced by the local environment.
3. **Why Is Popular Culture Widely Distributed?** Popular culture diffuses rapidly across Earth, facilitated by modern communications, especially television. Differences in popular culture are more likely to be observed in one place at different points in time than among different places at one point in time.
4. **Why Does Globalization of Popular Culture Cause Problems?** Geographers observe two kinds of problems from diffusion of popular culture across the landscape. First, popular culture—generally originating in Western MDCs—may cause elimination of some folk culture. Second, popular culture may adversely affect the environment.

CASE STUDY REVISITED / Food and Geography

Food taboos and preferences derived from environmental conditions have been shown in this chapter to be a feature of folk culture. In Thailand and Myanmar (Burma), for example, giant water bugs are deep-fried as a snack food or ground up in sauces. Mixing insects with rice provides lysine, an amino acid that is often deficient in the diet of people in folk cultures where rice is the staple food.

Though less likely to be related to environmental conditions, food taboos are also characteristic of popular culture. Despite the nutritional value of insects, Americans avoid eating them. The aversion of most Americans to eating insects is contradicted by consumption of such foods as canned mushrooms and tomato paste, which contain insects, although that is not commonly acknowledged.

Global food preferences are influenced more by cultural values than by environmental features. For example, why do Coca-Cola and Pepsi have different global sales patterns (Figure 4-26)? The two beverages are similar, and many people are unable to taste the difference. Yet consumers prefer Coke in some countries and Pepsi in others.

Coca-Cola accounts for more than one-half of the world's cola shares, and Pepsi for another one-fourth. Coca-Cola is the sales leader in most of the Western Hemisphere. The principal exception is Canada's French-speaking province of Québec, where Pepsi is preferred. Pepsi won over the Québécois with advertising that tied Pepsi to elements of uniquely French Canadian culture. The major indoor arena in Québec City is named the Colisée Pepsi (Pepsi Coliseum).

Cola preferences are influenced by politics in Russia. Under communism, government officials made a deal with Pepsi to allow that cola to be sold in the Soviet Union. With the breakup of the Soviet Union and the end of communism, Coke entered the Russian market. Russians quickly switched their preference to Coke, because Pepsi was associated with the discredited Communist government.

In the Middle East, it's religion that influences cola preferences. At one time, the region's predominantly Muslim countries boycotted products that were sold in predominantly Jewish Israel. Because Coke but not Pepsi was sold in Israel, in most of Israel's neighbors Pepsi was preferred. ■

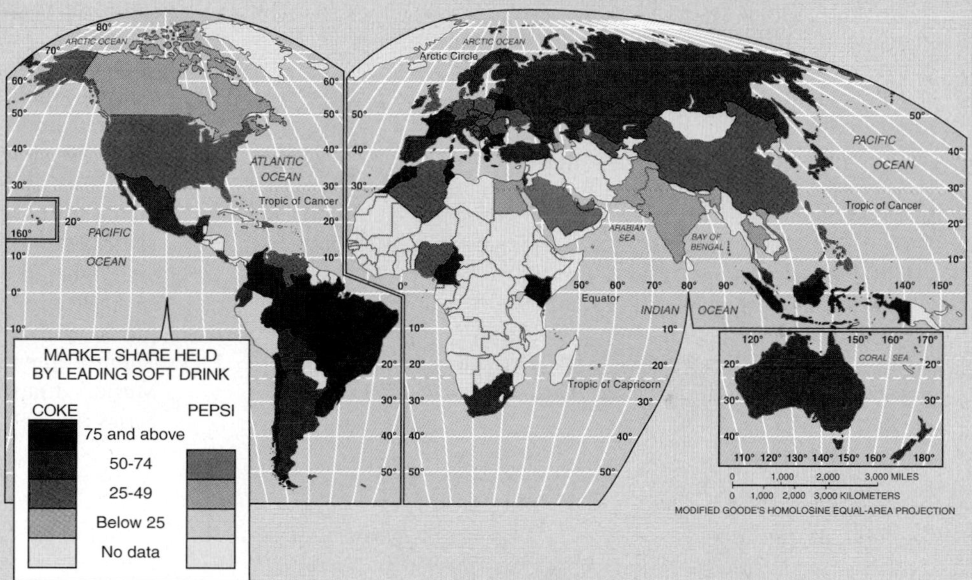

FIGURE 4-26 Cola preferences. Coca-Cola leads in sales in the United States, Latin America, Europe, and Russia. Pepsi leads in Canada and Asia.

KEY TERMS

Custom (p. 106) The frequent repetition of an act, to the extent that it becomes characteristic of the group of people performing the act.

Folk culture (p. 106) Culture traditionally practiced by a small, homogeneous, rural group living in relative isolation from other groups.

Habit (p. 106) A repetitive act performed by a particular individual.

Popular culture (p. 106) Culture found in a large, heterogeneous society that shares certain habits despite differences in other personal characteristics.

Taboo (p. 112) A restriction on behavior imposed by social custom.

Terroir (p. 114) The contribution of a location's distinctive physical features to the way food tastes.

THINKING GEOGRAPHICALLY

1. Should geographers regard culture and social customs as meaningful generalizations about a group of people, or should they concentrate on understanding how specific individuals interact with the physical environment? Why?
2. In what ways might gender affect the distribution of social customs in a community?
3. Are there examples of groups, either in more developed countries or in less developed countries, that have successfully resisted the diffusion of popular customs? Describe such a group and tell how it has succeeded in preserving its culture.
4. Which elements of the physical environment are emphasized in the portrayal of various places on television?
5. Which images of social customs do countries depict in campaigns to promote tourism? To what extent do these images reflect local social customs realistically?

RESOURCES

Some recent and classic books and articles on cultural geography:

Atkinson, David, ed. *Cultural Geography: A Critical Dictionary of Key Concepts*. London and New York: I.B. Tauris, 2005.

Bale, John. *Sports Geography*, 2nd ed. London and New York: Routledge, 2003.

Bennett, Merril K. *The World's Foods*. New York: Arno Press, 1954.

Blacking, John, and Joann W. Kealiinohomoku. *The Performing Arts: Music and Dance*. The Hague: Mouton, 1979.

Carlson, Alvar W. "The Contributions of Cultural Geographers to the Study of Popular Culture." *Journal of Popular Culture* 11 (1978): 830–31.

Chubb, Michael, and Holly R. Chubb. *One Third of Our Time*. New York: Macmillan, 1985.

Cohen, Ronald, ed. *Alan Lomax: Selected Writings, 1934–1997*. New York: Routledge, 2003.

Crouch, David, ed. *Leisure/Tourism Geographies: Practices and Geographical Knowledge*. London and New York: Routledge, 1999.

Crowley, William K. "Old Order Amish Settlement: Diffusion and Growth." *Annals of the Association of American Geographers* 68 (1978): 249–65.

DeBlij, Harm J. *A Geography of Viticulture*. Miami: University of Miami Geographical Society, 1981.

Denevan, William E. "The Pristine Myth: The Landscape of the Americas in 1492." *Annals of the Association of American Geographers* 82 (1992): 367–85.

Gade, Daniel W. "Tradition, Territory, and Terroir in French Viniculture: Cassis, France, and Appellation Contrôlée." *Annals of the Association of American Geographers* 94 (2004): 848–67.

Hall, C. Michael, and Stephen J. Page. *The Geography of Tourism and Recreation*, 2nd ed. London and New York: Routledge, 2002.

Jakle, John A., and Keith A. Sculle. *Fast Food: Roadside Restaurants in the Automobile Age*. Baltimore: Johns Hopkins University Press, 1999.

———. *The Gas Station in America*. Baltimore: Johns Hopkins University Press, 1994.

Jakle, John A., Keith A. Sculle, and Jefferson S. Rogers. *The Motel in America*. Baltimore: Johns Hopkins University Press, 1996.

Jakle, John A., Robert W. Bastian, and Douglas K. Meyer. *Common Houses in America's Small Towns: The Atlantic Seaboard to the Mississippi Valley*. Athens, GA: The University of Georgia Press, 1989.

Kaldjian, Paul J. "Istanbul's Bostans: A Millennium of Market Gardens." *Geographical Review* 94 (2004) 284–304.

Karan, Pradyumna P., and Cotton Mather. "Art and Geography: Patterns in the Himalayas." *Annals of the Association of American Geographers* 66 (1976): 487–515.

Kniffen, Fred B. "Folk-Housing: Key to Diffusion." *Annals of the Association of American Geographers* 55 (1965): 549–77.

Lewis, Peirce F., Yi-Fu Tuan, and David Lowenthal. *Visual Blight in America*. Washington, DC: Association of American Geographers, 1973.

Leyshon, Andrew, David Matless, and George Revill. "The Place of Music." *Transactions of the Institute of British Geographers New Series* 20 (1995): 423–33.

Lomax, Alan. *The Folk Songs of North America.* Garden City, NY: Doubleday, 1960.

Lornell, Christopher, and W. Theodore Mealor, Jr. "Traditions and Research Opportunities in Folk Geography." *Professional Geographer* 35 (1983): 51–56.

McAlester, Virginia, and Lee McAlester. *A Field Guide to American Houses.* New York: Alfred A. Knopf, 1984.

McColl, Robert W. "By Their Dwellings Shall We Know Them: Home and Setting Among China's Inner Asian Ethnic Groups." *Focus* 39 (1989): 1–6.

Rooney, John F., Jr. *A Geography of American Sport.* Reading, MA: Addison-Wesley, 1974.

Rooney, John F., Jr., and Paul L. Butt. "Beer, Bourbon and Boone's Farm: A Geographical Examination of Alcoholic Drink in the United States." *Journal of Popular Culture* 11 (1978): 832–56.

Rooney, John F., Jr., Wilbur Zelinsky, and Dean R. Louder, eds. *This Remarkable Continent: An Atlas of United States and Canadian Society and Culture.* College Station: Texas A & M University Press for the Society for the North American Cultural Survey, 1982.

Shortridge, Barbara G. "Cultural Geography of American Foodways: An Annotated Bibliography." *Journal of Cultural Geography* 15, no. 2 (1995): 79–108.

Shortridge, Barbara G., and James R. Shortridge. "Consumption of Fresh Produce in the Metropolitan United States." *Geographical Review* 79 (1989): 79–98.

Sommers, Brian J. *The Geography of Wine: How Landscapes, Cultures, Terroir, and the Weather Make a Good Drop.* London and New York: Penguin Group, 2008.

van Elteren, Mel. "Conceptualizing the Impact of US Popular Culture Globally." *Journal of Popular Culture* 30 (1996): 47–90.

Zelinsky, Wilbur. "North America's Vernacular Regions." *Annals of the Association of American Geographers* 70 (1980): 1–16.

Journals featuring cultural geography:

International Folk Music Council Journal, Journal of American Culture, Journal of American Folklore, Journal of American Studies, Journal of Cultural Geography, Journal of Leisure Research, Journal of Popular Culture, Journal of Sport History, Landscape, and *Leisure Science.*

Key Internet sites:

www.cybergeography.org The *Atlas of Cyberspace* includes items that depict information on familiar maps, such as the growth in Internet usage by country or U.S. state. Other "maps" in the atlas are actually diagrams or graphic representations, such as frequency of visits to web sites or chat rooms, as well as other versions of the popular music "map" on page 108.

www.internetworldstats.com Statistics concerning Internet users are published by country.

Log in to www.mygeoscienceplace.com for videos, interactive maps, RSS feeds, case studies, and self-study quizzes to enhance your study of Folk and Popular Culture.

CHAPTER 5

Language

How many languages do you speak? If you are Dutch, you were required to learn at least two foreign languages in high school. For those of you who do not happen to be Dutch, the number is probably a bit lower.

In fact, most people in the United States know only English. Fewer than one-half of American high school students have studied a foreign language. In contrast, nearly two-thirds of graduates from Dutch high schools have learned at least three foreign languages.

Even in other English-speaking countries, foreign languages are studied more frequently than in the United

KEY ISSUES

1 **Where Are English-Language Speakers Distributed?**

2 **Why Is English Related to Other Languages?**

3 **Where Are Other Language Families Distributed?**

4 **Why Do People Preserve Local Languages?**

States. For example, two-thirds of 10-year-olds in the United Kingdom are learning a foreign language in school.

Earth's heterogeneous collection of languages is one of its most obvious examples of cultural diversity. *Ethnologue*, one of the most authoritative sources of languages (SIL International, www.ethnologue.com), estimates that the world has 6,909 languages. Only 11 of these languages, including English, are spoken by at least 100 million people. Four of these are relatively familiar to North Americans (German, Portuguese, Spanish, and Russian), but others are less familiar (Arabic, Bengali, Hindi, Japanese, Lahnda, and Mandarin).

Approximately 153 languages are spoken by at least 3 million people, including the 11 largest ones. The remaining 6,756 languages are spoken by fewer than 3 million people. The distribution of some of these languages is easy for geographers to document, whereas others—especially in Africa and Asia—are difficult, if not impossible.

Language diversity. Signs in Korean and English in the Myeong-dong commercial district of Seoul, South Korea.

CASE STUDY / French and Spanish in the United States and Canada

The Tremblay family lives in a suburb of Montréal, Québec. The parents and two young children speak French at home, work, school, and shops. The Lopez family—also two parents and two children—lives in San Antonio, Texas, and speaks Spanish in their household.

The Tremblay and Lopez families share a common condition: They live in countries with an English-speaking majority, but English is not their native language. The French-speaking inhabitants of Canada and the Spanish-speaking residents of the United States continue to speak their languages, although English dominates the political, economic, and cultural life of their countries.

The two families use languages other than English because they believe that language is important in retaining and enhancing their cultural heritage. At the same time, both families recognize that knowledge of English is essential for career advancement and economic success.

French is one of Canada's two official languages, along with English. French speakers comprise one-fourth of the country's population. Most French-speaking Canadians are clustered in Québec, where they comprise more than three-fourths of the province's speakers. Colonized by the French in the seventeenth century, Québec was captured by the British in 1763, and in 1867 became one of the provinces in the Confederation of Canada.

In the United States, Spanish has become an increasingly important language in recent years because of large-scale immigration from Latin America. In some communities, public notices, government documents, and advertisements are printed in Spanish. Several hundred Spanish-language newspapers and radio and television stations operate in the United States, especially in southern Florida, the Southwest, and large northern cities, where most of the 34 million Spanish-speaking people live. ■

The two introductory case study examples—French-speaking residents of Canada and Spanish-speaking residents of the United States—illustrate the "where" and "why" questions that concern geographers who study languages. Where are different languages spoken? English, French, Spanish, and other languages are spoken in distinct locations around the world, and geographers can document the distribution of this important element of cultural identity.

Why in some cases are two different languages spoken in two locations, whereas in other cases the same language is spoken in two locations? The geography of language displays especially clearly this book's overall theme of interplay between forces of globalization and local diversity.

Language is a system of communication through speech, a collection of sounds that a group of people understands to have the same meaning. Many languages also have a **literary tradition**, or a system of written communication. However, hundreds of spoken languages lack a literary tradition. The lack of written records makes it difficult to document the distribution of many languages.

Many countries designate at least one language as their **official language**, which is the one used by the government for laws, reports, and public objects, such as road signs, money, and stamps. A country with more than one official language may require all public documents to be in all languages. Logically, an official language would be understood by most if not all of the country's citizens, but some countries that were once British colonies designate English as an official language, even though few of their citizens can speak it.

Language is part of culture, which, as shown in Chapter 1, has two main meanings—people's values and their tangible artifacts. Chapter 4 looked at the material objects of culture. This chapter and the next two discuss the three traits that best distinguish cultural values—language, religion, and ethnicity. We start our study of the geographic elements of cultural values with language in part because it is the means through which other cultural values, such as religion and ethnicity, are communicated.

Consistent with this book's where and why approach, this chapter first looks at ***where*** different languages are used and how these languages can be logically grouped in ***space***. The second and third key issues examine ***why*** languages have distinctive distributions. The study of language follows logically from migration, because the contemporary distribution of languages around the world results largely from past migrations of peoples.

Language is like luggage: People carry it with them when they move from place to place. They incorporate new words into their own language when they reach new ***places***, and they contribute words brought with them to the existing language at the new location. Geographers look at the similarities among languages to understand the diffusion and interaction of people around the world.

The final section of the chapter discusses contradictory trends of ***scale*** in language. On the one hand, English has achieved an unprecedented ***globalization*** because people around the world are learning it to participate in a global economy and culture. On the other hand, people are trying to preserve ***local diversity*** in language because language is one of the basic elements of cultural identity and a major feature of a region's uniqueness. Language is a source of pride to a people, a symbol of cultural unity. As a culture develops, language is both a cause of that development and a consequence.

The global distribution of languages results from a combination of two geographic processes—interaction and isolation. People in two locations speak the same language because of migration from one of the locations to another. If the two groups have few ***connections*** with each other after the migration, the language spoken by each will begin to differ. After a long period without contact, the two groups will speak languages that are so different they are classified as separate languages.

The interplay between interaction and isolation helps to explain ***regions*** of individual languages and entire language families. The difference is that individual languages emerged in the recent past as a result of historically documented events, whereas language families emerged several thousand years before recorded history.

KEY ISSUE 1
Where Are English-Language Speakers Distributed?

- **Origin and Diffusion of English**
- **Dialects of English**

The location of English-language speakers serves as a case study for understanding the process by which any language is distributed around the world. A language originates at a particular place and diffuses to other locations through the migration of its speakers. ■

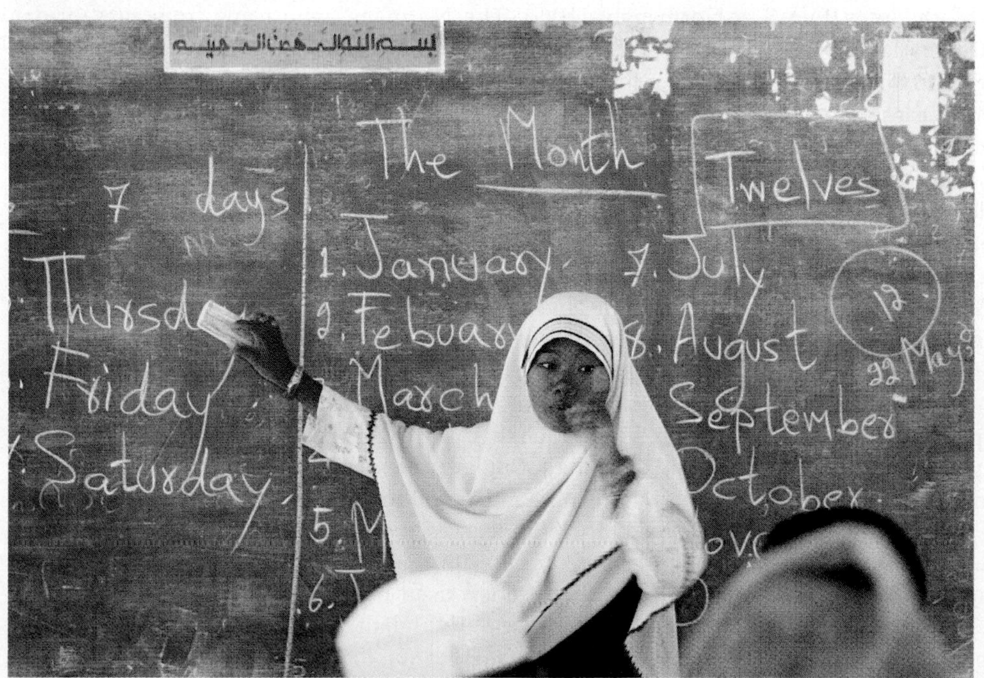

FIGURE 5-1 Teaching English. English is widely taught around the world, including this school in Thailand.

Origin and Diffusion of English

English is the first language of 328 million people and is spoken fluently by another one-half to one billion people (Figure 5-1). English is an official language in 57 countries, more than any other language, and is the predominant language in two more (Australia and the United States). Two billion people—one-third of the world—live in a country where English is an official language, even if they cannot speak it (Figure 5-2).

English Colonies

The contemporary distribution of English speakers around the world exists because the people of England migrated with their language when they established colonies during the past four centuries. Compare Figure 5-2 with Figure 8-8, which shows the location of former British colonies. English is an official language in most of the former British colonies.

English first diffused west from England to North America in the seventeenth century. The first English colonies were built in North America, beginning with Jamestown, Virginia, in 1607, and Plymouth, Massachusetts, in 1620. After England defeated France in a battle to dominate the North American colonies during the eighteenth century, the position of English as the principal language of North America was assured, even after the United States and Canada became independent countries.

Similarly, the British took control of Ireland in the seventeenth century, South Asia in the mid-eighteenth century, the South Pacific in the late eighteenth and early nineteenth centuries, and southern Africa in the late nineteenth century. In each case, English became an official language, even if only the colonial rulers and a handful of elite local residents could speak it.

More recently, the United States has been responsible for diffusing English to several places, most notably the Philippines, which Spain ceded to the United States in 1899, a year after losing the Spanish-American War. After gaining full independence in 1946, the Philippines retained English as one of its official languages along with Filipino.

Origin of English in England

The global distribution of English may be a function primarily of migration from England since the seventeenth century, but that does not explain how English came to be the principal language of the British Isles in the first place, or why English is classified as a Germanic language.

The British Isles had been inhabited for thousands of years, but we know nothing of their early languages until tribes called the Celts arrived around

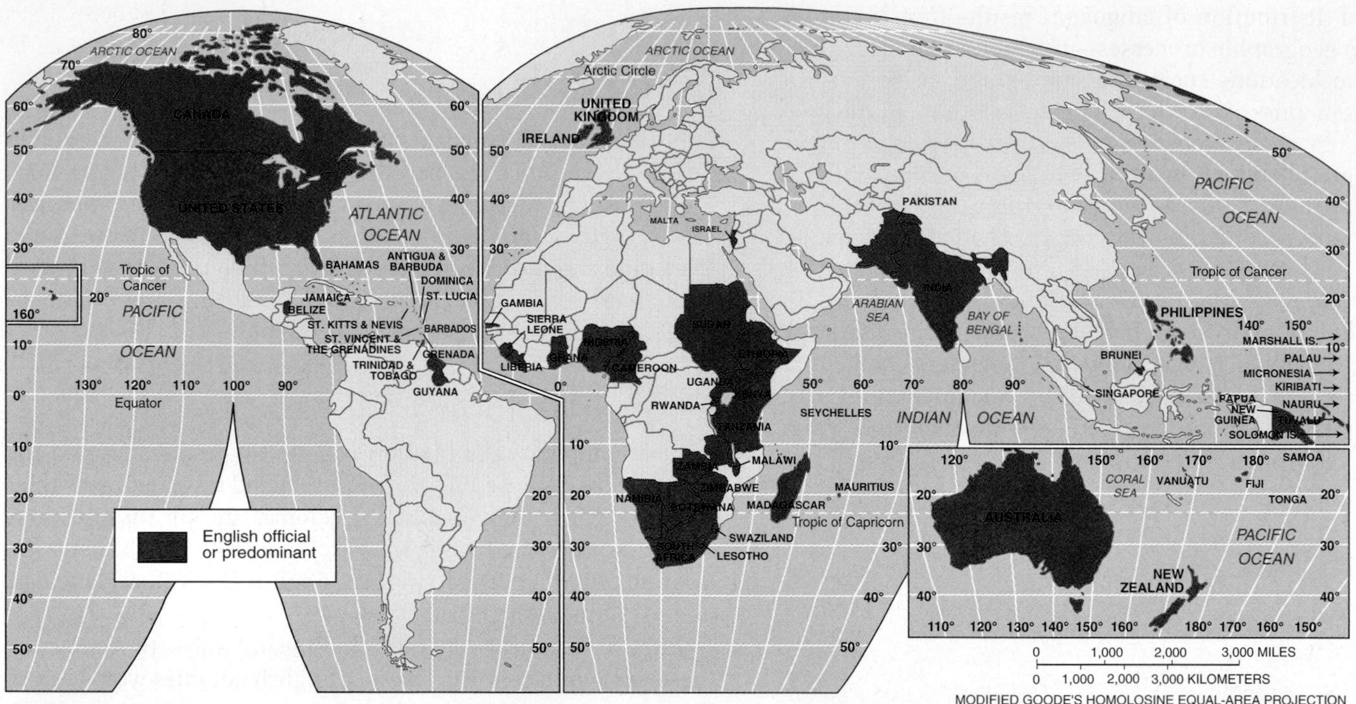

FIGURE 5-2 English-speaking countries. English is an official language in 57 countries. English is also the predominant language in the United States and Australia, although neither country has declared it to be the official language.

2000 B.C., speaking languages we call Celtic. Then, around A.D. 450, tribes from mainland Europe invaded, pushing the Celts into the remote northern and western parts of Britain, including Cornwall and the highlands of Scotland and Wales.

GERMAN INVASION. The invading tribes were the Angles, Jutes, and Saxons. All three were Germanic tribes—the Jutes from northern Denmark, the Angles from southern Denmark, and the Saxons from northwestern Germany (Figure 5-3). The three tribes who brought the beginnings of English to the British Isles shared a language similar to that of other peoples in the region from which they came. Today, English people and others who trace their cultural heritage back to England are often called Anglo-Saxons, after the two larger tribes. Modern English has evolved primarily from the language spoken by the Angles, Jutes, and Saxons. The name *England* comes from *Angles' land.* In Old English, *Angles* was spelled *Engles,* and the Angles' language was known as *englisc.* The Angles came from a corner, or *angle,* of Germany known as Schleswig-Holstein.

At some time in history, all Germanic people spoke a common language, but that time predates written records. The common origin of English with other Germanic languages can be reconstructed by analyzing language differences that emerged after Germanic groups migrated to separate territories and lived in isolation from each other, allowing their languages to continue evolving independently.

Other peoples subsequently invaded England and added their languages to the basic English. Vikings from present-day Norway landed on the northeast coast of England in the ninth century. Although defeated in their effort to conquer the islands, many Vikings remained in the country to enrich the language with new words.

NORMAN INVASION. English is a good bit different from German today primarily because England was conquered by the Normans in 1066. The Normans, who came from present-day Normandy in France, spoke French, which they established as England's official language for the next 300 years. The leaders of England, including the royal family, nobles, judges, and clergy, therefore spoke French. However, the majority of the people, who had little education, did not know French, so they continued to speak English to each other.

England lost control of Normandy in 1204, during the reign of King John, and entered a long period of conflict with France. As a result, fewer people in England wished to speak French, and English again became the country's unchallenged dominant language. Recognizing that nearly everyone in England was speaking English, Parliament enacted the Statute of Pleading in 1362 to change the official language of court business from French to English. However, Parliament continued to conduct business in French until 1489.

During the 300-year period that French was the official language of England, the Germanic language used by the common people and the French used by the leaders mingled to form a new language. Modern English owes its simpler, straightforward words, such as *sky, horse, man,* and *woman,* to its Germanic roots, and fancy, more elegant words, such as *celestial, equestrian, masculine,* and *feminine,* to its French invaders.

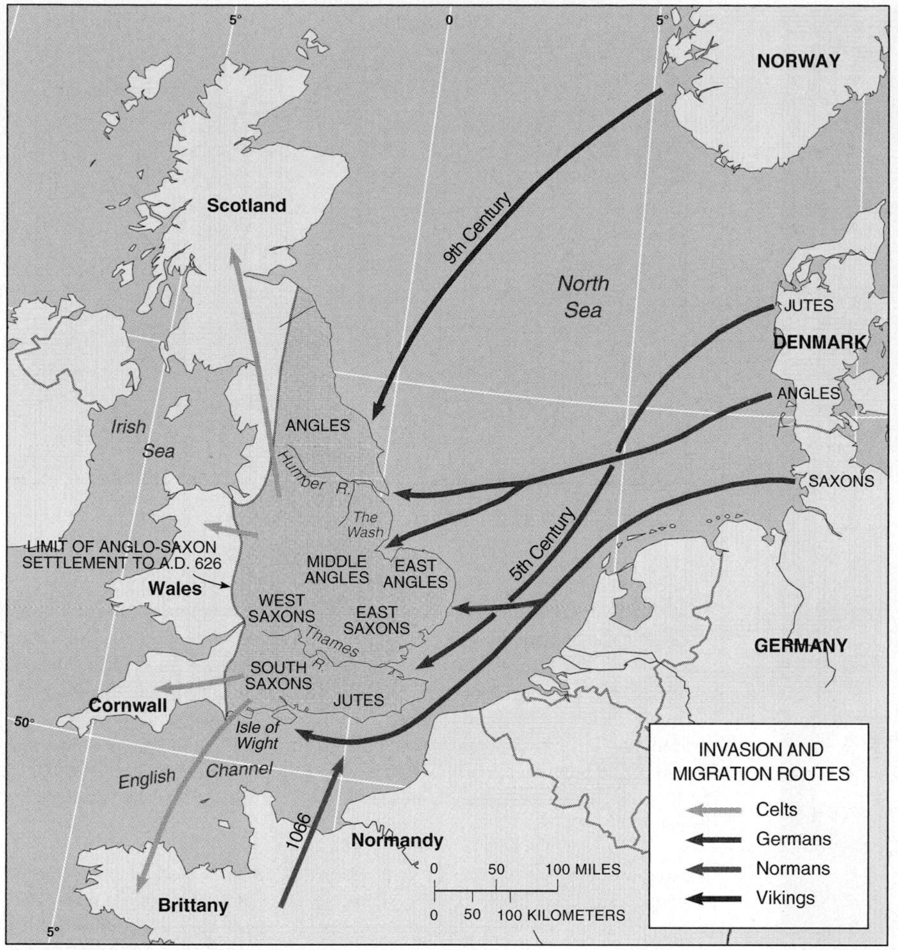

FIGURE 5-3 Invasions of England. The first speakers of the language that became known as English were tribes that lived in present-day Germany and Denmark. They invaded England in the fifth century. The Jutes settled primarily in southeastern England, the Saxons in the south and west, and the Angles in the north, eventually giving the country its name—Angles' Land, or England. From this original spatial separation, the first major regional differences in English dialect developed, as shown in Figure 5-5. Invasions by Vikings in the ninth century and Normans in the eleventh century brought new words to the language spoken in the British Isles. The Normans were the last successful invaders of England.

(*Source:* From Albert C. Baugh and Thomas Cable, *A History of the English Language*, 3rd ed., 1978, p. 47. Reprinted by permission of Prentice Hall, Englewood Cliffs, NJ.)

Dialects of English

A **dialect** is a regional variation of a language distinguished by distinctive vocabulary, spelling, and pronunciation. Generally, speakers of one dialect can understand speakers of another dialect. Geographers are especially interested in differences in dialects, because they reflect distinctive features of the environments in which groups live.

The distribution of dialects is documented through the study of particular words. Every word that is not used nationally has some geographic extent within the country and therefore has boundaries. Such a word-usage boundary, known as an **isogloss**, can be constructed for each word. Isoglosses are determined by collecting data directly from people, particularly natives of rural areas. People are shown pictures to identify or are given sentences to complete with a particular word. Although every word has a unique isogloss, boundary lines of different words coalesce in some locations to form regions.

When speakers of a language migrate to other locations, various dialects of that language may develop. This was the case with the migration of English speakers to North America several hundred years ago. Because of its large number of speakers and widespread distribution, English has an especially large number of dialects. North Americans are well aware that they speak English differently from the British, not to mention people living in India, Pakistan, Australia, and other English-speaking countries. Further, English varies by regions within individual countries. In both the United States and England, northerners sound different from southerners.

In a language with multiple dialects, one dialect may be recognized as the **standard language**, which is a dialect that is well established and widely recognized as the most acceptable for government, business, education, and mass communication. One particular dialect of English, the one associated with upper-class Britons living in the London area, is recognized in much of the English-speaking world as the standard form of British speech. This speech, known as **British Received Pronunciation (BRP)**, is well known because it is commonly used by politicians, broadcasters, and actors. Why don't Americans or, for that matter, other British people speak that way?

Dialects in England

"If you use proper English, you're regarded as a freak; why can't the English learn to speak?" asked Professor Henry Higgins in the Broadway musical *My Fair Lady* (Figure 5-4). He was referring to the Cockney-speaking Eliza Doolittle, who pronounced *rain* like "rine" and dropped the /h/ sound from the beginning of words like *happy*. Eliza Doolittle's speech illustrates that English, like other languages, has a wide variety of dialects that use different pronunciations, spellings, and meanings for particular words.

As already discussed, English originated with three invading groups from Northern Europe who settled in different parts of Britain—the Angles in the north, the Jutes in the southeast, and the Saxons in the south and west. The language each spoke was the basis of distinct regional dialects of Old English—Kentish in the southeast, West Saxon in the southwest, Mercian in the center of the island, and Northumbrian in the north (Figure 5-5, left).

French replaced English as the language of the government and aristocracy following the Norman invasion of 1066. After several hundred years of living in isolation in rural settlements under the control of a French-speaking government, five major regional dialects had emerged—Northern, East

FIGURE 5-4 *My Fair Lady*. In the 1950s Broadway musical *My Fair Lady* by Alan J. Lerner and Frederick Loewe, based on George Bernard Shaw's play *Pygmalion*, language expert Professor Henry Higgins (played by Rex Harrison) transforms Eliza Doolittle, a Cockney from the poor East End of London (played by Julie Andrews), into an upper-class woman by teaching her to speak with the accent used by upper-class Britons.

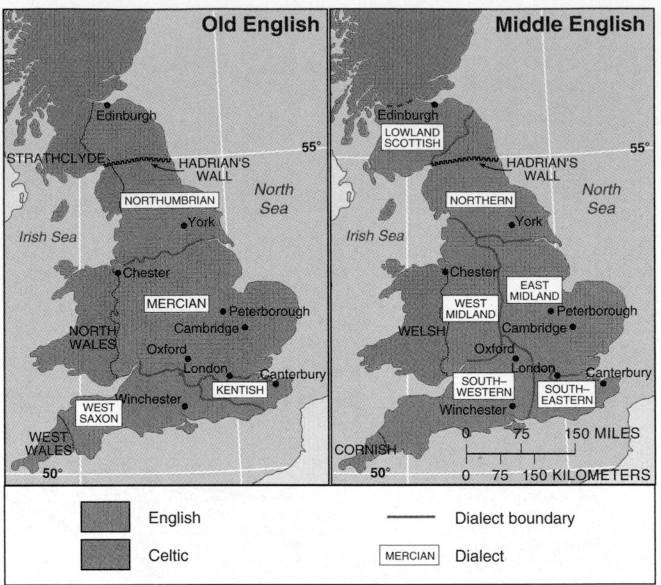

FIGURE 5-5 (Left) Old English dialects, before the Norman invasion of 1066. (Right) Middle English dialects (1150–1500). The two maps show that important dialects of Middle English corresponded closely to those of Old English. The Old English Northumbrian dialect, spoken by the Angles, split into Scottish and Northern dialects. The Old English Mercian dialect, spoken by the Saxons, divided into East Midland and West Midland, and the West Saxon dialect became known as the Southwestern dialect. The Old English Kentish dialect, spoken by the Jutes, extended considerably in area and became known as the Southeastern dialect.

(*Source:* From Albert C. Baugh and Thomas Cable, *A History of the English Language*, 3rd ed., 1978, p. 53. Reprinted by permission of Prentice Hall, Englewood Cliffs, NJ.)

Midland, West Midland, Southwestern, and Southeastern or Kentish (Figure 5-5, right).

From a collection of local dialects, one often emerges as the standard language for writing and speech. In the case of England, it was the dialect used by upper-class residents in the capital city of London and the two important university cities of Cambridge and Oxford. The diffusion of the upper-class London and university dialects was encouraged by the introduction of the printing press to England in 1476. Grammar books and dictionaries printed in the eighteenth century established rules for spelling and grammar that were based on the London dialect. These frequently arbitrary rules were then taught in schools throughout the country.

Despite the current dominance of BRP, strong regional differences persist in English dialects spoken in the United Kingdom, especially in rural areas. They can be grouped into three main ones—Northern, Midland, and Southern. For example,

- Southerners pronounce words like *grass* and *path* with an /*ah*/ sound; Northerners and people in the Midlands use a short /*a*/, as do most people in the United States.
- Northerners and people in the Midlands pronounce *butter* and *Sunday* with the /*oo*/ sound of words like *boot*.

The main dialects can be subdivided. For example, distinctive southwestern and southeastern accents occur within the Southern dialect.

- Southwesterners pronounce *thatch* and *thing* with the /*th*/ sound of *then*, rather than *thin*. *Fresh* and *eggs* have an /*ai*/ sound.
- Southeasterners pronounce the /*a*/ in *apple* and *cat* like the short /*e*/ in *bet*.

Local dialects can be further distinguished. Some words have distinctive pronunciations and meanings in each county of the United Kingdom.

Differences Between British and American English

The English language was brought to the North American continent by colonists from England who settled along the Atlantic

Coast beginning in the seventeenth century. The early colonists naturally spoke the language used in England at the time and established seventeenth-century English as the dominant form of European speech in colonial America.

Later immigrants from other countries found English already implanted here. Although they made significant contributions to American English, they became acculturated into a society that already spoke English. Therefore, the earliest colonists were most responsible for the dominant language patterns that exist today in the English-speaking part of the Western Hemisphere.

Why is the English language in the United States so different from that in England? As is so often the case with languages, the answer is isolation. Separated by the Atlantic Ocean, English in the United States and England evolved independently during the eighteenth and nineteenth centuries, with little influence on one another. Few residents of one country could visit the other, and the means to transmit the human voice over long distances would not become available until the twentieth century.

U.S. English differs from that of England in three significant ways:

- **Vocabulary.** The vocabulary is different largely because settlers in America encountered many new objects and experiences (Figure 5-6). The new continent contained physical features, such as large forests and mountains, that had to be given new names.

  New animals were encountered, including the *moose, raccoon,* and *chipmunk,* all of which were given names borrowed from Native Americans. Indigenous American "Indians" also enriched American English with names for objects such as *canoe, moccasin,* and *squash.*

  As new inventions appeared, they acquired different names on either side of the Atlantic. For example, the elevator is called a *lift* in England, and the flashlight is known as a *torch.* The British call the hood of a car the *bonnet* and the trunk the *boot.*
- **Spelling.** American spelling diverged from the British standard because of a strong national feeling in the United States for an independent identity. Noah Webster, the creator of the first comprehensive American dictionary and grammar books, was not just a documenter of usage, he had an agenda.

  Webster was determined to develop a uniquely American dialect of English. He either ignored or was unaware of recently created rules of grammar and spelling developed in England. Webster argued that spelling and grammar reforms would help establish a national language, reduce cultural dependence on England, and inspire national pride. The spelling differences between British and American English, such as the elimination of the "u" from the British spelling of words like *honour* and *colour* and the substitution of "s" for "c" in "*defence,*" are due primarily to the diffusion of Webster's ideas inside the United States.
- From the time of their arrival in North America, colonists began to pronounce words differently from the British. Such divergence is normal, for interaction between the two groups was largely confined to exchange of letters and other printed matter rather than direct speech.

  Such words as *fast, path,* and *half* are pronounced in England like the */ah/* in *father* rather than the */a/* in *man.* The British also eliminate the *r* sound from pronunciation except before vowels. Thus *lord* in British pronunciation sounds like *laud.*

  Americans pronounce unaccented syllables with more clarity. The words *secretary* and *necessary* have four syllables in American English but only three in British (*secret'ry* and *necess'ry*).

  Surprisingly, pronunciation has changed more in England than in the United States. The letters *a* and *r* are pronounced in the United States closer to the way they were pronounced in Britain in the seventeenth century when the first colonists arrived. A single dialect of Southern English did not emerge as the British national standard until the late eighteenth century, after the American colonies had declared independence and were politically as well as physically isolated from England. Thus people in the United States do not speak "proper" English because when the colonists left England, "proper" English was not what it is today. Furthermore, few colonists were drawn from the English upper classes.

FIGURE 5-6 Differences between British and American. In Britain, a circus is a place where several roads come together. The station sign is over an entrance to what Americans call the "subway," and Britons the "underground" or the "tube."

Dialects in the United States

Major differences in U.S. dialects originated because of differences in dialects among the original settlers (Figure 5-7). The English dialect spoken by the first colonists, who arrived in the seventeenth century, determined the future speech patterns for their communities because later immigrants adopted the language used in their new homes when they arrived. The language may have been modified somewhat by the new arrivals, but the distinctive elements brought over by the original settlers continued to dominate.

SETTLEMENT IN THE EAST. The original American settlements stretched along the Atlantic Coast in 13 separate colonies. The settlements can be grouped into three areas:

- **New England.** These colonies were established and inhabited almost entirely by settlers from England. Two-thirds of the New England colonists were Puritans from East Anglia in southeastern England, and only a few came from the north of England.
- **Southeastern.** About half came from southeast England, although they represented a diversity of social-class backgrounds, including deported prisoners, indentured servants, and political and religious refugees.
- **Middle Atlantic.** These immigrants were more diverse. The early settlers of Pennsylvania were predominantly Quakers from the north of England. Scots and Irish also went to Pennsylvania, as well as to New Jersey and Delaware. The Middle Atlantic colonies also attracted many German, Dutch, and Swedish immigrants who learned their English from the English-speaking settlers in the area.

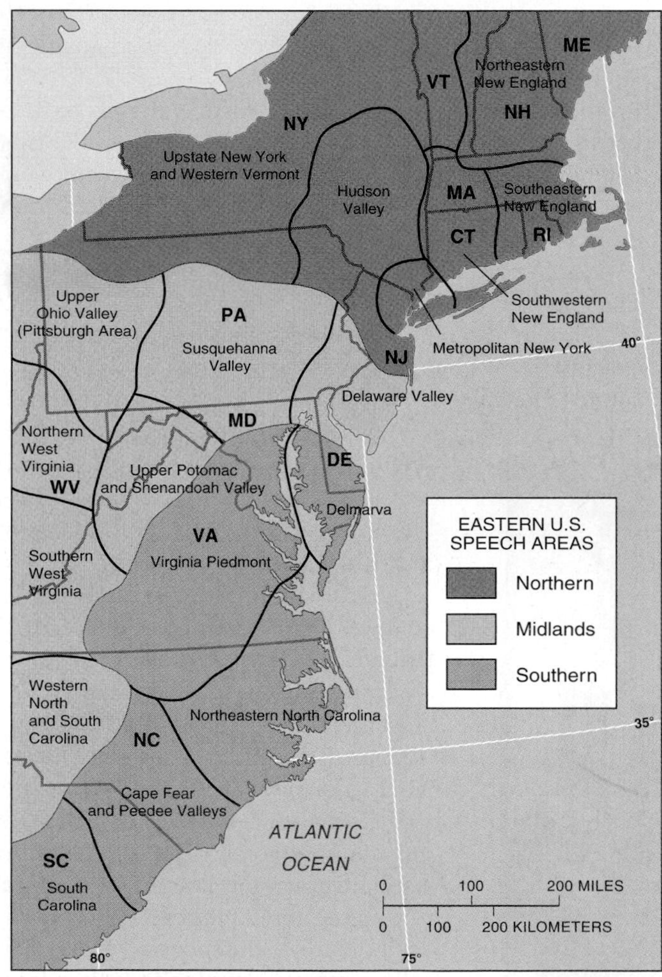

FIGURE 5-7 Dialects in eastern United States. The most comprehensive classification of dialects in the United States was made by Hans Kurath in 1949. He found the greatest diversity of dialects in the eastern part of the country, especially in vocabulary used on farms. Kurath divided the eastern United States into three major dialect regions—Northern, Midlands, and Southern—each of which contained a number of important subareas. Compare to the map of source areas of U.S. house types (Figure 4-12). As Americans migrated west they took with them distinctive house types as well as distinctive dialects

The English dialects now spoken in the U.S. Southeast and New England are easily recognizable. Current distinctions result from the establishment of independent and isolated colonies in the seventeenth century. The dialect spoken in the Middle Atlantic colonies differs significantly from those spoken farther north and south, because most of the settlers came from the north rather than the south of England or from other countries.

CURRENT DIALECT DIFFERENCES IN THE EAST. Major dialect differences continue to exist within the United States, primarily on the East Coast, although some distinctions can be found elsewhere in the country. Two important isoglosses separate the eastern United States into three major dialect regions, known as Northern, Midlands, and Southern (Figure 5-7). Some words are commonly used within one of the three major dialect areas but rarely in the other two. In most instances, these words relate to rural life, food, and objects from daily activities. Language differences tend to be greater in rural areas than in cities, because farmers are relatively isolated from interaction with people from other dialect regions.

Many words that were once regionally distinctive are now national in distribution. Mass media, especially television and radio, influence the adoption of the same words throughout the country. Nonetheless, regional dialect differences persist in the United States (Figure 5-8). For example, the word for soft drink varies. Most people in the Northeast and Southwest, as well as the St. Louis area, use *soda* to describe a soft drink. Most people in the Midwest, Great Plains, and Northwest prefer *pop*. Southerners refer to all soft drinks as *coke*.

PRONUNCIATION DIFFERENCES. Regional pronunciation differences are more familiar to us than word differences, although it is harder to draw precise isoglosses for them.

- The Southern dialect includes making such words as *half* and *mine* into two syllables ("ha-af" and "mi-yen"), pronouncing *poor* as "po-ur," and pronouncing *Tuesday* and *due* with a /y/ sound ("Tyuesday" and "dyue").
- The New England dialect is well known for dropping the /r/ sound, so that *heart* and *lark* are pronounced "hot" and "lock." Also, *ear* and *care* are pronounced with /ah/ substituted for the /r/ endings. This characteristic dropping of the /r/ sound is shared with speakers from the south of England

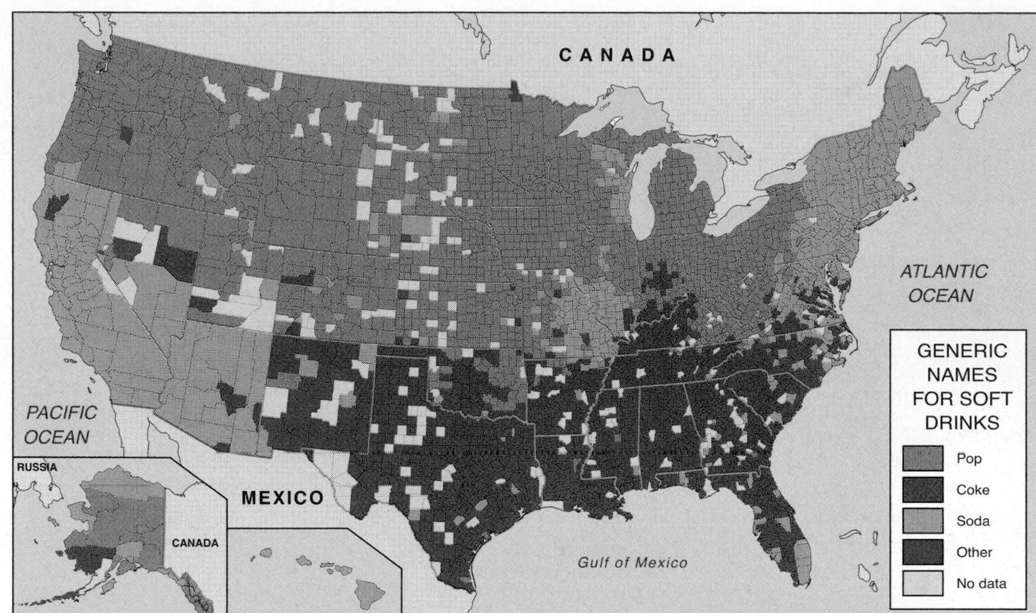

FIGURE 5-8 Soft-drink dialects. Soft drinks are called "soda" in the Northeast and Southwest, "pop" in the Midwest and Northwest, and "Coke" in the South. Map reflects voting at www.popvssoda.com.

and reflects the place of origin of most New England colonists.

It also reflects the relatively high degree of contact between the two groups. Residents of Boston, New England's main port city, maintained especially close ties to the important ports of southern England, such as London, Plymouth, and Bristol. Compared to other colonists, New Englanders received more exposure to changes in pronunciation that occurred in Britain during the eighteenth century.

The New England and southern accents sound unusual to the majority of Americans because the standard pronunciation throughout the American West comes from the Middle Atlantic states rather than the New England and Southern regions. This pattern occurred because most western settlers came from the Middle Atlantic states.

The diffusion of particular English dialects into the middle and western parts of the United States is a result of the westward movement of colonists from the three dialect regions of the East. The area of the Midwest south of the Ohio River was settled first by colonists from Virginia and the other southern areas. The Middle Atlantic colonies sent most of the early settlers north of the Ohio River, although some New Englanders moved to the Great Lakes area. The pattern by which dialects diffused westward resembles the diffusion of East Coast house types discussed in Chapter 4 (compare Figure 5-7 with Figure 4-12).

As more of the West was opened to settlement during the nineteenth century, people migrated from all parts of the East Coast. The California gold rush attracted people from throughout the East, many of whom subsequently moved to other parts of the West. The mobility of Americans has been a major reason for the relatively uniform language that exists throughout much of the West.

KEY ISSUE 2
Why Is English Related to Other Languages?

- Indo-European Branches
- Origin and Diffusion of Indo-European

English is part of the Indo-European language family. A **language family** is a collection of languages related through a common ancestral language that existed long before recorded history. Indo-European is the world's most extensively spoken language family by a wide margin. ■

Indo-European Branches

Within a language family, a **language branch** is a collection of languages related through a common ancestral language that existed several thousand years ago. Differences are not as extensive or as old as with language families, and archaeological evidence can confirm that the branches derived from the same family.

Indo-European is divided into eight branches (Figure 5-9). Four of the branches—Indo-Iranian, Romance, Germanic, and Balto-Slavic—are spoken by large numbers of people. Indo-Iranian languages are clustered in South Asia, Romance languages in southwestern Europe and Latin America, Germanic languages in northwestern Europe and North America, and Balto-Slavic languages in Eastern Europe. The four less extensively used Indo-European language branches are Albanian, Armenian, Greek, and Celtic.

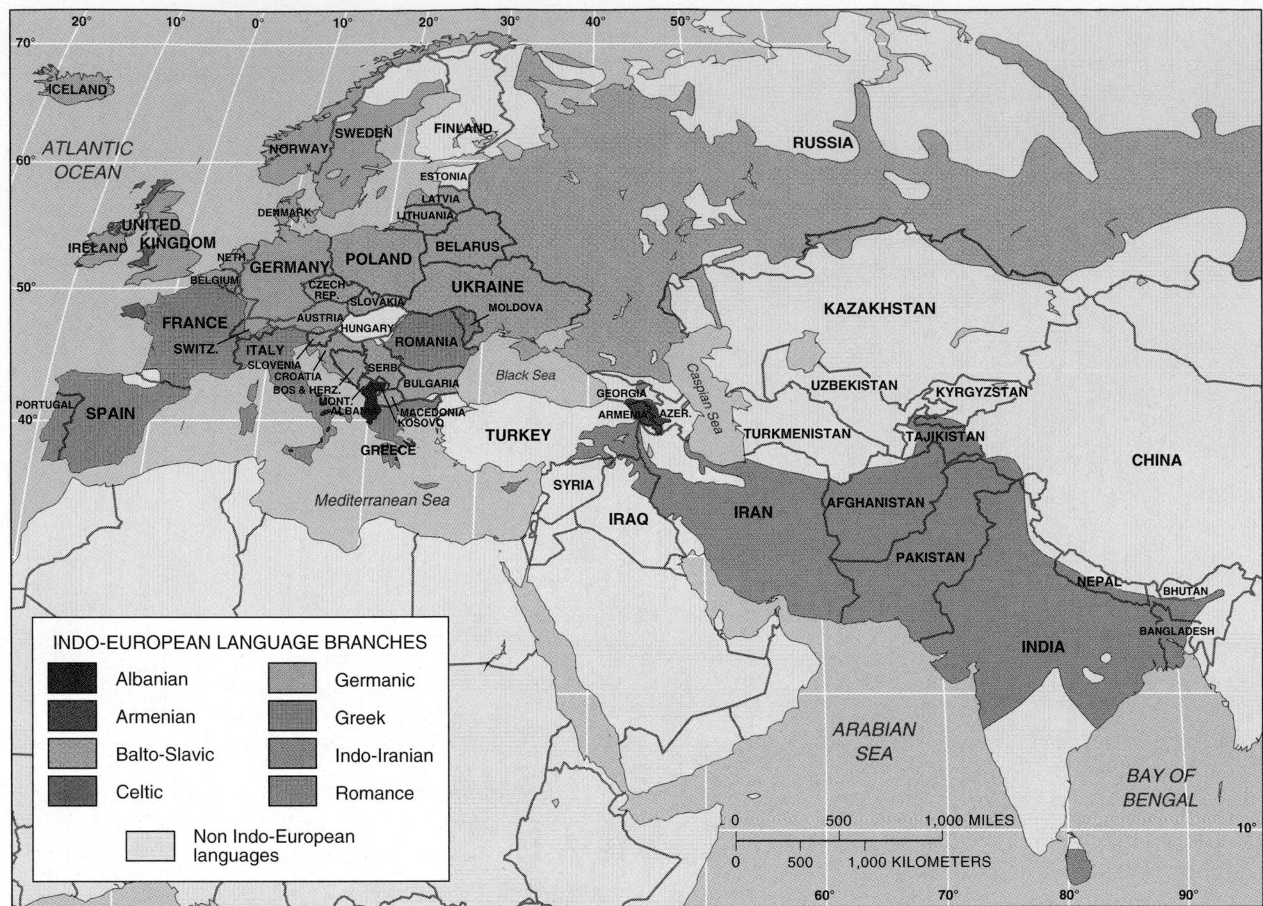

FIGURE 5-9 Branches of the Indo-European language family. Most Europeans speak languages from the Indo-European language family. In Europe, the three most important branches are Germanic (north and west), Romance (south and west), and Slavic (east). The fourth major branch, Indo-Iranian, clustered in southern and western Asia, has more than 1 billion speakers, the greatest number of any Indo-European branch.

Germanic Branch of Indo-European

German may seem a difficult language for many English speakers to learn, but the two languages are actually closely related. Both belong to the Germanic language branch of Indo-European. English is part of the Germanic branch of the Indo-European family due to the language spoken by the Germanic tribes that invaded England 1,500 years ago.

A **language group** is a collection of languages within a branch that share a common origin in the relatively recent past and display relatively few differences in grammar and vocabulary. West Germanic is the group within the Germanic branch of Indo-European to which English belongs. Although they sound very different, English and German are both languages in the West Germanic group because they are structurally similar and have many words in common (Figure 5-10).

West Germanic is further divided into High Germanic and Low Germanic subgroups, so named because they are found in high and low elevations within present-day Germany. High German, spoken in the southern mountains of Germany, is the basis for the modern standard German language. English is classified in the Low Germanic subgroup of the West Germanic group. Other Low Germanic languages include Dutch, which is spoken in the Netherlands, as well as Flemish, which is generally considered a dialect of Dutch spoken in northern Belgium. Afrikaans, a language of South Africa, is similar to Dutch, because Dutch settlers migrated to South Africa 300 years ago. Frisian is spoken by a few residents in northeastern Netherlands. A dialect of German spoken in the northern lowlands of Germany is also classified as Low Germanic.

The Germanic language branch also includes North Germanic languages, spoken in Scandinavia. The four Scandinavian languages—Swedish, Danish, Norwegian, and Icelandic—all derive from Old Norse, which was the principal language spoken throughout Scandinavia before A.D. 1000. Four distinct languages emerged after that time because of migration and the political organization of the region into four independent and isolated countries.

Indo-Iranian Branch of Indo-European

The branch of the Indo-European language family with the most speakers is Indo-Iranian. This branch includes more than 100 individual languages. The branch is divided into an eastern group (Indic) and a western group (Iranian).

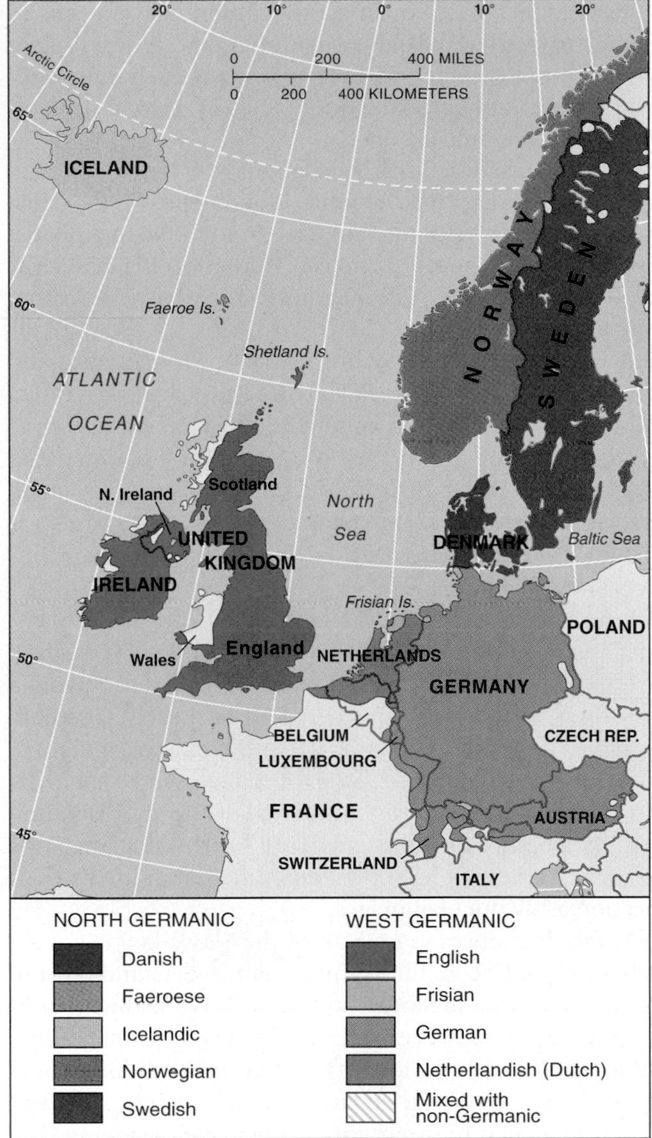

FIGURE 5-10 Germanic branch of the Indo-European language family. Germanic languages predominate in Northern and Western Europe.

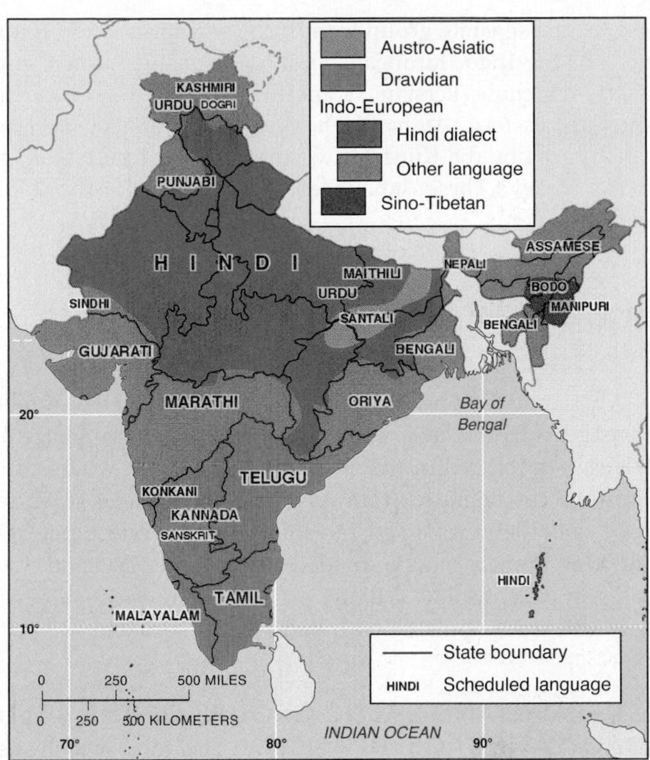

FIGURE 5-11 Languages and language families in India. India's official language is Hindi, which has many dialects. The country has 22 scheduled languages that the government is required to protect.

INDIC (EASTERN) GROUP OF INDO-IRANIAN LANGUAGE BRANCH. The most widely used languages in India, as well as in the neighboring countries of Pakistan and Bangladesh, belong to the Indo-European language family and, more specifically, to the Indic group of the Indo-Iranian branch of Indo-European.

One of the main elements of cultural diversity among the 1 billion plus residents of India is language (Figure 5-11). *Ethnologue* identifies 438 languages currently spoken in India, including 29 by at least one million people.

The official language of India is Hindi, which is an Indo-European language. Originally a variety of Hindustani spoken in the area of New Delhi, Hindi grew into a national language in the nineteenth century when the British encouraged its use in government.

After India became an independent state in 1947, Hindi was proposed as the official language, but speakers of other languages strongly objected. Consequently, English—the language of the British colonial rulers—has been retained as an official language. Speakers of different Indian languages who wish to communicate with each other sometimes are forced to turn to English as a common language.

India also recognizes 22 so-called scheduled languages, including 15 Indo-European (Assamese, Bengali, Dogri, Gujarati, Hindi, Kashmiri, Konkani, Maithili, Marathi, Nepali, Oriya, Panjabi, Sanskrit, Sindhi, and Urdu), four Dravidian (Kannada, Malayalam, Tamil, and Telugu), two Sino-Tibetan (Bodo and Manipuri), and one Austro-Asiatic (Santali). The government of India is obligated to encourage the use of these languages.

Hindi is spoken many different ways—and therefore could be regarded as a collection of many individual languages. But there is only one official way to write Hindi, using a script called Devanagari, which has been used in India since the seventh century A.D. For example, the word for *sun* is written in Hindi as सूरज pronounced "surya." Local differences arose in the spoken forms of Hindi but not in the written form because until recently few speakers of that language could read or write it.

Adding to the complexity, Urdu is spoken very much like Hindi, but it is recognized as a distinct language. Urdu is written with the Arabic alphabet, a legacy of the fact that most of its speakers are Muslims and their holiest book (the Quran) is written in Arabic.

IRANIAN (WESTERN) GROUP OF INDO-IRANIAN LANGUAGE BRANCH. Indo-Iranian languages are also spoken in Iran and neighboring countries in southwestern Asia.

These form a separate group from Indic within the Indo-Iranian branch of the Indo-European family. The major Iranian group languages include Persian (sometimes called Farsi) in Iran, Pashto in eastern Afghanistan and western Pakistan, and Kurdish, used by the Kurds of western Iran, northern Iraq, and eastern Turkey. These languages are written in the Arabic alphabet.

Balto-Slavic Branch of Indo-European

The other Indo-European language branch with large numbers of speakers is Balto-Slavic. Slavic was once a single language, but differences developed in the seventh century A.D. when several groups of Slavs migrated from Asia to different areas of Eastern Europe and thereafter lived in isolation from one other. As a result, this branch can be divided into East, West, and South Slavic groups as well as a Baltic group. Figure 7–30 shows the widespread area populated with Balto-Slavic speakers.

EAST SLAVIC AND BALTIC GROUPS OF THE BALTO-SLAVIC LANGUAGE BRANCH. The most widely used Slavic languages are the eastern ones, primarily Russian, which is spoken by more than 80 percent of Russian people. Russian is one of the six official languages of the United Nations.

The importance of Russian increased with the Soviet Union's rise to power after the end of World War II in 1945. Soviet officials forced native speakers of other languages to learn Russian as a way of fostering cultural unity among the country's diverse peoples. In Eastern European countries that were dominated politically and economically by the Soviet Union, Russian was taught as the second language. The presence of so many non-Russian speakers was a measure of cultural diversity in the Soviet Union, and the desire to use languages other than Russian was a major drive in its breakup. With the demise of the Soviet Union, the newly independent republics adopted official languages other than Russian, although Russian remains the language for communications among officials in the countries that were formerly part of the Soviet Union.

After Russian, Ukrainian and Belarusan are the two most important East Slavic languages and are the official languages in Ukraine and Belarus. *Ukraine* is a Slavic word meaning "border," and *Bela-* is translated as "white."

WEST AND SOUTH SLAVIC GROUPS OF THE BALTO-SLAVIC LANGUAGE BRANCH. The most spoken West Slavic language is Polish, followed by Czech and Slovak. The latter two are quite similar, and speakers of one can understand the other.

The government of the former state of Czechoslovakia tried to balance the use of the two languages, even though the country contained twice as many Czechs as Slovaks. For example, the announcers on televised sports events used one of the languages during the first half and switched to the other for the second half. These balancing measures were effective in promoting national unity during the Communist era, but in 1993, four years after the fall of communism, Slovakia split from the Czech Republic. Slovaks rekindled their long-suppressed resentment of perceived dominance of the national culture by the Czech ethnic group.

The most important South Slavic language is the one spoken in Bosnia and Herzegovina, Croatia, Montenegro, and Serbia. Bosnians and Croats write the language in the Roman alphabet (what you are reading now), whereas Montenegrans and Serbs use the Cyrillic alphabet (for example, *Yugoslavia* is written Југославија.

When Bosnia and Herzegovina, Croatia, Montenegro, and Serbia were all part of Yugoslavia, the language was called Serbo-Croatian. This name now offends Bosnians and Croatians because it recalls when they were once in a country that was dominated by Serbs. Instead, the names *Bosnian, Croatian,* and *Serbian* are preferred by people in these countries, to demonstrate that each language is unique, even though linguists consider them one.

Differences have crept into the language of the South Slavs. Bosnian Muslims have introduced Arabic words used in their religion, and Croats have replaced words regarded as having a Serbian origin with words considered to be purely Croatian. For example, the Serbo-Croatian word for martyr or hero—*junak*—has been changed to *heroj* by Croats and *shahid* by Bosnian Muslims. In the future, after a generation of isolation and hostility among Bosnians, Croats, and Serbs, the languages spoken by the three may be sufficiently different to justify their classification as distinct languages.

In general, differences among all of the Slavic languages are relatively small. A Czech, for example, can understand most of what is said or written in Slovak and could become fluent without much difficulty. However, because language is a major element in a people's cultural identity, relatively small differences among Slavic as well as other languages are being preserved and even accentuated in recent independence movements.

Romance Branch of Indo-European

The Romance language branch evolved from the Latin language spoken by the Romans 2,000 years ago. The four most widely used contemporary Romance languages are Spanish, Portuguese, French, and Italian (Figure 5-12). Spanish and French are two of the six official languages of the United Nations.

The European regions in which these four languages are spoken correspond somewhat to the boundaries of the modern states of Spain, Portugal, France, and Italy. Rugged mountains serve as boundaries among these four countries. France is separated from Italy by the Alps and from Spain by the Pyrenees, and several mountain ranges mark the border between Spain and Portugal. Physical boundaries such as mountains are strong intervening obstacles, creating barriers to communication between people living on opposite sides.

The fifth most important Romance language, Romanian, is the principal language of Romania and Moldova. It is separated

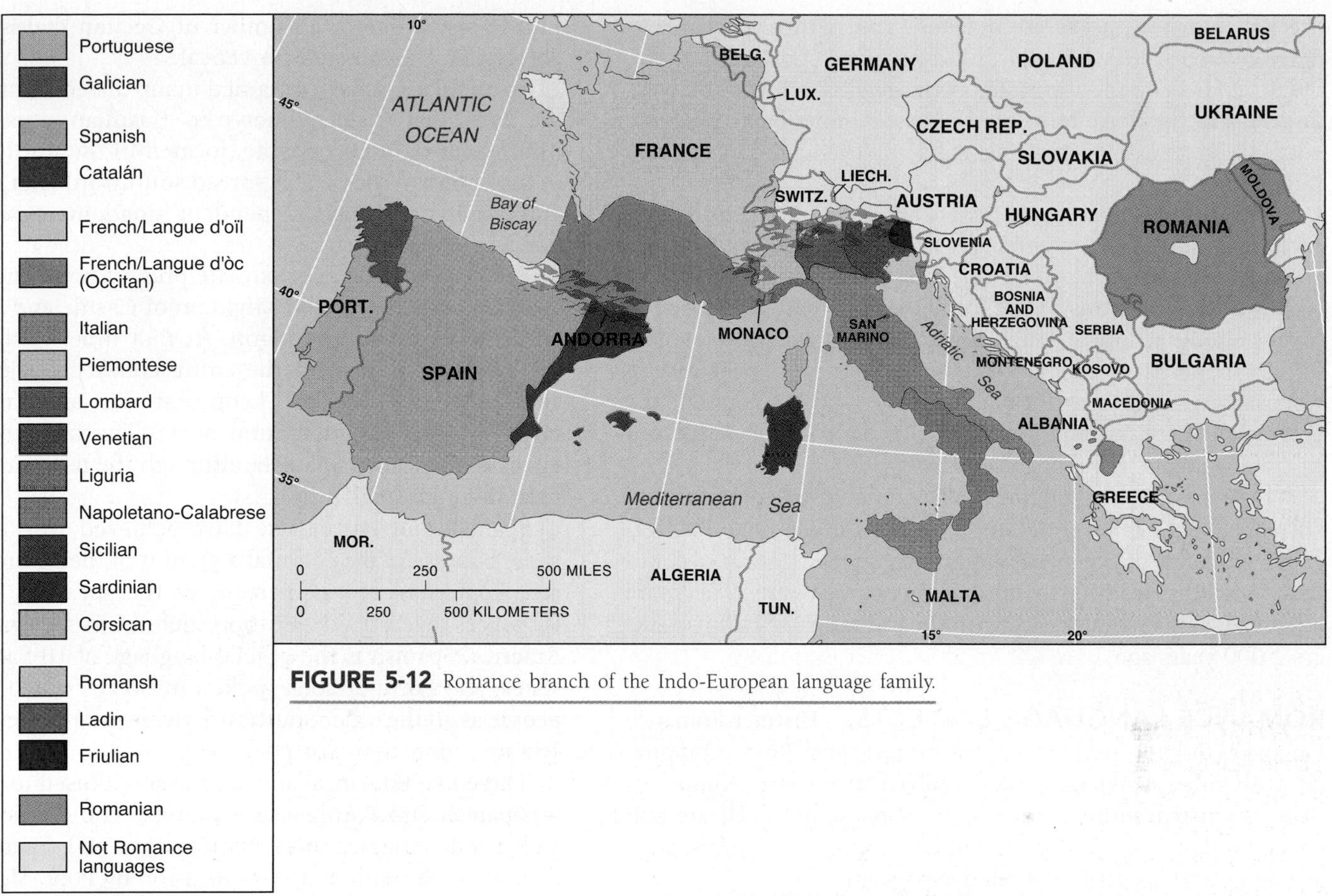

FIGURE 5-12 Romance branch of the Indo-European language family.

from the other Romance-speaking European countries by Slavic-speaking peoples.

The distribution of Romance languages shows the difficulty in trying to establish the number of distinct languages in the world. In addition to the five languages already mentioned, two other official Romance languages are Romansh and Catalán. Romansh is one of four official languages of Switzerland, although it is spoken by only 40,000 people. Catalán is the official language of Andorra, a tiny country of 70,000 inhabitants situated in the Pyrenees Mountains between Spain and France. Catalán is also spoken by 6 million people in eastern Spain and is the official language of Spain's highly autonomous Catalonia province, centered on the city of Barcelona. A third Romance language, Sardinian—a mixture of Italian, Spanish, and Arabic—was once the official language of the Mediterranean island of Sardinia.

In addition to these official languages, several other Romance languages have individual literary traditions. In Italy, Ladin (not Latin) is spoken by 30,000 people living in the South Tyrol, and Friulian is spoken by 800,000 people in the northeast. Ladin and Friulian (along with the official Romansh) are dialects of Rhaeto-Romanic.

A Romance tongue called Ladino—a mixture of Spanish, Greek, Turkish, and Hebrew—is spoken by 100,000 Sephardic Jews, most of whom now live in Israel. None of these languages have an official status in any country, although they are used in literature.

ORIGIN AND DIFFUSION OF ROMANCE LANGUAGES.

The Romance languages, including Spanish, Portuguese, French, Italian, and Romanian, are part of the same branch because they all developed from Latin, the "Romans' language." The rise in importance of the city of Rome 2,000 years ago brought a diffusion of its Latin language.

At its height in the second century A.D., the Roman Empire extended from the Atlantic Ocean on the west to the Black Sea on the east and encompassed all lands bordering the Mediterranean Sea (the empire's boundary is shown in Figure 6-9). As the conquering Roman armies occupied the provinces of this vast empire, they brought the Latin language with them. In the process, the languages spoken by the natives of the provinces were either extinguished or suppressed in favor of the language of the conquerors.

Even during the period of the Roman Empire, Latin varied to some extent from one province to another. The empire grew over a period of several hundred years, so the Latin used in each province was based on that spoken by the Roman army at the time of occupation. The Latin spoken in each province also integrated words from the language formerly spoken in the area.

The Latin that people in the provinces learned was not the standard literary form but a spoken form, known as **Vulgar Latin** from the Latin word referring to "the masses" of the populace. Vulgar Latin was introduced to the provinces by the soldiers stationed throughout the empire. For example, the literary term for "horse" was *equus,* from which English has derived such words as *equine* and *equestrian.* The Vulgar term, used by the common people, was *caballus,* from which are derived the modern terms for "horse" in Italian (*cavallor*), Spanish (*caballo*), Portuguese (*cavalo*), French (*cheval*), and Romanian (*cal*).

Following the collapse of the Roman Empire in the fifth century, communication among the former provinces declined, creating still greater regional variation in spoken Latin. By the eighth century, regions of the former empire had been isolated from each other long enough for distinct languages to evolve. But Latin persisted in parts of the former empire. People in some areas reverted to former languages; others adopted the languages of conquering groups of people from the north and east who spoke Germanic and Slavic.

In the past, when migrants were unable to communicate with speakers of the same language back home, major differences emerged between the languages spoken in the old and new locations, leading to the emergence of distinct, separate languages. This was the case with the migration of Latin speakers 2,000 years ago.

ROMANCE LANGUAGE DIALECTS. Distinct Romance languages did not suddenly appear in the former Roman Empire. As with other languages, they evolved over time. Numerous dialects existed within each province, many of which are still spoken today. The creation of standard national languages, such as French and Spanish, was relatively recent.

The dialect of the Île-de-France region, known as *Francien,* became the standard form of French because the region included Paris, which became the capital and largest city of the country. *Francien* French became the country's official language in the sixteenth century, and local dialects tended to disappear as a result of the capital's longtime dominance over French political, economic, and social life.

The most important surviving dialect difference within France is between the north and the south (refer to Figure 5-11). The northern dialect is known as *langue d'oïl* and the southern as *langue d'òc.* It is worth exploring these names, for they provide insight into how languages evolve.

These terms derive from different ways in which the word for "yes" was said. One Roman term for "yes" was *hoc illud est,* meaning "that is so." In the south, the phrase was shortened to *hoc,* or *òc,* because the /h/ sound was generally dropped, just as we drop it on the word *honor* today. Northerners shortened the phrase to *o-il* after the first sound in the first two words of the phrase, again with the initial /h/ suppressed. If the two syllables of *o-il* are spoken very rapidly, they are combined into a sound like the English word *wheel.* Eventually, the final consonant was eliminated, as in many French words, giving a sound for "yes" like the English *we,* spelled in French *oui.*

A province where the southern dialect is spoken in southwestern France is known as Languedoc. The southern French dialect is now sometimes called Occitan, derived from the French region of Aquitaine, which in French has a similar pronunciation to Occitan. About 2 million people in southern France speak one of a number of Occitan dialects, including Auvergnat, Gascon, and Provençal.

Spain, like France, contained many dialects during the Middle Ages. One dialect, known as Castilian, arose during the ninth century in Old Castile, located in the north-central part of the country. The dialect spread southward over the next several hundred years as independent kingdoms were unified into one large country.

Spain grew to its approximate present boundaries in the fifteenth century, when the Kingdom of Castile and Léon merged with the Kingdom of Aragón. At that time, Castilian became the official language for the entire country. Regional dialects, such as Aragón, Navarre, Léon, Asturias, and Santander, survived only in secluded rural areas. The official language of Spain is now called Spanish, although the term *Castilian* is still used in Latin America.

Spanish and Portuguese have achieved worldwide importance because of the colonial activities of their European speakers. Approximately 90 percent of the speakers of these two languages live outside Europe, mainly in Central and South America. Spanish is the official language of 18 Latin American states, and Portuguese is spoken in Brazil, which has as many people as all the other South American countries combined and 18 times more than Portugal itself.

These two Romance languages were diffused to the Americas by Spanish and Portuguese explorers. The division of Central and South America into Portuguese- and Spanish-speaking regions is the result of a 1493 decision by Pope Alexander VI to give the western portion of the New World to Spain and the eastern part to Portugal. The Treaty of Tordesillas, signed a year later, carried out the papal decision.

The Portuguese and Spanish languages spoken in the Western Hemisphere differ somewhat from their European versions, as is the case with English. The members of the Spanish Royal Academy meet every week in a mansion in Madrid to clarify rules for the vocabulary, spelling, and pronunciation of the Spanish language around the world. The Academy's official dictionary, published in 1992, has added hundreds of "Spanish" words that originated either in the regional dialects of Spain or the Indian languages of Latin America.

Brazil, Portugal, and several Portuguese-speaking countries in Africa agreed in 1994 to standardize the way their common language is written. Many people in Portugal are upset that the new standard language more closely resembles the Brazilian version, which eliminates most of the accent marks—such as tildes (São Paulo), cedillas (Alcobaça), circumflexes (Estância), and hyphens—and the agreement recognizes as standard thousands of words that Brazilians have added to the language.

The standardization of Portuguese is a reflection of the level of interaction that is possible in the modern world between groups of people who live tens of thousands of kilometers apart. Books and television programs produced in one country diffuse rapidly to other countries where the same language is used.

DISTINGUISHING BETWEEN DIALECTS AND LANGUAGES. Difficulties arise in determining whether two languages are distinct or whether they are merely two dialects of the same language:

- Galician, spoken in northwestern Spain and northeastern Portugal, is as distinct from Portuguese as, say, Catalán is from Spanish. However, Catalán is generally classified as a distinct language, and Galician is classified as a dialect of Portuguese.
- Moldovan is the official language of Moldova yet is generally classified as a dialect of Romanian.
- Flemish, the official language of northern Belgium, is generally considered a dialect of Dutch.

Several languages of Italy are viewed as different enough to merit consideration as languages distinct from Italian according to *Ethnologue*. In southern Italy, the most widespread of those possible distinct languages are Napoletano-Calebrese, spoken by 7 million people, and Sicilian, spoken by 5 million. In the north, the most widespread are Lombard, spoken by 9 million people; Piemontese, spoken by 3 million; and Emiliano-Romagnolo, Liguria, and Venetian, spoken by 2 million each. Distinguishing individual languages from dialects is difficult, because many speakers choose to regard their languages as distinct.

Romance languages spoken in some former colonies can also be classified as separate languages because they differ substantially from the original introduced by European colonizers. Examples include French Creole in Haiti, Papiamento (creolized Spanish) in Netherlands Antilles (West Indies), and Portuguese Creole in the Cape Verde Islands off the African coast.

A **creole or creolized language** is defined as a language that results from the mixing of the colonizer's language with the indigenous language of the people being dominated (Figure 5-13). A creolized language forms when the colonized group adopts the language of the dominant group but makes some changes, such as simplifying the grammar and adding words from their former language. The word *creole* derives from a word in several Romance languages for a slave who is born in the master's house.

FIGURE 5-13 Creole. A mix of French and English adorns this public bus in Port-au-Prince, capital of Haiti. French and Haitian Creole, a dialect of French, are both official languages in Haiti, although English is the lingua franca in Haiti because of the country's proximity to the United States.

Origin and Diffusion of Indo-European

If Germanic, Romance, Balto-Slavic, and Indo-Iranian languages are all part of the same Indo-European language family, then they must be descended from a single common ancestral language. Unfortunately, the existence of a single ancestor—which can be called Proto-Indo-European—cannot be proved with certainty, because it would have existed thousands of years before the invention of writing or recorded history.

The evidence that Proto-Indo-European once existed is "internal," derived from the physical attributes of words themselves in various Indo-European languages. For example, the words for some animals and trees in modern Indo-European languages have common roots, including *beech*, *oak*, *bear*, *deer*, *pheasant*, and *bee*. Because all Indo-European languages share these similar words, linguists believe the words must represent things experienced in the daily lives of the original Proto-Indo-European speakers. In contrast, words for other features, such as *elephant*, *camel*, *rice*, and *bamboo*, have different roots in the various Indo-European languages. Such words therefore cannot be traced back to a common Proto-Indo-European ancestor and must have been added later, after the root language split into many branches. Individual Indo-European languages share common root words for *winter* and *snow* but not for *ocean*. Therefore, linguists conclude that original Proto-Indo-European speakers probably lived in a cold climate, or one that had a winter season, but did not come in contact with oceans.

Linguists and anthropologists generally accept that Proto-Indo-European must have existed, but they disagree on when and where the language originated and the process and routes by which it diffused. The debate over place of origin and paths of diffusion is significant, because one theory argues that language diffused primarily through warfare and conquest, and the other theory argues that the diffusion resulted from peaceful sharing of food. So where did Indo-European originate? Not surprisingly, scholars disagree on where and when the first speakers of Proto-Indo-European lived.

- **Nomadic Warrior Thesis.** One influential hypothesis, espoused by Marija Gimbutas, is that the first Proto-Indo-European speakers were the Kurgan people, whose homeland was in the steppes near the border between present-day Russia and Kazakhstan. The earliest archaeological evidence of the Kurgans dates to around 4300 B.C.

The Kurgans were nomadic herders. Among the first to domesticate horses and cattle, they migrated in search of grasslands for their animals. This took them westward through Europe, eastward to Siberia, and southeastward to Iran and South Asia. Between 3500 and 2500 B.C., Kurgan warriors, using their domesticated horses as weapons, conquered much of Europe and South Asia (Figure 5-14).

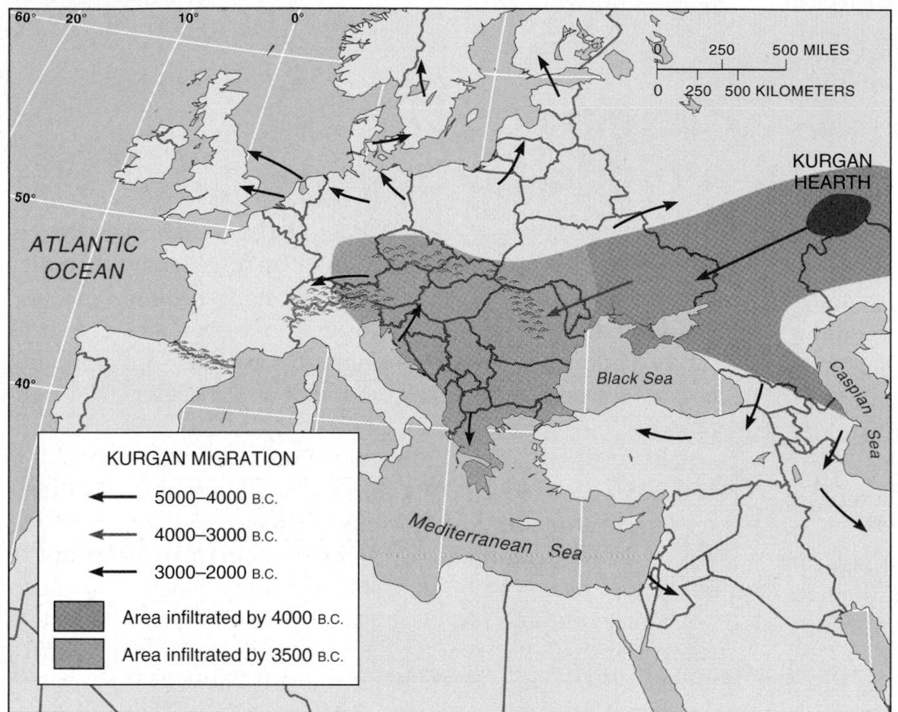

FIGURE 5-14 Origin and diffusion of Indo-European (nomadic warrior theory). The Kurgan homeland was north of the Caspian Sea, near the present-day border between Russia and Kazakhstan. According to this theory, the Kurgans may have infiltrated into Eastern Europe beginning around 4000 B.C. and into central Europe and southwestern Asia beginning around 2500 B.C.

- **Sedentary Farmer Thesis.** Archaeologist Colin Renfrew argues that the first speakers of Proto-Indo-European lived 2,000 years before the Kurgans, in eastern Anatolia, part of present-day Turkey (Figure 5-15). Biologist Russell D. Gray supports the Renfrew position but dates the first speakers even earlier, at around 6700 B.C.

  Renfrew believes they diffused from Anatolia westward to Greece (the origin of the Greek language branch) and from Greece westward toward Italy, Sicily, Corsica, the Mediterranean coast of France, Spain, and Portugal (the origin of the Romance language branch). From the Mediterranean coast, the speakers migrated northward toward central and northern France and on to the British Isles (perhaps the origin of the Celtic language branch).

  Indo-European is also said to have diffused northward from Greece toward the Danube River (Romania) and westward to central Europe, according to Renfrew. From there the language diffused northward toward the Baltic Sea (the origin of the Germanic language branch) and eastward toward the Dnestr River near Ukraine (the origin of the Slavic language branch). From the Dnestr River, speakers migrated eastward to the Dnepr River (the homeland of the Kurgans).

  The Indo-Iranian branch of the Indo-European language family originated either directly through migration from Anatolia along the south shores of the Black and Caspian seas by way of Iran and Pakistan, or indirectly by way of Russia north of the Black and Caspian seas.

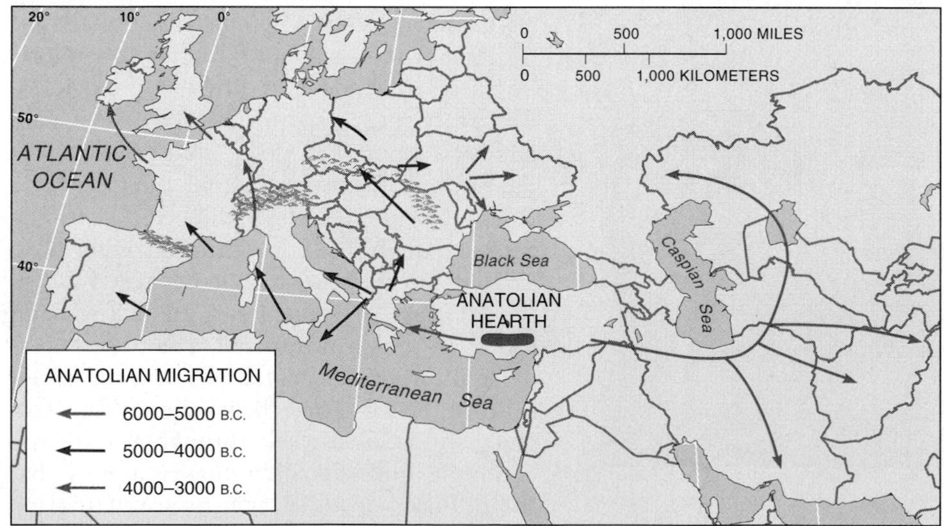

FIGURE 5-15 Origin and diffusion of Indo-European (sedentary farmer theory). Indo-European may have originated in present-day Turkey 2,000 years before the Kurgans. According to this theory, the language diffused along with agricultural innovations west into Europe and east into Asia.

Renfrew argues that Indo-European diffused into Europe and South Asia along with agricultural practices rather than by military conquest. The language triumphed because its speakers became more numerous and prosperous by growing their own food instead of relying on hunting.

Regardless of how Indo-European diffused, communication was poor among different peoples, whether warriors or farmers. After many generations of complete isolation, individual groups evolved increasingly distinct languages.

KEY ISSUE 3

Where Are Other Language Families Distributed?

- **Classification of Languages**
- **Distribution of Language Families**

This section describes where different languages are found around the world. The several thousand spoken languages can be organized logically into a small number of language families. Larger language families can be further divided into language branches and language groups. ■

Classification of Languages

Figure 5-16 shows the world's language families:

- A language in the ***Indo-European family***, such as English, is spoken by 46 percent of the world's people.
- A language in the ***Sino-Tibetan family***, such as Mandarin, is spoken by 21 percent of the world, mostly in China.
- A language in the ***Afro-Asiatic family***, including Arabic, is spoken by 6 percent, mostly in the Middle East
- A language in the ***Austronesian family*** is spoken by 6 percent, mostly in Southeast Asia.
- A language in the ***Niger-Congo family*** is spoken by 6 percent, mostly in Africa.
- A language in the ***Dravidian family*** is spoken by 4 percent, mostly in India.
- A language in the ***Altaic family*** is spoken by 2 percent, mostly in Asia.
- A language in the ***Austro-Asiatic family*** is spoken by 2 percent, mostly in Southeast Asia.
- ***Japanese***, a separate language family, is spoken by 2 percent.
- The remaining 5 percent of the world's people speak a language belonging to one of 100 smaller families.

Figure 5-17 attempts to depict differences among language families, branches, and groups. Language families form the trunks of the trees, whereas individual languages are displayed as leaves. The larger the trunks and leaves are, the greater the number of speakers of those families and languages. Some trunks divide into several branches, which logically represent language branches. The branches representing Germanic, Balto-Slavic, and Indo-Iranian in Figure 5-17 divide a second time into language groups.

Figure 5-17 displays each language family as a separate tree at ground level because differences among families predate recorded history. Linguists speculate that language families were joined together as a handful of superfamilies tens of thousands of years ago. Superfamilies are shown as roots below the surface because their existence is highly controversial and speculative.

Distribution of Language Families

Nearly one-half the people in the world speak an Indo-European language. The second-largest family is Sino-Tibetan, spoken by one-fifth of the world. Another half-dozen families account for most of the remainder.

Sino-Tibetan Family

The Sino-Tibetan family encompasses languages spoken in the People's Republic of China—the world's most populous state at more than 1 billion—as well as several smaller countries in Southeast Asia. The languages of China generally belong to the Sinitic branch of the Sino-Tibetan family.

There is no single Chinese language. Rather, the most important is Mandarin (or, as the Chinese call it, *pu tong hua*—"common speech"). Spoken by approximately three-fourths of the Chinese people, Mandarin is by a wide margin the most used language in the world. Once the language of emperors in Beijing, Mandarin is now the official language of both the People's Republic of China and Taiwan, as well as one of the six official languages of the United Nations. Other Sinitic branch languages are spoken by tens of millions of people in China, mostly in the southern and eastern parts of the country—Wu, Yue (also known as Cantonese), Min, Jinyu, Xiang, Hakka, and Gan. However, the Chinese government is imposing Mandarin countrywide.

The relatively small number of languages in China (compared to India, for example) is a source of national strength and unity. Unity is also fostered by a consistent written form for all Chinese languages. Although the words are pronounced differently in each language, they are written the same way.

You already know the general structure of Indo-European quite well because you are a fluent speaker of at least one Indo-European language. But the structure of Chinese languages is quite different (Figure 5-18). They are based on 420 one-syllable words. This number far exceeds the possible one-syllable sounds that humans can make, so Chinese languages use each sound to denote more than one thing. The sound *shi,* for example, may mean "lion," "corpse," "house," "poetry," "ten," "swear," or "die." The sound *jian* has more than 20 meanings, including "to see." The listener must infer the meaning from the context in the sentence and the intonation the speaker uses. In addition, two one-syllable words can be combined into two syllables, forming a new word. For example, the two-syllable word "Shanghai" is a

FIGURE 5-16 Language families. Most language can be classified into one of a handful of language families. The pie chart shows the percentage of people who speak a language from each major family. Languages that have more than 50 million speakers are identified on the map.

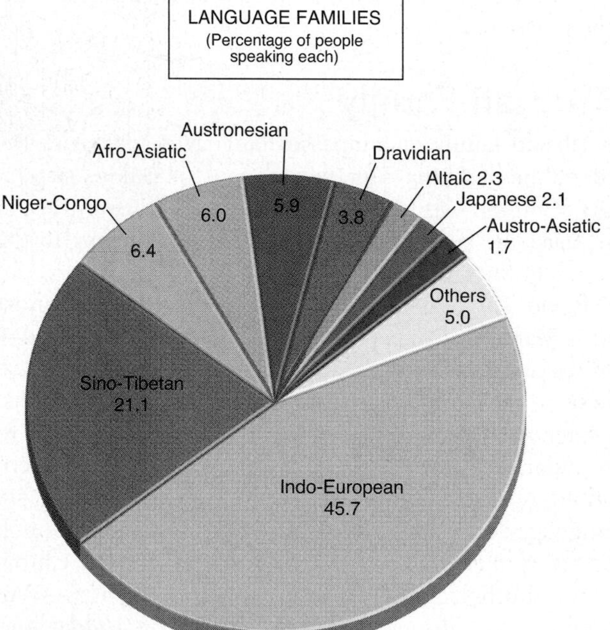

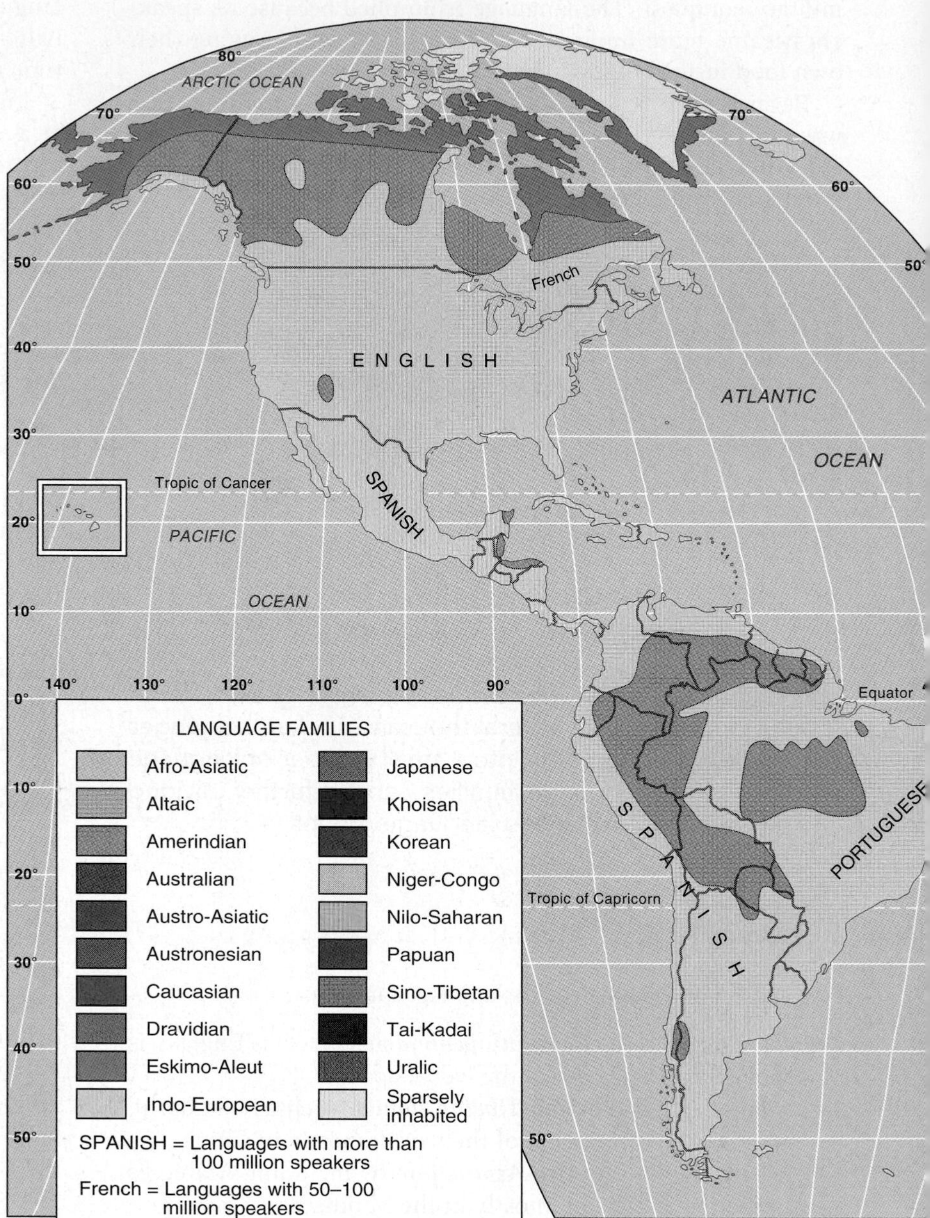

combination of words that mean "above" and "sea." *Kan jian*—a combination of the words for "look" and "see," which would be redundant in English—clarifies that "to see" is the intended meaning for the multiple meanings of *jian*.

The other distinctive characteristic of the Chinese languages is the method of writing. The Chinese languages are written with a collection of thousands of characters. Some of the characters represent sounds pronounced in speaking, as in English. However, most are **ideograms**, which represent ideas or concepts, not specific pronunciations. The system is intricate and mature, having developed over 4,000 years. The main language problem for the Chinese is the difficulty in learning to write because of the large number of characters. The Chinese government reports that 16 percent of the population over age 16 is unable to read or write more than a few characters.

Other East and Southeast Asian Language Families

In addition to Sino-Tibetan, several other language families spoken by large numbers of people can be found in East and Southeast Asia. If you look at their distribution in Figure 5-16, you can see a physical reason for their independent development: These language families are clustered either on islands or peninsulas.

- **Austronesian.** Spoken by about 6 percent of the world's people, speakers of Austronesian languages are mostly in Indonesia, the world's fourth most populous country. With its inhabitants dispersed among thousands of islands, Indonesia has an extremely large number of distinct

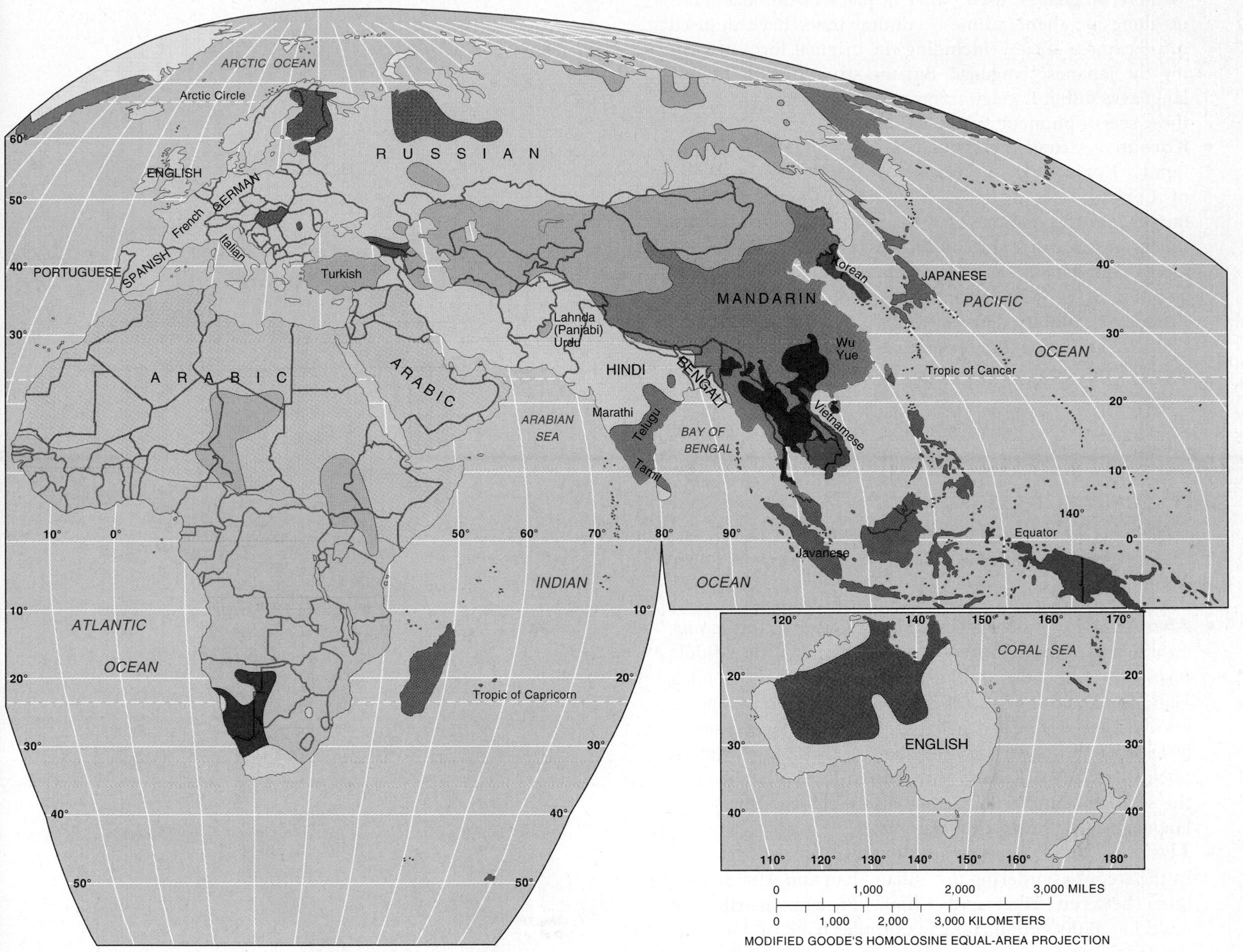

languages and dialects; 722 actively used languages are identified by *Ethnologue*. Indonesia's most widely used first language is Javanese, spoken by 85 million people, mostly on the island of Java, where two-thirds of the country's population is clustered.

Language maps show a striking oddity: The people of Madagascar, the large island off the east coast of Africa, speak Malagasy, which belongs to the Austronesian family, even though the island is 3,000 kilometers (1,900 miles) distant from any other Austronesian-speaking country. This is strong evidence of migration to Madagascar from present-day Indonesia. Malayo-Polynesian people apparently sailed in small boats across the Indian Ocean to reach Madagascar approximately 2,000 years ago.

- **Austro-Asiatic.** Spoken by about 2 percent of the world's population, Austro-Asiatic is based in Southeast Asia. Vietnamese, the most spoken tongue of the Austro-Asiatic language family, is written with our familiar Roman alphabet, with the addition of a large number of diacritical marks above the vowels. The Vietnamese alphabet was devised in the seventh century by Roman Catholic missionaries.
- **Tai Kadai.** Once classified as a branch of Sino-Tibetan, the principal languages of this family are spoken in Thailand and neighboring portions of China. Similarities with the Austronesian family lead some linguistic scholars to speculate that people speaking these languages may have migrated from the Philippines.
- **Japanese.** Written in part with Chinese ideograms, Japanese also uses two systems of phonetic symbols, like

Western languages, used either in place of the ideograms or alongside them. Chinese cultural traits have diffused into Japanese society, including the original form of writing the Japanese language. But the structures of the two languages differ. Foreign terms may be written with one of these sets of phonetic symbols.

- **Korean.** Usually classified as a separate language family, Korean may be related to the Altaic languages of Central Asia or to Japanese. Unlike Sino-Tibetan languages and Japanese, Korean is written not with ideograms but in a system known as *hankul* (also called *hangul* and *onmun*). In this system, each letter represents a sound, as in Western languages. More than half of the Korean vocabulary derives from Chinese words. In fact, Chinese and Japanese words are the principal sources for creating new words to describe new technology and concepts.

Languages of the Middle East and Central Asia

Major language families in the Middle East and Central Asia include Afro-Asiatic and Altaic. Uralic languages were once classified with Altaic.

- **Afro-Asiatic.** Arabic is the major language of this family, an official language in two dozen countries of the Middle East, and one of six official languages of the United Nations. In addition to the 200-million-plus native speakers of Arabic, a large percentage of the world's Muslims have at least some knowledge of Arabic because Islam's holiest book, the Quran (Koran), was written in that language in the seventh century. This family also includes Hebrew, the language of the Judeo-Christian Bible.
- **Altaic.** These languages are thought to have originated in the steppes bordering the Qilian Shan and Altai mountains between Tibet and China. Present distribution covers an 8,000-kilometer (5,000-mile) band of Asia. The Altaic language with by far the most speakers is Turkish.

Turkish was once written with Arabic letters. But in 1928 the Turkish government, led by Kemal Ataturk, ordered that the language be written with the Roman alphabet instead. Ataturk believed that switching to Roman letters would help modernize the economy and culture of Turkey through increased communications with European countries.

When the Soviet Union governed most of the Altaic-speaking region of Central Asia, use of Altaic languages was suppressed to create a homogeneous national culture. One element of Soviet policy was to force everyone to write with the Russian Cyrillic alphabet, even though some had traditionally employed Arabic letters. With the dissolution of the Soviet Union in the early 1990s, Altaic languages became official in several newly independent countries, including Azerbaijan, Kazakhstan, Kyrgyzstan, Turkmenistan, and Uzbekistan. People in these countries are no longer forced to learn Russian and write with Cyrillic letters.

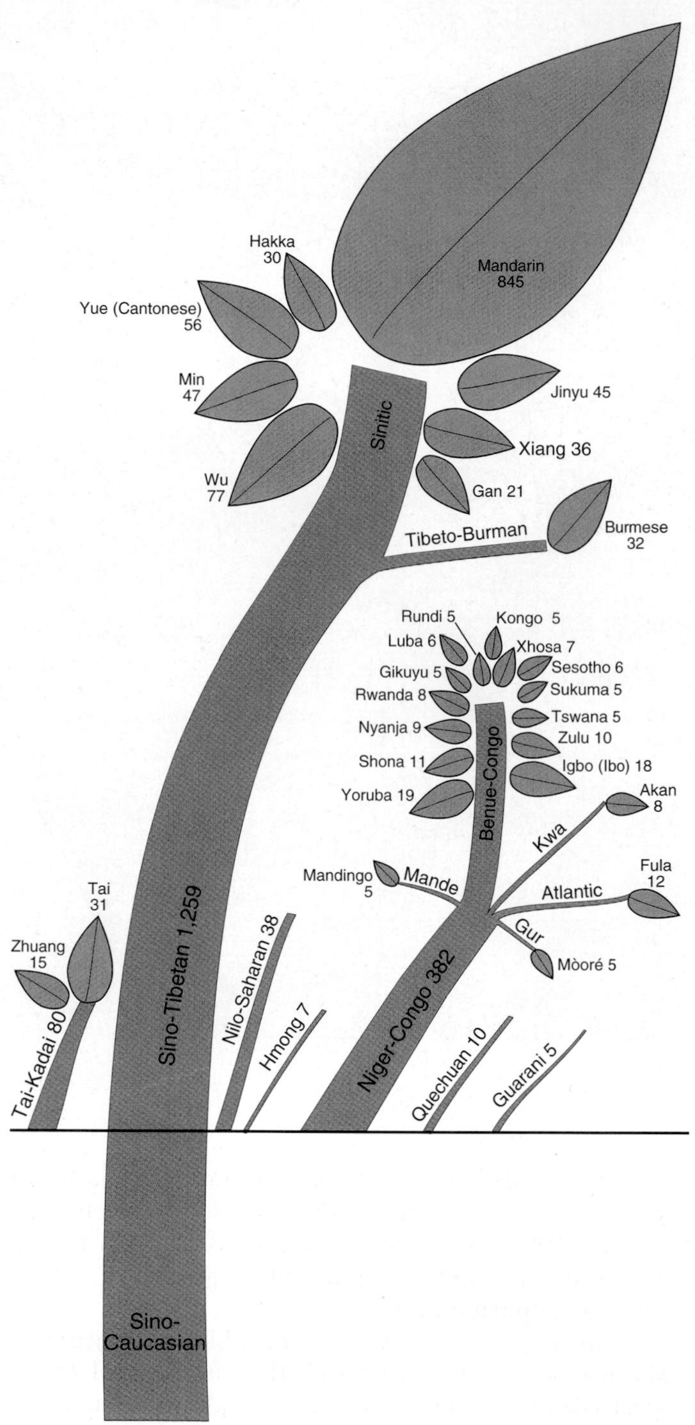

FIGURE 5-17 Language family tree. Language families are divided into branches and groups. Shown here are language families and individual languages that have more than 5 million speakers. Numbers on the tree are in millions of speakers. Below ground level, the language tree's "roots" are shown. However, the theory that several language families had common origins tens of thousands of years ago is a highly controversial speculation advocated by some linguists and rejected by others.

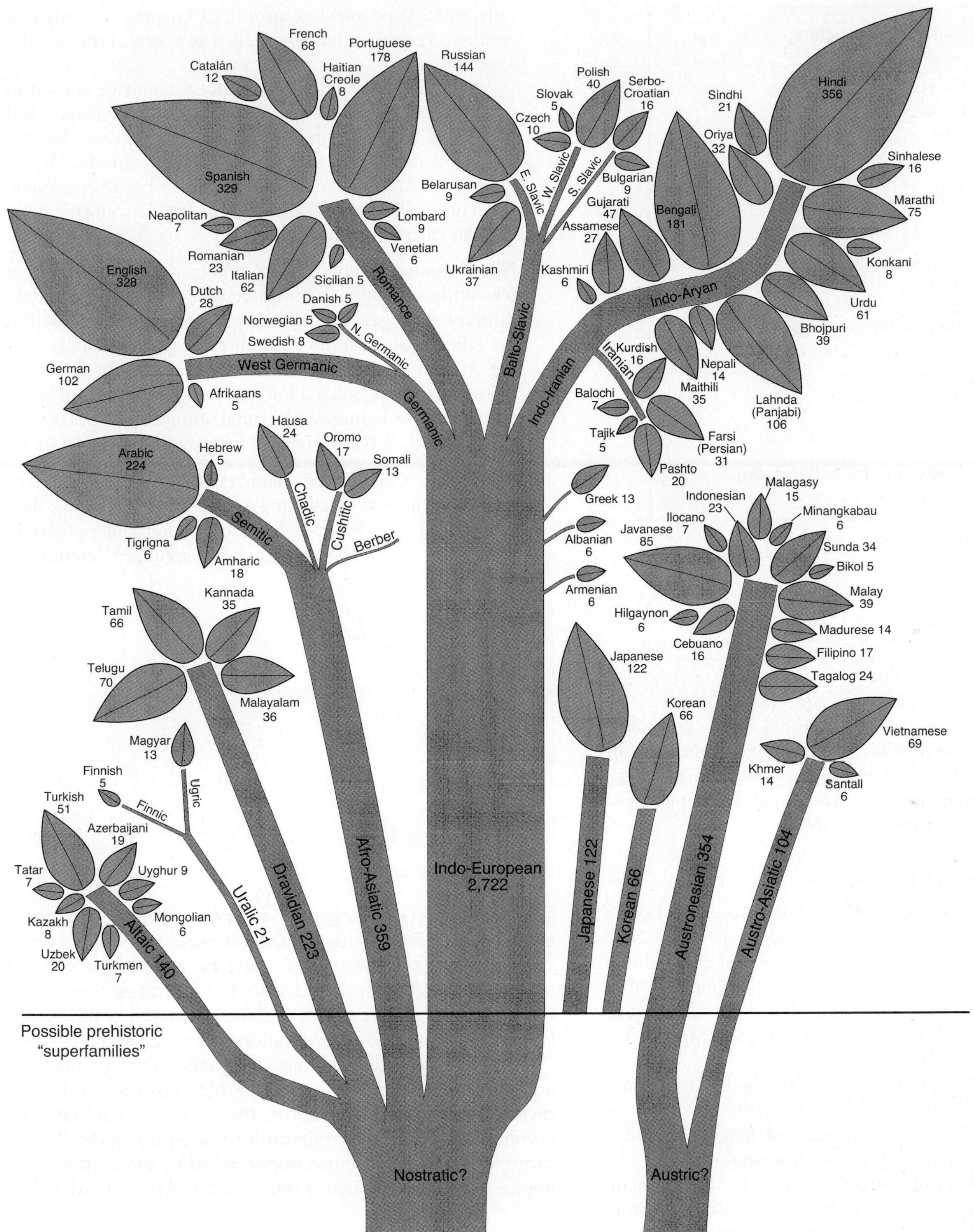

- **Uralic.** Every European country is dominated by Indo-European speakers, except for three—Estonia, Finland, and Hungary (refer to Figure 5-9). The Estonians, Finns, and Hungarians speak languages that belong to the Uralic family.

The Altaic and Uralic language families were once thought to be linked as one family, but recent studies point to geographically distinct origins. Uralic languages are traceable back to a common language, Proto-Uralic, first used 7,000 years ago by people living in the Ural Mountains

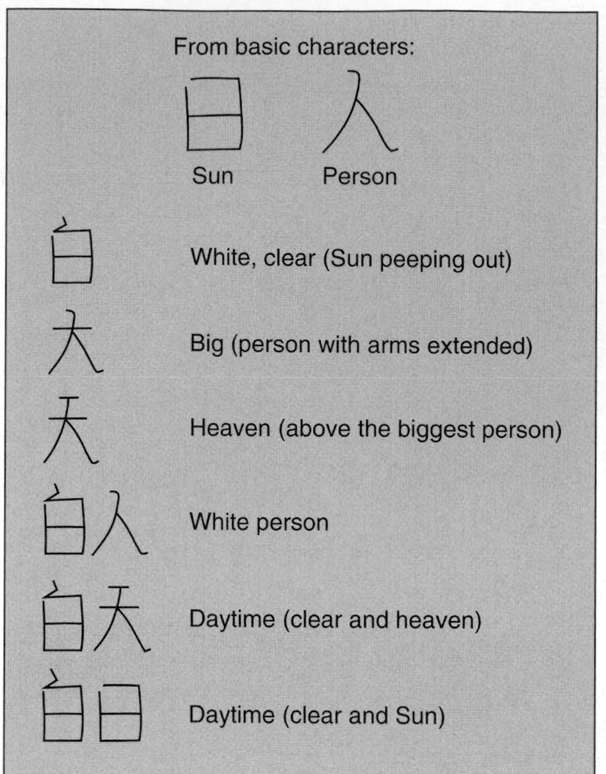

From basic characters:
日 人
Sun
Person
白
White, clear (Sun peeping out)
大
Big (person with arms extended)
天
Heaven (above the biggest person)
White person
Daytime (clear and heaven)
Daytime (clear and Sun)

FIGURE 5-18 Chinese language ideograms. The Chinese languages are written with ideograms, most of which represent ideas or concepts rather than sounds.

of present-day Russia, north of the Kurgan homeland. Migrants carried the Uralic languages to Europe, carving out homelands for themselves in the midst of Germanic- and Slavic-speaking peoples and retaining their language as a major element of cultural identity.

African Language Families

No one knows the precise number of languages spoken in Africa, and scholars disagree on classifying those known into families. In the 1800s, European missionaries and colonial officers began to record African languages using the Roman or Arabic alphabet. More than 1,000 distinct languages and several thousand named dialects have been documented. Most lack a written tradition.

Figure 5-19 shows the broad view of African language families, and Figure 5-20 hints at the complex pattern of multiple tongues of Nigeria. This great number of languages results from at least 5,000 years of minimal interaction among the thousands of cultural groups inhabiting the African continent. Each group developed its own language, religion, and other cultural traditions in isolation from other groups.

In northern Africa the language pattern is relatively clear, because Arabic, an Afro-Asiatic language, dominates, although in a variety of dialects. In sub-Saharan Africa, however, languages grow far more complex.

- **Niger-Congo.** More than 95 percent of the people in sub-Saharan Africa speak languages of the Niger-Congo family. One of these languages—Swahili—is the first language of only 800,000 people and an official language in only one country (Tanzania), but it is spoken as a second language by approximately 30 million Africans.

  Especially in rural areas, the local language is used to communicate with others from the same village, and Swahili is used to communicate with outsiders. Swahili originally developed through interaction among African groups and Arab traders, so its vocabulary has strong Arabic influences. Also, Swahili is one of the few African languages with an extensive literature.
- **Nilo-Saharan.** Languages of this family are spoken by a few million people in north-central Africa, immediately north of the Niger-Congo language region. Divisions within the Nilo-Saharan family exemplify the problem of classifying African languages. Despite fewer speakers, the Nilo-Saharan family is divided into six branches, plus numerous groups and subgroups. The total number of speakers of each individual Nilo-Saharan language is extremely small.
- **Khoisan.** A distinctive characteristic of the Khoisan languages is the use of clicking sounds. Upon hearing this, whites in southern Africa derisively and onomatopoeically named the most important Khoisan language Hottentot.

KEY ISSUE 4
Why Do People Preserve Local Languages?

- **Preserving Language Diversity**
- **Global Dominance of English**

The distribution of a language is a measure of the fate of an ethnic group. English has diffused around the world from a small island in northwestern Europe because of the cultural dominance of England and the United States over other territory on Earth's surface. Icelandic remains a little-used language because of the isolation of the Icelandic people.

As in other cultural traits, language displays the two competing geographic trends of globalization and local diversity. English has become the principal language of communication and interaction for the entire world. At the same time, local languages endangered by the global dominance of English are being protected and preserved. ■

Preserving Language Diversity

Thousands of languages are **extinct languages** once in use—even in the recent past—but no longer spoken or read in daily activities by anyone in the world. *Ethnologue* considers 473 languages as nearly extinct because only a few older speakers are

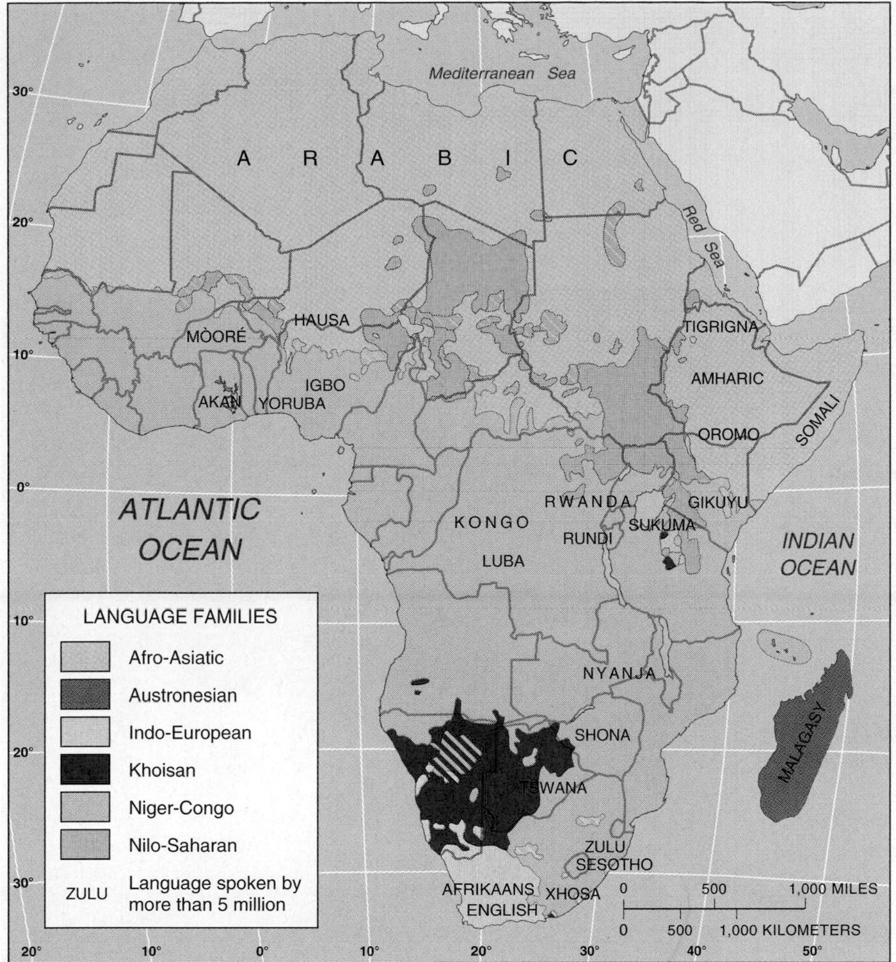

FIGURE 5-19 Africa's language families. More than 1,000 languages have been identified in Africa, and experts do not agree on how to classify them into families, especially languages in central Africa. Languages with more than 5 million speakers are named on the map.

still living, and they are not teaching the languages to their children. According to *Ethnologue*, 46 of these nearly extinct languages are in Africa, 182 in the Americas, 84 in Asia, 9 in Europe, and 152 in the Pacific.

When Spanish missionaries reached the eastern Amazon region of Peru in the sixteenth century, they found more than 500 languages. Only 92 survive today, according to *Ethnologue*, and 14 of these face immediate extinction because fewer than 100 speakers remain. Of Peru's 92 surviving indigenous languages, only Cusco, a Quechuan language, is currently used by more than 1 million people.

Gothic was widely spoken by people in Eastern and Northern Europe in the third century. Not only is Gothic extinct but so is the entire language group to which it belonged, the East Germanic group of the Germanic branch of Indo-European. The last speakers of Gothic lived in the Crimea in Russia in the sixteenth century. The Gothic language died because the descendants of the Goths were converted to other languages through processes of integration, such as political dominance and cultural preference. For example, many Gothic people switched to speaking the Latin language after their conversion to Christianity. Similarly, indigenous languages are disappearing in Peru as speakers switch to Spanish.

Some endangered languages are being preserved. The European Union has established the European Bureau for Lesser Used Languages (EBLUL), based in Dublin, Ireland, to provide financial support for the preservation of several dozen indigenous, regional, and minority languages spoken by 46 million Europeans. Nonetheless, linguists expect that hundreds of languages will become extinct during the twenty-first century and that only about 300 languages are clearly safe from extinction because they have sufficient speakers and official government support.

Hebrew: Reviving Extinct Languages

Hebrew is a rare case of an extinct language that has been revived (Figure 5-21). Most of the Jewish Bible (Christian Old Testament) was written in Hebrew (a small part of it was written in another Afro-Asiatic language, Aramaic). A language of daily activity in biblical times, Hebrew diminished in use in the fourth century B.C. and was thereafter retained only for Jewish religious services. At the time of Jesus, people in present-day Israel generally spoke Aramaic, which in turn was replaced by Arabic.

When Israel was established as an independent country in 1948, Hebrew became one of the new country's two official languages, along with Arabic. Hebrew was chosen because the Jewish population of Israel consisted of refugees and migrants from many countries who spoke many languages. Because Hebrew was still used in Jewish prayers, no other language could so symbolically unify the disparate cultural groups in the new country.

The task of reviving Hebrew as a living language was formidable. Words had to be created for thousands of objects and inventions unknown in biblical times, such as telephones, cars, and electricity. The revival effort was initiated by Eliezer Ben-Yehuda, who lived in Palestine before the creation of the state of Israel and who refused to speak any language other than Hebrew. Ben-Yehuda is credited with the invention of 4,000 new Hebrew words—related when possible to ancient ones—and the creation of the first modern Hebrew dictionary.

Celtic: Preserving Endangered Languages

The Celtic branch of Indo-European is of particular interest to English speakers because it was the major language in the British Isles before the Germanic Angles, Jutes, and Saxons invaded. Two thousand years ago, Celtic languages were spoken in much of present-day Germany, France, and northern Italy, as well as in

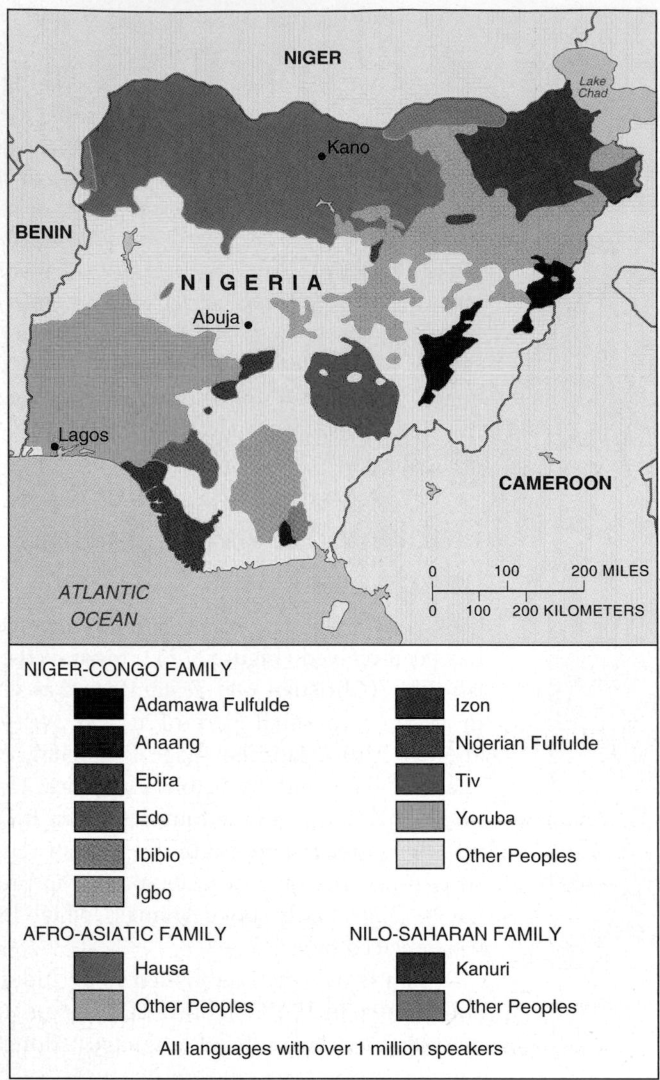

FIGURE 5-20 Nigeria's main languages. Africa's most populous country, Nigeria, displays problems that can arise from the presence of many speakers of many languages. Nigeria has 514 distinct languages, according to *Ethnologue*, only a few of which have widespread use. National unity is severely strained by the lack of a common language that a large percentage of the population can understand. Groups living in different regions of Nigeria have often battled. To reduce these regional tensions, the government moved the capital from Lagos in the Yoruba-dominated southwest to Abuja in the center of Nigeria. This central and "neutral" location was selected to avoid existing concentrations of the major rival cultural groups. Nigeria reflects the problems that can arise when great cultural diversity—and therefore language diversity—is packed into a relatively small region. Nigeria also illustrates the importance of language in identifying distinct cultural groups on a local scale. Speakers of one language are unlikely to understand any of the others in the same family, let alone languages from other families.

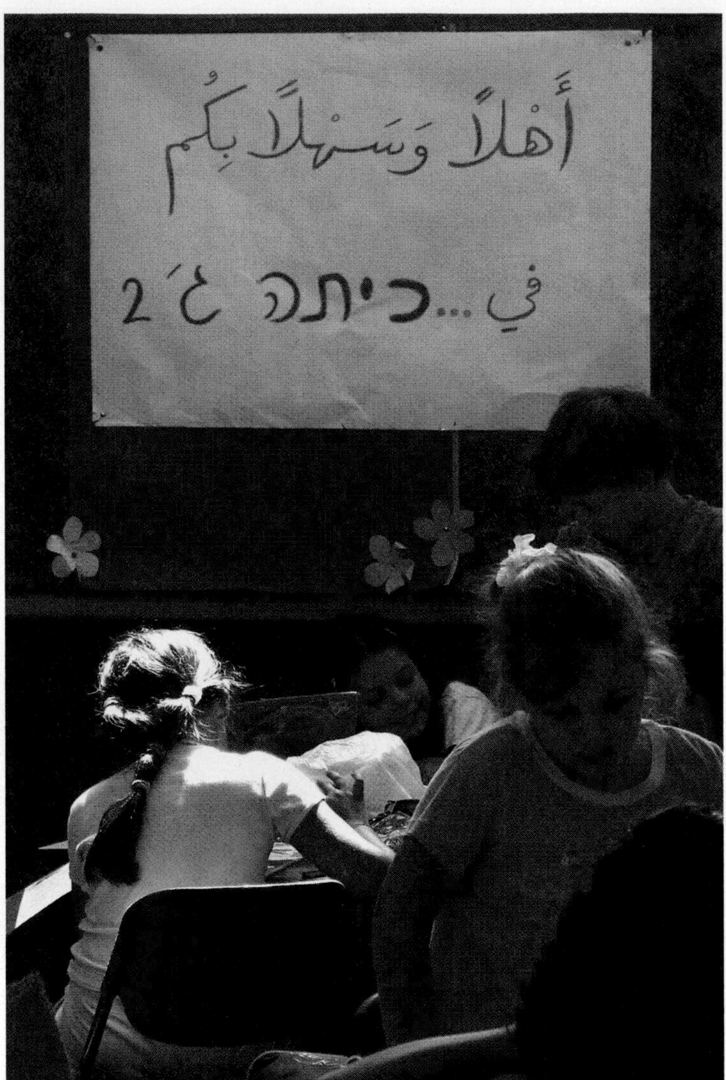

FIGURE 5-21 Revival of Hebrew. Hebrew and Arabic are both official languages in Israel. A third-grade class in Jerusalem is taught simultaneously in Arabic (in blue) and Hebrew (in red) by Arab and Jewish teachers.

the British Isles. Today, Celtic languages survive only in remoter parts of Scotland, Wales, and Ireland and on the Brittany peninsula of France (Figure 5-22).

The Celtic language branch is divided into Goidelic (Gaelic) and Brythonic groups. Two Goidelic languages survive—Irish Gaelic and Scottish Gaelic. Speakers of Brythonic (also called Cymric or Britannic) fled westward during the Germanic invasions to Wales, southwestward to Cornwall, or southward across the English Channel to the Brittany peninsula of France.

- **Irish Gaelic.** Irish Gaelic and English are the Republic of Ireland's two official languages. Irish is spoken by 350,000 people on a daily basis, and 1.5 million say that they can speak it.
- **Scottish Gaelic.** In Scotland 59,000, or 1 percent of the people, speak Scottish Gaelic. An extensive body of literature exists in Gaelic languages, including the Robert Burns poem *Auld Lang Syne* ("old long since"), the basis for the popular New Year's Eve song. Gaelic was carried from Ireland to Scotland about 1,500 years ago.
- **Brythonic (Welsh).** Wales—the name derived from the Germanic invaders' word for *foreign*—was conquered by the English in 1283. Welsh remained dominant in Wales until the nineteenth century, when many English speakers migrated there to work in coal mines and factories. A 2004 survey found 611,000 Welsh speakers in Wales, 22 percent of the population. In some isolated communities in the northwest, especially in the county of Gwynedd, two-thirds speak Welsh.

FIGURE 5-22 Celtic language branch. Direction sign on the Beara Peninsula in the southwest of County Cork, Ireland, indicates the distance to New York in Gaelic and English.

- **Cornish.** Cornish became extinct in 1777, with the death of the language's last known native speaker, Dolly Pentreath, who lived in Mousehole (pronounced "muzzle"). Before Pentreath died, an English historian recorded as much of her speech as possible so that future generations could study the Cornish language. One of her last utterances was later translated as "I will not speak English . . . you ugly, black toad!"
- **Breton.** In Brittany—like Cornwall, an isolated peninsula that juts out into the Atlantic Ocean—around 250,000 speak Breton regularly. Breton differs from the other Celtic languages in that it has more French words.

The survival of any language depends on the political and military strength of its speakers. The Celtic languages declined because the Celts lost most of the territory they once controlled to speakers of other languages. In the 1300s, the Irish were forbidden to speak their own language in the presence of their English masters. By the nineteenth century, Irish children were required to wear "tally sticks" around their necks at school. The teacher carved a notch in the stick every day the child used an Irish word, and at the end of the day meted out punishment based on the number of tallies. Parents encouraged their children to learn English so that they could compete for jobs. Most remaining Celtic speakers also know the language of their English or French conquerors.

Recent efforts have prevented the disappearance of Celtic languages. In Wales, the *Cymdeithas yr Iaith Gymraeg* (Welsh Language Society) has been instrumental in preserving the language. Britain's 1988 Education Act made Welsh language training a compulsory subject in all schools in Wales, and Welsh history and music have been added to the curriculum. All local governments and utility companies are now obliged to provide services in Welsh. Welsh-language road signs have been posted throughout Wales, and the British Broadcasting Corporation (BBC) produces Welsh-language television and radio programs. Knowledge of Welsh is now required for many jobs, especially in public service, media, culture, and sports.

An Irish-language TV station began broadcasting in 1996. English road signs were banned from portions of western Ireland in 2005. The revival is being led by young Irish living in other countries who wish to distinguish themselves from the English (in much the same way that Canadians traveling abroad often make efforts to distinguish themselves from U.S. citizens). Irish singers, including many rock groups (although not U2), have begun to record and perform in Gaelic.

A few hundred people have become fluent in the formerly extinct Cornish language, which was revived in the 1920s. Cornish is taught in grade schools and adult evening courses and is used in some church services. Some banks accept checks written in Cornish. EBLUL granted Cornish minority language status in 2002. After years of dispute over how to spell the revived language, various groups advocating for the revival of Cornish reached an agreement in 2008 on a standard written version of the language. Because the language became extinct, it is impossible to know precisely how to pronounce Cornish words.

The long-term decline of languages such as Celtic provides an excellent example of the precarious struggle for survival that many languages experience. Faced with the diffusion of alternatives used by people with greater political and economic strength, speakers of Celtic and other languages must work hard to preserve their linguistic cultural identity.

Multilingual States

Difficulties can arise at the boundary between two languages. Note in Figures 5-9 (Indo-European languages) and 5-10 (Germanic languages) that the boundary between the Romance and Germanic branches runs through the middle of two small European countries, Belgium and Switzerland. Belgium has had more difficulty than Switzerland in reconciling the interests of the different language speakers.

Southern Belgians (known as Walloons) speak French, whereas northern Belgians (known as Flemings) speak a dialect of the Germanic language, Dutch, called Flemish (Figure 5-23). The language boundary sharply divides the

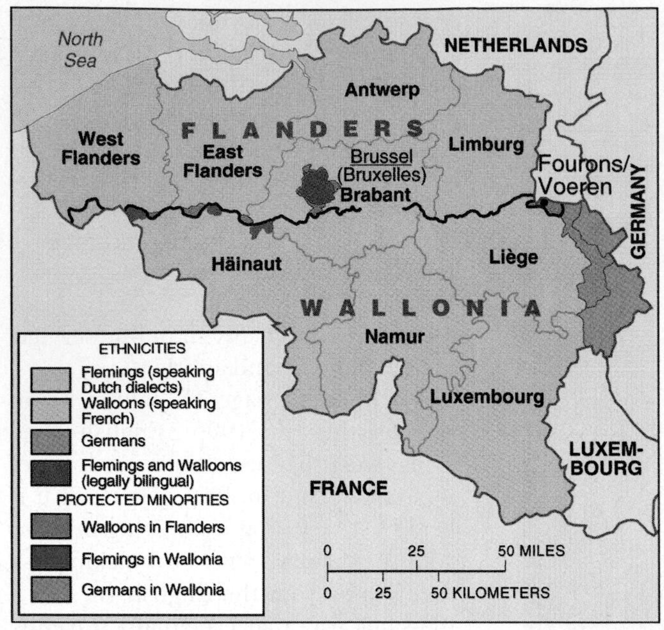

FIGURE 5-23 Languages in Belgium. Flemings in the north speak Flemish, a Dutch dialect. Walloons in the south speak French. The two groups have had difficulty sharing national power. Flemish activists seeking a division of Belgium into two countries spray-painted over the French names on this road sign, leaving the Flemish names intact.

country into two regions. Antagonism between the Flemings and Walloons is aggravated by economic and political differences. Historically, the Walloons dominated Belgium's economy and politics, and French was the official state language.

Motorists in Belgium clearly see the language boundary on expressways. Heading north, the highway signs suddenly change from French to Flemish at the boundary between Wallonia and Flanders. Brussels, the capital city, is an exception. Although located in Flanders, Brussels is officially bilingual and signs are in both French and Flemish. As an example, some stations on the subway map of Brussels are identified by two names—one French and one Flemish (for instance, Porte de Hal and Halle Poort—see Figure 13-28).

In response to pressure from Flemish speakers, Belgium has been divided into two independent regions, Flanders and Wallonia. Each elects an assembly that controls cultural affairs, public health, road construction, and urban development in its region. But for many in Flanders, regional autonomy is not enough. They want to see Belgium divided into two independent countries. Were that to occur, Flanders would be one of Europe's richest countries and Wallonia one of the poorest.

In contrast with Belgium, Switzerland peacefully exists with multiple languages. The key is a decentralized government, in which local authorities hold most of the power, and decisions are frequently made by voter referenda. Switzerland has four official languages—German (used by 65 percent of the population), French (18 percent), Italian (10 percent), and Romansh (1 percent). Swiss voters made Romansh an official language in a 1938 referendum, despite the small percentage of people who use the language.

Switzerland is divided into four main linguistic regions, as shown in Figure 5-24, but people living in individual communities, especially in the mountains, may use a language other than the prevailing local one. The Swiss, relatively tolerant of speakers of other languages, have institutionalized cultural diversity by creating a form of government that places considerable power in small communities.

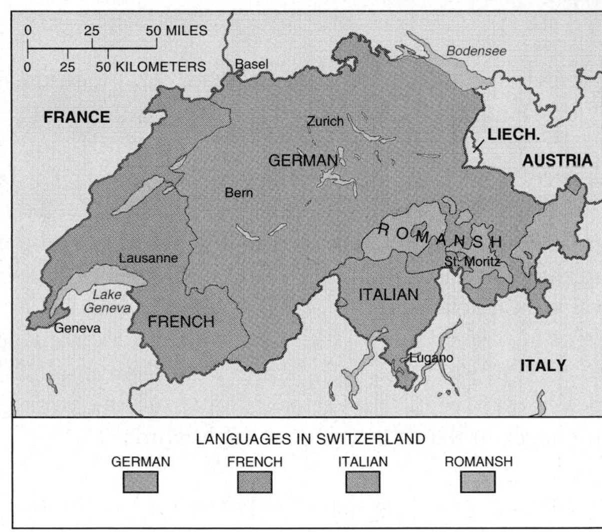

FIGURE 5-24 Languages in Switzerland. Switzerland lives peacefully with four official languages, including Romansh, which is used by only 1 percent of the population.

Isolated Languages

An **isolated language** is a language unrelated to any other and therefore not attached to any language family. Similarities and differences between languages—our main form of communication—are a measure of the degree of interaction among groups of people.

GLOBAL FORCES, LOCAL IMPACTS
Language Policy in Australia and New Zealand

English is the most widely used language in Australia and New Zealand as a result of British colonization during the early nineteenth century. Settlers in Australia and New Zealand established and maintained outposts of British culture, including use of the English language.

An essential element in maintaining British culture was restriction of immigration from non-English-speaking places during the nineteenth and early twentieth centuries. Fear of immigration was especially strong in Australia because of its proximity to other Asian countries. Under a "White Australia" policy, every prospective immigrant was required to write 50 words of a European language dictated by an immigration officer. The dictation test was not eliminated until 1957. The Australian government now merely requires that immigrants learn English.

New Zealand's language requirement is more stringent: Immigrants must already be fluent in English, although free English lessons are available to immigrants. More remote from Asian landmasses, New Zealand has attracted fewer Asian immigrants.

Though English remains the dominant language of Australia and New Zealand, the languages that predate British settlement survive in both countries. However, the two countries have adopted different policies with regard to indigenous languages. Australia regards English as a tool for promoting cultural diversity, whereas New Zealand regards linguistic diversity as an important element of cultural diversity (Figure 5-25).

In Australia, 1 percent of the population is Aboriginal. Many elements of Aboriginal culture are now being preserved. But education is oriented toward teaching English rather than maintaining local languages. English is the language of instruction throughout Australia, and others are relegated to the status of second language.

In New Zealand, more than 10 percent of the population is Maori, descendents of Polynesian people who migrated there around 1,000 years ago. In contrast with Australia, New Zealand has adopted policies to preserve the Maori language. Most notably, Maori has became one of New Zealand's three official languages, along with English and sign language. A Maori Language Commission was established to preserve the language. Despite official policies, only 1 percent of New Zealanders are fluent in Maori, most of whom are over age 50. Preserving the language requires skilled teachers and the willingness to endure inconvenience compared to using the world's lingua franca, English. ■

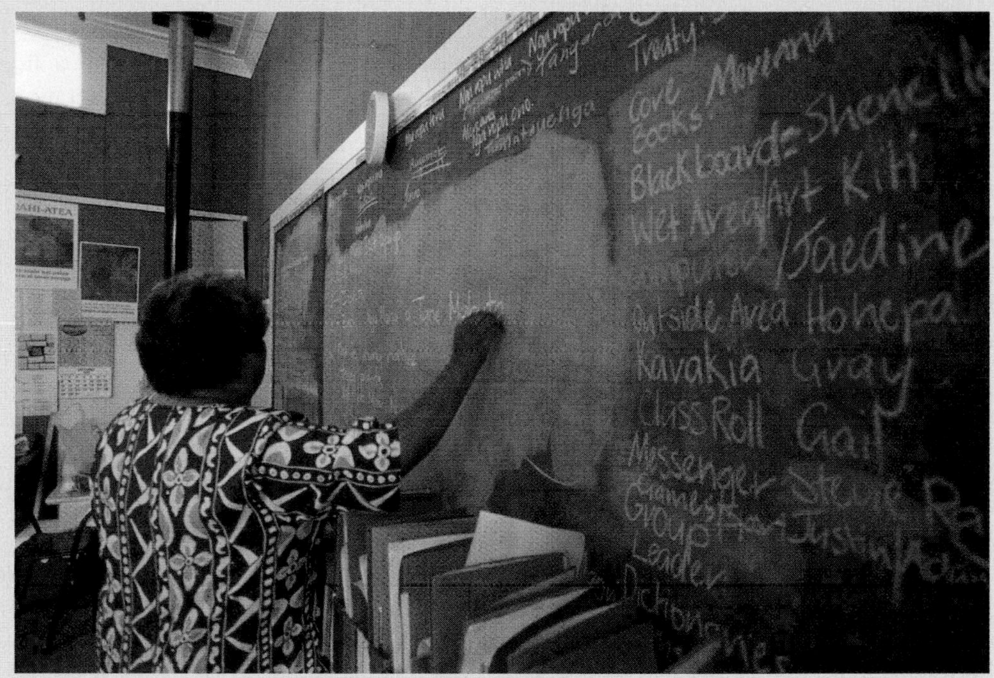

FIGURE 5-25 Preserving language diversity in New Zealand. Maori language is being taught at an elementary school in a predominantly Maori village.

The diffusion of Indo-European languages demonstrates that a common ancestor dominated much of Europe before recorded history. Similarly, the diffusion of Indo-European languages to the Western Hemisphere is a result of conquests by Indo-European speakers in more recent times. In contrast, isolated languages arise through lack of interaction with speakers of other languages.

A PRE-INDO-EUROPEAN SURVIVOR: BASQUE. The best example of an isolated language in Europe is Basque, apparently the only language currently spoken in Europe that survives from the period before the arrival of Indo-European speakers. No attempt to link Basque to the common origin of the other European languages has been successful.

Basque was probably once spoken over a wider area but was abandoned where its speakers came in contact with Indo-Europeans. It is now the first language of 666,000 people in the Pyrenees Mountains of northern Spain and southwestern France (refer to Figure 5-12, the gray area in northern Spain). Basque's lack of connection to other languages reflects the isolation of the Basque people in their mountainous homeland.

This isolation has helped them preserve their language in the face of the wide diffusion of Indo-European languages.

AN UNCHANGING LANGUAGE: ICELANDIC. Icelandic is related to other languages in the North Germanic group of the Germanic branch of the Indo-European family. Icelandic's significance is that over the past thousand years it has changed less than any other in the Germanic branch. As was the case with England, people in Iceland speak a Germanic language because their ancestors migrated to the island from the east, in this case from Norway. Norwegian settlers colonized Iceland in A.D. 874.

When an ethnic group migrates to a new location, it takes along the language spoken in the former home. The language spoken by most migrants—such as the Germanic invaders of England—changes in part through interaction with speakers of other languages. But in the case of Iceland, the Norwegian immigrants had little contact with speakers of other languages when they arrived in Iceland, and they did not have contact with speakers of their language back in Norway. After centuries of interaction with other Scandinavians, Norwegian and other North Germanic languages had adopted new words and pronunciation, whereas the isolated people of Iceland had less opportunity to learn new words and no reason to change their language.

Global Dominance of English

One of the most fundamental needs in a global society is a common language for communication. Increasingly in the modern world, the language of international communication is English (see Global Forces, Local Impacts box). A Polish airline pilot who flies over Spain speaks to the traffic controller on the ground in English. Swiss bankers speak a dialect of German among themselves, but with German bankers they prefer to speak English rather than German. English is the official language at an aircraft factory in France and an appliance company in Italy.

English: An Example of a Lingua Franca

A language of international communication, such as English, is known as a **lingua franca**. To facilitate trade, speakers of two different languages would create a lingua franca by mixing elements of the two languages into a simple common language. The term, which means *language of the Franks*, was originally applied by Arab traders during the Middle Ages to describe the language they used to communicate with Europeans, whom they called *Franks*.

A group that learns English or another lingua franca may learn a simplified form, called a **pidgin language**. To communicate with speakers of another language, two groups construct a pidgin language by learning a few of the grammar rules and words of a lingua franca, while mixing in some elements of their own languages. A pidgin language has no native speakers—it is always spoken in addition to one's native language.

Other than English, modern lingua franca languages include Swahili in East Africa, Hindi in South Asia, Indonesian in Southeast Asia, and Russian in the former Soviet Union. A number of African and Asian countries that became independent in the twentieth century adopted English or Swahili as an official language for government business, as well as for commerce, even if the majority of the people couldn't speak it.

The rapid growth in importance of English is reflected in the percentage of students learning English as a second language in school. More than 90 percent of students in the European Union learn English in middle or high school, not just in smaller countries like Denmark and the Netherlands but also in populous countries such as France, Germany, and Spain. The Japanese government, having determined that fluency in English is mandatory in a global economy, has even considered adding English as a second official language.

Foreign students increasingly seek admission to universities in countries that teach in English rather than in German, French, or Russian. Students around the world want to learn in English because they believe it is the most effective way to work in a global economy and participate in a global culture.

Expansion Diffusion of English

In the past, a lingua franca achieved widespread distribution through migration and conquest. Two thousand years ago, use of Latin spread through Europe along with the Roman Empire, and in recent centuries use of English spread around the world primarily through the British Empire.

In contrast, the recent growth in the use of English is an example of expansion diffusion, the spread of a trait through the snowballing effect of an idea rather than through the relocation of people. Expansion diffusion has occurred in two ways with English. First, English is changing through diffusion of new vocabulary, spelling, and pronunciation. Second, English words are fusing with other languages. For a language to remain vibrant, new words and usage must always be coined to deal with new situations. Unlike most examples of expansion diffusion, recent changes in English have percolated up from common usage and ethnic dialects rather than being directed down to the masses by elite people. Examples include dialects spoken by African Americans and residents of Appalachia.

Some African Americans speak a dialect of English heavily influenced by the group's distinctive heritage of forced migration from Africa during the eighteenth century to be slaves in the southern colonies. African American slaves preserved a distinctive dialect in part to communicate in a code not understood by their white masters. Black dialect words such as *gumbo* and *jazz* have long since diffused into the standard English language.

In the twentieth century, many African Americans migrated from the South to the large cities in the Northeast and Midwest (see Chapter 7). Living in racially segregated neighborhoods within northern cities and attending segregated schools, many African Americans preserved their distinctive dialect. That dialect has been termed **Ebonics**, a combination of *ebony* and *phonics*. The American Speech, Language and Hearing Association classified Ebonics as a distinct dialect, with a recognized vocabulary, grammar, and word meaning. Among the distinctive elements of Ebonics are the use of double negatives, such

CONTEMPORARY GEOGRAPHIC TOOLS
English on the Internet

English was the dominant language of the Internet during the 1990s. In 1998, 71 percent of people online were using English (Figure 5-26). The early dominance of English on the Internet was partly a reflection of the fact that the most populous English-speaking country, the United States, had a head start on the rest of the world in making the Internet available to most of its citizens (refer to Figure 4-19).

English continued as the leading Internet language in the first years of the twenty-first century, but it was far less dominant. The percentage of English-language online users declined from 71 percent in 1998 to 29 percent in 2008. Chinese (Mandarin) language online users increased from 2 percent of the world total in 1998 to 20 percent in 2008. English may be less dominant as the language of the Internet in the twenty-first century. But the United States remains the Internet leader in key respects—and with it the English language.

The United States created the English-language nomenclature for the Internet that the rest of the world has followed. The designation "www," which English speakers recognize as an abbreviation of "World Wide Web," is awkward in other languages, most of which do not have an equivalent sound to the English "w." In French, for example, "w" is pronounced "doo-blah-vay."

The U.S.-based Internet Corporation for Assigned Names and Numbers (ICANN) has been responsible for assigning domain names and for the suffixes following the dot, such as "com" and "edu." Domain names in the rest of the world include a two-letter suffix for the country, such as "fr" for France and "jp" for Japan, whereas U.S.-based domain names don't need the suffix.

U.S.-based companies provide the principal search engines for Internet users everywhere. In 2007, 67 percent of all searches worldwide used Google. Second place was another U.S.-based company, Yahoo, with 15 percent. These companies offer search engines in languages other than English. Google was heavily criticized when its Mandarin-language Google.cn was designed to block web sites deemed unsuitable by China's government.

Reflecting the globalization of the languages of the Internet, ICANN agreed in 2009 to permit domain names in characters other than Latin. Arabic, Chinese, and other characters may now be used. ■

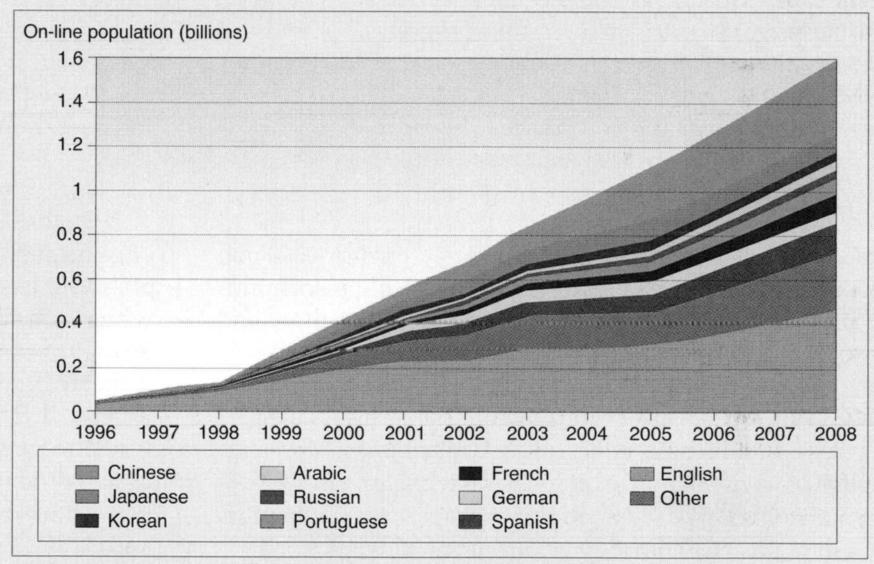

FIGURE 5-26 Languages of online speakers 1996–2008. English remains the most widely used language on the Internet, but Chinese is growing more rapidly.

as "I ain't going there no more," and such sentences as "She be at home" instead of "She is usually at home."

Natives of Appalachian communities, such as in rural West Virginia, also have a distinctive dialect, pronouncing *hollow* as "holler," and *creek* as "crick." Distinctive grammatical practices include the use of the double negative as in Ebonics and adding "a" in front of verbs ending in "ing," such as *a-sitting*.

Use of Ebonics is controversial within the African American community. On the one hand, some regard it as substandard, a measure of poor education, and an obstacle to success in the United States. Others see Ebonics as a means for preserving a distinctive element of African American culture and an effective way to teach African Americans who otherwise perform poorly in school.

Similarly, speaking an Appalachian dialect produces both pride and problems. An Appalachian dialect is a source of regional identity but has long been regarded by other Americans as a sign of poor education and an obstacle to obtaining employment in other regions of the United States. Some Appalachian residents are "bidialectic"—they speak "standard" English outside Appalachia and slip back into their regional dialect at home.

Diffusion to Other Languages

English words have become increasingly integrated into other languages. Many French speakers regard the invasion of English words with alarm, but Spanish speakers may find the mixing of the two languages stimulating.

FRANGLAIS. Traditionally, language has been an especially important source of national pride and identity in France. The French are particularly upset with the increasing worldwide domination of English, especially the invasion of their language by English words and the substitution of English for French as the most important language of international communications.

French is an official language in 29 countries and for hundreds of years served as the lingua franca for international diplomats. Many French are upset that English words such as *cowboy*, *hamburger*, *jeans*, and *T-shirt* were allowed to diffuse into the French language and destroy the language's purity. The widespread use of English in the French language is called **Franglais**, a combination of *français* and *anglais*, the French words for *French* and *English*.

Since 1635, the French Academy has been the supreme arbiter of the French language. In modern times, it has promoted the use of French terms in France, such as *stationnement* rather than *parking*, *fin du semaine* rather than *le weekend*, *logiciel* rather than *software*, and *arrosage* rather than *spam*. France's highest court, however, ruled in 1994 that most of the country's laws banning franglais were illegal.

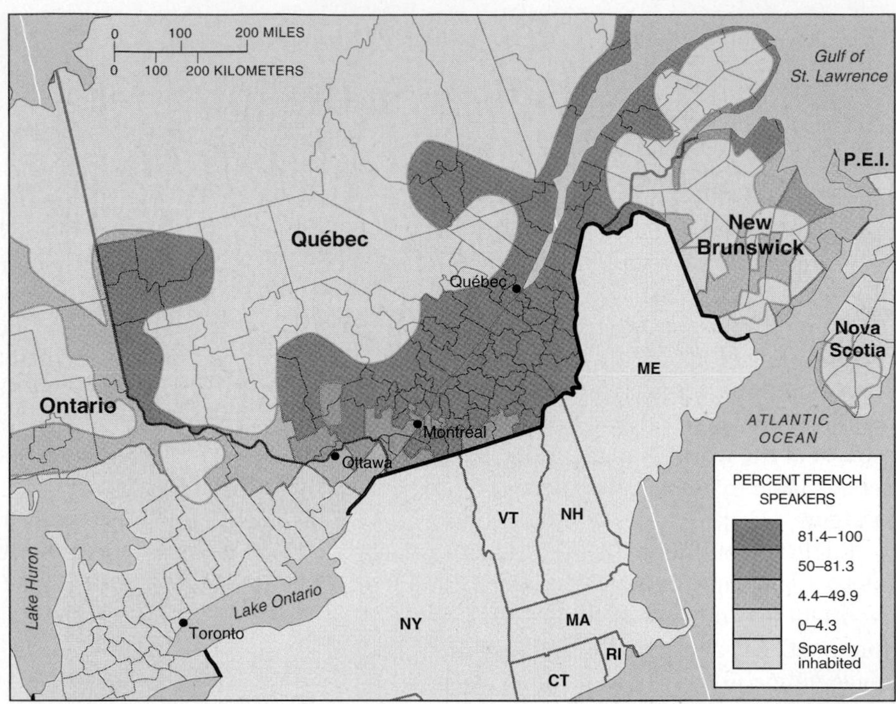

FIGURE 5-27 English/French language boundary in Canada. More than 80 percent of Québec's residents speak French, compared to approximately 6 percent for the rest of Canada.

Protection of the French language is even more extreme in Québec, which is completely surrounded by English-speaking provinces and U.S. states (Figure 5-27). Québécois are committed to preserving their distinctive French-language culture and to do so, they may secede from Canada.

SPANGLISH. English is diffusing into the Spanish language spoken by 34 million Hispanics in the United States, to create **Spanglish**, a combination of Spanish and English. In Miami's large Cuban-American community, Spanglish is sometimes called Cubonics, a combination of Cuban and phonetics.

As with franglais, Spanglish involves converting English words to Spanish forms. Some of the changes modify the spelling of English words to conform to Spanish preferences and pronunciations, such as dropping final consonants and replacing *v* with *b*. For example, *shorts* (pants) becomes *chores*, and *vacuum cleaner* becomes *bacuncliner.* In other cases, awkward Spanish words or phrases are dropped in favor of English words. For example, *parquin* is used rather than *estacionamiento* for *parking*, and *taipear* is used instead of *escribir a máquina* for *to type*.

Spanglish is a richer integration of English with Spanish than the mere borrowing of English words. New words have been invented in Spanglish that do not exist in English but would be useful if they did. For example, *bipiar* is a verb derived from the English *beeper* that means to "beep someone on a pager," and *i-meiliar* is a verb that means to "e-mail someone." Spanglish also mixes English and Spanish words in the same phrase. For example, a magazine article is titled "When he says me voy . . . what does he really mean?" (*me voy* means "I'm leaving").

Spanglish has become especially widespread in popular culture, such as song lyrics, television, and magazines aimed at young Hispanic women, but it has also been adopted by writers of serious literature. Inevitably, critics charge that Spanglish is a substitute for rigorously learning the rules of standard English and Spanish. And Spanglish has not been promoted for use in schools, as has Ebonics. Rather than a threat to existing languages, Spanglish is generally regarded as an enriching of both English and Spanish by adopting the best elements of each—English's ability to invent new words and Spanish's ability to convey nuances of emotion. Many Hispanic Americans like being able to say *Hablo un mix de los dos languages*.

DENGLISH. The diffusion of English words into German is called **Denglish**, with the "D" for Deutsch, the German word for *German*. For many Germans, wishing someone "happy birthday" sounds more melodic than the German *Herzlichen Gluuuckwunsch zum Geburtstag*.

The German telephone company Deutsche Telekom, uses the German word *Deutschlandverbindungen* for long distance and the Denglish word *Cityverbindungen* for local (rather than the German word *Ortsverbindungen*. The telephone company originally wanted to use the English "German calls" and "city calls" to describe its long-distance and local services, but the Institute for the German Language, which defines rules for the use of German, protested, so Deutsche Telekom compromised with one German word and one Denglish word.

English has diffused into other languages as well. The Japanese, for example, refer to *beisboru* ("baseball"), *naifu* ("knife"), and *sutoroberi keki* ("strawberry cake").

SUMMARY

The emergence of the Internet as an important means of communication has further strengthened the dominance of English. Because a majority of the material on the Internet is in English, knowledge of English is essential for Internet users around the world.

Some e-mail systems and interactive Internet programs do not accept accent marks used in other languages such as French. Languages that are not written in Latin letters, such as Japanese and Russian, are extremely cumbersome if not impossible to write on the Web.

The dominance of English as an international language has facilitated the diffusion of popular culture and science and the growth of international trade. In Germany, for example, airlines, car dealers, and telephone companies use English slogans in advertising. However, people who forsake their native language must weigh the benefits of using English against the cost of losing a fundamental element of local cultural identity.

People in smaller countries need to learn English to participate more fully in a global economy and culture. All children learn English in the schools of countries such as the Netherlands and Sweden to facilitate international communication. This may seem culturally unfair, but obviously it is more likely that several million Dutch people will learn English than that a half-billion English speakers around the world will learn Dutch.

In view of the global dominance of English, many U.S. citizens do not recognize the importance of learning other languages. One of the best ways to learn about the beliefs, traits, and values of people living in other regions is to learn their language. The lack of effort by Americans to learn other languages is a source of resentment among people elsewhere in the world, especially when Americans visit or work in other countries. The inability to speak other languages is also a handicap for Americans who try to conduct international business. Successful entry into new overseas markets requires knowledge of local culture, and officials who can speak the local language are better able to obtain important information. Japanese businesses that wish to expand in the United States send English-speaking officials, but American businesses that wish to sell products to the Japanese are rarely able to send a Japanese-speaking employee.

Here again are the key issues raised by the geography of languages:

1. **Where Are English-Language Speakers Distributed?** English can be traced to invasions of England by Germanic tribes 1,500 years ago. From England, the language diffused around the world when English speakers established colonies. Americans and British speak different dialects of English because of the relative isolation of the two groups.
2. **Why Is English Related to Other Languages?** English is in the Germanic branch of the Indo-European language family. Nearly one-half of the world speaks a language in the Indo-European family. All Indo-European languages can be traced to a common ancestor. Individual languages developed from this single ancestor through migration, followed by the isolation of one group from others who formerly spoke the same language.
3. **Where Are Other Language Families Distributed?** One-fifth of the world speaks a language in the Sino-Tibetan family. Seven other language families encompass most of the remainder. Each has a distinctive distribution, as with Indo-European, which is a result of a combination of migration and isolation.
4. **Why Do People Preserve Local Languages?** English has become the most important language for international communication in popular arts, science, and business. In the face of the global dominance of a lingua franca such as English, less widely used languages can face extinction, but recent efforts have been made to preserve and revive local languages because of the importance of language as an element of cultural identity.

CASE STUDY REVISITED / The Future of French and Spanish in Anglo-America

The French-speaking people of Canada and the Spanish-speaking people of the United States both live on a continent dominated by English speakers. Both languages will continue to play important roles in the region.

French Canada

Until recently, Québec was one of Canada's poorest and least developed provinces. Its economic and political activities were dominated by an English-speaking minority, and the province suffered from cultural isolation and lack of French-speaking leaders (Figure 5-28).

When French President Charles de Gaulle visited Québec in 1967, he encouraged the development of an independent Québec by shouting in his speech, "Vive le Québec libre!" ("Long live free Québec!") Voters in Québec have thus far rejected separation from Canada, but by a slim majority.

The Québec government has made the use of French mandatory in many daily activities. Québec's Commission de Toponymie has renamed towns, rivers, and mountains that have names with English-language origins. French must be the predominant language on all commercial signs, and the legislature passed a law banning non-French outdoor signs altogether (ruled unconstitutional by the Canadian Supreme Court).

Confrontation during the 1970s and 1980s has been replaced in Québec by increased cooperation between French and English speakers. The neighborhoods of Montréal, Québec's largest city, were once highly segregated between French-speaking residents on the east and English-speaking residents on the west, but in recent years they have become more linguistically mixed. One-third of Quebec's native English speakers have married French speakers in recent years. Children of English speakers are increasingly likely to be bilingual.

Although French dominates over English, Québec faces a fresh challenge of integrating a large number of immigrants from Europe, Asia, and Latin America who don't speak French. Many immigrants would prefer to use English rather than French as their lingua franca but are prohibited from doing so by the Québec government. Even immigrants who learn to speak French charge that they face discrimination because of their accents.

Hispanic America

Linguistic unity is an apparent feature of the United States, a nation of immigrants who learn English to become Americans. However, the diversity of languages in the United States is greater than it first appears.

(Continued)

CASE STUDY REVISITED (*Continued*)

FIGURE 5-28 Languages other than English in Canada. French dominates in the province of Québec. The word *Stop* has been replaced by *Arrêt* on the red octagonal road signs, even though *Stop* is used throughout the world, even in France and other French-speaking countries. In the village of Nutashkuan Québec, the French *arrêt* sign is also in Innu, the local native American language.

FIGURE 5-29 Languages other than English in the United States. Spanish dominates in the Little Havana area of Miami.

In 2006, a language other than English was spoken at home by 55 million Americans over age 5, 20 percent of the population. Spanish was spoken at home by 34 million people in the United States. More than 2 million spoke Chinese; at least 1 million each spoke French, German, Korean, Tagalog, and Vietnamese.

In reaction against the increasing use of Spanish in the United States, 30 states and a number of localities have laws making English the official language. (Hawaii has two official languages: English and Hawaiian, in the Austronesian language family.) Some courts have judged these laws to be unconstitutional restrictions on free speech. The U.S. Congress has debated enacting similar legislation. For a state such as Montana, the law is symbolic, because it has few non-English speakers. But for states such as California and Florida, with large Hispanic populations, the debate affects access to jobs, education, and social services.

Promoting the use of English symbolizes that language is the chief cultural bond in the United States in an otherwise heterogeneous society. With the growing dominance of the English language in the global economy and culture, knowledge of English is important for people around the world, not just inside the United States. At the same time, the increasing use of other languages in the United States is a reminder of the importance that groups place on preserving cultural identity and the central role that language plays in maintaining that identity. ■

KEY TERMS

British Received Pronunciation (BRP) (p. 139) The dialect of English associated with upper-class Britons living in London and now considered standard in the United Kingdom.

Creole or creolized language (p. 149) A language that results from the mixing of a colonizer's language with the indigenous language of the people being dominated.

Denglish (p. 164) Combination of German and English.

Dialect (p. 139) A regional variety of a language distinguished by vocabulary, spelling, and pronunciation.

Ebonics (p. 162) Dialect spoken by some African Americans.

Extinct language (p. 156) A language that was once used by people in daily activities but is no longer used.

Franglais (p. 164) A term used by the French for English words that have entered the French language; a combination of *français* and *anglais*, the French words for "French" and "English," respectively.

Ideograms (p. 152) The system of writing used in China and other East Asian countries in which each symbol represents an idea or a concept rather than a specific sound, as is the case with letters in English.

Isogloss (p. 139) A boundary that separates regions in which different language usages predominate.

Isolated language (p. 160) A language that is unrelated to any other languages and therefore not attached to any language family.

Language (p. 136) A system of communication through the use of speech, a collection of sounds understood by a group of people to have the same meaning.

Language branch (p. 143) A collection of languages related through a common ancestor that existed several thousand years ago. Differences are not as extensive or as old as with language families, and archaeological evidence can confirm that the branches derived from the same family.

Language family (p. 143) A collection of languages related to each other through a common ancestor long before recorded history.

Language group (p. 144) A collection of languages within a branch that share a common origin in the relatively recent past and display relatively few differences in grammar and vocabulary.

Lingua franca (p. 162) A language mutually understood and commonly used in trade by people who have different native languages.

Literary tradition (p. 136) A language that is written as well as spoken.

Official language (p. 136) The language adopted for use by the government for the conduct of business and publication of documents.

Pidgin language (p. 162) A form of speech that adopts a simplified grammar and limited vocabulary of a lingua franca; used for communications among speakers of two different languages.

Spanglish (p. 164) Combination of Spanish and English, spoken by Hispanic Americans.

Standard language (p. 139) The form of a language used for official government business, education, and mass communications.

Vulgar Latin (p. 147) A form of Latin used in daily conversation by ancient Romans, as opposed to the standard dialect, which was used for official documents.

THINKING GEOGRAPHICALLY

1. Thirty U.S. states have passed laws mandating English as the language of all government functions. Should the use of English be encouraged in the United States to foster cultural integration, or should bilingualism be encouraged to foster cultural diversity? Why?
2. Does the province of Québec possess the resources, economy, political institutions, and social structures to be a viable, healthy country? What would be the impact of Québec's independence on the remainder of Canada, on the United States, and on France?
3. How is American English different from British English as a result of contributions by African Americans and immigrants who speak languages other than English?
4. The southern portion of Belgium (Wallonia) suffers from higher rates of unemployment, industrial decline, and other economic problems compared to Flanders, in the north. How do differences in language exacerbate Belgium's regional economic differences?
5. Many countries now receive Cable News Network (CNN) broadcasts that originate in the United States, but even English-speaking viewers in other countries have difficulty understanding some American English. A recent business program on CNN created a stir outside the United States when it reported that McDonald's was a major IRA contributor. Viewers in the United Kingdom thought that the American hamburger chain was financing the purchase of weapons by the Irish Republican Army, which sometimes resorts to violence in its attempt to achieve unification of Ireland. However, McDonald's, in fact, was contributing to Individual Retirement Accounts for its employees. Can you think of other examples where the use of a word could cause a British–American misunderstanding?

RESOURCES

Some recent and classic books and articles on migration geography:

Aitchison, John, and Harold Carter. *Language, Economy, and Society: The Changing Fortunes of the Welsh Language in the Twentieth Century.* Cardiff: University of Wales Press, 2000.

Asher, R. E., and Christopher Moseley, eds. *Atlas of the World's Languages*, 2nd ed. London, New York: Routledge, 2007.

Baugh, Albert C., and Thomas Cable. *A History of the English Language*, 5th ed. London and New York: Routledge, 2002.

Cardona, George, Henry M. Hoeningswald, and Alfred Senn, eds. *Indo-European and Indo-Europeans.* Philadelphia: University of Pennsylvania Press, 1970.

Delgado de Carvalho, C. M. "The Geography of Languages." In *Readings in Cultural Geography,* eds. Philip L. Wagner and Marvin W. Mikesell. Chicago: University of Chicago Press, 1962.

Gordon, Raymond G., Jr., ed. *Ethnologue: Languages of the World*, 15th ed. Dallas: SIL International, 2005.

Harrison, K. David. *When Languages Die: The Extinction of the World's Languages and the Erosion of Human Knowledge.* Oxford, New York: Oxford University Press, 2007.

Katzner, Kenneth. *The Languages of the World*, 3rd ed. New York: Routledge, 2002.

Kirk, John M., Stewart Sanderson, and J. D. A. Widdowson, eds. *Studies in Linguistic Geography: The Dialects of English in Britain and Ireland.* London: Croom Helm, 1985.

Krantz, Grover S. *Geographical Development of European Languages.* New York: Peter Lang, 1988.

Kurath, Hans. *Word Geography of the Eastern United States.* Ann Arbor: University of Michigan Press, 1949.

Renfrew, Colin. *Archaeology and Language.* Cambridge, UK: Cambridge University Press, 1988.

Trudgill, Peter. "Linguistic Geography and Geographical Linguistics." *Progress in Geography* 7 (1975): 227–52.

Williams, Colin H., ed. *Language in Geographic Context.* Clevedon, UK: Multilingual Matters, 1988.

Key Internet sites:

www.ethnologue.com. The book *Ethnologue: Languages of the World*, listed in the Further Readings, is available online. For every country there is a map showing the distribution of languages within the country. Detailed information is also provided for every language of the world.

Log in to www.mygeoscienceplace.com for videos, interactive maps, RSS feeds, case studies, and self-study quizzes to enhance your study of Language.

CHAPTER 6

Religion

And He shall judge between the nations,
And shall decide for many peoples;
And they shall beat their swords into ploughshares,
And their spears into pruning-hooks:
Nation shall not lift up sword against nation,
Neither shall they learn war any more.

Isaiah 2:4

This passage from the Bible, the holiest book of Christianity and Judaism, is one of the most eloquent pleas for peace among the nations of the world. For many religious people, especially in the Western Hemisphere and Europe, Isaiah evokes a highly attractive image of the ideal future landscape.

KEY ISSUES

1 Where Are Religions Distributed?

2 Why Do Religions Have Different Distributions?

3 Why Do Religions Organize Space in Distinctive Patterns?

4 Why Do Territorial Conflicts Arise Among Religious Groups?

Islam's holiest book, the Quran (sometimes spelled Koran), also evokes powerful images of a peaceful landscape:

> *He it is who sends down water from the sky, whence ye have drink, and whence the trees grow whereby ye feed your flocks.*
>
> *He makes the corn to grow, and the olives, and the palms, and the grapes, and some of every fruit; verily, in that is a sign unto a people who reflect.*
>
> Sûrah (Chapter) of the Bee XVI.9

Most religious people pray for peace, but religious groups may not share the same vision of how peace will be achieved. Geographers see that the process by which one religion diffuses across the landscape may conflict with the distribution of others. Geographers are concerned with the regional distribution of different religions and the resulting potential for conflict. Geographers also observe that religions are derived in part from elements of the physical environment, and that religions, in turn, modify the landscape. As evidence of this, note the rich agricultural images in the passages just quoted from the Bible and the Quran.

Muslims in Karachi, Pakistan, pray on the holiday of Eid ul-Fitr, which marks the end of the holy month of Ramadan.

CASE STUDY / The Dalai Lama Versus the People's Republic of China

The Dalai Lama, the spiritual leader of Tibetan Buddhists, is as important to that religion as the Pope is to Roman Catholics. Traditionally, the Dalai Lama—which translates as "oceanic teacher"—was not only the spiritual leader of Tibetan Buddhism but also the head of the government of Tibet. The photograph on page 204 shows the Dalai Lama's former palace in Tibet's capital Lhasa, situated in the Himalaya Mountains.

China, which had ruled Tibet from 1720 until its independence in 1911, invaded the rugged, isolated country in 1950, turned it into a province named Xizang in 1951, and installed a Communist government in Tibet in 1953. After crushing a rebellion in 1959, China executed or imprisoned tens of thousands and forced another 100,000, including the Dalai Lama, to emigrate. Buddhist temples were closed and demolished, and religious artifacts and scriptures were destroyed.

Why did the Chinese try to dismantle the religious institutions of a poor, remote country? At issue was the fact that the presence of strong religious feelings among the Tibetan people conflicted with the aims of the Chinese government.

The conflict between traditional Buddhism and the Chinese government is one of many examples of the impact of religion. In the modern world of global economics and culture, local religious belief continues to play a strong role in people's lives. ■

Religion interests geographers because it is essential for understanding how humans occupy Earth. As always, human geographers start by asking "where?" and "why?"

The predominant religion varies among ***regions*** of the world, as well as among regions within North America. Geographers document the places ***where*** various religions are located in the world and offer explanations for why some religions have widespread distributions and others are highly clustered in particular ***places***.

To understand ***why*** some religions occupy more ***space*** than others, geographers must look at differences among practices of various faiths. Geographers, though, are not theologians, so they stay focused on those elements of religions that are geographically significant. Geographers study spatial ***connections*** in religion: the distinctive place of origin of religions, the extent of diffusion of religions from their places of origin, the processes by which religions diffused to other locations, and the religious practices and beliefs that lead some religions to have more widespread distributions.

Geographers find the tension in ***scale*** between ***globalization*** and ***local diversity*** especially acute in religion for a number of reasons:

- People care deeply about their religion and draw from religion their core values and beliefs, an essential element of the definition of culture.
- Some religions are actually *designed* to appeal to people throughout the world, whereas other religions are designed to appeal primarily to people in geographically limited areas.
- Religious values are important in understanding not only how people identify themselves, as was the case with language, but also the meaningful ways that they organize the landscape.
- Most (though not all) religions require exclusive adherence, so adopting a global religion usually requires turning away from a traditional local religion. In contrast, people can learn a globally important language such as English and at the same time still speak the language of their local culture.
- Like language, migrants take their religion with them to new locations, but although migrants typically learn the language of the new location, they retain their religion.

This chapter starts by describing the distribution of major religions, then in the second section explains why some religions have diffused widely and others have not. As a major facet of culture, religion leaves a strong imprint on the physical environment, discussed in the third section of the chapter.

Religion, like other cultural characteristics, can be a source of pride and a means of identification with a distinct culture. Unfortunately, intense identification with one religion can lead adherents into conflict with followers of other religions, discussed in the fourth key issue of the chapter.

KEY ISSUE 1
Where Are Religions Distributed?

- Universalizing Religions
- Ethnic Religions

Only a few religions can claim the adherence of large numbers of people. Each of these faiths has a distinctive distribution across Earth's surface. ■

Geographers distinguish two types of religions: universalizing and ethnic. A **universalizing religion** attempts to be global, to appeal to all people, wherever they may live in the world, not just to those of one culture or location. An **ethnic religion** appeals primarily to one group of people living in one place.

This section examines the world's three main universalizing religions and some representative ethnic religions.

Universalizing Religions

According to *Adherents.com*, about 58 percent of the world's population practice a universalizing religion, 26 percent an ethnic religion, and 16 percent no religion. The three main universalizing religions are Christianity, Islam, and Buddhism.

Each of the three is divided into branches, denominations, and sects. A **branch** is a large and fundamental division within a religion. A **denomination** is a division of a branch that unites a number of local congregations in a single legal and administrative body. A **sect** is a relatively small group that has broken away from an established denomination.

Statistics on the number of followers of religions, branches, and denominations can be controversial. No official count of religious membership is taken in the United States and in many other countries. Most statistics in this chapter come from *Adherents.com*, an organization not affiliated with any religion.

Christianity

Christianity has more than 2 billion adherents, far more than any other world religion, and has the most widespread distribution. It is the predominant religion in North America, South America, Europe, and Australia, and countries with a Christian majority exist in Africa and Asia as well (Figure 6-1).

BRANCHES OF CHRISTIANITY. Christianity has three major branches—Roman Catholic, Protestant, and Orthodox. According to the *Encyclopaedia Britannica*, Roman Catholics comprise 51 percent of the world's Christians, Protestants 24 percent, and Orthodox 11 percent. In addition, 14 percent of Christians belong to churches that do not consider themselves within one of these three branches.

Within Europe, Roman Catholicism is the dominant Christian branch in the southwest and east, Protestantism in the northwest, and Orthodoxy in the east and southeast. The regions of Roman Catholic and Protestant majorities frequently have sharp boundaries, even when they run through the middle of countries. For example, the Netherlands and Switzerland have approximately equal percentages of Roman Catholics and Protestants, but the Roman Catholic populations are concentrated in the south of these countries and the Protestant populations in the north.

The Orthodox branch of Christianity (often called Eastern Orthodox) is a collection of 14 self-governing churches in Eastern Europe and the Middle East. More than 40 percent of all Orthodox Christians belong to one of these 14—the Russian Orthodox Church. Christianity came to Russia in the tenth century, and the Russian Orthodox Church was established in the sixteenth century.

FIGURE 6-1 Branches of Christianity in Europe. In the United Kingdom, Germany, and Scandinavia, the majority adhere to a Protestant denomination. In Eastern and Southeastern Europe, Orthodoxy dominates. Roman Catholicism is dominant in Southern, Central, and Southwestern Europe.

Nine of the other 13 self-governing churches were established in the nineteenth or twentieth century. The largest of these 9, the Romanian Church, includes 20 percent of all Eastern Orthodox Christians. The Bulgarian, Greek, and Serbian Orthodox churches have approximately 10 percent each. The other 5 recently established Orthodox churches—Albania, Cyprus, Georgia, Poland, and Sinai—combined have about 2 percent of all Orthodox Christians. The remaining 4 of the 14 Eastern Orthodox churches—Constantinople, Alexandria, Antioch, and Jerusalem—trace their origins to the earliest days of Christianity. They have a combined membership of about 3 percent of all Orthodox Christians.

CHRISTIANITY IN THE WESTERN HEMISPHERE. The overwhelming percentage of people living in the Western Hemisphere—nearly 90 percent—are Christian. About 5 percent belong to other religions, and the remaining 6 percent profess adherence to no religion.

A fairly sharp boundary exists within the Western Hemisphere in the predominant branches of Christianity. Roman Catholics comprise 93 percent of Christians in Latin America, compared with 40 percent in North America. Within North America, Roman Catholics are clustered in the southwestern and northeastern United States and the Canadian province of Québec (Figure 6-2).

Protestant churches have approximately 82 million members, or about 28 percent of the U.S. population over age 5 (Table 6-1). Baptist churches have the largest number of adherents in the United States, about 37 million combined over age 5. Membership in some Protestant churches varies by region of the United States. Baptists, for example, are highly clustered in the southeast, and Lutherans in the upper Midwest. Other Christian denominations are more evenly distributed around the country.

SMALLER BRANCHES OF CHRISTIANITY. Several other Christian churches developed independently of the three main branches. Many of these Christian communities were isolated from others at an early point in the development of Christianity, partly because of differences in doctrine and partly as a result of Islamic control of intervening territory in Southwest Asia and North Africa.

Two small Christian churches survive in northeast Africa—the Coptic Church of Egypt and the Ethiopian Church. The

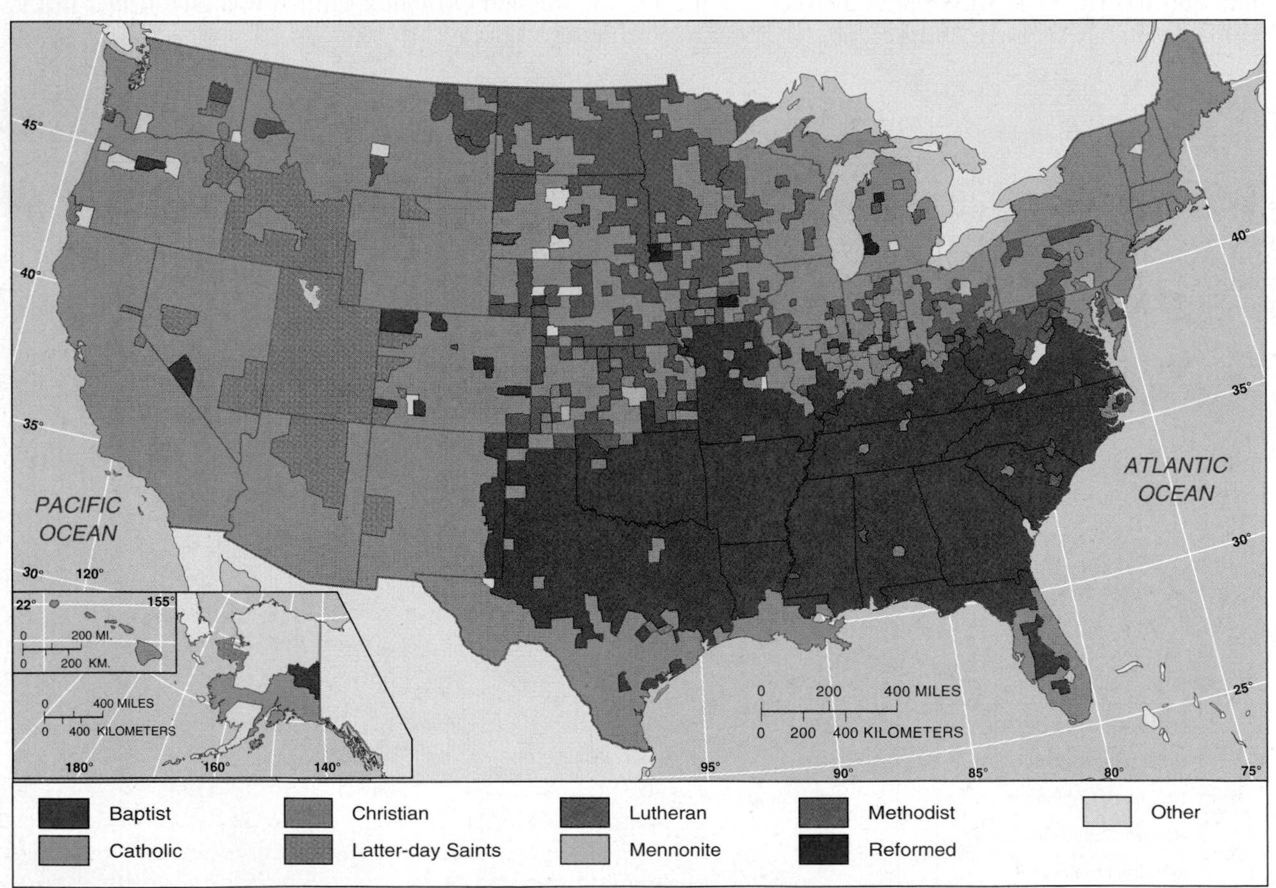

FIGURE 6-2 Distribution of Christians in the United States. The shaded areas are U.S. counties in which more than 50 percent of church membership is concentrated in either Roman Catholicism or one Protestant denomination. The distinctive distribution of religious groups within the United States results from patterns of migration, especially from Europe in the nineteenth century and from Latin America in recent years.

TABLE 6-1 RELIGIONS OF THE UNITED STATES

- 30 million nonreligious or atheist
- 1 million Buddhists
- 1 million Hindus
- 3 million Jews
- 1 million Muslims
- 6 million other faiths
- 161 million Christians
    - 66 million Roman Catholics
    - 3 million Orthodox
        - 2 million a church of the Greek Orthodox Archdiocese of America
        - 1 million another Orthodox church
    - 82 million Protestants
        - 37 million a Baptist church
            - 17 million a Southern Baptist Convention church
            - 8 million a National Baptist Convention, U.S.A., church
            - 4 million a National Baptist Convention of America church
            - 3 million a National Missionary Baptist Convention of America church
            - 3 million a Progressive National Baptist Convention church
            - 2 million an American Baptist Church, USA
            - 3 million another Baptist church
        - 13 million a Methodist church
            - 8 million a United Methodist church
            - 4 million an African Methodist Episcopal or Episcopal Zion church
        - 11 million a Pentecostal church
            - 6 million a Church of God in Christ
            - 3 million one of the Assemblies of God churches
            - 2 million one of the Pentecostal Assemblies of the world churches
        - 8 million a Lutheran church
            - 5 million an Evangelical Lutheran Church in America
            - 3 million one of the Lutheran Church Missouri Synod churches
        - 4 million a Presbyterian Church U.S.A.
        - 2 million a Reformed church
            - 1 million a United Church of Christ
            - 1 million another Reformed Church
        - 2 million an Episcopal church
        - 3 million one of the Churches of Christ
        - 1 million a Christian Church (Disciples of Christ)
        - 1 million a Seventh Day Adventist church
    - 10 million other Christians
        - 6 million a Church of Jesus Christ of Latter-Day Saints
        - 1 million a Jehovah's Witness church
        - 3 million other Christians

Ethiopian Church, with perhaps 10 million adherents, split from the Egyptian Coptic Church in 1948, although it traces its roots to the fourth century, when two shipwrecked Christians, who were taken as slaves, ultimately converted the Ethiopian king to Christianity.

The Armenian Church originated in Antioch, Syria, and was important in diffusing Christianity to South and East Asia between the seventh and thirteenth centuries. The church's few present-day adherents are concentrated in Lebanon and Armenia, as well as in northeastern Turkey and western Azerbaijan. Despite the small number of adherents, the Armenian Church, like other small sects, plays a significant role in regional conflicts. For example, Armenian Christians have fought for the independence of Nagorno-Karabakh, a portion of Azerbaijan, because Nagorno-Karabakh is predominantly Armenian, whereas the remainder of Azerbaijan is overwhelmingly Shiite Muslim (see Chapter 7).

The Maronites are another example of a small Christian sect that plays a disproportionately prominent role in political unrest. They are clustered in Lebanon, which has suffered through a long civil war fought among religious groups (see Chapter 7).

In the United States, members of The Church of Jesus Christ of Latter-day Saints (Mormons) regard their church as a branch of Christianity separate from other branches. About 3 percent of Americans are members of the Latter-day Saints, and a large percentage is clustered in Utah and surrounding states.

Islam

Islam, the religion of 1.3 billion people, is the predominant religion of the Middle East from North Africa to Central Asia (Figure 6-3). Half of the world's Muslims live in four countries outside the Middle East—Indonesia, Pakistan, Bangladesh, and India.

The word *Islam* in Arabic means "submission to the will of God," and it has a similar root to the Arabic word for *peace*. An adherent of the religion of Islam is known as a Muslim, which in Arabic means "one who surrenders to God." The core of Islamic belief is represented by five pillars of faith:

1. There is no god worthy of worship except the one God, the source of all creation, and Muhammad is the messenger of God.
2. Five times daily, a Muslim prays, facing the city of Makkah (Mecca), as a direct link to God.
3. A Muslim gives generously to charity as an act of purification and growth.
4. A Muslim fasts during the month of Ramadan as an act of self-purification.
5. If physically and financially able, a Muslim makes a pilgrimage to Makkah.

FIGURE 6-3 World distribution of religions. These are the religions of the world with the most adherents:

- Christianity: 31 percent, especially in Europe and the Western Hemisphere
- Islam: 22 percent, especially in northern Africa and Southwest and Southeast Asia
- Hinduism: 13 percent, virtually all in India
- Buddhism: 6 percent, especially in East and Southeast Asia
- Other ethnic religions: 13 percent, especially in Asia and Africa
- Nonreligious or atheist: 16 percent

The small pie charts on the map show the overall proportion of the world's religions in each world region. The large pie chart below shows the worldwide percentage of people adhering to the various religions.

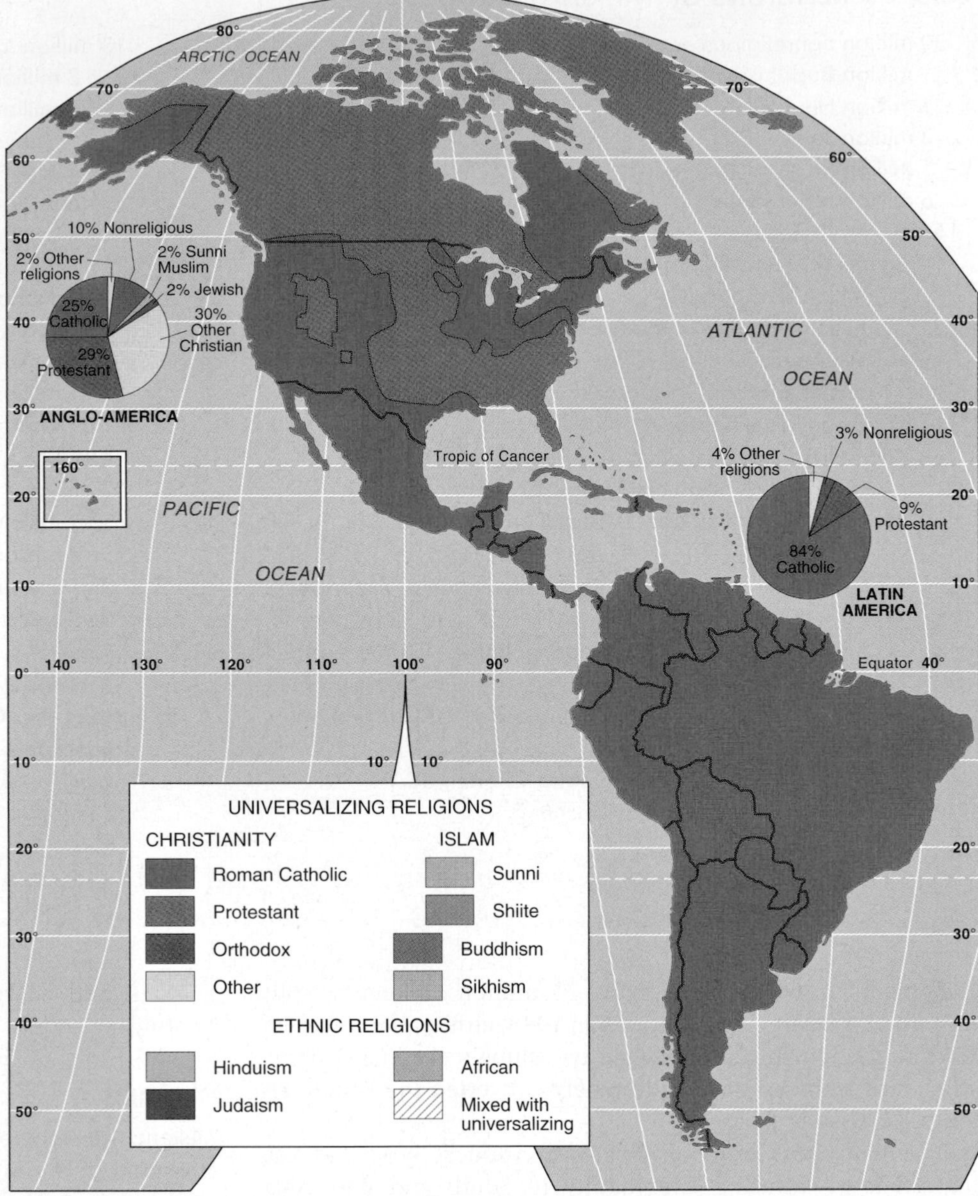

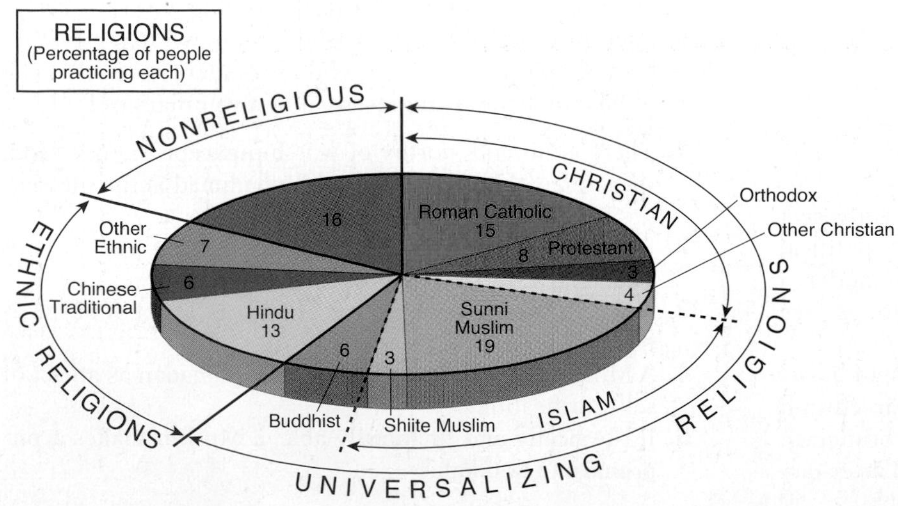

BRANCHES OF ISLAM. Islam is divided into two important branches: *Sunni* and *Shiite* (sometimes written *Shia* in English).

- Sunnis comprise 83 percent of Muslims and are the largest branch in most Muslim countries in the Middle East and Asia. The word Sunni comes from the Arabic for "people following the example of Muhammad."
- Shiites comprise 16 percent of Muslims, clustered in a handful of countries. Nearly 30 percent of all Shiites live in Iran, 15 percent in Pakistan, and 10 percent in Iraq. Shiites comprise nearly 90 percent of the population in Iran and more than half of the population in Azerbaijan, Iraq, and the less populous countries of Oman and Bahrain. The word *Shiite* comes from the Arabic word for "sectarian."

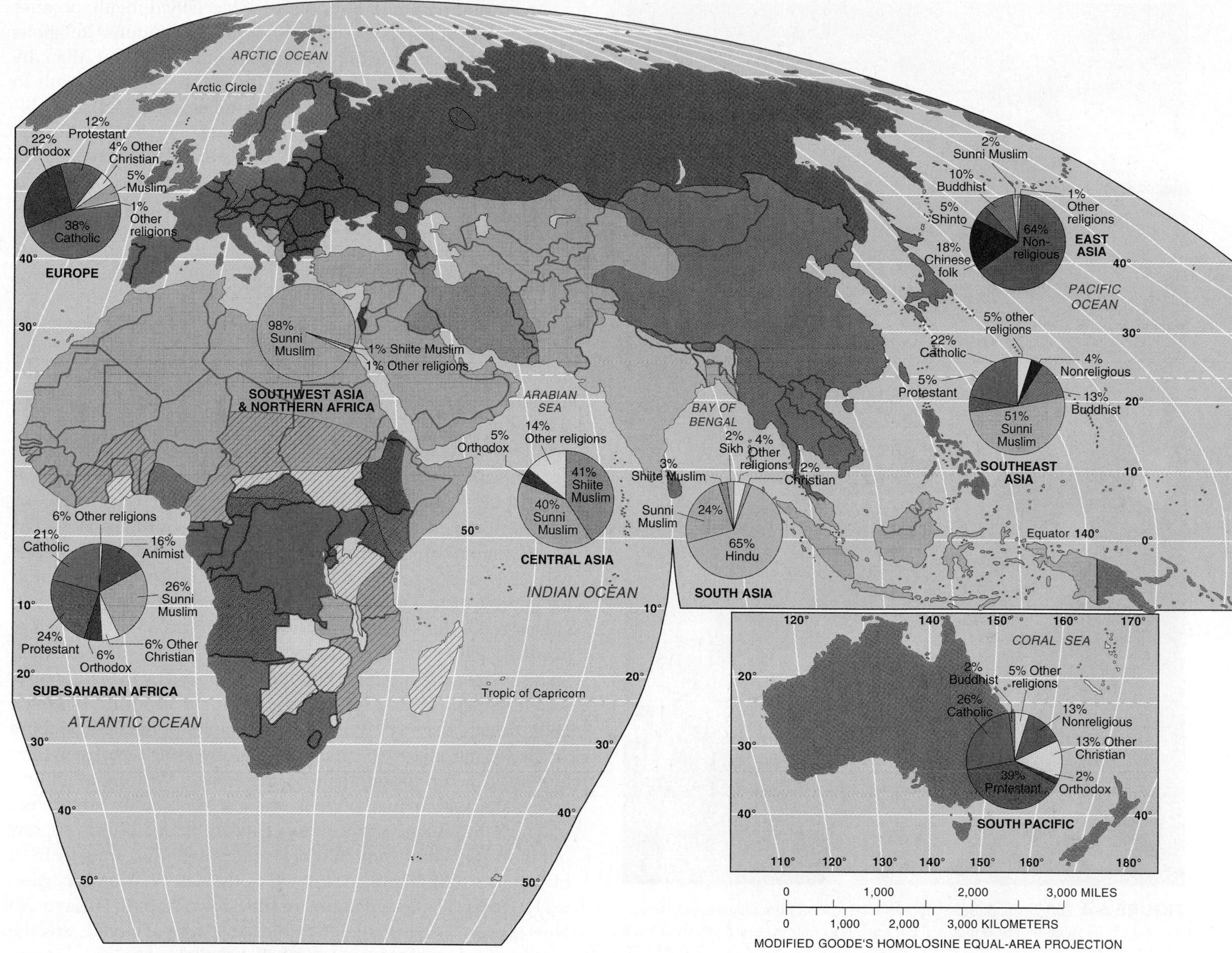

ISLAM IN NORTH AMERICA AND EUROPE. The Muslim population of North America and Europe has increased rapidly in recent years. Estimates of the number of Muslims in North America vary widely, from 1 million to 5 million, but in any event it has increased from only a few hundred thousand in 1990.

In Europe, Muslims account for 5 percent of the population. France has the largest Muslim population, about 4 million, a legacy of immigration from predominantly Muslim former colonies in North Africa. Germany has about 3 million Muslims, also a legacy of immigration, in Germany's case primarily from Turkey. In Southeast Europe, Albania, Bosnia, and Serbia each have about 2 million Muslims.

Islam also has a presence in the United States through the Nation of Islam, also known as Black Muslims, founded in Detroit in 1930 and led for more than 40 years by Elijah Muhammad, who called himself "the messenger of Allah." Black Muslims lived austerely and advocated a separate autonomous nation within the United States for their adherents. Tension between Muhammad and a Black Muslim minister, Malcolm X, divided the sect during the 1960s. After a pilgrimage to Makkah in 1963, Malcolm X converted to orthodox Islam and founded the Organization of Afro-American Unity. He was assassinated in 1965. After Muhammad's death, in 1975, his son Wallace D. Muhammad led the Black Muslims closer to the principles of orthodox Islam, and the organization's name was changed to the American Muslim Mission. A splinter group adopted the original name, Nation of Islam, and continues to follow the separatist teachings of Elijah Muhammad.

Buddhism

Buddhism, the third of the world's major universalizing religions, has nearly 400 million adherents, who are mainly found in China and Southeast Asia (Figure 6-4). The foundation of Buddhism is represented by these concepts, known as the Four Noble Truths:

1. All living beings must endure suffering.
2. Suffering, which is caused by a desire to live, leads to reincarnation (repeated rebirth in new bodies or forms of life).

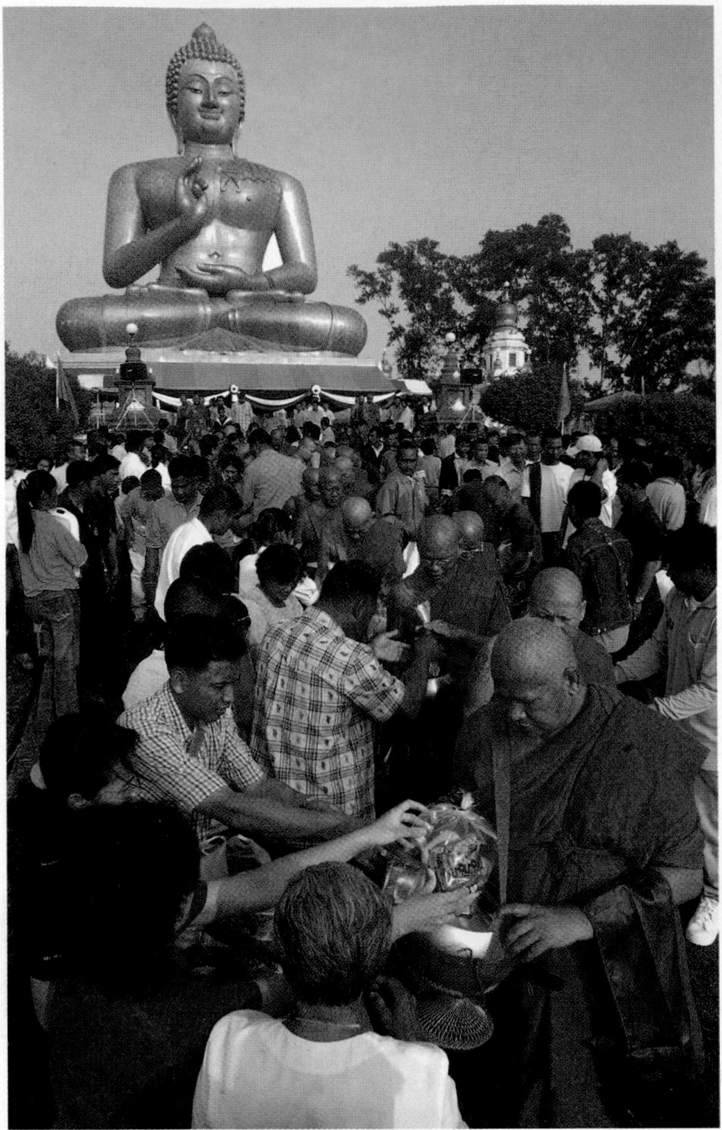

FIGURE 6-4 Buddhism. Buddhist monks receive food from villagers to celebrate the end of the three-month Buddhist Lent festival. Narathiwat province, Thailand.

3. The goal of all existence is to escape from suffering and the endless cycle of reincarnation into Nirvana (a state of complete redemption), which is achieved through mental and moral self-purification.
4. Nirvana is attained through an Eightfold Path, which includes rightness of belief, resolve, speech, action, livelihood, effort, thought, and meditation.

Like the other two universalizing religions, Buddhism split into more than one branch, as followers disagreed on interpreting statements by the founder, Siddhartha Gautama. The three main branches are Mahayana, Theravada, and Tantrayana.

Mahayanists account for about 56 percent of Buddhists, primarily in China, Japan, and Korea. Theravadists comprise about 38 percent of Buddhists, especially in Cambodia, Laos, Myanmar, Sri Lanka, and Thailand. The remaining 6 percent are Tantrayanists, found primarily in Tibet and Mongolia.

An accurate count of Buddhists is especially difficult, because only a few people participate in Buddhist institutions. Religious functions are performed primarily by monks rather than by the general public. The number of Buddhists is also difficult to count because Buddhism, although a universalizing religion, differs in significant respects from the Western concept of a formal religious system. Someone can be both a Buddhist and a believer in other Eastern religions, whereas Christianity and Islam both require exclusive adherence. Most Buddhists in China and Japan, in particular, believe at the same time in an ethnic religion.

Other Universalizing Religions

Sikhism and Bahá'í are the two universalizing religions other than Christianity, Islam, and Buddhism with the largest numbers of adherents. There are an estimated 23 million Sikhs and 7 million Bahá'ís. All but 3 million Sikhs are clustered in the Punjab region of India; Bahá'ís are dispersed among many countries, primarily in Africa and Asia.

Sikhism's first guru (religious teacher or enlightener) was Nanak (1469–1538), who lived in a village near the city of Lahore, in present-day Pakistan. God was revealed to Guru Nanak as The One Supreme Being, or Creator, who rules the universe by divine will. Only God is perfect, but people have the capacity for continual improvement and movement toward perfection by taking individual responsibility for their deeds and actions on Earth, such as heartfelt adoration, devotion, and surrender to the one God. Sikhism's most important ceremony, introduced by the tenth guru, Gobind Singh (1666–1708), is the Amrit (or Baptism), in which Sikhs declare they will uphold the principles of the faith. Gobind Singh also introduced the practice of men wearing turbans on their heads and never cutting their beards or hair. Wearing a uniform gave Sikhs a disciplined outlook and a sense of unity of purpose.

The Bahá'í religion is even more recent than Sikhism. It grew out of the Bábi faith, which was founded in Shíráz, Iran, in 1844 by Siyyid 'Ali Muhammad, known as the Báb (Persian for "gateway"). Bahá'ís believe that one of the Báb's disciples, Husayn 'Ali Nuri, known as Bahá'u'lláh (Arabic for "Glory of God"), was the prophet and messenger of God. Bahá'u'lláh's function was to overcome the disunity of religions and establish a universal faith through abolition of racial, class, and religious prejudices.

Ethnic Religions

The ethnic religion with by far the largest number of followers is Hinduism. With 900 million adherents, Hinduism is the world's third-largest religion, behind Christianity and Islam. Ethnic religions in Asia and Africa comprise most of the remainder.

Hinduism

Ethnic religions typically have much more clustered distributions than do universalizing religions. Hinduism is the world's third-largest religion, but 97 percent of Hindus are concentrated in one country, India, and most of the remainder can be found in India's neighbor Nepal. Hindus comprise more than 80 percent of the population of these two countries and a small minority in every other country (Figure 6-5).

FIGURE 6-5 Hindus bathe in the Ganges River. The river attracts Hindu pilgrims from all over India because they believe that the Ganges springs from the hair of Siva, one of the main deities. Hindus achieve purification by bathing in the Ganges, and bodies of the dead are washed with water from it before being cremated.

A rigid approach to theological matters is not central to Hinduism. Hindus believe that it is up to the individual to decide the best way to worship God. Various paths to reach God include the path of knowledge, the path of renunciation, the path of devotion, and the path of action. You can pursue your own path and follow your own convictions, as long as they are in harmony with your true nature. You are responsible for your own actions and you alone suffer the consequences. Because people start from different backgrounds and experiences, the appropriate form of worship for any two individuals may not be the same. Hinduism does not have a central authority or a single holy book, so each individual selects suitable rituals. If one person practices Hinduism in a particular way, other Hindus will not think that the individual has made a mistake or strayed from orthodox doctrine.

The average Hindu has allegiance to a particular god or concept within a broad range of possibilities. The manifestation of God with the largest number of adherents—an estimated 70 percent—is Vaishnavism, which worships the god Vishnu, a loving god incarnated as Krishna. An estimated 26 percent adhere to Sivaism, dedicated to Siva, a protective and destructive god. Shaktism is a form of worship dedicated to the female consorts of Vishnu and Siva. Although these and other deities and approaches are supported throughout India, some geographic concentration exists: Siva and Shakti are concentrated in the north; Shakti and Vishnu in the east; Vishnu in the west; and Siva, along with some Vishnu, in the south. However, holy places for Siva and Vishnu are dispersed throughout India.

Other Ethnic Religions

Several hundred million people practice ethnic religions in East Asia, especially in China and Japan. The coexistence of Buddhism with these ethnic religions in East Asia differs from the Western concept of exclusive religious belief.

Confucianism and Daoism (sometimes spelled Taoism) are often distinguished as separate ethnic religions in China, but many Chinese consider themselves both Buddhists and either Confucian, Daoist, or some other Chinese ethnic religion. Buddhism does not compete for adherents with Confucianism, Daoism, and other ethnic religions in China because many Chinese accept the teachings of both universalizing and ethnic religions.

Such commingling of diverse philosophies is not totally foreign to Americans. The tenets of Christianity or Judaism, the wisdom of the ancient Greek philosophers, and the ideals of the Declaration of Independence can all be held dear without doing grave injustice to the others.

CONFUCIANISM. Confucius (551–479 B.C.) was a philosopher and teacher in the Chinese province of Lu. His sayings, which were recorded by his students, emphasized the importance of the ancient Chinese tradition of *li*, which can be translated roughly as "propriety" or "correct behavior." Confucianism is an ethnic religion because of its especially strong rooting in traditional values of special importance to Chinese people. Confucianism prescribed a series of ethical principles for the orderly conduct of daily life in China, such as following traditions, fulfilling obligations, and treating others with sympathy and respect. These rules applied to China's rulers as well as to their subjects.

DAOISM (TAOISM). Lao-Zi (604– 531? B.C., also spelled Lao Tse), a contemporary of Confucius, organized Daoism. Although a government administrator by profession, Lao-Zi's writings emphasized the mystical and magical aspects of life rather than the importance of public service, which Confucius had emphasized.

Daoists seek *dao* (or *tao*), which means the "way" or "path." A virtuous person draws power (*de* or *te*) from being absorbed in *dao*. *Dao* cannot be comprehended by reason and knowledge because not everything is knowable. Because the universe is not ultimately subject to rational analysis, myths and legends develop to explain events. Only by avoidance of daily activities and introspection can a person live in harmony with the principles that underlie and govern the universe.

Daoism split into many sects, some acting like secret societies, and followers embraced elements of magic. The religion was officially banned by the Communists after they took control of China in 1949, but it is still practiced in China, and it is legal in Taiwan.

SHINTOISM. Since ancient times, Shintoism has been the distinctive ethnic religion of Japan. Ancient Shintoists considered forces of nature to be divine, especially the Sun and Moon, as well as rivers, trees, rocks, mountains, and certain animals. The religion was transmitted from one generation to the next orally until the fifth century A.D., when the introduction of Chinese writing facilitated the recording of ancient rituals and prayers. Gradually, deceased emperors and other ancestors became more important deities for Shintoists than natural features.

Under the reign of the Emperor Meiji (1868–1912), Shintoism became the official state religion, and the emperor was regarded as divine. Shintoism therefore was as much a political cult as a religion, and in a cultural sense all Japanese were Shintoists. After defeating Japan in World War II, the victorious Allies ordered Emperor Hirohito to renounce his divinity in a speech to the Japanese people, but he was allowed to retain ceremonial powers.

Shintoism still thrives in Japan, although no longer as the official state religion. Prayers are recited to show reverence for ancestors, and pilgrimages are made to shrines believed to house deities. More than 80,000 shrines serve as places for neighbors to meet or for children to play.

JUDAISM. Around one-third of the world's 14 million Jews live in the United States, one-third in Israel, and one-third in the rest of the world. Within the United States, Jews are heavily concentrated in the large cities, especially in the New York metropolitan area. Jews constitute a majority in Israel, where for the first time since the biblical era an independent state has had a Jewish majority.

Judaism plays a more substantial role in Western civilization than its number of adherents would suggest because two of the three main universalizing religions—Christianity and Islam—find some of their roots in Judaism. Jesus was born a Jew, and Muhammad traced his ancestry to Abraham.

Judaism is an ethnic religion based in the lands bordering the eastern end of the Mediterranean Sea, called Canaan in the Bible, Palestine by the Romans, and the state of Israel since 1948. About 4,000 years ago Abraham, considered the patriarch or father of Judaism, migrated from present-day Iraq to Canaan, along a route known as the Fertile Crescent (see discussion of the Fertile Crescent in Chapter 8 and Figure 8-6). The Bible recounts the ancient history of the Jewish people.

Fundamental to Judaism was belief in one all-powerful God. It was the first recorded religion to espouse **monotheism**, belief that there is only one God. Judaism offered a sharp contrast to the **polytheism** practiced by neighboring people, who worshipped a collection of gods. Jews considered themselves the "chosen" people because God had selected them to live according to God's ethical and moral principles, such as the Ten Commandments.

The name *Judaism* derives from *Judah*, one of the patriarch Jacob's 12 sons; *Israel* is another biblical name for Jacob. Descendants of 10 of Jacob's sons, plus 2 of his grandsons, constituted the 12 tribes of Hebrews who emigrated from Egypt in the Exodus narrative. Each received a portion of Canaan. Judah is one of the surviving tribes of the Hebrews; 10 of the tribes were considered lost after they were conquered and forced to migrate to Assyria in 721 B.C.

ETHNIC AFRICAN RELIGIONS. Approximately 100 million Africans, 12 percent of the continent's people, follow traditional ethnic religions, sometimes called **animism**. Animists believe that such inanimate objects as plants and stones, or such natural events as thunderstorms and earthquakes, are "animated," or have discrete spirits and conscious life. Relatively little is known about African religions because few holy books or other written documents have come down from ancestors. Religious rituals are passed from one generation to the next by word of mouth. African animist religions are apparently based on monotheistic concepts, although below the supreme god there is a hierarchy of divinities. These divinities may be assistants to the supreme god or personifications of natural phenomena, such as trees or rivers.

As recently as 1980, some 200 million Africans—half the population of the region at the time—were classified as animists. Some atlases and textbooks persist in classifying Africa as predominantly animist, even though the actual percentage is small and declining. Followers of traditional African religions now constitute a clear majority of the population only in Botswana. The rapid decline in animists in Africa has been caused by increases in the numbers of Christians and Muslims. Africa is now 46 percent Christian—split about evenly among Roman Catholic, Protestant, and other—and another 40 percent are Muslims. The growth in the two universalizing religions at the expense of ethnic religions reflects fundamental geographical differences between the two types of religions, discussed in the next key issue.

KEY ISSUE 2
Why Do Religions Have Different Distributions?

- **Origin of Religions**
- **Diffusion of Religions**
- **Holy Places**
- **The Calendar**

We can identify several major geographical differences between universalizing and ethnic religions. These differences include the locations where the religions originated, the processes by which they diffused from their place of origin to other regions, the types of places that are considered holy, the calendar dates identified as important holidays, and attitudes toward modifying the physical environment. ■

Origin of Religions

Universalizing religions have precise places of origin based on events in the life of a man. Ethnic religions have unknown or unclear origins, not tied to single historical individuals.

Origin of Universalizing Religions

Each of the three universalizing religions can be traced to the actions and teachings of a man who lived since the start of recorded history. The beginnings of Buddhism go back about 2,500 years, Christianity 2,000 years, and Islam 1,500 years. Specific events also led to the division of the universalizing religions into branches.

ORIGIN OF CHRISTIANITY. Christianity was founded upon the teachings of Jesus, who was born in Bethlehem between 8 and 4 B.C. and died on a cross in Jerusalem about A.D. 30. Raised as a Jew, Jesus gathered a small band of disciples and preached the coming of the Kingdom of God. The four Gospels of the Christian Bible—Matthew, Mark, Luke, and John—documented miracles and extraordinary deeds that Jesus performed. He was referred to as *Christ*, from the Greek word for the Hebrew word *messiah*, which means "anointed."

In the third year of his mission, Jesus was betrayed to the authorities by one of his companions, Judas Iscariot. After sharing the Last Supper (the Jewish Passover seder) with his disciples in Jerusalem, Jesus was arrested and put to death as an agitator. On the third day after his death, his tomb was found empty (Figure 6-6). Christians believe that Jesus died to atone for human sins, that he was raised from the dead by God, and that his Resurrection from the dead provides people with hope for salvation.

Roman Catholics accept the teachings of the Bible, as well as the interpretation of those teachings by the Church hierarchy, headed by the Pope. According to Roman Catholic belief, God conveys His grace directly to humanity through seven sacraments, including Baptism, Confirmation, Penance, Anointing the sick, Matrimony, Holy Orders, and the Eucharist (the partaking of bread and wine that repeats the actions of Jesus at the Last Supper). Roman Catholics believe that the Eucharist literally and miraculously become the body and blood of Jesus while keeping only the appearances of bread and wine, an act known as transubstantiation.

Orthodoxy comprises the faith and practices of a collection of churches that arose in the eastern part of the Roman Empire. The split between the Roman and Eastern churches dates to the fifth century, as a result of rivalry between the Pope of Rome and the Patriarchy of Constantinople, which was especially intense after the collapse of the Roman Empire. The split between the two churches became final in 1054, when Pope Leo IX condemned the Patriarch of Constantinople. Orthodox Christians accepted the seven sacraments but rejected doctrines that the Roman Catholic Church had added since the eighth century.

Protestantism originated with the principles of the Reformation in the sixteenth century. The Reformation movement is regarded as beginning when Martin Luther (1483–1546) posted 95 theses on the door of the church at Wittenberg on October 31, 1517. According to Luther, individuals had primary responsibility for achieving personal salvation through direct communication with God. Grace is achieved through faith rather than through sacraments performed by the Church.

ORIGIN OF ISLAM. Islam traces its origin to the same narrative as Judaism and Christianity. All three religions consider Adam to have been the first man and Abraham to have been one of his descendants.

According to the Biblical narrative, Abraham married Sarah, who did not bear children. As polygamy was a custom of the culture, Abraham then married Hagar, who bore a son, Ishmael. However, Sarah's fortunes changed, and she bore a son, Isaac. Sarah then successfully prevailed upon Abraham to banish Hagar and Ishmael. Jews and Christians trace their story through Abraham's original wife Sarah and her son Isaac. Muslims trace their story through his second wife Hagar and her son Ishmael. After their banishment, Ishmael and Hagar wandered through the Arabian desert, eventually reaching Makkah (spelled Mecca on many English-language maps), in present-day Saudi Arabia. Centuries later, one of Ishmael's descendants, Muhammad, became the Prophet of Islam.

Muhammad was born in Makkah about 570. At age 40, while engaged in a meditative retreat, Muhammad received his first revelation from God through the Angel Gabriel. The Quran, the holiest book in Islam, is a record of God's words

FIGURE 6-6 Origin of Christianity. This tomb in the center of the Church of the Holy Sepulchre in Jerusalem was erected on the site where Jesus is thought to have been buried and resurrected. Orthodox Christians observe Holy Saturday, the day before Easter, with a Holy Fire ceremony at the tomb.

FIGURE 6-7 Origin of Islam. Muhammad is buried in the Mosque of the Prophet in Madinah, Saudi Arabia. The mosque, built on the site of Muhammad's house, is the second holiest in Islam and the second largest mosque in the world.

as revealed to the Prophet Muhammad through Gabriel. Arabic is the lingua franca, or language of communication, within the Muslim world, because it is the language in which the Quran is written.

As he began to preach the truth that God had revealed to him, Muhammad suffered persecution, and in 622 he was commanded by God to emigrate. His migration from Makkah to the city of Yathrib—an event known as the *Hijra* (from the Arabic word for "migration," sometimes spelled *hegira*)—marks the beginning of the Muslim calendar. Yathrib was subsequently renamed Madinah, Arabic for "the City of the Prophet" (Figure 6-7). After several years, Muhammad and his followers returned to Makkah and established Islam as the city's religion. By Muhammad's death, in 632 at about age 63, the armies of Islam controlled most of present-day Saudi Arabia.

Differences between the two main branches—Shiites and Sunnis—go back to the earliest days of Islam and basically reflect disagreement over the line of succession in Islamic leadership. Muhammad had no surviving son and no follower of comparable leadership ability. His successor was his father-in-law Abu Bakr (573–634), an early supporter from Makkah, who became known as *caliph* ("successor of the prophet"). The next two caliphs, Umar (634–644) and Uthman (644–656), expanded the territory under Muslim influence to Egypt and Persia.

Uthman was a member of a powerful Makkah clan that had initially opposed Muhammad before the clan's conversion to Islam. More zealous Muslims criticized Uthman for seeking compromises with other formerly pagan families in Makkah. Uthman's opponents found a leader in Ali (600?– 661), a cousin and son-in-law of Muhammad, and thus Muhammad's nearest male heir. When Uthman was murdered, in 656, Ali became caliph, although five years later he, too, was assassinated.

Ali's descendants claim leadership of Islam, and Shiites support this claim. But Shiites disagree among themselves about the precise line of succession from Ali to modern times. They acknowledge that the chain of leadership was broken, but they dispute the date and events surrounding the disruption.

During the 1970s both the shah (king) of Iran and an ayatollah (religious scholar) named Khomeini claimed to be the divinely appointed interpreter of Islam for the Shiites. The allegiance of the Iranian Shiites switched from the shah to the ayatollah largely because the ayatollah made a more convincing case that he was more faithfully adhering to the rigid laws laid down by Muhammad in the Quran.

ORIGIN OF BUDDHISM. The founder of Buddhism, Siddhartha Gautama, was born about 563 B.C. in Lumbinī in present-day Nepal, near the border with India, about 160 kilometers (100 miles) from Vārā nasi (Benares). The son of a lord, he led a privileged existence sheltered from life's hardships. Gautama had a beautiful wife, palaces, and servants.

According to Buddhist legend, Gautama's life changed after a series of four trips. He encountered a decrepit old man on the first trip, a disease-ridden man on the second trip, and a corpse on the third trip. After witnessing these scenes of pain and suffering, Gautama began to feel he could no longer enjoy his life of comfort and security. Then, on a fourth trip, Gautama saw a monk, who taught him about withdrawal from the world.

At age 29 Gautama left his palace one night and lived in a forest for the next 6 years, thinking and experimenting with forms of meditation. Gautama emerged as the *Buddha*, the "awakened or enlightened one," and spent 45 years preaching his views across India. In the process, he trained monks, established orders, and preached to the public.

Theravada is the older of the two largest branches of Buddhism. The word means "the way of the elders," indicating the Theravada Buddhists' belief that they are closer to Buddha's original approach. Theravadists believe that Buddhism is a full-time occupation, so to become a good Buddhist, one must renounce worldly goods and become a monk.

Mahayana split from Theravada Buddhism about 2,000 years ago. *Mahayana* is translated as "the bigger ferry" or "raft," and Mahayanists call Theravada Buddhism by the name *Hinayana*, or "the little raft." Mahayanists claim that their approach to Buddhism can help more people because it is less demanding and all-encompassing. Theravadists emphasize Buddha's life of self-help and years of solitary introspection, and Mahayanists emphasize Buddha's later years of teaching and helping others. Theravadists cite Buddha's wisdom, and Mahayanists his compassion.

ORIGIN OF OTHER UNIVERSALIZING RELIGIONS. Sikhism and Bahá'í were founded more recently than the three large universalizing religions. The founder of Sikhism, Guru Nanak, traveled widely through South Asia around 500 years ago preaching his new faith, and many people became his *Sikhs*, which is the Hindi word for "disciples." Nine other gurus succeeded Guru Nanak. Arjan, the fifth guru, compiled and edited in 1604 the *Guru Granth Sahib* (the Holy Granth of Enlightenment), which became the book of Sikh holy scriptures.

When it was established in Iran during the nineteenth century, Bahá'í provoked strong opposition from Shiite Muslims. The Báb was executed in 1850, as were 20,000 of his followers. Bahá'u'lláh, the prophet of Bahá'í, was also arrested but was released in 1853 and exiled to Baghdad. In 1863, his claim that he was the messenger of God anticipated by the Báb was accepted by other followers. Before he died in 1892, Bahá'u'lláh appointed his eldest son 'Abdu'l-Bahá (1844–1921) to be the leader of the Bahá'í community and the authorized interpreter of his teachings.

Origin of Hinduism, an Ethnic Religion

Unlike the three universalizing religions, Hinduism did not originate with a specific founder. The origins of Christianity, Islam, and Buddhism are recorded in the relatively recent past, but Hinduism existed prior to recorded history.

The word *Hinduism* originated in the sixth century B.C. to refer to people living in what is now India. The earliest surviving Hindu documents were written around 1500 B.C., although archaeological explorations have unearthed objects relating to the religion from 2500 B.C. Aryan tribes from Central Asia invaded India about 1400 B.C. and brought with them Indo-European languages, as discussed in Chapter 5. In addition to their language, the Aryans brought their religion. The Aryans first settled in the area now called the Punjab in northwestern India and later migrated east to the Ganges River valley, as far as Bengal. Centuries of intermingling with the Dravidians already living in the area modified their religious beliefs.

Diffusion of Religions

The three universalizing religions diffused from specific hearths, or places of origin, to other regions of the world. In contrast, ethnic religions typically remain clustered in one location.

Diffusion of Universalizing Religions

The hearths where each of the three largest universalizing religions originated are based on the events in the lives of the three key individuals (Figure 6-8). All three hearths are in Asia (Christianity and Islam in Southwest Asia, Buddhism in South Asia). Followers transmitted the messages preached in the hearths to people elsewhere, diffusing them across Earth's surface along distinctive paths, as shown in Figure 6-8. Today, these three universalizing religions together have several billion adherents distributed across wide areas of the world.

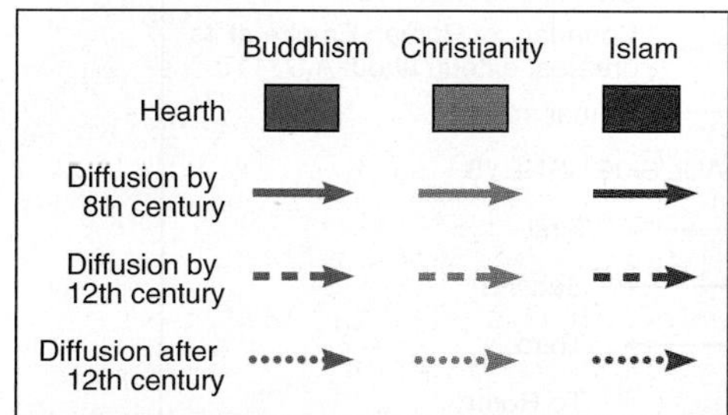

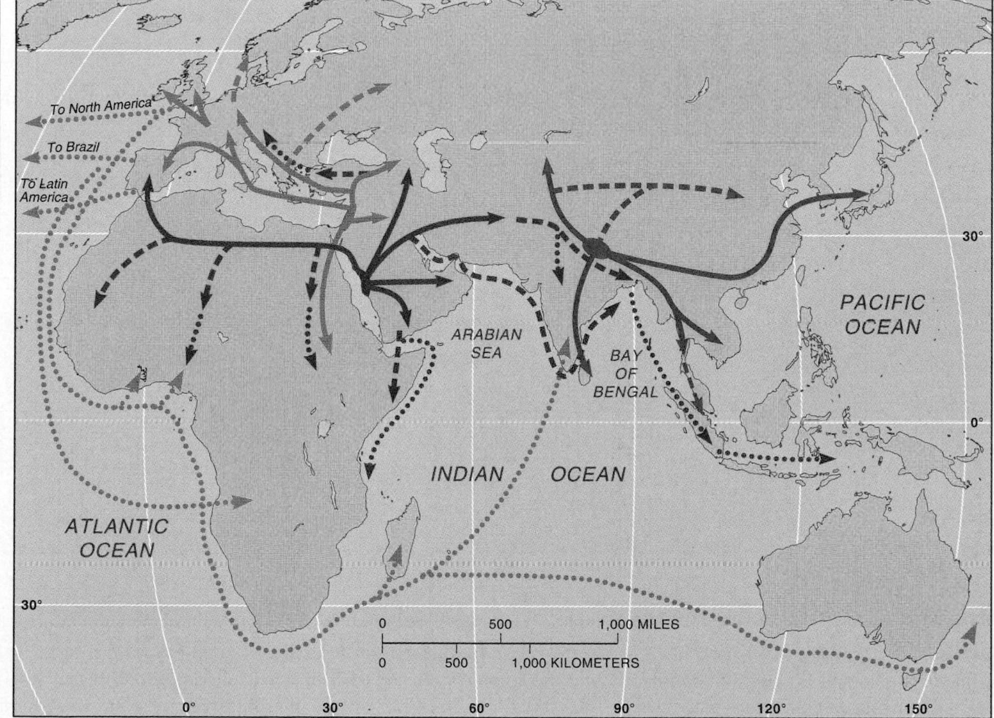

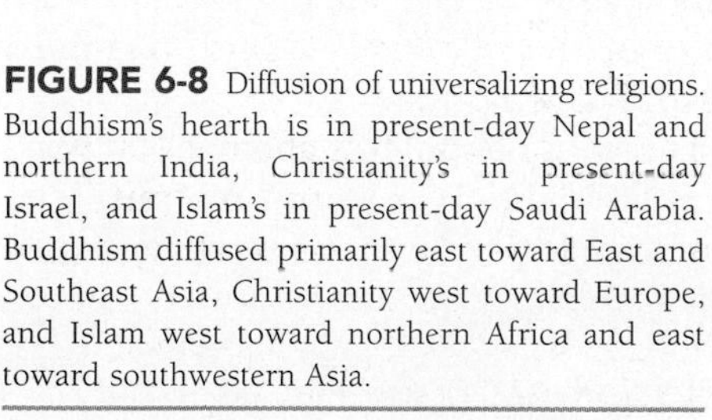

FIGURE 6-8 Diffusion of universalizing religions. Buddhism's hearth is in present-day Nepal and northern India, Christianity's in present-day Israel, and Islam's in present-day Saudi Arabia. Buddhism diffused primarily east toward East and Southeast Asia, Christianity west toward Europe, and Islam west toward northern Africa and east toward southwestern Asia.

DIFFUSION OF CHRISTIANITY. Christianity's diffusion has been rather clearly recorded since Jesus first set forth its tenets in the Roman province of Palestine. Consequently, geographers can examine its diffusion by reconstructing patterns of communications, interaction, and migration.

In Chapter 1 two processes of diffusion were identified—*relocation* (diffusion through migration) and *expansion* (diffusion through a snowballing effect). Within expansion diffusion we distinguished between *hierarchical* (diffusion through key leaders) and *contagious* (widespread diffusion). Christianity diffused through a combination of all of these forms of diffusion.

Christianity first diffused from its hearth in Palestine through *relocation diffusion*. **Missionaries**—individuals who help to transmit a universalizing religion through relocation diffusion—carried the teachings of Jesus along the Roman Empire's protected sea routes and excellent road network to people in other locations. Paul of Tarsus, a disciple of Jesus, traveled especially extensively through the Roman Empire as a missionary. The outline of the empire and spread of Christianity are shown in Figure 6-9.

People in commercial towns and military settlements that were directly linked by the communications network received the message first from Paul and other missionaries. But Christianity spread widely within the Roman Empire through *contagious diffusion*—daily contact between believers in the towns and nonbelievers in the surrounding countryside. **Pagan**, the word for a follower of a polytheistic religion in ancient times, derives from the Latin word for "countryside."

The dominance of Christianity throughout the Roman Empire was assured during the fourth century through *hierarchical diffusion*—acceptance of the religion by the empire's key elite figure, the emperor. Emperor Constantine (274?–337) encouraged the spread of Christianity by embracing it in 313, and Emperor Theodosius proclaimed it the empire's official religion in 380. In subsequent centuries, Christianity further diffused into Eastern Europe through conversion of kings or other elite figures.

Migration and missionary activity by Europeans since the year 1500 has extended Christianity to other regions of the world, as shown in Figure 6-3. Through permanent resettlement of Europeans, Christianity became the dominant religion in North and South America, Australia, and New Zealand. Christianity's dominance was further achieved by conversion of indigenous populations and by intermarriage. In recent decades, Christianity has further diffused to Africa, where it is now the most widely practiced religion.

Latin Americans are predominantly Roman Catholic because their territory was colonized by the Spanish and Portuguese, who brought with them to the Western Hemisphere their religion as well as their languages. Canada (except Québec) and the United States have Protestant majorities because their early colonists came primarily from Protestant England. Some regions and localities within the United States and Canada are predominantly

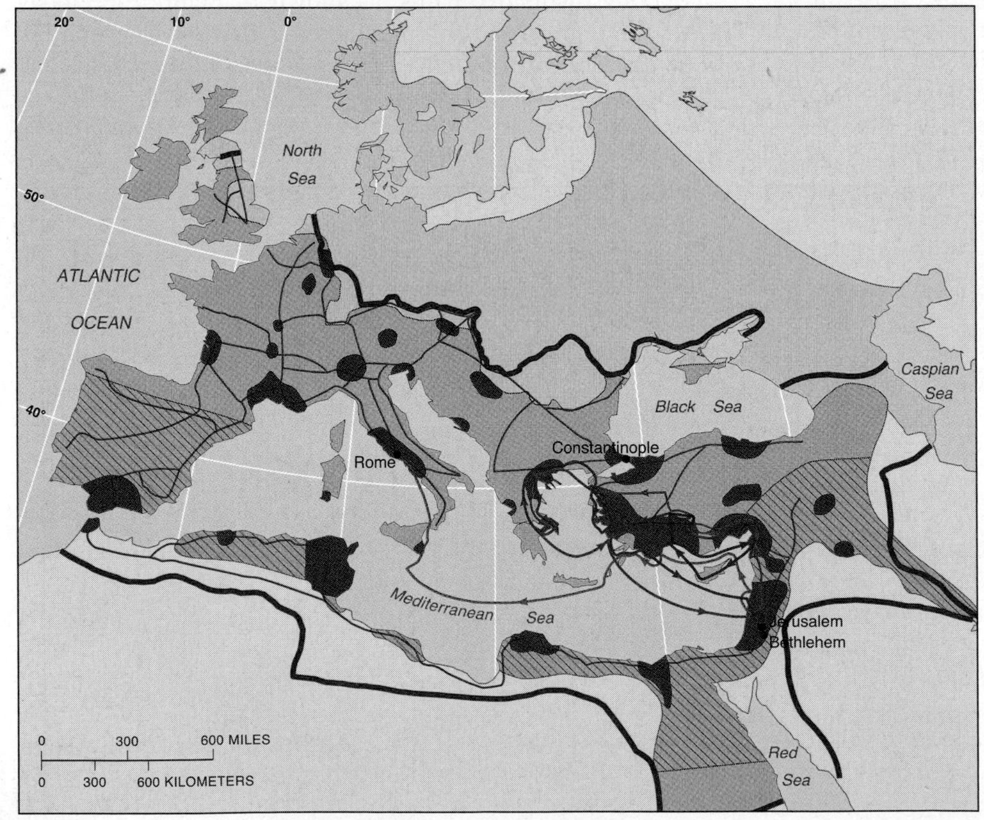

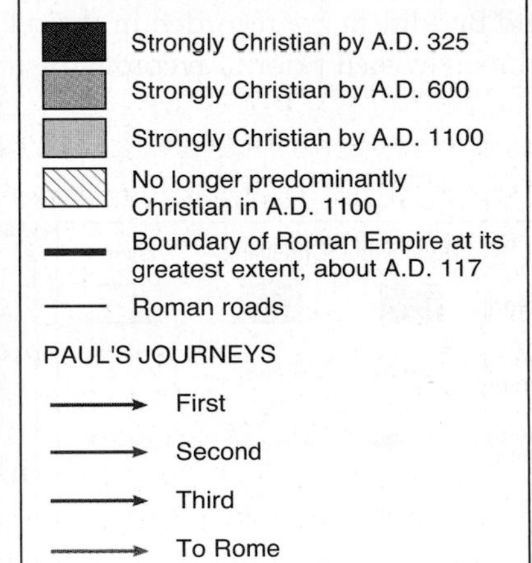

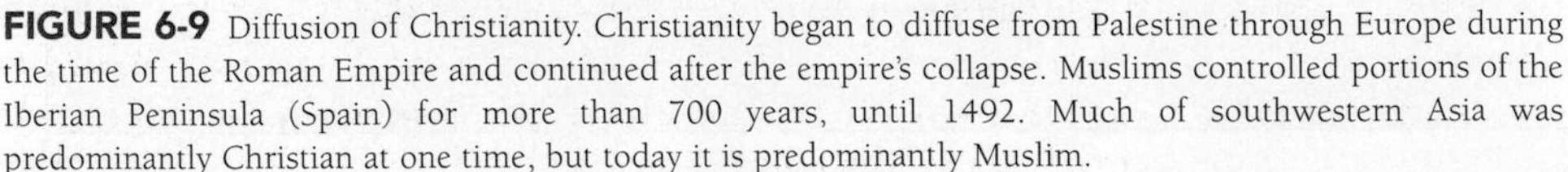
FIGURE 6-9 Diffusion of Christianity. Christianity began to diffuse from Palestine through Europe during the time of the Roman Empire and continued after the empire's collapse. Muslims controlled portions of the Iberian Peninsula (Spain) for more than 700 years, until 1492. Much of southwestern Asia was predominantly Christian at one time, but today it is predominantly Muslim.

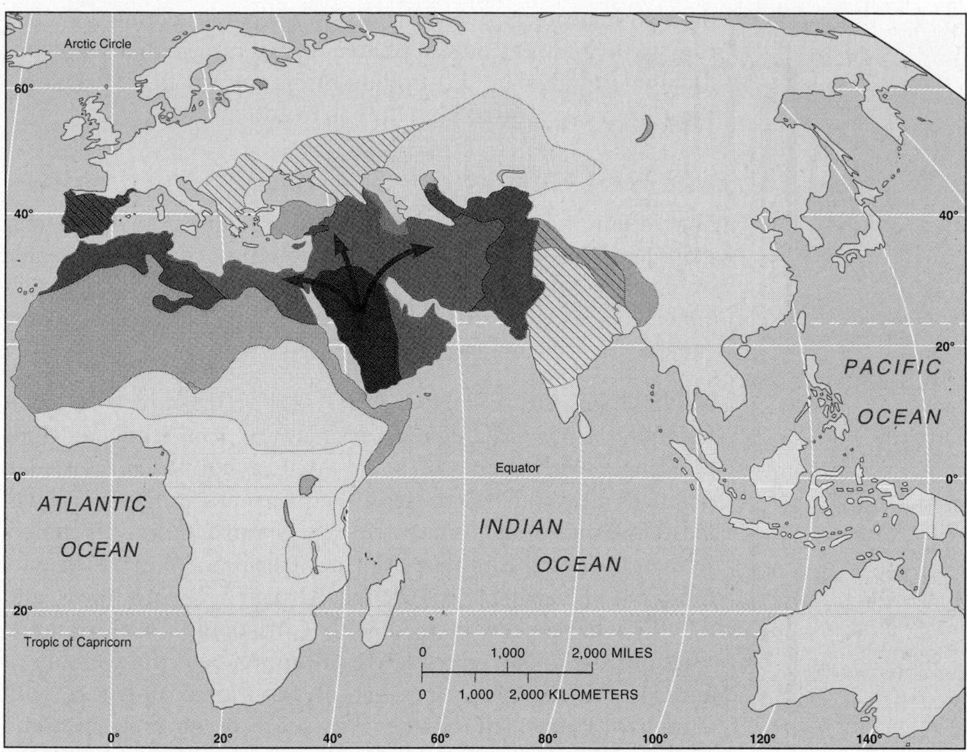

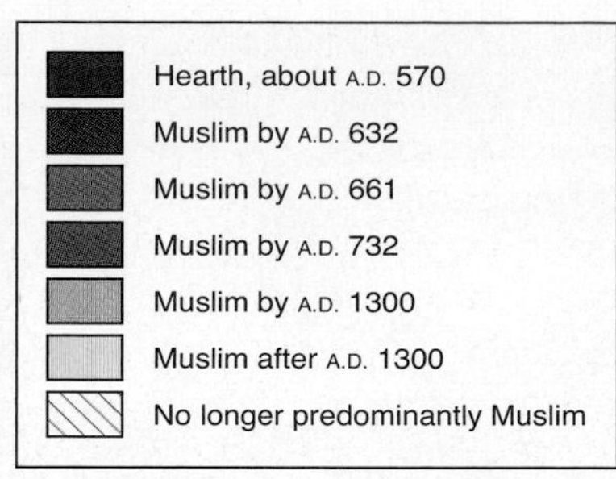

FIGURE 6-10 Diffusion of Islam. Islam diffused rapidly from its point of origin in present-day Saudi Arabia. Within 200 years, Islamic armies controlled much of North Africa, southwestern Europe, and southwestern Asia. Subsequently, Islam became the predominant religion as far east as Indonesia.

Roman Catholic because of immigration from Roman Catholic countries (refer to Figure 6-2). New England and large Midwestern cities such as Cleveland, Chicago, Detroit, and Milwaukee have concentrations of Roman Catholics because of immigration from Ireland, Italy, and Eastern Europe, especially in the late nineteenth and early twentieth centuries. Immigration from Mexico and other Latin American countries has concentrated Roman Catholics in the Southwest, whereas French settlement from the seventeenth century, as well as recent immigration, has produced a predominantly Roman Catholic Québec.

Similarly, geographers trace the distribution of other Christian denominations within the United States to the fact that migrants came from different parts of Europe, especially during the nineteenth century. Followers of The Church of Jesus Christ of Latter-day Saints, popularly known as Mormons, settled at Fayette, New York, near the hometown of their founder Joseph Smith. During Smith's life, the group moved several times in search of religious freedom. Eventually, under the leadership of Brigham Young, they migrated to the sparsely inhabited Salt Lake Valley in the present-day state of Utah.

DIFFUSION OF ISLAM. Muhammad's successors organized followers into armies that extended the region of Muslim control over an extensive area of Africa, Asia, and Europe (Figure 6-10). Within a century of Muhammad's death, Muslim armies conquered Palestine, the Persian Empire, and much of India, resulting in the conversion of many non-Arabs to Islam, often through intermarriage.

To the west, Muslims captured North Africa, crossed the Strait of Gibraltar, and retained part of Western Europe, particularly much of present-day Spain, until 1492. During the same century in which the Christians regained all of Western Europe, Muslims took control of much of southeastern Europe and Turkey.

As was the case with Christianity, Islam, as a universalizing religion, diffused well beyond its hearth in Southwest Asia through relocation diffusion of missionaries to portions of sub-Saharan Africa and Southeast Asia. Although it is spatially isolated in Southeast Asia from the Islamic core region, Indonesia, the world's fourth most populous country, is predominantly Muslim because Arab traders brought the religion there in the thirteenth century.

DIFFUSION OF BUDDHISM. Buddhism did not diffuse rapidly from its point of origin in northeastern India (Figure 6-11). Most responsible for the spread of Buddhism was Asoka, emperor of the Magadhan Empire from about 273 to 232 B.C.

The Magadhan Empire formed the nucleus of several powerful kingdoms in South Asia between the sixth century B.C. and the eighth century A.D. About 257 B.C., at the height of the Magadhan Empire's power, Asoka became a Buddhist and thereafter attempted to put into practice Buddha's social principles. A council organized by Asoka at Pataliputra decided to send missionaries to territories neighboring the Magadhan Empire. Emperor Asoka's son, Mahinda, led a mission to the island of Ceylon (now Sri Lanka), where the king and his subjects were converted to Buddhism. As a result, Sri Lanka is the country that

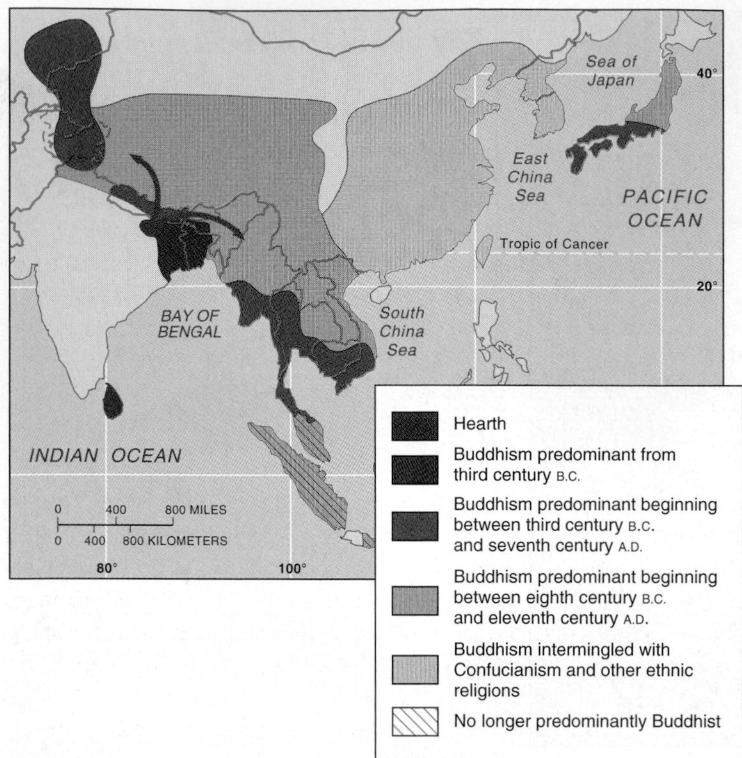

FIGURE 6-11 Diffusion of Buddhism. Buddhism diffused slowly from its core in northeastern India. Buddhism was not well established in China until 800 years after Buddha's death.

claims the longest continuous tradition of practicing Buddhism. Missionaries were also sent in the third century B.C. to Kashmir, the Himalayas, Burma (Myanmar), and elsewhere in India.

In the first century A.D., merchants along the trading routes from northeastern India introduced Buddhism to China. Many Chinese were receptive to the ideas brought by Buddhist missionaries, and Buddhist texts were translated into Chinese languages. Chinese rulers allowed their people to become Buddhist monks during the fourth century A.D., and in the following centuries Buddhism turned into a genuinely Chinese religion. Buddhism further diffused from China to Korea in the fourth century and from Korea to Japan two centuries later. During the same era, Buddhism lost its original base of support in India.

DIFFUSION OF OTHER UNIVERSALIZING RELIGIONS. The Bahá'í religion diffused to other regions in the late nineteenth and early twentieth centuries, under the leadership of 'Abdu'l-Bahá, son of the prophet Bahá'u'lláh. Bahá'í also spread rapidly during the late twentieth century, when a temple was constructed on every continent.

Sikhism remained relatively clustered in the Punjab, where the religion originated. Sikhs fought with the Muslims to gain control of the Punjab region, and they achieved their ambition in 1802 when they created an independent state in the Punjab. The British took over the Punjab in 1849 as part of its India colony but granted the Sikhs a privileged position and let them fight in the British army.

When the British government created the independent states of India and Pakistan in 1947, it divided the Punjab between the two instead of giving the Sikhs a separate country. Preferring to live in Hindu-dominated India rather than Muslim-dominated Pakistan, 2.5 million Sikhs moved from Pakistan's West Punjab region to East Punjab in India.

Lack of Diffusion of Ethnic Religions

Most ethnic religions have limited, if any, diffusion. These religions lack missionaries who are devoted to converting people from other religions. Thus, the diffusion of universalizing religions, especially Christianity and Islam, typically comes at the expense of ethnic religions.

MINGLING OF ETHNIC AND UNIVERSALIZING RELIGIONS. Universalizing religions may supplant ethnic religions or mingle with them. In some African countries, Christian practices are similar to those of their former European colonial masters. Equatorial Guinea, a former Spanish colony, is mostly Roman Catholic; Namibia, a former German colony, is heavily Lutheran. Elsewhere, traditional African religious ideas and practices have been merged with Christianity. For example, African rituals may give relative prominence to the worship of ancestors. Desire for a merger of traditional practices with Christianity has led to the formation of several thousand churches in Africa not affiliated with established churches elsewhere in the world.

In East Asia, Buddhism is the universalizing religion that has most mingled with ethnic religions, such as Shintoism in Japan. Shintoists first resisted Buddhism when it first diffused to Japan from Korea in the ninth century. Later, Shintoists embraced Buddhism and amalgamated elements of the two religions. Buddhist priests took over most of the Shinto shrines, but Buddhist deities came to be regarded by the Japanese as Shintoist deities instead.

The current situation in Japan offers a strong caution to anyone attempting to document the number of adherents of any religion. Although Japan is a wealthy country with excellent record-keeping, the number of Shintoists in the country is currently estimated at either 4 million or 100 million. When responding to questionnaires, around 4 million, or 3 percent, of the Japanese state they are Shintoist, but Shinto organizations in Japan place the number at 100 million, or 80 percent, based on record-keeping and participating in major Shinto holidays. Meanwhile, around 100 million Japanese say they are Buddhists. So if the higher number for Shintoists is correct, then most of the 122 million inhabitants of Japan profess to follow both religions.

Ethnic religions can diffuse if adherents migrate to new locations for economic reasons and are not forced to adopt a strongly entrenched universalizing religion. For example, the 1.3 million inhabitants of Mauritius include 52 percent Hindu, 28 percent Christian (26 percent Roman Catholic and 2 percent Anglican), and 17 percent Muslim. The religious diversity is a function of the country's history of immigration.

A 2,040-square-kilometer (788-square-mile) island located in the Indian Ocean 800 kilometers (500 miles) east of Madagascar, Mauritius was uninhabited until 1638, so it had no traditional ethnic religion. That year, Dutch settlers arrived to plant sugarcane and naturally brought their religion—Christianity—with them. France gained control in 1721 and imported African slaves to work on the sugarcane plantations. Then the British took over in 1810 and brought workers from India. Mauritius became independent in 1992. Hinduism on Mauritius traces back to the Indian immigrants, Islam to the African immigrants, and Christianity to the European immigrants.

JUDAISM, AN EXCEPTION. The spatial distribution of Jews differs from that of other ethnic religions because Judaism is practiced in many countries, not just its place of origin. Only since the creation of the state of Israel in 1948 has a significant percentage of the world's Jews lived in their Eastern Mediterranean homeland (see Global Forces, Local Impact box).

Most Jews have not lived in the Eastern Mediterranean since A.D. 70, when the Romans forced them to disperse throughout the world, an action known as the *diaspora*, from the Greek word for "dispersion." The Romans forced the diaspora after crushing an attempt by the Jews to rebel against Roman rule.

Most Jews migrated from the Eastern Mediterranean to Europe, although some went to North Africa and Asia. Having been exiled from the home of their ethnic religion, Jews lived among other nationalities, retaining separate religious practices but adopting other cultural characteristics of the host country, such as language.

Other nationalities often persecuted the Jews living in their midst. Historically, the Jews of many European countries were forced to live in **ghettos**, defined as a city neighborhood set up by law to be inhabited only by Jews. The term *ghetto* originated during the sixteenth century in Venice, Italy, as a reference to the city's foundry or metal-casting district, where Jews were forced to live. Ghettos were frequently surrounded by walls, and the gates were locked at night to prevent escape.

Beginning in the 1930s, but especially during World War II (1939–1945), the Nazis systematically rounded up a large percentage of European Jews, transported them to concentration camps, and exterminated them. About 4 million Jews died in the camps and 2 million in other ways. Many of the survivors migrated to Israel. Today, less than 15 percent of the world's 15 million Jews live in Europe, compared to 90 percent a century ago.

Holy Places

Religions may elevate particular places to a holy position. Universalizing and ethnic religions differ on the types of places that are considered holy:

- An ethnic religion typically has a less widespread distribution than a universalizing one in part because its holy places derive from the distinctive physical environment of its hearth, such as mountains, rivers, or rock formations.
- A universalizing religion endows with holiness cities and other places associated with the founder's life. Its holy places do not necessarily have to be near each other, and they do not need to be related to any particular physical environment.

Making a **pilgrimage** to these holy places—a journey for religious purposes to a place considered sacred—is incorporated into the rituals of some universalizing and ethnic religions. Hindus and Muslims are especially encouraged to make pilgrimages to visit holy places in accordance with recommended itineraries, and Shintoists are encouraged to visit holy places in Japan.

Holy Places in Universalizing Religions

Buddhism and Islam are the universalizing religions that place the most emphasis on identifying shrines. Places are holy because they are the locations of important events in the life of Buddha or Muhammad.

BUDDHIST SHRINES. Eight places are holy to Buddhists because they were the locations of important events in Buddha's life (Figure 6-12). The four most important of the eight places are concentrated in a small area of northeastern India and southern Nepal:

1. Lumbinī in southern Nepal, where Buddha was born around 563 B.C,. is most important. Many sanctuaries and monuments were built there, but all are in ruins today.

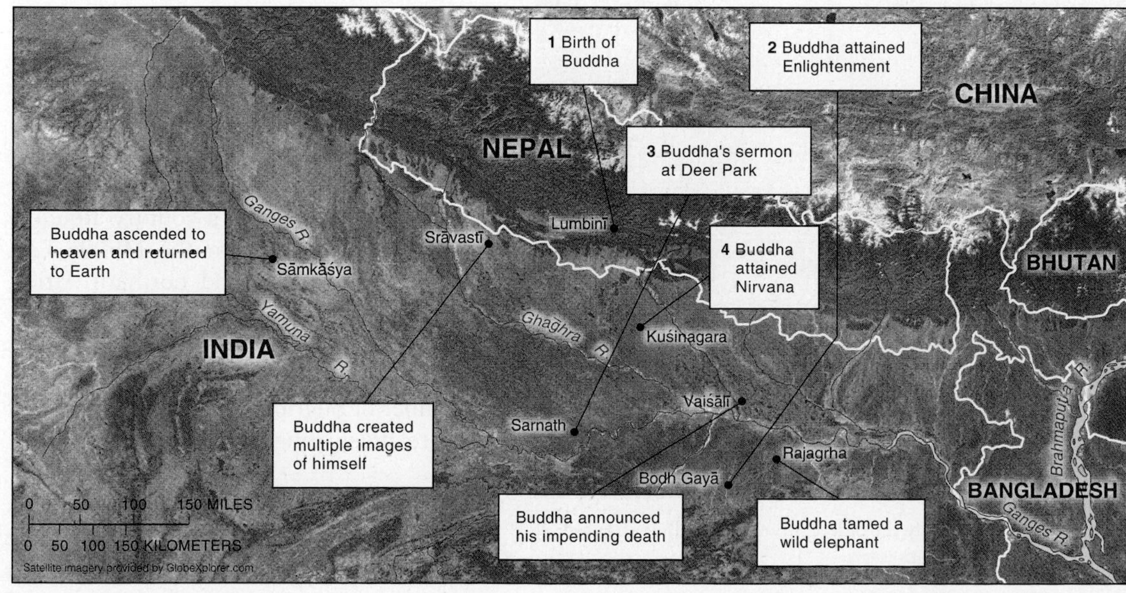

FIGURE 6-12 Holy places in Buddhism. Most are clustered in northeastern India and southern Nepal because they were the locations of important events in Buddha's life. Most of the sites are in ruins today.

2. Bodh Gayā, 250 kilometers (150 miles) southeast of Buddha's birthplace, is the site of the second great event in his life, where he reached perfect wisdom. A temple has stood near the site since the third century B.C., and part of the surrounding railing built in the first century A.D. still stands. Because Buddha reached perfect enlightenment while sitting under a bo tree, that tree has become a holy object as well. To honor Buddha, the bo tree has been diffused to other Buddhist countries, such as China and Japan.
3. Deer Park in Sarnath, where Buddha gave his first sermon, is the third important location. The Dhamek pagoda at Sarnath, built in the third century B.C., is probably the oldest surviving structure in India (Figure 6-13). Nearby is an important library of Buddhist literature, including many works removed from Tibet when Tibet's Buddhist leader, the Dalai Lama, went into exile.
4. Kuśinagara, the fourth holy place, is where Buddha died at age 80 and passed into nirvana, a state of peaceful extinction. Temples built at the site are currently in ruins.

Four other sites in northeastern India are particularly sacred because they were the locations of Buddha's principal miracles:

- Srāvastī is where Buddha performed his greatest miracle. Before an assembled audience of competing religious leaders, Buddha created multiple images of himself and visited heaven. Srāvastī became an active center of Buddhism, and one of the most important monasteries was established there.
- Sāmkāśya, the second miracle site, is where Buddha is said to have ascended to heaven, preached to his mother, and returned to Earth.
- Rajagrha, the third site, is holy because Buddha tamed a wild elephant there, and shortly after Buddha's death, it became the site of the first Buddhist Council.
- Vaisālī, the fourth location, is the site of Buddha's announcement of his impending death and the second Buddhist Council.

All four miracle sites are in ruins today, although excavation activity is under way.

FIGURE 6-13 Buddhist shrine. The Dhamek pagoda is probably the oldest surviving Buddhist structure. It was built where Buddha gave his first sermon.

HOLY PLACES IN ISLAM. The holiest locations in Islam are in cities associated with the life of the Prophet Muhammad. The holiest city for Muslims is Makkah (Mecca), the birthplace of Muhammad (Figure 6-14). The word *mecca* now has a general meaning in the English language as a goal sought or a center of activity.

Now a city of 1.3 million inhabitants, Makkah contains the holiest object in the Islamic landscape, namely al–Ka'ba, a cube-like structure encased in silk, which stands at the center of the Great Mosque, Masjid al-Haram, Islam's largest mosque (Figure 6-15). The Ka'ba, thought to have been built by Abraham and Ishmael, contains a black stone given to Abraham by Gabriel as a sign of the covenant with Ishmael and the Muslim people.

The Ka'ba had been a religious shrine in Makkah for centuries before the origin of Islam. After Muhammad defeated the local people, he captured the Ka'ba, cleared it of idols, and rededicated it to the all-powerful Allah (God). The Masjid al-Haram mosque also contains the well of Zamzam, considered to have the same water source as that used by Ishmael and Hagar when they were wandering in the desert after their exile from Canaan.

The second most holy geographic location in Islam is Madinah (Medina), a city of 1.3 million inhabitants, 350 kilometers (220 miles) north of Makkah. Muhammad received his first support from the people of Madinah and became the city's chief administrator. Muhammad's tomb is at Madinah, inside Islam's second-largest mosque (refer to Figure 6-7).

Every healthy Muslim who has adequate financial resources is expected to undertake a pilgrimage, called a *hajj*, to Makkah (Mecca). Regardless of nationality and economic background, all pilgrims dress alike in plain white robes to emphasize common loyalty to Islam and the equality of people in the eyes of Allah. A precise set of rituals is practiced, culminating in a visit to the Ka'ba. The *hajj* attracts millions of Muslims annually to Makkah. *Hajj* visas are issued by the government of Saudi Arabia according to a formula of 1 per 1,000 Muslims in a country. Roughly 80 percent come from the Middle East (southwestern Asia and northern Africa, and 20 percent from Asia. Although Indonesia is the world's most populous Muslim country, it has not sent the largest number of pilgrims to Makkah because of the relatively long travel distance.

HOLY PLACES IN SIKHISM. Sikhism's most holy structure, the Darbar Sahib, or Golden Temple, was built at Amritsar, in

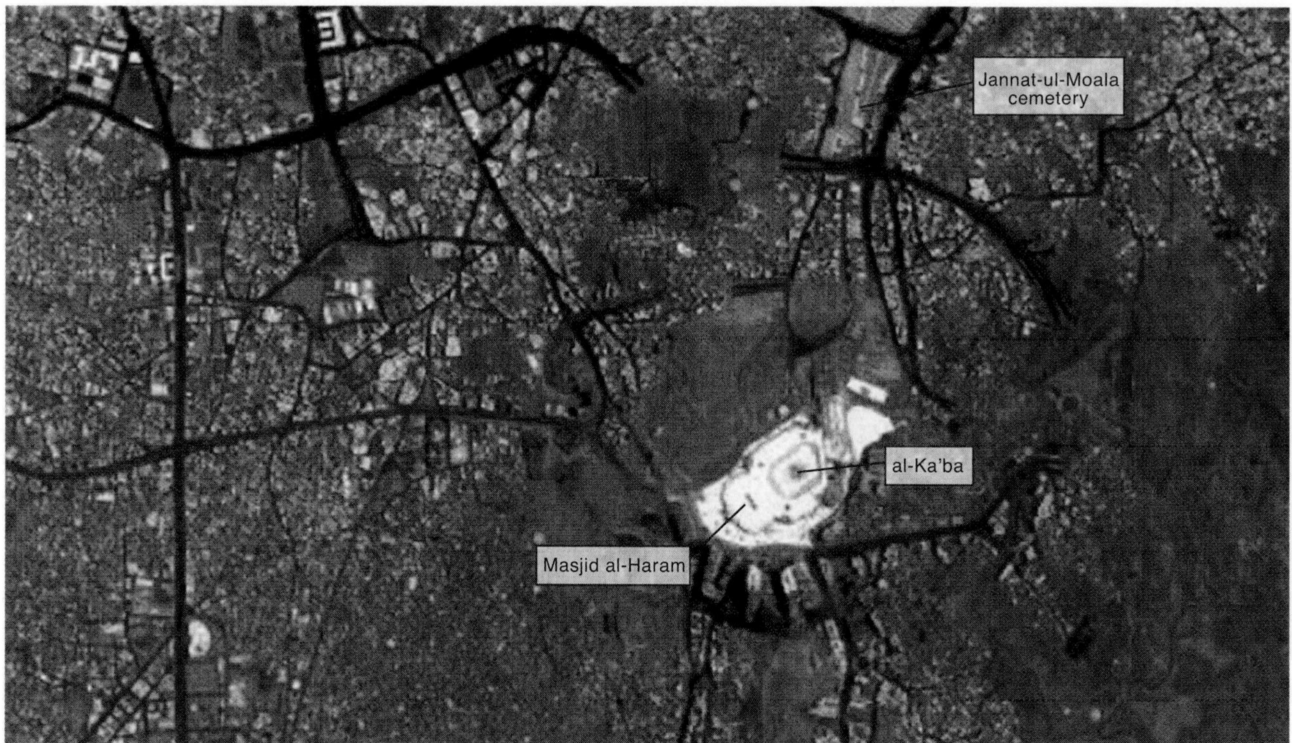

FIGURE 6-14 Makkah (Mecca), in Saudi Arabia, is the holiest city for Muslims because Muhammad was born there. Millions of Muslims make a pilgrimage to Makkah each year and gather at Masjid al-Haram, Islam's largest mosque, in the center of a city of 1.3 million inhabitants.

FIGURE 6-15 The black, cubelike structure in the center of Masjid al-Haram, called al-Ka'ba, once had been a shrine to tribal idols until Muhammad rededicated it to Allah. Muslims believe that Abraham and Ishmael originally built the Ka'ba.

the Punjab, by Arjan, the fifth guru, during the sixteenth century (Figure 6-16). The holiest book in Sikhism, the *Guru Granth Sahib*, is kept there.

Militant Sikhs used the Golden Temple at Amritsar as a base for launching attacks in support of greater autonomy for the Punjab during the 1980s. In 1984, the Indian army attacked the Golden Temple at Amritsar and killed approximately a thousand Sikhs defending the temple. In retaliation later that year, India's Prime Minister Indira Gandhi was assassinated by two of her guards, who were Sikhs.

Holy Places in Ethnic Religions

One of the principal reasons that ethnic religions are highly clustered is that they are closely tied to the physical geography of a particular place. Pilgrimages are undertaken to view these physical features.

HOLY PLACES IN HINDUISM. As an ethnic religion of India, Hinduism is closely tied to the physical geography of India. According to a survey conducted by the geographer Surinder Bhardwaj, the natural features most likely to rank among the holiest shrines in India are riverbanks or coastlines (Figure 6-17). Hindus consider a pilgrimage, known as a *tirtha*, to be an act of purification. Although not a substitute for meditation, the pilgrimage is an important act in achieving redemption.

Hindu holy places are organized into a hierarchy. Particularly sacred places attract Hindus from all over India, despite the relatively remote locations of some; less important shrines attract primarily local pilgrims. Because Hinduism has no central authority, the relative importance of shrines is established by tradition, not by doctrine. For example, many Hindus make long-distance pilgrimages to Mt. Kailās, located at the source of the Ganges in the Himalayas, which is holy because Siva lives there. Other mountains may attract only local pilgrims: Local residents may consider a nearby mountain to be holy if Siva is thought to have visited it at one time.

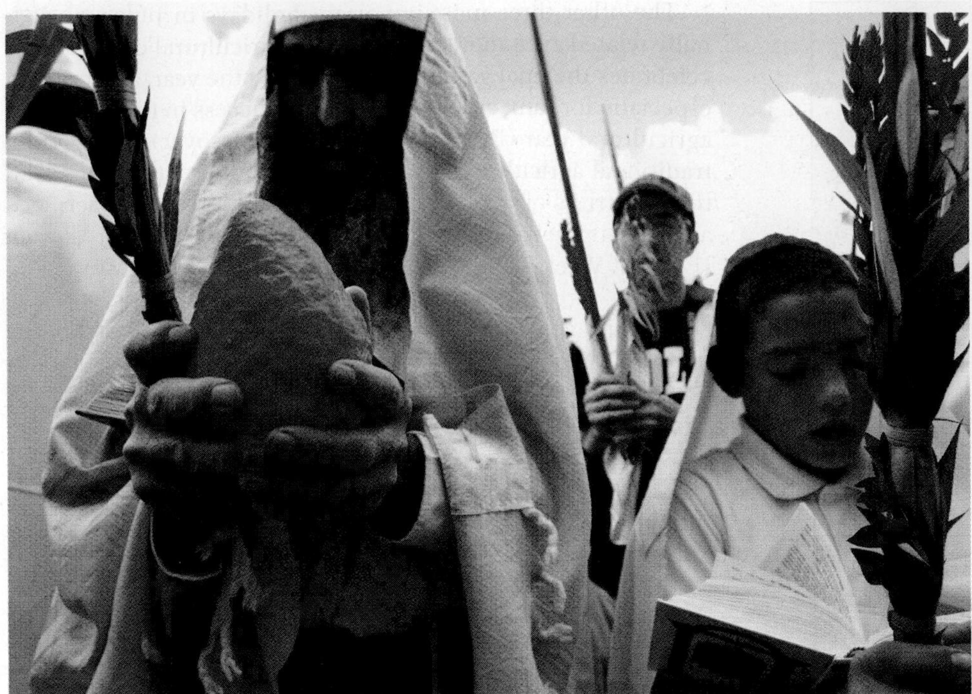

FIGURE 6-18 Ethnic religious holiday. On the holiday of Sukkoth, Jews carry a *lulav* (branches of date palm entwined with myrtle and willow) and an *etrog* (yellow citron) to symbolize gratitude for the many agricultural bounties offered by God.

Capitol down Maryland Avenue. Will archaeologists of the distant future think we erected the Capitol Building and aligned the streets as a religious ritual? Did the planner of Washington, Pierre L'Enfant, create the pattern accidentally or deliberately, and if so, why?

The Calendar in Universalizing Religions

The principal purpose of the holidays in universalizing religions is to commemorate events in the founder's life. Christians in particular associate their holidays with seasonal variations in the calendar, but climate and the agricultural cycle are not central to the liturgy and rituals.

ISLAMIC AND BAHÁ'Í CALENDARS. Islam, like Judaism, uses a lunar calendar. Whereas the Jewish calendar inserts an extra month every few years to match the agricultural and solar calendars, Islam as a universalizing religion retains a strict lunar calendar. In a 30-year cycle, the Islamic calendar has 19 years with 354 days and 11 years with 355 days.

As a result of using a lunar calendar, Muslim holidays arrive in different seasons from generation to generation. For example, during the holy month of Ramadan, Muslims fast during daylight every day and try to make a pilgrimage to the holy city of Makkah. At the moment, the start of Ramadan is occurring in the Northern Hemisphere summer—for example, August 11, 2010, on the western Gregorian calendar. In A.D. 1990, Ramadan fell on March 28, and in A.D. 2020 Ramadan will start April 24. Because Ramadan occurs at different times of the solar year in different generations, the number of hours of the daily fast varies widely because the amount of daylight varies by season and by location on Earth's surface.

Observance of Ramadan can be a hardship by interfering with critical agricultural activities, depending on the season. However, as a universalizing religion with more than 1 billion adherents worldwide, Islam is practiced in various climates and latitudes. If Ramadan were fixed at the same time of the Middle East's agricultural year, Muslims in various places of the world would need to make different adjustments to observe Ramadan.

The Bahá'ís use a calendar established by the Báb and confirmed by Bahá'u'lláh, in which the year is divided into 19 months of 19 days each, with the addition of four intercalary days (five in leap years). The year begins on the first day of spring, March 21, which is one of several holy days in the Bahá'í calendar. Bahá'ís are supposed to attend the Nineteen Day Feast, held on the first day of each month of the Bahá'í calendar, to pray, read scriptures, and discuss community activities.

CHRISTIAN, BUDDHIST, AND SIKH HOLIDAYS. Christians commemorate the resurrection of Jesus on Easter, observed on the first Sunday after the first full Moon following the spring equinox in late March. But not all Christians observe Easter on the same day, because Protestant and Roman Catholic branches calculate the date on the Gregorian calendar, but Orthodox churches use the Julian calendar.

Christians may relate Easter to the agricultural cycle, but that relationship differs depending on where they live. In Southern Europe, Easter is a joyous time of harvest. Northern Europe and North America do not have a major Christian holiday at harvest time, which would be placed in the fall. Instead, Easter in Northern Europe and North America is a time of anxiety over planting new crops, as well as a celebration of spring's arrival after a harsh winter. In the United States and Canada, Thanksgiving has been endowed with Christian prayers to play the role of harvest festival.

Most Northern Europeans and North Americans associate Christmas, the birthday of Jesus, with winter conditions, such as low temperatures, snow cover, and the absence of vegetation except for needleleaf evergreens. But for Christians in the Southern Hemisphere, December 25 is the height of the summer, with warm days and abundant sunlight.

All Buddhists celebrate as major holidays Buddha's birth, Enlightenment, and death. However, not all Buddhists observe them on the same days. Japanese Buddhists celebrate Buddha's birth on April 8, his Enlightenment on December 8, and his death on February 15; Theravadist Buddhists observe all three events on the same day, usually in April.

The major holidays in Sikhism are the births and deaths of the religion's ten gurus. The tenth guru, Gobind Singh, declared that after his death, instead of an eleventh guru, Sikhism's highest spiritual authority would be the holy scriptures of the Guru Granth Sahib. A major holiday in Sikhism is the day when the Holy Granth was installed as the religion's spiritual guide. Commemorating historical events distinguishes Sikhism as a universalizing religion, in contrast to India's major ethnic religion, Hinduism, which glorifies the physical geography of India.

KEY ISSUE 3

Why Do Religions Organize Space in Distinctive Patterns?

- Places of Worship
- Sacred Space
- Administration of Space

Geographers study the major impact on the landscape made by all religions, regardless of whether they are universalizing or ethnic. In large cities and small villages around the world, regardless of the region's prevailing religion, the tallest, most elaborate buildings are often religious structures.

The distribution of religious elements on the landscape reflects the importance of religion in people's values. The impact of religion on the landscape is particularly profound, for many religious people believe that their life on Earth ought to be spent in service to God. ■

Places of Worship

Church, basilica, mosque, temple, pagoda, and synagogue are familiar names that identify places of worship in various religions. Sacred structures are physical "anchors" of religion. All major religions have structures, but the functions of the buildings influence the arrangement of the structures across the landscape. They may house shrines or be places where people assemble for worship. Some religions require a relatively large number of elaborate structures, whereas others have more modest needs.

Christian Churches

The Christian landscape is dominated by a high density of churches. The word *church* derives from a Greek term meaning "lord," "master," and "power." *Church* also refers to a gathering of believers, as well as the building at which the gathering occurs.

The church plays a more critical role in Christianity than buildings in other religions, in part because the structure is an expression of religious principles, an environment in the image of God. The church is also more prominent in Christianity because attendance at a collective service of worship is considered extremely important.

The prominence of churches on the landscape also stems from their style of construction and location. In some communities, the church was traditionally the largest and tallest building and was placed at an important square or other prominent location. Although such characteristics may no longer apply in large cities, they are frequently still true for small towns and neighborhoods within cities.

Since Christianity split into many denominations, no single style of church construction has dominated. Churches reflect both the cultural values of the denomination and the region's architectural heritage. Orthodox churches follow an architectural style that developed in the Byzantine Empire during the fifth century. Byzantine-style Orthodox churches tend to be highly ornate, topped by prominent domes. Many Protestant churches in North America, on the other hand, are simple, with little ornamentation. This austerity is a reflection of the Protestant conception of a church as an assembly hall for the congregation.

Availability of building materials also influences church appearance. In the United States early churches were most frequently built of wood in the Northeast, brick in the Southeast, and adobe in the Southwest. Stucco and stone predominated in Latin America. This diversity reflected differences in the most common building materials found by early settlers.

Places of Worship in Other Religions

Religious buildings are highly visible and important features of the landscapes in regions dominated by religions other than Christianity. But unlike Christianity, other major religions do not consider their important buildings a sanctified place of worship.

MUSLIM MOSQUES. Muslims consider the mosque as a space for community assembly. Unlike a church, a mosque is not viewed as a sanctified place, but rather as a location for the community to gather together for worship. Mosques are found primarily in larger cities of the Muslim world; simple structures may serve as places of prayer in rural villages.

The mosque is organized around a central courtyard—traditionally open-air, although it may be enclosed in harsher climates. The pulpit is placed at the end of the courtyard facing Makkah, the direction toward which all Muslims pray. Surrounding the courtyard is a cloister used for schools and nonreligious activities. A distinctive feature of the mosque is the *minaret*, a tower where a man known as a *muzzan* summons people to worship. Two minarets are shown in Figure 6-7.

HINDU TEMPLES. Sacred structures for collective worship are relatively unimportant in Asian ethnic and universalizing religions. Instead, important religious functions are more likely to take place at home within the family. Temples are built to house shrines for particular gods rather than for congregational worship. The Hindu temple serves as a home to one or more gods, although a particular god may have more than one temple.

The typical Hindu temple contains a small, dimly lit interior room where a symbolic artifact or some other image of the god rests. Because congregational worship is not part of Hinduism, the temple does not need a large closed interior space filled with seats. The site of the temple, usually demarcated by a wall, may also contain a structure for a caretaker and a pool for ritual baths. Space may be devoted for ritual processions.

Wealthy individuals or groups usually maintain local temples. Size and frequency of temples are determined by local preferences and commitment of resources rather than standards imposed by religious doctrine.

BUDDHIST AND SHINTOIST PAGODAS. The pagoda is a prominent and visually attractive element of the Buddhist and Shintoist landscapes. Frequently elaborate and delicate in appearance, pagodas typically include tall, many-sided towers arranged in a series of tiers, balconies, and slanting roofs. Pagodas contain relics that Buddhists believe to be a portion of Buddha's body or clothing. After Buddha's death, his followers scrambled to obtain these relics. As part of the process of diffusing the religion, Buddhists carried these relics to other countries and built pagodas for them. Pagodas are not designed for congregational worship. Individual prayer or meditation is more likely to be undertaken at an adjacent temple, a remote monastery, or in a home.

BAHÁ'Í HOUSES OF WORSHIP. Bahá'ís have built Houses of Worship in Wilmette, Illinois, in 1953; Sydney, Australia, and Kampala, Uganda, both in 1961; Lagenhain, near Frankfurt, Germany, in 1964; Panama City, Panama, in 1972; Tiapapata, near Apia, Samoa, in 1984; and New Delhi, India, in 1986 (Figure 6-19). The first Bahá'í House of Worship, built in 1908 in Ashgabat, Russia, now the capital of Turkmenistan, was turned into a museum by the Soviet Union and demolished in 1962 after a severe earthquake. Additional ones are planned in Tehran, Iran; Santiago, Chile; and Haifa, Israel.

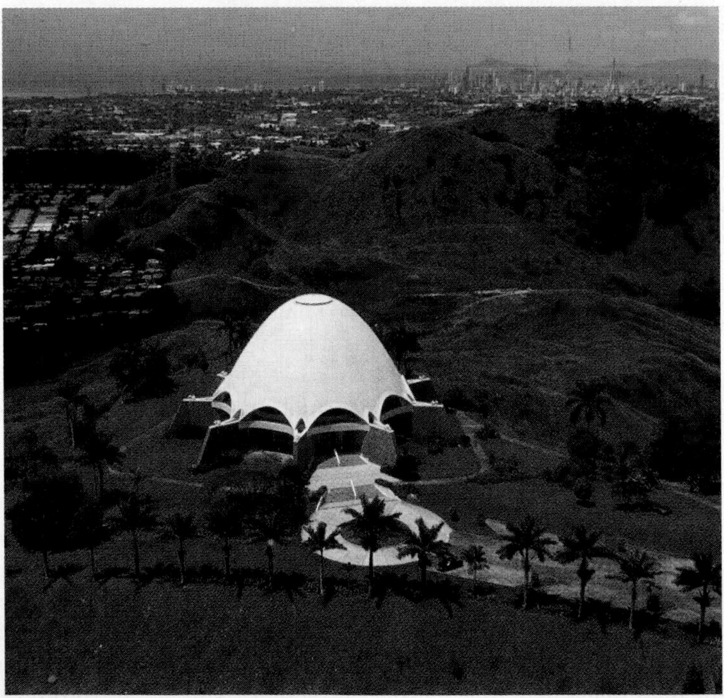

FIGURE 6-19 Bahá'í Temple, Panama City, Panama. Bahá'ís have built a temple in each continent to demonstrate the religion's status as a universalizing religion, with adherents around the world.

The locations have not been selected because of proximity to clusters of Bahá'ís. Instead, the Houses of Worship have been dispersed to different continents to dramatize Bahá'í as a universalizing religion with adherents all over the world. The Houses of Worship are open to adherents of all religions, and services include reciting the scriptures of various religions.

Sacred Space

The impact of religion is clearly seen in the arrangement of human activities on the landscape at several scales, from relatively small parcels of land to entire communities. How each religion distributes its elements on the landscape depends on its beliefs. Important religious land uses include burial of the dead and religious settlements.

Disposing of the Dead

A prominent example of religiously inspired arrangement of land at a smaller scale is burial practices. Climate, topography, and religious doctrine combine to create differences in practices to shelter the dead.

BURIAL. Christians, Muslims, and Jews usually bury their dead in a specially designated area called a cemetery. The Christian burial practice can be traced to the early years of the religion. In ancient Rome, underground passages known as *catacombs* were used to bury early Christians (and to protect the faithful when the religion was still illegal). After Christianity became legal, Christians buried their dead in the yard around the church. As these burial places became overcrowded, separate burial grounds had to be established outside the city walls. Public health and sanitation considerations in the nineteenth century led to public management of many cemeteries. Some cemeteries are still operated by religious organizations. The remains of the dead are customarily aligned in some traditional direction. Some Christians bury the dead with the feet toward Jerusalem so that they may meet Christ there on the Day of Judgment.

Cemeteries may consume significant space in a community, increasing the competition for scarce space. In congested urban areas, Christians and Muslims have traditionally used cemeteries as public open space. Before the widespread development of public parks in the nineteenth century, cemeteries were frequently the only green space in rapidly growing cities. Cemeteries are still used as parks in Muslim countries, where the idea faces less opposition than in Christian societies.

Traditional burial practices in China have put pressure on agricultural land. By burying dead relatives, rural residents have removed as much as 10 percent of the land from productive agriculture. The government in China has ordered the practice discontinued, even urging farmers to plow over old burial mounds. Cremation is encouraged instead.

OTHER METHODS OF DISPOSING OF BODIES. Not all faiths bury their dead. Hindus generally practice cremation rather than burial. The body is washed with water from the Ganges River and then burned with a slow fire on a funeral pyre (Figure 6-20). Burial is reserved for children, ascetics, and people with certain diseases. Cremation is considered an act of purification, although it tends to strain India's wood supply.

Motivation for cremation may have originated from unwillingness on the part of nomads to leave their dead behind, possibly because of fear that the body could be attacked by wild beasts or evil spirits, or even return to life. Cremation could also free the soul from the body for departure to the afterworld and provide warmth and comfort for the soul as it embarked on the journey to the afterworld. Cremation was the principal form of disposing of bodies in Europe before Christianity. It is still practiced in parts of Southeast Asia, possibly because of Hindu influence.

To strip away unclean portions of the body, Parsis (Zoroastrians) expose the dead to scavenging birds and animals. The ancient Zoroastrians did not want the body to contaminate the sacred elements of fire, earth, or water. Tibetan Buddhists also practice exposure for some dead, with cremation reserved for the most exalted priests.

Disposal of bodies at sea is used in some parts of Micronesia, but the practice is much less common than in the past. The bodies of lower-class people would be flung into the sea; elites could be set adrift on a raft or boat. Water burial was regarded as a safeguard against being contaminated by the dead.

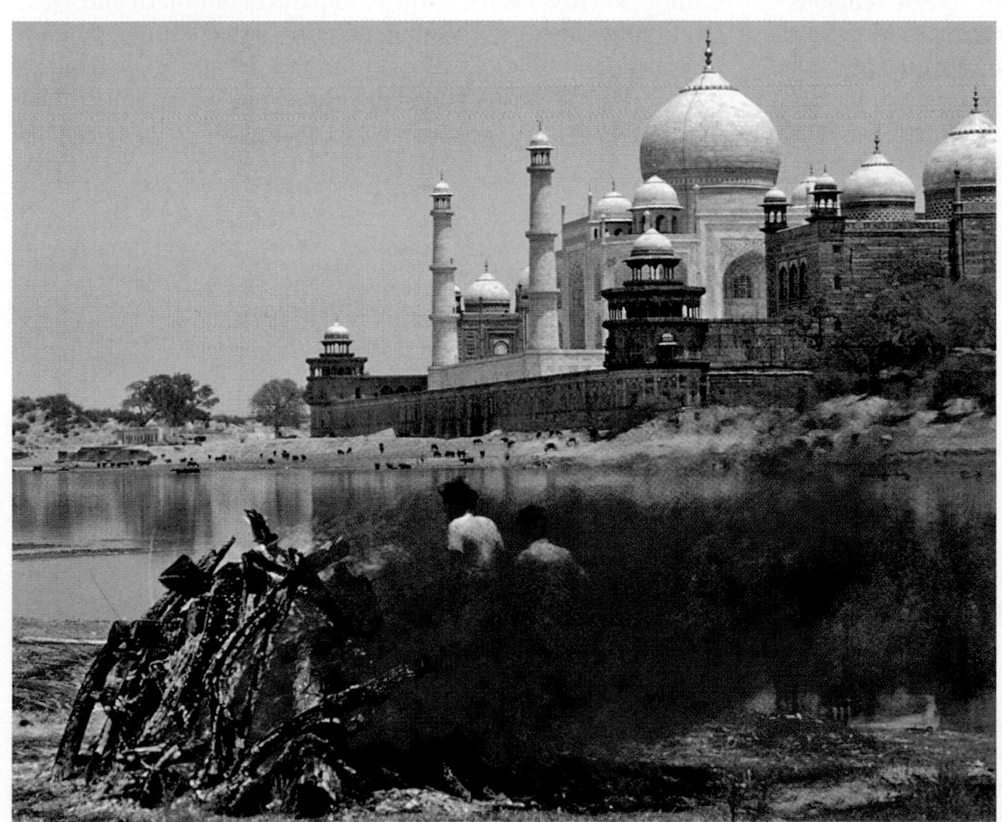

FIGURE 6-20 Cremation. The most common form of disposal of bodies in India is cremation. In middle-class families, bodies are more likely to be cremated in an electric oven at a crematorium. A poor person may be cremated in an open fire, such as this one within sight of the Taj Mahal. High-ranking officials and strong believers in traditional religious practices may also be cremated on an outdoor fire.

Religious Settlements

Buildings for worship and burial places are smaller-scale manifestations of religion on the landscape, but there are larger-scale examples—entire settlements. Most human settlements serve an economic purpose (see Chapter 12), but some are established primarily for religious reasons.

A utopian settlement is an ideal community built around a religious way of life. Buildings are sited and economic activities organized to integrate religious principles into all aspects of daily life. An early utopian settlement in the United States was Bethlehem, Pennsylvania, founded in 1741 by Moravians, Christians who had emigrated from the present-day Czech Republic. By 1858, some 130 different utopian settlements had begun in the United States in conformance with a group's distinctive religious beliefs. Examples include Oneida, New York; Ephrata, Pennsylvania; Nauvoo, Illinois; and New Harmony, Indiana.

The culmination of the utopian movement in the United States was the construction of Salt Lake City by the Mormons, beginning in 1848. The layout of Salt Lake City is based on a plan of the city of Zion given to the church elders in 1833 by the Mormon prophet Joseph Smith. The city has a regular grid pattern, unusually broad boulevards, and church-related buildings situated at strategic points.

Most utopian communities declined in importance or disappeared altogether. Some disappeared because the inhabitants were celibate and could not attract immigrants; in other cases residents moved away in search of better economic conditions. The utopian communities that have not been demolished are now inhabited by people who are not members of the original religious sect, although a few have been preserved as museums.

Although most colonial settlements were not planned primarily for religious purposes, religious principles affected many of the designs. Most early New England settlers were members of a Puritan Protestant denomination. The Puritans generally migrated together from England and preferred to live near each other in clustered settlements rather than on dispersed, isolated farms. Reflecting the importance of religion in their lives, New England settlers placed the church at the most prominent location in the center of the settlement, usually adjacent to a public open space known as a common,

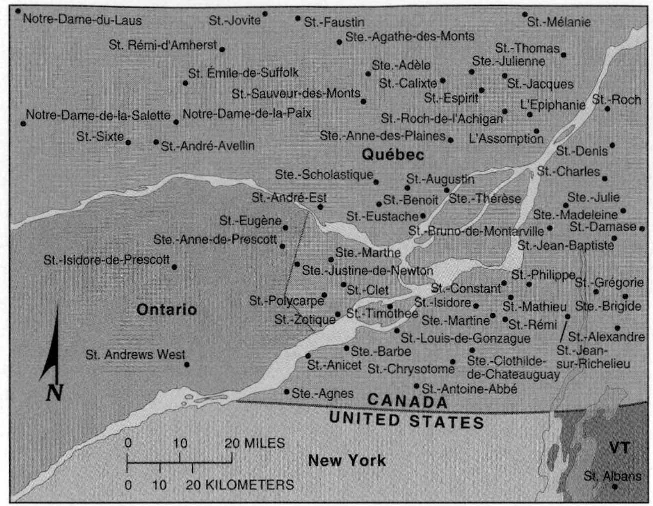

FIGURE 6-21 Religious toponyms. Place names near Québec's boundaries with Ontario and the United States show the impact of religion on the landscape. In Québec, a province with a predominantly Roman Catholic population, a large number of settlements are named for saints, whereas relatively few religious toponyms are found in predominantly Protestant Ontario, New York, and Vermont.

because it was for common use by everyone (see an example in Figure 12-13).

Religious Place Names

Roman Catholic immigrants have frequently given religious place names, or toponyms, to their settlements in the New World, particularly in Québec and the U.S. Southwest. Québec's boundaries with Ontario and the United States clearly illustrate the difference between toponyms selected by Roman Catholic and Protestant settlers. Religious place names are common in Québec but rare in the two neighbors (Figure 6-21).

Administration of Space

Followers of a universalizing religion must be connected so as to ensure communication and consistency of doctrine. The method of interaction varies among universalizing religions, branches, and denominations. Ethnic religions tend not to have organized, central authorities.

Hierarchical Religions

A **hierarchical religion** has a well-defined geographic structure and organizes territory into local administrative units. Roman Catholicism provides a good example of a hierarchical religion.

LATTER-DAY SAINTS. Latter-day Saints (Mormons) exercise strong organization of the landscape. The territory occupied by Mormons, primarily Utah and portions of surrounding states, is organized into *wards*, with populations of approximately 750 each. Several wards are combined into a *stake* of approximately 5,000 people. The highest authority in the Church—the board and president—frequently redraws ward and stake boundaries in rapidly growing areas to reflect the ideal population standards.

ROMAN CATHOLIC HIERARCHY. The Roman Catholic Church has organized much of Earth's inhabited land into an administrative structure ultimately accountable to the Pope in Rome (Figure 6-22). Here is the top hierarchy of Roman Catholicism:

- ***The Pope*** (he is also the bishop of the Diocese of Rome).
- ***Archbishops*** report to the Pope. Each heads a ***province***, which is a group of several dioceses. The archbishop also is bishop of one diocese within the province, and some distinguished archbishops are elevated to the rank of cardinal.
- ***Bishops*** report to an archbishop. Each administers a ***diocese***, which is the basic unit of geographic organization in the Roman Catholic Church. The bishop's headquarters, called a "see," is typically the largest city in the diocese.
- ***Priests*** report to Bishops. A diocese is spatially divided into ***parishes***, each headed by a priest.

The area and population of parishes and dioceses vary according to historical factors and the distribution of Roman Catholics across Earth's surface. In parts of Southern and Western Europe the overwhelming majority of the dense population is Roman Catholic. Consequently, the density of parishes is high. A typical parish may encompass only a few square kilometers and fewer than 1,000 people. At the other extreme,

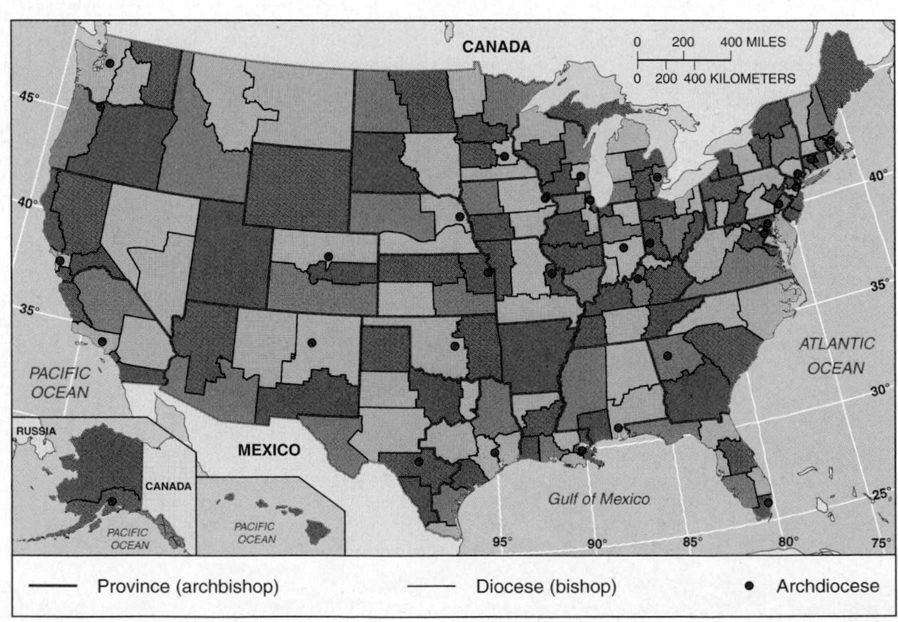

FIGURE 6-22 Roman Catholic hierarchy in the United States. The Roman Catholic Church divides the United States into provinces, each headed by an archbishop. Provinces are subdivided into dioceses, each headed by a bishop. The archbishop of a province also serves as the bishop of a diocese. Dioceses that are headed by archbishops are called archdioceses.

Latin American parishes may encompass several hundred square kilometers and 5,000 people. The more dispersed Latin American distribution is attributable partly to a lower population density than in Europe.

Because Roman Catholicism is a hierarchical religion, individual parishes must work closely with centrally located officials concerning rituals and procedures. If Latin America followed the European model of small parishes, many would be too remote for the priest to communicate with others in the hierarchy. The less intensive network of Roman Catholic institutions also results in part from colonial traditions, for both Portuguese and Spanish rulers discouraged parish development in Latin America.

The Roman Catholic population is growing rapidly in the U.S. Southwest and suburbs of some large North American and European cities. Some of these areas have a low density of parishes and dioceses compared to the population, so the Church must adjust its territorial organization. New local administrative units can be created, although funds to provide the desired number of churches, schools, and other religious structures might be scarce. Conversely, the Roman Catholic population is declining in inner cities and rural areas. Maintaining services in these areas is expensive, but the process of combining parishes and closing schools is very difficult.

Locally Autonomous Religions

Some universalizing religions are highly **autonomous religions**, or self-sufficient, and interaction among communities is confined to little more than loose cooperation and shared ideas. Islam and some Protestant denominations are good examples.

LOCAL AUTONOMY IN ISLAM. Among the three large universalizing religions, Islam provides the most local autonomy. Like other locally autonomous religions, Islam has neither a religious hierarchy nor a formal territorial organization. A mosque is a place for public ceremony, and a leader calls the faithful to prayer, but everyone is expected to participate equally in the rituals and is encouraged to pray privately.

In the absence of a hierarchy, the only formal organization of territory in Islam is through the coincidence of religious territory with secular states. Governments in some predominantly Islamic countries include in their bureaucracy people who administer Islamic institutions. These administrators interpret Islamic law and run welfare programs.

Strong unity within the Islamic world is maintained by a relatively high degree of communication and migration, such as the pilgrimage to Makkah. In addition, uniformity is fostered by Islamic doctrine, which offers more explicit commands than other religions.

PROTESTANT DENOMINATIONS. Protestant Christian denominations vary in geographic structure from extremely autonomous to somewhat hierarchical. The Episcopalian, Lutheran, and most Methodist churches have hierarchical structures, somewhat comparable to the Roman Catholic Church. Extremely autonomous denominations such as Baptists and United Church of Christ are organized into self-governing congregations. Each congregation establishes the precise form of worship and selects the leadership.

Presbyterian churches represent an intermediate degree of autonomy. Individual churches are united in a presbytery, several of which in turn are governed by a synod, with a general assembly as ultimate authority over all churches. Each Presbyterian church is governed by an elected board of directors with lay members.

ETHNIC RELIGIONS. Judaism and Hinduism also have no centralized structure of religious control. To conduct a full service, Judaism merely requires the presence of ten adult males. (Females count in some Jewish communities.)

Hinduism is even more autonomous, because worship is usually done alone or with others in the household. Hindus share ideas primarily through undertaking pilgrimages and reading traditional writings.

KEY ISSUE 4
Why Do Territorial Conflicts Arise Among Religious Groups?

- **Religion Versus Government Policies**
- **Religion Versus Religion**

The twentieth century was a century of global conflict—two world wars during the first half of the century and the Cold War between supporters of democracy and Communism during the second half. With the end of the Cold War, the threat of global conflict has receded in the twenty-first century, but local conflicts have increased in areas of cultural diversity, as will be discussed in Chapters 7 and 8.

The element of cultural diversity that has led to conflict in many localities is religion. The attempt by intense adherents of one religion to organize Earth's surface can conflict with the spatial expression of other religious or nonreligious ideas. ■

Religion Versus Government Policies

Religious groups may oppose government policies seen as promoting social change conflicting with traditional religious values. The role of religion in organizing Earth's surface has diminished in some societies because of political and economic change.

Islam has been particularly affected by a perceived conflict between religious values and modernization of the economy.

Hinduism also has been forced to react to new nonreligious ideas from the West. Buddhism, Christianity, and Islam have all been challenged by Communist governments that diminish the importance of religion in society. Yet, in recent years, religious principles have become increasingly important in the political organization of countries, especially where a branch of Christianity or Islam is the prevailing religion.

Religion Versus Social Change

In LDCs, participation in the global economy and culture can expose local residents to values and beliefs originating in MDCs of North America and Western Europe. North Americans and Western Europeans may not view economic development as incompatible with religious values, but many religious adherents in LDCs do, especially where Christianity is not the predominant religion.

TALIBAN VERSUS WESTERN VALUES. When the Taliban gained power in Afghanistan in 1996, many Afghans welcomed them as preferable to the corrupt and brutal warlords who had been running the country. U.S. and other Western officials also welcomed them as strong defenders against a possible new invasion by Russia.

The Taliban (which means "religious students") had run Islamic Knowledge Movement religious schools, mosques, shrines, and other religious and social services since the seventh century A.D., shortly after the arrival of Islam in Afghanistan. Once in control of Afghanistan's government in the late 1990s, the Taliban imposed very strict laws inspired by Islamic values as the Taliban interpreted them. They banned "Western, non-Islamic" leisure activities, such as playing music, flying kites, watching television, and surfing the Internet; and they converted soccer stadiums to settings for executions and floggings. Men were beaten for shaving their beards and stoned for committing adultery. Homosexuals were buried alive, and prostitutes were hanged in front of large audiences. Thieves had their hands cut off, and women wearing nail polish had their fingers cut off. Western values were not the only targets: Enormous Buddhist statues as old as the second century A.D. were destroyed in 2001 because they were worshipped as "graven images" in violation of Islam. The Ministry for the Promotion of Virtue and the Prevention of Vice enforced the laws. The Taliban believed that they had been called by Allah to purge Afghanistan of sin and violence and make it a pure Islamic state. Islamic scholars criticized the Taliban as poorly educated in Islamic law and history and for misreading the Quran.

A U.S.-led coalition overthrew the Taliban in 2001 and replaced it with a democratically elected government. However, the Taliban was able to regroup and resume its fight to regain control of Afghanistan and Pakistan (see Chapter 8).

HINDUISM VERSUS SOCIAL EQUALITY. Hinduism has been strongly challenged since the 1800s, when British colonial administrators introduced their social and moral concepts to India. The most vulnerable aspect of the Hindu religion was its rigid **caste** system, which was the class or distinct hereditary order into which a Hindu was assigned according to religious law.

The caste system apparently originated around 1500 B.C. when Aryans invaded India from the west. The Aryans divided themselves into four castes that developed strong differences in social and economic position—Brahmans, the priests and top administrators; Kshatriyas, or warriors; Vaisyas, or merchants; and Shudras, or agricultural workers and artisans. The Shudras occupied a distinctly lower status than the other three castes. Below the four castes were the outcasts, or untouchables, who did work considered too dirty for other castes. In theory, the untouchables were descended from the indigenous people who dwelled in India prior to the Aryan conquest.

Over the centuries, these original castes split into thousands of subcastes. Until recently, social relations among the castes were limited, and the rights of non-Brahmans, especially untouchables, were restricted. In Hinduism, because everyone was different, it was natural that each individual should belong to a particular caste or position in the social order. British administrators and Christian missionaries pointed out the shortcomings of the caste system, such as neglect of the untouchables' health and economic problems.

The type of Hinduism practiced will depend in part on the individual's caste. A high-caste Brahman may practice a form of Hinduism based on knowledge of relatively obscure historical texts. At the other end of the caste system, a low-caste illiterate in a rural village may perform religious rituals without a highly developed set of written explanations for them.

The rigid caste system has been considerably relaxed in recent years. The Indian government legally abolished the untouchable caste, and the people formerly in that caste now have equal rights with other Indians. But consciousness of caste persists: A government plan to devise a quota system designed to give untouchables more places in the country's universities generated strong opposition.

Religion Versus Communism

Organized religion was challenged in the twentieth century by the rise of Communism in Eastern Europe and Asia. The three religions most affected were Orthodox Christianity, Islam, and Buddhism.

ORTHODOX CHRISTIANITY AND ISLAM VERSUS THE SOVIET UNION. In 1721, Czar Peter the Great made the Russian Orthodox Church a part of the Russian government. The patriarch of the Russian Orthodox Church was replaced by a 12-member committee, known as the Holy Synod, nominated by the czar.

Following the 1917 Bolshevik revolution, which overthrew the czar, the Communist government of the Soviet Union pursued antireligious programs. Karl Marx had called religion "the opium of the people," a view shared by V. I. Lenin and other early Communist leaders. Marxism became the official doctrine of the Soviet Union, so religious doctrine was a potential threat to the success of the revolution.

The Soviet government in 1918 eliminated the official church–state connection that Peter the Great had forged. All church buildings and property were nationalized and could be used only with local government permission. People's religious beliefs could not be destroyed overnight, but the role of organized religion in Soviet life could be reduced, and was. The Orthodox religion retained adherents in the Soviet Union, especially among the elderly, but younger people generally had little contact with the church beyond attending a service perhaps once a year. With religious organizations prevented from conducting social and cultural work, religion dwindled in daily life.

The end of Communist rule in the late twentieth century brought a religious revival in Eastern Europe, especially where Roman Catholicism is the most prevalent branch of Christianity, including Croatia, the Czech Republic, Hungary, Lithuania, Poland, Slovakia, and Slovenia. Property confiscated by the Communist governments reverted to Church ownership, and attendance at church services increased.

In Central Asian countries that were former parts of the Soviet Union—Kazakhstan, Kyrgyzstan, Tajikistan, Turkmenistan, and Uzbekistan—most people are Muslims. These newly independent countries are struggling to determine the extent to which laws should be rewritten to conform to Islamic custom rather than to the secular tradition inherited from the Soviet Union.

BUDDHISM VERSUS SOUTHEAST ASIAN COUNTRIES. In Southeast Asia, Buddhists were hurt by the long Vietnam War—waged between the French and later by the Americans, on one side, and Communist groups on the other. Neither antagonist was particularly sympathetic to Buddhists. U.S. air raids in Laos and Cambodia destroyed many Buddhist shrines, and others were vandalized by Vietnamese and by the Khmer Rouge Cambodian Communists. On a number of occasions, Buddhists immolated (burned) themselves to protest policies of the South Vietnamese government.

The current Communist governments in Southeast Asia have discouraged religious activities and permitted monuments to decay, most notably the Angkor Wat complex in Cambodia, considered one of the world's most beautiful Buddhist structures. In any event, these countries do not have the funds necessary to restore the structures.

Religion Versus Religion

Refer back to the map of world religions (Figure 6-3) near the beginning of this chapter. Conflicts are most likely to occur where colors change, indicating a boundary between two religious groups.

Contributing to more intense religious conflict has been a resurgence of religious **fundamentalism**, which is a literal interpretation and a strict and intense adherence to basic principles of a religion (or a religious branch, denomination, or sect). In a world increasingly dominated by a global culture and economy, religious fundamentalism is one of the most important ways in which a group can maintain a distinctive cultural identity. A group convinced that its religious view is *the* correct one may spatially intrude upon the territory controlled by other religious groups. Two long-standing conflicts involving religious groups are in Northern Ireland and the Middle East.

Religious Wars in Ireland

The most troublesome religious boundary in Western Europe lies on the island of Eire (Ireland). The Republic of Ireland, which occupies five-sixths of the island, is 87 percent Roman Catholic, but the island's northern one-sixth, which is part of the United Kingdom rather than Ireland, is 46 percent Protestant and 40 percent Roman Catholic, according to the 2001 census (the remaining 14 percent stated no religion or did not respond).

The entire island was an English colony for many centuries and was made part of the United Kingdom in 1801. Agitation for independence from Britain increased in Ireland during the nineteenth century, especially after poor economic conditions and famine in the 1840s led to mass emigration. Following a succession of bloody confrontations, Ireland became a self-governing dominion within the British Empire in 1921. Complete independence was declared in 1937, and a republic was created in 1949. When most of Ireland became independent, a majority in six northern counties voted to remain in the United Kingdom. Protestants, who comprised the majority in Northern Ireland, preferred to be part of the predominantly Protestant United Kingdom rather than join the predominantly Roman Catholic Republic of Ireland (Figure 6-23).

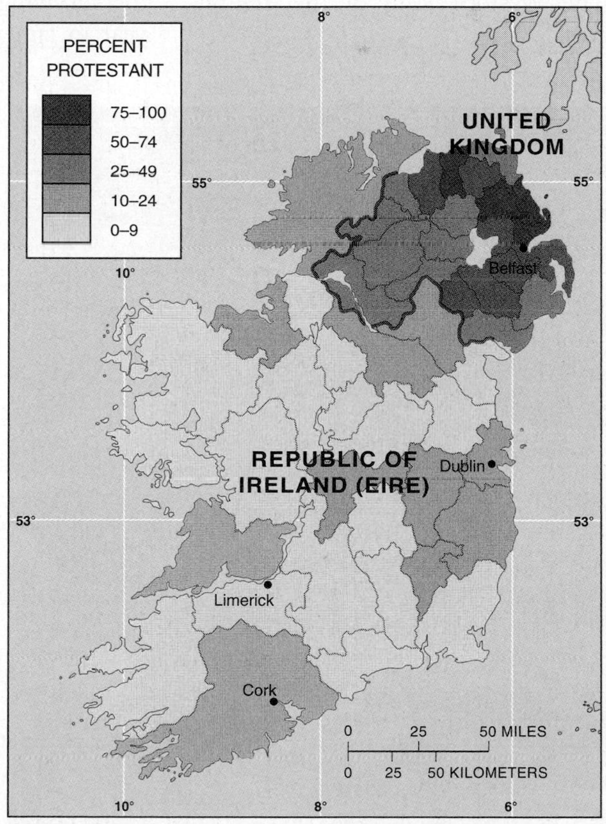

FIGURE 6-23 Distribution of Protestants in Ireland, 1911. Long a colony of England, Ireland became a self-governing dominion within the British Empire in 1921. In 1937, it became a completely independent country, but 26 districts in the north of Ireland chose to remain part of the United Kingdom. The Republic of Ireland today is 87 percent Roman Catholic, whereas Northern Ireland has a Protestant majority. The boundary between Roman Catholics and Protestants does not coincide precisely with the international border, so Northern Ireland includes some communities that are predominantly Roman Catholic. This is the root of a religious conflict that continues today.

GLOBAL FORCES, LOCAL IMPACTS
Jerusalem: Contested Geography

The geography of Jerusalem makes it difficult if not impossible to settle the long-standing religious conflicts. The difficulty is that the most sacred space in Jerusalem for Muslims was literally built on top of the most sacred space for Jews (Figure 6-24).

Jerusalem is especially holy to Jews as the location of the Temple, their center of worship in ancient times. The First Temple, built by King Solomon in approximately 960 B.C,. was destroyed by the Babylonians in 586 B.C. After the Persian Empire, led by Cyrus the Great, gained control of Jerusalem in 614 BC, Jews were allowed to build a Second Temple in 516 BC. The Romans destroyed the Jewish Second Temple in A.D. 70. The Western Wall of the Temple survives.

The most important Muslim structure in Jerusalem is the Dome of the Rock, built in 691. Muslims believe that the large rock beneath the building's dome is the place from which Muhammad ascended to heaven, as well as the altar on which Abraham prepared to sacrifice his son Isaac. Immediately south of the Dome of the Rock is the al-Aqsa Mosque. The challenge facing Jews and Muslims is that al-Aqsa Mosque was built on the site of the ruins of the Jewish Second Temple. Thus, the surviving Western Wall of the Jewish Temple is situated immediately beneath holy Muslim structures.

Christians and Muslims call the Western Wall the Wailing Wall, because for many centuries Jews were allowed to visit the surviving Western Wall only once a year to lament the Temple's destruction. After Israel captured the entire city of Jerusalem during the 1967 Six-Day War, it removed the barriers that had prevented Jews from visiting and living in the Old City of Jerusalem, including the Western Wall. The Western Wall soon became a site for daily prayers by observant Jews.

Israel allows Muslims unlimited access to that religion's holy structures in Jerusalem and some control over them. Ramps and passages patrolled by Palestinian guards provide Muslims access to the Dome of the Rock and the al-Aqsa Mosque without having to walk in front of the Western Wall where Jews are praying. However, because the holy Muslim structures sit literally on top of the holy Jewish structure, the two sets of holy structures cannot be logically divided by a line on a map. ■

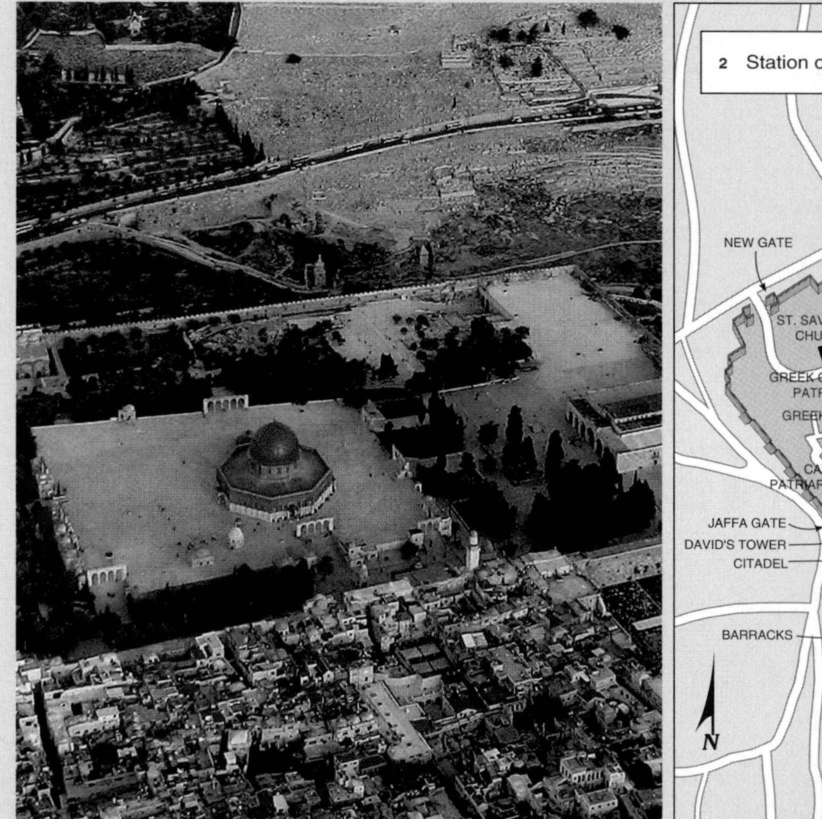

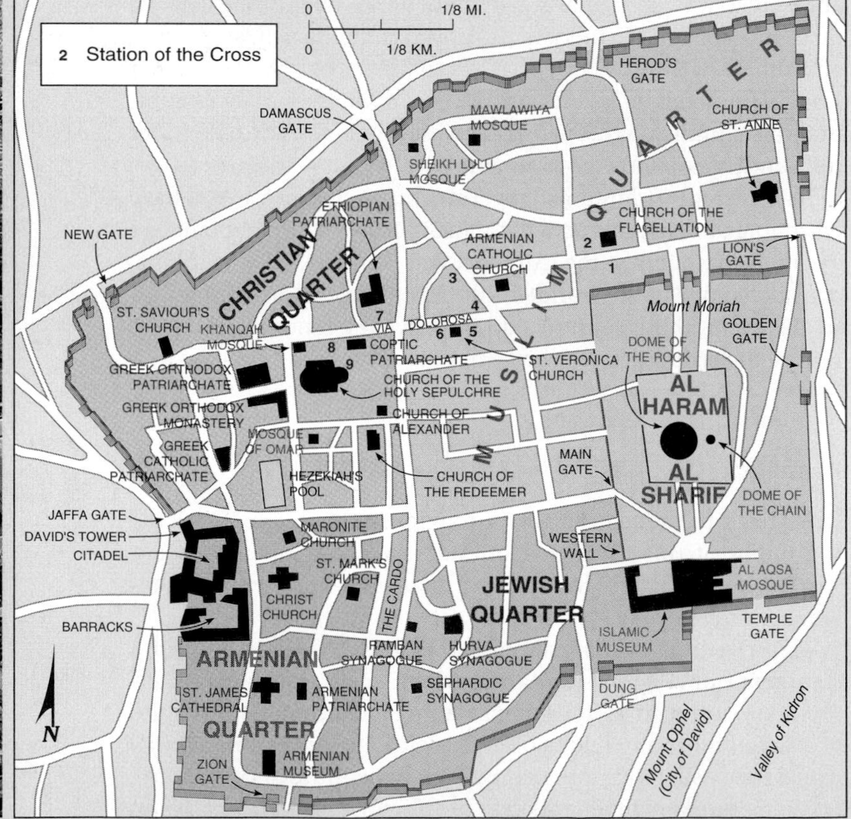

FIGURE 6-24 Jerusalem's contested space. The Old City of Jerusalem contains holy places for three religions. The flattened hill on the eastern side of the Old City is the site of two structures holy to Muslims, the Dome of the Rock (the golden dome in the photograph) and the al-Aqsa Mosque. The west side of the Old City contains the most important Christian shrines, including the Church of the Holy Sepulchre. In front of the Dome of the Rock is the Western Wall of the ancient Jewish Temple.

Roman Catholics in Northern Ireland have been victimized by discriminatory practices, such as exclusion from higher-paying jobs and better schools. Demonstrations by Roman Catholics protesting discrimination began in 1968. Since then, more than 3,000 have been killed in Northern Ireland—both Protestants and Roman Catholics—in a never-ending cycle of demonstrations and protests.

A small number of Roman Catholics in both Northern Ireland and the Republic of Ireland joined the Irish Republican Army (IRA), a militant organization dedicated to achieving Irish national unity by whatever means available, including violence. Similarly, a scattering of Protestants created extremist organizations to fight the IRA, including the Ulster Defense Force (UDF).

Although the overwhelming majority of Northern Ireland's Roman Catholics and Protestants are willing to live peacefully with the other religious group, extremists disrupt daily life for everyone and do well in elections. As long as most Protestants are firmly committed to remaining in the United Kingdom and most Roman Catholics are equally committed to union with the Republic of Ireland, peaceful settlement appears difficult.

Religious Wars in the Middle East

Conflict in the Middle East is among the world's longest standing and most intractable. Jews, Christians, and Muslims have fought for 2,000 years to control the same small strip of land in the Eastern Mediterranean.

To some extent the hostility among Christians, Muslims, and Jews in the Middle East stems from their similar heritage. All three groups trace their origins to Abraham in the Hebrew Bible narrative, but the religions diverged in ways that have made it difficult for them to share the same territory.

- ***Judaism***, an ethnic religion, makes a special claim to the territory it calls the Promised Land. The major events in the development of Judaism took place there, and the religion's customs and rituals acquired meaning from the agricultural life of the ancient Hebrew tribe. After the Romans gained control of the area, which they called the province of Palestine, they dispersed the Jews from Palestine, and only a handful were permitted to live in the region until the twentieth century.
- ***Islam*** became the most widely practiced religion in Palestine after the Muslim army conquered it in the seventh century A.D. Muslims regard Jerusalem as their third holiest city, after Makkah and Madinah, because it is the place from which Muhammad is thought to have ascended to heaven (see Global Forces, Local Impacts box).
- ***Christianity*** considers Palestine the Holy Land and Jerusalem the Holy City because the major events in Jesus's life, death, and Resurrection were concentrated there. Most inhabitants of Palestine accepted Christianity, after the religion was officially adopted by the Roman Empire and before the Muslim army conquest in the seventh century.

CRUSADES BETWEEN CHRISTIANS AND MUSLIMS. In the seventh century, Muslims, now also called Arabs because they came from the Arabian peninsula, captured most of the Middle East, including Palestine and Jerusalem. The Arab army diffused the Arabic language across the Middle East and converted most of the people from Christianity to Islam.

The Arab army moved west across North Africa and invaded Europe at Gibraltar in AD 711 (see Figure 6-10). The army conquered most of the Iberian Peninsula, crossed the Pyrenees Mountains a few years later, and for a time occupied much of present-day France. Its initial advance in Europe was halted by the Franks (a West Germanic people), led by Charles Martel, at Poitiers, France, in 732. The Arab army made further gains in Europe in subsequent years and continued to control portions of present-day Spain until 1492, but Martel's victory ensured that Christianity rather than Islam would be Europe's dominant religion.

To the east, Ottoman Turks captured Eastern Orthodox Christianity's most important city, Constantinople (present-day Istanbul in Turkey), in 1453 and advanced a few years later into Southeast Europe, as far north as present-day Bosnia and Herzegovina. The recent civil war in that country is a legacy of the fifteenth-century Muslim invasion (see Chapter 7).

To recapture the Holy Land from its Muslim conquerors, European Christians launched a series of military campaigns, known as Crusades, over a 150-year period. Crusaders captured Jerusalem from the Muslims in 1099 during the First Crusade, lost it in 1187 (which led to the Third Crusade), regained it in 1229 as part of a treaty ending the Sixth Crusade, and lost it again in 1244.

JEWS VERSUS MUSLIMS IN PALESTINE. The Muslim Ottoman Empire controlled Palestine for most of the four centuries between 1516 and 1917. Upon the empire's defeat in World War I, Great Britain took over Palestine under a mandate from the League of Nations, and later from the United Nations.

For a few years the British allowed some Jews to return to Palestine, but immigration was restricted again during the 1930s in response to intense pressure by Arabs in the region. As violence initiated by both Jewish and Muslim settlers escalated after World War II, the British announced their intention to withdraw from Palestine.

The United Nations voted in 1947 to partition Palestine into two independent states, one Jewish and one Muslim (Figure 6-25). Jerusalem was to be an international city, open to all religions, and run by the United Nations. When the British withdrew in 1948, Jews declared an independent state of Israel within the boundaries prescribed by the UN resolution. The next day its neighboring Arab Muslim states declared war.

The combatants signed an armistice in 1949 that divided control of Jerusalem. The Old City of Jerusalem, which contained the famous religious shrines, became part of the Muslim country of Jordan. The newer, western portion of Jerusalem became part of Israel, but Jews were still not allowed to visit the historic shrines in the Old City.

Israel won three more wars with its neighbors, in 1956, 1967, and 1973. Especially important was the 1967 Six-Day War, when Israel captured territory from its neighbors. From Jordan, Israel captured the West Bank (the territory west of the Jordan River taken by Jordan in the 1948–1949 war).

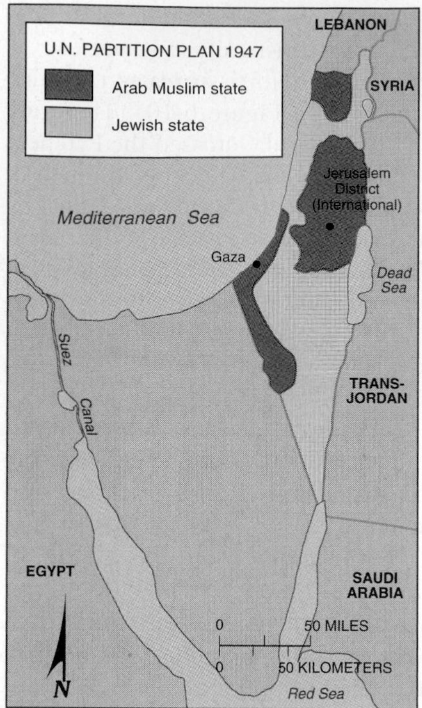

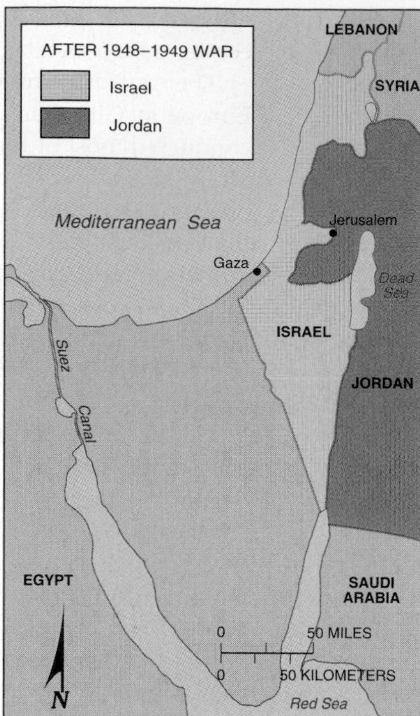

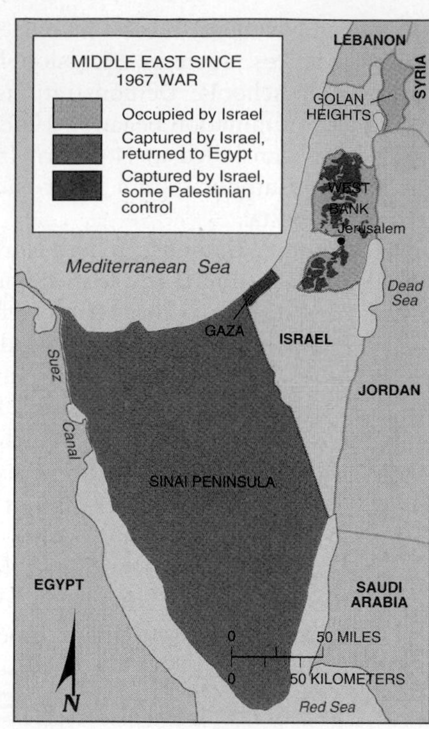

FIGURE 6-25 Boundary changes in Palestine/Israel. (left) The 1947 UN plan to partition Palestine. The plan was to create two countries, with the boundaries drawn to separate the predominantly Jewish areas from the predominantly Arab Muslim areas. Jerusalem was intended to be an international city, run by the United Nations. (center) Israel after the 1948–49 war. The day after Israel declared its independence, several neighboring states began a war, which ended in an armistice. Israel's boundaries were extended beyond the UN partition to include the western suburbs of Jerusalem. Jordan gained control of the West Bank and East Jerusalem, including the Old City, where holy places are clustered. (right) The Middle East since the 1967 war. Israel captured the Golan Heights from Syria, the West Bank and East Jerusalem from Jordan, and the Sinai Peninsula and Gaza Strip from Egypt. Israel returned Sinai to Egypt in 1979 and turned over Gaza and a portion of the West Bank to the Palestinians in 1994. Israel still controls the Golan Heights, most of the West Bank, and East Jerusalem.

From Jordan, Israel also gained control of the entire city of Jerusalem, including the Old City. From Syria, Israel acquired the Golan Heights. From Egypt came the Gaza Strip and Sinai Peninsula.

Israel returned the Sinai Peninsula to Egypt, and in return Egypt recognized Israel's right to exist. Egypt's President Anwar Sadat and Israel's Prime Minister Menachem Begin signed a peace treaty including these terms in 1979, following a series of meetings with U.S. President Jimmy Carter at Camp David, Maryland. Sadat was assassinated by Egyptian soldiers, who were extremist Muslims opposed to compromising with Israel, but his successor Hosni Mubarak carried out the terms of the treaty. Four decades after the Six-Day War, the status of the other territories occupied by Israel has still not been settled.

CONFLICT OVER THE HOLY LAND: PALESTINIAN PERSPECTIVES. After the 1973 war, the Palestinians emerged as Israel's principal opponent. Egypt and Jordan renounced their claims to the Gaza Strip and the West Bank, respectively, and recognized the Palestinians as the legitimate rulers of these territories. The Palestinians in turn also saw themselves as the legitimate rulers of Israel.

Five groups of people consider themselves Palestinians:

- People living in the West Bank, Gaza, and East Jerusalem territories captured by Israel in 1967
- Citizens of Israel who are Muslims rather than Jews
- People who fled from Israel to other countries after the 1948–49 war
- People who fled from the West Bank or Gaza to other countries after the 1967 war
- Citizens of other countries, especially Jordan, Lebanon, Syria, Kuwait, and Saudi Arabia, who identify themselves as Palestinians

After capturing the West Bank from Jordan in 1967, Israel permitted Jewish settlers to construct more than 100 settlements in the territory (Figure 6-26, left). Some Israelis built settlements in the West Bank because they regarded the territory as an integral part of the biblical Jewish homeland, known as Judea and Samaria. Others migrated to the settlements because of a shortage of affordable housing inside Israel's pre-1967 borders. Jewish settlers comprise about 10 percent of the West Bank population, and Palestinians see their immigration as a hostile act. To protect the settlers, Israel has military control over most of the West Bank.

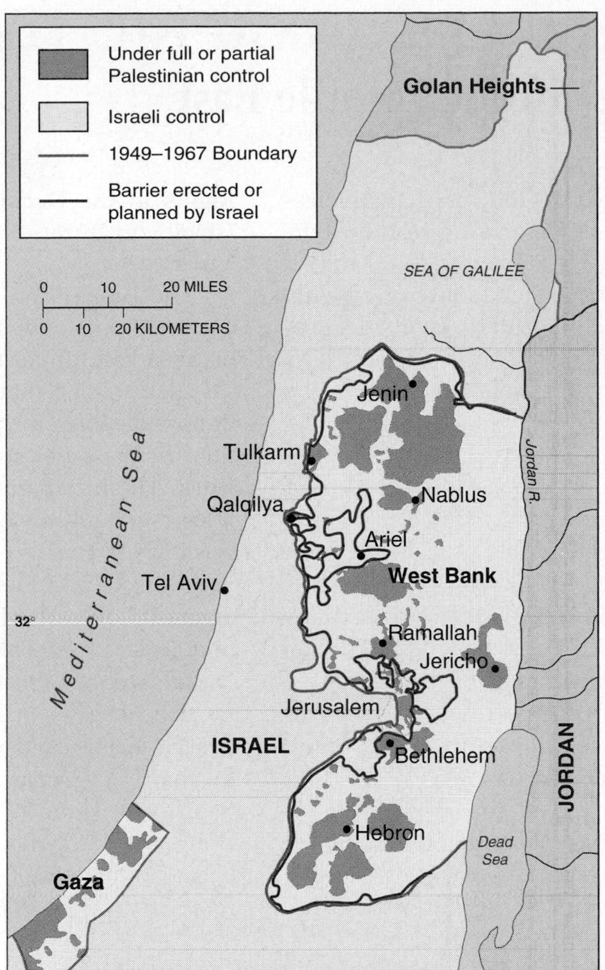

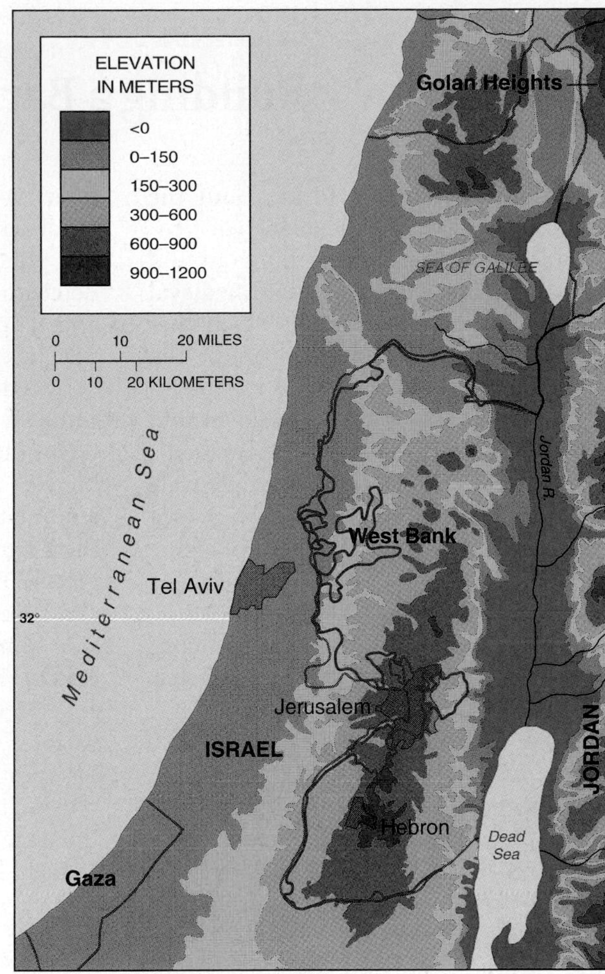

FIGURE 6-26 Two perspectives on Palestine/Israel. (left) A Palestinian perspective. Since Israel captured the West Bank in 1967, Jewish settlers have constructed more than 100 settlements in the territory. (right) A Jewish perspective. Between 1948 and 1967, Israel's boundaries encompassed primarily the coastal lowlands, whereas Jordan and Syria controlled the highlands. During the 1967 war, Israel captured these highlands and retained them to stop attacks on population concentrations in the Jordan River valley and the coastal plain.

The Palestinian fight against Israel was coordinated by the Palestine Liberation Organization (PLO) under the longtime leadership of Yassir Arafat until his death in 2004. Israel has permitted the organization of a limited form of government in much of the West Bank and Gaza, called the Palestinian Authority, but Palestinians are not satisfied with either the territory or the power they have received thus far.

The Palestinians have been divided by sharp differences, reflected in a struggle for power between the Fatah and Hamas parties. Some Palestinians, especially those aligned with the Fatah Party, are willing to recognize the state of Israel with its Jewish majority in exchange for return of all territory taken by Israel in the 1967 war. Other Palestinians, especially those aligned with the Hamas Party, do not recognize the right of Israel to exist and want to continue fighting for control of the entire territory between the Jordan River and the Mediterranean Sea. The United States, European countries, and Israel consider Hamas to be a terrorist organization.

CONFLICT OVER THE HOLY LAND: ISRAELI PERSPECTIVE. Israel sees itself as a very small country—20,000 square kilometers (8,000 square miles)—with a Jewish majority, surrounded by a region of hostile Muslim Arabs encompassing more than 25 million square kilometers (10 million square miles). In dealing with its neighbors, Israel considers two elements of the local landscape especially meaningful.

First, the country's major population centers are quite close to international borders, making them vulnerable to surprise attack. The country's two largest cities, Tel Aviv and Haifa, are only 20 and 60 kilometers (12 and 37 miles), respectively, from Palestinian-controlled territory, and its third-largest city, Jerusalem, is adjacent to the border.

The second geographical problem from Israel's perspective derives from local landforms. The northern half of Israel is a strip of land 80 kilometers (50 miles) wide between the Mediterranean Sea and the Jordan River. It is divided into three roughly parallel physical regions (Figure 6-26, right):

- A coastal plain along the Mediterranean, extending inland as much as 25 kilometers (15 miles) and as little as a few meters
- A series of hills reaching elevations above 1,000 meters (3,300 feet)
- The Jordan River valley, much of which is below sea level

CONTEMPORARY GEOGRAPHIC TOOLS
Building a Barrier in the Middle East

Constructing a barrier to keep out the unwanted is one of the oldest of geographic tools. Walls were built around cities from ancient Ur through medieval Paris to modern Québec (see Chapter 12). The longest structure ever built, the Great Wall of China, is a 6,700-kilometer-(4,200-mile-) long barrier started around 688 B.C. In the twentieth century, walls were built across Cyprus to separate warring Greek and Turkish ethnicities (see Chapter 8) and around West Berlin by the Communists to prevent East Germans from escaping.

The most ambitious barrier constructed in the twenty-first century has been the one Israel has placed between it and the West Bank. The government of Israel started building the barrier in 2002, with the support of most of its citizens, as a way to deter Palestinian suicide bombers from crossing into Israel (Figure 6-27). Israel had already built a fence along its border with the Gaza Strip after turning over some of that territory to Palestinian control in 1994.

The West Bank barrier is 670 kilometers (420 miles) in length. About 20 percent of it follows the Green Line, which was the boundary between Israel and Jordan between 1949 and 1967. The remaining 80 percent is 20 meters (65 feet) and 20 kilometers (12 miles) inside the West Bank. Israel's separation barrier is actually a wall for only 5 percent of the route, mostly in dense urban areas such as suburbs of Jerusalem. The wall is typically concrete slabs 8 meters (26 feet) high and 3 meters (10 feet) thick. Most of the barrier is actually a wide area, averaging 60 meters (200 feet) in width with several obstacles. Beginning on the Palestinian side, the first obstacle is a 1.8-meter (6-foot) pyramid-shaped stack of six barbed wire coils, then a 2-meter (7-foot) trench, then a dirt road suitable for military vehicles, then a 3-meter (10-foot) fence with electrical sensors, then a paved road for border police, then a strip of fine sand to detect footprints, then more barbed wire, and then finally on the Israeli side, surveillance cameras.

According to Israel's government, the route of the barrier was selected for two technical reasons. First, the area had to be wide enough to make construction of a 60-meter-wide barrier feasible. Second, the route was designed to place the high ground on the Israeli side. The barrier is controversial because it places on Israel's side some of the West Bank, 7 to 12 percent of the land, home to between 10,000 and 50,000 Palestinians, according to various sources. In addition, the route has put three-quarters of the 240,000 of Israeli settlers in the West Bank on the Israeli side, according to B'Tselem (Israeli Information Center for Human Rights in the Occupied Territories), an Israeli organization that opposes the barrier.

The Israel Supreme Court has twice declared portions of the route illegal because Palestinian rights were violated. The court ruled that the barrier made it impossible for some Palestinians to reach their fields, water sources, and places of work. The International Court of Justice also issued an advisory that the barrier was illegal.

Ultimately, Israel and international organizations call the barrier the "separation fence," but Palestinians call it the "racial segregation wall" in Arabic, or "apartheid wall" in English. Meanwhile, Israeli officials have been providing advice to the U.S. Department of Homeland Security on how to construct a barrier along the U.S.–Mexico border to deter illegal immigration (see Chapter 3). ■

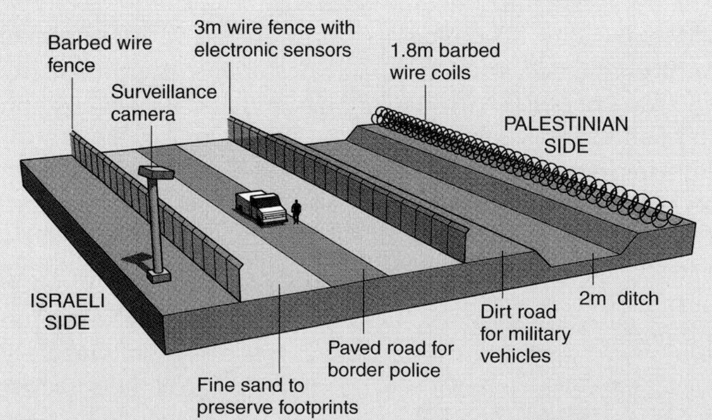

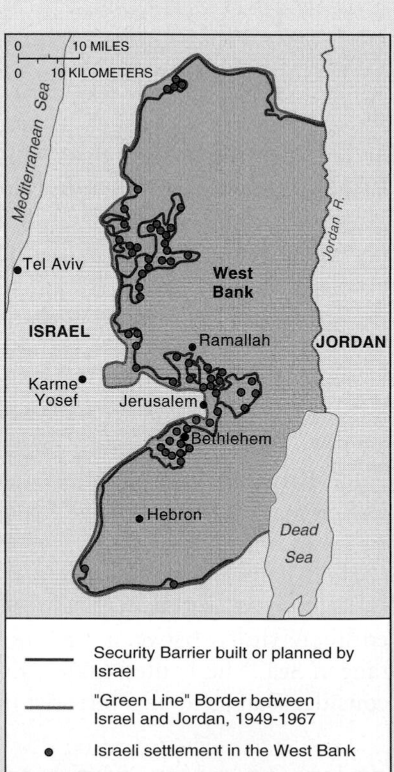

FIGURE 6-27 Israel's "separation fence." (left) A typical cross section of the barrier built by Israel in the West Bank. (right) Route of the barrier built by Israel in the West Bank.

The UN plan for the partition of Palestine in 1947, as modified by the armistice ending the 1948–49 war, allocated most of the coastal plain to Israel, whereas Jordan took most of the hills between the coastal plain and the Jordan River valley, a region generally called the West Bank (of the Jordan River). Farther north, Israel's territory extended eastward to the Jordan River valley, but Syria controlled the highlands east of the valley, known as the Golan Heights.

Jordan and Syria used the hills between 1948 and 1967 as staging areas to attack Israeli settlements on the adjacent coastal plain and in the Jordan River valley. Israel captured these highlands during the 1967 war to stop attacks on the lowland population concentrations. Israel still has military control over the Golan Heights and West Bank a generation later, yet attacks by Palestinians against Israeli citizens have continued.

Israeli Jews were divided for many years between those who wished to retain the occupied territories and those who wished to make compromises with the Palestinians. In recent years, a large majority of Israelis have supported construction of a barrier to deter Palestinian attacks (see Contemporary Geographic Tools box).

The ultimate obstacle to comprehensive peace in the Middle East is the status of Jerusalem. As long as any one religion—Jewish, Muslim, or Christian—maintains exclusive political control over Jerusalem, the other religious groups will not be satisfied. But Israelis have no intention of giving up control of the Old City of Jerusalem, and Palestinians have no intention of giving up their claim to it.

SUMMARY

North Americans pride themselves on tolerance of religious diversity. Most North Americans are Christian, but they practice Christianity in many ways, including Roman Catholicism, many denominations of Protestantism, and other Christian faiths. In addition, North America is home to millions of Muslims, Jews, Buddhists, Bahá'ís, Hindus, and other faiths. And tens of millions practice no religion. The freedom to establish a religion is a protected right.

The religious landscape looks different outside North America. One-third of the world's people are Christian, but that leaves two-thirds who are not. Around the world, people care deeply about their religion and are willing to fight other religious groups and governments to protect their right to worship as they choose. The growth of Islam in Europe and of Christianity in Africa shows that the religious landscape can change through migration and conversion.

Almost all religions preach a doctrine of peace and love, yet religion has been at the center of conflicts throughout history. For geographers, religion represents a critical factor in explaining cultural differences among locations as well as interrelationships between the environment and culture. Given the importance of religion to people everywhere, geographers are sensitive to the importance of accurately understanding global similarities and local diversity among religions.

The key issues of this chapter demonstrate the impact of religion on the cultural landscape. Here again are the key issues for Chapter 6:

1. **Where Are Religions Distributed?** The world has three large universalizing religions—Christianity, Islam, and Buddhism, each of which is divided into branches and denominations. Hinduism is the largest ethnic religion.
2. **Why Do Religions Have Different Distributions?** A universalizing religion has a known origin and clear patterns of diffusion, whereas ethnic religions typically have unknown origins and little diffusion. Holy places and holidays in a universalizing religion are related to events in the life of its founder or prophet and are related to the local physical geography in an ethnic religion. Some religions encourage pilgrimages to holy places.
3. **Why Do Religions Organize Space in Distinctive Patterns?** Some religions have elaborate places of worship. Religions affect the landscape in other ways: Religious communities are built, religious toponyms mark the landscape, and extensive tracts are reserved for burying the dead. Some but not all universalizing religions organize their territory into a rigid administrative structure to disseminate religious doctrine.
4. **Why Do Territorial Conflicts Arise Among Religious Groups?** With Earth's surface dominated by four large religions, expansion of the territory occupied by one religion may reduce the territory of another. In addition, religions must compete for control of territory with nonreligious ideas, notably communism and economic modernization.

CASE STUDY REVISITED / Future of Buddhism in Tibet

When the Dalai Lama dies, Tibetan Buddhists believe that his spirit enters the body of a child. In 1937, a group of priests located and recognized a two-year-old child named Tenzin Gyatso as the fourteenth Dalai Lama, the incarnation of the deceased thirteenth Dalai Lama, Bodhisattva Avalokiteshvara.

The child was brought to Lhasa in 1939 when he was 4 and enthroned a year later (Figure 6-28). Priests trained the young Dalai Lama to assume leadership and sent him to college when he was 16.

Daily life in Tibet was traditionally dominated by Buddhist rites. As recently as the 1950s, one-fourth of all males were monks, and polygamy was encouraged among other males to produce enough children to prevent the population from declining.

After taking control of Tibet in 1950, the Chinese Communists sought to reduce the domination of Buddhist monks in the country's daily life by destroying monasteries and temples. Farmers were

(Continued)

CASE STUDY REVISITED (*Continued*)

required to join agricultural communes unsuitable for their nomadic style of raising livestock, especially yaks.

In recent years, the Chinese have built new roads and power plants to help raise the low standard of living in Tibet. The Chinese argue that they have brought modern conveniences to Tibet, including paved roads, hospitals, schools, and agricultural practices. Some monasteries have been rebuilt, but no new monks are being trained. At the same time, the Chinese have secured their hold on Tibet by encouraging immigration from other parts of China. The Chinese government opposes efforts by other countries and international organizations to encourage greater autonomy for Tibet.

The Dalai Lama has become an articulate spokesperson for religious freedom, and in 1989 he was awarded the world's most prestigious award for peace, the Nobel Prize. Despite the efforts of the Dalai Lama and other Buddhists, though, when the current generation of priests dies, many Buddhist traditions in Tibet may be lost forever. ■

FIGURE 6-28 Dalai Lama's Palace in Lhasa, Tibet.

KEY TERMS

Animism (p. 178) Belief that objects, such as plants and stones, or natural events, like thunderstorms and earthquakes, have a discrete spirit and conscious life.

Autonomous religion (p. 195) A religion that does not have a central authority but shares ideas and cooperates informally.

Branch (p. 171) A large and fundamental division within a religion.

Caste (p. 196) The class or distinct hereditary order into which a Hindu is assigned according to religious law.

Cosmogony (p. 188) A set of religious beliefs concerning the origin of the universe.

Denomination (p. 171) A division of a branch that unites a number of local congregations into a single legal and administrative body.

Ethnic religion (p. 170) A religion with a relatively concentrated spatial distribution whose principles are likely to be based on the physical characteristics of the particular location in which its adherents are concentrated.

Fundamentalism (p. 197) Literal interpretation and strict adherence to basic principles of a religion (or a religious branch, denomination, or sect).

Ghetto (p. 185) During the Middle Ages, a neighborhood in a city set up by law to be inhabited only by Jews; now used to denote a section of a city in which members of any minority group live because of social, legal, or economic pressure.

Hierarchical religion (p. 194) A religion in which a central authority exercises a high degree of control.

Missionary (p. 182) An individual who helps to diffuse a universalizing religion.

Monotheism (p. 178) The doctrine or belief of the existence of only one god.

Pagan (p. 182) A follower of a polytheistic religion in ancient times.

Pilgrimage (p. 185) A journey to a place considered sacred for religious purposes.

Polytheism (p. 178) Belief in or worship of more than one god.

Sect (p. 171) A relatively small group that has broken away from an established denomination.

Solstice (p. 189) Astronomical event that happens twice each year, when the tilt of Earth's axis is most inclined toward or away from the Sun, causing the Sun's apparent position in the sky to reach it most northernmost or southernmost extreme, and resulting in the shortest and longest days of the year.

Universalizing religion (p. 170) A religion that attempts to appeal to all people, not just those living in a particular location.

THINKING GEOGRAPHICALLY

1. Sharp differences in demographic characteristics, such as natural increase, crude birth, and migration rates, can be seen among Jews, Christians, and Muslims in the Middle East and between Roman Catholics and Protestants in Northern Ireland. How might demographic differences affect future relationships among the groups in these two regions?
2. People carry their religious beliefs with them when they migrate. Over time, change occurs in the regions from which most U.S. immigrants originate and in the U.S. regions where they settle. How has the distribution of U.S. religious groups been affected by these changes?
3. To what extent have increased interest in religion and ability to practice religious rites served as forces for unification in Eastern Europe and the countries that were formerly part of the Soviet Union? Has the growing role of religion in the region fostered political instability? Explain.
4. Why does Islam seem strange and threatening to some people in predominantly Christian countries? To what extent is this attitude shaped by knowledge of the teachings of Muhammad and the Quran, and to what extent is it based on lack of knowledge of the religion?
5. Some Christians believe that they should be prepared to carry the word of God and the teachings of Jesus Christ to people who have not been exposed to them, at any time and at any place. Are evangelical activities equally likely to occur at any time and at any place, or are some places more suited than others? Why?

RESOURCES

Some recent and classic books and articles on ethnic geography:

Aharoni, Yohanan, Michael Avi-Yonah, Anson F. Rainey, and Ze'ev Safrai. *The Macmillan Bible Atlas*, 3rd ed. New York: Macmillan, 1993.

Al Faruqi, Isma'il R., and David E. Sopher. *Historical Atlas of the Religions of the World*. New York: Macmillan, 1974.

Al Faruqi, Isma'il R., and Lois Lamaya' Al Faruqi. *The Cultural Atlas of Islam*. New York: Macmillan, 1986.

Bhardwaj, Surinder M. *Hindu Places of Pilgrimage in India*. Berkeley: University of California Press, 1973.

Cooper, Adrian. "New Directions in the Geography of Religion." *Area* 24 (1992): 123–29.

Falah, Ghazi-Walid, and Caroline Nagel, eds. *Geographies of Muslim Women: Gender, Religion, and Space*. New York: Guilford Press, 2005.

Fickeler, Paul. "Fundamental Questions in the Geography of Religions." In *Readings in Cultural Geography*, eds. Philip L. Wagner and Marvin W. Mikesell. Chicago: University of Chicago Press, 1962.

Gaustad, Edwin Scott, and Philip L. Barlow. *New Historical Atlas of Religion in America*. New York: Oxford University Press, 2001.

Ivakhiv, Adrian. "Toward a Geography of 'Religion': Mapping the Distribution of an Unstable Signifier." *Annals of the Association of American Geographers* 96 (2006): 169–75.

Kay, Jeanne. "Human Dominion over Nature in the Hebrew Bible." *Annals of the Association of American Geographers* 79 (1989): 214–32.

Levine, Gregory J. "On the Geography of Religion." *Transactions of the Institute of British Geographers*, New Series 11, no. 4 (1987): 428–40.

Morin, Karen M., and Jeanne Kay Guelke, eds. *Women, Religion, & Space: Global Perspectives on Gender and Faith*. Syracuse, NY: Syracuse University Press, 2007.

Pacione, Michael. "The Relevance of Religion for a Relevant Human Geography." *Scottish Geographical Journal* 115 (1999): 117–31.

Park, Chris C. *Sacred Worlds: An Introduction to Geography and Religion*. London and New York: Routledge, 1994.

Pritchard, James B., ed. *HarperCollins Atlas of Bible History*. New York: Harper, 2008.

Shortridge, James R. "Patterns of Religion in the United States." *Geographical Review* 66 (1976): 420–34.

Smart, Ninian, and Frederick W. Denny, eds. *Atlas of the World's Religions*, 2nd ed. Oxford and New York: Oxford University Press, 2007.

Sopher, David E. "Geography and Religions." *Progress in Human Geography* 5 (1981): 510–24.

Stump, Roger W. *The Geography of Religion: Faith, Place, and Space*. Lanham, MD: Rowman and Littlefield, 2008.

Zelinsky, Wilbur. "An Approach to the Religious Geography of the United States: Patterns of Church Membership in 1952." *Annals of the Association of American Geographers* 51 (1961): 139–67.

Key Internet sites

www.adherents.com Statistics on the number of adherents to religions, branches, and denominations can be controversial. Adherents.com maintains an authoritative nondenominational source of data. Statistics are provided by religion and by location. The site also notes when different sources sharply disagree about the numbers.

www.glenmary.org Glenmary Research Center is the principal source of information about adherents within the United States. The Center, which is affiliated with the Roman Catholic Church, provides maps of the largest branch or denomination by county.

Log in to www.mygeoscienceplace.com for videos, interactive maps, RSS feeds, case studies, and self-study quizzes to enhance your study of Religion.

CHAPTER 7

Ethnicity

Few humans live in total isolation. People are members of groups with which they share important attributes. If you are a citizen of the United States of America, you are identified as an American, which is a type of nationality.

Many Americans further identify themselves as belonging to an ethnicity, a group with which they share cultural background. One-third of Americans identify their ethnicity as African American, Hispanic, or Asian American. Other Americans identify with ethnicities tracing back to Europe.

KEY ISSUES

1 Where Are Ethnicities Distributed?

2 Why Have Ethnicities Been Transformed into Nationalities?

3 Why Do Ethnicities Clash?

4 What Is Ethnic Cleansing?

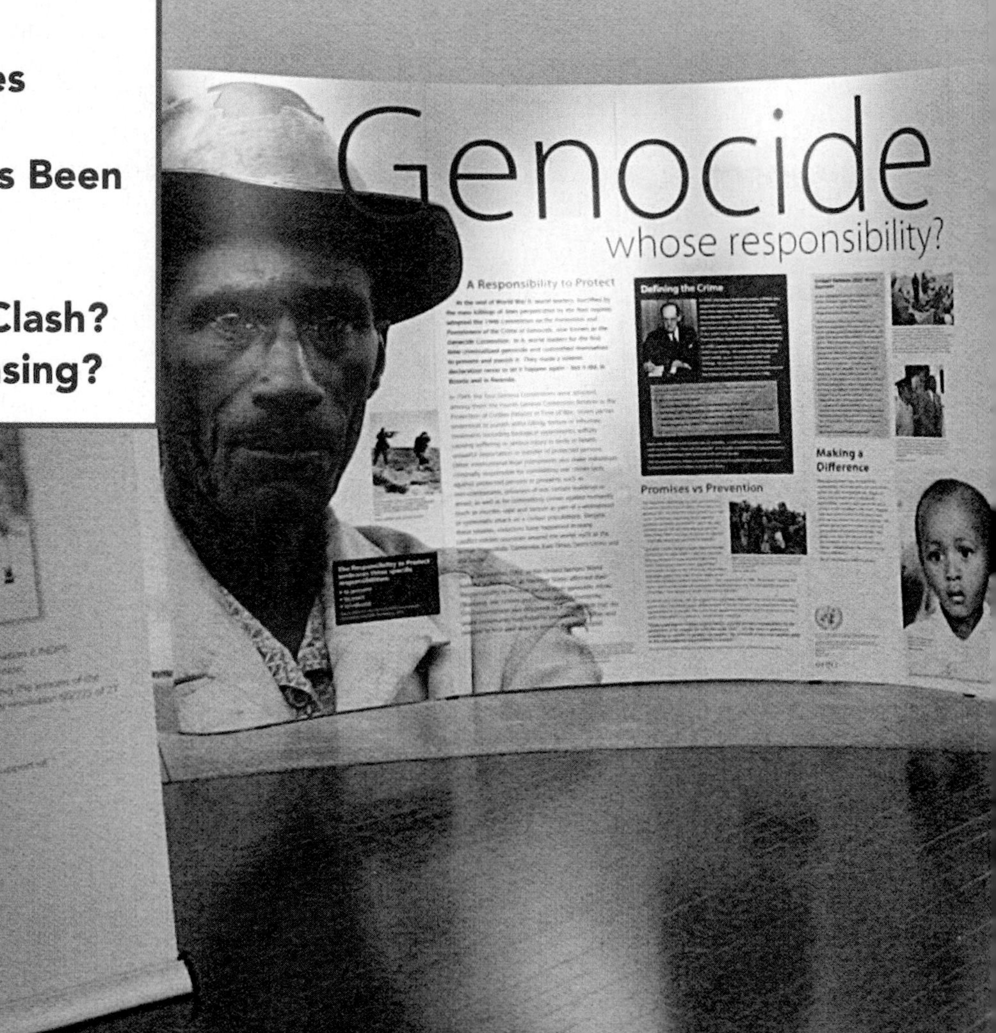

Ethnicity is a source of pride to people, a link to the experiences of ancestors and to cultural traditions, such as food and music preferences. The ethnic group to which one belongs has important measurable differences, such as average income, life expectancy, and infant mortality rate. Ethnicity also matters in places with a history of discrimination by one ethnic group against another.

The significance of ethnic diversity is controversial in the United States:

- To what extent does discrimination persist against minority ethnicities, especially African Americans and Hispanics?
- Should preferences be given to minority ethnicities to correct past patterns of discrimination?
- To what extent should the distinct cultural identity of ethnicities be encouraged or protected?

Exhibition on ethnic cleansing in Rwanda during the 1990s organized by the United Nations and on display in Johannesburg, South Africa, in 2008.

CASE STUDY / Ethnic Diversity in America

The United States is a country of ethnic diversity. The complexity of ethnic identity in the United States is clearly illustrated by Barack Obama: the country's first black president, son of a white mother and black father.

- President Obama's father, Barack Obama, Senior, was born in the village of Kanyadhiang, Kenya. He was a member of Kenya's third-largest ethnic group, known as the Luo.
- President Obama's mother, Ann Dunham, was born in Kansas. Most of her ancestors migrated to the United States from England in the nineteenth century.
- President Obama's step-father—his mother's second husband, Lolo Soetoro—was born in the village of Yogyakarta, Indonesia. He was a member of Indonesia's largest ethnic group, known as the Javanese.

Race and ethnicity are often confused. Ethnicity, such as the president's Luo ancestry through his father, is important to geographers because its characteristics derive from the distinctive features of particular places on Earth, such as rural eastern Kenya.

Features of race, such as skin color, hair type and color, blood traits, and shape of body, head, and facial features, were once thought to be scientifically classifiable. Contemporary geographers reject the entire biological basis of classifying humans into a handful of races because these features are not rooted in specific places.

However, one feature of race does matter to geographers—the color of skin. President Obama's race is black because of the color of his skin. The distribution of persons of color matters to geographers because it is the fundamental basis by which people in many societies sort out where they reside, attend school, recreate, and perform many other activities of daily life. ■

Ethnicity is identity with a group of people who share the cultural traditions of a particular homeland or hearth. Ethnicity comes from the Greek word *ethnikos*, which means "national." Ethnicity is distinct from **race**, which is identity with a group of people who share a biological ancestor. Race comes from a middle-French word for *generation*.

Geographers are interested in ***where*** ethnicities are distributed across ***space***, like other elements of culture. An ethnic group is tied to a particular ***place***, because members of the group—or their ancestors—were born and raised there. The cultural traits displayed by an ethnicity derive from particular conditions and practices in the group's homeland.

The reason ***why*** ethnicities have distinctive traits should by now be familiar. Like other cultural elements, ethnic identity derives from the interplay of ***connections*** with other groups and isolation from them.

Ethnicity is an especially important cultural element of ***local diversity*** because our ethnic identity is immutable. We can deny or suppress our ethnicity, but we cannot choose to change it in the same way we can choose to speak a different language or practice a different religion. If our parents come from two ethnic groups or our grandparents from four, our ethnic identity may be extremely diluted, but it never completely disappears.

The study of ethnicity lacks the tension in ***scale*** between preservation of local diversity and ***globalization*** observed in other cultural elements. Despite efforts to preserve local languages, it is not far-fetched to envision a world in which virtually all educated people speak English. And universalizing religions continue to gain adherents around the world. But no ethnicity is attempting or even aspiring to achieve global dominance, although ethnic groups are fighting with each other to control specific areas of the world.

Ethnicity is especially important to geographers because in the face of globalization trends in culture and economy, ethnicity stands as the strongest bulwark for the preservation of local diversity. Even if globalization engulfs language, religion, and other cultural elements, ***regions*** of distinct ethnic identity will remain.

KEY ISSUE 1
Where Are Ethnicities Distributed?

- **Distribution of Ethnicities in the United States**
- **Differentiating Ethnicity and Race**

An ethnicity may be clustered in specific areas within a country, or the area it inhabits may match closely the boundaries of a country. This section of the chapter examines the clustering of ethnicities within countries, and the next key issue looks at ethnicities on the national scale. ■

Distribution of Ethnicities in the United States

The two most numerous ethnicities in the United States are Hispanics (or Latinos), at 15 percent of the total population, and African Americans, at 13 percent. In addition, about 4 percent are Asian American and 1 percent American Indian.

Clustering of Ethnicities

Within a country, clustering of ethnicities can occur on two scales. Ethnic groups may live in particular regions of the country, and they may live in particular neighborhoods within cities. Within the United States, ethnicities are clustered at both scales.

REGIONAL CONCENTRATIONS OF ETHNICITIES. On a regional scale, ethnicities have distinctive distributions within the United States:

- **Hispanic or Latino/Latina.** Clustered in the Southwest, Hispanics exceed one-third of the population of Arizona, New Mexico, and Texas, and one-quarter of California (Figure 7-1). California is home to one-third of all Hispanics, Texas one-fifth, and Florida and New York one-sixth each.

  Hispanic or *Hispanic American* is a term that the U.S. government chose in 1973 to describe the group because it was an inoffensive label that could be applied to all people from Spanish-speaking countries. Some Americans of Latin American descent have instead adopted the terms *Latino* (males) and *Latina* (females). A 1995 U.S. Census Bureau survey found that 58 percent of Americans of Latin American descent preferred the term *Hispanic* and 12 percent *Latino/Latina*.

  Most Hispanics identify with a more specific ethnic or national origin. Around two-thirds come from Mexico and are sometimes called Chicanos (males) or Chicanas (females). Originally the term was considered insulting, but in the 1960s Mexican American youths in Los Angeles began to call themselves Chicanos and Chicanas with pride.
- **African Americans.** Clustered in the Southeast, African Americans comprise at least one-fourth of the population in Alabama, Georgia, Louisiana, Maryland, and South Carolina, and more than one-third in Mississippi (Figure 7-2). Concentrations are even higher in selected counties. At the other extreme, nine states in upper New England and the West have less than 1 percent African Americans.
- **Asian Americans.** Clustered in the West, Asian Americans comprise more than 40 percent of the population of Hawaii (Figure 7-3). One-half of all Asian Americans live in California, where they comprise 12 percent of the population. Chinese account for one-fourth of Asian Americans, Indians and Filipinos one-fifth each, and Korean and Vietnamese one-tenth each.
- **American Indians and Alaska Natives.** Within the 48 continental United States, American Indians are most numerous in the Southwest and the Plains states (Figure 7-4).

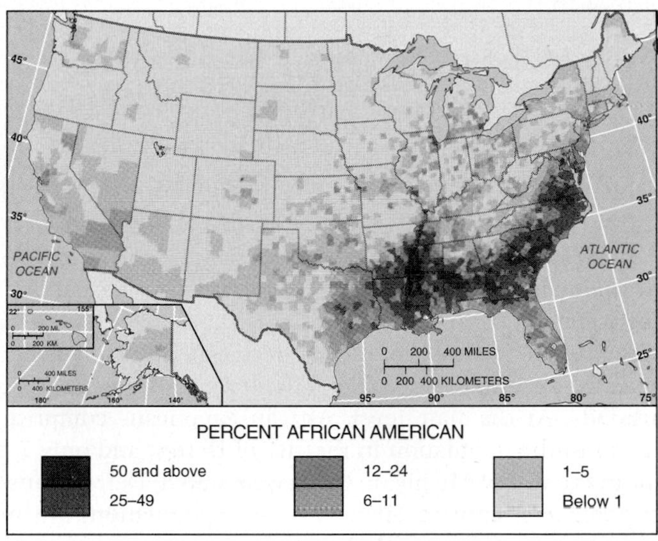

FIGURE 7-2 Distribution of African Americans in the United States. The highest percentages of African Americans are in the rural South and in northern cities.

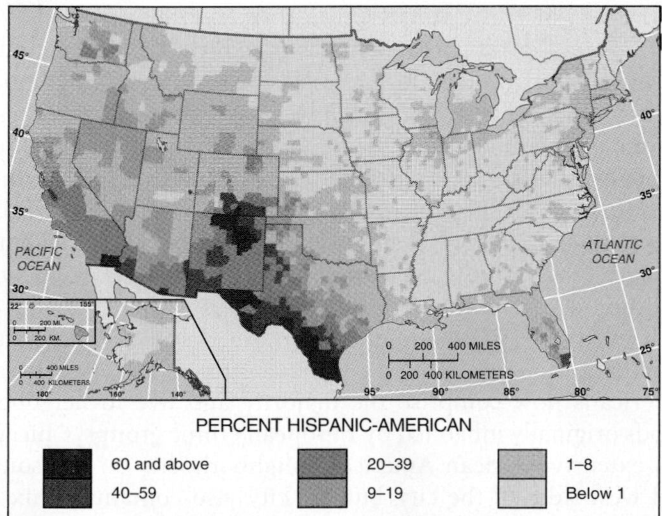

FIGURE 7-1 Distribution of Hispanic Americans in the United States. The highest percentages are in the Southwest, near the Mexican border, and in northern cities.

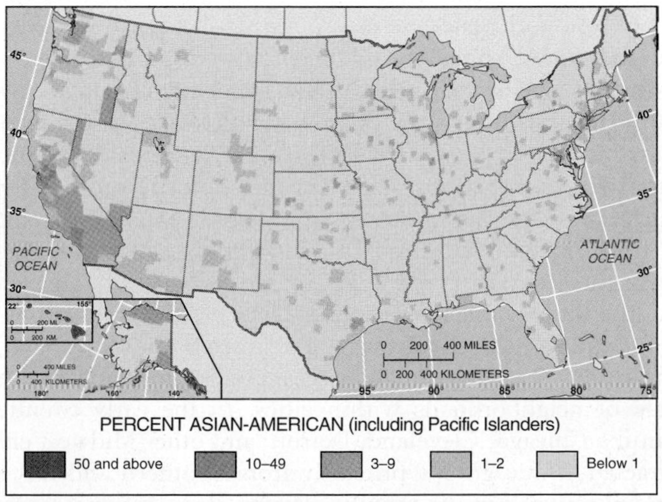

FIGURE 7-3 Distribution of Asian Americans in the United States. The highest percentages are in Hawaii and California.

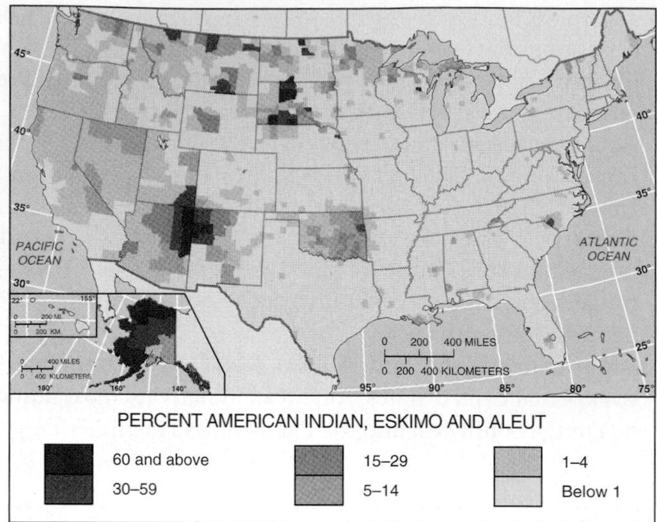

FIGURE 7-4 Distribution of American Indians in the United States. The highest percentages are in Alaska and the Plains states.

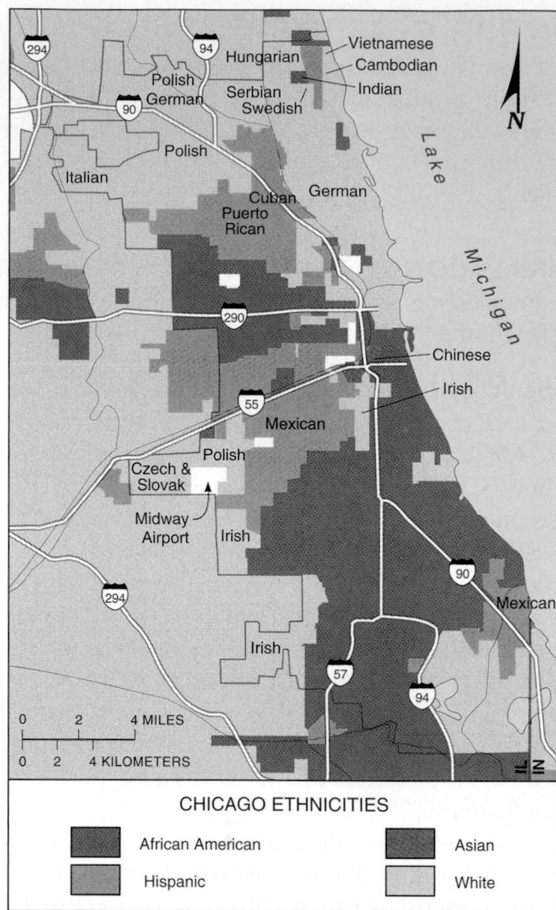

FIGURE 7-5 Distribution of ethnicities in Chicago. African Americans occupy extensive areas on the south and west sides. Hispanic Americans are clustered in several neighborhoods on the west side. European ethnic groups are located to the northwest, southwest, and far south side. Asian ethnic groups are clustered in the far north side.

CONCENTRATION OF ETHNICITIES IN CITIES. African Americans and Hispanics are highly clustered in urban areas. Around 90 percent of these ethnicities live in metropolitan areas, compared to around 75 percent for all Americans.

The distinctive distribution of African Americans and Hispanics is especially noticeable at the levels of states and neighborhoods. At the state level, African Americans comprise 85 percent of the population in the city of Detroit and only 7 percent in the rest of Michigan. Otherwise stated, Detroit contains less than one-tenth of Michigan's total population, but more than one-half of the state's African American population. Similarly, Chicago is more than one-third African American, compared to one-twelfth in the rest of Illinois. Chicago has less than one-fourth of Illinois' total population and more than one-half of the state's African Americans.

The distribution of Hispanics is similar to that of African Americans in large northern cities. For example, New York City is more than one-fourth Hispanic, compared to one-sixteenth in the rest of New York State, and New York City contains two-fifths of the state's total population and three-fourths of its Hispanics.

In the states with the largest Hispanic populations—California and Texas—the distribution is mixed. In California, Hispanics comprise nearly half of Los Angeles's population, but the percentage of Hispanics in California's other large cities is less than or about equal to the overall state average. In Texas, El Paso and San Antonio—the two large cities closest to the Mexican border—are more than one-half Hispanic, but the state's other large cities have percentages below or about equal to the state's average of around one-third.

The clustering of ethnicities is especially pronounced on the scale of neighborhoods within cities. In the early twentieth century, Chicago, Cleveland, Detroit, and other Midwest cities attracted ethnic groups primarily from Southern and Eastern Europe to work in the rapidly growing steel, automotive, and related industries. For example, in 1910, when Detroit's auto production was expanding, three-fourths of the city's residents were immigrants and children of immigrants. Southern and Eastern European ethnic groups clustered in newly constructed neighborhoods that were often named for their predominant ethnicities, such as Detroit's Greektown and Poletown.

The children and grandchildren of European immigrants moved out of most of the original inner-city neighborhoods during the twentieth century. For descendants of European immigrants, ethnic identity is more likely to be retained through religion, food, and other cultural traditions rather than through location of residence. A visible remnant of early twentieth-century European ethnic neighborhoods is the clustering of restaurants in such areas as Little Italy and Greektown.

Ethnic concentrations in U.S. cities increasingly consist of African Americans who migrate from the South or immigrants from Latin America and Asia. In cities such as Detroit, African Americans now comprise the majority and live in neighborhoods originally inhabited by European ethnic groups. Chicago has extensive African American neighborhoods on the south and west sides of the city, but the city also contains a mix of neighborhoods inhabited by European, Latin American, and Asian ethnicities (Figure 7-5).

In Los Angeles, which contains large percentages of African Americans, Hispanics, and Asian Americans, the major ethnic

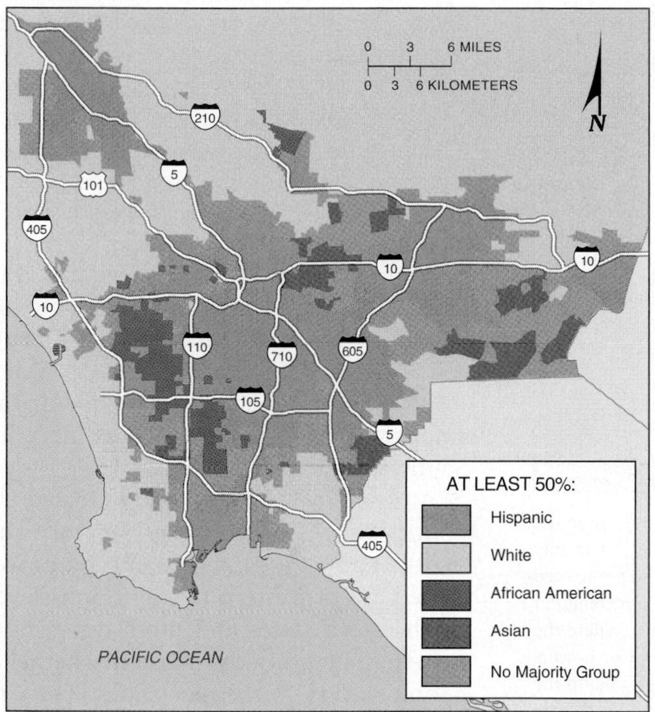

FIGURE 7-6 Distribution of ethnicities in Los Angeles. African Americans are clustered to the south of downtown Los Angeles, and Hispanics to the east. Asian American neighborhoods are contiguous to the African American and Hispanic areas.

groups are clustered in different areas (Figure 7-6). African Americans are located in south-central Los Angeles and Hispanics in the east. Asian Americans are located to the south and west, contiguous to the African American and Hispanic areas.

African American Migration Patterns

The clustering of ethnicities within the United States is partly a function of the same process that helps geographers to explain the regular distribution of other cultural factors, such as language and religion—namely migration. The migration patterns of African Americans have been especially distinctive.

Three major migration flows have shaped the current distribution of African Americans within the United States:

- Forced migration from Africa to the American colonies in the eighteenth century.
- Immigration from the U.S. South to northern cities during the first half of the twentieth century.
- Immigration from inner-city ghettos to other urban neighborhoods during the second half of the twentieth and first decade of the twenty-first centuries.

FORCED MIGRATION FROM AFRICA. Most African Americans are descended from Africans forced to migrate to the Western Hemisphere as slaves. Slavery is a system whereby one person owns another person as a piece of property and can force that slave to work for the owner's benefit.

The first Africans brought to the American colonies as slaves arrived at Jamestown, Virginia, on a Dutch ship in 1619

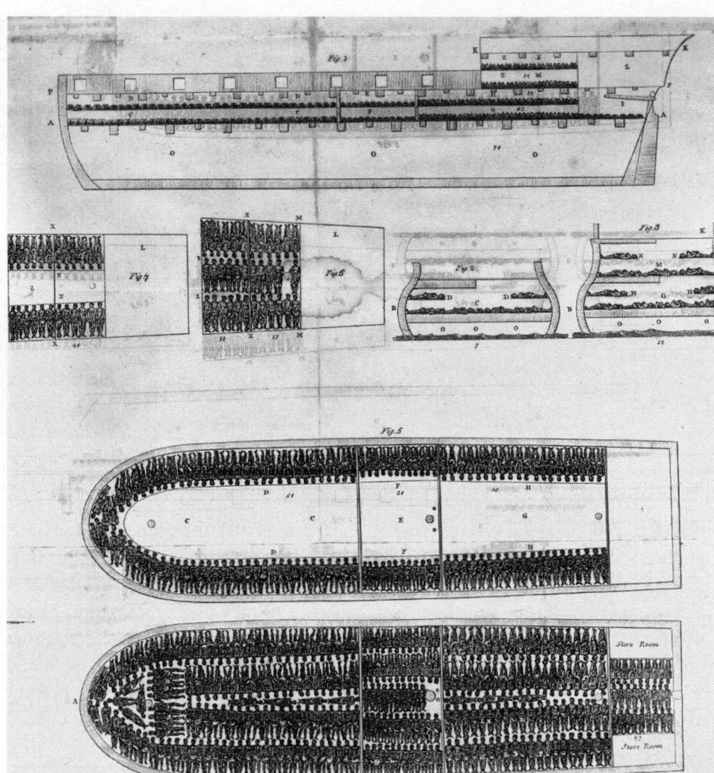

FIGURE 7-7 Slave ship. The diagrams show the extremely high density by which Africans were transported to the Americas to be sold as slaves. The bottom two images show human figures packed into the hold of the ship lying next to each other with no room to move.

(Figure 7-7). During the eighteenth century, the British shipped about 400,000 Africans to the 13 colonies that later formed the United States. In 1808, the United States banned bringing in additional Africans as slaves, but an estimated 250,000 were illegally imported during the next half-century.

Slavery was widespread during the time of the Roman Empire, about 2,000 years ago. During the Middle Ages, slavery was replaced in Europe by a feudal system, in which laborers working the land (known as serfs) were bound to the land and not free to migrate elsewhere. Serfs had to turn over a portion of their crops to the lord and provide other services as demanded by the lord.

Although slavery was rare in Europe, Europeans were responsible for diffusing the practice to the Western Hemisphere. Europeans who owned large plantations in the Americas turned to African slaves as an abundant source of labor that cost less than paying wages to other Europeans.

At the height of the slave trade between 1710 and 1810, at least 10 million Africans were uprooted from their homes and sent on European ships to the Western Hemisphere for sale in the slave market. During that period, the British and Portuguese each shipped about 2 million slaves to the Western Hemisphere, with most of the British slaves going to Caribbean islands and the Portuguese slaves to Brazil.

The forced migration began when people living along the east and west coasts of Africa, taking advantage of their superior weapons, captured members of other groups living farther

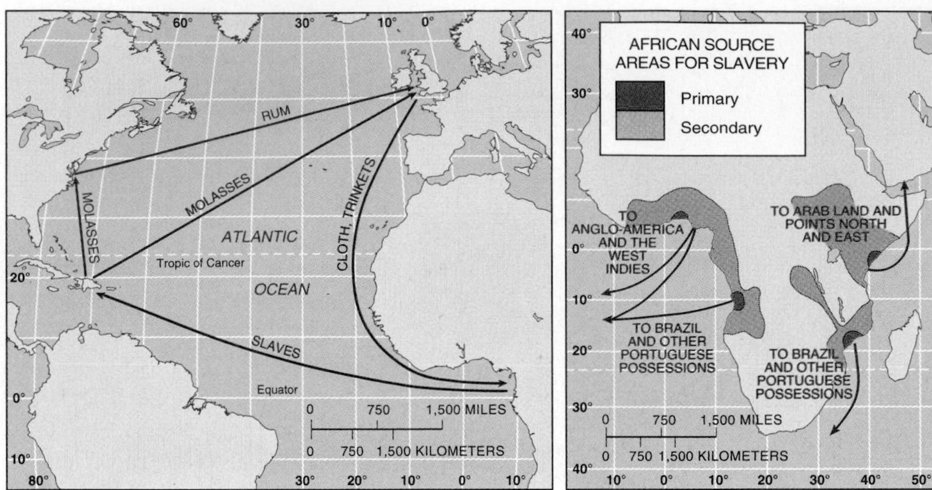

FIGURE 7-8 Triangular slave pattern. (left) The British initiated a triangular slave trading pattern in the eighteenth century. Cloth, iron bars, and other goods were carried by ship from Britain to Africa to buy slaves. The same ships transported slaves from Africa to the Caribbean islands. The ships then completed the triangle by returning to Britain with molasses to make rum. Sometimes the ships formed a rectangular pattern by carrying the molasses from the Caribbean islands to the North American colonies, where the rum was distilled and shipped to Britain. (right) Slave sources. The British and other European powers obtained slaves primarily from a narrow strip along the west coast of Africa, from Liberia to Angola. In the early days of colonization, Europeans secured territory along the Atlantic Coast and rarely ventured more than 160 kilometers (100 miles) into the interior of the continent.

inland and sold the captives to Europeans. Europeans in turn shipped the captured Africans to the Americas, selling them as slaves either on consignment or through auctions. The Spanish and Portuguese first participated in the slave trade in the early sixteenth century, and the British, Dutch, and French joined in during the next century.

Different European countries operated in various regions of Africa, each sending slaves to different destinations in the Americas (Figure 7-8, right). The Portuguese shipped slaves primarily from their principal African colonies—Angola and Mozambique—to their major American colony, Brazil. Other European countries took slaves primarily from a coastal strip of West Africa between Liberia and the Congo, 4,000 kilometers (2,500 miles) long and 160 kilometers (100 miles) wide. The majority of these slaves went to Caribbean islands and most of the remainder to Central and South America. Fewer than 5 percent of the slaves ended up in the United States.

At the height of the eighteenth-century slave demand, a number of European countries adopted the **triangular slave trade**, an efficient triangular trading pattern (Figure 7-8, left):

- Ships left Europe for Africa with cloth and other trade goods, used to buy the slaves.
- They then transported slaves and gold from Africa to the Western Hemisphere, primarily to the Caribbean islands.
- To complete the triangle, the same ships then carried sugar and molasses from the Caribbean on their return trip to Europe.

Some ships added another step, making a rectangular trading pattern, in which molasses was carried from the Caribbean to the North American colonies, and rum from the colonies to Europe.

The large-scale forced migration of Africans obviously caused them unimaginable hardship, separating families and destroying villages. Traders generally seized the stronger and younger villagers, who could be sold as slaves for the highest price. The Africans were packed onto ships at extremely high density, kept in chains, and provided with minimal food and sanitary facilities. Approximately one-fourth died crossing the Atlantic.

In the 13 colonies that later formed the United States, most of the large plantations in need of labor were located in the South, primarily those growing cotton as well as tobacco. Consequently, nearly all Africans shipped to the 13 colonies ended up in the Southeast.

Attitudes toward slavery dominated U.S. politics during the nineteenth century. During the early 1800s, when new states were carved out of western territory, anti-slavery northeastern states and pro-slavery southeastern states bitterly debated whether to permit slavery in the new states. The Civil War (1861–1865) was fought to prevent 11 pro-slavery Southern states from seceding from the Union. In 1863, during the Civil War, Abraham Lincoln issued the Emancipation Proclamation, freeing the slaves in the 11 Confederate states. The Thirteenth Amendment to the Constitution, adopted 8 months after the South surrendered, outlawed slavery.

Freed as slaves, most African Americans remained in the rural South during the late nineteenth century working as sharecroppers (Figure 7-9). A **sharecropper** works fields rented from a landowner and pays the rent by turning over to the landowner a share of the crops. To obtain seed, tools, food, and living quarters, a sharecropper gets a line of credit from the landowner and repays the debt with yet more crops. The sharecropper system burdened poor African Americans with high interest rates and heavy debts. Instead of growing food that they could eat, sharecroppers were forced by landowners to plant extensive areas of crops such as cotton that could be sold for cash.

IMMIGRATION TO THE NORTH. Sharecropping became less common into the twentieth century as the introduction of farm machinery and a decline in land devoted to cotton reduced demand for labor. At the same time sharecroppers were being pushed off the farms, they were being pulled by the prospect of jobs in the booming industrial cities of the North.

African Americans migrated out of the South along several clearly defined channels (Figure 7-10). Most traveled by bus and car along the major two-lane long-distance U.S. roads that were paved and signposted in the early decades of the twentieth century and have since been replaced by interstate highways:

- East coast: From the Carolinas and other South Atlantic states north to Baltimore, Philadelphia, New York, and other northeastern cities, along U.S. Route 1 (parallel to present-day I-95).

FIGURE 7-9 Sharecroppers. Many African Americans became sharecroppers after slavery was abolished. Fields were rented from a landowner, and rent was paid in crops, in this case cotton from a Georgia farm in 1898.

- East central: From Alabama and eastern Tennessee north to either Detroit, along U.S. Route 25 (present-day I-75), or Cleveland, along U.S. Route 21 (present-day I-77).
- West central: From Mississippi and western Tennessee north to St. Louis and Chicago, along U.S. routes 61 and 66 (present-day I-55)
- Southwest: From Texas west to California, along U.S. routes 80 and 90 (present-day I-10 and I-20).

Southern African Americans migrated north and west in two main waves, the first in the 1910s and 1920s before and after World War I and the second in the 1940s and 1950s before and after World War II. The world wars stimulated expansion of factories in the 1910s and 1940s to produce war materiel, while the demands of the armed forces created shortages of factory workers. After the wars, during the 1920s and 1950s, factories produced steel, motor vehicles, and other goods demanded in civilian society.

In 1910, only 5,741 of Detroit's 465,766 inhabitants were African American. With the expansion of the auto industry during the 1910s and 1920s, the African American population increased to 120,000 in 1930, 300,000 in 1950, and 500,000 in 1960.

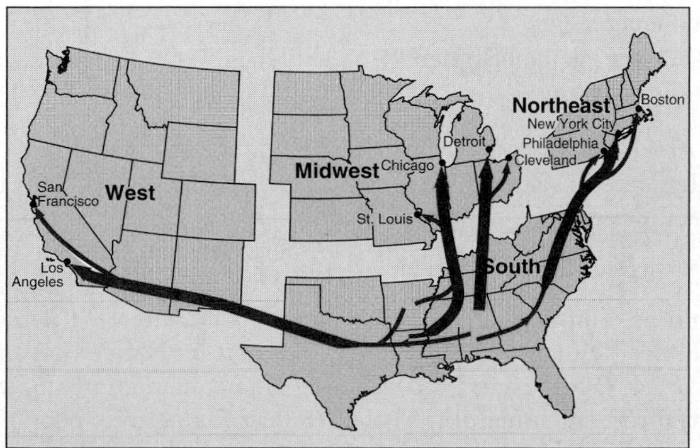

FIGURE 7-10 African American twentieth-century migration within the United States. Migration followed four distinctive channels along the east coast, east central, west central, and southwest regions of the country.

EXPANSION OF THE GHETTO. When they reached the big cities, African American immigrants clustered in the one or two neighborhoods where the small numbers who had arrived in the nineteenth century were already living. These areas became known as ghettos, after the term for neighborhoods in which Jews were forced to live in the Middle Ages (see Chapter 6).

In 1950, most of Baltimore's quarter-million African Americans lived in a 3-square-kilometer (1-square-mile) neighborhood northwest of downtown (Figure 7-11). The remainder were clustered east of downtown or in a large isolated housing project on the south side built for black wartime workers in port industries. Densities in the ghettos were high, with 40,000 inhabitants per square kilometer (100,000 per square mile) common. Contrast that density with the current level found in typical American suburbs of 2,000 inhabitants per square kilometer (5,000 per square mile). Because of the shortage of housing in the ghettos, families were forced to live in one room. Many dwellings lacked bathrooms, kitchens, hot water, and heat.

African Americans moved from the tight ghettos into immediately adjacent neighborhoods during the 1950s and 1960s. In Baltimore, the west side African American area expanded from 3 square kilometers (1 square mile) in 1950 to 25 square kilometers (10 square miles) in 1970, and a 5-square-kilometer (2-square-mile) area on the east side became mainly populated by African Americans. Expansion of the ghetto continued to follow major avenues to the northwest and northeast in subsequent decades.

Differentiating Ethnicity and Race

Race and ethnicity are often confused. In the United States, consider three prominent ethnic groups—Asian Americans, African Americans, and Hispanic Americans. All three ethnicities display

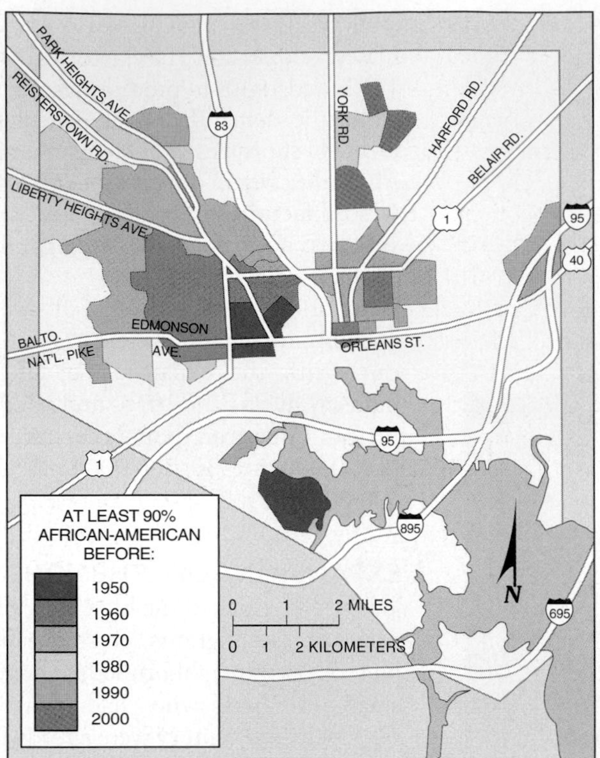

FIGURE 7-11 Expansion of African American ghetto in Baltimore, Maryland. In 1950, most African Americans in Baltimore lived in a small area northwest of downtown. During the 1950s and 1960s, the African American area expanded to the northwest, along major radial roads, and a second node opened on the east side. The south-side African American area was an isolated public housing complex built for wartime workers in the nearby port industries.

distinct cultural traditions that originate at particular hearths, but the three are regarded in different ways:

- Asian is recognized as a distinct race by the U.S. Bureau of the Census, so Asian as a race and Asian American as an ethnicity encompass basically the same group of people. However, the Asian American ethnicity lumps together people with ties to many countries in Asia.
- African American and black are different groups, although the 2000 census combined the two. Most black Americans are descended from African immigrants and therefore also belong to an African American ethnicity. Some American blacks, however, trace their cultural heritage to regions other than Africa, including Latin America, Asia, and Pacific islands.
- The term *African American* identifies a group with an extensive cultural tradition, whereas the term *black* in principle denotes nothing more than dark skin. Because many Americans make judgments about the values and behavior of others simply by observing skin color, *black* is substituted for *African American* in daily language.
- Hispanic or Latino is not considered a race, so on the census form members of the Hispanic or Latino ethnicity select any race they wish—white, black, or other.

The traits that characterize race are those that can be transmitted genetically from parents to children. For example, lactose intolerance affects 95 percent of Asian Americans, 65 percent of African Americans and Native Americans, and 50 percent of Hispanics, compared to only 15 percent of Americans of European ancestry. Nearly everyone is born with the ability to produce lactase, which enables infants to digest the large amount of lactose in milk. Lactase production typically slackens during childhood, leaving some with difficulty in absorbing a large amount of lactose as adults. A large percentage of persons of Northern European descent have a genetic mutation that results in lifelong production of lactase.

At best, biological features are so highly variable among members of a race that any prejudged classification is meaningless. Perhaps many tens or hundreds of thousands of years ago, early "humans" (however they emerged as a distinct species) lived in such isolation from other early "humans" that they were truly distinct genetically. But the degree of isolation needed to keep biological features distinct genetically vanished when the first human crossed a river or climbed a hill.

At worst, biological classification by race is the basis for **racism**, which is the belief that race is the primary determinant of human traits and capacities and that racial differences produce an inherent superiority of a particular race. A **racist** is a person who subscribes to the beliefs of racism.

Race in the United States

Every 10 years, the U.S. Bureau of the Census asks people to classify themselves according to the race with which they most closely identify. Americans are asked to identify themselves by checking the box next to one of the following fourteen races:

- White
- Black, African American, or Negro
- American Indian or Alaska Native
- Asian Indian
- Chinese
- Filipino
- Japanese
- Korean
- Vietnamese
- Other Asian
- Native Hawaiian
- Guamanian or Chamorro
- Samoan
- Other Pacific Islander
- Other race

If American Indian, Other Pacific Islander, Other Asian, or Other race is selected, the respondent is asked to write in the specific name.

In 2000 about 75 percent of Americans checked that they were white, 12 percent black, 4 percent Asian (Asian Indian, Chinese, Filipino, Japanese, Korean, or Vietnamese), 1 percent American Indian or Alaska Native, 0.1 percent Native Hawaiian or other Pacific Islander (including Guamanian and Samoan), and 6 percent some other race. The census permits people to check more than one box, and 7 million Americans (2 percent) of the respondents did that in 2000. President Obama is an example of an American of more than one race. His father was a black native of Africa and his mother was white.

"SEPARATE BUT EQUAL" DOCTRINE. In explaining spatial regularities, geographers look for patterns of spatial interaction. A distinctive feature of race relations in the United States has been the strong discouragement of spatial interaction—in the past through legal means, today through cultural preferences or discrimination.

The U.S. Supreme Court in 1896 upheld a Louisiana law that required black and white passengers to ride in separate railway cars. In *Plessy v. Ferguson*, the Supreme Court stated that Louisiana's law was constitutional because it provided *separate, but equal*, treatment of blacks and whites, and equality did not mean that whites had to mix socially with blacks.

Once the Supreme Court permitted "separate but equal" treatment of the races, southern states enacted a comprehensive set of laws to segregate blacks from whites as much as possible (Figure 7-12). These were called "Jim Crow" laws, named for a nineteenth-century song-and-dance act that depicted blacks offensively. Blacks had to sit in the back of buses, and shops, restaurants, and hotels could choose to serve only whites. Separate schools were established for blacks and whites. After all, white southerners argued, the bus got blacks sitting in the rear to the destination at the same time as the whites in the front, some commercial establishments served only blacks, and all of the schools had teachers and classrooms.

Throughout the country, not just in the South, house deeds contained restrictive covenants that prevented the owners from selling to blacks, as well as to Roman Catholics or Jews in some places. Restrictive covenants kept blacks from moving into an all-white neighborhood. And because schools, especially at the elementary level, were located to serve individual neighborhoods, most were segregated in practice, even if not legally mandated.

FIGURE 7-12 Segregation in the United States. Until the 1960s in the U.S. South, whites and blacks had to use separate drinking fountains, as well as separate restrooms, bus seats, hotel rooms, and other public facilities.

"WHITE FLIGHT." Segregation laws were eliminated during the 1950s and 1960s. The landmark Supreme Court decision *Brown v. Board of Education of Topeka, Kansas*, in 1954, found that having separate schools for blacks and whites was unconstitutional, because no matter how equivalent the facilities, racial separation branded minority children as inferior and therefore was inherently unequal. A year later, the Supreme Court further ruled that schools had to be desegregated "with all deliberate speed."

Rather than integrate, whites fled. The expansion of the black ghettos in American cities was made possible by "white flight," the emigration of whites from an area in anticipation of blacks immigrating into the area. Detroit provides a clear example. Black immigration into Detroit from the South subsided during the 1950s, but as legal barriers to integration crumbled, whites began to emigrate out of Detroit. Detroit's white population dropped by about 1 million between 1950 and 1975 and by another half million between 1975 and 2000. While whites fled, Detroit's black population continued to grow, but at a more modest rate, as a result of natural increase. In sum, Detroit in 1950 contained about 1.7 million whites and 300,000 blacks. The black population increased to 500,000 in 1960, 700,000 in 1970, and 800,000 in both 1990 and 2000, while the white population declined from 1.7 million in 1950 to 1.3 million in 1960, 900,000 in 1970, 500,000 in 1980, 300,000 in 1990, and 200,000 in 2000.

White flight was encouraged by unscrupulous real estate practices, especially blockbusting. Under **blockbusting**, real estate agents convinced white homeowners living near a black area to sell their houses at low prices, preying on their fears that black families would soon move into the neighborhood and cause property values to decline. The agents then sold the houses at much higher prices to black families desperate to escape the overcrowded ghettos. Through blockbusting, a neighborhood could change from all-white to all-black in a matter of months, and real estate agents could start the process all over again in the next white area.

The National Advisory Commission on Civil Disorders, known as the Kerner Commission, wrote in 1968 that U.S. cities were divided into two separate and unequal societies, one black and one white. Four decades later, despite serious efforts to integrate and equalize the two, segregation and inequality persist.

Division by Race in South Africa

Discrimination by race reached its peak in the late twentieth century in South Africa. While the United States was repealing laws that segregated people by race, South Africa was enacting them. The cornerstone of the South African policy was the creation of a legal system called apartheid (Figure 7-13). **Apartheid** was the physical separation of different races into different

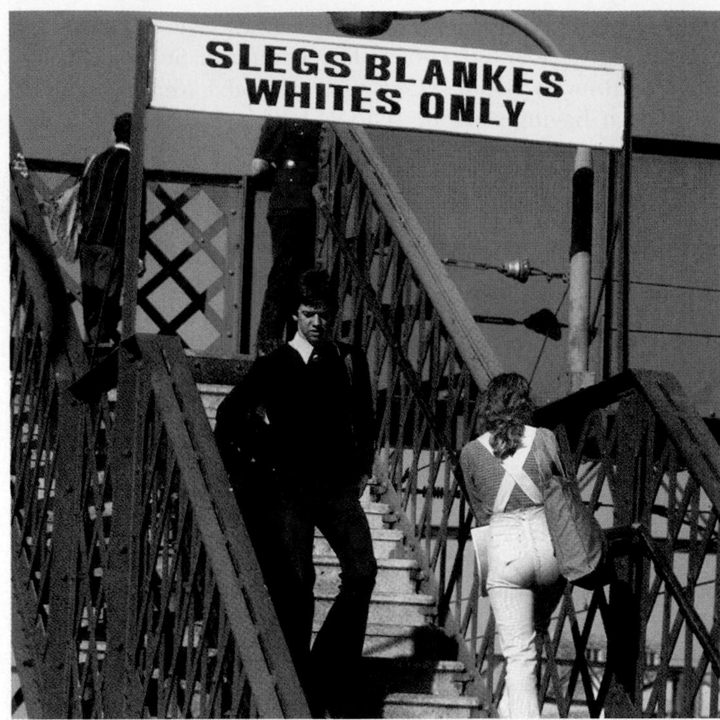

FIGURE 7-13 Apartheid. South Africa's apartheid laws were designed to spatially segregate races as much as possible. Blacks and whites reached the platform at this train station in Johannesburg by walking up separate stairs. Whites waited at the front of the platform to get into cars at the head of the train, while blacks waited at the rear.

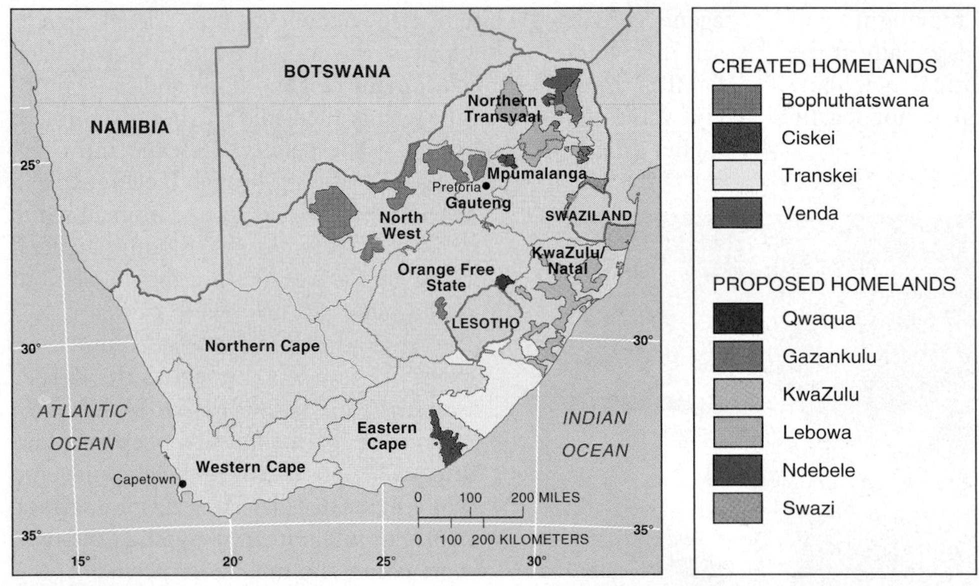

FIGURE 7-14 Apartheid in South Africa. As part of its apartheid system, the government of South Africa designated ten homelands, expecting that ultimately every black would become a citizen of one of them. South Africa declared four of these homelands to be independent states, but no other country recognized the action. With the end of apartheid and the election of a black majority government, the homelands were abolished, and South Africa was reorganized into nine provinces.

geographic areas. Although South Africa's apartheid laws were repealed during the 1990s, it will take many years to erase the impact of past policies.

In South Africa, under apartheid, a newborn baby was classified as being one of four races—black, white, colored (mixed white and black), or Asian. Under apartheid, each of the four races had a different legal status in South Africa. The apartheid laws determined where different races could live, attend school, work, shop, and own land. Blacks were restricted to certain occupations and were paid far lower wages than were whites for similar work. Blacks could not vote or run for political office in national elections. The apartheid system was created by descendants of whites who arrived in South Africa from Holland in 1652 and settled in Cape Town, at the southern tip of the territory. They were known either as *Boers*, from the Dutch word for "farmer," or *Afrikaners*, from the word "Afrikaans," the name of their language, which is a dialect of Dutch.

The British seized the Dutch colony in 1795, and controlled South Africa's government until 1948, when the Afrikaner-dominated Nationalist Party won elections. The Afrikaners gained power at a time when colonial rule was being replaced in the rest of Africa by a collection of independent states run by the local black population. The Afrikaners vowed to resist pressures to turn over South Africa's government to blacks, and the Nationalist Party created the apartheid laws in the next few years to perpetuate white dominance of the country. To ensure geographic isolation of different races, the South African government designated ten so-called homelands for blacks (Figure 7-14). The white minority government expected every black to become a citizen of one of the homelands and to move there. More than 99 percent of the population in the ten homelands was black.

The white-dominated government of South Africa repealed the apartheid laws in 1991. The principal antiapartheid organization, the African National Congress, was legalized, and its leader, Nelson Mandela, was released from jail after more than 27 years of imprisonment. When all South Africans were permitted to vote in national elections for the first time, in 1994, Mandela was overwhelmingly elected the country's first black president.

Now that South Africa's apartheid laws have been dismantled and the country is governed by its black majority, other countries have reestablished economic and cultural ties. However, the legacy of apartheid will linger for many years: South Africa's blacks have achieved political equality, but they are much poorer than white South Africans. Average income among white South Africans is about ten times higher than that of blacks.

KEY ISSUE 2

Why Have Ethnicities Been Transformed into Nationalities?

- **Rise of Nationalities**
- **Multinational States**
- **Revival of Ethnic Identity**

Ethnicity and race are distinct from nationality, another term commonly used to describe a group of people with shared traits. **Nationality** is identity with a group of people who share legal attachment and personal allegiance to a particular country. It comes from the Latin word *nasci*, which means "to have been born."

Nationality and ethnicity are similar concepts in that membership in both is defined through shared cultural values. In principle, the cultural values shared with others of the same ethnicity derive from religion, language, and material culture, whereas those shared with others of the same nationality derive from voting, obtaining a passport, and performing civic duties. ■

Rise of Nationalities

In the United States, nationality is generally kept reasonably distinct from ethnicity and race in common usage:

- *Nationality* identifies citizens of the United States of America, including those born in the country and those who immigrated and became citizens.
- *Ethnicity* identifies groups with distinct ancestry and cultural traditions, such as African Americans, Hispanic Americans, Chinese Americans, or Polish Americans.
- *Race* distinguishes blacks and other persons of color from whites.

The United States forged a nationality in the late eighteenth century out of a collection of ethnic groups gathered primarily from Europe and Africa, not through traditional means of issuing passports (African Americans weren't considered citizens then) or voting (women and African Americans couldn't vote then), but through sharing the values expressed in the Declaration of Independence, the U.S. Constitution, and the Bill of Rights. To be an American meant believing in the "unalienable rights" of "life, liberty, and the pursuit of happiness."

In Canada, the Québécois are clearly distinct from other Canadians in language, religion, and other cultural traditions. But do the Québécois form a distinct ethnicity within the Canadian nationality or a second nationality separate altogether from Anglo-Canadian? The distinction is critical, because if Québécois is recognized as a separate nationality from Anglo-Canadian, the Québec government would have a much stronger justification for breaking away from Canada to form an independent country (refer to Figure 5-27).

Outside North America, distinctions between ethnicity and nationality are even muddier. We have already seen in this chapter that identification with ethnicity and race can lead to discrimination and segregation. Confusion between ethnicity and nationality can lead to violent conflicts.

Nation-States

To preserve and enhance distinctive cultural characteristics, ethnicities seek to govern themselves without interference. A **nation-state** is a state whose territory corresponds to that occupied by a particular ethnicity that has been transformed into a nationality.

Ethnic groups have been transformed into nationalities because desire for self-rule is a very important shared attitude for many of them. The concept that ethnicities have the right to govern themselves is known as **self-determination**.

DENMARK: THERE ARE NO PERFECT NATION-STATES. Denmark is a fairly good example of a nation-state, because the territory occupied by the Danish ethnicity closely corresponds to the state of Denmark. The Danes have a strong sense of unity that derives from shared cultural characteristics and attitudes and a recorded history that extends back more than 1,000 years. Nearly all Danes speak the same language—Danish—and nearly all the world's speakers of Danish live in Denmark.

But even Denmark is not a perfect example of a nation-state. Ten percent of Denmark's population consists of ethnic minorities. The two largest groups are guest workers from Turkey and refugees from ethnic cleansing in the former Yugoslavia (see Chapter 3).

To dilute the concept of a nation-state further, Denmark controls two territories in the Atlantic Ocean that do not share Danish cultural characteristics. One is the Faeroe Islands, a group of 21 islands ruled by Denmark for more than 600 years. The nearly 50,000 inhabitants of the Faeroe Islands speak

Faeroese (see red area in Figure 5-10). Denmark also controls Greenland, the world's largest island, which is 50 times larger than Denmark proper. Only 12 percent of Greenland's 58,000 residents are considered Danish; the remainder are native-born Greenlanders, primarily Inuit. Greenlanders have received authority from Denmark to control their own domestic affairs. One decision was to change all place names in Greenland from Danish to the local Inuit language. Greenland is now officially known as Kalaallit Nunaat, and the capital city was changed from Godthaab to Nuuk. In 2009, Greenlandic became the official language of Greenland.

NATION-STATES IN EUROPE. Ethnicities were transformed into nationalities throughout Europe during the nineteenth century (Figure 7-15, upper left). Most of Western Europe was made up of nation-states by the early twentieth century.

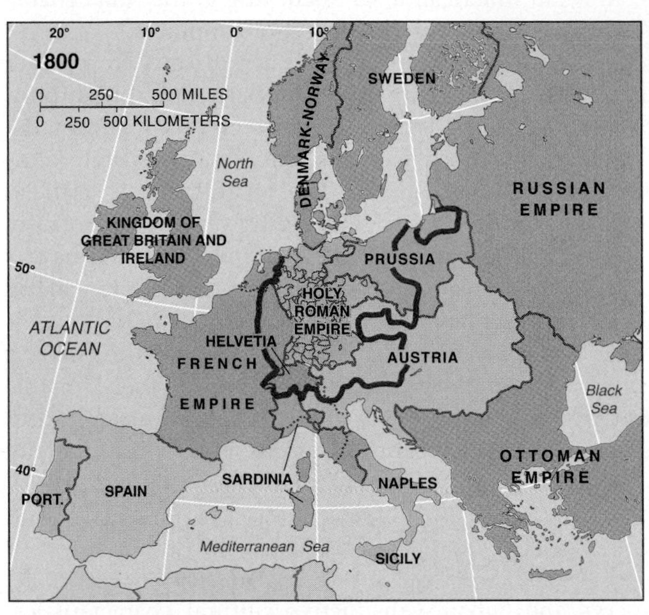

German-speaking territory in 1914

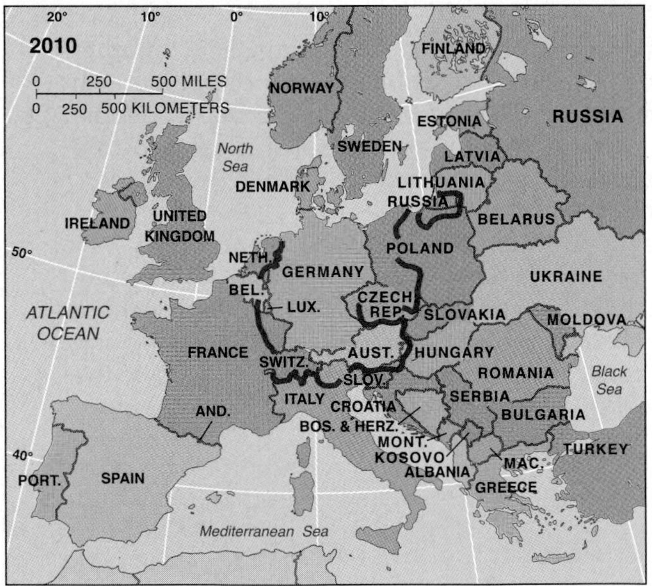

FIGURE 7-15 Nation-states in Europe. (upper left) In 1800, Europe's German-speaking territory was divided into a large number of principalities. (upper right) After losing World War I, Germany was divided into two discontinuous areas, separated by the Danzig Corridor, part of the newly created state of Poland. (lower left) Germany's boundaries changed again after World War II, as eastern portions of the country were taken by Poland and the Soviet Union. (lower right) With the collapse of Communism in Eastern Europe, East Germany and West Germany were united. Because of forced migration of Germans (as well as other peoples) after World War II, the territory occupied by German speakers today is much farther west than the location a century ago.

Germany did not emerge as a nation-state until 1871, more recently than its neighbors. Prior to that time, the map of the central European area now called Germany was a patchwork of small states—more than 300 during the seventeenth century, for example. In 1871, Prussia—the most powerful German state—forced most its neighbors to join a Prussian-dominated German Empire. Germany lost much of its territory after World War I (Figure 7-15, upper right). Although the boundaries of states in Southern and Eastern Europe were fixed to conform when possible to those of ethnicities, Germany's new boundaries were arbitrary.

During the 1930s, German National Socialists (Nazis) claimed that all German-speaking parts of Europe constituted one nationality and should be unified into one state. They pursued this goal forcefully, and other European powers did not attempt to stop the Germans from taking over Austria and the German-speaking portion of Czechoslovakia, known as the Sudetenland. Not until the Germans invaded Poland (clearly not a German-speaking country) in 1939 did England and France try to stop them, marking the start of World War II.

After it was defeated in World War II, Germany was divided into two countries (Figure 7-15, lower left). Two Germanys existed from 1949 until 1990. With the end of communism, the German Democratic Republic ceased to exist, and its territory became part of the German Federal Republic (Figure 7-15, lower right).

Nationalism

A nationality, once established, must hold the loyalty of its citizens to survive (Figure 7-16). Politicians and governments try to instill loyalty through **nationalism**, which is loyalty and devotion to a nationality. Nationalism typically promotes a sense of national consciousness that exalts one nation above all others and emphasizes its culture and interests as opposed to those of other nations. People display nationalism by supporting a state that preserves and enhances the culture and attitudes of their nationality.

States foster nationalism by promoting symbols of the nation-state, such as flags and songs. The symbol of the hammer and sickle on a field of red was long synonymous with the beliefs of communism. After the fall of communism, one of the first acts in a number of Eastern European countries was to redesign flags without the hammer and sickle. Legal holidays were changed from dates associated with Communist victories to those associated with historical events that preceded Communist takeovers.

Nationalism can have a negative impact. The sense of unity within a nation-state is sometimes achieved through the creation of negative images of other nation-states. Travelers in southeastern Europe during the 1970s and 1980s found that jokes directed by one nationality against another recurred in the same form throughout the region, with only the name of the target changed. For example, "How many [fill in the name of a nationality] are needed to change a lightbulb?" Such jokes seemed harmless, but in hindsight reflected the intense dislike for other nationalities that led to conflict in the 1990s.

Nationalism is an important example of a **centripetal force**, which is an attitude that tends to unify people and enhance support for a state. (The word *centripetal* means "directed toward the center"; it is the opposite of *centrifugal*, which means "to spread out from the center.") Most nation-states find that the best way to achieve citizen support is to emphasize shared attitudes that unify the people.

Multinational States

A state that contains more than one ethnicity is a **multiethnic state**. In some multiethnic states, ethnicities all contribute cultural features to the formation of a single nationality. The United States has numerous ethnic groups, all of whom consider themselves as belonging to the American nationality.

Other multiethnic states, known as **multinational states**, contain two ethnic groups with traditions of self-determination that agree to coexist peacefully by recognizing each other as distinct nationalities. A multinational state contains two or more nationalities with traditions of self-determination. Relationships among nationalities vary in different multinational states. In some states, one nationality tries to dominate another, especially if one of the nationalities is much more numerous than the other, whereas in other states nationalities coexist peacefully. The people of one nation may be assimilated into the cultural characteristics of another nation, but in other cases, the two nationalities remain culturally distinct.

FIGURE 7-16 Nationalism. Ukrainians celebrate independence day on August 24 by waving flags while walking along Khreshchatyk Street in the capital, Kiev. Ukraine declared its independence from the former Soviet Union on August 24, 1991.

One example of a multinational state is the United Kingdom, which contains four main nationalities—England, Scotland, Wales, and Northern Ireland. The four display some ethnic differences, but the main reason for considering them as distinct nationalities is that each had very different historical experiences.

- *Wales* was conquered by England in 1282 and formally united with England through the Act of Union of 1536. Welsh laws were abolished, and Wales became a local government unit.
- *Scotland* was an independent country for nearly a thousand years, until 1603 when Scotland's King James VI also became King James I of England, thereby uniting the two countries. The Act of Union in 1707 formally merged the two governments, although Scotland was allowed to retain its own systems of education and local laws. England, Wales, and Scotland together comprise Great Britain, and the term British refers to the combined nationality of the three groups.
- *Northern Ireland*, along with the rest of Ireland, was ruled by the British until the 1920s. The 1801 Act of Union created the United Kingdom of Great Britain and Ireland. During the 1920s most of Ireland became a separate country, but the northern portion—with a majority of Protestants—remained under British control. The official name of the country was changed to the United Kingdom of Great Britain and Northern Ireland.

Today, the strongest element of national identity comes from sports. England, Scotland, Wales, and Northern Ireland field their own national soccer teams and compete separately in major international tournaments, such as the World Cup. The most important annual rugby tournament, known as the Six Nations' Championship, includes teams from England, Scotland, and Wales, as well as Ireland, Italy, and France. Given the history of English conquest, the other nationalities often root against England when it is playing teams from other countries.

FIGURE 7-17 Union of Soviet Socialist Republics. The former Soviet Union included 15 republics, named for the country's largest ethnicities. With the breakup of the Soviet Union, the 15 republics became independent states.

Former Soviet Union: The Largest Multinational State

The Soviet Union was an especially prominent example of a multinational state until its collapse in the early 1990s (Figure 7-17). When the Soviet Union existed, its 15 republics were based on the 15 largest ethnicities. Less numerous ethnicities were not given the same level of recognition.

With the breakup of the Soviet Union into 15 independent countries, a number of these less numerous ethnicities are now divided among more than one state. The 15 republics that once constituted the Soviet Union are now independent countries. These 15 newly independent states consist of five groups:

- Three Baltic: Estonia, Latvia, and Lithuania
- Three European: Belarus, Moldova, and Ukraine
- Five Central Asian: Kazakhstan, Kyrgyzstan, Tajikistan, Turkmenistan, and Uzbekistan
- Three Caucasus: Azerbaijan, Armenia, and Georgia
- Russia

Reasonably good examples of nation-states have been carved out of the Baltic, European, and some Central Asian states. On the other hand, peaceful nation-states have not been created in any of the small Caucasus states, and Russia is an especially prominent example of a state with major difficulties in keeping all of its ethnicities contented.

NEW BALTIC NATION-STATES. Estonia, Latvia, and Lithuania are known as the Baltic states for their location on the Baltic Sea. They had been independent countries between the end of World War I in 1918 and 1940, when the former Soviet Union annexed them under an agreement with Nazi Germany. Of the three Baltic states, Lithuania most closely fits the definition of a nation-state because ethnic Lithuanians comprise 85 percent of its population. In Estonia, ethnic Estonians comprise only 69 percent of the population; in Latvia, only 59 percent are ethnic Latvians.

These three small neighboring Baltic countries have clear cultural differences and distinct historical traditions. Most Estonians are Protestant (Lutherans), most Lithuanians are Roman Catholics, and Latvians are predominantly Lutheran with a substantial Roman Catholic minority. Estonians speak a Uralic language related to Finnish, whereas Latvians and Lithuanians speak languages of the Baltic group within the Balto-Slavic branch of the Indo-European language family.

NEW EUROPEAN NATION-STATES. To some extent, the former Soviet republics of Belarus, Moldova, and Ukraine now qualify as nation-states. Belarusians comprise 81 percent of the population of Belarus, Moldovans comprise 78 percent of the population of Moldova, and Ukrainians comprise 78 percent of the population of Ukraine. The ethnic distinctions among Belarusians, Ukrainians, and Russians are somewhat blurred. The three groups speak similar East Slavic languages, and all are predominantly Orthodox Christians (some western Ukrainians are Roman Catholics).

Belarusians and Ukrainians became distinct ethnicities because they were isolated from the main body of Eastern Slavs—the Russians—during the thirteenth and fourteenth centuries. This was the consequence of Mongolian invasions and conquests by Poles and Lithuanians. Russians conquered the Belarusian and Ukrainian homelands in the late 1700s, but after five centuries of exposure to non-Slavic influences, the three Eastern Slavic groups displayed sufficient cultural diversity to consider themselves as three distinct ethnicities.

The situation is different in Moldova. Moldovans are ethnically indistinguishable from Romanians, and Moldova (then called Moldavia) was part of Romania until the Soviet Union seized it in 1940. When Moldova changed from a Soviet republic back to an independent country in 1992, many Moldovans pushed for reunification with Romania, both to reunify the ethnic group and to improve the region's prospects for economic development.

But it was not to be that simple. When Moldova became a Soviet republic in 1940, its eastern boundary was the Dniester River. The Soviet government increased the size of Moldova by about 10 percent, transferring from Ukraine a 3,000-square-kilometer (1,200-square-mile) sliver of land on the east bank of the Dniester. The majority of the inhabitants of this area, known as Trans-Dniestria, are Ukrainian and Russian. They, of course, oppose Moldova's reunification with Romania.

NEW CENTRAL ASIAN STATES. The five states in Central Asia carved out of the former Soviet Union display varying degrees of conformance to the principles of a nation-state. Together the five provide an important reminder that multinational states can be more peaceful than nation-states.

In Turkmenistan and Uzbekistan, the leading ethnic group has an overwhelming majority—85 percent Turkmen and 80 percent Uzbek, respectively. Both ethnic groups are Muslims who speak an Altaic language; they were conquered by Russia in the nineteenth century. Turkmen and Uzbeks are examples of ethnicities split into more than one country—the Turkmen between Turkmenistan and Russia, and Uzbeks among Kyrgyzstan, Tajikistan, and Uzbekistan.

Kyrgyzstan is 69 percent Kyrgyz, 15 percent Uzbek, and 9 percent Russian. The Kyrgyz—also Muslims who speak an Altaic language—resent the Russians for seizing the best farmland when they colonized this mountainous country early in the twentieth century.

In principle, Kazakhstan, twice as large as the other four Central Asian countries combined, is a recipe for ethnic conflict. The country is divided between Kazakhs, who comprise 67 percent of the population, and Russians, at 18 percent. Kazakhs are Muslims who speak an Altaic language similar to Turkish, whereas the Russians are Orthodox Christians who speak an Indo-European language. Tensions exist between the two groups, but Kazakhstan has been peaceful, in part because it has a somewhat less depressed economy than its neighbors.

In contrast, Tajikistan—80 percent Tajik, 15 percent Uzbek, and only 1 percent Russian—would appear to be a stable country, but it suffers from a civil war among the Tajik people, Muslims who speak a language in the Indic group of the Indo-Iranian branch of Indo-European language. The civil war has been between Tajiks, who are former Communists, and an unusual alliance of Muslim fundamentalists and Western-oriented intellectuals. Fifteen percent of the population has been made homeless by the fighting.

Russia: Now the Largest Multinational State

Russia officially recognizes the existence of 39 nationalities, many of which are eager for independence. Russia's ethnicities are clustered in two principal locations (Figure 7-18). Some are located along borders with neighboring states, including Buryats and Tuvinian near Mongolia, and Chechens, Dagestani, Kabardins, and Ossetians near the two former Soviet republics of Azerbaijan and Georgia. Overall, 20 percent of the country's population is non-Russian.

Other ethnicities are clustered in the center of Russia, especially between the Volga River basin and the Ural Mountains. Among the more numerous in this region are Bashkirs, Chuvash, and Tatars, who speak Altaic languages similar to Turkish, and Mordvins and Udmurts, who speak Uralic languages similar to Finnish. Most of these groups were conquered by the Russians in the sixteenth century under the leadership of Ivan IV (Ivan the Terrible).

Independence movements are flourishing because Russia is less willing to suppress these movements forcibly than the Soviet Union had once been. Particularly troublesome for the Russians are the Chechens, a group of Sunni Muslims who speak a Caucasian language and practice distinctive social customs.

Chechnya was brought under Russian control in the nineteenth century only after a 50-year fight. When the Soviet Union broke up into 15 independent states in 1991, the Chechens declared their independence and refused to join the newly created country of Russia. Russian leaders ignored the declaration of independence for 3 years but then sent in the Russian army in an attempt to regain control of the territory. Russia fought hard to prevent Chechnya from gaining independence because it feared that other ethnicities would follow suit. Chechnya was also important to Russia because the region contained deposits of petroleum. Russia viewed political stability in the area as essential for promoting economic development and investment by foreign petroleum companies.

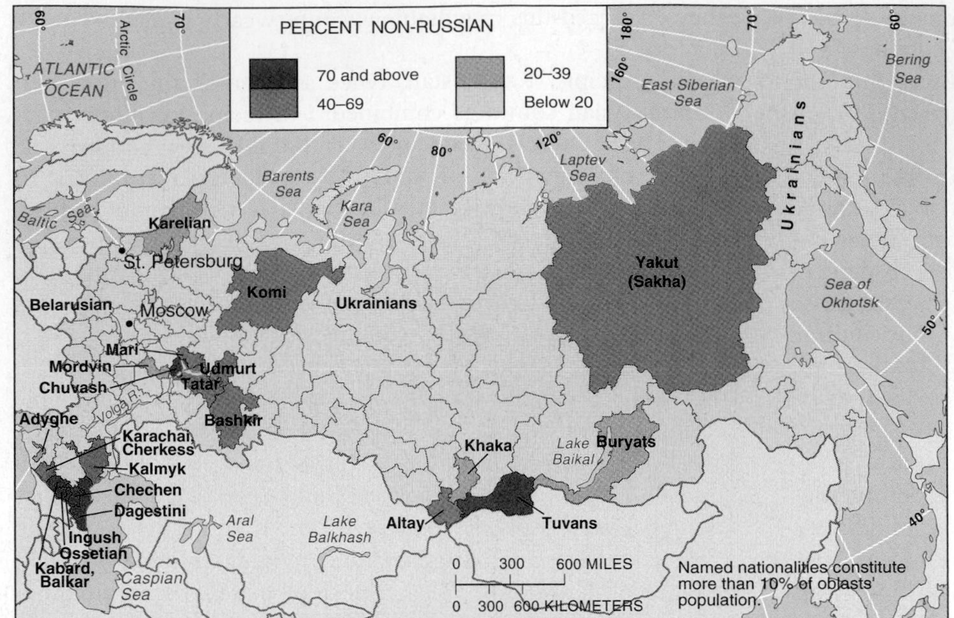

FIGURE 7-18 Ethnicities in Russia. Russians are clustered in the western portion of Russia, and the percentage declines to the south and east. The largest numbers of non-Russians are found in the center of the country between the Volga River and the Ural Mountains and near the southern boundaries.

Turmoil in the Caucasus

The Caucasus region, an area about the size of Colorado, is situated between the Black and Caspian seas and gets its name from the mountains that separate Russia from Azerbaijan and Georgia. The region is home to several ethnicities, with Azeris, Armenians, and Georgians the most numerous (Figure 7-19). Other important ethnicities include Abkhazians, Chechens, Ingush, and Ossetians. Kurds and Russians—two ethnicities that are more numerous in other regions—are also represented in the Caucasus.

When the entire Caucasus region was part of the Soviet Union, the Soviet government promoted allegiance to communism and the Soviet state and quelled disputes among ethnicities, by force if necessary. With the breakup of the region into several independent countries, long-simmering conflicts among ethnicities have erupted into armed conflicts. Each ethnicity has a long-standing and complex set of grievances against others in the region. But from a political geography perspective, every ethnicity in the Caucasus has the same aspiration—to carve out a sovereign nation-state. The region's ethnicities have had varying success in achieving this objective, but none have fully achieved it.

AZERBAIJAN. Azeris (or Azerbaijanis) trace their roots to Turkish invaders who migrated from Central Asia in the eighth and ninth centuries and merged with the existing Persian population. An 1828 treaty allocated northern Azeri territory to Russia and southern Azeri territory to Persia (now Iran). In 1923, the Russian portion became the Azerbaijan Soviet Socialist Republic within the Soviet Union.

With the Soviet Union's breakup in 1991, Azerbaijan again became an independent country. The western part of the country,

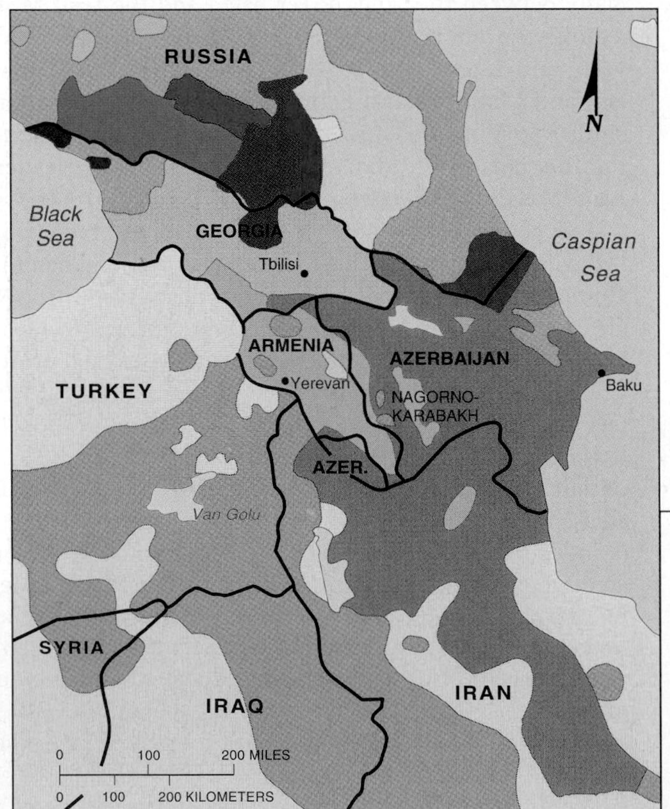

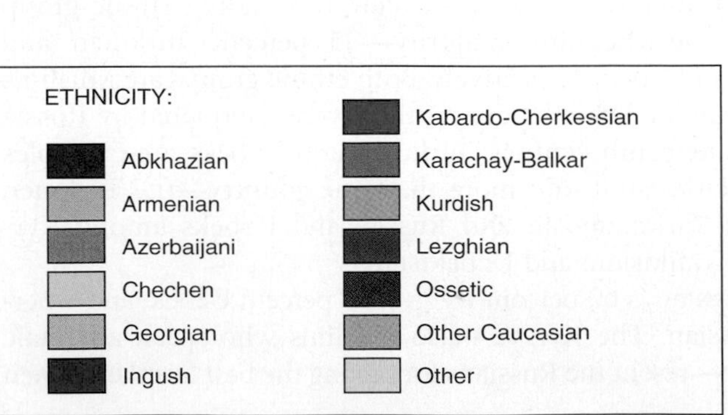

FIGURE 7-19 Ethnicities in the Caucasus. Armenians, Azeris, and Georgians are examples of ethnicities that were able to dominate new states during the 1990s following the breakup of the Soviet Union. But the boundaries of the states of Armenia, Azerbaijan, and Georgia do not match the territories occupied by the Armenian, Azeri, and Georgian ethnicities. The Abkhazians, Chechens, Kurds, and Ossetians are examples of ethnicities in this region that have not been able to organize nation-states.

Nakhichevan (named for the area's largest city), is separated from the rest of Azerbaijan by a 40-kilometer (25-mile) corridor belonging to Armenia.

More than 7 million Azeris now live in Azerbaijan, 91 percent of the country's total population. Another 16 million Azeris are clustered in northwestern Iran, where they constitute 24 percent of that country's population. Azeris hold positions of responsibility in Iran's government and economy, but Iran restricts teaching of the Azeri language.

ARMENIA. More than 3,000 years ago Armenians controlled an independent kingdom in the Caucasus. Converted to Christianity in 303, they lived for many centuries as an isolated Christian enclave under the rule of Turkish Muslims.

During the late nineteenth and early twentieth centuries, hundreds of thousands of Armenians were killed in a series of massacres organized by the Turks. Others were forced to migrate to Russia, which had gained possession of eastern Armenia in 1828.

After World War I the allies created an independent state of Armenia, but it was soon swallowed by its neighbors. In 1921, Turkey and the Soviet Union agreed to divide Armenia between them. The Soviet portion became the Armenian Soviet Socialist Republic and then an independent country in 1991. Armenians comprise 98 percent of the population in Armenia, making it the most ethnically homogeneous country in the region.

Armenians and Azeris both have achieved long-held aspirations of forming nation-states, but after their independence from the Soviet Union the two went to war over the boundaries between them. The war concerned possession of Nagorno-Karabakh, a 5,000-square-kilometer (2,000-square-mile) enclave within Azerbaijan that is inhabited primarily by Armenians but placed under Azerbaijan's control by the Soviet Union during the 1920s. A 1994 cease-fire has left Nagorno-Karabakh technically part of Azerbaijan, but in reality it acts as an independent republic.

FIGURE 7-20 Turmoil in the Caucasus. Georgian people demonstrate against deployment of Russian troops in Abkhazia and South Ossetia during 2008.

GEORGIANS. The population of Georgia is more diverse than that in Armenia and Azerbaijan. Ethnic Georgians comprise 84 percent of the population. The country also includes about 7 percent Azeri, 6 percent Armenian, 2 percent Russian, and 3 percent Abkhazian, Ajar, and Ossetian.

Georgia's cultural diversity has been a source of unrest, especially among the Ossetians and Abkhazians (Figure 7-20). During the 1990s, the Abkhazians fought for control of the northwestern portion of Georgia and have declared Abkhazia to be an independent state. In 2008, the Ossetians fought a war with the Georgians that resulted in the Ossetians declaring the South Ossetia portion of Georgia to be independent.

Russia has recognized Abkhazia and South Ossetia as independent countries and has sent troops there. Only a handful of other countries recognize the independence of Abkazia and South Ossetia, although the two operate as if they were independent of Georgia.

Revival of Ethnic Identity

Europeans thought that ethnicity had been left behind as an insignificant relic, such as wearing quaint costumes to amuse tourists. Karl Marx wrote that nationalism was a means for the dominant social classes to maintain power over workers, and he believed that workers would identify with other working-class people instead of with an ethnicity.

Until they lost power around 1990, Communist leaders in Eastern Europe and the former Soviet Union used centripetal forces to discourage ethnicities from expressing their cultural uniqueness. Writers and artists were pressured to conform to a style known as "socialist realism," which emphasized Communist economic and political values. Use of the Russian language was promoted as a centripetal device throughout the former Soviet Union. It was taught as the second language in other Eastern European countries. The role of organized religion was minimized, suppressing a cultural force that competed with the government.

In the twenty-first century, ethnic identity has once again become more important than nationality, even in much of Europe. In Eastern Europe the breakup of the Soviet Union, Yugoslavia, and Czechoslovakia during the 1990s gave more numerous ethnicities the opportunity to organize nation-states. But the less numerous ethnicities found themselves existing as minorities in multinational states or divided among more than one of the new states. Especially severe problems have occurred in the Balkans, a rugged, mountainous region where nation-states could not be delineated peacefully.

The Soviet Union, Yugoslavia, and Czechoslovakia were dismantled largely because minority ethnicities opposed the long-standing dominance of the most numerous ones in each country—Russians in the Soviet Union, Serbs in Yugoslavia, and Czechs in

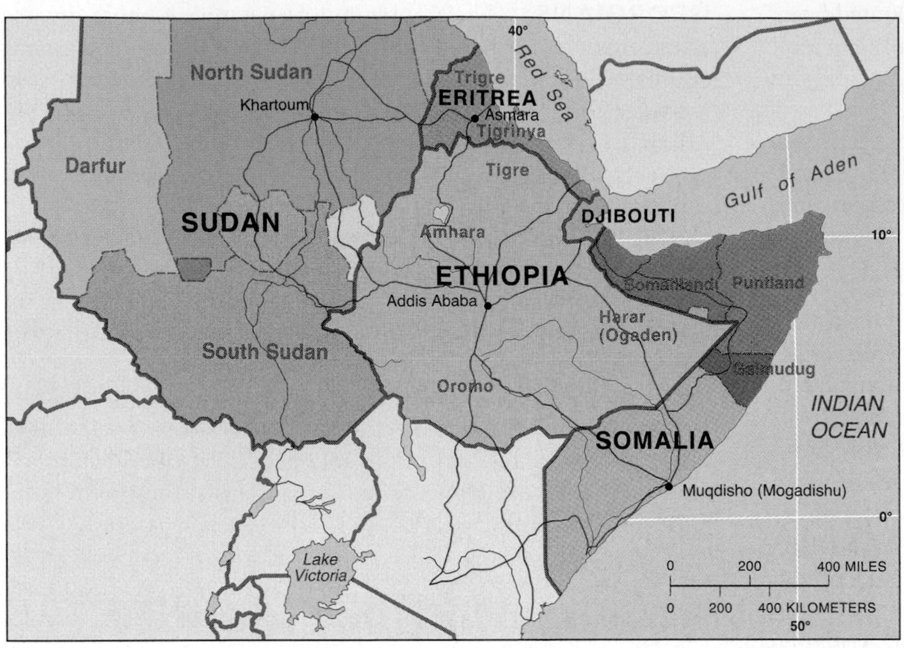

FIGURE 7-21 Ethnic diversity in the Horn of Africa. Conflicts have been widespread within the East African countries in the area known as the Horn of Africa because each contains numerous ethnicities.

Czechoslovakia. The dominance was pervasive, including economic, political, and cultural institutions. No longer content to control a province or some other local government unit, ethnicities sought to be the majority in completely independent nation-states. Republics that once constituted local government units within the Soviet Union, Yugoslavia, and Czechoslovakia generally made peaceful transitions into independent countries—as long as their boundaries corresponded reasonably well with the territory occupied by a clearly defined ethnicity.

Slovenia is a good example of a nation-state that was carved from the former Yugoslavia in the 1990s. Slovenes comprise 83 percent of the population of Slovenia, and nearly all the world's 2 million Slovenes live in Slovenia. The relatively close coincidence between the boundaries of the Slovene ethnic group and the country of Slovenia has promoted the country's relative peace and stability, compared to other former Yugoslavian republics.

KEY ISSUE 3

Why Do Ethnicities Clash?

- **Ethnic Competition to Dominate Nationality**
- **Dividing Ethnicities Among More Than One State**

Ethnicities do not always find ways to live together peacefully. In some cases, ethnicities compete in civil wars to dominate the national identity. In other cases, problems result from division of ethnicities among more than one state. ■

Ethnic Competition to Dominate Nationality

Sub-Saharan Africa has been a region especially plagued by conflicts among ethnic groups competing to become dominant within the various countries. The Horn of Africa and central Africa are the two areas within sub-Saharan Africa where conflicts among ethnic groups have been particularly complex and brutal.

Ethnic Competition in the Horn of Africa

The Horn of Africa encompasses the countries of Djibouti, Ethiopia, Eritrea, and Somalia. Especially severe problems have occurred in Ethiopia, Eritrea, and Somalia, as well as in the neighboring country of Sudan.

ETHIOPIA AND ERITREA. Eritrea, located along the Red Sea, became an Italian colony in 1890. Ethiopia, an independent country for more than 2,000 years, was captured by Italy during the 1930s. After World War II, Ethiopia regained its independence, and the United Nations awarded Eritrea to Ethiopia (Figure 7-21). The United Nations expected Ethiopia to permit Eritrea considerable authority to run its own affairs, but Ethiopia dissolved the Eritrean legislature and banned the use of Tigrinya, Eritrea's major local language. The Eritreans rebelled, beginning a 30-year fight for independence (1961–1991). During this civil war, an estimated 665,000 Eritrean refugees fled to neighboring Sudan.

Eritrean rebels defeated the Ethiopian army in 1991, and 2 years later Eritrea became an independent state. But war between Ethiopia and Eritrea flared up again in 1998 because of disputes over the location of the border. Eritrea justified its claim through a 1900 treaty between Ethiopia and Italy, which then controlled Eritrea, but Ethiopia cited a 1902 treaty with Italy. Ethiopia defeated Eritrea in 2000 and took possession of the disputed areas.

A country of 5 million people split evenly between Christian and Muslim, Eritrea has two principal ethnic groups: Tigrinya and Tigre. At least in the first years of independence, a strong sense of national identity united Eritrea's ethnicities as a result of shared experiences during the 30-year war to break free of Ethiopia.

Even with the loss of Eritrea, Ethiopia remained a complex multiethnic state. From the late nineteenth century until the 1990s, Ethiopia was controlled by the Amharas, who are Christians. After the government defeat in the early 1990s, power passed to a combination of ethnic groups. The Oromo, who are Muslim fundamentalists from the south, are the largest ethnicity in Ethiopia, at 34 percent of the population. The Amhara, who comprise 27 percent of the population, had banned the use of languages other than Amharic, including Oromo.

SUDAN. In Sudan, a country of 41 million, several civil wars have raged since the 1980s between the Arab-Muslim dominated government in the north and other ethnicities in the south, west, and east (Figure 7-22):

- South: Black Christian and animist ethnicities resisted government attempts to convert the country from a multiethnic society to one nationality tied to Muslim traditions. A north-south war between 1983 and 2005 resulted in the death of an estimated 1.9 million Sudanese, mostly civilians. The war ended with the establishment of Southern Sudan as an autonomous region scheduled to have a referendum on independence in 2011. Three bordering districts—Abyei, Nuba Mountains, and Blue Nile—may also become part of Southern Sudan in 2011.
- West: Black Muslim ethnic groups in the Darfur region of western Sudan fought against the government of Sudan beginning in 2003. The United Nations estimates that 400,000 died in Darfur and 2 million became refugees. The United States considers the mass murders and rape of civilians conducted by Sudanese troops to be genocide.
- East: Ethnicities along the Eastern Front fought the government of Sudan between 2004 and 2006 with the support of neighboring Eritrea. At issue was disbursement of profits from oil.

SOMALIA. On the surface, Somalia should face fewer ethnic divisions than its neighbors in the Horn of Africa. Somalis are overwhelmingly Sunni Muslims and speak Somali. Most share a sense that Somalia is a nation-state, with a national history and culture.

FIGURE 7-22 Ethnic cleansing in Darfur, Sudan. Several million blacks in the Darfur region of Sudan have been killed or forced to live in refugee camps as a result of attacks supported by the government of Sudan.

Somalia's 9 million inhabitants are divided among several ethnic groups known as clans, each of which is divided into a large number of subclans. Traditionally, the major clans occupied different portions of Somalia. In 1991, a dictatorship that ran the country collapsed, and various clans and subclans claimed control over portions of the country. Clans have declared independent states of Somaliland in the north, Puntland in the northeast, Galmudug in the center, and Southwestern Somalia in the south

The United States sent several thousand troops to Somalia in 1992, after an estimated 300,000 people, mostly women and children, died from famine and from warfare among clans. The purpose of the mission was to protect delivery of food by international relief organizations to starving Somali refugees and to reduce the number of weapons in the hands of the clan and subclan armies. After peace talks among the clans collapsed in 1994, U.S. troops withdrew.

Islamist militias took control of much of Somalia between 2004 and 2006. Neighboring countries were drawn into the conflict, Eritrea on the side of the Islamists and Ethiopia against them. Claiming that some of the leaders were terrorists, the United States also opposed the Islamists, and launched air strikes in 2007. The fighting generated several hundred thousand refugees. Islamist militias withdrew from most of Somalia in 2006, but have since returned and again control much of the country.

Ethnic Competition in Lebanon

Lebanon has 4 million people in an area of 10,000 square kilometers (4,000 square miles), a bit smaller and more populous than Connecticut. Once known as a financial and recreational center in the Middle East, Lebanon has been severely damaged by fighting among ethnicities since the 1970s.

Lebanon is divided between around 60 percent Muslims and 39 percent Christians (Figure 7-23). The precise distribution of religions in Lebanon is unknown, because no census has been taken since 1932. Lebanon's most numerous Christian sect is Maronite, which split from the Roman Catholic Church in the seventh century. Maronites, ruled by the patriarch of Antioch, perform the liturgy in the ancient Syrian language. The second-largest are Greek Orthodox, the Orthodox church that uses a Byzantine liturgy.

Most of Lebanon's Muslims belong to one of several Shiite sects. Sunnis, who are much more numerous than Shiites in the world, account for a minority of Lebanon's Muslims. Lebanon also has an important community of Druze, who were once considered a separate religion but now consider themselves Muslim. Many Druze rituals are kept secret from outsiders.

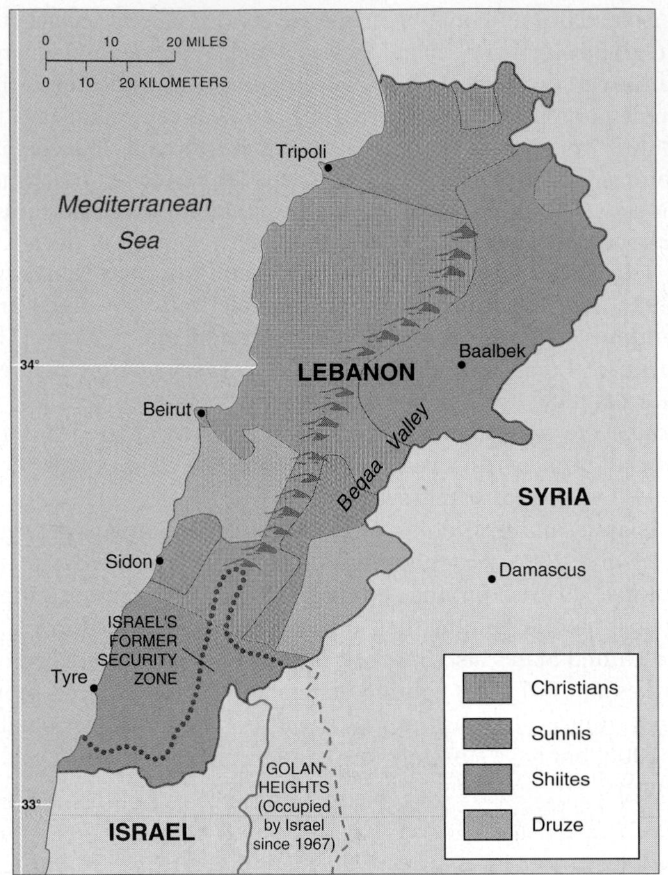

FIGURE 7-23 Ethnicities in Lebanon. Christians dominate in the south and the northwest, Sunni Muslims in the far north, Shiite Muslims in the northeast and south, and Druze in the south-central and southeast.

Lebanon's diversity appears to be religious not ethnic. But most of Lebanon's Christians consider themselves ethnically descended from the ancient Phoenicians who once occupied present-day Lebanon. In this way, Lebanon's Christians differentiate themselves from the country's Muslims, who are considered Arabs.

When Lebanon became independent in 1943, the constitution required that each religion be represented in the Chamber of Deputies according to its percentage in the 1932 census. By unwritten convention, the president of Lebanon was a Maronite Christian, the premier a Sunni Muslim, the speaker of the Chamber of Deputies a Shiite Muslim, and the foreign minister a Greek Orthodox Christian. Other cabinet members and civil servants were similarly apportioned among the various faiths.

Lebanon's religious groups have tended to live in different regions of the country. Maronites are concentrated in the west central part, Sunnis in the northwest, and Shiites in the south and east. Beirut, the capital and largest city, has been divided between an Christian eastern zone and a Muslim western zone. During a civil war between 1975 and 1990, each religious group formed a private army or militia to guard its territory. The territory controlled by each militia changed according to results of battles with other religious groups.

When the governmental system was created, Christians constituted a majority and controlled the country's main businesses, but as the Muslims became the majority, they demanded political and economic equality. The agreement ending the civil war in 1990 gave each religion one-half of the 128 seats in Parliament. Israel and the United States sent troops into Lebanon at various points in failed efforts to restore peace. The United States pulled out after 241 U.S. marines died in their barracks from a truck bomb in 1983. Lebanon was left under the control of neighboring Syria, which had a historical claim over the territory until it, too, was forced to withdraw its troops in 2005.

Dividing Ethnicities Among More Than One State

Newly independent countries are often created to separate two ethnicities. However, two ethnicities can rarely be segregated completely. Conflicts arise when an ethnicity is split among more than one country (see Global Forces, Local Impacts box).

South Asia provides vivid examples of what happens when independence comes to colonies that contain two major ethnicities. Several major ethnic conflicts have ensued in the region.

India and Pakistan

When the British ended their colonial rule of the Indian subcontinent in 1947, they divided the colony into two irregularly shaped countries—India and Pakistan (Figure 7-24). Pakistan

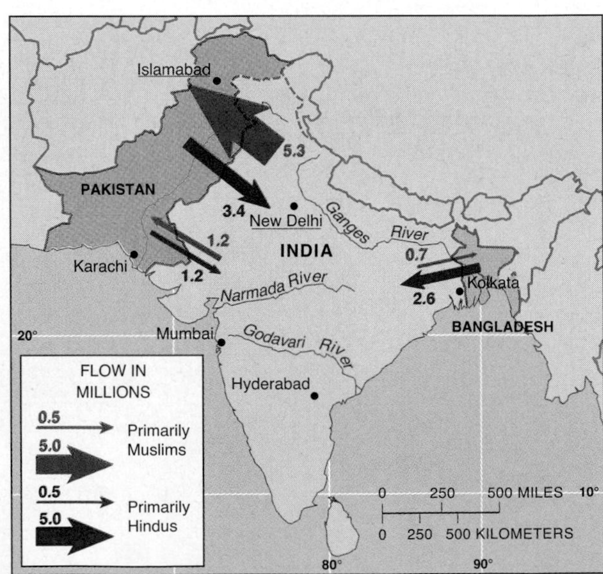

FIGURE 7-24 Ethnic division of South Asia. In 1947, British India was partitioned into two independent states, India and Pakistan, which resulted in the migration of an estimated 17 million people. The creation of Pakistan as two territories nearly 1,600 kilometers (1,000 miles) apart proved unstable, and in 1971 East Pakistan became the independent country of Bangladesh.

GLOBAL FORCES, LOCAL IMPACTS
Dividing the Kurds

An example of an ethnicity divided among several states is the Kurds, who live in the Caucasus south of the Armenians and Azeris (Figure 7-25). The Kurds are Sunni Muslims who speak a language in the Iranian group of the Indo-Iranian branch of Indo-European and have distinctive literature, dress, and other cultural traditions.

Kurds lived in an independent nation-state called Kurdistan during the 1920s, but today 30 million Kurds are split among several countries. Fifteen million live in eastern Turkey, 6 million in northern Iraq, 5 million in western Iran, 2 million in Syria, and the rest in other countries (refer to Figure 7-19). Kurds comprise 20 percent of the population in Turkey, 15-20 percent in Iraq, 8 percent in Syria, and 7 percent in Iran.

When the victorious European allies carved up the Ottoman Empire after World War I, they created an independent state of Kurdistan to the south and west of Van Gölü (Lake Van) under the 1920 Treaty of Sèvres. Before the treaty was ratified, however, the Turks, under the leadership of Mustafa Kemal (later known as Kemal Ataturk), fought successfully to expand the territory under their control beyond the small area the allies had allocated to them. The Treaty of Lausanne in 1923 established the modern state of Turkey, with boundaries nearly identical to the current ones. Kurdistan became part of Turkey and disappeared as an independent state.

To foster the development of Turkish nationalism, the Turks have tried repeatedly to suppress Kurdish culture. Use of the Kurdish language was illegal in Turkey until 1991, and laws banning its use in broadcasts and classrooms remain in force. Kurdish nationalists, for their part, have waged a guerrilla war since 1984 against the Turkish army. Kurds in other countries have fared just as poorly as those in Turkey. Iran's Kurds secured an independent republic in 1946, but it lasted less than a year. Iraq's Kurds have made several unsuccessful attempts to gain independence, including in the 1930s, 1940s, and 1970s.

A few days after Iraq was defeated in the 1991 Gulf War, the country's Kurds launched another unsuccessful rebellion. The United States and its allies decided not to resume their recently concluded fight against Iraq on behalf of the Kurdish rebels, but after the revolt was crushed, they sent troops to protect the Kurds from further attacks by the Iraqi army. After the United States attacked Iraq and deposed Saddam Hussein in 2003, Iraqi Kurds achieved even more autonomy, but still not independence. Thus, despite their numbers, the Kurds are an ethnicity with no corresponding Kurdish state today. Instead, they are forced to live under the control of the region's more powerful nationalities. ■

FIGURE 7-25 Kurdish refugees, escaping from attacks by Saddam Hussein's army during the 1991 war in Iraq, head for the Turkish border on foot.

comprised two noncontiguous areas, West Pakistan and East Pakistan—1,600 kilometers (1,000 miles) apart, separated by India. East Pakistan became the independent state of Bangladesh in 1971. An eastern region of India was also practically cut off from the rest of the country, attached only by a narrow corridor north of Bangladesh that is less than 13 kilometers (8 miles) wide in some places.

The basis for separating West and East Pakistan from India was ethnicity. The people living in the two areas of Pakistan were predominantly Muslim; those in India were predominantly Hindu. Antagonism between the two religious groups was so great that the British decided to place the Hindus and Muslims in separate states. Hinduism has become a great source of national unity in India. In modern India, with its hundreds of languages and ethnic groups, Hinduism has become the cultural trait shared by the largest percentage of the population.

Muslims have long fought with Hindus for control of territory, especially in South Asia. After the British took over India in the early 1800s, a three-way struggle began, with the Hindus and Muslims fighting each other as well as the British rulers. Mahatma Gandhi, the leading Hindu advocate of nonviolence and reconciliation with Muslims, was assassinated in 1948, ending the possibility of creating a single state in which Muslims and Hindus could live together peacefully.

The partition of South Asia into two states resulted in massive migration because the two boundaries did not correspond precisely to the territory inhabited by the two ethnicities. Approximately 17 million people caught on the wrong side of a boundary felt compelled to migrate during the late 1940s. Some 6 million Muslims moved from India to West Pakistan and about 1 million from India to East Pakistan. Hindus who migrated to India included approximately 6 million from West Pakistan and 3.5 million from East Pakistan. As they attempted to reach the other side of the new border, Hindus in Pakistan and Muslims in India were killed by people from the rival religion. Extremists attacked small groups of refugees traveling by road and halted trains to massacre the passengers.

Pakistan and India never agreed on the location of the boundary separating the two countries in the northern region of Kashmir (Figure 7-26). Since 1972, the two countries have maintained a "line of control" through the region, with Pakistan administering the northwestern portion and India the southeastern portion. Muslims, who comprise a majority in both portions, have fought a guerrilla war to secure reunification of Kashmir, either as part of Pakistan or as an independent country. India blames Pakistan for the unrest and vows to retain its portion of Kashmir. Pakistan argues that Kashmiris on both sides of the border should choose their own future in a vote, confident that the majority Muslim population would break away from India.

India's religious unrest is further complicated by the presence of 25 million Sikhs, who have long resented that they were not given their own independent country when India was partitioned (see Chapter 6). Although they constitute only 2 percent of India's total population, Sikhs comprise a majority in the Indian state of Punjab, situated south of Kashmir along the border with Pakistan. Sikh extremists have fought for more control over the Punjab or even complete independence from India.

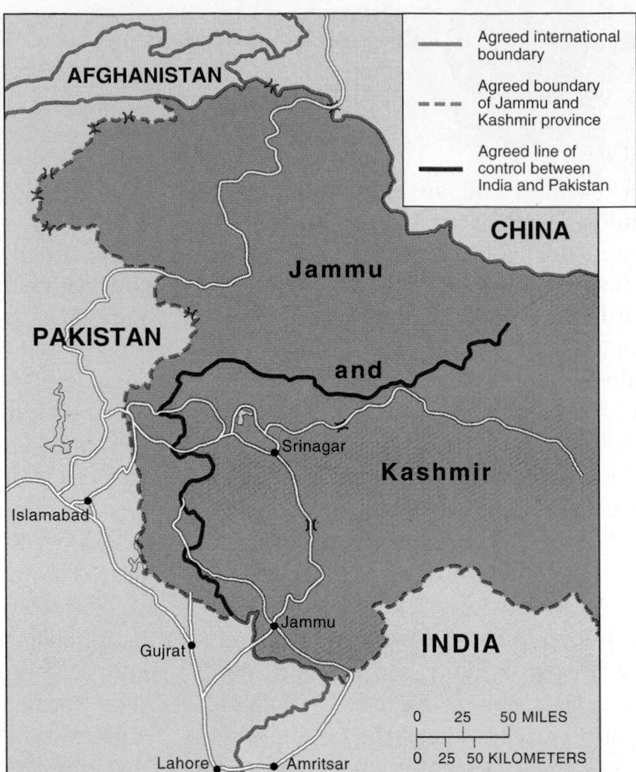

FIGURE 7-26 Kashmir, India, and Pakistan dispute the location of their border. India claims Kashmir, in northernmost Pakistan, and India accuses Pakistan of encouraging unrest in India's state of Jammu and Kashmir, where the majority is Muslim.

Sinhalese and Tamil in Sri Lanka

Sri Lanka, an island country of 20 million inhabitants off the Indian coast, is inhabited by two principal ethnicities known as Sinhalese and Tamil (Figure 7-27). War between the two ethnicities erupted in 1983 and continued until 2009. During that period, 80,000 died in the conflict between the Sinhalese and Tamil.

Sinhalese, who comprise 82 percent of Sri Lanka's population, migrated from northern India in the fifth century B.C., occupying the southern portion of the island. Three hundred years later, the Sinhalese were converted to Buddhism, and Sri Lanka became one of that religion's world centers. Sinhalese is an Indo-European language, in the Indo-Iranian branch.

Tamils—14 percent of Sri Lanka's population—migrated across the narrow 80-kilometer-wide (50-mile-wide) Palk Strait from India beginning in the third century B.C. and occupied the northern part of the island. Tamils are Hindus, and the Tamil language, in the Dravidian family, is also spoken by 60 million people in India.

The dispute between Sri Lanka's two ethnicities extends back more than 2,000 years but was suppressed during 300 years of European control. Since independence in 1948, Sinhalese have dominated the government, military, and most of the commerce. Tamils feel that they suffer from discrimination at the hands of

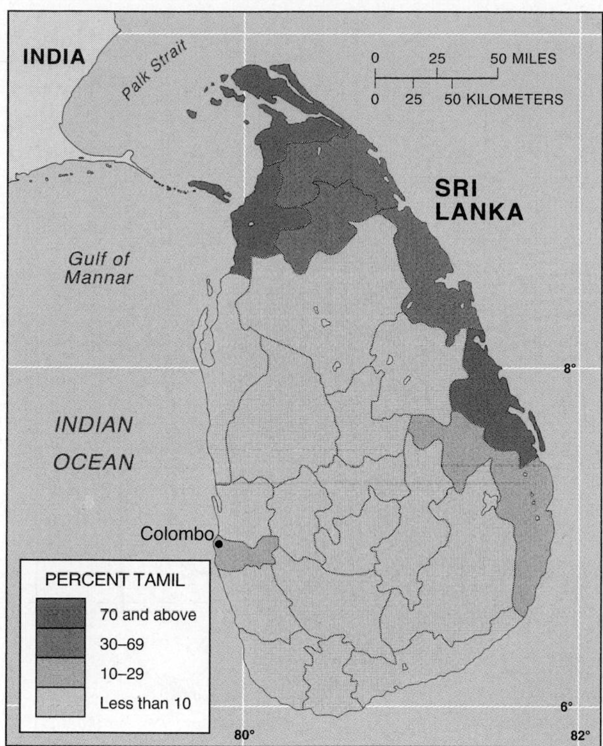

FIGURE 7-27 Ethnicities in Sri Lanka 1976. The Sinhalese are Buddhists who speak an Indo-European language, whereas the Tamils are Hindus who speak a Dravidian language.

the Sinhalese-dominated government and have received support for a rebellion that began in 1983 from Tamils living in other countries.

The long war between the ethnicities ended in 2009 with the defeat of the Tamil. With their defeat, the Tamil fear that the future of Sri Lanka as a multinational state is jeopardized. Back in 1956, Sinhalese leaders made Buddhism the sole official religion and Sinhala the sole official language of Sri Lanka. The Tamil fear that their military defeat jeopardizes their ethnic identity again.

KEY ISSUE 4
What Is Ethnic Cleansing?

- **Ethnic Cleansing in Europe**
- **Ethnic Cleansing in Central Africa**

Throughout history, ethnic groups have been forced to flee from other ethnic groups' more powerful armies. **Ethnic cleansing** is a process in which a more powerful ethnic group forcibly removes a less powerful one in order to create an ethnically homogeneous region. In recent years, ethnic cleansing has been carried out primarily in Europe and Africa.

Ethnic cleansing is undertaken to rid an area of an entire ethnicity so that the surviving ethnic group can be the sole inhabitants. The point of ethnic cleansing is not simply to defeat an enemy or to subjugate them, as was the case in traditional wars. Rather than a clash between armies of male soldiers, ethnic cleansing involves the removal of every member of the less powerful ethnicity—women as well as men, children as well as adults, the frail elderly as well as the strong youth. ■

Ethnic Cleansing in Europe

The largest forced migration came during World War II (1939–1945) because of events leading up to the war, the war itself, and postwar adjustments (Figure 7-28). Especially notorious was the deportation by the German Nazis of millions of Jews, gypsies, and other ethnic groups to the infamous concentration camps, where they exterminated most of them.

After World War II ended, millions of ethnic Germans, Poles, Russians, and other groups were forced to migrate as a result of boundary changes. For example, when a portion of eastern Germany became part of Poland, the Germans living in the region were forced to move west to Germany and Poles were allowed to move into the area. Similarly, Poles were forced to move when the eastern portion of Poland was turned over to the Soviet Union.

The scale of forced migration during World War II has not been repeated, but in recent years ethnic cleansing within Europe has occurred in portions of former Yugoslavia, especially Bosnia and Herzegovina and Kosovo. Ethnic cleansing in the former Yugoslavia is part of a complex pattern of ethnic diversity in the region of southeastern Europe known as the Balkan Peninsula. The region, about the size of Texas, is named for the Balkan Mountains (known in Slavic languages as Stara Planina), which extend east–west across the region. The Balkans includes Albania, Bulgaria, Greece, and Romania, as well as several countries that once comprised Yugoslavia.

Creation of Multiethnic Yugoslavia

The Balkan Peninsula, a complex assemblage of ethnicities, has long been a hotbed of unrest (Figure 7-29). Northern portions were incorporated into the Austro-Hungary Empire; southern portions were ruled by the Ottomans. Austria-Hungary extended its rule farther south in 1878 to include Bosnia and Herzegovina, where the majority of the people had been converted to Islam by the Ottomans.

In June 1914, the heir to the throne of Austria-Hungary was assassinated in Sarajevo by a Serb who sought independence for Bosnia. The incident sparked World War I. After World War I, the allies created a new country, Yugoslavia, to unite several Balkan ethnicities that spoke similar South Slavic languages (Figure 7-30). The prefix "Yugo" in the country's name derives from the Slavic word for "south."

Under the long leadership of Josip Broz Tito, who governed Yugoslavia from 1953 until his death in 1980, Yugoslavs liked to repeat a refrain that roughly translates as follows: "Yugoslavia

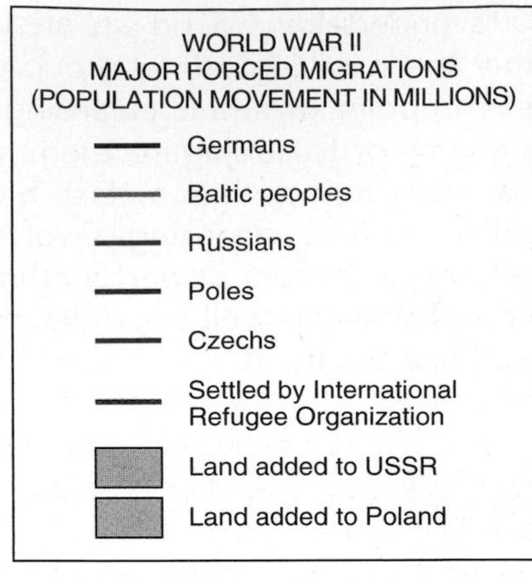

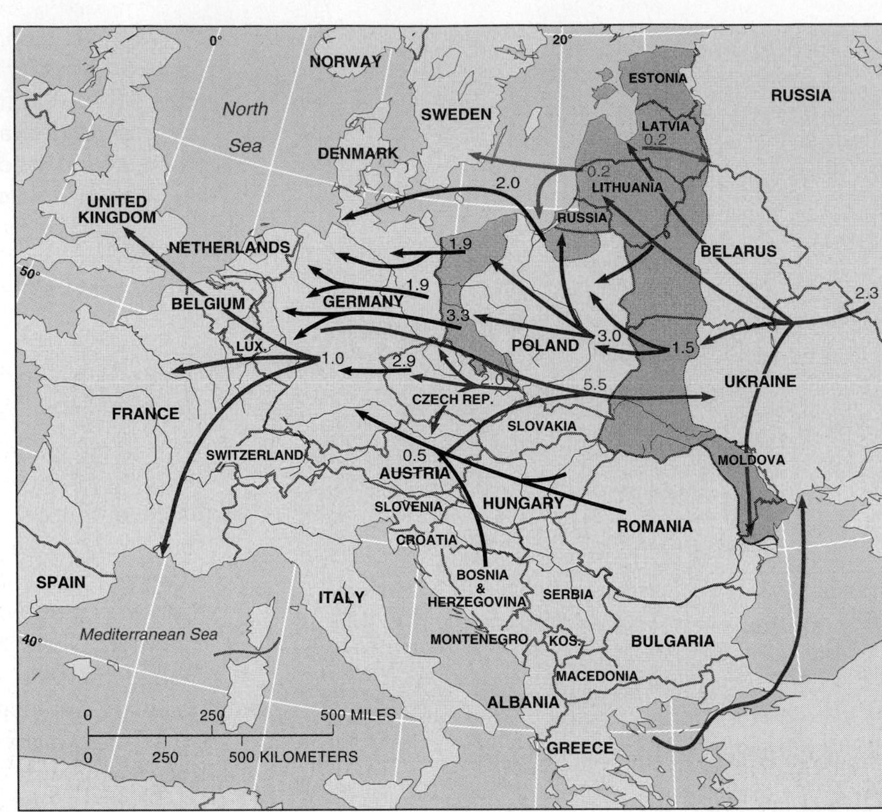

FIGURE 7-28 Forced migration of ethnicities as a result of territorial changes after World War II. The largest number were Poles forced to move from territory occupied by the Soviet Union, Germans forced to migrate from territory taken over by Poland and the Soviet Union, and Russians forced to return to the Soviet Union from Western Europe

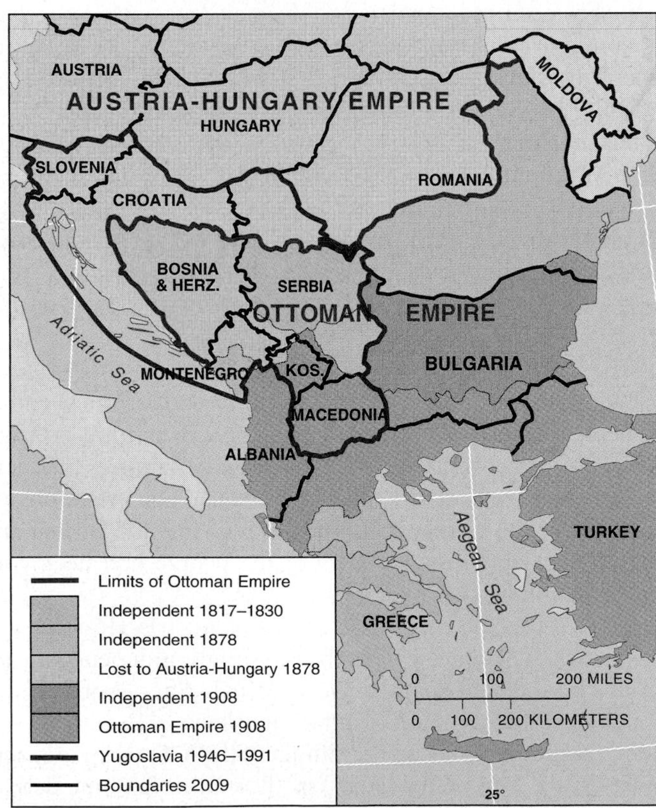

FIGURE 7-29 The Balkans in 1914. At the outbreak of World War I, Austria-Hungary controlled the northern part of the region, including all or part of Croatia, Slovenia, and Romania. The Ottoman Empire controlled some of the south, although during the nineteenth century it had lost control of Albania, Bosnia & Herzegovina, Greece, Romania, and Serbia.

has seven neighbors, six republics, five nationalities, four languages, three religions, two alphabets, and one dinar" (Figure 7-31). Specifically:

- ***Seven*** neighbors of Yugoslavia included three longtime democracies (Austria, Greece, and Italy) and four states then governed by Communists (Albania, Bulgaria, Hungary, and Romania). The diversity of neighbors reflected Yugoslavia's strategic location between the Western democracies and Communist Eastern Europe. Although a socialist country, Yugoslavia was militarily neutral after it had been expelled in 1948 from the Soviet-dominated military alliance for being too independent-minded. Yugoslavia's Communists permitted more communication and interaction with Western democracies than did other Eastern European countries.
- ***Six*** republics within Yugoslavia—Bosnia & Herzegovina, Croatia, Macedonia, Montenegro, Serbia, and Slovenia—had more autonomy from the national government to run their own affairs than was the case in other Eastern European countries.
- ***Five*** of the republics were named for the country's five recognized nationalities—Croats, Macedonians, Montenegrens, Serbs, and Slovenes. Bosnia & Herzegovina contained a mix of Serbs, Croats, and Muslims.
- ***Four*** official languages were recognized—Croatian, Macedonian, Serbian, and Slovene. (Montenegrens spoke Serbian.)
- ***Three*** major religions included Roman Catholic in the north, Orthodox in the east, and Islam in the south. Croats and Slovenes were predominantly Roman Catholic, Serbs

FIGURE 7-30 Languages in Southern and Eastern Europe. After World War I, world leaders created several new states and realigned the boundaries of existing ones so that the boundaries of states matched language boundaries as closely as possible. These state boundaries proved to be relatively stable for much of the twentieth century. In the late twentieth century, the region became a center of conflict among speakers of different languages.

and Macedonians predominantly Orthodox, and the Bosnians and Montenegrens predominantly Muslim.

- *Two* of the four official languages—Croatian and Slovene—were written in the Roman alphabet; Macedonian and Serbian were written in Cyrillic. Most linguists outside Yugoslavia considered Serbian and Croatian to be the same language except for different alphabets.
- *One*, the refrain concluded, was the dinar, the national unit of currency. This meant that despite cultural diversity, common economic interests kept Yugoslavia's nationalities unified.

The creation of Yugoslavia brought stability that lasted for most of the twentieth century. Old animosities among ethnic groups were submerged, and younger people began to identify themselves as Yugoslavs rather than as Serbs, Croats, or Montenegrens.

Destruction of Multiethnic Yugoslavia

Rivalries among ethnicities resurfaced in Yugoslavia during the 1980s after Tito's death, leading to the breakup of the country. Breaking away to form independent countries were Bosnia & Herzegovina, Croatia, Macedonia, and Slovenia during the 1990s, and Montenegro in 2006. The breakup left Serbia standing on its own as well.

As long as Yugoslavia comprised one country, ethnic groups were not especially troubled by the division of the country into six republics. But when Yugoslavia's republics were transformed from local government units into five separate countries, ethnicities fought to redefine the boundaries (Figure 7-31). Not only did the boundaries of Yugoslavia's six republics fail to match the territory occupied by the five major nationalities, but the country contained other important ethnic groups that had not received official recognition as nationalities.

ETHNIC CLEANSING IN BOSNIA. The creation of a viable country proved especially difficult in the case of Bosnia and Herzegovina. The population of Bosnia & Herzegovina consisted of 48 percent Bosnian Muslim, 37 percent Serb, and 14 percent Croat. Bosnian Muslim was considered an ethnicity rather than a nationality. Rather than live in an independent multiethnic country with a Muslim plurality, Bosnia & Herzegovina's Serbs and Croats fought to unite the portions of the republic that they inhabited with Serbia and Croatia, respectively.

To strengthen their cases for breaking away from Bosnia & Herzegovina, Serbs and Croats engaged in ethnic cleansing of Bosnian Muslims (Figure 7-32). Ethnic cleansing ensured that areas did not merely have majorities of Bosnian Serbs and Bosnian Croats, but were ethnically homogeneous and therefore better candidates for union with Serbia and Croatia. Ethnic cleansing by Bosnian Serbs against Bosnian Muslims was especially severe because much of the territory inhabited by Bosnian Serbs was separated from Serbia by areas with Bosnian Muslim majorities. By ethnically cleansing Bosnian Muslims from intervening areas, Bosnian Serbs created one continuous area of Bosnian Serb domination rather than several discontinuous ones.

Accords reached in Dayton, Ohio, in 1996 by leaders of the various ethnicities divided Bosnia & Herzegovina into three regions, one each dominated, respectively, by the Bosnian Croats, Muslims, and Serbs. The Bosnian Croat and Muslim regions were combined into a federation, with some cooperation between the two groups, but the Serb region has operated with almost complete independence in all but name from the

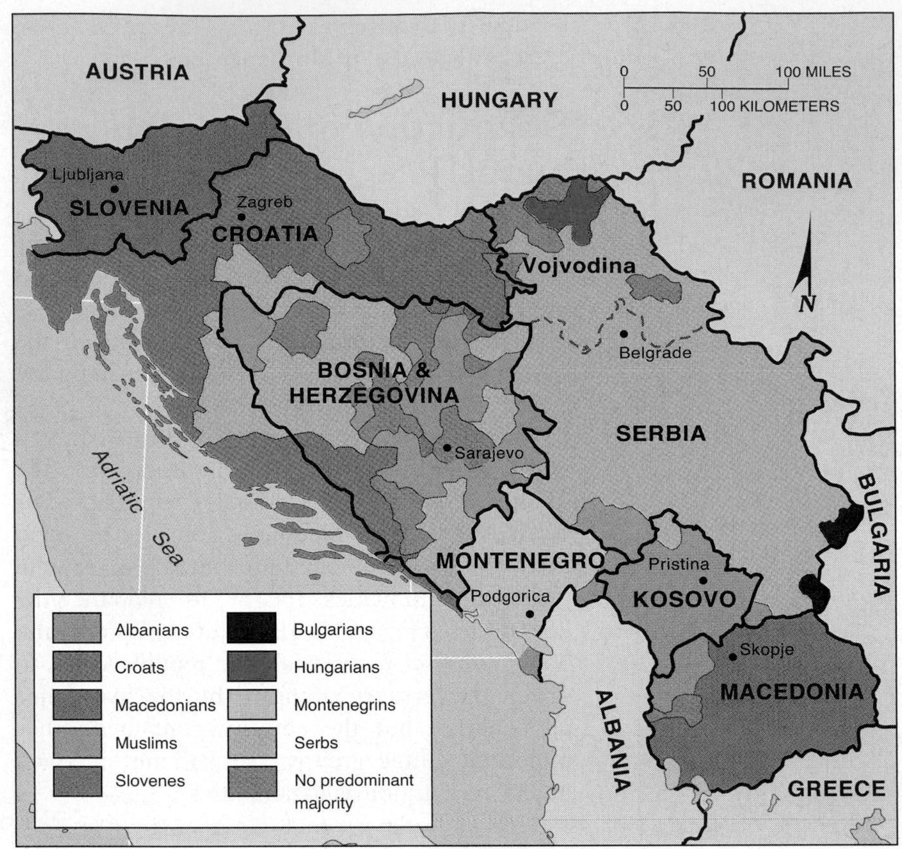

FIGURE 7-31 Yugoslavia, until its breakup in 1992. Yugoslavia comprised six republics (plus Kosovo and Vojvodina, autonomous regions within the Republic of Serbia).

others. In recognition of the success of their ethnic cleansing, Bosnian Serbs received nearly half of the country, although they comprised one-third of the population, and Bosnian Croats got one-fourth of the land, although they comprised one-sixth of the population. Bosnian Muslims, one-half of the population before the ethnic cleansing, got one-fourth of the land.

ETHNIC CLEANSING IN KOSOVO. After the breakup of Yugoslavia, Serbia remained a multiethnic country. Particularly troubling was the province of Kosovo, where ethnic Albanians comprised 90 percent of the population. Under Tito, ethnic Albanians in Kosovo received administrative autonomy and national identity.

Serbia had a historical claim to Kosovo, having controlled it between the twelfth and fourteenth centuries. Serbs fought an important—though losing—battle in Kosovo against the Ottomon Empire in 1389. In recognition of its role in forming the Serb ethnicity, Serbia was given control of Kosovo when Yugoslavia was created in the early twentieth century.

With the breakup of Yugoslavia, Serbia took direct control of Kosovo and launched a campaign of ethnic cleansing of the Albanian majority (see Figure 7-33 in Contemporary Geographic Tools box). At its peak in 1999, Serb ethnic cleansing had forced 750,000 of Kosovo's 2 million ethnic Albanian residents from their homes, mostly to camps in Albania. Outraged by the ethnic cleansing, the United States and Western European countries, operating through the North Atlantic Treaty Organization (NATO), launched an air attack against Serbia. The bombing campaign ended when Serbia agreed to withdraw all of its soldiers and police from Kosovo. Kosovo declared its independence from Serbia in 2008. Around 60 countries, including the United States, recognize Kosovo as an independent country, but Serbia and Russia oppose it.

BALKANIZATION. A century ago, the term **Balkanized** was widely used to describe a small geographic area that could not successfully be organized into one or more stable states because it was inhabited by many ethnicities with complex, long-standing antagonisms toward each other. World leaders at the time regarded **Balkanization**—the process by which a state breaks down through conflicts among its ethnicities—as a threat to peace throughout the world, not just in a small area. They were right: Balkanization led directly to World War I, because the various nationalities in the Balkans dragged into the war the larger powers with which they had alliances.

After two world wars and the rise and fall of communism during the twentieth century, the Balkans have once again become Balkanized in the twenty-first century. Will the United States, Western Europe, and Russia once again be drawn reluctantly into conflict through entangled alliances in the Balkans? If peace comes to the Balkans, it will be because in a tragic way ethnic cleansing "worked." Millions of people were rounded up and killed or forced to migrate because they constituted ethnic minorities. Ethnic homogeneity may be the price of peace in areas that once were multiethnic.

Ethnic Cleansing in Central Africa

Ethnic conflict is widespread in Africa largely because the present-day boundaries of states do not match the boundaries of ethnic groups (Figure 7-34). During the late nineteenth and early twentieth centuries, European countries carved up the continent into a collection of colonies with little regard for the distribution of ethnicities.

Traditionally, the most important unit of African society was the tribe rather than independent states with political and economic self-determination. Africa contains several thousand ethnicities (usually referred to as tribes) with a common sense of language, religion, and social customs (refer to Figure 5-19 for a map of African languages). The precise number of tribes is impossible to determine, because boundaries separating them are not usually defined clearly. Further, it is hard to determine

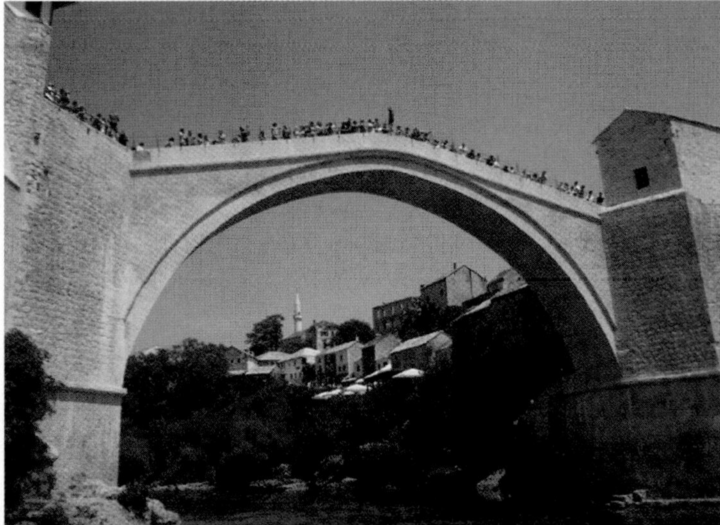

FIGURE 7-32 Ethnic cleansing in Bosnia & Herzegovina. The Stari Most (old bridge), built by the Turks in 1566 across the Neretva River, was an important symbol and tourist attraction in the city of Mostar (top). The bridge was blown up by Serbs in 1993 (middle), in an attempt to demoralize Bosnian Muslims as part of ethnic cleansing. With the end of the war in Bosnia & Herzegovina, the bridge was rebuilt in 2004 (bottom).

whether a particular group forms a distinct tribe or is part of a larger collection of similar groups.

When the European colonies in Africa became independent states, especially during the 1950s and 1960s, the boundaries of the new states typically matched the colonial administrative units imposed by the Europeans. As a result, some tribes were divided among more than one modern state, and others were grouped with dissimilar tribes.

Long-standing conflicts between two ethnic groups, the Hutus and Tutsis, lie at the heart of a series of wars in central Africa. The Hutus were settled farmers, growing crops in the fertile hills and valleys of present-day Rwanda and Burundi, known as the Great Lakes region of central Africa. The Tutsi were cattle herders who migrated to present-day Rwanda and Burundi from the Rift Valley of western Kenya beginning 400 years ago. Relations between settled farmers and herders are often uneasy—this is also an element of the ethnic cleansing in Darfur described earlier in the chapter. The Tutsi took control of the kingdom of Rwanda and turned the Hutu into their serfs, although Tutsi comprised only about 15 percent of the population.

Rwanda, as well as Burundi, became a colony of Germany in 1899, and after the Germans were defeated in World War I, the League of Nations gave a mandate over the two small colonies to Belgium. Colonial administrators permitted a few Tutsis to attend university and hold responsible government positions, while excluding the Hutu altogether.

Shortly before Rwanda gained its independence in 1962, Hutus killed or ethnically cleansed most of the Tutsis out of fear that the Tutsis would seize control of the newly independent country. Those fears were realized in 1994 after the airplane carrying the presidents of Rwanda and Burundi back from peace talks was shot down, probably by a Tutsi. Descendents of the ethnically cleansed Tutsis, most of whom lived in neighboring Uganda, poured back into Rwanda, defeated the Hutu army, and killed a half-million Hutus, while suffering a half-million casualties of their own. Through ethnic cleansing, 3 million of the country's 7 million Hutus fled to Zaire, Tanzania, Uganda, and Burundi.

The conflict between Hutus and Tutsis spilled into neighboring countries, especially the Democratic Republic of Congo. The region's largest and most populous country, the Congo is thought to have had the world's deadliest war since the end of World War II in 1945. An estimated 5.4 million have died in Congo civil wars as of 2009.

Tutsis were instrumental in the successful overthrow of the Congo's longtime president, Joseph Mobutu, in 1997. Mobutu had amassed a several-billion-dollar personal fortune from the sale of minerals while impoverishing the rest of the country. After succeeding Mobutu as president, Laurent Kabila relied heavily on Tutsis and permitted them to kill some of the Hutus who had been responsible for atrocities against Tutsis back in the early 1990s. But Kabila soon split with the Tutsis, and the Tutsis once again found themselves offering support to rebels seeking to overthrow Congo's government.

Kabila turned for support to Hutus, as well as to Mayi Mayi, another ethnic group in the Congo that also hated Tutsis. Armies from Angola, Namibia, Zimbabwe, and other neighboring countries came to Kabila's aid. Kabila was assassinated in 2001 and succeeded by his son, who negotiated an accord with rebels the following year.

CONTEMPORARY GEOGRAPHIC TOOLS

Documenting Ethnic Cleansing

Early reports of ethnic cleansing by Serbs in the former Yugoslavia were so shocking that many people dismissed them as journalistic exaggeration or partisan propaganda. It took one of geography's most important analytic tools, aerial-photography interpretation, to provide irrefutable evidence of the process, as well as the magnitude, of ethnic cleansing.

The process of ethnic cleansing involved four steps. A series of three photographs taken by NATO air reconnaissance over the village of Glodane, in western Kosovo, illustrated the four steps. The first step was to move a large amount of military equipment and personnel into a village with no strategic value.

Figure 7-33 shows the village's houses and farm buildings clustered on the left side, with fields on the outskirts of the village, including the center and right portions of the photograph. As discussed in Chapter 12, rural settlements in most of the world have houses and farm buildings clustered together and surrounded by fields rather than in isolated, individual farms typical of North America. The red circles in Figure 7-33 show the location of Serb armored vehicles along the main street of the village.

The second step in ethnic cleansing was to round up all the people in the village. In Bosnia, Serbs often segregated men from women, children, and old people. The men were placed in detention camps or "disappeared"—undoubtedly killed—and the others were forced to leave the village. In Kosovo, men were herded together with the others rather than killed.

In the photograph of Glodane, the farm field immediately to the east of the main north–south road is filled with the villagers. At the scale that the photograph is reproduced in this book, the people appear as a dark mass. The white rectangles to the north of the people are civilian cars and trucks.

The third step in ethnic cleansing was to force the people to leave the village. This step appeared dramatically in the second photograph of the sequence, depicting the same location a short time later. The second photograph showed one major change: The people and vehicles massed in the field in the first photograph were gone—no people and no vehicles. The villagers were forced into a convoy—some in the vehicles, others on foot—heading for the Albanian border 16 kilometers (10 miles) to the west.

The fourth step in ethnic cleansing was to destroy the vacated village. The third photograph of the sequence showed that the buildings in the village had been set on fire.

Aerial photographs such as these not only "proved" that ethnic cleansing was occurring but also provided critical evidence to prosecute Serb leaders for war crimes. ■

FIGURE 7-33 Evidence of ethnic cleansing. Ethnic cleansing by Serbs forced Albanians living in Kosovo to flee in 1999. The village of Glodane is on the west (left) side of the road. The villagers and their vehicles have been rounded up and placed in the field east of the road. The red circles show the locations of Serb armored vehicles.

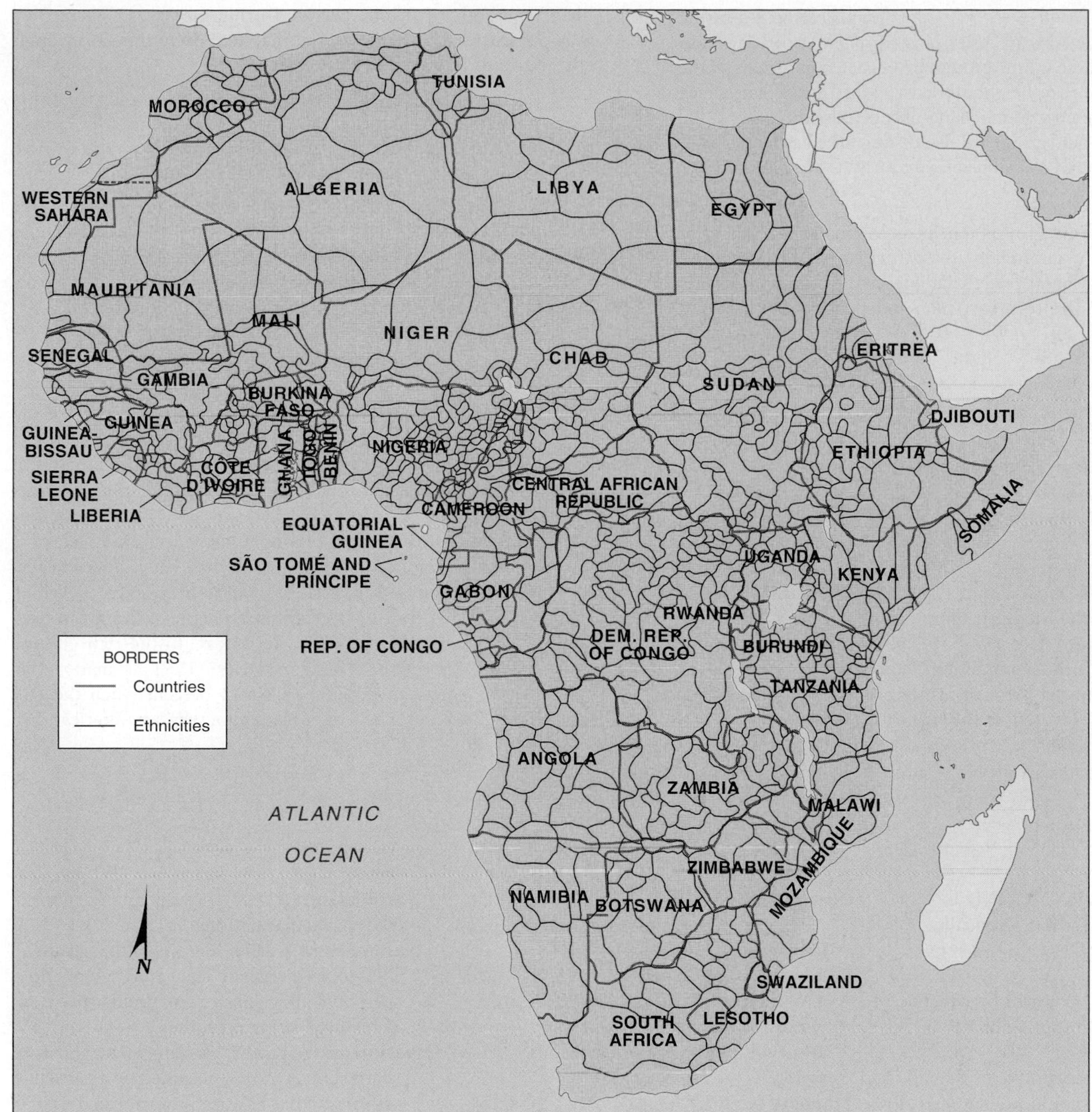

FIGURE 7-34 Ethnicities in Africa. The boundaries of modern African states do not match the territories long occupied by thousands of ethnic groups. State boundaries derive from the administrative units imposed by European colonial powers a century ago.

SUMMARY

Line up five Hutus and five Tutsis, and the ethnic origin of perhaps half of them would be plain. The two ethnicities speak the same language, hold similar beliefs, and practice similar social customs, and intermarriage has lessened the physical differences between the two. Yet Hutus and Tutsis have engaged in large-scale ethnic cleansing.

Line up five Sinhalese and five Tamils, and their ethnic origin would not be visible. When they open their mouths they speak different languages, and when they pray they adhere to different beliefs. They too have engaged in large-scale ethnic cleansing as they try to share an island nation.

For many ethnicities, sharing space with other ethnicities is difficult, if not impossible. Grievances real and imagined, extending back hundreds of years, prevent peaceful coexistence. Here again are the key issues for Chapter 7:

1. **Where Are Ethnicities Distributed?** Major ethnicities in the United States include African Americans, Hispanic Americans, and Asian Americans. These ethnic groups are clustered in regions of the country and within urban neighborhoods. In the United States, race and ethnicity are often used interchangeably because members of the African American ethnic group are also distinguished as members of the black race (although not all blacks are African Americans).
2. **Why Have Ethnicities Been Transformed into Nationalities?** Nationalities are ethnic groups that possess among their cultural traditions attachment and loyalty to a particular country. A nationality combines an ethnic group's language, religion, and artistic expressions with a country's particular independence movement, history, and other patriotic events. During the past two centuries, many countries have been created that attempt to transform single ethnic groups into single nationalities.
3. **Why Do Ethnicities Clash?** Conflicts can arise when a country contains several ethnicities competing with each other for control or dominance. Conflicts also arise when an ethnicity is divided among more than one country.
4. **What Is Ethnic Cleansing?** Ethnic cleansing is an attempt by a more powerful ethnic group to create an ethnically homogeneous region by forcibly evicting all members of another ethnic group. The practice has been especially widespread in the countries that comprise the former country of Yugoslavia.

CASE STUDY REVISITED / Ethnic Diversity in America

Two major museums standing one block apart in Detroit illustrate the challenges of encouraging respect for different ethnic identities in the United States. One of the museums, the Detroit Institute of Arts (DIA), contains a major collection of paintings by medieval European artists, many of which were donated a century ago by rich Detroit industrialists. The DIA's most famous work is an enormous mural completed in 1932 by the Mexican muralist Diego Rivera, glorifying workers in Detroit's auto factories. The 80-year-old building, the country's fifth-largest art museum, looks like a Greek temple.

The nearby Museum of African American History houses the nation's largest exhibit devoted to the history and culture of African Americans. Founded in 1965, the museum has moved twice to larger buildings, including the current one opened in 1997. The building is designed to reflect the cultural heritage of Africa, including an entry with large bronze doors topped by 14-carat gold-plate decorative masks. The exhibits are primarily photographs, videos, and text.

The financially strapped city of Detroit has had difficulty adequately funding both museums, so it has had to make choices. Which museum should take priority—a crumbling temple of European masterpieces or an emotionally powerful testimony to the rich cultural traditions of an important ethnic minority? Does it matter that Detroit's African American population was 5 percent when the DIA was built and 75 percent when the Museum of African American history was built? ■

KEY TERMS

Apartheid (p. 215) Laws (no longer in effect) in South Africa that physically separated different races into different geographic areas.

Balkanization (p. 232) Process by which a state breaks down through conflicts among its ethnicities.

Balkanized (p. 232) Descriptive of a small geographic area that could not successfully be organized into one or more stable states because it was inhabited by many ethnicities with complex, long-standing antagonisms toward each other.

Blockbusting (p. 215) A process by which real estate agents convince white property owners to sell their houses at low prices because of fear that persons of color will soon move into the neighborhood.

Centripetal force (p. 219) An attitude that tends to unify people and enhance support for a state.

Ethnic cleansing (p. 229) Process in which a more powerful ethnic group forcibly removes a less powerful one in order to create an ethnically homogeneous region.

Ethnicity (p. 208) Identity with a group of people that share distinct physical and mental traits as a product of common heredity and cultural traditions.

Multiethnic state (p. 219) State that contains more than one ethnicity.

Multinational state (p. 219) State that contains two or more ethnic groups with traditions of self-determination that agree to coexist peacefully by recognizing each other as distinct nationalities.

Nationalism (p. 219) Loyalty and devotion to a particular nationality.

Nationality (p. 217) Identity with a group of people that share legal attachment and personal allegiance to a particular place as a result of being born there.

Nation-state (p. 217) A state whose territory corresponds to that occupied by a particular ethnicity that has been transformed into a nationality.

Race (p. 208) Identity with a group of people descended from a common ancestor.

Racism (p. 214) Belief that race is the primary determinant of human traits and capacities and that racial differences produce an inherent superiority of a particular race.

Racist (p. 214) A person who subscribes to the beliefs of racism.

Self-determination (p. 217) Concept that ethnicities have the right to govern themselves.

Sharecropper (p. 212) A person who works fields rented from a landowner and pays the rent and repays loans by turning over to the landowner a share of the crops.

Triangular slave trade (p. 212) A practice, primarily during the eighteenth century, in which European ships transported slaves from Africa to Caribbean islands, molasses from the Caribbean to Europe, and trade goods from Europe to Africa.

THINKING GEOGRAPHICALLY

1. The U.S. Census permits people to identify themselves as being of more than one race, in recognition that several million American children have parents of two races. Discuss the merits and difficulties of permitting people to choose more than one race.
2. Sarajevo, capital of Bosnia & Herzegovina, once contained concentrations of many ethnic groups. In retaliation for ethnic cleansing by the Serbs and Croats, the Bosnian Muslims now in control of Sarajevo have been forcing other ethnic groups to leave the city, and Sarajevo is now inhabited overwhelmingly by Bosnian Muslims. Discuss the challenges in restoring Sarajevo as a multiethnic city.
3. Despite the 1954 U.S. Supreme Court decision that racially segregated school systems are inherently unequal, most schools remain segregated, with virtually none or virtually all African American or Hispanic pupils. As long as most neighborhoods are segregated, how can racial integration in the schools be achieved?
4. A century ago European immigrants to the United States had much stronger ethnic ties than today, including clustering in specific neighborhoods. Discuss the rationale for retaining strong ethnic identity in the United States as opposed to full assimilation into the American nationality identity.
5. With the removal of the apartheid laws, South Africa now offers legal equality to all races in principle. Discuss obstacles that South Africa's blacks face in achieving cultural and economic equality.

RESOURCES

Some recent and classic books and articles on ethnic geography:

Alberts, Heike C. "Changes in Ethnic Solidarity in Cuban Miami." *Geographical Review* 95 (2005): 231–48.

Allen, James P., and Eugene Turner. "Ethnic Residential Concentrations in United States Metropolitan Areas." *Geographical Review* 95 (2005): 267–85.

Arreola, Daniel D., ed. *Hispanic Spaces, Latino Places: Community and Cultural Diversity in Contemporary America.* Austin: University of Texas Press, 2004.

Christopher, A. J. *The Atlas of Changing South Africa,* 2nd ed. London and New York: Routledge, 2001.

———."'To Define the Undefinable': Population Classification and the Census in South Africa." *Area* 34 (2002): 401–8.

Delaney, David. "The Space That Race Makes." *Professional Geographer* 54 (2002): 6–14.

Dwyer, Claire, and Caroline Bressey, eds. *New Geographies of Race and Racism.* Aldershot, England, and Burlington, Vermont: Ashgate, 2008.

Fawcett, Liz. *Religion, Ethnicity, and Social Change.* New York: St. Martin's Press, 2000.

Marriott, Alan. "Nationalism and Nationality in India and Pakistan." *Geography* 85 (2000): 173–77.

Miyares, Ines M., and Christopher A. Airriess, eds. *Contemporary Ethnic Geographies in America.* Lanham, MD: Rowman & Littlefield, 2007.

Murphy, Alexander B. "Territorial Policies in Multiethnic States." *Geographical Review* 79 (1989): 410–21.

Peake, Linda, and Audrey Kobayashi. "Policies and Practices for an Antiracist Geography at the Millennium." *Professional Geographer* 54 (2002): 50–61.

Skop, Emily, and Wei Li. "Asians in America's Suburbs: Patterns and Consequences of Settlement." *Geographical Review* 95 (2005): 167–88.

Key Internet sites:

www.GlobalSecurity.org Images of ethnic cleansing in Kosovo and elsewhere are collected by GlobalSecurity.org, which is an organization that tries to be a reliable source of background information on conflicts by providing facts and figures.

www.defenselink.mil/photos/ The U.S. Department of Defense DefenseLINK web site posts photos of conflicts.

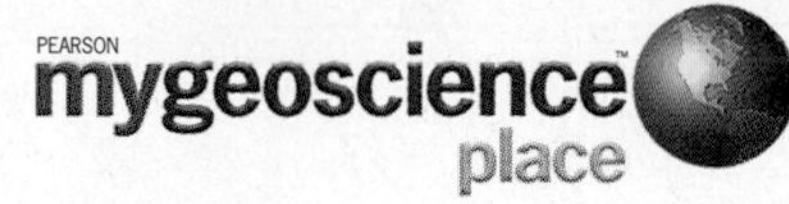

Log in to www.mygeoscienceplace.com for videos, interactive maps, RSS feeds, case studies, and self-study quizzes to enhance your study of Ethnicity.

CHAPTER 8

Political Geography

How many countries can you name? Old-style geography sometimes required memorization of countries and their capitals. Human geographers now emphasize a thematic approach. We are concerned with the location of activities in the world, the reasons for particular spatial distributions, and the significance of the arrangements.

Despite this change in emphasis, you still need to know the locations of countries. Without such knowledge, you lack a basic frame of reference—knowing where things are. It would be like translating an article from a foreign language by looking up each word in a dictionary.

In recent years, we have repeatedly experienced military conflicts and revolutionary changes in once

KEY ISSUES

1 Where Are States Located?

2 Why Do Boundaries Between States Cause Problems?

3 Why Do States Cooperate with Each Other?

4 Why Has Terrorism Increased?

obscure places. No one can predict where the next war will erupt, but political geography helps to explain the cultural and physical factors that underlie political unrest in the world. Political geographers study how people have organized Earth's land surface into countries and alliances, reasons underlying the observed arrangements, and the conflicts that result from the organization.

Political conflicts during the twentieth century were dominated by wars between states or collections of allied states. For the United States, World War I, World War II, the Korean War, and the Vietnam War were the bloodiest of these conflicts with other states. In contrast, the attack against the United States on September 11, 2001, was initiated not by a hostile state, but by a group of individual terrorists. For political geography, the challenge is to explain "why" terrorism occurs if "where" facts about the terrorists, such as countries of birth and current places of residence, are not particularly important factors in the explanation. Even so, reasons for terrorist attacks may relate to the political geography of particular regions of the world.

North Korean girls celebrate April 15, a major holiday in North Korea, that observes the birthday of Kim Il Sung, the country's leader between 1948 and1994.

CASE STUDY / Changing Borders in Europe

Daniel Lenig lives in the village of Rittershoffen and works at a Mercedes-Benz truck factory in the town of Worth, about 50 kilometers (30 miles) away. Lenig's journey to work takes him across an international border, because Rittershoffen is in France, whereas Worth is in Germany. As a citizen of France, Lenig has no legal difficulty crossing the German–French border twice a day; no guards ask him to show his passport or require him to pay customs duties on goods he purchases on the other side. If he is delayed, the cause is heavy traffic on the bridge that spans the Rhine River, which serves as the border between the two countries.

The boundary between France and Germany has not always been so easy to cross peacefully. The French have long argued that the Rhine River forms the logical physical boundary between France and Germany. But the Germans once claimed that they should control the Rhine, including the lowlands on the French side between the west bank of the river and the Vosges Mountains, an area known as Alsace.

Alsace was initially inhabited by Germanic tribes but was annexed by France in 1670. Two centuries later, in 1870, Alsace and its neighboring province of Lorraine were captured by Prussia (which a year later formed the core of the newly proclaimed German Empire). France regained Alsace and Lorraine after Germany was defeated in World War I and has possessed them ever since, except between 1940 and 1945 when Germany controlled them during World War II.

With the end of the Cold War and the demise of communism in Eastern Europe, France and Germany now lie at the core of the world's wealthiest market area. Most French and German people consider the pursuit of higher standards of living to be more important than rehashing centuries-old boundary disputes.

Although old boundaries between France and Germany have been virtually eliminated, new ones have been erected elsewhere in Europe. Travelers between Ljubljana and Zagreb now must show their passports and convert their cash into a different currency. These two cities were once part of the same country—Yugoslavia—but now they are the capitals of two separate countries, Slovenia and Croatia. Similarly, travelers between Vilnius and Moscow—both once part of the Soviet Union—now must show their passports and change money when they cross the international boundary between Lithuania and Russia. ■

For several decades during the Cold War, many countries belonged to one of two ***regions***, one allied with the former Soviet Union and the other allied with the United States. With the end of the Cold War in the 1990s, the global political landscape changed fundamentally.

Geographic concepts help us to understand this changing political organization of Earth's surface. We can also use geographic methods to examine the causes of political change and instability and to anticipate potential trouble spots around the world.

When looking at satellite images of Earth, we easily distinguish ***places***—landmasses and water bodies, mountains and rivers, deserts and fertile agricultural land, urban areas and forests. What we cannot see are ***where*** boundaries are located between countries. Boundary lines are not painted on Earth, but they might as well be, for these national divisions are very real.

To many, national boundaries are more meaningful than natural features. One of Earth's most fundamental cultural characteristics—one that we take for granted—is the division of our planet's surface into a collection of ***spaces*** occupied by individual countries.

In the post–Cold War era, the familiar division of the world into countries or states is crumbling. Geographers observe ***why*** this familiar division of the world is changing. Between the mid-1940s and the late 1980s two superpowers—the United States and the Soviet Union—essentially "ruled" the world. As on superpowers, they competed at a global ***scale***. But the United States is less dominant in the political landscape of the twenty-first century, and the Soviet Union no longer exists.

Today, ***globalization*** means more ***connections*** among states. Individual countries have transferred military, economic, and political authority to regional and worldwide collections of states. Power is exercised through connections among states created primarily for economic cooperation.

Despite (or perhaps because of) greater global political cooperation, ***local diversity*** has increased in political affairs, as individual cultural groups demand more control over the territory they inhabit. States have transferred power to local governments, but this does not placate cultural groups who seek complete independence.

Wars have broken out in recent years—both between small neighboring states and among cultural groups within countries—over political control of territory. Old countries have been broken up in a collection of smaller ones, some barely visible on world maps.

KEY ISSUE 1
Where Are States Located?

- **Problems of Defining States**
- **Varying Size of States**
- **Development of the State Concept**

The question posed in this key issue may seem self-evident, because a map of the world shows that virtually all habitable land belongs to a country. But for most of history, until recently, this was not so. As recently as the 1940s, the world contained only about 50 countries, compared to 192 members of the United Nations as of 2009. (Refer ahead to Figure 8-5.) ■

Problems of Defining States

A **state** is an area organized into a political unit and ruled by an established government that has control over its internal and foreign affairs. It occupies a defined territory on Earth's surface and contains a permanent population. The term *country* is a synonym for *state*. A state has **sovereignty**, which means independence from control of its internal affairs by other states. Because the entire area of a state is managed by its national government, laws, army, and leaders, it is a good example of a formal or uniform region. The term *state*, as used in political geography, does not refer to the 50 regional governments inside the United States. The 50 states of the United States are subdivisions within a single state—the United States of America.

There is some disagreement about the actual number of sovereign states. Among places that test the definition of a state are Korea, China, and Western Sahara (Sahrawi Republic).

Korea: One State or Two?

A colony of Japan for many years, Korea was divided into two occupation zones by the United States and the former Soviet Union after they defeated Japan in World War II (Figure 8-1). The country was divided into northern and southern sections along 38° north latitude. The division of these zones became permanent in the late 1940s, when the two superpowers established separate governments and withdrew their armies. The new government of the Democratic People's Republic of Korea (North Korea) then invaded the Republic of Korea (South Korea) in 1950, touching off a 3-year war that ended with a cease-fire line near the 38th parallel.

Both Korean governments are committed to reuniting the country into one sovereign state. Leaders of the two countries agreed in 2000 to allow exchange visits of families separated for a half century by the division and to increase economic cooperation. However, progress toward reconciliation was halted by North Korea's decision to build nuclear weapons, even though the country lacked the ability to provide its citizens with food, electricity, and other basic needs. Meanwhile, in 1992, North Korea and South Korea were admitted to the United Nations as separate countries.

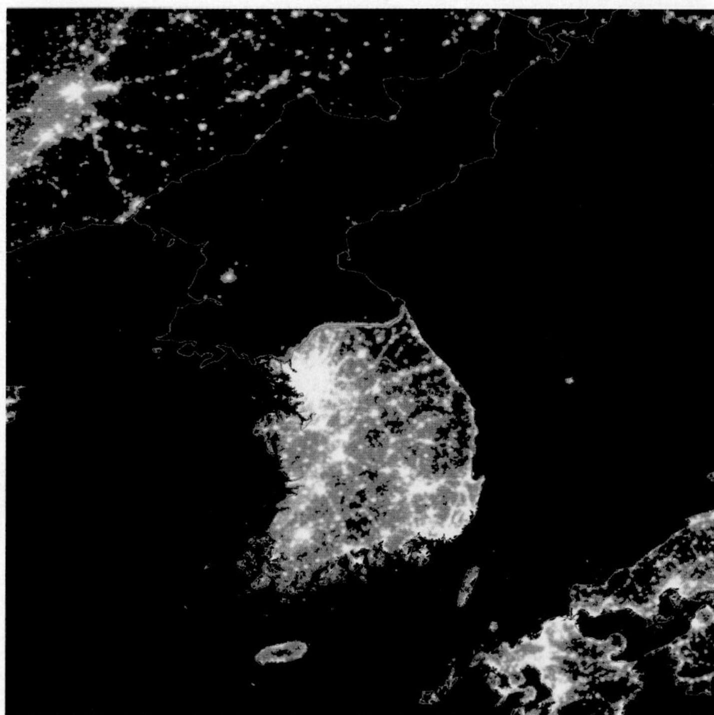

FIGURE 8-1 North and South Korea. A nighttime satellite image recorded by the U.S. Air Force Defense Meteorological Satellite Program shows the illumination of electric lights in South Korea, whereas North Korea has virtually no electric lights, a measure of its poverty and limited economic activity.

China and Taiwan: One State or Two?

Are China and the island of Taiwan two sovereign states or one? Most other countries consider China (officially, the People's Republic of China) and Taiwan (officially, the Republic of China) as separate and sovereign states. According to China's government, Taiwan is not sovereign, but a part of China. This confusing situation arose from a civil war in China during the late 1940s between the Nationalists and the Communists. After losing, nationalist leaders in 1949 fled to Taiwan, 200 kilometers (120 miles) off the Chinese coast.

The Nationalists proclaimed that they were still the legitimate rulers of the entire country of China. Until some future occasion when they could defeat the Communists and recapture all of China, the Nationalists argued, at least they could continue to govern one island of the country. In 1999, Taiwan's president announced that Taiwan would regard itself as a sovereign independent state, but the government of China viewed that announcement as a dangerous departure from the long-standing arrangement between the two.

The question of who constituted the legitimate government of China plagued U.S. officials during the 1950s and 1960s. The United States had supported the Nationalists during the civil war, so many Americans opposed acknowledging that China was firmly under the control of the Communists. Consequently, the United States continued to regard the Nationalists as the official government of China until 1971, when U.S. policy finally changed and the United Nations voted to transfer China's seat from the Nationalists to the Communists. Taiwan is now the most populous state not in the United Nations.

Western Sahara (Sahrawi Republic)

The Sahrawi Arab Democratic Republic, also known as Western Sahara, is considered by most African countries as a sovereign state. Morocco, however, claims the territory and to prove it has built a 2,700-kilometer wall around the territory to keep out rebels (Figure 8-2).

FIGURE 8-2 Western Sahara. A French soldier attached to a United Nations missions patrols a portion of the sand walls built by Morocco during the 1980s to isolate Polisario Front rebels fighting for independence.

Spain controlled the territory on the continent's west coast between Morocco and Mauritania until withdrawing in 1976. An independent Sahrawi Republic was declared by the Polisario Front and recognized by most African countries, but Morocco and Mauritania annexed the northern and southern portions, respectively. Three years later Mauritania withdrew, and Morocco claimed the entire territory.

Morocco controls most of the populated area, but the Polisario Front operates in the vast, sparsely inhabited deserts, especially the one-fifth of the territory that lies east of Morocco's wall. The United Nations has tried but failed to reach a resolution among the parties.

Polar Regions: Many Claims

The South polar region contains the only large landmasses on Earth's surface that are not part of a state. Several states claim portions of the region, and some claims are overlapping and conflicting.

Several states, including Argentina, Australia, Chile, France, New Zealand, Norway, and the United Kingdom, claim portions of Antarctica (Figure 8-3). Argentina, Chile, and the United Kingdom have made conflicting, overlapping claims. The United States, Russia, and a number of other states do not recognize the claims of any country to Antarctica. The Antarctic Treaty, signed in 1959, provides a legal framework for managing Antarctica. States may establish research stations there for scientific investigations, but no military activities are permitted. The Treaty has been signed by 47 states.

As for the Arctic, the 1982 United Nations Convention on the Law of the Sea permitted countries to submit claims inside the Arctic Circle by 2009 (Figure 8-4). The Arctic region is thought to be rich in energy resources.

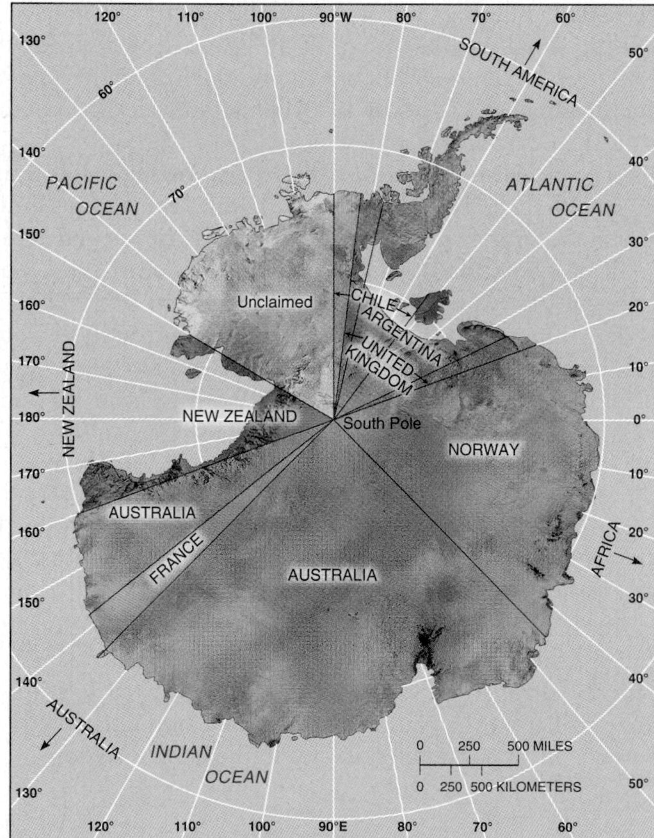

FIGURE 8-3 National claims to Antarctica. Antarctica is the only large landmass in the world that is not part of a sovereign state. It comprises 14 million square kilometers (5.4 million square miles), which makes it 50 percent larger than Canada. Portions are claimed by Argentina, Australia, Chile, France, New Zealand, Norway, and the United Kingdom; claims by Argentina, Chile, and the United Kingdom are conflicting.

Varying Size of States

The land area occupied by the states of the world varies considerably. The largest state is Russia, which encompasses 17.1 million square kilometers (6.6 million square miles), or 11 percent of the world's entire land area (Figure 8-5). Other states with more than 5 million square kilometers (2 million square miles) include Canada, the United States, China, Brazil, and Australia.

At the other extreme are about two dozen **microstates**, which are states with very small land areas. If Russia were the size of this page, the a microstate would be the size of a single letter. The smallest microstate in the United Nations—Monaco— encompasses only 1.5 square kilometers (0.6 square miles). See Figure 8-17 for an image of Monaco.

Other UN member states that are smaller than 1,000 square kilometers include Andorra, Antigua and Barbuda, Bahrain,

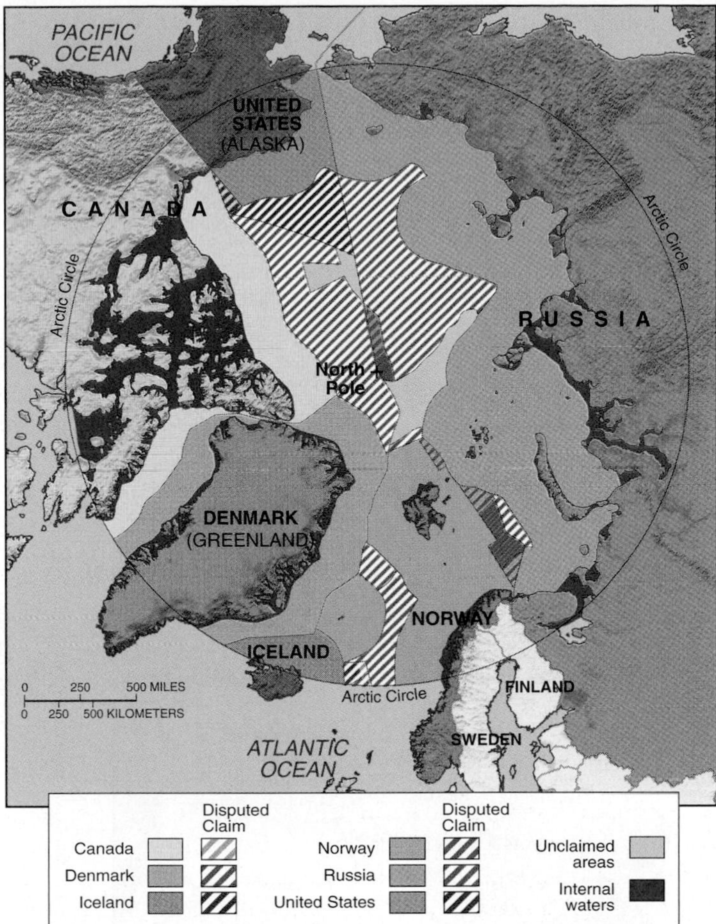

FIGURE 8-4 National claims to the Arctic. Under The Law of the Sea Treaty of 1982 countries had until 2009 to submit territory claims inside the Arctic Circle. Some of these claims overlap.

Barbados, Dominica, Grenada, Kiribati, Liechtenstein, Maldives, Malta, Micronesia, Nauru, Palau, St. Kitts & Nevis, St. Lucia, St. Vincent & the Grenadines, San Marino, São Tomé e Príncipe, the Seychelles, Singapore, Tonga, and Tuvalu. Many of the microstates are islands, which explains both their small size and sovereignty.

Development of the State Concept

The concept of dividing the world into a collection of independent states is recent. Prior to the 1800s, Earth's surface was organized in other ways, such as city-states, empires, and tribes. Much of Earth's surface consisted of unorganized territory.

Ancient and Medieval States

The development of states can be traced to the ancient Middle East, in an area known as the Fertile Crescent. The modern movement to divide the world into states originated in Europe.

ANCIENT STATES. The ancient Fertile Crescent formed an arc between the Persian Gulf and the Mediterranean Sea (Figure 8-6). The eastern end, Mesopotamia, was centered in the valley formed by the Tigris and Euphrates rivers, in present-day Iraq. The Fertile Crescent then curved westward over the desert, turning southward to encompass the Mediterranean coast through present-day Syria, Lebanon, and Israel. The Nile River valley of Egypt is sometimes regarded as an extension of the Fertile Crescent. Situated at the crossroads of Europe, Asia, and Africa, the Fertile Crescent was a center for land and sea communications in ancient times.

The first states to evolve in Mesopotamia were known as city-states. A **city-state** is a sovereign state that comprises a town and the surrounding countryside. Walls clearly delineated the boundaries of the city, and outside the walls the city controlled agricultural land to produce food for urban residents. The countryside also provided the city with an outer line of defense against attack by other city-states. Periodically, one city or tribe in Mesopotamia would gain military dominance over the others and form an empire. Mesopotamia was organized into a succession of empires by the Sumerians, Assyrians, Babylonians, and Persians.

Meanwhile, the state of Egypt emerged as a separate empire to the west of the Fertile Crescent. Egypt controlled a long, narrow region along the banks of the Nile River, extending from the Nile Delta at the Mediterranean Sea southward for several hundred kilometers. Egypt's empire lasted from approximately 3000 B.C. until the fourth century B.C.

EARLY EUROPEAN STATES. Political unity in the ancient world reached its height with the establishment of the Roman Empire, which controlled most of Europe, North Africa, and Southwest Asia, from modern-day Spain to Iran and from Egypt to England. At its maximum extent, the empire comprised 38 provinces, each using the same set of laws that had been created in Rome. Massive walls helped the Roman army defend many of the empire's frontiers.

The Roman Empire collapsed in the fifth century after a series of attacks by people living on its frontiers and because of internal disputes. The European portion of the Roman Empire was fragmented into a large number of estates owned by competing kings, dukes, barons, and other nobles. Beginning about the year 1100, a handful of powerful kings emerged as rulers over large numbers of these European estates. The consolidation of neighboring estates under the unified control of a king formed the basis for the development of such modern Western European states as England, France, and Spain. Much of central Europe, however—notably present-day Germany and Italy—remained fragmented into a large number of estates that were not consolidated into states until the nineteenth century.

Colonies

A **colony** is a territory that is legally tied to a sovereign state rather than being completely independent. In some cases, a sovereign state runs only the colony's military and foreign policy. In others, it also controls the colony's internal affairs.

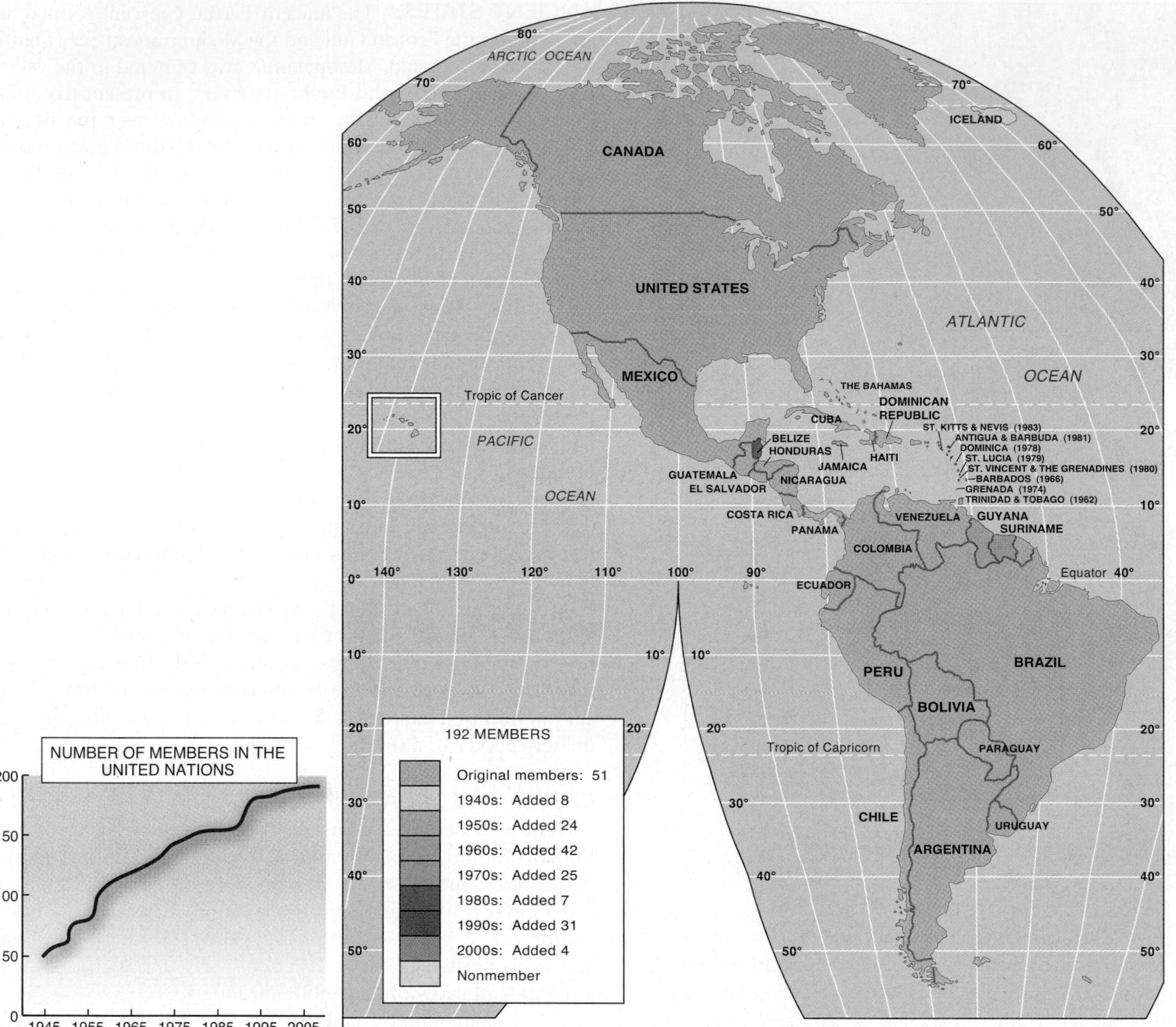

FIGURE 8-5 U.N. members. When it was organized in 1945, the United Nations had only 51 members, including 49 sovereign states plus Byelorussia (now Belarus) and Ukraine, then part of the Soviet Union. The number increased to 192 in 2006. The greatest increase in sovereign states has occurred in Africa. Only 4 African states were original members of the United Nations—Egypt, Ethiopia, Liberia, and South Africa—and only 6 more joined during the 1950s. Beginning in 1960, however, a collection of independent states was carved from most of the remainder of the region. In 1960 alone, 16 newly independent African states became UN members. Creation of new sovereign states slowed during the 1980s. The breakup of the Soviet Union and Yugoslavia stimulated the formation of more new states during the early 1990s, and several microstates in the Pacific Ocean joined during the late 1990s.

COLONIALISM. European states came to control much of the world through **colonialism**, which is the effort by one country to establish settlements in a territory and to impose its political, economic, and cultural principles on that territory (Figure 8-7). European states established colonies elsewhere in the world for three basic reasons:

- To promote Christianity
- To extract useful resources and to serve as captive markets for their products
- To establish relative power through the number of their colonies.

The three motives could be summarized as God, gold, and glory.

The colonial era began in the 1400s, when European explorers sailed westward for Asia but encountered and settled in the Western Hemisphere instead. Eventually, the European states lost most of their Western Hemisphere colonies: Independence was declared by the United States in 1776 and by most Latin American states between 1800 and 1824. European states then turned their attention to Africa and Asia (Figure 8-8). This European colonization of Africa and Asia is often termed **imperialism**, which is control of territory already occupied and organized by an indigenous society, whereas colonialism is control of previously uninhabited or sparsely inhabited land.

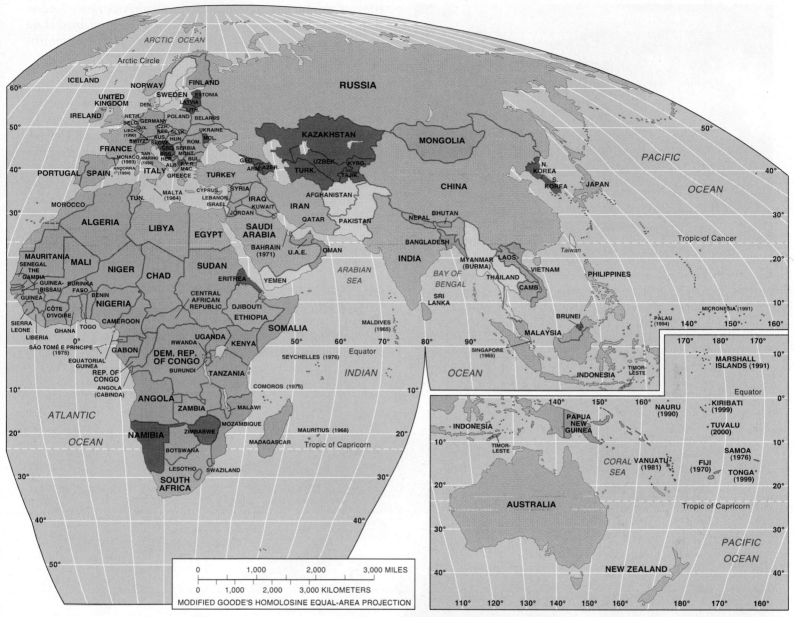

The British planted colonies on every continent, including much of eastern and southern Africa, South Asia, the Middle East, Australia, and Canada. With by far the largest colonial empire, the British proclaimed that the "Sun never set" on their empire. France had the second-largest overseas territory, primarily in West Africa and Southeast Asia. The colonial practices of European states varied. France attempted to assimilate its colonies into French culture and

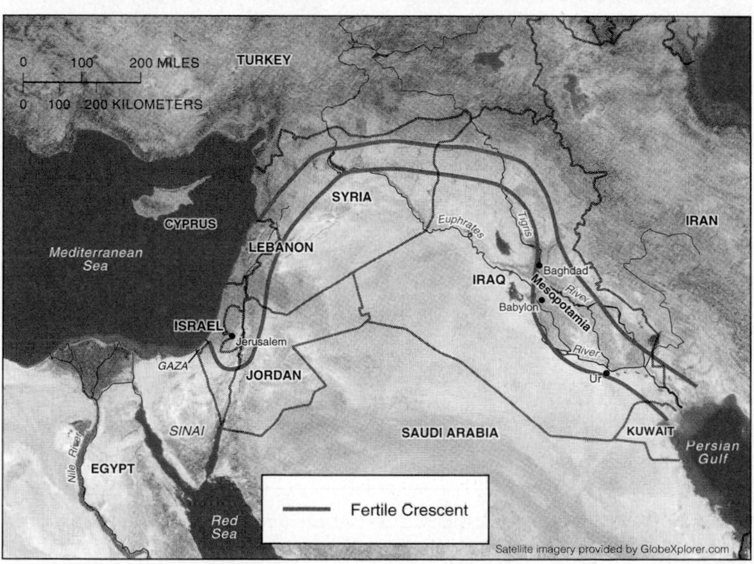

FIGURE 8-6 The Fertile Crescent. The crescent-shaped area of relatively fertile land was organized into a succession of empires starting several thousand years ago.

FIGURE 8-7 European colonialism. European countries carved up much of Africa and Asia into colonies during the nineteenth century. The British assembled the largest collection. In this 1893 photograph, Britain's Queen Victoria is writing at the desk while her Indian servant holds her walking stick and awaits orders.

educate an elite group to provide local administrative leadership. After independence, most of these leaders retained close ties with France. The British created different government structures and policies for various territories of their empire. This decentralized approach helped to protect the diverse cultures, local customs, and educational systems in their extensive empire. British colonies generally made peaceful transitions to independence, although exceptions can be found in the Middle East, Southern Africa, and Ireland, where recent conflicts can be traced in part to the legacy of British rule.

Most African and Asian colonies became independent after World War II. Only 15 African and Asian states were members of the United Nations when it was established in 1945, compared to 106 in 2010. The boundaries of the new states frequently coincide with former colonial provinces, although not always.

THE FEW REMAINING COLONIES. At one time, colonies were widespread over Earth's surface, but only a handful remain. The U.S. Department of State lists 43 colonies with indigenous populations (Figure 8-9).

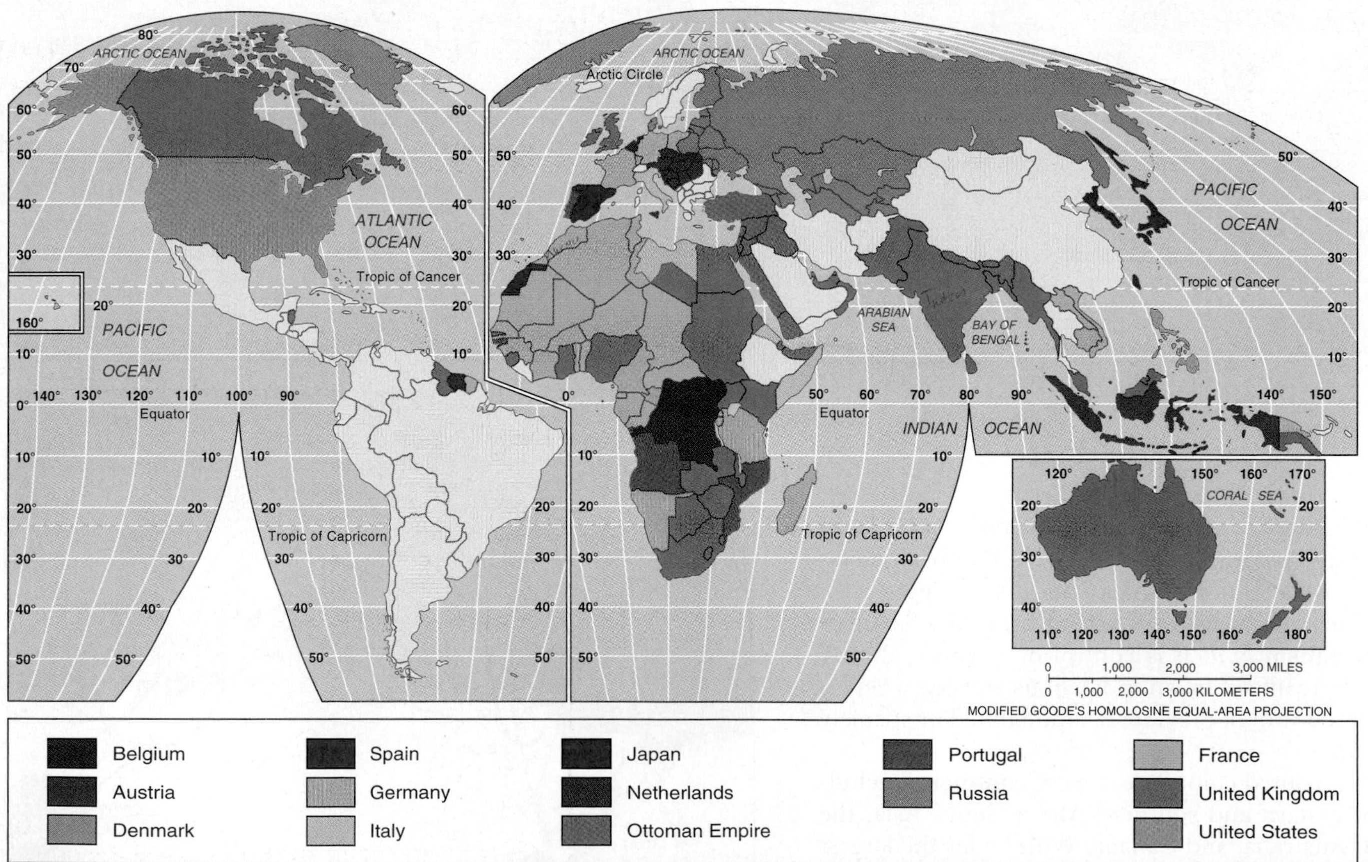

FIGURE 8-8 Colonial possessions, 1914. At the outbreak of World War I in 1914, European states held colonies in much of the world, especially in Africa and Asia. Most of the countries in the Western Hemisphere at one time had been colonized by Europeans but gained their independence in the eighteenth or nineteenth centuries.

Most current colonies are islands in the Pacific Ocean or Caribbean Sea. The most populous is Puerto Rico, a Commonwealth of the United States, with 4 million residents on an island of 8,870 square kilometers (3,500 square miles). Puerto Ricans are citizens of the United States, but do not participate in U.S. elections, nor have a voting member of Congress.

One of the world's least populated colonies is Pitcairn Island, a 47-square-kilometer (18-square-mile) possession of the United Kingdom. The island in the South Pacific was settled in 1790 by British mutineers from the ship *Bounty*, commanded by Captain William Bligh. Its 48 islanders survive by selling fish, as well as postage stamps to collectors.

The State Department list does not include several inhabited islands considered by other sources to be colonies, including Australia's Lord Howe Island, Britain's Ascension Island, and Chile's Easter Island. On the other hand, the State Department list includes several entities that others do not classify as colonies, including Greenland, Hong Kong, and Macao. Greenland regards the Queen of Denmark as its head of state. But it has a high degree of autonomy and self-rule and makes even foreign policy decisions independently of Denmark. Hong Kong and Macao, attached to the mainland of China, were colonies of the United Kingdom and Portugal, respectively. The British returned Hong Kong to China in 1997 and the Portuguese returned Macao to China 2 years later. These two areas are classified as Special Administrative Regions with autonomy from the rest of China in economic matters but not in foreign and military affairs.

KEY ISSUE 2

Why Do Boundaries Between States Cause Problems?

- **Shapes of States**
- **Types of Boundaries**
- **Boundaries Inside States**

A state is separated from its neighbors by a **boundary**, an invisible line marking the extent of a state's territory. Boundaries completely surround an individual state to mark the outer limits of its territorial control and to give it a distinctive shape. Boundaries interest geographers because the process of selecting their location is frequently difficult. ■

Shapes of States

The shape of a state controls the length of its boundaries with other states. The shape therefore affects the potential for communication and conflict with neighbors. The shape also, as in the outline of the United States or Canada, is part of its unique identity. Beyond its value as a centripetal force, the shape of a state can influence the ease or difficulty of internal administration and can affect social unity.

Five Basic Shapes

Countries have one of five basic shapes—compact, prorupted, elongated, fragmented, or perforated—examples of each can be seen in southern Africa (Figure 8-10). Each shape displays distinctive characteristics and challenges.

COMPACT STATES: EFFICIENT. In a **compact state**, the distance from the center to any boundary does not vary significantly. The ideal theoretical compact state would be shaped like a circle, with the capital at the center and with the shortest possible boundaries to defend.

Compactness can be a beneficial characteristic for smaller states, because good communications can be more easily established to all regions, especially if the capital is located near the center. However, compactness does not necessarily mean peacefulness, as compact states are just as likely as others to experience civil wars and ethnic rivalries.

ELONGATED STATES: POTENTIAL ISOLATION. A handful of **elongated states** have a long and narrow shape. Examples include:

- Malawi, which measures about 850 kilometers (530 miles) north–south but only 100 kilometers (60 miles) east–west (refer to Figure 8–9).
- Chile, which stretches north–south for more than 4,000 kilometers (2,500 miles) but rarely exceeds an east–west distance of 150 kilometers (90 miles); Chile is wedged between the Pacific Coast of South America and the rugged Andes Mountains, which rise more than 6,700 meters (20,000 feet).
- Italy, which extends more than 1,100 kilometers (700 miles) from northwest to southeast but is only approximately 200 kilometers (120 miles) wide in most places.
- Gambia, which extends along the banks of the Gambia River about 500 kilometers (300 miles) east–west but is only about 25 kilometers (15 miles) north–south.

Elongated states may suffer from poor internal communications. A region located at an extreme end of the elongation might be isolated from the capital, which is usually placed near the center.

PRORUPTED STATES: ACCESS OR DISRUPTION. An otherwise compact state with a large projecting extension is a **prorupted state**. Proruptions are created for two principal reasons:

1. **To provide a state with access to a resource, such as water.** For example, in southern Africa, Congo has a 500-kilometer (300-mile) proruption to the west along the Zaire (Congo) River. The Belgians created the proruption to give their colony access to the Atlantic.
2. **To separate two states that otherwise would share a boundary.** For example, in southern Africa, Namibia has a 500-kilometer (300-mile) proruption to the east

called the Caprivi Strip. When Namibia was a colony of Germany, the proruption disrupted communications among the British colonies of southern Africa. It also provided the Germans with access to the Zambezi, one of Africa's most important rivers.

Elsewhere in the world, the otherwise compact state of Afghanistan has a proruption approximately 300 kilometers (200 miles) long and as narrow as 20 kilometers (12 miles) wide. The British created the proruption to prevent Russia from sharing a border with Pakistan (refer ahead to Figure 8-25 later in this chapter).

PERFORATED STATES: SOUTH AFRICA. A state that completely surrounds another one is a **perforated state**. The one good example of a perforated state is South Africa, which completely surrounds the state of Lesotho. Lesotho must depend almost entirely on South Africa for the import and export of goods. Dependency on South Africa was especially difficult for Lesotho when South Africa had a government controlled by whites who discriminated against the black majority population.

Gambia, described above as an elongated state, is completely surrounded by Senegal except for a short coastline along the Atlantic Ocean. The shapes of Gambia and Senegal are a legacy of competition among European countries to establish colonies during the nineteenth century. Gambia became a British colony, whereas Senegal was French. The border between the two countries divided families and ethnic groups but was never precisely delineated, so people trade and move across the border with little concern for its location.

FRAGMENTED STATES: PROBLEMATIC. A **fragmented state** includes several discontinuous pieces of territory. Technically, all states that have offshore islands as part of their territory are fragmented. However, fragmentation is particularly significant for some states. There are two kinds of fragmented states:

1. **Fragmented states separated by water.** Examples include:
    - Tanzania, which was created in 1964 as a union of the island of Zanzibar with the mainland territory of Tanganyika. Although home to different ethnic groups, the two entities agreed to join together because they shared common development goals and political priorities.
    - Indonesia, which comprises 13,677 islands that extend more than 5,000 kilometers (3,000 miles) between the Indian Ocean and Pacific oceans. Although more than 80 percent of the country's population live on two of the islands—Java and Sumatra—the fragmentation hinders communications and makes integration of people living on remote islands nearly impossible. To foster national integration, the Indonesian government has encouraged migration from the more densely populated islands to some of the sparsely inhabited ones.

Not all of the fragments joined Indonesia voluntarily. A few days after Timor-Leste (East Timor) gained its independence from Portugal in 1975, Indonesia invaded. A long struggle against Indonesia culminated in independence in 2002. West Papua, another fragment of Indonesia (the western portion of the island shared with Papua New Guinea), also claims that it should be an independent country. However, West Papua's attempt to break away from Indonesia gained less support from the international community.

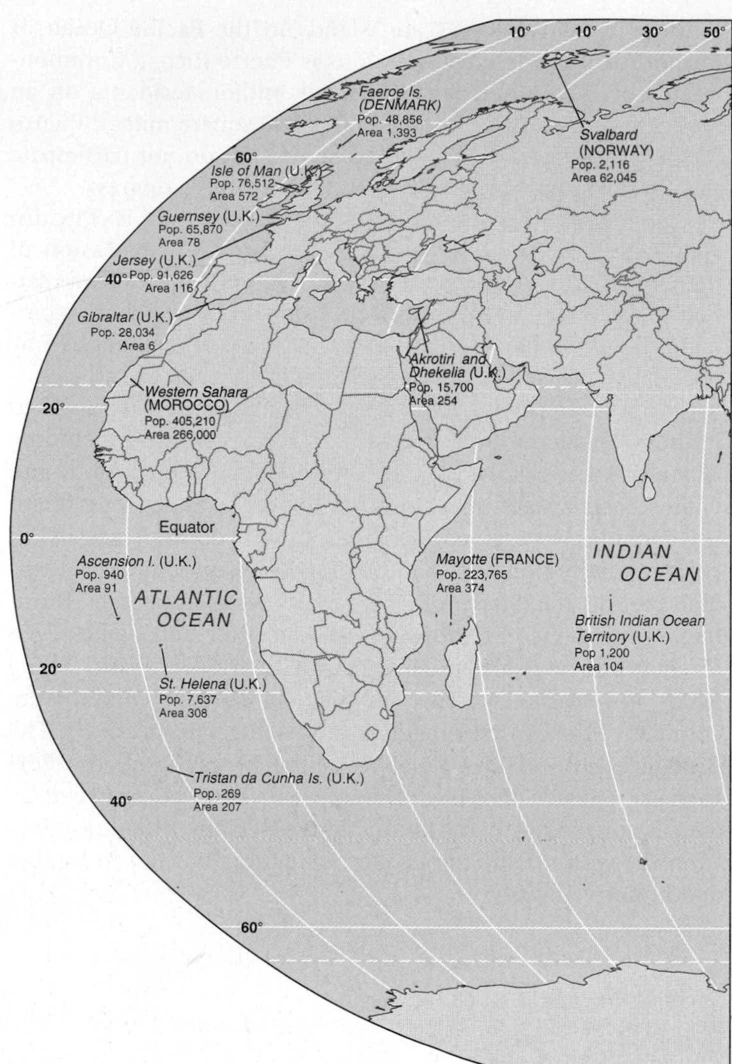

FIGURE 8-9 Colonial possessions, 2006. Most remaining colonies are tiny specks in the Pacific Ocean or the Caribbean Sea, too small to appear on the map. Svalbard, which belongs to Norway, is the only remaining colony with a land area greater than 10,000 square kilometers.

2. **Fragmented states separated by an intervening state.** Examples include:
    - Angola, which is divided into two fragments by the Congo proruption described above. An independence movement is trying to detach Cabinda as a separate state from Angola, with the justification that its population belongs to distinct ethnic groups.
    - Russia, which has a fragment called Kaliningrad (Konigsberg), a 16,000-square-kilometer (6,000-square-mile) entity 400 kilometers (250 miles) west of the remainder of Russia, separated by the states of Lithuania and Belarus (refer to Figure 7–17). The area was part of

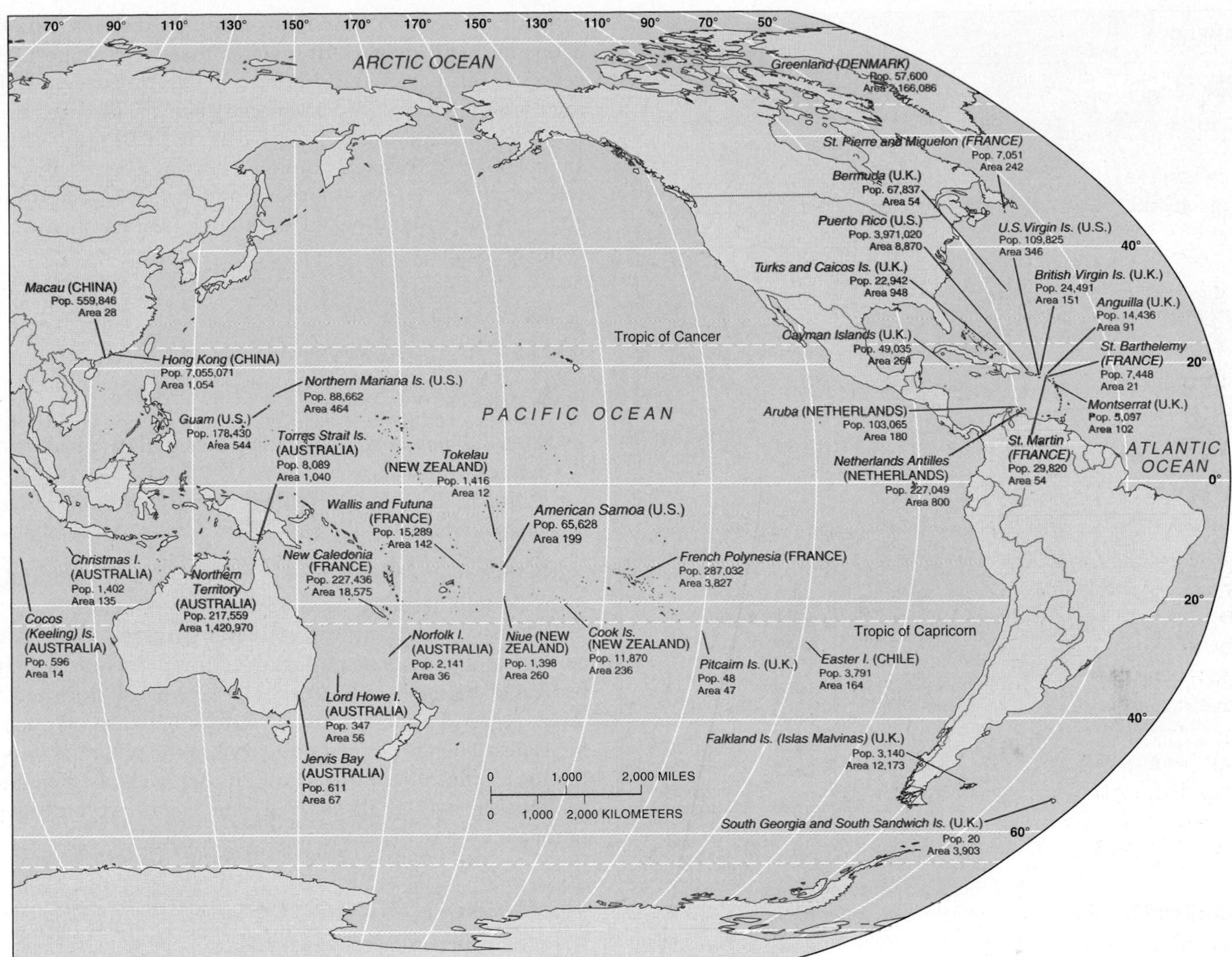

Germany until the end of World War II when the Soviet Union seized it after the German defeat. The German population fled westward after the war, and virtually all of the area's 430,000 residents are Russians. Russia wants Kaliningrad because it has the country's largest naval base on the Baltic Sea.

- Panama, which was a fragmented state for most of the twentieth century, divided in two parts by the canal, built in 1914 by the United States. After the United States withdrew from the Canal Zone in 1999, Panama became an elongated state, 700 kilometers (450 miles) long and 80 kilometers (50 miles) wide.
- India's Tin Bigha corridor, which is a tiny strip of land only 178 meters (about 600 feet) by 85 meters (about 300 feet). It fragments Dahagram and Angarpota from the rest of Bangladesh (Figure 8-11). It is a legacy of the late 1940s when the British divided the region according to religion, allocating predominantly Hindu enclaves to India and predominantly Muslim ones to Bangladesh (see Figure 7-26).

India and Bangladesh reached a novel agreement that opens the Tin Bigha to citizens of both countries every day between dawn and dusk. Bangladeshis may travel between Dahagram and Angarpota and the rest of Bangladesh, and Indians may travel between Cooch Behār and the rest of India without submitting to passport inspection, customs declarations, and other international border controls.

Landlocked States

A **landlocked state** lacks a direct outlet to the sea because it is completely surrounded by several other countries (only one country in the case of Lesotho). Landlocked states are most common in Africa, where 14 of the continent's 54 states have no direct ocean access. The prevalence of landlocked states in Africa is a remnant of the colonial era, when Britain and France controlled extensive regions. The European powers built railroads, mostly in the early twentieth century, to connect the interior of Africa with the sea. Railroads moved minerals from interior mines to seaports, and in the opposite direction, rail lines carried mining equipment and supplies from seaports to the interior.

Now that the British and French empires are gone, and former colonies have become independent states, some important colonial railroad lines pass through several independent

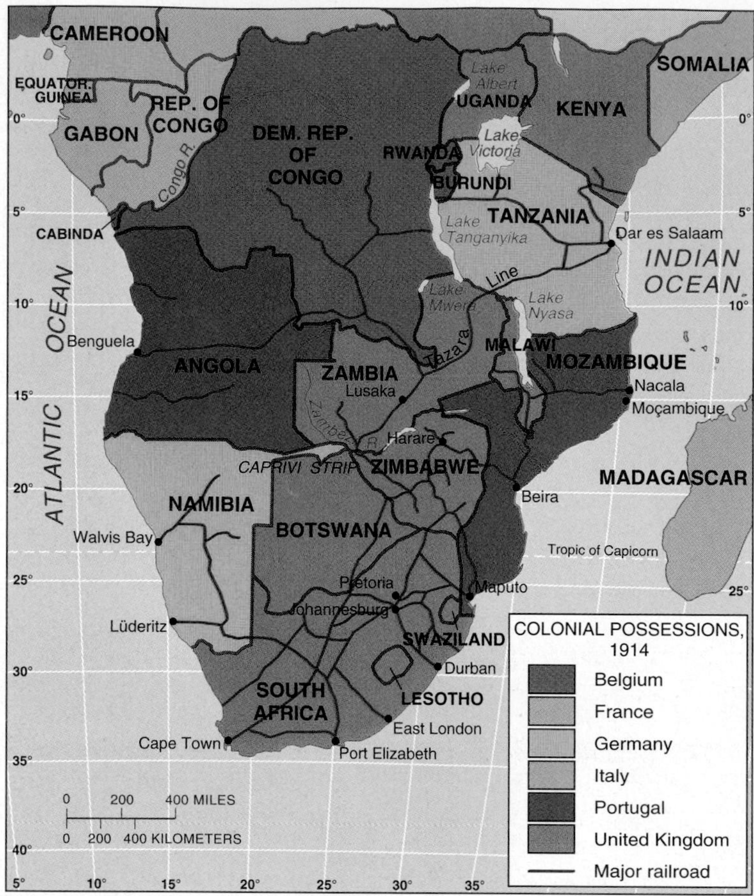

FIGURE 8-10 Shapes of states in Southern Africa. Burundi, Kenya, Rwanda, and Uganda are examples of compact states. Malawi and Mozambique are elongated states. Namibia and the Democratic Republic of Congo are prorupted states. Angola and Tanzania are fragmented states. South Africa is a perforated state. Also shown are landlocked African states, which must import and export goods by land-based transportation, primarily rail lines, to reach ocean ports in cooperating neighbor states. Colors show the European colonial rulers in 1914.

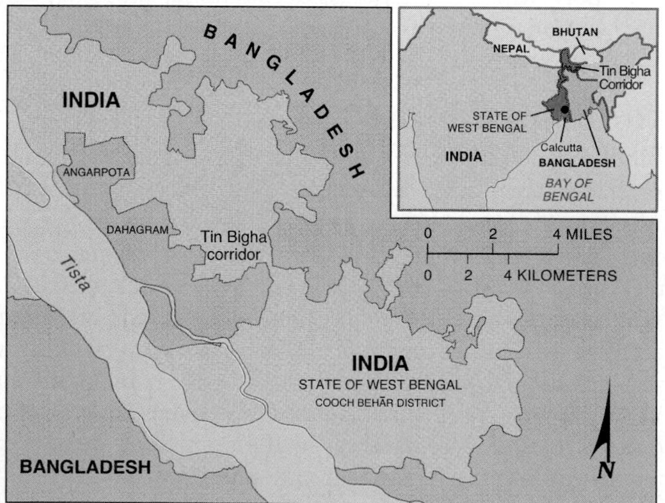

FIGURE 8-11 The Tin Bigha corridor. Less than 300 meters (900 feet), the Tin Bigha corridor is a part of India that fragments Dahagram and Angarpota from the rest of Bangladesh.

countries. This has created new landlocked states, which must cooperate with neighboring states that have seaports. Direct access to an ocean is critical to states because it facilitates international trade. Bulky goods, such as petroleum, grain, ore, and vehicles, are normally transported long distances by ship. This means that a country needs a seaport where goods can be transferred between land and sea. To send and receive goods by sea, a landlocked state must arrange to use another country's seaport.

Types of Boundaries

Boundaries are of two types:

- ***Physical boundaries*** coincide with significant features of the natural landscape.
- ***Cultural boundaries*** follow the distribution of cultural characteristics.

Neither type of boundary is better or more "natural," and many boundaries are a combination of both types.

Boundary locations can generate conflict, both within a country and with its neighbors. The boundary line, which must be shared by more than one state, is the only location where direct physical contact must take place between two neighboring states. Therefore, the boundary has the potential to become the focal point of conflict between them. The best boundaries are those to which all affected states agree, regardless of the rationale used to draw the line.

Physical Boundaries

Important physical features on Earth's surface can make good boundaries because they are easily seen, both on a map and on the ground. Three types of physical elements serve as boundaries between states—deserts, mountains, and water.

DESERT BOUNDARIES. A boundary drawn in a desert can effectively divide two states. Like mountains, deserts are hard to cross and sparsely inhabited. Desert boundaries are common in Africa and Asia. In North Africa, the Sahara has generally proved to be a stable boundary separating Algeria, Libya, and Egypt on the north from Mauritania, Mali, Niger, Chad, and the Sudan on the south.

MOUNTAIN BOUNDARIES. Mountains can be effective boundaries if they are difficult to cross (Figure 8-12). Contact between nationalities living on opposite sides may be limited, or completely impossible if passes are closed by winter storms. Mountains are also useful boundaries because they are rather permanent and are usually sparsely inhabited.

Mountains do not always provide for the amicable separation of neighbors. Argentina and Chile agreed to be divided by the crest of the Andes Mountains but could not decide on the precise location of the crest. Was the crest a jagged line, connecting mountain peak to mountain peak? Or was it a curving line following the continental divide (the continuous ridge that divides rainfall and snowmelt between flow toward the

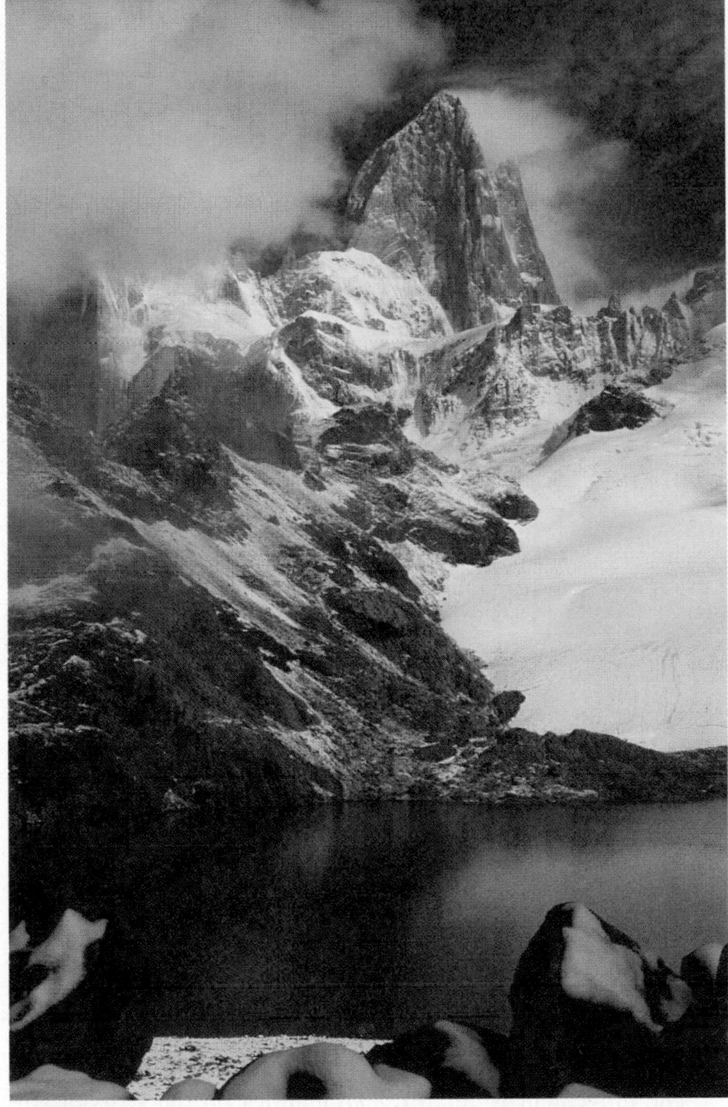

FIGURE 8-12 Mountain boundary: Andes Mountains. The Andes serve as the boundary between Chile and Argentina.

Atlantic or Pacific)? The two countries almost fought a war over the boundary line. But with the help of U.S. mediators, they finally decided on the line connecting adjacent mountain peaks.

WATER BOUNDARIES. Rivers, lakes, and oceans are the physical features most commonly used as boundaries. Water boundaries are readily visible on maps and aerial imagery. Historically, water boundaries offered good protection against attack from another state, because an invading state had to transport its troops by air or ship and secure a landing spot in the country being attacked. The state being invaded could concentrate its defense at the landing point.

Water boundaries are especially common in East Africa:

- The boundary between Democratic Republic of the Congo and Uganda runs through Lake Albert.
- The boundary separating Kenya, Tanzania, and Uganda runs through Lake Victoria (Figure 8-13).
- The boundary separating Burundi, Democratic Republic of Congo, Tanzania, and Zambia runs through Lake Tanganyika.
- The boundary between Democratic Republic of Congo and Zambia runs through Lake Mwera.
- The boundary between Malawi and Mozambique runs through Lake Malawi (Lake Nyasa).

Water boundaries may seem to be set permanently, but the precise position of the water may change over time. Rivers, in particular, can slowly change their course. The Rio Grande, the river separating the United States and Mexico, has frequently meandered from its previous course since it became part of the boundary in 1848. Land that had once been on the U.S. side of the boundary came to be on the Mexican side, and vice versa. The United States and Mexico have concluded treaties that restore land affected by the shifting course of the river to the country in control at the time of the original nineteenth-century delineation. The International Boundary and Water Commission, jointly staffed by the United States and Mexico, oversees the border treaties and settles differences.

Ocean boundaries also cause problems because states generally claim that the boundary lies not at the coastline but out at sea. The reasons are for defense and for control of valuable fishing industries. Beginning in the late eighteenth century, some states recognized a boundary, known as the territorial limit, which extended 3 nautical miles (about 5.5 kilometers or 3.5 land miles) from the shore into the ocean. Some states claimed more extensive territorial limits, and others identified a contiguous zone of influence that extended beyond the territorial limits.

The Law of the Sea, signed by 158 countries, has standardized the territorial limits for most countries at 12 nautical miles (about 22 kilometers or 14 land miles). Under the Law of the Sea, states also have exclusive rights to the fish and other marine life within 200 miles (320 kilometers). Countries separated by less than 400 miles of sea must negotiate the location of the boundary between exclusive fishing rights. Disputes can be taken to a Tribunal for the Law of the Sea or to the International Court of Justice.

Cultural Boundaries

Two types of cultural boundaries are common—geometric and ethnic. Geometric boundaries are simply straight lines drawn on a map. Other boundaries between states coincide with differences in ethnicity, especially language and religion.

GEOMETRIC BOUNDARIES. Part of the northern U.S. boundary with Canada is a 2,100-kilometer (1,300-mile) straight line (more precisely, an arc) along 49° north latitude, running from Lake of the Woods between Minnesota and Manitoba to the Strait of Georgia between Washington State and British Columbia. This boundary was established in 1846 by a treaty between the United States and Great Britain, which still controlled Canada. The two countries share an additional 1,100-kilometer (700-mile) geometric boundary between Alaska and the Yukon Territory along the north–south arc of 141° west longitude.

FIGURE 8-13 Water boundary: Lake Victoria. The boundary between Kenya, Tanzania, and Uganda runs through Lake Victoria.

The 1,000-kilometer (600-mile) boundary between Chad and Libya is a straight line drawn across the desert in 1899 by the French and British to set the northern limit of French colonies in Africa (Figure 8-14). Libya claimed that the straight line should be 100 kilometers (60 miles) to the south. Citing an agreement between France and Italy in 1935, Libya seized the territory in 1973. In 1987, Chad expelled the Libyan army with the help of French forces and regained control of the strip.

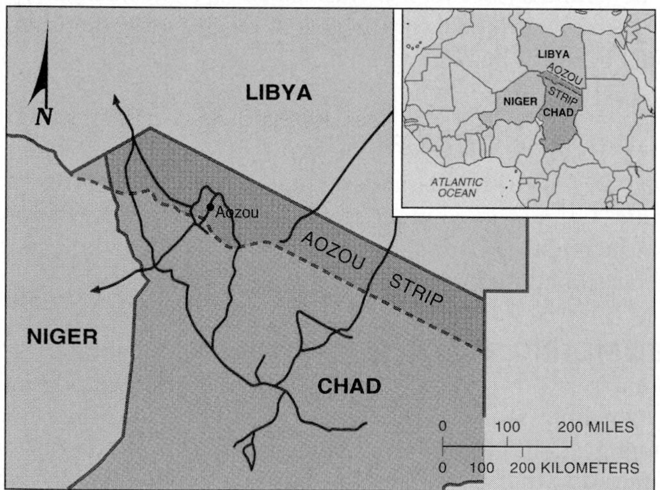

FIGURE 8-14 Geometric boundary: Aozou Strip. The boundary between Libya and Chad is a straight line, drawn by European countries early in the twentieth century when the area comprised a series of colonies. Libya, however, claims that the boundary should be located 100 kilometers to the south and that it should have sovereignty over the Aozou Strip.

RELIGIOUS BOUNDARIES. Boundaries between countries have been placed where possible to separate speakers of different languages or followers of different religions. Religious differences often coincide with boundaries between states, but in only a few cases has religion been used to select the actual boundary line.

The most notable example was in South Asia, when the British partitioned India into two states on the basis of religion. The predominantly Muslim portions were allocated to Pakistan, whereas the predominantly Hindu portions became the independent state of India (see Figure 7-26). Religion was also used to some extent to draw the boundary between two states on the island of Eire (Ireland). Most of the island became an independent country, but the northeast—now known as Northern Ireland—remained part of the United Kingdom. Roman Catholics comprise approximately 95 percent of the population in the 26 counties that joined the Republic of Ireland, whereas Protestants constitute the majority in the six counties of Northern Ireland (see Figure 6-23).

LANGUAGE BOUNDARIES. Language is an important cultural characteristic for drawing boundaries, especially in Europe. England, France, Portugal, and Spain are examples of European states that coalesced around distinctive languages before the nineteenth century. Germany and Italy emerged in the nineteenth century as states unified by language.

The movement to identify nationalities on the basis of language spread elsewhere in Europe during the twentieth century. After World War I, leaders of the victorious countries met at the Versailles Peace Conference to redraw the map of Europe. One of the chief advisers to President Woodrow Wilson, the geographer Isaiah Bowman, played a major role in the decisions. Language was the most important criterion the allied leaders used to create new states in Europe and to adjust the boundaries of existing ones.

The Versailles conference was particularly concerned with Eastern and Southern Europe, regions long troubled by political instability and conflict. Boundaries were drawn around the states of Bulgaria, Hungary, Poland, and Romania to conform closely to the distribution of Bulgarian, Hungarian (Magyar), Polish, and Romanian speakers. Speakers of several similar South Slavic languages were placed together in the new country of Yugoslavia. Czechoslovakia was created by combining the speakers of Czech and Slovak, mutually intelligible West Slavic languages (refer to Figure 7-30).

The nation-states created by the Versailles conference on the basis of language lasted with minor adjustment through most of the twentieth century. As discussed in Chapter 7, a nation-state exists when the boundaries of a state match the boundaries of the territory inhabited by an ethnic group. Problems exist when the boundaries do not match. However, during the 1990s, the boundaries on the map of Europe drawn at Versailles in 1919 collapsed. Despite speaking similar languages, Czechs and Slovaks found that they could no longer live together peacefully in the same state. Croats, Macedonians, Serbs, and Slovenes realized the same.

Cyprus's "Green Line" Boundary

Cyprus, the third-largest island in the Mediterranean Sea, contains two nationalities—Greek and Turkish (Figure 8-15). Although the island is physically closer to Turkey, Turks comprise only 18 percent of the country's population, whereas Greeks account for 78 percent. When Cyprus gained independence from Britain in 1960, its constitution guaranteed the Turkish minority a substantial share of elected offices and control over its own education, religion, and culture. But Cyprus has never peacefully integrated the Greek and Turkish nationalities.

Several Greek Cypriot military officers who favored unification of Cyprus with Greece seized control of the government in 1974. Shortly after the coup, Turkey invaded Cyprus to protect the Turkish Cypriot minority. The Greek coup leaders were removed within a few months, and an elected government was restored, but the Turkish army remained on Cyprus. The northern 36 percent of the island controlled by Turkey declared itself the independent Turkish Republic of Northern Cyprus in 1983, but only Turkey recognizes it as a separate state.

A wall was constructed between the two areas, and a buffer zone patrolled by the United Nations was delineated across the entire island. Traditionally, the Greek and Turkish Cypriots had mingled, but after the wall and buffer zone were established, the two nationalities became geographically isolated. The northern part of the island is now overwhelmingly Turkish, whereas the southern part is overwhelmingly Greek. Approximately one-third of the island's Greeks were forced to move from the region controlled by the Turkish army, whereas nearly one-fourth of the Turks moved from the region now considered to be the Greek side.

The two sides have been brought closer in recent years. A portion of the wall was demolished, and after three decades the two nationalities could again cross to the other side. The European Union accepted the entire island of Cyprus as a member in 2004. A UN Peace Plan for reunification was accepted by the Turkish side but rejected by the Greek side.

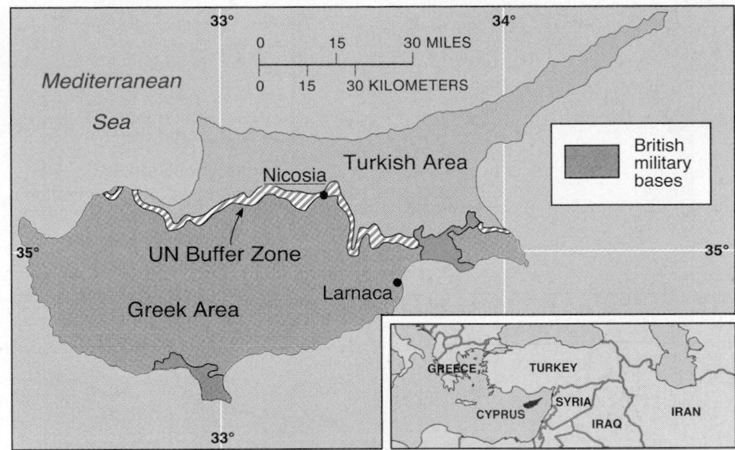

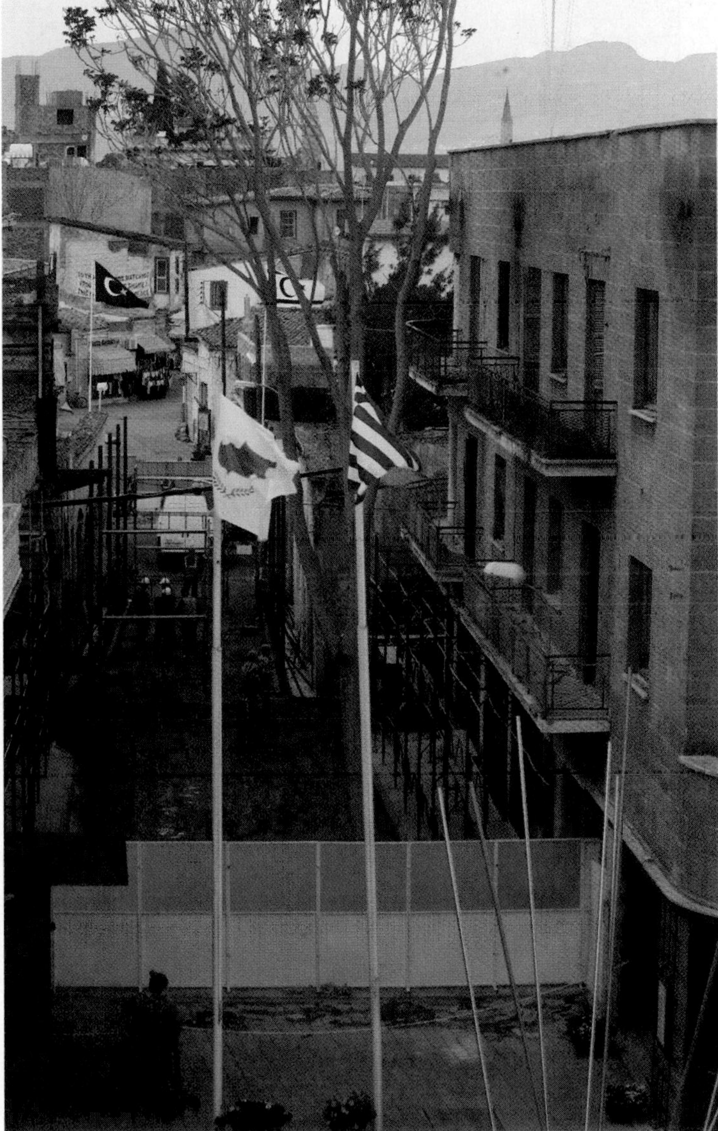

FIGURE 8-15 Cultural boundary: Cyprus. Since 1974, Cyprus has been divided into Greek and Turkish areas, separated by a United Nations Buffer Zone. The photo shows a crossing between the Greek side (foreground) and Turkish side (background), through the UN Buffer Zone (middle).

Frontiers

Historically, frontiers rather than boundaries separated states (Figure 8-16). A **frontier** is a zone where no state exercises complete political control. It is a tangible geographic area, whereas a boundary is an infinitely thin, invisible, imaginary line. A frontier provides an area of separation, often kilometers in width, but a boundary brings two neighboring states into direct contact, increasing the potential for violent face-to-face meetings. A frontier area is either uninhabited or sparsely

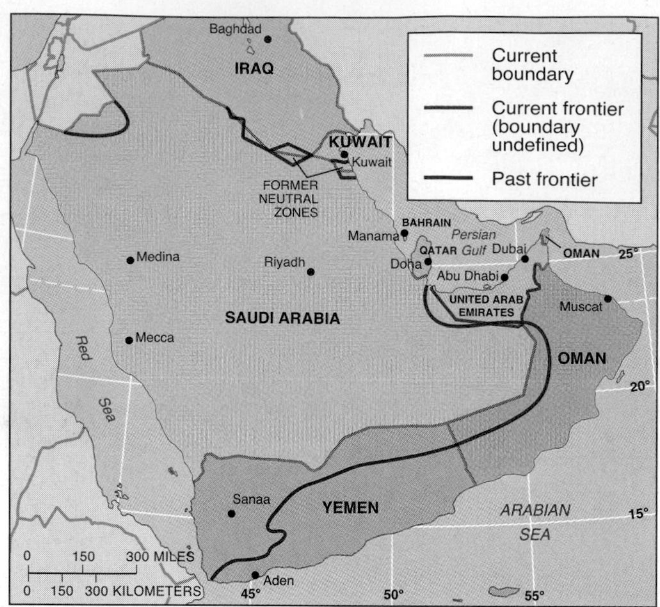

FIGURE 8-16 Frontiers in the Arabian Peninsula. Frontiers rather than boundaries separated Saudi Arabia from its neighbors. In the late twentieth century, most of these frontiers were converted to boundaries, but frontiers remain between Saudi Arabia and Yemen and UAE.

settled by a few isolated pioneers seeking to live outside organized society.

Almost universally, frontiers between states have been replaced by boundaries. Modern communications systems permit countries to monitor and guard boundaries effectively, even in previously inaccessible locations. Once-remote frontier regions have become more attractive for agriculture and mining. Most of the borders between Saudi Arabia and its neighbors, for example, remained frontiers until recently. Saudi Arabia was separated from Kuwait by a diamond-shaped frontier called a Neutral Zone until 1965, and another diamond-shaped Neutral Zone separated Saudi Arabia from Iraq until 1981. Saudi Arabia converted its frontiers with Oman to boundaries in 1990, and with Yemen in 2000. In all three cases, Saudi Arabia agreed with its neighbors to share the former frontier regions' resources, such as water, oil, and grazing land, and to permit nomads to wander freely in the frontier.

Boundaries Inside States

Within countries, local government boundaries are sometimes drawn to separate different nationalities or ethnicities. They are also drawn sometimes to provide advantage to a political party.

Unitary and Federal States

In the face of increasing demands by ethnicities for more self-determination, states have restructured their governments to transfer some authority from the national government to local government units. An ethnicity that is not sufficiently numerous to gain control of the national government may be content with control of a regional or local unit of government.

The governments of states are organized according to one of two approaches—the unitary system or the federal system. The **unitary state** places most power in the hands of central government officials, whereas the **federal state** allocates strong power to units of local government within the country.

A country's cultural and physical characteristics influence the evolution of its governmental system. In principle, the unitary government system works best in nation-states characterized by few internal cultural differences and a strong sense of national unity. Because the unitary system requires effective communications with all regions of the country, smaller states are more likely to adopt it (Figure 8-17). Unitary states are especially common in Europe.

In a federal state, such as the United States, local governments possess more authority to adopt their own laws. Multinational states may adopt a federal system of government to empower different nationalities, especially if they live in separate regions of the country. Under a federal system, local government boundaries can be drawn to correspond with regions inhabited by different ethnicities.

The federal system is also more suitable for very large states because the national capital may be too remote to provide effective control over isolated regions. Most of the world's largest states are federal, including Russia, Canada, the United States, Brazil, and India. However, the size of the state is not always an accurate predictor of the form of government: Tiny Belgium is a federal state (to accommodate the two main cultural groups, the Flemish and the Waloons, as discussed in Chapter 5), whereas China is a unitary state (to promote Communist values).

Some multinational states have adopted unitary systems, so that the values of one nationality can be imposed on others. In Kenya and Rwanda, for instance, the mechanisms of a unitary state have enabled one ethnic group to extend dominance over weaker groups.

Trend Toward Federal Government

In recent years there has been a strong global trend toward federal government. Unitary systems have been sharply curtailed in a number of countries and scrapped altogether in others.

FRANCE: CURBING A UNITARY GOVERNMENT. A good example of a nation-state, France has a long tradition of unitary government in which a very strong national government dominates local government decisions. Their basic local government unit is the *département* (department). Each of the 96 departments has an elected general council, but its administrative head is a powerful *préfet* appointed by the national government rather than directly elected by the people.

A second tier of local government in France is the *commune*. Each of the 36,686 communes has a locally elected mayor and council, but the mayor can be a member of the national parliament at the same time. Further, the median size of a commune is 380 inhabitants, too small to govern effectively, with the possible exception of the largest ones, such as those in Paris, Lyon, Lille, and Marseille.

FIGURE 8-17 Monaco, a unitary microstate. The smallest microstate in the United Nations, Monaco is a principality, ruled by a prince.

The French government has granted additional legal powers to the departments and communes in recent years. Local governments can borrow money freely to finance new projects without explicit national government approval, which was formerly required. The national government gives a block of funds to localities with no strings attached. In addition, 22 regional councils that previously held minimal authority have been converted into full-fledged local government units, with elected councils and the power to levy taxes.

POLAND: A NEW FEDERAL GOVERNMENT. Poland switched from a unitary to a federal system after control of the national government was wrested from the Communists. The federal system was adopted to dismantle legal structures by which Communists had maintained unchallenged power for more than 40 years. Under the Communists' unitary system, local governments held no legal authority. The national government appointed local officials and owned public property. This system led to deteriorated buildings, roads, and water systems, because the national government did not allocate sufficient funds to maintain property and no one had clear responsibility for keeping property in good condition. In 1999, Poland adopted a three-tier system of local government. The country was divided into 16 *województwa* (provinces). Each *województwo* was divided into between 12 and 42 *powiats* (counties), and each *powiat* was divided into between 3 and 19 *gmina* (municipalities).

The transition to a federal system of government proved difficult in Poland and other Eastern European countries. Given the absence of local government for a half century, elected officials had no experience in governing a community. The first task for many newly elected councilors was to attend a training course in how to govern. To compound the problem of adopting a federal system, Poland's newly elected local government officials had to find thousands of qualified people to fill appointed positions, such as directors of education, public works, and planning.

Electoral Geography

The boundaries separating legislative districts within the United States and other countries are redrawn periodically to ensure that each district has approximately the same population. Boundaries must be redrawn because migration inevitably results in some districts gaining population, whereas others are losing. The districts of the U.S. House of Representatives are redrawn every 10 years following the release of official population figures by the Census Bureau.

The job of redrawing boundaries in most European countries is entrusted to independent commissions. Commissions typically try to create compact homogeneous districts without regard for voting preferences or incumbents. A couple of U.S. states, including Iowa and Washington, also use independent or bipartisan commissions, but in most U.S. states the job of redrawing boundaries is entrusted to the state legislature. The political party in control of the state legislature naturally attempts to redraw boundaries to improve the chances of its supporters to win seats. The process of redrawing legislative boundaries for the purpose of benefiting the party in power is called **gerrymandering**.

The term *gerrymandering* was named for Elbridge Gerry (1744–1814), governor of Massachusetts (1810–1812) and vice president of the United States (1813–1814). As governor, Gerry signed a bill that redistricted the state to benefit his party. An opponent observed that an oddly shaped new district looked like a "salamander," whereupon another opponent responded that it was a "gerrymander." A newspaper subsequently printed an editorial cartoon of a monster named "gerrymander" with a body shaped like the district.

Gerrymandering takes three forms (Figure 8-18). "Wasted vote" spreads opposition supporters across many districts but in the minority. "Excess vote" concentrates opposition supporters into a few districts. "Stacked vote" links distant areas of like-minded voters through oddly shaped boundaries. "Stacked vote" gerrymandering has been especially attractive for creating districts inclined to elect ethnic minorities. Because the two largest ethnic groups in the United States (African Americans and most Hispanics other than Cubans) tend to vote Democratic—in some elections more than 90 percent of African Americans vote Democratic—creating a majority African American district virtually guarantees election of a Democrat. Republicans support a "stacked" Democratic district because they are better able to draw boundaries that are favorable to their candidates in the rest of the state.

The U.S. Supreme Court ruled gerrymandering illegal in 1985 but did not require dismantling of existing oddly shaped districts, and a 2001 ruling allowed North Carolina to add another oddly

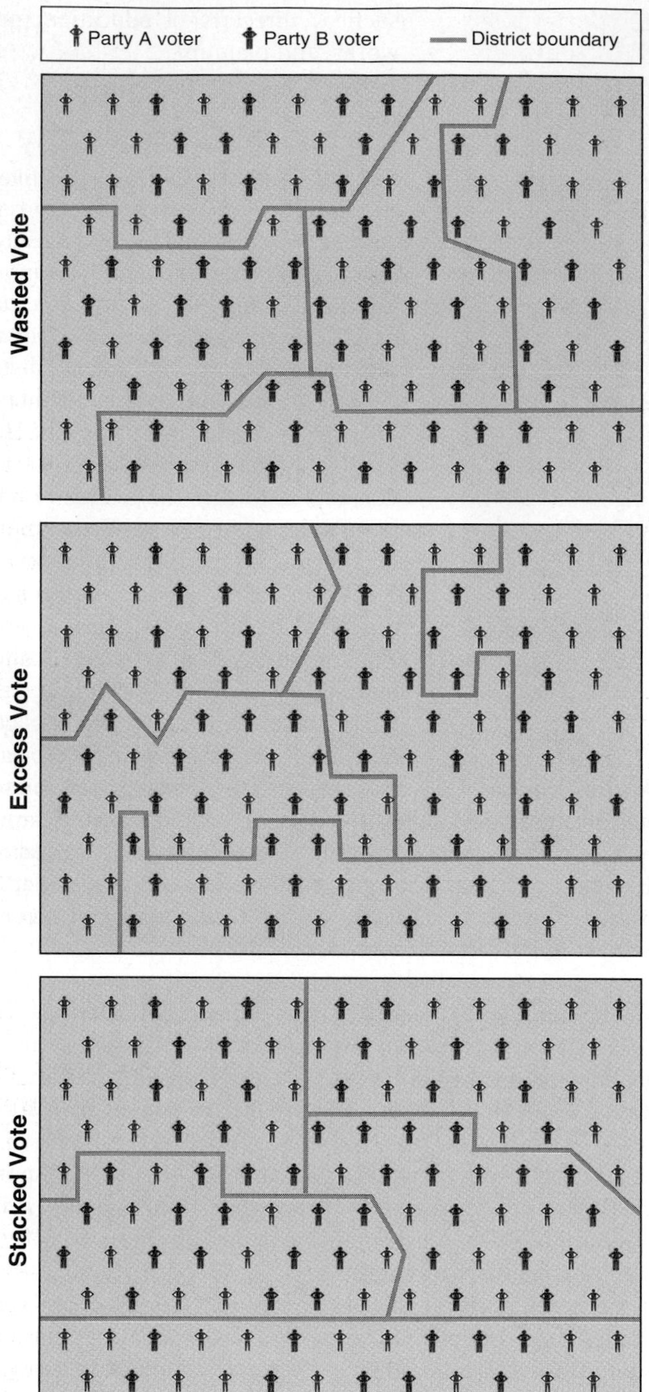

FIGURE 8-18 Three forms of gerrymandering. In all cases, Party A has 52 percent and Party B 48 percent of the overall vote. (top) "Wasted vote" spreads opposition supporters across many districts as a minority. If Party A controls the redistricting process, it could do a "wasted vote" gerrymander by putting in each of the five districts 13 of its voters and 12 of Party B voters, thereby giving Party A the opportunity to win all five districts. (middle) "Excess vote" concentrates opposition supporters into a few districts. If Party B controls the redistricting process, it could do an "excess vote" gerrymander by putting 13 of its voters and 12 of Party A voters in four of the five districts and concentrating 17 Party A voters and only 8 Party B voters in the fifth district, thereby giving Party B the likelihood of winning four of five districts. (bottom) A "stacked vote" links distant areas of like-minded voters through oddly shaped boundaries. If Party A controls the redistricting process, the trend is for state legislatures to create three districts each with 15 of its voters and 10 of Party B's voters and two districts both with 10 of its voters and 15 of Party B's voters. That way, all five districts are safely in possession of one party, with a majority of three for Party A and two for Party B.

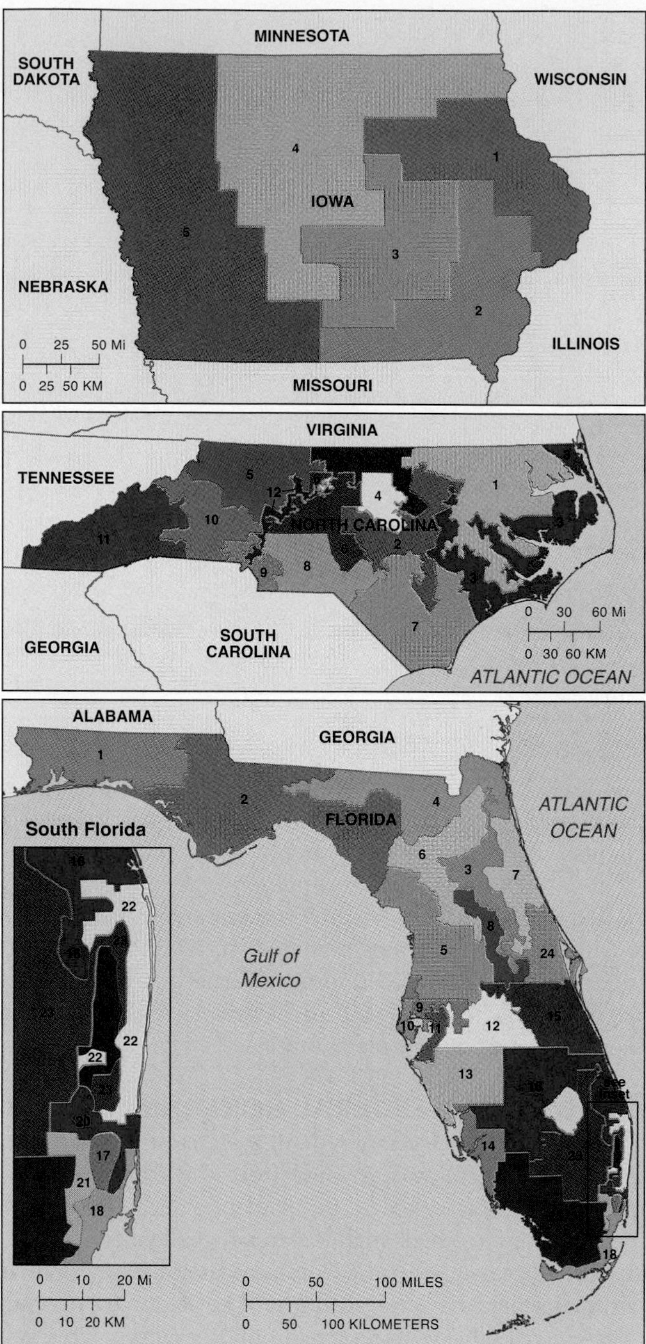

FIGURE 8-19 Gerrymandering examples. (top) Iowa is a state that does not have gerrymandered congressional districts. Each district is relatively compact, and boundaries coincide with county boundaries. (middle) In North Carolina, Democrats gerrymandered congressional districts to concentrate Republican voters. Democrats won 8 of North Carolina's 13 seats in 2008, although Democrat Barack Obama won North Carolina by only 14,000 votes (0.4 percent). (bottom) Meanwhile in Florida, Republicans gerrymandered congressional district boundaries to concentrate Democratic voters. Republicans won 15 of Florida's 25 seats in 2008, although President Obama carried the state by a relatively comfortable margin of 205,000 votes (2.5 percent).

shaped district that ensured the election of an African American Democrat (Figure 8-19). Through gerrymandering, only about one-tenth of congressional seats are competitive, making a shift of more than a few seats unlikely from one election to another in the United States except in unusual circumstances.

KEY ISSUE 3

Why Do States Cooperate with Each Other?

- Political and Military Cooperation
- Economic Cooperation

Chapter 7 illustrated examples of challenges to the survival of states from the trend toward local diversity. Ethnicities seek the right of self-determination as an expression of unique cultural identity. The inability to accommodate the diverse aspirations of ethnicities has led to the breakup of states into smaller ones, especially in Eastern Europe.

The future of the world's current collection of sovereign states is also challenged by the trend toward globalization. States are willingly transferring authority to regional organizations, established primarily for economic cooperation. Although it has limited authority, the United Nations includes all but a handful of states. ■

Political and Military Cooperation

During the Cold War era (late 1940s until early 1990s), global and regional organizations were established primarily to prevent a third world war in the twentieth century and to protect countries from a foreign attack. With the end of the Cold War, some of these organizations have flourished and found new roles, whereas others have withered.

The United Nations

The most important global organization is the United Nations, created at the end of World War II by the victorious Allies. When established in 1945, the United Nations comprised 49 states, but membership grew to 192 in 2006, making it a truly global institution (refer to Figure 8-5).

The number of countries in the United Nations increased rapidly on three occasions—1955, 1960, and the early 1990s. Sixteen countries joined in 1955, mostly European countries that had been liberated from Nazi Germany during World War II. Seventeen new members were added in 1960, all but one a former African colony of Britain or France. Twenty-six countries were added between 1990 and 1993, primarily from the breakup of the Soviet Union and Yugoslavia. UN membership also increased in the 1990s because of the admission of several microstates.

The United Nations was not the world's first attempt at international peacemaking. It replaced an earlier organization known as the League of Nations, which was established after World War I. The League was never an effective peacekeeping organization. The United States did not join, despite the fact that President Woodrow Wilson initiated the idea, because the U.S. Senate refused to ratify the membership treaty. By the 1930s, Germany, Italy, Japan, and the Soviet Union had all withdrawn, and the League could not stop aggression by these states against neighboring countries.

UN members can vote to establish a peacekeeping force and request states to contribute military forces. The United Nations is playing an important role in trying to separate warring groups in a number of regions, especially in Eastern Europe, the Middle East, and sub-Saharan Africa. However, any one of the five permanent members of the Security Council—China, France, Russia (formerly the Soviet Union), the United Kingdom, and the United States—can veto a peacekeeping operation. During the Cold War era, the United States and the Soviet Union used the veto to prevent undesired UN intervention, and it was only after the Soviet Union's delegate walked out of a Security Council meeting in 1950 that the UN voted to send troops to support South Korea. More recently, the opposition of China, France, and Russia prevented the United Kingdom and the United States from securing support from the United Nations for the 2003 attack on Iraq to overthrow Saddam Hussein.

Because it must rely on individual countries to supply troops, the UN often lacks enough of them to keep peace effectively. The UN tries to maintain strict neutrality in separating warring factions, but this has proved difficult in places such as Bosnia & Herzegovina, where most of the world sees one ethnicity (Bosnian Serbs) as a stronger aggressor and another (Bosnian Muslims) as a weaker victim. Despite its shortcomings, though, the UN represents a forum where, for the first time in history, virtually all states of the world can meet and vote on issues without resorting to war.

Regional Military Alliances

In addition to joining the United Nations, many states joined regional military alliances after World War II. The division of the world into military alliances resulted from the emergence of two states as superpowers—the United States and the Soviet Union.

ERA OF TWO SUPERPOWERS. During the Cold War era, the United States and the Soviet Union were the world's two superpowers. As very large states, both superpowers could quickly deploy armed forces in different regions of the world (Figure 8-20). To maintain strength in regions that were not contiguous to their own territory, the United States and the Soviet Union established military bases in other countries. From these bases, ground and air support were in proximity to local areas of conflict. Naval fleets patrolled the major bodies of water.

Before the Cold War, the world typically contained more than two superpowers. For example, before the outbreak of World War I in the early twentieth century, there were eight great powers: Austria, France, Germany, Italy, Japan, Russia, the United Kingdom, and the United States. When a large number of states ranked as great powers of approximately equal strength, no single state could dominate. Instead, major powers joined together to form temporary alliances.

A condition of roughly equal strength between opposing alliances is known as a **balance of power**. In contrast, the post–World War II balance of power was bipolar between the

FIGURE 8-20 The Cold War: 1962 Cuban Missile Crisis. A major confrontation during the Cold War between the United States and Soviet Union came in 1962 when the Soviet Union secretly began to construct missile launching sites in Cuba, less than 150 kilometers (90 miles) from U.S. territory. President Kennedy demanded that the missiles be removed and ordered a naval blockade to prevent further Soviet material from reaching Cuba. The crisis ended when the Soviet Union agreed to dismantle the sites. The U.S. Department of Defense took aerial photographs to show the Soviet buildup in Cuba. (Top) Three Soviet ships with missile equipment are being unloaded at Mariel naval port in Cuba. Within the outline box (enlarged below and rotated 90° clockwise) are Soviet missile transporters, fuel trailers, and oxider trailers (used to support the combustion of missile fuel).

United States and the Soviet Union. Because the power of these two states was so much greater than all others, the world comprised two camps, each under the influence of one of the superpowers. Other states lost the ability to tip the scales significantly in favor of one or the other superpower. They were relegated to a new role, that of ally or satellite.

Both superpowers repeatedly demonstrated that they would use military force if necessary to prevent an ally from becoming too independent. The Soviet Union sent its armies into Hungary in 1956 and Czechoslovakia in 1968 to install more sympathetic governments. Because these states were clearly within the orbit of the Soviet Union, the United States chose not to intervene militarily. Similarly, the United States sent troops to the Dominican Republic in 1965, Grenada in 1983, and Panama in 1989 to ensure that they would remain allies.

MILITARY COOPERATION IN EUROPE. After World War II, most European states joined one of two military alliances dominated by the superpowers—NATO (North Atlantic Treaty Organization) or the Warsaw Pact (Figure 8-21, left). NATO was a military alliance among 16 democratic states, including the United States and Canada plus 14 European states. The Warsaw Pact was a military agreement among Communist Eastern European countries to defend each other in case of attack. Seven members joined the Warsaw Pact when it was founded in 1955. Some of Hungary's leaders in 1956 asked for the help of Warsaw Pact troops to crush an uprising that threatened Communist control of the government. Warsaw Pact troops also invaded Czechoslovakia in 1968 to depose a government committed to reforms.

NATO and the Warsaw Pact were designed to maintain a bipolar balance of power in Europe. For NATO allies, the principal objective was to prevent the Soviet Union from overrunning West Germany and other smaller countries. The Warsaw Pact provided the Soviet Union with a buffer of allied states between it and Germany to discourage a third German invasion of the Soviet Union in the twentieth century.

In a Europe no longer dominated by military confrontation between two blocs, the Warsaw Pact was disbanded, and the number of troops under NATO command was sharply reduced. NATO expanded its membership to include most of the former Warsaw Pact countries. Membership in NATO offered Eastern European countries an important sense of security against any future Russian threat, no matter how remote that appears at the moment, as well as participation in a common united Europe (Figure 8-21, right).

OTHER REGIONAL ORGANIZATIONS. Other prominent regional organizations include:

- **The Organization on Security and Cooperation in Europe (OSCE).** Its 56 members include the United States, Canada, and Russia, as well as most European countries. When founded in 1975, the Organization on Security and Cooperation was composed primarily of Western European countries and played only a limited role. With the end of the Cold War in the 1990s, the renamed OSCE expanded to include Warsaw Pact countries and became a more active forum for countries concerned with ending conflicts in Europe, especially in the Balkans and Caucasus. Although the OSCE does not directly command armed forces, it can call upon member states to supply troops if necessary.
- **The Organization of American States (OAS).** All 35 states in the Western Hemisphere are members. Cuba is a member but was suspended from most OAS activities in 1962. The organization's headquarters, including the permanent council and general assembly, are located in Washington, D.C. The OAS promotes social, cultural, political, and economic links among member states.
- **The African Union (AU).** Established in 2002, it encompasses 53 countries in Africa. The AU replaced an earlier organization called the Organization of African Unity, founded in 1963 primarily to seek an end to colonialism and

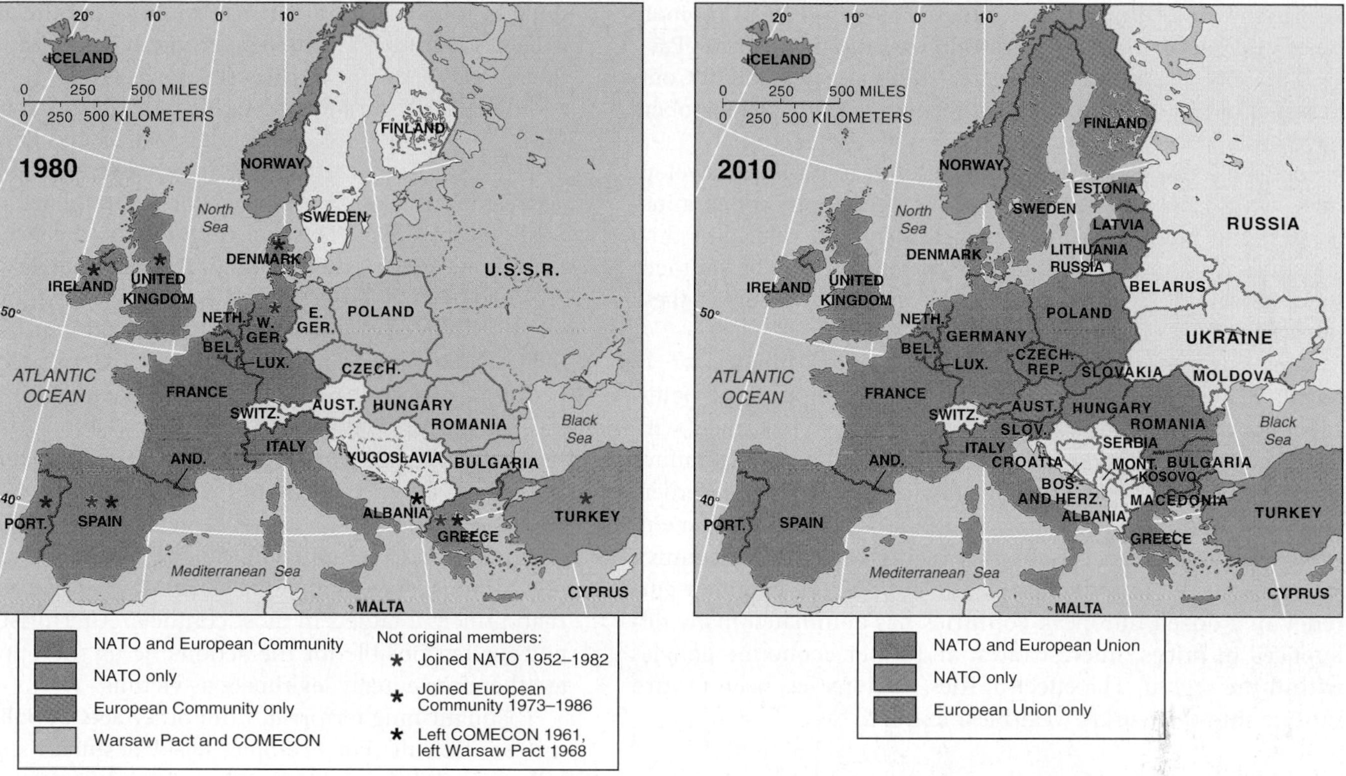

FIGURE 8-21 (left) Economic and military alliances in Europe during the Cold War. Western European countries joined the European Union and the North Atlantic Treaty Organization (NATO), whereas Eastern European countries joined COMECON and the Warsaw Pact. (right) Post–Cold War economic and military alliances in Europe. COMECON and the Warsaw Pact have been disbanded, whereas the European Union and NATO have accepted or plan to accept new members.

apartheid in Africa. The new organization has placed more emphasis on promoting economic integration in Africa.

- **The Commonwealth.** It includes the United Kingdom and 52 other states that were once British colonies, including Australia, Bangladesh, Canada, India, Nigeria, and Pakistan. Most other members are African states or island countries in the Caribbean or Pacific. Commonwealth members seek economic and cultural cooperation.

Economic Cooperation

The era of a bipolar balance of power formally ended when the Soviet Union was disbanded in 1992, and the world has returned to the pattern of more than two superpowers that predominated before World War II. The contemporary pattern of global power displays two key differences, however:

1. The most important elements of state power are increasingly economic rather than military. China, Germany, and Japan have joined the ranks of superpowers on the basis of their economic success, whereas Russia has slipped in strength because of economic problems (see Chapter 9).
2. The leading superpower is not a single state, such as the United States or Russia, but an economic union of European states.

With the decline in the military-oriented alliances, European states increasingly have turned to economic cooperation. Western Europe's most important economic organization is the European Union (formerly known as the European Economic Community, the Common Market, and the European Community). When it was established in 1958, the predecessor to the European Union included six countries—Belgium, France, Italy, Luxembourg, the Netherlands, and the Federal Republic of Germany (West Germany). The union was designed to heal Western Europe's scars from World War II (which had ended only 13 years earlier) when Nazi Germany, in alliance with Italy, conquered the other four countries. The European Union has expanded from 6 countries during the 1950s and 1960s to 12 countries during the 1980s and 27 countries during the first decade of the twenty-first century. A European Parliament is elected by the people in each of the member states simultaneously.

Croatia and Turkey have begun negotiations to join, but the European Union has not yet set a timetable. Macedonia is a candidate to join but negotiations have not started. Albania, Bosnia and Herzegovina, Iceland, Kosovo, Montenegro, and Serbia have been designated by the European Union as potential candidates.

In 1949, during the Cold War, the seven Eastern European Communist states in the Warsaw Pact formed an organization for economic cooperation, the Council for Mutual Economic Assistance (COMECON). Cuba, Mongolia, and Vietnam were also members of the alliance, which was designed to promote trade and sharing of natural resources. Like the Warsaw Pact, COMECON disbanded in the early 1990s after the fall of Communism in Eastern Europe. Most former COMECON members have joined the European Union.

The main task of the European Union is to promote development within the member states through economic cooperation. At first the European Union played a limited role, providing subsidies to farmers and to depressed regions such as southern Italy. Most of the European Union's budget still serves these purposes.

The European Union has taken on more importance in recent years as member states seek greater economic and political cooperation. It has removed most barriers to free trade; with a few exceptions, goods, services, capital, and people can move freely through Europe. Trucks can carry goods across borders without stopping, and a bank can open branches in any member country with supervision only by the bank's home country. In addition, the introduction of the euro as the common currency in a dozen European countries has eliminated many differences in prices, interest rates, and other economic policies within the region. The effect of these actions has been to turn Europe into the world's wealthiest market.

KEY ISSUE 4
Why Has Terrorism Increased?

- **Terrorism by Individuals and Organizations**
- **State Support for Terrorism**

Terrorism is the systematic use of violence by a group in order to intimidate a population or coerce a government into granting its demands. Terrorists attempt to achieve their objectives through organized acts that spread fear and anxiety among the population, such as bombing, kidnapping, hijacking, taking of hostages, and assassination. Terrorists consider violence necessary as a means of bringing widespread publicity to goals and grievances that are not being addressed through peaceful means. Terrorists' belief in their cause is so strong that they do not hesitate to strike despite knowing they will probably die in the act. ■

Terrorism by Individuals and Organizations

The term *terror* (from the Latin "to frighten") was first applied to the period of the French Revolution between March 1793 and July 1794, known as the Reign of Terror. In the name of protecting the principles of the revolution, the Committee of Public Safety, headed by Maximilien Robespierre, guillotined several thousand of its political opponents. In modern times, the term *terrorism* has been applied to actions by groups operating outside government rather than to those of official government agencies, although some governments provide military and financial support for terrorists.

Four U.S. presidents have been assassinated—Lincoln (1865), Garfield (1881), McKinley (1901), and Kennedy (1963). The Roman Emperor Julius Caesar's assassination 2,000 years ago has been vividly re-created for future generations through Shakespeare's play. The assassination of the Archduke Franz Ferdinand, heir to the throne of Austria-Hungary, by a Serb in Sarajevo (capital of present-day Bosnia & Herzegovina) June 28, 1914, led directly to the outbreak of World War I. But terrorism differs from assassinations and other acts of political violence because attacks are aimed at ordinary people rather than at military targets or political leaders. Victims of terrorism are a cross section of citizens who happen to be at the target at the time of the attack. Other types of military action can result in civilian deaths—bombs can go astray, targets can be misidentified, and enemy's military equipment can be hidden in civilian buildings—but average individuals are unintended victims rather than principal targets in most conflicts. A terrorist considers all citizens responsible for the actions he or she opposes, so they are therefore equally legitimate as victims.

Distinguishing terrorism from other acts of political violence can be difficult. For example, if a Palestinian suicide bomber kills several dozen Israeli teenagers in a Jerusalem restaurant, is that an act of terrorism or wartime retaliation against Israeli government policies and army actions? Competing arguments are made: Israel's sympathizers denounce the act as a terrorist threat to the country's existence, whereas advocates of the Palestinian cause argue that long-standing injustices and Israeli army attacks on ordinary Palestinian civilians provoked the act.

Terrorism against Americans

The United States suffered several terrorist attacks during the late twentieth century:

- December 21, 1988: A terrorist bomb destroyed Pan Am Flight 103 over Lockerbie, Scotland, killing all 259 aboard, plus 11 on the ground.
- February 26, 1993: A car bomb parked in the underground garage damaged New York's World Trade Center, killing 6 and injuring about 1,000.
- April 19, 1995: A car bomb killed 168 people in the Alfred P. Murrah Federal Building in Oklahoma City.
- June 25, 1996: A truck bomb blew up an apartment complex in Dhahran, Saudi Arabia, killing 19 U.S. soldiers who lived there and injuring more than 100 people.
- August 7, 1998: U.S. embassies in Kenya and Tanzania were bombed, killing 190 and wounding nearly 5,000.
- October 12, 2000: The USS *Cole* was bombed while in the port of Aden, Yemen, killing 17 U.S. service personnel.

With the exception of the Oklahoma City bombing, Americans generally paid little attention to the attacks and had only a vague notion of who had committed them. It took the attack on the World Trade Center and Pentagon on September 11, 2001, for most Americans to feel threatened by terrorism.

Some of the terrorists during the 1990s were American citizens operating alone or with a handful of others. Theodore

J. Kaczynski, known as the Unabomber, was convicted of killing 3 people and injuring 23 others by sending bombs through the mail during a 17-year period. His targets were mainly academics in technological disciplines and executives in businesses whose actions he considered to be adversely affecting the environment. Timothy J. McVeigh was convicted and executed for the Oklahoma City bombing, and for assisting him Terry I. Nichols was convicted of conspiracy and involuntary manslaughter but not executed. McVeigh claimed his terrorist act was provoked by rage against the U.S. government for such actions as the Federal Bureau of Investigation's 51-day siege of the Branch Davidian religious compound near Waco, Texas, culminating with an attack on April 19, 1993, that resulted in 80 deaths.

September 11, 2001, Attacks

The most dramatic terrorist attack against the United States came on September 11, 2001 (Figure 8-22). The tallest buildings in the United States, the 110-story twin towers of the World Trade Center in New York City, were destroyed, and the Pentagon, near Washington, D.C., was damaged (Figure 8-23). The attacks resulted in nearly 3,000 fatalities:

- 93 (5 terrorists, 77 other passengers, and 11 crew members) on American Airlines Flight 11, which crashed into World Trade Center Tower 1 (North Tower)
- 65 (5 terrorists, 51 other passengers, and 9 crew members) on United Airlines Flight 175, which crashed into World Trade Center Tower 2 (South Tower)
- 2,605 on the ground at the World Trade Center
- 64 (5 terrorists, 53 other passengers, and 6 crew members) on American Airlines Flight 77, which crashed into the Pentagon
- 125 on the ground at the Pentagon
- 44 (4 terrorists, 33 other passengers, and 7 crew members) on United Airlines Flight 93, which crashed near Shanksville, Pennsylvania, after passengers fought with terrorists on board, preventing an attack on another Washington, D.C., target.

Al-Qaeda

Responsible or implicated in most of the anti-U.S. terrorism during the 1990s, as well as the September 11, 2001, attack, was the al-Qaeda network, founded by Osama bin Laden (see Global Forces, Local Impacts box). His father, Mohammed bin Laden, a native of Yemen, established a construction company in Saudi Arabia and became a billionaire through close connections to the royal family. Osama bin Laden, one of about 50 children fathered by Mohammed with several wives, used his several hundred million dollar inheritance to fund al-Qaeda (an Arabic word meaning "the foundation" or "the base") around 1990 to unite *jihad* fighters in Afghanistan, as well as supporters of bin Laden elsewhere in the Middle East.

Bin Laden moved to Afghanistan during the mid-1980s to support the fight against the Soviet army and the country's Soviet-installed government. Calling the anti-Soviet fight a holy

FIGURE 8-22 Terrorist attack on the World Trade Center. On September 11, 2001, at 9:03 A.M., United Flight 175 approaches World Trade Center Tower 2 (left) and crashes into it (right). Tower 1 is already burning from the crash of American Flight 11 at 8:45 A.M.

FIGURE 8-23 Aftermath of World Trade Center attack. Laser technology was used to create a topographic map of the World Trade Center site on September 19, 2001, 8 days after the attack. Colors represent elevation above sea level (in green) or below sea level (in red) of the destroyed buildings. Rubble was piled more than 60 feet high where the twin towers once stood. The top of the image faces northeast. West St. runs across the foreground, and Liberty St. runs between the bottom center and the upper right. Tower 1 rubble is the square-shaped pile in the middle of the block facing West St. The remains of Tower 2 face Liberty St.

war or *jihad*, bin Laden recruited militant Muslims from Arab countries to join the cause. After the Soviet Union withdrew from Afghanistan in 1989, bin Laden returned to Saudi Arabia, but he was expelled in 1991 for opposing the Saudi government's decision permitting the United States to station troops there during the 1991 war against Iraq. Bin Laden moved to Sudan but was expelled in 1994 for instigating attacks against U.S. troops in Yemen and Somalia, so he returned to Afghanistan, where he lived as a "guest" of the Taliban-controlled government.

Bin Laden issued a declaration of war against the United States in 1996 because of U.S. support for Saudi Arabia and Israel. In a 1998 *fatwa* ("religious decree"), bin Laden argued that Muslims had a duty to wage a holy war against U.S. citizens because the United States was responsible for maintaining the Saud royal family as rulers of Saudi Arabia and a state of Israel dominated by Jews. Destruction of the Saudi monarchy and the Jewish state of Israel would liberate from their control Islam's three holiest sites of Makkah (Mecca), Madinah, and Jerusalem.

Al-Qaeda has been implicated in several bombings since 9/11:

- May 12, 2003: 35 died (including 9 terrorists) in car bomb detonations at two apartment complexes in Riyadh, Saudi Arabia.
- November 15, 2003: Truck bombs killed 29 (including 2 terrorists) at two synagogues in Istanbul, Turkey.
- November 20, 2003: 32 (including 2 terrorists) were killed at the British consulate and British-owned HSBC Bank in Istanbul.
- May 29, 2004: 22 died in attacks on oil company offices in Khobar, Saudi Arabia.
- July 7, 2005: 56 died (including 4 terrorists) when several subway trains and buses were bombed in London, England.
- July 23, 2005: 88 died in bombings of resort hotels in Sharm-el-Sheikh, Egypt.
- November 9, 2005: 60 died in the bombing of three American-owned hotels in Amman, Jordan.

Al-Qaeda is not a single unified organization, and the number involved is unknown. Bin Laden is advised by a small leadership council, which has several committees that specialize in such areas as finance, military, media, and religious policy. In addition to the original organization founded by Osama bin Laden responsible for the World Trade Center attack, al-Qaeda also encompasses local franchises concerned with country-specific issues, as well as imitators and emulators ideologically aligned with al-Qaeda but not financially tied to it.

Jemaah Islamiyah is an example of an al-Qaeda franchise with local concerns, specifically with establishing fundamentalist Islamic governments in Southeast Asia. Jemaah Islamiyah terrorist activities have been concentrated in the world's most populous Muslim country, Indonesia:

- October 12, 2002: A nightclub in the resort town of Kuta on the island of Bali was bombed, killing 202.
- August 5, 2003: Car bombs killed 12 at a Marriott hotel in the capital Jakarta.
- September 9, 2004: Car bombs killed 9 or 11 at the Australian embassy, also in Jakarta.

GLOBAL FORCES, LOCAL IMPACTS

Where Is Osama bin Laden?

Osama bin Laden, founder of al-Qaeda, was last seen in public in Jalalabad, Afghanistan, 2 months after al-Qaeda's September 11, 2001, attacks on the World Trade Center and Pentagon. He last sent a radio transmission on November 28, 2001, from the Tora Bora cave complex in eastern Afghanistan (Figure 8-24). When U.S. forces captured the Tora Bora complex in December 2001, though, bin Laden was not there. Because his corpse was not found and he was not seen or heard live since then, the assumption is that bin Laden has been spending the years since the battle of Tora Bora hiding somewhere in the region. Was he living in a cave? Was he in Afghanistan, or in Pakistan?

A team of geographers at UCLA led by professors Tom Gillespie and John Agnew applied geographic techniques to pinpoint the likely whereabouts of bin Laden. In 2009, they reported that bin Laden was likely to be hiding in one of three buildings in the town of Parachinar in western Pakistan. The team reached its conclusion by applying various geographic concepts at three scales—global, regional, and local.

- **Global scale.** The UCLA geography team used the concept of distance-decay (see Chapter 1) to calculate the probability that bin Laden traveled various distances from Tora Bora to other places in the world. The closer to Tora Bora, the more likely bin Laden's location was to be. The probability was 98 percent that bin Laden was hiding in the Kurram Valley, along Pakistan's border with Afghanistan.
- **Regional scale.** A physical geography theory states that relatively few species with relatively high extinction rates are found on small isolated islands than on large islands near other habitats. Applying this theory to bin Laden's hideout, the UCLA geography team concluded that bin Laden was most likely to be hiding in a larger settlement relatively close to Tora Bora. Supporting this conclusion was bin Laden's need for daily dialysis and thus a source of electricity, most likely to be found in a larger town. Also, bin Laden had several bodyguards, easier to hide in a larger community. The team concluded that the most likely region was Parachinar, a town of 20,000 inhabitants.
- **Local scale.** Bin Laden's life history was applied to finding structures in Parachinar most suitable for his needs. Using air photos, the team found three structures—essentially fortified compounds—that fit best with bin Laden's behavior and needs.

Critics have pointed out that Parachinar is an unlikely hideout for bin Laden because it is a predominantly Shiite community, whereas bin Laden and most of his al-Qaeda followers are Sunnis. ■

FIGURE 8-24 Tora Bora, Afghanistan. This cave was Osama bin Laden's last confirmed residence.

- October 1, 2005: Attacks on a downtown square in Kuta as well as a food court in Jimbaran, also on Bali, killed 23 (including 3 terrorists).

Other terrorist groups have been loosely associated with al-Qaeda. For example:

- November 28, 2002: A Somali terrorist group killed 10 Kenyan dancers and 3 Israeli tourists at a resort in Mombasa, Kenya, and fired two missiles at an Israeli airplane taking off from the Mombasa airport.
- March 11, 2004: A local terrorist group blew up several commuter trains in Madrid, Spain, killing 192.

Al-Qaeda's use of religion to justify attacks has posed challenges to Muslims and non-Muslims alike. For many Muslims, the challenge has been to express disagreement with the policies of governments in the United States and Europe yet disavow the use of terrorism. For many Americans and Europeans, the challenge has been to distinguish between the peaceful but unfamiliar principles and practices of the world's 1.3 billion Muslims and the misuse and abuse of Islam by a handful of terrorists.

State Support for Terrorism

Several states in the Middle East have provided support for terrorism in recent years, at three increasing levels of involvement:

- Providing sanctuary for terrorists wanted by other countries
- Supplying weapons, money, and intelligence to terrorists
- Planning attacks using terrorists

Libya

The government of Libya was accused of sponsoring a 1986 bombing of a nightclub in Berlin, Germany, popular with U.S. military personnel then stationed there, killing three (including one U.S. soldier). U.S. relations with Libya had been poor since 1981 when U.S. aircraft shot down attacking Libyan warplanes while conducting exercises over waters that the United States considered international but that Libya considered inside its territory. In response to the Berlin bombing, U.S. bombers attacked the Libyan cities of Tripoli and Benghazi in a failed attempt to kill Colonel Muammar el-Qaddafi.

Libyan agents were found to have planted bombs that killed 270 people on Pan Am Flight 103 over Lockerbie, Scotland, in 1988, as well as 170 people on UTA Flight 772 over Niger in 1989. Following 8 years of UN economic sanctions, Qaddafi turned over suspects in the Lockerbie bombing for a trial that was held in the Netherlands under Scottish law. One of the two was acquitted; the other, Abdel Basset Ali al-Megrahi, was convicted and sentenced to life imprisonment, but he was released in 2009 after he was diagnosed with terminal cancer. Libya renounced terrorism in 2003, and has provided compensation for victims of Pan Am 103. UN sanctions have been lifted, and it is no longer considered a state sponsor of terrorism.

Afghanistan

U.S. accusations of state-sponsored terrorism escalated after 9/11. The governments of first Afghanistan, then Iraq, and then Iran were accused of providing at least one of the three levels of state support for terrorists. As part of its war against terrorism, the U.S. government with the cooperation of some other countries attacked Afghanistan in 2001 and Iraq in 2003 to depose those country's government leaders, who were considered supporters of terrorism.

The United States attacked Afghanistan in 2001 when its leaders, known as the Taliban, sheltered Osama bin Laden and other al-Qaeda terrorists. The Taliban (Pashto for "students") had gained power in Afghanistan in 1995 and had imposed strict Islamic fundamentalist law on the population. Afghanistan's Taliban leadership treated women especially harshly. Women were prohibited from attending school, working outside the home, seeking health care, or driving a car. They were permitted to leave home only if fully covered by clothing and escorted by a male relative.

The 6 years of Taliban rule temporarily suppressed a civil war that has raged in Afghanistan on and off since the 1970s. The civil war began in 1973 when the king was overthrown in a bloodless coup led by Mohammed Daoud Khan. Daoud was murdered 5 years later and replaced by a government led by military officers sympathetic to the Soviet Union. The Soviet Union sent 115,000 troops to Afghanistan beginning in 1979 after fundamentalist Muslims, known as *mujahedeen*, or "holy warriors," started a rebellion against the pro-Soviet government.

Although heavily outnumbered by Soviet troops and possessing much less sophisticated equipment, the *mujahedeen* offset the Soviet advantage by waging a guerrilla war in the country's rugged mountains, where they were more comfortable than the Soviet troops and where Soviet air superiority was ineffective. Unable to subdue the *mujahedeen*, the Soviet Union withdrew its troops in 1989; the Soviet-installed government in Afghanistan collapsed in 1992. After several years of infighting among the factions that had defeated the Soviet Union, the Taliban gained control over most of the country.

Six years of Taliban rule came to an end in 2001 following the U.S. invasion. Destroying the Taliban was necessary in order for the United States to go after al-Qaeda leaders, including Osama bin Laden, who were living in Afghanistan as guests of the Taliban. Removal of the Taliban unleashed a new struggle for control of Afghanistan among the country's many ethnic groups (Figure 8-25). When U.S. attention shifted to Iraq and Iran, the Taliban were able to regroup and resume an insurgency against the U.S.-backed Afghanistan government.

Iraq

U.S. claims of state-sponsored terrorism proved more controversial with regard to Iraq than to Afghanistan. The United States led an attack against Iraq in 2003 in order to depose Saddam Hussein, the country's longtime president. U.S. officials' justification for removing Hussein was that he had created biological and chemical weapons of mass destruction. These weapons could fall into the hands of terrorists, the U.S. government charged, because close links were said to exist between Iraq's government and al-Qaeda. The United Kingdom and a few other countries joined in the 2003 attack, but most countries did not offer support.

U.S. confrontation with Iraq predated the war on terror. From the time he became president of Iraq in 1979, Hussein's behavior had raised concern around the world. War with neighbor Iran, begun in 1980, ended 8 years later in stalemate. A nuclear reactor near Baghdad, where nuclear weapons to attack Israel were allegedly being developed, was destroyed in 1981 by Israeli planes. Hussein ordered the use of poison gas in 1988 against Iraqi Kurds, killing 5,000. Iraq's 1990 invasion of neighboring Kuwait, which Hussein claimed was part of Iraq, was opposed by the international community.

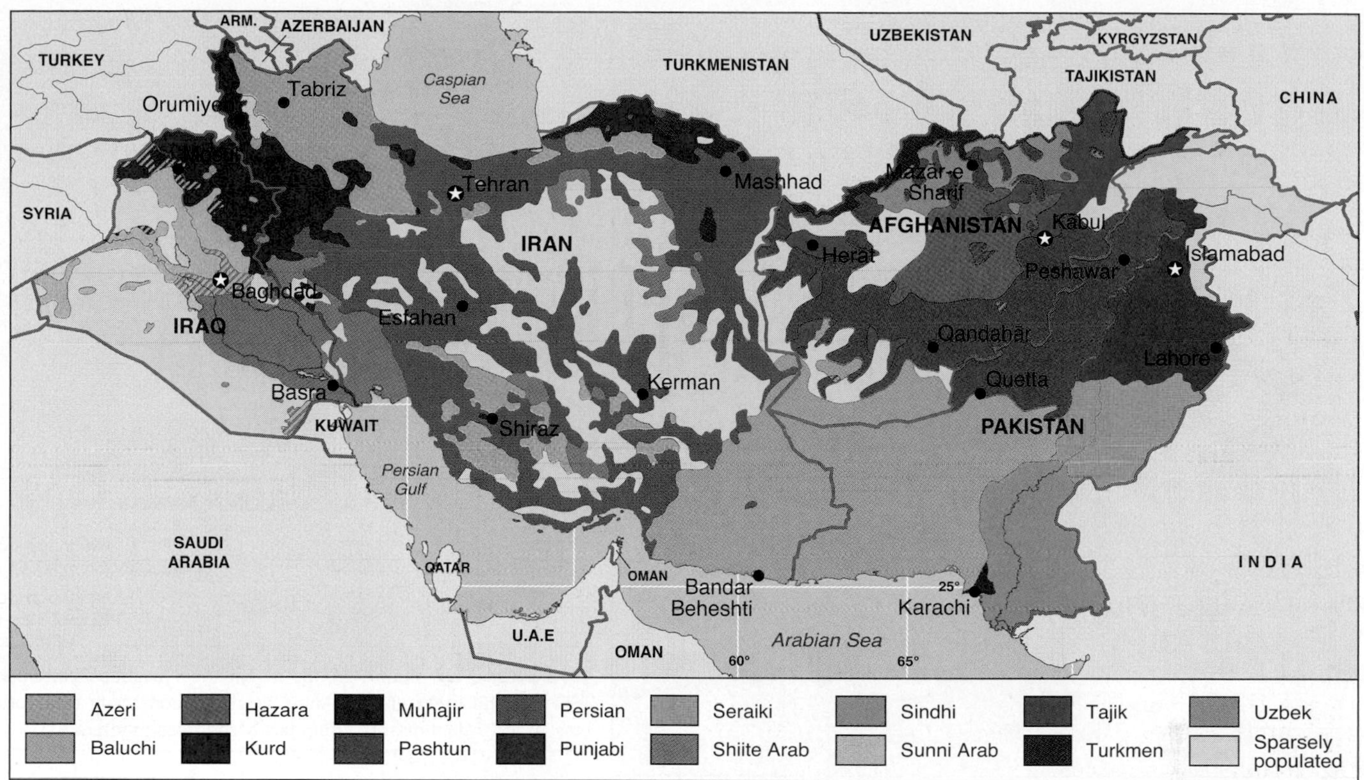

FIGURE 8-25 Ethnic groups in Southwest Asia. During the Cold War, the United States and the Soviet Union viewed conflicts in this region as part of a global struggle in support of or against the spread of communism. More recently, U.S. officials have regarded conflicts in Southwest Asia as part of the global war on terrorism. As the map shows, boundaries between ethnic groups do not match boundaries between countries in Southwest Asia, especially Iraq, Iran, Afghanistan, and Pakistan. This mismatch plays a critical role in the region's many wars.

The 1991 U.S.-led Gulf War, known as Operation Desert Storm, drove Iraq out of Kuwait, but it failed to remove Hussein from power. Desert Storm was supported by nearly every country in the United Nations because the purpose was to end one country's unjustified invasion and attempted annexation of another. In contrast, few countries supported the U.S.-led attack in 2003; most did not agree with the U.S. assessment that Iraq still possessed weapons of mass destruction or intended to use them.

Inspectors sent by the United Nations had found the following evidence of weapons of mass destruction in Iraq during the 1980s:

- A nuclear radiation weapon program, including 40 nuclear-research facilities and 3 uranium-enrichment programs
- A program for making weapons from the VX nerve agent
- A biological weapons program, including production of botulinum, anthrax, aflatoxin, and clostridium, and bombs to deliver toxic agents

UN experts concluded that Iraq had destroyed these weapons in 1991 after its Desert Storm defeat. U.S. officials believed instead that Iraq still had the weapons hidden, though they were never able to find them and their judgment may have been based on faulty intelligence.

The U.S. assertion that Hussein had close links with al-Qaeda was also challenged by most other countries, as well as ultimately by U.S. intelligence agencies. Hussein's Ba'ath Party, which ruled Iraq between 1968 and 2003, espoused different principles than the al-Qaeda terrorists. The guiding principle of the Ba'ath Party was Pan-Arab nationalism, which was the belief that the several hundred million Arabs living in the vast territory between North Africa and Central Asia should be joined together into one powerful nation-state, with financial strength garnered by sharing the region's extensive oil wealth. Whereas al-Qaeda terrorists justified their attacks on the basis of their interpretation of Islam, Ba'ath Party leaders were not observant Muslims and did not derive Pan-Arab philosophy from religious principles.

Lacking evidence of weapons of mass destruction and ties to al-Qaeda, the United States argued instead that Iraq needed a "regime change." Hussein's quarter-century record of brutality justified replacing him with a democratically elected government, according to U.S. officials. The U.S. position drew little international support—sovereign states are reluctant to invade another sovereign state just because they dislike its leader, no matter how odious.

Iraq is divided into around 150 tribes (Figure 8-26). After Hussein was toppled, tribes stepped into the political vacuum,

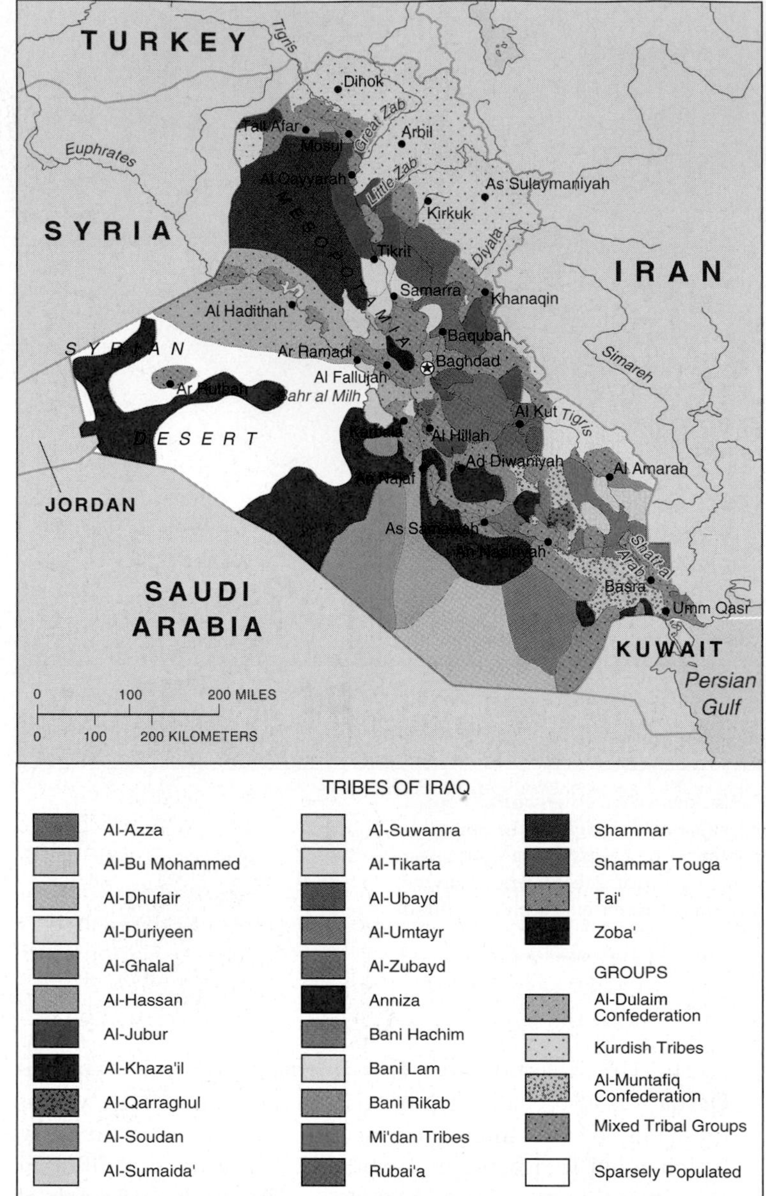

FIGURE 8-26 Major tribes in Iraq. Iraq is home to around 150 distinct tribes. Some of the larger ones are shown on this map.

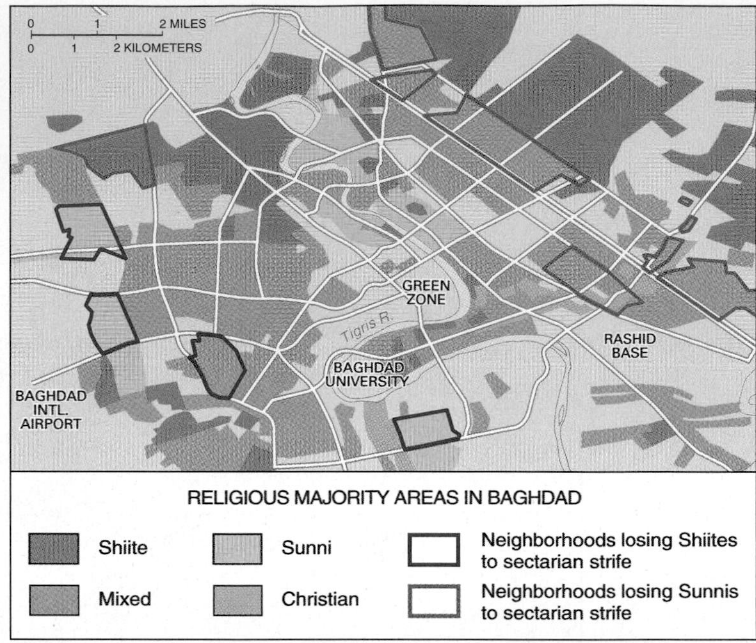

FIGURE 8-27 Ethnic groups in Baghdad. Baghdad contains a mix of Sunnis, Shi'ites, and other groups. Many neighborhoods were traditionally mixed, but in recent years the minority group has been forced to migrate.

establishing control over their local territories. A tribe (*'ashira*) is divided into several clans (*fukhdhs*), which in turn encompass several houses (*beit*), which in turn include several extended families (*kham*). Tribes are grouped into more than a dozen federations (*qabila*). Most Iraqis have stronger loyalty to a tribe or clan than to a national government.

In addition, Iraq's principal ethnic groups are split into regions, with Kurds in the north, Sunnis in the center, and Shi'ites in the south. The Kurds welcomed the United States because they gained more security and autonomy than they had under Hussein. Sunni Muslims opposed the U.S.-led attack because they feared loss of power and privilege given to them by Hussein, who was a Sunni. Shiite Muslims also opposed the U.S. presence. Although they had been treated poorly by Hussein and came to control Iraq's post-Hussein government, Shiites shared a long-standing hostility toward the United States with their neighbors in Shiite-controlled Iran. The capital, Baghdad, where one-fourth of the Iraqi people live, has some neighborhoods where virtually all residents are of one ethnicity, but most areas are mixed. In many of these historically mixed neighborhoods, the less numerous ethnicity has been forced to move away (Figure 8-27).

Having invaded Iraq and removed Hussein from power, the United States expected an enthusiastic welcome from the Iraqi people. Instead, the United States became embroiled in a complex and violent struggle among these various religious sects and tribes. A quarter-century of Hussein's dictatorial rule had suppressed long-standing tensions among these groups, and violence erupted among them after his removal.

Iran

Hostility between the United States and Iran dates from 1979, when a revolution forced abdication of Iran's pro-U.S. Shah Mohammad Reza Pahlavi. Iran's majority Shiite population had demanded more democratic rule and opposed the Shah's economic modernization program that generated social unrest. Supporters of exiled fundamentalist Shiite Muslim leader Ayatollah Ruholiah Khomeini then proclaimed Iran an Islamic republic and rewrote the constitution to place final authority with the ayatollah. Militant supporters of the ayatollah seized the U.S. embassy on November 4, 1979, and held 62 Americans hostage until January 20, 1981.

Iran and Iraq fought a war between 1980 and 1988 over control of the Shatt al-Arab waterway, formed by the confluence of

CONTEMPORARY GEOGRAPHIC TOOLS

Air Photos in War and Peace

Photographs taken by reconnaissance aircraft and satellites have long been an important tool in guiding military operations during conflicts, such as pinpointing targets for air strikes and the deployment of opposition armies. Air photos have also occasionally played a critical role on the diplomatic front.

A major confrontation during the Cold War between the United States and Soviet Union came in 1962 when the Soviet Union secretly began to construct missile-launching sites in Cuba, less than 150 kilometers (90 miles) from U.S. territory. President Kennedy went on national television to demand that the missiles be removed, and he ordered a naval blockade to prevent additional Soviet material from reaching Cuba.

At the United Nations, immediately after Soviet Ambassador Valerian Zorin denied that his country had placed missiles in Cuba, U.S. Ambassador Adlai Stevenson dramatically revealed aerial photographs taken by the U.S. Department of Defense clearly showing preparations for them (see examples in Figure 8-20). Faced with irrefutable evidence that the missiles existed, the Soviet Union ended the crisis by dismantling them.

As the United States moved toward war with Iraq in 2003, Secretary of State Colin Powell scheduled a speech at the United Nations. The speech was supposed to present irrefutable evidence to the world justifying military action against Iraq. Adding credibility to the presentation, Powell was known to be the senior U.S. diplomat most reluctant to go to war. Recalling the Cuban missile crisis, Powell displayed a series of air photos designed to prove that Iraq possessed weapons of mass destruction. He introduced the photos with these words: "Let me say a word about satellite images before I show a couple. The photos that I am about to show you are sometimes hard for the average person to interpret, hard for me. The painstaking work of photo analysis takes experts with years and years of experience, pouring for hours and hours over light tables. But as I show you these images, I will try to capture and explain what they mean, what they indicate to our imagery specialists."

Powell first showed an image of 15 munitions bunkers at Taji, Iraq (Figure 8-28). "We know that this one has housed chemical munitions," Powell stated. "How do I know that? How can I say that? Let me give you a closer look. Look at the image on the left [in Figure 8-29]. On the left is a close-up of one of the four chemical bunkers. The two arrows indicate the presence of sure signs that the bunkers are storing chemical munitions. The arrow at the top that says security points to a facility that is the signature item for this kind of bunker. Inside that facility are special guards and special equipment to monitor any leakage that might come out of the bunker. The truck you also see is a signature item. It's a decontamination vehicle in case something goes wrong." Subsequent close-ups of the bunkers showed them being cleaned immediately before UN inspectors arrived. Powell also showed a ballistic missile facility being cleaned immediately before the arrival of UN inspectors.

Unlike Stevenson in 1962, Powell could not through air photos make a convincing case at the United Nations for the U.S. position. As a result, the United States went to war with Iraq without the support of the United Nations. A subsequent U.S. State Department analysis found many inaccuracies in the interpretation of the photos presented by Powell. For example, the "decontamination vehicle" in Figure 8-29 turned out to be a water truck. Two years later, Powell himself said that the 2003 speech had been a "blot" on his record. ■

FIGURE 8-28 U.S. satellite image purporting to show munitions bunkers in Taji, Iraq.

FIGURE 8-29 (left) Close-up of alleged munitions bunker outlined in red near the bottom of Figure 8-28. (right) Close-up of the two bunkers, outlined in red in the middle of Figure 8-28, allegedly sanitized.

the Tigris and Euphrates rivers flowing into the Persian Gulf. Forced to cede control of the waterway to Iran in 1975, Iraq took advantage of Iran's revolution to seize the waterway in 1980, but Iran was not defeated outright, so an 8-year war began that neither side was able to win. An estimated 1.5 million died in the war, which ended when the two countries accepted a UN peace plan.

When the United States launched its war on terrorism after 9/11, Afghanistan was the immediate target, followed by Iraq. But after the election of Mahmoud Ahmadinejad as president in 2005, relations between the United States and Iran deteriorated. The United States accused Iran of harboring al-Qaeda members and of trying to gain influence in Iraq, where, as in Iran, the majority of the people were Shiites. More troubling to the international community was Iran's aggressive development of a nuclear program. Iran claimed that its nuclear program was for civilian purposes, but other countries believed that it was intended to develop weapons. Prolonged negotiations were undertaken to dismantle Iran's nuclear capabilities without resorting to yet another war in the Middle East.

Pakistan

Pakistan along with India was created in 1947 when South Asia was partitioned into predominantly Muslim and Hindu states. The war on terrorism has spilled over from Pakistan's western neighbors Afghanistan and Iran. Although the overwhelming majority of Pakistanis are Muslim, Pakistan is a multiethnic state. Punjabis comprise around 45 percent of the population and, combined, Pashtuns, Sindhis, and Seraikis around 40 percent; the remaining 15 percent are other ethnicities. Around 70 percent of Pakistanis are Sunni Muslims and 30 percent are Shiite. Western Pakistan, along the border with Afghanistan, is a rugged, mountainous region inhabited by several ethnic minorities where the Taliban have been largely in control. Osama bin Laden is thought to have hidden in Pakistan after escaping from Tora Bora (see Contemporary Geographic Tools box).

SUMMARY

The political geography of the second half of the twentieth century was dominated by the Cold War between two superpowers—the United States and the former Soviet Union. In the twenty-first century, military alliances among states have become less important than patterns of global and regional economic cooperation and competition among states.

At the same time, with the end of the Cold War, the world has entered a period characterized by an unprecedented increase in the number of new states created to satisfy the desire of nationalities for self-determination as an expression of cultural distinctiveness. Turmoil has resulted in many places where the boundaries of the new states do not match the territories occupied by distinct nationalities. Terrorism led by ethnic groups has replaced direct military confrontation between states as a leading source of political unrest.

Here is a review of issues raised at the beginning of the chapter:

1. **Where Are States Located?** A state is a political unit, with an organized government and sovereignty, whereas a nation is a group of people with a strong sense of cultural unity. Most of Earth's surface is allocated to states, and only a handful of colonies and tracts of unorganized territory remain.
2. **Why Do Boundaries Between States Cause Problems?** Boundaries between states, where possible, are drawn to coincide either with physical features, such as mountains, deserts, and bodies of water, or with such cultural characteristics as geometry, religion, and language. Boundaries affect the shape of countries and affect the ability of a country to live peacefully with its neighbors. Problems arise when the boundaries of states do not coincide with the boundaries of ethnicities.
3. **Why Do States Cooperate with Each Other?** Following World War II, the United States and the Soviet Union, as the world's two superpowers, formed military alliances with other countries. With the end of the Cold War, nationalities now are cooperating with each other, especially in Western Europe, primarily to promote economic growth rather than to provide military protection.
4. **Why Has Terrorism Increased?** Terrorism initiated by individuals, organizations, and states has increased, especially against the United States. Terrorists consider all U.S. citizens justifiable targets because they hold all U.S. citizens responsible for U.S. government policies and cultural practices.

CASE STUDY REVISITED / Future of the Nation-State in Europe

In the twenty-first century, the importance of the nation-state has diminished in Western Europe, the world region most closely associated with development of the concept during the previous two centuries (Figure 8-30). Western Europeans carry European Union rather than national passports, which they don't need to show when traveling within Western Europe.

More importantly, European nation-states have put aside their centuries-old rivalries to forge the world's most powerful economic union. France's franc, Germany's mark, and Italy's lira—powerful symbols of sovereign nation-states—have disappeared, replaced by a single currency, the euro. European leaders have bet that every country in the region will be stronger economically by replacing national currencies with the euro.

(Continued)

CASE STUDY REVISITED (*Continued*)

Cultural differences persist at borders. For example, highways in the Netherlands are more likely than those in neighboring Belgium to be flanked by well-manicured vegetation and paths reserved for bicycles. But boundaries where hundreds of thousands of soldiers once stood guard now have little more economic significance in Europe than boundaries between states inside the United States.

Rather than national boundaries, the most fundamental obstacle to Western European integration is the multiplicity of languages. Although English has rapidly become the principal language of business in the European Union, much of the European Union's budget is spent translating documents into other languages. Businesses must figure out how to effectively advertise their products in several languages.

At the same time that residents of Western European countries are displaying increased tolerance for the cultural values of their immediate neighbors, opposition has increased to the immigration of people from the south and east, especially those who have darker skin and adhere to Islam. Immigrants from poorer regions of Europe, Africa, and Asia fill low-paying jobs (such as cleaning streets and operating buses) that Western Europeans are not willing to perform. Nonetheless, many Western Europeans fear that large-scale immigration will transform their nation-states into multiethnic societies.

Underlying this fear of immigration is recognition that natural increase rates are higher in most African and Asian countries than in Western Europe as a result of higher crude birth rates. Many Western Europeans believe that Africans and Asians who immigrate to their countries will continue to maintain relatively high crude birth rates and consequently will constitute even higher percentages of the population in Western Europe in the future. ■

FIGURE 8-30 Rhine River. The Rhine River is the boundary between France and Germany. The Passarelle Mimram Pedestrian Bridge (known informally as the Two Banks Bridge) connects Strasbourg, France, and Kehl, Germany. To mark the sixtieth anniversary of NATO in 2009, world leaders including U.S. President Barack Obama (front right) and Poland's President Lech Kaczynski (front left) walked across the bridge.

KEY TERMS

Balance of power (p. 257) Condition of roughly equal strength between opposing countries or alliances of countries.

Boundary (p. 247) Invisible line that marks the extent of a state's territory.

City-state (p. 243) A sovereign state comprising a city and its immediate hinterland.

Colonialism (p. 244) Attempt by one country to establish settlements and to impose its political, economic, and cultural principles in another territory.

Colony (p. 243) A territory that is legally tied to a sovereign state rather than completely independent.

Compact state (p. 247) A state in which the distance from the center to any boundary does not vary significantly.

Elongated state (p. 247) A state with a long, narrow shape.

Federal state (p. 254) An internal organization of a state that allocates most powers to units of local government.

Fragmented state (p. 248) A state that includes several discontinuous pieces of territory.

Frontier (p. 253) A zone separating two states in which neither state exercises political control.

Gerrymandering (p. 255) Process of redrawing legislative boundaries for the purpose of benefiting the party in power.

Imperialism (p. 245) Control of territory already occupied and organized by an indigenous group.

Landlocked state (p. 249) A state that does not have a direct outlet to the sea.

Microstate (p. 242) A state that encompasses a very small land area.

Perforated state (p. 248) A state that completely surrounds another one.

Prorupted state (p. 247) An otherwise compact state with a large projecting extension.

Sovereignty (p. 241) Ability of a state to govern its territory free from control of its internal affairs by other states.

State (p. 241) An area organized into a political unit and ruled by an established government with control over its internal and foreign affairs.

Unitary state (p. 254) An internal organization of a state that places most power in the hands of central government officials.

THINKING GEOGRAPHICALLY

1. In his book *1984*, George Orwell envisioned the division of the world into three large unified states, held together through technological controls. To what extent has Orwell's vision of a global political arrangement been realized?
2. Gerald Helman and Steven Ratner have identified countries that they call "failed nation-states," including Cambodia, Liberia, Somalia, and Sudan. Helman and Ratner argue that the governments of these countries were maintained in power during the Cold War era through massive military and economic aid from the United States or the Soviet Union. With the end of the Cold War, these failed nation-states have sunk into civil wars, fought among groups who share language, religion, and other cultural characteristics. What obligations do other countries have to restore order in failed nation-states?
3. Given the movement toward increased local government autonomy on the one hand and increased authority for international organizations on the other, what is the future of the nation-state? Have political and economic trends since the 1990s strengthened the concept of nation-state or weakened it?
4. The world has been divided into a collection of countries on the basis of the principle that ethnicities have the right of self-determination. National identity, however, derives from economic interests as well as from such cultural characteristics as language and religion. To what extent should a country's ability to provide its citizens with food, jobs, economic security, and material wealth, rather than the principle of self-determination, become the basis for dividing the world into independent countries?
5. A century ago the British geographer Halford J. Mackinder identified a heartland in the interior of Eurasia (Europe and Asia) that was isolated by mountain ranges and the Arctic Ocean. Surrounding the heartland was a series of fringe areas, which the geographer Nicholas Spykman later called the *rimland*, oriented toward the oceans. Mackinder argued that whoever controlled the heartland would control Eurasia and hence the entire world. To what extent has Mackinder's theory been validated during the twentieth century by the creation and then the dismantling of the Soviet Union?

RESOURCES

Some recent and classic books and articles on political geography:

Agnew, John. *Making Political Geography*. London: Arnold, 2002.

———. "Sovereignty Regimes: Territoriality and State Authority in Contemporary World Politics." *Annals of the Association of American Geographers* 95 (2005): 437–61.

Allen, John. *Lost Geographies of Power*. Malden, MA: Blackwell, 2003.

Cohen, Saul B. "Geopolitical Realities and United States Foreign Policy." *Political Geography* 22 (2003): 1–33.

Cox, Kevin R., Murray Low, and Jennifer Robinson, eds. *The SAGE Handbook of Political Geography*. Los Angeles: SAGE, 2008.

Cutter, Susan L., Douglas B. Richardson, and Thomas J. Wilbanks, eds. *The Geographical Dimensions of Terrorism*. New York: Routledge, 2003.

Dale, E. H. "Some Geographical Aspects of African Land-Locked States." *Annals of the Association of American Geographers* 58 (1968): 485–505.

Demko, George J., and William B. Wood, eds. *Reordering the World: Geopolitical Perspectives on the Twenty-First Century*. Boulder, CO: Westview Press, 1999.

Falah, Gazi-Walid, Colin Flint, and Virginie Mamadouh. "Just War and Extraterritoriality: The Popular Geopolitics of the United States' War on Iraq as Reflected in Newspapers of the Arab World." *Annals of the Association of American Geographers* 96 (2006): 142–64.

Gillespie, Thomas W., John A. Agnew, Erika Mariano, Scott Mossler, Nolan Jones, Matt Braughton, and Jorge Gonzalez. "Finding Osama bin Laden: An Application of Biogeographic Theories and Satellite Imagery." *MIT International Review*. Web-published February 17, 2009. Accessed June 10, 2009 at http://web.mit.edu/mitir/.

Helman, Gerald B., and Steven R. Ratner. "Saving Failed States." *Foreign Policy* 89 (1992): 3–20.

Johnston, R. J., Peter J. Taylor, and Michael J. Watts. *Geographies of Global Change*, 2nd ed. Cambridge, MA: Blackwell, 2002.

Kliot, N., and Y. Mansfield. "The Political Landscape of Partition: The Case of Cyprus." *Political Geography* 16 (1997): 495–521.

MacLaughlin, Jim. *Reimaging the Nation-State: The Contested Terrains of Nation-Building*. London: Pluto Press, 2001.

Murphy, Alexander B. "Historical Justifications for Territorial Claims." *Annals of the Association of American Geographers* 80 (1990): 531–48.

Nijman, Jan. "The Limits of Superpower: The United States and the Soviet Union Since World War II." *Annals of the Association of American Geographers* 82 (1992): 681–95.

Ó Tuathail, Gearóid. "The Postmodern Geopolitical Condition: States, Statecraft, and Security at the Millennium." *Annals of the Association of American Geographers* 90 (2000): 166–78.

Painter, Joe, and Alex Jeffrey. *Political Geography: An Introduction To Space and Power*, 2nd ed. Los Angeles: SAGE, 2009.

Parker, W. H. *Mackinder: Geography as an Aid to Statecraft*. Oxford: Clarendon Press, 1982.

Prescott, J. R. V. *Political Frontiers and Boundaries*. London: Unwin Hyman, 1990.

Smith, Dan. *The Penguin Atlas of War and Peace*, 4th ed. London and New York: Penguin, 2003.

———. *Penguin State of the World Atlas*, 7th ed. New York: Penguin, 2003.

Taylor, Peter J., and Colin Flint. *Political Geography: World Economy, Nation-State and Locality*, 5th ed. Upper Saddle River, NJ: Prentice Hall, 2007.

Journals featuring political geography:

American Journal of Political Science; American Political Science Review; Foreign Affairs; Foreign Policy; International Affairs; International Journal of Middle East Studies; Political Geography; Post-Soviet Geography.

Key Internet site:

https://www.cia.gov/library/publications/the-world-factbook/ The U.S. Central Intelligence Agency publishes a World Factbook. Select a country from the drop-down list to find background information, as well as facts and figures about the country's demography, economy, physical geography, government, and military. Maps are also available.

Log in to www.mygeoscienceplace.com for videos, interactive maps, RSS feeds, case studies, and self-study quizzes to enhance your study of Political Geography.

CHAPTER 9

Development

Have you ever traveled to a Caribbean island? Even if you haven't, you have probably seen advertisements for resorts featuring a bronzed couple sipping exotic drinks, lying on a deserted beach surrounded by palm trees.

Beyond this paradise is another world, fleetingly glimpsed by tourists traveling between the resort and the airport. The permanent residents of the islands may live in poverty, earning less money in a year than a night's hotel bill. They are ill fed, ill clothed, and underemployed.

KEY ISSUES

1 Why Does Development Vary Among Countries?

2 Where Are MDCs and LDCs Distributed?

3 Where Does Level of Development Vary by Gender?

4 Why Do LDCs Face Obstacles to Development?

This depressing view of conditions on the islands is shielded from tourists, of course. They do not travel hundreds of kilometers to encounter misery on their vacation or honeymoon. Tourists bring money to the islands and in the process help pay for whatever improvements can be made to the squalid living conditions.

But can you imagine the feelings of the local residents? What would you think if a very expensive and exclusive resort were built in your neighborhood, and you and your family, who were economically disadvantaged, were expected to work there (for good wages, perhaps) to serve the needs of the vacationers? You might welcome the money, but would you resent the wealthy tourists?

The world is divided between relatively rich and relatively poor countries. Geographers try to understand the reasons for this division and learn what can be done about it.

Workers polish diamonds at a factory in Gaborone, Botswana.

CASE STUDY / Bangladesh's Development Challenges

Rabea Rahman lives in the village of Bathoimuri, Bangladesh, with her three children—a son, 18, and two daughters, ages 10 and 7. Rahman's two other children died in infancy. Her husband died of tuberculosis.

Rahman's husband was a tenant farmer, or sharecropper. Under this arrangement, he shared a portion of his crops with the landowner instead of paying rent. After he died, Rahman went to work as a domestic servant and water carrier, working from 7 A.M. to 4 P.M. and from 6 P.M. to 11 P.M., seven days a week. Her son sells bread and prepares a midday meal for his two sisters. Total household income is $16 per month (compared to a monthly household average of around $4,000 in the United States).

Their house has a dirt floor and leaky roof, but the rent is only $2 per month, plus $3 per month for fuel. The remaining $11 a month goes for food. The sum is sufficient to provide each member of the household with 100 grams (about a quarter pound) of rice per day, but little else. The diet is supplemented by leftover food that Rahman receives from her employer. After paying for rent, fuel, and food, the family has no money left for other necessities. Because they cannot afford shoes, the family members often go barefoot. Rahman suffers from a gastric ulcer but cannot afford treatment.

Underlying the impoverished condition of the Rahman household is the role of women in a predominantly Muslim country such as Bangladesh. In rural villages, fewer than 10 percent of the women can read and write. Typically, a woman is married as a teenager and bears six babies in her lifetime, although on average one of the six does not survive infancy.

A woman like Rahman who is forced to find a job is limited to working as a servant or farm laborer. The condition of women—poor, illiterate, overburdened with children—is one of the most important factors holding back economic development in South Asian countries such as Bangladesh. ■

Previous chapters examined global demographic and cultural patterns. Birth, death, and natural increase rates vary among regions of the world, and people in different regions also have different social customs, languages, religions, and ethnic identities. Political problems arise when the distribution of cultural characteristics does not match the boundaries between states. Chapter 8 pointed out that in the contemporary world, global military confrontation and alliances have been replaced by global economic competition and cooperation.

The second half of this book concentrates on economic elements of human geography. This chapter examines the most fundamental global economic pattern—the division of the world into relatively wealthy ***regions*** and relatively poor ones. Subsequent chapters look at the three basic ways that humans earn their living—growing food, manufacturing products, and providing services.

Earth's nearly 200 countries can be classified according to their level of **development**, which is the process of improving the material conditions of people through diffusion of knowledge and technology. The development process is continuous, involving never-ending actions to constantly improve the health and prosperity of the people. Every ***place*** lies at some point along a continuum of development.

Because many countries cluster at the high or low end of the continuum of development, they can be divided into two groups. A **more developed country (MDC)**, also known as a relatively developed country or simply as a developed country, has progressed further along the development continuum. A country in an earlier stage of development is frequently called a **less developed country (LDC)**, although many analysts prefer the term developing country or emerging country. *"Developing"* or *"emerging"* implies that the country has already made some progress and expects to continue.

The first geographic task is to identify ***where*** MDCs and LDCs are located. Geographers observe that MDCs cluster in some ***spaces*** and LDCs cluster in others. Next, geographers are concerned with ***why*** some regions are more developed than others. A number of economic, social, and demographic indicators distinguish regions of MDCs from regions of LDCs.

The ***scale*** of the severe economic downturn that began in 2008 has illustrated the ***globalization*** of the economy in the twenty-first century. In the recent recession, individual countries have seen their economies severely buffeted by close ***connections*** to the global economy. A return to economic growth has necessitated taking advantage of ***local diversity*** in skills and resources.

KEY ISSUE 1
Why Does Development Vary Among Countries?

- **Economic Indicators of Development**
- **Social Indicators of Development**
- **Demographic Indicators of Development**

A country's level of development can be distinguished according to three factors—*economic, social,* and *demographic.* The **Human Development Index (HDI)**, created by the United Nations, recognizes that a country's level of development is a function of all three of these factors (Figure 9-1). This key issue examines the three sets of development indicators. ■

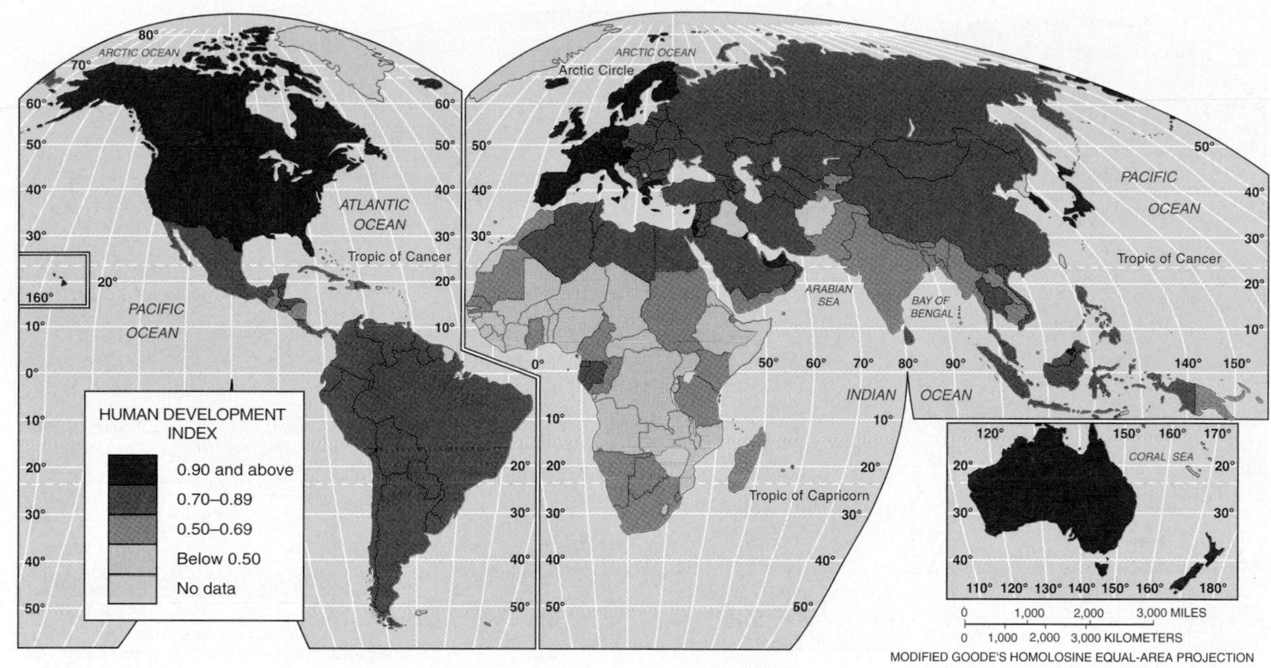

FIGURE 9-1 Human Development Index (HDI). The United Nations updated HDI scores in October 2009 based on 2006 data. It will be several years before lower HDI scores reflect the severe recession.

Economic Indicators of Development

To create the HDI, the United Nations selects one economic factor, two social factors, and one demographic factor that in the opinion of an international team of analysts best reveal a country's level of development:

- The economic factor is gross domestic product (GDP) per capita.
- The social factors are the literacy rate and amount of education.
- The demographic factor is life expectancy.

The four factors are combined to produce a country's HDI. The highest HDI possible is 1.0, or 100 percent. The UN has computed HDIs for countries every year since 1990, although it has tinkered a few times with the method of computation. The highest-ranking countries are typically in Europe and include Canada. The highest HDI in most recent years has been Norway's, at 0.971 in 2009. The lowest-ranked country in 2009 was Niger, with an HDI of 0.340. Thirty of the thirty-two lowest-ranking countries were in sub-Saharan Africa.

Gross Domestic Product Per Capita

The average individual earns a much higher income in an MDC than in an LDC. Per capita income is a difficult figure to obtain in many countries, so to get a sense of average incomes in various countries, geographers substitute per capita gross domestic product, a more readily available indicator.

The **gross domestic product (GDP)** is the value of the total output of goods and services produced in a country, normally during a year. Dividing the GDP by total population measures the contribution made by the average individual toward generating a country's wealth in a year. For example, GDP in the United States was $14 trillion in 2009 and its population was 307 million, so GDP per capita was about $45,600.

In 2008, per capita GDP exceeded $30,000 in MDCs, compared with less than $3,000 in most LDCs (Figure 9-2). And the gap has widened: Since 1980 GDP per capita has increased from around $15,000 to $30,000 in MDCs and from around $1,000 to $4,000 in LDCs. Per capita GDP—or, for that matter, any other single indicator—cannot measure perfectly the level of a country's development. Few people may be starving in LDCs with per capita GDPs of a few thousand dollars. And not everyone is wealthy in MDCs with per capita GDP of more than $40,000. Per capita GDP measures average (mean) wealth, not its distribution. If only a few people receive much of the GDP, then the standard of living for the majority may be lower than the average figure implies. The higher the per capita GDP, the greater the potential for ensuring that all citizens enjoy a comfortable life.

Types of Jobs

In addition to GDP per capita, three other economic indicators are especially useful in distinguishing between MDCs and LDCs—types of jobs, worker productivity, and availability of consumer goods.

Average per capita income is higher in MDCs because people typically earn their living by different means than in LDCs (Figure 9-3). Jobs fall into three types:

- Primary (including agriculture)
- Secondary (including manufacturing)
- Tertiary (including services)

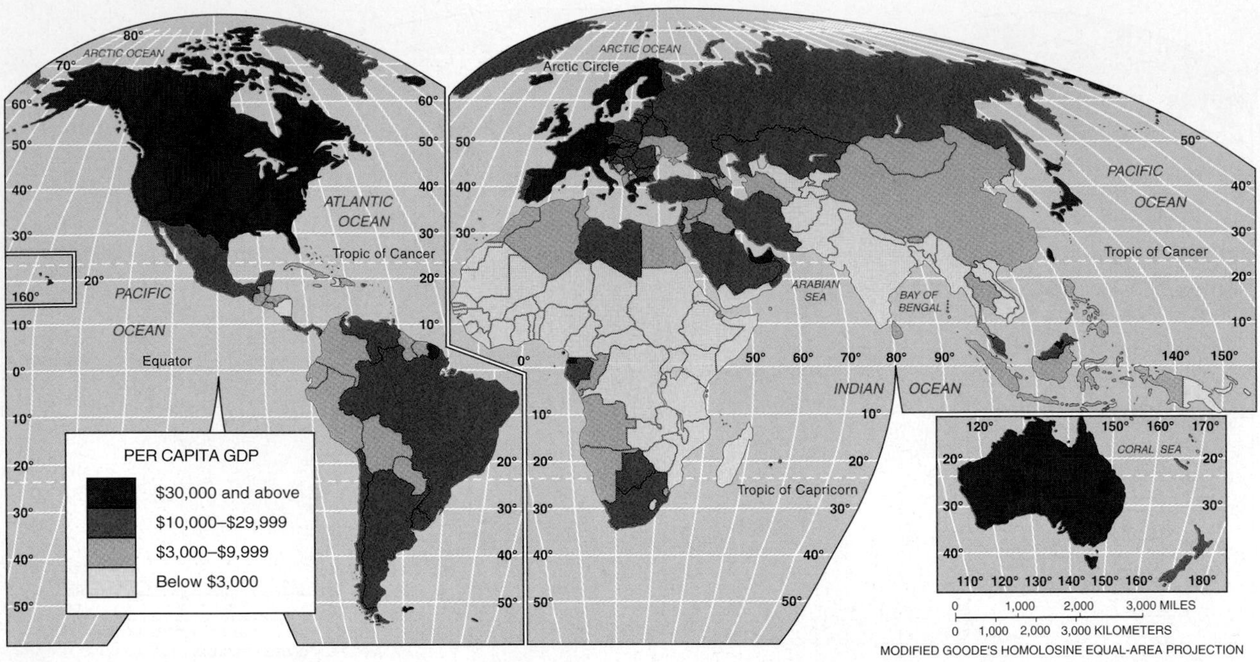

FIGURE 9-2 Annual gross domestic product (GDP) per capita. This measure exceeds $30,000 in most MDCs, compared to less than $10,000 in most LDCs. Figures are for "purchasing power parity," which is a method for comparing living standards based on the price for equivalent products in different local currencies. Figures are latest estimates by the CIA, mostly from 2008.

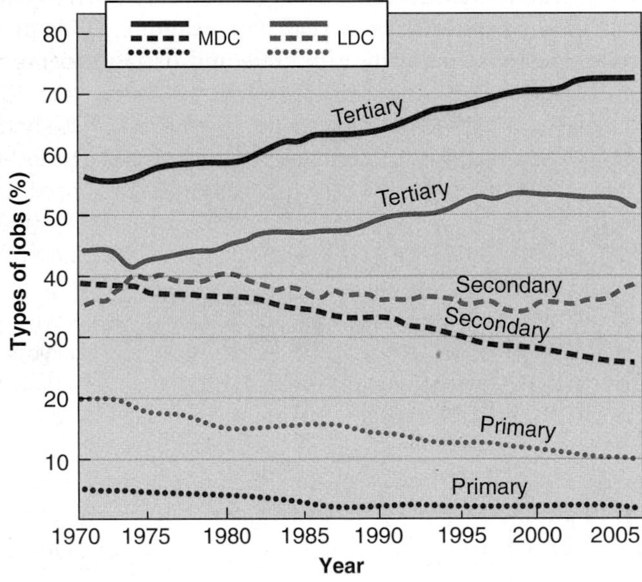

FIGURE 9-3 Percent GDP contributed by type of job. The tertiary sector contributes a greater share to GDP in MDCs than in LDCs. In MDCs, the tertiary sector contributes an increasing share to GDP, whereas the secondary sector contributes a decreasing share.

Workers in the **primary sector** directly extract materials from Earth through agriculture, and sometimes by mining, fishing, and forestry. The **secondary sector** includes manufacturers that process, transform, and assemble raw materials into useful products. Other secondary-sector industries take manufactured goods and fabricate them into finished consumer goods. The **tertiary sector** involves the provision of goods and services to people in exchange for payment. Tertiary-sector activities include retailing, banking, law, education, and government.

To compare the types of economic activities found in MDCs and LDCs, we can compute the contribution to GDP from each of these three sectors. The contribution to GDP among primary, secondary, and tertiary sectors varies between MDCs and LDCs.

- The share of GDP accounted for by the primary sector has decreased in LDCs, but it remains higher than in MDCs.
- The share of GDP accounted for by the secondary sector has decreased sharply in MDCs and is now less than in LDCs.
- The share of GDP accounted for by the tertiary sector is relatively large in MDCs, and it continues to grow.

Productivity

Workers in MDCs are more productive than those in LDCs. **Productivity** is the value of a particular product compared to the amount of labor needed to make it. Productivity can be measured by the value added per capita. The **value added** in manufacturing is the gross value of the product minus the costs of raw materials and energy. The value added per capita exceeds $5,000 in the United States and $7,000 in Japan, compared to around $500 in China and $100 in India.

Workers in MDCs produce more with less effort because they have access to more machines, tools, and equipment to perform much of the work. On the other hand, production in LDCs must rely more on human and animal power. The larger per capita GDP in MDCs in part pays for the manufacture and purchase of machinery, which in turn makes workers more productive and generates more wealth.

Consumer Goods

Part of the wealth generated in MDCs is used to purchase goods and services. Especially important are goods and services related to transportation and communications, including motor vehicles, telephones, and computers. Motor vehicles provide individuals with access to jobs and services and permit businesses to distribute their products (Figure 9-4). Telephones enhance interaction with providers of raw materials and customers for goods and services (Figure 9-5). Computers facilitate the sharing of information with other buyers and suppliers (see Figure 4-20).

Products that promote better transportation and communications are accessible to virtually all residents in MDCs and are vital to the economy's functioning and growth. In contrast, in LDCs these products do not play a central role in daily life for many people. Motor vehicles, computers, and telephones are not essential to people who live in the same village as their friends and relatives and work all day growing food in nearby fields. In many LDCs, those who have these products are concentrated in urban areas; those who do not live in the countryside. Technological innovations tend to diffuse from urban to rural areas. Access to these goods is more important in urban areas because of the dispersion of homes, factories, offices, and shops.

In MDCs, the number of telephones is around 800 per 1,000 inhabitants, motor vehicles 400, and Internet users 400. In LDCs, the figures are around 200 telephones per 1,000 inhabitants, motor vehicles 20, and Internet users 100. Lower numbers indicate that people in LDCs are much less likely to have access to these products. Most people in LDCs are familiar with these goods, even if they cannot afford them, and may desire them as symbols of development. Because possession of consumer goods is not universal in LDCs, a gap can emerge between the "haves" and the "have-nots." The minority of people who have these goods may include government officials, business owners, and other elites, whereas their lack among the majority who are denied access may provoke political unrest.

Technological change is helping to reduce the gap between MDCs and LDCs in access to communications. Cell phone ownership, for example, is expanding rapidly in LDCs because these phones do not require the costly investment of connecting wires to each individual building and more individuals can obtain service from a single tower or satellite.

Social Indicators of Development

MDCs use part of their greater wealth to provide schools, hospitals, and welfare services. As a result, their people are better educated, healthier, and better protected from hardships. Infants are more likely to survive, and adults are more likely to live longer. In turn, this well-educated, healthy, and secure population can be more economically productive.

Education and Literacy

In general, the higher the level of development, the greater are both the quantity and the quality of a country's educational services. Two measures of education for which data are regularly collected for most countries of the world are student/teacher ratio and literacy rate.

In elementary or primary school, the number of students per teacher exceeds 30 in most LDCs, whereas it is less than 20 in most MDCs. The fewer pupils a teacher has, the more likely that each student will receive personalized instruction (Figure 9-6).

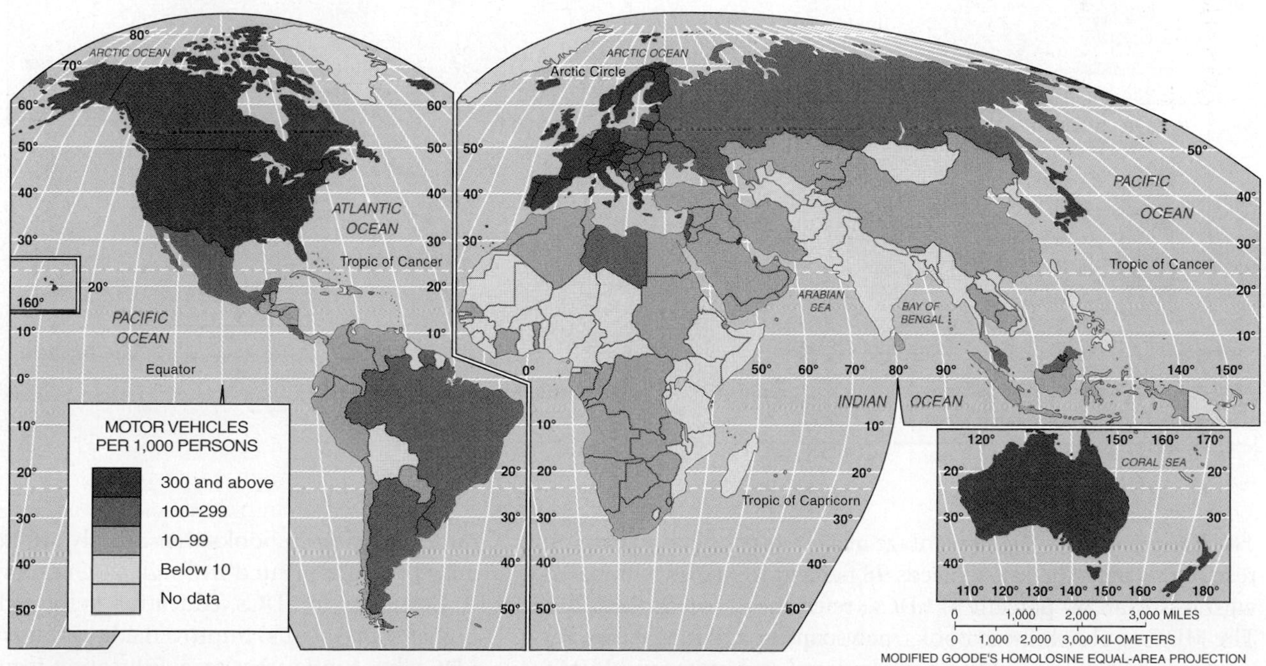

FIGURE 9-4 Motor vehicles per 1,000 persons. MDCs have several hundred vehicles per 1,000 persons, compared with less than 100 in most LDCs.

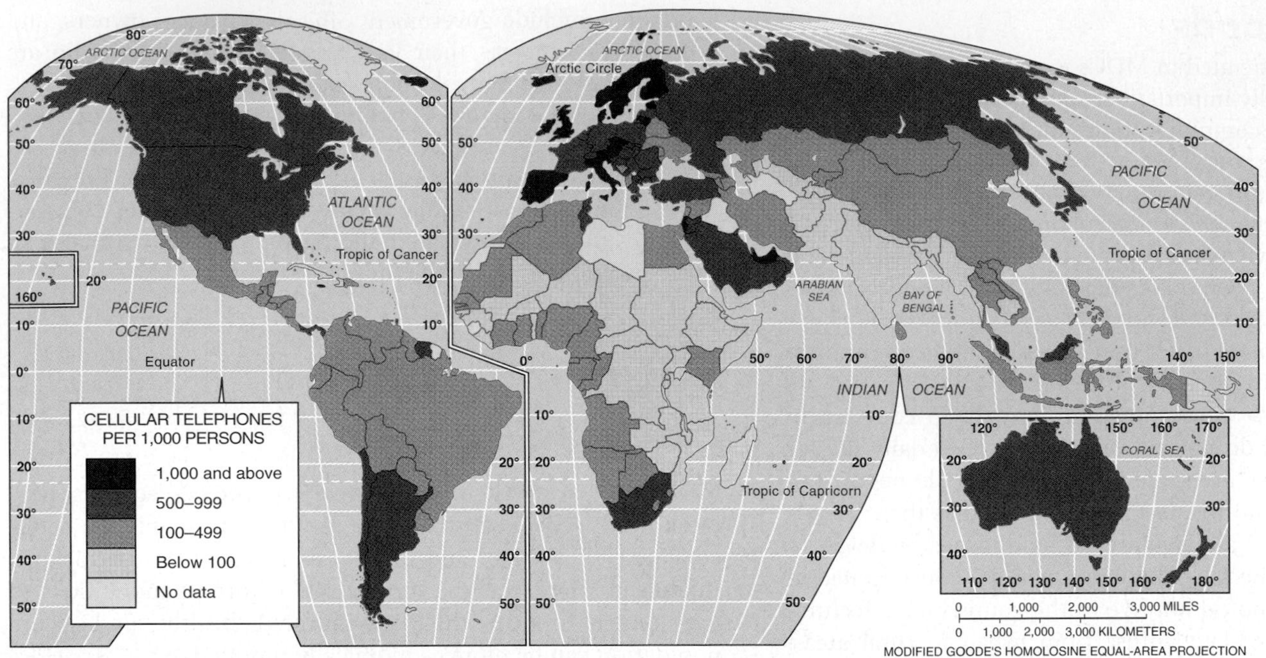

FIGURE 9-5 Cellular telephone lines per 1,000 persons. MDCs have nearly as many cell phones as inhabitants.

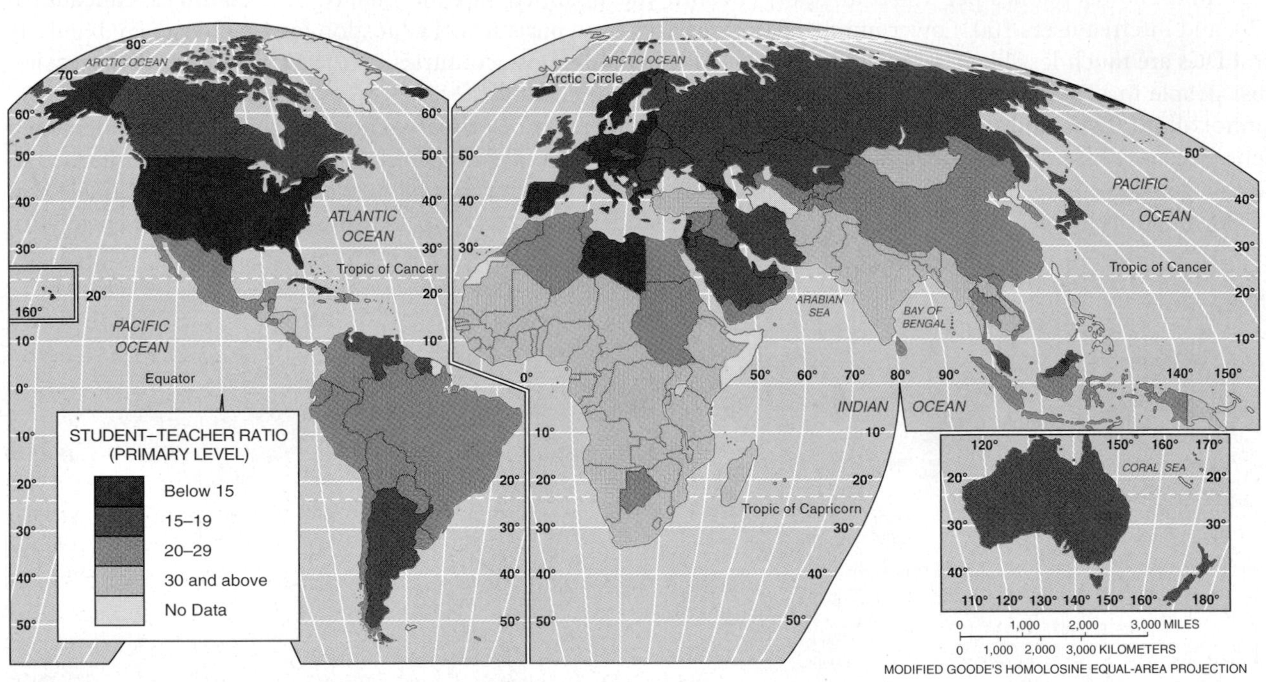

FIGURE 9-6 Students per teacher, primary school. Primary-school teachers must deal with much larger average class sizes in LDCs than in MDCs.

The **literacy rate** is the percentage of a country's people who can read and write. The rate exceeds 98 percent in MDCs, compared with less than 60 percent in LDCs (refer ahead to Figure 9-20). The MDCs publish more books, newspapers, and magazines per person because more of their citizens read and write, and MDCs dominate scientific and nonfiction publishing worldwide—this textbook is an example. Students in LDCs must learn technical information from books that usually are not in their native language but are printed in English, German, Russian, or French.

For many in LDCs, education is the ticket to better jobs and higher social status. Improved education is a major goal of many LDCs, but funds are scarce. Education may receive a higher percentage of the GDP in LDCs, but their GDP is far lower to begin with, so they spend far less per pupil than do MDCs.

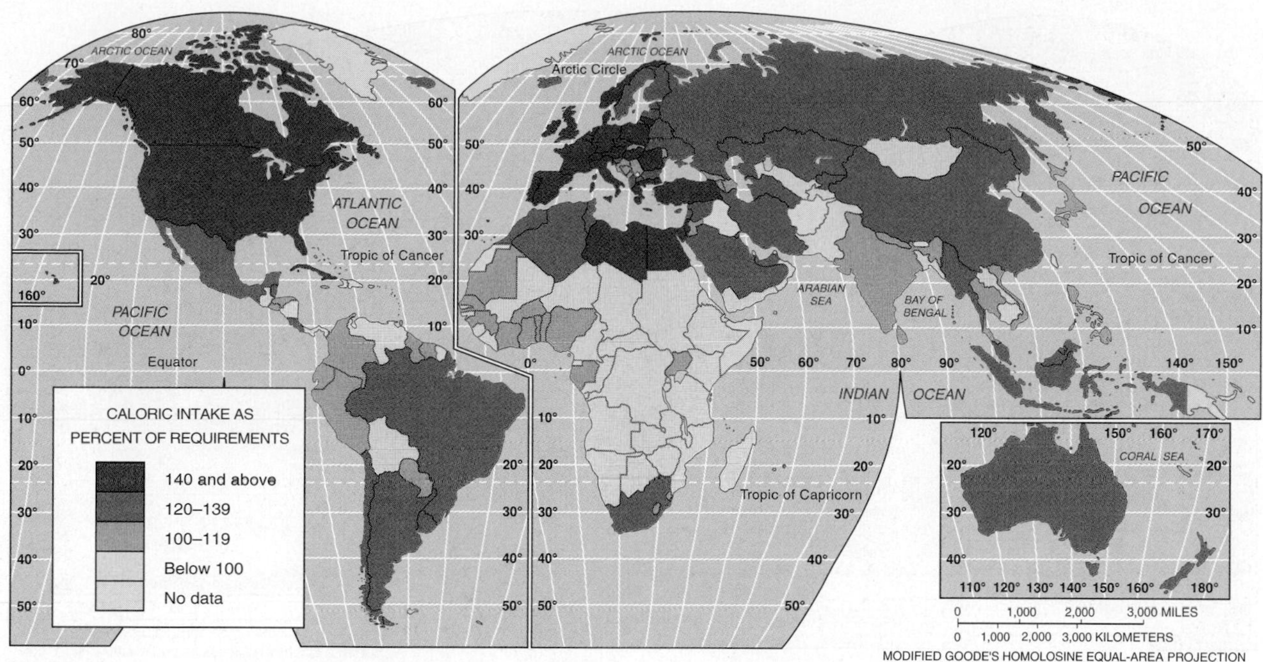

FIGURE 9-7 Daily available calories per capita as a percentage of requirements. Daily available calories per capita (food supply) is the domestic agricultural production plus imports, minus exports and nonfood uses. To maintain a moderate level of physical activity, an average individual requires at least 2,350 calories a day, according to the United Nations Food and Agricultural Organization. The figure must be adjusted for age, sex, and region of the world. In more developed countries, the average citizen consumes about one-third more calories than the minimum needed. The typical resident of a less developed country receives almost precisely the minimum number of calories needed to maintain moderate physical activity—on average. At first glance, this does not reveal a serious problem. However, because these figures are means, a substantial proportion of the population must be receiving less than the necessary daily minimum. The problem is especially severe in Africa, where most people consume less than the needed minimum.

Health and Welfare

People are healthier in MDCs than in LDCs. The health of a population is influenced by diet. On average, people in MDCs receive more calories and proteins daily than they need. But in the LDCs of Africa and Asia, most people receive less than the daily minimum allowance of calories and proteins recommended by the United Nations (Figure 9-7).

When people get sick, MDCs possess the resources to care for them. Total expenditures on health care exceed 8 percent of GDP in MDCs, compared to less than 6 percent in LDCs (Figure 9-8). So not only do MDCs have much higher GDP per capita than LDCs, they spend a higher percentage of that GDP on health care. Some of that additional expenditure on health in MDCs is reflected in more hospitals, doctors, and nurses per capita (Figure 9-9).

In most MDCs, health care is a public service that is available at little or no cost. Government programs pay more than 70 percent of health-care costs in most European countries, and private individuals pay less than 30 percent. In LDCs, private individuals must pay more than half of the cost of health care (Figure 9-10). An exception is the United States, where private individuals are required to pay an average of 55 percent of health care, more closely resembling the pattern in LDCs.

The MDCs use part of their wealth to protect people who, for various reasons, are unable to work. In these countries, some public assistance is offered to those who are sick, elderly, poor, disabled, orphaned, veterans of wars, widows, unemployed, or single parents. Countries in northwestern Europe, such as Denmark, Norway, and Sweden, typically provide the highest level of public-assistance payments. However, MDCs are hard-pressed to maintain their current levels of public assistance. In the past, rapid economic growth permitted these states to finance generous programs with little difficulty. But in recent years economic growth has slowed, whereas the percentage of people needing public assistance has increased. Governments have faced a choice between reducing benefits or increasing taxes to pay for them.

Demographic Indicators of Development

MDCs display many demographic differences from LDCs. The UN's HDI utilizes life expectancy as a measure of development. Other demographic characteristics described in Chapter 2 that distinguish between more and less developed countries include infant mortality, natural increase, and crude birth rates.

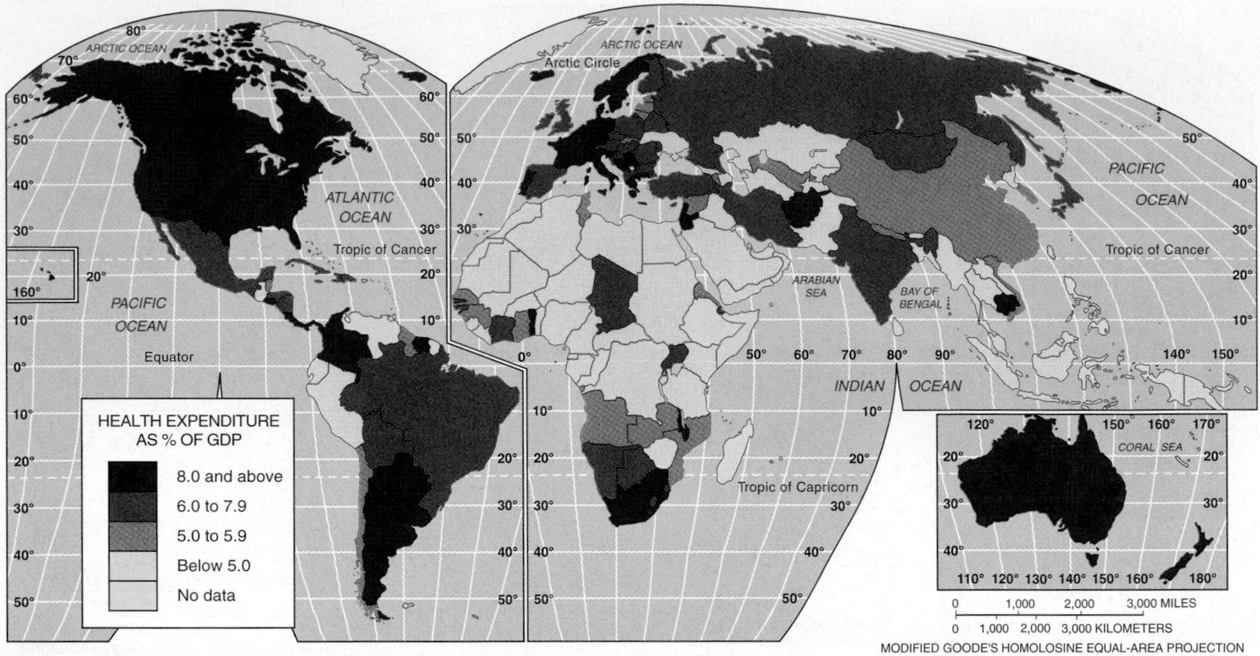

FIGURE 9-8 Expenditure on health care as percent of GDP. MDCs have a much higher gross domestic product (GDP) than LDCs, and they spend a higher percentage of that GDP on health care.

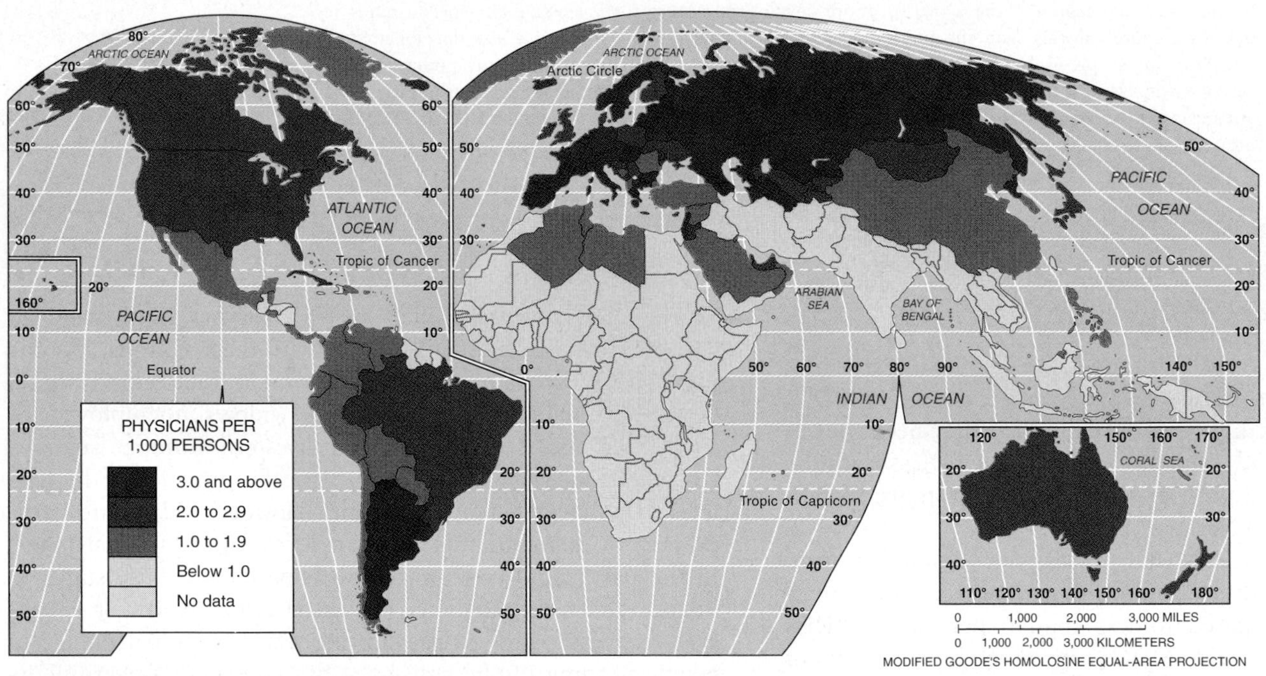

FIGURE 9-9 Physicians per 1,000 persons. MDCs have three or more physicians per 1,000 persons, compared with less than one in most LDCs

Life Expectancy

Better health and welfare in MDCs permit people to live longer. Life expectancy at birth was defined in Chapter 2 as the average number of years a newborn infant can expect to live at current mortality levels. Babies born today can expect to live into their sixties in LDCs compared to their seventies in MDCs (see Figure 2-13). The gap in life expectancy is greater for females than for males. Males can expect to live 10 years longer in MDCs than in LDCs, whereas females can expect to live 13 years longer in MDCs.

With longer life expectancies, MDCs have a higher percentage of older people who have retired and receive public support and a lower percentage of children under age 15 who are too young to work and must also be supported by employed adults and government programs. The number of young people is six times higher than the number of older

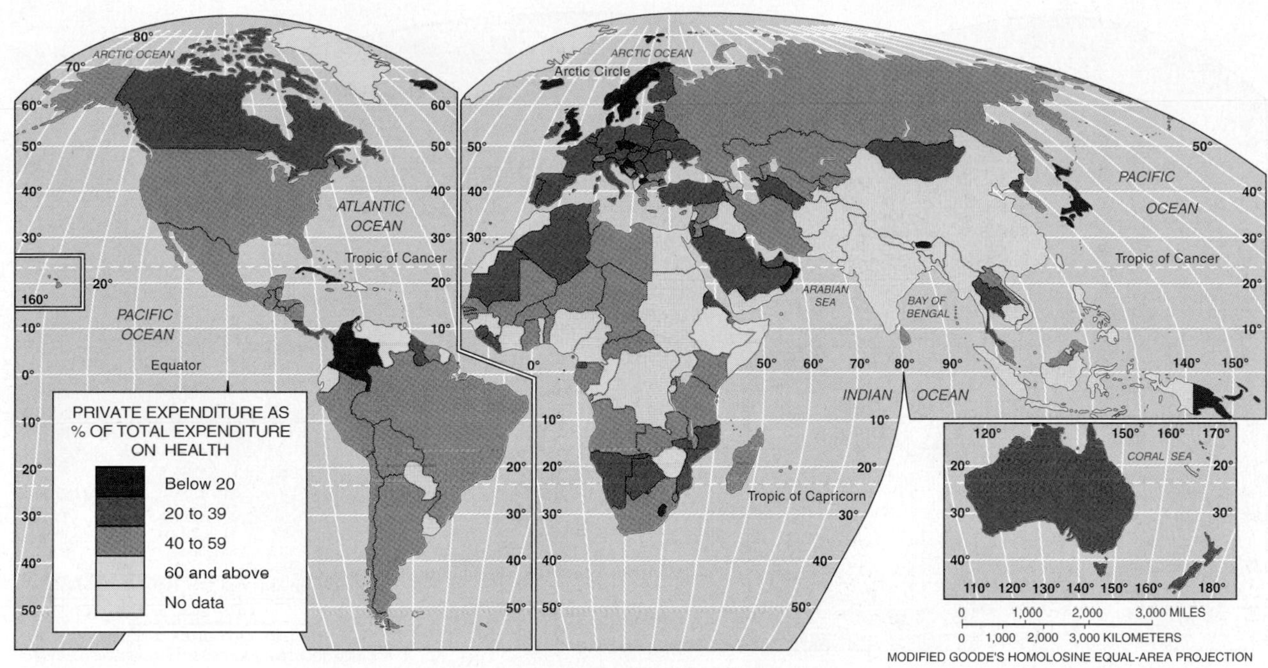

FIGURE 9-10 Private expenditure on health care as percent of total health-care expenditure, 2005. Health care is considered a public service in most MDCs, except for the United States, where—like in most LDCs—private individuals must pay most health-care costs.

people in LDCs, whereas the two are nearly the same in MDCs (see Figure 2-15).

Infant Mortality Rate

Better health and welfare also permit more babies to survive infancy in MDCs. About 94 percent of infants survive and 6 percent die in LDCs, whereas in MDCs more than 99.5 percent survive and fewer than one-half of 1 percent perish (see Figure 2-12). The infant mortality rate is greater in LDCs for several reasons. Babies may die from malnutrition or lack of medicine needed to survive illness, such as dehydration from diarrhea. They may also die from poor medical practices that arise from lack of education.

Natural Increase Rate

The natural increase rate averages 1.5 percent annually in LDCs compared to only 0.2 percent in MDCs. Greater natural increase strains a country's ability to provide hospitals, schools, jobs, and other services that can make its people healthier and more productive. Many LDCs must allocate increasing percentages of their GDPs just to care for the rapidly expanding population rather than to improve care for the current population (see Figure 2-9).

Crude Birth Rate

LDCs have higher natural increase rates because they have higher crude birth rates. The annual crude birth rate is 23 per 1,000 in LDCs, compared to 12 per 1,000 in MDCs. Women in MDCs choose to have fewer babies for various economic and social reasons, and they have access to various birth-control devices to achieve this goal (see Figure 2-10).

The crude death rate (CDR) does not indicate a society's level of development. The CDR is lower in LDCs than in MDCs, 8 per 1,000 compared to 10 per 1,000. Two reasons account for the lower rate in LDCs. First, diffusion of medical technology from MDCs has eliminated or sharply reduced the incidence of several diseases in LDCs. Second, MDCs have higher percentages of older people, who have high mortality rates, as well as lower percentages of children, who have low mortality rates once they survive infancy.

KEY ISSUE 2

Where Are MDCs and LDCs Distributed?

- **More Developed Regions**
- **Less Developed Regions**

The countries of the world can be categorized into nine major regions according to their level of development—North America, Europe, Latin America, East Asia, Southwest Asia (with North Africa), Southeast Asia, Central Asia, South Asia, and sub-Saharan Africa (Figure 9-11). In addition to these nine major regions, three other distinctive areas can be identified—Japan, Oceania, and Russia. These

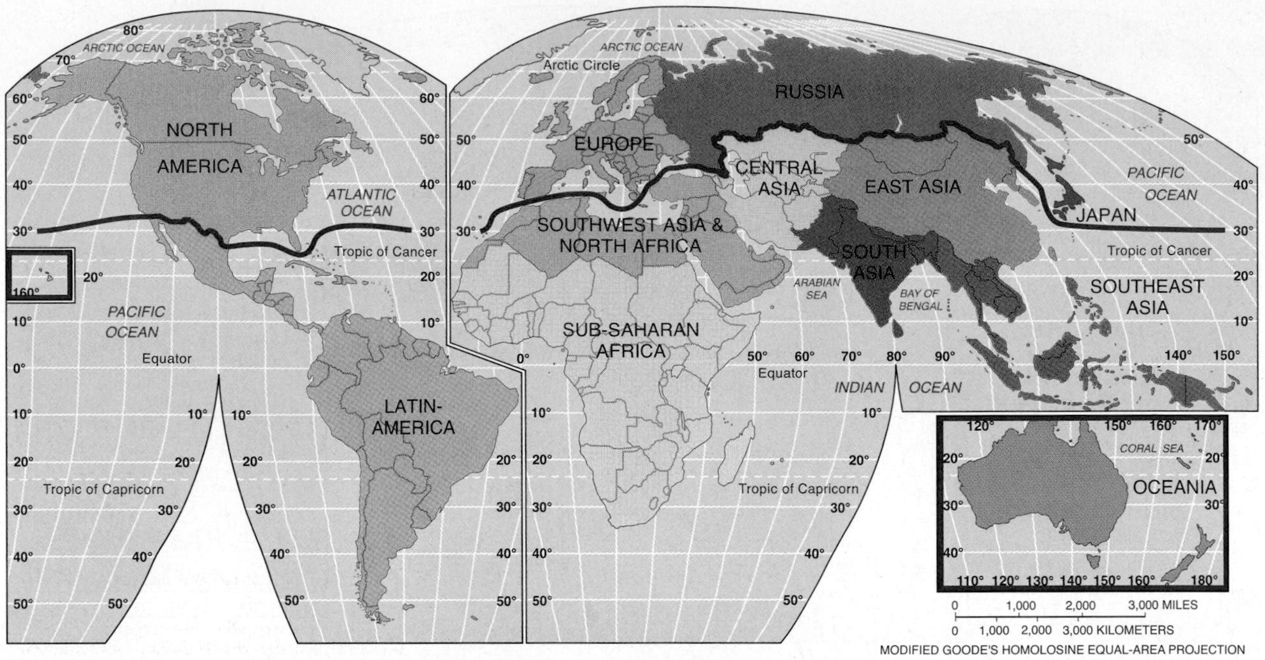

FIGURE 9-11 More and less developed regions. With the exception of Oceania, the more developed regions are located north of the red line on the map.

regions have distinctive demographic and cultural characteristics that have been discussed in earlier chapters. Subsequent chapters will show that the nine major regions also differ in how people earn their living, how the societies use their wealth, and other economic characteristics. In a global economy, geographers are increasingly concerned with both the similarities and the differences in the economic patterns of the various regions. ■

More Developed Regions

Two of the nine major cultural regions—North America and Europe—are considered more developed. The other seven regions are considered less developed. This section examines the more developed regions.

The distribution of more and less developed countries reflects a clear global pattern. If we draw a circle around the world at about 30° north latitude, we find that nearly all of the MDCs are situated to the north, whereas nearly all of the LDCs lie south of the circle. This division of the world between more and less developed and developing countries is known as the *north–south split*. The north–south split between MDCs and LDCs shows up clearly in world maps of measures of development, such as the HDI created by the United Nations (Figure 9-1). MDCs in the north have relatively high HDIs, whereas southern countries have lower indexes.

North America: HDI 0.95

The United States ranked only thirteenth in HDI in 2009. The United States was near the top in two of the four indicators—GDP per capita and literacy rate—but lower than a number of other countries in education and life expectancy. The education indicator suffered because of a relatively high school dropout rate, and life expectancy was lower because many households have inadequate health-care coverage.

North America was once the world's major manufacturer of steel, automobiles, and other goods, but in the past three decades, Japan and Europe as well as LDCs led by China have eroded the region's dominance. Americans remain the leading consumers and world's largest market for many of these products. The region adapted to the loss of manufacturing in the global economy by holding the world's highest percentage of tertiary-sector employment, especially health care, leisure, and financial services.

The relatively large number of health-care providers is a result of the service being provided primarily by the private sector in the United States (see Figure 9-10). The region also provides entertainment, mass media, sports, recreation equipment, and other services that promote use of leisure time. North America's financial institutions played a leading role in precipitating the recent deep recession. So-called *subprime* loans were made at high interest rates to businesses and individuals who were unable to repay them. Financial institutions also profited by selling insurance to enterprises with poor prospects and spreading the risk associated with holding the insurance among many financial institutions.

North America is also the world's leading food exporter. Few Americans are farmers, but a large percentage of the region's workforce is engaged in some aspect of producing or serving food.

Europe: HDI 0.93

During the Cold War era between the 1940s and 1990s, Europe was regarded as two regions—a democratic West closely linked economically and militarily with the United States and a

Communist East closely linked to the Soviet Union. With the fall of communism and the breakup of many of the states in Eastern Europe, the two parts of Europe have become much closer, and are now treated as a single world region.

The elimination of most economic barriers within the European Union makes Europe the world's largest and richest market. European countries hold 15 of the 19 highest HDI rankings. Within Europe the level of development is the world's highest in a core area that includes western Germany, northeastern France, northern Italy, Switzerland, southern Scandinavia, southeastern United Kingdom, Belgium, the Netherlands, and Luxembourg. Southern and Eastern European countries lag in level of development, resulting in an overall HDI for Europe lower than that for North America.

Europe is especially dependent on international trade, both among countries within Europe and increasingly with other regions of the world. To pay for their imports, Western Europeans have provided high-value goods and services, such as insurance, banking, and luxury motor vehicles, including BMW and Mercedes-Benz. The recent severe recession has exacerbated regional and national differences within Europe. Countries most dependent on international trade have been especially hard hit.

Government officials representing the region's wealthiest core area have been accused of protecting jobs in their individual countries rather than in the European Union as a whole. For example, cutbacks at major European carmakers, such as Opel and Renault, have been viewed as less severe in their home countries (Germany and France, respectively) than elsewhere in Europe. In Southern and Eastern Europe, unemployment rates have been double the regional average. European governments have also disagreed on optimal strategies for fighting the recession. In the United Kingdom, as in North America, hundreds of billions of dollars have been spent on government projects, loans, and grants to stimulate the economy. Most European governments have limited government spending because they fear high inflation once the economy recovers.

Russia: HDI 0.73

Under communism, the Soviet Union had a centrally planned economy. Five-year plans prescribed production goals for the entire country by economic sector and region. They specified the type and quantity of minerals, manufactured goods, and agricultural commodities to be produced and the factories, railways, roads, canals, and houses to be built in each part of the country.

After the dissolution of the Soviet Union in 1991, Russia rapidly converted to a market economy. The transition proved painful. Unemployment soared as inefficient Communist-era businesses were either streamlined or closed. A handful of Russians—some of them gangsters—became very rich, but most Russians saw their standard of living decline sharply. Reflecting the deteriorating standard of living in Russia, the HDI declined from more than 0.9 in the 1980s under communism to below 0.9 in the 1990s and below 0.8 after 2000. In the first years of the twenty-first century, Russia experienced economic growth, fueled in large measure by escalating production of oil. The severe worldwide recession caused a sharp drop in demand, however, and with it the possibility of a renewed decline in the HDI.

Japan: HDI 0.96

North America and Europe share many cultural characteristics. North America was colonized by European immigrants, so the regions share language, religion, and other political, economic, and cultural traditions. From the perspective of LDCs, the economic influence wielded by these two regions is closely intertwined with the global influence of European and American culture. Japan, the third area of high HDI, has a different cultural tradition.

Japan's development is especially remarkable because it has an extremely unfavorable ratio of population to resources. Japan became an industrial power by taking advantage of the country's one asset, an abundant supply of people willing to work hard for low wages. The Japanese government encouraged manufacturers to sell their products in other countries at prices lower than domestic competitors. Having gained a foothold in the global economy by selling low-cost products, Japan then specialized in high-quality, high-value products, such as electronics, motor vehicles, and cameras.

Japan's eminence was achieved in part by concentrating resources in rigorous educational systems and training programs to create a skilled labor force. Japanese companies spend twice as much as U.S. firms on research and development, and the government provides further assistance in developing new products and manufacturing processes.

Oceania: HDI 0.90

Oceania is relatively marginal in the global economy because of its small number of inhabitants and peripheral location. Although the HDIs of Australia and New Zealand are comparable to those of other MDCs, the area's remaining people are scattered among sparsely inhabited islands that are generally less developed.

As former British colonies, Australia and New Zealand share many cultural characteristics with the United Kingdom. Over 90 percent of the residents are descendants of nineteenth-century British settlers, although indigenous populations remain. Australia plays an increasingly important role in the global economy because it is a leader in mining numerous important minerals, including iron ore, lead, manganese, nickel, titanium, and zinc. Australia and New Zealand are also net exporters of food and other resources, especially to the United Kingdom. Increasingly, their economies are tied to Japan and other Asian countries.

Less Developed Regions

Seven regions are classified as less developed. The level of development varies widely among them. Latin America has the highest HDI among the seven regions. Behind Latin America, four of the five Asian regions—East Asia, Southwest Asia (with North Africa), Southeast Asia, and Central Asia—have similar HDIs. South Asia and sub-Saharan Africa lag behind the others.

GLOBAL FORCES, LOCAL IMPACTS
Regional Variations Within Countries

Brazil, China, and Mexico are among the world's largest and most populous countries. At the national scale, the three countries fall somewhere in the middle of the pack in GDP per capita and most other HDI indicators—well above sub-Saharan Africa and South Asia but well behind Europe and North America.

Hidden in nationwide statistics are substantial variations at the regional scale within all three countries (Figure 9-12). All three countries have GDP per capita greater than 150 percent of the national average in some provinces or states and less than 75 percent of the national average in other regions. MDCs also have regional variations in GDP per capita, but they are less extreme. In the United States, for example, the GDP per capita is 122 percent of the national average in the wealthiest region (New England) and 90 percent of the national average in the poorest region (Southeast).

Regional variations can be traced to distinctive features of each country:

- Brazil: Wealth is highest along the Atlantic coast and lowest in the interior Amazon tropical rain forest.
- China: As in Brazil, wealth is highest along the east coast and lowest in the remote and inhospitable mountain and desert environments of the interior.
- Mexico: Wealth is relatively high in the region bordering its even wealthier neighbor to the north and in the principal tourist region on the Yucatan Peninsula.

At a local scale, wealth in these intermediate-development countries is concentrated in large urban areas, such as Rio de Janeiro and São Paulo in Brazil, Beijing and Shanghai in China, and Mexico City. These cities contain a large share of the national services and manufacturing sectors and are where many leaders of the public and private sectors live. They also contain extensive areas of poverty and slum conditions, as discussed in Chapter 13. ■

FIGURE 9-12 GDP per capita as percent of national average in three large countries: (center) states of Brazil, (top) provinces of China, (bottom) states of Mexico.

Latin America: HDI 0.82

Latin America's population is highly concentrated along the South Atlantic Coast between Curitiba, Brazil, and Buenos Aires, Argentina, especially in large urban areas, including Rio de Janeiro and São Paulo, Brazil, as well as Buenos Aires (Figure 9-13, left). Mexico City also ranks among the world's largest cities. Overall, Latin Americans are more likely to live in urban areas than people in other LDCs.

The level of development varies sharply within Latin America. Neighborhoods within the large cities enjoy a level of development comparable to that of MDCs. The coastal area as a whole has a relatively high GDP per capita (see Global Forces, Local Impacts box). Outside the coastal area, development is lower in Central America, several Caribbean islands, and the interior of South America. Large areas of interior tropical rain forest are being destroyed to sell the timber or to clear the land for settled agriculture.

Overall development in Latin America is hindered by inequitable income distribution. In many countries, a handful of wealthy families control much of the land and rent parcels to individual farmers. Many tenant farmers grow coffee, tea, and fruits for export to relatively developed countries rather than food for domestic consumption. Latin American governments encourage redistribution of land to peasants but do not wish to alienate the large property owners, who generate much of the national wealth.

Latin America's economy is closely linked to that of the United States. Mexico is especially dependent on trade with the United States. As a result, the severe global recession has hit Latin America especially hard.

East Asia: HDI 0.77

The economy of East Asia—and the entire world, for that matter—is in the twenty-first century being driven increasingly by China. Now the world's second-largest economy, behind only the United States, China has become the world's largest market and manufacturer (see Contemporary Geographic Tools box on 289).

China has been the world's most populous country throughout recorded history. It was the world's wealthiest country from ancient times until it was passed by Europe in the sixteenth century. As recently as the early nineteenth century, China still accounted for one-third of world GDP. But after a century of civil wars and foreign invasions, China had fallen far behind the level of development achieved in Europe and North America in the twentieth century.

China's watershed year was 1949, when the Communist party won a civil war and created the People's Republic of China. The old Nationalist government fled to the island of Taiwan, setting up a government in exile. Since then, dramatic changes have occurred in China's economy. At first, priority was given to rural areas, where two-thirds of the Chinese people

FIGURE 9-13 Developing regions with higher HDI: Latin America and East Asia. Brazil (left) and China (right) are leading producers of motor vehicles.

live. Before communism, most Chinese farmers had been tenants, forced to pay high rents and turn over a percentage of their crops to property owners. Most years, farmers produced just enough food to survive, but they frequently suffered from famines, epidemics, floods, and wars.

Under communism, the government took control of most agricultural land. In some villages, officials assigned specific tasks to each farmer, distributed food to each family according to individual needs, and sold any remaining food to urban residents. In other cases, farmers rented land from the local government, received orders to grow specific amounts of particular crops, and sold for their own profit any crops above the minimum production targets. The system assured the production and distribution of enough food to support China's one-billion-plus population. In recent years, farmers have been permitted to hold long-term leases on land and control their own production.

In the twenty-first century, manufacturing has been increasing dramatically in China (Figure 9-13, right). With rising wealth, the world's largest population has been transformed into the world's largest market for consumer products like detergent, shampoo, and toothpaste. And with its factories paying much lower wages than those in MDCs, China is producing two-thirds of the world's DVD players, microwaves, photocopiers, and shoes for export to other countries as well as for domestic consumption.

In partnership with the world's largest retailer Wal-Mart, China's manufacturing might is pushing down prices for consumer goods throughout the world. At the same time, the low wages being paid to China's factory workers are driving down factory pay around the world. The severe recession has slowed China's economic growth because of declining global demand for manufactured goods.

Weaknesses remain in China's economic performance. Middle management is weak, quality control is minimal, banking is primitive, and legal protection is inadequate. Rapid development is straining resources, as China has become the world's largest consumer of steel, copper, coal, and cement and the second-largest consumer of petroleum behind the United States. China is also responsible for an increasing share of the world's pollution.

Southwest Asia and North Africa: HDI 0.74

Much of Southwest Asia and North Africa is desert that can sustain only sparse concentrations of plant and animal life. This region—once more commonly called the Middle East—must import most products; however, it possesses one major economic asset: a large percentage of the world's petroleum reserves.

Saudi Arabia, the United Arab Emirates, and other oil-rich states in the region, most of them concentrated in states that border the Persian (Arabian) Gulf, have used the billions of dollars generated from petroleum sales to finance development (Figure 9-14, center). But not every country in the region has abundant petroleum reserves.. Development possibilities are limited in countries that lack significant reserves—Egypt, Jordan, Syria, and others. The large gap in per capita income between the petroleum-rich countries and those that lack resources causes tension in the region.

Islam, the religion of more than 95 percent of the region's population, dominates the culture of the region. It professes some religious principles that conflict with business practices in MDCs. In some countries, all business halts several times a day when Muslims are called to prayers. Shops close their checkout lines and permit people to unwrap their prayer rugs and prostrate themselves on the floor. Women are excluded from holding most jobs and from visiting public places, such as restaurants and swimming pools. In some places, they are expected to wear traditional black clothes, a shroud, and a veil. The low level of literacy among women is the main reason the United Nations considers development among these petroleum-rich states to be lower than the region's wealth would support. The challenge for many Middle Eastern states is to promote development without abandoning the traditional cultural values of Islam.

FIGURE 9-14 Developing regions with middle HDI. (left) Central Asia: A bed of cotton is being weeded by hand in Uzbekistan. (center) Southwest Asia: An oil valve is reopened at a refinery in Baiji, Iraq, in 2009. (right) Southeast Asia: Packets are being checked at a herbal medicine factory in Ungaran, Semarang, Indonesia.

The region also suffers from serious internal cultural disputes, as discussed in Chapters 6 through 8. Iraq's long war with Iran and attempted annexation of Kuwait split the Arab world. Lack of resolution of the long-standing conflict between Israel and its neighbors has diverted resources from development to military conflict. Southwest Asia has also struggled with terrorism. The attitude of most people in the region toward terrorism is ambivalent. On the one hand, very few endorse acts of violence against Americans, Israelis, and other civilians not directly involved in combat or the interpretation of Islam used to justify the attacks. On the other hand, few supported the U.S.-led invasion of Iraq, and alternatives are sought to U.S.-influenced culture and development.

Southeast Asia: HDI 0.73

Southeast Asia's most populous country, Indonesia, includes 13,667 islands. Southeast Asia's other most populous countries are Vietnam and Thailand (situated on the Asian mainland) and the Philippines (situated like Indonesia on a series of islands).

The region's tropical climate limits intensive cultivation of most grains (Figure 9-14, right). The heat is nearly continuous, the rainfall abundant, and the vegetation dense. Soils are generally poor because the heat and humidity rapidly destroy nutrients when land is cleared for cultivation. Development is also limited in Southeast Asia by several mountain ranges, active volcanoes, frequent typhoons, and occasional tsunamis. This inhospitable environment traditionally kept population growth low in much of the region. Nearly two-thirds of the population live on the island of Java, which has one of the world's highest arithmetic densities. People have concentrated on Java partly because the island's soil, derived from volcanic ash, is more fertile than elsewhere in the region and partly because the Dutch established their colonial headquarters there.

Because of distinctive vegetation and climate, farmers in Southeast Asia concentrate on harvesting products that are used in manufacturing. The region produces a large percentage of the world's supply of palm oil and copra (coconut oil), natural rubber, kapok (fibers from the ceiba tree used for insulation and filling), and abaca (fibers from banana leafstalks used in fabrics and ropes). Southeast Asia also contains a large percentage of the world's tin as well as some petroleum reserves. Rice, the region's most important food, is now exported in large quantities from India, Malaysia, and Thailand.

The region has suffered from a half-century of nearly continuous warfare. Japan, the Netherlands, France, and the United Kingdom were all forced to withdraw from colonies they had established in the region. In addition, France and the United States both fought unsuccessfully to prevent Communists from controlling Vietnam during the Vietnam War, which ran from the 1950s to 1975. Wars have also devastated neighboring Laos and Cambodia. In some Southeast Asian countries, however, notably Thailand, Singapore, Malaysia, and the Philippines, development has been rapid. The region has become a major manufacturer of textiles and clothing, taking advantage of cheap labor. Thailand has become the region's center for the manufacturing of motor vehicles and other consumer goods.

But economic growth in the region has slowed since the last years of the twentieth century. Earlier economic growth had been achieved through very close cooperation among manufacturers, financial institutions, and government agencies. In the absence of independent watchdogs and regulators, funds for development were sometimes invested unwisely or stolen by corrupt officials. To restore economic confidence among international investors, Southeast Asian countries have been forced to undertake painful reforms that reduce the people's standard of living.

Central Asia: HDI 0.70

Most of the countries in Central Asia were once part of the Soviet Union. With the breakup of the Soviet Union in the 1990s, regional geographers now prefer to identify a distinct region in Central Asia that encompasses eight of the fifteen former Soviet republics. Iran and Afghanistan are also included in the Central Asia region, although these two countries are closely tied to those of Southwest Asia, discussed in the next section.

Within Central Asia, the level of development is relatively high in Kazakhstan and Iran (Figure 9-14, left). Not by coincidence, these two countries are the region's leading producers of petroleum. In Kazakhstan, rising oil revenues are being used to promote carefully managed improvements in overall development. In Iran, a large share of the rising oil revenues has been used to maintain low consumer prices rather than to promote development. Since coming to power in a 1979 revolution, the fundamentalist Shiite leaders in control of Iran have also used oil revenues to promote revolutions elsewhere in the region and to sweep away elements of development and social customs they perceive to be influenced by Europe or North America.

The level of development is lower in this region's other "stan" republics. Minerals and agricultural products are their principal economic resources. Afghanistan probably has one of the world's lowest HDI's, but the United Nations hasn't calculated it for many years because of the extended war.

South Asia: HDI 0.61

South Asia includes India, Pakistan, Bangladesh, Sri Lanka, and the small Himalayan states of Nepal and Bhutan. The region has the world's second-highest population and second-lowest per capita income. Population density is very high, and the natural increase rate is among the world's highest. The overall ratio of population to resources in the region is unfavorable because of the huge population.

South Asia was a principal beneficiary of the Green Revolution, a series of developments beginning in the 1960s that dramatically increased agricultural productivity. As a result of the Green Revolution, "miracle" rice and wheat seeds were widely diffused throughout South Asia. But agricultural productivity in South Asia also depends on climate. The region receives nearly all its precipitation from rain that falls during the monsoon season between May and August. Agricultural output declines sharply if the monsoon rains fail to arrive. In a typical year, farmers in South Asia produce a grain surplus that is stored for distribution during dry years. However, several consecutive years without monsoon rains produce widespread hardship.

FIGURE 9-15 Developing regions with low HDIs: South Asia and sub-Saharan Africa. (left) Sugarcane is transported by rickshaw to a wholesale market in Hyderabad, India. (right) Family in Kenya hoe a field to plant tomatoes.

India, South Asia's largest country, has become the world's fourth-largest economy, behind the United States, China, and Japan, and the rate of growth of its economy is second only to China's. India is the world's leading producer of jute (used to make burlap and twine), peanuts, sugarcane, and tea and has mineral reserves, including uranium, bauxite (aluminum ore), coal, manganese, iron ore, and chromite (chromium ore). It is also one of the world's leading rice and wheat producers (Figure 9-15, left). The country has become a major manufacturer, though not as rapidly as China. In addition, India has become a major service provider. When you phone an airline, a help desk, or a credit card company, chances are your call will be answered by someone actually located in India.

Sub-Saharan Africa: HDI 0.51

Africa has been divided into two regions: Countries north of the Sahara share economic and cultural characteristics with Southwest Asia; south of the desert is sub-Saharan Africa.

Among the countries of this region, South Africa is a major source of minerals, including chromium, diamonds, manganese, and platinum. Other countries in the region also contain resources important for development, including bauxite in Guinea, cobalt in the Democratic Republic of Congo and Zambia, diamonds in Botswana and Congo, manganese in Gabon, petroleum in Nigeria, and uranium in Niger. Regional wealth is comparable to levels found in other LDCs.

Despite these assets, sub-Saharan Africa offers the least favorable prospect for development (Figure 9-15, right). The region has the world's highest percentage of people living in poverty and suffering from poor health and low education levels. And economic conditions in sub-Saharan Africa have deteriorated in recent years: The average African consumes less today than three decades ago. Some of the region's economic problems are a legacy of the colonial era. Mining companies and other businesses were established to supply European industries with needed raw materials rather than to promote overall economic development in sub-Saharan Africa. Africa's many landlocked states have difficulties shipping out raw materials through neighboring countries. And in recent years, African countries have suffered because world prices for their resources have fallen.

Political problems have also plagued sub-Saharan Africa. European colonies were converted to states without regard for the distribution of ethnicities (see Figure 7-34). After independence, leaders of many countries in the region pursued personal economic gain and local wars rather than policies to promote development of their national economies. These frequent internal wars, as well as those between countries in sub-Saharan Africa, have retarded development.

But the fundamental problem in many countries of sub-Saharan Africa is a dramatic imbalance between the number of inhabitants and the capacity of the land to feed the population. Nearly the entire region has either a tropical or a dry climate. Both climate regions can support some people, but not large concentrations. Yet, because sub-Saharan Africa has by far the world's highest rate of natural increase, its land is more and more overworked, and agricultural output has declined.

KEY ISSUE 3

Where Does Level of Development Vary by Gender?

- **Gender-Related Development Index**
- **Gender Empowerment**

A country's overall level of development masks inequalities in the status of men and women. Gender inequality

CONTEMPORARY GEOGRAPHIC TOOLS
Wal-Mart and China

No corporation exposes the effects of globalization on the world's economy more effectively than Wal-Mart. Wal-Mart Stores, Inc., was the world's largest corporation in 2008, with revenues of $379 billion. The company operated 7,200 stores worldwide in 2008, including 4,200 in the United States, and employed 2.1 million.

Geographic research has documented the key role played by geographic tools in Wal-Mart's growth. Most important has been application of the distance-decay concept, which was defined in Chapter 1 as the diminishing in importance and eventual disappearance of a phenomenon with increasing distance from its origin.

Wal-Mart was founded by Sam Walton in 1962 with a single store in Bentonville, Arkansas. It was not until 1995 that each of the 50 states had at least one Wal-Mart. Maps of opening dates of each Wal-Mart store show a diffusion pattern (Figure 9-16). In the 1970s, Wal-Mart was a small company confined largely to Arkansas and Missouri. In the 1980s, it was concentrated in the south-central regions of the United States. In the 1990s, Wal-Mart reached the northeast, north-central, and west coast regions. The diffusion pattern resulted from a deliberate application of distance-decay principles. Rather than sprinkling stores around the country, the company preferred to saturate communities with stores before moving to new territory. New stores were opened near existing ones so that they could share the same advertising and central management control and, most importantly, the same distribution center.

Distribution centers are very large facilities of more than 1 million square feet typically serving several dozen stores. State-of-the-art inventory controls move merchandise from the distribution center to the stores. To locate new stores, Wal-Mart mapped a one-day driving distance around a distribution center. New stores were placed first at the outer edge of the one-day ring around the distribution center; additional stores were added closer in until the market area was saturated. New distribution centers were opened at ever-increasing distances from Arkansas until the company finally reached California in 1990 and New England in 1991.

Wal-Mart's U.S. retail stores are part of a global network. Most of Wal-Mart's merchandise is made in China; moved by truck from factories to Wal-Mart's Global Procurement Center in Shenzhen, China; loaded on container ships for the two-week passage across the Pacific Ocean; unloaded in the port of Long Beach, California; and transported primarily by truck across the United States to the distribution centers. ■

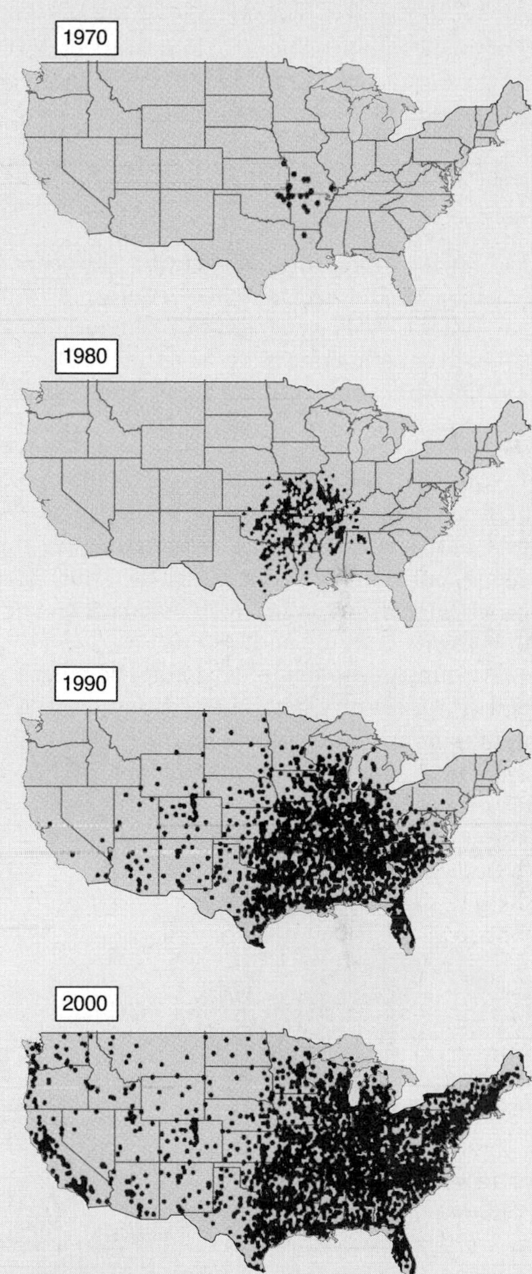

FIGURE 9-16 Growth of Wal-Mart stores. Wal-Mart diffused from Arkansas through the United States using a distance-decay model of store location.

exists in every country of the world, according to the United Nations. In some countries women have achieved near equality with men, whereas in other countries women lag far behind the level of development for men. The United Nations has not found a single country in the world where women are treated as well as men.

To measure the extent of each country's gender inequality, the United Nations has created two indexes. The **Gender-Related Development Index (GDI)** compares the level of women's development with that of both sexes. The **Gender Empowerment Measure (GEM)** compares the ability of women and men to participate in economic and political decision making. ■

Gender-Related Development Index

The GDI is constructed in a manner similar to the HDI, discussed in the first two sections of this chapter (Figure 9-17). The GDI combines the same indicators of development used in the HDI, adjusted to reflect differences in the accomplishments and conditions of men and women:

- **Economic indicator of gender differences:** Per capita female income as a percentage of per capita male income (Figure 9-18)
- **Social indicators of gender differences:** Number of females enrolled in school compared to number of males (Figure 9-19) and percent of literate females compared to percent of literate males (Figure 9-20)
- **Demographic indicator of gender differences:** Life expectancy of females compared to males (Figure 9-21)

The GDI penalizes a country for having a large disparity between the well-being of men and women. For example, Hungary and Saudi Arabia have approximately the same GDP per capita, but Hungary has a higher GDI than Saudi Arabia in part because the disparity between female and male income is lower in Hungary than in Saudi Arabia.

A country with complete gender equality would have a GDI of 1.0. No country has achieved that level. A high GDI means that both men and women have achieved a high level of development, though women may have a slightly lower level than men. A low GDI means that women have a low level of development and the level is substantially below that of men.

Gender Empowerment

The GEM measures the ability of women to participate in the process of achieving improvements in their status, that is, to achieve economic and political power. In every country of the world, both MDCs and LDCs, fewer women than men hold positions of economic and political power, according to the United Nations GEM scoring system.

The GEM is calculated by combining two indicators of economic power and two indicators of political power (Figure 9-22):

- **Economic indicators of empowerment:** Per capita female income as a percentage of per capita male income (Figure 9-18) and percentage of professional and technical jobs held by women (Figure 9-23)
- **Political indicators of empowerment:** Percentage of administrative jobs held by women (Figure 9-24) and percentage of members of the national parliament who are women (Figure 9-25)

A country with complete equality of power between men and women would have a score of 1.0. As with the GDI, countries with the highest GEMs are MDCs, especially in North America, Northern Europe, and Oceania. The lowest scores are in Africa and Asia, though lack of data prevents calculating scores for many LDCs. Every country has a lower GEM than GDI. A higher

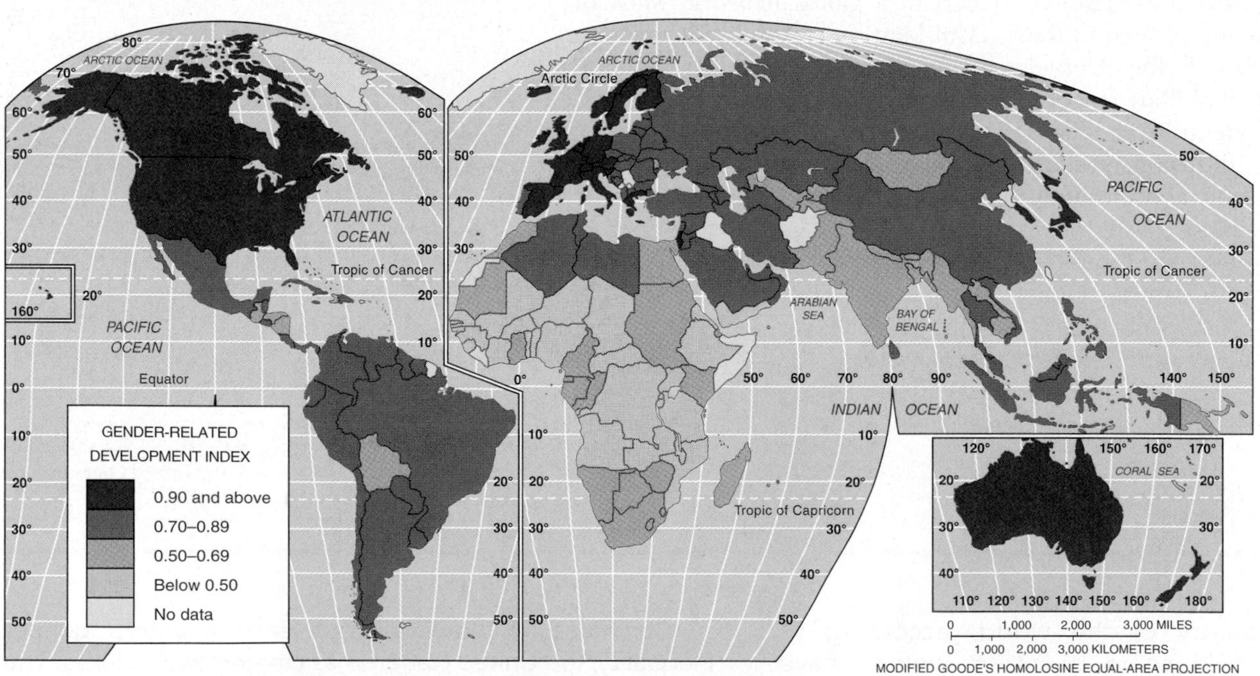

FIGURE 9-17 Gender-Related Development Index (GDI). Similar to the Human Development Index (HDI), the GDI combines four measures of development, lowered by the amount of disparity between males and females. A high GDI means that men and women have both achieved high levels of development, though women have a slightly lower level. A low GDI means that women have a low level of development and a level substantially lower than that for men.

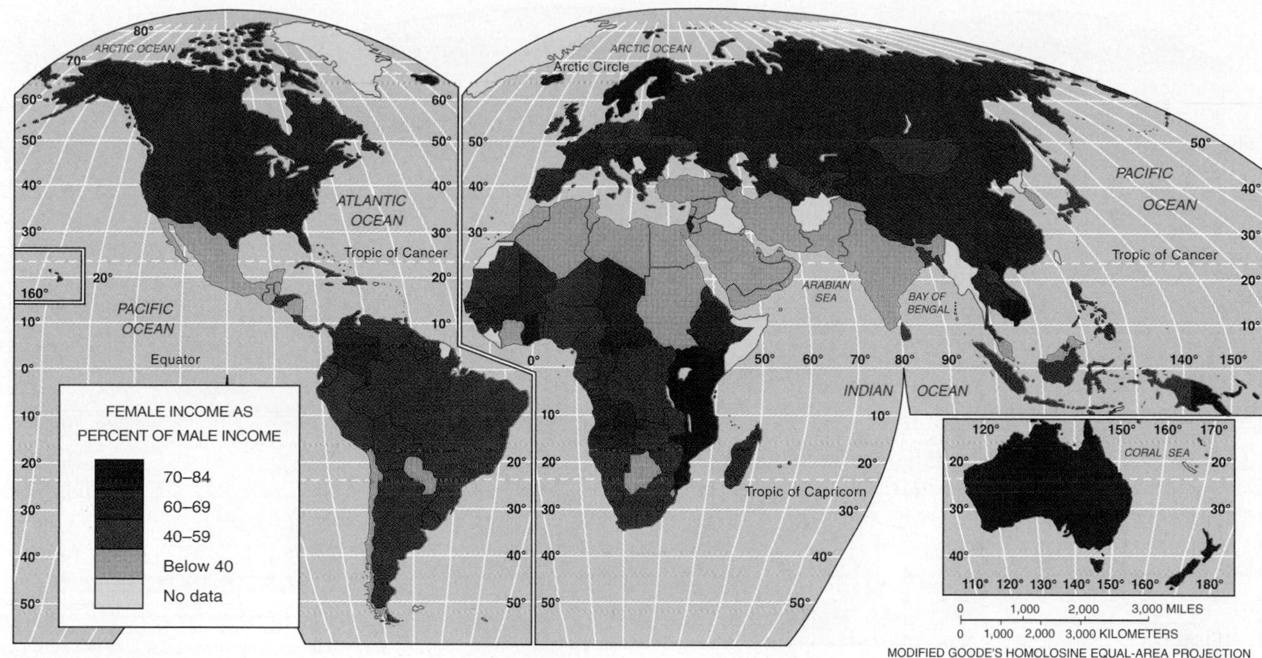

FIGURE 9-18 Economic indicator of gender difference: Income. The average income of women is lower than that of men in every country of the world, both in MDCs and LDCs. Women on average have two-thirds of the income of men in MDCs. This translates into an income gap of $12,000. In LDCs, the disparity between male and female income is relatively low in dollar terms but high on a percentage basis. Earnings for women lag far behind those of men in LDCs, although both figures are much lower than those found in MDCs.

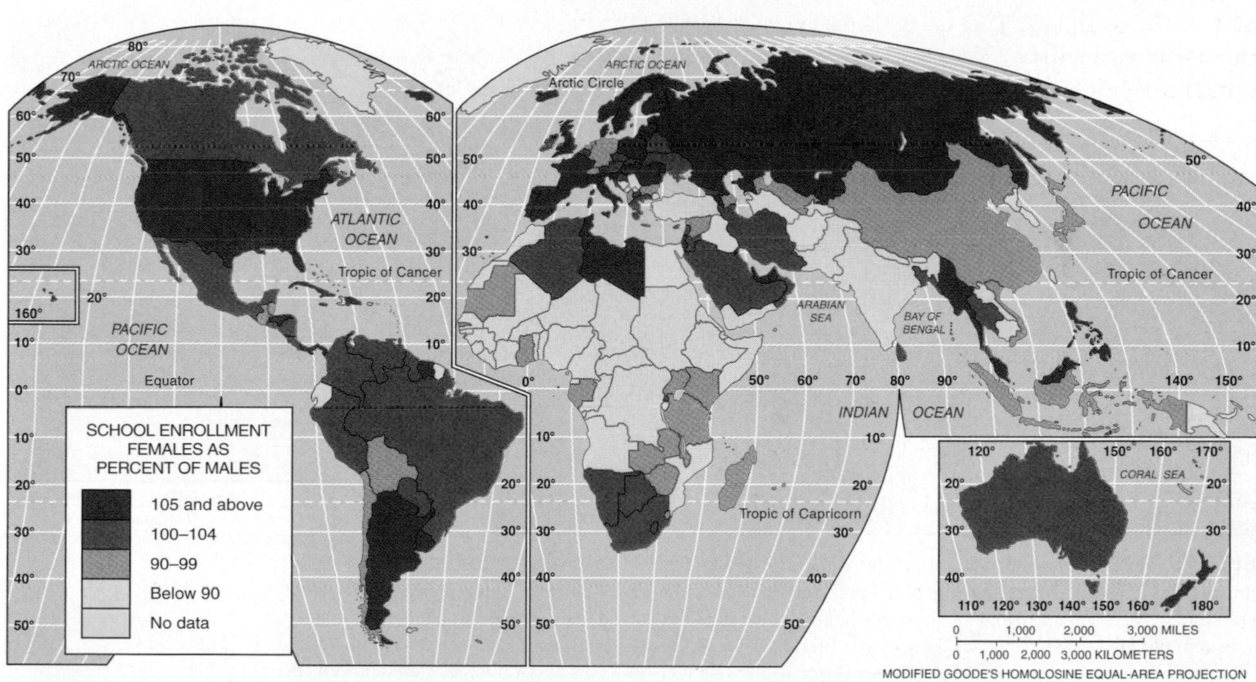

FIGURE 9-19 Social indicator of gender difference (one of two): school enrollment. Girls are more likely than boys to be enrolled in school in MDCs, but less likely in LDCs. The percentage of females attending school is a key measure of gender disparity in sub-Saharan Africa and Southwest Asia. In Latin America and much of Asia, boys and girls are equally likely to attend school, but attendance is lower than in MDCs.

GDI compared to GEM means that women possess a greater share of a country's resources than they do power over allocation of those resources.

The indicators presented in the previous key issues reflect sharp differences in the levels of development of MDCs and LDCs. To promote development, LDCs seek improvements in these indicators. Progress has been mixed (Figure 9-26). On the one hand, key indicators look better for LDCs now than they did a generation ago. On the other hand, the gap in key development indicators between LDCs and MDCs remains wide.

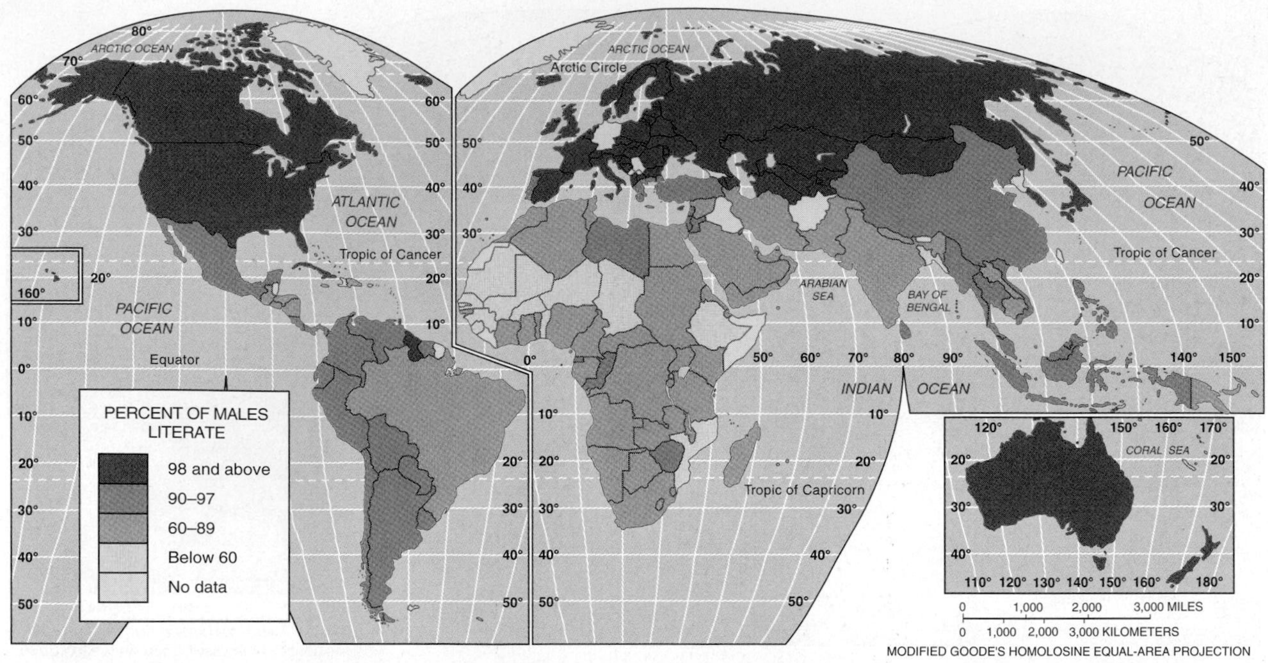

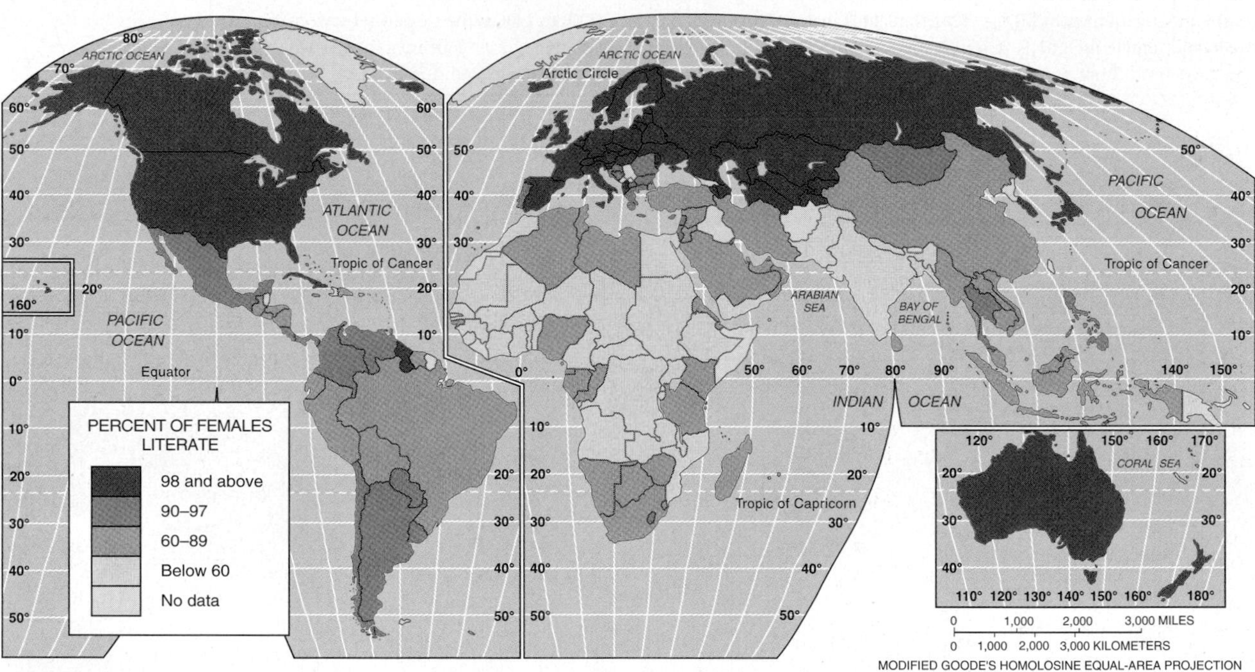

FIGURE 9-20 Social indicator of gender difference (two of two): literacy. In MDCs, literacy is nearly universal among both men and women. In Latin America and much of Asia, literacy is not universal, but rates are similar for men and women. In sub-Saharan Africa and Southwest Asia, female literacy is low, and substantially lower than for males. Low female literacy is an especially important obstacle to development in these regions. It is both a cause and a consequence of the relatively low contribution females are allowed to make to the economy and culture of these regions.

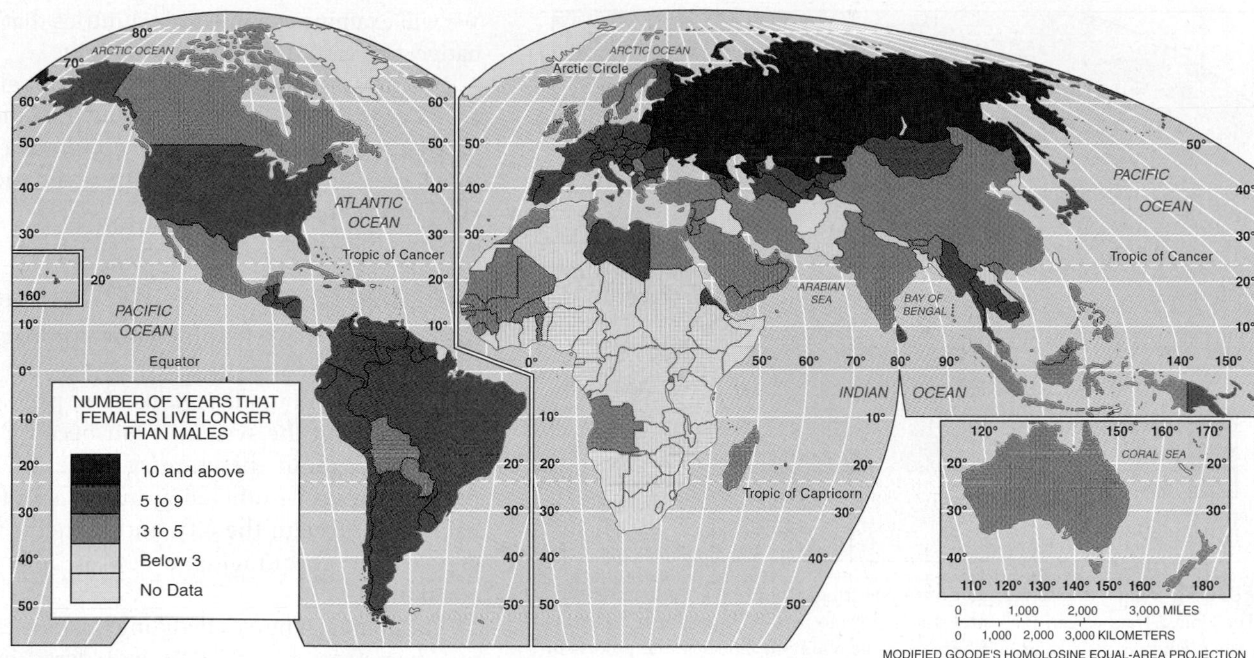

FIGURE 9-21 Demographic indicator of gender difference: life expectancy. The gender gap in life expectancy is greater in MDCs than in LDCs. In MDCs, a female baby born today is expected to live several years longer than a male baby, whereas in most LDCs, the gap in life expectancy between females and males is only a year or two. The inability of women to outlive men in LDCs derives primarily from the hazards of childbearing. Women in LDCs bear more children than in MDCs, often under poor medical conditions.

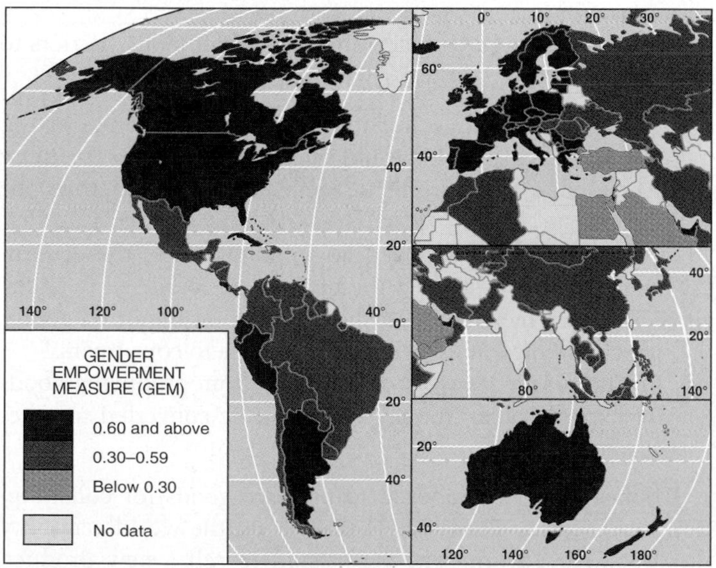

FIGURE 9-22 Gender Empowerment Measure (GEM). The GEM combines two measures of the economic power of women and two measures of their political power. Information was not available to calculate the GEM for most LDCs. Compare to Figure 9-17: A country with a much lower GEM than GDI offers women less power than economic resources.

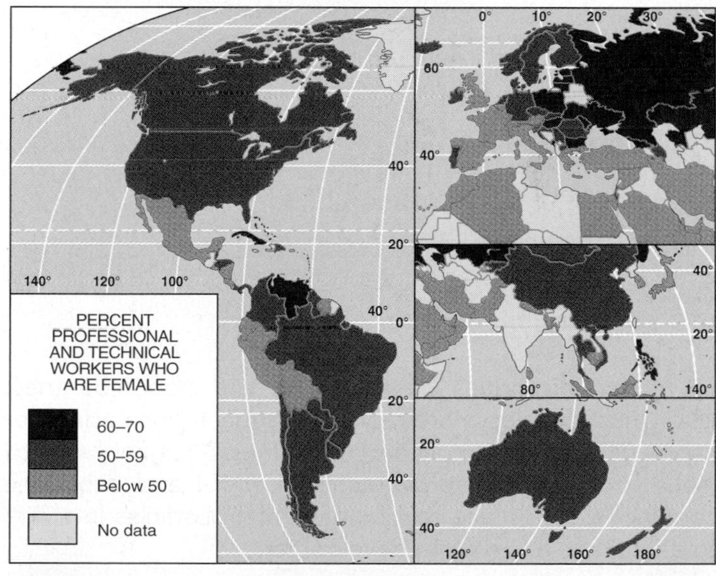

FIGURE 9-23 Economic indicator of empowerment: professionals. The percentage of women occupying professional and technical jobs is considered an important measure of the economic power held by women in a country. Professional and technical jobs are regarded by the United Nations as those offering women the greatest opportunities for advancement to positions of influence in a country's economy. Cultural barriers may restrict the ability of women to obtain these jobs in the first place or to secure promotions to top-level decision-making positions.

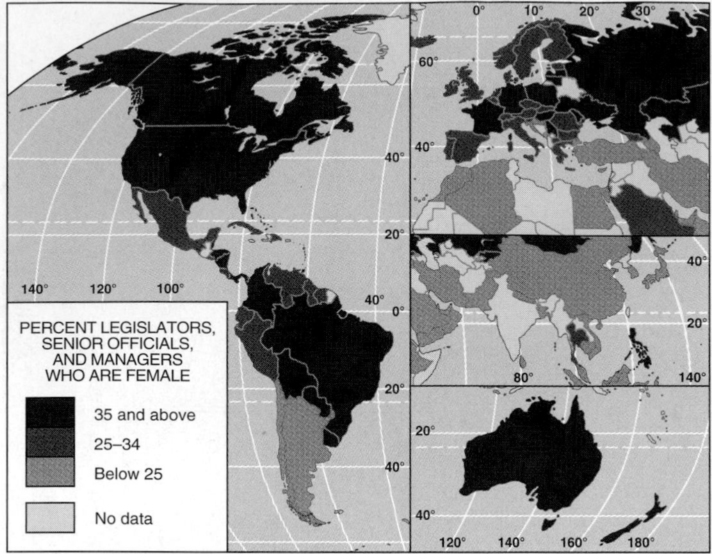

FIGURE 9-24 Political indicator of empowerment (one of two): administrators. The United Nations considers professional jobs to be a key measure of economic power, whereas managerial jobs represent the ability to influence the process of decision making.

KEY ISSUE 4
Why Do LDCs Face Obstacles to Development?

- Development Through Self-Sufficiency
- Development Through International Trade
- International Trade Approach Triumphs
- Financing Development
- Fair Trade

To reduce disparities between rich and poor countries, LDCs must develop more rapidly. This means increasing per capita GDP more rapidly and using the additional funds to make more rapid improvements in social and economic conditions. LDCs face two fundamental obstacles in trying to encourage more rapid development:

- Adopting policies that successfully promote development
- Finding funds to pay for development ■

Development Through Self-Sufficiency

To promote development, LDCs choose one of two models: One emphasizes international trade; the other advocates self-sufficiency. Each has important advantages and serious problems. We will examine examples of countries that have tried each alternative, successfully and unsuccessfully.

For most of the twentieth century, self-sufficiency, or balanced growth, was the more popular of the development alternatives. The world's two most populous countries, China and India, once adopted this strategy, as did most African and Eastern European countries.

Elements of Self-Sufficiency Approach

According to the self-sufficiency approach, a country should spread investment as equally as possible across all sectors of its economy and in all regions. The pace of development may be modest, but the system is fair because residents and enterprises throughout the country share the benefits of development. Under self-sufficiency, incomes in the countryside keep pace with those in the city, and reducing poverty takes precedence over encouraging a few people to become wealthy consumers.

The approach nurses fledgling businesses in an LDC by isolating them from competition with large international corporations. Such insulation from the potentially adverse impacts of decisions made by businesses and governments in the MDCs encourages a country's fragile businesses to achieve independence. Countries promote such self-sufficiency by setting barriers that limit the import of goods from other places. Three widely used barriers include setting high taxes (tariffs) on imported goods to make them more expensive than domestic goods, fixing quotas to limit the quantity of imported goods, and requiring licenses in order to restrict the number of legal importers. The approach also restricts local businesses from exporting to other countries.

For many years India made effective use of many barriers to trade. For example:

- To import goods into India, most foreign companies had to secure a license, a long and cumbersome process because several dozen government agencies had to approve the request.
- Once a company received an import license, the government severely restricted the quantity it could sell in India.
- The government imposed heavy taxes on imported goods, which doubled or even tripled the price to consumers.
- Indian businesses were discouraged from producing goods for export; Indian money could not be converted to other currencies.

Businesses were supposed to produce goods for consumption inside India. Effectively cut off from the world economy, they required government permission to sell a new product, modernize a factory, expand production, set prices, hire or fire workers, and change the job classification of existing workers. If private companies were unable to make a profit selling goods only inside India, the government provided subsidies, such as cheap electricity, or wiped out debts. The government owned not just communications, transportation, and power companies, a common feature around the world, but also businesses such as insurance companies and automakers, left to the private sector in most countries.

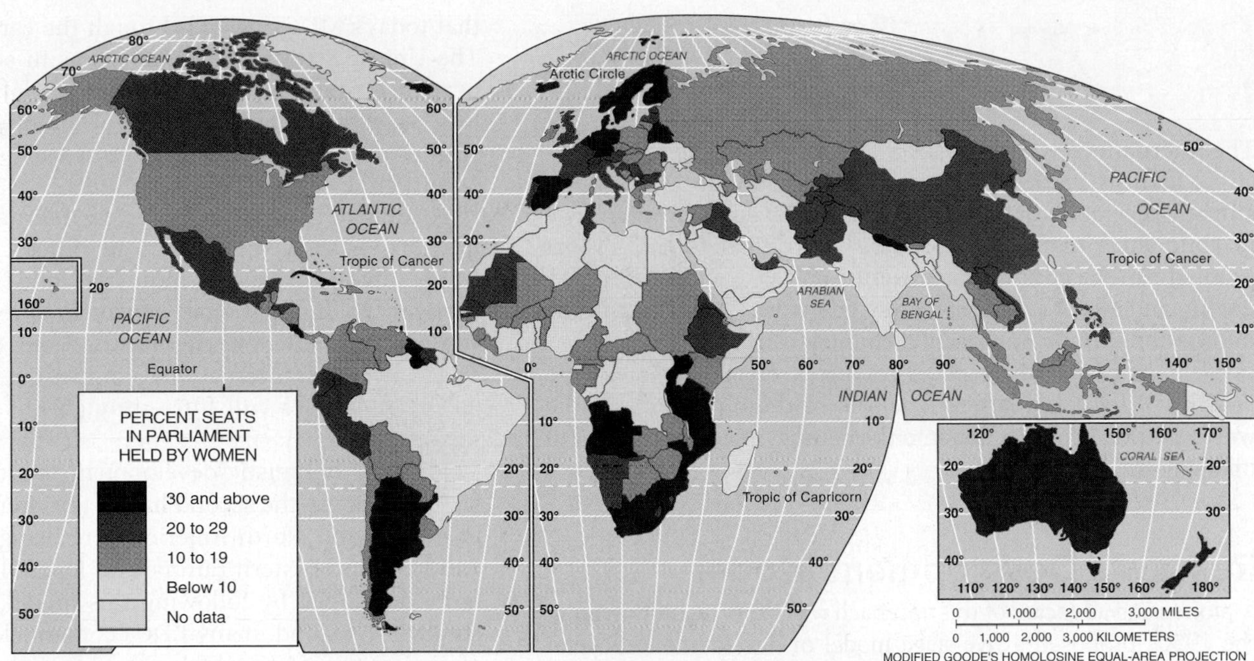

FIGURE 9-25 Political indicator of empowerment (two of two): elected officials. No particular gender-specific skills are required to be elected as a representative and to serve effectively. Although more women than men vote in most places, no country has a national parliament or congress with a majority of women. The highest percentages are in Northern Europe, where women comprise approximately one-third of members of national parliaments. In the United States, 15 percent of the U.S. Senate and House of Representatives are women.

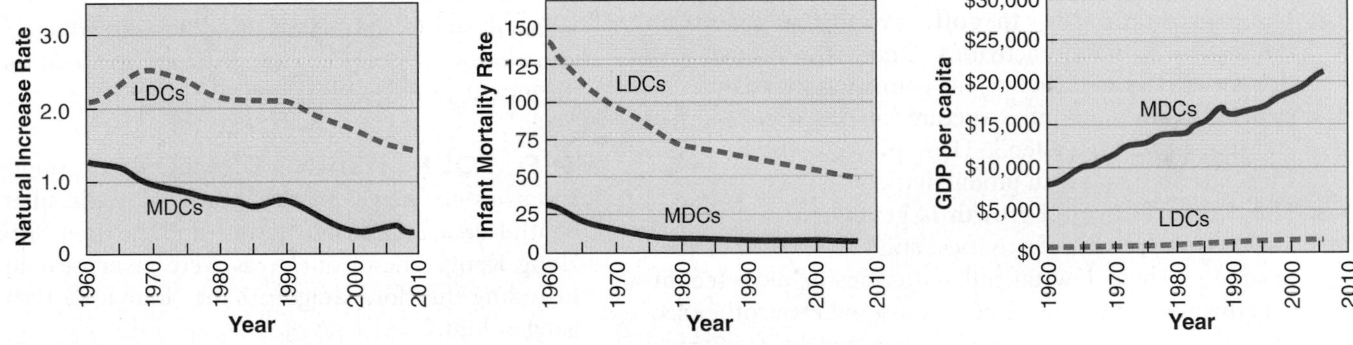

FIGURE 9-26 Progress toward development. Key three indicators of development show that the gap between MDCs and LDCs remains wide. The natural increase rate has declined at about the same rate in MDCs and LDCs since 1970, and the infant mortality rate has declined at about the same rate since 1990. GDP per capita has increased more rapidly in MDCs than in LDCs.

Problems with the Self-Sufficiency Alternative

The experience of India and other LDCs with self-sufficiency revealed two major problems:

1. **Protection of inefficient businesses.** Businesses could sell all they made, at high government-controlled prices, to customers culled from long waiting lists, so they had little incentive to improve quality, lower production costs, reduce prices, or increase production. Companies protected from international competition were not pressured to keep abreast of rapid technological changes.
2. **Need for large bureaucracy.** The complex administrative system needed to administer the controls encouraged abuse and corruption. Potential entrepreneurs found that struggling to produce goods or offer services was less rewarding financially than advising others how to get around the complex government regulations. Other potential entrepreneurs earned more money by illegally importing goods and selling them at inflated prices on the black market.

Development Through International Trade

The international trade model of development calls for a country to identify its distinctive or unique economic assets. What animal, vegetable, or mineral resources does the country have in abundance that other countries are willing to buy? What product can the country manufacture and distribute at a higher quality and a lower cost than other countries? According to the international trade approach, a country can develop economically by concentrating scarce resources on expansion of its distinctive local industries. The sale of these products in the world market brings funds into the country that can be used to finance other development.

Rostow's Development Model

A pioneering advocate of this approach was W. W. Rostow, who in the 1950s proposed a five-stage model of development. Several countries adopted this approach during the 1960s, although most continued to follow the self-sufficiency approach. The five stages were as follows:

1. **The traditional society.** A traditional society has not yet started a process of development. It contains a very high percentage of people engaged in agriculture and a high percentage of national wealth allocated to what Rostow called "nonproductive" activities, such as the military and religion.
2. **The preconditions for takeoff.** An elite group initiates innovative economic activities. Under the influence of these well-educated leaders, the country starts to invest in new technology and infrastructure, such as water supplies and transportation systems. These projects will ultimately stimulate an increase in productivity.
3. **The takeoff.** Rapid growth is generated in a limited number of economic activities, such as textiles or food products. These few takeoff industries achieve technical advances and become productive, whereas other sectors of the economy remain dominated by traditional practices.
4. **The drive to maturity.** Modern technology, previously confined to a few takeoff industries, diffuses to a wide variety of industries, which then experience rapid growth comparable to the takeoff industries. Workers become more skilled and specialized.
5. **The age of mass consumption.** The economy shifts from production of heavy industry, such as steel and energy, to consumer goods, such as motor vehicles and refrigerators.

According to the international trade model, each country is in one of these five stages of development. MDCs are in stage 4 or 5, whereas LDCs are in one of the three earlier stages. The model assumes that LDCs will achieve development by moving along from an earlier to a later stage. The model also asserts that today's MDCs passed through the early stages in the past. The United States, for example, was in stage 1 prior to independence, stage 2 during the first half of the nineteenth century, stage 3 during the middle of the nineteenth century, and stage 4 during the late nineteenth century, before entering stage 5 during the early twentieth century.

A country that concentrates on international trade benefits from exposure to consumers in other countries. To remain competitive, the takeoff industries must constantly evaluate changes in international consumer preferences, marketing strategies, production engineering, and design technologies. This concern for international competitiveness in the exporting takeoff industries will filter through less advanced economic sectors.

Rostow's optimistic development model was based on two factors. First, in the second half of the twentieth century MDCs in Europe and North America were being joined by others in Southern and Eastern Europe and Japan. If they could become more developed by following this model, why couldn't other countries? Second, many LDCs contained an abundant supply of raw materials sought by manufacturers and producers in MDCs. In the past, European colonial powers extracted many of these resources without paying compensation to the colonies. In a global economy, the sale of these raw materials could generate funds for LDCs with which they could promote development.

Examples of the International Trade Approach

When most LDCs were following the self-sufficiency approach, two groups of countries chose the international trade approach during the mid-twentieth century.

THE FOUR ASIAN DRAGONS. Among the first countries to adopt the international trade alternative were South Korea, Singapore, Taiwan, and the then-British colony of Hong Kong. These four areas were given several nicknames, including the "four dragons," the "four little tigers," and "the gang of four."

Singapore and Hong Kong, British colonies until 1965 and 1997, respectively, have virtually no natural resources. Both comprise large cities surrounded by very small amounts of rural land. South Korea and Taiwan have traditionally taken their lead from Japan, which occupied both countries until after World War II. Their adoption of the international trade approach was strongly influenced by Japan's success. Lacking many natural resources, the four dragons promoted development by concentrating on producing a handful of manufactured goods, especially clothing and electronics. Low labor costs enabled these countries to sell products inexpensively in MDCs.

PETROLEUM-RICH ARABIAN PENINSULA STATES. The Arabian Peninsula includes Saudi Arabia, the region's largest and most populous country, plus Kuwait, Bahrain, Oman, and the United Arab Emirates. Once among the world's least developed countries, they were transformed overnight into some of the wealthiest thanks to escalating petroleum prices during the 1970s.

Arabian Peninsula countries have used petroleum revenues to finance large-scale projects, such as housing, highways, airports, universities, and telecommunications networks. Their steel, aluminum, and petrochemical factories compete on world markets with the help of government subsidies. The landscape has been further changed by the diffusion of consumer goods. Large motor vehicles, color TVs, audio equipment, and motorcycles are readily available and affordable. Supermarkets are stocked with food imported from Europe and North America.

Problems with the International Trade Alternative

Three problems have hindered countries outside the four Asian dragons and the Arabian Peninsula from developing through the international trade approach:

1. **Uneven resource distribution.** Arabian Peninsula countries achieved successful development by means of rising petroleum prices. Other countries found that the prices of their commodities did not increase and in some cases actually decreased. LDCs that depended on the sale of one product suffered because the price of their leading commodity did not rise as rapidly as the cost of the products they needed to buy. For example, Zambia has extensive copper reserves, but it has been unable to use this asset to promote development because of declining world prices for copper.
2. **Increased dependence on MDCs.** Building up a handful of takeoff industries that sell to people in MDCs may force LDCs to cut back on production of food, clothing, and other necessities for their own people. Rather than finance new development, funds generated from the sale of products to other countries may have to be used to buy these necessities from MDCs for the employees of the takeoff industries.
3. **Market decline.** Countries that depend on selling low-cost manufactured goods find that the world market for many products has declined sharply in recent years. Even before the recent severe recession, MDCs had limited growth in population and market size.

International Trade Approach Triumphs

In the late twentieth century, most countries embraced the international trade approach as the preferred alternative for stimulating development. Trade has increased more rapidly than wealth (as measured by GDP), a measure of the growing importance of the international trade approach, especially in LDCs (Figure 9-27).

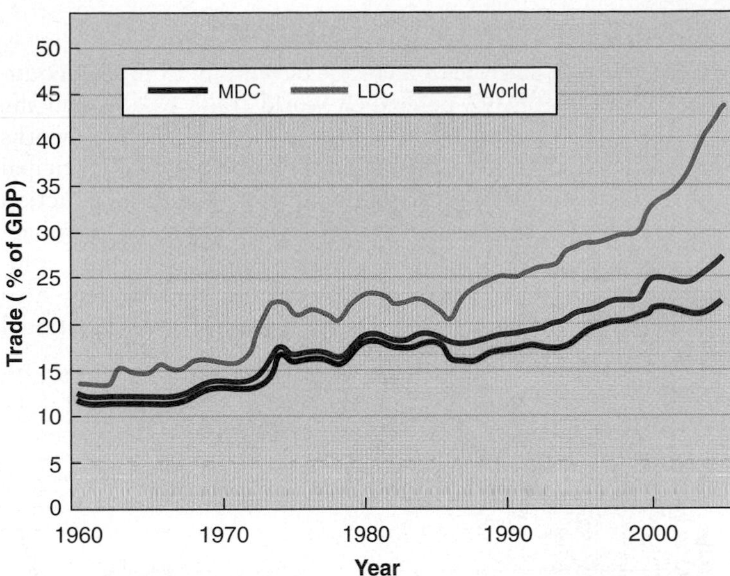

FIGURE 9-27 Trade as percent of GDP. Trade has grown much more rapidly than GDP, especially in LDCs after 1990, a measure of the conversion of many LDC economies from self-sufficiency to international trade.

Longtime advocates of the self-sufficiency approach converted to international trade during the 1990s. India, for example, dismantled its formidable collection of barriers to international trade:

- Foreign companies were allowed to set up factories and sell in India.
- Tariffs and restrictions on the import and export of goods were reduced or eliminated.
- Monopolies in communications, insurance, and other industries were eliminated.
- With increased competition, Indian companies have improved the quality of their products.

During the self-sufficiency era, India's auto industry was dominated by Maruti-Udyog Ltd., which was controlled by the Indian government. Nursed by import duties that rose from 15 percent in 1984 to 66 percent in 1991, Maruti captured more than 80 percent of the Indian market selling cars that would be considered out-of-date in other countries. In the international trade era, the government sold control of Maruti to the Japanese company Suzuki, which now holds only 40 percent of India's market.

Countries like India converted from self-sufficiency to international trade during the 1990s for one simple reason—overwhelming evidence that international trade better promoted development (Figure 9-28). The World Bank found that between 1990 and 2005 per capita GDP increased more than 4 percent annually in countries strongly oriented toward international trade, compared with less than 1 percent for countries strongly oriented toward self-sufficiency.

World Trade Organization

To promote the international trade development model, countries representing 97 percent of world trade established the World Trade Organization (WTO) in 1995. The WTO works to reduce barriers to international trade in two principal ways. First, through the WTO, countries negotiate reduction or elimination of international trade restrictions on manufactured goods, such as government subsidies for exports, quotas for imports, and tariffs on both imports and exports. Also reduced or eliminated are restrictions on the international movement of money by banks, corporations, and wealthy individuals.

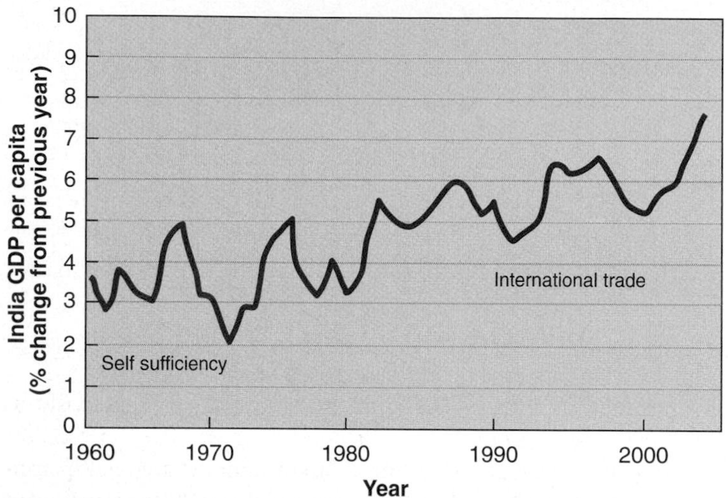

FIGURE 9-28 GDP change in India. After conversion from self-sufficiency to international trade around 1990, India's GDP increased more rapidly.

The WTO also promotes international trade by enforcing agreements. One country can bring to the WTO an accusation that another country has violated a WTO agreement. The WTO is authorized to rule on the validity of the charge and order remedies. The WTO also protects intellectual property in the age of the Internet. An individual or corporation can also bring charges to the WTO that someone in another country has violated their copyright or patent, and the WTO can order illegal actions to stop.

The WTO has been sharply attacked by critics. Protesters routinely gather in the streets outside high-level meetings of the WTO (Figure 9-29). Progressive critics charge that the WTO is antidemocratic, because decisions made behind closed doors promote the interests of large corporations rather than the poor. Conservatives charge that the WTO compromises the power and sovereignty of individual countries because it can order changes in taxes and laws that it considers unfair trading practices.

Foreign Direct Investment

International trade requires corporations based in a particular country to invest in other countries. Investment made by a foreign company in the economy of another country is known as **foreign direct investment (FDI)**.

Foreign direct investment grew rapidly during the 1990s, from $130 billion in 1990 to $1.5 trillion in 2000. The level declined to $647 billion in 2003 in the wake of the 9/11 al-Qaeda attacks on the United States, before returning to $1.5 trillion later in the decade. Foreign direct investment does not

FIGURE 9-29 Anti-World Trade Organization (WTO) protest. Protesters oppose WTO meeting in Jakarta, Indonesia, in 2009.

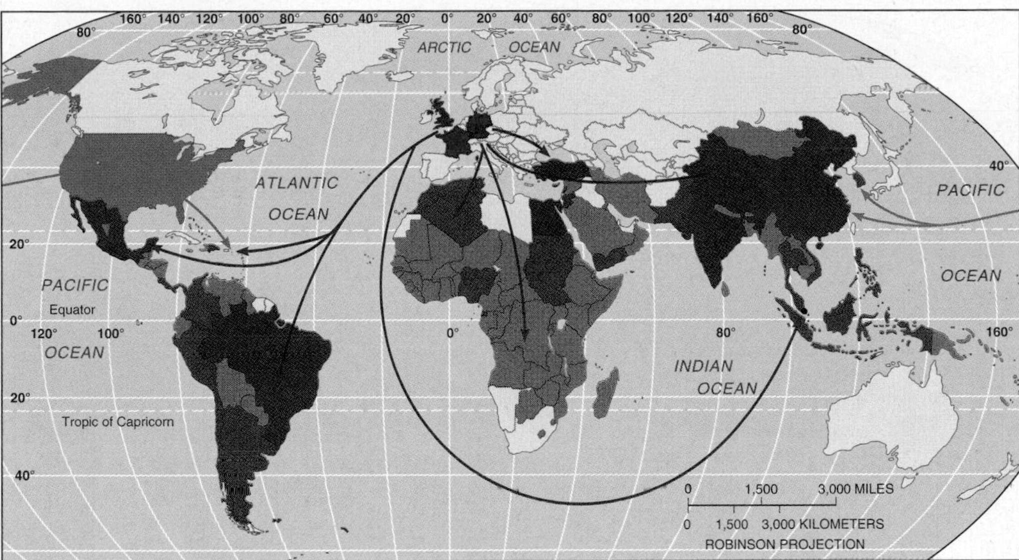

FIGURE 9-30 Foreign direct investment in LDCs. Most transnational companies invest in the three core areas—North America, Europe, and Japan. Outside the core regions, the largest amount of investment by transnational corporations is in China.

flow equally around the world (Figure 9-30). Only one-fourth of foreign investment in 2007 went from an MDC to a LDC, whereas the other three-fourths went from one MDC to another MDC. And FDI is not evenly distributed among LDCs. More than one-third of all FDI destined for LDCs went to China in 2007, one-third to all other Asian countries, one-fifth to all Latin American countries, and one-tenth to all African countries.

The major sources of FDI are transnational corporations (TNCs). A **transnational corporation** invests and operates in countries other than the one in which its headquarters are located. Of the 500 largest TNCs in 2008, 140 had headquarters in the United States and 163 in Europe.

Financing Development

LDCs lack money to fund development, so they obtain financial support from MDCs. Finance comes from two primary sources—loans from banks and international organizations and direct investment by transnational corporations.

Loans

The two major lenders to LDCs are the World Bank and the International Monetary Fund (IMF):

- **The World Bank:** Includes the International Bank for Reconstruction and Development (IBRD) and the International Development Association (IDA). The IBRD provides loans to countries to reform public administration and legal institutions, develop and strengthen financial institutions, and implement transportation and social service projects. The IDA provides support to poor countries considered too risky to qualify for IBRD loans. The IBRD has loaned about $400 billion since 1945, primarily in Europe and Latin America, and the IDA about $150 billion since 1960, primarily in Asia and Africa. The IBRD lends money raised from sales of bonds to private investors; the IDA from government contributions.
- **The IMF:** Provides loans to countries experiencing balance-of-payments problems that threaten expansion of international trade. IMF assistance is designed to help a country rebuild international reserves, stabilize currency exchange rates, and pay for imports without having to impose harsh trade restrictions or capital controls that could hamper the growth of world trade. Unlike the development banks, the IMF does not lend for specific projects. Funding of the IMF is based on each member country's relative size in the world economy.

The World Bank and IMF were conceived at a 1944 United Nations Monetary and Financial Conference in Bretton Woods, New Hampshire, to promote economic development and stability after the devastation of World War II and to avoid a repetition of the disastrous economic policies contributing to the Great Depression of the 1930s. The IMF and World Bank became specialized agencies of the United Nations when it was established in 1945.

LDCs borrow money to build new infrastructure, such as hydroelectric dams, electric transmission lines, flood-protection systems, water supplies, roads, and hotels. The theory is that new infrastructure will make conditions more favorable for domestic and foreign businesses to open or expand. After all, no business wants to be located in a place that lacks paved roads, running water, and electricity.

In principle, new or expanded businesses are attracted to an area because improved infrastructure will contribute additional

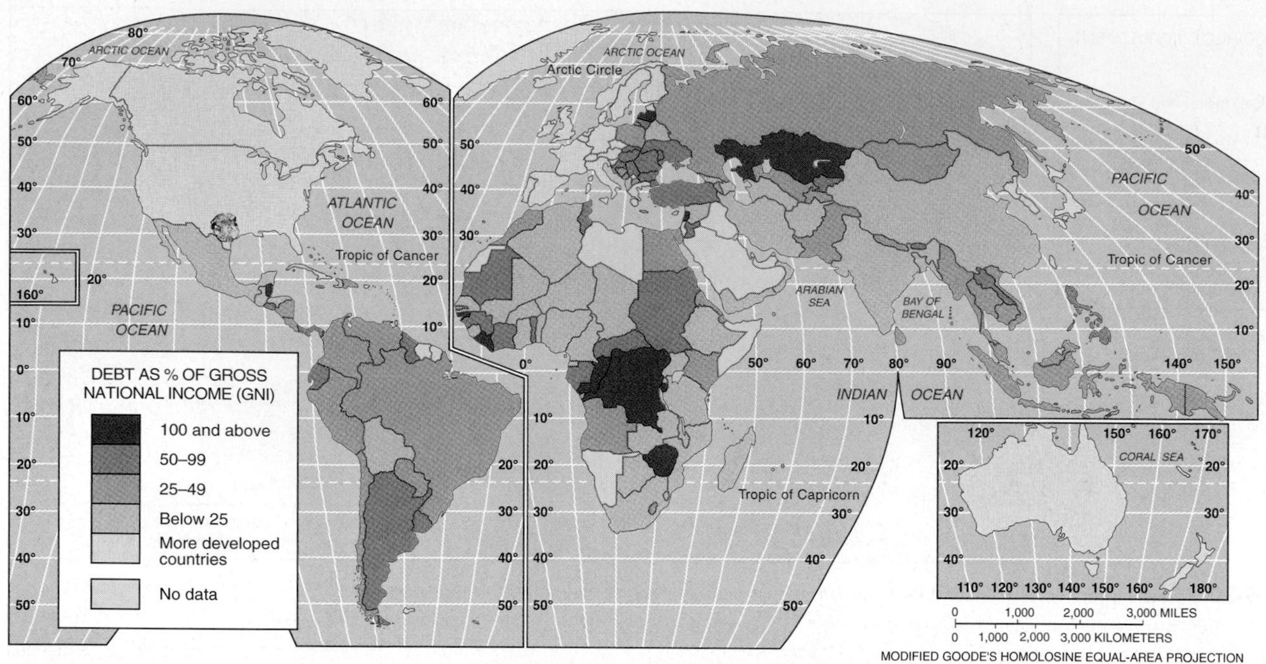

FIGURE 9-31 Debt as percentage of income. To finance development, some LDCs have accumulated large foreign debts relative to their GDP. As a result, a large percentage of their national budgets must be used to repay loans. When LDCs cannot repay their debts, financial institutions in MDCs suffer because they were a major source of the loans.

taxes that the LDC will use in part to repay the loans and in part to improve its citizens' living conditions. In reality, the World Bank itself has judged half of the projects it has funded in Africa to be failures. Common reasons include:

- Projects don't function as intended because of faulty engineering.
- Aid is squandered, stolen, or spent on armaments by recipient nations.
- New infrastructure does not attract other investment.

Many LDCs have been unable to repay the interest on their loans, let alone the principal (Figure 9-31). Debt actually exceeds annual income in a dozen countries. When these countries cannot repay their debts, financial institutions in MDCs refuse to make further loans, so construction of needed infrastructure stops. The inability of many LDCs to repay loans also damages the financial stability of banks in the MDCs.

Structural Adjustment Programs

The IMF, World Bank, and MDCs fear that granting, canceling, or refinancing debts without strings attached will perpetuate bad habits in LDCs. Therefore, before granting debt relief, an LDC is required to prepare a Policy Framework Paper (PFP) outlining a **structural adjustment program**, which includes economic goals, strategies for achieving the objectives, and external financing requirements.

A structural adjustment program includes economic "reforms" or "adjustments." Requirements placed on an LDC typically include:

- Spend only what it can afford
- Direct benefits to the poor, not just the elite
- Divert investment from military to health and education spending
- Invest scarce resources where they would have the most impact
- Encourage a more productive private sector
- Reform the government, including a more efficient civil service, more accountable fiscal management, more predictable rules and regulations, and more dissemination of information to the public

Critics charge that poverty worsens under structural adjustment programs. By placing priority on reducing government spending and inflation, structural adjustment programs may result in:

- Cuts in health, education, and social services that benefit the poor

- Higher unemployment
- Loss of jobs in state enterprises and the civil service
- Less support for those most in need, such as poor pregnant women, nursing mothers, young children, and elderly people

In short, structural reforms allegedly punish Earth's poorest people for actions they did not commit—waste, corruption, misappropriation, and military buildups.

International organizations respond that the poor suffer more when a country does not undertake reforms. Economic growth is what benefits the poor the most in the long run. Nevertheless, in response to criticisms, the IMF and the World Bank now encourage innovative programs to reduce poverty and corruption and consult more with average citizens. A safety net must be included to ease short-term pain experienced by poor people.

Fair Trade

Fair trade has been proposed as a variation of the international trade model of development. **Fair trade** means that products are made and traded according to standards that protect workers and small businesses in LDCs. Standards for fair trade are set internationally by Fairtrade Labelling Organisations International (FLO). A nonprofit organization, TransFair USA, certifies the products sold in the United States that are fair trade.

In North America, fair trade products have been primarily craft products such as decorative home accessories, jewelry, textiles, and ceramics. Ten Thousand Villages is the largest fair trade organization in North America, specializing in handicrafts. In Europe, most fair trade sales are in food, including coffee, tea, banana, chocolate, cocoa, juice, sugar, and honey products.

Two sets of standards distinguish fair trade: one set applies to workers on farms and in factories and the other to producers.

Fair Trade Producer Standards

Fair trade advocates work with small businesses, especially worker-owned and democratically run cooperatives. Small-scale farmers and artisans in LDCs are unable to borrow from banks the money they need to invest in their businesses. By banding together, they can get credit, reduce their raw material costs, and maintain higher and fairer prices for their products. Cooperatives thus benefit the local farmers and artisans who are members, rather than absentee corporate owners interested only in maximizing profits. Because cooperatives are managed democratically, farmers and artisans learn leadership and organizational skills. The people who grew or made the products thereby have a say in how local resources are utilized and sold. Safe and healthy working conditions can be protected.

Consumers pay higher prices for fair trade coffee than for grocery store brands, but prices are comparable to those charged for gourmet brands. However, fair trade coffee producers receive a significantly higher price per pound than traditional coffee producers. North American consumers pay $4 to $11 a pound for coffee bought from growers for about 80 cents a pound. Growers who sell to fair trade organizations earn $1.12 to $1.26 a pound. Because fair trade organizations bypass exploitative middlemen and work directly with producers, they are able to cut costs and return a greater percentage of the retail price to the producers. In some cases, the quality is higher because fair traders factor in the environmental cost of production. For instance, in the case of coffee, fairly traded coffee is usually organic and shade grown, which results in a higher-quality coffee.

Fair Trade Worker Standards

Critics of international trade charge that only a tiny percentage of the price a consumer pays for a good reaches the individual in the LDC responsible for making or growing it. A Haitian sewing clothing for the U.S. market, for example, earns less than 1 percent of the retail price, according to the National Labor Committee. In contrast, fair trade returns on average one-third of the price to the producer in the LDC. The rest goes to the wholesaler who imports the item and for the retailer's rent, wages, and other expenses.

Protection of workers' rights is not a high priority in the international trade development approach, according to its critics. With minimal oversight by governments and international lending agencies, workers in LDCs allegedly work long hours in poor conditions for low pay. The workforce may include children or forced labor. Health problems may result from poor sanitation and injuries from inadequate safety precautions. Injured, ill, or laid-off workers are not compensated.

In contrast, fair trade requires employers to pay workers fair wages, permit union organizing, and comply with minimum environmental and safety standards. Under fair trade, workers are paid at least the country's minimum wage. Sixty to seventy percent of the artisans providing fair trade hand-crafted products are women. Often these women are mothers and the sole wage earners in the home. Because the minimum wage is often not enough for basic survival, whenever feasible, workers are paid enough to cover food, shelter, education, health care, and other basic needs. Cooperatives are encouraged to reinvest profits back into the community, such as by providing health clinics, child care, and training.

Paying fair wages does not necessarily mean that products cost the consumer more. Because fair trade organizations bypass exploitative middlemen and work directly with producers, they are able to cut costs and return a greater percentage of the retail price to the producers. The cost remains the same as traditionally traded goods, but the distribution of the cost of the product is different, because the large percentage taken by middlemen is removed from the equation.

SUMMARY

The relationship between MDCs and LDCs—described at the beginning of the chapter as a north–south split—appears somewhat different on a north polar projection. MDCs form a triangular-shaped inner-core area, whereas LDCs occupy peripheral locations (Figure 9-32). This unorthodox world map projection emphasizes the central role played by MDCs in the world economy and the secondary role of LDCs.

In an increasingly unified world economy, the MDCs clustered in the core play dominant roles in forming the economies of the LDCs on the periphery. North America, Europe, and Japan account for a high percentage of the world's economic activity and wealth. The LDCs in the periphery have less access to the world centers of consumption, communications, wealth, and power, which are clustered in the core. The development prospects of Latin America are tied to governments and businesses in North America, those of Africa and Eastern Europe to Western Europe, and those of Asia to Japan and to a lesser extent Europe and North America.

To reduce disparities between MDCs and LDCs, the United Nations has set eight so-called **Millennium Development Goals**:

1. End poverty and hunger. Extreme poverty has been cut substantially in the world, primarily because of success in Asia, but it has not declined in sub-Saharan Africa.
2. Achieve universal primary (elementary school) education. The percentage of children not enrolled in school remains relatively high in South Asia and sub-Saharan Africa.
3. Promote gender equality and empower women. Gender disparities remain in all regions, as discussed in Key Issue 3 of this chapter.
4. Reduce child mortality. Infant mortality rates have declined in most LDCs, but not in most countries in sub-Saharan Africa.
5. Improve maternal health. One-half million women die from complications during pregnancy; 99 percent of these women live in LDCs.
6. Combat HIV/AIDS, malaria, and other diseases. The number of people living with HIV continues to rise, as discussed in Chapter 2.
7. Ensure environmental sustainability. Water scarcity and quality, deforestation, and overfishing are especially critical environmental issues, according to the United Nations (see Chapter 14).
8. Develop a global partnership for development. Aid from MDCs to LDCs has been declining.

Here again are the key issues concerning development:

1. **Why Does Development Vary Among Countries?** Development is the process by which the material conditions of a country's people are improved. An MDC has a higher level of per capita GDP, achieved through a transformation in the structure of the economy from a predominantly agricultural to a service-providing society. MDCs use their wealth in part to provide better health, education, and welfare services. Conversely, LDCs must use their additional wealth primarily to meet the needs of a rapidly growing population.
2. **Where Are MDCs and LDCs Distributed?** We can identify two regions of MDCs—North America and Europe—plus three other developed areas—Japan, Oceania, and Russia. Seven regions of LDCs include Latin America, East Asia, Southwest Asia, Southeast Asia, Central Asia, South Asia, and sub-Saharan Africa. These less developed regions have varying prospects for promoting development.
3. **Where Does Level of Development Vary by Gender?** The United Nations has found evidence of gender inequality in every country of the world. Women have lower levels of income, literacy, and education than men. Even in countries where women have achieved near equality with men in living conditions, they still have much less economic and political power.
4. **Why Do LDCs Face Obstacles to Development?** LDCs choose between the international trade and the self-sufficiency paths toward development. In either case, LDCs may need to borrow considerable sums of money to promote development. The inability of many LDCs to pay back these loans is a source of considerable tension between them and MDCs.

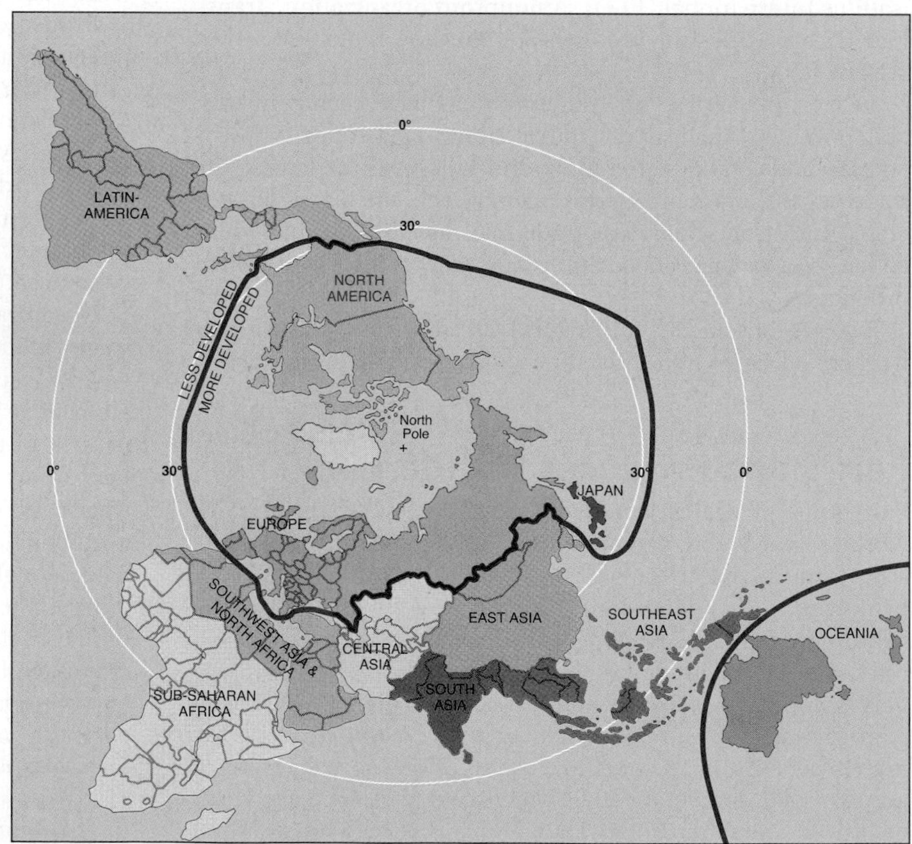

FIGURE 9-32 Core and periphery. Most of the countries that have achieved relatively high levels of development are located above 30° north latitude. Viewed from this north polar projection, more developed countries appear clustered in an inner core, whereas less developed countries are generally relegated to a peripheral or outer-ring location.

CASE STUDY REVISITED / Future Prospects for Development

The most fundamental obstacle to development in many LDCs is gender inequality. A precondition for effective nurturing of take-off industries and effective use of loans is ensuring an effective role for women in the development process. Excluding women is not merely unfair, it wastes a major economic asset.

One organization trying to do something about the legacy of gender inequality in South Asia is the Grameen Bank (Figure 9-33). Based in Bangladesh, Grameen specializes in making loans to women, who make up three-fourths of the borrowers since the bank was established in 1977. For founding the bank, Muhammad Yunus was awarded the Nobel Peace Prize in 2006.

The Grameen Bank has made several hundred thousand loans to women in Bangladesh and neighboring South Asian countries, and only 1 percent of the borrowers have failed to make their weekly loan repayments, an extraordinarily low percentage for a bank. Several million loans have also been provided to women by the Bangladesh Rural Advancement Committee.

FIGURE 9-33 The Grameen Bank. In the village of Sharifun Begeum, Bangladesh, women are paying back their loans to the Grameen Bank.

Rabea Rahman borrowed $90 from the Grameen Bank to buy a cow. Earnings from selling the cow's milk enabled her to buy her son an $85 rickshaw bicycle so that he could make a living. The smallest loan the bank has made was $1, to a woman who wanted to sell plastic bangles door to door. Other women have borrowed money to make perfume, bind books, and sell matches, mirrors, and bananas. The average loan is about $60 ■.

KEY TERMS

Development (p. 274) A process of improvement in the material conditions of people through diffusion of knowledge and technology.

Fair trade (p. 301) Alternative to international trade that emphasizes small businesses and worker-owned and democratically run cooperatives and requires employers to pay workers fair wages, permit union organizing, and comply with minimum environmental and safety standards.

Foreign direct investment (FDI) (p. 298) Investment made by a foreign company in the economy of another country.

Gender Empowerment Measure (GEM) (p. 289) Compares the ability of women and men to participate in economic and political decision making.

Gender-Related Development Index (GDI) (p. 289) Compares the level of development of women with that of both sexes.

Gross domestic product (GDP) (p. 275) The value of the total output of goods and services produced in a country in a given time period (normally 1 year).

Human Development Index (HDI) (p. 274) Indicator of level of development for each country, constructed by the United Nations, combining income, literacy, education, and life expectancy.

Less developed country (LDC) (p. 274) A country that is at a relatively early stage in the process of economic development.

Literacy rate (p. 278) The percentage of a country's people who can read and write.

Millennium Development Goals (p. 302) Eight international development goals that all members of the United Nations have agreed to achieve by 2015.

More developed country (MDC) (p. 274) A country that has progressed relatively far along a continuum of development.

Primary sector (p. 276) The portion of the economy concerned with the direct extraction of materials from Earth's surface, generally through agriculture, although sometimes by mining, fishing, and forestry.

Productivity (p. 276) The value of a particular product compared to the amount of labor needed to make it.

Secondary sector (p. 276) The portion of the economy concerned with manufacturing useful products through processing, transforming, and assembling raw materials.

Structural adjustment program (p. 300) Economic policies imposed on less developed countries by international agencies to create conditions encouraging international trade, such as raising taxes, reducing government spending, controlling inflation, selling publicly owned utilities to private corporations, and charging citizens more for services.

Tertiary sector (p. 276) The portion of the economy concerned with transportation, communications, and utilities, sometimes extended to the provision of all goods and services to people, in exchange for payment.

Transnational corporation (p. 299) A company that conducts research, operates factories, and sells products in many countries, not just where its headquarters or shareholders are located.

Value added (p. 276) The gross value of the product minus the costs of raw materials and energy.

THINKING GEOGRAPHICALLY

1. Review the major economic, social, and demographic characteristics that contribute to a country's level of development. Which indicators can vary significantly by gender within countries and between countries at various levels of development? Why?
2. Some geographers have been attracted to the concepts of Immanuel Wallerstein, who argued that the modern world consists of a single entity, the capitalist world economy that is divided into three regions: the core, semiperiphery, and periphery. How have the boundaries among these three regions changed?
3. China historically relied on self-sufficiency to promote development, whereas Hong Kong was a prominent practitioner of international trade. Explain how these two approaches have been reconciled since Hong Kong became part of China in 1997.
4. Some LDCs claim that the requirements placed on them by lending organizations such as the World Bank impede rather than promote development. Should LDCs be given a greater role in deciding how much the international organizations should spend and how such funds should be spent? Why or why not?
5. In what ways has the severe recession encouraged countries to switch from international trade to self-sufficiency? What are the advantages and challenges of returning to self-sufficiency in poor economic conditions?

RESOURCES

Some recent and classic books and articles on development geography:

Barnes, Trevor J. "Retheorizing Economic Geography: From the Quantitative Revolution to the 'Cultural Turn'". *Annals of the Association of American Geographers* 91 (2001): 546–65.

———, Jamie Peck, Eric Sheppard, and Adam Tickell, eds. *Reading Economic Geography*. Malden, MA: Blackwell, 2000.

Brakman, Steven, Harry Garretsen, and Charles van Marrewijk. *An Introduction to Geographical Economics: Trade, Location and Growth*. New York: Cambridge University Press, 2001.

Buvinic, Mayra, and Geeta Rao Gupta. "Female-Headed Households and Female-Maintained Families: Are They Worth Targeting to Reduce Poverty in Developing Countries?" *Economic Development and Cultural Change* 45 (1997): 259–80.

Clark, Gordon L., Maryann P. Feldman, and Meric S. Gertler, eds. *The Oxford Handbook of Economic Geography*. New York: Oxford University Press, 2000.

Coe, Neil M., Martin Hess, Henry Wai-Chung Yeung, Peter Dicken, and Jeffrey Henderson. "'Globalizing' Regional Development: A Global Production Networks Perspective." *Transactions of the Institute of British Geographers New Series* 29 (2004): 468–84.

Dicken, Peter. *Global Shift: Mapping the Changing Contours of the World Economy.* 5th ed. New York: Guilford, 2007.

Hayter, Roger, Trevor J. Barnes, and Michael J. Bradshaw. "Relocating Resource Peripheries to the Core of Economic Geography's Theorizing: Rationale and Agenda." *Area* 35 (2003): 15–23.

Holmes, Thomas J. "The Diffusion of Wal-Mart and Economies of Density." NBER Working Paper Series 13783. Cambridge, MA: National Bureau of Economic Research, 2008.

Khan, A. U. "A Decade of Indian Economic Reforms and the Inflow of Foreign Investment." *Regional Studies* 21 (2003): 63–83.

Mabogunje, Akin L. *The Development Process: A Spatial Perspective*, 2nd ed. London: Unwin Hyman, 1989.

Martin, R., and P. Sunley. "Rethinking the 'Economic' in Economic Geography: Broadening Our Vision or Losing Our Focus?" *Antipode* 33 (2001): 148–61.

Meier, Gerald M., and James E. Rauch. *Leading Issues in Economic Development*, 8th ed. New York: Oxford University Press, 2005.

Neumark, David, Junfu Zhang, and Stephen Ciccarella. "The Effects of Wal-Mart on Local Labor Markets." NBER Working Paper 11782. Cambridge, MA: National Bureau of Economic Research, 2005.

Peet, Richard. *Theories of Development*. New York: Guilford, 1999.

Rostow, Walter W. *The Stages of Economic Growth*. Cambridge: Cambridge University Press, 1960.

Samers, M. "What Is the Point of Economic Geography?" *Antipode* 33 (2001): 183–93.

Sheppard, Eric, and Trevor Barnes. *A Companion to Economic Geography*. Malden, MA: Blackwell, 2000.

Storper, Michael. *The Regional World: Territorial Development in a Global Economy*. New York: Guilford Press, 1997.

Wallerstein, Immanuel. *The Capitalist World-Economy*. Cambridge: Cambridge University Press, 1979.

Yapa, Lakshman. "What Causes Poverty?: A Postmodern View." *Annals of the Association of American Geographers* 86 (1996): 707–28.

Journals featuring development geography:

Economic Development and Cultural Change; Economic Geography; International Development Review; International Economic Review; International Journal of Political Economy; Journal of Developing Areas; Netherlands Journal of Economic and Social Geography; Regional Studies.

Key Internet sites:

http://hdr.undp.org. The complete Human Development Index Report includes numerous indicators that can be viewed for every country. The web site also includes an interactive calculator that permits the user to see the impact on the HDI from changing the values of one or more of the variables.

http://earthtrends.wri.org/. The indicators cited in this chapter that are not part of the Human Development Index Report can be found through the Earth Trends portion of the World Resources Institute (WRI) web site.

www.NationMaster.com. Several data sources, including the United Nations and the CIA, are brought together on this web site.

Log in to www.mygeoscienceplace.com for videos, interactive maps, RSS feeds, case studies, and self-study quizzes to enhance your study of Development.

CHAPTER 10

Agriculture

When you buy food in the supermarket, are you reminded of a farm? Not likely. The meat is carved into pieces that no longer resemble an animal and is wrapped in paper or plastic film. Often the vegetables are canned or frozen. The milk and eggs are in cartons.

The food industry in the United States and Canada is vast, but only a few people are full-time farmers, and they

KEY ISSUES

1 Where Did Agriculture Originate?

2 Where Are Agricultural Regions in LDCs?

3 Where Are Agricultural Regions in MDCs?

4 Why Do Farmers Face Economic Difficulties?

may be more familiar with the operation of computers and advanced machinery than the typical factory or office worker. The mechanized, highly productive American or Canadian farm contrasts with the subsistence farm found in much of the world. The most "typical" human—if there is such a person—is an Asian farmer who grows enough food to survive, with little surplus. This sharp contrast in agricultural practices constitutes one of the most fundamental differences between the more developed and less developed countries of the world.

Market, Douz, Tunisia

CASE STUDY / Wheat Farmers in Kansas and Pakistan

The Iqbel family grows wheat on its 1-hectare (2.5-acre) plot of land in the Punjab province of Pakistan in a manner similar to that of their ancestors. They perform most tasks by hand or with the help of animals. To irrigate the land, for example, they lift water from a 20-meter (65-foot) well by pushing a water wheel. More prosperous farmers in Pakistan use bullocks to turn the wheel.

The farm produces about 1,500 kilograms (3,300 pounds) of wheat per year—enough to feed the Iqbel family. Some years they produce a small surplus, which they can sell. They can then use that money to buy other types of food or household items. In drought years, however, the crop yield is lower, and the Iqbel family must receive food from government and international relief organizations.

A world away, in Kansas, the McKinleys farm the prairie sod. Like the Iqbels, they grow wheat in a climate that receives little rain. Otherwise, the two farm families lead very different lives. The McKinley family's farm is 200 times as large—200 hectares (500 acres). The McKinleys derive several hundred times more income from the sale of wheat than do the Iqbels.

The wheat grown on the McKinleys farm is not consumed directly by them. Instead, it is sold to a processing company and ultimately turned into bread wrapped in plastic and sold in a supermarket hundreds of kilometers away. Most of the wheat from the Iqbels' farm is consumed in the village where it is grown. ■

Approximately one-half of the people in less developed countries are farmers. The overwhelming majority of them are like the Iqbels, growing enough food to feed themselves, but little more. LDCs are home to 97 percent of the world's farmers. In contrast, fewer than 2 percent of the people in the United States are farmers. Yet the advanced technology used by these farmers allows them to produce enough food for people in the United States at a very high standard, plus food for many people elsewhere in the world.

The previous chapter divided economic activities into primary, secondary, and tertiary sectors. This chapter is concerned with the principal form of primary-sector economic activity—agriculture. The next two chapters look at the secondary and tertiary sectors.

Geographers study ***where*** agriculture is distributed across Earth. The most important distinction is what happens to farm products. In less developed ***regions***, the farm products are most often consumed on or near the farm where they are produced, whereas in MDCs farmers sell what they produce.

Geographers observe a wide variety of agricultural practices. The reason ***why*** farming varies around the world relates to the distribution of cultural and environmental factors across ***space***. Elements of the physical environment, such as climate, soil, and topography, set broad limits on agricultural practices, and farmers make choices to modify the environment in a variety of ways.

Farming is an economic activity that still depends very much on the ***local diversity*** of environmental and cultural conditions in each ***place***. Despite increased knowledge of alternatives, farmers practice distinctive agriculture in different regions and, in fact, on neighboring farms. Broad climate patterns influence the crops planted in a region, and local soil conditions influence the crops planted on an individual farm.

In each society, farmers possess very specific knowledge of their environmental conditions and certain technology for modifying the landscape. Within the limits of their technology, farmers choose from a variety of agricultural practices, based on their perception of the value of each alternative. These values are partly economic and partly cultural. How farmers deal with their physical environment varies according to dietary preferences, availability of technology, and other cultural traditions. Farmers select agricultural practices based on cultural perceptions, because a society may hold some foods in high esteem while avoiding others.

Although individual farmers may make specific decisions on a very local ***scale***, agriculture is as caught up in the ***globalization*** of the economy as other industries. Agriculture is big business in MDCs and a major component of international trade ***connections*** in LDCs.

After examining the origins and diffusion of agriculture, we will consider the agricultural practices used in LDCs and MDCs. We will also examine the problems farmers face in each type of region. Although each farm has a unique set of physical conditions and choice of crops, geographers group farms into several types by their distinctive environmental and cultural characteristics.

KEY ISSUE 1
Where Did Agriculture Originate?

- **Origins of Agriculture**
- **Subsistence and Commercial Agriculture**

The origins of agriculture cannot be documented with certainty because it began before recorded history. Scholars try to reconstruct a logical sequence of events based on fragments of information about ancient agricultural practices and historical environmental conditions. Improvements in cultivating plants and domesticating animals evolved over thousands of years. This section offers an explanation for the origin and diffusion of agriculture. ■

Origins of Agriculture

Agriculture is deliberate modification of Earth's surface through cultivation of plants and rearing of animals to obtain sustenance or economic gain. Agriculture originated when humans domesticated plants and animals for their use. The word *cultivate* means "to care for," and a **crop** is any plant cultivated by people.

Hunters and Gatherers

Before the invention of agriculture, all humans probably obtained the food they needed for survival through hunting for animals, fishing, or gathering plants (including berries, nuts, fruits, and roots). Hunters and gatherers lived in small groups, of usually fewer than 50 persons, because a larger number would quickly exhaust the available resources within walking distance (Figure 10-1). The men hunted game or fished, and the women collected berries, nuts, and roots. This division of labor sounds like a stereotype but is based on evidence from archaeology and anthropology. They collected food often, perhaps daily. The food search might take only a short time or much of the day, depending on local conditions.

The group traveled frequently, establishing new home bases or camps. The direction and frequency of migration depended on the movement of game and the seasonal growth of plants at various locations. We can assume that groups communicated with each other concerning hunting rights, intermarriage, and other specific subjects. For the most part, they kept the peace by steering clear of each other's territory.

Today, perhaps a quarter-million people, or less than 0.005 percent of the world's population, still survive by hunting and gathering rather than by agriculture. Examples include the Spinifex (also known as Pila Nguru) people, who live in Australia's Great Victorian Desert; the Sentinelese people, who live in India's Andaman Islands; and the Bushmen, who live in Botswana and Namibia. Contemporary hunting and gathering societies are isolated groups living on the periphery of world settlement, but they provide insight into human customs that prevailed in prehistoric times, before the invention of agriculture.

FIGURE 10-1 Hunting and gathering. Botswana Bushmen dig up wild onions called kjon.

Invention of Agriculture

Why did most nomadic groups convert from hunting, gathering, and fishing to agriculture? Geographers and other scientists agree that agriculture originated in multiple hearths around the world. They do not agree on when agriculture originated and diffused, or why.

Southwest Asia was an early center of crop domestication (Figure 10-2). The earliest crops domesticated in Southwest Asia are thought to have been barley and wheat, around 10,000 years ago. Lentil and olive were also early domestications in Southwest Asia. From this hearth, cultivation diffused west to Europe and east to Central Asia. Rice is now thought to have been domesticated in East Asia more than 10,000 years ago, along the Yangtze River in eastern China. Millet was cultivated at an early date along the Yellow River. Sorghum was domesticated in central Africa around 8,000 years ago. Yams may have been domesticated even earlier. Millet and rice may have been domesticated in sub-Saharan Africa independently of the hearth in East Asia. From central Africa, domestication of crops probably diffused further south in Africa.

In Latin America, two important hearths of crop domestication are thought to have emerged in Mexico and Peru around 4,000 to 5,000 years ago. Mexico is considered a hearth for beans and cotton, and Peru for potato. Squashes may have been first domesticated in a third hearth in the Americas, in southeastern present-day United States, as well as in Mexico. The most important contribution of the Americas to crop domestication, maize (corn), may have emerged in the two hearths independently around the same time. From these two hearths, cultivation of maize and other crops diffused northward into North America and southward into tropical South America.

Animals were also domesticated in multiple hearths at various dates. Southwest Asia is thought to have been the hearth for the domestication of the largest number of animals that would prove to be most important for agriculture, including cattle, goats, pigs, and sheep, between 8,000 and 9,000 years ago (Figure 10-3). Domestication of the dog is thought to date from around 12,000 years ago, also in Southwest Asia. The horse is considered to have been domesticated in Central Asia; diffusion of the domesticated horse is thought to be associated with the diffusion of the Indo-European language, as discussed in Chapter 5.

Inhabitants of Southwest Asia may have been the first to integrate cultivation of crops with domestication of herd animals such as cattle, sheep, and goats. These animals were used to prepare the land before planting seeds and, in turn, were fed part of the harvested crop. Other animal products, such as milk, meat, and skins, may have been exploited at a later date. This integration of plants and animals is a fundamental element of modern agriculture.

Scientists do not agree on whether agriculture originated primarily because of environmental factors or cultural factors. Probably a combination of both factors contributed. Those favoring environmental reasons point to the coinciding of the first domestication of crops and animals with climate change

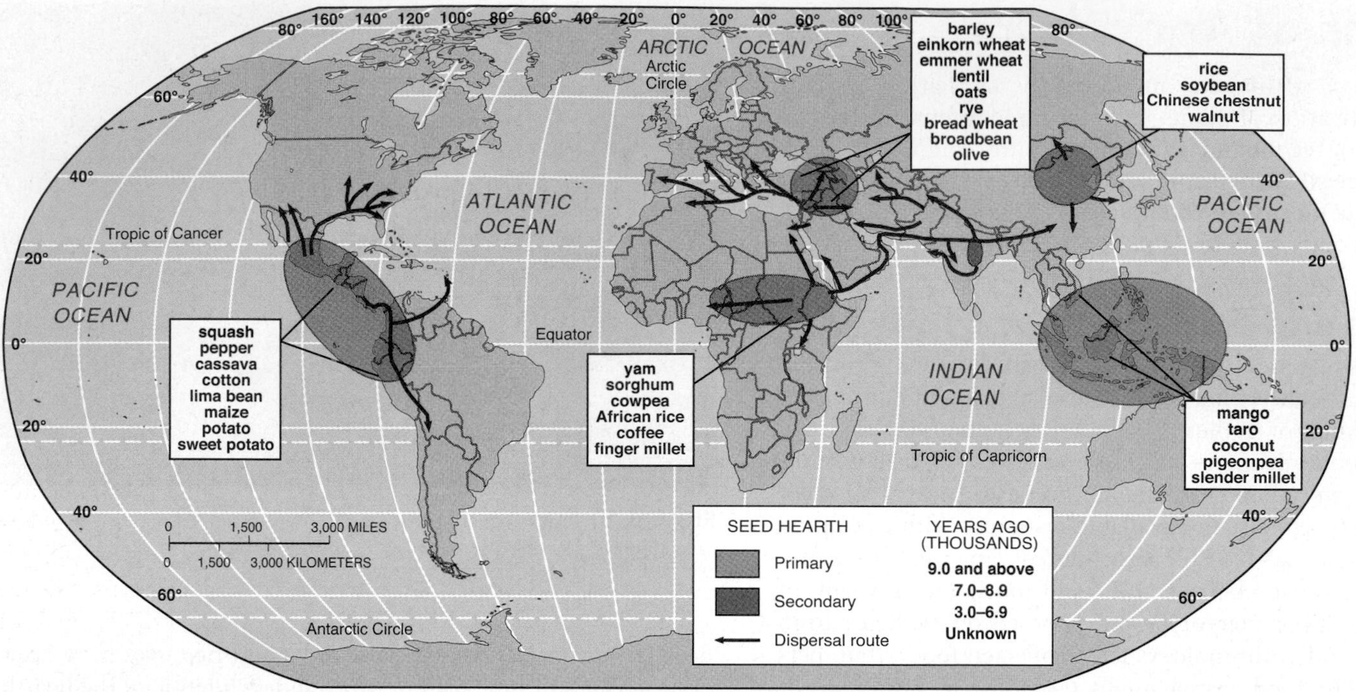

FIGURE 10-2 Crop hearths. Agriculture originated in multiple hearths. Domestication of some crops can be dated back more than 10,000 years.

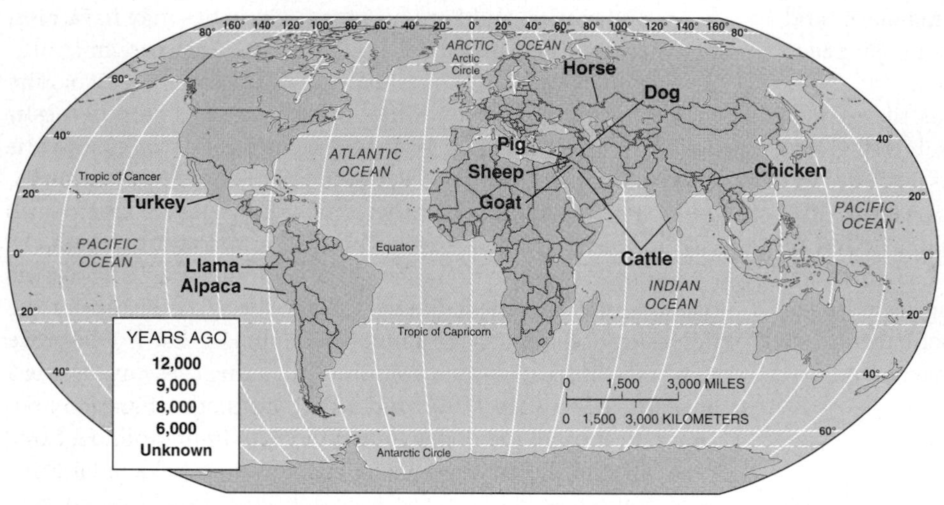

FIGURE 10-3 Animal hearths. Animal domestication also originated in multiple hearths.

around 10,000 years ago. This marked the end of the last ice age, when permanent ice cover receded from Earth's midlatitudes to polar regions, resulting in a massive redistribution of humans, other animals, and plants at that time. Alternatively, human behavior may be primarily responsible for the origin of agriculture. A preference for living in a fixed place rather than as nomads may have led hunters and gatherers to build permanent settlements and to store surplus vegetation there.

In gathering wild vegetation, people inevitably cut plants and dropped berries, fruits, and seeds. These hunters probably observed that, over time, damaged or discarded food produced new plants. They may have deliberately cut plants or dropped berries on the ground to see if they would produce new plants. Subsequent generations learned to pour water over the site and to introduce manure and other soil improvements. Over thousands of years, plant cultivation apparently evolved from a combination of accident and deliberate experiment.

That agriculture had multiple origins means that, from earliest times, people have produced food in distinctive ways in different regions. This diversity derives from a unique legacy of wild plants, climatic conditions, and cultural preferences in each region. Improved communications in recent centuries have encouraged the diffusion of some plants to varied locations around the world. Many plants and animals thrive across a wide portion of Earth's surface, not just in their place of original domestication. Only after 1500, for example, were wheat, oats, and barley introduced to the Western Hemisphere and maize to the Eastern Hemisphere.

Subsistence and Commercial Agriculture

The most fundamental differences in agricultural practices are between those in LDCs and those in MDCs. Farmers in LDCs generally practice subsistence agriculture, whereas farmers in MDCs practice commercial agriculture. **Subsistence agriculture**, found in LDCs, is the production of food primarily for consumption by the farmer's family. **Commercial agriculture**, found in MDCs, is the production of food primarily for sale off the farm.

The most widely used map of world agricultural regions is based on work done by geographer Derwent Whittlesey in 1936. Whittlesey identified 11 main agricultural regions, plus an area where agriculture was nonexistent. Whittlesey's 11 regions are divided between 5 that are important in LDCs and 6 that are important in MDCs (Figure 10-4). Figure 10-4 also includes a small, simplified version of the world climate map (see Figure 1-19).

Similarities between the agriculture and climate maps are striking. For example, pastoral nomadism is the predominant type of agriculture in the Middle East, which has a dry climate, whereas shifting cultivation is the predominant type of agriculture in central Africa, which has a tropical climate. Note the division between southeastern China (warm midlatitude climate, intensive subsistence agriculture with wet rice dominant) and northeastern China (cold midlatitude climate, intensive subsistence agriculture with wet rice not dominant). In the United States, much of the West is distinguished from the rest of the country according to climate (dry) and agriculture (livestock ranching). Thus, agriculture varies between the dry lands and the tropics within LDCs—as well as between the dry lands of LDCs and MDCs.

Because of the problems involved with the concept of environmental determinism, discussed in Chapter 1, geographers are wary of placing too much emphasis on the role of climate. Cultural preferences (discussed in Chapter 4) also explain agricultural differences in areas of similar climate. Hog production is virtually nonexistent in predominantly Muslim regions because of that religion's taboo against consuming pork products (Figure 4-8). Wine production is relatively low in Africa and Asia, even where the climate is favorable for growing grapes, because of alcohol avoidance in predominantly non-Christian countries (Figure 4-15).

Five principal features distinguish commercial agriculture from subsistence agriculture:

- Purpose of farming
- Percentage of farmers in the labor force
- Use of machinery
- Farm size
- Relationship of farming to other businesses

Purpose of Farming

Subsistence and commercial agriculture are undertaken for different purposes. In LDCs, most people produce food for their own consumption. Some surplus may be sold to the government or to private firms, but the surplus product is not the farmer's primary purpose and may not even exist some years because of growing conditions.

In commercial farming, farmers grow crops and raise animals primarily for sale off the farm rather than for their own consumption. Agricultural products are not sold directly to consumers but to food-processing companies. Large processors, such as General Mills and Kraft, typically sign contracts with commercial farmers to buy their grain, chickens, cattle, and other output. Farmers may have contracts to sell sugar beets to sugar refineries, potatoes to distilleries, and oranges to manufacturers of concentrated juices.

Percentage of Farmers in the Labor Force

In MDCs, around 5 percent of workers are engaged directly in farming, compared to around 50 percent in LDCs (Figure 10-5). The percentage of farmers is even lower in North America—only around 2 percent. Yet the small percentage of farmers in the United States and Canada produces not only enough food for themselves and the rest of the region but also a surplus to feed people elsewhere.

The number of farmers declined dramatically in MDCs during the twentieth century. The United States had about 6 million farms in 1940 and 4 million in 1960; the number has stabilized during the past quarter-century at around 2 million. Both push and pull migration factors have been responsible for the decline: People were pushed away from farms by lack of opportunity to earn a decent income, and at the same time they were pulled to higher-paying jobs in urban areas.

Use of Machinery

In MDCs, a small number of farmers can feed many people because they rely on machinery to perform work, rather than relying on people or animals (Figure 10-6). In LDCs, farmers do much of the work with hand tools and animal power.

Traditionally, the farmer or local craftspeople made equipment from wood, but beginning in the late eighteenth century, factories produced farm machinery. The first all-iron plow was made in the 1770s and was followed in the nineteenth and twentieth centuries by inventions that made farming less dependent on human or animal power. Tractors, combines, corn pickers, planters, and other factory-made farm machines have replaced or supplemented manual labor.

Transportation improvements have also aided commercial farmers. The building of railroads in the nineteenth century, and highways and trucks in the twentieth century, have enabled farmers to transport crops and livestock farther and faster. Cattle arrive at market heavier and in better condition when transported by truck or train than when driven on hoof. Crops reach markets without spoiling.

Commercial farmers use scientific advances to increase productivity. Experiments conducted in university laboratories, industry, and research organizations generate new fertilizers, herbicides, hybrid plants, animal breeds, and farming practices, which produce higher crop yields and healthier animals. Access to other scientific information has enabled farmers to make more intelligent decisions concerning proper agricultural practices. Some farmers conduct their own on-farm research.

Electronics also help commercial farmers. Global positioning systems (GPS) determine the precise coordinates for spreading different types and amounts of fertilizers. On large ranches, GPS is also used to monitor the location of cattle. Satellite imagery monitors crop progress. Yield monitors attached to combines determine the precise number of bushels being harvested.

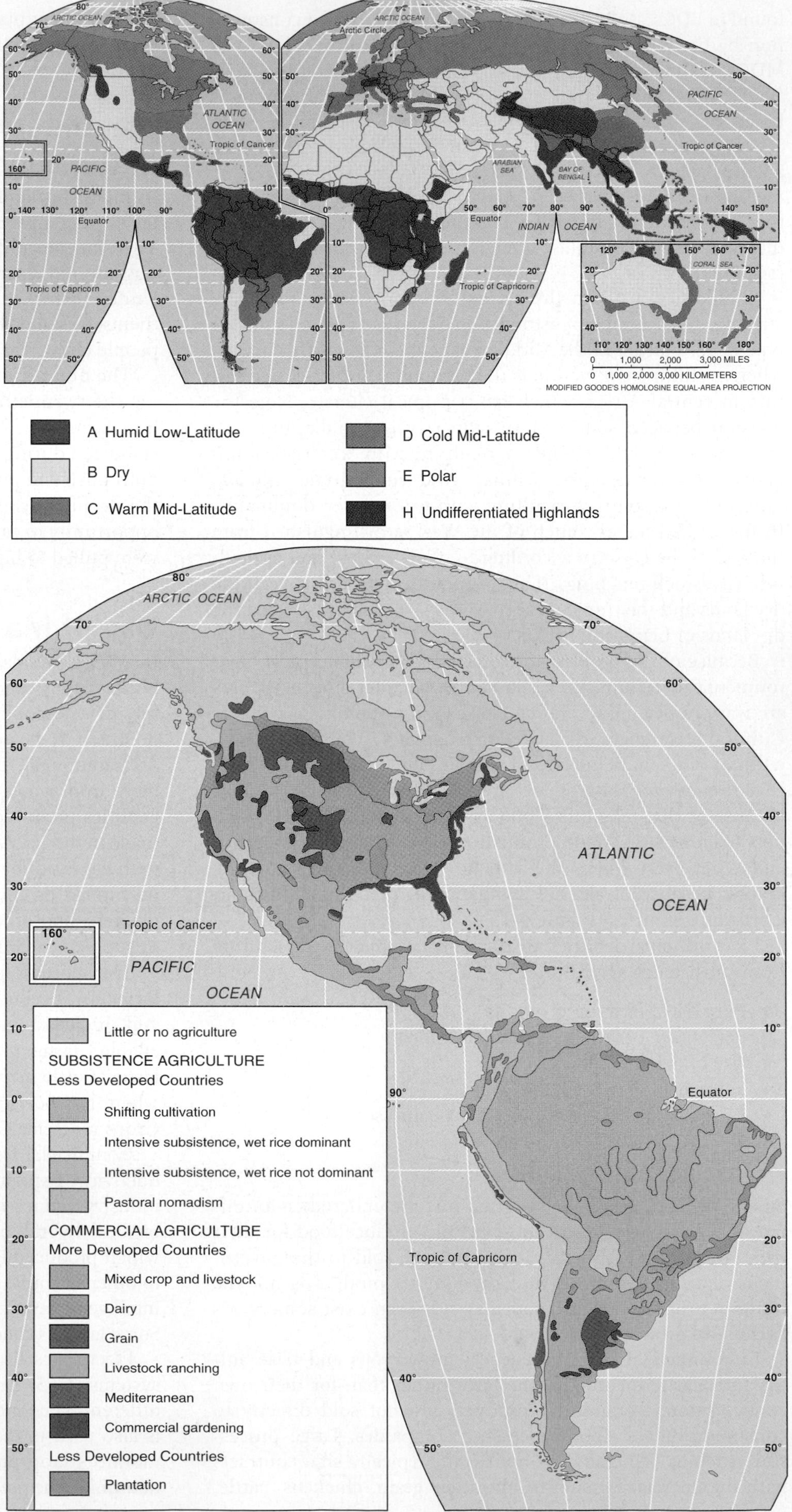

FIGURE 10-4 Agriculture and climate. (Top) Simplified climate regions. Compare the broad distribution of the major climate regions with the distinctive types of agriculture in MDCs and LDCs. (Bottom) Agricultural regions. The major agricultural practices of the world can be divided into subsistence and commercial regions. Subsistence regions include the following:

- Shifting cultivation—primarily the tropical regions of South America, Africa, and Southeast Asia
- Pastoral nomadism—primarily the dry lands of North Africa and Asia
- Intensive subsistence, wet rice dominant—primarily the large population concentrations of East and South Asia
- Intensive subsistence, crops other than rice dominant—primarily the large population concentrations of East and South Asia where growing rice is difficult

Commercial regions include the following:

- Mixed crop and livestock—primarily U.S. Midwest and central Europe
- Dairying—primarily near population clusters in the northeastern United States, southeastern Canada, and northwestern Europe
- Grain—primarily north-central United States and Eastern Europe
- Ranching—primarily the drylands of the western United States, southeastern South America, Central Asia, southern Africa, and Australia
- Mediterranean—primarily lands surrounding the Mediterranean Sea, western United States, and Chile
- Commercial gardening—primarily the southeastern United States and southeastern Australia
- Plantation—primarily the tropical and subtropical regions of Latin America, Africa, and Asia

Farm Size

The average farm size is relatively large in commercial agriculture, especially in the United States and Canada, with U.S. farms averaging about 180 hectares (449 acres). Despite their size, most commercial farms in MDCs are family owned and operated—98 percent in the United States. Commercial farmers frequently expand their holdings by renting nearby fields.

Commercial agriculture is increasingly dominated by a handful of large farms. In the United States, the largest 5 percent of farms produced 75 percent of the country's total agriculture. Large size is partly a consequence of mechanization. Combines, pickers, and other machinery perform most efficiently at very large scales, and their considerable expense cannot be justified on a small farm. As a result of the large size and the high level of mechanization, commercial agriculture is an expensive business. Farmers spend hundreds of thousands of dollars to buy or rent land and machinery before beginning operations. This money is frequently borrowed from a bank and repaid after the output is sold.

Although the United States currently has fewer farms and farmers than in 1900, the amount of land devoted to agriculture has increased. The United States had 60 percent fewer farms and 85 percent fewer farmers in 2000 than in 1900, but 13 percent more farmland, primarily because of irrigation and reclamation. However, the amount of U.S. farmland has declined from its all-time peak around 1960. Primarily because of the expansion of urban areas, the United States has been losing 500,000 hectares (1.2 million acres) per year from its 400 million hectares (1 billion acres) of farmland. A more serious problem in the United States has been the loss of 200,000 hectares (500,000 acres) of the most productive farmland, known as **prime agricultural land**, as urban areas sprawl into the surrounding countryside (see Contemporary Geographic Tools box).

Relationship of Farming to Other Businesses

Commercial farming is closely tied to other businesses. The system of commercial farming found in the United States and other MDCs has been called **agribusiness** because the family farm is not an isolated activity but is integrated into a large

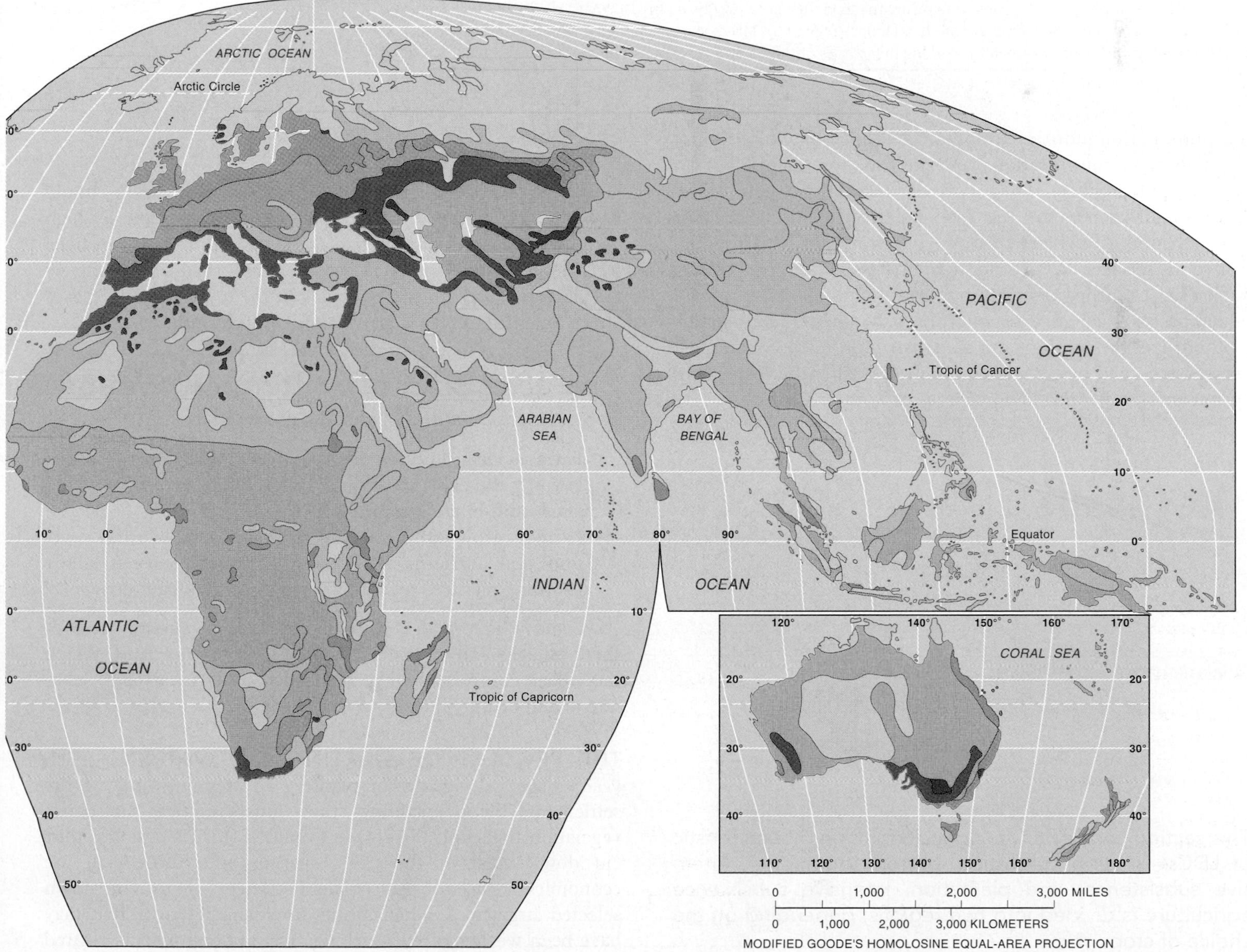

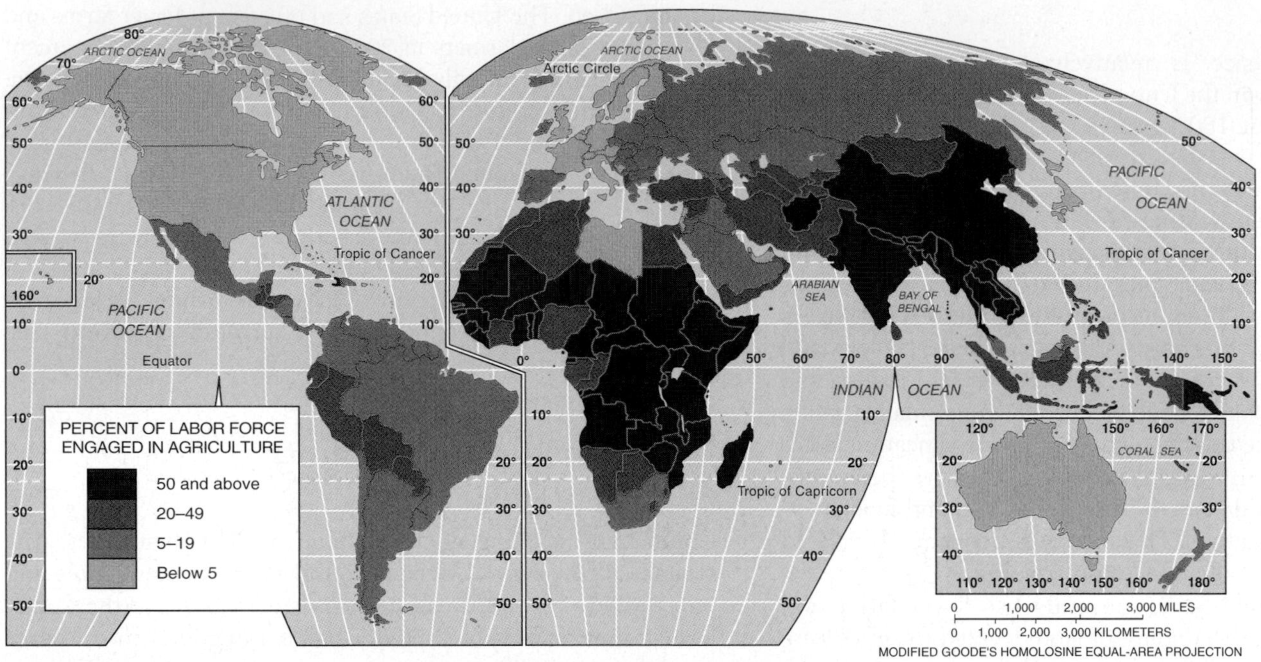

FIGURE 10-5 Agricultural workers. The percent of the workforce engaged in agriculture is higher in LDCs than in MDCs. A priority for all people is to secure the food they need to survive. In LDCs most people work in agriculture to produce the food they and their families require. In MDCs few people are farmers, and most people buy food with money earned by working in factories or offices or by performing other services.

food-production industry. Commercial farmers make heavy use of modern communications and information technology to stay in touch and keep track of prices, yields, and expenditures.

Although farmers are less than 2 percent of the U.S. labor force, around 20 percent of U.S. labor works in food production and services related to agribusiness—food processing, packaging, storing, distributing, and retailing. Agribusiness encompasses such diverse enterprises as tractor manufacturing, fertilizer production, and seed distribution. Although most farms are owned by individual families, many other aspects of agribusiness are controlled by large corporations.

KEY ISSUE 2
Where Are Agricultural Regions in LDCs?

- **Shifting Cultivation**
- **Pastoral Nomadism**
- **Intensive Subsistence Agriculture**
- **Plantation Farming**

This section considers four agricultural types characteristic of LDCs—shifting cultivation, pastoral nomadism, intensive subsistence, and plantation. Intensive subsistence agriculture is divided into two regions, depending on the choice of crop. ■

Shifting Cultivation

Shifting cultivation is practiced in much of the world's Humid Low-Latitude, or A, climate regions, which have relatively high temperatures and abundant rainfall (Figure 10-8). It is practiced by roughly 250 million people across 36 million square kilometers (14 million square miles), especially in the tropical rainforests of South America, Central and West Africa, and Southeast Asia.

Characteristics of Shifting Cultivation

Two distinctive features of **shifting cultivation** are:

- Farmers clear land for planting by slashing vegetation and burning the debris (shifting cultivation is sometimes called **slash-and-burn agriculture**).
- Farmers grow crops on a cleared field for only a few years until soil nutrients are depleted and then leave it fallow (nothing planted) for many years so the soil can recover.

People who practice shifting cultivation generally live in small villages and grow food on the surrounding land, which the village controls. Well-recognized boundaries usually separate neighboring villages.

THE PROCESS OF SHIFTING CULTIVATION. Each year villagers designate for planting an area surrounding the settlement. Before planting, they must remove the dense vegetation that typically covers tropical land. Using axes, they cut down most of the trees, sparing only those that are economically useful. An efficient strategy is to cut down selected large trees, which bring down smaller trees that may have been weakened by notching. The undergrowth is cleared

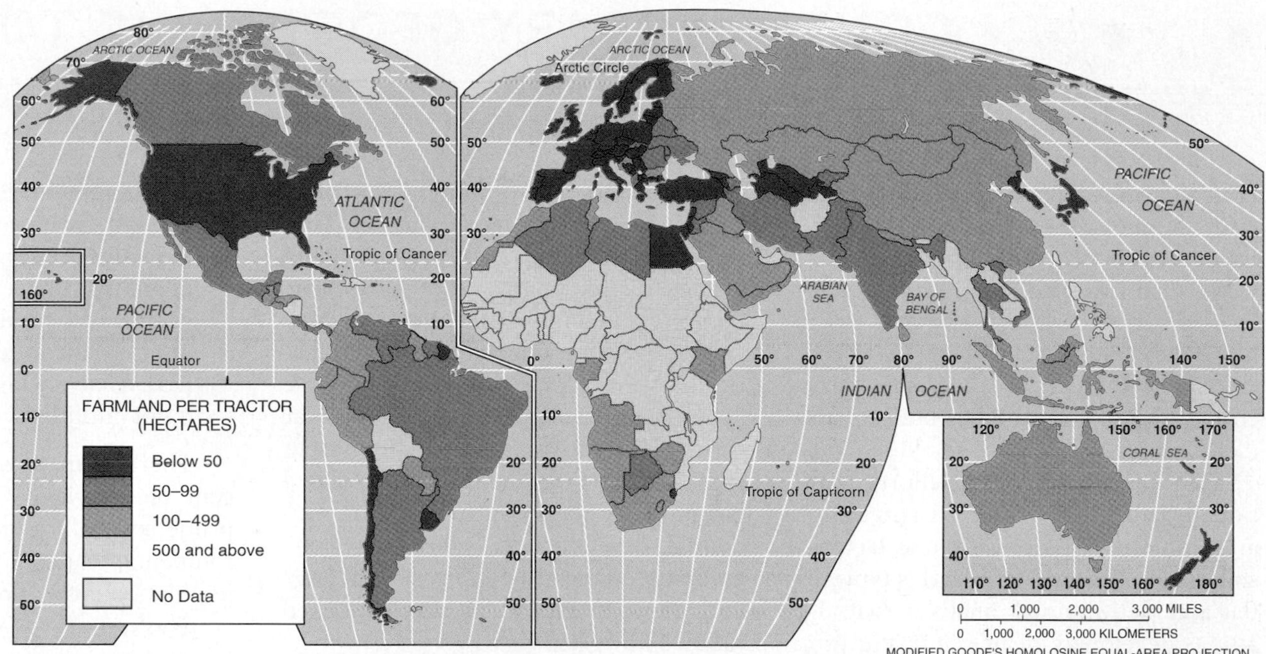

FIGURE 10-6 Area of farmland per tractor. Farmers in MDCs have more tractors per hectare or acre of land than do farmers in LDCs. The machinery makes it possible for commercial farmers to farm extensive areas, a practice necessary to pay for the expensive machinery.

away with a machete or other long knife. On a windless day the debris is burned under carefully controlled conditions. The rains wash the fresh ashes into the soil, providing needed nutrients.

Before planting, the cleared area, known by a variety of names in different regions, including **swidden**, *lading*, *milpa*, *chena*, and *kaingin*, is prepared by hand, perhaps with the help of a simple implement such as a hoe; plows and animals are rarely used. The only fertilizer generally available is potash (potassium) from burning the debris when the site is cleared. Little weeding is done the first year that a cleared patch of land is farmed; weeds may be cleared with a hoe in subsequent years.

The cleared land can support crops only briefly, usually 3 years or less. In many regions, the most productive harvest comes in the second year after burning. Thereafter, soil nutrients are rapidly depleted and the land becomes too infertile to nourish crops. Rapid weed growth also contributes to the abandonment of a swidden after a few years. When the swidden is no longer fertile, villagers identify a new site and begin clearing it. They leave the old site uncropped for many years, allowing it to become overrun again by natural vegetation. The field is not actually abandoned; the villagers will return to the site someday, perhaps as few as 6 years or as many as 20 years later, to begin the process of clearing the land again. In the meantime, they may still care for fruit-bearing trees on the site.

If a cleared area outside a village is too small to provide food for the population, then some of the people may establish a new village and practice shifting cultivation there. Some farmers may move temporarily to another settlement if the field they are clearing that year is distant.

CROPS OF SHIFTING CULTIVATION. The crops grown by each village vary by local custom and taste. The predominant crops include upland rice in Southeast Asia, maize (corn) and manioc (cassava) in South America, and millet and sorghum in Africa. Yams, sugarcane, plantain, and vegetables are also grown in some regions. These crops have originated in one region of shifting cultivation and have diffused to other areas in recent years.

The Kayapo people of Brazil's Amazon tropical rain forest do not arrange crops in the rectangular fields and rows that are familiar to us. They plant in concentric rings. At first they plant sweet potatoes and yams in the inner area. In successive rings go corn and rice, manioc, and more yams. In subsequent years the inner area of potatoes and yams expands to replace corn and rice. The outermost ring contains plants that require more nutrients, including papaya, banana, pineapple, mango, cotton, and beans.

CONTEMPORARY GEOGRAPHIC TOOLS
Protecting Farmland

Loss of farmland to urban growth is especially severe at the edge of the string of large metropolitan areas along the East Coast of the United States. Some of the most threatened agricultural land lies in Maryland, a small state where two major cities—Washington and Baltimore—have coalesced into a continuous built-up area (see Chapter 13).

Farmland preservation efforts traditionally identify "prime" agricultural areas on the basis of only one factor—soil quality. "Prime" farmland is typically flat and well drained, qualities that also attract developers of new housing projects. In Maryland, a geographic information system (GIS) was used to identify which farms should be preserved. Through GIS, the distribution of Maryland's most productive soils could be compared to the distribution of other factors.

Maps generated through GIS were essential in identifying agricultural land to protect, because the most appropriate farms to preserve were not necessarily those with the highest-quality soil. Why should the state and nonprofit organizations spend scarce funds to preserve "prime" farmland that is nowhere near the path of urban sprawl? Conversely, why purchase an expensive, isolated farm already totally surrounded by residential developments, when the same amount of money could buy several large contiguous farms that effectively blocked urban sprawl elsewhere?

To identify the "best" lands according to several economic and environmental factors, not just soil quality, GIS consultants produced a series of maps at the state and county levels.

- Environmental maps identified farmland in need of preservation because of water quality, flood control, species habitats, historic sites, or especially attractive scenery.
- Economic maps included the market value of the products grown or raised on the land and areas projected to have relatively high population growth if not curtailed.

The GIS maps showed that 63 percent of Maryland's farmland contained prime soils, 32 percent important environmental features, and 23 percent high population growth pressures.

The various soil quality, environmental, and economic maps were combined through the GIS to produce a single composite map of all three sets of important factors (Figure 10-7). The map shows that 4 percent of the state's farmland had prime soils, significant environmental features, and high projected population growth, and 25 percent had two of the three factors (such as prime soils and significant environmental features but not high population growth).

GIS cannot rank the relative importance of the various physical, environmental, and economic features. Land with one important physical, environmental, or economic feature may be as important to preserve as land with three features. Still, Maryland officials are making use of the results of the GIS as part of an overall strategy to minimize sprawl and keep new developments as tightly packed around existing urban areas as possible. For example, state highway money is allocated to improving roads in existing built-up areas rather than extend new roads through rural areas. ■

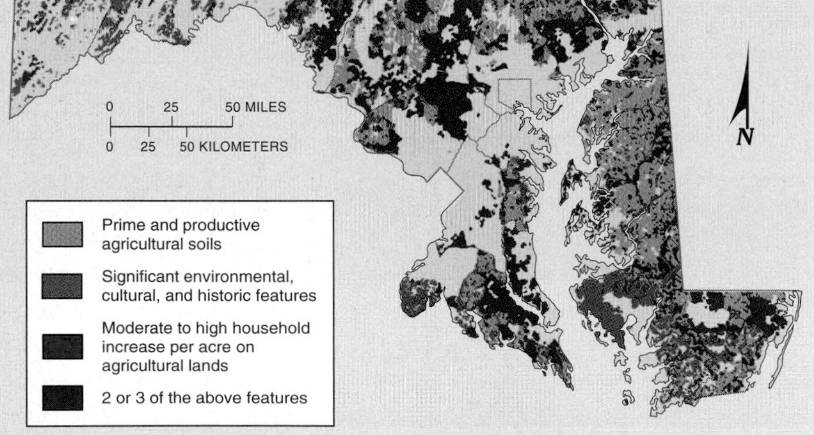

FIGURE 10-7 Maryland soil quality, environmental conditions, and population growth.

It is here that the leafy crowns of cut trees fall when the field is cleared, and their rotting releases more nutrients into the soil.

Most families grow only for their own needs, so one swidden may contain a large variety of intermingled crops, which are harvested individually at the best time. In shifting cultivation a "farm field" appears much more chaotic than do fields in MDCs, where a single crop such as corn or wheat may grow over an extensive area. In some cases, families may specialize in a few crops and trade with villagers who have a surplus of others.

OWNERSHIP AND USE OF LAND IN SHIFTING CULTIVATION. Traditionally, land was owned by the village as a whole rather than separately by each resident. The chief or ruling council allocated a patch of land to each family and allowed it to retain the output. Individuals may also have had the right to own or protect specific trees surrounding the village. Today, private individuals now own the land in some communities, especially in Latin America.

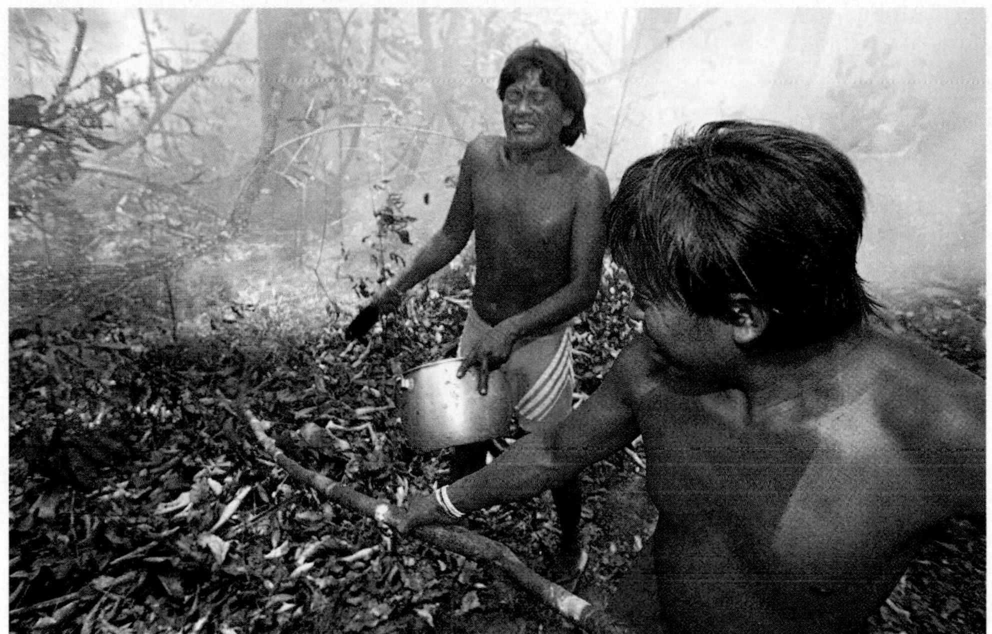

FIGURE 10-8 Shifting cultivation farmer. The Brazilian Yanomami people use shifting cultivation to prepare a field for planting by slashing and burning the vegetation. The dense vegetation is chopped down, and the debris is burned in order to provide the soil with needed nutrients.

FIGURE 10-9 Deforestation in the Amazon tropical rainforest. Cash crops such as soybeans are replacing forests in Brazil.

Shifting cultivation occupies approximately one-fourth of the world's land area, a higher percentage than any other type of agriculture. However, less than 5 percent of the world's people engage in shifting cultivation. The gap between the percentage of people and land area is not surprising, because the practice of moving from one field to another every couple of years requires more land per person than do other types of agriculture.

Future of Shifting Cultivation

Land devoted to shifting cultivation is declining in the tropics at the rate of about 75,000 square kilometers (30,000 square miles), or 0.2 percent, per year according to the United Nations (Figure 10-9). The amount of Earth's surface allocated to tropical rain forests has already been reduced to less than half of its original area, for until recent years the World Bank supported deforestation with loans to finance development schemes that required clearing forests. Shifting cultivation is being replaced by logging, cattle ranching, and the cultivation of cash crops. Selling timber to builders or raising beef cattle for fast-food restaurants are more effective development strategies than maintaining shifting cultivation. LDCs also see shifting cultivation as an inefficient way to grow food in a hungry world. Indeed, compared to other forms of agriculture, shifting cultivation can support only a small population in an area without causing environmental damage.

To its critics, shifting cultivation is at best a preliminary step in economic development. Pioneers use shifting cultivation to clear forests in the tropics and to open land for development where permanent agriculture never existed. People unable to find agricultural land elsewhere can migrate to the tropical forests and initially practice shifting cultivation. Critics say it then should be replaced by more sophisticated agricultural techniques that yield more per land area. Defenders of shifting cultivation consider it the most environmentally sound approach for the tropics. Practices used in other forms of agriculture, such as fertilizers and pesticides and permanently clearing fields, may damage the soil, cause severe erosion, and upset balanced ecosystems.

Large-scale destruction of the rain forests also may contribute to global warming. When large numbers of trees are cut, their burning and decay release large volumes of carbon dioxide. This gas can build up in the atmosphere, acting like the window glass in a greenhouse to trap solar energy in the atmosphere, resulting in the "greenhouse effect," discussed in Chapter 14. Elimination of shifting cultivation could also upset the traditional local diversity of cultures in the tropics. The activities of shifting cultivation are intertwined with other social, religious, political, and various folk customs. A drastic change in the agricultural economy could disrupt other activities of daily life.

As the importance of tropical rain forests to the global environment has become recognized, LDCs have been pressured to restrict further destruction of them. In one innovative strategy, Bolivia agreed to set aside 1.5 million hectares (3.7 million acres) in a forest reserve in exchange for cancellation of $650,000,000 of its debt to developed countries. Meanwhile, in Brazil's Amazon rain forest, deforestation has increased from 2.7 million hectares (7 million acres) per year during the 1990s to 3.1 million hectares (8 million acres) since 2000.

Pastoral Nomadism

Pastoral nomadism is a form of subsistence agriculture based on the herding of domesticated animals. The word *pastoral* refers to sheepherding. It is adapted to dry climates, where planting crops is impossible. Pastoral nomads live primarily in the large belt of arid and semiarid land that includes Central and Southwest Asia and North Africa (Figure 10-10). The Bedouins of Saudi Arabia and North Africa and the Masai of East Africa are examples of nomadic groups. Only about 15 million people are pastoral nomads, but they sparsely occupy about 20 percent of Earth's land area.

Characteristics of Pastoral Nomadism

Unlike other subsistence farmers, pastoral nomads depend primarily on animals rather than crops for survival. The animals provide milk, and their skins and hair are used for clothing and tents. Like other subsistence farmers, though, pastoral nomads consume mostly grain rather than meat. Their animals are usually not slaughtered, although dead ones may be consumed. To nomads, the size of their herd is both an important measure of power and prestige and their main security during adverse environmental conditions.

Some pastoral nomads obtain grain from sedentary subsistence farmers in exchange for animal products. More often, part of a nomadic group—perhaps the women and children—may plant crops at a fixed location while the rest of the group wanders with the herd. Nomads might hire workers to practice sedentary agriculture in return for grain and protection. Other nomads might sow grain in recently flooded areas and return later in the year to harvest the crop. Yet another strategy is to remain in one place and cultivate the land when rainfall is abundant; then, during periods that are too dry to grow crops, the group can increase the size of the herd and migrate in search of food and water.

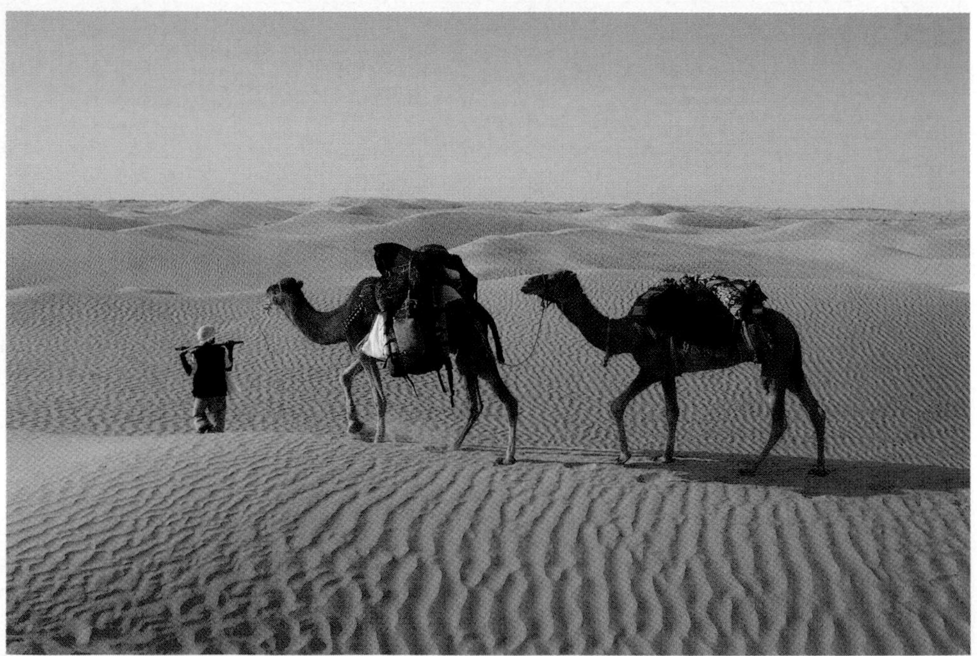

FIGURE 10-10 Pastoral nomad. Camels are led across the Sahara desert.

CHOICE OF ANIMALS. Nomads select the type and number of animals for the herd according to local cultural and physical characteristics. The choice depends on the relative prestige of animals and the ability of species to adapt to a particular climate and vegetation. The camel is the most highly desired animal in North Africa and Southwest Asia, along with sheep and goats. The horse is particularly important in Central Asia.

- Camels are well suited to arid climates because they can go long periods without water, carry heavy baggage, and move rapidly, but they are particularly bothered by flies and sleeping sickness and have a relatively long gestation period—12 months from conception to birth.
- Goats need more water than do camels but are tough and agile and can survive on virtually any vegetation, no matter how poor.
- Sheep are relatively slow moving and affected by climatic changes; they require more water and are more selective as to which plants they will eat.

The minimum number of animals necessary to support each family adequately varies according to the particular group and animal. The typical nomadic family needs 25 to 60 goats or sheep or 10 to 25 camels.

MOVEMENTS OF PASTORAL NOMADS. Pastoral nomads do not wander randomly across the landscape but have a strong sense of territoriality. Every group controls a piece of territory and will invade another group's territory only in an emergency or if war is declared. The goal of each group is to control a territory large enough to contain the forage and water needed for survival. The actual amount of land a group controls depends on its wealth and power.

The precise migration patterns evolve from intimate knowledge of the area's physical and cultural characteristics. Groups frequently divide into herding units of five or six families and choose routes based on the most likely water sources during the various seasons of the year. The selection of routes varies in unusually wet or dry years and is influenced by the condition of their animals and the area's political stability.

Some pastoral nomads practice **transhumance**, which is seasonal migration of livestock between mountains and lowland pasture areas. **Pasture** is grass or other plants grown for feeding grazing animals, as well as land used for grazing. Sheep or other animals may pasture in alpine meadows in the summer and be herded back down into valleys for winter pasture.

The Future of Pastoral Nomadism

Agricultural experts once regarded pastoral nomadism as a stage in the evolution of agriculture—between the hunters and gatherers who migrated across Earth's surface in search of food and sedentary farmers who cultivated grain in one place. Because they had domesticated animals but not plants, pastoral nomads were considered more advanced than hunters and gatherers but less advanced than settled farmers.

Pastoral nomadism is now generally recognized as an offshoot of sedentary agriculture, not as a primitive precursor of it. It is simply a practical way of surviving on land that receives too little rain for cultivation of crops. The domestication of animals—the basis for pastoral nomadism—probably was achieved originally by sedentary farmers, not by nomadic hunters. Pastoral nomads therefore had to be familiar with sedentary farming, and in many cases they practiced it.

Today, pastoral nomadism is a declining form of agriculture, partly a victim of modern technology. Before recent transportation and communications inventions, pastoral nomads played an important role as carriers of goods and information across the sparsely inhabited dry lands. They used to be the most powerful inhabitants of the dry lands, but now, with modern weapons, national governments can control the nomadic population more effectively.

Government efforts to resettle nomads have been particularly vigorous in China, Kazakhstan, and several Southwest Asia countries, including Israel, Saudi Arabia, and Syria. Nomads are reluctant to cooperate, so these countries have experienced difficulty in trying to force their settlement in collectives and cooperatives. Governments force groups to give up pastoral nomadism because they want the land for other uses. Land that can be irrigated is converted from nomadic to sedentary agriculture. In some instances, the mining and petroleum industries now operate in dry lands formerly occupied by pastoral nomads. Some nomads are encouraged to try sedentary agriculture or to work for mining or petroleum companies. Others are still allowed to move about, but only within ranches of fixed boundaries. In the future, pastoral nomadism will be increasingly confined to areas that cannot be irrigated or that lack valuable raw materials.

Intensive Subsistence Agriculture

Shifting cultivation and pastoral nomadism are forms of subsistence agriculture found in regions of low density. But three-fourths of the world's people live in LDCs, and the form of subsistence agriculture that feeds most of them is **intensive subsistence agriculture**. The term *intensive* implies that farmers must work intensively to subsist on a parcel of land. In densely populated East, South, and Southeast Asia, most farmers practice intensive subsistence agriculture. The typical farm in Asia's intensive subsistence agriculture regions is much smaller than elsewhere in the world. Many Asian farmers own several fragmented plots, frequently a result of dividing individual holdings among several children over several centuries. Because the agricultural density—the ratio of farmers to arable land—is so high in parts of East and South Asia, families must produce enough food for their survival from a very small area of land. They do this through careful agricultural practices, refined over thousands of years in response to local environmental and cultural patterns. Most of the work is done by hand or with animals rather than with machines, in part due to abundant labor, but largely from lack of funds to buy equipment.

To maximize food production, intensive subsistence farmers waste virtually no land. Corners of fields and irregularly shaped pieces of land are planted rather than left idle. Paths and roads are kept as narrow as possible to minimize the loss of arable land. Livestock are rarely permitted to graze on land that could be used to plant crops, and little grain is grown to feed the animals.

Intensive Subsistence with Wet Rice Dominant

The intensive agriculture region of Asia can be divided between areas where wet rice dominates and areas where it does not (refer to Figure 10-4). The term **wet rice** refers to the practice of planting rice on dry land in a nursery and then moving the seedlings to a flooded field to promote growth (Figure 10-11). Wet rice occupies a relatively small percentage of Asia's agricultural land but is the region's most important source of food. Intensive wet-rice farming is the dominant type of agriculture in Southeast China, East India, and much of Southeast Asia (Figure 10-12). Successful production of large yields of rice is an elaborate process that is time-consuming and done mostly by hand. The consumers of the rice also perform the work, and all family members, including children, contribute to the effort.

Growing rice involves several steps. First, a farmer prepares the field for planting, using a plow drawn by water buffalo or oxen. The use of a plow and animal power is one

FIGURE 10-11 Wet rice farmer, Laomeng, China. Growing wet rice is labor-intensive. It needs to be grown on flat land, so hillsides are terraced to increase the area of production.

characteristic that distinguishes subsistence agriculture from shifting cultivation. The plowed land is then flooded with water. The water is collected from rainfall, river overflow, or irrigation. Too much or too little can damage the crop—a particular problem for farmers in South Asia who depend on monsoon rains, which do not always arrive at the same time each summer. Before planting, dikes and canals are repaired to ensure the right quantity of water in the field. The flooded field is called a **sawah** in the Austronesian language widely spoken in Indonesia, including Java. Europeans and North Americans frequently, but incorrectly, call it a **paddy**, the Malay word for wet rice.

The customary way to plant rice is to grow seedlings on dry land in a nursery and then transplant the seedlings into the flooded field. Typically, one-tenth of a sawah is devoted to the cultivation of seedlings. After about a month they are transferred to the rest of the field. Rice plants grow submerged in water for approximately three-fourths of the growing period. Another method of planting rice is to broadcast dry seeds by scattering them through the field, a method used to some extent in South Asia.

Rice plants are harvested by hand, usually with knives. To separate the husks, known as **chaff**, from the seeds, the heads are **threshed** by beating them on the ground or treading on them barefoot. The threshed rice is placed in a tray, and the lighter chaff is **winnowed**—that is, allowed to be blown away by the wind. If the rice is to be consumed directly by the farmer, the **hull**, or outer covering, is removed by mortar and pestle. Rice that is sold commercially is frequently whitened and polished, a

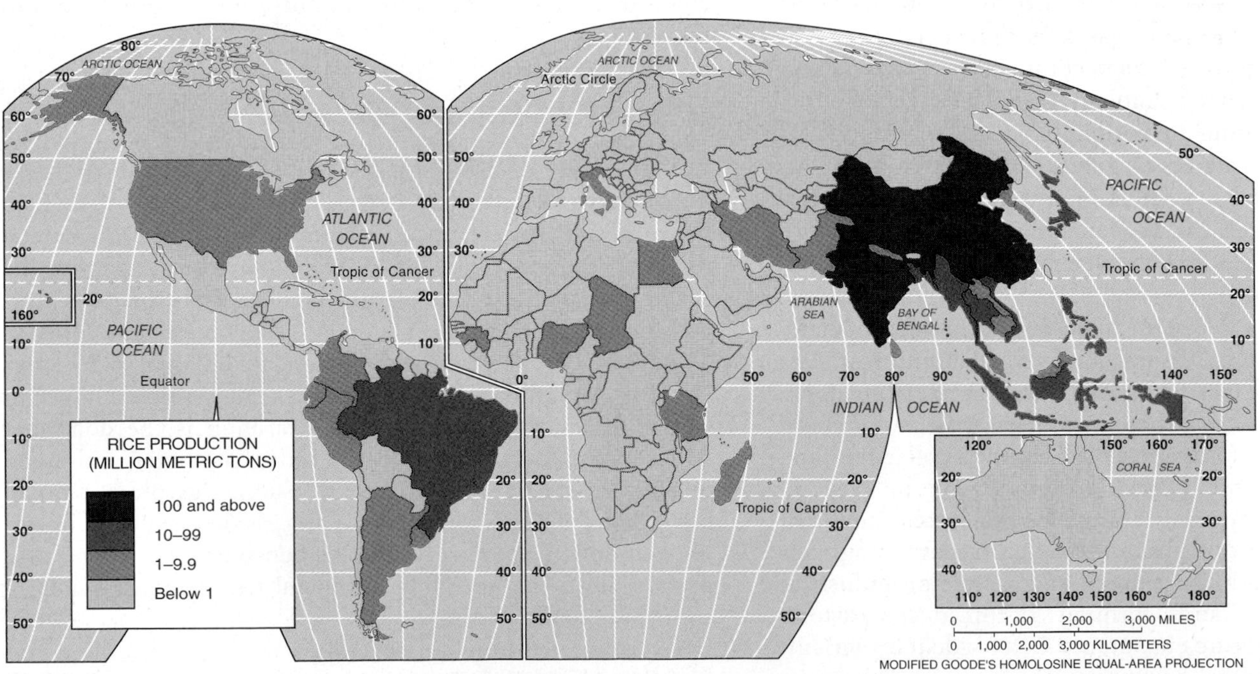

FIGURE 10-12 Rice production. Rice is the most important crop in the large population concentrations of East, South, and Southeast Asia. China and India account for more than half of world production.

process that removes some nutrients but leaves rice more pleasing in appearance and taste to many consumers.

Wet rice is most easily grown on flat land, because the plants are submerged in water much of the time. Thus most wet-rice cultivation takes place in river valleys and deltas. But the pressure of population growth in parts of East Asia has forced expansion of areas under rice cultivation. One method of developing additional land suitable for growing rice is to terrace the hillsides of river valleys.

Land is used even more intensively in parts of Asia by obtaining two harvests per year from one field, a process known as **double cropping**. Double cropping is common in places that have warm winters, such as South China and Taiwan, but is relatively rare in India, where most areas have dry winters. Normally, double cropping involves alternating between wet rice, grown in the summer when precipitation is higher, and wheat, barley, or another dry crop, grown in the drier winter season. Crops other than rice may be grown in the wet-rice region in the summer on nonirrigated land.

Intensive Subsistence with Wet Rice Not Dominant

Climate prevents farmers from growing wet rice in portions of Asia, especially where summer precipitation levels are too low and winters are too harsh (refer to Figure 10-4). Agriculture in much of interior India and northeast China is devoted to crops other than wet rice. Wheat is the most important crop, followed by barley. Various other grains and legumes are grown for household consumption, including millet, oats, corn, kaoliang, sorghum, and soybeans. Also grown are some crops sold for cash, such as cotton, flax, hemp, and tobacco.

Aside from what is grown, this region shares most of the characteristics of intensive subsistence agriculture with the wet-rice region. Land is used intensively and worked primarily by human power with the assistance of some hand implements and animals. In milder parts of the region where wet rice does not dominate, more than one harvest can be obtained some years through skilled use of **crop rotation**, which is the practice of rotating use of different fields from crop to crop each year to avoid exhausting the soil. In colder climates, wheat or another crop is planted in the spring and harvested in the fall, but no crops can be sown through the winter.

Since the Communist Revolution in 1949, private individuals have owned little agricultural land in China. Instead, the Communist government organized agricultural producer communes, which typically consisted of several villages of several hundred people. By combining several small fields into a single large unit, China's government hoped to promote agricultural efficiency—scarce equipment and animals and larger improvement projects, such as flood control, water storage, and terracing, could be shared. In reality, productivity did not increase as much as the government had expected because people worked less efficiently for the commune than when working for themselves.

China has therefore dismantled the agricultural communes. The communes still hold legal title to agricultural land, but villagers sign contracts entitling them to farm portions of the land as private individuals. Chinese farmers may sell to others the right to use the land and to pass on the right to their children. Reorganization has been difficult because irrigation systems, equipment, and other infrastructure were developed to serve large communal farms rather than small individually managed ones, which cannot afford to operate and maintain the machinery. But production has increased greatly.

Plantation Farming

The plantation is a form of commercial agriculture found in the tropics and subtropics, especially in Latin America, Africa, and Asia (Figure 10-13). Although generally situated in LDCs, plantations are often owned or operated by Europeans or North Americans and grow crops for sale primarily to MDCs. Crops

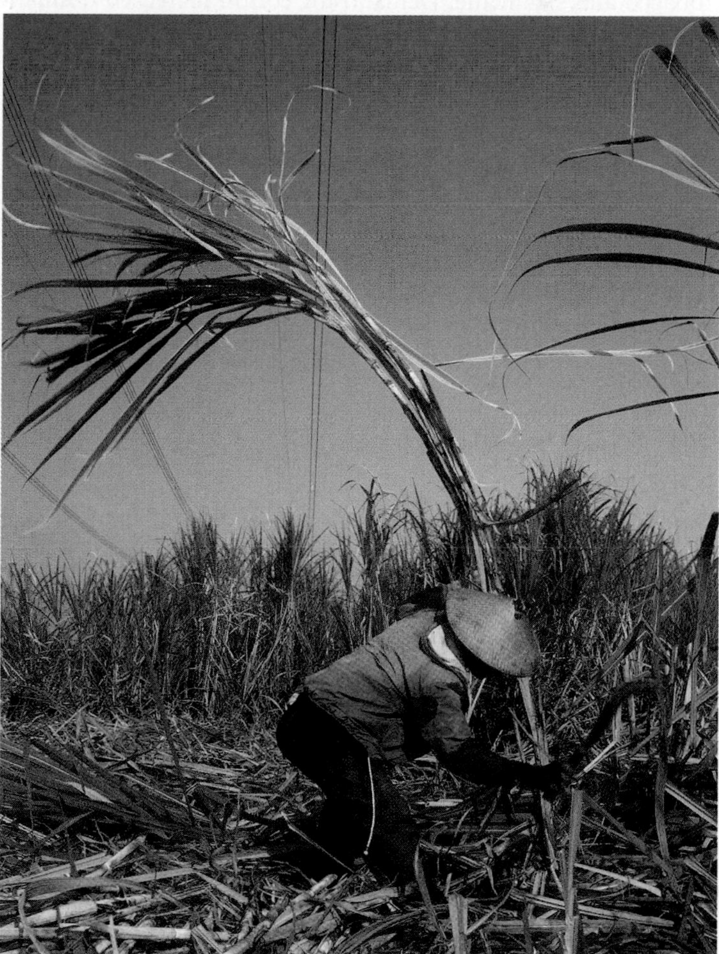

FIGURE 10-13 Plantation farm workers, Bogor, Indonesia. Workers are harvesting palm fruit, to be processed into palm oil. Indonesia is the world's largest palm oil producer.

are normally processed at the plantation before shipping because processed goods are less bulky and are therefore cheaper to ship long distances to the North American and European markets.

A **plantation** is a large farm that specializes in one or two crops. Among the most important crops grown on plantations are cotton, sugarcane, coffee, rubber, and tobacco. Also produced in large quantities are cocoa, jute, bananas, tea, coconuts, and palm oil. Latin American plantations are more likely to grow coffee, sugarcane, and bananas, whereas Asian plantations may provide rubber and palm oil. Crops such as tobacco, cotton, and sugarcane, which can be planted only once a year, are less likely to be grown on large plantations today than in the past.

Because plantations are usually situated in sparsely settled locations, they must import workers and provide them with food, housing, and social services. Plantation managers try to spread the work as evenly as possible throughout the year to make full use of the large labor force. Where the climate permits, more than one crop is planted and harvested annually. Rubber-tree plantations try to spread the task of tapping the trees throughout the year.

Until the Civil War, plantations were important in the U.S. South, where the principal crop was cotton, followed by tobacco and sugarcane. Demand for cotton increased dramatically after the establishment of textile factories in England at the start of the Industrial Revolution in the late eighteenth century. Cotton production was stimulated by the improvement of the cotton gin by Eli Whitney in 1793 and the development of new varieties of cotton that were hardier and easier to pick. Slaves brought from Africa performed most of the labor until the abolition of slavery and the defeat of the South in the Civil War. Thereafter, plantations declined in the United States; they were subdivided and either sold to individual farmers or worked by tenant farmers.

KEY ISSUE 3

Where Are Agricultural Regions in MDCs?

- Mixed Crop and Livestock Farming
- Dairy Farming
- Grain Farming
- Livestock Ranching
- Mediterranean Agriculture
- Commercial Gardening and Fruit Farming

Commercial agriculture in MDCs can be divided into six main types, as listed above. Each type is predominant in distinctive regions within MDCs, depending largely on climate. ■

Mixed Crop and Livestock Farming

Mixed crop and livestock farming is the most common form of commercial agriculture in the United States west of the Appalachians and east of 98° west longitude and in much of Europe from France to Russia (refer to Figure 10-4).

Characteristics of Mixed Crop and Livestock Farming

The most distinctive characteristic of mixed crop and livestock farming is its integration of crops and livestock (Figure 10-14). Most of the crops are fed to animals rather than consumed directly by humans. In turn, the livestock supply manure to improve soil fertility to grow more crops. A typical mixed crop and livestock farm devotes nearly all land area to growing crops but derives more than three-fourths of its income from the sale of animal products, such as beef, milk, and eggs. In the United States pigs are often bred directly on the farms, whereas cattle may be brought in to be fattened on corn.

Mixing crops and livestock permits farmers to distribute the workload more evenly through the year. Fields require less attention in the winter than in the spring, when crops are planted, and in the fall, when they are harvested. Meanwhile, livestock require year-long attention. A mix of crops and livestock also reduces seasonal variations in income; most income from crops comes during the harvest season, but livestock products can be sold throughout the year.

In the United States, corn (maize) is the crop most frequently planted in the mixed crop and livestock region because it generates higher yields per area than other crops (Figure 10-15). Some of the corn is consumed by people as oil, margarine, and other food products, but most is fed to pigs and cattle. The most important mixed crop and livestock farming region in the United States—extending from Ohio to the Dakotas, with its center in Iowa—is frequently called the Corn Belt, because around half of the cropland is planted in corn. Soybeans have become the second most important crop in the U.S. mixed commercial farming region. Like corn, soybeans are mostly used to make animal feed. Tofu (made from soybean milk) is a major food source, especially for people in China and Japan. Soybean oil is widely used in U.S. foods, but as a hidden ingredient.

Crop Rotation

Mixed crop and livestock farming typically involves crop rotation. The farm is divided into a number of fields, and each field is planted on a planned cycle, often of several years. The crop planted changes from one year to the next, typically going through a cycle of two or more crops, and perhaps a year of fallow before the cycle is repeated. Crop rotation helps maintain the fertility of a field because various crops deplete the soil of certain nutrients but restore others. Crop rotation contrasts with shifting cultivation, in which nutrients depleted from a field are restored only by leaving the field fallow (uncropped) for many years. In any given year, crops cannot be planted in

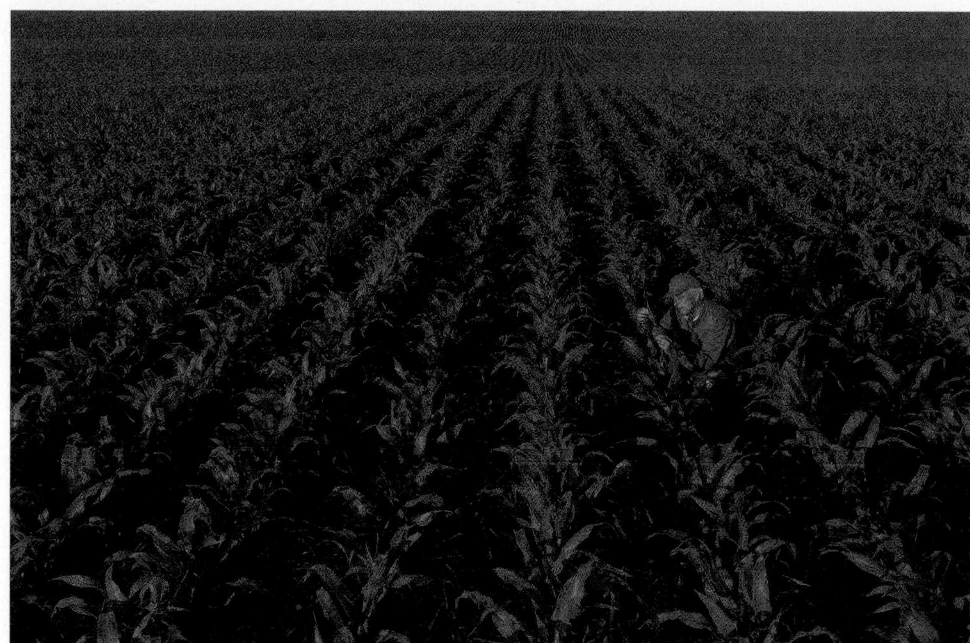

FIGURE 10-14 Mixed crop and livestock farm, Iowa.

most of an area's fields, so overall production in shifting cultivation is much lower than in mixed commercial farming.

A two-field crop-rotation system was developed in Northern Europe as early as the fifth century. A **cereal grain**, such as oats, wheat, rye, or barley, was planted in Field A one year, while Field B was left fallow. The following year Field B was planted but A left fallow, and so forth. Beginning in the eighth century, a three-field system was introduced. The first field was planted with a winter cereal, the second with a spring cereal, and the third was left fallow. As a result, each field yielded four harvests every 6 years, compared to three every 6 years under the two-field system.

A four-field system was introduced in Europe during the eighteenth century. The first year, the farmer could plant a root crop (such as turnips) in Field A, a cereal in Field B, a "rest" crop (such as clover, which helps restore the field) in Field C, and a cereal in Field D. The second year, the farmer might select a cereal for Field A, a rest crop for Field B, a cereal for Field C, and a root for Field D. The rotation would continue for two more years before the cycle would start again. Each field thus passed through a cycle of four crops—root, cereal, rest crop, and another cereal.

Cereals such as wheat and barley were sold for flour and beer production, and straw (the stalks remaining after the heads of wheat are threshed) was retained for animal bedding. Root crops such as turnips were fed to the animals during the winter. Clover and other "rest" crops were used for cattle grazing and restoration of nitrogen to the soil.

Dairy Farming

Dairy farming is the most important commercial agriculture practiced on farms near the large urban areas of the Northeast United States, Southeast Canada, and Northwest Europe (Figure 10-16). Dairying has also become an important type

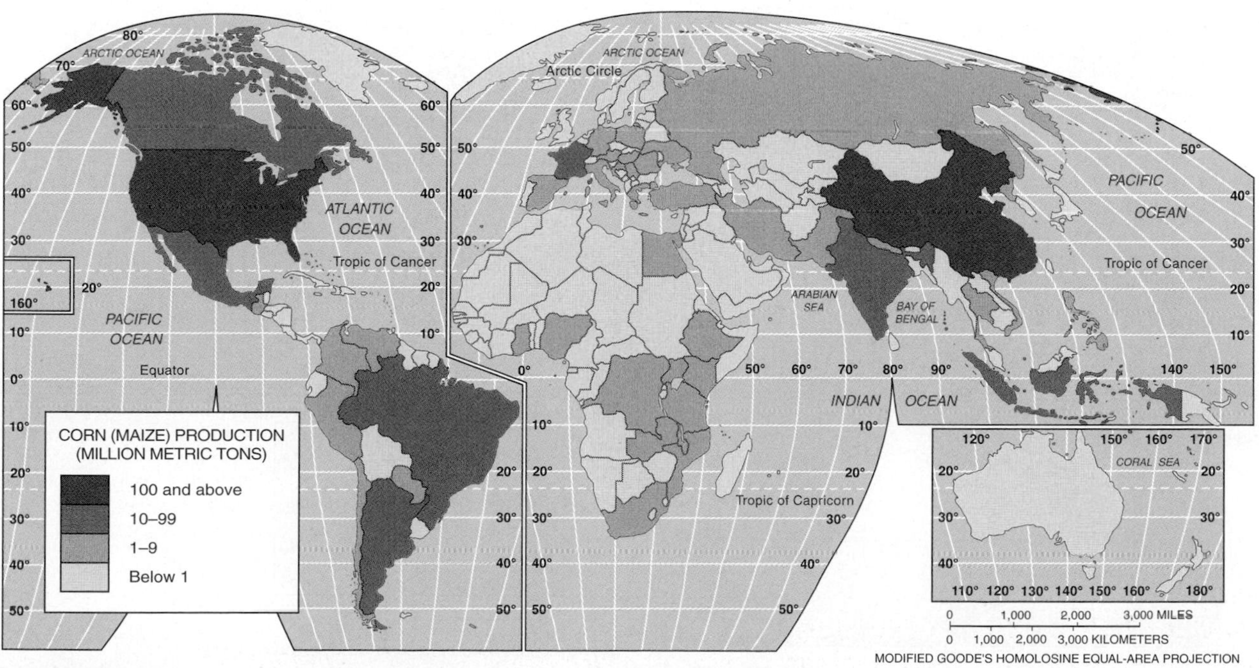

FIGURE 10-15 Corn (maize) production. The United States accounts for more than 40 percent and China nearly 20 percent of the world's corn production. Outside North America, corn is called maize.

of farming in South and East Asia. Traditionally, fresh milk was rarely consumed except directly on the farm or in nearby villages. With the rapid growth of cities in MDCs during the nineteenth century, demand for the sale of milk to urban residents increased. Rising incomes permitted urban residents to buy milk products, which were once considered luxuries.

FIGURE 10-16 Dairy farmer, Vermont. Dairying is a major type of agriculture in the northeastern United States, a region that is within the milksheds of several major metropolitan areas.

Regional Distribution of Dairying

For most of the twentieth century, the world's milk production was clustered in a handful of MDCs (Figure 10-17). However, the share of the world's dairy farming conducted in LDCs has risen dramatically, from 26 percent in 1980 to 51 percent in 2007. In the twenty-first century, India has become the world's largest milk producer, ahead of the United States, the traditional leader, and China and Pakistan have passed Russia as third and fourth largest.

In MDCs, dairying is the most important type of commercial agriculture in the first ring outside large cities because of transportation factors. Dairy farms must be closer to their market than other types of farms because their products are highly perishable. The ring surrounding a city from which milk can be supplied without spoiling is known as the **milkshed**. Improvements in transportation have permitted dairying to be undertaken farther from the market. Until the 1840s, when railroads were first used for transporting dairy products, milksheds rarely had a radius beyond 50 kilometers (30 miles). Today, refrigerated railcars and trucks enable farmers to ship milk more than 500 kilometers (300 miles). As a result, nearly every farm in the U.S. Northeast and Northwest Europe is within the milkshed of at least one urban area.

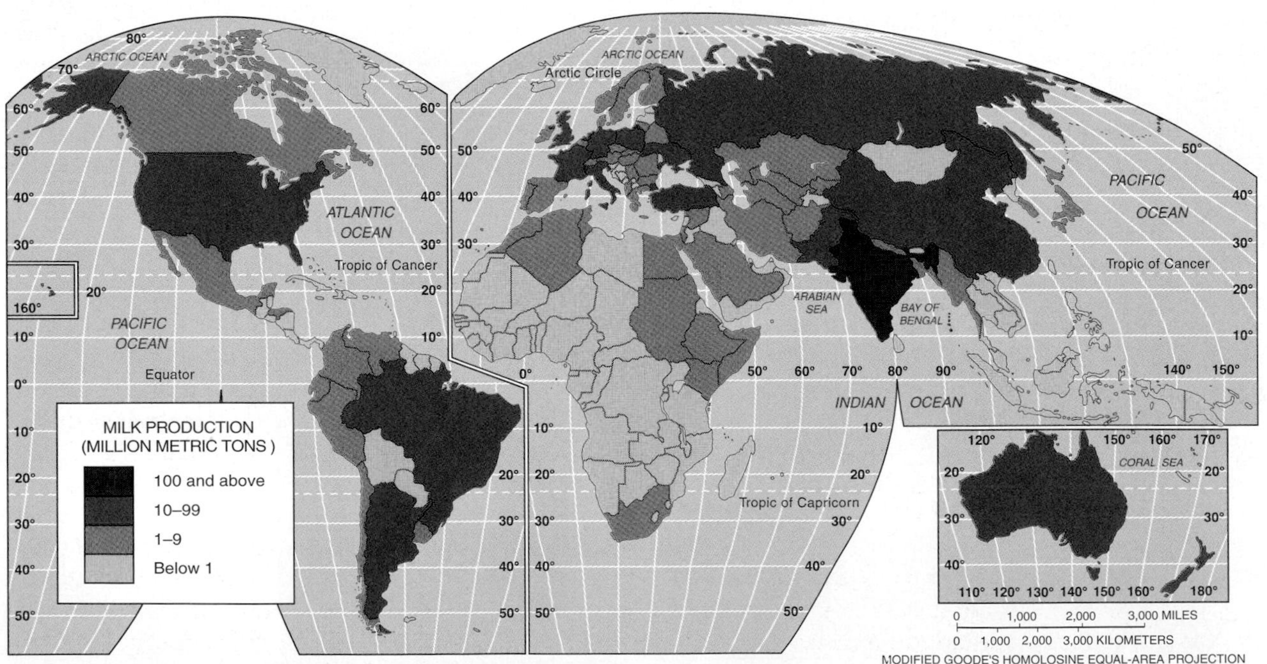

FIGURE 10-17 Milk production. MDCs have one-fourth of the world's population but produce more than one-half of the world's milk. However, production has expanded rapidly in recent years in LDCs.

Dairy farmers, like other commercial farmers, usually do not sell their products directly to consumers. Instead, they generally sell milk to wholesalers, who distribute it in turn to retailers. Retailers then sell milk to consumers in shops or at home. Farmers also sell milk to butter and cheese manufacturers. The choice of product varies within the U.S. dairy region, depending on whether the farms are within the milkshed of a large urban area. In general, the farther the farm is from large urban concentrations, the smaller is the percentage of output devoted to fresh milk. Farms located farther from consumers are more likely to sell their output to processors who make butter, cheese, or dried, evaporated, and condensed milk. The reason is that these products keep fresh longer than milk does and therefore can be safely shipped from remote farms.

In the East, virtually all milk is sold to consumers living in New York, Philadelphia, Boston, and the other large urban areas. Farther west, most milk is processed into cheese and butter. Most of the milk in Wisconsin is processed, for example, compared to only 5 percent in Pennsylvania. The proximity of northeastern farmers to several large markets accounts for these regional differences.

Countries likewise tend to specialize in certain products. New Zealand, the world's largest per capita producer of dairy products, devotes about 5 percent to liquid milk, compared to more than 50 percent in the United Kingdom. New Zealand farmers do not sell much liquid milk, because the country is too far from North America and Northwest Europe, the two largest relatively wealthy population concentrations.

Challenges for Dairy Farmers

Like other commercial farmers, dairy farmers face economic difficulties because of declining revenues and rising costs. Dairy farmers who have quit most often cite lack of profitability and excessive workload as reasons for getting out of the business. Distinctive features of dairy farming have exacerbated the economic difficulties:

- **Labor-intensive.** Cows must be milked twice a day, every day; although the actual milking can be done by machines, dairy farming nonetheless requires constant attention throughout the year.
- **Winter Feed.** Dairy farmers face the expense of feeding the cows in the winter, when they may be unable to graze on grass. In Northwest Europe and in the Northeastern United States, farmers generally purchase hay or grain for winter feed. In the western part of the U.S. dairy region, crops are more likely to be grown in the summer and stored for winter feed on the same farm.

Grain Farming

Some form of grain is the major crop on most farms. **Grain** is the seed from various grasses, like wheat, corn, oats, barley, rice, millet, and others. Commercial grain agriculture is distinguished from mixed crop and livestock farming because crops on a grain farm are grown primarily for consumption by humans rather than by livestock (Figure 10-18). Farms in LDCs also grow crops for human consumption, but the output is directly consumed by the farmers. Commercial grain farms sell their output to manufacturers of food products, such as breakfast cereals and snack-food makers.

FIGURE 10-18 Wheat farmer, Palouse, Washington. Eastern Washington is one of three major wheat-growing regions in the United States, along with the winter-wheat belt of Kansas, Colorado, and Oklahoma, and the spring-wheat belt of the Dakotas and Montana.

The most important crop grown is wheat, used to make bread flour. Wheat generally can be sold for a higher price than other grains, such as rye, oats, and barley, and it has more uses as human food. It can be stored relatively easily without spoiling and can be transported a long distance. Because wheat has a relatively high value per unit weight, it can be shipped profitably from remote farms to markets.

Wheat's significance extends beyond the amount of land or number of people involved in growing it. Unlike other agricultural products, wheat is grown to a considerable extent for international trade and is the world's leading export crop. The United States and Canada account for about half of the world's wheat exports; consequently, the North American prairies are accurately labeled the world's "breadbasket." The ability to provide food for many people elsewhere in the world is a major source of economic and political strength for these two countries.

The largest commercial producer of grain by far is the United States (Figure 10-19). Large-scale commercial grain production is found in only a few other countries, including Canada, Argentina, Australia, France, and the United Kingdom. Commercial grain farms are generally located in regions that are too dry for mixed crop and livestock agriculture. Within North America, large-scale grain production is concentrated in three areas:

- The **winter-wheat** belt through Kansas, Colorado, and Oklahoma. The crop is planted in the autumn and develops a strong root system before growth stops for the winter. The wheat survives the winter, especially if it is insulated beneath a snow blanket, and is ripe by the beginning of summer.
- The **spring-wheat** belt through the Dakotas, Montana, and southern Saskatchewan in Canada. Winters are usually too

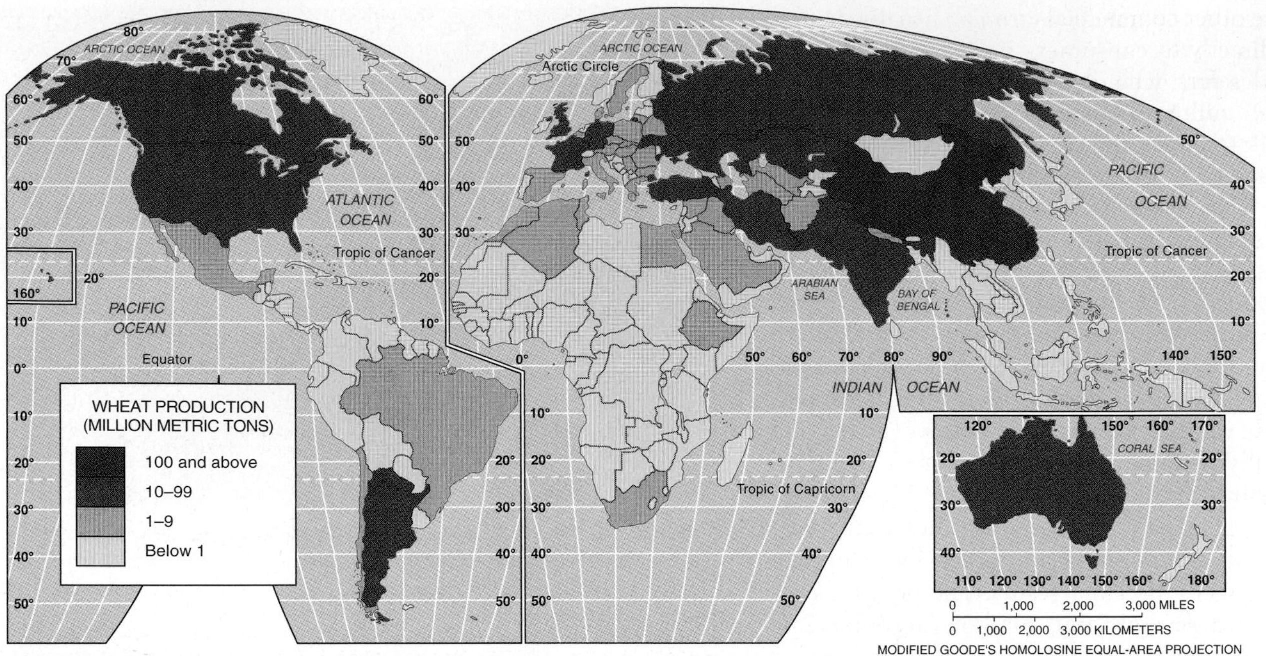

FIGURE 10-19 Wheat production. China and India have passed the United States as the leading producer of wheat. The United States remains the leading producer of wheat for commercial sale off the farm.

severe for winter wheat in this region, so spring wheat is planted in the spring and harvested in the late summer.

- The Palouse region of Washington State.

Large-scale grain production, like other commercial farming ventures in MDCs, is heavily mechanized, conducted on large farms, and oriented to consumer preferences. The McCormick **reaper** (a machine that cuts grain standing in the field), invented in the 1830s, first permitted large-scale wheat production. Today the **combine** machine performs in one operation the three tasks of reaping, threshing, and cleaning.

Unlike work on a mixed crop and livestock farm, the effort required to grow wheat is not uniform throughout the year. Some individuals or firms may therefore have two sets of fields—one in the spring-wheat belt and one in the winter-wheat belt. Because the planting and harvesting in the two regions occur at different times of the year, the workload can be distributed throughout the year. In addition, the same machinery can be used in the two regions, thus spreading the cost of the expensive equipment. Combine companies start working in Oklahoma in early summer and work their way northward.

Livestock Ranching

Ranching is the commercial grazing of livestock over an extensive area (Figure 10-20). This form of agriculture is adapted to semiarid or arid land and is practiced in MDCs where the vegetation is too sparse and the soil too poor to support crops.

The importance of ranching in the United States extends beyond the people who choose this form of commercial farming. Its prominence in popular culture, especially in Hollywood films and television, has not only helped to draw attention to this form of commercial farming but has also served to illustrate, albeit in sometimes romanticized ways, the crucial role that ranching played in the history and settlement of areas of the United States. Cattle ranching in Texas, as glamorized in popular culture, did actually dominate commercial agriculture, but only for a short period—from 1867 to 1885.

Cattle ranching expanded in the United States during the 1860s because of the demand for beef in the East Coast cities. If they could get their cattle to Chicago, ranchers were paid $30 to $40 per head, compared to only $3 or $4 per head in Texas. Once in Chicago, the cattle could be slaughtered and processed by meat-packing companies and shipped in packages to consumers in the East. To reach Chicago, cattle were driven on hoof by cowboys over trails from Texas to the nearest railhead. There the cattle were driven into cattle cars for the rest of their journey. The western terminus of the rail line reached Abilene, Kansas, in 1867. Wichita, Caldwell, Dodge City, and other towns in Kansas took their turns as the main destination for cattle driven north on trails from Texas. The most famous route from Texas northward to the rail line was the Chisholm Trail, which began near Brownsville at the Mexican border and extended northward through Texas.

Cattle ranching declined in importance during the 1880s after it came into conflict with sedentary agriculture. Most early U.S. ranchers adhered to "the Code of the West," although the system had no official legal status. Under the code, ranchers had range rights—that is, their cattle could graze on any open land and had access to scarce water sources and grasslands. The early cattle ranchers in the West owned little land, only cattle. The U.S. government, which owned most of the land used for open grazing, began to sell it to farmers to grow crops, leaving cattle ranchers with no legal claim to it. For a few years the ranchers tried to drive out the farmers by cutting fences and then illegally erecting their own fences on public land, and "range wars" flared. The farmers' most potent weapon proved to be barbed wire, first commercially produced in 1873. The farmers eventually won the battle, and ranchers were compelled to buy or lease land to accommodate their cattle. Large cattle ranches were established, primarily on land that was too dry to support crops. Ironically, 60 percent of cattle grazing today takes place on land leased from the U.S. government.

With the spread of irrigation techniques and hardier crops, land in the United States has been converted from ranching to crop growing. Ranching generates lower income per area of land, although it has lower operating costs. Cattle are still raised on ranches but are frequently sent for fattening to farms or to local feed lots along major railroad and highway routes rather than directly to meat processors.

Commercial ranching is conducted in several other MDCs (Figure 10-21). The interior of Australia was opened for grazing in the nineteenth century, although sheep are more common than cattle. Ranching is rare in Europe, except in Spain and Portugal. In South America, a large portion of the pampas of Argentina, southern Brazil, and Uruguay are devoted to grazing cattle and sheep. The cattle industry grew rapidly in Argentina in part because the land devoted to ranching was relatively accessible to the ocean, making it possible for meat to be transported to overseas markets.

Ranching has followed similar stages around the world. First was the herding of animals over open ranges, in a seminomadic style. Then ranching was transformed into fixed farming by dividing the open land into ranches. When many of the farms converted to growing crops, ranching was confined to the drier lands. To survive, the remaining ranches experimented with new methods of breeding and sources of water and feed. Ranching has become part of the meat-processing industry rather than an economic activity carried out on isolated farms. In this way,

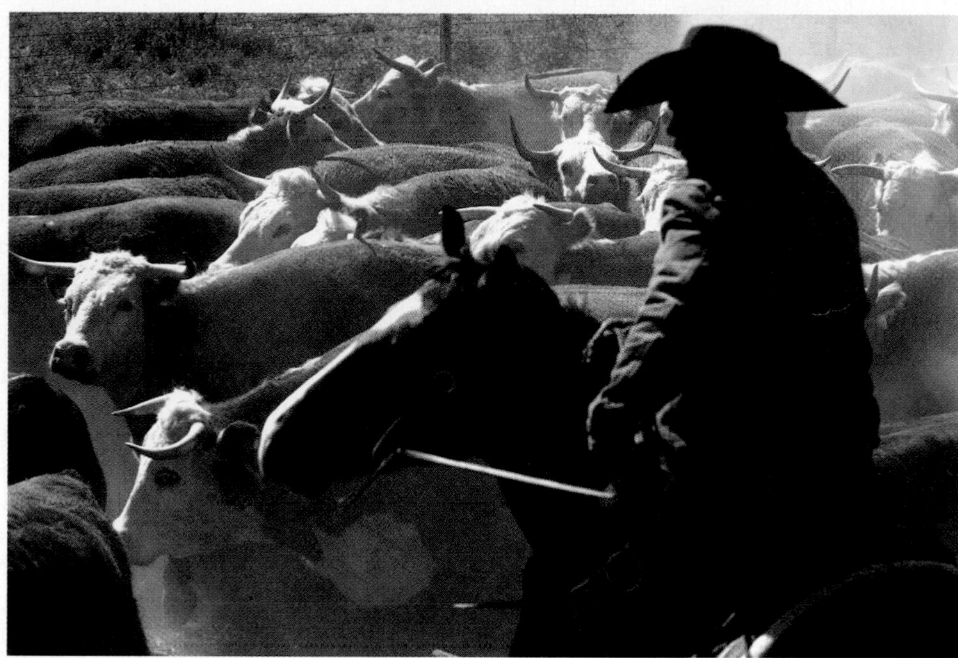

FIGURE 10-20 Rancher, North Dakota. Cattle are being branded.

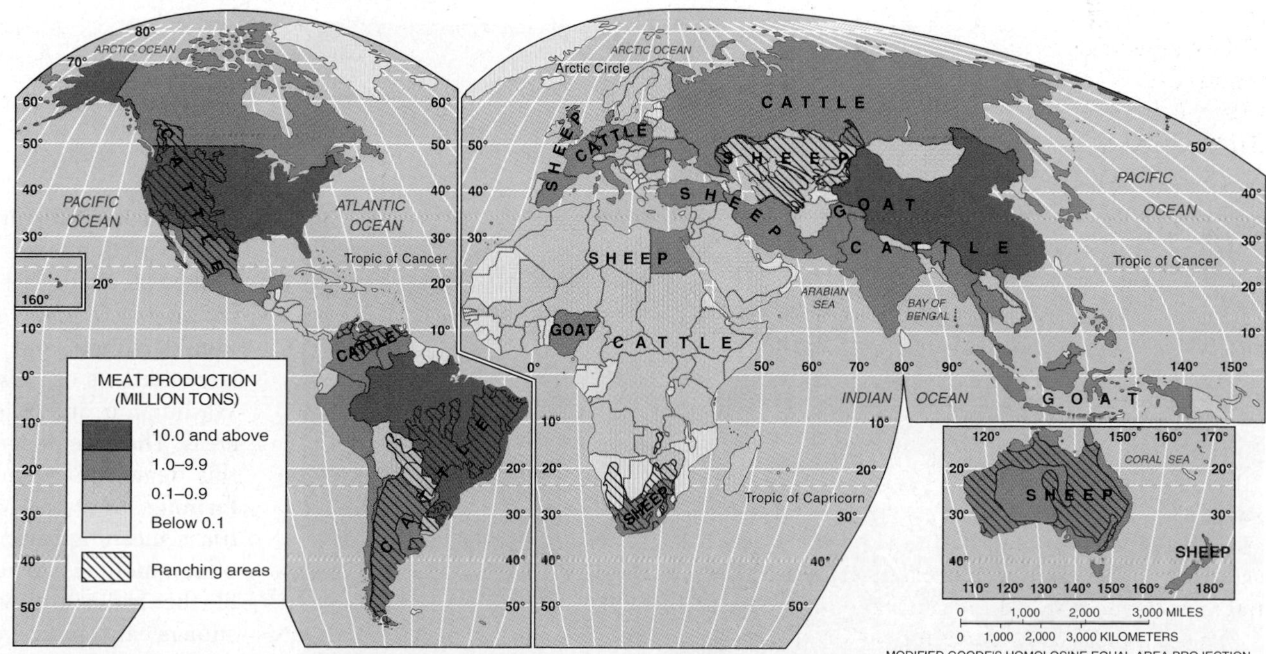

FIGURE 10-21 Meat production. Cattle, sheep, and goats are the three animals most commonly found on ranches. Cattle are found on ranches in the Western Hemisphere, sheep in Australia, and goats in Central Asia.

commercial ranching differs from pastoral nomadism, the form of animal herding practiced in less developed regions.

Mediterranean Agriculture

Mediterranean agriculture exists primarily on the lands that border the Mediterranean Sea in Southern Europe, North Africa, and western Asia (Figure 10-22). Farmers in California, central Chile, the southwestern part of South Africa, and southwestern Australia practice Mediterranean agriculture as well.

These Mediterranean areas share a similar physical environment (refer to Figure 10-4). Every Mediterranean area borders a sea and most are on west coasts of continents (except for some lands surrounding the Mediterranean Sea). Prevailing sea winds provide moisture and moderate the winter temperatures. Summers are hot and dry, but sea breezes provide some relief. The land is very hilly, and mountains frequently plunge directly to the sea, leaving very narrow strips of flat land along the coast.

Farmers derive a smaller percentage of income from animal products in the Mediterranean region than in the mixed crop and livestock region. Livestock production is hindered during the summer by the lack of water and good grazing land. Some farmers living along the Mediterranean Sea traditionally used transhumance to raise animals, although the practice is now less common. Under transhumance, animals—primarily sheep and goats—are kept on the coastal plains in the winter and transferred to the hills in the summer.

Most crops in Mediterranean lands are grown for human consumption rather than for animal feed. **Horticulture**—which is the growing of fruits, vegetables, and flowers—and tree crops form the commercial base of Mediterranean farming. A combination of local physical and cultural characteristics determines which crops are grown in each area. The hilly landscape encourages farmers to plant a variety of crops within one farming area.

In the lands bordering the Mediterranean Sea, the two most important cash crops are olives and grapes. Two-thirds of the world's wine is produced in countries that border the Mediterranean, especially Italy, France, and Spain. Mediterranean agricultural regions elsewhere in the world produce most of the remaining one-third. The lands near the Mediterranean Sea are also responsible for a large percentage of the world's supply of olives, an important source of cooking oil. Despite the importance of olives and grapes to commercial farms bordering the Mediterranean Sea, approximately half of the land is devoted to growing cereals, especially wheat for pasta and bread. As in the U.S. winter-wheat belt, the seeds are sown in the fall and harvested in early summer. After cultivation, cash crops are planted on some of the land, whereas the remainder is left fallow for a year or two to conserve moisture in the soil.

Cereals occupy a much lower percentage of the cultivated land in California than in other Mediterranean climates. Instead, a large portion of California farmland is devoted to fruit and vegetable horticulture, which supplies much of the citrus fruits, tree nuts, and deciduous fruits consumed in the United States. Horticulture is practiced in other Mediterranean climates, but not to the extent found in California. The rapid growth of urban areas in California, especially Los Angeles, has converted high-quality agricultural land into housing developments. Thus far, the loss of farmland has been offset by the expansion of agriculture into arid lands. However, farming in dry lands requires massive irrigation to provide water. In the future, agriculture may face stiffer competition for the Southwest's increasingly scarce water supply.

FIGURE 10-22 Mediterranean agriculture farmer, Spain. Most of the world's olives are grown in the Mediterranean agriculture region.

Commercial Gardening and Fruit Farming

Commercial gardening and fruit farming is the predominant type of agriculture in the U.S. Southeast (Figure 10-23). The region has a long growing season and humid climate and is accessible to the large markets of New York, Philadelphia, Washington, and other eastern U.S. urban areas. The type of agriculture practiced in this region is frequently called **truck farming**, from the Middle English word *truck*, meaning bartering or the exchange of commodities. Truck farms grow many of the fruits and vegetables that consumers in more developed societies demand, such as apples, asparagus, cherries, lettuce, mushrooms, and tomatoes. Some of these fruits and vegetables are

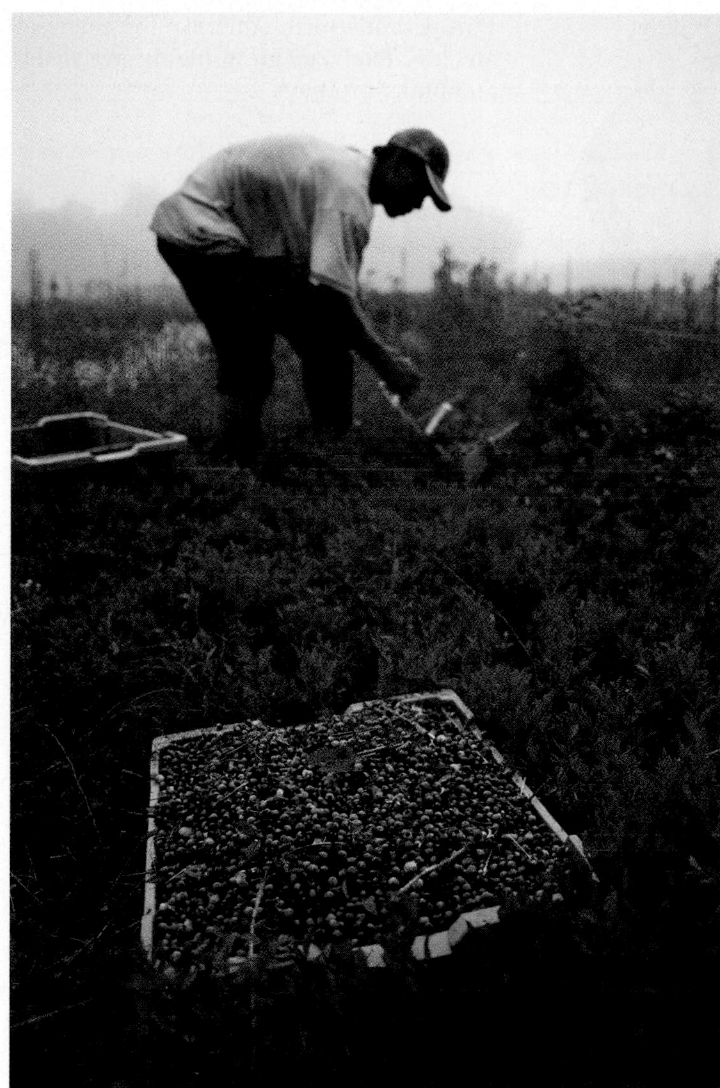

FIGURE 10-23 Truck farmer, Maine. Specialty fruits such as blueberries are grown in the commercial gardening and fruit farming region along the East Coast of the United States.

sold fresh to consumers, but most are sold to large processors for canning or freezing.

Truck farms are highly efficient large-scale operations that take full advantage of machines at every stage of the growing process. Truck farmers are willing to experiment with new varieties, seeds, fertilizers, and other inputs to maximize efficiency. Labor costs are kept down by hiring migrant farm workers, some of whom are undocumented immigrants from Mexico who work for very low wages. Farms tend to specialize in a few crops, and a handful of farms may dominate national output of some fruits and vegetables.

A form of truck farming called specialty farming has spread to New England. Farmers are profitably growing crops that have limited but increasing demand among affluent consumers, such as asparagus, peppers, mushrooms, strawberries, and nursery plants. Specialty farming represents a profitable alternative for New England farmers, at a time when dairy farming is declining because of relatively high operating costs and low milk prices.

KEY ISSUE 4
Why Do Farmers Face Economic Difficulties?

- Challenges for Commercial Farmers
- Challenges for Subsistence Farmers
- Strategies to Increase Food Supply

Commercial and subsistence farmers face comparable challenges. Both commercial and subsistence farmers have difficulty generating enough income to continue farming. The underlying reasons, though, are different. Commercial farmers can produce a surplus of food, whereas many subsistence farmers are barely able to produce enough food to survive. ■

Challenges for Commercial Farmers

Commercial farmers are in some ways victims of their own success. Having figured out how to produce large quantities of food, they face low prices for their output. Government subsidies help prop up farm income, but many believe that the future health of commercial farming rests with embracing more sustainable practices.

Importance of Access to Markets

Because the purpose of commercial farming is to sell produce off the farm, the distance from the farm to the market influences the farmer's choice of crop to plant. Geographers use the von Thünen model to help explain the importance of proximity to market in the choice of crops on commercial farms.

Johann Heinrich von Thünen, an estate owner in northern Germany, first proposed the model in 1826 in a book titled *The Isolated State* (Figure 10-24). According to the model, which was later modified by geographers, a commercial farmer initially considers which crops to cultivate and which animals to raise based on market location. In choosing an enterprise, the farmer compares two costs—the cost of the land versus the cost of transporting products to market.

Von Thünen based his general model of the spatial arrangement of different crops on his experiences as owner of a large estate in northern Germany during the early nineteenth century. He found that specific crops were grown in different rings around the cities in the area. Market-oriented gardens and milk producers were located in the first ring out from the cities. These products are expensive to deliver and must reach the market quickly because they are perishable. The next ring out from the cities contained wood lots, where timber was cut for construction and fuel; closeness to market is important for this commodity because of its weight. The

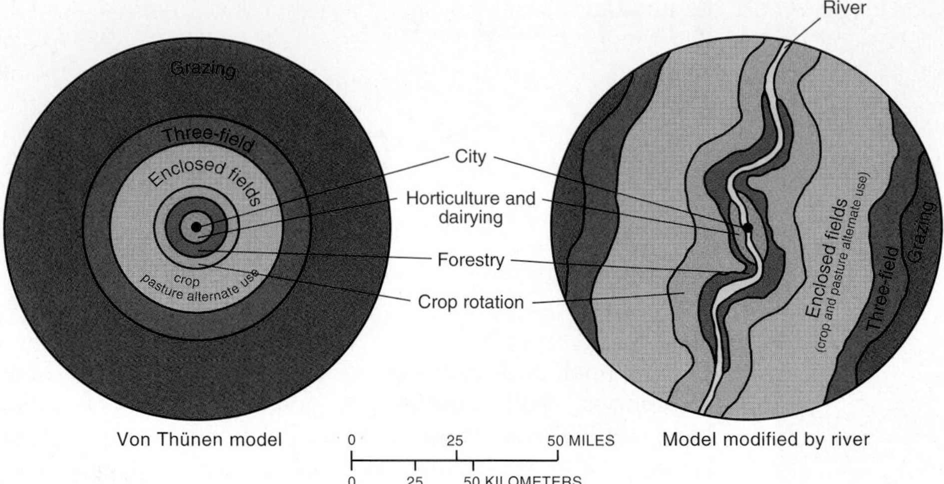

FIGURE 10-24 (left) Von Thünen model of the role of situation factors in choice of crop. According to the von Thünen model, in the absence of topographic factors, different types of farming are conducted at different distances from a city, depending on the cost of transportation and the value of the product. (right) Von Thünen recognized that his model would be modified by site factors, such as a river in this sketch, which changes the accessibility of different land parcels to the market center. Agricultural uses that seek highly accessible locations need to locate nearer the river. The following example illustrates the influence of transportation cost on the profitability of growing wheat:

1. Gross profit from sale of wheat grown on 1 hectare of land not including transportation costs:
   a. Wheat can be grown for $0.25 per kilogram.
   b. Yield per hectare of wheat is 1,000 kilograms.
   c. Gross profit is $250 per hectare ($0.25 per kilogram × 1,000 kilograms per hectare).
2. Net profit from sale of wheat grown on 1 hectare of land *including* transportation costs:
   a. Cost of transporting 1,000 kilograms of wheat to the market is $62.50 per kilometer.
   b. Net profit from the sale of 1,000 kilograms of wheat grown on a farm located 1 kilometer from the market is $187.50 ($250 gross profit – $62.50 per kilometer transport costs).
   c. Net profit from sale of 1,000 kilograms of wheat grown on a farm located 4 kilometers from the market is $0 ($250 gross profit – $62.50 per kilometer × 4 kilometers).

The example shows that a farmer would make a profit growing wheat on land located less than 4 kilometers from the market. Beyond 4 kilometers, wheat is not profitable because the cost of transporting it exceeds the gross profit. These calculations demonstrate that farms located closer to the market tend to select crops with higher transportation costs per hectare of output, whereas more distant farms are more likely to select crops that can be transported less expensively.

next rings were used for various crops and for pasture; the specific commodity was rotated from one year to the next. The outermost ring was devoted exclusively to animal grazing, which requires lots of space.

The model assumed that all land in a study area had similar site characteristics and was of uniform quality, although von Thünen recognized that the model could vary according to topography and other distinctive physical conditions. For example, a river might modify the shape of the rings because transportation costs change when products are shipped by water routes rather than over roads. The model also failed to consider that social customs and government policies influence the attractiveness of plants and animals for a commercial farmer.

Although von Thünen developed the model for a small region with a single market center, it is also applicable on a national or global scale. Farmers in relatively remote locations who wish to sell their output in the major markets of Western Europe and North America, for example, are less likely to grow highly perishable and bulky products.

Overproduction in Commercial Farming

Commercial farmers suffer from low incomes because they are capable of producing much more food than is demanded by consumers in MDCs. A surplus of food can be produced because of widespread adoption of efficient agricultural practices. New seeds, fertilizers, pesticides, mechanical equipment, and management practices have enabled farmers to obtain greatly increased yields per area of land. The experience of dairy farming in the United States demonstrates the growth in productivity. The number of milk cows in the United States decreased from 10.8 million to 9.3 million between 1980 and 2008. But milk production increased from 128 billion to 190 billion pounds—yield per cow thus nearly doubled in the period.

Although the food supply has increased in MDCs, demand has remained constant, because the market for most products is already saturated. In MDCs, consumption of a particular commodity may not change significantly if the price changes. Americans, for example, do not switch from Wheaties to Corn Flakes if the price of corn falls more rapidly than wheat. Demand is also stagnant for most agricultural products in MDCs because of low population growth.

The U.S. government has three policies that are supposed to address the problem of excess productive capacity:

1. **Farmers are encouraged to avoid producing crops that are in excess supply.** Because soil erosion is a constant threat, the government encourages planting fallow crops, such as clover, to restore nutrients to the soil and to help hold the soil in place. These crops can be used for hay, forage for pigs, or to produce seeds for sale.
2. **The government pays farmers when certain commodity prices are low.** The government sets a target price for the commodity and pays farmers the difference between the price they receive in the market and a target price set by the government as a fair level for the commodity. The target prices are calculated to give farmers the same price for the commodity today as in the past, when compared to other consumer goods and services.
3. **The government buys surplus production and sells or donates it to foreign governments.** In addition, low-income Americans receive food stamps in part to stimulate their purchase of additional food.

The United States has averaged about $16 billion a year on farm subsidies in recent years. Annual spending varies considerably from one year to the next: Subsidy payments are lower in years when market prices rise and production is down, typically as a result of poor weather conditions in the United States or political problems in other countries. Farming in Europe is subsidized even more than in the United States. More farmers receive subsidies in Europe, and they receive more than American farmers. The high subsidies are a legacy of a long-standing commitment by the European Union to maintain agriculture in its member states, especially in France. Supporters point to the preservation of rural village life in parts of Europe, while critics charge that Europeans pay needlessly high prices for food as a result of the subsidies.

Government policies in MDCs point out a fundamental irony in worldwide agricultural patterns. In an MDC such as the United States, farmers are encouraged to grow less food, whereas LDCs struggle to increase food production to match the rate of growth in the population.

Sustainable Agriculture

Some commercial farmers are converting their operations to **sustainable agriculture**, an agricultural practice that preserves and enhances environmental quality (Figure 10-25). Farmers practicing sustainable agriculture typically generate lower revenues than do conventional farmers, but they also have lower costs.

An increasingly popular form of sustainable agriculture is organic farming. However, some organic farms, especially the larger ones, may rely in part on nonsustainable practices, such as use of fossil fuels to operate tractors. Worldwide, 32.2 million hectares (79 million acres), or 0.24 percent of farmland, was classified as organic in 2007. Australia was the leader, with 12 million of the hectares, or 37 percent of the worldwide total. Argentina, Brazil, the United States, China, Italy, India, Spain, Uruguay, and Germany together accounted for 40 percent of the worldwide total.

Three principal practices distinguish sustainable agriculture (and at its best, organic farming) from conventional agriculture:

- Sensitive land management
- Limited use of chemicals
- Better integration of crops and livestock

SENSITIVE LAND MANAGEMENT. Sustainable agriculture protects soil in part through **ridge tillage**, which is a system of planting crops on ridge tops. Crops are planted on 10-to 20-centimeter (4- to 8-inch) ridges that are formed during cultivation or after harvest. The crop is planted on the same ridges, in the same rows, year after year. Ridge tillage is attractive for two main reasons—lower production costs and greater soil conservation.

Production costs are lower with ridge tillage in part because it requires less investment in tractors and other machinery than conventional planting. An area that would be prepared for planting under conventional farming with three to five tractors can be prepared for ridge tillage with only one or two tractors. The primary tillage tool is a row-crop cultivator that can form ridges. There is no need for a plow, or field cultivator, or a 300-horsepower four-wheel-drive tractor. With ridge tillage, the space between rows needs to match the distance between wheels of the machinery. If 75 centimeters (30 inches) are left between rows, tractor tires will typically be on 150-centimeter (60-inch) centers and combine wheels on 300-centimeter (120-inch) centers. Wheel spacers are available from most manufacturers to fit the required spacing.

Ridge tillage features a minimum of soil disturbance from harvest to the next planting. A compaction-free zone is created under each ridge and in some row middles. Keeping the trafficked area separate from the crop-growing area improves soil properties. Over several years the soil will tend to have increased organic matter, greater water-holding capacity and more earthworms. The channels left by earthworms and decaying roots enhance drainage.

Ridge tillage compares favorably with conventional farming for yields while lowering the cost of production. Although more labor-intensive than other systems, it is profitable on a per-acre basis. In Iowa, for example, ridge tillage has gained favor for production of organic and herbicide-free soybeans, which sell for more than regular soybeans.

FIGURE 10-25 Organic farmer, Iowa. Pigs on this organic farm are fed a mixture of wheat, oats, and tofu (soybeans) grown on the farm.

LIMITED USE OF CHEMICALS. In conventional agriculture, seeds are often genetically modified to survive when herbicides and insecticides are sprayed on

GLOBAL FORCES, LOCAL IMPACTS

Genetically Modified Foods and Sub-Saharan Africa

Sub-Saharan African countries have been urged by the United States to increase their food supply in part through increased use of genetic modification (GM) of crops and livestock (Figure 10-26). Africans are divided on whether to accept GM organisms.

Farmers have been manipulating crops and livestock for thousands of years: The very nature of agriculture is to deliberately manipulate nature. Humans control selective reproduction of plants and animals in order to produce a larger number of stronger, hardier survivors. Beginning in the nineteenth century, the science of genetics expanded understanding of how to manipulate plants and animals to secure dominance of the most favorable traits. However, GM, which became widespread in the late twentieth century, marks a sharp break with the agricultural practices of the past several thousand years. Under GM the genetic composition of an organism is not merely studied, it is actually altered, for GM involves mixing genetic material of two or more species that would not otherwise mix in nature.

GM is especially widespread in the United States: 89 percent of soybeans, 83 percent of cotton, and 61 percent of maize; three-fourths of the processed food that Americans consume has at least one GM ingredient. Worldwide, 102 million hectares of farmland were devoted to GM in 2006, three times more land than was devoted to organic farming. The United States was responsible for 53 percent of the world's GM crops in 2006, and Canada another 6 percent. Argentina was second to the United States, accounting for 18 percent of the world's GM crops.

Africans must weigh arguments both for and against adoption of GM. The positives of GM are higher yields, increased nutrition, and more resistance to pests. GM foods are also better tasting, at least to some palates. Despite these benefits, opposition to GM is strong in Africa for several reasons:

- **Health problems.** Consuming large quantities of GM may reduce the effectiveness of antibiotics and could destroy long-standing ecological balances in local agriculture.
- **Export problems.** European countries, the main markets for Africa's agricultural exports, require GM foods to be labeled. Europeans are especially strongly opposed to GM because they believe the food is not as nutritious as that from traditionally bred crops and livestock. Because European consumers shun GM food, African farmers fear that if they are no longer able to certify their exports as GM-free, European customers will stop buying them.
- **Increased dependence on the United States.** U.S.-based transnational corporations, such as Monsanto, manufacture most of the GM seeds. Africans fear that the biotech companies could—and would—introduce a so-called "terminator" gene in the GM seeds, to prevent farmers from replanting them after harvest and require them to continue to purchase seeds year after year from the transnational corporations.

"We don't want to create a habit of using genetically modified maize that the country cannot maintain," explained Mozambique's prime minister. If agriculture is regarded as a way of life, not just a food production business, GM represents for many Africans an unhealthy level of dependency on MDCs. ■

FIGURE 10-26 Genetically modified potatoes may not be sold in Japan.

the fields to kill weeds and insects. These are known as "Roundup-Ready" seeds, because its creator Monsanto Corp. sells its weedkillers under the brand name "Roundup." "Roundup-Ready" seeds were planted in 80 percent of all soybean acreage, 54 percent of all cotton acreage, and 12 percent of all corn acreage in the United States in 2003. Aside from adverse impacts on soil and water quality, widespread use of "Roundup-Ready" seeds is causing some weeds to become resistant to the herbicide.

Sustainable agriculture, on the other hand, involves application of limited if any herbicides to control weeds. In principle, farmers can control weeds without chemicals, although it requires additional time and expense that few farmers can afford. Researchers have found that combining mechanical weed control with some chemicals yields higher returns per acre than relying solely on one of the two methods.

Ridge tilling also promotes decreased use of chemicals, which can be applied only to the ridges and not the entire field. Combining herbicide banding—which applies chemicals in narrow bands over crop rows—with cultivating may be the best option for many farmers.

INTEGRATED CROP AND LIVESTOCK. Sustainable agriculture attempts to integrate the growing of crops and the raising of livestock as much as possible at the level of the individual farm. Animals consume crops grown on the farm and are not confined to small pens. In conventional farming, integration between crops and livestock generally takes place through intermediaries rather than inside an individual farm. As discussed earlier in the chapter, mixed crop and livestock is a common form of farming in many LDCs and in the Corn Belt of the United States. But many farmers in the mixed crop and livestock region actually choose to only grow crops or raise more animals than the crops they grow can feed. They sell their crops off the farm or purchase feed for their animals from outside suppliers. Integration of crops and livestock reflects a return to the historical practice of mixed crop and livestock farming, in which growing crops and raising animals were regarded as complementary activities on the farm. This was the common practice for centuries until the mid-1900s when technology, government policy, and economics encouraged farmers to become more specialized.

Sustainable agriculture is sensitive to the complexities of biological and economic interdependencies between crops and livestock:

1. **Number of livestock.** The correct number, as well as the distribution, of livestock for an area is determined based on the landscape and forage sources. Prolonged concentration of livestock in a specific location can result in permanent loss of vegetative cover, so the farmer needs to move the animals to reduce overuse in some areas. Growing row crops on the more level land while confining pastures to steeper slopes will reduce soil erosion, so it may be necessary to tolerate some loss of vegetation in specific locations. The farmer may need to balance the need to secure livestock inside fences with the convenience of tilling large unfenced fields through the use of temporary fencing.
2. **Animal confinement.** The moral and ethical debate over animal welfare is particularly intense regarding confined livestock production systems. Confined livestock are a source of surface and ground water pollutants, particularly where the density of animals is high. Expensive waste management facilities are a necessary cost of confined production systems. If animals are not confined, manure can contribute to soil fertility. However, quality of life in nearby communities may be adversely affected by the smell.
3. **Management of extreme weather conditions.** Herd size may need to be reduced during periods of short- and long-term droughts. On the other hand, livestock can buffer the negative impacts of low rainfall periods by consuming crops that in conventional farming would be left as failures. Especially in Mediterranean climates such as California's, properly managed grazing significantly reduces fire hazards by reducing fuel buildup in grasslands and brushlands.
4. **Flexible feeding and marketing.** This can help cushion farmers against trade and price fluctuations and, in conjunction with cropping operations, make more efficient use of farm labor. Feed costs are the largest single variable cost in any livestock operation. Most of the feed may come from other enterprises on the ranch, though some is usually purchased off the farm. Feed costs can be kept to a minimum by monitoring animal condition and performance and understanding seasonal variations in feed and forage quality on the farm.

Challenges for Subsistence Farmers

Two issues discussed in earlier chapters influence the choice of crops planted by subsistence farmers:

- Subsistence farmers must feed an increasing number of people because of rapid population growth in LDCs (discussed in Chapter 2).
- Subsistence farmers must grow food for export instead of for direct consumption due to the adoption of the international trade approach to development (discussed in Chapter 9).

Subsistence Farming and Population Growth

Population growth influences the distribution of types of subsistence farming, according to economist Ester Boserup. It compels subsistence farmers to consider new farming approaches that produce enough food to take care of the additional people.

For hundreds if not thousands of years, subsistence farming in LDCs yielded enough food for people living in rural villages to survive, assuming no drought, flood, or other natural disaster occurred. Suddenly in the late twentieth century, the LDCs needed to provide enough food for a rapidly increasing population as well as for the growing number of urban residents who cannot grow their own food. According to Boserup, subsistence farmers increase the supply of food through intensification of production, achieved in two ways:

1. **Adoption of new farming methods.** Plows replace axes and sticks. More weeding is done, more manure applied, more terraces carved out of hillsides, and more irrigation ditches dug. The additional labor needed to perform these operations comes from the population growth. The farmland yields more food per area of land, but with the growing population, output per person remains about the same.

2. **Land is left fallow for shorter periods.** This expands the amount of land area devoted to growing crops at any given time. Boserup identified five basic stages in the intensification of farmland:
   - Forest Fallow. Fields are cleared and utilized for up to 2 years and left fallow for more than 20 years, long enough for the forest to grow back.
   - Bush Fallow. Fields are cleared and utilized for up to 8 years and left fallow for up to 10 years, long enough for small trees and bushes to grow back.
   - Short Fallow. Fields are cleared and utilized for perhaps 2 years (Boserup was uncertain) and left fallow for up to 2 years, long enough for wild grasses to grow back.
   - Annual Cropping. Fields are used every year and rotated for a few months with planting legumes and roots.
   - Multicropping. Fields are used several times a year and never left fallow.

Contrast shifting cultivation, practiced in regions of low population density, such as central Africa, with intensive subsistence agriculture, practiced in regions of high population density, such as East Asia. Under shifting cultivation, cleared fields are utilized for a couple of years, then left fallow for 20 years or more. This type of agriculture supports a small population living at low density. As the number of people living in an area increases (that is, the population density increases) and more food must be grown, fields will be left fallow for shorter periods of time. Eventually, farmers achieve the very intensive use of farmland characteristic of areas of high population density.

Subsistence Farming and International Trade

To expand production, subsistence farmers need higher-yield seeds, fertilizer, pesticides, and machinery. Some needed supplies can be secured by trading food with urban dwellers. For many African and Asian countries, though, the main way to obtain agricultural supplies is to import them from other countries. However, they lack the money to buy agricultural equipment and materials from MDCs.

To generate the funds they need to buy agricultural supplies, LDCs must produce something they can sell in MDCs. The LDCs sell some manufactured goods (see Chapter 11), but most raise funds through the sale of crops in MDCs. Consumers in MDCs are willing to pay high prices for fruits and vegetables that would otherwise be out of season or for crops such as coffee and tea that cannot be grown in MDCs because of the climate.

In an LDC such as Kenya, families may divide by gender between traditional subsistence agriculture and contributing to international trade. Women practice most of the subsistence agriculture—that is, growing food for their families to consume—in addition to the tasks of cooking, cleaning, and carrying water from wells. Men may work for wages, either growing crops for export or at jobs in distant cities. Because men in Kenya frequently do not share the wages with their families, many women try to generate income for the household by making clothes, jewelry, baked goods, and other objects for sale in local markets.

The sale of export crops brings an LDC foreign currency, a portion of which can be used to buy agricultural supplies. But governments in LDCs face a dilemma: The more land that is devoted to growing export crops, the less that is available to grow crops for domestic consumption. Rather than helping to increase productivity, the funds generated through the sale of export crops may be needed to feed the people who switched from subsistence farming to growing export crops.

Drug Crops

The export crops chosen in some LDCs, especially in Latin America and Asia, are those that can be converted to drugs. Marijuana, the most popular drug, is estimated to be used by 140 million worldwide. Cocaine and heroin, the two leading, especially dangerous drugs, are abused by 15 million and 14 million people, respectively, worldwide. The United Nations estimated that in 1998 the incomes of 4 million people, primarily in Asia and Latin America, were dependent on cultivation of the opium poppy or coca leaf (Figure 10-27).

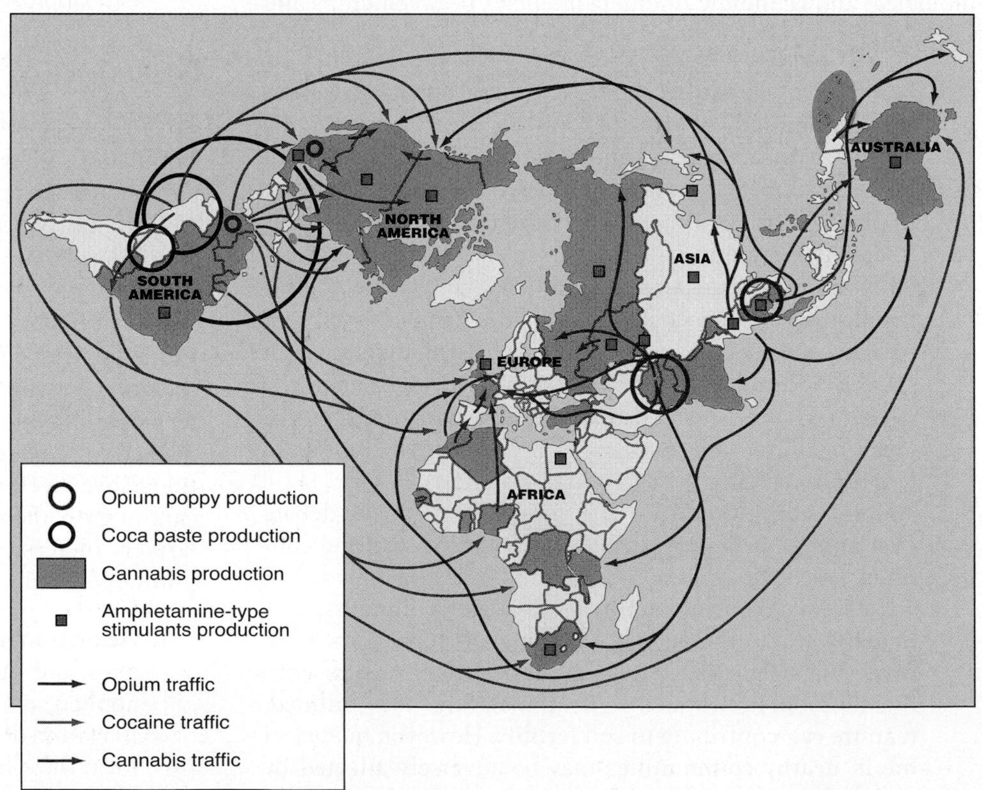

FIGURE 10-27 Drug trade. Most of the world's opium comes from Afghanistan, and most of the world's cocaine originates in Colombia.

Heroin is derived from raw opium gum, which is produced by the opium poppy plant. Afghanistan is the source of around 80 percent of the world's opium; most of the remainder is grown in Myanmar (Burma). Most consumers are located in Central Asia. One-half of the world's coca leaf is grown in Columbia, and most of the remainder in neighboring Peru and Bolivia. Most of the processing of cocaine, as well as its distribution to the United States and other MDCs, is based in Colombia. Marijuana, produced from the *Cannabis sativa* plant, is cultivated widely around the world. The overwhelming majority of the marijuana that reaches the United States is grown in Mexico. Cultivation of *C. sativa* is not thought to be expanding worldwide, whereas opium poppies and coca leaf are.

Strategies to Increase the Food Supply

Four strategies are being employed to increase the world's food supply:

- Expanding the land area used for agriculture
- Increasing the productivity of land now used for agriculture
- Identifying new food sources
- Increasing exports from other countries

Challenges underlie each of these strategies.

Expanding Agricultural Land

Historically, world food production has increased primarily by expanding the amount of land devoted to agriculture (Figure 10-28). When the world's population began to increase more rapidly in the late eighteenth and early nineteenth centuries, during the Industrial Revolution, pioneers could migrate to uninhabited territory and cultivate the land. Sparsely inhabited land suitable for agriculture was available in western North America, central Russia, and Argentina's pampas.

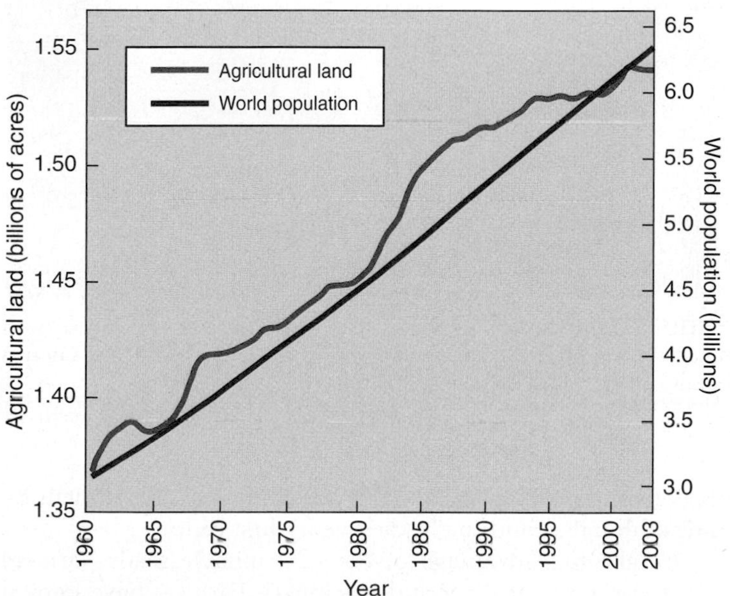

FIGURE 10-28 Agricultural land and population. The world's amount of agricultural land increased more rapidly than population through the late twentieth century. In the first years of the twenty-first century, agricultural land is expanding more slowly than population.

Two centuries ago people believed that good agricultural land would always be available for willing pioneers. Today few scientists believe that further expansion of agricultural land can feed the growing world population. At first glance, new agricultural land appears to be available because only 11 percent of the world's land area is currently cultivated. In fact, cultivated land has been expanding in Africa at a rate of 1 percent per year. But population in Africa is increasing more than 2 percent per year. Worldwide, despite the recent decline in the natural increase, agricultural land is expanding more slowly than population.

In some regions, farmland is abandoned for lack of water (Figure 10-29). Especially in semiarid regions, human actions are causing land to deteriorate to a desertlike condition, a process called **desertification** (more precisely, semiarid land degradation). Semiarid lands that can support only a handful of pastoral nomads are overused because of rapid population growth. Excessive crop planting, animal grazing, and tree cutting exhaust the soil's nutrients and preclude agriculture. The Earth Policy Institute estimates that 2 billion hectares (5 million acres) of land have been degraded around the world. Overgrazing is thought to be responsible for 34 percent of the total, deforestation for 30 percent, and agricultural use for 28 percent. The United Nations estimates that desertification removes 27 million hectares (70 million acres) of land from agricultural production each year, an area roughly equivalent to Colorado.

Excessive water threatens other agricultural areas, especially drier lands that receive water from human-built irrigation systems. If the irrigated land has inadequate drainage, the underground water level rises to the point where roots become waterlogged. The United Nations estimates that 10 percent of all irrigated land is waterlogged, mostly in Asia and South America. If the water is salty, it can damage plants. The ancient civilization of Mesopotamia may have collapsed in part because of waterlogging and excessive salinity in its agricultural lands near the Tigris and Euphrates rivers.

Urbanization can also contribute to reducing agricultural land. As urban areas grow in population and land area, farms on the periphery are replaced by homes, roads, shops, and other urban land uses. In North America, farms outside urban areas are left idle until the speculators who own them can sell them at a profit to builders and developers, who convert the land to urban uses.

Increasing Productivity

Population grew at the fastest rate in human history during the second half of the twentieth century, as discussed in Chapter 2. Many experts forecast massive global famine, but these dire predictions did not come true. New agricultural practices have permitted farmers worldwide to achieve

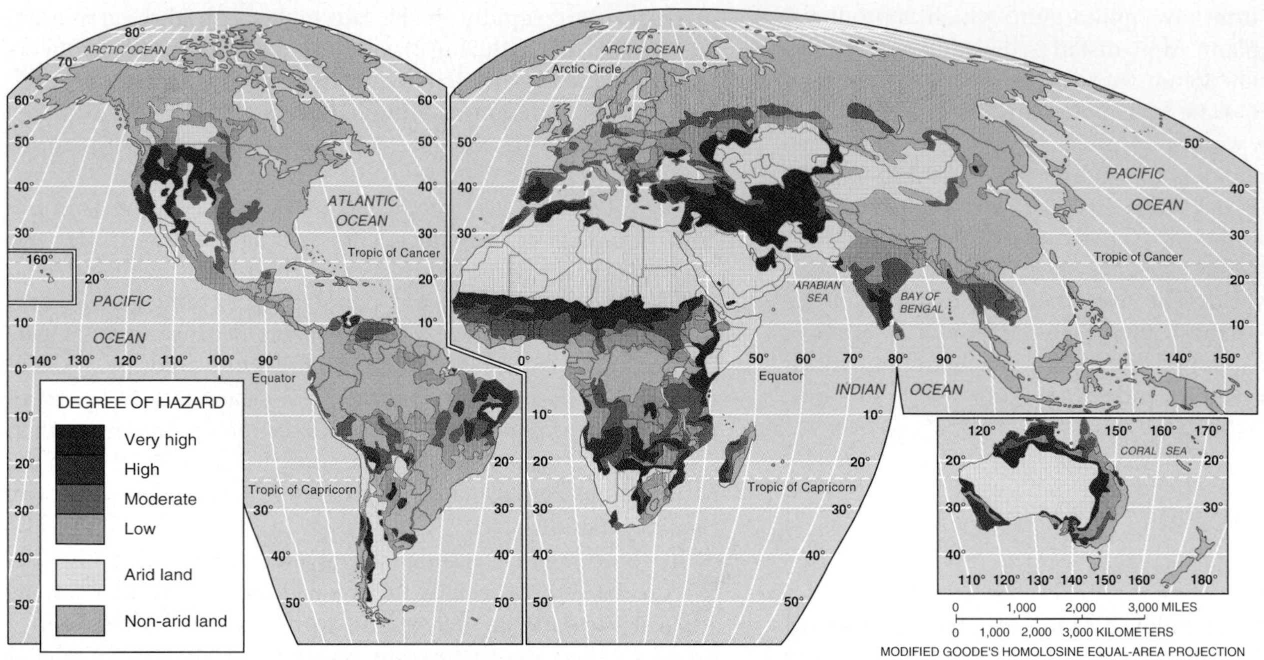

FIGURE 10-29 Desertification (semiarid land degradation). The most severe problems are in northern Africa, central Australia, and the southwestern parts of Africa, Asia, North America, and South America.

much greater yields from the same amount of land. The invention and rapid diffusion of more productive agricultural techniques during the 1970s and 1980s is called the **green revolution**. The green revolution involves two main practices—the introduction of new higher-yield seeds and the expanded use of fertilizers. Because of the green revolution, agricultural productivity at a global scale has increased faster than population growth.

Scientists began an intensive series of experiments during the 1950s to develop a higher-yield form of wheat (Figure 10-30). A decade later, the "miracle wheat seed" was ready. Shorter and stiffer than traditional breeds, the new wheat was less sensitive to variation in day length, responded better to fertilizers, and matured faster. The Rockefeller and Ford foundations sponsored many of the studies, and the program's director, Dr. Norman Borlaug, won the Nobel Peace Prize in 1970. The International Rice Research Institute, established in the Philippines by the Rockefeller and Ford foundations, worked to create a miracle rice seed. During the 1960s, their scientists introduced a hybrid of Indonesian rice and Taiwan dwarf rice that was hardier and that increased yields. More recently, scientists have developed new high-yield maize (corn).

FIGURE 10-30 Green revolution. "Miracle" high-yield seeds have been produced through laboratory experiments at the International Rice Research Institute (IRRI). The IRRI is testing rice varieties in the Philippines.

The new miracle seeds were diffused rapidly around the world. India's wheat production, for example, more than doubled in 5 years. After importing 10 million tons of wheat annually in the mid-1960s, India by 1971 had a surplus of several million tons. Other Asian and Latin American countries recorded similar productivity increases. The green revolution was largely responsible for preventing a food crisis in these regions during the 1970s and 1980s. But will these scientific breakthroughs continue in the twenty-first century?

To take full advantage of the new miracle seeds, farmers must use more fertilizer and machinery. Farmers have known for thousands of years that application of manure, bones, and ashes somehow increases, or at least maintains, the fertility of the land. Not until the nineteenth century did scientists identify nitrogen, phosphorus, and potassium (potash) as the critical elements in these substances that improved fertility. Today these three elements form the basis for fertilizers—products that farmers apply to their fields to enrich the soil by restoring lost nutrients.

Nitrogen, the most important fertilizer, is a ubiquitous substance. China is the leading producer of nitrogen fertilizer. Europeans most commonly produce a fertilizer known as urea, which contains 46 percent nitrogen. In North America, nitrogen is available as ammonia gas, which is 82 percent nitrogen but more awkward than urea to transport and store. Both urea and ammonia gas combine nitrogen and hydrogen. The problem is that the cheapest way to produce both types of nitrogen-based fertilizers is to obtain hydrogen from natural gas or petroleum. As fossil-fuel prices increase, so do the prices for nitrogen-based fertilizers, which then become too expensive for many farmers in LDCs. In contrast to nitrogen, phosphorus and potash reserves are not distributed uniformly across Earth's surface. Phosphate rock reserves are clustered in China, Morocco, and the United States. Proven potash reserves are concentrated in Canada, Russia, and Ukraine.

Farmers need tractors, irrigation pumps, and other machinery to make the most effective use of the new miracle seeds. In LDCs, farmers cannot afford such equipment and cannot, in view of high energy costs, buy fuel to operate the equipment. To maintain the green revolution, governments in LDCs must allocate scarce funds to subsidize the cost of seeds, fertilizers, and machinery.

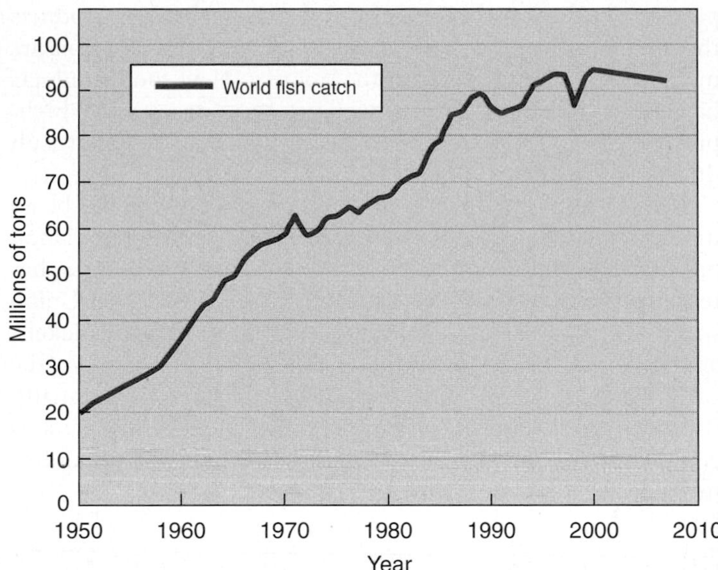

FIGURE 10-31 Fish. The world fish catch has not increased since around 1990.

Identifying New Food Sources

A third alternative for increasing the world's food supply is to develop new food sources. Three strategies being considered are to cultivate the oceans, to develop higher-protein cereals, and to improve palatability of rarely consumed foods.

CULTIVATING OCEANS. At first glance, increased use of food from the sea is attractive. Oceans are vast, covering nearly three-fourths of Earth's surface and lying near most population concentrations. Historically the sea has provided only a small percentage of the world food supply. About two-thirds of the fish caught from the ocean is consumed directly, whereas the remainder is converted to fish meal and fed to poultry and hogs.

Hope grew during the mid-twentieth century that increased fish consumption could meet the needs of a rapidly growing global population. Indeed, the world's annual fish catch increased from around 30 million tons in 1950 to 100 million tons in 1990. However, the population of some fish species declined because they were harvested faster than they could reproduce. Overfishing has been particularly acute in the North Atlantic and Pacific oceans. Because of overfishing, the population of large predatory fish, such as tuna and swordfish, declined by 90 percent in the past half-century. The United Nations estimates that one-quarter of fish stocks have been overfished and one-half fully exploited, leaving only one-fourth underfished. Consequently, the total world fish catch has remained relatively constant since the 1980s despite population growth (Figure 10-31).

To protect fishing areas, many countries have claimed control of the oceans within 200 nautical miles of the coast. These countries have the right to seize foreign fishing vessels that venture into the so-called exclusive economic zone.

DEVELOPING HIGHER-PROTEIN CEREALS. A second possible new food source is higher-protein cereal grains. People in MDCs obtain protein by consuming meat, but people in LDCs generally rely on wheat, corn, and rice, which lack certain proteins. Scientists are experimenting with hybrids of the world's major cereals that have higher protein content. People can also obtain needed nutrition by consuming foods that are fortified during processing with vitamins, minerals, and protein-carrying amino acids. This approach achieves better nutrition without changing food-consumption habits. However, fortification has limited application in LDCs, where most people grow their own food rather than buy processed food.

IMPROVING PALATABILITY OF RARELY CONSUMED FOODS. To fulfill basic nutritional needs, people consume types of food adapted to their community's climate, soil, and other physical characteristics. People also select foods on the basis of religious values, taboos, and other social customs that are unrelated to nutritional or environmental factors. A third way to make more effective use of existing global resources is to encourage consumption of foods that are avoided for social reasons.

A prominent example of an underused food resource in North America is the soybean. Although the soybean is one of the region's leading crops, most of the output is processed into animal feed, in part because many North Americans avoid consuming tofu, sprouts, and other recognizable soybean products. However, burgers, franks, oils, and other products that are made from soybeans but do not look like soybeans are more widely accepted in North America. New food products have been created in LDCs as well. In Asia, for example, high-protein beverages made from seeds resemble popular soft drinks.

Krill (a term for a group of small crustaceans) could be an important source of food from the oceans. The krill population has increased rapidly in recent years, because overhunting has reduced the number of whales that eat krill. The Soviet Union was a major harvester of krill, used primarily to feed chickens and livestock. Since the breakup of the Soviet Union in the early 1990s, the world krill harvest has declined substantially. Because krill deteriorates rapidly new processing methods could substantially increase the harvest for human food; unfortunately, krill does not taste very good.

Increasing Trade

A fourth alternative for increasing the world's food supply is to export more food from countries that produce surpluses (Figure 10-32). The three top export grains are wheat, maize (corn), and rice. Few countries are major exporters of food, but increased production in these countries could cover the gap elsewhere.

Before World War II, Western Europe was the only major grain-importing region. Prior to their independence, colonies of Western European countries supplied food to their parent states. Asia became a net grain importer in the 1950s, Africa and Eastern Europe in the 1960s, and Latin America in the 1970s. Population increases in these regions largely accounted for the need to import grain. By 1980 North America was the only major exporting region in the world. In response to the increasing global demand for food imports, the United States passed Public Law 480, the Agricultural, Trade, and Assistance Act of 1954 (frequently referred to as "P.L.-480"). Title I of the act provided for the sale of grain at low interest rates, and Title II gave grants to needy groups of people. The United States remains the world's leading exporter of grain by a wide margin, accounting for one-third of the total exports of the three leading grains, including more than one-half of all maize exports and more than one-fourth of all wheat exports.

Elsewhere in the world the picture has changed in the twenty-first century. From net importers of grain, South Asia and Southeast Asia have now become net exporters. Thailand has replaced the United States as the leading exporter of rice, accounting for one-third of the world total, followed by India in second place with one-sixth. Vietnam and Pakistan ranked fourth and fifth, respectively, in rice exports in 2004, behind the United States in third place. Japan is by far the world's leading grain importing country, followed by China. Japan is the leading importer of maize and China of wheat, and both rank among leading rice importers. On a regional scale, Southwest Africa (with Northern Africa) has become the leading net importer of all three major grains, and Saudi Arabia was the world's leading importer of rice in 2007. Sub-Saharan Africa also ranks among the leaders in net imports of all three grains.

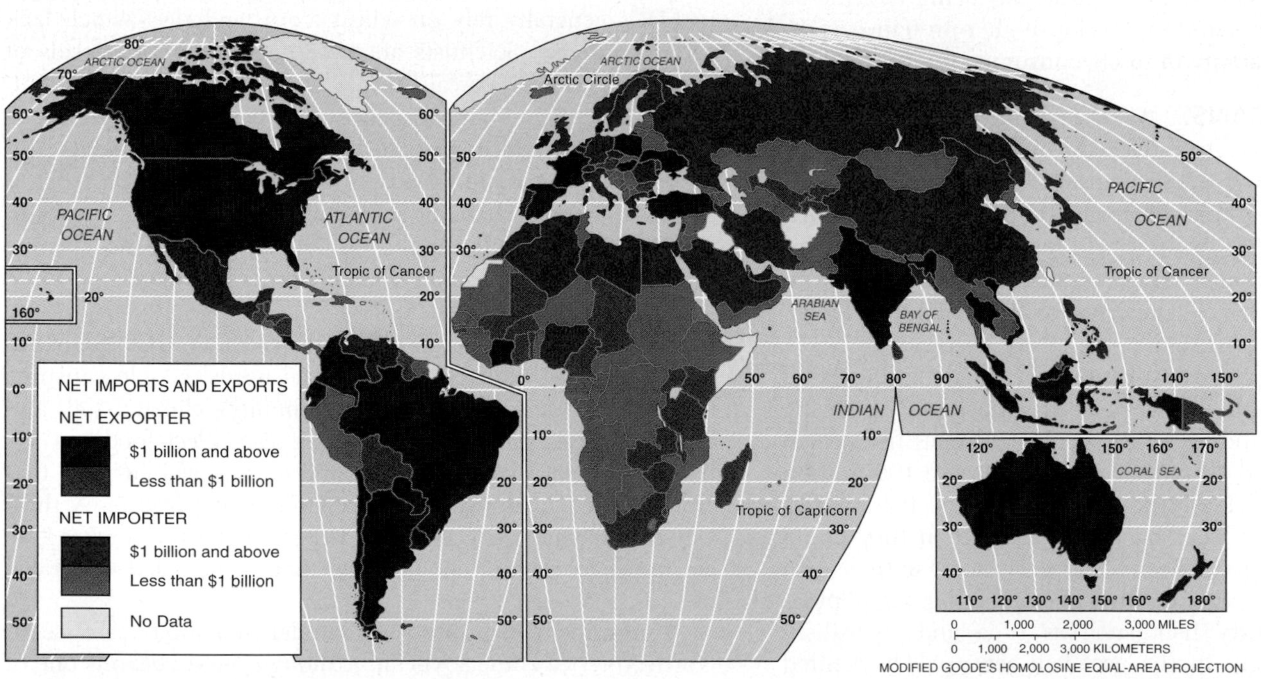

FIGURE 10-32 Grain imports and exports. Most countries must import more food than they export. The United States has by far the largest excess of food exports compared to imports. European countries are also leading food exporters, though they are also leading food importers.

SUMMARY

A country's agricultural system remains one of the best measures of its level of development and standard of material comfort. Despite major changes, agriculture in LDCs still employs the majority of the population, and producing food for local survival is still paramount.

Farming in MDCs directly employs few people, but when manufacturers of food products, supermarkets, restaurants, and other businesses that handle food are considered, then the food industry is actually the largest employer. The production and distribution of food are not primary-sector or agricultural activities, though; they are part of the industrial and service sectors of the economies in MDCs.

Even farming may one day no longer be considered a distinct primary-sector activity in MDCs. True, farmers still deliberately modify the land by planting seeds or grazing animals, but they spend more time sitting at computers, operating sophisticated machinery, and reviewing finances and devising marketing strategies.

Here again are the key questions concerning agricultural geography:

1. **Where Did Agriculture Originate?** Prior to the development of agriculture, people survived by hunting animals, gathering wild vegetation, or fishing. Agriculture was not simply invented, but was the product of thousands of years of experiments and accidents. Current agricultural practices vary between MDCs and LDCs.
2. **Where Are Agricultural Regions in LDCs?** Most people in the world, especially those in LDCs, are subsistence farmers, growing crops primarily to feed themselves. Important types of subsistence agriculture include shifting cultivation, pastoral nomadism, and intensive farming. Regions where subsistence agriculture is practiced are characterized by a large percentage of the labor force engaged in agriculture, with few mechanical aids.
3. **Where Are Agricultural Regions in MDCs?** The most common type of farm found in MDCs is mixed crop and livestock. Where mixed crop and livestock farming is not suitable, commercial farmers practice a variety of other types of agriculture, including dairying, commercial grain, and ranching.
4. **Why Do Farmers Face Economic Difficulties?** Agriculture in LDCs faces distinctive economic problems resulting from rapid population growth and pressure to adopt international trade strategies to promote development. Agriculture in MDCs faces problems resulting from access to markets and overproduction.

CASE STUDY REVISITED / Africa's Food-Supply Crisis

As discussed in Chapter 2, Thomas Malthus argued in 1798 that population would increase more rapidly than food supply. At a global scale, the Malthus thesis has not come true.

Population grew at the most rapid rate in human history during the second half of the twentieth century, but food supply increased more rapidly. The percentage of people suffering from undernourishment declined in all LDCs combined from 37 percent in 1970 to 17 percent in 1995 (Figure 10-33). The exception has been sub-Saharan Africa. Over the past half-century, undernourishment in the sub-Saharan region has remained constant at just under one-third of the entire population. Sub-Saharan Africa is losing the race to keep food production ahead of population growth (Figure 10-34). Food production in the

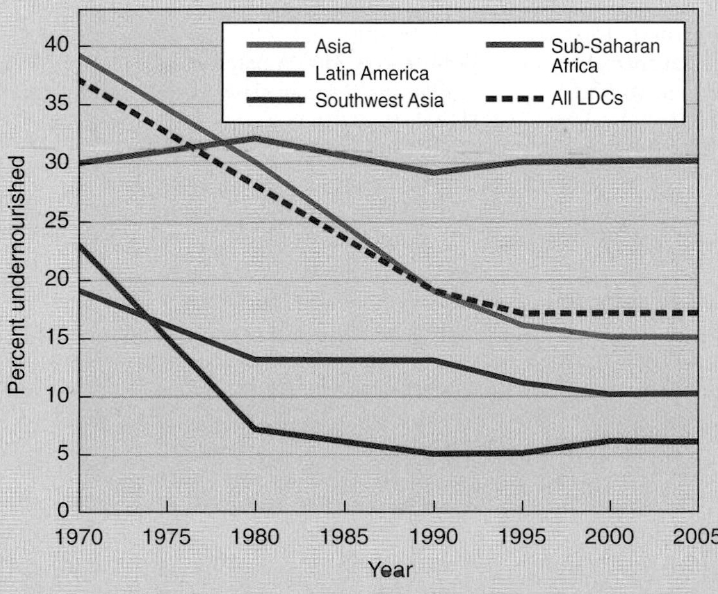

FIGURE 10-33 Percent of population undernourished. One-third of Africans and one-sixth of all people living in LDCs are considered undernourished. Progress in reducing undernourishment has been substantial in Asia but not in other LDCs.

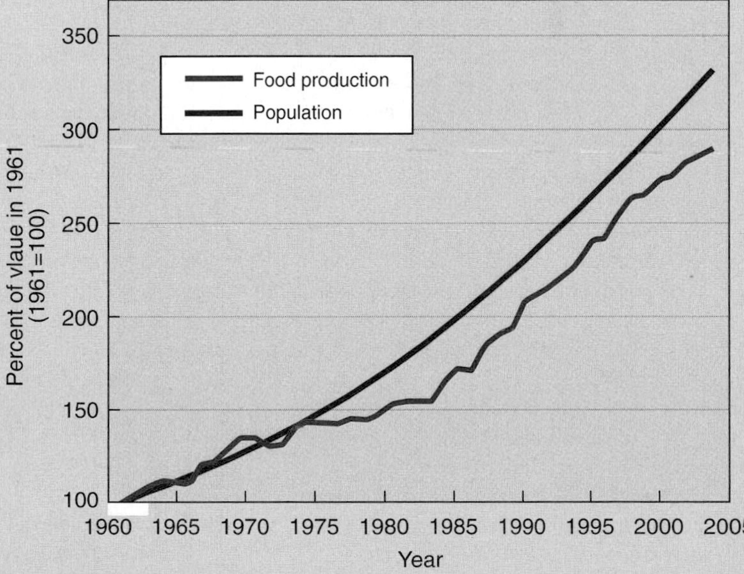

FIGURE 10-34 Population and food production in sub-Saharan Africa. The levels of population and food production in 1961 were set at 100 in the chart. Since the 1980s, population has increased more rapidly than food production in sub-Saharan Africa.

(Continued)

CASE STUDY REVISITED (Continued)

region has tripled over the past half-century, but population has quadrupled.

The problem is particularly severe in the Horn of Africa, including Somalia, Ethiopia, and Sudan. Also facing severe food shortages are countries in the Sahel region, a 400- to 550-kilometer (250- to 350-mile) belt in West Africa that marks the southern border of the Sahara. The most severely affected countries in the Sahel are Gambia, Senegal, Mali, Mauritania, Burkina Faso, Niger, and Chad. Traditionally, this region supported limited agriculture. Pastoral nomads moved their herds frequently, permitting vegetation to regenerate. Farmers grew groundnuts for export and used the receipts to import rice. With rapid population growth, herd size increased beyond the capacity of the land to support the animals. Animals overgrazed the limited vegetation and clustered at scarce water sources. Many died of hunger. Farmers overplanted, exhausting soil nutrients, and reduced fallow time, during which unplanted fields can recover. Soil erosion increased after most of the remaining trees were cut for wood and charcoal, used for urban cooking and heating.

Government policies have aggravated the food-shortage crisis. To make food affordable for urban residents, governments keep agricultural prices low. Constrained by price controls, farmers are unable to sell their commodities at a profit and therefore have little incentive to increase productivity. ■

KEY TERMS

Agribusiness (p. 313) Commercial agriculture characterized by the integration of different steps in the food-processing industry, usually through ownership by large corporations.

Agriculture (p. 309) The deliberate effort to modify a portion of Earth's surface through the cultivation of crops and the raising of livestock for sustenance or economic gain.

Cereal grain (p. 323) A grass yielding grain for food.

Chaff (p. 320) Husks of grain separated from the seed by threshing.

Combine (p. 326) A machine that reaps, threshes, and cleans grain while moving over a field.

Commercial agriculture (p. 311) Agriculture undertaken primarily to generate products for sale off the farm.

Crop (p. 309) Grain or fruit gathered from a field as a harvest during a particular season.

Crop rotation (p. 321) The practice of rotating use of different fields from crop to crop each year, to avoid exhausting the soil.

Desertification (p. 335) Degradation of land, especially in semiarid areas, primarily because of human actions like excessive crop planting, animal grazing, and tree cutting.

Double cropping (p. 321) Harvesting twice a year from the same field.

Grain (p. 325) Seed of a cereal grass.

Green revolution (p. 336) Rapid diffusion of new agricultural technology, especially new high-yield seeds and fertilizers.

Horticulture (p. 328) The growing of fruits, vegetables, and flowers.

Hull (p. 320) The outer covering of a seed.

Intensive subsistence agriculture (p. 319) A form of subsistence agriculture in which farmers must expend a relatively large amount of effort to produce the maximum feasible yield from a parcel of land.

Milkshed (p. 324) The area surrounding a city from which milk is supplied.

Paddy (p. 320) Malay word for wet rice, commonly but incorrectly used to describe a sawah.

Pastoral nomadism (p. 318) A form of subsistence agriculture based on herding domesticated animals.

Pasture (p. 319) Grass or other plants grown for feeding grazing animals, as well as land used for grazing.

Plantation (p. 322) A large farm in tropical and subtropical climates that specializes in the production of one or two crops for sale, usually to a more developed country.

Prime agricultural land (p. 313) The most productive farmland.

Ranching (p. 326) A form of commercial agriculture in which livestock graze over an extensive area.

Reaper (p. 326) A machine that cuts cereal grain standing in the field.

Ridge tillage (p. 331) System of planting crops on ridge tops in order to reduce farm production costs and promote greater soil conservation.

Sawah (p. 320) A flooded field for growing rice.

Shifting cultivation (p. 314) A form of subsistence agriculture in which people shift activity from one field to another; each field is used for crops for a relatively few years and left fallow for a relatively long period.

Slash-and-burn agriculture (p. 314) Another name for shifting cultivation, so named because fields are cleared by slashing the vegetation and burning the debris.

Spring wheat (p. 325) Wheat planted in the spring and harvested in the late summer.

Subsistence agriculture (p. 310) Agriculture designed primarily to provide food for direct consumption by the farmer and the farmer's family.

Sustainable agriculture (p. 331) Farming methods that preserve long-term productivity of land and minimize pollution, typically by rotating soil-restoring crops with cash crops and reducing inputs of fertilizer and pesticides.

Swidden (p. 315) A patch of land cleared for planting through slashing and burning.

Thresh (p. 320) To beat out grain from stalks by trampling it.

Transhumance (p. 319) The seasonal migration of livestock between mountains and lowland pastures.

Truck farming (p. 328) Commercial gardening and fruit farming, so named because *truck* was a Middle English word meaning *bartering* or the exchange of commodities.

Wet rice (p. 319) Rice planted on dryland in a nursery and then moved to a deliberately flooded field to promote growth.

Winnow (p. 320) To remove chaff by allowing it to be blown away by the wind.

Winter wheat (p. 325) Wheat planted in the autumn and harvested in the early summer.

THINKING GEOGRAPHICALLY

1. Assume that the United States constitutes one agricultural market, centered around New York City, the largest metropolitan area. To what extent can the major agricultural regions of the United States be viewed as irregularly shaped rings around the market center, as von Thünen applied to southern Germany?
2. New Zealand once sold nearly all its dairy products to the British, but since the United Kingdom joined the European Union in 1973, New Zealand has been forced to find other markets. What are some other examples of countries that have restructured their agricultural production in the face of increased global interdependence and regional cooperation?
3. Review the concept of overpopulation (the number of people in an area exceeding the capacity of the environment to support life at a decent standard of living). What agricultural regions have relatively limited capacities to support intensive food production? Which of these regions face rapid population growth?
4. Compare world distributions of corn, wheat, and rice production. To what extent do differences derive from environmental conditions and to what extent from food preferences and other social customs?
5. How might the loss of farmland on the edge of rapidly growing cities alter the choice of crops that other farmers make in a commercial agricultural society?

RESOURCES

Some recent and classic books and articles on agricultural geography:

Chakravarti, A. K. "Green Revolution in India." *Annals of the Association of American Geographers* 63 (1973): 319–30.

Durand, Loyal, Jr. "The Major Milksheds of the Northeastern Quarter of the United States." *Economic Geography* 40 (1964): 9–33.

Ebeling, Walter. *The Fruited Plain: The Story of American Agriculture.* Berkeley: University of California Press, 1979.

Grigg, David B. *The Agricultural Systems of the World: An Evolutionary Approach*. London: Cambridge University Press, 1974.

———. "Ester Boserup's Theory of Agrarian Change: A Critical Review." *Progress in Human Geography* 3 (1979): 64–84.

———. "Food Imports, Food Exports and Their Role in National Food Consumption." *Geography* 86 (2001): 171–76.

Hart, John Fraser. *The Changing Scale of American Agriculture*. Charlottesville: University of Virginia Press, 2003.

Ilbery, Brian W., Quentin Pablo Chiotti, and Timothy Rickard, eds. *Agricultural Restructuring and Sustainability: A Geographical Perspective*. Wallingford, England: CAB International: 1997.

Mannion, A. M. "Domestication and the Origins of Agriculture: An Appraisal." *Progress in Physical Geography* 23 (1999): 37–56.

Morgan, W. B. *Agriculture in the Third World: A Spatial Analysis*. Boulder, CO: Westview Press, 1978.

Newbury, Paul A. R. *A Geography of Agriculture*. Estover, England: Macdonald and Evans, 1980.

Pacione, Michael, ed. *Progress in Agricultural Geography*. London and Dover, NH: Croom Healm, 1986.

Rumney, Thomas A. *The Study of Agricultural Geography*. Lanham, MD: Rowman and Littlefield, 2005.

Sauer, Carl O. *Agricultural Origins and Dispersals*, 2nd ed. Cambridge, MA: M.I.T. Press, 1969.

Symons, Leslie. *Agricultural Geography*, rev. ed. London: G. Bell, 1979.

Tarrant, John R. *Agricultural Geography*. New York: John Wiley, 1974.

Turner, B. L., II, and Stephen B. Brush, eds. *Comparative Farming Systems*. New York: Guilford Press, 1987.

United Nations Office on Drugs and Crime. *World Drug Report 2009*. Vienna: UNDOC, 2009.

von Thünen, Johann Heinrich. *Von Thünen's Isolated State: An English Edition of "Der Isolierte Staat."* Trans. by Carla M. Wartenberg. Elmsford, NY: Pergamon Press, 1966.

Whittlesey, Derwent. "Major Agricultural Regions of the Earth." *Annals of the Association of American Geographers* 26 (1936): 199–240.

Willer, Helga, and Lukas Kilcher, eds. *The World of Organic Agriculture—Statistics and Emerging Trends*. Bonn, Germany: IFOAM, published annually.

Journal featuring agricultural geography:

Journal of Rural Studies

Key Internet sites:

www.fao.org. Agricultural statistics can be found on the Internet at the United Nation's Food and Agriculture Organization's web site. The FAO maintains a database known as FAOSTAT, with information on crops, food, and land use.

www.nass.usda.gov. The principal source of data about U.S. agriculture is the National Agricultural Statistical Service, in the U.S. Department of Agriculture.

www.organic-world.net. This web site is maintained by the Research Institute of Organic Agriculture funded under a project of the International Trade Centre and the Swiss State Secretariat of Economic Affairs.

www.sare.org. Information about sustainable agriculture can be found through the Sustainable Agriculture Research and Education web site.

Log in to www.mygeoscienceplace.com for videos, interactive maps, RSS feeds, case studies, and self-study quizzes to enhance your study of Agriculture.

CHAPTER 11

Industry

Huffy bicycles were manufactured in Ohio for more than a century, first in Dayton, where George P. Huffman founded the company's predecessor Davis Sewing Machine Co. in 1892, and beginning in 1954 at the world's largest bicycle factory in Celina. Huffy's two main U.S. competitors, Murray and Roadmaster, also manufactured bicycles in Ohio beginning in the 1930s, both in Cleveland.

In the 1950s, Murray and Roadmaster both moved bicycle production out of Ohio, Murray to Lawrenceburg, Tennessee, and then in the 1990s to Mississippi. Roadmaster moved to Little Rock, Arkansas, and then in the 1960s to Olney, Illinois. Huffy held out in Celina

KEY ISSUES

1 Where Is Industry Distributed?

2 Why Are Situation Factors Important?

3 Why Are Site Factors Important?

4 Why Are Location Factors Changing?

until 1998, when it moved production to Missouri for a year and then to Mexico.

In 1999, all three companies, as well as other U.S.-based bicycle manufacturers, including Schwinn, were sold to Pacific Cycle, a division of a Canadian company Dorel Industries. Pacific Cycle "designs, markets, and distributes" bicycles, according to its web site. What Pacific Cycle doesn't do is actually manufacture bicycles. All of the U.S. factories it had acquired were closed. Instead, Pacific Cycle contracts with companies in China to make the bicycles.

The bicycle story is not isolated. Ohio Art Co. moved production of its *Etch A Sketch* toy from Bryan, Ohio, to Shenzhen, China, in 2000. Maytag moved production of its refrigerators from Galesburg, Illinois, to Reynosa, Mexico, in 2002. Carrier Corp. moved production of its refrigeration units from Syracuse, New York, to Singapore in 2003. Hoover vacuum cleaners, once made in North Canton, Ohio, have been produced since 2007 by a Chinese company, Techtronic Industries Co.

Ultimately responsible for the changing geography of manufacturing is the American consumer. When Huffy bicycles were made in Celina, they sold for $80. After production was shifted overseas, the price came down to $40. For nearly all consumers, low price is much more important than place of origin.

Postcard from 1916, titled "The Ford Motor Plant [in Highland Park, Michigan] and 1,000 cars, a single day's output."

CASE STUDY / Maquiladoras in Mexico

Edi Bencomo is a factory worker in Chihuahua, Mexico. Her job is to clip together several color-coded wires for Alambrados y Circuitos Eléctricos, a factory that is owned by Delphi Automotive Systems. Bencomo migrated to Chihuahua 4 years ago, at age 16, from Madera, a village in the Sierra Madre Occidental, a mountain range 250 kilometers (150 miles) to the west. One of seven children, Bencomo saw no future for herself in remaining on her parents' corn farm. Had she remained in Madera, Bencomo probably would have been unemployed, along with 25 percent of the villagers.

In Chihuahua, Bencomo lives with her husband in a two-room shack more than an hour from the plant. They can afford to rent a somewhat better dwelling, but none are available in this rapidly growing city. She leaves home each weekday at 4 A.M. to battle hordes of workers who crowd onto buses that serve the factory area.

Bencomo earns about $4 an hour. She also receives two important benefits by working for Alambrados—a bus pass so that she can reach the plant at no cost and two meals in the cafeteria, paid for almost entirely by the company. She considers her job to be superior to that of her husband, who makes piñatas; both are paid minimum wage, but he receives no benefits.

Delphi's Chihuahua plant is known as a **maquiladora**, from the Spanish verb *maquilar*, which means to receive payment for grinding or processing corn. The term originally applied to a tax when Mexico was a Spanish colony. Under U.S. and Mexican laws, companies receive tax breaks if they ship materials from the United States, assemble components at a *maquiladora* plant in Mexico, and export the finished product back to the United States. More than 1 million Mexicans are employed at over 3,000 *maquiladoras*. Delphi has more than 50 *maquiladoras* employing 75,000 people and is one of Mexico's largest employers. ■

The title of this chapter, "Industry," refers to the manufacturing of goods in a factory. The word is appropriate because it also means persistence or diligence in creating value. A factory utilizes a large number of people, machinery, and money to turn out valuable products.

In the previous chapter, we looked at agriculture, practiced throughout the inhabited world because the need for food is universal. Industry is much more highly clustered in ***space*** than is agriculture. In this chapter, we look at the ***regions*** where factories are located and ***why***. A particular ***place*** may be well suited or poorly suited for industry, depending on the distinctive characteristics of land, labor, and capital there.

Geographers also recognize that ***connections*** with the rest of the world are critical in determining whether a particular place is suitable for industry. Two connections are critical in determining the best location for a factory—*where* the markets for the product are located and where the resources needed to make the product are located.

A generation ago, industry was highly clustered in a handful of communities within a handful of MDCs, but industry has diffused to many communities in many LDCs. The United States lost one-third of its manufacturing jobs during the first decade of the twenty-first century. Americans alarmed by this loss heard "a giant sucking sound" of manufacturing jobs being "sucked" into other countries from recently closed U.S. factories. The future of manufacturing in the United States was "now in jeopardy," according to the National Association of Manufacturers, a leading industry group.

Government officials everywhere recognize the powerful role of industry in the economic health of a community. Manufacturing jobs are viewed as a special asset by communities around the world and they mourn when factories close and rejoice when they open. To attract and retain them, officials offer financial support that, when scrutinized by independent analysts, is considered excessive.

Americans' fears of manufacturing job losses were echoed elsewhere in the world. A former president of the European Union warned against the "deindustrialization of Europe." Japan's loss of manufacturing jobs to overseas locations was called a "hollowing out" by Japanese politicians. In Mexico, the loss of manufacturing jobs during the early twenty-first century led to "a wave of soul-searching."

Transnational corporations operate at a global ***scale*** of concern for the distribution of markets and resources. Raw materials may be collected from many places, sent to factories located in several other places for a succession of specialized manufacturing procedures, and shipped to consumers located in yet other places.

With ***globalization*** of competition to attract new industries—or, in many places, to retain existing ones—each place possesses distinctive location characteristics. Geographers identify the ***local diversity*** in assets that enable some communities to compete successfully for industries, as well as handicaps that must be overcome to retain older companies.

KEY ISSUE 1
Where Is Industry Distributed?

- **Origin of Industry**
- **Industrial Regions**

The modern concept of industry—meaning the manufacturing of goods in a factory—originated in northern England and southern Scotland during the second half of the eighteenth century. From there, industry diffused in the nineteenth century to Europe and to North America and in the twentieth century to other regions. ■

Origin of Industry

The **Industrial Revolution** was a series of improvements in industrial technology that transformed the process of manufacturing goods. Prior to the Industrial Revolution, industry was geographically dispersed across the landscape. People made household tools and agricultural equipment in their own homes or obtained them in the local village. Home-based manufacturing was known as the **cottage industry** system.

The term *Industrial Revolution* is somewhat misleading, because the transformation was far more than industrial, and it didn't happen overnight. The Industrial Revolution resulted in new social, economic, and political inventions, not just industrial ones. The changes involved a gradual diffusion of new ideas and techniques over decades, rather than an instantaneous revolution. Nonetheless, the term is commonly used to define the process that began in the United Kingdom in the late 1700s.

The root of the Industrial Revolution was technology, involving several inventions that transformed the way in which goods were manufactured. The revolution in industrial technology created an unprecedented expansion in productivity, resulting in substantially higher standards of living. In Chapter 2, the Industrial Revolution was cited as a principal cause of population growth in stage 2 of the demographic transition.

The invention most important to the development of factories was the steam engine, patented in 1769 by James Watt (1736–1819), a maker of mathematical instruments in Glasgow, Scotland (Figure 11-1). Watt built the first useful steam engine, which could supply power far more efficiently than the watermills then in common use, let alone human or animal power. The large supply of steam power available from James Watt's steam engines induced firms to concentrate all steps in one building attached to a single power source.

FIGURE 11-1 James Watt's steam engine. Steam injected in a cylinder (left side of engine) pushes a piston attached to a crankshaft that drives machinery (right side of engine).

Industries impacted by the Industrial Revolution included:

- **Iron:** The first industry to benefit from Watt's steam engine. The usefulness of iron had been known for centuries, but the scale of production was small. The process demanded constant heating and cooling of the iron, a time-consuming and skilled operation because energy could not be generated to keep the ovens hot for a sufficiently long period of time. The Watt steam engine provided a practical way to keep the ovens constantly heated.

  Henry Cort, a navy agent, established an iron forge near Fareham, England, to shape iron into useful objects. The combination of Watt's engine and Cort's iron purification process increased iron-manufacturing capability.
- **Coal:** The source of energy to operate the ovens and the steam engines. Iron production requires a source of heat to smelt the iron ore as well as to run the furnaces, forges, and steam engines. Wood, the main energy source prior to the Industrial Revolution, was becoming scarce in England because it was in heavy demand for construction of ships, buildings, and furniture, as well as for heat. Manufacturers turned to coal, which was then plentiful in England.

  Abraham Darby of Coalbrookdale in Shropshire, England, produced high-quality iron smelted with purified carbon made from coal, known as coke. Coke is richer in carbon and more combustible than coal, so it is a better source for the heat and gases needed to smelt iron ore.
- **Transportation:** Critical for diffusing the Industrial Revolution. First canals and then railroads enabled factories to bring in bulky raw materials such as iron ore and coal and ship finished goods to consumers (Figure 11-2).

  Europe's political problems retarded the diffusion of the railroad. Cooperation among small neighboring states was essential to build an efficient rail network and to raise money for constructing and operating the system. Because such cooperation could not be attained, railroads in some parts of Europe were delayed 50 years after their debut in Britain.
- **Textiles:** Transformed from a dispersed cottage industry to a concentrated factory system during the late eighteenth century. Prior to the Industrial Revolution, thread was spun at home on spinning wheels operated by hand and foot. People known as putters-out were hired by merchants to drop off cotton or wool at homes, where women and children sorted, cleaned, and spun it into thread. The putters-out then picked up the finished work and paid according to the number of pieces that were completed ("piece-rate").

  In 1768, Richard Arkwright, a barber and wigmaker in Preston, England, invented machines to untangle cotton prior to spinning. Too large to fit inside a cottage, spinning frames were placed inside factories near sources of rapidly flowing water, which supplied the power. Because the buildings resembled large watermills, they were known as mills.

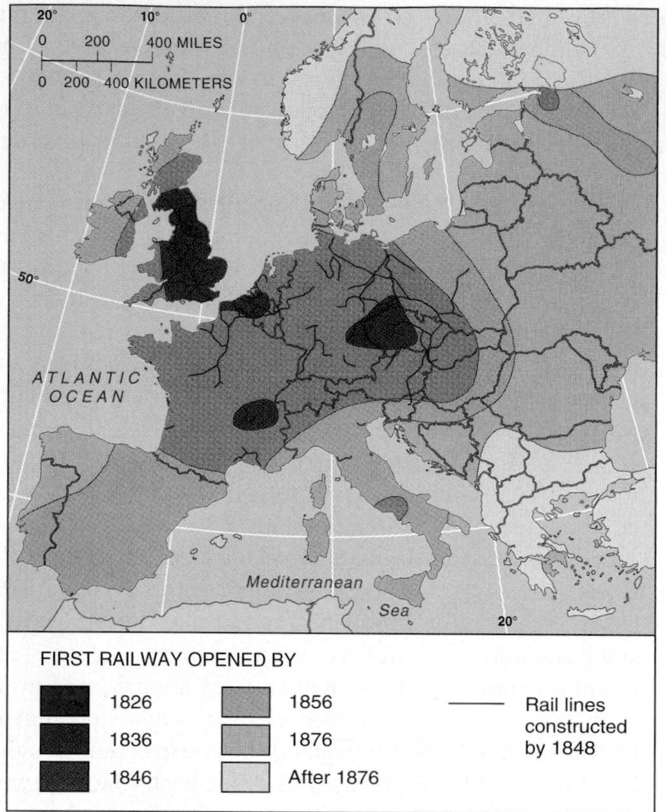

FIGURE 11-2 Diffusion of Industrial Revolution. The construction of railroads from the United Kingdom to the European continent reflects the diffusion of the Industrial Revolution. More than 50 years passed between the construction of the first railroads in Britain and the first ones in Eastern Europe.

- **Chemicals:** An industry created to bleach and dye cloth. In 1746, John Roebuck and Samuel Garbett established a factory to bleach cotton with sulfuric acid obtained from burning coal. When combined with various metals, sulfuric acid produced another acid called vitriol, useful for dying clothing. Sulfuric acid produced a blue vitriol when combined with copper, green with iron, and white with zinc.
- **Food processing:** Essential to feed the factory workers no longer living on farms. In 1810, French confectioner Nicholas Appert started canning food in glass bottles sterilized in boiling water.

Industrial Regions

Industry is concentrated in three of the nine world regions discussed in Chapter 9: Europe, North America, and East Asia (Figure 11-3). Each of the three regions accounts for roughly one-fourth of the world's total industrial output. Outside these three regions the leading industrial producers are Brazil and India. The three industrial regions are discussed in this section, beginning with the oldest.

Europe's Industrial Areas

Numerous industrial areas emerged in Europe during the nineteenth and early twentieth centuries (Figure 11-4). These include several clustered in Western Europe centered on western Germany and extending north to the United Kingdom and south to Italy and Spain, and several in Eastern Europe, primarily in the former Soviet Union.

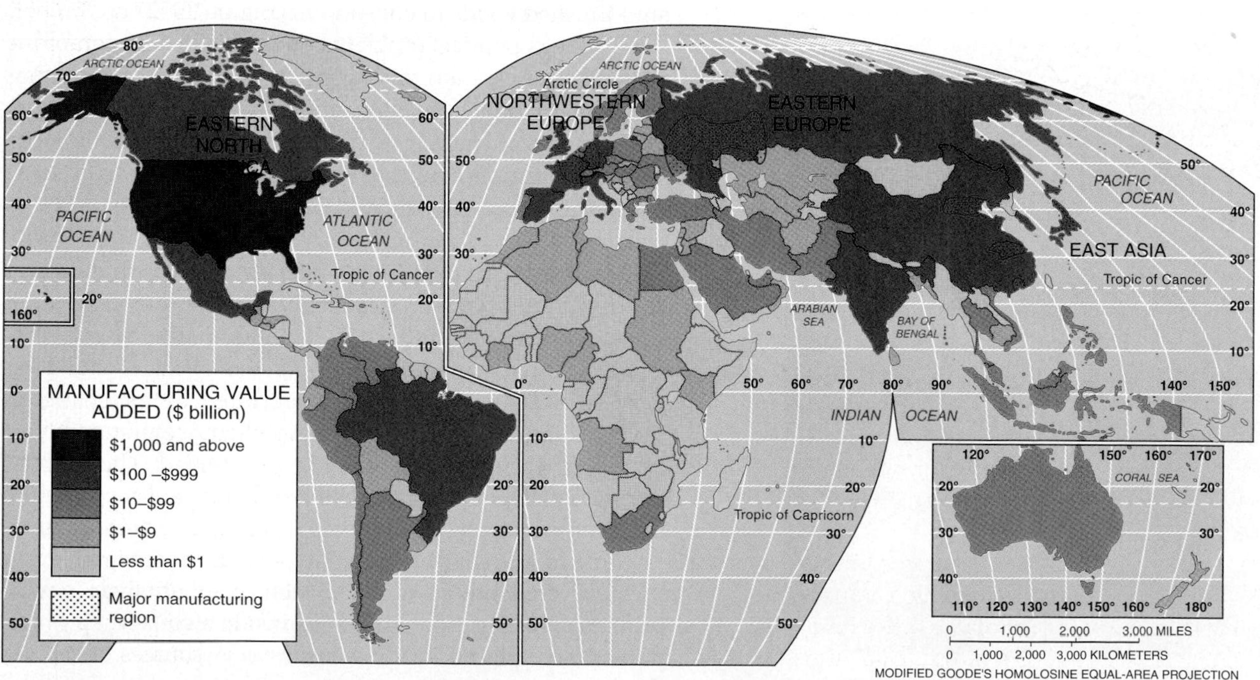

FIGURE 11-3 Manufacturing value added. Three-fourths of the world's manufacturing takes place in North America, Europe, and East Asia.

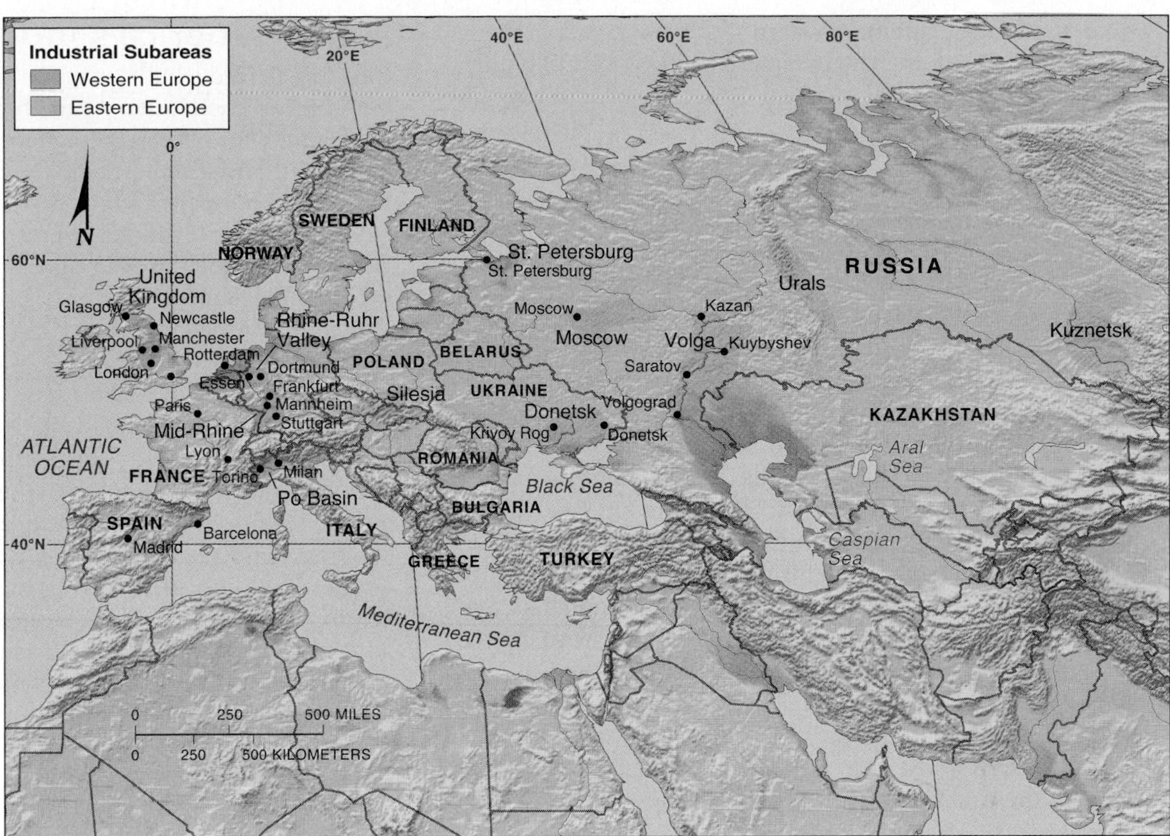

FIGURE 11-4 Industrial areas in Europe. In Western Europe, manufacturing is clustered on a north–south axis between the North Sea and the Mediterranean Sea. In Eastern Europe, industrial areas are dispersed on an east–west axis across the former Soviet Union.

- **United Kingdom:** Dominated world production of steel and textiles during the nineteenth century. As the first country to enter the Industrial Revolution, Britain was saddled in the twentieth century with what became outmoded and deteriorating factories and support services.

  Although no longer a world leader in steel, textiles, and other early Industrial Revolution industries, the United Kingdom expanded industrial production in the late twentieth century by attracting new high-tech industries that serve the European market. Japanese companies have built more factories in the United Kingdom than in any other European country. The British have done more than the other major European countries to lower taxes on businesses, reduce government regulations, convert government monopolies to private ownership, and utilize computers.
- **Rhine–Ruhr Valley:** Western Europe's most important and most centrally located industrial area. Within the area, industry is dispersed rather than concentrated in one or two cities.

  Iron and steel manufacturing concentrated in the Rhine–Ruhr Valley because of proximity to large coalfields. Access to iron and steel production stimulated other heavy-metal industries, such as railroad, machinery, and armaments, to locate in the area. Rotterdam, the world's largest port, lies at the mouth of several branches of the Rhine River as it flows into the North Sea.
- **Mid-Rhine:** Western Europe's second most important industrial area. The German portion of the Mid-Rhine area lacks abundant raw materials but lies at the center of Europe's most important consumer market. The French portion of the Mid-Rhine region—Alsace and Lorraine—contains Europe's largest iron-ore field and is the production center for two-thirds of France's steel.

  The three largest cities in the German portion are Frankfurt, Stuttgart, and Mannheim. When Germany was divided into eastern and western portions during the Cold War, Frankfurt was West Germany's most important financial and commercial center and the hub of its road, rail, and air networks. Stuttgart's industries specialize in high-value goods and require skilled labor; Mercedes-Benz and Audi automobiles are among the city's best-known products. Mannheim, an inland port along the Rhine, has a large chemical industry that manufactures synthetic fibers, dyes, and pharmaceuticals.
- **Po Basin:** Southern Europe's oldest and most important industrial area. The Po Valley contains about two-thirds of Italy's manufacturing in one-fifth of its land area.

  Modern industrial development in the Po Basin began with establishment of textile manufacturing during the nineteenth century. The area had two key assets: inexpensive hydroelectricity from the nearby Alps and a large labor supply, especially from Italy's poorer south, willing to work for relatively low wages.

- **Northeastern Spain:** Western Europe's fastest-growing industrial area in the late twentieth century. Spain's leading industrial area, Catalonia, is centered on the city of Barcelona. The area is the center of Spain's textile industry and the location of the country's largest motor-vehicle plant. Spain's motor-vehicle industry has grown into the second largest in Europe, behind only Germany's, although it is entirely foreign-owned.
- **Moscow:** Russia's oldest industrial area, centered around the country's capital and largest market. Moscow specializes in fabrics and products that require skilled labor.
- **St. Petersburg:** Eastern Europe's second largest city, specializing in shipbuilding and other industries serving Russia's navy and ports in the Baltic Sea.
- **Volga:** Russia's largest petroleum and natural gas fields. The motor-vehicle industry is concentrated in Togliatti, oil refining in Kuybyshev, chemicals in Saratov, metallurgy in Volgograd, and leather and fur in Kazan.
- **Urals:** Contains more than 1,000 types of minerals, the most varied collection found in any mining area in the world. Proximity to these inputs encouraged the Communists to locate iron and steel, chemicals, machinery, and metal fabricating in this area.
- **Kuznetsk:** Russia's most important manufacturing district east of the Ural Mountains. Soviet planners took advantage of the area's coal and iron ore to invest in iron and steel factories there.
- **Donetsk:** In Eastern Ukraine, an area of coal, iron ore, manganese, and natural gas. These assets make the area Eastern Europe's largest producer of iron and steel. Major plants are located at Krivoy Rog, near iron-ore fields, and at Donetsk, near coalfields.
- **Silesia:** Eastern Europe's leading industrial area outside the former Soviet Union. Silesia, which includes southern Poland and northern Czech Republic, is an important steel production center, near coalfields.

North America's Industrial Areas

Industry arrived a bit later in the United States than in Europe, but it grew much faster. At the time of independence in 1776, the United States was a predominantly agricultural society, dependent on the import of manufactured goods from England. The first U.S. textile mill opened in Pawtucket, Rhode Island, in 1791. The textile industry grew rapidly after 1808, when the U.S. government imposed an embargo on European trade to avoid entanglement in the Napoleonic Wars. By 1860, the United States had become a major industrial nation, second only to the United Kingdom.

Manufacturing in North America concentrated in the northeastern quadrant of the United States and in southeastern Canada (Figure 11-5). This industrial area has achieved its dominance through a combination of historical and environmental factors. As the first area of European settlement in the Western Hemisphere, the U.S. East Coast was tied to European markets and industries during the first half of the nineteenth century. The early date of settlement gave eastern cities an advantage in creating the infrastructure needed to become the country's dominant industrial center.

The Northeast also had essential raw materials, including iron and coal. Good transportation moved raw materials to factories and manufactured goods to markets. The Great Lakes and major rivers (Mississippi, Ohio, St. Lawrence) were supplemented in the 1800s by canals, railways, and highways. All helped to connect the westward-migrating frontier with manufacturing centers.

Within the North American manufacturing region, several specialized areas developed:

- **New England:** The oldest industrial area in the northeastern United States. It developed as an industrial center in the early nineteenth century, beginning with cotton textiles. Cotton was imported from southern states, where it was grown, and finished cotton products were shipped to Europe.
- **Middle Atlantic:** The largest U.S. market. It attracts industries that need proximity to a large number of consumers and that depend on foreign trade through one of this region's large ports. Other firms seek locations near the financial, communications, and entertainment industries, which are highly concentrated in New York.
- **Mohawk Valley:** A linear industrial belt in upper New York State along the Hudson River and Erie Canal. Buffalo, near the confluence of the Erie Canal and Lake Erie, was the region's most important industrial center, especially for steel and food processing. Inexpensive, abundant electricity, generated at nearby Niagara Falls, has attracted aluminum, paper, and electrochemical industries to the region.

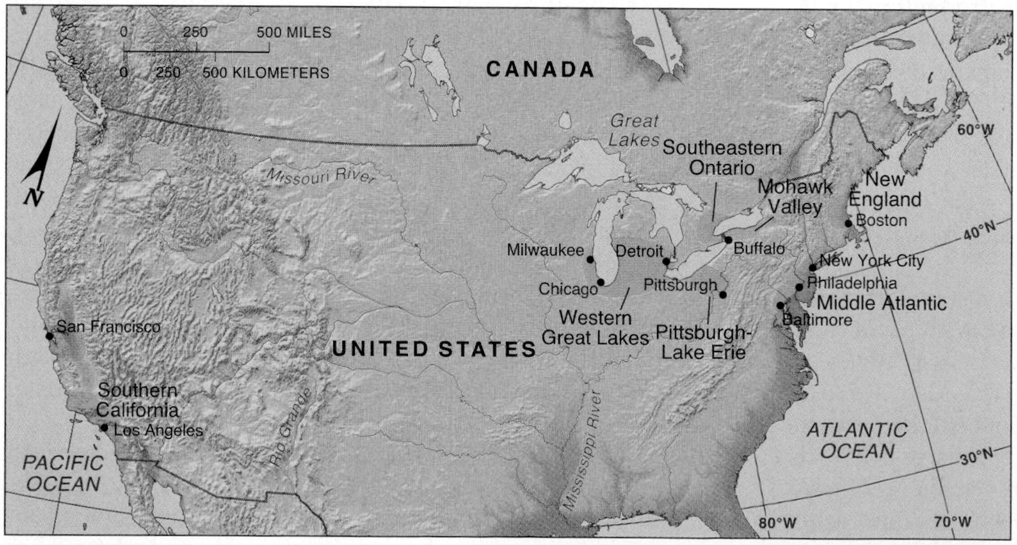

FIGURE 11-5 Industrial areas in North America. Manufacturing in North America was traditionally highly clustered in several regions within the northeastern United States and southeastern Canada.

- **Pittsburgh–Lake Erie:** The leading steel-producing area in the nineteenth century because of proximity to Appalachian coal and iron ore. Steel manufacturing originally concentrated in the area between Pittsburgh and Cleveland because of its proximity to Appalachian coal and iron ore. Proximity to steel makers attracted other manufacturers that made heavy use of steel in their own products.
- **Western Great Lakes:** Centered on Chicago, the hub of the nation's transportation network, now the center of steel production. Motor-vehicle manufacturers and other industries that have a national market locate in the western Great Lakes area to take advantage of this convergence of transportation routes. The area supplies machine tools, transportation equipment, clothing, furniture, agricultural machinery, and food products to people living in the interior of the country.
- **Southern California:** The leading industrial area outside of the Northeast. When the United States entered World War II in 1941, more than one-third of Los Angeles' manufacturing was in the aircraft industry, attracted by clear skies, light winds, and mild winters. More recently, Los Angeles has become the country's largest area of clothing and textile production, the second-largest furniture producer, and a major food-processing center. Immigrants from Latin America and Asia provide a large pool of low-wage workers.
- **Southeastern Ontario:** Canada's most important industrial area, central to the Canadian and U.S. markets and near the Great Lakes and Niagara Falls. Most of Canada's steel production is concentrated in Hamilton, Ontario, and motor-vehicle assembly in the Toronto area. Inexpensive electricity has attracted aluminum manufacturing, paper making, flour mills, textile manufacturing, and sugar refining.

East Asia's Industrial Areas

Faced with isolation from world markets and a shortage of nearly all essential resources, East Asia has taken advantage of its most abundant resource—people. The region's two leading industrial countries—Japan and China—rank second and third in manufacturing value behind only the United States (Figure 11-6).

- **Japan:** Became an industrial power in the 1950s and 1960s initially by producing goods that could be sold in large quantity at cut-rate prices to consumers in other countries. Prices were kept low, despite high shipping costs, because workers received much lower wages in Japan than in North America or Europe. The country became the world's leading manufacturer of automobiles, ships, cameras, stereos, and televisions.

  Aware that South Korea, Taiwan, and other Asian countries were building industries based on even lower-cost labor, Japan started training workers for highly skilled jobs. "Made in Japan," a phrase once synonymous with cheap, poorly made goods, now refers to high-quality motor vehicles, electronics, and precision instruments. Japan's manufacturing is concentrated in the central region between Tokyo and Nagasaki, especially in the two large urban areas of Tokyo–Yokohama and Osaka–Kobe–Kyoto.
- **China:** The world's largest supply of low-cost labor and the world's largest market for many consumer products. China is the largest manufacturer of textiles and apparel, steel, and many household products.

  Policy changes in the 1990s opened China's market and labor force to transnational corporations. Foreign-owned firms seeking low-cost labor were permitted to open factories in China to manufacture labor-intensive products such as apparel for export. Rapid economic expansion put money in the pockets of enough of China's 1.3 billion people to encourage more manufacturing for domestic consumption.

  China's manufacturers have clustered in three areas along the east coast—near Guangdong and Hong Kong, the Yangtze River valley between Shanghai and Wuhan, and along the Gulf of Bo Hai from Tianjin and Beijing to Shenyang. These three areas contain only one-fourth of China's population but one-half of its wealth, three-fourths of its foreign investment, and five-sixths of its foreign trade. The clustering of investment has produced large and increasing gaps in wealth within China.

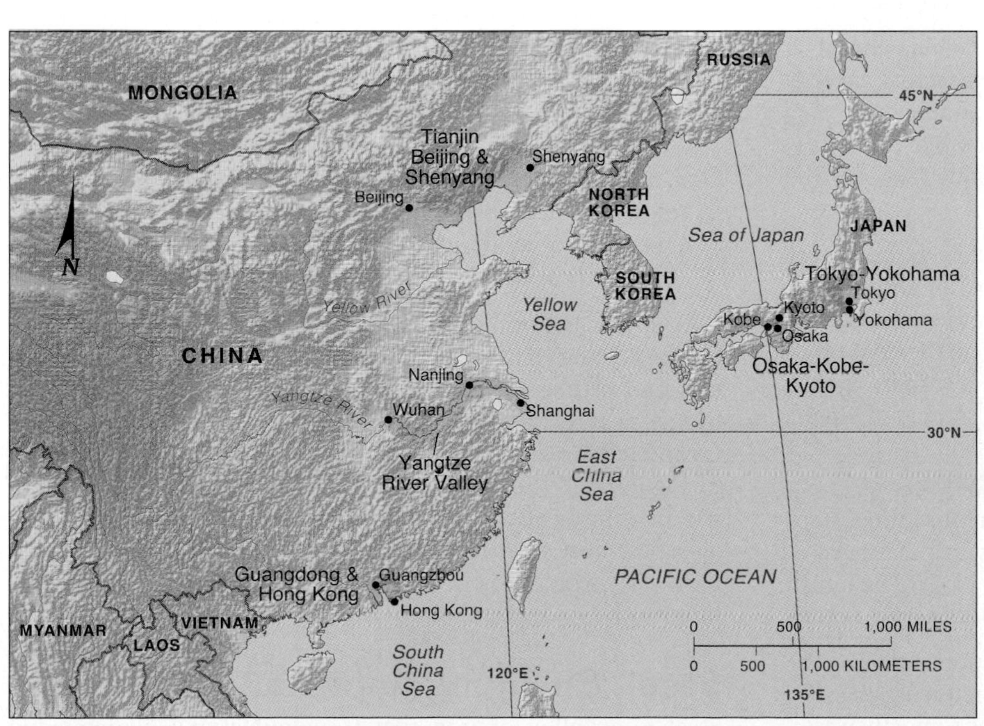

FIGURE 11-6 Industrial areas in East Asia. Within Japan, production is clustered along the southeast coast. Within China, a large percentage of industries are clustered in three centers along the east coast.

KEY ISSUE 2

Why Are Situation Factors Important?

- Proximity to Inputs
- Proximity to Markets
- Ship, Rail, Truck, or Air?

Having looked at the "where" question for industrial location, we can next consider the "why" question: Why are industries located where they are? Geographers try to explain why one location may prove more profitable for a factory than others.

A company ordinarily faces two geographical costs—situation and site. Situation factors are discussed in this section and site factors in the next section. **Situation factors** involve transporting materials to and from a factory. A firm seeks a location that minimizes the cost of transporting inputs to the factory and finished goods to consumers. ■

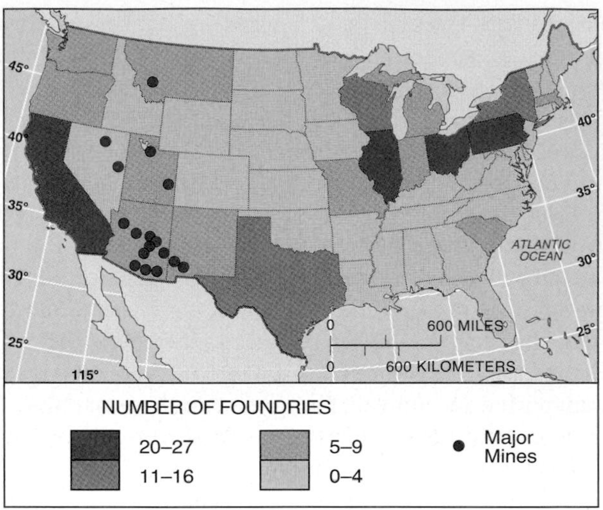

FIGURE 11-7 Bulk-reducing industries. Copper mining, concentrating, and smelting are examples of bulk-reducing industries. In the United States, most plants that concentrate, smelt, and refine copper are in or near Arizona, where most copper mines are located. In contrast, most foundries, where copper products are manufactured, are located near markets in the east and west coasts.

Proximity to Inputs

Manufacturers buy from companies and individuals who supply inputs, such as materials, energy, machinery, and supporting services. They sell to companies and individuals who purchase the product. All manufacturers try to minimize the aggregate cost of transporting inputs to their factories and transporting finished products from their plants to consumers. The farther something is transported, the higher the cost, so a manufacturer tries to locate its factory as close as possible to both buyers and sellers.

- The optimal plant location is as close as possible to inputs if the cost of transporting raw materials to the factory exceeds the cost of transporting the product to consumers.
- The optimal plant location is as close as possible to the customer if the cost of transporting the product exceeds the cost of transporting inputs.

Every industry uses some inputs. These may be resources from the physical environment (minerals, wood, or animals), or they may be parts or materials made by other companies. An industry in which the inputs weigh more than the final products is a **bulk-reducing industry**. To minimize transport costs, a bulk-reducing industry needs to locate near its sources of inputs.

Copper: A Bulk-Reducing Industry

Copper production involves several steps. The first three steps provide good examples of bulk-reducing activities that need to be located near their sources of inputs (Figure 11-7). The fourth step is not bulk-reducing, so does not need to be near inputs.

1. **Mining.** The first step in copper production is mining the copper ore. Mining in general is bulk-reducing because the heavy, bulky ore extracted from mines is mostly waste, known as *gangue*. Copper ore mined in North America is especially low-grade, less than 0.7 percent copper.
2. **Concentration.** Concentration mills crush and grind the ore into fine particles, mix them with water and chemicals, and filter and dry them. Copper concentrate is about 25 percent copper. Concentration mills are always near the mines because concentration transforms the heavy, bulky copper ore into a product of much higher value per weight.
3. **Smelting.** The concentrated copper becomes the input for smelters, which remove more impurities. Smelters produce copper matte (about 60 percent copper), blister copper (about 97 percent copper), and anode copper (about 99 percent copper). As another bulk-reducing industry, smelters are built near their main inputs—the concentration mills—again to minimize transportation cost.
4. **Refining.** The purified copper produced by smelters is treated at refineries to produce copper cathodes, about 99.99 percent pure copper. Little further weight loss occurs, so proximity to the mines, mills, and smelters is a less critical factor in determining the location of refineries.

Another important locational consideration is the source of energy to power these energy-demanding operations. In general, metal processors such as the copper industry also try to locate near economical electrical sources and to negotiate favorable rates from power companies.

Figure 11-7 shows the distribution of the U.S. copper industry. Two-thirds of U.S. copper is mined in Arizona, so the state also has most of the concentration mills and smelters. Most foundries, where copper is manufactured, are located near markets on the east and west coasts.

Steel: Changing Importance of Inputs

Steel is an alloy of iron that is manufactured by removing impurities in iron, such as silicon, phosphorus, sulfur, and oxygen, and adding desirable elements, such as manganese and chromium. The two principal inputs in steel production are iron ore and coal. Steelmaking is an example of a bulk-reducing industry that has located to minimize the cost of transporting these two inputs.

Steel was a luxury item until Henry Bessemer (1813–1898) patented an efficient process for casting steel in 1855. The Bessemer process remained the most common method of manufacturing steel until the mid-twentieth century. Because of the need for large quantities of bulky, heavy iron ore and coal, steelmaking has clustered near sources of the two key raw materials.

Steelmaking demonstrates that when the source of inputs or the relative importance of inputs changes, the optimal location for the industry changes. In the United States, the distribution of steel production has changed several times because of changing inputs (Figure 11-8).

- **Mid-nineteenth century:** The U.S. steel industry concentrated around Pittsburgh in southwestern Pennsylvania, where iron ore and coal were both mined. The area no longer has steel mills, but it remains the center for research and administration.
- **Late nineteenth century:** Steel mills were built around Lake Erie, in the Ohio cities of Cleveland, Youngstown, and Toledo, and near Detroit. The locational shift was largely influenced by the discovery of rich iron ore in the Mesabi Range, a series of low mountains in northern Minnesota. This area soon became the source for virtually all iron ore used in the U.S. steel industry. The ore was transported by way of Lake Superior, Lake Huron, and Lake Erie. Coal was shipped from Appalachia by train.
- **Early twentieth century:** Most new steel mills were located near the southern end of Lake Michigan—Gary in Indiana, Chicago, and other communities. The main raw materials continued to be iron ore and coal, but changes in steelmaking required more iron ore in proportion to coal. Thus, new steel mills were built closer to the Mesabi Range to minimize transportation cost. Coal was available from nearby southern Illinois, as well as from Appalachia.
- **Mid-twentieth century:** Most new U.S. steel mills were located in communities near the East and West coasts, including Baltimore, Los Angeles, and Trenton, New Jersey. These coastal locations partly reflected further changes in transportation cost. Iron ore increasingly came from other countries, especially Canada and Venezuela, and locations near the Atlantic and Pacific oceans were more accessible to those foreign sources. Further, scrap iron and steel—widely available in the large metropolitan areas of the East and West coasts—became an important input in the steel-production process.
- **Late twentieth century:** Most steel mills in the United States closed. Most of the survivors were around southern Lake Michigan and along the East Coast.

Thus, for surviving steel mills in the United States, proximity to markets has become more important than the traditional situation factor of proximity to inputs. Coastal plants provide steel to large East Coast population centers, and southern Lake Michigan plants are centrally located to distribute their products countrywide.

The increasing importance of proximity to markets is also demonstrated by the recent growth of steel minimills, which have captured one-fourth of the U.S. steel market (Figure 11-9).

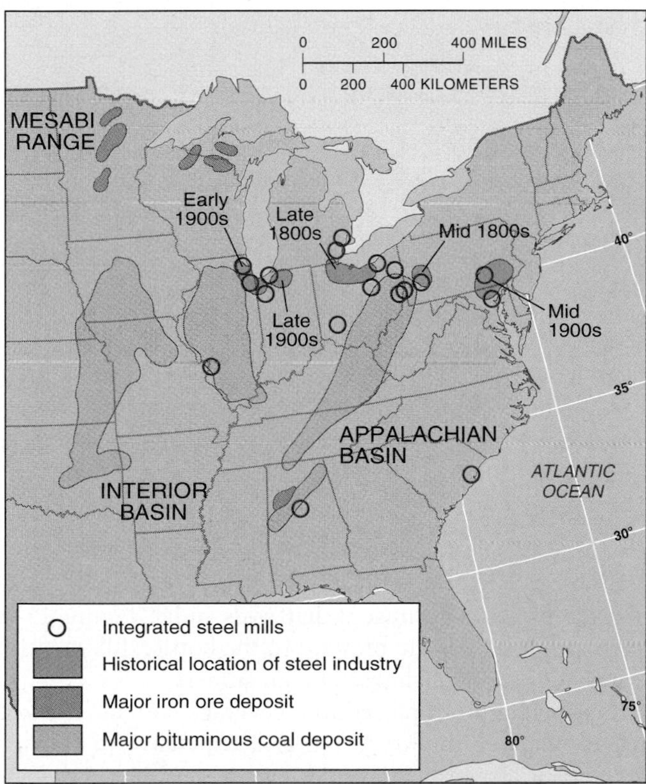

FIGURE 11-8 Impact of changing situation factors. Integrated steel mills are highly clustered near the southern Great Lakes, especially Lake Erie and Lake Michigan. Historically, the most critical factor in siting a steel mill was to minimize transportation cost for raw materials, especially heavy, bulky iron ore and coal. In recent years, many integrated steel mills have closed. Most surviving mills are in the Midwest to maximize access to consumers.

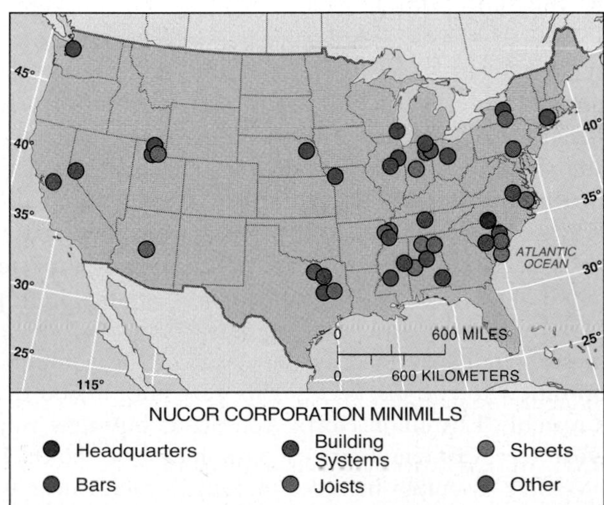

FIGURE 11-9 Minimills for steel production. Shown are the plants of Nucor, the largest minimill operator in the United States. Minimills, which produce steel from scrap metal, are more numerous than integrated steel mills, and they are distributed around the country near local markets.

Rather than iron ore and coal, the main input into minimill production is scrap metal. In the past, most steel was produced at large integrated mill complexes. They processed iron ore, converted coal into coke, converted the iron into steel, and formed the steel into sheets, beams, rods, or other shapes. Minimills, generally limited to one step in the process—steel production—are less expensive than integrated mills to build and operate, and they can locate near their markets because their main input—scrap metal—is widely available.

Proximity to Markets

For many firms, the optimal location is close to customers. Proximity to markets is a critical locational factor for three types of industries—bulk-gaining, single market, and perishable.

Bulk-Gaining Industries

A **bulk-gaining industry** makes something that gains volume or weight during production. To minimize transport costs, a bulk-gaining industry needs to locate near where the product is sold.

FABRICATED METALS. A prominent example of a bulk-gaining industry is the fabrication of parts and machinery from steel and other metals. A fabricated-metal factory brings together metals such as steel and previously manufactured parts as the main inputs and transforms them into a more complex product. Fabricators shape individual pieces of metal using such processes as bending, forging (hammering or rolling metal between two dies), stamping (pressing metal between two dies), and forming (pressing metal against one die). Separate parts are joined together through welding, bonding, and fastening with bolts and rivets.

Because fabricated and machined products typically occupy a larger volume than the sum of their individual parts and metals, the cost of shipping the final product to consumers is usually the most critical factor. Whereas steelmakers have traditionally located near raw materials, steel fabricators have traditionally located near markets. Machinery is fabricated for use in farms, factories, offices, and homes. Common fabricated goods include televisions, refrigerators, and air conditioners. Machine shops also transform metal into useful products such as structural metal for buildings and bridges.

The largest market for fabricated metal and machinery manufacturers is motor vehicles. Motor vehicles are fabricated in the United States at about 40 final assembly plants, from parts made at several thousand other plants (Figure 11-10). As a bulk-gaining industry, the critical location factor is minimizing transportation to the market, in this case the 20,000 dealers where roughly 12 million North Americans buy new vehicles each year (see Contemporary Geographic Tools box). Thus, motor-vehicle assembly involves making vehicles near where they are to be sold:

- At a global scale: Three-fourths of vehicles sold in the United States are assembled in the United States, and most of the remainder are assembled in Canada and Mexico.

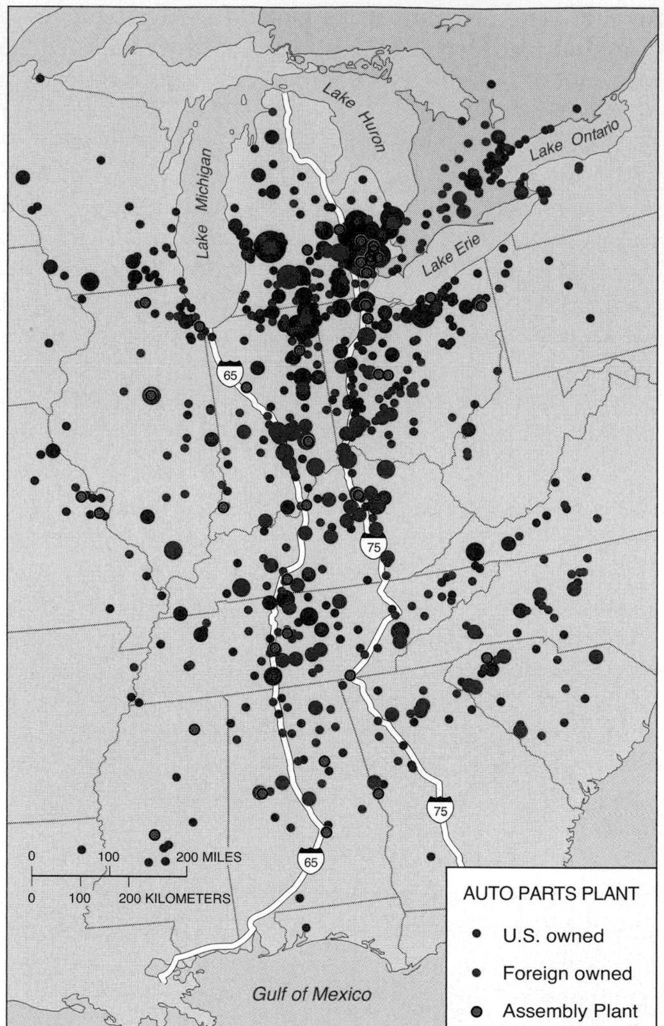

FIGURE 11-10 Auto alley. U.S.- and foreign-owned motor-vehicle parts plants. Plants are clustered in the interior of the country, near the major customers, the final assembly plants. Foreign-owned plants are more likely to be farther south, where workers are less likely to join a union. Circles are proportional to the number of plants in a particular ZIP code.

- At a national scale: Most assembly plants are located in the interior of the United States, between Michigan and Alabama, centered in a corridor known as "auto alley," formed by north–south interstate highways 65 and 75.

BEVERAGE PRODUCTION. Beverage bottling is another good example of an industry that adds bulk (Figure 11-12). Empty cans or bottles are brought to the bottler, filled with the soft drink or beer, and shipped to consumers.

The principal input placed in a beverage container is water, which is relatively bulky, heavy, and expensive to transport. Major soft-drink companies add syrups, and beer companies add barley, hops, and yeast, according to proprietary recipes. These added ingredients are much less bulky than the water and much easier to transport.

If water were only available in a few locations around the country, then bottlers might cluster near the source of such a scarce, bulky input. But because water is available where

CONTEMPORARY GEOGRAPHIC TOOLS
Honda Selects a Factory Location

When Honda decided that it needed another assembly plant in the United States, it applied situation and site factors to select a location for the factory. Situation factors were considered first in the decision-making process, then site factors.

The most critical situation factor for Honda was minimizing the cost of shipping finished vehicles to its customers around North America. That led Honda to look for locations within auto alley, where its other U.S. assembly plants, as well as nearly all of its competitors, are located (Figure 11-11).

The other situation factor, minimizing the cost of shipping its inputs, was also important. Honda's most important inputs were the engine and transmission, which were to come from existing factories in western Ohio. Other parts would come from factories already shipping to Honda's two assembly plants in central Ohio. That guided Honda to the portion of auto alley encompassing Illinois, Indiana, and Ohio.

Site factors helped Honda find specific locations within auto alley. Principal site factors were land and labor, though these pointed Honda to different locations. The land site factor suggested a rural location. Honda wanted a large tract of land in order to construct a spread-out one-story factory. It needed to be near at least one interstate highway because most parts would arrive and finished vehicles would leave by truck. It also needed to be next to a rail line because some parts would come from Japan by boat and train, and finished vehicles would be shipped to the west coast by rail. An assembly plant hires several thousand workers, so Honda needed a large labor supply within a 1-hour commuting range. But it didn't want to compete for workers with existing assembly plants. That could lead to a shortage of skilled workers and push up wages. So Honda looked for areas outside the 1-hour commuting range around existing assembly plants.

Honda's short list of locations included Decatur in eastern Illinois, Greensburg in southwestern Indiana, and unnamed communities in west-central Ohio. The third site factor, capital, helped Honda make its final pick. The state governments of Illinois, Indiana, and Ohio were all willing to provide Honda with financial support for roads, utilities, and worker training. But Honda considered Indiana the safest choice: The governors of the other two states at the time were involved in financial scandals. ■

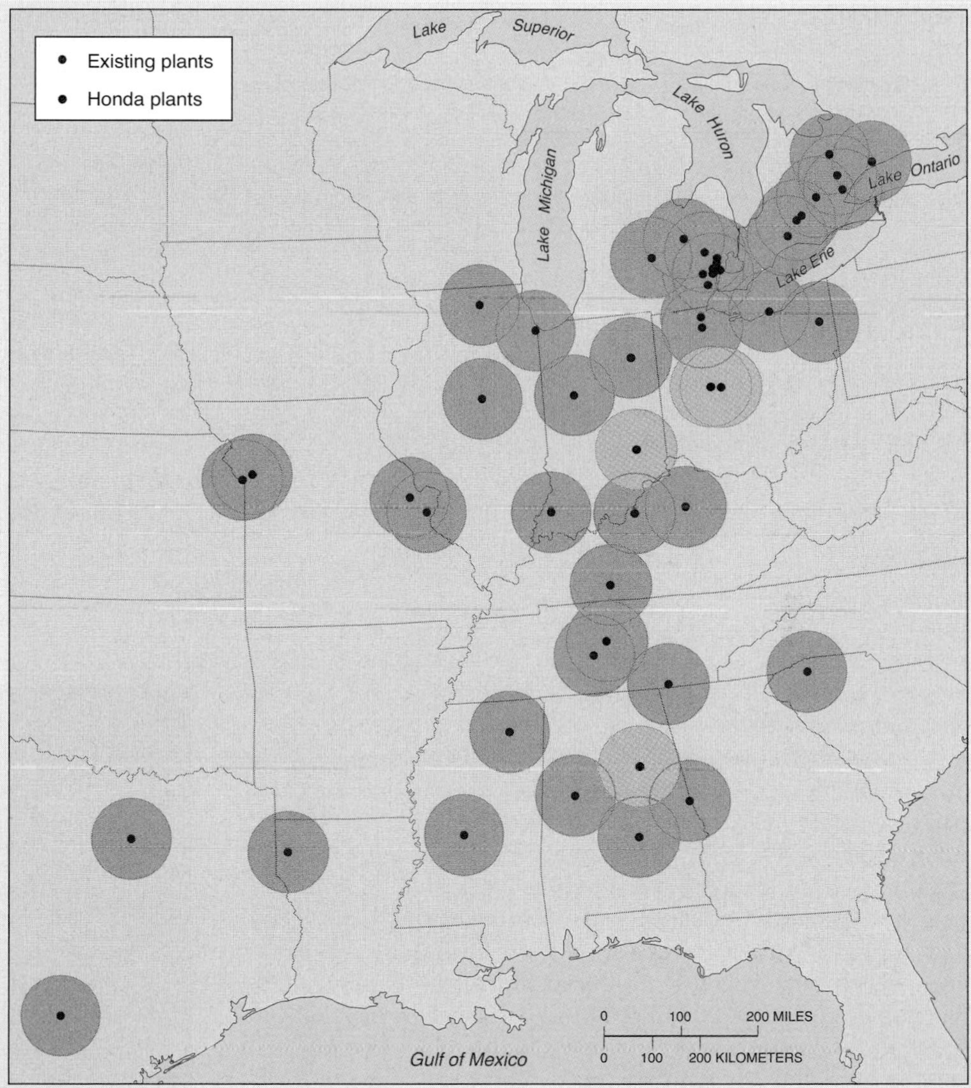

FIGURE 11-11 Labor markets around motor-vehicle assembly plants. Assembly plants draw their workforce from within a roughly 1-hour radius. New plants have been located outside the labor market areas of existing plants.

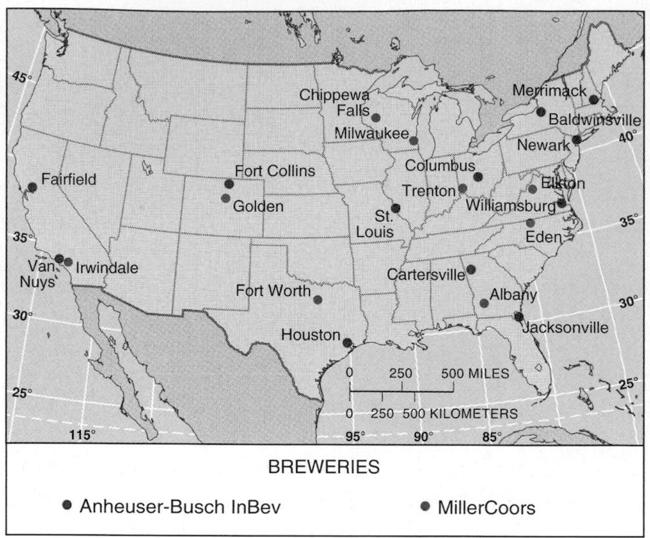

FIGURE 11-12 A bulk-gaining industry: beer-making. The two best-selling brewing companies locate their plants near major population concentrations. Most breweries are clustered in the heavily populated Northeast.

people live, bottlers can minimize costs by producing beverages near their consumers instead of shipping water (their heaviest and bulkiest input) long distances. A filled container has the same volume as an empty one, but it is much heavier. Because they are heavier, the filled containers are more expensive to ship than the empty ones, and bottlers locate near their customers rather than the manufacturers of the containers.

Single-Market Manufacturers

Single-market manufacturers are specialized manufacturers with only one or two customers. The optimal location for these factories is often in close proximity to the customer.

An example of a single-market manufacturer is a producer of parts for motor vehicles (Figure 11-13). The typical passenger car weighs about 3,300 pounds and contains about 54 percent steel, 11 percent iron, 8 percent plastic, 7 percent aluminum, 6 percent fluids and lubricants, 4 percent rubber, 3 percent glass, and 6 percent other materials. The total value of the parts attached to new vehicles produced annually in the United States is more than $200 billion. Most parts makers have only a handful of customers—the major motor-vehicle producers such as Ford and Toyota. Parts makers now ship most of their products directly to the carmaker's assembly plants clustered in "auto alley."

Proximity to the assembly plant is increasingly important for parts producers because of the diffusion of "just-in-time" delivery (see Key Issue 4). Under "just-in-time," parts are delivered to the assembly plant just in time to be used, often within minutes, rather than weeks or months in advance. For some parts makers, just-in-time delivery dictates that they build their factories as close as possible to their customers, the final assembly plants. Seats, for example, are invariably manufactured at a location within an hour of the final assembly plant. The seat is an especially large and bulky object, and carmakers do not want to waste valuable space in their assembly plants by piling up an inventory of them. Most engines, transmissions, and metal body parts are also produced at locations only a couple of hours away from an assembly plant.

On the other hand, many parts do not need to be manufactured close to the customer. For them, changing site factors are more important. Some locate in Mexico to take advantage of lower labor costs. Others come from China, where labor costs are even lower but shipping costs are higher.

Perishable Products

To deliver their products to consumers as rapidly as possible, perishable-product industries must be located near their markets. Because few people want stale bread or sour milk, food producers such as bakers and milk bottlers must locate near their customers to assure rapid delivery (Figure 11-14). Processors of fresh food into frozen, canned, and preserved products can, however, locate far from their customers. Cheese and butter, for example, are manufactured in Wisconsin because rapid delivery to the urban markets is not critical for products with a long shelf life, and the area is well suited agriculturally for raising dairy cows.

The daily newspaper is an example of a product other than food that is highly perishable, because it contains dated information. People demand their newspaper as soon after its printing as possible. Therefore, newspaper publishers must locate near markets to minimize transportation cost.

FIGURE 11-13 A single-market manufacturer: seat-making. Faurecia, a French company, makes seats for the BMW 3 Series at a factory in Leipzig, Germany, adjacent to BMW's assembly plant.

FIGURE 11-14 A perishable product: newspapers. Delivery of afternoon newspapers was once commonly handled by youths after school. Most afternoon newspapers have ceased production.

Difficulty with timely delivery is one of the main factors in the decline of newspapers. Electronic devices—computers and handheld devices—can deliver news more quickly than a newspaper. Little wonder that during the first decade of the twenty-first century, print publishing jobs declined from 1 million to 800,000 in the United States, whereas Internet publishing jobs increased from 70,000 to 80,000.

Ship, Rail, Truck, or Air?

Inputs and products are transported in one of four ways—via ship, rail, truck, or air. Firms seek the lowest-cost mode of transport, but which of the four alternatives is cheapest changes with the distance that goods are being sent.

The farther something is transported, the lower is the cost per kilometer (or mile). Longer-distance transportation is cheaper per kilometer in part because firms must pay workers to load goods on and off vehicles, whether the material travels 10 kilometers or 10,000. The cost per kilometer decreases at different rates for each of the four modes because the loading and unloading expenses differ for each mode.

- **Trucks.** Most often used for short-distance delivery, because they can be loaded and unloaded quickly and cheaply. Truck delivery is especially advantageous if the driver can reach the destination within one day, before having to stop for an extended rest.
- **Trains.** Often used to ship to destinations that take longer than one day to reach, such as between the east and west coasts of the United States. Trains take longer than trucks to load, but once underway aren't required to make daily rest stops like truck drivers.
- **Ships.** Attractive for very long distances because the cost per kilometer is very low. Slower than land-based transportation, but used when shipping cannot be done by train or truck, such as to North America from Europe or Asia.
- **Air.** Most expensive for all distances, so is usually reserved for speedy delivery of small-bulk, high-value packages.

Modes of delivery are often mixed. For example, air-freight companies pick up packages in the afternoon and transport them by truck to the nearest airport. Late at night, planes filled with packages are flown to a central hub airport in the interior of the country, such as Memphis, Tennessee, and Louisville, Kentucky. The packages are transferred to other planes, flown to airports nearest their destination, transferred to trucks, and delivered the next morning.

Containerization has facilitated transfer of packages between modes. Containers may be packed into a rail car, transferred quickly to a container ship to cross the ocean, and unloaded into trucks at the other end. Large ships have been specially built to accommodate large numbers of rectangular box-like containers.

Regardless of transportation mode, cost rises each time that inputs or products are transferred from one mode to another. For example, workers must unload goods from a truck and then reload them onto a plane. The company may need to build or rent a warehouse to store goods temporarily after unloading from one mode and before loading to another mode. Some companies may calculate that the cost of one mode is lower for some inputs and products, whereas another mode may be cheaper for other goods. Many companies that use multiple transport modes locate at a **break-of-bulk point**, which is a location where transfer among transportation modes is possible (Figure 11-15). Important break-of-bulk points include seaports and airports. For example, a steel mill near the port of Baltimore receives iron ore by ship from South America and coal by train from Appalachia.

FIGURE 11-15 Break-of-bulk point: Port of Long Beach, California. Most goods shipped across the ocean are packed in uniformly sized containers, which can be quickly transferred between ships and trucks or trains.

KEY ISSUE 3

Why Are Site Factors Important?

- Labor
- Land
- Capital

Site factors result from the unique characteristics of a location. Land, labor, and capital are the three traditional production factors that may vary among locations. ■

Labor

The most important site factor at a global scale is labor. Minimizing labor costs is important for some industries, and the variation of labor costs around the world is large. Worldwide, around one-half billion workers are engaged in industry, according to the UN International Labor Organization (ILO). China has around one-fourth of the world's manufacturing workers, India around one-fifth, and all MDCs combined around one-fifth.

Labor-Intensive Industries

A **labor-intensive industry** is one in which wages and other compensation paid to employees constitute a high percentage of expenses. Labor constitutes an average of 11 percent of overall manufacturing costs in the United States, so a labor-intensive industry in the United States would have a much higher percentage than that. The reverse case, an industry with a much lower than average percentage of expenditures on labor, is considered capital-intensive.

The average wage paid to manufacturing workers exceeds $20 per hour in North America, Western Europe, and other MDCs. Health care, retirement pensions, and other benefits add substantially to the compensation. In LDCs, average wages are less than $5 per hour and include limited additional benefits. For some manufacturers—but not all—the difference between paying workers $5 and $20 per hour is critical.

A labor-intensive industry is not the same as a high-wage industry. "Labor-intensive" is measured as a percentage, whereas "high-wage" is measured in dollars or other currencies. For example, motor-vehicle workers are paid much higher hourly wages than textile workers, yet the textile industry is labor-intensive and the auto industry is not. Although auto workers earn relatively high wages, most of the value of a car is accounted for by the parts and the machinery needed to put the parts together. On the other hand, labor accounts for a large percentage of the cost of producing a towel or shirt when compared with materials and machinery.

Textiles: Labor-Intensive

Production of apparel and **textiles**, which are woven fabrics, is a prominent example of an industry that generally requires less-skilled, low-cost workers. Textile and apparel production involves three principal steps:

- Spinning of fibers and other preparatory work to make yarn from natural or human-made materials
- Weaving or knitting of yarn into fabric (as well as finishing of fabric by bleaching or dyeing)
- Cutting and sewing of fabric for assembling into clothing and other products.

The textile and apparel industry accounts for 6 percent of the dollar value of world manufacturing but a much higher 14 percent of world manufacturing employment, an indicator that it is a labor-intensive industry. The percentage of the world's women employed in this type of manufacturing is even higher.

Spinning, weaving, and sewing are all labor-intensive compared to other industries, but the importance of labor varies somewhat among them. As a result, their global distributions are not identical, because the three steps are not equally labor-intensive.

TEXTILE AND APPAREL SPINNING. Fibers can be spun from natural or synthetic elements. The principal natural fiber is cotton. Synthetics now account for three-fourths and natural fibers only one-fourth of world thread production.

Before the Industrial Revolution, spinning of cotton was a job for women, often an unmarried daughter still living at home, called a spinster, a term that came to be applied to any unmarried woman. Children usually performed carding, which involved preparing the fibers for spinning by untangling them onto rolls called cards.

Because it is still a labor-intensive industry, spinning is done primarily in low-wage countries (Figure 11-16). China produces two-thirds of the world's cotton thread.

Synthetic fibers include regenerated synthetics and true synthetics. Regenerated synthetic fibers are produced from natural raw materials modified to produce fibers suitable for weaving. The first commercially successful regenerated synthetic was rayon, made by processing the cellulose in wood pulp. True synthetic fibers are produced from substances like petrochemicals that do not naturally form fibers. The first true synthetic fiber, nylon, was developed from petroleum in 1937. Polyester is now the leading true synthetic, accounting for one-third of synthetic fiber production.

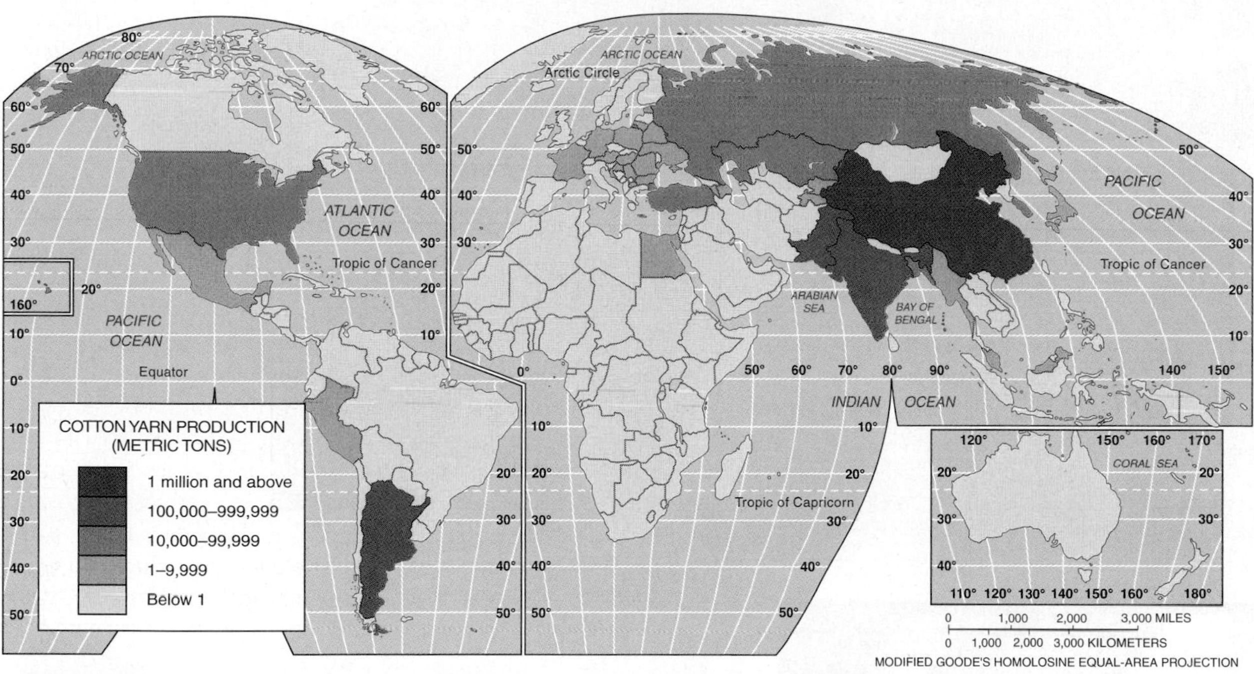

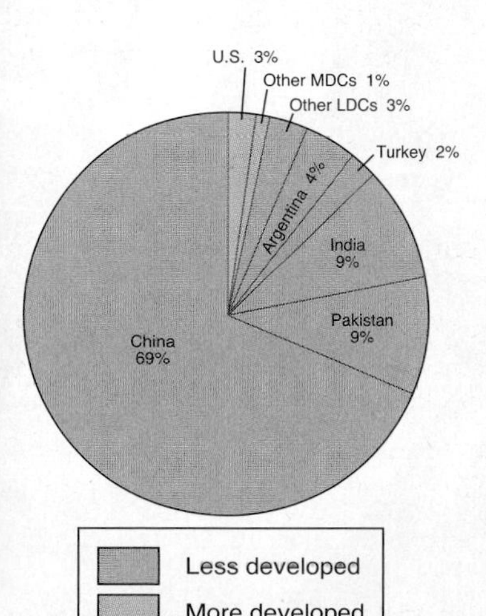

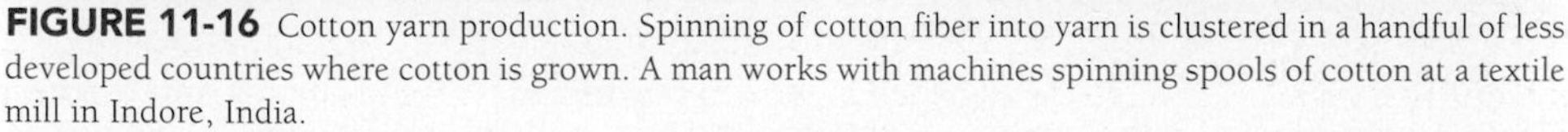
FIGURE 11-16 Cotton yarn production. Spinning of cotton fiber into yarn is clustered in a handful of less developed countries where cotton is grown. A man works with machines spinning spools of cotton at a textile mill in Indore, India.

TEXTILE AND APPAREL WEAVING. For thousands of years, fabric has been woven or laced together by hand on a loom, which is a frame on which two sets of threads are placed at right angles to each other. One set of threads, called a warp, is strung lengthwise. A second set of threads, called a weft, is carried in a shuttle that is inserted over and under the warp. As the process of weaving was physically hard work, weavers were traditionally men.

For mechanized weaving, labor constitutes a high percentage of the total production cost. Consequently, weaving especially is highly clustered in low-wage countries: 93 percent of the world's woven cotton fabric is produced in LDCs (Figure 11-17). Despite their remoteness from European and North American markets, China and India have become the dominant fabric producers because lower labor costs offset the expense of shipping inputs and products long distances. China alone accounts for

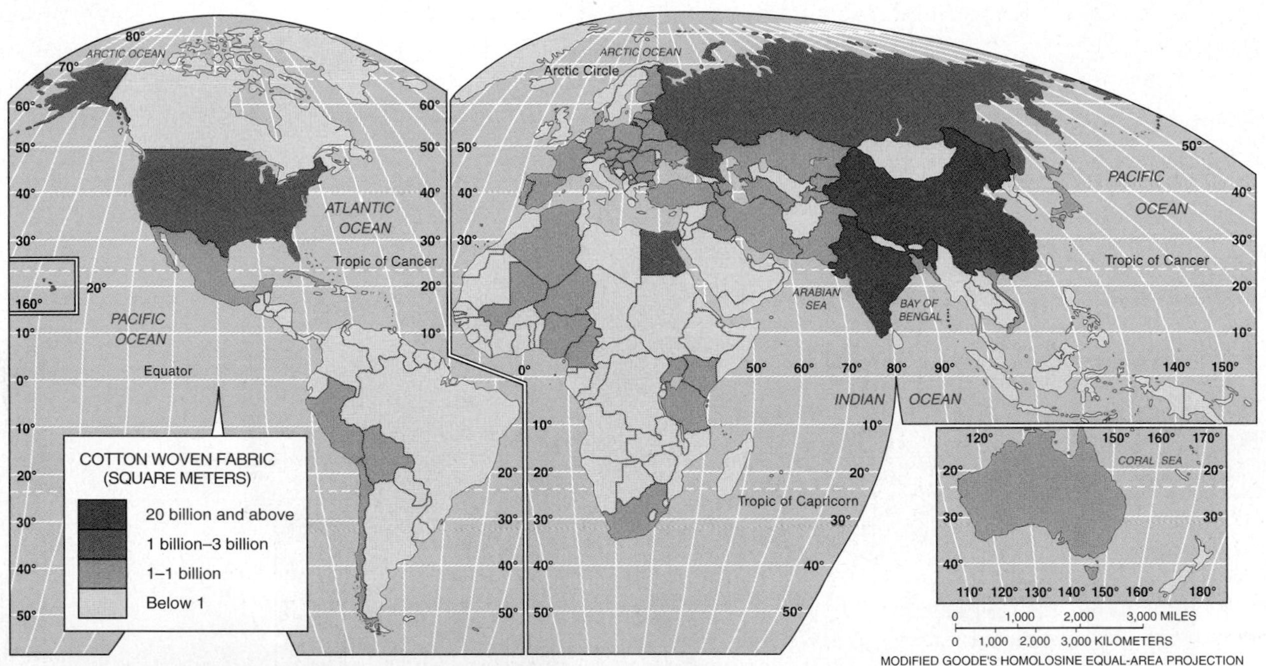

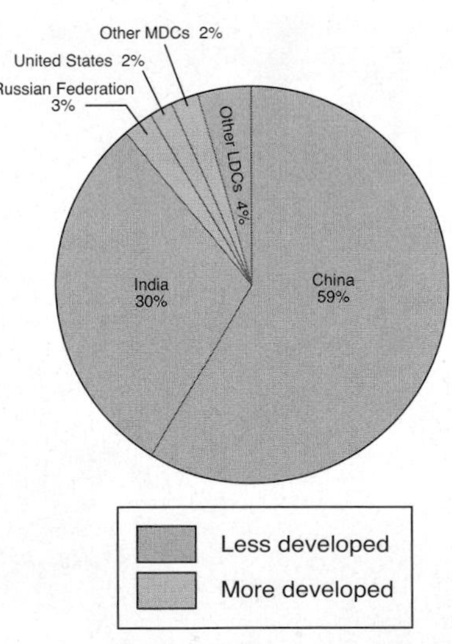

FIGURE 11-17 Woven cotton fabric production. Woven cotton fabric is likely to be produced in LDCs because the process is more labor-intensive than the other major processes in textile and clothing manufacturing. The cotton looms in the photograph are from a factory in North Carolina during the 1990s before it closed.

nearly 60 percent of the world's woven cotton fabric production, and India another 30 percent.

TEXTILE AND APPAREL ASSEMBLY. Sewing is probably an even older human activity than spinning and weaving. Needles made from animal horns or bones date back tens of thousands of years, and iron needles date from the fourteenth century.

The first functional sewing machine was invented by French tailor Barthelemy Thimonnier in 1830. In 1841, Thimonnier installed 80 sewing machines in a factory in St.-Etienne, France, to sew uniforms for the French army. However, Parisian tailors, fearing the machines would put them out of work, stormed the factory and destroyed the machines. Isaac Singer manufactured the first commercially successful sewing machine in the United States during the 1850s, but he was convicted of infringing a patent filed by Elias Howe in 1846.

Textiles are assembled into four main types of products—garments, carpets, home products such as bed linens and curtains, and industrial items such as headliners inside motor vehicles. MDCs play a larger role in assembly than in spinning and weaving because most of the consumers of assembled products are located in MDCs. For example, two-thirds of the women's blouses sold worldwide in a year are sewn in MDCs (Figure 11-18).

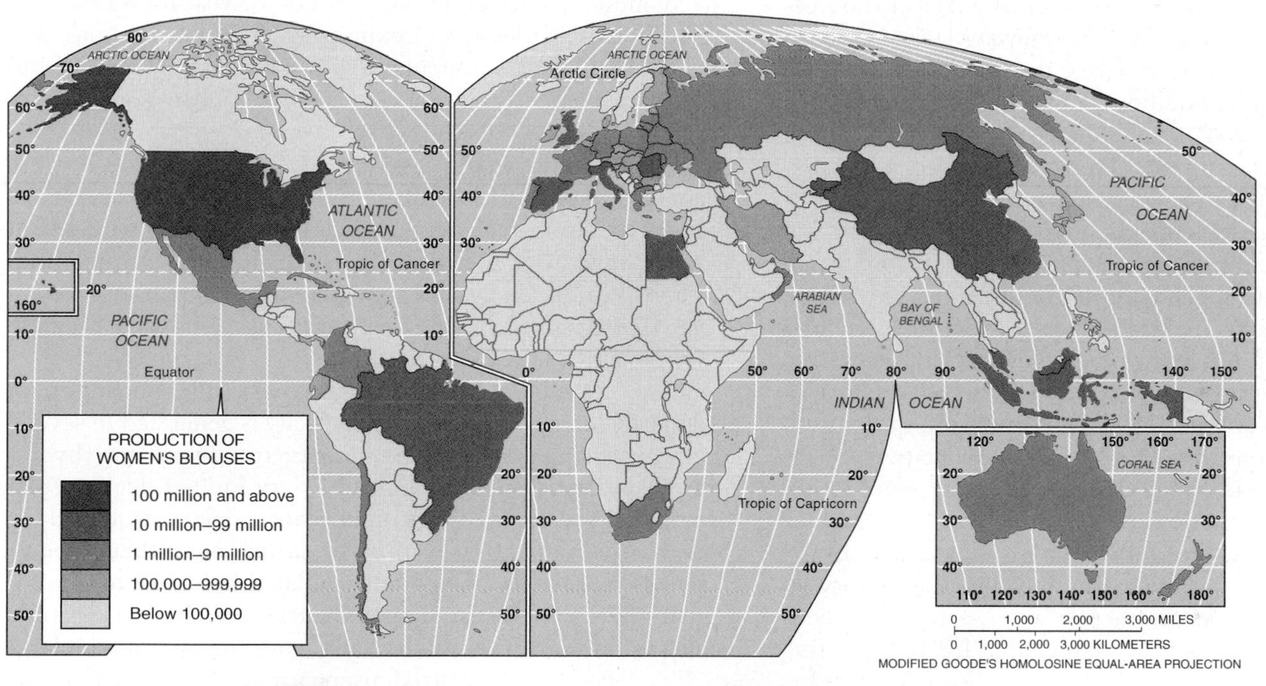

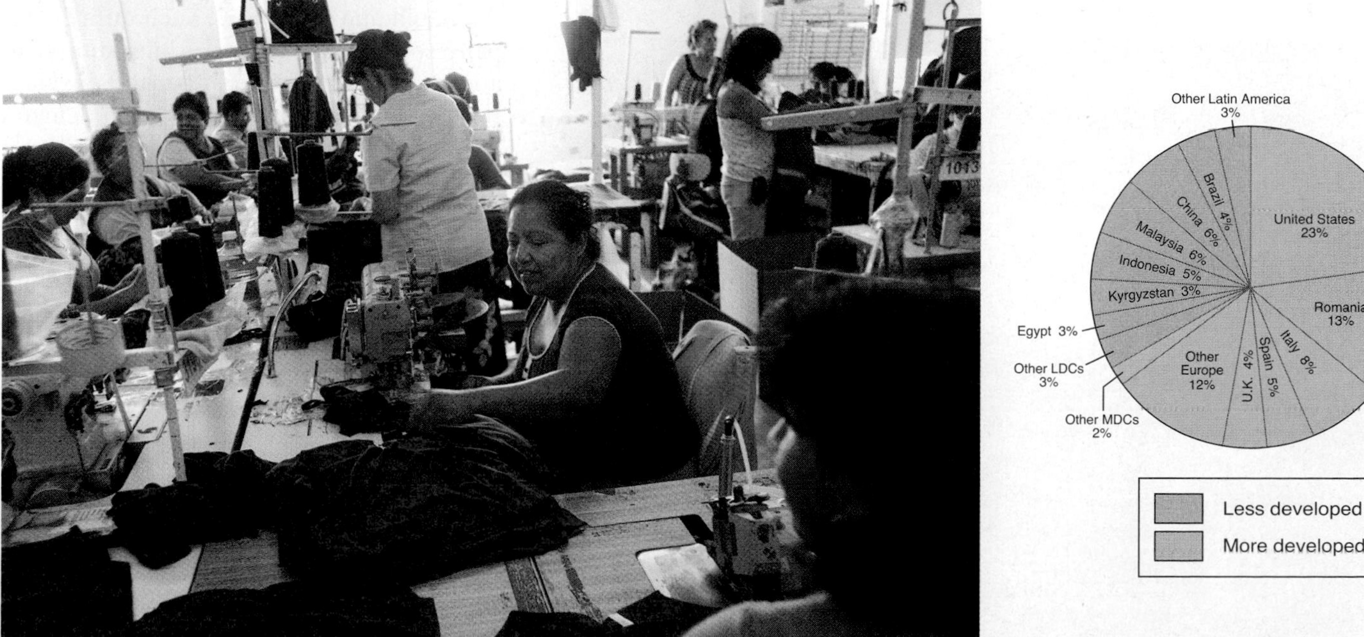

FIGURE 11-18 Production of women's blouses. Sewing of cotton fabric into blouses is more likely to take place in MDCs. Clothing producers must balance the need for low-wage workers against the need for proximity to customers. These women are sewing garments in Los Angeles.

Land

Land suitable for constructing a factory can be found in many places. If considered to encompass natural and human resources in addition to terra firma, "land" is a critical site factor.

Rural Sites

Early factories located inside cities due to a combination of situation and site factors. A city offered an attractive situation—proximity to a large local market and convenience in shipping to a national market by rail. A city also offered an attractive site—proximity to a large supply of labor as well as to sources of capital. The site factor that cities have always lacked is abundant land. To get the necessary space in cities, early factories were typically multistory buildings. Raw materials were hoisted to the upper floors to make smaller parts, which were then sent downstairs on chutes and pulleys for final assembly and shipment. Water was stored in tanks on the roof.

Contemporary factories operate more efficiently when laid out in one-story buildings (Figure 11-19). Raw materials are typically delivered at one end and moved through the factory on conveyors or forklift trucks. Products are assembled in logical order and shipped out at the other end. The land needed to build one-story factories is now more likely to be available in suburban or rural locations. Also, land is much cheaper in suburban or rural locations than near the center of a city.

In addition to providing enough space for one-story buildings, locations outside cities are also attractive because they facilitate delivery of inputs and shipment of products. In the past, when most material moved in and out of a factory by rail, a central location was attractive because rail lines converged there. With trucks now responsible for transporting most inputs and products, proximity to major highways is more important for a factory. Especially attractive is the proximity to the junction of a long-distance route and the beltway or ring road that encircles most cities. Thus, factories cluster in industrial parks located near suburban highway junctions.

FIGURE 11-19 Land as a site factor: rural location. South Korean-owned Samick Musical Instruments Company located its piano-making factory in Gallatin, Tennessee.

Environmental Factors

Not every location has the same climate, topography, recreational opportunities, cultural facilities, and cost of living. Some executives select locations because they are attracted to the distinctive amenities of a site. Attractions could be relatively mild climates and opportunities for year-round outdoor recreation activities, or proximity to cultural facilities and major-league sports franchises. Industries may be attracted to specific parcels of land that are accessible to low-cost energy sources. Prior to the Industrial Revolution, many economic activities were located near rivers and close to forests because running water and the burning of wood were the two most important sources of energy. When coal became the dominant form of industrial energy in the late eighteenth century, location near coalfields became more important. Because coalfields were less ubiquitous than streams or forests, industry began to concentrate in fewer locations.

In the twentieth century, electricity became an important source of energy for industry. Electricity is generated in several ways, by using coal, oil, natural gas, running water (hydroelectricity), nuclear fuel, and, to a very limited degree, solar energy and wind. In the United States, electricity usually is purchased from utility companies, which are either publicly owned or privately owned but regulated by the state government. Like home consumers, industries are charged a certain rate per kilowatt hour of electricity consumed, although large industrial users usually pay a lower rate than do home consumers. Each utility company sets its own rate schedule, subject to approval by its state's regulatory agency. Industries with a particularly high demand for energy may select a location with lower electrical rates.

The aluminum industry, for example, requires a large amount of electricity to separate pure aluminum from bauxite ore (Figure 11-20). Aluminum producers locate near dams to take advantage of the large amount of cheap hydroelectric power generated there. The oldest continuously operating aluminum production and fabricating plant in the United States at Massena, New York, was established in 1902 by the Pittsburgh Reduction Co. (now Alcoa, Inc.) near a dam constructed by the St. Lawrence River Power Co. as part of a three-mile canal linking the St. Lawrence and Grasse rivers. Alcoa, the world's largest aluminum producer, also makes aluminum near other sources of inexpensive hydroelectric power.

FIGURE 11-20 Land as a site factor: aluminum industry. Low-cost electricity is a critical site factor for aluminum producers. To obtain electricity, Alcoa constructed the Cheoah Dam across the Little Tennessee River in 1919.

As an indication of the importance of inexpensive electricity for aluminum production, a subsidiary of Alcoa even owns dams that generate power along the Cheoah, Little Tennessee, and Yadkin rivers in eastern Tennessee and western North Carolina.

Capital

Manufacturers typically borrow funds to establish new factories or expand existing ones. The U.S. motor-vehicle industry concentrated in Michigan early in the twentieth century largely because this region's financial institutions were more willing than eastern banks to lend money to the industry's pioneers. The most important factor in the clustering of high-tech industries in California's Silicon Valley—even more important than proximity to skilled labor—was the availability of capital. Banks in Silicon Valley have long been willing to provide money for new software and communications firms even though lenders elsewhere have hesitated. High-tech industries have been risky propositions—roughly two-thirds of them fail—but Silicon Valley financial institutions have continued to lend money to engineers with good ideas so that they can buy the software, communications, and networks they need to get started. One-fourth of all capital in the United States is spent on new industries in the Silicon Valley.

The ability to borrow money has become a critical factor in the distribution of industry in LDCs. Financial institutions in many LDCs are short of funds, so new industries must seek loans from banks in MDCs. But enterprises may not get loans if they are located in a country that is perceived to have an unstable political system, a high debt level, or ill-advised economic policies.

KEY ISSUE 4
Why Are Location Factors Changing?

- **Attraction of New Industrial Regions**
- **Renewed Attraction of Traditional Industrial Regions**

Industry is on the move around the world. Changing site factors have been especially important in stimulating industrial growth in new regions, internationally and within MDCs. At the same time, some industries remain in the traditional regions, primarily because of changing situation factors. ■

Attraction of New Industrial Regions

Labor is the site factor that is changing especially dramatically in the twenty-first century. To minimize labor costs, some manufacturers are locating in places where prevailing wage rates are lower than in traditional industrial regions.

Changing Industrial Distribution Within MDCs

Within MDCs, industry is shifting away from the traditional industrial areas of northwestern Europe and northeastern United States. In the United States, industry has shifted from the northeast toward the south and west. In Europe, government policies have encouraged relocation toward economically distressed peripheral areas.

Interregional Shift in the United States

The northeastern United States lost 6 million jobs in manufacturing between 1950 and 2009 (Figure 11-21). Especially large declines were recorded by New York State and Pennsylvania, states that once served as centers for clothing, textile, steel, and fabricated metal manufacturing. Meanwhile, 2 million manufacturing jobs were added in the South and West between 1950 and 2009. California and Texas each added one-half million manufacturing jobs.

Industrialization during the late nineteenth and early twentieth centuries largely bypassed the South, which had not recovered from losing the Civil War. The South lacked infrastructure needed for industrial development: Road and rail networks were less intensively developed in the South, and electricity was less common. As a result, the South was the poorest region of the United States. Industrial growth in the South since the 1930s has been stimulated in part by government policies to reduce historical disparities. The Tennessee Valley Authority brought electricity to much of the rural South, and roads were constructed in previously inaccessible sections of the Appalachians, Piedmont, and Ozarks. Air-conditioning made living and working in the South more tolerable during the summer.

RIGHT-TO-WORK LAWS. The principal lure for many manufacturers was enactment by Southern states of **right-to-work laws**. A right-to-work law requires a factory to maintain a so-called "open shop" and prohibits a "closed shop." In a "closed shop," a company and a union agree that everyone must join the union to work in the factory. In an "open shop," a union and a company may not negotiate a contract that requires workers to join a union as a condition of employment.

By enacting right-to-work laws, Southern states made it much more difficult for unions to organize factory workers, collect dues, and bargain with employers from a position of strength. More importantly, the region was especially attractive for companies working hard to keep out a union altogether. The right-to-work laws sent a powerful signal that antiunion attitudes would be tolerated, even actively supported. As a result, the percentage of workers who are members of a union is much lower in the South than elsewhere in the United States.

Steel, textiles, tobacco products, and furniture industries have become dispersed through smaller communities in the South, many in search of a labor force willing to work for less pay than in the North and forgo joining a union. The Gulf Coast has become an important industrial area because of access to oil and natural gas. Along the Gulf Coast are oil refining, petrochemical manufacturing, food processing, and aerospace product manufacturing.

TEXTILE PRODUCTION. The textile and apparel industry has been especially prominent in opening production in lower-wage locations while shutting down production in higher-wage locations. The U.S. textile and apparel industry was heavily concentrated in the Northeast during the early twentieth century, then shifted to the South and West.

New York's Garment District, near Pennsylvania Station at 7th Avenue and 33rd Street, once housed a large percentage of the nation's textile and apparel manufacturers. Its major attraction was a large supply of European immigrants willing to weave and sew long hours in sweatshops for low pay. Buyers from around the country arrived in New York, mostly by train, twice a year to select the clothing for sale in their stores during the next season and to place orders for making the clothes with Garment District manufacturers.

Most textile and apparel production in the United States moved from the Northeast to the Southeast during the mid-twentieth century. Favored sites were small towns of the Appalachian, Piedmont, and Ozark mountains, especially western North and South Carolina and northern Georgia and Alabama. The area is home to 99 percent of U.S. hosiery and sock producers, half of them in North Carolina (Figure 11-22).

Prevailing wage rates were much lower in the Southeast. Even more important for manufacturers, workers in the Southeast showed little interest in joining one of the unions established by

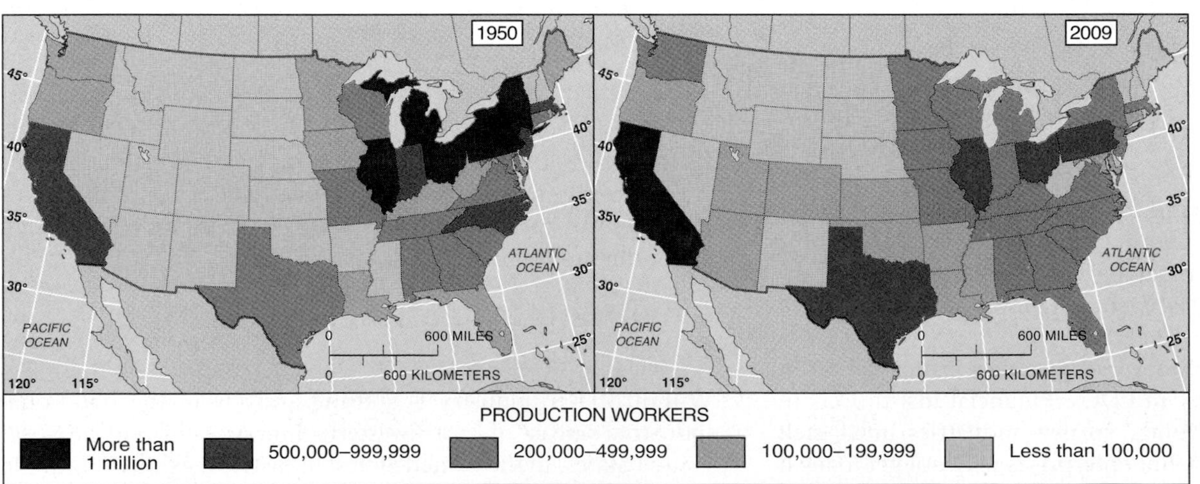

FIGURE 11-21 Changing U.S. manufacturing. States traditionally associated with manufacturing in the Northeast accounted for two-thirds of the country's manufacturing in 1950, compared to only two-fifths in 2009.

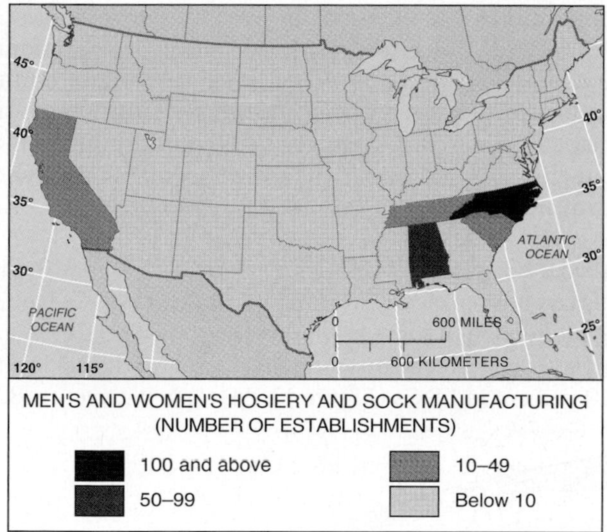

FIGURE 11-22 Men's and women's socks and hosiery manufacturers. To support their labor-intensive industry, hosiery manufacturers locate where a low-cost workforce exists. In the United States, the lowest-cost labor is concentrated in the Southeast. The U.S. Bureau of the Census classifies these manufacturers as North American Industry Classification System (NAICS) 31511.

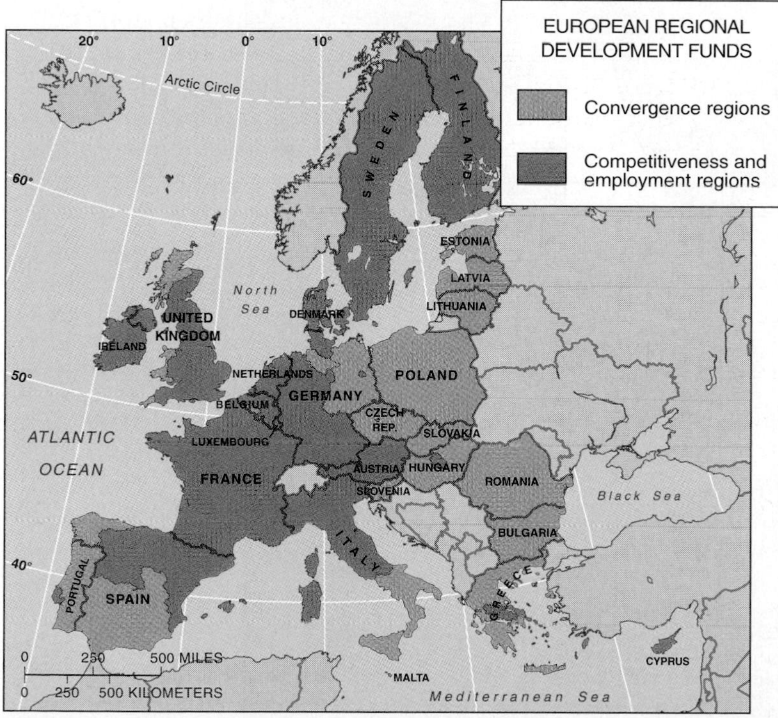

FIGURE 11-23 European Union Structural Funds. The European Union provides subsidies in regions with economic difficulties because of declining industries, as well as to regions that have lower-than-average incomes.

Northeastern textile and apparel workers to bargain for higher wages and safer working conditions. Although located farther from Northeastern population centers, Southeastern mills were able to reach markets easily after the opening of the interstate highway system beginning in the 1950s.

INTERREGIONAL SHIFTS IN EUROPE. Manufacturing has diffused from traditional industrial centers in northwestern Europe toward southern and eastern Europe. In contrast to the United States, European government policies have explicitly encouraged this industrial relocation (Figure 11-23). The European Union provides assistance to what it calls convergence regions and competitive and employment regions:

- **Convergence Regions:** Primarily Eastern and Southern Europe, where incomes lag behind Europe's average.
- **Competitive and Employment Regions:** Primarily Western Europe's traditional core industrial areas, which have experienced substantial manufacturing job losses in recent years.

The Western European country with the most rapid manufacturing growth since the late twentieth century has been Spain, especially since its admission to the European Union in 1986. Until then, Spain's manufacturing growth had been retarded by physical and political isolation. Spain's motor-vehicle industry has grown into the second largest in Europe, behind only Germany's, although it is entirely foreign-owned. Spain's leading industrial area is Catalonia, in the northeast, centered on the city of Barcelona. The region has the country's largest motor-vehicle plant and is the center of Spain's textile industry as well.

Several European countries situated east of Germany and west of Russia have become major centers of industrial investment since the fall of communism in the early 1990s. Poland, Czech Republic, and Hungary have had the most industrial development, though other countries in the region have shared in the growth. The region prefers to be called *Central Europe*, reverting to a common pre-Cold War term, to signify its more central location in Europe's changing economy. Central Europe offers manufacturers an attractive combination of two important site and situation factors—labor and market proximity. Central Europe's workers offer manufacturers good value for money—they are less skilled but much cheaper than in Western Europe, more expensive but much more skilled than in Asia and Latin America. At the same time, the region offers closer proximity to the wealthy markets of Western Europe than other emerging industrial centers.

International Shifts in Industry

In 1970, nearly one-half of world industry was in Europe and nearly one-third was in North America; now these two regions account for only one-fourth each. The share of world industry in other regions has increased from one-sixth in 1970 to one-half in 2010.

Increasingly important industrial areas outside of North America and Europe include:

- **East Asia.** Already one of the world's three major industrial regions, as discussed in Key Issue 2. Rapid industrial growth in China means that East Asia likely will account for an increasing share of world industrial production, pulling well ahead of Europe and North America. In addition to China and Japan, East Asia also includes South Korea, which is the world's leading producer of large container ships that play an important role in international trade. South Korea is a leading producer of steel and fabricated metal products, including motor vehicles.

- **South Asia.** Led by India, with one of the fastest-growing economies among large countries. Textiles are India's dominant industrial sector, but motor-vehicle production is growing rapidly. India is now an important center for business services, as discussed in the next chapter. India's GDP is expected to match that of the United States by 2050.
- **Latin America.** The nearest low-wage region to the United States. The cost of shipping from Mexico to the United States is lower than from other LDCs. *Maquiladora* plants have located in Mexico's far north to be as close as possible to the United States. Mexico City, the country's largest market, is the center for industrial production for domestic consumption. Brazil is the leading industrial country in Latin America. Its industries serve primarily the domestic market, which is also the region's largest. Industry is clustered in the southeast of the country, especially around the two largest cities São Paulo and Rio de Janeiro.

CHANGING DISTRIBUTIONS. The shift to new industrial regions can be seen clearly in steel and clothing. MDCs have been losing production of these key industries to LDCs.

In 1980, 80 percent of world steel was produced in MDCs and 20 percent in LDCs (Figure 11-24, top). Between 1980 and

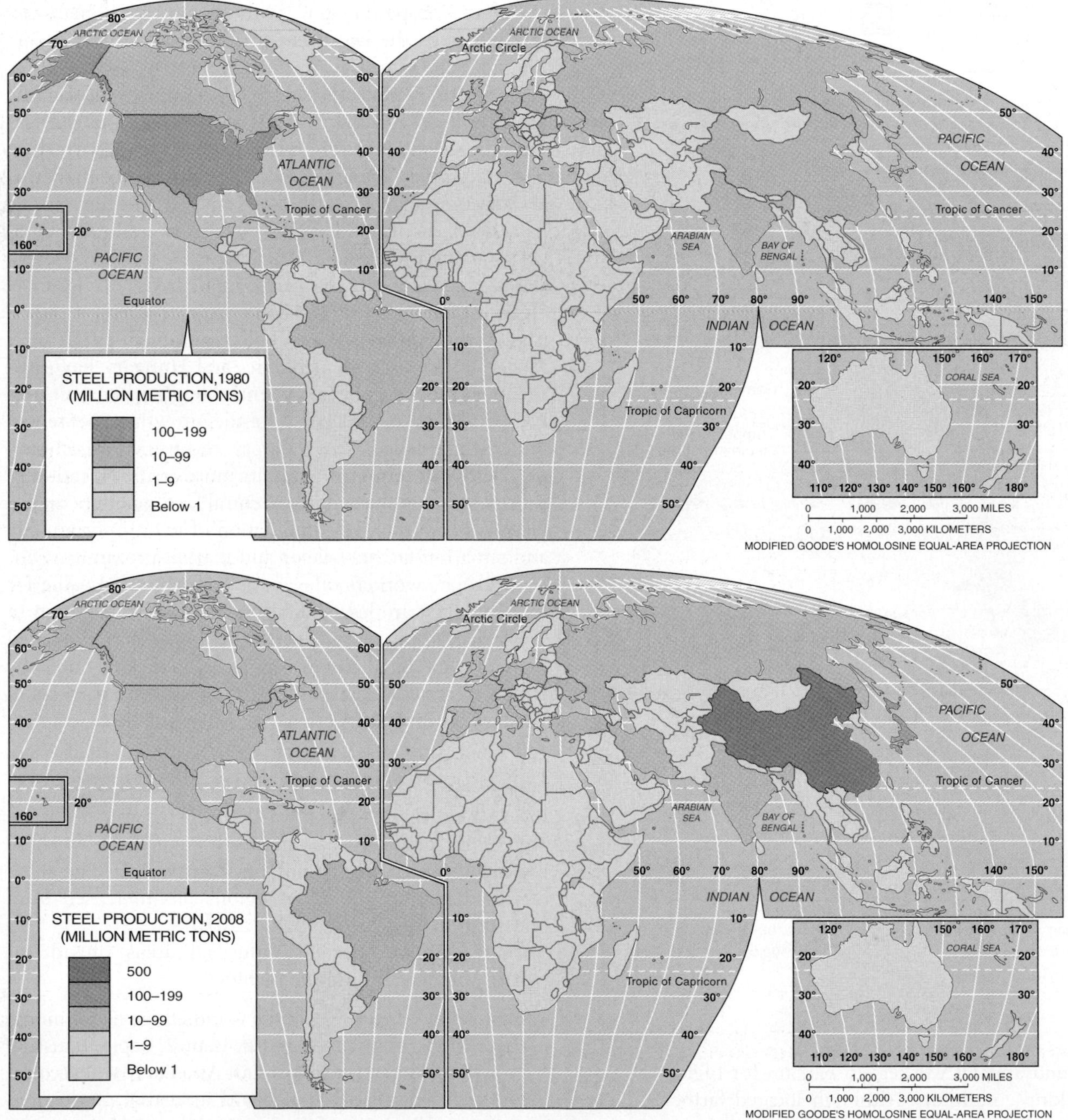

FIGURE 11-24 World steel production, 1980 and 2008. All of the world's increase in steel production has been in LDCs, especially China.

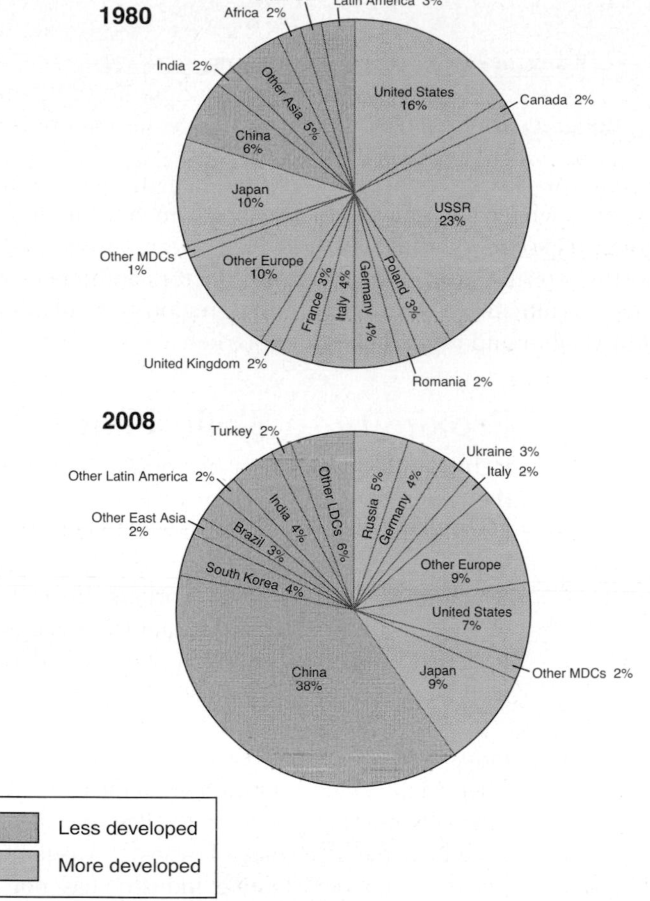

FIGURE 11-25 MDCs accounted for 80 percent of global steel production in 1980, compared to only 40 percent in 2008.

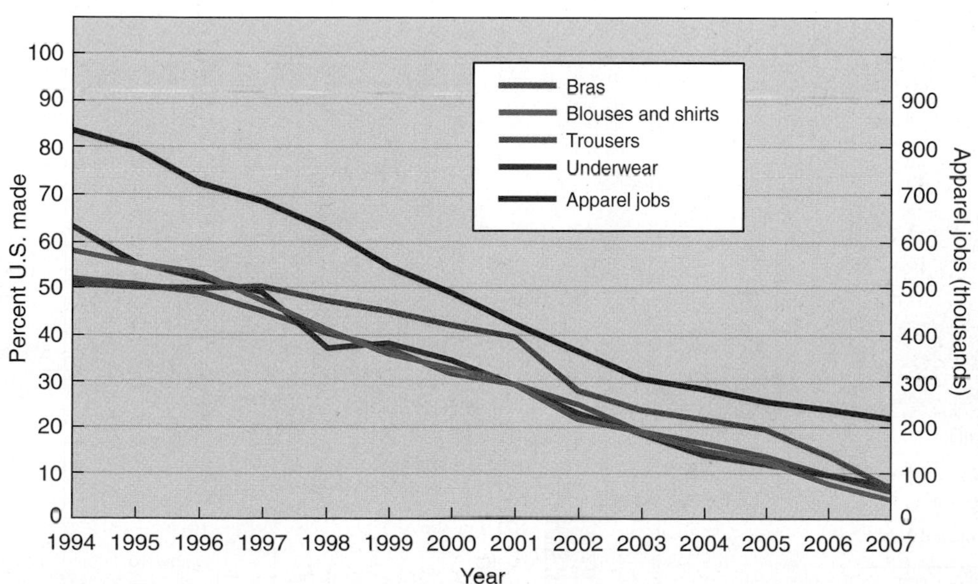

FIGURE 11-26 Apparel production and jobs in the United States. The number of jobs in the apparel industry has declined sharply in the United States since the 1990s. Not by coincidence, the percentage of everyday clothing accounted for by domestic production has decreased sharply, replaced with imports.

2008, the share of world steel production declined to 40 percent in MDCs and increased to 60 percent in LDCs (Figure 11-24, bottom). During this quarter-century, the share of world steel production declined from 50 percent to 23 percent in Europe and from 20 to 8 percent in North America. China, now the world's largest steel producer, accounted for 38 percent of world steel output in 2008, nearly as much as all MDCs combined (Figure 11-25). Otherwise stated, world steel production increased from around 700 million metric tons in 1980 to around 1,300 million metric tons in 2008. Production in MDCs remained about the same at around 500 million metric tons. Meanwhile, production increased during the period by around 500 million metric tons in China and by around 300 million metric tons in other LDCs.

Labor-intensive industries have been especially attracted to LDCs. The number of apparel workers in the United States declined from 900,000 in 1990 to 500,000 in 2000 and to 150,000 in 2009. During this period, most apparel sold in the United States switched from domestic-made to foreign-made (Figure 11-26). As apparel from other countries has become less expensive and less complicated to import into the United States, mills in the Southeast paying $10 to $15 per hour wages have been unable to compete with manufacturers in countries paying less than $1 per hour (Figure 11-27). European countries have been even harder hit by international competition. Compensation for manufacturing employees exceeds $30 per hour in much of Europe.

OUTSOURCING. Transnational corporations have been especially aggressive in using low-cost labor in LDCs. To remain competitive in the global economy, they carefully review their production processes to identify steps that can be performed by low-paid, low-skilled workers in LDCs. Despite greater transportation cost, transnational corporations can profitably transfer some work to LDCs, given the substantial difference in wages between MDCs and LDCs. At the same time, operations that require highly skilled workers remain in factories in MDCs. This selective transfer of some jobs to LDCs is known as the **new international division of labor**.

Transnational corporations allocate production to low-wage countries through **outsourcing**, which is turning over much of the responsibility for production to independent suppliers. Outsourcing contrasts with the approach typical of traditional mass production, called vertical integration, in which a company controls all phases of a highly complex production process. Vertical integration was traditionally regarded as a source of strength for manufacturers because it gave them the ability to do and control everything. Carmakers once made nearly all of their own parts, for example, but now most of this operation is outsourced to other companies able to make the parts cheaper and better. As a result of outsourcing, though, carmakers account for only around 30 percent of the value of the vehicles that bear their names. The rest of the value is tied up in the thousands of parts that go into the vehicles. The makers of these parts must also decide where to locate their factories.

Outsourcing has had a major impact on the distribution of manufacturing because each step in the production process is now scrutinized closely in order to determine the optimal location. For example, carmakers have outsourced production of seats to independent companies. The seats installed in U.S. vehicles are invariably put together in the United States, but many of the parts in the seats are made in other countries (see Global Forces, Local Impacts box).

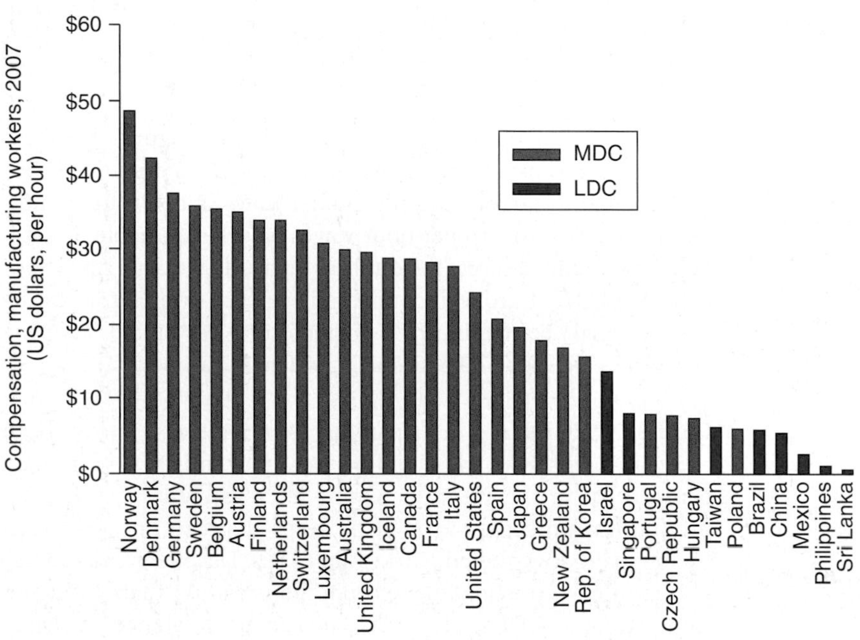

FIGURE 11-27 Manufacturing compensation. Compensation including wages and benefits is much higher in MDCs, especially in Europe, than in LDCs.

Renewed Attraction of Traditional Industrial Regions

Given the strong lure of low-cost labor in new industrial regions, why would any industry locate in one of the traditional regions, especially in the northeastern United States or northwestern Europe? Two location factors influence industries to remain in these traditional regions—availability of skilled labor and rapid delivery to market.

Proximity to Skilled Labor

Henry Ford boasted that he could take people off the street and put them to work with only a few minutes of training. That has changed for many industries, including motor vehicle assembly, which now want skilled workers instead. The search for skilled labor has important geographic implications because it is an asset found principally in the traditional industrial regions.

Computer manufacturing is an example of an industry that has concentrated in relatively high-wage, high-skilled communities of the United States, especially near universities in the Bay Area of California and at the University of Texas at Austin (Figure 11-28). Even the clothing industry has not completely abandoned the Northeast. Dresses, woolens, and other "high-end" clothing products are still made in the region. They require more skill in cutting and assembling the material, and skilled textile workers are more plentiful in the Northeast (Figure 11-29).

Traditionally, factories assigned each worker one specific task to perform repeatedly. Some

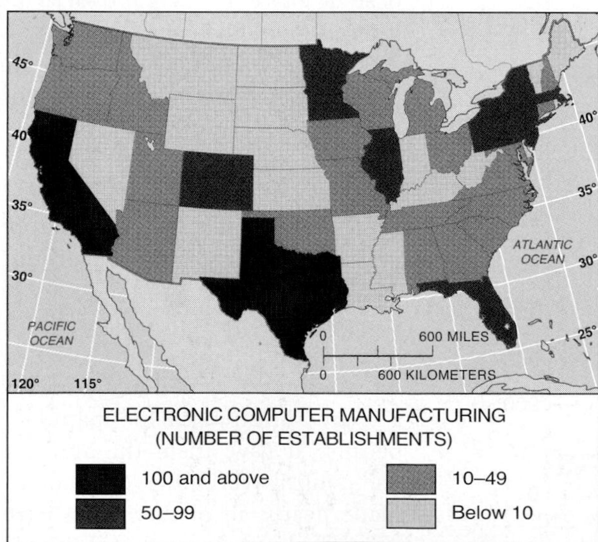

FIGURE 11-28 Electronic computing manufacturing (NAICS 3341). Manufacturers of computing equipment need access to highly skilled workers to perform precision tasks. They are willing to pay relatively high wages to attract the workers. The largest clusters of skilled workers are in the Northeast and on the West Coast.

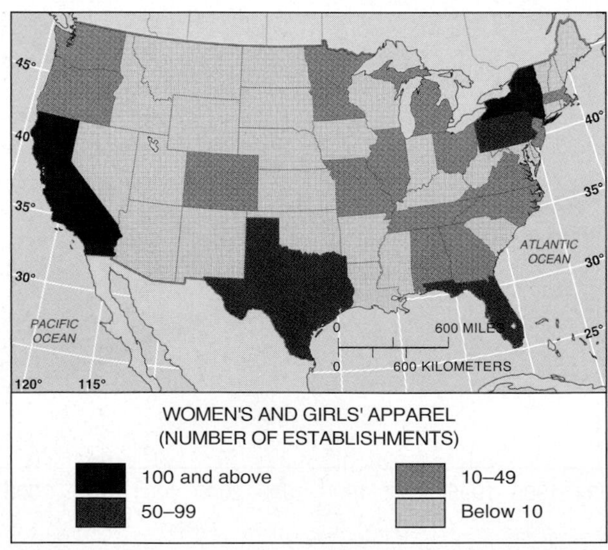

FIGURE 11-29 Women's and girls' cut and sew apparel manufacturing (NAICS 31523). Products that require more skilled workers, such as dresses and knit outerwear, are still produced primarily in or near New York City.

GLOBAL FORCES, LOCAL IMPACTS
What Is an American Car?

Distinctions between "American" and "foreign" motor vehicles have been blurred for the past three decades. Popular media have delighted in showcasing examples of "American" vehicles produced by the Detroit 3 (Chrysler, Ford, and General Motors) that have lower U.S. content than those produced by "Japanese" carmakers such as Honda and Toyota. The U.S. government distinguishes between domestic and foreign vehicles in three ways:

- For measuring fuel efficiency, the U.S. Environmental Protection Agency considers a vehicle domestic if at least 75 percent of its content comes from North America, originally defined as the United States and Canada, and, after enactment of the North American Free Trade Agreement (NAFTA), including Mexico.
- For setting import tariffs, the U.S. Department of Treasury Customs Service considers as domestic a vehicle having at least 50 percent U.S. and Canadian content.
- For informing consumers, the American Automobile Labeling Act of 1992 considers a vehicle domestic if at least 85 percent of the parts originate in the United States and Canada; a part is counted as domestic if at least 70 percent of its overall content comes from the United States and Canada.

According to data derived from Labeling Act reports, vehicles built by foreign-owned carmakers at assembly plants located in the United States had around 60 percent domestic content in 2008. Domestic content for the Detroit 3 in 2008 was 76 percent (Figure 11-30, top). The lower domestic content for foreign carmakers masks differences among individual companies. Honda and Toyota have a level of U.S. content comparable to that of the Detroit 3. German-owned carmakers such as BMW and Daimler have much lower percentages.

The gap in domestic content between the two sets of carmakers narrowed during the 1990s primarily because the foreign-owned companies bought more North American parts. After opening assembly plants in the United States during the 1980s, Japanese-owned carmakers convinced many of their Japanese-owned suppliers to build factories in the United States. During the first decade of the twenty-first century, the gap in domestic content narrowed further because the Detroit 3 bought more foreign parts. More than one-fourth of all new vehicle parts are imported. Mexico has become the leading source of imported parts, and China has been increasing its share rapidly (Figure 11-30, bottom).

In the United States, one-half of vehicle parts are made in the United States by U.S.-owned companies, one-fourth are made in the United States by foreign-owned transnational corporations, and one-fourth are made overseas and imported into the United States. As variations in situation and site costs continually shift from one country to another, these percentages are bound to change. ■

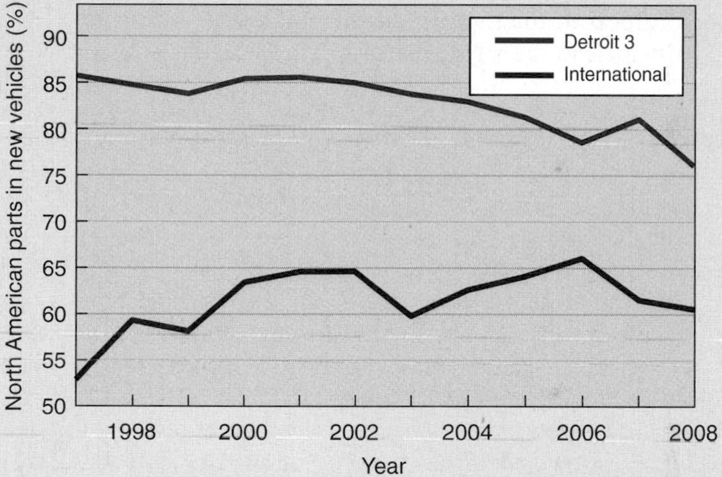

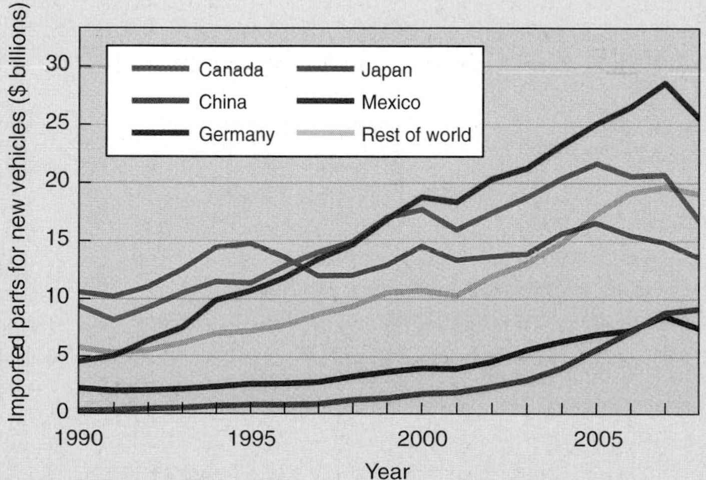

FIGURE 11-30 (top) Motor vehicles sold in the United States by the Detroit 3 contain a higher percentage of U.S.-made parts than do vehicles sold by the internationals (carmakers with headquarters in Japan, Korea, and Germany). (bottom) Mexico is the leading source of imported motor vehicle parts.

geographers call this approach **Fordist** or mass production because the Ford Motor Company was one of the first to organize its production this way early in the twentieth century. At its peak, Ford's factory complex along the River Rouge in Dearborn, Michigan, near Detroit, employed more than 100,000. Most of these workers did not need education or skills to do their jobs, and many were immigrants from Europe or the southern United States. Many industries now follow a lean or flexible production approach. The term **post-Fordist** is sometimes used to describe lean production, to contrast with Fordist production. Again, a carmaker is best known for pioneering lean production—in this case, Toyota.

Three types of work rules distinguish post-Fordist lean production:

1. **Teams.** Workers are placed in teams and told to figure out for themselves how to perform a variety of tasks.
2. **Problem solving.** A problem is addressed through consensus after consulting with all affected parties rather than through filing a complaint or grievance.
3. **Leveling.** Factory workers are treated alike and managers and veterans do not get special treatment; they wear the same uniform, eat in the same cafeteria, park in the same lot, and participate in the same athletic and social activities.

Just-in-Time Delivery

Proximity to market has long been important for many types of manufacturers, as discussed earlier in this chapter. This factor has become even more important in recent years because of the rise of just-in-time delivery. As the name implies, just-in-time is shipment of parts and materials to arrive at a factory moments before they are needed. Just-in-time delivery is especially important for delivery of inputs, such as parts and raw materials, to manufacturers of fabricated products, such as cars and computers.

Under just-in-time, parts and materials arrive at a factory frequently, in many cases daily if not hourly. Suppliers of the parts and materials are told a few days in advance how much will be needed over the next week or two, and first thing each morning they are told exactly what will be needed at precisely what time that day. To meet a tight timetable, a supplier of parts and materials must locate factories near its customers. If given only an hour or two notice, a supplier has no choice but to locate a factory within 50 miles or so of the customer.

Just-in-time delivery reduces the money that a manufacturer must tie up in wasteful inventory. In fact, the percentage of the U.S. economy tied up in inventory has been cut in half during the past quarter-century. Manufacturers also save money through just-in-time delivery by reducing the size of the factory, because space does not have to be wasted on piling up a mountain of inventory. Leading computer manufacturers have eliminated inventory altogether. They build computers only in response to customer orders placed primarily by telephone or over the Internet. In some cases, just-in-time delivery merely shifts the burden of maintaining inventory to suppliers. Wal-Mart, for example, holds low inventories but tells its suppliers to hold high inventories "just in case" a sudden surge in demand requires restocking on short notice.

Just-in-time delivery means that producers have less inventory to cushion against disruptions in the arrival of needed parts. Two kinds of disruptions can result from reliance on just-in-time delivery:

- **Labor unrest.** A strike at one supplier plant can shut down the entire production within a couple of days. A strike in the logistics industry, such as a strike by truckers or dockworkers, could also disrupt deliveries.
- **"Acts of God."** Most common are weather-related incidents, such as blizzards that close highways or floods that damage factories. A notable non-weather-related disruption followed the September 11, 2001, terrorist attacks on the United States. Suppliers in Canada and Mexico were unable to maintain just-in-time deliveries to manufacturers in the United States because the border crossings were closed. The grounding of all civilian aircraft for several days after the attacks prevented delivery of small high-value parts.

SUMMARY

Three recent changes in the structure of manufacturing have geographic consequences:

- Factories have become more productive through introduction of new machinery and processes. A factory may continue to operate at the same location but require fewer workers to produce the same output. Faced with meager prospects of getting another job in the same community, workers laid off at these factories migrate to other regions.
- Companies are locating production in communities where workers are willing to adopt more flexible work rules. Firms are especially attracted to smaller towns where low levels of union membership reduce vulnerability to work stoppages, even if wages are kept low and layoffs become necessary.
- By spreading production among many countries, or among many communities within one country, large corporations have increased their bargaining power with local governments and labor forces. Production can be allocated to locations where the local government is especially helpful and generous in subsidizing the costs of expansion, and the local residents are especially eager to work in the plant.

These, again, are the key issues in the geography of industry:

1. **Where Is Industry Distributed?** Industry is highly concentrated. Three regions where industry clustered during the twentieth century are Europe, North America, and East Asia.
2. **Why Are Situation Factors Important?** Factories try to identify a location where production cost is minimized. Critical industrial location costs include situation factors for some firms and site factors for others. Situation factors involve the cost of transporting both inputs into the factory and products from the factory to consumers.
3. **Why Are Site Factors Important?** Three site factors—land, labor, and capital—control the cost of doing business at a location.
4. **Why Are Location Factors Changing?** New industrial regions are able to attract some industries, especially because of low wage rates. For their part, traditional industrial regions have been able to offer manufacturers skilled workers and proximity to customers demanding just-in-time delivery.

CASE STUDY REVISITED / Throwing BRIC at NAFTA

NAFTA has joined the United States with its neighbors to its immediate north and south to form one of the world's three main industrial regions. Motor vehicles sold in the United States may be assembled in Canada with many Mexican parts.

Integration of North American industry has generated fear in the United States and Canada:

- Labor leaders fear that more manufacturers will relocate production to Mexico to take advantage of lower wage rates. Such labor-intensive industries as food processing and textile manufacturing may be especially attracted to a region where prevailing wage rates are lower.
- Environmentalists fear that NAFTA encourages firms to move production to Mexico because laws governing air- and water-quality standards are less stringent than in the United States and Canada. Mexico has adopted regulations to reduce air pollution in Mexico City; catalytic converters have been required on Mexican automobiles since 1991. But environmentalists charge that environmental protection laws are still not strictly enforced in Mexico.

Mexico faces its own challenges: It has lost a quarter million *maquiladora* jobs since 2000. Electronics firms are especially likely to pull out of Mexico. The reason: At $2 an hour, Mexican wages are higher than in other LDCs, although much lower than in the United States. Many firms are moving to China, where wages are only $1 an hour.

Meanwhile, some analysts believe that industry in North America, as well as in Europe, will be challenged in the coming decades by a new industrial alliance called BRIC. This is an acronym for four countries—Brazil, Russia, India, and China. The four BRIC countries together currently control one-fourth of the world's land and two-fifths of the world's population, but the four combined account for only one-sixth of world GDP (Figure 11-31). In alphabetical order, their economies rank tenth, eighth, twelfth, and third in the world.

The BRIC concept is that if the four giants work together, they will become the world's dominant industrial bloc in the twenty-first century. China and India have the two largest labor forces, whereas Russia and Brazil are especially rich in inputs critical for industry. The four BRIC countries could possess four of the six largest economies in the world by the mid-twenty-first century. However, as an industrial region, BRIC has the obvious drawback of Brazil's being on the other side of the planet from the other three. China, India, and Russia could form a contiguous region, but long-standing animosity among them has limited economic interaction so far. Still, a generation ago, few would have predicted that industry in Mexico would be highly integrated with industry in the United States and Canada. ■

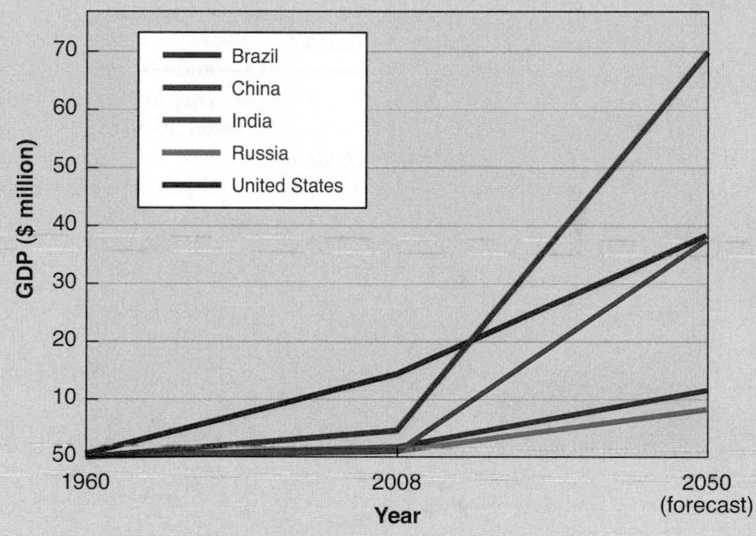

FIGURE 11-31 GDP for BRIC countries and the United States.

KEY TERMS

Break-of-bulk point (p. 355) A location where transfer is possible from one mode of transportation to another.

Bulk-gaining industry (p. 352) An industry in which the final product weighs more or comprises a greater volume than the inputs.

Bulk-reducing industry (p. 350) An industry in which the final product weighs less or comprises a lower volume than the inputs.

Cottage industry (p. 345) Manufacturing based in homes rather than in a factory, commonly found prior to the Industrial Revolution.

Fordist production (p. 368) Form of mass production in which each worker is assigned one specific task to perform repeatedly.

Industrial Revolution (p. 345) A series of improvements in industrial technology that transformed the process of manufacturing goods.

Labor-intensive industry (p. 356) An industry for which labor costs comprise a high percentage of total expenses.

Maquiladora (p. 344) Factories built by U.S. companies in Mexico near the U.S. border, to take advantage of much lower labor costs in Mexico.

New international division of labor (p. 365) Transfer of some types of jobs, especially those requiring low-paid, less skilled workers, from more developed to less developed countries.

Outsourcing (p. 365) A decision by a corporation to turn over much of the responsibility for production to independent suppliers.

Post-Fordist production (p. 368) Adoption by companies of flexible work rules, such as the allocation of workers to teams that perform a variety of tasks.

Right-to-work state (p. 362) A U.S. state that has passed a law preventing a union and company from negotiating a contract that requires workers to join a union as a condition of employment.

Site factors (p. 356) Location factors related to the costs of factors of production inside the plant, such as land, labor, and capital.

Situation factors (p. 350) Location factors related to the transportation of materials into and from a factory.

Textile (p. 356) A fabric made by weaving, used in making clothing.

THINKING GEOGRAPHICALLY

1. What have been the benefits and costs to Canada, Mexico, and the United States as a result of NAFTA?
2. To induce Kia to build its U.S. production facility in Georgia, the state spent $36 million to buy the site; $25 million to prepare the site, including grading; $30 million to provide road improvements, including an interchange off I-85; $6 million to build a rail spur; $20 million to construct a training center; $6 million to operate the center for 5 years; $6 million to develop a training course; $76 million in tax credits; $14 million in sales tax exemptions; and $41 million in training equipment. Did Georgia overpay to win the Kia factory? Explain.
3. Foreign cars account for one-fourth of the sales in the midwestern United States, compared to half in California and other West Coast states. What factors might account for this regional difference?
4. Draw a large triangle on a map of Russia, with one point near Moscow, one point in the Ural Mountains, and one point in Central Asia. What are the principal economic assets of the three regions at each side of the triangle? How do the distributions of markets, resources, and surplus labor vary within Russia?
5. What are the principal manufacturers in your community or area? How have they been affected by increasing global competition?

RESOURCES

Some recent and classic books and articles on industrial geography:

Ashton, Thomas S. *The Industrial Revolution*. London and New York: Oxford University Press, 1997.

Bluestone, Barry, and Bennett Harrison. *The Deindustrialization of America: Plant Closings, Community Abandonment, and the Dismantling of Basic Industry*. New York: Basic Books, 1982.

Dicken, Peter. "Transnational Corporations and Nation-States." *International Social Science Journal* 49 (1997): 77–90.

Essletzbichler, Jürgen. "The Geography of Job Creation and Destruction in the U.S. Manufacturing Sector, 1967–1997." *Annals of the Association of American Geographers* 94 (2004): 602–19.

Harner, John. "Place Identity and Copper Mining in Sonora, Mexico." *Annals of the Association of American Geographers* 91 (2001): 660–80.

Hogan, William T. *Minimills and Integrated Mills: A Comparison of Steelmaking in the United States*. Lexington, MA: Lexington Books, 1987.

———. "What Do They Make, Where, and Does It Matter Any More? Regional Industrial Structures in Britain Since the Great War." *Geography* 7 (1986): 289–304.

Hughes, Alex, and Suzanne Reimer, eds. *Geographies of Commodity Chains*. New York: Routledge, 2004.

Klier, Thomas, and James M. Rubenstein. *Who Really Made Your Car? Restructuring and Geographic Change in the Auto Industry*. Kalamazoo, MI: Upjohn Institute, 2008.

Labrianidis, Lois, ed. *The Moving Frontier: The Changing Geography of Production in Labour-intensive Industries*. Aldershot, England, and Burlington, VT: Ashgate Publishing, 2008.

Langton, John. "The Industrial Revolution and the Regional Geography of England." *Transactions of the Institute of British Geographers New Series* 9 (1984): 145–67.

Lugo, Alejandro. *Fragmented Lives, Assembled Parts: Culture, Capitalism, and Conquest at the U.S.–Mexico Border.* Austin: University of Texas Press, 2008.

Massey, Doreen, and Richard Meegan, eds. *Politics and Method: Contrasting Studies in Industrial Geography*. New York: Methuen, 1986.

Phelps, N. A. "When Was Post-Fordism? The Uneven Institution of New Work Practices in a Multinational." *Antipode* 34 (2002): 205–26.

Rubenstein, James M. *The Changing US Auto Industry*. London: Routledge, 1992.

———. *Making and Selling Cars: Innovation and Change in the U.S. Automotive Industry*. Baltimore: Johns Hopkins University Press, 2001.

Salzinger, Leslie. *Genders in Production: Making Workers in Mexico's Global Factories*. Berkeley: University of California Press, 2003.

Scott, Allen J., and Michael Storper, eds. *Production, Work, Territory*. Boston: Allen and Unwin, 1986.

South, Robert B. "Transnational 'Maquiladora' Location." *Annals of the Association of American Geographers* 80 (1990): 529–70.

Storper, Michael, and Richard Walker. *The Capitalist Imperative: Territory, Technology, and Industrial Growth*. New York: Basil Blackwell, 1989.

Thrift, Nigel, and Kris Olds. "Refiguring the Economic in Economic Geography." *Progress in Human Geography* 20 (1996): 311–37.

Toyne, Brian, Jeffrey S. Arpan, David A. Ricks, Terence A. Shimp, and Andy Barnett. *The Global Textile Industry*. London: Allen and Unwin, 1984.

Yeung, H. W. "Industrial Geography: Industrial Restructuring and Labour Markets." *Progress in Human Geography* 26 (2002): 367–80.

Journals featuring industrial geography:

Journal of Industrial Economics, Journal of International Economics, Journal of Marketing, Journal of Transport Economics and Policy, Journal of Transport History, and Journal of Urban Economics

Key Internet sites:

www.bls.gov. Statistics on employment in manufacturing, as well as other sectors of the U.S. economy, are compiled by the U.S. Department of Labor Bureau of Labor Statistics. Some international labor statistics are also supplied.

www.ilo.org. International statistics concerning wages and working conditions are found at the web site maintained by the International Labour organization.

Log in to www.mygeoscienceplace.com for videos, interactive maps, RSS feeds, case studies, and self-study quizzes to enhance your study of Industry.

CHAPTER 12

Services

Flying across the United States on a clear night, you look down on the lights of settlements, large and small. You see small clusters of lights from villages and towns, and large, brightly lit metropolitan areas. It may appear that the light clusters are random, but geographers discern a regular pattern in them. These regularities have been documented, and concepts from economic geography can be applied to understand why this pattern exists.

The regular distribution observed over North America and over other MDCs is not seen in LDCs. Geographers

KEY ISSUES

1 Where Did Services Originate?

2 Where Are Contemporary Services Located?

3 Why Are Consumer Services Distributed in a Regular Pattern?

4 Why Do Business Services Cluster in Large settlements?

explain this difference and why the absence of a regular pattern is significant. The regular pattern of settlement in MDCs reflects where services are provided. In MDCs the majority of the workers are employed in the tertiary sector of the economy, defined in Chapter 9 as the provision of goods and services to people in exchange for payment. In contrast, less than 10 percent of the labor force in LDCs provides services.

Everyone needs food for survival. In LDCs, most people work in the primary sector, growing food. In MDCs, people purchase food at supermarkets or restaurants. The people employed at the supermarkets and restaurants are examples of service-sector workers, and the customers pay for the food with money earned in other service-sector jobs, such as retailing, banking, law, education, and government.

Retailing in Causeway Bay district of Hong Kong, China.

CASE STUDY / Phoning the Help Desk

Need to have your computer fixed? Correct a mistake on your credit card bill? Change your plane reservation? Relief is just a single toll-free call away, the company assures you. You punch in the company's "800" number, and after several loops through "press 1 for X, press 2 for Y," you actually reach a live human who offers to help you.

The human you have reached on the phone could be in India. The company whose name is on the computer, credit card, or airplane may not actually employ the person who answered your call. Instead, the call-answering job may have been contracted out to another company known as a call center. Leading call centers are located in India.

Call centers are one of the fastest-growing activities in the global economy. They take orders and provide customer service at the other end of the "800" numbers. They are also the source of many of those "annoying" calls that interrupt your dinner to ask you questions or sell you something. Kalldesk is one such call center, located in Chandigarh, a city of nearly 1 million in northern India. Started by 27-year-old Anuj Mahajan in 2002, Kalldesk grew within 1 year to 80 employees, mostly local college students.

The attraction of an LDC is, naturally, low wages, about $200 a month at Kalldesk, for example. India—rather than other LDCs—has attracted call centers for another reason. A call-center employee must be able to understand what a customer located in North America is trying to say and must be able to respond clearly in language understood by a "typical" North American. In India, English is understood and spoken among educated people, such as the college students working at Kalldesk. And Kalldesk trains employees in what it calls "accent neutralization." In other words, Indians are taught to alter their accents to sound like a "typical North American." And they adopt "typical North American" names, which they use when they place or answer a call.

Call centers in India can "pretend" that they are located in North America and are employing Americans. But in one respect, they can't escape the "tyranny" of geography. Refer to Figure 1-13, the map of world time zones. In the middle of the day, when most Americans are placing calls, it is the middle of the night in India. So call center employees in India typically work all night. ■

In MDCs, most people work in such places as shops, offices, restaurants, universities, and hospitals. These are examples of the tertiary or service sector of the economy. A **service** is any activity that fulfills a human want or need and returns money to those who provide it. A smaller number of people work in factories or farms, the primary and secondary sectors.

In sorting out ***where*** services are distributed in ***space***, geographers see a close link between services and settlements, because services are located in settlements. A **settlement** is a permanent collection of buildings where people reside, work, and obtain services. Settlements range in size from tiny rural villages with barely a hundred inhabitants to teeming cities with 20 million people. They occupy a very small percentage of Earth's surface, well under 1 percent, but settlements are home to nearly all humans, because few people live in isolation.

Explaining ***why*** services are clustered in settlements is at one level straightforward for geographers. In geographic terms, only one locational factor is critical for a service—proximity to the market. The optimal location of industry, described in the last chapter, requires balancing a number of site and situation factors, but the optimal location for a service is simply near its customers.

On the other hand, locating a service calls for far more precise geographic skills than locating a factory. The optimal location for a factory may be an area of several hundred square kilometers—such as Honda's factory, described in the Contemporary Geographic Tools box in Chapter 11—whereas the optimal location for a service may be a very specific ***place***, such as a street corner.

Service providers often say that the three critical factors in selecting a suitable site are "location, location, and location." Although geographically imprecise, the expression is a way for nongeographers to appreciate that a successful service must carefully select its precise location. Industries can locate in remote areas, confident that workers, water, and highways will be brought to the location if necessary. The distribution of services must follow to a large extent the distribution of where people live, within a city, country, or world ***region***.

However, if services were located merely where people lived, then China and India would have the most, rather than the United States and other MDCs. Services cluster in MDCs because more people able to buy services live there. Within MDCs, larger cities offer a larger ***scale*** of services than do small towns, because more customers reside there.

As in other economic and cultural features, geographers observe trends toward both globalization and local diversity in the distribution of services. In terms of ***globalization***, the provision of services is increasingly uniform from one urban settlement to another, especially within MDCs.

Every urban settlement in the United States above a certain size has a branch of a large retail chain, such as a McDonald's restaurant, and the larger cities have several. In England, every city above a certain size has a Tesco supermarket, and the larger cities have several. In an MDC, the demand for many types of services produces regular ***connections*** among settlements.

Despite the strong globalization trend so clearly visible on the landscape, ***local diversity*** is alive and well in the provision of services. Within MDCs, fast-food restaurants may be located in every settlement, but other services cluster in particular locations. A settlement may offer a service such as a medical clinic, an advertising agency, or a film studio not found in other settlements of comparable size. And every place—MDCs and LDCs alike—offers distinctive services that attract tourists and visitors.

KEY ISSUE 1
Where Did Services Originate?

- Three Types of Services
- Services in Early Rural Settlements
- Services in Early Urban Settlements

Services are provided in all societies, but in MDCs a majority of workers are engaged in the provision of services. In North America, three-fourths of workers are in services. The percentage of service workers varies widely in LDCs but is typically less than one-fourth. One reason for the wide variation is that in a number of LDCs, workers engaged in agriculture or manufacturing are counted in the service sector because they are employed by the government. ■

Three Types of Services

Services generate more than two-thirds of GDP in most MDCs, compared to less than one-half in most LDCs (Figure 12-1). Logically, the distribution of service workers is opposite that of the percentage of primary workers (see Figure 10-5). The service sector of the economy is subdivided into three types—consumer services, business services, and public services. Each of these sectors is divided into several major subsectors (Figure 12-2).

Consumer Services

The principal purpose of **consumer services** is to provide services to individual consumers who desire them and can afford to pay for them. Around 44 percent of all jobs in the United States are in consumer services. Four main types of consumer services are retail, education, health, and leisure.

- **Retail and Wholesale Services.** About 15 percent of all U.S. jobs. Department stores, grocers, and motor vehicle sales and service account for nearly one-half of these jobs; another one-fourth are wholesalers who provide merchandise to retailers.
- **Education Services.** About 10 percent of all U.S. jobs. Two-thirds of educators are employed in public schools, the other one-third in private schools. In Figure 12-2, educators at public schools are counted in public-sector employment.
- **Health Services.** About 12 percent of all U.S. jobs, primarily hospitals, doctors' offices, and nursing homes.
- **Leisure and Hospitality Services.** About 10 percent of all U.S. jobs. Around 70 percent of these jobs are in restaurants and bars; the other 30 percent is divided evenly between lodging and entertainment.

Business Services

The principal purpose of **business services** is to facilitate other businesses. Around 24 percent of all jobs in the United States are in business services. Professional services, financial services, and transportation services are the three main types of business services.

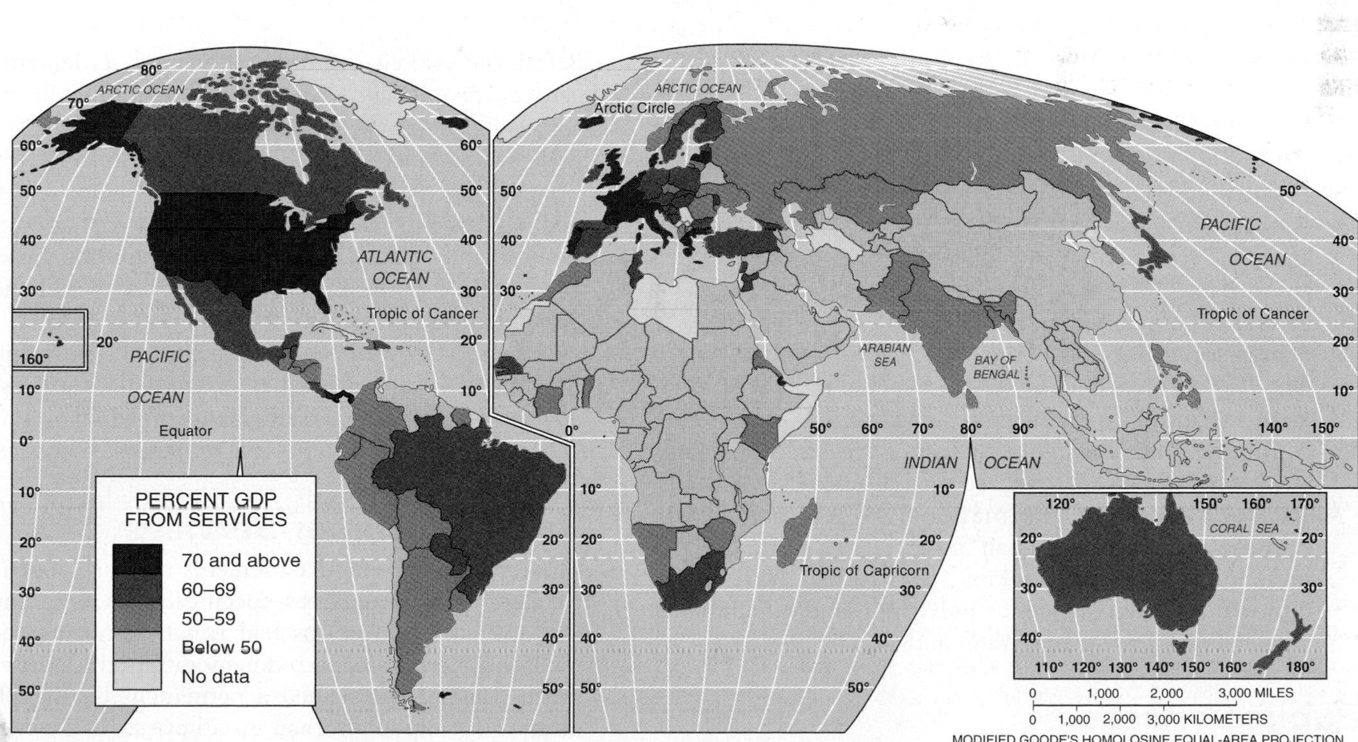

FIGURE 12-1 Percentage of gross domestic product (GDP) from services, 2006. Services contribute more than two-thirds of GDP in MDCs, compared to less than one-half in LDCs.

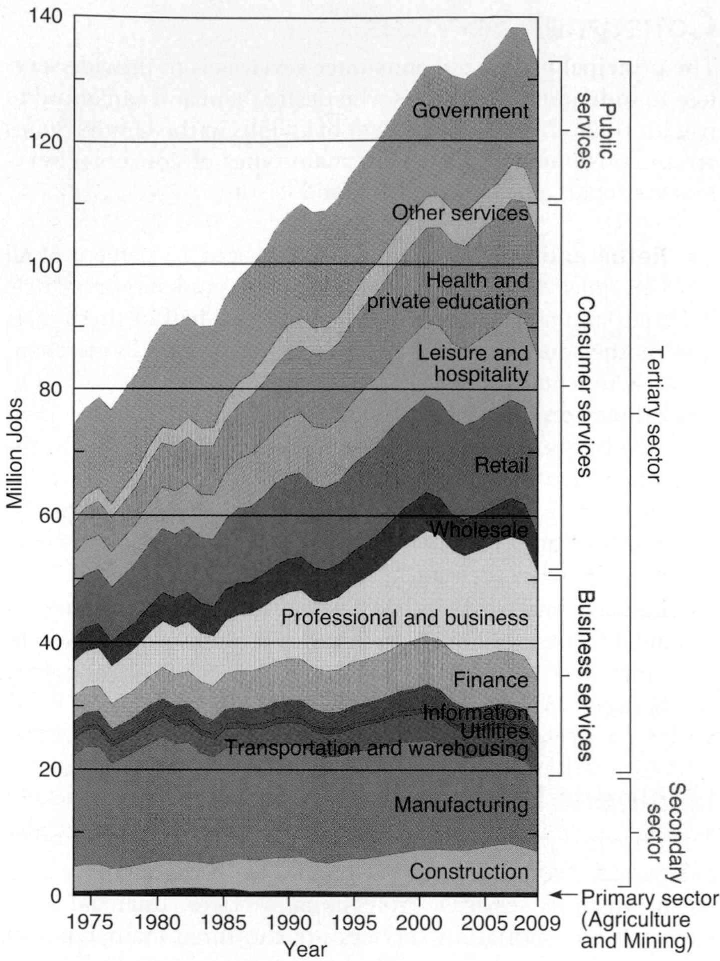

FIGURE 12-2 Employment change in the United States by sector. All of the growth in employment since 1970 has been in the tertiary sector, whereas employment has remained unchanged in the primary and secondary sectors. Within the tertiary sector the greatest increase has occurred in professional services.

- **Financial Services.** About 6 percent of all U.S. jobs. This sector is often called "FIRE," an acronym for finance, insurance, and real estate. One-half of the financial services jobs are in banks and other financial institutions, one-third in insurance companies, and the remainder in real estate.
- **Professional Services.** About 12 percent of all U.S. jobs. One-half are in technical services, including law, management, accounting, architecture, engineering, design, and consulting. The remaining one-half of this sector is in support services, such as clerical, secretarial, and custodial work.
- **Transportation and Information Services.** About 6 percent of all U.S. jobs. One-half of these services are in transportation, primarily trucking. The other half are in information services such as publishing and broadcasting, as well as utilities such as water and electricity.

Public Services

The purpose of **public services** is to provide security and protection for citizens and businesses. About 17 percent of all U.S. jobs are in the public sector. Nine percent of public school employees are excluded from this total and counted instead under education (consumer) services. Excluding educators, one-fourth of public-sector employees work for the federal government, one-fourth for one of the 50 state governments, and one-half for one of the tens of thousands of local governments.

Changes in Number of Employees

Figure 12-2 shows changes in employment in the United States between 1972 and 2009. All of the growth in employment in the United States has been in services, whereas employment in primary- and secondary-sector activities has declined.

Within business services, jobs expanded most rapidly in professional services (such as engineering, management, and law), data processing, advertising, and temporary employment agencies. Jobs grew more slowly in finance and transportation services because of improved efficiency—fewer workers are needed to run trains and answer phones, for example.

On the consumer services side, the most rapid increase has been in the provision of health care, including hospital staff, clinics, nursing homes, and home health-care programs. Other large increases have been recorded in education, entertainment, and recreation. The share of jobs in retailing has not increased—more stores are opening all the time, but they don't need as many employees as in the past. The Global Forces, Local Impacts box discusses the impact on services of the severe recession in the early twenty-first century.

Services in Early Rural Settlements

Before the establishment of permanent settlements as service centers, people lived as nomads, migrating in small groups across the landscape in search of food and water. They gathered wild berries and roots or killed wild animals for food (see Chapter 10). At some point, groups decided to build permanent settlements. Several families clustered together in a rural location and obtained food in the surrounding area. What services would these nomads require? Why would they establish permanent settlements to provide these services?

No one knows the precise sequence of events through which settlements were established to provide services. Based on archaeological research, settlements probably originated to provide consumer and public services. Business services came later.

Early Consumer Services

The earliest permanent settlements may have been established to offer consumer services, specifically places to bury the dead. Perhaps nomadic groups had rituals honoring the deceased, including ceremonies commemorating the anniversary of a death. Having established a permanent resting place for the dead, the group might then install priests at the site to perform the service of saying prayers for the deceased. This would have encouraged the building of structures—places for ceremonies and dwellings. By the time recorded history began about 5,000

GLOBAL FORCES, LOCAL IMPACTS

Services in the Recession

The service sector of the economy has been the engine of growth in the economy of MDCs, even as industry and agriculture have declined. But it was the service sector that triggered the severe economic recession that began in 2008. Principal contributors to the recession were some of the practices involved in financial services and real estate services, including:

- A rapid rise in real estate prices, encouraging speculators to acquire properties for the purpose of reselling them quickly at even higher prices
- Poor judgment in lending by financial institutions, especially by offering so-called "subprime" mortgages to individuals who were unable to repay them
- Invention of new financial services practices, such as derivatives, in which investors bought and sold risky assets, with the expectation that the value of the assets would continually rise
- Decisions by government agencies to reduce or eliminate regulation of the practices of new financial institutions
- Unwillingness of financial institutions to make loans once the recession started.

The early twenty-first century recession was also distinctive because it rapidly diffused to every region of the world. At the same time, the impact of the global recession varied by region and locality.

At a global scale, the early twenty-first century recession resulted in an absolute decline in world GDP, for the first time since the 1930s (Figure 12-3). GDP grew by an annual average of 3.7 percent between 1960 and the start of the recession in 2008. Only twice in the past 50 years did GDP grow at a rate of less than 1 percent per year.

At a regional scale, MDCs were more affected by the global recession. GDP grew slowly in LDCs, but it declined sharply in MDCs. The countries least affected by the global recession were the poorest countries of sub-Saharan Africa. These countries are the most peripheral to the global economy.

At a local scale, the recession hit some communities harder than others (Figure 12-4). In the United States, some of the hardest-hit communities were industrial centers in the Midwest, where bankrupt carmakers Chrysler and GM were based. But most of the hardest-hit communities were in the South and West, regions that had been the most prosperous. These communities were especially affected by declines in services, especially real estate and finance. ■

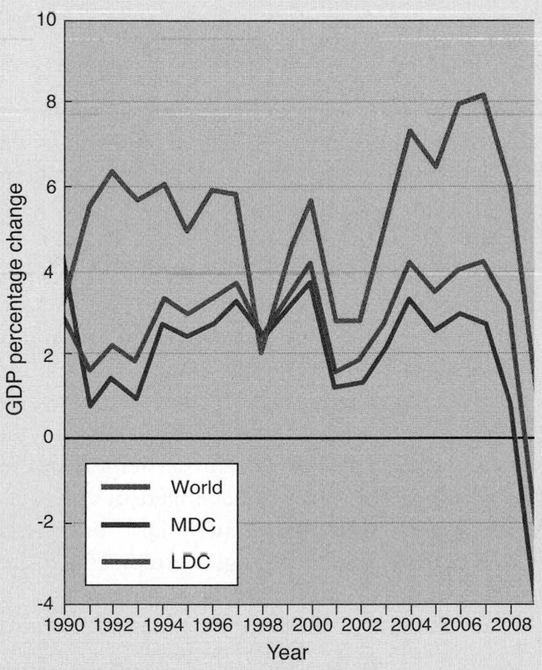

FIGURE 12-3 GDP change from prior year.

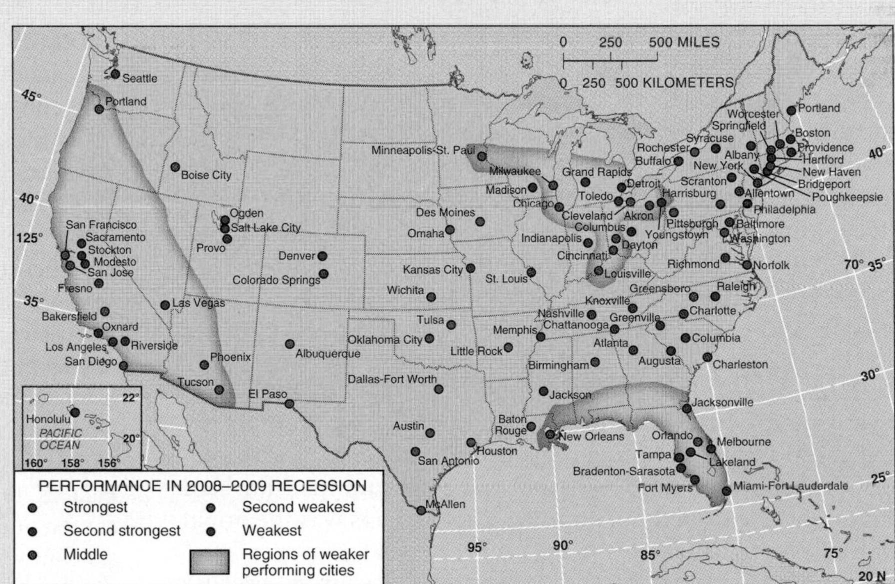

FIGURE 12-4 Impact of recession on 100 largest U.S. metropolitan areas.

years ago, many settlements existed, and some featured a temple. In fact, until the invention of skyscrapers in the late nineteenth century, religious buildings were often the tallest structures in a community.

Settlements also may have been places to house families, permitting unburdened males to travel farther and faster in their search for food. Women kept "home and hearth," making household objects, such as pots, tools, and clothing, as well as educating the children. These household-based services evolved over thousands of years into schools, libraries, theaters, museums, and other institutions that create and store a group's values and heritage and transmit them from one generation to the next.

People also needed tools, clothing, shelter, containers, fuel, and other material goods. Settlements therefore became manufacturing centers. Men gathered the materials needed to make a variety of objects, including stones for tools and weapons, grass for containers and matting, animal hair for clothing, and wood for shelter and heat. Women used these materials to manufacture household objects and maintain their dwellings. The variety of consumer services expanded as people began to specialize. One person could be skilled at repairing tools, another at training horses. People could then trade these services with one another. Settlements took on a retail-service function.

Early Public Services

Public services probably followed religious activities into the early permanent settlements. The group's political leaders also chose to live permanently in the settlement, which may have been located for strategic reasons, to protect the group's land claims.

Everyone in a settlement was vulnerable to attack from other groups, so for protection, some members became soldiers, stationed in the settlement. The settlement likely was a good base from which the group could defend nearby food sources against competitors. For defense, the group might surround the settlement with a wall. Defenders were stationed at small openings or atop the wall, giving them a great advantage over attackers. Thus settlements became citadels—centers of military power. Walls proved an extremely effective defense for thousands of years, until warfare was revolutionized by the introduction of gunpowder in Europe in the fourteenth century.

Early Business Services

Everyone in settlements needed food, which was supplied by the group through hunting or gathering. At some point, someone probably wondered, why not bring in extra food for hard times, such as drought or conflict? This perhaps was the origin of transportation services.

Not every group had access to the same resources, because of the varied distribution of vegetation, animals, fuel wood, and mineral resources across the landscape. People brought objects and materials they collected or produced into the settlement and exchanged them for items brought by others. Settlements became warehousing centers to store the extra food. The settlement served as neutral ground where several groups could safely come together to trade goods and services. To facilitate this trade, officials in the settlement provided producer services, such as regulating the terms of transactions, setting fair prices, keeping records, and creating a currency system.

Through centuries of experiments and accidents, residents of early settlements realized that some of the wild vegetation they had gathered could generate food if deliberately placed in the ground and nursed to maturity—in other words, agriculture, as described in Chapter 10. Over time, settlements became surrounded by fields, where people produced most of their food by planting seeds and raising animals rather than by hunting and gathering.

Services in Early Urban Settlements

Urban settlements date from the beginning of documented history in the Middle East and Asia. In ancient times, a handful of urban settlements provided business and public services, as well as some consumer services with large market areas. Virtually all settlements were rural, though, because the economy was based on the agriculture of the surrounding fields.

Services in Ancient Cities

Urban settlements may have originated in Mesopotamia, part of the Fertile Crescent of the Middle East (see Figure 8-6), and diffused at an early date to Egypt, China, and South Asia's Indus Valley. Or they may have originated independently in each of the four hearths. In any case, from these four hearths, the concept of urban settlements diffused to the rest of the world.

EARLIEST URBAN SETTLEMENTS. Among the oldest well-documented urban settlements is Ur in Mesopotamia (present-day Iraq). Ur, which means "fire," was where Abraham lived prior to his journey to Canaan in approximately 1900 B.C., according to the Bible. Archaeologists have unearthed ruins in Ur that date from approximately 3000 B.C (Figure 12-5).

Titris Höyük, in present-day Turkey, occupied a 50-hectare (125-acre) site and apparently had a population of about 10,000. The site is especially well preserved today because after 300 years the settlement was abandoned and never covered by newer buildings. Recent evidence unearthed at Titris Höyük from about 2500 B.C. suggests that early urban settlements were well-planned communities. Houses were arranged in a regular pattern—walls and streets were apparently laid out first. Palaces, temples, and other public buildings were placed at the center, and cemeteries were beyond the walls. Houses varied in size but were of similar design, built around a central courtyard that contained a crypt where some of the family members were buried. The several cooking areas within the houses indicate that they were apparently occupied by an extended family, and evidence of wine production and weaving has been found in the houses.

ANCIENT ATHENS. Settlements were first established in the eastern Mediterranean about 2500 B.C. The oldest include

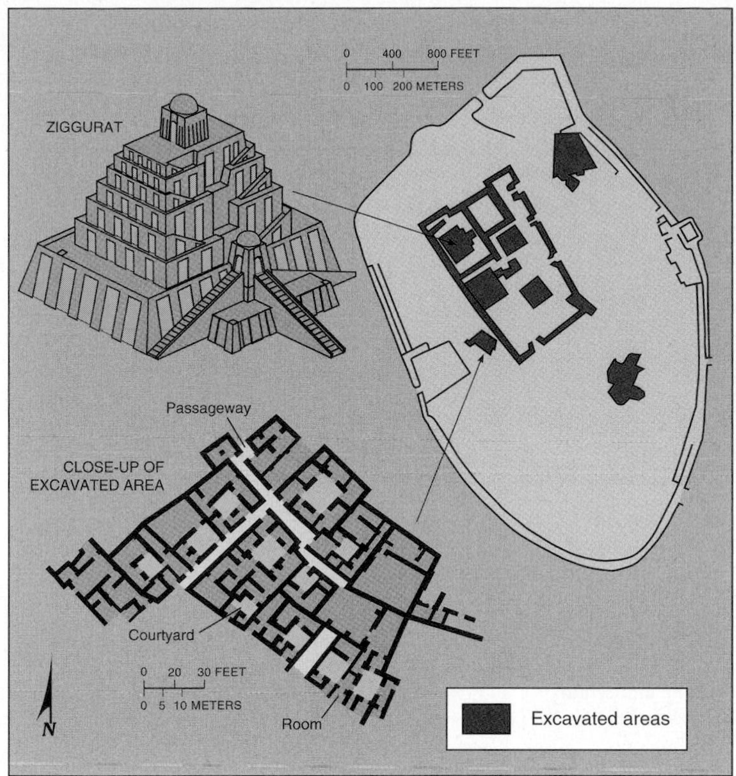

FIGURE 12-5 Ancient Ur. The remains of Ur, in present-day Iraq, provide evidence of early urban civilization. Ancient Ur was compact, perhaps covering 100 hectares (250 acres), and was surrounded by a wall. The most prominent building was the stepped temple, called a *ziggurat*, shown in the photo. The ziggurat was originally a three-story structure with a base that was 64 by 46 meters (210 by 150 feet) and the upper stories stepped back. Four more stories were added in the sixth century B.C. Surrounding the ziggurat was a dense network of small residences built around courtyards and opening onto narrow passageways. The excavation site was damaged during the two wars in Iraq.

Knossos on the island of Crete, Troy in Asia Minor (Turkey), and Mycenae in Greece. These settlements were trading centers for the thousands of islands dotting the Aegean Sea and the eastern Mediterranean and provided the government, military protection, and other public services for their surrounding hinterlands. They were organized into **city-states**—independent self-governing communities that included the settlement and nearby countryside.

Athens, the largest city-state in ancient Greece, was probably the first city to attain a population of 100,000 (Figure 12-6). Athens made substantial contributions to the development of culture, philosophy, and other elements of Western civilization, an example of the traditional distinction between urban settlements and rural. The urban settlements provided not only public services but also a concentration of consumer services, notably cultural activities, not found in smaller settlements.

ANCIENT ROME. The rise of the Roman Empire encouraged urban settlement. With much of Europe, North Africa, and Southwest Asia under Roman rule, settlements were established as centers of administrative, military, and other public services, as well as retail and other consumer services. Trade was encouraged through transportation and utility services, notably construction of many roads and aqueducts, and the security the Roman legions provided.

The city of Rome—the empire's center for administration, commerce, culture, and all other services—grew to at least 250,000 inhabitants, although some claim that the population may have reached a million. The city's centrality in the empire's communications network was reflected in the old saying "All roads lead to Rome" (see Figure 6-9).

With the fall of the Roman Empire in the fifth century, urban settlements declined. Their prosperity had rested on trading in the secure environment of imperial Rome. But with the empire fragmented under hundreds of rulers, trade diminished. Large urban settlements shrank or were abandoned. For several hundred years Europe's cultural heritage was preserved largely in monasteries and isolated rural areas.

Services in Medieval Cities

Urban life began to revive in Europe in the eleventh century as feudal lords established new urban settlements (Figure 12-7). The lords gave residents charters of rights with which to establish independent cities in exchange for their military service. Both the lord and the urban residents benefited from this arrangement. The lord obtained people to defend his territory at less cost than maintaining a standing army. For their part, urban residents preferred periodic military service to the burden faced by rural serfs, who farmed the lord's land and could keep only a small portion of their own agricultural output.

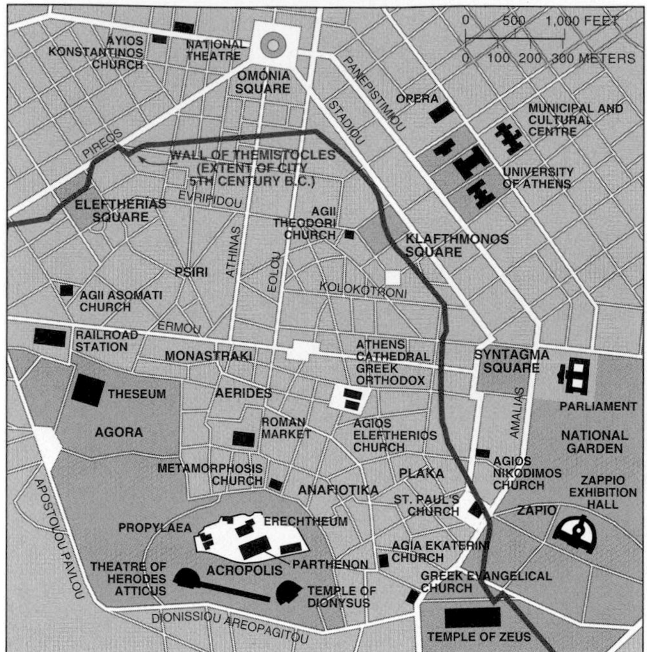

FIGURE 12-6 Ancient city: Athens, Greece. Dominating the skyline of modern Athens is the ancient hilltop site of the city, the Acropolis. Ancient Greeks selected this high place because it was defensible, and they chose it as a place to erect shrines to their gods. The most prominent structure on the Acropolis is the Parthenon, built in the fifth century B.C. to honor the goddess Athena. The structure to the right of the Parthenon in the photograph is the Propylaea, which was the entrance gate to the Acropolis. Greek fighter helicopters fly over the Acropolis to mark the anniversary of the Greek rebellion against the Ottomans on March 25, 1821.

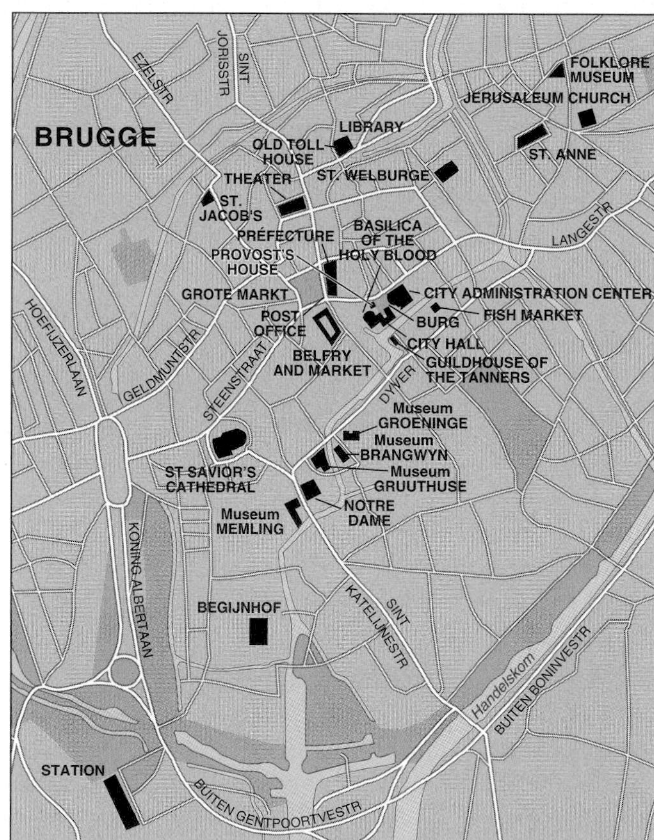

FIGURE 12-7 Medieval city: Brugge, Belgium. Modern Brugge (Bruges, in French) is a town of more than 100,000 in the western part of Belgium, near the North Sea coast. Beginning in the twelfth century, Brugge was the most important port in northwestern Europe and a major center for manufacturing wool. However, three events forced the city's decline during the fifteenth century: Foreign competitors captured much of the wool industry; the Belgian city of Antwerpen developed a better port; and the River Zwin silted, stranding the town 13 kilometers (8 miles) inland from the North Sea. Typical of medieval towns, the center of Brugge is dominated by squares surrounded by public buildings, churches, and markets. The photo shows the Grote Markt.

With their newly won freedom from the relentless burden of rural serfdom, the urban dwellers set about expanding trade. Surplus from the countryside was brought into the city for sale or exchange, and markets were expanded through trade with other free cities. The trade among different urban settlements was enhanced by new roads and more use of rivers. By the fourteenth century, Europe was covered by a dense network of small market towns serving the needs of particular lords.

The largest medieval European urban settlements served as power centers for the lords and church leaders, as well as major market centers. The most important public services occupied palaces, churches, and other prominent buildings arranged around a central market square. The tallest and most elaborate structures were usually churches, many of which still dominate the landscape of smaller European towns. In medieval times, European urban settlements were usually surrounded by walls even though by then cannonballs could destroy them (Figure 12-8). Dense and compact within the walls, medieval urban settlements lacked space for construction, so ordinary shops and houses nestled into the side of the walls and the large buildings. Most of these modest medieval shops and homes, as well as the walls, have been demolished in modern times, with only the massive churches and palaces surviving. Modern tourists can appreciate the architectural beauty of these medieval churches and palaces, but they do not receive an accurate image of a densely built medieval town.

Most of the world's largest cities were in Asia, not Europe, however, from the collapse of the Roman Empire until the diffusion of the Industrial Revolution across Europe during the nineteenth century. The five most populous cities in 900 are thought to have included Baghdad (in present-day Iraq), Constantinople (now called Istanbul, in Turkey), Kyoto (in Japan), and Changan and Hangchow (in China). Beijing (China) competed with Constantinople as the world's most populous city for several hundred years, until London claimed the distinction during the early 1800s. Agra (India), Cairo (Egypt), Canton (China), Isfahan (Iran), and Osaka (Japan) also ranked among the world's most populous cities prior to the Industrial Revolution.

FIGURE 12-8 Medieval city: Carcassonne, France. Medieval European cities, such as Carcassonne, in southwestern France, were often surrounded by walls for protection. The walls have been demolished in most places, but they still stand around the medieval center of Carcassonne.

KEY ISSUE 2
Where Are Contemporary Services Located?

- **Services in Rural Settlements**
- **Services in Urban Settlements**

Services are clustered in settlements. Rural settlements are centers for agriculture and provide a small number of services; urban settlements are centers for consumer and business services. One-half of the people in the world currently live in a rural settlement, and the other half in an urban settlement. ■

Services in Rural Settlements

A **clustered rural settlement** is a place where a number of families live in close proximity to each other, with fields surrounding the collection of houses and farm buildings. A **dispersed rural settlement**, typical of the North American rural landscape, is characterized by farmers living on individual farms isolated from neighbors rather than alongside other farmers in settlements.

Clustered Rural Settlements

A clustered rural settlement typically includes homes, barns, tool sheds, and other farm structures, plus consumer services, such as religious structures, schools, and shops. A handful of public and business services may also be present in the clustered rural settlement. In common language such a settlement is called a hamlet or village.

Each person living in a clustered rural settlement is allocated strips of land in the surrounding fields. The fields must be accessible to the farmers and are thus generally limited to a radius of 1 or 2 kilometers (½ to 1 mile) from the buildings. The strips of land are allocated in different ways. In some places, individual farmers own or rent the land. In other places, the land is owned collectively by the settlement or by a lord, and farmers do not control the choice of crops or use of the output.

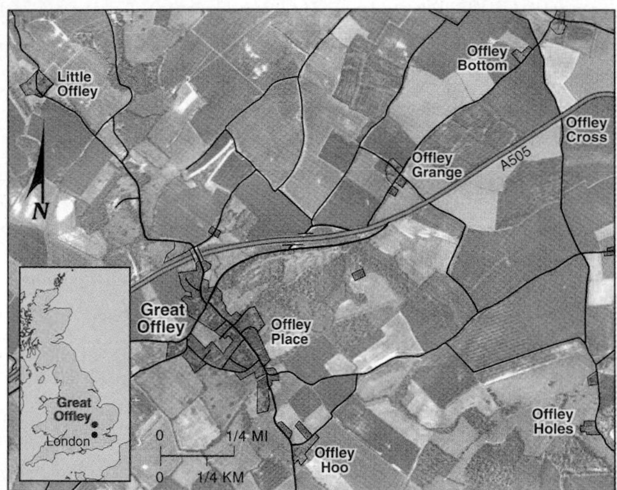

FIGURE 12-9 Clustered rural settlements. The English rural landscape reflects the historical pattern of growth through establishment of satellite settlements. On the map, note the numerous places with "Offley" in their name: Great Offley (the largest and the original settlement), Little Offley, Offley Grange (barn), Offley Cross, Offley Bottom, Offley Place, Offley Hoo (house), and Offley Hole. These are satellite rural settlements in the parish of Offley, in Hertfordshire. The name "Offley" means the wooded clearing of Offa, who was a ruler of Mercia (see Figure 5-3) during the eighth century and is said to have died at the site of the settlement.

Farmers typically own, or have responsibility for, a collection of scattered parcels in several fields. This pattern of controlling several fragmented parcels of land has encouraged living in a clustered rural settlement to minimize travel time to the various fields. Traditionally, when the population of a settlement grew too large for the capacity of the surrounding fields, new settlements were established nearby. This was possible because not all land was under cultivation (Figure 12-9).

Homes, public buildings, and fields in a clustered rural settlement are arranged according to local cultural and physical characteristics. Clustered rural settlements are often arranged in one of two types of patterns—circular or linear (Figure 12-10).

CIRCULAR RURAL SETTLEMENTS. These comprise a central open space surrounded by structures. Examples include:

- Kraal villages in southern Africa, which have enclosures for livestock in the center, surrounded by a ring of houses (Figure 12-11).
- Gewandorf settlements, once found in rural Germany, which consisted of a core of houses, barns, and churches, encircled by different types of agricultural activities. Small garden plots were located in the first ring surrounding the village, with cultivated land, pastures, and woodlands in successive rings. Von Thünen observed this circular rural pattern in his landmark agricultural studies of the early nineteenth century (Figure 10-24).

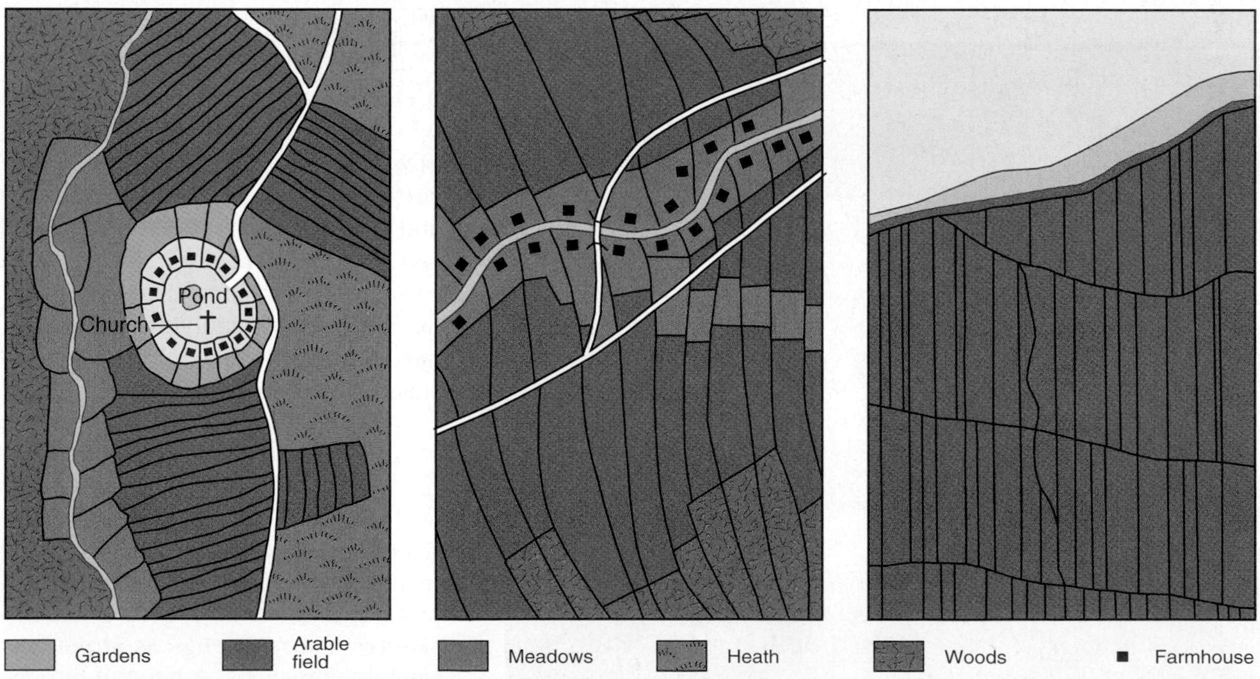

FIGURE 12-10 Rural settlement patterns. (Left) Circular rural settlement once common in Germany. (Center) Linear rural settlement, called "long-lot," once common in France, which gives everyone access to the river. (Right) Long-lot rural settlement established by French settlers in Québec, especially along the St. Lawrence River.

FIGURE 12-11 Kraal circular-shaped village, named Oshakatik, created by the Ovambo people in northern Namibia.

FIGURE 12-12 Québec long-lot rural settlement. Along the St. Lawrence River, farm fields were traditionally arranged in narrow strips.

LINEAR RURAL SETTLEMENTS. These comprise buildings clustered along a road, river, or dike to facilitate communications. The fields extend behind the buildings in long, narrow strips. Long-lot farms can be seen today along the St. Lawrence River in Québec (Figure 12-12).

In the French long-lot system, houses were erected along a river, which was the principal water source and means of communication. Narrow lots from 5 to 100 kilometers deep (3 to 60 miles) were established perpendicular to the river, so that each original settler had river access. This created a linear settlement along the river. These long, narrow lots eventually were subdivided. French law required that each son inherit an equal portion of an estate, so the heirs established separate farms in each division. Roads were constructed inland parallel to the river for access to inland farms. In this way, a new linear settlement emerged along each road, parallel to the original riverfront settlement.

Clustered Settlements in Colonial America

New England colonists built clustered settlements centered on an open area called a common (Figure 12-13). Settlers grouped their homes and public buildings, such as the church and school, around the common. In addition to their houses, each settler had a home lot of 1 to 5 acres (1/2 to 2 hectares), which contained a barn, garden, and enclosures for feeding livestock. Clustered settlements were favored by New England colonists for several reasons:

- They typically traveled to the New World in a group. The English government granted an area of land, in New England perhaps 4 to 10 square miles (10 to 25 square kilometers). Members of the group then traveled to America to settle the land, and usually built the settlement near the center of the land grant.
- The colonists wanted to live close together to reinforce common cultural and religious values. Most came from the same English village and belonged to the same church. Many of them left England in the 1600s to gain religious freedom. The settlement's leader was often an official of the Puritan Church, and the church played a central role in daily activities.
- They clustered their settlements for defense against Indian attacks.

Each villager owned several discontinuous parcels on the periphery of the settlement, to provide the variety of land types needed for different crops. Beyond the fields the town held pastures and woodland for the common use of all residents. Outsiders could obtain land in the settlement only by permission of the town's residents. Land was not sold, but rather was awarded to an individual after the town's residents felt confident that the recipient would work hard. Settlements accommodated a growing population by establishing new settlements nearby. As in the older settlements, the newer ones contained central commons surrounded by houses and public buildings, home lots, and outer fields.

The contemporary New England landscape contains remnants of the old clustered rural settlement pattern. Many New England towns still have a central common surrounded by the church, school, and various houses. However, quaint New England towns are little more than picturesque shells of clustered rural settlements, because today's residents work in shops and offices rather than on farms.

FIGURE 12-13 Newfane, Vermont, a New England clustered settlement. Public buildings are grouped around a common, including Windham County Courthouse (foreground), Congregational Church (right), and Newfane Village Union Hall (background, opposite courthouse).

Dispersed Rural Settlements

Outside of New England, dispersed rural settlements were more common in the American colonies. Meanwhile, in New England and in Great Britain clustered rural settlements were converted to a dispersed pattern. Owning several discontinuous fields around a clustered rural settlement had several disadvantages: Farmers lost time moving between fields, villagers had to build more roads to connect the small lots, and farmers were restricted in what they could plant. With the introduction of farm machinery, farms operated more efficiently at a larger scale.

DISPERSED RURAL SETTLEMENTS IN THE UNITED STATES. The Middle Atlantic colonies were settled by more heterogeneous groups than those in New England. Colonists came from Germany, Holland, Ireland, Scotland, and Sweden, as well as from England. Most arrived in Middle Atlantic colonies individually rather than as a member of a cohesive religious or cultural group. Some bought tracts of land from speculators. Others acquired land directly from individuals who had been given large land grants by the English government, including William Penn (Pennsylvania), Lord Baltimore (Maryland), and Sir George Carteret (the Carolinas).

Dispersed settlement patterns dominated in the American Midwest in part because the early settlers came primarily from the Middle Atlantic colonies. The pioneers crossed the Appalachian Mountains and established dispersed farms on the frontier. Land was plentiful and cheap, and people bought as much as they could manage. In New England a dispersed distribution began to replace clustered settlements in the eighteenth century. Eventually people bought, sold, and exchanged land to create large, continuous holdings instead of several isolated pieces.

The clustered rural settlement pattern worked when the population was low, but settlements had no spare land to meet the needs of a population that was growing through natural increase and net in-migration. A shortage of land eventually forced immigrants and children to strike out alone and claim farmland on the frontier. In addition, the cultural bonds that had created clustered rural settlements were weakened. Descendants of the original settlers were less interested in the religious and cultural values that had unified the original immigrants.

DISPERSED RURAL SETTLEMENTS IN GREAT BRITAIN. To improve agricultural production, a number of European countries converted their rural landscapes from clustered settlements to dispersed patterns. Dispersed settlements were considered more efficient for agriculture than clustered settlements. A prominent example was the **enclosure movement** in Great Britain, between 1750 and 1850. The British government transformed the rural landscape by consolidating individually owned strips of land surrounding a village into a single large farm, owned by an individual. When necessary, the government forced people to give up their former holdings.

The enclosure movement brought greater agricultural efficiency, but it destroyed the self-contained world of village life. Village populations declined drastically as displaced farmers moved to urban settlements. Because the enclosure movement coincided with the Industrial Revolution, villagers who were displaced from farming moved to urban settlements and became workers in factories and services. Some villages became the centers of the new, larger farms, but villages that were not centrally located to a new farm's extensive land holdings were abandoned and replaced with entirely new farmsteads at more strategic locations. As a result, the isolated, dispersed farmstead, unknown in medieval England, is now a common feature of that country's rural landscape.

As recently as 1800, only 3 percent of Earth's population lived in cities, and only one city in the world—Beijing—had more than 1 million inhabitants. Two centuries later, one-half of the world's people live in cities, and more than 400 of them have at least 1 million inhabitants. This rapid growth has made it difficult to define the boundaries of cities (see Chapter 13).

Services in Urban Settlements

The population of urban settlements exceeded that of rural settlements for the first time in human history in 2008 (Figure 12-14). The percentage of people living in urban settlements had increased from 3 percent in 1800 to 6 percent in 1850, 14 percent in 1900, 30 percent in 1950, and 47 percent in 2000.

Differences Between Urban and Rural Settlements

A century ago, social scientists observed striking differences between urban and rural residents. Louis Wirth argued during the 1930s that an urban dweller follows a different way of life than does a rural dweller. Thus Wirth defined a city as a permanent settlement that has three characteristics—large size, high population density, and socially heterogeneous people. These characteristics produced differences in the social behavior of urban and rural residents.

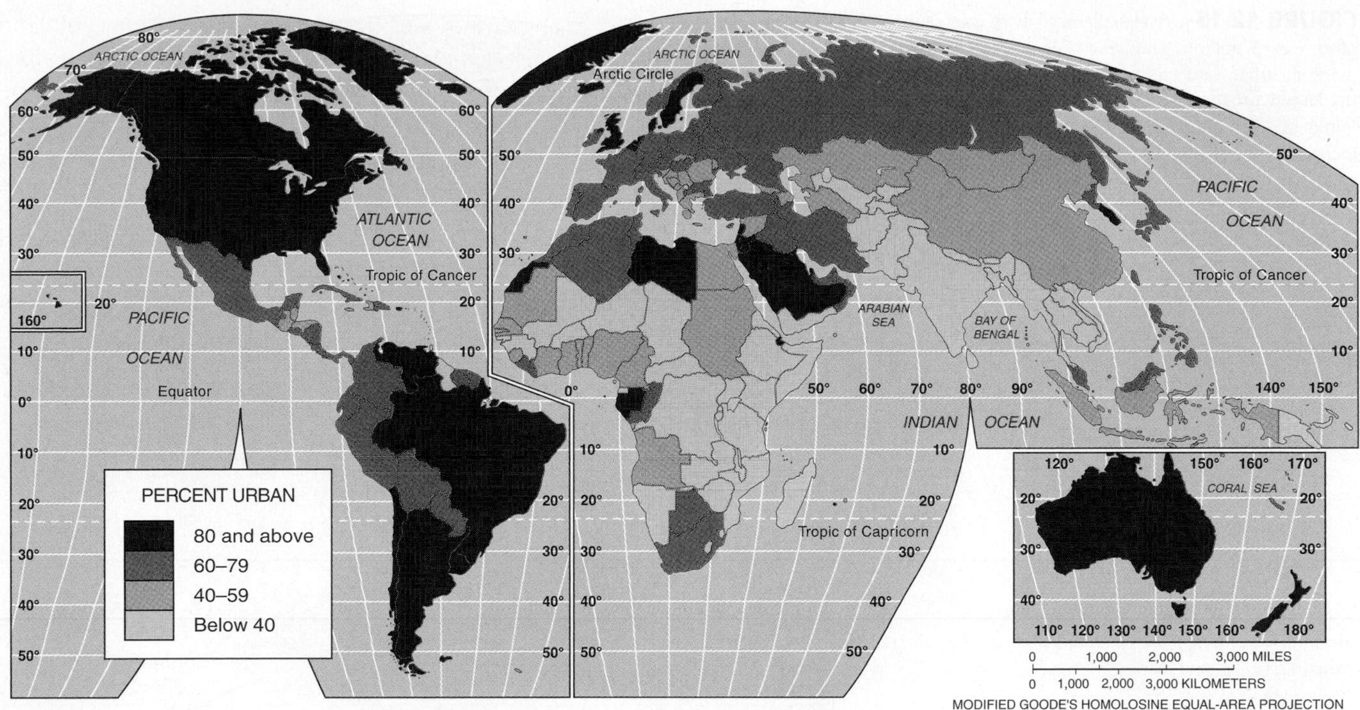

FIGURE 12-14 Percent living in urban settlements. MDCs have a higher percentage of people living in urban settlements.

LARGE SIZE. If you live in a rural settlement, you know most of the other inhabitants and may even be related to many of them. The people with whom you relax are probably the same ones you see in local shops and at church.

In contrast, if you live in an urban settlement, you can know only a small percentage of the other residents. You meet most of them in specific roles—your supervisor, your lawyer, your supermarket cashier, your electrician. Most of these relationships are contractual: You are paid wages according to a contract, and you pay others for goods and services. Consequently, the large size of an urban settlement produces different social relationships than those formed in rural settlements.

HIGH DENSITY. High density also produces social consequences for urban residents, according to Wirth. The only way that a large number of people can be supported in a small area is through specialization. Each person in an urban settlement plays a special role or performs a specific task to allow the complex urban system to function smoothly. At the same time, high density also encourages people to compete for survival in limited space. Social groups compete to occupy the same territory, and the stronger group dominates. This behavior distinguishes an urban settlement from a rural one.

SOCIAL HETEROGENEITY. The larger the settlement, the greater the variety of people. A person has greater freedom in an urban settlement than in a rural settlement to pursue an unusual profession, sexual orientation, or cultural interest. In a rural settlement, unusual actions might be noticed and scorned, but urban residents are more tolerant of diverse social behavior. Regardless of values and preferences, in a large urban settlement individuals can find people with similar interests. But despite the freedom and independence of an urban settlement, people may also feel lonely and isolated. Residents of a crowded urban settlement often feel that they are surrounded by people who are indifferent and reserved.

Wirth's three-part distinction between urban and rural settlements may still apply in LDCs. But in MDCs social distinctions between urban and rural residents have blurred. According to Wirth's definition, nearly everyone in an MDC now is urban. All but 1 percent of workers in developed societies hold "urban" types of jobs. Nearly universal ownership of automobiles, telephones, televisions, and other modern communications and transportation has also reduced the differences between urban and rural lifestyles in MDCs. Almost regardless of where you live in an MDC you have access to urban jobs, services, culture, and recreation.

Increasing Percentage of People in Cities

The process by which the population of urban settlements grows, known as **urbanization**, has two dimensions—an increase in the *number* of people living in cities and an increase in the *percentage* of people living in cities. The distinction between the two factors is important because they occur for different reasons and have different global distributions.

A large percentage of people living in urban settlements reflects a country's level of development. In MDCs, about three-fourths of the people live in urban areas, compared to about two-fifths in LDCs. The major exception to the global pattern is Latin America, where the urban percentage is comparable to the

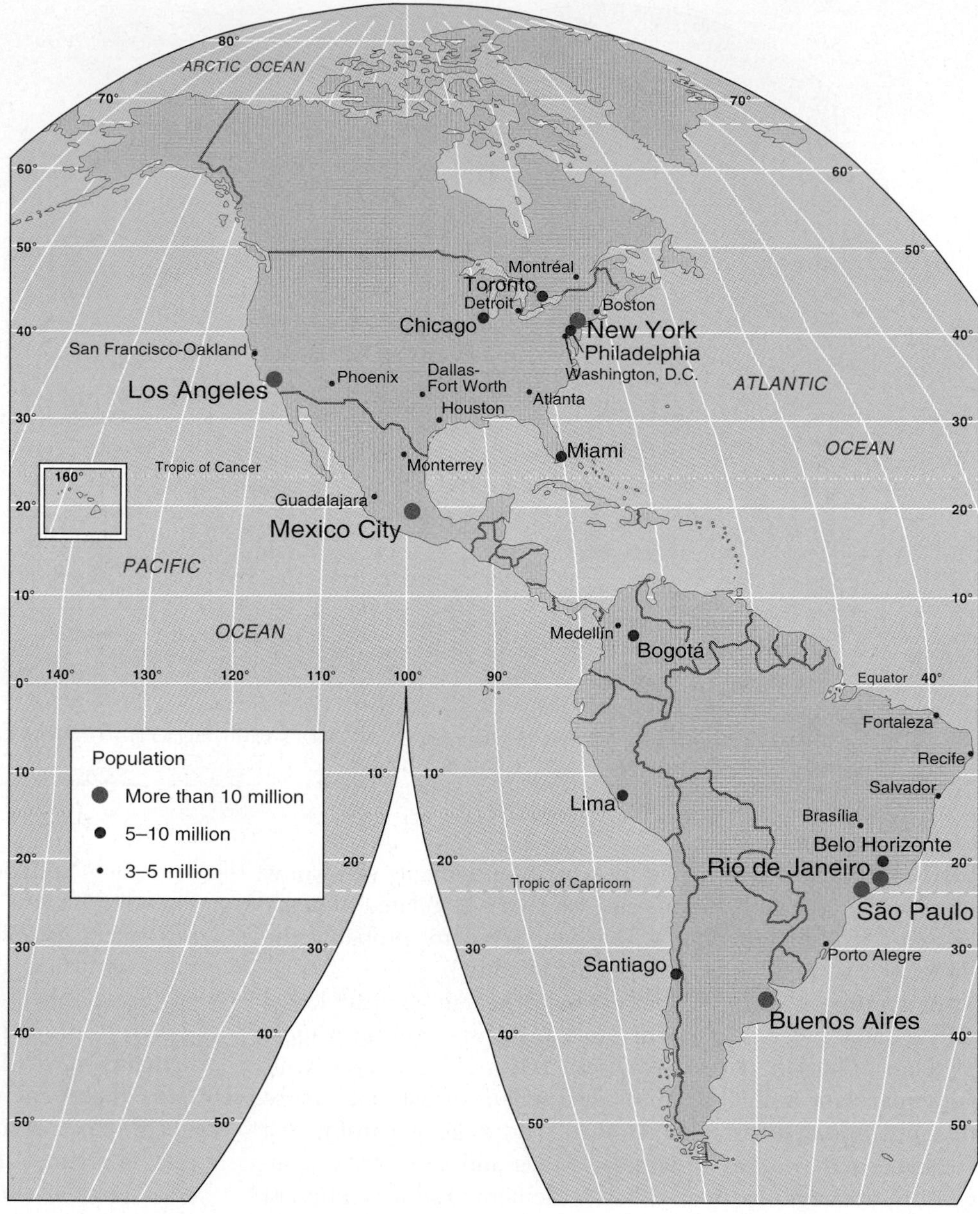

FIGURE 12-15 Urban settlements with populations of at least 3 million. Though the percentage of people living in urban settlements is greater in MDCs, most of the largest urban settlements are now located in LDCs. Rapid growth of urban settlements in LDCs reflects increasing overall population plus migration from rural areas.

level of MDCs. The higher percentage of urban residents in MDCs is a consequence of changes in economic structure during the past two centuries—first the Industrial Revolution in the nineteenth century and then the growth of services in the twentieth. The world map of urban percentages looks very much like the world map of percentage of GDP derived from services (Figure 12-1).

The percentage of urban dwellers is high in MDCs because over the past 200 years rural residents have migrated from the countryside to work in the factories and services that are concentrated in cities. The need for fewer farm workers has pushed people out of rural areas, and rising employment opportunities in manufacturing and services have lured them into urban areas. Because everyone resides either in an urban settlement or a rural settlement, an increase in the percentage living in urban areas has produced a corresponding decrease in the percentage living in rural areas.

Because the percentage living in urban areas simply cannot increase much more in MDCs, the process of urbanization that began around 1800 has largely ended. Nearly everyone interested in migrating from rural to urban areas has already done so, leaving those who choose to live in rural areas. We can now speak of MDCs as being fully urbanized, because the percentage of urban residents is so high. In recent years in LDCs, the percentage living in cities has risen rapidly because of the migration of rural residents to the cities in search of jobs in manufacturing or services. As in MDCs, people in LDCs are pushed off the farms by declining opportunities. However, urban jobs are by no means assured in LDCs experiencing rapid overall population growth.

Increasing Number of People in Cities

MDCs have a higher percentage of urban residents, but LDCs have more of the very large urban settlements (Figure 12-15).

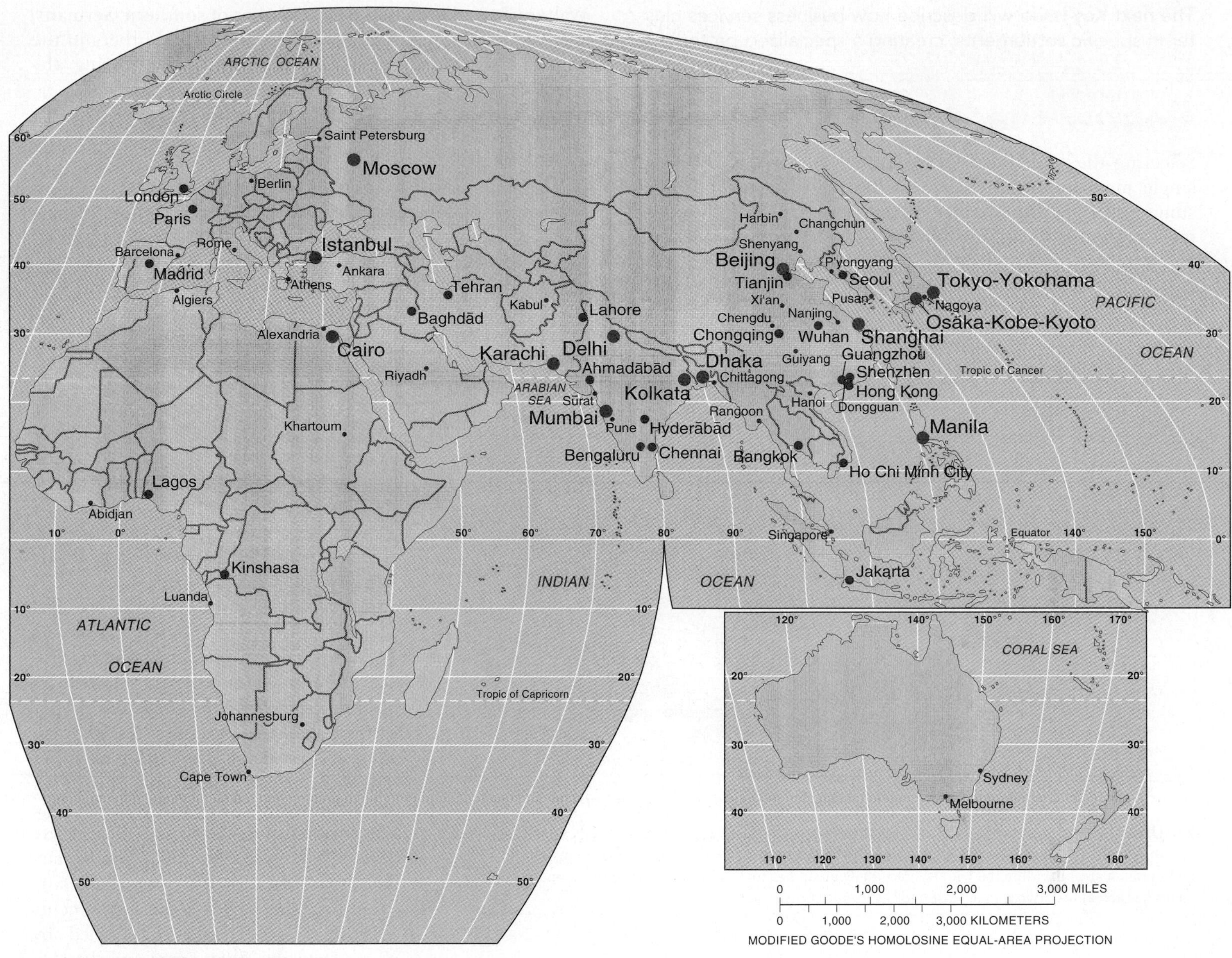

Eight of the ten most populous cities are currently in LDCs—Buenos Aires, Delhi, Dhaka, Kolkata (Calcutta), Mexico City, Mumbai (Bombay), São Paulo, and Shanghai. New York and Tokyo are the two large cities in MDCs. That LDCs dominate the list of largest urban settlements is remarkable because urbanization was once associated with economic development. In 1800, seven of the world's ten largest cities were in Asia. In 1900, after diffusion of the Industrial Revolution from Great Britain to today's MDCs, all ten of the world's largest cities were in Europe and North America.

In LDCs, migration from the countryside is fueling half of the increase in population in urban settlements, even though job opportunities may not be available. The other half results from high natural increase rates; in Africa, the natural increase rate accounts for three-fourths of urban growth.

KEY ISSUE 3

Why Are Consumer Services Distributed in a Regular Pattern?

- **Central Place Theory**
- **Market-Area Analysis**
- **Hierarchy of Services and Settlements**

Consumer services and business services do not have the same distributions. Consumer services generally follow a regular pattern based on size of settlements, with larger settlements offering more consumer services than smaller ones.

The next Key Issue will describe how business services cluster in specific settlements, creating a specialized pattern. ■

Central Place Theory

Selecting the right location for a new shop is probably the single most important factor in the profitability of a consumer service. **Central place theory** helps to explain how the most profitable location can be identified. Central place theory was first proposed in the 1930s by German geographer Walter Christaller, based on his studies of southern Germany. August Lösch in Germany and Brian Berry and others in the United States further developed the concept during the 1950s.

A **central place** is a market center for the exchange of goods and services by people attracted from the surrounding area. The central place is so called because it is centrally located to maximize accessibility. Central places compete against each other to serve as markets for goods and services for the surrounding region. According to central place theory, this competition creates a regular pattern of settlements.

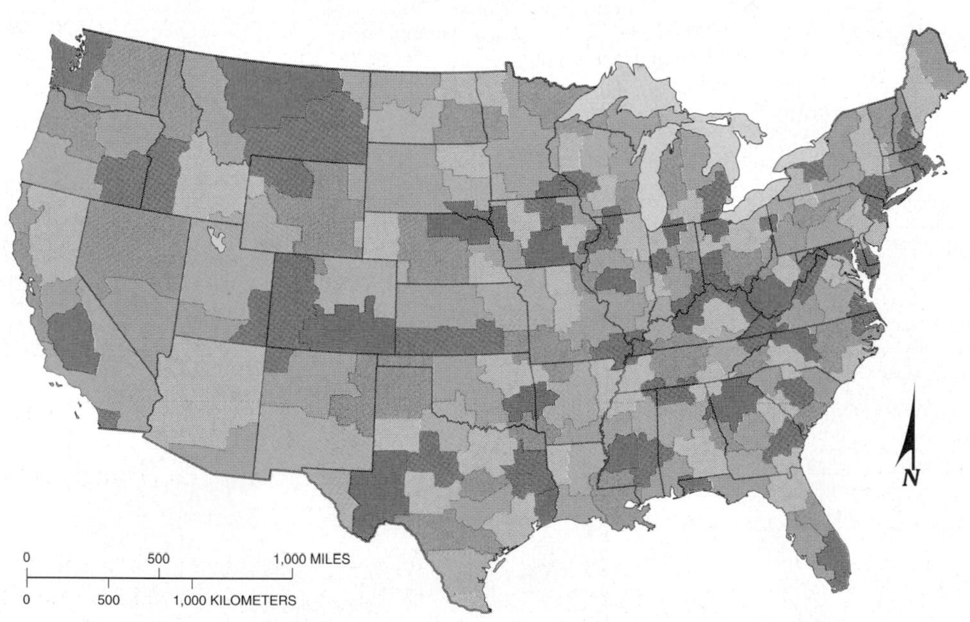

FIGURE 12-16 "Daily urban systems." The U.S. Department of Commerce divided the 48 contiguous states into "daily urban systems," delineated by functional ties, especially commuting, to the nearest metropolitan area. Dividing the country into "daily urban systems" demonstrates that everyone in the United States has access to services in at least one large settlement.

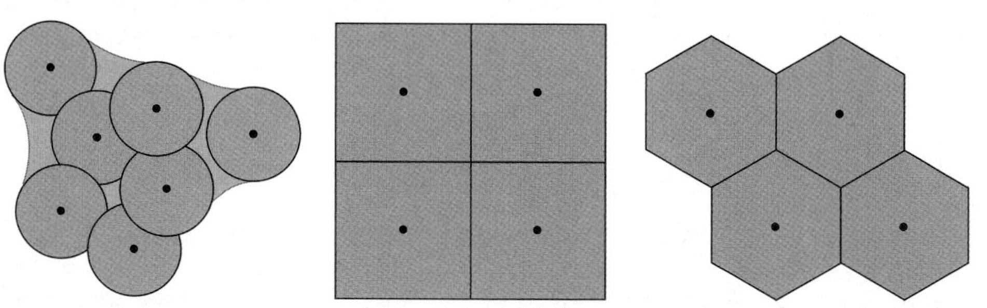

FIGURE 12-17 Why central place theory uses hexagons to delineate market areas. (left) The problem with circles. Circles are equidistant from center to edge, but they overlap or leave gaps. An arrangement of circles that leaves gaps indicates that people living in the gaps are outside the market area of any service, which is obviously not true. Overlapping circles are also unsatisfactory, for one service or another will be closer, and people will tend to patronize it. (center) The problem with squares. Squares nest together without gaps, but their sides are not equidistant from the center. If the market area is a circle, the radius—the distance from the center to the edge—can be measured, because every point around a circle is the same distance from the center. But in a square the distance from the center varies among points along a square. (right) The hexagon compromise. Geographers use hexagons to depict the market area of a good or service because hexagons offer a compromise between the geometric properties of circles and squares.

Market Area of a Service

The area surrounding a service from which customers are attracted is the **market area** or **hinterland**. A market area is a good example of a nodal region—a region with a core where the characteristic is most intense. To establish the market area, a circle is drawn around the node of service on a map. The territory inside the circle is its market area.

Because most people prefer to get services from the nearest location, consumers near the center of the circle obtain services from local establishments. The closer to the periphery of the circle, the greater is the percentage of consumers who will choose to obtain services from other nodes. People on the circumference of the market-area circle are equally likely to use the service, or go elsewhere. The entire United States can be divided into market areas based on the hinterland surrounding the largest urban settlements (Figure 12-16). Studies conducted by C. A. Doxiadis, Brian Berry, and the U.S. Department of Commerce allocated the 48 contiguous states to 171 functional regions centered around commuting hubs, which they called "daily urban systems."

To represent market areas in central place theory, geographers draw hexagons around settlements (Figure 12-17). (Hexagons represent a compromise between circles and squares.) Like squares, hexagons nest without gaps. Although all points along the hexagon are not the same distance from the center, the variation is less than with a square.

Size of Market Area

The market area of every service varies. To determine the extent of a market area, geographers need two pieces of information about a service—its range and its threshold.

RANGE OF A SERVICE. How far are you willing to drive for a pizza? To see a doctor for a serious problem? To watch a ball game? The **range** is the maximum distance people are willing to travel to use a service. The range is the radius of the circle (or hexagon) drawn to delineate a service's market area.

People are willing to go only a short distance for everyday consumer services, like groceries. But they will travel a long distance for other services, such as a concert or ball game. Thus a convenience store has a small range, whereas an arena has a large range. In a large urban settlement, for example, the range of a fast-food franchise like McDonald's is roughly 5 kilometers (3 miles); the range of a casual dining chain like Steak'n'Shake is roughly 8 kilometers (5 miles), and an arena has a range of 100 kilometers (60 miles) or more.

As a rule, people tend to go to the nearest available service: someone in the mood for a McDonald's hamburger is likely to go to the nearest McDonald's. Therefore, the range of a service is irregularly shaped to take in only the area for which the site is closer than competitors' sites. For example, irregularly shaped circles can be drawn around the location of all Kroger supermarkets in Dayton, Ohio (Figure 12-18). The radius of each irregular circle is the range for each store. The median radius for Kroger supermarkets in Dayton is approximately 2 kilometers (1.2 miles).

Most people go to the nearest provider of a service, but some travel to a more distant location. Consequently, retailers typically define their range as the maximum distance that three-fourths of their customers are willing to travel (see Contemporary Geographic Tools Box). The range must be modified further because most people think of distance in terms of time, rather than in terms of a linear measure like kilometers or miles. If you ask people how far they are willing to travel to a restaurant or a baseball game, they are more likely to answer in minutes or hours than in distance. If the range of a good or service is expressed in travel time, then the irregularly shaped circle must be drawn to acknowledge that travel time varies with road conditions. "One hour" may translate into traveling 90 kilometers (60 miles) while driving on an expressway but only 50 kilometers (30 miles) driving congested city streets.

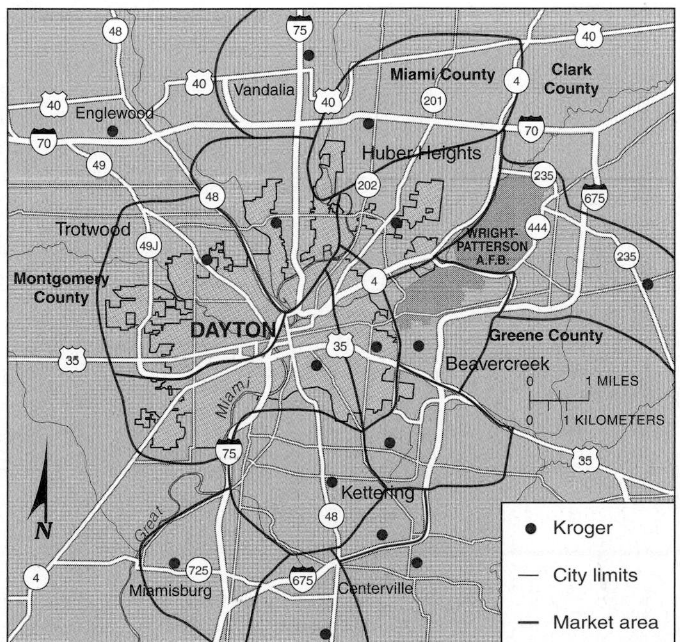

FIGURE 12-18 Market area, range, and threshold for Kroger supermarkets, in the Dayton, Ohio, metropolitan area. Fewer stores are in the southwest and northeast, which are predominantly industrial areas, and in the west, which contains lower-income residents.

THRESHOLD OF A SERVICE. The second piece of geographic information needed to compute a market area is the **threshold**, which is the minimum number of people needed to support the service. Every enterprise has a minimum number of customers required to generate enough sales to make a profit. So once the range has been determined, a service provider must determine whether a location is suitable by counting the potential customers inside the irregularly shaped circle. For example, the median threshold needed to support a Kroger supermarket in Dayton is about 30,000 people. Census data help to determine the population within the circle.

How potential consumers inside the range are counted depends on the product. Convenience stores and fast-food restaurants appeal to nearly everyone, whereas other goods and services appeal primarily to certain consumer groups.

- Movie theaters attract younger people; chiropractors attract older folks.
- Poorer people are drawn to thrift stores; wealthier ones might frequent upscale department stores.
- Amusement parks attract families with children; nightclubs appeal to singles.

If a good or service appeals to certain customers, then only the type of good or service that appeals to them should be counted inside the range.

Developers of shopping malls, department stores, and large supermarkets may count only higher-income people, perhaps those whose annual incomes exceed $50,000. Even though the stores may attract individuals of all incomes, higher-income people are likely to spend more and purchase items that carry higher profit margins for the retailer. Hence, in the Dayton area, Kroger operates more supermarkets in the south, where higher-income people are clustered, and fewer in the west, a lower-income area.

Market-Area Analysis

Would a convenience store be profitable in your community? Retailers and other providers of consumer services use market-area analysis to determine if locating in the area would be profitable and where the best location would be within the market area.

Profitability of a Location

The range and threshold together determine whether a good or service can be profitable in a particular location. Here's how:

1. **Compute the range.** You might survey local residents and determine that people are generally willing to travel up to 15 minutes to reach a convenience store.

CONTEMPORARY GEOGRAPHIC TOOLS

Locating a New Department Store

Macy's opened a new department store in Gilbert, Arizona, in 2009. Major U.S. department store chains, mall developers, and other large retailers employ geographers to determine the best locations to build new stores (Figure 12-19). A large retailer has many locations to choose from when deciding to build new stores. A suitable site is one with the potential for generating enough sales to justify using the company's scarce capital to build it. The role of the geographer is to forecast the sales expected at a proposed new store.

The first step in forecasting sales for a proposed new retail outlet is to define the market or trade area where the store would derive most of its sales. Analysis relies heavily on the company's records of their customers' credit card transactions at existing stores. The ZIP codes of customers who paid by credit card are used to determine the market area of a department store, which is typically defined as the ZIP codes where two-thirds to three-fourths of the customers live. Based on the ZIP codes of credit card customers, geographers estimate that the range for a typical mid-priced department store such as Macy's is about a 15-minute driving time.

In terms of population, the threshold for a typical department store is about 250,000. In other words, a typical department store needs about 250,000 people living within the 15-minute range. But in determining the threshold, the amount of money available in the area to spend in department stores is more important than simply the number of people.

If a potential site has enough customers with enough money within the market area, then the geographic analysis proceeds to the second step in estimating sales—market share. The proposed new department store will have to share customers with competitors' department stores. Geographers typically predict market share through the so-called analog method. One or more existing stores are identified in locations that the geographer judges to be comparable to the location of the proposed store. The market share of the comparable stores is then applied to the proposed new store.

Information about the viability of a proposed new store is depicted through Geographic Information Systems (GIS). One layer of the GIS depicts the trade area of the proposed store. Other layers display characteristics of the people living in the area, such as distribution of households, average income, and competitors' stores. A simplified example in Figure 12-19 shows the location of Macy's in the Phoenix area compared to average income. The market areas are smaller in higher-income areas and larger in lower-income areas.

The ability of the retail geographer is judged on the accuracy of the forecasts. After a new store is open for several years, how close to the actual sales were the forecasts that the geographer made several years earlier? ■

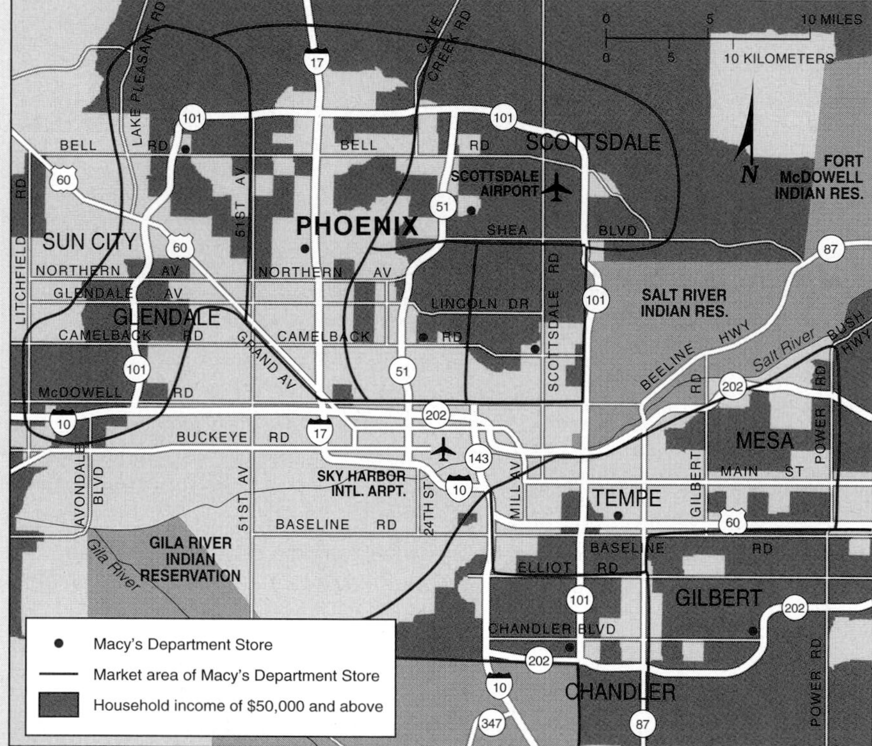

FIGURE 12-19 Market areas, ranges, and thresholds for Macy's department stores in the Phoenix, Arizona, metropolitan area. Stores are closer together in areas with higher incomes, including the newest location in Gilbert.

2. **Compute the threshold.** Suppose a convenience store must sell at least $10,000 worth of goods per week to make a profit, and the average customer spends $2 a week. The store needs at least 5,000 customers each week, spending $2 each, to achieve the break-even sales level of $10,000. If the average customer goes to a convenience store once a week, the threshold in this example would be 5,000.
3. **Draw the market area.** For a proposed location, draw an irregular circle with a 15-minute travel radius, adjusting the boundaries to account for any competitors. Count the number of people within the irregularly shaped circle. If more than 5,000 people are within the radius, then the threshold may be high enough to justify locating the new convenience store in your community. However, your store may need a larger threshold and range to attract some of the available customers if competitors are located nearby.

Optimal Location Within a Market

If the threshold and range justify the service, the next geographic question is: Where should the service be located within the market area to maximize profit? According to geographers, the best location is the one that minimizes the distance to the service for the largest number of people.

BEST LOCATION IN A LINEAR SETTLEMENT. Suppose that you want to establish your hot business idea, *Geographers' Pizza*, in your community. Where is the best place to build it?

Assume that you are seeking the optimal location for your business in an elongated community such as Miami Beach, Florida; Atlantic City, New Jersey; or Ocean City, Maryland. The community has only one major north–south street and a number of short east–west streets that are numbered consecutively.

The best location will be the one that minimizes the distance your van must travel to deliver to all potential customers. It corresponds to the median, which mathematically is the middle point in any series of observations. In a linear community such as an Atlantic Ocean resort, the service should be located where half of the customers are to the north and half are to the south (Figure 12-20).

What if a different number of customers live in each block of the city? What if the buildings are apartments, each housing a different number of families? To compute the optimal location in these cases, geographers have adapted the gravity model from physics. The **gravity model** predicts that the optimal location of a service is directly related to the number of people in the area and inversely related to the distance people must travel to access it. According to the gravity model, consumer behavior reflects two patterns:

1. The greater the number of people living in a particular place, the greater is the number of potential customers for a service. A city block or apartment building that contains 100 families will generate more customers than a house containing only one family.
2. The farther people are from a particular service, the less likely they are to use it. People who live 1 kilometer from a store are more likely to patronize it than people who live 10 kilometers away.

BEST LOCATION IN A NONLINEAR SETTLEMENT. Most settlements are more complex than a single main street. Geographers still apply the gravity model to find the best location, following these steps:

1. Identify a possible site for a new service.
2. Within the range of the service, identify where every potential user lives.

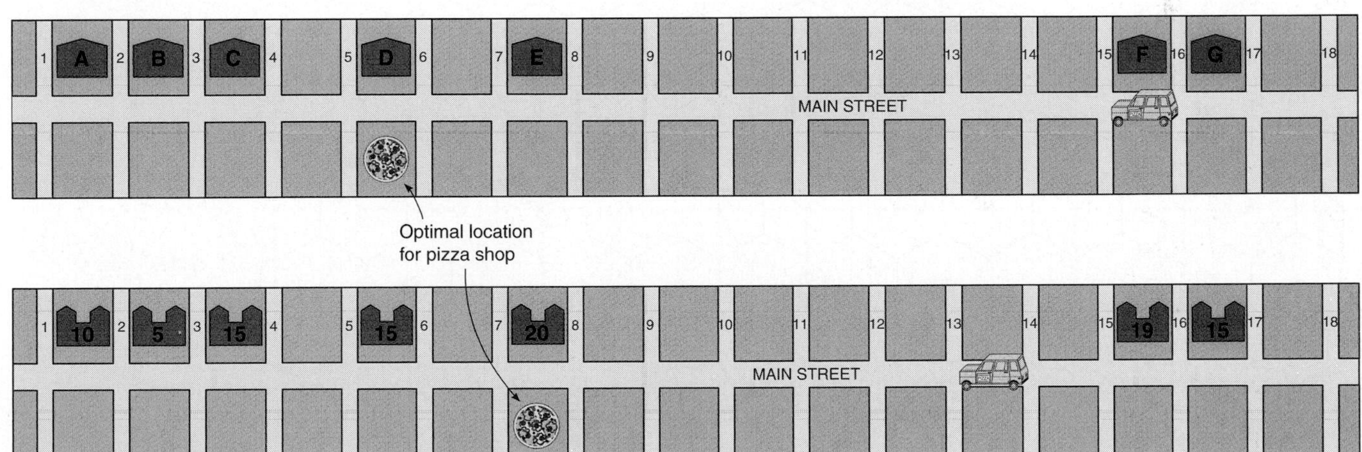

FIGURE 12-20 (top) Best location for a pizza-delivery service in a linear settlement with seven potential customers, families A through G. The optimal location for the shop is between 5th and 6th streets, the median location. The delivery van would travel four blocks to deliver a pizza to Family A (between 1st and 2nd streets), three blocks to Family B (between 2nd and 3rd), two blocks to Family C (between 3rd and 4th), zero blocks to Family D (between 5th and 6th), two to Family E (between 7th and 8th), ten to Family F (between 15th and 16th), and 11 to Family G (between 16th and 17th). The van would have to travel a total of 32 blocks to deliver a pizza to each of the seven customers, three located to the west and three to the east. (bottom) Best location for a pizza-delivery service in a linear settlement with 99 families. Numbers in each apartment building represent families at each location. The median location is the middle observation among these 99 families, the place where 49 families live to the west and 49 families live to the east. *Geo Pizza* should locate between 7th and 8th streets.

3. Measure the distance from the possible site of the new service to every potential user.
4. Divide each potential user by the distance to the potential site for the service.
5. Sum all of the results of potential users divided by distances.
6. Select a second possible location for the new service, and repeat steps 2, 3, 4, and 5.
7. Compare the results of step 5 for all possible sites. The site with the highest score has the highest potential number of users and is therefore the optimal location for the service.

Hierarchy of Services and Settlements

Small settlements are limited to consumer services that have small thresholds, short ranges, and small market areas, because too few people live in small settlements to support many services. A large department store or specialty store cannot survive in a small settlement because the minimum number of people needed exceeds the population within range of the settlement.

Larger settlements provide consumer services having larger thresholds, ranges, and market areas. Neighborhoods within large settlements also provide services having small thresholds and ranges. Services patronized by a small number of locals can coexist in a neighborhood ("mom-and-pop stores") along with services that attract many from throughout the settlement. This difference is vividly demonstrated by comparing the yellow pages for a small settlement with those for a major city. The major city's yellow pages are thick with more services, and diverse headings show widely varied services unavailable in small settlements.

We spend as little time and effort as possible in obtaining consumer services and thus go to the nearest place that fulfills our needs. There is no point in traveling to a distant department store if the same merchandise is available at a nearby one. We travel greater distances only if the price is much lower or if the item is unavailable locally.

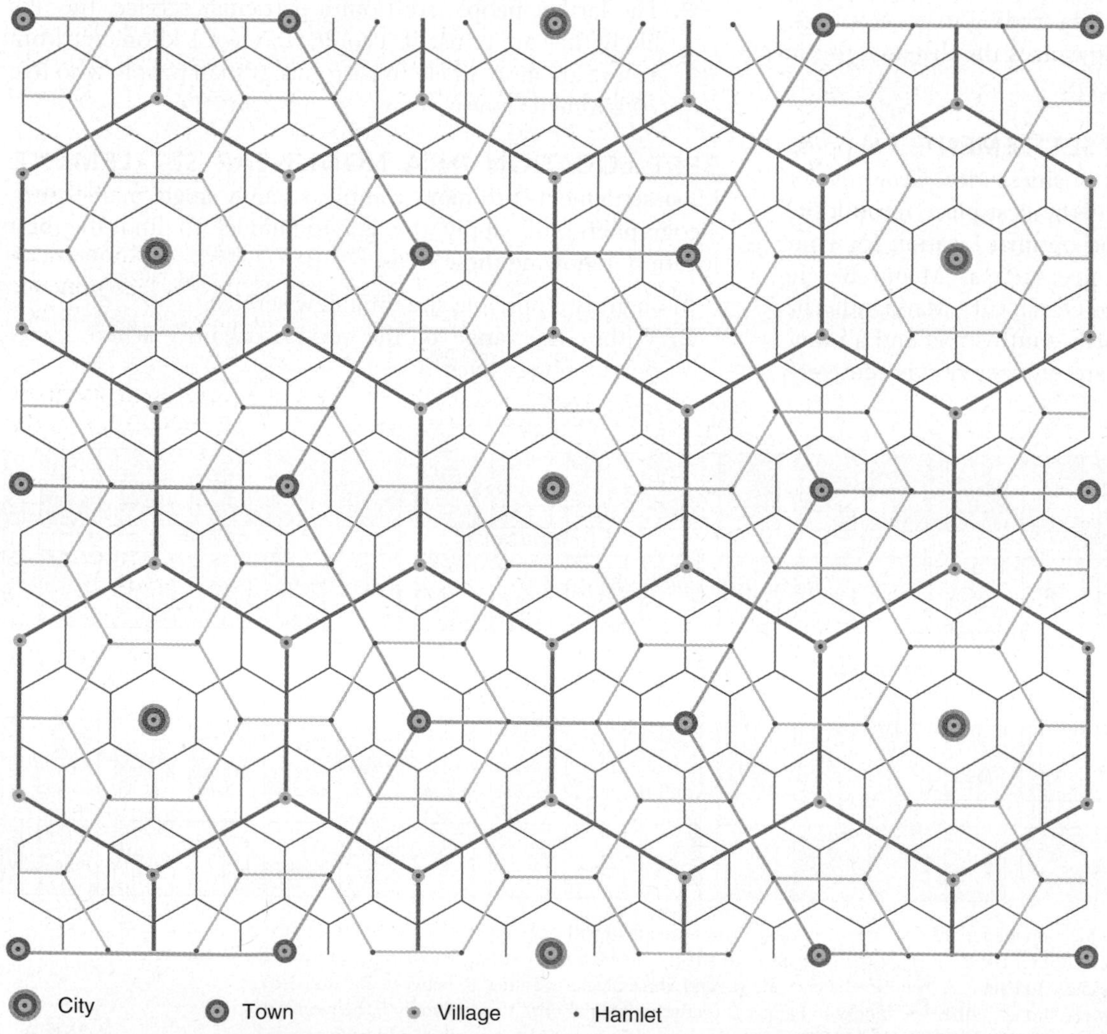

FIGURE 12-21 Central place theory. According to central place theory, market areas are arranged in a regular pattern. Larger market areas, based in larger settlements, are fewer in number and farther apart from each other than smaller market areas and settlements. However, larger settlements also provide goods and services with smaller market areas; consequently, larger settlements have both larger and smaller market areas drawn around them.

Nesting of Services and Settlements

According to central place theory, market areas across an MDC would be a series of hexagons of various sizes, unless interrupted by physical features such as mountains and bodies of water (Figure 12-21). MDCs have numerous small settlements with small thresholds and ranges, and far fewer large settlements with large thresholds and ranges.

The nesting pattern can be illustrated with overlapping hexagons of different sizes. Four different levels of market area—for hamlet, village, town, and city—are shown in Figure 12-21. Hamlets with very small market areas are represented by the smallest contiguous hexagons. Larger hexagons represent the market areas of larger settlements and are overlaid on the smaller hexagons, because consumers from smaller settlements shop for some goods and services in larger settlements.

In his original study, Walter Christaller showed that the distances between settlements in southern Germany followed a regular pattern. He identified seven sizes of settlements (market hamlet, township center, county seat, district city, small state capital, provincial head capital, and regional capital city). In southern Germany, the smallest (market hamlet) had an average population of 800 and a market area of 45 square kilometers (17 square miles). The average distance between market hamlets was 7 kilometers (4.4 miles). The figures were higher for the average settlement at each increasing level in the hierarchy. Brian Berry has documented a similar hierarchy of settlements in parts of the U.S. Midwest.

The principle of nesting market areas also works at the scale of services within cities. For example, compare the market areas within Dayton of United Dairy Farmers (UDF) in Figure 12-22 with those of Kroger in Figure 12-18. The UDF convenience stores are more numerous than Kroger stores and have smaller thresholds, ranges, and market areas.

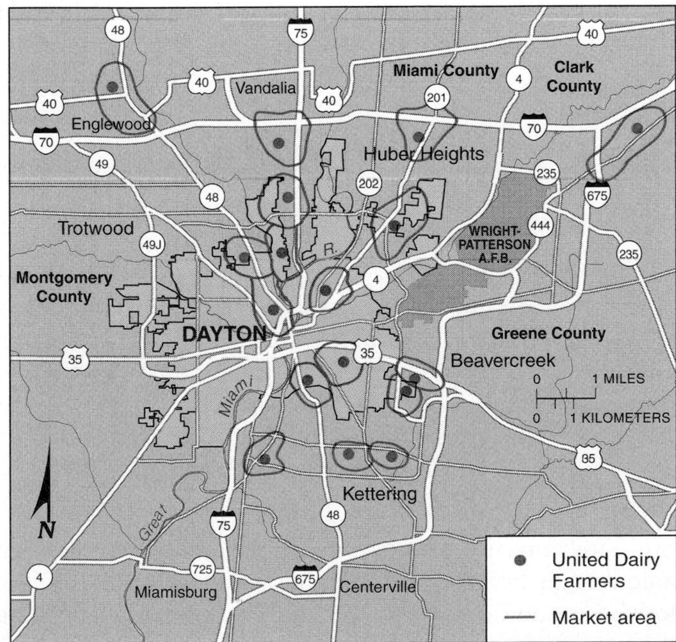

FIGURE 12-22 Market area, range, and threshold for UDF convenience stores in the Dayton, Ohio, metropolitan area. Compared to Kroger supermarkets, shown in Figure 12-18, UDF stores are more numerous and have smaller market areas, ranges, and thresholds.

Rank-Size Distribution of Settlements

In many MDCs, geographers observe that ranking settlements from largest to smallest (population) produces a regular pattern or hierarchy. This is the **rank-size rule**, in which the country's *nth*-largest settlement is 1/*n* the population of the largest settlement. In other words, the second-largest city is one-half the size of the largest, the fourth-largest city is one-fourth the size of the largest, and so on. When plotted on logarithmic paper, the rank-size distribution forms a fairly straight line. In the United States and a handful of other countries (Figure 12-23), the distribution of settlements closely follows the rank-size rule.

If the settlement hierarchy does not graph as a straight line, then the country does not have a rank-size distribution of settlements. Instead, it may follow the **primate city rule**, in which the largest settlement has more than twice as many people as the second-ranking settlement. In this distribution, the country's largest city is called the **primate city**. Several primate city distributions exist in Europe:

- In Denmark, København (Copenhagen) is a primate city with 1 million inhabitants, whereas the second-largest urban area, Århus, has only 200,000, instead of the 500,000 that the rank-size rule predicts.
- In the United Kingdom, London has 8 million inhabitants, whereas Birmingham—the second-largest—has only 2 million.
- In Romania, the largest city, Bucharest, has 1.9 million inhabitants, and the second-largest, Iasi, has 315,000. Romania also has fewer settlements with population between 1,000 and 10,000 than expected.

The existence of a rank-size distribution of settlements is not merely a mathematical curiosity. It has a real impact on the

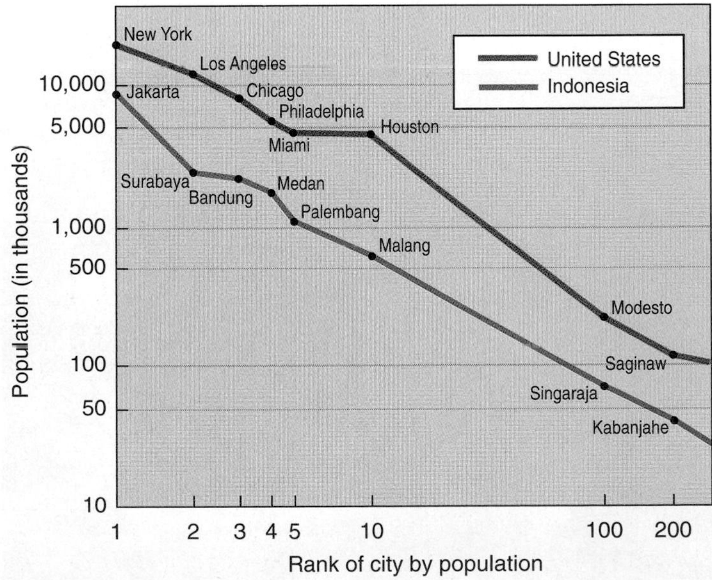

FIGURE 12-23 Rank-size distribution of settlements in the United States and Indonesia.

quality of life for a country's inhabitants. A regular hierarchy—as in the United States—indicates that the society is sufficiently wealthy to justify the provision of goods and services to consumers throughout the country. Conversely, the absence of the rank-size distribution in an LDC indicates that there is not enough wealth in the society to pay for a full variety of services. The absence of a rank-size distribution constitutes a hardship for people who must travel long distances to reach an urban settlement with shops and such services as hospitals. Because most people in LDCs do not have cars, buses must be provided to reach larger towns. A trip to a shop or a doctor that takes a few minutes in the United States could take several hours in an LDC.

Periodic Markets

Services at the lower end of the central place hierarchy may be provided at a periodic market, which is a collection of individual vendors who come together to offer goods and services in a location on specified days. The periodic market typically is set up in a street or other public space early in the morning, taken down at the end of the day, and set up in another location the next day (Figure 12-24).

A periodic market provides goods to residents of LDCs, as well as rural areas in MDCs, where sparse populations and low incomes produce purchasing power too low to support full-time retailing. A periodic market makes services available in more villages than would otherwise be possible, at least on a part-time basis. In urban areas, periodic markets offer residents fresh food brought in that morning from the countryside.

Many of the vendors in periodic markets are mobile, driving their trucks from farm to market, back to the farm to restock, then to another market. Other vendors, especially local residents who cannot or prefer not to travel to other villages, operate on a part-time basis, perhaps only a few times a year. Other part-time vendors are individuals who are capable of producing only a small quantity of food or handicrafts.

FIGURE 12-24 Periodic market. Women are selling herbs and spices at the periodic market in Kuch, Ethiopia.

The frequency of periodic markets varies by culture.

- Muslim countries: Typically conform to the weekly calendar—once a week in each of six cities and no market on Friday, the Muslim day of rest.
- Rural China: A three-city, 10-day cycle of periodic markets, according to G. William Skinner. The market operates in a central market on days 1, 4, and 7; in a second location on days 2, 5, and 8; in a third location on days 3, 6, and 9; and no market on the tenth day. Three 10-day cycles fit in a lunar month.
- Korea: Two 15-day cycles in a lunar month.
- Africa: Varies from 3 to 7 days. Variations in the cycle stem from ethnic differences.

KEY ISSUE 4
Why Do Business Services Cluster in Large Settlements?

- **Hierarchy of Business Services**
- **Business Services in LDCs**
- **Economic Base of Settlements**

Every urban settlement provides consumer services to people in a surrounding area, but not every settlement of a given size has the same number and types of business services. Business services disproportionately cluster in a handful of urban settlements, and individual settlements specialize in particular business services. ■

Hierarchy of Business Services

Geographers distinguish four levels of urban settlements according to their importance in the provision of business services. At the top are a handful of urban settlements known as world cities that play an especially important role in global business services (Figure 12-25). World cities, as well as the other three levels, can be further subdivided (Figure 12-26).

Services in World Cities

World cities are most closely integrated into the global economic system because they are at the center of the flow of information and capital. Business services, including law, banking, insurance, accounting, and advertising, concentrate in disproportionately large numbers in world cities.

New forms of transportation and communications were expected to reduce the need for clustering of services in large cities.

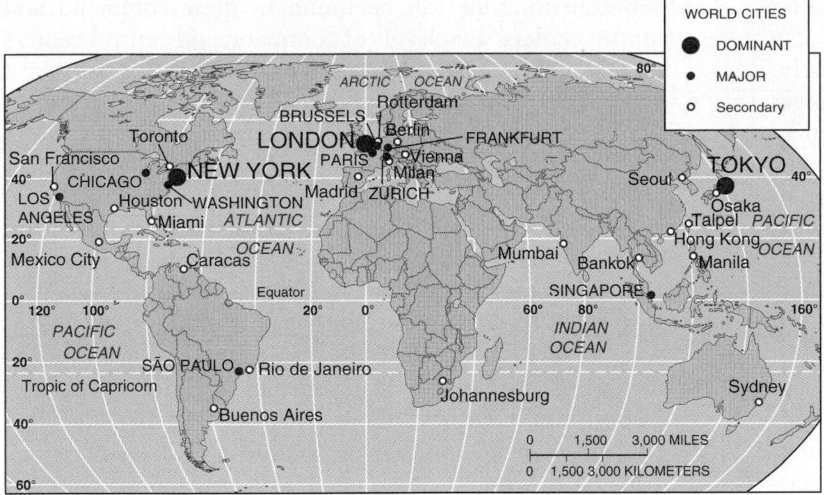

FIGURE 12-25 World cities. World cities are centers for the provision of services in the global economy. London, New York, and Tokyo are the three dominant world cities. Major and secondary world cities play somewhat less central roles in the provision of services than the three dominant world cities.

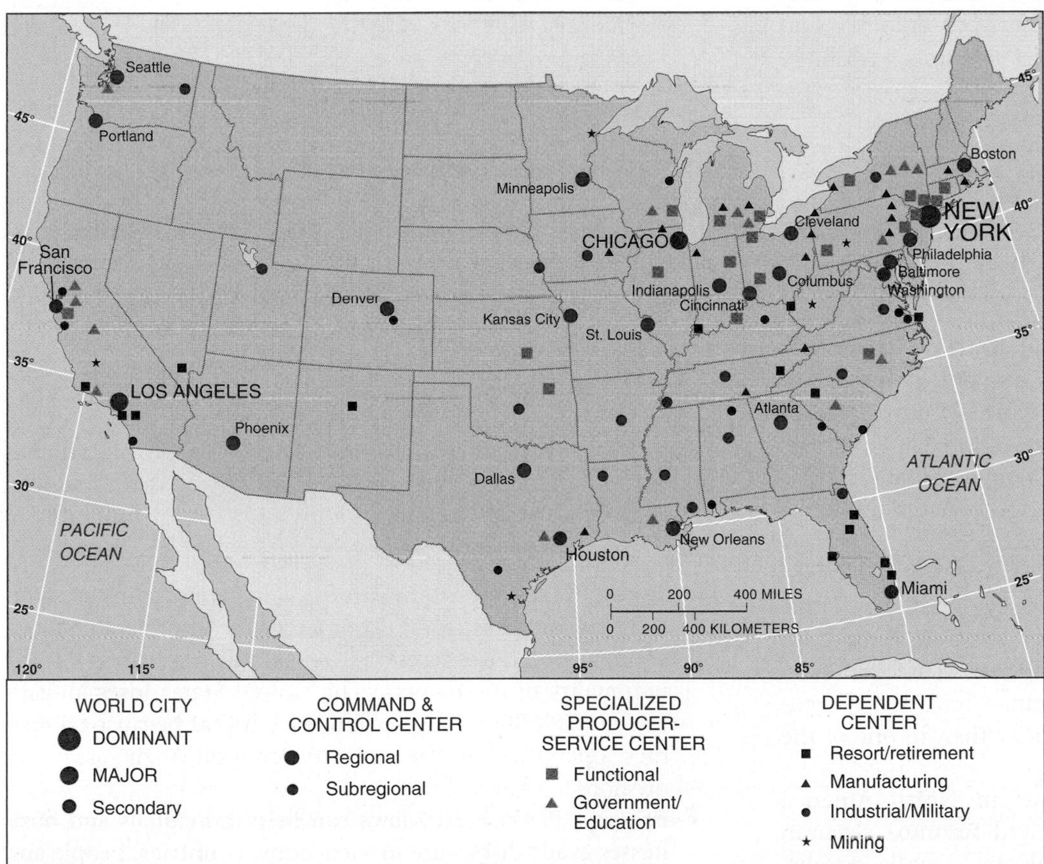

FIGURE 12-26 Business-service cities in the United States. Atop the hierarchy are New York and the major and secondary world cities. Below the world cities in the hierarchy are regional command and control centers, specialized producer-service centers, and dependent centers.

- The telegraph and telephone in the nineteenth century and the computer in the twentieth century made it possible to communicate immediately with coworkers, clients, and customers around the world.
- The railroad in the nineteenth century and the motor vehicle and airplane in the twentieth century made it possible to deliver people, inputs, and products quickly.

To some extent, economic activities have decentralized, especially manufacturing, but modern transportation and communications reinforce rather than diminish the primacy of world cities in the global economy. Transportation services converge on world cities. They tend to have busy harbors and airports and lie at the junction of rail and highway networks.

BUSINESS SERVICES IN WORLD CITIES. The clustering of business services in the modern world city is a product of the Industrial Revolution. Modern industry is managed by large corporations formed to minimize the liability to any individual owner. A board of directors located far from the factory building makes key decisions concerning what to make, how much to produce, and what prices to charge. Support staff also far from the factory accounts for the flow of money and materials to and from the factories. This work is done in offices in world cities.

World cities offer many financial services to these businesses. As centers for finance, world cities attract the headquarters of the major banks, insurance companies, and specialized financial institutions where corporations obtain and store funds for expansion of production. Shares of major corporations are bought and sold on the stock exchanges, which are located in world cities. Obtaining information in a timely manner is essential in order to buy and sell shares at attractive prices.

Lawyers, accountants, and other professionals cluster in world cities to provide advice to major corporations and financial institutions. Advertising agencies, marketing firms, and other services concerned with style and fashion locate in world cities to help corporations anticipate changes in taste and to help shape those changes.

CONSUMER SERVICES IN WORLD CITIES. Because of their large size, world cities have retail services with extensive market areas, but they may have even more retailers than large size alone would predict. A disproportionately large number of wealthy people live in world cities, so luxury and highly specialized products are especially likely to be sold there.

Leisure services of national significance are especially likely to cluster in world cities, in part because they require large thresholds and large ranges and in part because of the presence of wealthy patrons. World cities typically offer the most plays, concerts, operas, night clubs, restaurants, bars, and professional sporting events. They contain the largest libraries, museums, and theaters. London presents more plays than the rest of the United Kingdom combined, and New York has nearly more theaters than the rest of the United States combined.

PUBLIC SERVICES IN WORLD CITIES. World cities may be centers of national or international political power. Most are national capitals, so they contain mansions or palaces for the head of state, imposing structures for the national legislature and courts, and offices for the government agencies. Also clustered in the world cities are offices for groups having business with the government, such as representatives of foreign countries, trade associations, labor unions, and professional organizations.

Unlike other world cities, New York is not a national capital. But as the home of the world's major international organization, the United Nations, it attracts thousands of diplomats and bureaucrats, as well as employees of organizations with business at the United Nations. Brussels is a world city because it is the most important center for European Union activities.

Four Levels of Business Services

According to the hierarchy of business services in urban settlements, cities can be divided into four levels of importance:

- **World Cities.** Subdivided into three tiers:
  - **Dominant World Cities.** London, New York, and Tokyo. Each is the largest city in one of the three main regions of the more developed world (Europe, North America, and East Asia), as discussed in Chapter 9. The world's most important stock exchanges operate in these three cities, and they contain large concentrations of financial and related business services.
  - **Major World Cities.** Chicago, Los Angeles, and Washington in North America, and Brussels, Frankfurt, Paris, and Zurich in Europe. Only two of the nine second-tier world cities—São Paulo and Singapore—are in LDCs. Some major corporations and banks have their headquarters in major world cities rather than in one of the three dominant ones.
  - **Secondary World Cities.** Four in North America (Houston, Miami, San Francisco, and Toronto), seven in Asia (Bangkok, Bombay, Hong Kong, Manila, Osaka, Seoul, and Taipei), five in Europe (Berlin, Madrid, Milan, Rotterdam, and Vienna), four in Latin America (Buenos Aires, Caracas, Mexico City, and Rio de Janeiro), and one each in Africa (Johannesburg) and the Oceania (Sydney).
- **Command and Control Centers.** These contain the headquarters of many large corporations, well-developed banking facilities, and concentrations of other business services, including insurance, accounting, advertising, law, and public relations. Important educational, medical, and public institutions can be found in these command and control centers. Two levels of command and control centers are regional centers and subregional centers.
- **Specialized Producer-Service Centers.** These centers offer a more narrow and highly specialized variety of services. One group of these cities specializes in the management and R&D (research and development) activities related to specific industries, such as motor vehicles in Detroit; steel in Pittsburgh; office equipment in Rochester, New York; and semiconductors in San Jose, California. A second group of these cities specializes as centers of government and education, notably state capitals that also have a major university, such as Albany, Lansing, Madison, and Raleigh-Durham.
- **Dependent Centers.** These provide relatively unskilled jobs and depend for their economic health on decisions made in the world cities, regional command and control centers, and specialized producer-service centers. Four subtypes of dependent centers in the United States include:
  - **Resort, Retirement, and Residential Centers.** Clustered in the South and West.
  - **Manufacturing Centers.** Clustered in the old northeastern manufacturing belt.
  - **Military Centers.** Clustered mostly in the South and West.
  - **Mining Centers.** Clustered in mining areas.

Business Services in LDCs

In the global economy, LDCs specialize in two distinctive types of business services:

- Offshore financial services
- Back-office functions

Offshore Financial Services

Small countries, usually islands and microstates, exploit niches in the circulation of global capital by offering offshore financial services. Offshore centers provide two important functions in the global circulation of capital:

- **Taxes.** Taxes on income, profits, and capital gains are typically low or nonexistent. Companies incorporated in an offshore center also have tax-free status regardless of the nationality of the owners. The United States loses an estimated $70 billion in tax revenue each year because companies operating in the country conceal their assets in offshore tax havens.
- **Privacy.** Bank secrecy laws can help individuals and businesses evade disclosure in their home countries. People and corporations in litigious professions, such as a doctor or lawyer accused of malpractice or the developer of a collapsed building, can protect some of their assets from lawsuits in offshore centers, as can a wealthy individual who wants to protect assets in a divorce. Creditors cannot reach such assets in bankruptcy hearings. Short statutes of limitation protect offshore accounts from long-term investigation.

The privacy laws and low tax rates in offshore centers can also provide havens to tax dodges and other illegal schemes. By definition, the extent of illegal activities is unknown and unknowable. A prominent example is the Cayman Islands, a British Crown Colony in the Caribbean near Cuba. The Caymans comprise three main islands and several smaller ones totaling around 260 square kilometers (100 square miles), with 40,000 inhabitants. Several hundred banks with assets of more than $1 trillion are legally based in the Caymans. Most of these banks have only a handful of people, if any, actually working in the Caymans.

In the Caymans, it is a crime to discuss confidential business—defined as matters learned on the job—in public. Assets placed in an offshore center by an individual or corporation in a trust are not covered by lawsuits originating in the United States, Britain, or other service centers. To get at those assets, additional lawsuits would have to be filed in the offshore centers, where privacy laws would shield the individual or corporation from undesired disclosures.

Other offshore centers include:

- **British Dependencies Other Than the Caymans.** Including Anguilla, Montserrat, and the British Virgin Islands in the Caribbean; Guernsey/Sark/Alderney, Isle of Man, and Jersey in the English Channel; and Gibralter, off Spain.
- **Dependencies of Other Countries.** Including Cook Island and Niue, controlled by New Zealand; Aruba and the Netherlands Antilles, controlled by the Netherlands; and the U.S. Virgin Islands.
- **Independent Island Countries.** Including Antigua & Barbuda, Bahamas, Barbados, Dominica, Grenada, St. Kitts & Nevis, St. Lucia, St. Vincent & the Grenadines, and Turks & Caicos in the Caribbean; the Marshall Islands Nauru, Samoa, Tonga, and Vanuatu in the Pacific Ocean; and the Maldives and Seychelles in the Indian Ocean.
- **Other Independent Countries.** Including Andorra, Liechtenstein, and Monaco in Europe; Belize and Panama in Central America; Bahrain in the Middle East; and Liberia in Africa.

Back Offices

The second distinctive type of business service found in peripheral regions is back-office functions, also known as business-process outsourcing (BPO). Typical back-office functions include processing insurance claims, payroll management, transcription work, and other routine clerical activities (Figure 12-27). Back-office work also includes centers for responding to billing inquiries related to credit cards, shipments, and claims, or technical inquiries related to installation, operation, and repair.

Traditionally, companies housed their back-office staff in the same office building downtown as their management staff, or at least in nearby buildings. A large percentage of the employees in a downtown bank building, for example, would be responsible for sorting paper checks and deposit slips. Proximity was considered important to assure close supervision of routine office workers and rapid turnaround of information.

Rising rents downtown have induced many business services to move routine work to lower-rent buildings outside the central business district (CBD). In most cases, sufficiently low rents can be obtained in buildings in the suburbs or nearby small towns.

FIGURE 12-27 Back office, India. A call center in India sends and receives phone calls primarily from the United States and other English-speaking MDCs.

However, for many business services, improved telecommunications have eliminated the need for spatial proximity.

Selective LDCs have attracted back offices for two reasons related to labor:

- **Low Wages.** Most back-office workers earn a few thousand dollars per year—higher than wages paid in most other sectors of the economy, but only one-tenth the wages paid for workers performing similar jobs in MDCs. As a result, what is regarded as menial and dead-end work in MDCs may be considered relatively high-status work in LDCs and therefore able to attract better-educated, more-motivated employees in the LDCs than would be possible in MDCs.
- **Ability to Speak English.** Many LDCs offer lower wages than MDCs, but only a handful of LDCs possess a large labor force fluent in English. In Asia, countries such as India, Malaysia, and the Philippines have substantial numbers of workers with English-language skills, a legacy of British and American colonial rule. Major multinational companies such as American Express and General Electric have extensive back-office facilities in these countries.

The ability to communicate in English over the telephone is a strategic advantage in competing for back offices with neighboring countries, such as Indonesia and Thailand, where English is less commonly used. Familiarity with English is an advantage not only for literally answering the telephone but also for gaining a better understanding of the preferences of American consumers through exposure to English-language music, movies, and television.

Workers in back offices are often forced to work late at night, when it's daytime in the United States, peak demand for inquiries. Many employees must arrive at work early and stay late because they lack their own transportation, so they depend on public transportation, which typically does not operate late at night. Sleeping and entertainment rooms are provided at work to fill the extra hours.

Economic Base of Settlements

A settlement's distinctive economic structure derives from its **basic industries**, which export primarily to consumers outside the settlement. **Nonbasic industries** are enterprises whose customers live in the same community—essentially, consumer services. A community's unique collection of basic industries defines its **economic base**.

A settlement's economic base is important, because exporting by the basic industries brings money into the local economy, thus stimulating the provision of more nonbasic consumer services for the settlement. New basic industries attract new workers to a settlement, and they bring their families with them. The settlement then attracts additional consumer services to meet the needs of the new workers and their families. Thus a new basic industry stimulates establishment of new supermarkets, laundromats, restaurants, and other consumer services. But a new nonbasic service, such as a supermarket, will not induce construction of new basic industries.

A community's basic industries can be identified by computing the percentage of the community's workers employed in different types of businesses. The percentage of workers employed in a particular industry in a settlement is then compared to the percentage of all workers in the country employed in that industry. If the percentage is much higher in the local community, then that type of business is a basic economic activity.

Specialization of Cities in Different Services

Settlements in the United States can be classified by their type of basic activity (Figure 12-28). Each type of basic activity has a different spatial distribution. The concept of basic industries originally referred to manufacturing. Some communities specialize in durable manufactured goods, such as steel and automobiles, others in nondurable manufactured goods, such as textiles, apparel, food, chemicals, and paper. Most communities that have an economic base of manufacturing durable goods are clustered between northern Ohio and southeastern Wisconsin, near the southern Great Lakes. Nondurable manufacturing industries, such as textiles, are clustered in the Southeast, especially in the Carolinas. Compared to the national average, some settlements have a very high percentage of workers employed in the primary sector, notably mining. Mining settlements are located near reserves of coal, petroleum, and other resources.

In a postindustrial society, such as the United States, increasingly the basic economic activities are in business, consumer, or public services. Geographers Ó hUallacháin and Reid have documented examples of settlements that specialize in particular types of services:

- Examples of settlements specializing in business services:
  - General business: Large metropolitan areas, especially Chicago, Los Angeles, New York, and San Francisco
  - Computing and data processing services: Boston and San Jose

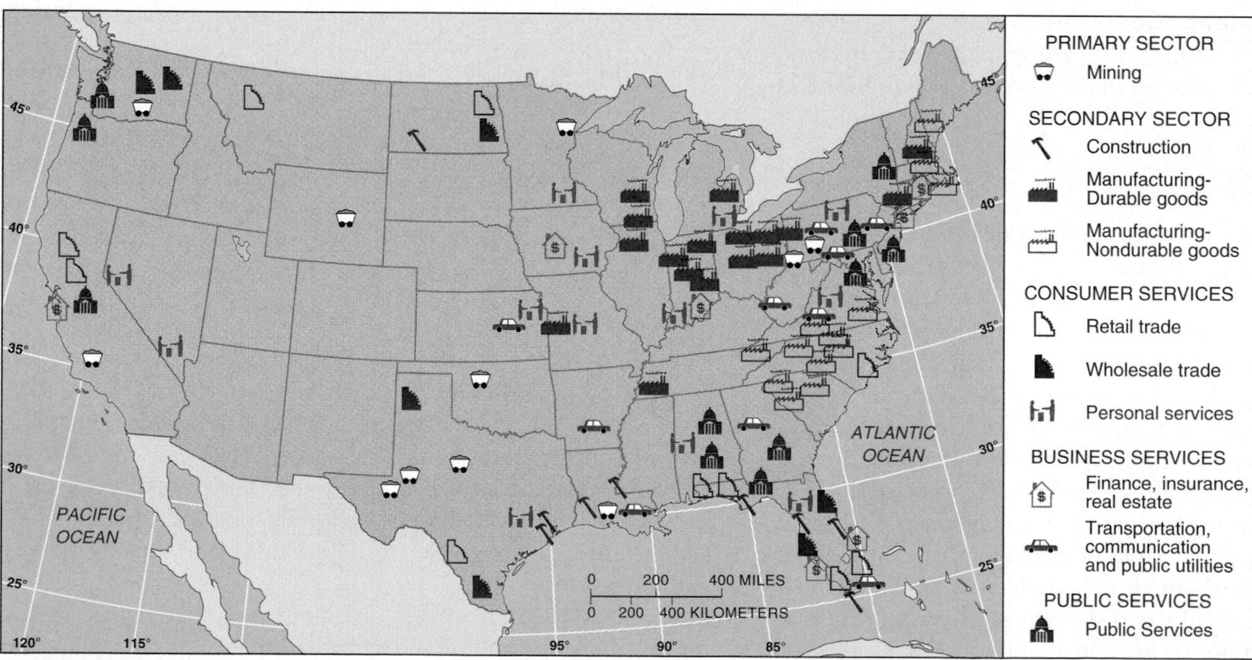

FIGURE 12-28 Economic base of U.S. cities. Symbols on this map represent cities that have a significantly higher percentage of their labor force engaged in the type of economic activity shown. Other cities also engage in such activities but are not shown because they specialize in multiple activities or are near the national average for all sectors. (Mathematically, a city was included if the percentage of its labor force in one sector was more than 2 standard deviations above the mean for all U.S. cities.)

- High-tech industries support services: Austin, Orlando, and Raleigh-Durham
- Military activity support services: Albuquerque, Colorado Springs, Huntsville, Knoxville, and Norfolk
- Management-consulting services: Washington, D.C.

- Examples of settlements specializing in consumer services:
  - Entertainment and recreation: Atlantic City, Las Vegas, and Reno
  - Medical services: Rochester, Minnesota
- Examples of settlements specializing in public services:
  - State capitals
  - Large universities
  - Military bases

Although the population of cities in the South and West has grown more rapidly in recent years, Ó hUallacháin and Reid found that cities in the North and East have expanded their provision of business services more rapidly. Northern and eastern cities that were once major manufacturing centers have been transformed into business service centers. These cities have moved more aggressively to restructure their economic bases to offset sharp declines in manufacturing jobs.

Steel was once the most important basic industry of Cleveland and Pittsburgh, but now health services such as hospitals and clinics and medical high-technology research are more important. Baltimore once depended for its economic base on manufacturers of fabricated steel products, such as Bethlehem Steel, General Motors, and Westinghouse. The city's principal economic asset was its port, through which raw materials and fabricated products passed. As these manufacturers declined, the city's economic base turned increasingly to services, taking advantage of its clustering of research-oriented universities, especially in medicine. The city is trying to become a center for the provision of services in biotechnology.

Distribution of Talent

Individuals possessing special talents are not distributed uniformly among cities. Some cities have a higher percentage of talented individuals than others. To some extent, talented

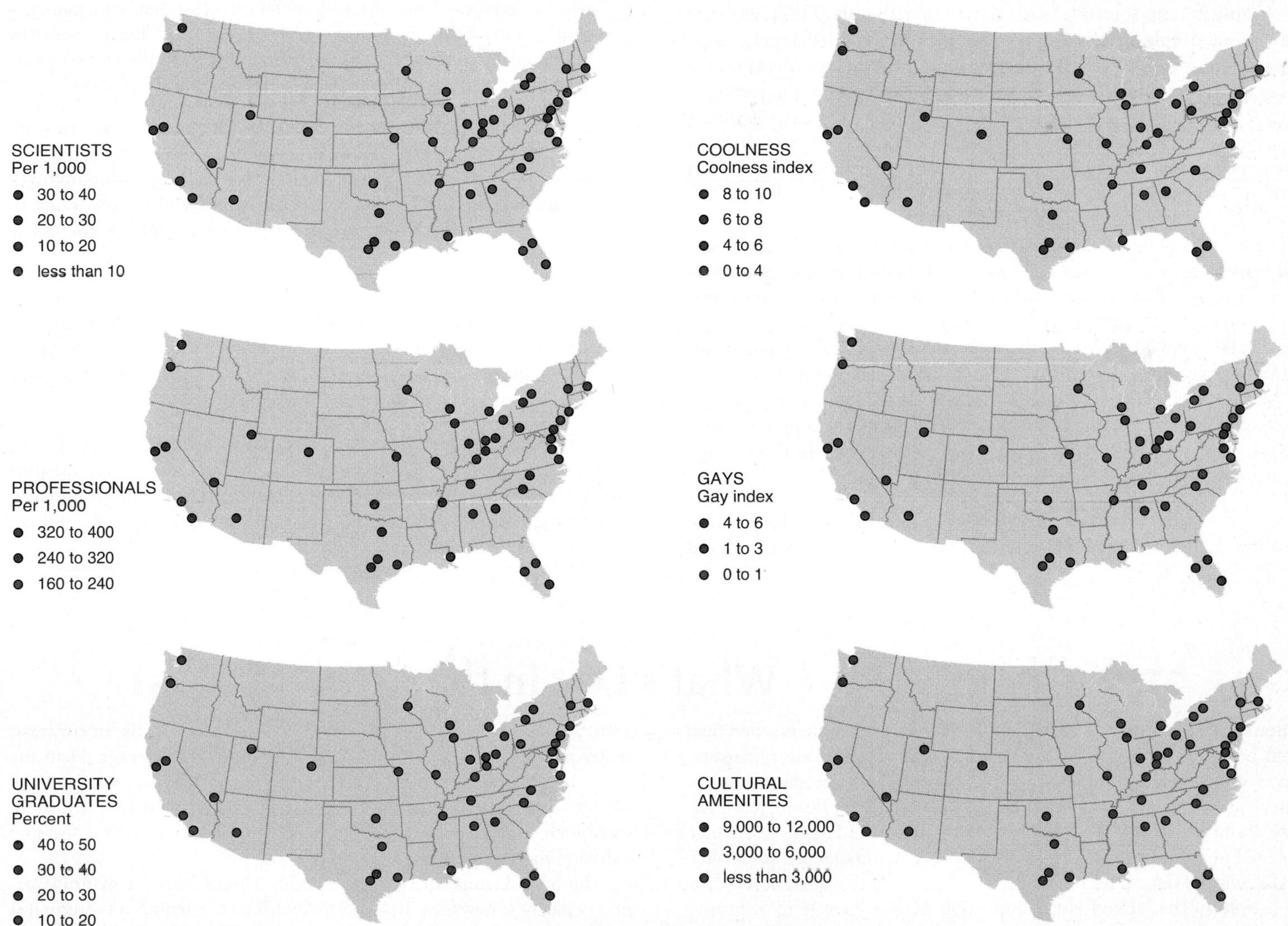

FIGURE 12-29 Geography of talent. People with relatively high levels of talent—scientists, professionals, and college-educated people, for example (left)—are attracted to cities with high levels of diversity, as measured by cultural facilities and gays per capita, as well as a "coolness" factor (right).

individuals are attracted to the cities with the most job opportunities and financial incentives. But the principal enticement for talented individuals to cluster in some cities more than others is cultural rather than economic, according to research conducted by Richard Florida. Individuals with special talents gravitate toward cities that offer more cultural diversity.

Talent was measured by Florida as a combination of the percentage of people in the city with college degrees, the percentage employed as scientists or engineers, and the percentage employed as professionals or technicians. Three measures of cultural diversity were used—the number of cultural facilities per capita, the percentage of gay men, and a "coolness" index. The "coolness" index, developed by *POV Magazine*, combined the percentage of population in their 20s, the number of bars and other nightlife places per capita, and the number of art galleries per capita. A city's gay population was based on census figures for the percentage of households consisting of two adult men. Two adult men who share a house may not be gay, but Florida assumed that the percentage of adult men living together who were gay did not vary from one city to another.

Florida found a significant positive relationship between the distribution of talent and the distribution of diversity in the largest U.S. cities (Figure 12-29, see previous page). In other words, cities with high cultural diversity tended to have relatively high percentages of talented individuals. Washington, San Francisco, Boston, and Seattle ranked among the top in both talent and diversity, whereas Las Vegas was near the bottom in both. Attracting talented individuals is important for a city, because these individuals are responsible for promoting economic innovation. They are likely to start new businesses and infuse the local economy with fresh ideas.

SUMMARY

Geographers do not merely observe the distribution of services; they play a major role in creating it. Shopping center developers, large department store and supermarket chains, and other retailers employ geographers to identify new sites for stores and assess the performance of existing stores. Geographers conduct statistical analyses based on the gravity model to delineate underserved market areas where new stores could be profitable, as well as to identify overserved market areas where poorly performing stores are candidates for closure.

Developers of new retail services obtain loans from banks and financial institutions to construct new stores and malls. Lending institutions want assurance that the proposed retail development has a market area with potential to generate sufficient profits to repay the loan. They employ geographers to make objective market-area analyses independent of the excessively optimistic forecasts submitted by the retailer.

Many service providers make location decisions on the basis of instinct, intuition, and tradition. In an increasingly competitive market, retailers and other services that place themselves in the optimal location secure a critical advantage.

1. **Where Did Services Originate?** Services are divided into three types—consumer (including retail, health, education, and leisure), business (including financial, professional, and management), and public (including federal, state, and local). Services originated in settlements. Early settlements, established to serve rural areas, provided primarily personal and public services.
2. **Where Are Contemporary Services Located?** Services are clustered in settlements. Rural settlements are centers for agriculture and provide a small number of services. Urbanization involves increases in the percentage and in the number of people living in urban settlements. Urban settlements are centers for consumer and business services.
3. **Why Are Consumer Services Distributed in a Regular Pattern?** Consumer services attract customers from market areas of varying size. Geographers calculate whether a service can be profitable within a market area. In MDCs, market areas form a regular hierarchy by size and distance from each other.
4. **Why Do Business Services Locate in Large Settlements?** Financial, professional, and other business services cluster disproportionately in large world cities to support the operations of major corporations. World cities also play major consumer- and public-service functions.

CASE STUDY REVISITED / What's Left in the United States?

China's manufacturing growth is highly visible to consumers when they open boxes stamped "made in China" or read about a factory closure. Much less visible is India's growing service sector. The person at the other end of a telephone could be anywhere in the world. Increasingly, they are in India. Just as China now has the world's largest labor force engaged in manufacturing, India now has the world's largest labor force engaged in services (Figure 12-30).

People in the United States and other MDCs have long been told that they needed a good education to get a good job. Reinforcing this view, the manufacturing jobs being added in China and other LDCs generally do not require high levels of education or skills. In contrast, many of the service jobs being added in India have been held in the United States by college graduates. Service jobs outsourced from the United States to India include financial analysis, patent research, insurance claims processing, architectural drafting, radiology, and software management.

The lure of India for service providers, like China for manufacturers, is partly low wages. India's universities are turning out graduates with MBAs willing to do the same job for $12,000 that Wall Street analysts have been doing in the United States for $100,000. Computer

(Continued)

CASE STUDY REVISITED (*Continued*)

FIGURE 12-30 Advertisement for call-center jobs in Bengalūru (Bangalore), India.

programmers paid $45 an hour in the United States can be replaced by college graduates in India willing to do the same job for $7 an hour. Adding to the appeal of India's labor force is the ability to work in English. Fluency in English is not a critical skill for manufacturers of goods exported to the United States, but it is a critical skill for providers of services to U.S. consumers and businesses, and it is a skill available in India.

Business services are moving from the United States to India at the rate of 300,000 jobs a year, according to Forrester Research. Another survey has estimated an annual relocation from the United States to India of 100,000 financial services jobs, 30,000 management jobs, and 20,000 architectural design jobs. The United States lost 4 percent of its jobs in the first year of the severe recession that started in 2008. The losses were not uniform: Jobs declined 12 percent in manufacturing, 7 percent in professional services, 5 percent in financial services, 4 percent in retail services, and 2 percent in leisure services. So what jobs are left in the United States? Even in the worst economy since the 1930s, jobs in education and in health care grew by 2 percent each. ■

KEY TERMS

Basic industries (p. 398) Industries that sell their products or services primarily to consumers outside the settlement.

Business services (p. 375) Services that primarily meet the needs of other businesses, including professional, financial, and transportation services.

Central place (p. 388) A market center for the exchange of services by people attracted from the surrounding area.

Central place theory (p. 388) A theory that explains the distribution of services based on the fact that settlements serve as centers of market areas for services; larger settlements are fewer and farther apart than smaller settlements and provide services for a larger number of people who are willing to travel farther.

City-state (p. 379) A sovereign state comprising a city and its immediate hinterland.

Clustered rural settlement (p. 381) A rural settlement in which the houses and farm buildings of each family are situated close to each other and fields surround the settlement.

Consumer services (p. 375) Businesses that provide services primarily to individual consumers, including retail services and education, health, and leisure services.

Dispersed rural settlement (p. 381) A rural settlement pattern characterized by isolated farms rather than clustered villages.

Economic base (p. 398) A community's collection of basic industries.

Enclosure movement (p. 384) The process of consolidating small landholdings into a smaller number of larger farms in England during the eighteenth century.

Gravity model (p. 391) A model that holds that the potential use of a service at a particular location is directly related to the number of people in a location and inversely related to the distance people must travel to reach the service.

Market area (or hinterland) (p. 388) The area surrounding a central place from which people are attracted to use the place's goods and services.

Nonbasic industries (p. 398) Industries that sell their products primarily to consumers in the community.

Primate city (p. 393) The largest settlement in a country, if it has more than twice as many people as the second-ranking settlement.

Primate city rule (p. 393) A pattern of settlements in a country such that the largest settlement has more than twice as many people as the second-ranking settlement.

Public services (p. 376) Services offered by the government to provide security and protection for citizens and businesses.

Range (of a service) (p. 389) The maximum distance people are willing to travel to use a service.

Rank-size rule (p. 393) A pattern of settlements in a country such that the *n*th largest settlement is 1/*n* the population of the largest settlement.

Service (p. 374) Any activity that fulfills a human want or need and returns money to those who provide it.

Settlement (p. 374) A permanent collection of buildings and inhabitants.

Threshold (p. 389) The minimum number of people needed to support the service.

Urbanization (p. 385) An increase in the percentage of the number of people living in urban settlements.

THINKING GEOGRAPHICALLY

1. Determine the economic base of your community. Consult the U.S. Census of Manufacturing or County Business Patterns. To make a rough approximation of your community's basic industries, compute the decimal fraction of the nation's population that lives in your community. It will be a small number, such as 0.0005. Then, find the total number of U.S. firms (or employees) in each industrial sector that is present in your community. Multiply these national figures by your local population fraction. Subtract the result from your community's actual number of firms (or employees) for that type of industry. If the difference is positive, you have identified one of your community's basic industries.
2. Your community's economy is expanding or contracting as a result of the performance of its basic employment. Two factors can explain the performance of your community's basic employment. One is that the sector is expanding or contracting nationally. The second is that the sector is performing much better or worse in the community than in the country as a whole. Which of the two factors better explains the performance of your community's basic employment?
3. Rural settlement patterns along the U.S. East Coast were influenced by migration during the Colonial era. To what extent do distinctive rural settlement patterns elsewhere in the United States result from international or internal migration?
4. Nearly all residents of MDCs lead urban lifestyles even if they live in rural areas. In contrast, many residents in LDCs lead rural lifestyles even though they live in large cities. They practice subsistence agriculture, raising animals or growing crops. Lacking electricity, they gather wood for fuel. Lacking running water and sewers, they dig latrines. Why do so many urban dwellers in LDCs lead rural lifestyles?
5. What evidence can you find in your community of economic ties to world cities located elsewhere in North America, Europe, or East Asia?

RESOURCES

Some recent and classic books and articles on industrial geography:

Archer, Clark J., and Ellen R. White. "A Service Classification of American Metropolitan Areas." *Urban Geography* 6 (1985): 122–51.

Bagchi-Sen, Sharmistha. "Service Employment in Large, Medium, and Small Metropolitan Areas in the United States." *Urban Geography* 18 (1997): 264–81.

Beaverstock, J. V., P. Taylor and R. G. Smith, "A Roster of World Cities," *Cities* 16 (1999): 445–58.

Benevolo, Leonardo. *The History of the City*, 2nd ed. Cambridge, MA: MIT Press, 1991.

Berry, Brian J. L. *The Geography of Market Centers and Retail Distribution.* Englewood Cliffs, NJ: Prentice Hall, 1967.

Bourne, L. S., and J. W. Simmons. *Systems of Cities*. New York: Oxford University Press, 1978.

Brunn, Stanley D., Maureen Hays-Mitchell, and Donald J. Zeigler, eds. *Cities of the World: World Regional Urban Development*, 4th ed. Lanham, MD: Rowman and Littlefield, 2008.

Chisholm, Michael. *Rural Settlement and Land Use*, 3rd ed. London: Hutchinson, 1979.

Christaller, Walter. *The Central Places of Southern Germany.* Englewood Cliffs, NJ: Prentice Hall, 1966.

Crewe, Louise. "Geographies of Retailing and Consumption." *Progress in Human Geography* 24 (2000): 275–90.

Daniels, P. W. *Service Industries: A Geographical Appraisal.* London: Methuen, 1986.

Davis, Kingsley. *Cities: Their Origin, Growth, and Human Impact.* San Francisco: W. H. Freeman, 1973.

Florida, Richard. "The Economic Geography of Talent." *Annals of the Association of American Geographers* 92 (2002): 743–55.

Harris, Chauncey D. "A Functional Classification of Cities in the United States." *Geographical Review* 33 (1943): 86–99.

Hauser, Philip M., and Leo F. Schnore, eds. *The Study of Urbanization.* New York: John Wiley, 1965.

Jacobs, Jane. *The Economy of Cities*. New York: Vintage Books, 1970.

Keeble, D., and L. Nachum. "Why Do Business Service Firms Cluster? Small Consultancies, Clustering and Decentralization in London and Southern England." *Transactions of the Institute of British Geographers, New Series* 27 (2002): 67–90.

King, Leslie J. *Central Place Theory.* Beverly Hills, CA: Sage, 1984.

Kirn, Thomas J. "Growth and Change in the Service Sector of the U.S.: A Spatial Perspective." *Annals of the Association of American Geographers* 77 (1987): 353–72.

Lord, J. Dennis. "Retail Saturation: Inevitable or Irrelevant?" *Urban Geography* 21 (2000): 342–60.

Lösch, August. *The Economics of Location.* New Haven, CT: Yale University Press, 1954.

Marshall, John U. "Beyond the Rank-Size Rule: A New Descriptive Model of City Sizes." *Urban Geography* 18 (1997): 36–55.

Ó hUallacháin, Breandan, and Neil Reid. "The Location and Growth of Business and Professional Services in American Metropolitan Areas, 1976–1986." *Annals of the Association of American Geographers* 81 (1991): 254–70.

Scott, Peter. *Geography and Retailing.* London: Hutchinson University Press, 1970.

Taylor, P. J., G. Catalano, and D. R. F. Walker. "Measurement of the World City Network." *Urban Studies* 39 (2002): 2367–76.

Trewartha, Glen T. "Types of Rural Settlements in North America." *Geographical Review* 36 (1946): 568–96.

Journals featuring settlement geography:

Journal of Historical Geography, Journal of Regional Science, Journal of Rural Studies, Journal of Urban Economics, Urban Geography.

Key Internet sites:

http://www.citypopulation.de/World.html. A number of organizations publish lists of the population of cities based on officially published national statistics. One such site has been created by Thomas Brinkoff, a professor at the University of Oldenburg in Germany.

www.creativeclass.com. Information on the geography of talent can be found on a web site maintained by Richard Florida.

www.oecd.org. The Organisation for Economic Co-operation and Development maintains a statistics service on economic performance of countries, including how each is faring in the severe recession that started in 2008.

Log in to www.mygeoscienceplace.com for videos, interactive maps, RSS feeds, case studies, and self-study quizzes to enhance your study of Services.

CHAPTER 13

Urban Patterns

Suppose as a geography class assignment you were dropped off on a street corner in a very large city and told to meet your instructor and classmates in 1 hour at city hall. How would you find it? In a small town you could simply ask for directions, but in an unfamiliar neighborhood of a large city would you hesitate to ask strangers?

Your destination is probably downtown, because that's where public services such as city hall cluster. Which direction is downtown? The skyscrapers far in the distance are probably a clue, and house numbers on major streets get lower as you head toward downtown.

KEY ISSUES

1 **Why Do Services Cluster Downtown?**

2 **Where Are People Distributed Within Urban Areas?**

3 **Why Do Inner Cities Face Distinctive Challenges?**

4 **Why Do Suburbs Face Distinctive Challenges?**

In a small town everything is within easy walking distance, but in a large city your destination is too far to walk. How would you get there without a car? Hitchhiking is dangerous, and you don't have enough money to hire a taxi. What about the bus? Where does the bus stop? What route does it follow? How much is the fare? Do you have the exact change or a prepaid fare card, as required on most big-city buses?

Once on the bus, you sit down next to another passenger. Is your neighbor of the same ethnicity as you? In fact, are you the only person on the bus of your ethnicity? Have you been in other large groups where you were the only person of your ethnicity? Do the other passengers smile at you and chat, or do they mind their own business?

A large city is stimulating and agitating, entertaining and frightening, welcoming and cold. A city has something for everyone, but a lot of those things are for people who are different from you. Urban geography helps to sort out the complexities of familiar and unfamiliar patterns in urban areas.

Waiting for the Paris Métro

CASE STUDY / Two Families in New Jersey

Ruth Merritt lives in the city of Camden, New Jersey. She is a 24-year-old single parent with three children (ages 7, 2, and 1). Her income, derived from the community's program of child support, is $250 per month. That works out to $3,000 a year.

The Merritt family lives in a four-room apartment in a row house that was divided some years ago into six dwelling units. The apartment has generally adequate plumbing and kitchen facilities, but the residents sometimes see rats in the building. The rent is $75 per month, plus an average of $50 per month for electricity and other utilities.

Ruth Merritt receives food stamps, but her monthly expenses for food, clothing, and shelter exceed her income. In cold weather she must sometimes reduce the food budget to pay for heat.

Just 10 kilometers away, east of Camden, the Johnson family lives in Cherry Hill, New Jersey. William Johnson is a lawyer. He commutes to downtown Philadelphia, across the Delaware River from Camden. Diane Johnson works for a nonprofit organization with offices in the suburban community where they live. Their two children attend a recently built school in the community.

The Johnson family's dwelling is a detached house with three bedrooms, a living room, dining room, family room, and kitchen. The attached garage contains two cars, one for each parent to get to work. The half-acre lawn surrounding the house provides ample space for the children to play. The Johnsons bought their house 10 years ago for $250,000. The monthly payments for mortgage and utilities are $3,000, but the family's combined annual income of $200,000 is more than adequate to pay the housing costs. The house is now worth a half-million dollars.

The Merritt and Johnson households illustrate the contrasts that exist today in U.S. urban areas. As you have seen throughout this book, dramatic differences in material standards exist around the world. However, the picture drawn here is based on families living in the same urban area, only a few kilometers apart.

Were these examples taken from an urban area elsewhere in the world, the spatial patterns might be reversed. In most of the world the higher-status Johnsons would live near the center of the city, whereas the lower-status Merritts would live in the suburbs. ■

When you stand at the corner of Fifth Avenue and 34th Street in New York City, staring up at the Empire State Building, you know that you are in a city. When you are standing in an Iowa cornfield, you have no doubt that you are in the country. Geographers help explain what makes city and countryside different ***places***.

Chapter 12 and this chapter are both concerned with urban geography, but at different ***scales***. The previous chapter examined the distribution of urban settlements at national and global scales. This chapter looks at ***where*** people and activities are distributed within urban ***spaces***. Models have been developed to explain ***why*** differences occur within urban areas.

We all experience the interplay between ***globalization*** and ***local diversity*** of urban settlements. If you were transported to the downtown of another city, you might be able to recognize the city from its skyline. Many downtowns have a collection of high-rise buildings, towers, and landmarks that are identifiable even to people who have never visited them.

On the other hand, if you were transported to a suburban residential neighborhood, you would have difficulty identifying the urban area. Suburban houses, streets, schools, and shopping centers look very much alike from one American city to another.

In ***regions*** of MDCs, people are increasingly likely to live in suburbs. This changing structure of cities is a response to conflicting desires. People wish to spread across the landscape to avoid urban problems, but at the same time, they want convenient ***connections*** to the city's jobs, shops, culture, and recreation.

In this chapter, the causes and consequences of today's evolving urban patterns are examined. Although different internal structures characterize urban areas in the United States and elsewhere, the problems arising from current spatial trends are similar. Geographers describe where different types of people live and try to explain the reasons for the observed patterns.

KEY ISSUE 1
Why Do Services Cluster Downtown?

- CBD Land Uses
- Competition for Land in the CBD
- CBDs Outside North America

Downtown is the best-known and the most visually distinctive area of most cities. It is usually one of the oldest districts in a city, often the site of the original settlement. The downtowns of most North American cities have different features than those in the rest of the world. ■

CBD Land Uses

Downtown is known to geographers by the more precise term **central business district (CBD)**. The CBD is compact—less than 1 percent of the urban land area—but contains a large percentage

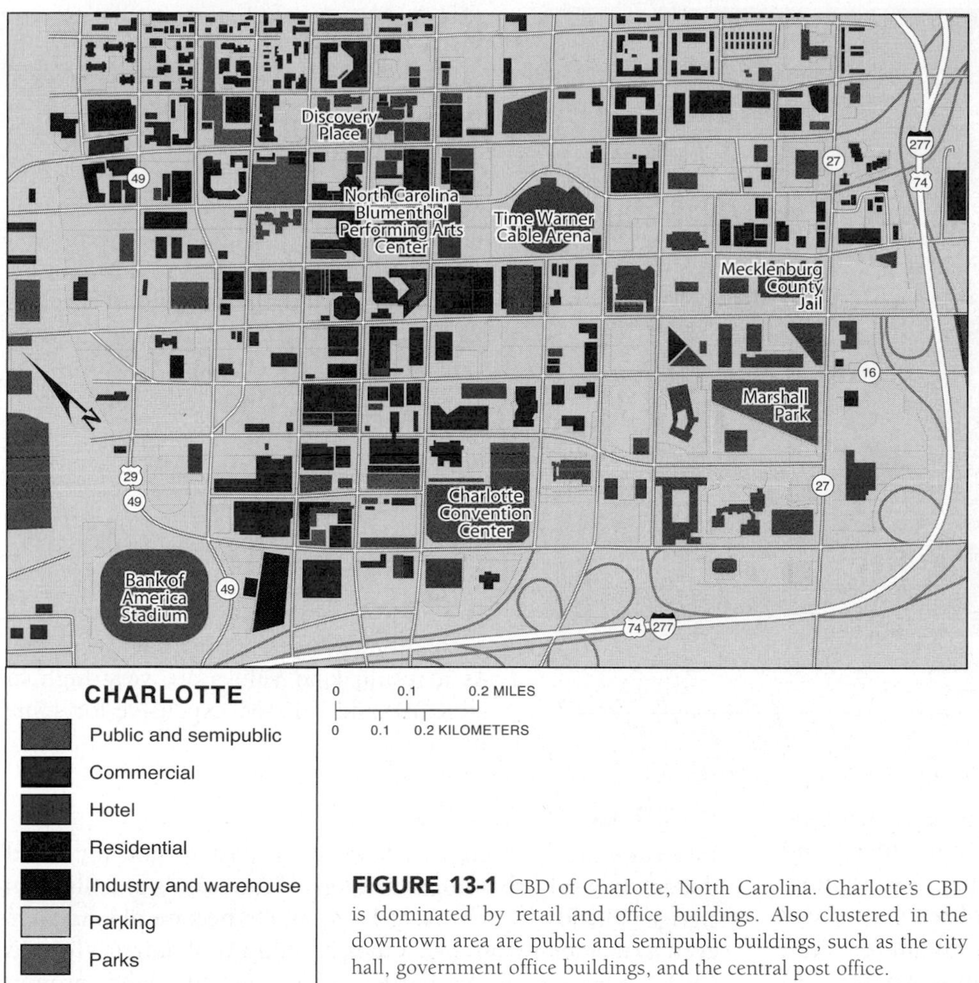

FIGURE 13-1 CBD of Charlotte, North Carolina. Charlotte's CBD is dominated by retail and office buildings. Also clustered in the downtown area are public and semipublic buildings, such as the city hall, government office buildings, and the central post office.

of the shops, offices, and public institutions (Figure 13-1). Consumer and business services are attracted to the CBD because of its accessibility. The center is the easiest part of the city to reach from the rest of the region and is the focal point of the region's transportation network.

Retail Services in the CBD

In the past, three types of retail services clustered in the CBD because they required accessibility to everyone in the region—retailers with a high threshold, those with a long range, and those that served people who worked in the CBD. Changing shopping habits and residential patterns have reduced the importance of retail services in the CBD.

RETAILERS WITH A HIGH THRESHOLD. Retailers with high thresholds, such as department stores, traditionally preferred a CBD location in order to be accessible to many people (Figure 13-2). Large department stores in the CBD would cluster near one intersection, which was known as the "100 percent corner." Rents were highest there because this location had the highest accessibility for the most customers.

In recent years, however, many high-threshold shops such as large department stores have closed their downtown branches. CBDs that once boasted three or four stores now have none, or perhaps one struggling survivor. The customers for downtown department stores now consist of downtown office workers, inner-city residents, and tourists. Department stores with high thresholds are now more likely to be in suburban malls.

RETAILERS WITH A HIGH RANGE. High-range retailers are often specialists, with customers who patronize them infrequently (Figure 13-3). These retailers once preferred CBD locations because their customers were scattered over a wide area. For example, a jewelry or clothing store attracted shoppers from all over the urban area, but each customer visited infrequently.

Like those with high thresholds, high-range retailers have moved with department stores to suburban locations. These retailers survive in some CBDs if they combine retailing with recreational activities. People are willing to make a special trip to a specific destination downtown for unusual shops in a dramatic setting, perhaps a central atrium with a fountain or a view of a harbor. New shopping areas that attract high-range retailers have been built in several North American CBDs:

- Boston: Faneuil Hall Marketplace, in renovated eighteenth-century buildings
- Baltimore: Harbor Place, built in the Inner Harbor, adjacent to waterfront museums, tourist attractions, hotels, and cultural facilities
- Philadelphia: Gallery at Market East, a suburban-style shopping center
- San Francisco Ferry Building: a gourmet food center where San Francisco Bay ferries dock

These downtown malls attract suburban shoppers as well as out-of-town tourists because in addition to shops they offer unique recreation and entertainment experiences.

Some CBDs have restored their food markets, with individual stalls operated by different merchants. They may have a high range because they attract customers who willingly travel far to find more exotic or higher-quality products. At the same time, inner-city residents may use these markets for their weekly grocery shopping.

RETAILERS SERVING DOWNTOWN WORKERS. A third type of retail activity in the center serves the many people

FIGURE 13-2 CBD retailer with high threshold. Shoppers flock to Macy's in Midtown Manhattan the day after Thanksgiving.

who work in the center and shop during lunch or working hours. These businesses sell office supplies, computers, and clothing, or offer shoe repair, rapid photocopying, dry cleaning, and so on. In contrast to the other two types of retailers, shops that appeal to nearby office workers are expanding in the CBD, in part because the number of downtown office workers has increased and in part because downtown offices require more services.

Patrons of downtown shops tend increasingly to be downtown employees who shop during the lunch hour. Thus, although the total volume of sales in downtown areas has been stable, the pattern of demand has changed. Large department stores have difficulty attracting their old customers, whereas smaller shops that cater to the special needs of the downtown labor force are expanding.

Business Services in the CBD

Offices cluster in the center for accessibility. People in such business services as advertising, banking, finance, journalism, and law particularly depend on proximity to professional colleagues. Lawyers, for example, choose locations near government offices and courts. Services such as temporary secretarial agencies and instant printers locate downtown to be near lawyers, forming a chain of interdependency that continues to draw offices to the center city.

Despite the diffusion of modern telecommunications, many professionals still exchange information with colleagues primarily through face-to-face contact. Financial analysts discuss attractive stocks or impending corporate takeovers. Lawyers meet to settle disputes out of court. Offices are centrally located to facilitate rapid communication of fast-breaking news through spatial proximity. Face-to-face contact also helps to establish a relationship of trust based on shared professional values.

A central location also helps businesses that employ workers from a variety of neighborhoods. Top executives may live in one neighborhood, junior executives in another, secretaries in another, and custodians in still another. Only a central location is readily accessible to all groups. Firms that need highly specialized employees are more likely to find them in the central area, perhaps currently working for another company downtown.

Competition for Land in the CBD

The center's accessibility produces extreme competition for the limited sites available. As a result, land values are very high in the CBD, and it is too expensive for some activities.

High Land Costs

In a rural area a hectare of land might cost several thousand dollars. In a suburb it might run tens of thousands of dollars. In a large CBD like New York or London, if a hectare of land were even available, it would cost tens of millions of dollars. Tokyo's CBD contains some of Earth's most expensive land, around $15,000 per square meter ($60,000,000 per acre). If this page were a parcel of land in Tokyo, it would sell for $1,000. Before the 2008 recession, prices were even higher.

Tokyo's high prices result from a severe shortage of buildable land. Buildings in most areas are legally restricted to less than 10 meters in height (normally three stories) for fear of earthquakes, even though recent earthquakes have demonstrated that modern, well-built skyscrapers are safer than older three-story structures. Two distinctive characteristics of the CBD follow from the high land cost. First, land is used more intensively in the center than elsewhere in the city. Second, some activities are excluded from the center because of the high cost of space.

INTENSIVE LAND USE. The intensive demand for space has given the CBD a three-dimensional character, pushing it vertically. Compared to other parts of the city, the CBD uses more space below and above ground level.

A vast underground network exists beneath most central cities. The typical "underground city" includes multistory parking garages, loading docks for deliveries to offices and shops, and utility lines (water, sewer, phone, electric, and some heating). Typically, telephone, electric, and cable television wires run beneath the surface in central areas. Not enough space is available in the center for the large number of telephone poles that would be needed for such a dense network, and the wires are unsightly and hazardous. Subways run beneath the streets of

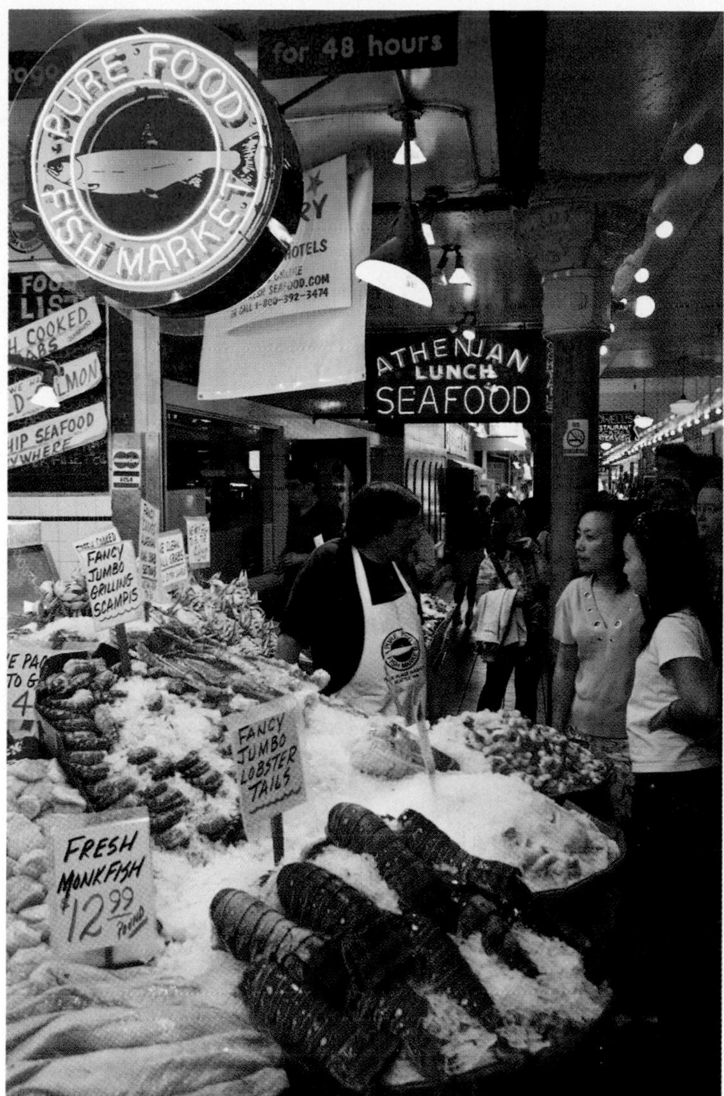

FIGURE 13-3 CBD retailer with high range. Pike Place Market in downtown Seattle sells food at individually owned stalls.

larger CBDs. And cities such as Minneapolis, Montreal, and Toronto have built extensive pedestrian passages and shops beneath the center. These underground areas segregate pedestrians from motor vehicles and shield them from harsh winter weather.

SKYSCRAPERS. Demand for space in the CBD has also made high-rise structures economically feasible. Downtown skyscrapers give a city one of its most distinctive images and unifying symbols. Suburban houses, shopping malls, and factories look much the same from one city to another, but each city has a unique downtown skyline, resulting from the particular arrangement and architectural styles of its high-rise buildings.

The first skyscrapers were built in Chicago in the 1880s, made possible by two inventions—the elevator and iron-frame building construction. The first high-rises caused great inconvenience to neighboring structures because they blocked light and air movement. Artificial lighting, ventilation, central heating, and air-conditioning have helped solve these problems. Most North American and European cities enacted zoning ordinances early in the twentieth century in part to control the location and height of skyscrapers.

Skyscrapers are an interesting example of "vertical geography." The nature of an activity influences which floor it occupies in a typical high-rise:

- Retailers pay high rents for street-level space to entice customers.
- Professional offices, less dependent on walk-in trade, occupy the middle levels at lower rents.
- Apartments in the upper floors take advantage of lower noise levels and panoramic views.

The one large U.S. CBD without skyscrapers is Washington, D.C., where no building is allowed to be higher than the U.S. Capitol dome. Consequently, offices in downtown Washington rise no more than 13 stories. As a result, the typical Washington office building uses more horizontal space—land area—than in other cities. Thus the city's CBD spreads over a much wider area than those in comparable cities.

Activities Excluded from the CBD

High rents and land shortage discourage two principal activities in the CBD—industrial and residential.

LACK OF INDUSTRY IN THE CBD. Modern factories require large parcels of land to spread operations among one-story buildings. Suitable land is generally available in suburbs. In the past, inner-city factories and retail establishments relied on waterfront CBDs that were once lined with piers for cargo ships to load and unload and warehouses to store the goods. Today's large oceangoing vessels are unable to maneuver in the tight, shallow waters of the old CBD harbors. Consequently, port activities have moved to more modern facilities downstream.

Port cities have transformed their waterfronts from industry to commercial and recreational activities. Derelict warehouses and rotting piers have been replaced with new offices, shops, parks, and museums. As a result, CBD waterfronts have become major tourist attractions in a number of North American cities, including Boston, Toronto, Baltimore, and San Francisco, as well as in European cities such as Barcelona and London. The cities took the lead in clearing the sites and constructing new parks, docks, walkways, museums, and parking lots. They also have built large convention centers to house professional meetings and trade shows. Private developers have added hotels, restaurants, boutiques, and entertainment centers to accommodate tourists and conventioneers.

LACK OF RESIDENTS IN CBDs. Many people used to live downtown. Poorer people jammed into tiny, overcrowded apartments, and richer people built mansions downtown. In the twentieth century, most residents abandoned downtown living because of a combination of pull and push factors. They were pulled to suburbs that offered larger homes with private yards and modern schools. And they were pushed from CBDs by high rents that business and retail services were willing to pay and by the dirt, crime, congestion, and poverty that they experienced by living downtown.

In the twenty-first century, however, the population of many U.S. CBDs has increased. New apartment buildings and townhouses have been constructed, and abandoned warehouses and outdated office buildings have been converted into residential lofts. Downtown living is especially attractive to people without school-age children, either "empty nesters" whose children have left home or young professionals who have not yet had children. These two groups are attracted by the entertainment, restaurants, museums, and nightlife that are clustered downtown, and they are not worried about the quality of neighborhood schools.

CBDs Outside North America

CBDs outside the United States are less dominated by commercial considerations. The most prominent structures may be churches and former royal palaces, situated on the most important public squares, at road junctions, or on hilltops. Parks in the center of European cities often were first laid out as private gardens for aristocratic families and later were opened to the public.

European cities display a legacy of low-rise structures and narrow streets, built as long ago as medieval times. Some European cities try to preserve their historic CBDs by limiting high-rise buildings and the number of cars. Several high-rise offices were built in Paris during the 1970s, including Europe's tallest office building (the 210-meter, or 688-foot, Tour Montparnasse). The public outcry over this disfigurement of the city's historic skyline was so great that officials reestablished lower height limits.

More people live downtown in cities outside North America. As a result, CBDs outside North America are more likely to contain supermarkets, bakeries, butchers, and other food stores. However, the 24-hour supermarket is rare outside North America because of shopkeeper preferences, government regulations, and longtime shopping habits. Many CBDs outside of North America ban motor vehicles from busy shopping streets, thus emulating one of the most attractive attributes of large shopping malls—pedestrian-only walkways. Shopping streets reserved for pedestrians are widespread in Northern Europe, including in the Netherlands, Germany, and Scandinavia. Rome periodically bans private vehicles from the CBD to reduce pollution and congestion and minimize damage to ancient monuments.

Although constructing large new buildings is difficult, many shops and offices still wish to be in the center of European cities. The alternative to new construction is renovation of older buildings. However, renovation is more expensive and does not always produce enough space to meet the demand. As a result, rents are much higher in the center of European cities than in U.S. cities of comparable size.

KEY ISSUE 2
Where Are People Distributed Within Urban Areas?

- **Models of Urban Structure**
- **Applying the Models Outside North America**

People are not distributed randomly within an urban area. They concentrate in particular neighborhoods, depending on their social characteristics. Geographers describe where people with particular characteristics are likely to live within an urban area, and they offer explanations for why these patterns occur. ■

Models of Urban Structure

Sociologists, economists, and geographers have developed three models to help explain where different types of people tend to live in an urban area—the concentric zone, sector, and multiple nuclei models.

The three models describing the internal social structure of cities were developed in Chicago, a city on a prairie. The three models were later applied to cities elsewhere in the United States and in other countries.

Except for Lake Michigan to the east, few physical features have interrupted the region's growth. Chicago includes a CBD known as the Loop because transportation lines (originally cable cars, now El trains) loop around it. Surrounding the Loop are residential suburbs to the south, west, and north.

Concentric Zone Model

The concentric zone model was the first to explain the distribution of different social groups within urban areas (Figure 13-4). It was created in 1923 by sociologist E. W. Burgess.

According to the **concentric zone model**, a city grows outward from a central area in a series of concentric rings, like the growth rings of a tree. The precise size and width of the rings vary from one city to another, but the same basic types of rings appear in all cities in the same order. Back in the 1920s, Burgess identified five rings:

1. CBD: The innermost ring, where nonresidential activities are concentrated.
2. A zone in transition, which contains industry and poorer-quality housing. Immigrants to the city first live in this zone in small dwelling units, frequently created by subdividing larger houses into apartments. The zone also contains rooming houses for single individuals.
3. A zone of working-class homes, which contains modest older houses occupied by stable, working-class families.

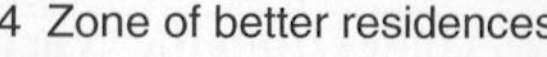

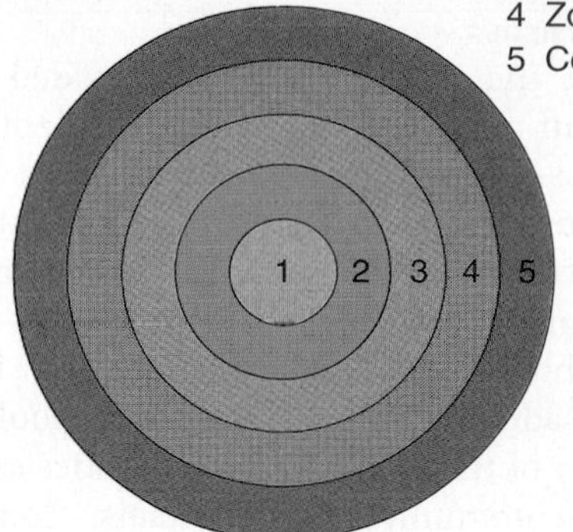

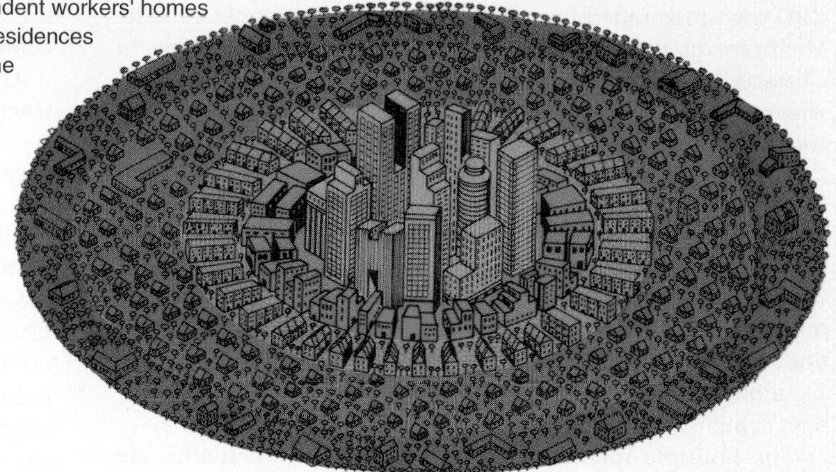

FIGURE 13-4 Concentric zone model. According to the model, a city grows in a series of rings that surround the central business district.

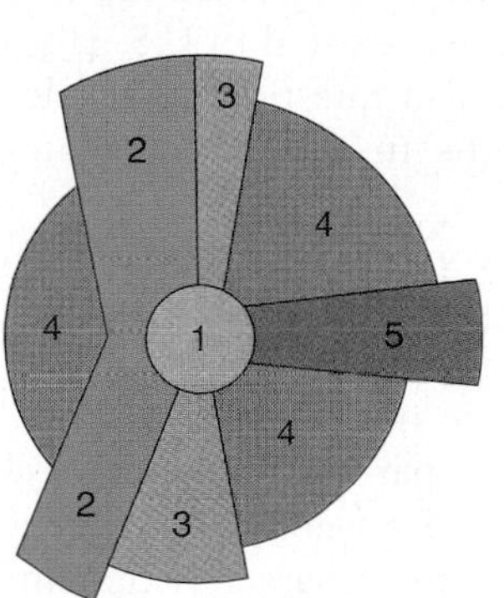

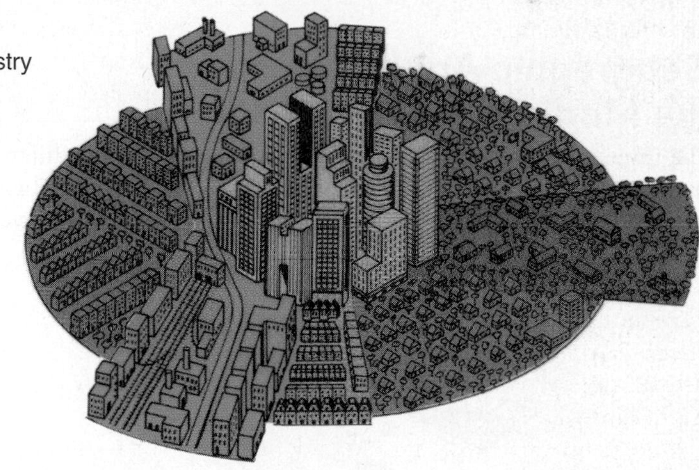

FIGURE 13-5 Sector model. According to the model, a city grows in a series of wedges or corridors, which extend out from the central business district.

4. A zone of better residences, which contains newer and more spacious houses for middle-class families.
5. A commuters' zone, beyond the continuous built-up area of the city. Some people who work in the center nonetheless choose to live in small villages that have become dormitory towns for commuters.

Sector Model

A second theory of urban structure, the **sector model**, was developed in 1939 by land economist Homer Hoyt (Figure 13-5). According to Hoyt, the city develops in a series of sectors, not rings. Certain areas of the city are more attractive for various activities, originally because of an environmental factor or even by mere chance. As a city grows, activities expand outward in a wedge, or sector, from the center.

Once a district with high-class housing is established, the most expensive new housing is built on the outer edge of that district, farther out from the center. The best housing is therefore found in a corridor extending from downtown to the outer edge of the city. Industrial and retailing activities develop in other sectors, usually along good transportation lines.

To some extent the sector model is a refinement of the concentric zone model rather than a radical restatement. Hoyt mapped the highest-rent areas for a number of U.S. cities at different times and showed that the highest social-class district usually remained in the same sector, although it moved farther out along that sector over time.

Hoyt and Burgess both claimed that social patterns in Chicago supported their model. According to Burgess, Chicago's CBD was surrounded by a series of rings, broken only by Lake Michigan on the east. Hoyt argued that the best housing in Chicago developed north from the CBD along Lake Michigan, whereas industry located along major rail lines and roads to the south, southwest, and northwest.

Multiple Nuclei Model

Geographers C. D. Harris and E. L. Ullman developed the multiple nuclei model in 1945. According to the **multiple nuclei model**, a city is a complex structure that includes more than one center around which activities revolve (Figure 13-6). Examples of these nodes include a port, neighborhood business center, university, airport, and park.

The multiple nuclei theory states that some activities are attracted to particular nodes, whereas others try to avoid them. For example, a university node may attract well-educated residents, pizzerias, and bookstores, whereas an airport may attract hotels and warehouses. On the other hand, incompatible land-use activities will avoid clustering in the same locations. Heavy industry and high-class housing, for example, rarely exist in the same neighborhood.

Geographic Applications of the Models

The three models help us understand where people with different social characteristics tend to live within an urban area. They can also help to explain why certain types of people tend to live in particular places. Effective use of the models depends on the availability of data at the scale of individual neighborhoods. In the United States and many other countries, that information comes from the census.

Urban areas in the United States are divided into **census tracts** that contain approximately 5,000 residents and correspond, where possible, to neighborhood boundaries. Every decade the U.S. Bureau of the Census publishes data summarizing the characteristics of the residents living in each tract. Examples of information the bureau publishes include the number of nonwhites, the median income of all families, and the percentage of adults who finished high school. The spatial distribution of any of these social characteristics can be plotted on a map of the community's census tracts. Computers have become invaluable in this task because they permit rapid creation of maps and storage of voluminous data about each census tract. Social scientists can compare the distributions of characteristics and create an overall picture of where various types of people tend to live. This kind of study is known as **social area analysis**.

None of the three models taken individually completely explains why different types of people live in distinctive parts of the city. Critics point out that the models are too simple and fail to consider the variety of reasons that lead people to select particular residential locations. Because the three models are all based on conditions that existed in U.S. cities between the two world wars, critics also question their relevance to contemporary urban patterns in the United States or in other countries.

But if the models are combined rather than considered independently, they help geographers explain where different types of people live in a city. People tend to reside in certain locations depending on their particular personal characteristics. This does not mean that everyone with the same characteristics must live in the same neighborhood, but the models say that most people prefer to live near others who have similar characteristics:

- **Applying the Concentric Zone Model.** Consider two families with the same income and ethnic background. One family owns its home, whereas the other rents. The owner-occupant is much more likely to live in an outer ring and the renter in an inner ring (Figure 13-7).
- **Applying the Sector Model.** Given two families who own their homes, the family with the higher income will not live in the same sector of the city as the family with the lower income (Figure 13-8).
- **Applying the Multiple Nuclei Model.** People with the same ethnic or racial background are likely to live near each other (Figure 13-9).

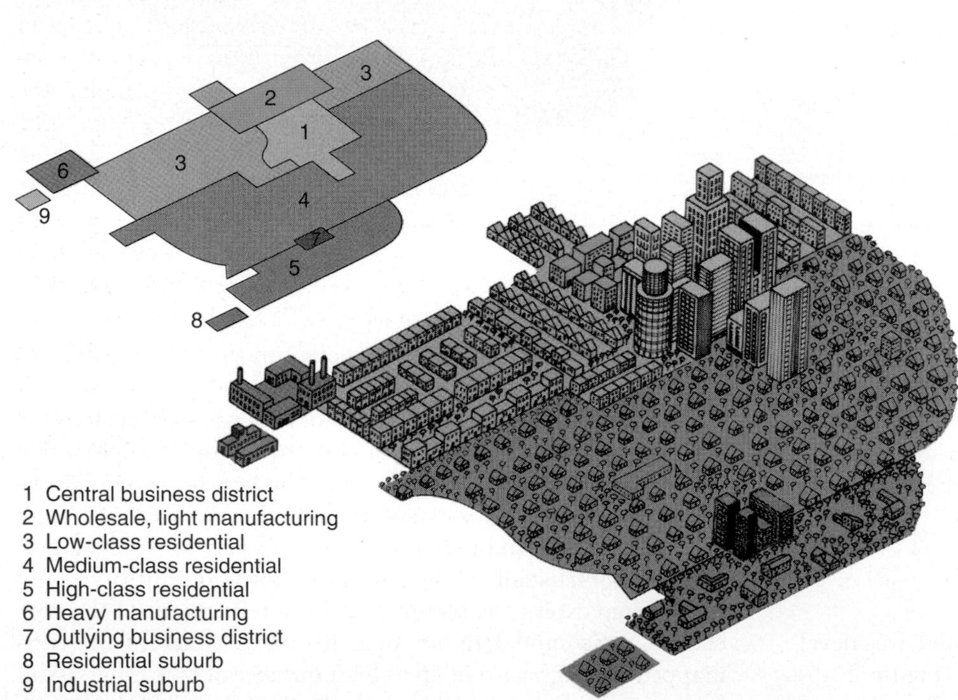

FIGURE 13-6 Multiple nuclei model. According to the model, a city consists of a collection of individual nodes, or centers, around which different types of people and activities cluster.

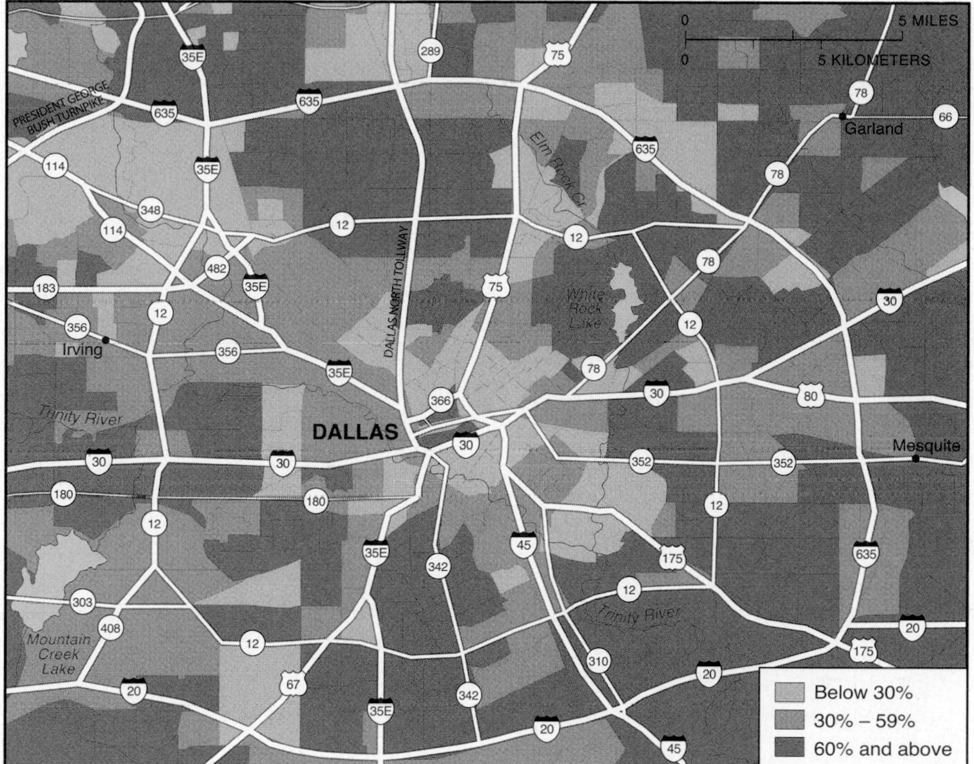

FIGURE 13-7 Example of concentric zone model in Dallas, the distribution of home owners. The percentage of households that own their home is greater in the outer rings of the city.

FIGURE 13-8 Example of sector model in Dallas, the distribution of high-income households. The median household income is the highest in a sector to the north.

Putting the three models together, we can identify, for example, the neighborhood in which a high-income, Asian American owner-occupant is most likely to live (see Contemporary Geographic Tools box).

Applying the Models Outside North America

The three models may describe the spatial distribution of social classes in the United States, but American urban areas differ from those elsewhere in the world. These differences do not invalidate the models, but they do point out that social groups in other countries may not have the same reasons for selecting particular neighborhoods within their cities.

European Cities

In contrast to most U.S. cities, wealthy Europeans still live in the inner rings of the upper-class sector, not just in the suburbs (Figure 13-10). A central location provides proximity to the region's best shops, restaurants, cafés, and cultural facilities. Wealthy people are also attracted by the opportunity to occupy elegant residences in carefully restored, beautiful old buildings.

As in the United States, though, wealthier people also cluster in European cities along a sector extending out from the CBD. In Paris, for example, the wealthy moved to the southwestern hills to be near the royal palace (the Louvre, beginning in the twelfth century, and the Palace of Versailles, from the sixteenth century until the French Revolution in 1789. The preference of Paris's wealthy to cluster in a southwest sector was reinforced in the nineteenth century during the Industrial Revolution. Factories were built to the south, east, and north along the Seine and Marne River valleys, but relatively few were built on the southwestern hills. Similar upper-class sectors emerged in other European cities, typically on higher elevations and near royal palaces.

In the past, low-income people also lived in the center of European cities. Before the invention of electricity in the nineteenth century, social segregation was vertical: Wealthier people lived on the first

FIGURE 13-9 Example of multiple nuclei model in Dallas, the distribution of minorities. African Americans and Hispanics occupy nodes to the south and west of downtown, respectively.

or second floors, whereas poorer people occupied the dark, dank basements or climbed many flights of stairs to reach the attics. As the city expanded during the Industrial Revolution, housing for these people was constructed in sectors near the factories and away from the wealthy. Today, low-income people are less likely to live in European inner-city neighborhoods. Poor-quality housing has been renovated for wealthy people or demolished and replaced by offices or luxury apartment buildings. Building and zoning codes prohibit anyone from living in basements, and upper floors are attractive to wealthy individuals once elevators are installed.

People with lower incomes have been relegated to the outskirts of European cities. Vast suburbs containing dozens of high-rise apartment buildings house these people who were displaced from the inner city. European suburban residents face the prospect of long commutes by public transportation to reach jobs and other downtown amenities. Shops, schools, and other services are worse in the suburbs than in inner neighborhoods; the suburbs are centers for crime, violence, and drug dealing; and people lack the American suburban amenity of large private yards. Many residents of these dreary suburbs are persons of color or recent immigrants from Africa or Asia who face discrimination and prejudice from "native" Europeans.

European officials encouraged the construction of high-density suburbs to help preserve the countryside from development and to avoid the inefficient sprawl that characterizes American suburbs, as discussed in the last section of this chapter. And tourists are attracted to the historic, lively centers of European cities. But these policies have resulted in the clustering of people with social and economic problems in remote suburbs rarely seen by wealthier individuals.

Less Developed Countries

In LDCs, as in Europe, the poor are accommodated in the suburbs, whereas the wealthy live near the center of cities as well as in a sector extending from the center. The similarity between European and LDC cities is not a coincidence: European colonial policies left a heavy mark on the development of cities in LDCs, many of which have passed through three stages of development—pre-European colonization, the European colonial period, and postcolonial independence.

PRECOLONIAL CITIES. Few cities existed in Africa, Asia, and Latin America before the Europeans established colonies. Most people lived in rural settlements. The principal cities in Latin America were located in Mexico and the Andean highlands of northwestern South America. In Africa, cities could be found along the western coast, Egypt's Nile River valley, and Islamic empires in the north and east (as well as in Southwest Asia). Cities were also built in South and East Asia, especially in India, China, and Japan.

Cities were often laid out surrounding a religious core, such as a mosque in Muslim regions. The center of Islamic cities also had a bazaar or marketplace, which served as the commercial core. Government buildings and the homes of wealthy families surrounded the mosque and bazaar. Narrow, winding streets led from the core to other quarters. Families with less wealth and lower status located farther from the core, and recent migrants to the city lived on the edge. Commercial activities were arranged in a concentric and hierarchical pattern:

- Higher-status businesses directly related to religious practices (such as selling religious books, incense, and candles) were located closest to the mosque.
- In the next ring were secular businesses, such as leather works, tailors, rug shops, and jewelers.
- Food products were sold in the next ring, then came blacksmiths, basket makers, and potters.
- A quarter would be reserved for Jews, a second for Christians, and a third for foreigners.

In Mexico, the Aztecs founded Mexico City—which they called Tenochtitlán—on a hill known as Chapultepec ("the hill of the grasshopper"). When forced by other people to leave the hill, they migrated a few kilometers south, near the present-day site of the University of Mexico, and then in 1325 to a marshy

FIGURE 13-10 Income distribution in the Paris region. Incomes are higher in the inner city of Paris than in the suburbs, with the exception of a high-income sector to the southwest. The inner city features sidewalk cafes and fancy housing. Suburbs, such as Le Courneuve, have high-rise apartments for low-income people.

10-square-kilometer (4-square-mile) island in Lake Texcoco (Figure 13-11).

The node of religious life was the Great Temple. Three causeways with drawbridges linked Tenochtitlán to the mainland and also helped to control flooding. An aqueduct brought fresh water from Chapultepec. Most food, merchandise, and building materials crossed from the mainland to the island by canoe, barge, or other type of boat, and the island was laced with canals to facilitate pickup and delivery of people and goods. Over the next two centuries the Aztecs conquered the neighboring peoples and extended their control through much of present-day Mexico. As their wealth and power grew, Tenochtitlán grew to a population of a half-million.

COLONIAL CITIES. When Europeans gained control of Africa, Asia, and Latin America, they expanded existing cities to provide colonial services, such as administration, military command, and international trade, as well as housing for Europeans who settled in the colony. Existing native towns were either left to one side or demolished because they were totally at variance with European ideas (Figure 13-12).

Colonial cities followed standardized plans. All Spanish cities in Latin America, for example, were built according to the Laws of the Indies, drafted in 1573. The laws explicitly outlined how colonial cities were to be constructed—a gridiron street plan centered on a church and central plaza, walls around individual houses, and neighborhoods built around central, smaller plazas with parish churches or monasteries. Compared to the existing cities, these European districts typically contain wider streets and public squares, larger houses surrounded by gardens, and much lower density. In contrast, the old quarters have narrow, winding streets, little open space, and cramped residences.

CONTEMPORARY GEOGRAPHIC TOOLS

Market Segmentation: You Are Where You Live

Marketing geographers identify sectors, rings, and nodes that come closest to matching customers preferred by a retailer. Companies use this information to understand, locate, and reach their customers better and to determine where to put new stores and where advertising should appear.

Segmentation is the process of partitioning markets into groups of potential customers with similar needs and characteristics who are likely to exhibit similar purchasing behavior. A prominent example of geographic segmentation is the Potential Rating Index for ZIP Markets (Prizm) clusters created by Nielsen Claritas. As Nielsen Claritas states, "birds of a feather flock together"—in other words, a person is likely to live near people who are similar.

Nielsen Claritas combines two types of geographic information—distribution of the social and economic characteristics of people obtained from the census and the addresses of purchasers of various products obtained from service providers. The variables are organized into 66 clusters that are given picturesque names. For each ZIP code in the United States, Nielsen Claritas determines the five clusters that are most prevalent. Nielsen Claritas calls this analysis "you are where you live."

We can compare Prizm clusters for two ZIP codes in the Dallas area. Refer to Figures 13-7, 13-8, and Figure 13-9 to see the close relationship between the Nielsen Claritas Prizm clusters and the models of urban structure. ZIP code 75215 is south of downtown Dallas. The five most common clusters (in alphabetical order) are as follows:

- City Roots: Older low-income ethnic minorities, living in older homes and apartments.
- Low-Rise Living: The lowest income of any Prizm cluster; many are single parents who rent their homes, travel by bus, and read *Ebony*.
- Multi-Culti Ethnic: Hispanics with modest incomes, who shop at Marshall's and read *Jet*.
- Urban Achievers: Young Hispanics, who watch BET and read *Latina*.
- Urban Elders: Elderly Hispanics who watch Spanish soaps and rarely eat out.

Compare the above to the five most common Prizm clusters in ZIP code 75230 in the northern suburbs of Dallas:

- Movers and Shakers: Wealthy, highly educated, suburban couples with two incomes.
- New Empty Nests: Older upper middle-class couples with grown children, an active lifestyle, and a preference for watching *60 Minutes*.
- Pools & Patios: High-income older couples who like to shop at Lord & Taylor and watch *Nova*.
- Upper Crust: The highest income of any Prizm cluster; many are empty nesters who shop at Saks and watch the Golf channel.
- Urban Achievers: A cluster shared with the south side of Dallas; two social areas that are seemingly very different can appear in different locations. ■

After the Spanish conquered Tenochtitlán in 1521 after a 2-year siege, they destroyed the city and dispersed or killed most of the inhabitants. The city, renamed Mexico City, was rebuilt around a main square, called the Zócalo, in the center of the island, on the site of the Aztecs' sacred precinct. The Spanish reconstructed the streets in a grid pattern extending from the Zócalo. A Roman Catholic cathedral was built on the north side of the square, near the site of the demolished Great Temple, and the National Palace was erected on the east side, on the site of the Aztec emperor Moctezuma's destroyed palace. The Spanish placed a church and monastery on the site of the Tlatelolco market.

In other examples, Fès (Fez), Morocco, now consists of two separate and distinct towns—one that existed before the French gained control and one built by the French colonialists (Figure 13-13). Similarly, the British built New Delhi near the existing city of Delhi, India. On the other hand, the French colonial city of Saigon, Vietnam (now Ho Chi Minh City), was built by completely demolishing the existing city without leaving a trace.

CITIES SINCE INDEPENDENCE. Following independence, cities have become the focal points of change in LDCs. Millions of people have migrated to the cities in search of work.

Geographers Ernest Griffin and Larry Ford show that in Latin American cities wealthy people push out from the center in a well-defined elite residential sector. The elite sector forms on either side of a narrow spine that contains offices, shops, and amenities attractive to wealthy people, such as restaurants, theaters, parks, and zoos (Figure 13-14). The wealthy are also attracted to the center and spine because services such as water and electricity are more readily available and reliable.

In Mexico City, Emperor Maximilian (1864–1867) designed a 14-lane, tree-lined boulevard patterned after the Champs-Elysées in Paris. The boulevard (now known as the Paseo de la Reforma) extended 3 kilometers southwest from the center to Chapultepec. The Reforma between downtown and Chapultepec became the spine of an elite sector. During the late nineteenth century, the wealthy built pretentious *palacios* (palaces) along it. Physical factors also influenced the movement of wealthy people toward the west along the Reforma. Because

FIGURE 13-11 Precolonial city (left). The Aztec city of Tenochtitlán was built on an island in Lake Texcoco. (right) The center of the city was dominated by the Templo Mayor. The twin shrines on the top of the temple were dedicated to the Aztec God of rain and agriculture (in blue) and to the Aztec God of war (in red).

FIGURE 13-12 Colonial city. The main square in downtown Mexico City, the Zócalo, was laid out by the Spanish. The Metropolitan Cathedral is in the center of this image, on the north side of the square. The National Palace on the west side, and City Hall on the south side. The site of the Templo Mayor is east of the cathedral and north of the palace.

elevation was higher than elsewhere in the city, sewage flowed eastward and northward away from Chapultepec. In 1903, most of Lake Texcoco was drained by a gigantic canal and tunnel project, allowing the city to expand to the north and east. The dried-up lakebed was a less desirable residential location than the west side because prevailing winds from the northeast stirred up dust storms. As Mexico City's population grew rapidly during the twentieth century, the social patterns inherited from the nineteenth century were reinforced.

Similarly, in Rio de Janeiro, Brazil, wealthy people are clustered in the center of the city along the west shore of Guanabara Bay, as well as in a sector to the south, including Ipanema and Copacabana, which offers spectacular views of the Atlantic Ocean and access to beaches (Figure 13-15). The poor live in northern suburbs, where steep mountains restrict construction of other types of buildings.

SQUATTER SETTLEMENTS. The LDCs are unable to house the rapidly growing number of poor people. Their cities are growing because of overall population increase and migration from rural areas for job opportunities. Because of the housing shortage, a large percentage of poor immigrants to urban areas in LDCs live in squatter settlements.

Squatter settlements are known by a variety of names, including *barriadas* and *favelas* in Latin America, *bidonvilles* in North Africa, *bastees* in India, *gecekondu* in Turkey, *kampongs* in Malaysia, and *barong-barong* in the Philippines. The United Nations estimated that 175 million people worldwide lived in squatter settlements in 2003. Squatter settlements have few services, because neither the city nor the residents can afford them. The settlements generally lack schools, paved roads, telephones, or sewers. Latrines are usually designated by the settlement's leaders, and water is carried from a central well or dispensed from a truck. Electricity service may be stolen by

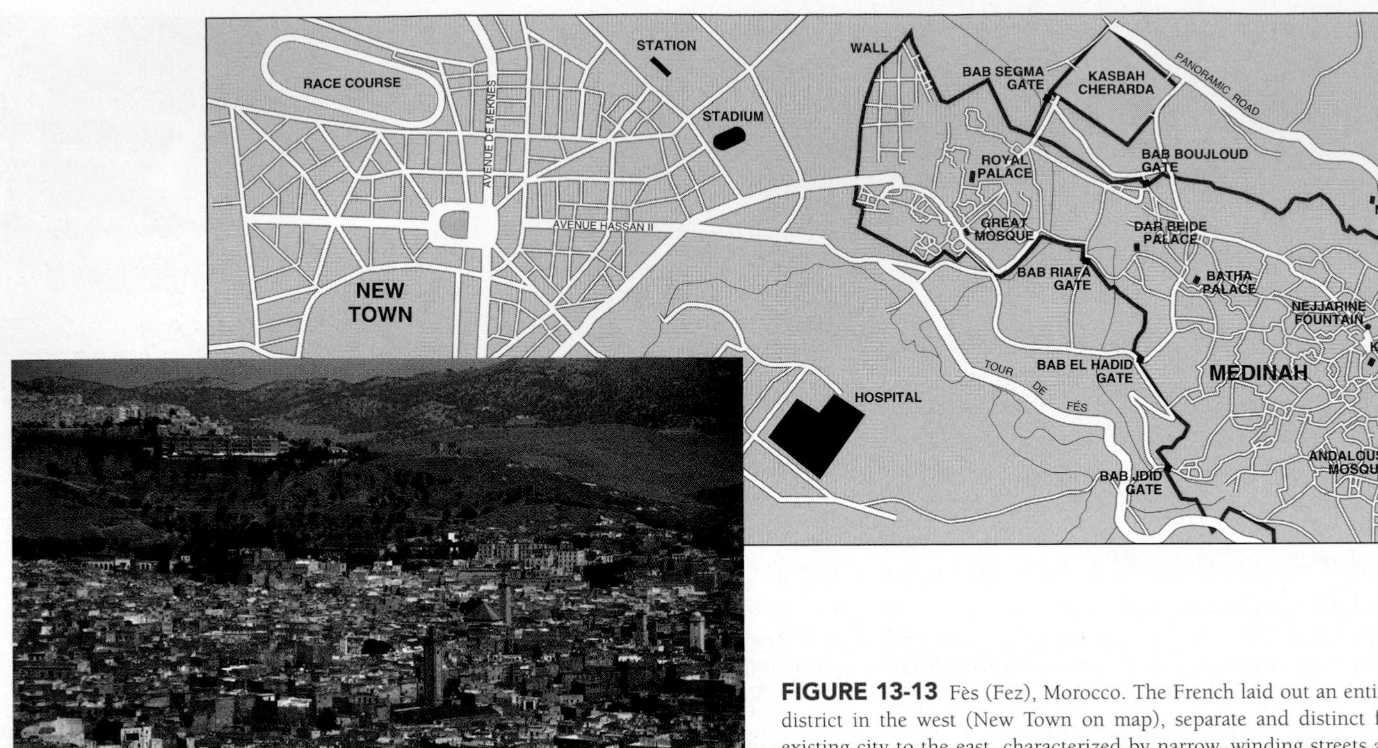

FIGURE 13-13 Fès (Fez), Morocco. The French laid out an entirely new district in the west (New Town on map), separate and distinct from the existing city to the east, characterized by narrow, winding streets and high density (Medinah on the map). In the photo, the Medinah is in the foreground. The tower in the foreground is the Karaouine Mosque. The New Town is in the background.

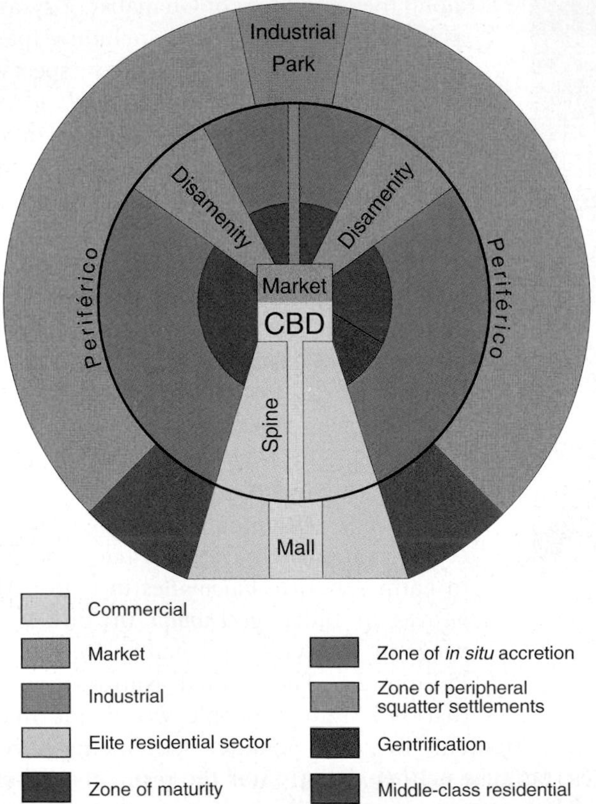

FIGURE 13-14 Model of a Latin American city. Wealthy people live in the inner city and a sector extending along a commercial spine. (Adapted from Larry R. Ford, "A New and Improved Model of Latin American City Structure," *Geographical Review* 86 (1996): 438. Used by permission of the publisher.)

running a wire from the nearest power line. In the absence of bus service or available private cars, a resident may have to walk 2 hours to reach a place of employment.

At first, squatters do little more than camp on the land or sleep in the street. In severe weather, they may take shelter in markets and warehouses. Families then erect primitive shelters with scavenged cardboard, wood boxes, sackcloth, and crushed beverage cans. As they find new bits of material, they add them to their shacks. After a few years they may build a tin roof and partition the space into rooms, and the structure acquires a more permanent appearance.

KEY ISSUE 3

Why Do Inner Cities Face Distinctive Challenges?

- **Inner-City Physical Issues**
- **Inner-City Social Issues**
- **Inner-City Economic Issues**

Most of the land in urban settlements is devoted to housing, where people live. Within U.S. urban areas, the most fundamental spatial distinction is between inner-city

FIGURE 13-15 (top) Favela in Rio de Janeiro, Brazil. A large percentage of people in the rapidly growing cities of less developed countries live in squatter settlements. In Rio, most of the squatter settlements, known as favelas, are on hillsides on the edge of the city. (bottom) Rio's highest income areas are near the CBD and in sectors along the ocean.

residential neighborhoods that surround the CBD and suburban residential neighborhoods on the periphery. Inner cities in the United States contain concentrations of low-income people who face a variety of physical, social, and economic problems very different from those faced by suburban residents. ■

Inner-City Physical Issues

The major physical problem faced by inner-city neighborhoods is the poor condition of the housing, most of which was built before 1940. Deteriorated housing can either be demolished and replaced with new housing or it can be rehabilitated.

Process of Deterioration

As the number of low-income residents increase in the city, the territory they occupy expands. Neighborhoods can shift from predominantly middle-class to low-income occupants within a few years. Middle-class families move out of a neighborhood to newer housing farther from the center and sell or rent their houses to lower-income families.

FILTERING. Large houses built by wealthy families in the nineteenth century are subdivided by absentee landlords into smaller dwellings for low-income families. This process of subdivision of houses and occupancy by successive waves of lower-income people is known as **filtering**.

Like a car, clothing, or any other object, the better a house is maintained, the longer it will last. Landlords stop maintaining houses when the rent they collect becomes less than the maintenance cost. In such a case, the building soon deteriorates and grows unfit for occupancy. Not even the poorest families will rent the dwelling. At this point in the filtering process, the owner may abandon the property, because the rents that can be collected are less than the costs of taxes and upkeep. Cities have codes that require owners to maintain houses in good condition. But governments that aggressively go after landlords to repair deteriorated properties may in fact hasten abandonment because landlords will not spend money on repairs that they are unable to recoup in rents. Thousands of vacant houses stand in the inner areas of U.S. cities because the landlords have abandoned them.

One hundred years ago, low-income inner-city neighborhoods in the United States teemed with throngs of recent immigrants from Europe. These neighborhoods that housed perhaps 100,000 a century ago contain less than 10,000 inhabitants today. Schools and shops close because they are no longer needed in inner-city neighborhoods with rapidly declining populations. Through the filtering process, many low-income families have moved to less deteriorated houses farther from the center.

REDLINING. Some banks engage in **redlining**—drawing lines on a map to identify areas in which they will refuse to loan money. As a result of redlining, families who try to fix up houses in the area have difficulty borrowing money. Although redlining is illegal, enforcement of laws against it is frequently difficult. The Community Reinvestment Act requires U.S. banks to document by census tract where they make loans. A bank must demonstrate that inner-city neighborhoods within its service area receive a fair share of its loans.

Urban Renewal

North American and European cities have demolished much of their substandard inner-city housing through urban renewal programs. Under **urban renewal**, cities identify blighted inner-city neighborhoods, acquire the properties from private owners, relocate the residents and businesses, clear the site, and build new roads and utilities. The land is then turned over to private developers or to public agencies, such as the board of education or the parks department, to construct new buildings or services. National government grants help cities pay for urban renewal.

Urban renewal has been criticized for destroying the social cohesion of older neighborhoods and reducing the supply of low-cost housing. Because African Americans comprised a large percentage of the displaced population in U.S. cities, urban renewal was often called "Negro Removal" during the 1960s. Most North American and European cities have turned away from urban renewal since the 1970s, and national governments, including that of the United States, have stopped funding it.

PUBLIC HOUSING. Many substandard inner-city houses have been demolished and replaced with public housing. In the United States, **public housing** is reserved for low-income households, who must pay 30 percent of their income for rent. A housing authority, established by the local government, manages the buildings, and the federal government pays the cost of construction and the maintenance, repair, and management that are not covered by rent. In the United States, public housing accounts for only 1 percent of all dwellings, compared to 14 percent in the United Kingdom. Elsewhere in Western Europe, governments typically subsidize construction cost and rent for a large percentage of the privately built housing.

Most of the high-rise public-housing projects built in the United States and Europe during the 1950s and early 1960s are now considered unsatisfactory environments for families with children. The elevators are frequently broken, juveniles terrorize other people in the hallways, and drug use and crime rates are high. Some observers claim that the high-rise buildings were responsible for the problem because too many low-income families were concentrated into a high-density environment. Because of poor conditions, high-rise public-housing projects have been demolished in many U.S. and European cities.

The U.S. government has stopped funding construction of new public housing. A federal program known as Hope VI supports renovation of older public housing, and the Housing Choice Voucher Program helps low-income households pay their rent in private housing. With the overall level of funding much lower, the supply of public housing and other government-subsidized housing in the United States diminished by approximately 1 million units between 1980 and 2000. But during the same period, the number of households that needed low-rent dwellings increased by more than 2 million.

In Britain, the supply of public housing, known as social housing (formerly council estates), has also declined because the government has forced local authorities to sell some of the dwellings to the residents. The British also expanded subsidies to nonprofit housing associations that build housing for groups with special needs, including single mothers, immigrants, the disabled, and the elderly as well as the poor.

RENOVATED HOUSING. An alternative to demolishing deteriorated inner-city houses is to renovate them. In some cases, nonprofit organizations renovate houses and sell or rent them to low-income people. But more often, the renovated housing attracts middle-class people. Most cities have at least one substantially renovated inner-city neighborhood where middle-class people live. In a few cases, inner-city neighborhoods never deteriorated because the community's social elite maintained them as enclaves of expensive property. In most cases, inner-city neighborhoods have only recently been renovated by the city and by private investors.

The process by which middle-class people move into deteriorated inner-city neighborhoods and renovate the housing is known as **gentrification**. Middle-class families are attracted to deteriorated inner-city housing because the houses may be larger, more substantially constructed, yet cheaper in the inner city than in the suburbs. Inner-city houses may also possess attractive architectural details such as ornate fireplaces, cornices, high ceilings, and wood trim. Gentrified inner-city neighborhoods also attract middle-class individuals who work downtown. Inner-city living eliminates the strain of commuting on crowded freeways or public transit. Others seek proximity to theaters, bars, restaurants, and other cultural and recreational facilities located downtown. Renovated inner-city housing appeals to single people and couples without children, who are not concerned with the quality of inner-city schools.

In cities where gentrification is especially strong, ethnic patterns are being altered. In Chicago, for example, the white population is increasing in inner-city neighborhoods and declining in the outer-city neighborhoods (Figure 13-16). Conversely, the population of African Americans and Hispanics is declining in the inner city and increasing in neighborhoods farther from the center.

Because renovating an old inner-city house can be nearly as expensive as buying a new one in the suburbs, cities encourage the process by providing low-cost loans and tax breaks. Public expenditures for renovation have been criticized as subsidies for the middle class at the expense of people with lower incomes, who are forced to move out of the gentrified neighborhoods because the rents in the area are suddenly too high for them. Cities try to reduce the hardship on poor families forced to move. U.S. law requires that they be reimbursed both for moving expenses and for rent increases over a 4-year period. Western European countries have similar laws. Cities are also renovating old houses specifically for lower-income families through public housing or other programs. By renting renovated houses, the city also helps to disperse low-income families throughout the city instead of concentrating them in large inner-city public-housing projects.

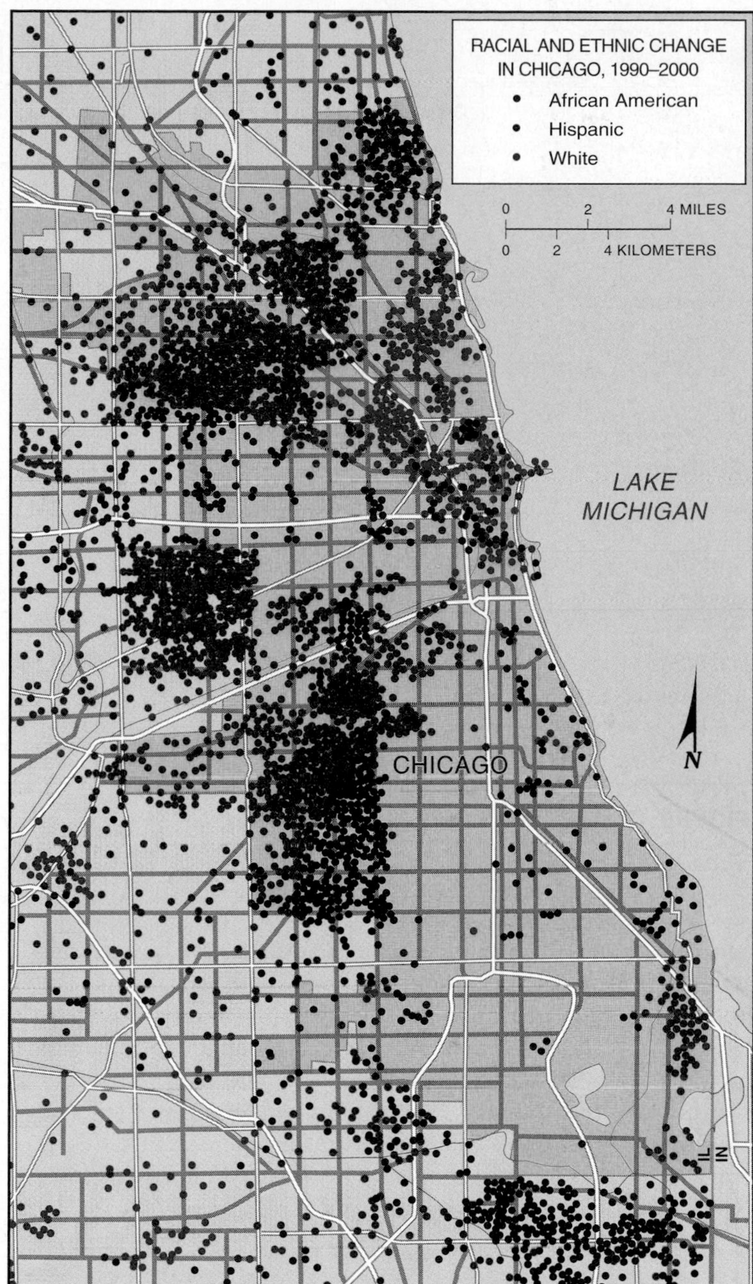

FIGURE 13-16 Racial change in Chicago. Dots represent where the population of each ethnicity increased between 1980 and 2000. Note growth of the white population in the inner city and North Side, while the African American and Hispanic populations have been increasing in the outer city and inner suburbs.

Inner-City Social Issues

Beyond the pockets of gentrified neighborhoods, inner cities contain primarily people with low incomes who face a variety of social problems. Inner-city residents constitute a permanent underclass who live in a culture of poverty.

Underclass

Inner-city residents are frequently referred to as a permanent **underclass** because they are trapped in an unending cycle of economic and social problems. The underclass suffers from relatively high rates of unemployment, alcoholism, drug addiction, illiteracy, juvenile delinquency, and crime.

The children of the underclass attend deteriorated schools, and affordable housing is increasingly difficult to find. Their neighborhoods lack adequate police and fire protection, shops, hospitals, clinics, or other health-care facilities. The future is especially bleak for the underclass because they are increasingly unable to compete for jobs. Inner-city residents lack the technical skills needed for most jobs because fewer than half complete high school. Despite the importance of education in obtaining employment, many in the underclass live in an atmosphere that ignores good learning habits, such as regular school attendance and completion of homework. The gap between skills demanded by employers and the training possessed by inner-city residents is widening. In the past, people with limited education could become factory workers or filing clerks, but today these jobs require skills in computing and handling electronics. Meanwhile, inner-city residents do not even have access to the remaining low-skilled jobs, such as custodians and fast-food servers, because these jobs are increasingly in the distant suburbs.

Some of the underclass are homeless. Accurate counts are impossible to obtain, but several surveys estimate that on a given night nearly 1 million Americans sleep in doorways, on heated street grates, and in bus and subway stations. Over the

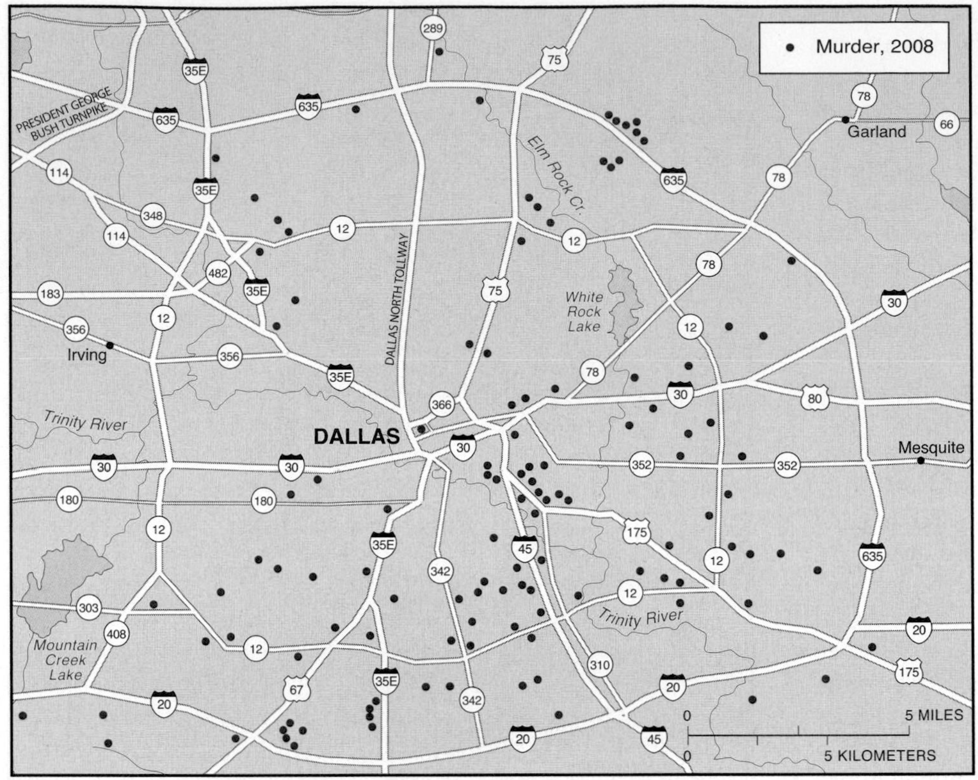

FIGURE 13-17 Inner-city social issues: Dallas murders 2008. Most murders were on the south and east sides of the city. Compare to Figures 13-7, 13-8, and 13-9.

course of a year, the number of Americans who are homeless at some time is estimated at more than 3 million. Most people are homeless because they cannot afford housing and have no regular income. Homelessness may have been sparked by family problems, job loss, or mental illness. Single men constitute two-fifths of the homeless, and the remainder are women and children. Homelessness is also a serious problem in LDCs. Several hundred thousand people in Kolkata (Calcutta), India, sleep, bathe, and eat on sidewalks and traffic islands.

Culture of Poverty

Inner-city residents are trapped as a permanent underclass because they live in a culture of poverty. Unwed mothers give birth to three-fourths of the babies in U.S. inner-city neighborhoods, and three-fourths of children in the inner city live with only one parent. Because of inadequate child-care services, single mothers may be forced to choose between working to generate income and staying at home to take care of the children.

In principle, government officials would like to see more fathers living with their wives and children, but they provide little incentive for them to do so. Only a small percentage of "deadbeat dads" are tracked down for failing to provide required child-care support. If the husband moves back home, his wife may lose welfare benefits, leaving the couple financially worse off together than apart.

Trapped in a hopeless environment, some inner-city residents turn to drugs. Although drug use is a problem in both the suburbs and rural areas, rates of use in recent years have increased most rapidly in the inner cities. Some drug users obtain money through criminal activities. Gangs form in inner-city neighborhoods to control lucrative drug distribution. Violence erupts when two gangs fight over the boundaries between their drug distribution areas. Compare Figure 13-17 with Figures 13-7, 13-8, and 13-9. Most of the murders in Dallas in 2008 occurred in low-income minority areas, and most victims, as well as those arrested for murder in Dallas, were minorities.

Many neighborhoods in the United States are segregated by ethnicity, as discussed in Chapter 7. African Americans and Hispanics concentrate in one or two large continuous areas of the inner city, whereas whites live in the suburbs. Even small cities display strong social distinctions among neighborhoods. A frequently noticed division is between the east and west sides of a city, or between the north and south sides, with one side attracting the higher-income residents and the other left to lower-status and minority families. A family seeking a new residence usually considers only a handful of districts, where the residents' social and financial characteristics match their own. Residential areas designed for wealthy families are developed in scenic, attractive areas, possibly on a hillside or near a water body, whereas flat, dull land closer to industry becomes built up with cheaper housing.

Inner-City Economic Issues

The concentration of low-income residents in inner-city neighborhoods of central cities has produced financial problems. The severe recession in recent years has aggravated those problems.

Eroding Tax Base

Low-income inner-city residents require public services, but they can pay very little of the taxes necessary to support those services. Central cities face a growing gap between the cost of needed services in inner-city neighborhoods and the availability of funds to pay for them. A city has two choices for closing the gap between the cost of services and the funding available from taxes:

- **Reduce Services.** For example, close libraries, eliminate bus routes, collect trash less frequently, delay replacement of outdated school equipment. Aside from the hardship imposed on individuals laid off from work, cutbacks in public services also encourage middle-class residents and industries to move from the city.

- **Raise Tax Revenues.** For example, provide tax breaks for downtown offices, luxury hotels, restaurants, and shops. Even with generous subsidies, these businesses pay more taxes than the buildings demolished to make way for them, and they provide minimum-wage personal-service jobs for low-income inner-city residents. But spending public money to increase the downtown tax base can take scarce funds away from projects in inner-city neighborhoods, such as subsidized housing and playgrounds.

During the mid-twentieth century, inner-city fiscal problems were alleviated by increasing contributions from the federal government. The percentage of the budgets of the 50 largest U.S. cities supplied by the federal government increased from 1 percent in 1950 to 18 percent in 1980. But the percentage shrank substantially during the 1980s, to 6 percent in 1990 and 2000. When adjusted for inflation, federal aid to U.S. cities has declined by two-thirds since the 1980s. To offset a portion of these lost federal funds, some state governments increased financial assistance to cities.

Impact of the Recession

One of the principal causes of the severe recession that began in 2008 was a collapse in the housing market, primarily in the inner city (Figure 13-18). To purchase a house, most people borrow money through a mortgage, which is repaid in monthly installments over many years. In the years leading up to the recession, financial institutions sharply increased the number of loans to low-income inner-city households buying their first homes. Despite having poor credit histories, first-time home buyers were approved for mortgages without background checks. These were known as subprime mortgages.

Financial institutions around the world were eager to invest in housing in the United States. Investing in housing was viewed as providing a higher rate of return at a lower risk than other investment options. Investors reasoned that their loans were safe: House prices had increased rapidly for many years, so even if a few home owners defaulted on their mortgages, investors would still recoup their investment. Inner-city residents were especially targeted for subprime mortgages. As the concentric zone model shows (Figures 13-4 and Figure 13-7), inner-city residents are more likely to be renters and therefore represent the best opportunity for financial institutions to increase the number of home owners.

When people are unable to repay their loans, lenders can take over the property in what is called a foreclosure. In the first year of the recession, 10 percent of all Americans with mortgages were behind in their mortgage payments or already in foreclosure. Compounding the problem, house prices have fallen in the United States and other MDCs since their peak in 2006. With falling house prices, houses are worth less than in earlier years. In many cases, the amount of the mortgage exceeded the value of the house once prices had fallen.

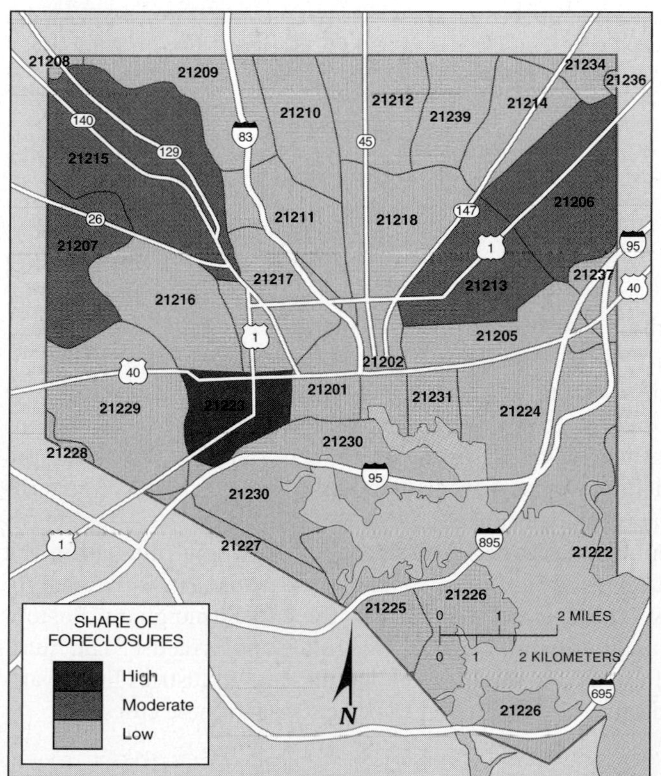

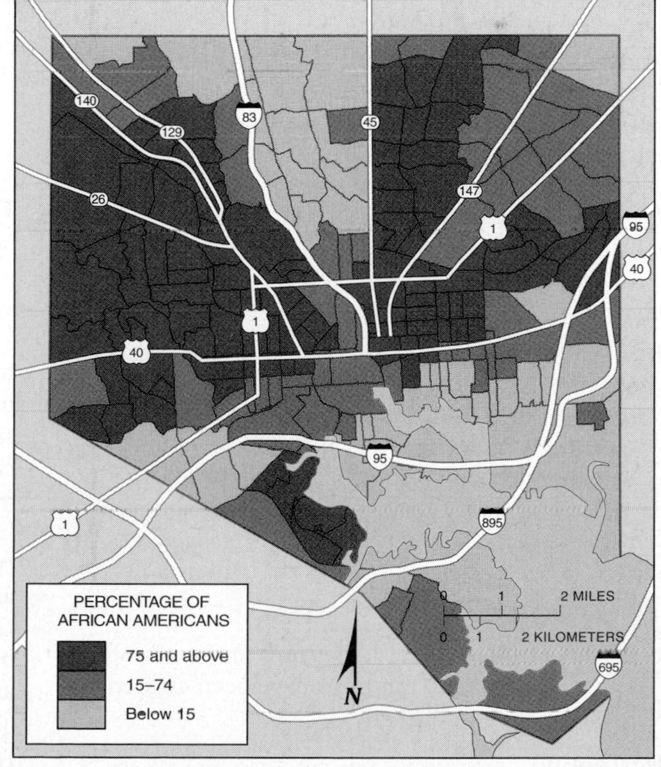

FIGURE 13-18 Foreclosures in Baltimore. Foreclosures in Baltimore are clustered in the inner city and in a sector to the northwest where the African American population has increased in recent years (see Figure 7-11).

KEY ISSUE 4

Why Do Suburbs Face Distinctive Challenges?

- Urban Expansion
- The Peripheral Model
- Suburban Segregation
- Transportation and Suburbanization

In 1950, only 20 percent of Americans lived in suburbs compared to 40 percent in cities and 40 percent in small towns and rural areas. In 2000, after a half-century of rapid suburban growth, 50 percent of Americans lived in suburbs compared to only 30 percent in cities and 20 percent in small towns and rural areas. ■

Urban Expansion

Until recently in the United States, as cities grew, they expanded by adding peripheral land. Now cities are surrounded by a collection of suburban jurisdictions whose residents prefer to remain legally independent of the large city.

Annexation

The process of legally adding land area to a city is **annexation**. Rules concerning annexation vary among states. Normally, land can be annexed to a city only if a majority of residents in the affected area vote in favor of doing so.

Peripheral residents generally desired annexation in the nineteenth century, because the city offered better services, such as water supply, sewage disposal, trash pickup, paved streets, public transportation, and police and fire protection. Thus, as U.S. cities grew rapidly in the nineteenth century, the legal boundaries frequently changed to accommodate newly developed areas. For example, the city of Chicago expanded from 26 square kilometers (10 square miles) in 1837 to 492 square kilometers (190 square miles) in 1900 (Figure 13-19).

Today, however, cities are less likely to annex peripheral land because the residents prefer to organize their own services rather than pay city taxes for them. Originally, some of these peripheral jurisdictions were small, isolated towns that had a tradition of independent local government before being swallowed up by urban growth. Others are newly created communities whose residents wish to live close to the large city but not be legally part of it.

Defining Urban Settlements

Instead of annexing peripheral areas, cities now are surrounded by suburbs. As a result, several definitions have been created to characterize cities and their suburbs:

- City: a legal entity
- Urbanized area: a continuously built-up area
- Metropolitan area: a functional area

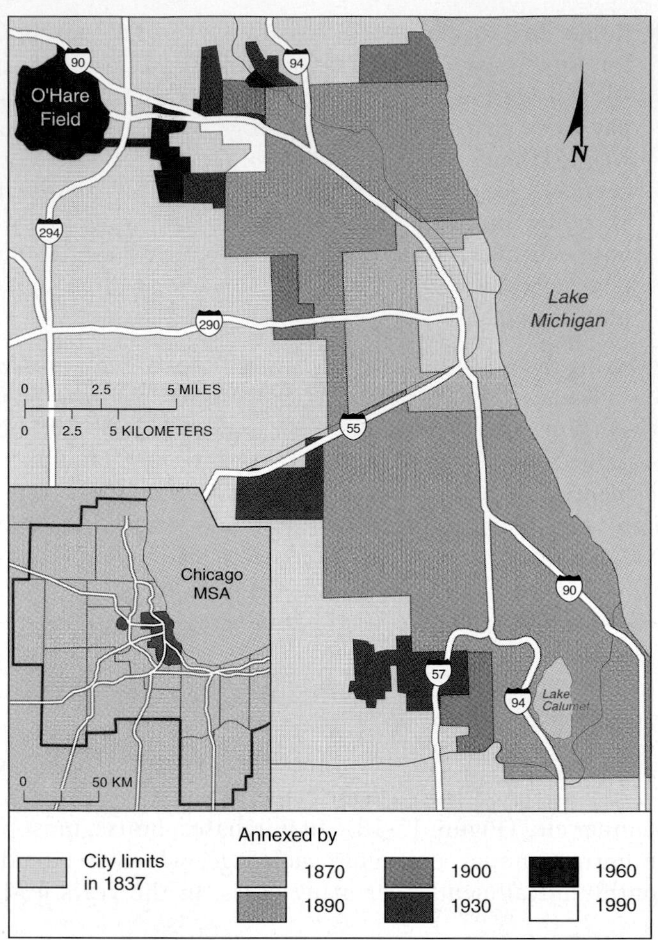

FIGURE 13-19 Annexation in Chicago. During the nineteenth century, the city of Chicago grew rapidly through annexation of peripheral land. Relatively little land was annexed during the twentieth century; the major annexation was on the northwest side for O'Hare Airport. The inset shows that the city of Chicago covers only a small portion of the Chicago metropolitan statistical area.

THE CITY. The term **city** defines an urban settlement that has been legally incorporated into an independent, self-governing unit (Figure 13-20). In the United States, a city surrounded by suburbs is sometimes called a **central city**.

Virtually all countries have a local government system that recognizes cities as legal entities with fixed boundaries. A city has locally elected officials, the ability to raise taxes, and responsibility for providing essential services. The boundaries of the city define the geographic area within which the local government has legal authority.

Population has declined since 1950 by about one-half in the central cities of Baltimore, Buffalo, Cleveland, Detroit, Pittsburgh, and St. Louis, and by about one-third in Birmingham, Boston, Cincinnati, Dayton, Newark, Rochester, and Syracuse. The number of tax-paying middle-class families and industries has invariably declined by much higher percentages in these cities.

URBANIZED AREA. In the United States, the central city and the surrounding built-up suburbs are called an **urbanized area**. More precisely, an urbanized area consists of a central city

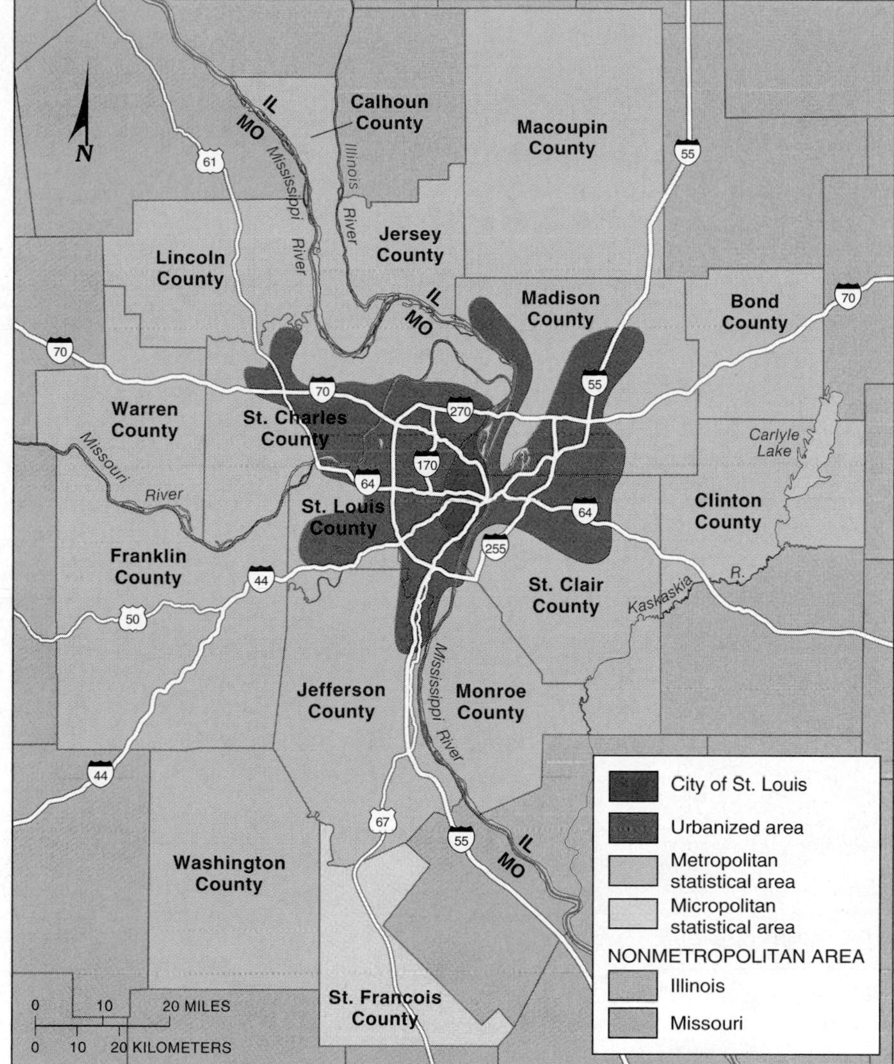

FIGURE 13-20 City, urbanized area, and metropolitan statistical area of St. Louis. Surrounding the city of St. Louis is an urbanized area that spreads westward into St. Louis County and eastward across the Mississippi River into Illinois. The St. Louis metropolitan statistical area includes seven Missouri counties and eight in Illinois, as well as the city of St. Louis. The St. Louis-St. Charles-Farmington combined statistical area includes the St. Louis MSA and the Farmington micropolitan statistical area (Farmington is the county seat and largest city in St. Francois County).

plus its contiguous built-up suburbs where population density exceeds 1,000 persons per square mile (400 persons per square kilometer). Approximately 70 percent of the U.S. population lives in urbanized areas, including about 30 percent in central cities and 40 percent in surrounding jurisdictions. Working with urbanized areas is difficult because few statistics are available about them. Most data in the United States and other countries are collected for cities, counties, and other local government units, but urbanized areas do not correspond to government boundaries.

METROPOLITAN STATISTICAL AREA. The concept of urbanized area also has limited applicability because it does not accurately reflect the full influence that an urban settlement has in contemporary society. The area of influence of a city extends beyond legal boundaries and adjacent built-up jurisdictions. For example, commuters may travel a long distance to work and shop in the city or built-up suburbs. People in a wide area watch the city's television stations, read the city's newspapers, and support the city's sports teams. Therefore, we need another definition of urban settlement to account for its more extensive zone of influence.

The U.S. Bureau of the Census has created a method of measuring the functional area of a city, known as the **metropolitan statistical area (MSA)**. An MSA includes the following:

- An urbanized area with a population of at least 50,000
- The county within which the city is located
- Adjacent counties with a high population density and a large percentage of residents working in the central city's county (e.g., a county with a density of 25 persons per square mile and at least 50 percent working in the central city's county)

Studies of metropolitan areas in the United States are usually based on information about MSAs. The MSAs are widely used because many statistics are published for counties, the basic MSA building block.

The census designated 366 MSAs as of 2009, encompassing 84 percent of the U.S. population. Older studies may refer to SMSAs, or standard metropolitan statistical areas, which the census used before 1983 to designate metropolitan areas in a manner similar to MSAs. An MSA is not the perfect tool for measuring the functional area of a city. One problem is that some MSAs include extensive land area that is not urban. For example, Great Smoky Mountains National Park is partly in the Knoxville, Tennessee, MSA; Sequoia National Park is in the Visalia-Porterville, California, MSA. The MSAs comprise some 20 percent of total U.S. land area, compared to only 2 percent for urbanized areas. The urbanized area typically occupies only 10 percent of an MSA land area but contains nearly 90 percent of its population.

The census has also designated smaller urban areas as **micropolitan statistical areas (μSAs)**. These include an urbanized area of between 10,000 and 50,000 inhabitants, the county in which it is found, and adjacent counties tied to the city. The United States had 574 micropolitan statistical areas as of 2008, for the most part found around southern and western communities previously considered rural in character. About 10 percent of Americans live in a micropolitan statistical area. The 366 MSAs and 574 μSAs together are known as **core based statistical areas (CBSAs)**.

Recognizing that many MSAs and μSAs have close ties, the census has combined some of them into 124 **combined statistical areas (CSAs).** A CSA is defined as two or more contiguous CBSAs tied together by commuting patterns. The 124 CSAs plus the remaining 187 MSAs and 406 μSAs not combined into CSAs together are known as **primary census statistical areas (PCSAs).**

Local Government Fragmentation

The fragmentation of local government in the United States makes it difficult to solve regional problems of traffic, solid-waste disposal, and the building of affordable housing. The number of local governments exceeds 1,400 in the New York area, 1,100 in the Chicago area, and 20,000 throughout the United States. Approximately 40 percent of these 20,000 local governments are general units, such as cities and counties. The remainder serve special purposes, such as schools, sanitation, transportation, water, and fire districts.

Long Island, which extends for 150 kilometers (90 miles) east of New York City and is approximately 25 kilometers (15 miles) wide, contains nearly 800 local governments. The island includes 2 counties, 2 cities, 13 towns, 95 villages, 127 school districts, and more than 500 special districts (such as for garbage collection). The multiplicity of local governments on Long Island leads to problems. When police or firefighters are summoned to the State University of New York at Old Westbury, two or three departments sometimes respond because the campus is in five districts. The boundary between the communities of Mineola and Garden City runs down the center of Old Country Road, a busy four-lane route. Mineola set a 40-mile-per-hour speed limit for the eastbound lanes, whereas Garden City set a 30-mile-per-hour speed limit for the westbound lanes.

The large number of local government units has led to calls for a metropolitan government that could coordinate—if not replace—the numerous local governments in an urban area. Most U.S. metropolitan areas have a **council of government**, which is a cooperative agency consisting of representatives of the various local governments in the region. The council of government may be empowered to do some overall planning for the area that local governments cannot logically do. Strong metropolitan-wide governments have been established in a few places in North America. Two kinds exist:

- **Consolidations of City and County Governments.** Examples include Indianapolis and Miami. The boundaries of Indianapolis were changed to match those of Marion County, Indiana. Government functions that were handled separately by city and county now are combined into a joint operation in the same office building. In Florida, the city of Miami and surrounding Dade County have combined some services, but the city boundaries have not been changed to match those of the county.
- **Federations.** Examples include Toronto and other large Canadian cities. Toronto's metropolitan government was created in 1953 through federation of 13 municipalities. A two-tier system of government existed until 1998, when the municipalities were amalgamated into a single government.

Overlapping Metropolitan Areas

Some adjacent MSAs overlap. A county between two central cities may send a large number of commuters to jobs in each. In the northeastern United States, large metropolitan areas are so close together that they now form one continuous urban

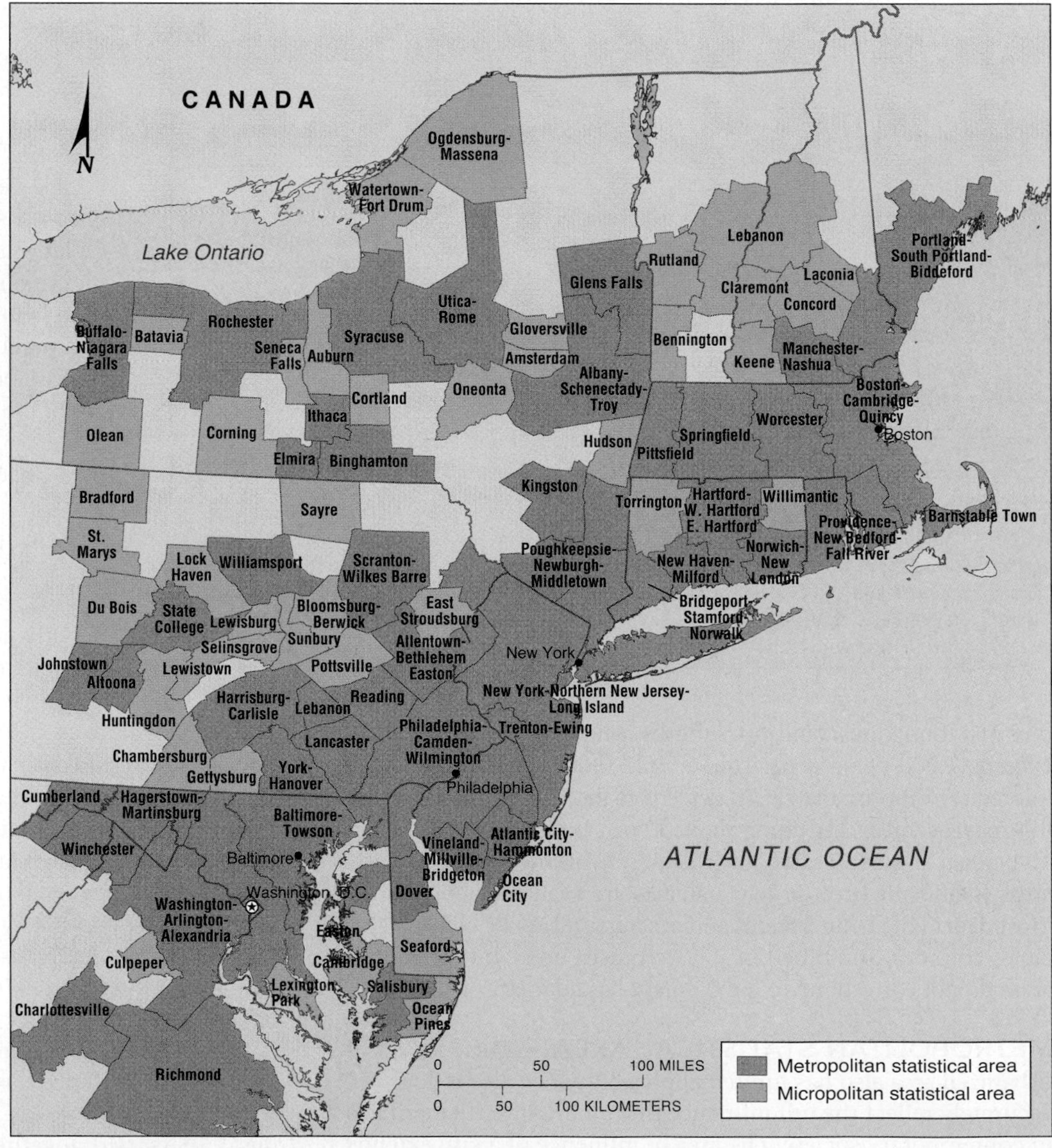

FIGURE 13-21 Megalopolis. Also known as the Boswash corridor, Megalopolis extends more than 700 kilometers (440 miles) from Boston on the northeast to Washington, D.C., on the southwest. Megalopolis contains one-fourth of the U.S. population on 2 percent of the country's total land area.

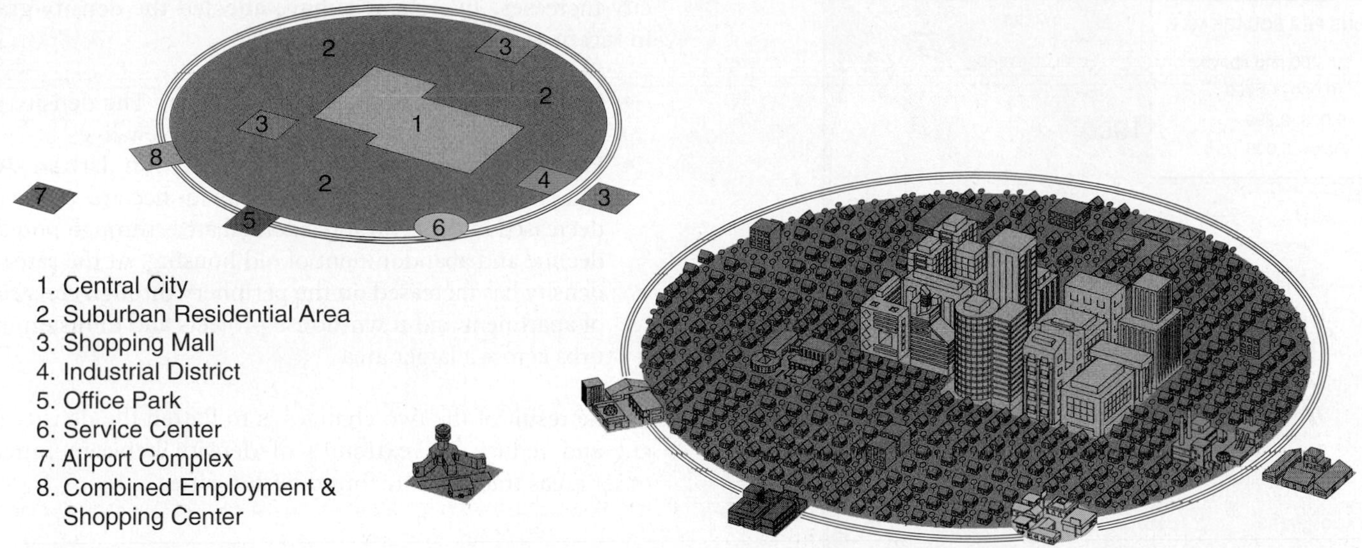

FIGURE 13-22 Peripheral model of urban areas. The central city is surrounded by a beltway or ring road. Around the beltway are suburban residential areas and nodes, or edge cities, where consumer and business services and manufacturing cluster. (Adapted from Chauncy D. Harris, "The Nature of Cities and Urban Geography in the Last Half Century." Reprinted with permission from *Urban Geography*, vol. 18, no. 1 (1997), p. 17. © V. H. Winston & Son, Inc., 360 South Ocean Blvd., Palm Beach, FL 33480. All rights reserved.)

complex, extending from north of Boston to south of Washington, D.C. Geographer Jean Gottmann named this region Megalopolis, a Greek word meaning "great city"; others have called it the Boswash corridor (Figure 13-21).

Other continuous urban complexes exist in the United States—the southern Great Lakes between Chicago and Milwaukee on the west and Pittsburgh on the east, and southern California from Los Angeles to Tijuana. Among important examples in other MDCs are the German Ruhr (including the cities of Dortmund, Düsseldorf, and Essen), Randstad in the Netherlands (including the cities of Amsterdam, the Hague, and Rotterdam), and Japan's Tokaido (including the cities of Tokyo and Yokohama).

Within Megalopolis, the downtown areas of individual cities such as Baltimore, New York, and Philadelphia retain distinctive identities, and the urban areas are visibly separated from each other by open space used as parks, military bases, and dairy or truck farms. But at the periphery of the urban areas, the boundaries overlap. Once considered two separate areas, Washington and Baltimore were combined into a single MSA after the 1990 census. Washingtonians visit the Inner Harbor in downtown Baltimore, and Baltimoreans attend major-league hockey and basketball games in downtown Washington. However, combining them into one MSA did not do justice to the distinctive character of the two cities, so the Census Bureau again divided them into two separate MSAs after the 2000 census but grouped them into one combined statistical area.

The Peripheral Model

North American urban areas follow what Chauncey Harris (creator of the multiple nuclei model) called the peripheral model. According to the **peripheral model**, an urban area consists of an inner city surrounded by large suburban residential and business areas tied together by a beltway or ring road (Figure 13-22). Peripheral areas lack the severe physical, social, and economic problems of inner-city neighborhoods. But the peripheral model points to problems of sprawl and segregation that characterize many suburbs.

Around the beltway are nodes of consumer and business services, called **edge cities**. Edge cities originated as suburban residences for people who worked in the central city, and then shopping malls were built to be near the residents. Now edge cities contain manufacturing centers spread out over a single story for more efficient operations and office parks where producer services cluster. Specialized nodes emerge in the edge cities—a collection of hotels and warehouses around an airport, a large theme park, a distribution center near the junction of the beltway, and a major long-distance interstate highway.

Density Gradient

As you travel outward from the center of a city, you can watch the decline in the density at which people live (Figure 13-23). Inner-city apartments or row houses may pack as many as 250 dwellings on a hectare of land (100 dwellings per acre). Older suburbs have larger row houses, semidetached houses, and individual houses on small lots, at a density of about 10 houses per hectare (4 houses per acre). A detached house typically sits on a lot of one-fourth to one-half hectare (0.6 to 1.2 acres) in new suburbs, and a lot of 1 hectare or greater (2.5 acres) on the fringe of the built-up area.

This density change in an urban area is called the **density gradient**. According to the density gradient, the number of houses per unit of land diminishes as distance from the center

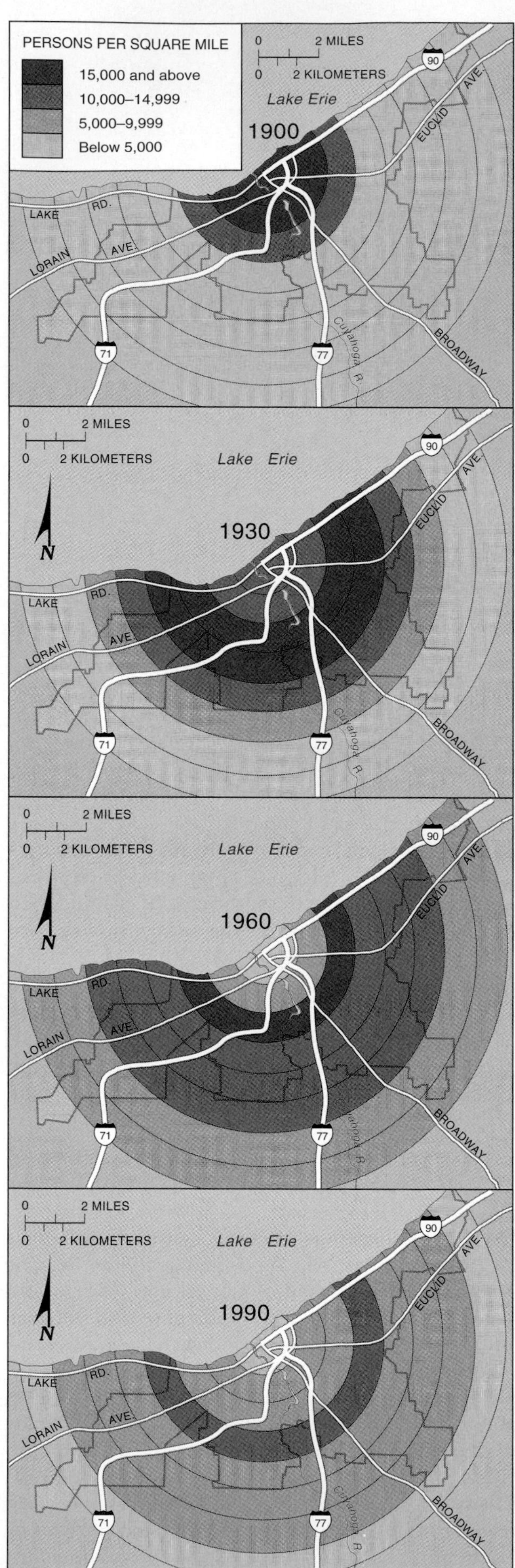

city increases. Two changes have affected the density gradient in recent years:

- **Fewer People Living in the Center.** The density gradient thus has a gap in the center, where few live.
- **Fewer Differences in Density Within Urban Areas.** The number of people living on a hectare of land has decreased in the central residential areas through population decline and abandonment of old housing. At the same time, density has increased on the periphery through construction of apartment and town-house projects and diffusion of suburbs across a larger area.

The result of the two changes is to flatten the density gradient and reduce the extremes of density between inner and outer areas traditionally found within cities.

Cost of Suburban Sprawl

U.S. suburbs are characterized by **sprawl**, which is the progressive spread of development over the landscape. When private developers select new housing sites, they seek cheap land that can easily be prepared for construction—land often not contiguous to the existing built-up area (Figure 13-24). Sprawl is also fostered by the desire of many families to own large tracts of land.

As long as demand for single-family detached houses remains high, land on the fringe of urbanized areas will be converted from open space to residential land use. Land is not transformed immediately from farms to housing developments. Instead, developers buy farms for future construction of houses by individual builders. Developers frequently reject land adjacent to built-up areas in favor of detached isolated sites, depending on the price and physical attributes of the alternatives. The peripheries of U.S. cities therefore look like Swiss cheese, with pockets of development and gaps of open space.

Urban sprawl has some undesirable traits. Roads and utilities must be extended to connect isolated new developments to nearby built-up areas. The cost of these new roads and utilities is funded by taxes or the services are installed by the developer, who passes on the cost to new residents through higher home prices. Sprawl also wastes land. Some prime agricultural land may be lost through construction of isolated housing developments. In the interim, other sites lie fallow while speculators await the most profitable time to build homes on them. In reality, sprawl has little impact on the total farmland in the United States, but it does reduce the ability of city dwellers to get to the country for recreation, and it can affect the supply of local dairy products and vegetables. The low-density suburb also wastes more energy, especially because motor vehicles are required for most trips.

FIGURE 13-23 Density gradient in Cleveland. In 1900, the population was highly clustered in and near the central business district (CBD). By 1930 and 1960, the population was spreading, leaving the original core less dense. By 1990, population was distributed over a much larger area, the variation in the density among different rings was much less, and the area's lowest densities existed in the rings near the CBD. The current boundary of the city of Cleveland is shown. (First three maps adapted from Avery M. Guest. "Population Suburbanization in American Metropolitan Areas, 1940–1970." *Geographical Analysis* 7 (1975): 267–83, table 4. Used by permission of the publisher.)

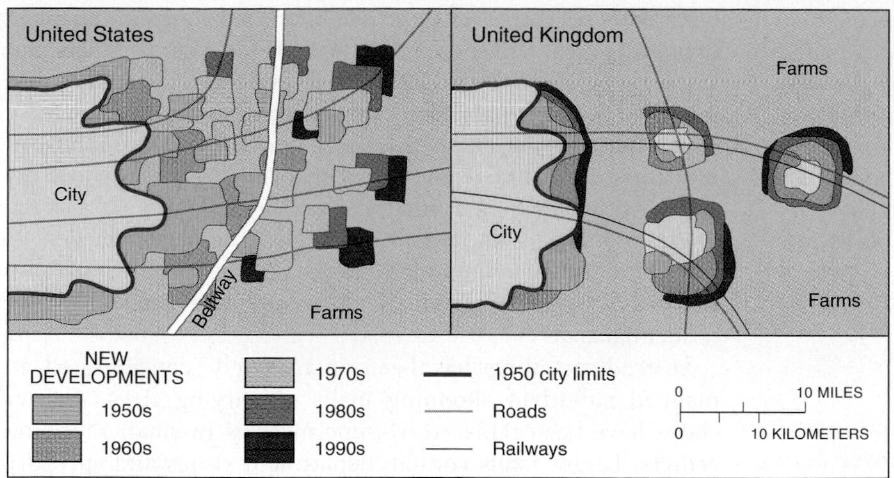

FIGURE 13-24 Suburban development patterns in the United Kingdom and the United States. The United States has much more sprawl than the United Kingdom. In the United Kingdom, new housing is more likely to be concentrated in new towns or planned extensions of existing small towns, whereas in the United States growth occurs in discontinuous developments.

The supply of land for the construction of new housing is more severely restricted in European urban areas. Officials attack sprawl by designating areas of mandatory open space. London, Birmingham, and several other British cities are surrounded by **greenbelts**, or rings of open space. New housing is built either in older suburbs inside the greenbelts or in planned extensions to small towns and new towns beyond the greenbelts. However, restriction of the supply of land on the urban periphery has driven up house prices in Europe.

Several U.S. states have taken strong steps in the past few years to curb sprawl, reduce traffic congestion, and reverse inner-city decline. The goal is to produce a pattern of compact and contiguous development, while protecting rural land for agriculture, recreation, and wildlife. Legislation and regulations to limit suburban sprawl and preserve farmland has been called **smart growth**. Oregon and Tennessee have defined growth boundaries within which new development must occur. Cities can annex only lands that have been included in the urban growth areas. New Jersey, Rhode Island, and Washington were also early leaders in enacting strong state-level smart-growth initiatives. Maryland enacted especially strong smart growth legislation in 1998. The Maryland smart-growth law prohibits the state from funding new highways and other projects that would extend suburban sprawl and destroy farmland. State money must be spent to "fill in" already urbanized areas.

Suburban Segregation

Public opinion polls in the United States show people's strong desire for suburban living. In most polls, more than 90 percent of respondents prefer the suburbs to the inner city. It is no surprise then that the suburban population has grown much faster than the overall population in the United States.

Suburbs offer varied attractions—a detached single-family dwelling rather than a row house or apartment, private land surrounding the house, space to park cars, and a greater opportunity for home ownership. The suburban house provides space and privacy, a daily retreat from the stress of urban living. Families with children are especially attracted to suburbs, which offer more space for play and protection from the high crime rates and heavy traffic that characterize inner-city life. As incomes rose in the twentieth century, first in the United States and more recently in other MDCs, more families were able to afford to buy suburban homes.

The modern residential suburb is segregated, and in two ways:

- **Segregated Social Classes.** Housing in a given suburban community is usually built for people of a single social class, with others excluded by virtue of the cost, size, or location of the housing.
- **Segregated Land Uses.** Residents are separated from commercial and manufacturing activities that are confined to compact, distinct areas.

Residential Segregation

The homogeneous suburb was a twentieth-century phenomenon. Before then, activities and classes in a city were more likely to be separated vertically rather than horizontally. In a typical urban building, shops were on the street level, with the shop owner or another well-to-do family living on one or two floors above the shop. Poorer people lived on the higher levels or in the basement, the least attractive parts of the building. The basement was dark and damp, and before the elevator was invented, the higher levels could be reached only by climbing many flights of stairs. Wealthy families lived in houses with space available in the basement or attic to accommodate servants. Once cities spread out over much larger areas, the old pattern of vertical separation was replaced by territorial segregation. Large sections of the city were developed with houses of similar interior dimension, lot size, and cost, appealing to people with similar incomes and lifestyles.

Zoning ordinances, developed in Europe and North America in the early decades of the twentieth century, encouraged spatial separation. They prevented the mixing of land uses within the same district. In particular, single-family houses, apartments, industry, and commerce were kept apart, because the location of one activity near another was considered unhealthy and inefficient. The strongest criticism of U.S. residential suburbs is that low-income people and minorities are unable to live in them because of the high cost of the housing and the unfriendliness of established residents. Suburban communities discourage the entry of those with lower incomes and minorities because of fear that property values will decline if the high-status composition of the neighborhood is altered. Legal devices, such as requiring each house to sit on a large lot and the prohibition of apartments, prevent low-income families from living in many suburbs.

In some metropolitan areas, the inner-city social and economic problems described earlier in this chapter are found in older suburbs immediately adjacent to the central city (Figure 13-25). As the central city is transformed into a vibrant community for higher-income people, inner suburbs become home to lower-income people displaced from gentrifying urban neighborhoods. Meanwhile, middle-class residents move from inner suburbs to newer homes on the periphery. Thus, the inner suburbs are unable to generate revenue to provide for the needs of a poorer population.

Suburbanization of Businesses

Businesses have moved to suburbs. Manufacturers have selected peripheral locations because land costs are lower. Service providers have moved to the suburbs because most of their customers are there.

SUBURBANIZATION OF RETAILING. Suburban residential growth has fostered change in traditional retailing patterns (Figure 13-26). Historically, urban residents bought food and other daily necessities at small neighborhood shops in the midst of housing areas and shopped in the CBD for other products. But since the end of World War II, downtown sales have not increased, whereas suburban sales have risen at an annual rate of 5 percent. Downtown sales have stagnated because suburban residents who live far from the CBD won't make the long journey there. At the same time, small corner shops do not exist in the midst of newer residential suburbs. The low density of residential construction discourages people from walking to stores, and restrictive zoning practices often exclude shops from residential areas.

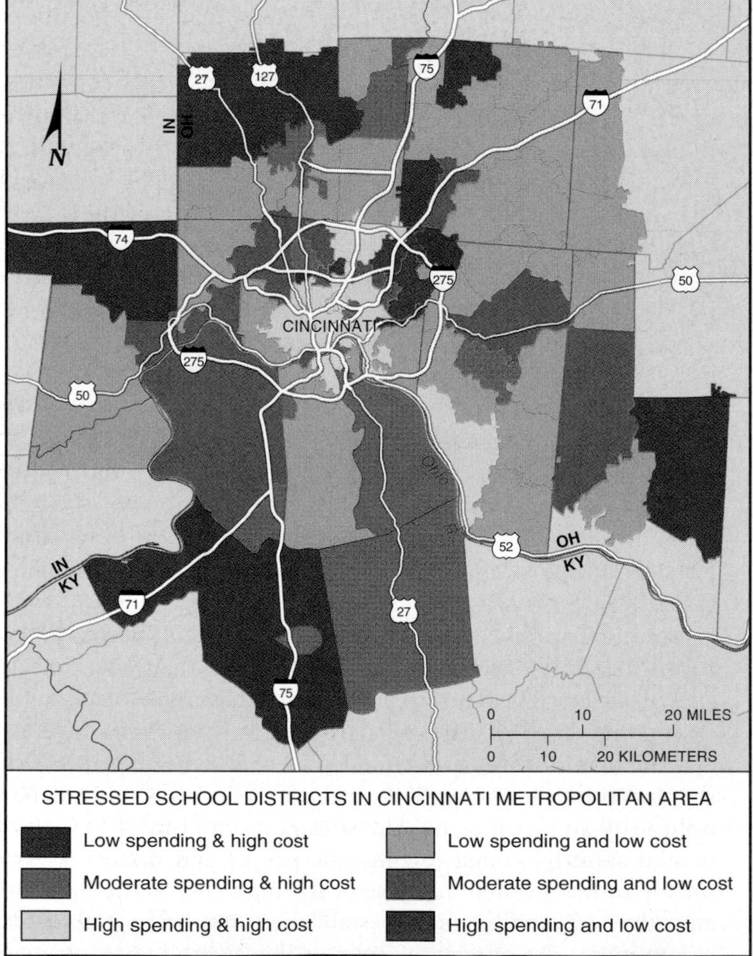

FIGURE 13-25 Suburban stress. In the Cincinnati MSA, the school districts considered high stress are mostly in the suburbs. A high-cost school district has either a rapidly growing or declining enrollment, or else a large percentage of students eligible for a free lunch program because of low income.

Instead, retailing has been increasingly concentrated in planned suburban shopping malls of varying sizes. Corner shops have been replaced by supermarkets in small shopping centers. Larger malls contain department stores and specialty shops traditionally reserved for the CBD. Generous parking lots surround the stores. A shopping mall is built by a developer, who buys the land, builds the structures, and leases space to individual merchants. Typically, a merchant's rent is a percentage of sales revenue.

Shopping malls require as many as 40 hectares (100 acres) of land and are frequently near key road junctions, such as the interchange of two interstate highways. Some shopping malls are elaborate multilevel structures exceeding 100,000 square meters (1 million square feet), with more than 100 stores arranged along covered walkways. The key to a successful large shopping mall is the inclusion of one or more anchors, usually large department stores. Most consumers go to a mall to shop at an anchor and, while there, patronize the smaller shops. In smaller shopping centers, the anchor is frequently a supermarket or discount store.

Malls have become centers for activities in suburban areas that lack other types of community facilities. Retired people go to malls for safe, vigorous walking exercise, or they sit on a bench to watch the passing scene. Teenagers arrive after school to meet their friends. Concerts and exhibitions are frequently set up in malls.

SUBURBANIZATION OF FACTORIES AND OFFICES. Factories and warehouses have migrated to suburbia for more space, cheaper land, and better truck access. Modern factories and warehouses demand more land because they spread their conveyor belts, forklift trucks, loading docks, and machinery over a single level for efficient operation. Suburban locations also facilitate truck shipments by providing good access to main highways and no central city traffic congestion, important because industries increasingly receive inputs and distribute products by truck.

Offices that do not require face-to-face contact are increasingly moving to suburbs where rents are lower than in the CBD. Executives can drive on uncongested roads to their offices from their homes in nearby suburbs and park their cars without charge. For other employees, though, suburban office locations can pose a hardship. Secretaries, custodians, and other lower-status office workers may not have cars, and public transportation may not serve the site. Other office workers might miss the stimulation and animation of a central location, particularly at lunchtime.

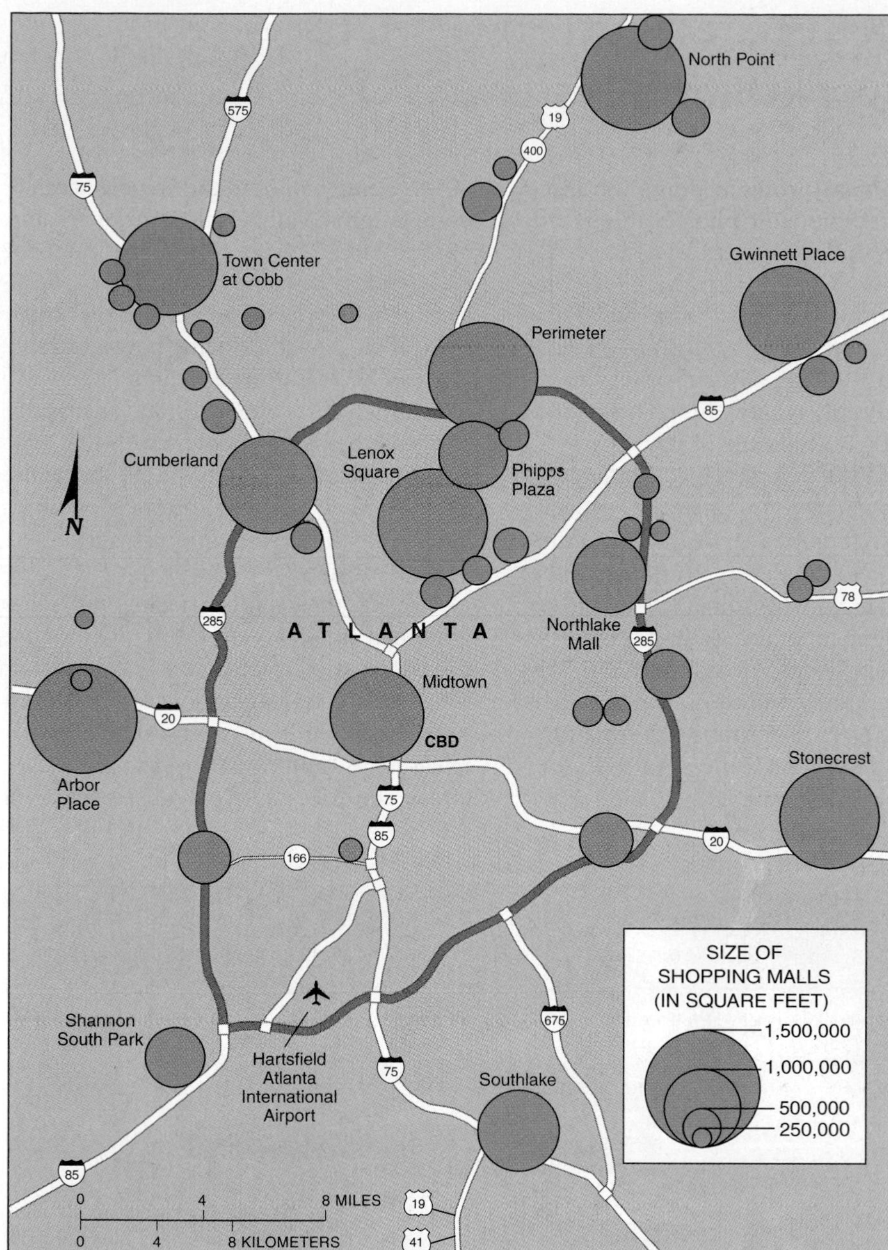

FIGURE 13-26 Major retail centers in Atlanta. Most shopping malls in the Atlanta metropolitan area, as elsewhere in North America, are in the suburbs, not the inner city. The optimal location for a large shopping mall is near an interchange on an interstate highway "beltway." These encircle many American cities, such as I-285 around Atlanta.

Transportation and Suburbanization

People do not travel aimlessly; their trips have a precise point of origin, destination, and purpose. More than half of all trips are work-related—commuting between work and home, business travel, or deliveries. Shopping or other personal business and social journeys each account for approximately one-fourth of all trips. Together, all of these trips produce congestion in urban areas. Congestion imposes costs on individuals and businesses by delaying arrival at destinations, and the high concentration of slowly moving vehicles produces increased air pollution (see Chapter 14).

Historically, the growth of suburbs was constrained by poor transportation. People lived in crowded cities because they had to be within walking distance of shops and places of employment. The invention of the railroad in the nineteenth century enabled people to live in suburbs and work in the central city. Cities then built railroads at street level (called trolleys, streetcars, or trams) and underground (subways) to accommodate commuters. Many so-called streetcar suburbs built in the nineteenth century still exist and retain unique visual identities. They consist of houses and shops clustered near a station or former streetcar stop at a much higher density than is found in newer suburbs.

Motor Vehicles

The suburban explosion in the twentieth century relied on motor vehicles rather than railroads, especially in the United States. Rail lines restricted nineteenth-century suburban development to narrow ribbons within walking distance of the stations. Cars and trucks permitted large-scale development of suburbs at greater distances from the center, in the gaps between the rail lines. Motor vehicle drivers have much greater flexibility in their choice of residence than was ever before possible.

Motor vehicle ownership is nearly universal among American households, with the exception of some poor families, older individuals, and people living in the centers of large cities such as New York. More than 95 percent of all trips within U.S. cities are made by car, compared to fewer than 5 percent by bus or rail. Outside the big cities, public transportation service is extremely rare or nonexistent. The U.S. government has encouraged the use of cars and trucks by paying 90 percent of the cost of limited-access, high-speed interstate highways, which stretch for 74,000 kilometers (46,000 miles) across the country. The use of motor vehicles is also supported by policies that keep the price of fuel below the level found in Europe.

The motor vehicle is an important user of land in the city. An average city allocates about one-fourth of its land to roads and parking lots. Multilane freeways cut a 23-meter (75-foot) path through the heart of cities, and elaborate interchanges consume even more space. Valuable land in the central city is devoted to parking cars and trucks, although expensive underground and

GLOBAL FORCES, LOCAL IMPACTS
Intelligent Transportation Systems

The future health of urban areas depends on relieving traffic congestion. Geographic tools, including global positioning systems (GPS) and electronic mapping, are playing central roles in the design of intelligent transportation systems, either through increasing road capacity or through reducing demand.

The current generation of innovative techniques to increase road capacity is aimed at providing drivers with information so that they can make intelligent decisions about avoiding congestion. Radio stations in urban areas have long broadcast reports to advise motorists of accidents or especially congested highways. Information about traffic congestion is now being transmitted through computers, handheld devices, and vehicle monitors. Traffic hot spots are displayed on electronic maps and images, using information collected through sensors in the roadbeds and cameras placed at strategic locations. An individual wishing to know about a particular route can program an electronic device to receive a congestion alert and to suggest alternatives.

The other current application of geographic tools is to reduce demand through "smart" highways. Toronto and several California cities charge motorists higher tolls to drive on freeways during congested times. A transponder attached to a vehicle records the time of day it is on the highway. A monthly bill sent to the vehicle's owner reflects the differential tolls. Singapore makes the most elaborate use of "smart" highway technology to minimize congestion. Every vehicle has a transponder that records tolls. To drive downtown during rush hour, a motorist must buy a license and demonstrate ownership of a parking space. The government limits the number of licenses and charges high tolls to drive downtown. Motorists must pay an £8 ($12) Congestion Charge to drive into Central London between 7 A.M. and 6.30 P.M. Monday through Friday (Figure 13-27). A similar system exists in Stockholm, where the charge varies depending on the time of day.

Future intelligent transportation systems are likely to increase capacity through hands-free driving. A motorist will drive to a freeway entrance, where the vehicle will be subjected to a thorough diagnostic (taking a half-second) to ensure that it has enough fuel and is in good operating condition. A menu offers a choice of predetermined destinations, such as "home" or "office," or a destination can be programmed by hand.

A release will send the vehicle accelerating automatically on the entrance ramp onto the freeway. Sensors in the bumpers and fenders, attached to radar or GPS, alert vehicle systems to accelerate, brake, or steer as needed. Spacing between vehicles can be as little as 2 meters.

While the vehicle is automatically controlled, the "driver" swivels the seat to a workstation to make phone calls, check e-mail, surf the Internet, or write letters. Or the driver can read, watch television, or nap.

When the vehicle nears the programmed freeway exit, a tone warns that the driver will have to take back control. The vehicle is halted on the exit ramp until the driver firmly presses the brake to release the "autodrive" system, much as cruise control is currently disengaged. ■

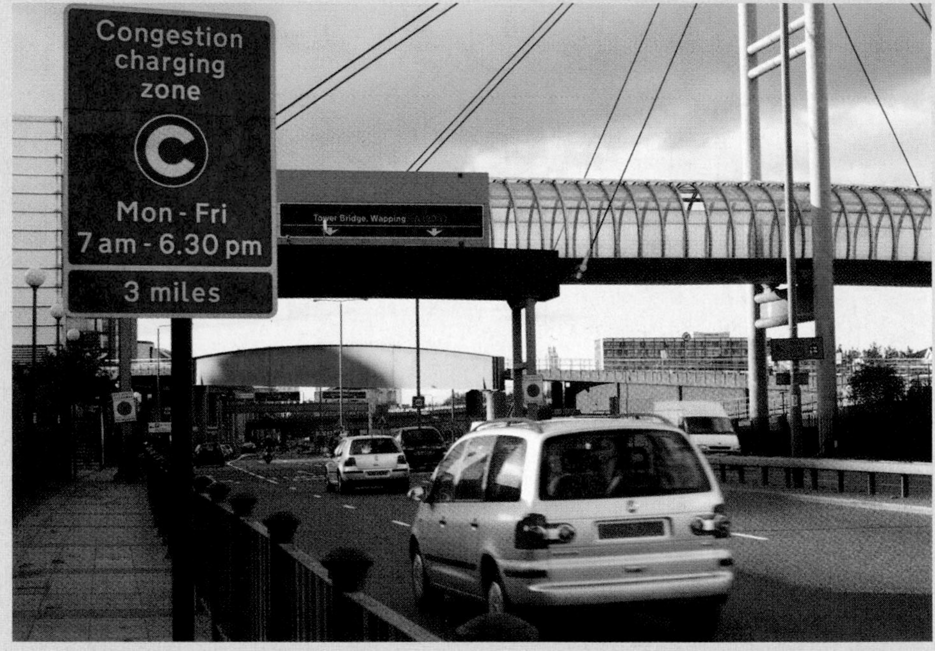

FIGURE 13-27 London Congestion Charge. The sign warns motorists that they are approaching the Congestion Zone. A charge of £8 is levied for driving a private vehicle into central London.

multistory parking structures can reduce the amount of ground-level space needed. European and Japanese cities have been especially disrupted by attempts to insert new roads and parking areas in or near the medieval central areas.

Technological improvements may help congestion (see Global Forces, Local Impacts box). In general, traffic flow can be improved by increasing the capacity of the roads or reducing demand to drive on them.

Public Transit

Because few people live within walking distance of their place of employment, urban areas are characterized by extensive commuting. The heaviest flow of commuters is into the CBD in the morning and out of it in the evening.

The intense concentration of people in the CBD during working hours strains transportation systems because a large number of people must reach a small area of land at the same time in the morning and disperse at the same time in the afternoon. As much as 40 percent of all trips made into or out of a CBD occur during four hours of the day—two in the morning and two in the afternoon. **Rush hour**, or peak hour, is the four consecutive 15-minute periods that have the heaviest traffic.

ADVANTAGES OF PUBLIC TRANSIT. In larger cities, public transit is better suited than motor vehicles to moving large numbers of people, because each transit traveler takes up far less space. Public transportation is cheaper, less polluting, and more energy efficient than the automobile. It also is particularly suited to rapidly bringing a large number of people into a small area. A bus can accommodate 30 people in the amount of space occupied by one automobile, whereas a double-track rapid transit line can transport the same number of people as 16 lanes of urban freeway.

Motor vehicles have costs beyond their purchase and operation: delays imposed on others, increased need for highway maintenance, construction of new highways, and pollution. One-third of the high-priced central land is devoted to streets and parking lots, although multistory and underground garages also are constructed.

In most cities around the world, extensive networks of bus, tram, and subway lines have been maintained, and funds for new construction have been provided in recent years (Figure 13-28). Since the late 1960s, London has opened 50 kilometers (35 miles) of subways, including two new lines, plus 25 kilometers (15 miles) in light-rail transit lines to serve the docklands area, which has been transformed from industrial to residential and office use. During the same period, Paris has added 400 kilometers (250 miles) of new subway lines, primarily in a new system known as the Réseau Express Régional (R.E.R.) to serve outer suburbs.

Smaller cities have shared the construction boom. In France alone, new subway lines have been built since the 1970s in Lille, Lyon, and Marseille, and hundreds of kilometers of entirely new tracks have been laid between the country's major cities to operate a high-speed train known as the TGV (Train à Grande Vitesse). Growth in the suburbs has stimulated nonresidential construction, including suburban shops, industry, and offices.

PUBLIC TRANSIT IN THE UNITED STATES. In the United States, public transit is used primarily for rush-hour commuting by workers into and out of the CBD. One-half of trips to work are by public transit in New York, one-third in Boston, San Francisco, and Washington, and one-fourth in Chicago and Philadelphia.

But in other cities, public transit service is minimal or nonexistent. Despite the obvious advantages of public transportation for commuting, only 5 percent of work trips are by public transit in the United States. Overall, public transit ridership in the United States declined from 23 billion per year in the 1940s to 10 billion in 2006. The average American loses 36 hours per year sitting in traffic jams and wastes 55 gallons of gasoline. In the United States, the total cost of congestion is valued at more than $87 billion per year. But most Americans still prefer to commute by vehicle. Most people overlook these costs because they place higher value on the car's privacy and flexibility of schedule.

Early in the twentieth century, U.S. cities had 50,000 kilometers (30,000 miles) of street railways and trolleys that carried 14 billion passengers a year, but only a few hundred kilometers of track remain. The number of U.S. and Canadian cities with trolley service declined from approximately fifty in 1950 to eight in the 1960s. General Motors acquired many of the privately owned streetcar companies and replaced the trolleys with buses that the company made. Buses offer a more flexible service than do trolleys because they are not restricted to fixed tracks. However, bus ridership in the United States declined from a peak of 11 billion riders annually in the late 1940s to 6 billion in 2006. Commuter railroad service, like trolleys and buses, has also been drastically reduced in most U.S. cities.

The one exception to the downward trend in public transit in the United States is rapid transit. It is known to transportation planners as either fixed heavy rail (such as subways) or fixed light rail (such as streetcars). Cities such as Boston and Chicago have attracted new passengers through construction of new subway lines and modernization of existing service. Chicago has been a pioneer in the construction of heavy-rail rapid transit lines in the median strips of expressways. Entirely new subway systems have been built in recent years in U.S. cities, including Atlanta, Baltimore, Miami, San Francisco, and Washington.

The federal government has permitted Boston, New York, and other cities to use funds originally allocated for interstate highways to modernize rapid transit service instead. New York's subway cars, once covered with graffiti spray-painted by gang members, have been cleaned so that passengers can ride in a more hospitable environment. As a result of these improvements, subway ridership in the United States increased from 2 billion in 1995 to 3 billion in 2006.

The trolley—now known by the more elegant term of fixed light-rail transit—is making a modest comeback in North America. Once relegated almost exclusively to a tourist attraction in New Orleans and San Francisco, new trolley lines have been built or are under construction in Baltimore, Buffalo, Calgary, Edmonton, Los Angeles, Portland (Oregon), Sacramento, St. Louis, San Diego, and San Jose. Ridership in all cities combined was 400 million in 2006.

California, the state that most symbolizes the automobile-oriented American culture, is the leader in construction of new fixed light-rail transit lines. San Diego has added more kilometers than any other city. One line that runs from the center south to the Mexican border has been irreverently dubbed the "Tijuana trolley" because it is heavily used by residents of nearby Tijuana, Mexico. Los Angeles—the city perhaps most associated with the motor vehicle—has planned the most

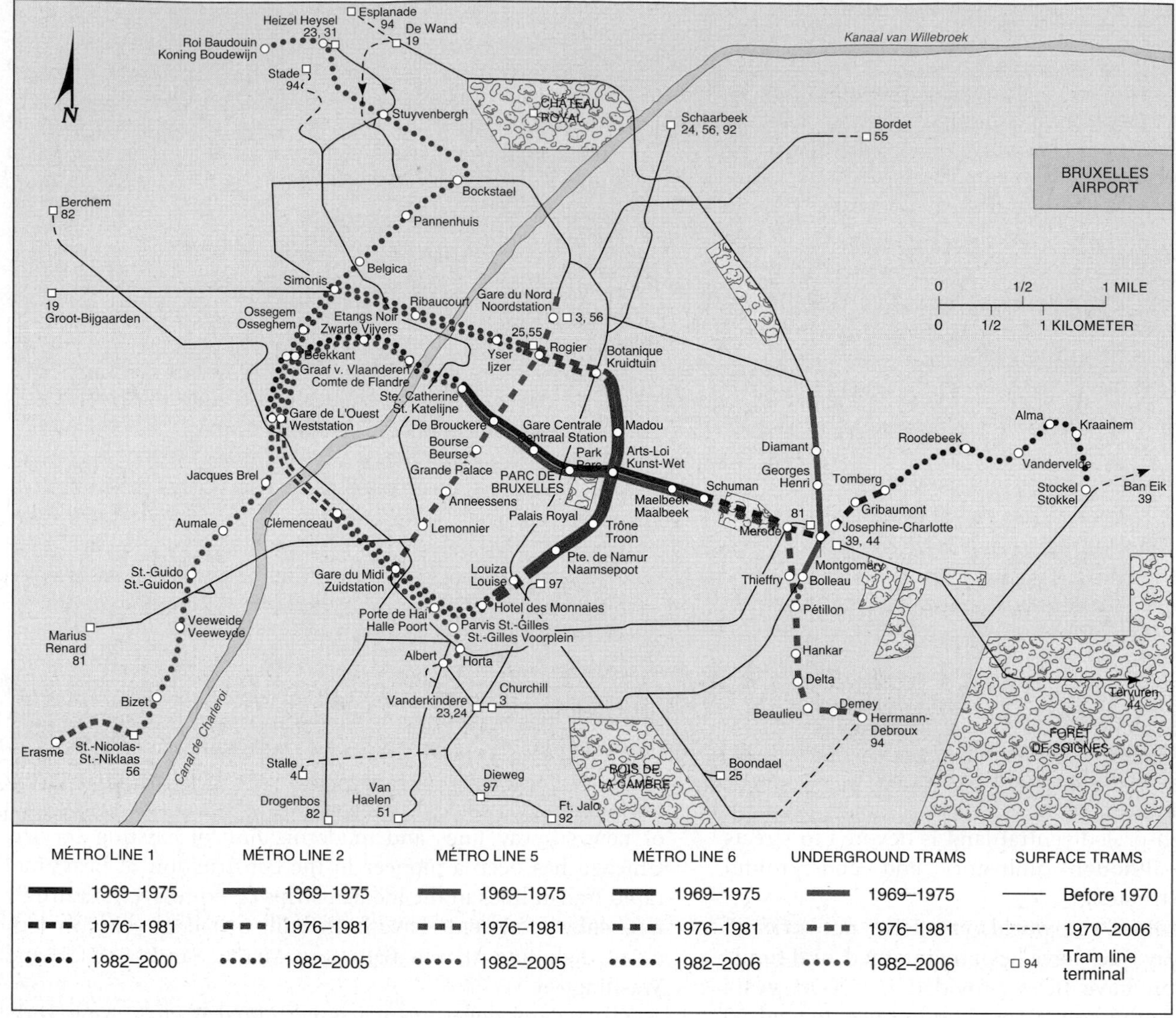

FIGURE 13-28 Brussels, Belgium, subway and tram lines. European cities such as Brussels have invested substantially to improve public transportation in recent years. Brussels provides a good example of a public transport system that integrates heavy rail (Métro) with light rail (trams). Trams initially used Métro tunnels, but the tunnels were large enough to convert to heavy-rail lines as funds became available.

extensive new light-rail system. The city had a rail network exceeding 1,600 kilometers (1,000 miles) as recently as the late 1940s, but the lines were abandoned when freeways were built to accommodate increasing automobile usage. Now Los Angeles wants to entice motorists out of their cars and trucks with new light-rail lines, but construction is very expensive and the lines serve only a tiny percentage of the region.

The minimal level of public transit service in most U.S. cities means that low-income people may not be able to reach places of employment. Low-income people tend to live in inner-city neighborhoods, but the job opportunities, especially those requiring minimal training and skill in personal services, are in suburban areas not well served by public transportation. Inner-city neighborhoods have high unemployment rates at the same time that suburban firms have difficulty attracting workers. In some cities, governments and employers subsidize vans to carry low-income inner-city residents to suburban jobs.

Despite modest recent successes, public transit in the United States is caught in a vicious circle, because fares do not cover operating costs. As patronage declines and expenses rise, the fares are increased, which drives away passengers and leads to service reduction and still higher fares. Public expenditures to subsidize construction and operating costs have increased, but the United States does not fully recognize that public transportation is a vital utility deserving of subsidy to the degree long assumed by European governments.

SUMMARY

Many people live in urban areas and never venture into inner-city neighborhoods or downtown. They live in suburbs, attend school in suburbs, work in suburbs, shop in suburbs, visit friends and family in suburbs, and attend movies and sports events in suburbs. Motor vehicles allow movement across urban areas without entering the center.

Conversely, inner-city residents may rarely venture out to suburbs. Lacking a motor vehicle, they have no access to most suburban locations. Lacking money, they do not shop in suburban malls or attend sporting events at suburban arenas. The spatial segregation of inner-city residents and suburbanites lies at the heart of the stark contrasts so immediately observed in any urban area.

Here is a review of the key issues raised at the beginning of the chapter:

1. **Why Do Services Cluster Downtown?** The central business district (CBD) contains a large percentage of a settlement's business services. Business services cluster downtown to facilitate face-to-face contact. Retailers with large thresholds or large ranges may also locate downtown.
2. **Where Are People Distributed Within Urban Areas?** Three models explain where various groups of people live in urban areas—the concentric zone, sector, and multiple nuclei models. Combined, the three models present a useful framework for understanding the distribution of social and economic groups within urban areas. With modifications, the models also apply to cities in Europe and in LDCs.
3. **Why Do Inner Cities Face Distinctive Challenges?** Inner-city residential areas have physical problems stemming from the high percentage of older deteriorated housing, social problems stemming from the high percentage of low-income households, and economic problems stemming from a gap between demand for services and supply of local tax revenue.
4. **Why Do Suburbs Face Distinctive Challenges?** The suburban lifestyle as exemplified by the detached single-family house with surrounding yard attracts most people. Transportation improvements, most notably the railroad in the nineteenth century and the automobile in the twentieth century, have facilitated the sprawl of urban areas. Among the negative consequences of large-scale sprawl are segregation and inefficiency.

CASE STUDY REVISITED / Contrasts in the City

What is the future for cities? As shown in this chapter, contradictory trends are at work simultaneously. Why does one inner-city neighborhood become a slum and another an upper-class district (Figure 13–29)? Why does one city attract new shoppers and visitors while another languishes?

The Camden, New Jersey, urban area displays the strong contrasts that characterize American urban areas. The central city of Camden houses an isolated underclass while suburban Camden County prospers.

- **Population Decline:** The population of the city of Camden has declined from 117,000 in 1960 to 70,000 in 2007.
- **Racial Change:** Camden's white, non-Hispanic population has declined from 90,000 in 1960 to 4,000 in 2007. African Americans comprise about 36,000 of the city's population; Hispanics about 30,000.
- **Demographic Stress:** More than one-fourth of Camden's residents are under age 15, closer to the level found in LDCs than to the rest of the United States. The infant mortality rate for the city's African American population is 27 per 1,000, about the level of Mexico, and four times higher than the rest of the United States.

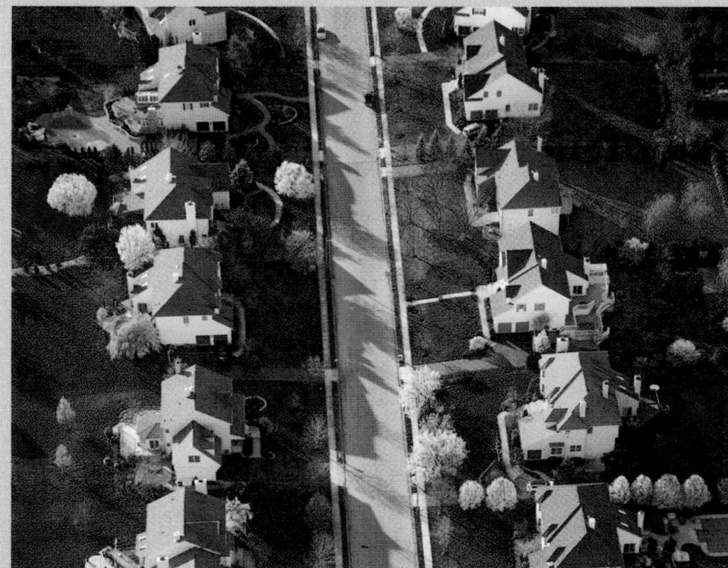

FIGURE 13-29 Urban contrasts. (left) Downtown Camden. (right) Suburban Camden County.

CASE STUDY REVISITED (*Continued*)

- **Low Income:** Median annual household income in Camden is $23,000, compared to $42,000 for the United States as a whole. More than half the population receives government assistance.

Job prospects are not promising for Camden's young people. In the past, they could find jobs in factories that produced Campbell's soups, Esterbrook pens, and RCA Victor records, radios, and televisions, but the city has lost 90 percent of its industrial jobs. The Esterbrook and Campbell factories in Camden are closed, although Campbell's corporate offices remain. The old RCA Victor building has been converted to apartments.

As Camden's population and industries decline, few shops have enough customers to remain open. The city once had 13 movie theaters, but none are left. The murder rate soared after gangs carved up the city into districts during the mid-1980s to control cocaine trafficking. Violent crimes such as murder, rape, and robbery are increasing in Camden while dropping nationally. New Jersey state troopers help the city's understaffed police force deal with crime.

Meanwhile, Camden County (excluding the city) grew from 275,000 in 1960 to about 443,000 in 2007. Cherry Hill had about 72,000 residents in 2007, compared to fewer than 10,000 in 1960. The population of Cherry Hill has increased modestly since 1990, as growth pushed east, much farther away from Camden, which is on the far western edge of the county.

Cherry Hill is an example of an edge city, a large node of office and retail activities on the edge of an urban area. Despite its rapid population growth and trained labor force, an edge city like Cherry Hill has become both a residential area that commuters leave and an employment center that attracts other commuters. Cherry Hill has attracted so many new jobs that a major obstacle to further economic growth is a shortage of qualified workers.

But many inner-city Camden residents lack transport to reach the jobs or the skills to hold the jobs. Camden's mismatch among locations of people, jobs, resources, and services exemplifies the urban crisis throughout the United States, as well as in other countries. Geographers help us understand why these patterns arise and what can be done about them. ■

KEY TERMS

Annexation (p. 424) Legally adding land area to a city in the United States.

Census tract (p. 412) An area delineated by the U.S. Bureau of the Census for which statistics are published; in urbanized areas, census tracts correspond roughly to neighborhoods.

Central business district (CBD) (p. 406) The area of a city where retail and office activities are clustered.

City (p. 424) An urban settlement that has been legally incorporated into an independent, self-governing unit.

Combined statistical area (CSA) (p. 425) In the United States, two or more contiguous core based statistical areas tied together by commuting patterns.

Concentric zone model (p. 410) A model of the internal structure of cities in which social groups are spatially arranged in a series of rings.

Core based statistical area (CBSA) (p. 425) In the United States, the combination of all metropolitan statistical areas and micropolitan statistical areas.

Council of government (p. 426) A cooperative agency consisting of representatives of local governments in a metropolitan area in the United States.

Density gradient (p. 427) The change in density in an urban area from the center to the periphery.

Edge city (p. 427) A large node of office and retail activities on the edge of an urban area.

Filtering (p. 419) A process of change in the use of a house, from single-family owner occupancy to abandonment.

Gentrification (p. 420) A process of converting an urban neighborhood from a predominantly low-income, renter-occupied area to a predominantly middle-class, owner-occupied area.

Greenbelt (p. 429) A ring of land maintained as parks, agriculture, or other types of open space to limit the sprawl of an urban area.

Metropolitan statistical area (MSA) (p. 425) In the United States, a central city of at least 50,000 population, the county within which the city is located, and adjacent counties meeting one of several tests indicating a functional connection to the central city.

Micropolitan statistical area (p. 425) An urbanized area of between 10,000 and 50,000 inhabitants, the county in which it is found, and adjacent counties tied to the city.

Multiple nuclei model (p. 412) A model of the internal structure of cities in which social groups are arranged around a collection of nodes of activities.

Peripheral model (p. 427) A model of North American urban areas consisting of an inner city surrounded by large suburban residential and business areas tied together by a beltway or ring road.

Primary census statistical area (PCSA) (p. 425) In the United States, all of the combined statistical areas plus all of the remaining metropolitan statistical areas and micropolitan statistical areas.

Public housing (p. 420) Housing owned by the government; in the United States, it is rented to residents with low incomes, and the rents are set at 30 percent of the families' incomes.

Redlining (p. 420) A process by which banks draw lines on a map and refuse to lend money to purchase or improve property within the boundaries.

Rush hour (p. 433) The four consecutive 15-minute periods in the morning and evening with the heaviest volumes of traffic.

Sector model (p. 411) A model of the internal structure of cities in which social groups are arranged around a series of sectors, or wedges, radiating out from the central business district (CBD).

Smart growth (p. 429) Legislation and regulations to limit suburban sprawl and preserve farmland.

Social area analysis (p. 412) Statistical analysis used to identify where people of similar living standards, ethnic background, and life style live within an urban area.

Sprawl (p. 428) Development of new housing sites at relatively low density and at locations that are not contiguous to the existing built-up area.

Squatter settlement (p. 417) An area within a city in a less developed country in which people illegally establish residences on land they do not own or rent and erect homemade structures.

Underclass (p. 421) A group in society prevented from participating in the material benefits of a more developed society because of a variety of social and economic characteristics.

Urban renewal (p. 420) Program in which cities identify blighted inner-city neighborhoods, acquire the properties from private owners, relocate the residents and businesses, clear the site, build new roads and utilities, and turn the land over to private developers.

Urbanized area (p. 424) In the United States, a central city plus its contiguous built-up suburbs.

Zoning ordinance (p. 429) A law that limits the permitted uses of land and maximum density of development in a community.

THINKING GEOGRAPHICALLY

1. Compare the CBDs of Toronto and Detroit. What might account for differences?
2. Draw a sketch of your community or neighborhood. In accordance with Kevin Lynch's *The Image of the City,* place five types of information on the map—districts (homogeneous areas), edges (boundaries that separate districts), paths (lines of communication), nodes (central points of interaction), and landmarks (prominent objects on the landscape). How clear an image does your community have for you?
3. Jane Jacobs wrote in *Death and Life of Great American Cities* that an attractive urban environment is one that is animated with an intermingling of a variety of people and activities, such as found in many New York City neighborhoods. What are the attractions and drawbacks to living in such environments?
4. Land-use activities in Communist cities were allocated by government rather than made by private market decisions. To what extent would the absence of a private-sector urban land market affect the form and structure of socialist cities? What impacts might Eastern European cities experience with the switch to market economies?
5. Officials of rapidly growing cities in LDCs discourage the building of houses that do not meet international standards for sanitation and construction methods. Also discouraged are privately owned transportation services, because the vehicles generally lack decent tires, brakes, and other safety features. Yet the residents prefer substandard housing to no housing, and they prefer unsafe transportation to no transportation. What would be the advantages and problems for a city if health and safety standards for housing, transportation, and other services were relaxed?

RESOURCES

Some recent and classic books and articles on industrial geography:

Berry, Brian J. L. *The Human Consequences of Urbanization.* New York: St. Martin's Press, 1973.

———, and John D. Kasarda. *Contemporary Urban Ecology.* New York: Macmillan, 1977.

———, and James O. Wheeler, eds. *Urban Geography in America, 1950–2000: Paradigms and Personalities.* New York: Routledge, 2005.

Bertaud, Alain, and Bertrand Renaud. "Socialist Cities Without Land Markets." *Journal of Urban Economics* 41 (1997): 137–51.

Bourne, Larry S., ed. *Internal Structure of the City,* 2nd ed. New York: Oxford University Press, 1982.

Brockerhoff, Martin P. "An Urbanizing World." *Population Bulletin* 55, no. 3. Washington, DC: Population Reference Bureau, 2000.

Clawson, Marion, and Peter Hall. *Planning and Urban Growth.* Baltimore: Johns Hopkins University Press, 1973.

Dear, Michael, and Steven Flusty. "Postmodern Urbanism." *Annals of the Association of American Geographers* 88 (1998): 50–72.

Ford, Larry R. "A New and Improved Model of Latin American City Structure." *Geographical Review* 86 (1996): 437–40.

Garreau, Joel. *Edge City: Life on the New Frontier.* New York: Doubleday, 1991.

Gottmann, Jean. *Megalopolis.* New York: Twentieth-Century Fund, 1961.

Griffin, Ernest, and Larry Ford, "A Model of Latin American City Structure." *Geographical Review* 70 (1980): 387–422.

Guest, Avery M. "Population Suburbanization in American Metropolitan Areas, 1940–1970." *Geographical Analysis* 7 (1976): 267–83.

Harris, Chauncy D. "Diffusion of Urban Models: A Case Study." *Urban Geography* 19 (1998): 49–67.

———. "The Nature of Cities and Urban Geography in the Last Half Century." *Urban Geography* 18 (1997): 15–35.

———, and Edward L. Ullman. "The Nature of Cities." *Annals of the American Academy of Political and Social Science* 143 (1945): 7–17.

Hodge, David C., Richard L. Morrill, and Kiril Stanilov. "Implications of Intelligent Transportation Systems for Metropolitan Form." *Urban Studies* 17 (1996): 714–39.

Hoyt, Homer. *The Structure and Growth of Residential Neighborhoods.* Washington, D.C: Federal Housing Administration, 1939.

Jacobs, Jane. *Death and Life of Great American Cities.* New York: Random House, 1961.

Knox, Paul L. *Metroburbia.* New Brunswick, NJ: Rutgers University Press, 2008.

Levy, John. *Contemporary Urban Planning,* 8th ed. Upper Saddle River, NJ: Prentice Hall, 2008

Lynch, Kevin. *The Image of the City.* Cambridge, MA: M.I.T. Press, 1960.

Mumford, Lewis. *The City in History.* New York: Harcourt, Brace, and World, 1961.

Park, Robert E., Ernest W. Burgess, and Roderick D. McKenzie, eds. *The City.* Chicago: University of Chicago Press, 1925.

Scott, Allen J. "Capitalism, Cities, and the Production of Symbolic Forms." *Transactions of the Institute of British Geographers,* New Series 26 (2001): 11–24.

Journals featuring urban geography:

Environment and Planning, Journal of Housing, Journal of the American Planning Association, Land Economics, Planning, Urban Geography, Urban Land, and *Urban Studies.*

Key Internet sites:

www.census.gov. Data concerning any urban area can be found at the U.S. Bureau of the Census web site. The American Factfinder service provides information from the most recent census as well as annual updates from the American Community Survey. Tables and maps can be generated for census tracts within urban areas as well as for entire urban areas. Access is also provided to data from earlier censuses.

www.socialexplorer.com. Social Explorer provides access to census data at all scales, including urban. An interactive map enables users to choose the area of interest and from among hundreds of census variables.

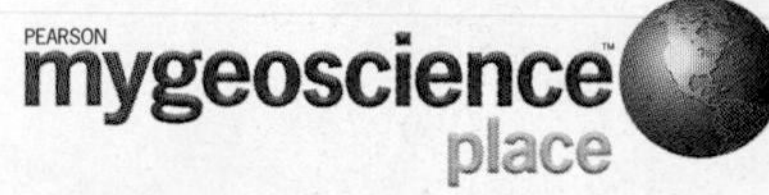

Log in to www.mygeoscienceplace.com for videos, interactive maps, RSS feeds, case studies, and self-study quizzes to enhance your study of Urban Patterns.

CHAPTER 14

Resource Issues

When you have finished drinking a soda, do you pitch the can in the trash or place it in a recycling bin? Do you drink bottled water or tap water? Do you normally eat with disposable plates, cups, and plastic utensils, or do you use washable ceramic and stainless steel products? When you leave a room, do you turn off the lights and electronics? When you buy a new motor vehicle, do you consider its fuel efficiency? Do you care that your family's sport-utility vehicle gets poorer gas mileage than your grandparents' big "old-fashioned" sedan?

KEY ISSUES

1 Why Are Resources Being Depleted?

2 Why Are Resources Being Polluted?

3 Why Are Resources Being Reused?

4 Why Should Resources Be Conserved?

People have always transformed Earth's land, water, and air for their benefit. But human actions in recent years have gone far beyond the impact of the past. The magnitude of transformations is disproportionately shared by North Americans; with only one-twentieth of Earth's population, North Americans consume one-fourth of the world's energy and generate one-fourth of many pollutants. Elsewhere in the world, 2 billion people live without clean water or sewers. One billion live in cities with unsafe sulfur dioxide levels.

Future generations will pay the price if we continue to mismanage Earth's resources. Our shortsightedness could lead to shortages of energy to heat homes and operate motor vehicles. Our carelessness has already led to unsafe drinking water and toxic air in some places.

Humans once believed Earth's resources to be infinite, or at least so vast that human actions could never harm or deplete them. But warnings from geographers and other scientists make it clear that resources are indeed a problem.

Trans-Alaska pipeline.

CASE STUDY / Pollution in Mexico City

Eight-year-old Carlos and nine-year-old Maria, residents of Mexico City, did not go to school today. Many of their classmates also did not attend. And many of their teachers failed to report for work. These people did not leave their homes because they feared that breathing outside air in Mexico City would be too dangerous.

For much of the year, a stationary cloud hangs over Mexico City, producing a gray-brown fog that irritates the eyes and burns the throat. Residents report frequent conjunctivitis and other eye disorders, skin rashes, bronchitis, other respiratory diseases, and increased susceptibility to heart attacks. The air is considered safe to breathe for only 31 days a year. The health benefits of outdoor sports such as soccer and running are outweighed by the health risks of breathing the air. Pregnant women are cautioned that living in Mexico City increases risk to fetal health.

This severe air pollution partly results from Mexico City's site: It rests in a basin some 2,250 meters (about 7,400 feet) above sea level, surrounded by a semicircle of volcanic peaks as high as 5,545 meters (16,900 feet). This giant bowl is open only to the north. Prevailing winds from the north enter the basin and back polluted air against the surrounding mountains. Most of the emissions come from burning fuels in 8 million motor vehicles and 50,000 factories. These emissions are trapped close to the ground in a stationary cloud, especially in the winter, when the climate is cool and dry and winds are calm. Because the city is at a high altitude, the level of available oxygen is low. Thus fossil fuels burn less completely than at lower altitudes, and burning them produces more carbon monoxide and ozone.

Air pollution is not Mexico City's only environmental problem. Inadequately treated sewage flows into nearby rivers, and one-fourth of the city's homes are not even connected to the sewer system. Solid waste is deposited at large municipal dumps, where 17,000 people known as *pepenadores*, or garbage pickers, survive by going through rubbish and, in many cases, actually live at the dump. Dust from fecal matter in unsewered areas increases skin and eye infections. ■

Plants and animals live in harmony with their environment, but people often do not. Geographers study the troubled relationship between human actions and the physical environment in which we live.

From the perspective of human geographers, Earth offers a large menu of resources available for people to use. A **resource** is a substance in the environment that is useful to people, is economically and technologically feasible to access, and is socially acceptable to use. Resources include food, water, soil, plants, animals, and minerals. A natural resource has little value in and of itself. Its value derives from its usefulness to humans, especially in production. Enterprises extract those resources for which humans are willing to pay a sufficiently high price to justify the investment. As the supply of a resource dwindles, consumers may be willing to pay higher prices, thus encouraging continued exploitation and further depletion of reserves rather than conservation for future generations.

The problem is that most resources are limited, and Earth has a tremendous number of consumers. Geographers observe two major misuses of resources:

- We deplete scarce resources, especially petroleum, natural gas, and coal, for energy production.
- We destroy resources through pollution of air, water, and soil.

These two misuses are the basic themes of this chapter.

As with other topics, geographers look first at ***where*** resources are distributed across ***space***. Both supply and demand of resources show pronounced ***local diversity***. Some ***regions*** are well endowed with minerals, water, and other resources, whereas other regions have limited supplies. The reason ***why*** problems arise from this uneven distribution is that resources are often located in ***places*** different from their users. Differences in demand may arise from the uneven distribution of people across Earth or from variations in development.

Nowhere is the ***globalization*** trend more pronounced than in the study of resources. The global economy depends on the availability of natural resources to produce the goods and services that people demand. Global uniformity in cultural preferences means that people in different places value similar natural resources, although not everyone has the same access to them. In a global environment, all places are ***connected***, so the misuse of a resource in one place affects the well-being of people everywhere.

To study resource problems, we also depend on our understanding of local ***scale***. As geographers, we understand that our energy problems derive from depletion of resources in particular regions and from differences in how consumers use resources in different places. We see that the pollution problem comes from the concentration of substances that harm the physical environment in particular regions.

KEY ISSUE 1
Why Are Resources Being Depleted?

- **Energy Resources**
- **Mineral Resources**

Two kinds of natural resources are especially valuable to humans—minerals and energy resources. We depend on abundant, low-cost energy and minerals to run our industries, transport ourselves, and keep our homes comfortable. But we are depleting the global supply of some resources. MDCs want to preserve current standards of living, and LDCs are struggling to attain a better standard. All this demands tremendous resources; so as we deplete our current sources, we must develop alternative ones. ■

Energy Resources

Historically, people relied primarily on power supplied by themselves or by animals, known as **animate power**. Energy from flowing water and burning biomass fuel supplemented animate power. **Biomass fuel**, such as wood, plant material, and animal waste, is burned directly or converted to charcoal, alcohol, or methane gas. Biomass remains the most important source of fuel in some LDCs, but during the Industrial Revolution MDCs converted to **inanimate power**, generated from machines.

Energy Supply and Demand

Around one-half of the world's energy is consumed in MDCs and one-half in LDCs. MDCs contain only around one-third of the population of LDCs, so per capita consumption of energy is thus around three times higher in MDCs than in LDCs (Figure 14-1). North Americans are the heaviest per capita consumers of energy. North America uses one-fourth of the world's energy but contains only one-twentieth of the world's people.

Three of Earth's substances provide five-sixths of the world's energy (Figure 14-2):

- **Coal.** Supplanted wood as the leading energy source in North America and Europe in the late 1800s.
- **Petroleum.** First pumped in 1859, but not an important resource until the diffusion of motor vehicles in the twentieth century.
- **Natural Gas.** Originally burned off as a waste product of oil drilling, but now used to heat homes.

In MDCs other energy comes primarily from nuclear and hydroelectric power. Burning wood and hydroelectric power provide much of the remaining energy in LDCs.

Energy is used in three principal places in the United States:

1. **Businesses.** The main energy resource is coal, followed by natural gas and oil. Some businesses directly burn coal in their own furnaces. Others rely on electricity, mostly generated at coal-burning power plants.
2. **Homes.** Energy is used primarily for the heating of living space and water. Natural gas is the most common source, followed by petroleum (heating oil and kerosene).
3. **Transportation.** Almost all transportation systems operate on petroleum products, including cars, trucks, buses, airplanes, and most railroads. Only subways, streetcars, and some trains run on coal-generated electricity.

Petroleum, natural gas, and coal are known as fossil fuels. A **fossil fuel** is the residue of plants and animals that were buried millions of years ago. As sediment accumulated over these remains, intense pressure and chemical reactions slowly converted them into the fossil fuels we use today. When we burn these substances, we are releasing energy originally stored in plants and animals millions of years ago.

Two characteristics of fossil fuels cause great concern for the future:

- **The Supply of Fossil Fuels Is Finite.** Once the present supply of fossil fuels is consumed, it is gone, and we must look to other resources for our energy. (Technically, fossil fuels are continually being formed, but the process takes millions of years, so humans must regard the current supply as essentially finite.)
- **Fossil Fuels Are Distributed Unevenly Around the Globe.** Some regions enjoy a generous supply of fossil fuels, whereas others have little, and fossil fuels are not consumed in the same regions where they are produced.

Finiteness of Fossil Fuels

Earth's resources are divided between those that are renewable and those that are not:

- **Renewable energy** has an essentially unlimited supply and is not depleted when used by people. Examples include hydroelectric, geothermal, fusion, wind, and solar energy.
- **Nonrenewable energy** forms so slowly that for practical purposes it cannot be renewed. The fossil fuels are examples.

As nonrenewable energy sources, the three main fossil fuels, once burned, are used up for all time. The world faces an energy problem in part because we are rapidly depleting the remaining supply of the three fossil fuels, especially petroleum.

Because of dwindling supplies of fossil fuels, most of the buildings in which we live, work, and study will have to be heated another way. Cars, trucks, and buses will have to operate on some other energy source. The many plastic objects that we use (because they are made from petroleum) must be made with other materials. We can use other resources for heat, fuel, and manufacturing, but they are likely to be more expensive and less convenient to use than fossil fuels. And converting from fossil fuels will likely disrupt our daily lives and cause us hardship.

PROVEN RESERVES. How much of the fossil-fuel supply remains? Despite the critical importance of this question for the future, no one can answer it precisely. Because petroleum, natural gas, and coal are deposited beneath Earth's surface, considerable technology and skill are required to locate these substances and estimate their volume. The amount of energy remaining in deposits that have been discovered is called a **proven reserve**. Proven reserves can be measured with reasonable accuracy—about 1.3 trillion barrels of petroleum, about 175 trillion cubic meters of natural gas, and about 1 quadrillion metric tons of coal.

To determine when remaining reserves of an energy source will be depleted, we must know the rate at which the resource is being consumed. At the current world petroleum consumption rate of about 31 billion barrels a year, Earth's proven petroleum reserves of 1.3 trillion barrels will last 43 years. Similarly,

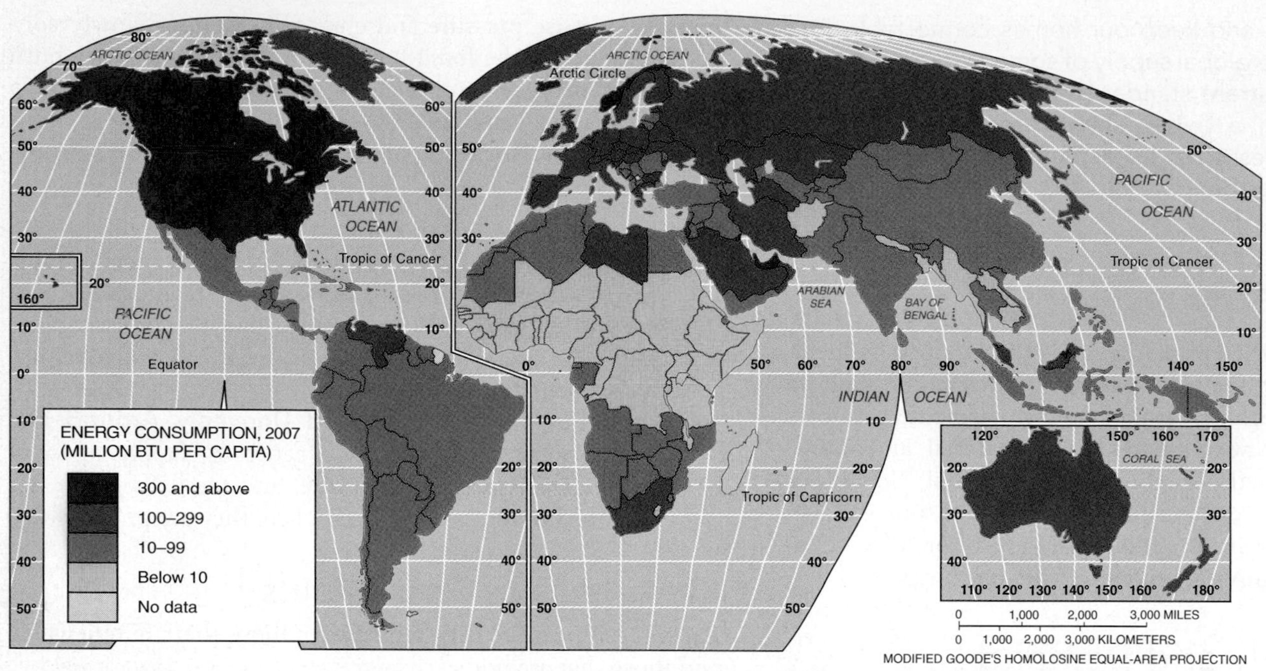

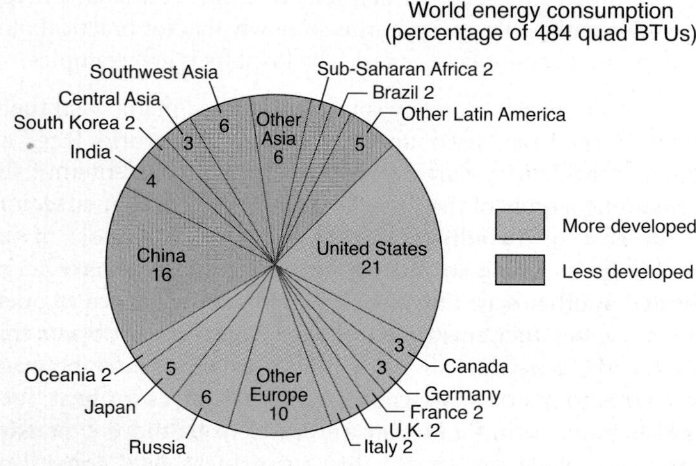

FIGURE 14-1 Per capita energy consumption. MDCs consume on average five times more energy per capita than do LDCs.

at current rates of use, the world's proven reserves of natural gas will last for about 49 years. For coal, the immediate future is less grim. At current consumption, proven coal reserves would last 131 years. One-half of U.S. electricity currently comes from power plants that burn coal.

POTENTIAL RESERVES. Some deposits in the world have not yet been discovered. The energy in deposits that are undiscovered but thought to exist is a **potential reserve**. When a potential reserve is actually discovered, it is reclassified as a proven reserve (Figure 14-3). Potential reserves can be converted to proven reserves in several ways:

- **Undiscovered Fields.** The largest, most accessible deposits of petroleum, natural gas, and coal have already been exploited. Newly discovered reserves are generally smaller and more remote, such as beneath the seafloor, where extraction is costly. Exploration costs have increased because methods are more elaborate and the probability of finding new reserves is less. But as energy prices climb, exploration costs may be justified.
- **Enhanced Recovery from Already Discovered Fields.** When it was first exploited, petroleum "gushed" from wells drilled into rock layers saturated with it. Coal was quarried in open pits. But now extraction is harder. Sometimes pumping is not sufficient to remove petroleum, so water or carbon dioxide may be forced into wells to push out the remaining resource. The problem of removing the last reserves from a proven field is comparable to wringing out a soaked towel. It is easy to quickly remove the main volume of water, but the last few percent require more time and patience and special technology.
- **Unconventional Sources.** They are called unconventional because methods currently used to extract resources won't work. Also we do not currently have economically feasible, environmentally sound technology with which to extract them. Examples include oil shale, which is a "rock that burns," and oil sands, which are saturated with a thick petroleum.

Abundant oil sands are found in Alberta, Canada, as well as in Venezuela and Russia. Oil shale is found in Utah, Wyoming, and Colorado in the United States. Native Americans used the tar to caulk canoes in the eighteenth century. The oil shale must be extracted through mining, which can be environmentally damaging, and current technology makes processing expensive. As with exploration, though, as energy prices rise, unconventional sources become economically feasible. Even then the adverse environmental impacts of using these sources may be high.

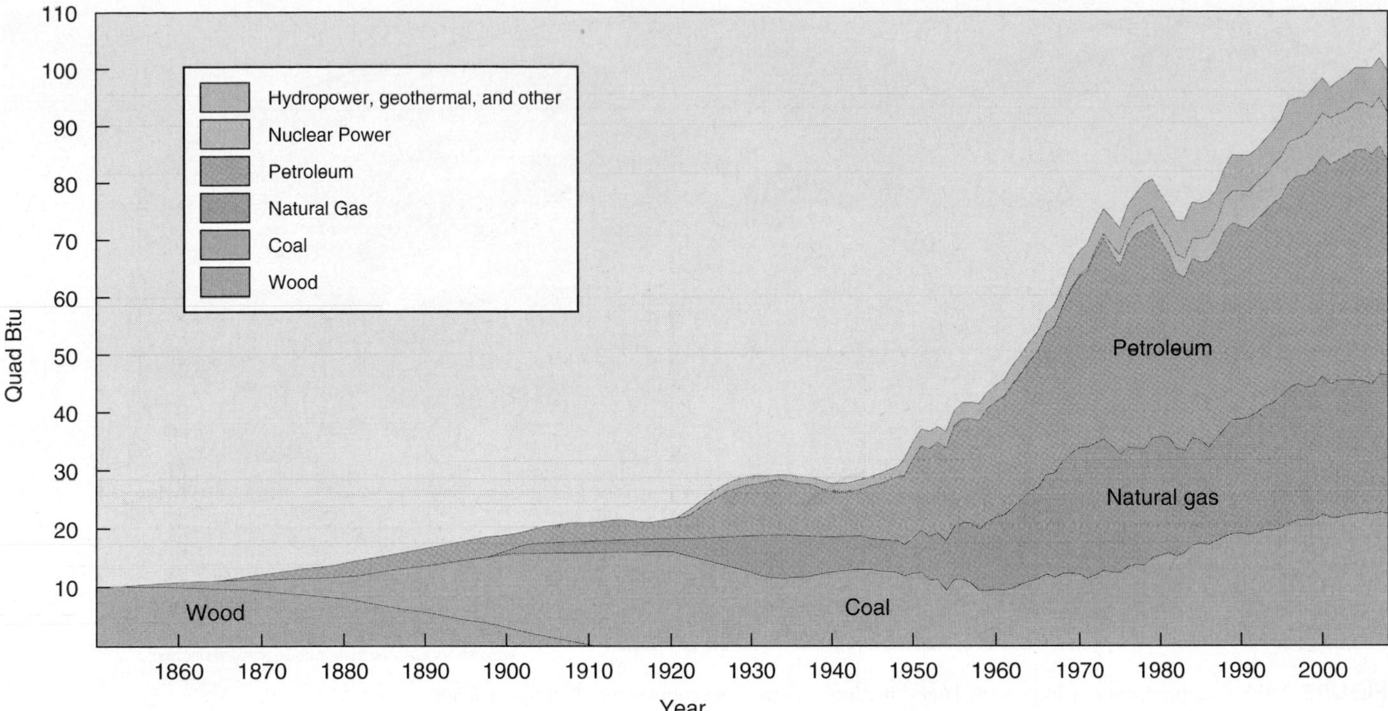

FIGURE 14-2 U.S. energy consumption. U.S. energy consumption increased rapidly during the 1960s, but since the early 1970s it has increased at a much slower rate. The amount of energy derived from petroleum and natural gas increased rapidly in the 1960s, when use of coal stagnated.

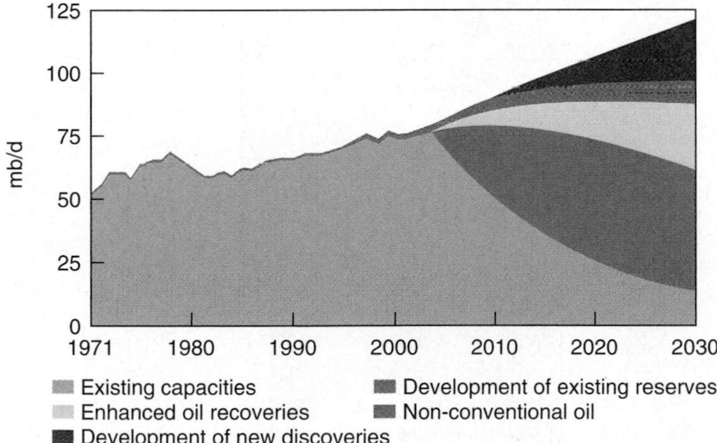

FIGURE 14-3 Outlook for world petroleum production. As production of proven reserves declines, new sources of petroleum will be developed. These include removal of more petroleum from existing fields, discovery of new fields, and exploitation of nonconventional sources. Source: World Energy Outlook 2004.

Uneven Distribution of Fossil Fuels

Geographers observe two important inequalities in the global distribution of fossil fuels: Some regions have abundant reserves, whereas others have little; the heaviest consumers of fossil fuel are in different regions than most of the reserves. Given the centrality of fossil fuels in contemporary economy and culture, unequal possession and consumption of fossil fuels have been major sources of global instability.

LOCATION OF RESERVES. Why do some regions have abundant reserves of one or more fossil fuels, but other regions have little? This partly reflects how fossil fuels form.

Coal forms in tropical locations, in lush, swampy areas rich in plants. Thanks to the slow movement of Earth's drifting continents, the tropical swamps of 250 million years ago have relocated to the midlatitudes. As a result, today's main reserves of coal are in midlatitude countries rather than in the tropics. China is responsible for extracting 39 percent of the world's coal, and the United States 16 percent (Figure 14-4). Three countries—the United States, Russia, and China—have nearly two-thirds of the world's proven coal reserves, and five other countries combined have most of the remainder (Figure 14-5).

Similarly, sources of petroleum and natural gas formed millions of years ago from sediment deposited on the seafloor. Some oil and natural gas reserves still lie beneath such seas as the Persian Gulf and the North Sea, but other reserves are located beneath land that had been under water millions of years ago. Southwest Asia produces 40 percent of the world's petroleum, Central Asia 15 percent, Russia 11 percent, and the United States 10 percent (Figure 14-6). Russia and the United States each account for 18 percent of current natural gas production (Figure 14-7).

Russia possesses more than one-fourth of the world's natural gas reserves. Southwest Asia and Central Asia together have more than one-half. Two regions—Southwest Asia and Central Asia—together account for nearly two-thirds of the world's proven petroleum reserves. Canada is now thought to have 13 percent of world petroleum reserves, second behind

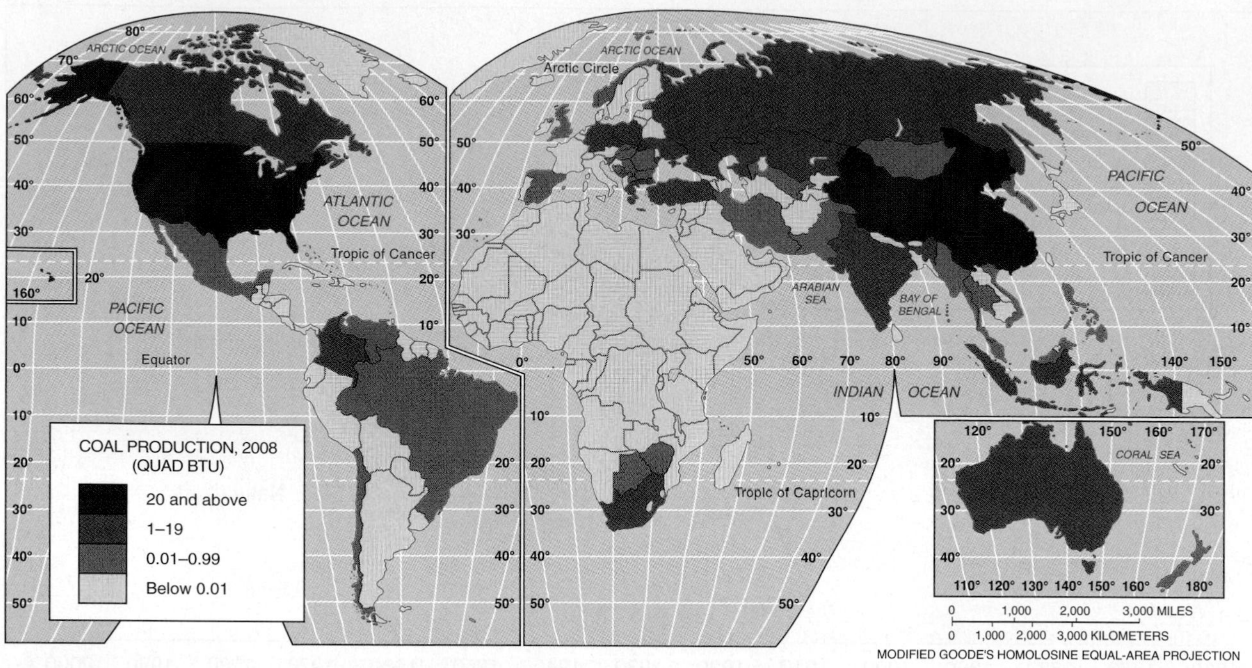

FIGURE 14-4 Coal production. China is the largest producer. To facilitate comparison, Figures 14-5 and 14-6 use the same scale as this map.

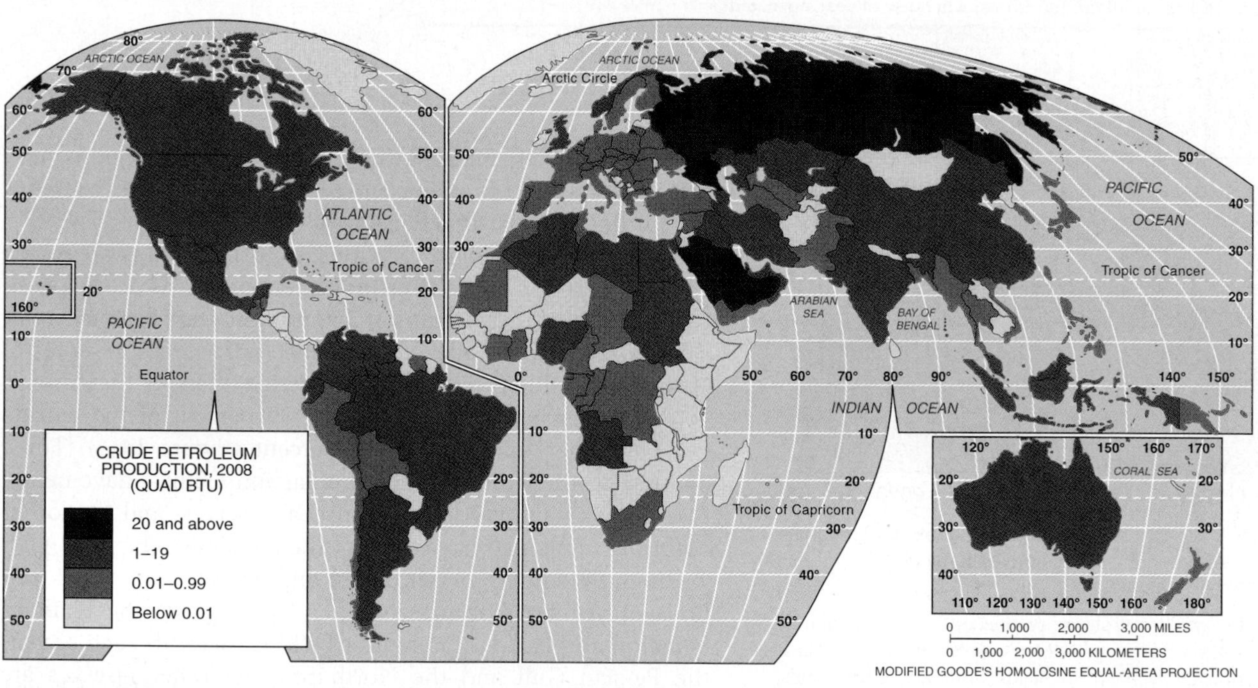

FIGURE 14-5 Petroleum production. Saudi Arabia and Russia are the largest producers.

Saudi Arabia. Despite challenges in extracting the petroleum, extensive deposits of oil in Alberta oil sands have been reclassified from potential to proven reserves in recent years because of rapidly escalating petroleum prices.

Taken as a group, MDCs have historically possessed a disproportionately high percentage of the world's proven fossil-fuel reserves. Europe's nineteenth-century industrial development depended on its abundant coalfields, and extensive coal and petroleum supplies helped the United States to become the leading industrial power of the twentieth century. But this dominance is ending in the twenty-first century. Many of Europe's coal mines have closed because either the coal was exhausted or the remaining supply was too expensive to extract, and the region's petroleum and natural gas (in the North Sea) account for only small percentages of worldwide reserves. Japan has never had significant fossil-fuel reserves. The United States still has extensive coal

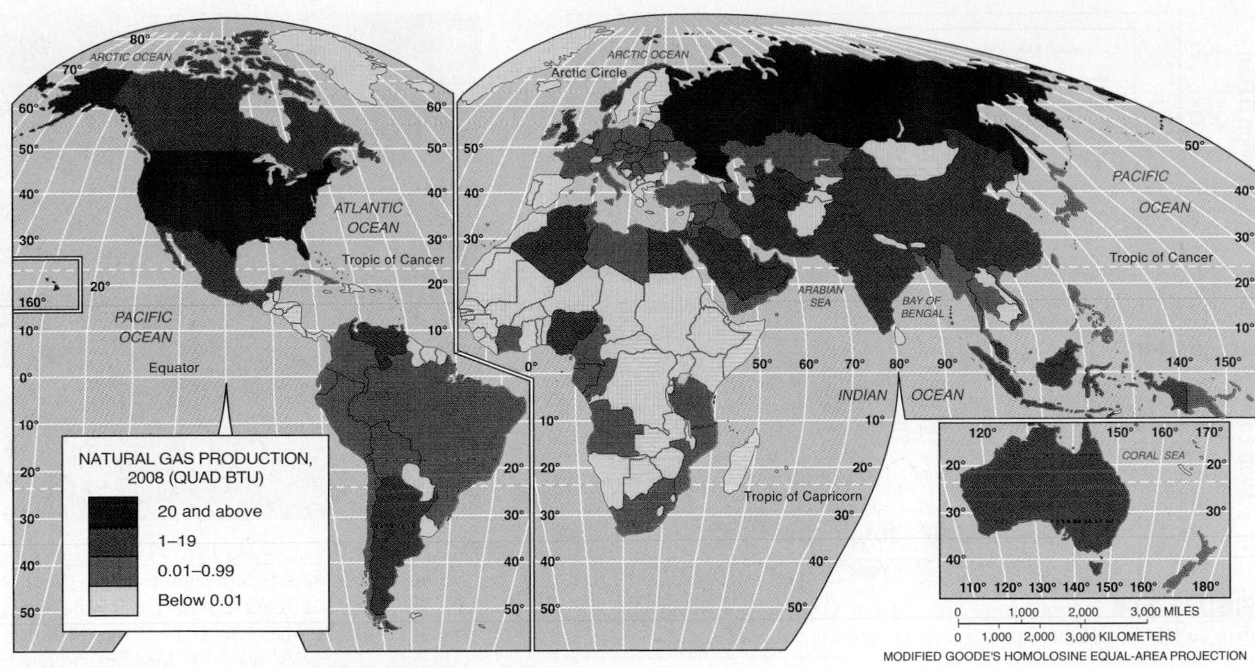

FIGURE 14-6 Natural gas production. The United States and Russia are the largest producers.

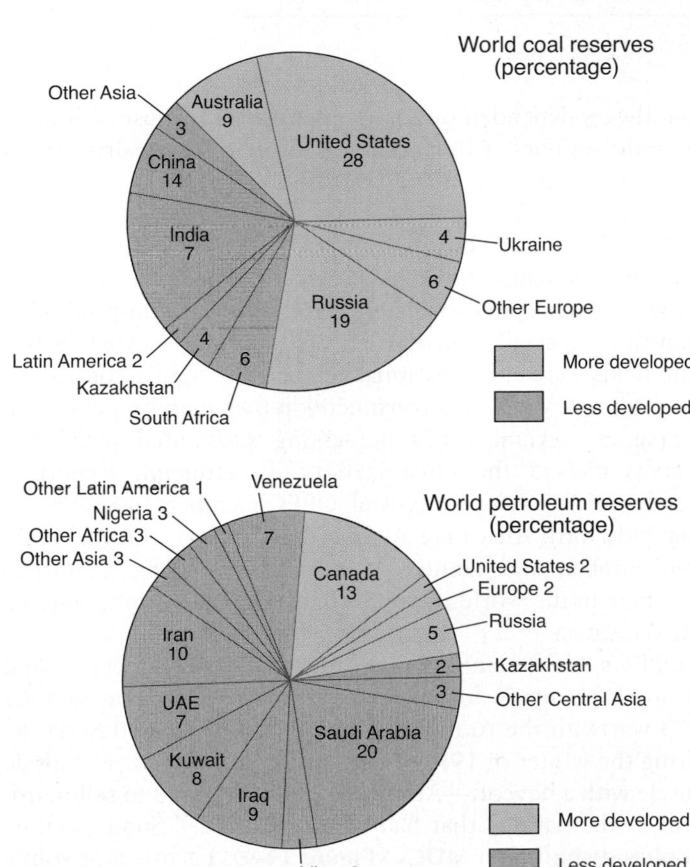

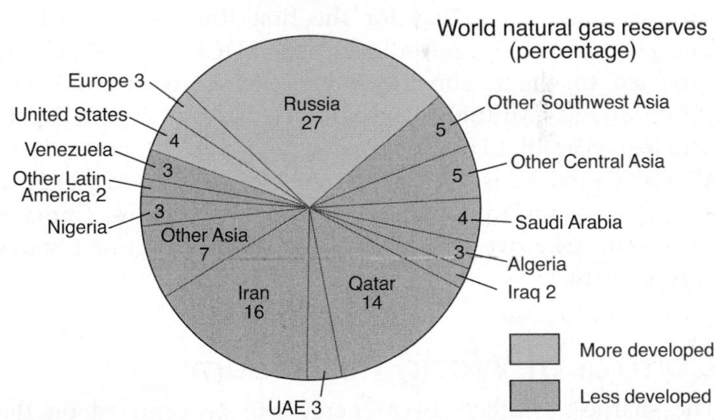

FIGURE 14-7 Fossil-fuel proven reserves. The largest proven reserves of coal are in the United States, Russia, and China; petroleum in Saudi Arabia, Canada, and Iran; and natural gas Russia, Iran, and Qatar.

reserves, but its petroleum and natural gas reserves are being depleted rapidly. It currently produces 18 percent of the world's natural gas but possesses only 4 percent of proven reserves, and it produces 10 percent of the world's petroleum but possesses only 2 percent of reserves.

CONSUMPTION OF FOSSIL FUELS. Because MDCs consume more energy than they produce, they must import fossil fuels, especially petroleum, from LDCs. The United States and Europe import more than half their petroleum, and Japan more than 90 percent.

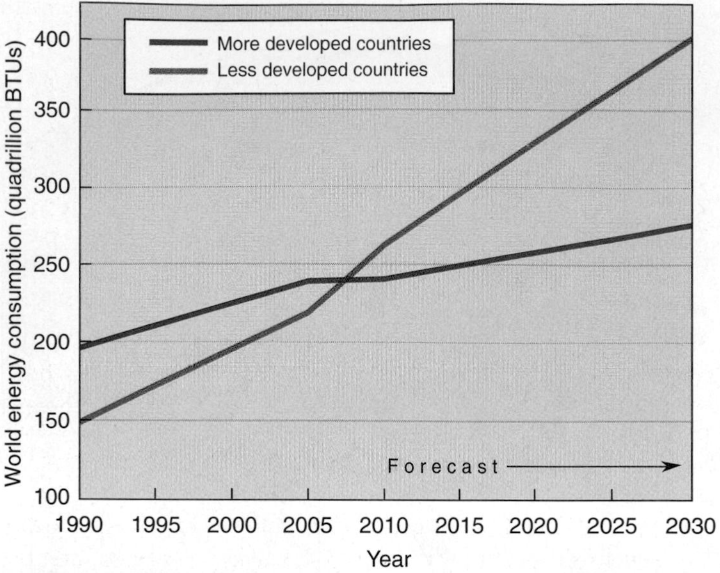

FIGURE 14-8 Future energy consumption. LDCs now consume more energy than do MDCs, and the gap is expected to grow in the years ahead.

FIGURE 14-9 Gas lines, 1973. The Organization of Petroleum Exporting Countries refused to sell petroleum to North American and European countries for a few months during the winter of 1973–74 to protest Western countries' support for Israel in the October 1973 war. Motorists in the United States, such as these in Los Angeles, waited in long lines to purchase gas. Note that the cars are in line to enter the Shell station in the background. The Exxon station in the foreground is closed because it already ran out of its day's allotment of fuel. Back in 1973, Americans regarded Exxon's posted price of 74¢ for a gallon of regular gas to be outrageously high.

Because of more rapid economic development in LDCs, the MDCs face greater competition in obtaining the world's remaining supplies of fossil fuels. During the first decade of the twenty-first century, fossil-fuel consumption in LDCs surpassed that of MDCs for the first time (Figure 14-8). The gap in consumption between LDCs and MDCs is expected to widen considerably in the years ahead. Consumption of fossil fuels has been increasing at around 3 percent per year in LDCs, compared to 1 percent per year in MDCs. China consumes 16 percent of the world's energy, currently second to the United States at 21 percent. China is forecast to take over the top ranking from the United States around 2015.

Control of World Petroleum

The sharpest conflicts over energy will be centered on the world's limited proven reserves of petroleum. The MDCs import most of their petroleum from Southwest and Central Asia, where most of the world's proven reserves are concentrated. Both U.S. and European transnational companies originally exploited Asian petroleum fields and sold the petroleum at a low price to consumers in MDCs.

The United States produced more petroleum than it consumed during the first half of the twentieth century. But beginning in the 1950s the handful of large transnational companies then in control of international petroleum distribution determined that extracting domestic petroleum was more expensive than importing it from Southwest and Central Asia. U.S. petroleum imports increased from 14 percent of total consumption in 1954 to 58 percent in 2009. European countries and Japan have always depended on foreign petroleum because of limited domestic supplies. China changed from a net exporter to an importer of petroleum during the 1990s.

OPEC. At first, Western companies set oil prices and paid Asian governments only a small percentage of their oil profits. But government policies changed in the petroleum-producing countries, especially during the 1970s. Foreign-owned petroleum fields were either nationalized or more tightly controlled, and prices were set by governments rather than by petroleum companies. Several LDCs possessing substantial petroleum reserves created the Organization of Petroleum Exporting Countries (OPEC) in 1960. Arab OPEC members in Southwest Asia and North Africa are Algeria, Iraq, Kuwait, Libya, Qatar, Saudi Arabia, and United Arab Emirates. OPEC countries elsewhere in the world include Angola, Ecuador, Iran, Nigeria, and Venezuela.

OPEC's Arab members were angry at North American and European countries for supporting Israel during that nation's 1973 war with the Arab states of Egypt, Jordan, and Syria. So during the winter of 1973–74, they flexed their new economic muscle with a boycott—Arab OPEC states refused to sell petroleum to the nations that had supported Israel. Soon gasoline supplies dwindled in MDCs (Figure 14-9). Each U.S. gasoline station received a small ration of fuel, which ran out early in the day. Gasoline was sold by license plate number (cars with licenses ending in an odd number could buy only on odd-numbered days). Long lines formed, and some motorists waited all night for fuel. European countries took more drastic action—the Netherlands, for example, banned all but emergency motor vehicle travel on Sundays.

OPEC lifted the boycott in 1974 but raised petroleum prices from $3 per barrel to more than $35 by 1981. Prices at U.S. gas pumps soared from an average of 39 cents in 1973 to $1.38 in 1981. The rapid escalation in petroleum prices during the 1970s caused severe economic problems in MDCs. Production of steel, motor vehicles, and other energy-dependent industries plummeted in the United States in the wake of the 1973–74 boycott and has never regained preboycott levels (recall Figure 11-24, which shows declining steel production in MDCs since the 1970s). Manufacturers were forced out of business by soaring energy costs, and the survivors were forced to restructure their operations to regain international competitiveness.

The LDCs were hurt even more. They depended on low-cost petroleum imports to spur economic development. Because many fertilizers are derived from petroleum, their fertilizer costs shot up. North American and European states cushioned themselves by creating a profitable return path for money that was going to OPEC: They encouraged OPEC countries to invest in American and European real estate, banks, and other safe and profitable investments. Comparable investment opportunities were limited in LDCs.

CHANGING SUPPLY AND DEMAND. The price of petroleum plummeted during the 1980s and settled during the 1990s at the lowest level in modern history, adjusting for inflation (Figure 14-10). The United States and other major consuming nations entered the twenty-first century optimistic that oil prices would remain low for some time.

The United States reduced its dependency on imported oil in the immediate wake of the 1970s shocks, and the share of imports from OPEC countries declined from two-thirds during the 1970s to one-third during the 1980s. Conservation measures also dampened demand for petroleum in most MDCs during the late twentieth century (Figure 14-11). The average vehicle driven in the United States, for example, got 14 miles per gallon in 1975, compared to 22 miles per gallon in 2000. However, with petroleum prices remaining low into the twenty-first century, consumption increased. Americans bought more gas-guzzling trucks and sport-utility vehicles and drove longer distances. Thus, once again, petroleum imports increased to record levels.

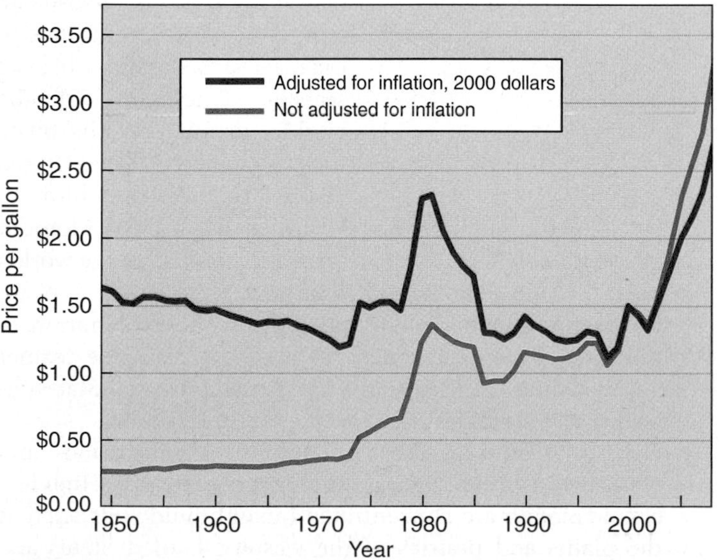

FIGURE 14-10 U.S. gasoline prices. The average price paid for a gallon of regular increased in the United States from around 25¢ in 1950 to around $3.25 in 2008. When adjusted for inflation, the price has been around $1.50 (in 2000 dollars) for most of the half-century, with the exception of the early 1980s and beginning in 2004.

As in the 1970s, Americans were unprepared for the shock of steep oil price rises in the twenty-first century when supplies were disrupted in the wake of terrorist attacks and several wars in the Middle East. Gas prices hit historic highs both in real terms and adjusted for inflation. Yet global demand continues to increase, especially demand from LDCs, led by China. Inadequate refinery capacity makes it difficult to expand supply.

The world will not literally "run out" of petroleum during the twenty-first century. However, at some point extracting the remaining petroleum reserves will prove so expensive and environmentally damaging that use of alternative energy sources will accelerate, and dependency on petroleum will diminish. The issues for the world are whether dwindling petroleum reserves are handled wisely and other energy sources are substituted peacefully. Given the massive growth in petroleum consumption expected in LDCs such as China and India, the United States and other MDCs may have little influence over when prices rise and supplies decline.

Mineral Resources

Earth has 92 natural elements, but about 99 percent of the crust is composed of 8 elements—oxygen, silicon, aluminum, iron, calcium, sodium, potassium, and magnesium. Oxygen alone accounts for nearly one-half of the crust and silicon more than one-fourth (Figure 14-12). The 8 most common elements combine with rare ones to form approximately 3,000 different minerals, all with their own properties of hardness, color, and density, as well as spatial distribution. Each mineral is potentially a resource, if people find a use for it.

Minerals are either metallic or nonmetallic. In weight, more than 90 percent of the minerals that humans use are nonmetallic, but metallic minerals are especially important for economic activities and so carry relatively high value. Because a mineral is valued primarily for its mechanical or chemical properties, the definition of which minerals constitute resources evolves as technology and economies change. When a new technological process or product is invented, demand can suddenly increase for a mineral that had little use in the past.

Mineral deposits are not uniformly distributed around the world. Most of the world's supply of particular minerals is found in a handful of countries (Figure 14-13). Countries such as Australia and China rank among leading producers of several minerals, whereas other countries have abundant supplies of only one mineral. Further, the leading producers at this time are not always the countries with the most extensive reserves, an indication that the relative fortunes of states may change in the future.

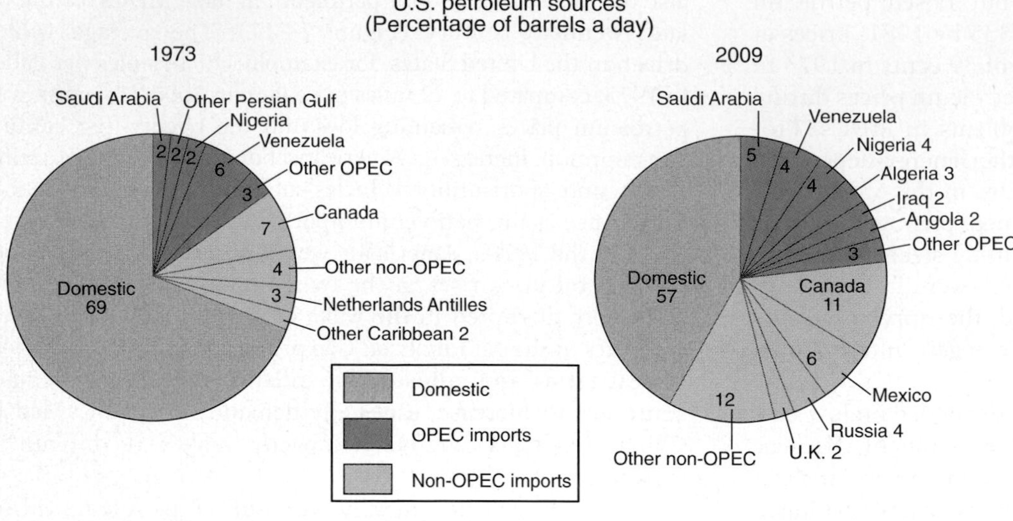

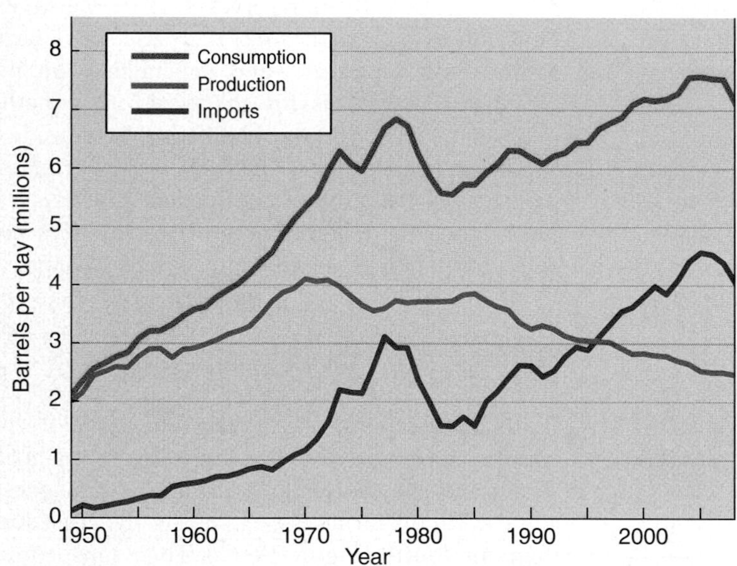

FIGURE 14-11 U.S. petroleum imports. The United States imported 58 percent of its crude oil in 2009, compared to only 31 percent in 1973. Since the 1980s, domestic production has declined 40 percent, whereas consumption has increased 30 percent. The gap has been covered by a 60 percent increase in imports.

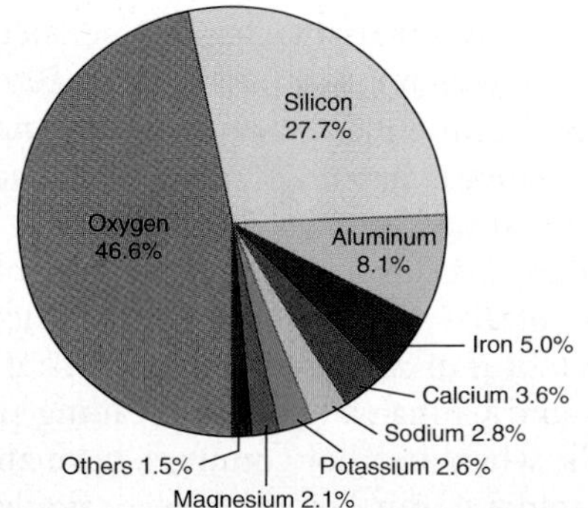

FIGURE 14-12 Elements in Earth's crust. Oxygen comprises nearly one-half of Earth's crust, silicon more than one-fourth, and six elements nearly all of the remainder.

Nonmetallic Minerals

Building stones, including large stones, coarse gravel, and fine sand, account for 90 percent of nonmetallic mineral extraction. These minerals are fashioned into structures, roads, monuments, tools, and many other objects of daily use. The rocks and earthen materials used for building purposes are so common that differences in distribution are of little consequence, at least at the international scale.

Nonmetallic minerals are also used for fertilizer. All crops must have at least some quantity of these minerals and obtain some of what they need from the soil. Because soils are often deficient in these minerals, farmers add them. Important nonmetallic mineral sources of fertilizers include phosphorus, potassium, calcium, and sulfur (Figure 14-14). All four are abundant elements in nature with wide distributions. However, mining is highly clustered where the minerals are most easily and cheaply extracted.

- **Phosphorus:** Essential to plant growth, but easily exhausted in cultivation. The chief source of phosphorus is phosphate rock (apatite), found among the marine sediments of old seabeds. One-fourth of the world's supply of phosphate rock is mined in the United States; another one-third in Morocco and China. Morocco possesses one-half of the world's reserves.
- **Potassium:** Obtained primarily from the evaporation of saltwater. Principal sources of potassium include former Soviet Union countries, Canada, and the United States, as well as the Dead Sea, shared by Israel and Jordan.
- **Calcium:** Essential for formation of strong bones and teeth, and especially important for growing corn. High levels of calcium are concentrated in subhumid soils such as the plains and prairies of the western United States and Canada, as well as Russia's steppes.
- **Sulfur:** Used to make insecticides and herbicides, as well as fertilizers. The United States and Canada are responsible for one-fourth of the world's sulfur production, with another one-fifth coming from China and Russia.

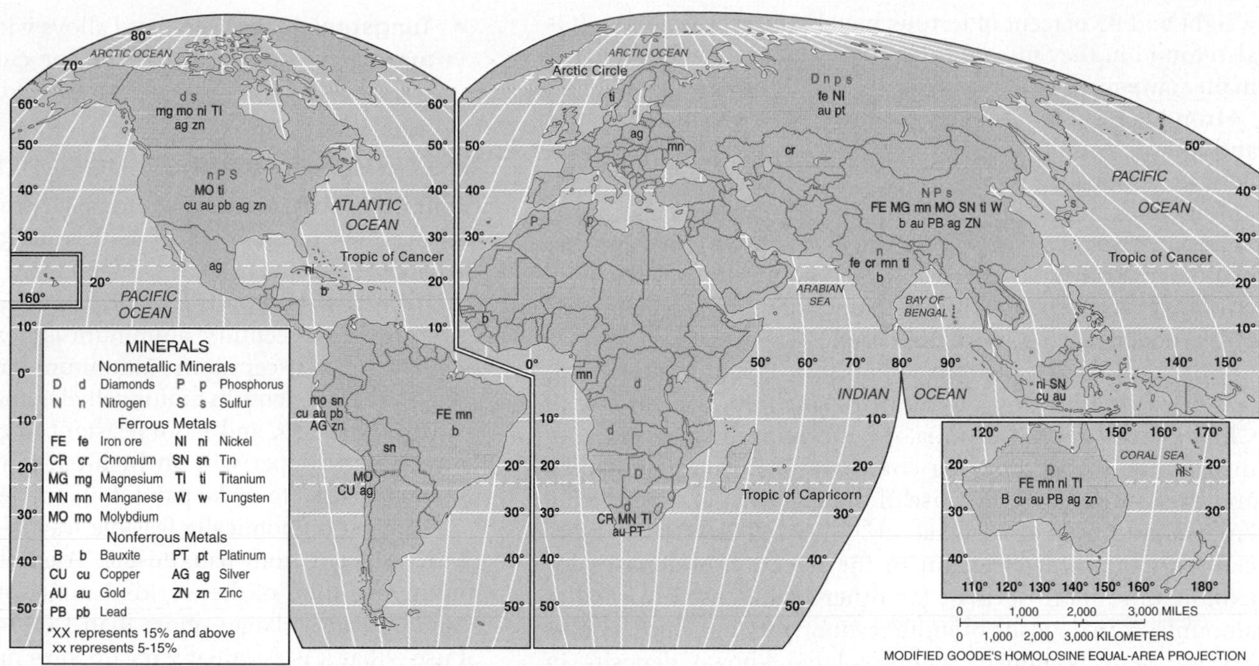

FIGURE 14-13 Production of important minerals. Australia and China are leading producers of several of the most important minerals.

FIGURE 14-14 Mining. Sulfur is being mined in Kawah Ijen, Indonesia. Miners carry chunks of sulfur by hand out of the bottom of an active volcano.

Nitrogen, obtained from the atmosphere, is an even more important fertilizer. Capturing it from the atmosphere utilizes a lot of energy, so its supply and demand are more closely associated with issues of energy resources rather than the other fertilizer elements.

Another group of nonmetallic minerals, gemstones, are valued highly for their color and brilliance when cut and polished. Diamonds are especially useful in manufacturing because they are the strongest and hardest known material and have the highest thermal conductivity of any material at room temperature. Two-thirds of the world's diamonds are currently mined in Australia, Botswana, and Russia.

Metallic Minerals: Ferrous

Metallic minerals have properties that are especially valuable for fashioning machinery, vehicles, and other essential components of an industrialized society. They are to varying degrees malleable (able to be hammered into thin plates) and ductile (able to be drawn into fine wire) and are conductors of heat and electricity. Many metals are also capable of combining with other metals to form alloys with yet other distinctive properties. A mineral bearing a metal such as aluminum or iron is known as an ore. Nearly all ore contains at least some metallic mineral, although the concentration is often too low to justify extracting it.

Metals are known as ferrous or nonferrous. **Ferrous** metals include iron ore and other alloys used in the production of iron and steel. The term "ferrous" refers to the Latin word for iron, and the symbol for iron in the periodic table of elements is Fe.

IRON ORE. By far the world's most widely used ferrous metal is iron, which accounts for 5 percent of Earth's crust by weight and 95 percent of ferrous metal mineral extraction. It is also found in the Sun and other stars and is thought to be the main component of Earth's core.

Iron is prized for its many assets: a good conductor of heat and electricity, able to be attracted by a magnet and to be magnetized, and malleable into useful shapes. Humans began fashioning tools and weapons from the silver-gray metal about 2000 B.C. The critical importance of iron to the past 4,000 years of human history is reflected by the application of the term "Iron Age" to the period. Iron remains an important element in every modern society, from least to most developed.

Mining of iron ore, from which iron is extracted, is concentrated in a handful of countries, including two-thirds in China, Brazil, and Australia. Major importers of iron ore include the steel-producing countries of Europe plus Japan and the United States. Because of the high cost of transporting large quantities of iron ore, accessibility to market is an especially important determinant in the selection of deposits for exploitation, more so than for other metals used in smaller amounts. Iron deposits of indifferent quality but close to market are actively mined, whereas large known deposits in remote areas are ignored for now, although they may become more important in the future once more accessible deposits are exhausted.

OTHER FERROUS METALS. Several less common ferrous metals are important for alloying with iron to produce steel:

- **Chromium:** A principal component of stainless steel, because it helps keep a sharp cutting edge even at high temperatures. Chromium is extracted from chromite ore, one-half of which is mined in South Africa.
- **Manganese:** An especially vital alloying metal for making steel because it imparts toughness and carries off undesirable sulfur and oxygen. Manganese ore is a relatively plentiful element in Earth's crust, so total world supply is not a problem. Responsible for one-half of world manganese production are Brazil, Gabon, and South Africa.
- **Molybdenum:** Imparts toughness and resilience to steel. Unlike the other rare metals discussed here, a leading role in providing this mineral is played by the United States, the leading producer, with one-third of world production.
- **Nickel:** Used primarily for stainless steel and high-temperature and electrical alloys. World reserves are around 100 years at current rates of use. Russia, Australia, and Canada are responsible for one-half of current production.
- **Tin:** Valued for its corrosion-resistant properties. Tin is used for plating iron and steel and has been used for more than 5,000 years as an alloy of copper for making bronze. China extracts two-fifths of the world's tin, Indonesia one-fourth, and Peru one-sixth. World reserves are estimated at only around 50 years.
- **Titanium:** A lightweight, high-strength, corrosion-resistant metal used as an alloy of steel, although its main use is as white pigment in paint. Titanium is extracted primarily from the mineral ilmenite. Sixty percent of world production is clustered in Australia, South Africa, and Canada.
- **Tungsten:** Makes very hard alloys with steel and is used to manufacture tungsten carbide for cutting tools. China is responsible for 90 percent of world production.

Metallic Minerals: Nonferrous

Nonferrous metals are utilized to make products other than iron and steel. The most abundant nonferrous metal is aluminum.

ALUMINUM (BAUXITE). Rarely used commercially prior to the twentieth century, aluminum is now in greater demand than any metal except iron. Aluminum has replaced some iron and steel components in motor vehicles and airplanes because it is lighter, stronger, and more resistant to corrosion. Aluminum has replaced copper wire in high-tension power transmission lines and is used to make paint, foil, and jewelry.

The most economically feasible way of obtaining aluminum is to extract it from bauxite ore. Australia is responsible for mining one-third of the world's bauxite ore. World reserves of aluminum are so large—more than 1,000 years at current rates of use—that it is essentially regarded as inexhaustible at realistic projections of future demand.

OTHER NONFERROUS METALS. Other especially important nonferrous include:

- **Copper:** Valued for its high ductility, malleability, thermal and electrical conductivity, and resistance to corrosion. Rated third in metal consumption behind iron and aluminum, copper is used primarily in electronics and constructing buildings. Chile is responsible for one-third of world production.
- **Lead:** Very corrosion-resistant, dense, ductile, and malleable. This blue-gray metal has been used for a variety of purposes for several thousand years, first in building materials and pipes, then in ammunition, brass, glass, and crystal, and now primarily in motor-vehicle batteries. Australia and China each supply one-fourth of the world's lead.
- **Magnesium:** Relatively light yet strong, so it is used to produce lightweight, corrosion-resistant alloys, especially with aluminum to make beverage cans. China supplies three-fourths of the world's magnesium.
- **Zinc:** Used primarily as a coating to protect iron and steel from corrosion and as an alloy to make bronze and brass. Again, China is the leading producer, with one-fourth of the world total. Australia and Peru together supply another one-fourth.

World supplies of some nonferrous metals are extremely limited—less than 60 years for copper, 25 years for lead, and 45 years for zinc. Supplies of magnesium are abundant because it can be removed from seawater brine.

PRECIOUS METALS. Nonferrous metals also include precious metals—silver, gold, and the platinum group. In addition to jewelry, both silver and gold are used in a variety of industrial applications, such as electrical and electronic products, and silver is a component of photographic film, whereas gold is important in dentistry. The principal use of the platinum group is in motor-vehicle catalytic converters to treat exhaust emissions, as well as in fuel cells.

- **Gold:** Prized since ancient times for its beauty and durability. In addition to jewelry, gold is used in a variety of industrial applications and in dentistry. One-third of the world's gold is mined in Australia, South Africa, and the United States.
- **Silver:** Like gold, also long prized for its beauty and durability, and used in photographic film as well as other industries. Associated with copper, lead, and zinc deposits, it is often mined at great depths. One-half of current production is in Peru, China, Mexico, and Australia.
- **Platinum group:** Includes six related and especially scarce metals that commonly occur together in nature—platinum, palladium, rhodium, ruthenium, iridium, and osmium. Platinum has the most highly clustered distribution of the major precious metals: South Africa is responsible for three-fourths of production.

World supply of most metals is high, including the most widely used ferrous metal (iron) and the most widely used nonferrous metal (aluminum). However, reserves of some metals are low, posing a challenge to manufacturers to find economically feasible substitutes.

KEY ISSUE 2
Why Are Resources Being Polluted?

- Air Pollution
- Water Pollution
- Land Pollution

In our consideration of resources, consumption is half of the equation—waste disposal is the other half. All of the resources we use are eventually returned to the atmosphere, bodies of water, or land surface, through burning, rinsing, or discarding. We rely on air, water, and land to remove and disperse our waste. **Pollution** occurs when more waste is added than a resource can accommodate.

In this section, we look at air, water, and land pollution. Each has distinctive features that illustrate the close connection between human activities and environmental quality. ■

Air Pollution

At ground level, Earth's average atmosphere is made up of about 78 percent nitrogen, 21 percent oxygen, and less than 1 percent argon. The remaining 0.04 percent includes several trace gases, some of which are critical. **Air pollution** is a concentration of trace substances at a greater level than occurs in average air.

The most common air pollutants are carbon monoxide, sulfur dioxide, nitrogen oxides, hydrocarbons, and solid particulates. Concentrations of these trace gases in the air can damage property and adversely affect the health of people, other animals, and plants. Three human activities generate most air pollution—motor vehicles, industry, and power plants. In all three cases, pollution results from the burning of fossil fuels. Burning gasoline or diesel oil in cars, trucks, buses, and motorcycles produces carbon monoxide, hydrocarbons, nitrogen oxides, and other pollutants. Factories and power plants produce sulfur dioxides and solid particulates, primarily from burning coal.

Global-Scale Air Pollution

Air pollution concerns geographers at three scales—global, regional, and local. At the global scale, air pollution may contribute to global warming. It also may be damaging the atmosphere's ozone layer.

GLOBAL WARMING. Human actions, especially the burning of fossil fuels, may be causing Earth's temperature to rise. The average temperature of Earth's surface has increased by 1° Celsius (2° Fahrenheit) during the past century (Figure 14-15).

Earth is warmed by sunlight that passes through the atmosphere, strikes the surface, and is converted to heat. When the heat tries to pass back through the atmosphere to space, some gets through and some is trapped. This process keeps Earth's temperatures moderate and allows life to flourish on the planet. A concentration of trace gases in the atmosphere can block or delay the return of some of the heat leaving the surface heading for space, thereby raising Earth's temperatures. When fossil fuels are burned, one of the trace gases, carbon dioxide, is discharged into the atmosphere.

Plants and oceans absorb much of the discharges, but increased fossil-fuel burning during the past 200 years has caused the level of carbon dioxide in the atmosphere to rise by more than one-fourth, according to the UN Intergovernmental Panel on Climate Change. Even if fossil-fuel burning is reduced immediately, the level will continue to increase because of lingering effects of past emissions. Carbon dioxide is also increasing in the atmosphere from the burning and rotting of trees cut in the rain forests. Contributing to the warming has been the buildup of carbon dioxide emissions at an annual rate of more than 1 percent, although scientists disagree on whether it caused most or only a small percentage of the warming. Unless carbon dioxide emissions are sharply curtailed in the near future, average temperatures at the surface of Earth will increase by several degrees over the next century.

The anticipated increase in Earth's temperature, caused by carbon dioxide trapping some of the radiation emitted by the surface, is called the **greenhouse effect**. The term is somewhat misleading, because a greenhouse does not work in the same way as do trace gases in the atmosphere. In a real greenhouse, the interior gets very warm when the windows remain closed on a sunny day. The Sun's light energy passes through the glass into the greenhouse and is converted to heat, while the heat trapped inside the building is unable to escape out through the glass. Although an imprecise analogy, "greenhouse effect" has been a widely adopted term to evoke the anticipated warming of Earth's surface when trace gases block some of the heat trying to escape into space.

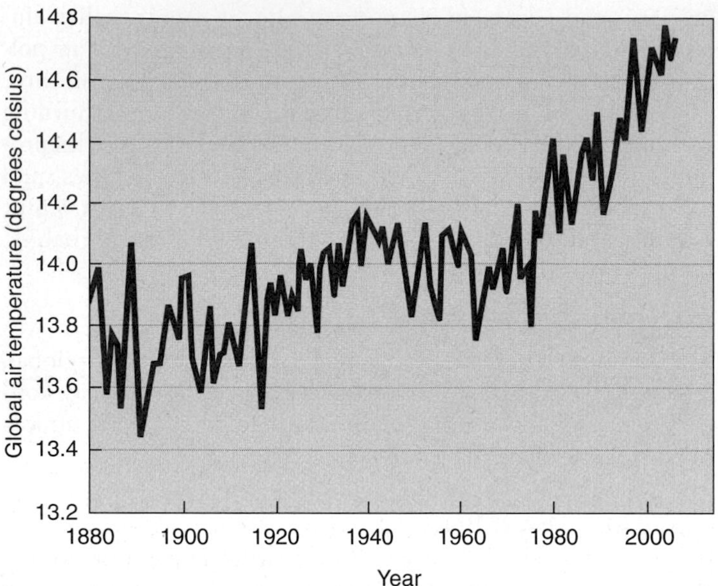

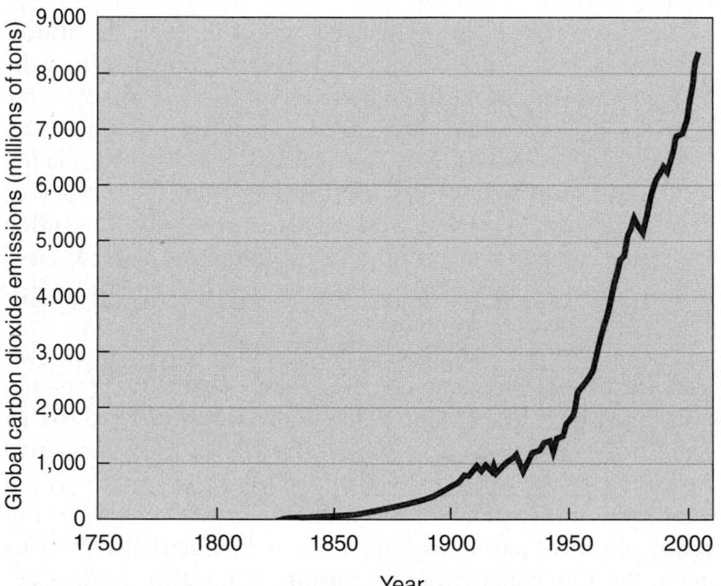

FIGURE 14-15 Global-scale air pollution: Global warming. (top) Earth's annual mean temperature has increased an average of 1°C per year between 1880 and 2007. (bottom) Contributing to global warming has been a rapid increase in carbon dioxide emissions, primarily from fossil-fuel burning.

Regardless of what it is called, global warming of only a few degrees could melt the polar ice caps and raise the level of the oceans many meters. Coastal cities such as New York, Los Angeles, Rio de Janeiro, and Hong Kong would flood (see Global Forces, Local Impacts box). Global patterns of precipitation could shift—some deserts could receive more rainfall, but currently productive agricultural regions, such as the U.S. Midwest, could become too dry for farming. Humans can adapt to a warmer planet, but the shifts in coastlines and precipitation patterns could require massive migration and be accompanied by political disputes.

GLOBAL-SCALE OZONE DAMAGE. Earth's atmosphere has zones with distinct characteristics. The stratosphere—the zone between 15 and 50 kilometers (9 to 30 miles) above Earth's surface—contains a concentration of **ozone** gas. The ozone layer absorbs dangerous ultraviolet (UV) rays from the Sun. Were it not for the ozone in the stratosphere, UV rays would damage plants, cause skin cancer, and disrupt food chains.

Earth's protective ozone layer is threatened by pollutants called **chlorofluorocarbons (CFCs)**. CFCs such as *freon* were once widely used as coolants in refrigerators and air conditioners. When they leak from these appliances, the CFCs are carried into the stratosphere, where they break down Earth's protective layer of ozone gas. In 2007, virtually all countries of the world agreed to cease using CFCs, by 2020 in MDCs and by 2030 in LDCs.

Regional-Scale Air Pollution

At the regional scale, air pollution may damage a region's vegetation and water supply through acid deposition. Industrialized, densely populated regions in Europe and eastern North America are especially affected by acid deposition.

Sulfur oxides and nitrogen oxides, emitted by burning fossil fuels, enter the atmosphere, where they combine with oxygen and water. Tiny droplets of sulfuric acid and nitric acid form and return to Earth's surface as **acid deposition** (Figure 14-18). When dissolved in water, the acids may fall as **acid precipitation**—rain, snow, or fog. The acids can also be deposited in dust. Before they reach the surface, these acidic droplets might be carried hundreds of kilometers.

Acid precipitation damages lakes, killing fish and plants. Aquatic life has been completely eliminated from 4 percent of the lakes in the eastern United States and Canada; another 5 percent of the lakes in the eastern United States and 20 percent in eastern Canada have acidity levels that threaten some species. On land, concentrations of acid in the soil can injure plants by depriving them of nutrients and can harm worms and insects. Acid precipitation has contributed to the decline of the red spruce tree at higher elevations. Buildings and monuments made of marble and limestone have suffered corrosion from acid rain; engravings on old marble tombstones, for example, may become illegible.

Geographers are particularly interested in the effects of acid precipitation because the worst damage is not experienced at the same location as the emission of the pollutants. Within the United States the major generators of acid deposition are in Ohio and other industrial states along the southern Great Lakes. However, the severest effects of acid rain are felt in several areas farther east.

The problem of acid precipitation is compounded by the fact that pollutants emitted in one country cause adverse impacts in another. Acid rain falling in Ontario, Canada, for example, can be traced to emissions from coal-burning power plants in the U.S. Great Lakes region. Government officials at the source of the pollution may be reluctant to impose strong controls on the offending factories because they fear damaging the local economy.

Europe has suffered especially severely from acid precipitation, a legacy of Communist policies that encouraged the construction of factories and power plants without pollution- control devices. Destruction of forests is especially widespread because of acid rain emitted in Silesia, Eastern Europe's major industrial region, in southern Poland and the northern Czech Republic. In the

GLOBAL FORCES, LOCAL IMPACTS
Climate Change in the South Pacific

One consequence of global warming is a rise in the level of the oceans. The large percentage of the world's population—including one-half of Americans—who live near the sea face increased threat of flooding. The threat is especially severe for island countries in the Pacific Ocean—they could be wiped off the map entirely.

One threatened Pacific island microstate is Tuvalu, the world's third-smallest country in land area and fourth-smallest in population (Figure 14-16). Others under threat include Kiribati, Micronesia, Nauru, Palau, Samoa, Solomon Islands, and Tonga. Tuvalu is one of the most isolated locations on Earth: in the middle of the South Pacific Ocean 4,000 kilometers (2,500 miles) from Australia. It consists of nine islands with a combined area of 26 square kilometers (10 square miles) spread across 600 kilometers (360 miles). Tuvalu's 12,000 inhabitants survive on fish, limited agriculture, and imported food. To raise money, it exports small quantities of copra (dried coconut meat); sells stamps, coins, and handcrafts; and leases fishing rights to U.S. and Japanese ships.

Despite its extreme isolation, global forces threaten Tuvalu's existence. Rising sea levels from global warming threaten Tuvalu because the highest elevation on its nine islands is only 4.5 meters (15 feet). Since 1978, the sea around Tuvalu has risen 1.2 millimeters (0.05 inches) per year. The capital, Funafuti, home to 5,000 of the country's 12,000 inhabitants, is at sea level.

Tuvalu and other Pacific island microstates are atolls, that is, made of coral reefs. A coral is a small sedentary marine animal having a horny or calcareous skeleton. Corals form colonies, and the skeletons build up to form coral reefs. Coral is very fragile. Humans are attracted to coral for its beauty and the diversity of species it supports, but handling coral can kill it. The threat of global warming to coral is especially severe: Coral stays alive in only a narrow range of ocean temperatures, between 23°C and 25°C (73°F and 77°F).

Tuvalu has an emergency response plan to rising sea levels: The 12,000 inhabitants will be evacuated to Australia and New Zealand. A small, isolated country, lacking in most resources, will disappear. ■

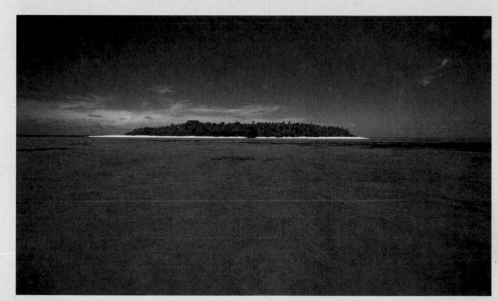

FIGURE 14-16 Tuvalu, a 26-square-kilometer (10-square-mile) Pacific island country of 12,000 inhabitants, fears it will disappear under rising sea levels caused by global warming.

Czech Republic, more than 80 percent of the Bohemian Forest has been affected by acid precipitation (Figure 14-17).

The destruction of trees has harmed Eastern Europe's seasonal water flow. In dense forests, snow used to melt slowly and trickle into rivers. Now, on the barren sites, it melts and drains quickly, causing flooding in the spring and water shortages in the summer. Perhaps the most severe impact is on human life. A 40-year-old man living in Poland's polluted southern industrial area has a life expectancy 10 years less than his father had at the same age. Poland is estimated to have between 20,000 and 50,000 additional deaths per year due to pollution.

The United States has reduced sulfur dioxide emissions significantly since the 1970s. Many European countries have also made substantial cuts, largely by reducing coal use. Despite this progress, acid precipitation continues to damage forests and lakes. Governments are reluctant to impose the high cost of controls on their industries and consumers.

Local-Scale Air Pollution

At the local scale, air pollution is especially severe in places where emission sources are concentrated, such as in urban areas. The air above urban areas may be polluted because a large number of factories, motor vehicles, and other polluters emit residuals in a concentrated area.

Urban air pollution has three basic components:

- **Carbon Monoxide:** Produced from incomplete burning in power plants and vehicles; proper combustion produces carbon dioxide. Breathing carbon monoxide reduces the

FIGURE 14-17 Regional-scale air pollution: Acid precipitation. A large percentage of the trees in the forests of the Czech Republic have died, including these spruce in Krkonose National Park. Emissions of sulfur dioxide and nitrogen oxides from factories and power plants built without pollution-control devices in the former Communist East Germany and Czechoslovakia caused this widespread death of trees.

oxygen level in blood, impairs vision and alertness, and threatens those with breathing problems.

- **Hydrocarbons:** Also result from incomplete fuel combustion, as well as evaporation of paint solvents. Hydrocarbons and nitrogen oxides in the presence of sunlight form **photochemical smog**, which causes respiratory problems, stinging in the eyes, and an ugly haze over cities.
- **Particulates:** Include dust and smoke particles. The dark plume of smoke from a factory stack and the exhaust of a diesel truck are examples of particulates being emitted.

The severity of air pollution resulting from emissions of carbon monoxide, hydrocarbons, and particulates depends on the weather. The worst urban air pollution occurs when winds are slight, skies are clear, and a temperature inversion exists. When the wind blows, it disperses pollutants; when it is calm, pollutants

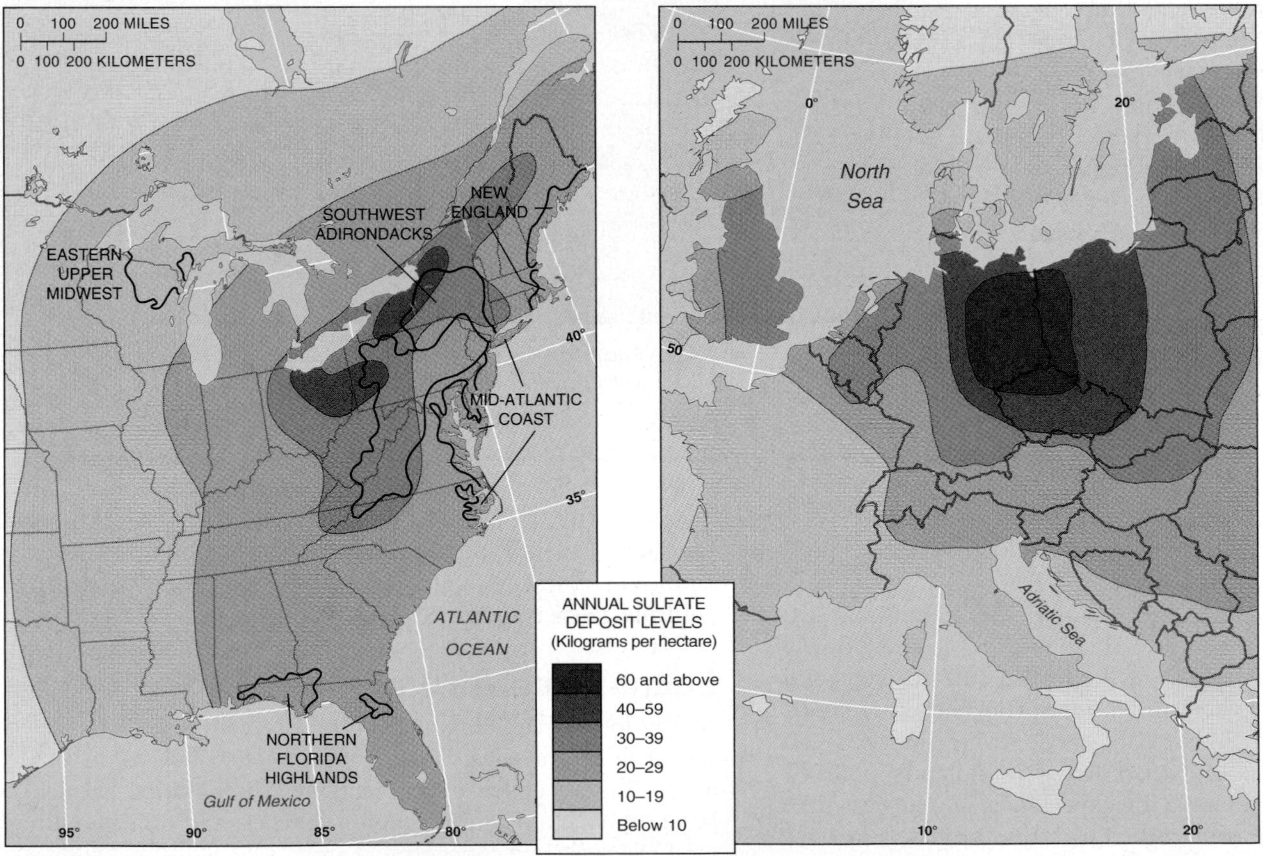

FIGURE 14-18 Regional-scale air pollution: Acid deposition in North America and Europe. Levels exceeding 20 kg/ha are considered threatening. (Left) Because of prevailing wind patterns across North America, damage is generally found to the east of the emissions. (Right) Deposition levels in eastern Germany are higher than anywhere in the United States, although elsewhere in central Europe levels are comparable to those in the eastern United States.

FIGURE 14-19 Local-scale air pollution: Smog in Mexico City. Mexico City suffers from a combination of circumstances that lead to significant air pollution problems. The city lies in a mountain basin that limits dispersion of pollutants, and motor-vehicle traffic is heavy. (left) Smog blankets Mexico City January 9, 1996, in this view looking south from Paseo de la Reforma. (right) Same view one week earlier on a day with clean air.

build. Sunlight provides the energy for the formation of smog. Air is normally cooler at higher elevations, but during temperature inversions—in which air is warmer at higher elevations—pollutants are trapped near the ground.

According to the American Lung Association, the two worst U.S. metropolitan areas for concentrations of particulates are Los Angeles and Pittsburgh. Mexico City may have the world's most serious air pollution problem, as discussed in the opening case study (Figure 14-19).

Progress in controlling urban air pollution is mixed. In MDCs, air has improved where strict clean-air regulations are enforced. Limited emission controls in LDCs are contributing to severe urban air pollution. Changes in automobile engines, manufacturing processes, and electric generation all have helped. For example, since the 1970s, when the U.S. government required catalytic converters on motor vehicles, carbon monoxide emissions have been reduced by more than three-fourths, and nitrogen oxide and hydrocarbon emissions have been reduced by more than 95 percent. But more people are driving, offsetting gains made by emission controls.

Water Pollution

Water serves many human purposes:

- It must be drunk to survive.
- It is used for cooking.
- It is used for bathing.
- It provides a location for boating, swimming, fishing, and other recreation activities.
- It is home to fish and other edible aquatic life.

When all of these uses are totaled, the average American consumes 5,300 liters (1,400 gallons) of water per day, including 680 liters (180 gallons) for drinking, cooking, and bathing. These uses depend on fresh, clean, unpolluted water.

But clean water is not always available, because people also use water for purposes that pollute it. Pollution is widespread because it is easy to dump waste into a river and let the water carry it downstream where it becomes someone else's problem. By polluting water, humans harm the health of aquatic life and the health of land-based life (including humans themselves).

Not all human actions harm the environment, for every resource, including water, can accept some waste. Water can decompose some waste without adversely impacting other activities; for example, when we send household cleaners and chemicals into a river, the river may dilute them until their concentration is insignificant. But the volume of discharge exceeds the capacity of many rivers and lakes to accommodate it.

Water Pollution Sources

Three main sources generate most water pollution:

- **Water-Using Industries.** Steel, chemicals, paper products, and food processing are major industrial polluters of water. Each requires a large amount of water in the manufacturing process and generates a lot of wastewater. Food processors, for example, wash pesticides and chemicals from fruit and vegetables. They also use water to remove skins, stems, and other parts. Water can also be polluted by industrial accidents, such as petroleum spills from ocean tankers and leaks from underground tanks at gasoline stations.
- **Municipal Sewage.** In MDCs, sewers carry wastewater from sinks, bathtubs, and toilets to a municipal treatment plant, where most—but not all—of the pollutants are removed. The treated wastewater is then typically dumped back into a river or lake. In LDCs, sewer systems are rare, and wastewater usually drains untreated into rivers and lakes (Figure 14-20).
- **Agriculture.** Fertilizers and pesticides spread on fields to increase agricultural productivity are carried into rivers and lakes by the irrigation system or natural runoff (see Contemporary Geographic Tools box). Expanded use of these products may help to avoid a global food crisis, yet they destroy aquatic life by polluting rivers and lakes.

FIGURE 14-20 Water quality in LDCs. Workers in Kolkata (Calcutta), India, construct a new sewer.

These sources of pollution can be divided into point sources and nonpoint sources. Point-source pollution enters a stream at a specific location, whereas nonpoint-source pollution comes from a large, diffuse area:

- Manufacturers and municipal sewage systems tend to pollute through point sources, such as a pipe from a wastewater treatment plant.
- Farmers tend to pollute through nonpoint sources, such as by permitting fertilizer to wash from a field during a storm.

Point-source pollutants are usually smaller in quantity and much easier to control. Nonpoint sources usually pollute in greater quantities and are much harder to control.

Impact on Aquatic Life

Polluted water can harm aquatic life. Aquatic plants and animals consume oxygen, but so does the decomposing organic waste that humans dump in the water. The oxygen consumed by the decomposing organic waste constitutes the **biochemical oxygen demand (BOD)**. If too much waste is discharged into the water, the water becomes oxygen starved and fish die.

This condition is typical when water becomes loaded with municipal sewage or industrial waste. The sewage and industrial pollutants consume so much oxygen that the water can become unlivable for normal plants and animals, creating a "dead" stream or lake. Similarly, when runoff carries fertilizer from farm fields into streams or lakes, the fertilizer nourishes excessive aquatic plant production—a "pond scum" of algae—that consumes too much oxygen. Either type of pollution reduces the normal oxygen level, threatening aquatic plants and animals. Some of the residuals may become concentrated in the fish, making them unsafe for human consumption. For example, salmon from the Great Lakes became unfit to eat because of high concentrations of the pesticide DDT, which washed into streams from farm fields.

Many factories and power plants use water for cooling and then discharge the warm water back into the river or lake. The warm water may not be polluted with chemicals, but it raises the temperature of the body of water it enters. Fish adapted to cold water, such as salmon and trout, might not be able to survive in the warmer water.

Wastewater and Disease

Since passage of the U.S. Clean Water Act and equivalent laws in other MDCs, most treatment plants meet high water-quality standards. Improved treatment procedures have resulted in cleaner rivers and lakes in MDCs. One dramatic example is the River Thames, which passes through London.

The Thames was once a major food source for Londoners, but during the Industrial Revolution it became the principal location for dumping waste. The fish died and the water grew unsafe to drink. The river became so dark, murky, and smelly that novelist Charles Dickens called the Thames "London's Styx," after the underworld river that the dead had to cross in Greek mythology. The British government began a massive cleanup during the 1960s to restore the Thames to health. Regulations prohibited industrial dumping, and sewage systems were modernized to improve treatment. A salmon was caught in the Thames just upstream from London in 1982, the first since 1833. Salmon are particularly sensitive to pollution, and for nearly 150 years the Thames was too polluted for salmon to survive.

Although LDCs generate less wastewater per person than do MDCs, they have less capacity to treat their wastewater. In LDCs, sewage often flows untreated directly into rivers. The drinking water, usually removed from the same rivers, may be inadequately treated as well. In squatter settlements on the edge of rapidly growing LDC cities, running water and sewers may be totally lacking. The combination of untreated water and poor sanitation makes drinking water deadly in LDCs. Waterborne diseases such as cholera, typhoid, and dysentery are major causes of death.

Some LDCs regard water pollution as a small price to pay for participating in a global economy. Industrialization may take a higher priority than clean water. In the past, MDCs caused most of the water pollution. Now they possess the wealth and technology to clean up polluted rivers and lakes.

Land Pollution

When we consume a product, we also consume an unwanted by-product—a glass, metal, paper, or plastic box, wrapper, or container in which the product is packaged. About 2 kilograms

CONTEMPORARY GEOGRAPHIC TOOLS
Monitoring the Disappearing Aral Sea

One of the world's most extreme instances of water pollution is the Aral Sea in the former Soviet Union, now divided between the countries of Kazakhstan and Uzbekistan. The world's fourth-largest lake in 1960, the Aral has been shrinking rapidly in area and volume and could disappear altogether by 2020.

The destruction of the Aral Sea over several decades was little known, and it was denied by the Soviet Union. Satellite imagery has enabled geographers and other scientists to document without question the extent of the destruction of the Aral and to monitor precisely the speed of destruction. A 1975 aerial photograph shows the Aral Sea in the early stages of destruction (Figure 14-21, upper left). It shows small islands barely visible in the center of the sea. Systematic monitoring of the Aral began in 1984 with images obtained by the National Oceanic and Atmospheric Administration (NOAA). Figure 14-21 (upper right), taken in 1989, shows a large island forming in the middle of the sea. The 1989 image was taken by a Landsat satellite, one of a series launched beginning in 1972 to support the Landsat Program jointly managed by National Aeronautic and Space Administration (NASA) and the U.S. Geological Survey (USGS).

NASA launched two satellites named *Terra* and *Aqua* as part of its Earth Observing System, which was begun in 1990 to collect even more precise images and data from Earth's surface, including the Aral Sea. Figure 14-21 (lower left) was captured by the Moderate Resolution Imaging Spectroradiometer (MODIS) on the *Aqua* satellite on August 12, 2003. It shows the Aral Sea now divided into two portions,western and eastern. Figure 14-21 (lower right) was also captured by MODIS on the *Terra* satellite on October 5, 2008. The western portion has not changed much, but the eastern portion has dried up. A small northern lake also remains.

Overall, the Aral declined from about 68,000 square kilometers in 1960 to 46,000 square kilometers in 1985 (when better-quality monitoring began), 29,000 square kilometers in 1998, and 7,000 square kilometers in 2007.

The Aral Sea died because beginning in 1954, the Soviet Union diverted its tributary rivers, the Amu Dar'ya and the Syr Dar'ya, to irrigate cotton fields. Ironically, the cotton now is withering because winds pick up salt from the exposed lakebed and deposit it on the cotton fields. Carp, sturgeon, and other fish species have disappeared, the last fish dying in 1983. Large ships lie aground in salt flats that were once the lakebed, outside of abandoned fishing villages that now lay tens of kilometers from the rapidly receding shore. ■

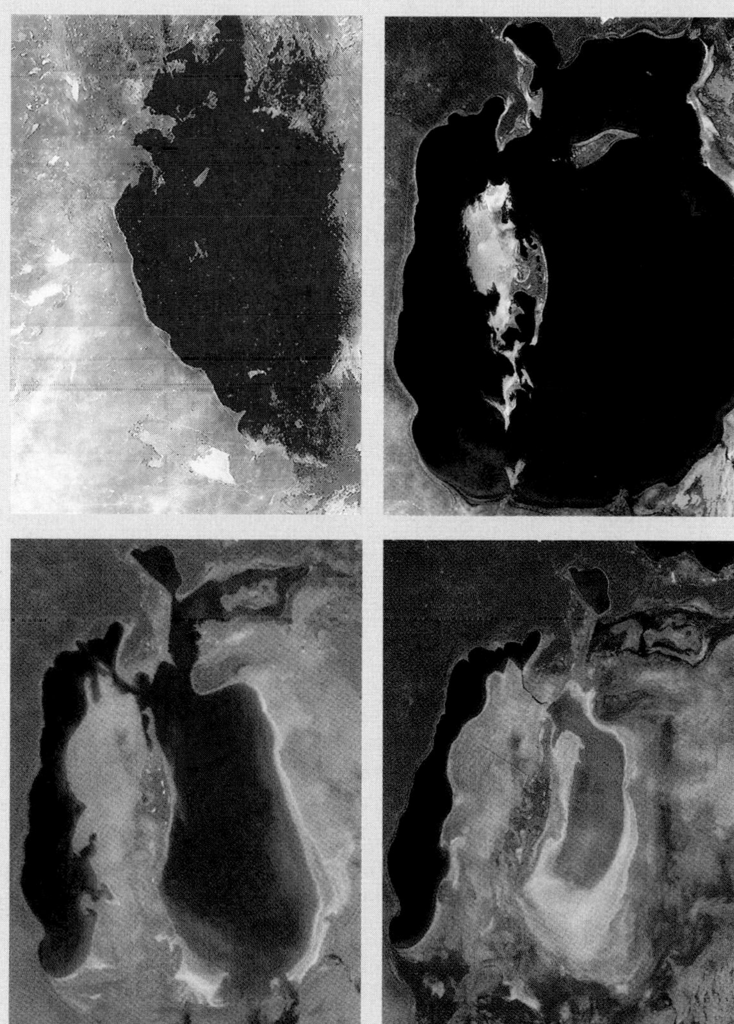

FIGURE 14-21 Disappearing Aral Sea. (upper left) 1975, (upper right) 1989, (lower left) 2003, (lower right) 2008.

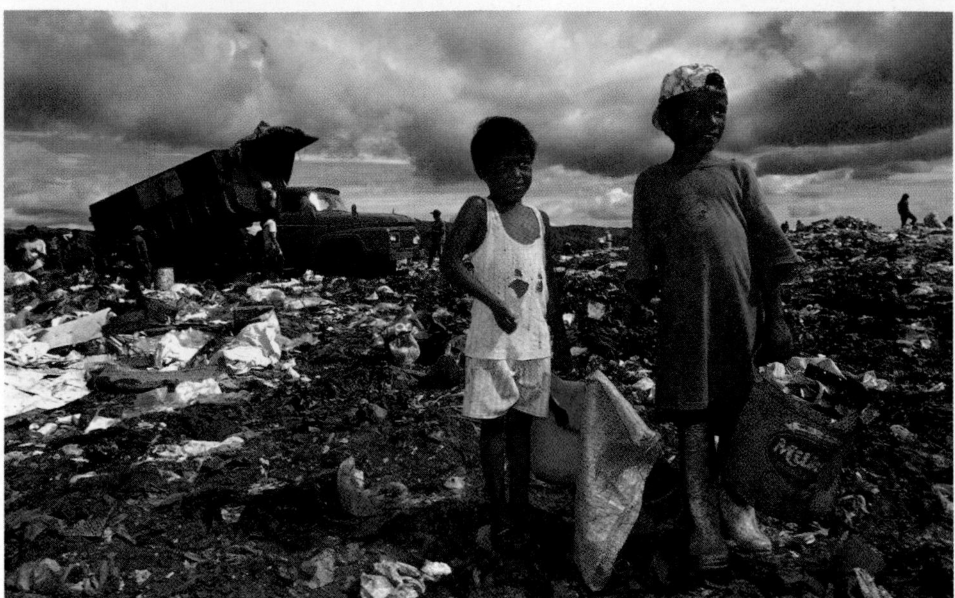

FIGURE 14-22 Solid waste. In Manila, Philippines, a child collects garbage that will be segregated and sold to a recycling center.

(4 pounds) of solid waste per person is generated daily in the United States, about 60 percent from residences and 40 percent from businesses.

Paper products, such as corrugated cardboard and newspapers, account for the largest percentage of solid waste in the United States, especially among residences and retailers (Figure 14-22). Food products, plastics, and rubbish cleanup from yards, such as grass clippings and leaves, are other important sources of solid waste. Manufacturers discard large quantities of metals as well as paper.

Some consumers demonstrate obvious unconcern for the environment by discarding waste along roadsides and sidewalks, where it causes visual pollution. But even consumers who carefully dispose of solid waste are contributing to a major pollution problem. A particularly severe threat is posed by the careless discharge of toxic waste.

Solid-Waste Disposal

The **sanitary landfill** is by far the most common strategy for disposal of solid waste in the United States: More than one-half of the country's waste is trucked to landfills and buried under soil. This strategy is the opposite of our disposal of gaseous and liquid wastes: We *disperse* air and water pollutants into the atmosphere, rivers, and eventually the ocean, but we *concentrate* solid waste in thousands of landfills. Concentration would seem to eliminate solid-waste pollution, but it may only hide it—temporarily. Chemicals released by the decomposing solid waste can leak from the landfill into groundwater. This can contaminate water wells, soil, and nearby streams.

The number of landfills in the United States has declined by three-fourths since 1990. Thousands of small-town "dumps" have been closed and replaced by a small number of large regional ones. Some communities now pay to use landfills elsewhere. New Jersey and New York are two states that regularly try to dispose of their solid waste by transporting it out of state. New York City exports 25,000 tons of trash a day to other communities. Passaic County, New Jersey, hauls waste 400 kilometers (250 miles) west to Johnstown, Pennsylvania. San Francisco trucks solid waste to Altamont, California, 100 kilometers (60 miles) away.

Better compaction methods, combined with expansion in the land area of some of the large regional dumps, have resulted in expanded landfill capacity. At the same time, the two principal alternatives to disposing solid waste in landfills—incineration and recycling—have both increased rapidly. Burning the trash reduces its bulk by about three-fourths and the remaining ash demands far less landfill space. Incineration also provides energy—the incinerator's heat can boil water to produce steam heat or to operate a turbine that generates electricity. Given the shortage of space in landfills, the percentage of solid waste that is burned has increased rapidly during the past three decades, to one-sixth of solid waste. However, solid waste, a mixture of many materials, may burn inefficiently. Burning releases some toxins into the air, and some remain in the ash. Thus solving one pollution problem may increase another.

Hazardous Waste

Disposing of hazardous waste is especially difficult. Hazardous wastes include heavy metals (including mercury, cadmium, and zinc), PCB oils from electrical equipment, cyanides, strong solvents, acids, and caustics. These may be unwanted by-products generated in manufacturing or waste to be discarded after usage.

If poisonous industrial residuals are not carefully placed in protective containers, the chemicals may leach into the soil and contaminate groundwater or escape into the atmosphere. Breathing air or consuming water contaminated with toxic wastes can cause cancer, mutations, chronic ailments, and even immediate death.

Companies in the United States that release chemicals classified as toxic by the Environmental Protection Agency (EPA) must report the amounts released. About 47 million tons of hazardous wastes were discharged in the United States in 2007, including 34 percent in Louisiana and 28 percent in Texas. Dow Chemical in Plaquemine, Louisiana, was responsible for 8 million tons out of the national total of 47 million tons in 2007. Nine other sites were responsible for at least 1 million tons of hazardous waste in 2007: Solutia in Alvin, Texas; Occidental Chemical in Hahnville, Lousiana; Rubicon in Geismer, Louisiana; Dow in Midland, Michigan; Shamrock Pipe Line in Sunray, Texas; DuPont in Pass Christian, Mississippi; Cytec Industries in Waggaman, Louisiana; BP Products in Texas City, Texas; and Ineos in Port Lavaca, Texas.

As toxic-waste disposal sites become increasingly hard to find, some European and North American firms have tried to transport their waste to West Africa, often unscrupulously. Some firms have signed contracts with West African countries; others have found isolated locations to dump waste without official consent.

KEY ISSUE 3

Why Are Resources Being Reused?

- Renewing Resources
- Recycling Resources

Depletion and destruction of resources can be reduced through reuse. Renewable resources can be substituted for nonrenewable ones. Recycling unwanted by-products into resources can replace the discharging of these products into the environment. ■

Renewing Resources

About 15 percent of energy consumed in the United States is generated by sources other than the three main fossil fuels. Nuclear power accounts for 8 percent of U.S. energy and renewable sources 7 percent. Energy poses an especially strong challenge in substituting renewable resources for nonrenewable ones. Although renewable resources can be harnessed for energy, continued reliance on the three main nonrenewable fossil fuels—petroleum, natural gas, and coal—continues to be the cheaper alternative.

Nuclear Energy

Nuclear power is not renewable, but it is viewed by some as an alternative to fossil fuels. The big advantage of nuclear power is the large amount of energy released from a small amount of material. One kilogram of enriched nuclear fuel contains more than 2 million times the energy in 1 kilogram of coal.

Nuclear power supplies about one-sixth of the world's electricity. Europe and North America are each responsible for generating one-third of the world's nuclear power. About 30 countries make some use of nuclear power. The countries most highly dependent on nuclear power are clustered in Europe (Figure 14-23), where it supplies 95 percent of all electricity in France and more than one-half in Armenia, Belgium, Bulgaria, Lithuania, Slovakia, and Ukraine. Dependency on nuclear power varies widely among U.S. states (Figure 14-24). Nuclear power accounts for 80 percent of electricity in Vermont and more than one-half in Connecticut, New Jersey, and South Carolina. At the other extreme, 20 states and the District of Columbia have no nuclear power plants.

Nuclear power presents serious problems, however: potential accidents, radioactive waste, generation of material for nuclear weapons, a limited uranium supply, geographic distribution, and cost (Figure 14-25).

POTENTIAL ACCIDENTS. A nuclear power plant produces electricity from energy released by splitting uranium atoms in a controlled environment, a process called **fission**. One product of all nuclear reactions is **radioactive waste**, certain types of which are lethal to people exposed to it. Elaborate safety precautions are taken to prevent the leaking of nuclear fuel from a power plant.

Nuclear power plants cannot explode, like a nuclear bomb, because the quantities of uranium are too small and cannot be brought together fast enough. However, it is possible to have a runaway reaction, which overheats the reactor, causing a meltdown, possible steam explosions, and scattering of radioactive material into the atmosphere. This happened in 1986 at Chernobyl, then in the Soviet Union and now in the north of Ukraine, near the Belarus border.

The Soviet Union reported at the time that the Chernobyl accident caused 28 deaths because of exposure to high radiation doses. Following the accident, the 135,000 people living within a 30-kilometer (18-mile) radius were forced to move to other homes. Despite the evacuation, in the first decade after the accident, cases of thyroid cancer were 10 times higher than normal in Ukraine and 84 times higher than normal in southern Belarus, where most of the fallout hit. The impact of the Chernobyl accident extended through Europe. Most European governments temporarily banned the sale of milk and fresh vegetables, which were contaminated with radioactive fallout. Half of the eventual victims may be residents of European countries other than Ukraine and Belarus.

American nuclear plants are designed with strong, thick containment buildings surrounding the reactors. But nuclear plants built by the former Soviet Union lack containment buildings and often have defective parts. At a Soviet-built plant in East Germany, 11 of 12 cooling pumps were disabled by a fire and power failure. Had the twelfth pump failed, a meltdown, with its inevitable release of strong radioactive materials, likely would have killed the 50,000 inhabitants of the nearby city of Greifswald. This 1975 accident went unreported for 15 years, making the case for all nuclear plants to be open for inspection.

RADIOACTIVE WASTE. When nuclear fuel fissions, the waste is highly radioactive and lethal and remains so for many years. Plutonium for making nuclear weapons can be harvested from it. Pipes, concrete, and water near the fissioning fuel also become "hot" with radioactivity.

No one has yet devised permanent storage for radioactive waste. The waste cannot be burned or chemically treated, and it must be isolated for several thousand years until it loses its radioactivity. Spent fuel in the United States is stored "temporarily" in cooling tanks at nuclear power plants, but these tanks are nearly full. The United States is Earth's third-largest country in land area, yet it has failed to find a suitable underground storage site because of worry about groundwater contamination. In 2002, the U.S. Department of Energy approved a plan to store the waste in Nevada's Yucca Mountains. But soon after taking office in 2009, the Obama administration reversed

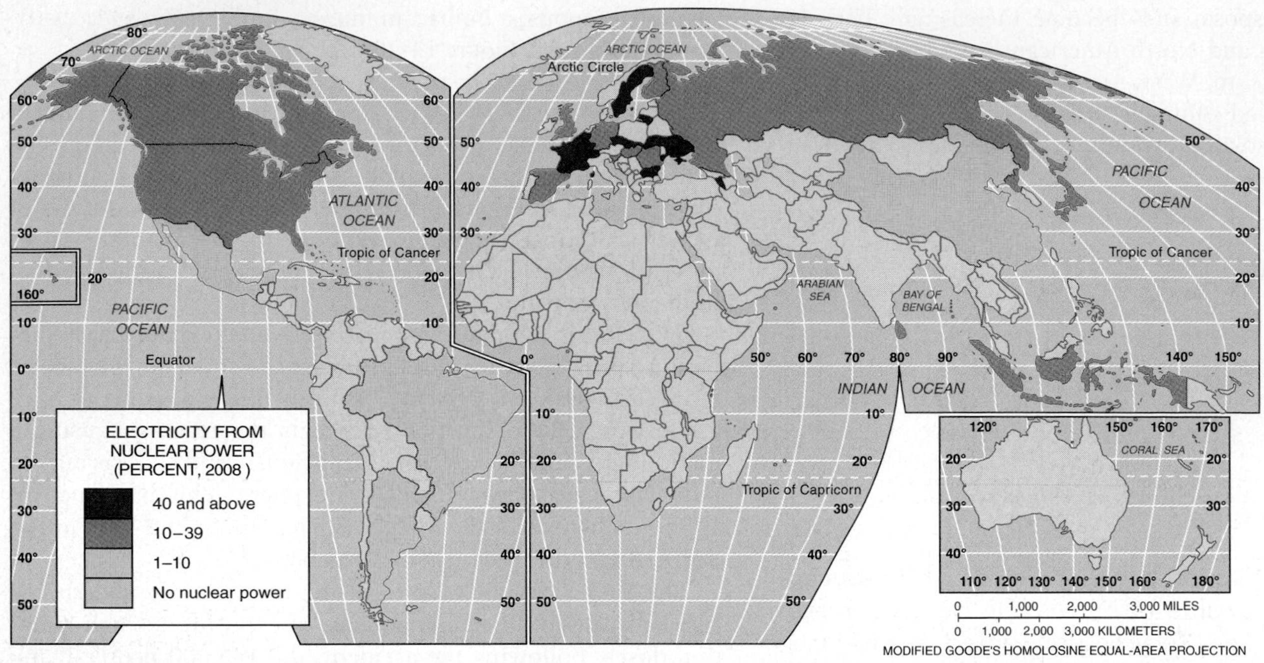

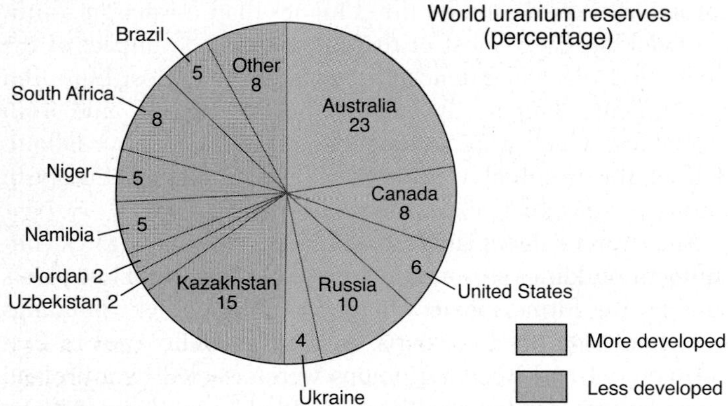

FIGURE 14-23 Nuclear power as percent of electricity. Nuclear power has been especially attractive to MDCs in Europe that lack abundant reserves of either petroleum or coal.

the decision and halted construction on the nearly complete repository.

BOMB MATERIAL. Nuclear power has been used in warfare twice, in August 1945, when the United States dropped an atomic bomb on first Hiroshima and then Nagasaki, Japan, ending World War II. No government has dared to use these bombs in a war since then, because leaders recognize that a full-scale nuclear conflict could terminate human civilization.

The United States and the Soviet Union (now Russia) each have several thousand nuclear weapons. China, France, and the United Kingdom have several hundred nuclear weapons each, India and Pakistan several dozen each, and North Korea a handful. Israel is suspected of possessing nuclear weapons but has not admitted to it, and Iran has been developing the capability. Other countries have initiated nuclear programs over the years but have not advanced to the weapons stage. The diffusion of nuclear programs to countries sympathetic to terrorists has been particularly worrying to the rest of the world.

LIMITED URANIUM RESERVES. Like fossil fuels, uranium is a nonrenewable resource. Proven uranium reserves are about 124 years at current rates of use. And they are not distributed uniformly around the world: one-fourth of the world's proven uranium reserves are in Australia and one-sixth are in Kazakhstan. The chemical composition of natural uranium further aggravates the scarcity problem. Uranium ore naturally contains only 0.7 percent U-235; a greater concentration is needed for power generation.

A **breeder reactor** turns uranium into a renewable resource by generating plutonium, also a nuclear fuel. However, plutonium is more lethal than uranium and could cause more deaths and injuries in an accident. It is also easier to fashion into a bomb. Because of these risks, few breeder reactors have been built, and none are in the United States.

HIGH COST. Nuclear power plants cost several billion dollars to build, primarily because of elaborate safety measures. Without double and triple backup systems, nuclear energy would be too dangerous to use. Uranium is mined in one place, refined in another, and used in still another. The complexities of safe transportation add to the cost. As a result, generating electricity from nuclear plants is much more expensive than from coal-burning plants.

The future of nuclear power has been seriously hurt by the combination of high risk and cost. Some countries in North America and Europe have curtailed construction of new plants. Italy closed its nuclear power plants in 1987. Sweden, which had been receiving half of its electricity from nuclear power, closed two nuclear reactors in 2005. On the other

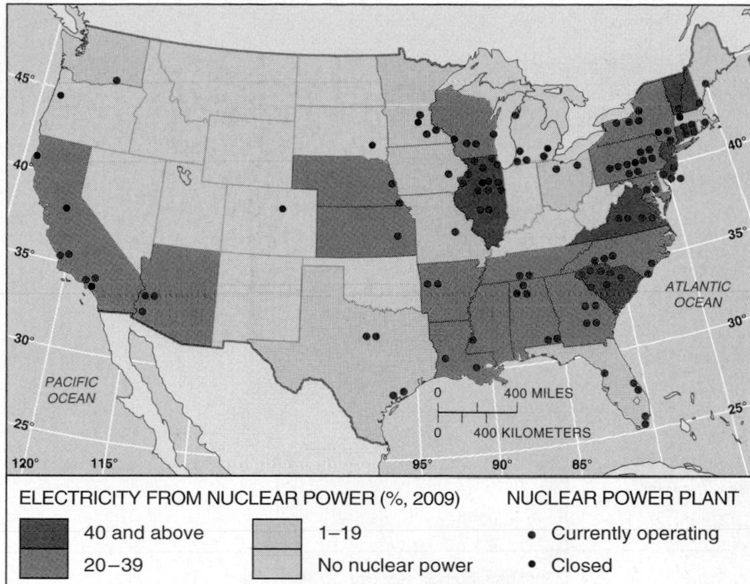

FIGURE 14-24 U.S. nuclear power plants and nuclear power as percent of electricity by state. Nuclear power is an important source of electricity in a number of northeastern and midwestern states. Each dot represents one reactor; more than one reactor is operating at many plants.

FIGURE 14-25 Nuclear waste disposal. This tunnel beneath the Yucca Mountains was constructed as a repository for nuclear waste in the United States, but the plan was halted in 2009.

hand, countries without nuclear power are moving toward introducing it, including Poland, Turkey, Indonesia, and Vietnam. Australia, with the most extensive reserves of uranium, is debating expansion of nuclear power. Advances in safety and reactor technology, combined with the high cost of other power sources, are driving the renewed interest.

NUCLEAR FUSION. Some nuclear power issues could be addressed through nuclear **fusion**, which is the fusing of hydrogen atoms to form helium. Fusion releases spectacular amounts of energy—a gnat-sized amount of hydrogen releases the energy of thousands of tons of coal. But fusion can occur only at very high temperatures (millions of degrees). Such high temperatures have been briefly achieved in hydrogen bomb tests but not on a sustained basis in a power-plant reactor, given present technology.

Alternatives such as fusion do not offer immediate solutions to energy shortages in the twenty-first century but may become more practical if the price of current energy sources substantially rises. Earth possesses a variety of energy resources, but the era of dependency on nonrenewable fossil fuels for energy will constitute a remarkably short period of human history.

Leading Renewable Energy Sources

The leading sources of renewable energy currently are biomass and hydroelectric. Geothermal, wind, and solar are also currently used but are less common.

BIOMASS. More than one-half of renewable energy comes from biomass, including wood and crops. When carefully harvested in forests, wood is a renewable resource that can be used to generate electricity and heat. The waste from processing wood, such as building construction and demolition, is also available. And crops such as sugarcane, corn, and soybeans can be processed into motor-vehicle fuels. Brazil in particular makes extensive use of biomass to fuel its cars and trucks.

The potential for increasing the use of biomass for fuel is limited, for several reasons:

- Burning biomass may be inefficient, because the energy used to produce the crops may be as much as the energy supplied by the crops.
- When wood is burned for fuel instead of being left in the forest, the fertility of the forest may be reduced.
- Biomass already serves essential purposes other than energy, such as providing much of Earth's food, clothing, and shelter.

HYDROELECTRIC POWER. Water has been a source of mechanical power since before recorded history. It turned water wheels and the rotational motion was used to grind grain, saw timber, pump water, and operate machines. Over the last hundred years, the energy of moving water has been used to generate electricity, called **hydroelectric power**. Hydroelectric power is the world's second-most popular source of electricity, after coal, supplying about one-fourth of worldwide demand. Many LDCs depend on hydroelectric power for most of their electricity (Figure 14-26). The most populous country to depend primarily on hydroelectric power is Brazil. Among MDCs, Canada gets two-thirds of its electricity from hydroelectric power, and although the United States is the fourth leading producer of hydroelectric power, it obtains only 8 percent of its electricity from that source. And this percentage may decline, because few acceptable sites to build new dams remain.

China is the world's leading producer of hydroelectric power. Unfortunately, its Three Gorges dam, under construction across the Yangtze River, is widely regarded as an environmental disaster (Figure 14-27) for the resulting decline in water quality in the river, endangerment of wildlife, including cranes and dolphins,

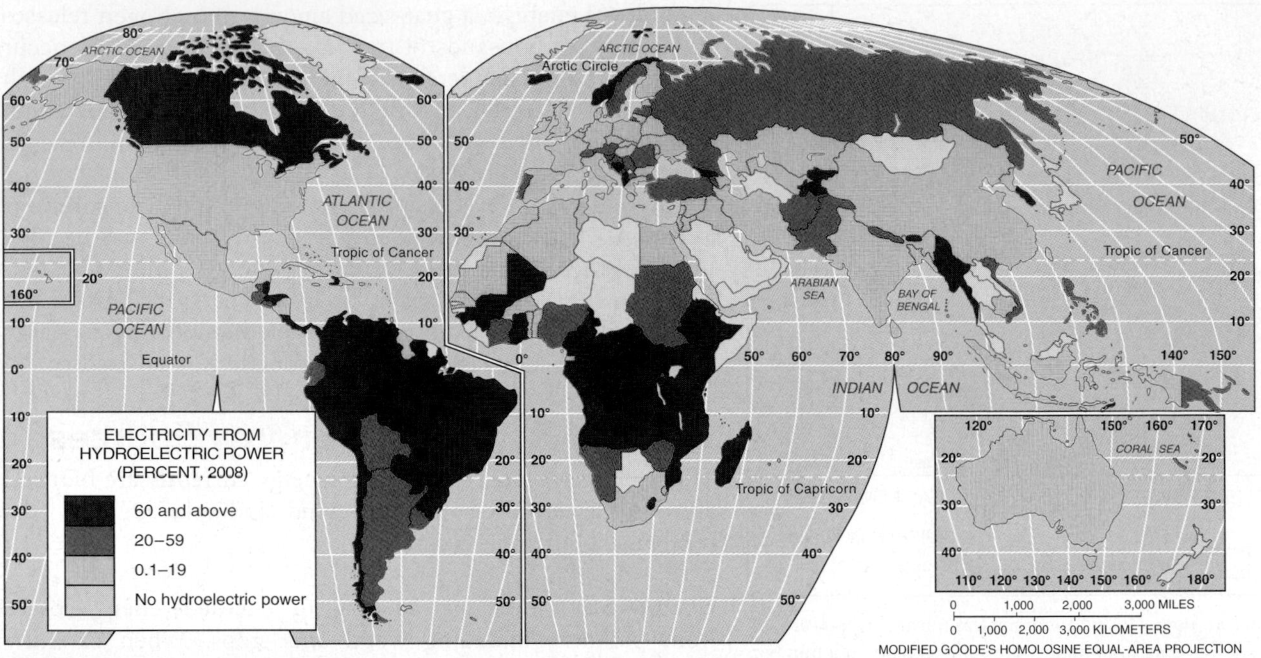

FIGURE 14-26 Electricity from hydroelectric power. Hydroelectricity provides a large percentage of electricity in a number of LDCs.

FIGURE 14-27 Hydroelectric power. Three Gorges Dam, on the Chang Jiang (Yangtze) River in China, is the world's largest hydroelectric dam. Although a renewable energy source and less polluting of the air than burning fossil fuels, the Three Gorges Dam has destroyed the habitat of several thousand species, forced the displacement of more than 1 million people in the way of the project, and flooded several-thousand-year-old archaeological sites.

and extinction of rare species of vegetation in a region known for biodiversity.

WIND POWER. Wind has also long been a source of energy, the most obvious examples of its uses being sailboats for travel and windmills for grinding grain. Like moving water, moving air can turn a turbine. Hundreds of wind "farms" consisting of dozens of windmills have been constructed across the United States, one-third of the country being considered windy enough to make wind power economically feasible (Figure 14-28), especially North Dakota, Texas, Kansas, South Dakota, and Montana. Twenty percent of Denmark's electricity is being generated through wind power.

The benefits of wind-generated power seem irresistible. Construction of a windmill modifies the environment much less severely than construction of a dam across a river. And wind power has a greater potential for increased use because only a small portion of the potential resource has been harnessed. However, wind power has divided the environmental community. Some oppose construction of windmills because they can be noisy and lethal for birds and bats. They also can constitute a visual blight when constructed on mountaintops or offshore in places of outstanding beauty.

GEOTHERMAL ENERGY. Natural nuclear reactions make the Earth's interior hot. Toward the surface, in volcanic areas, this heat is especially pronounced. The hot rocks can encounter groundwater, producing heated water or steam that can be tapped by wells. Energy from this hot water or steam is called **geothermal energy**.

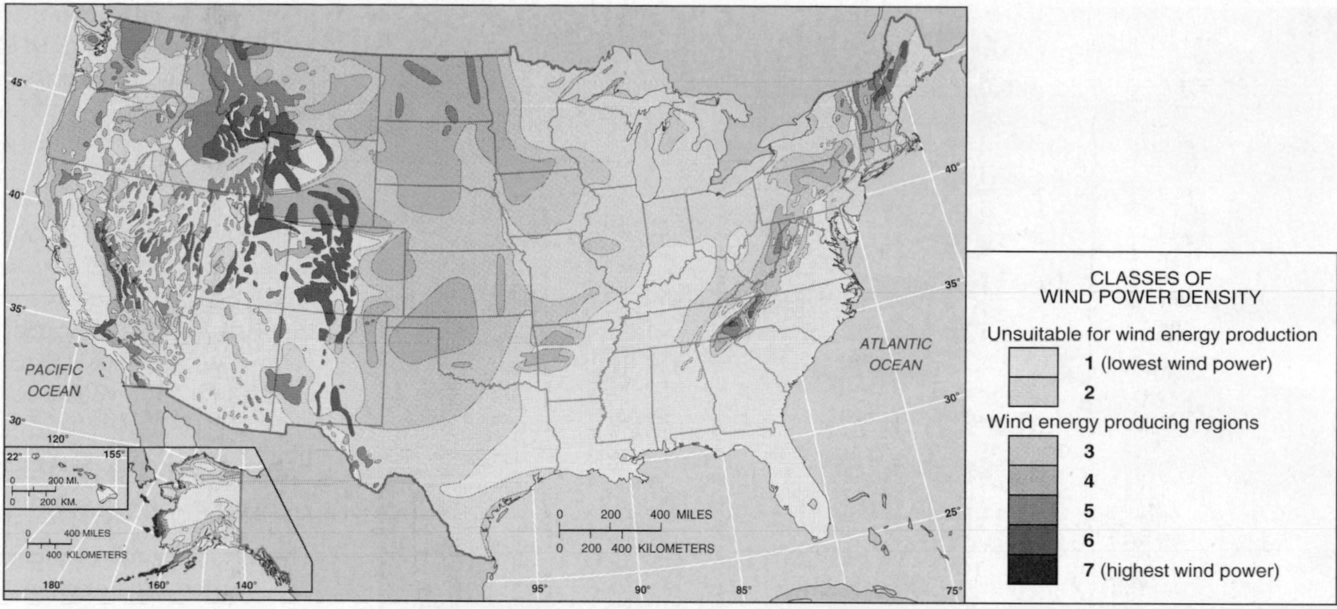

FIGURE 14-28 Wind power. Winds are especially strong enough to support generation of power in the U.S. Plains states, between the Dakotas and Texas.

Harnessing geothermal energy is most feasible at the sites along Earth's surface where crustal plates meet, which are also the sites of many earthquakes and volcanoes. Geothermal energy is being tapped in several locations, including California, Italy, New Zealand, and Japan, and other plate boundary sites are being explored. Iceland and Indonesia make extensive use of geothermal energy. Ironically, in Iceland, an island named for its glaciers, nearly all homes and businesses in the capital of Reykjavik are heated with geothermal steam.

Solar Energy

The ultimate renewable resource is solar energy supplied by the Sun. Solar sources currently supply the United States with only 1 percent of electricity, but the potential for growth is limitless. The Sun's remaining life is estimated at 5 billion years, and humans appear incapable of destroying or depleting that resource. The Sun's energy is free and ubiquitous and cannot be exclusively owned, bought, or sold by any particular individual or enterprise. Utilizing the Sun as a resource does not damage the environment or cause pollution, as does the extraction and burning of nonrenewable fossil fuels.

PASSIVE SOLAR ENERGY. Solar energy is harnessed through either passive or active means. **Passive solar energy systems** capture energy without special devices. These systems use south-facing windows and dark surfaces to heat and light buildings on sunny days. The Sun's rays penetrate the windows and are converted to heat. Humans act as passive solar energy collectors when they are warmed by sunlight. And since dark objects absorb more energy, wearing dark clothing warms a person even more when exposed to sunlight.

Reliance on passive solar energy increased during the nineteenth century when construction innovations first permitted the hanging of massive glass "curtains" on a thin steel frame. Greenhouses enabled people to grow and view vegetation that required more warmth to flourish than the local climate permitted. Early skyscrapers made effective use of passive solar energy. During World War II when fossil fuels were rationed, consumers looked for alternative energy sources. A major glass manufacturer, Libbey-Owens-Ford Glass Co., responded by publishing a book in 1947 entitled *Your Solar House*. But with electricity and petroleum cheap and abundant after World War II and through most of the twentieth century, passive solar energy rarely played a major role in construction of homes and commercial buildings.

In recent years building construction and remodeling have made more use of passive solar energy through advances in glass technology. Double- and triple-pane windows have higher insulating values, and low-E (low emissivity) glass can be coated to let heat in but not out. Window panes made with this glass are filled with argon or other gases that increase their insulating values beyond that of windows that have just air between the panes. Phase-change technologies can also switch the glass from opaque to translucent when a voltage is applied.

ACTIVE SOLAR ENERGY. **Active solar energy systems** collect solar energy and convert it either to heat energy or to electricity (Figure 14-29). The conversion can be accomplished either directly or indirectly.

In direct electric conversion, solar radiation is captured with **photovoltaic cells**, which convert light energy to electrical energy. Bell Laboratories invented the photovoltaic cell in 1954. Each cell generates only a small electric current, but large numbers of them wired together produce significant electricity. These cells are made primarily of silicon (also used in computers), the second most abundant element in Earth's crust. When the silicon is combined with one or more other materials, it exhibits distinctive electrical properties in the

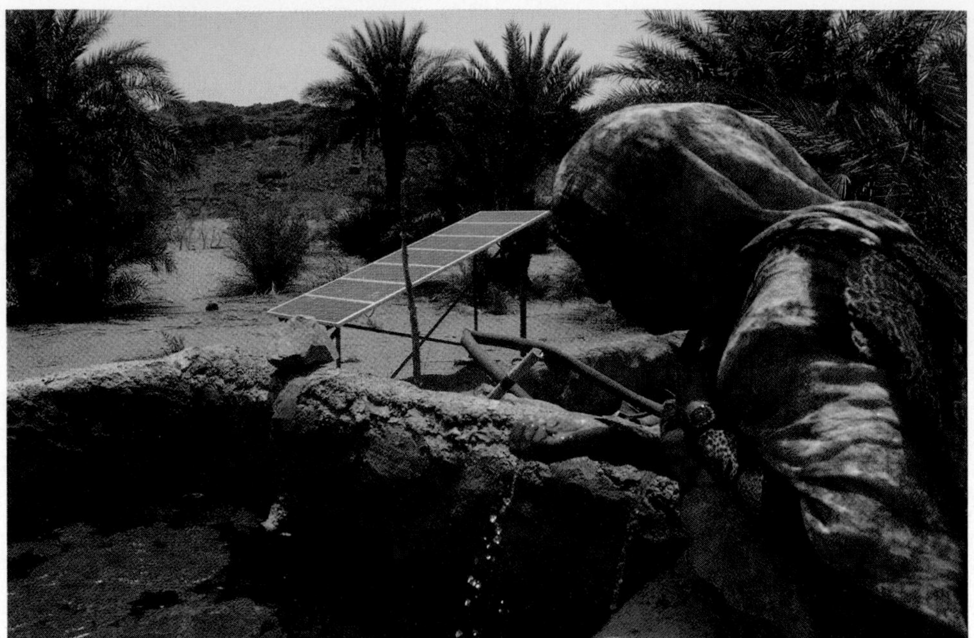

FIGURE 14-29 Solar energy. Solar energy is increasingly used in Africa, where it is not economical to string wires for electricity. Solar panels provide the power to pump water from this well in Mauritania.

presence of sunlight, known as the photovoltaic effect. Electrons excited by the light move through the silicon, producing direct current (DC) electricity.

In indirect electric conversion, solar radiation is first converted to heat, then to electricity. The Sun's rays are concentrated by reflectors onto a pipe filled with synthetic oil. The heat from the oil-filled pipe generates steam to run turbines. In heat conversion, solar radiation is concentrated with large reflectors and lenses to heat water or rocks. These store the energy for use at night and on cloudy days. A place that receives relatively little sunlight can still use solar energy by using more reflectors and lenses and larger storage containers.

GENERATING ELECTRICITY THROUGH SOLAR. Solar power can be produced at a central station and distributed by an electric company, as coal- and nuclear-generated electricity is now supplied. However, with coal still relatively cheap and investment in nuclear facilities already substantial, public and private utility companies have had little interest in solar technology.

In MDCs, solar-generated electricity is used in spacecraft, light-powered calculators, and at remote sites where conventional power is unavailable, such as California's Mojave Desert. Solar energy is used primarily as a substitute for electricity in heating water. Rooftop devices collect, heat, and store water for apartment buildings in Israel and Japan and individual homes in the United States. The initial cost of installing a solar water heater is higher than hooking into the central system but may be justified if an individual plans to stay in the same house for a long time. In LDCs, the largest and fastest-growing market for photovoltaic cells includes the 2 billion people who lack electricity, especially residents of remote villages. For example, in Kenya, more homes have been electrified in recent years using photovoltaic cells than by hooking up to the central power grid. In Morocco, solar panels are sold in bazaars and open markets, next to carpets and tinware.

The cost of cells must drop and their efficiency must improve for solar power to expand rapidly, with or without government support. Solar energy will become more attractive as other energy sources become more expensive. A bright future for solar energy is indicated, for petroleum companies now own the major U.S. manufacturers of photovoltaic cells.

Renewable Energy in Motor Vehicles

The most serious challenge to reducing dependence on nonrenewable energy is its importance as motor-vehicle fuel. As Americans become more concerned with the high cost, pollution, and increasing dependency on imported petroleum, several alternative sources of fuel are becoming available.

BATTERIES. Battery power was popular in early motor vehicles. Of the 4,000 cars sold in the United States in 1900, 38 percent were powered by electricity, 40 percent by steam, and only 22 percent by gasoline. The electric car was especially popular in 1900 in large cities of the Northeast, such as New York and Philadelphia, where their relative quietness and cleanliness made them popular as taxicabs. Women also preferred electric cars because they were easier to start than gasoline- or steam-powered ones.

The main shortcomings of the electric car in the early 1900s remain unchanged a century later. Compared to gasoline power, the electric-powered vehicle has a more limited range and costs more to operate. Recharging the battery can take several hours. To address these issues, carmakers offer a variety of vehicles that combine electric and gasoline power. Hybrid vehicles conserve gasoline by running on electricity at low speeds. Other vehicles operate exclusively on battery-powered electricity and use the gasoline engine to recharge the battery.

BIOFUELS. Motor-vehicle fuel, known as ethanol, can be produced from biomass material, specifically the sugars found in many crops. All motor vehicles can run on E10 fuel, which is a mix of 10 percent ethanol and 90 percent gasoline. A few vehicles on the market can run on E85, which is a mix of 85 percent ethanol and 15 percent gasoline. Corn is the principal source of ethanol in the United States. It is cheaper to produce ethanol from corn than from other grains, and corn is the nation's leading crop.

However, to grow and process the corn takes a lot of energy, supplied primarily by fossil fuels, so using ethanol may not actually reduce dependency on fossil fuels. And using corn for ethanol takes it away from its principal use as food for people and animals. Because of the limitations of using corn, engineers are looking at other sources of biofuels. Brazil, for example, makes use of sugarcane. In the United States, research is directed towards cellulosic sources such as stems and stalks, as well as switchgrasses. Some vehicles with diesel engines can be run on biodiesel, which is a fuel made from vegetable oils and fats. A principal source of biodiesel fuel is grease discarded by restaurants after they fry potatoes.

HYDROGEN FUEL CELLS. Hydrogen fuel cells convert hydrogen and oxygen into water, producing electricity and heat in the process. The electricity can be used to power motors or other electrical devices. The oxygen for the fuel cell reaction comes from the air, so it is free and ubiquitous. Obtaining the hydrogen is more problematic. Most hydrogen is currently produced by separating it from the carbon in methane (a form of natural gas), through a process called steam reforming. But methane is a finite fossil fuel and a polluting greenhouse gas.

Hydrogen fuel is currently used to lift space shuttles into orbit, and hydrogen fuel cells power the space shuttles' electrical systems. For motor vehicles, getting tanks of liquid or gaseous hydrogen to motorists will require a new distribution system. The United States had 63 hydrogen fueling stations as of 2008, primarily in Los Angeles, New York, and Washington.

Increasing and wildly fluctuating petroleum prices have stimulated interest in alternative-fuel vehicles. Another disruption of petroleum supplies as occurred in the 1970s would trigger a rapid switch to alternative-fuel vehicles. If that happens, by the middle of the twenty-first century large gasoline-powered vehicles would be limited to specialized tasks—or consigned to museums.

Recycling Resources

Unwanted by-products are usually "thrown away," perhaps in a "trash can." **Recycling** is the separation, collection, processing, marketing, and reuse of the unwanted material. Recycling increased in the United States from 7 percent of all solid waste in 1970 to 10 percent in 1980, 17 percent in 1990, and 33 percent in 2007.

As a result of recycling, about 85 million of the 254 million tons of solid waste generated in the United States in 2007 did not have to go to landfills and incinerators, compared to 34 million of the 200 million tons generated in 1990 (Figure 14-30). In other words, the amount of solid waste generated by Americans increased by 54 million tons between 1990 and 2007 and the amount recycled increased by 51 million tons, so about the same amount went into landfills or incinerators over the period. The percentage of materials recovered by recycling varies widely by product: Two-thirds of yard trimmings and one-half of paper products are recycled, compared to only 7 percent of plastic and 3 percent of food scraps.

Recycling Collection

Recycling involves two main series of activities. First, materials that would otherwise be "thrown away" are collected and

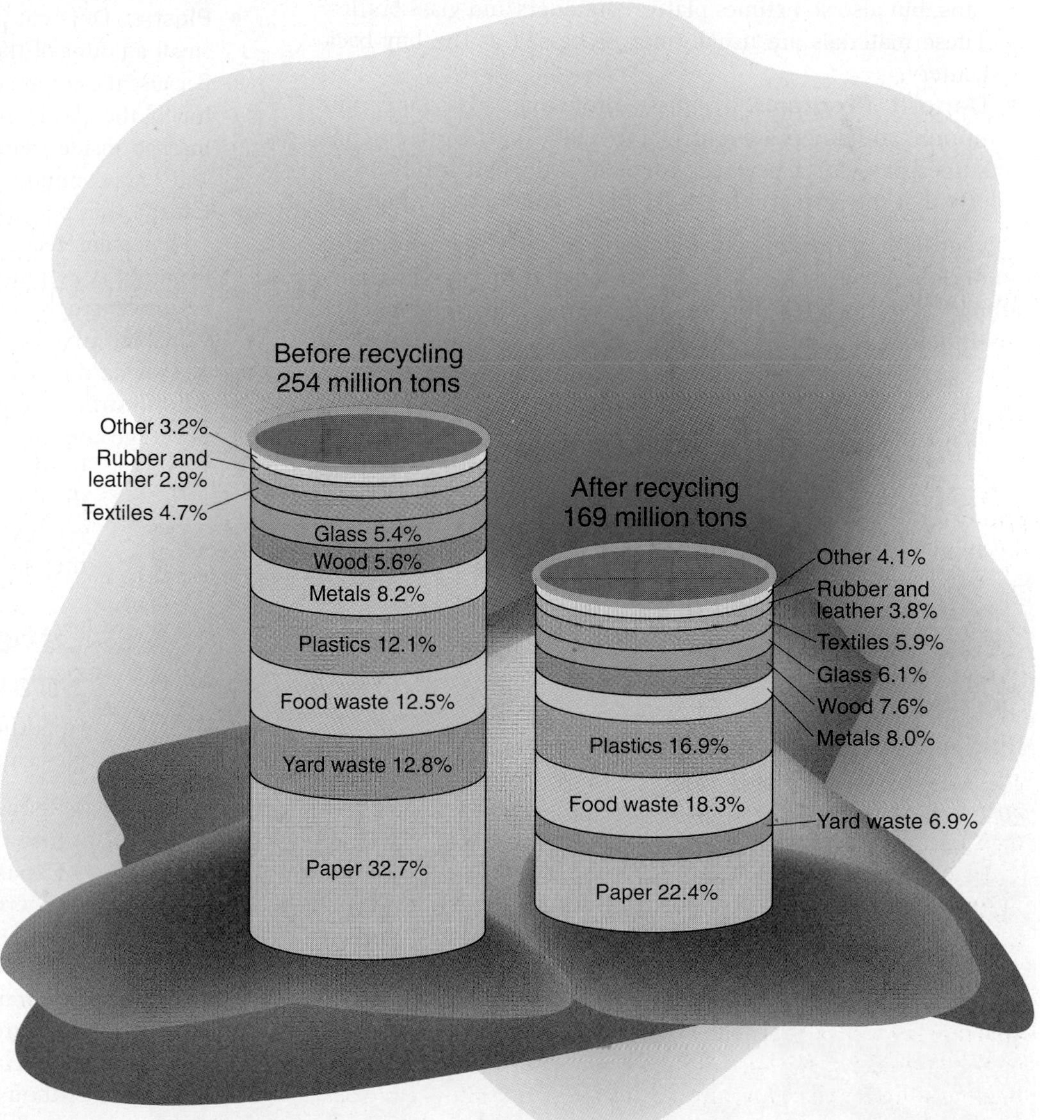

FIGURE 14-30 Sources of solid waste. Paper products account for the largest percentage of U.S. solid waste, followed by yard rubbish and food waste. One-third of the solid waste is recycled, especially paper and yard waste.

sorted (Figure 14-31). Then the materials are manufactured into new products for which a market exists.

PICK-UP AND PROCESSING. Recyclables are collected primarily in four ways:

- **Curbside Programs.** Recyclables are required to be placed at the curb in a container separate from the nonrecyclable trash at a specified time each week, either at the same or different time as the other trash. The trash collector usually supplies homes with specially marked containers for the recyclable items.
- **Drop-off Centers.** These are sites, typically several large containers placed at a central location, for individuals to leave recyclable materials. A separate container is designated for each type of recyclable material, and the containers are periodically emptied by a processor or recycler but are otherwise left unattended.
- **Buy-back Centers.** These are commercial operations that pay consumers for recyclable materials, especially aluminum cans, but also sometimes plastic containers and glass bottles. These materials are usually not processed at the buy-back center.
- **Deposit Programs.** In these programs, glass and aluminum containers are returned to retailers. The price a consumer pays for a beverage includes a deposit fee of 5¢ or 10¢ that the retailer refunds when the container is returned.

Regardless of the collection method, recyclables are sent to a materials recovery facility to be sorted and prepared as marketable commodities for manufacturing. Recyclables are bought and sold just like any other commodity; typical prices in recent years have been 30¢ per pound for plastic, $30 per ton for clear glass, and $90 per ton for corrugated paper. Prices for the materials change and fluctuate with the market.

FIGURE 14-31 Recycling and remanufacturing. Plastic waste imported from the United States is being sorted in China for use in a plastic factory in Zhan Long.

MANUFACTURING. Once cleaned and separated, the recyclables are ready to be manufactured into a marketable product. Four major manufacturing sectors accounted for more than half of the recycling activity—paper mills, steel mills, plastic converters, and iron and steel foundries.

Common household items that contain recycled materials include newspapers and paper towels; aluminum, plastic, and glass soft-drink containers; steel cans; and plastic laundry detergent bottles. Recycled materials are also used in such industrial applications as recovered glass in roadway asphalt ("glassphalt") or recovered plastic in carpeting, park benches, and pedestrian bridges.

- **Paper:** Most types can be recycled. Newspapers have been recycled profitably for decades, and recycling of other paper is growing, especially computer paper. Rapid increases in virgin paper pulp prices have stimulated construction of more plants capable of using waste paper. The key to recycling is collecting large quantities of clean, well-sorted, uncontaminated, and dry paper.
- **Plastic:** Different plastic types must not be mixed, as even a small amount of the wrong type of plastic can ruin the melt. Because it is impossible to tell one type from another by sight or touch, the plastic industry has developed a system of numbers marked inside triangles on the bottom of containers. Types 1 and 2 are commonly recycled, and the others generally are not.
- **Glass:** Can be used repeatedly with no loss in quality and is 100 percent recyclable. The process of creating new glass from old is extremely efficient, producing virtually no waste or unwanted by-products. Though unbroken clear glass is valuable, mixed-color glass is nearly worthless, and broken glass is hard to sort.
- **Aluminum:** The principal source of recycled aluminum is beverage containers. Aluminum cans began to replace glass bottles for beer during the 1950s and for soft drinks during the 1960s. Aluminum scrap is readily accepted for recycling, although other metals are rarely accepted.

Other Pollution Reduction Strategies

In addition to recycling, two other basic strategies can reduce pollution:

- Reducing the amount of waste discharged into the environment
- Expanding the capacity of the environment to accept discharges.

REDUCING DISCHARGES. Pollution can be prevented if the amount of waste being discharged into the environment is reduced to a level that the environment can assimilate. Although consumers purchase more "throw-away" packages than in the past, the packaging material is much less bulky. Glass bottles weigh less today than a generation ago, as do plastic jugs. Higher manufacturing and shipping costs following the 1973 energy crisis induced companies to cut costs by reducing the bulk of their packaging.

The mix of various inputs can be adjusted to produce a higher ratio of product to waste. For example, gasoline for motor vehicles once contained lead, most of which was discharged through the exhaust pipe, contributing to air pollution. To reduce the generation of lead—once a significant waste—carmakers modified engines so that they operate on unleaded instead of leaded gasoline.

The amount of waste can also be reduced if the production system produces less of the product—or if production ceases altogether—because of lower consumer demand. The creation of fewer products would result in the production of less waste as well. The amount of pollution generated by motor vehicles declined in 2008 and 2009 compared to earlier years primarily because consumers drove less during the severe recession.

Emissions-trading systems can reduce discharges, especially into the atmosphere. To reduce sulfur dioxide discharges, the United States introduced a market through an amendment to the 1990 Clean Air Act. Power companies can buy and sell allowances to emit sulfur dioxide. Dirty power companies have found it cheaper to install pollution-control devices to reduce pollution and sell some of their allowances. In Canada, Ontario's emissions-reduction trading sets emissions caps and creates a market for trading allowances and credits within the caps.

INCREASING ENVIRONMENTAL CAPACITY. The second way to handle pollution is to increase the capacity of the environment to accept the discharges. The capacity of air, water, and land to accept waste is not fixed but varies among places and at different times.

Adding a particular amount of wastewater to a stream may or may not constitute pollution, depending on the flow of the water. A deep, fast-flowing river has a greater capacity to absorb wastewater than a shallow, slow-moving one. Wastewater can be stored when the river level is low and released when the river is high. Similarly, exhaust released into stagnant air irritates, whereas exhaust released in windy conditions is quickly dispersed. Industries and utilities reduce local air pollution by building taller smokestacks, which better disperse gases at greater heights.

Environmental capacity can also be increased by transforming the waste so that it is discharged into a resource that has the capacity to assimilate it. Matter can be transformed among gaseous, liquid, and solid states and discharged into air, water, or land. For example, a coal-burning power plant can discharge gases into the atmosphere, causing air pollution. To reduce air pollution, wet scrubbers are installed to wash particulates from the gas before it is released to the atmosphere. These scrubbers capture the particulates in water, which then can be discharged into a stream. If the stream is polluted by the discharge, then the wastewater can be cleaned in a settling basin where the particulates drop out. This transforms the residue into a solid waste for disposal on land.

Comparing Pollution Reduction Strategies

Relying on an increase in the capacity of the environment to accept discharges is risky (Figure 14-32). Because we do not always know the environment's capacity to assimilate a particular waste, we are likely to exceed it at times.

Dispersed wastes may remain harmful. A pollutant like sulfur dioxide might exist at tolerable levels in the air, but it damages trees when it accumulates in the soil. Recent history is filled with examples of wastes discharged into the environment in the belief that they would be dispersed or isolated safely—CFCs in the stratosphere, garbage offshore, and toxic chemicals beneath Love Canal. Many pollutants are mobile. They often travel from air to soil, or soil to water. Tall smokestacks built to reduce sulfur dioxide discharges around coal-burning industries and metal smelters were successful at dispersing sulfur over a larger area. But the result of the dispersal was that acid precipitation (containing sulfur) fell hundreds of kilometers away, polluting vegetation and lakes over a wide area.

In view of the many uncertainties associated with increasing environmental capacity, reducing discharges into the environment (by either changing the production process or recycling) is usually the preferred alternative. Although the environment has the capacity to accept some discharges, consumers must learn to use this environmental capacity most efficiently. At the same time, consumers must learn to waste less, either by reducing the consumption of products that result in waste or by recycling more. With careful management, we can enjoy the benefits of both industrial development and a cleaner, safer environment.

KEY ISSUE 4
Why Should Resources Be Conserved?

- **Sustainable Development**
- **Biodiversity**

Because it is one part natural science and one part social science, geography is especially sensitive to the importance of protecting the natural environment while meeting human needs. "Conservation" is a concept that reflects balance between nature and society. ■

Sustainable Development

Sustainable development is "development that meets the needs of the present without compromising the ability of future generations to meet their own needs," according to the United Nations. Through sustainable development, humans can improve their quality of life while protecting Earth's resources for the benefit of future generations.

Conservation, Preservation, and Sustainability

Conservation is the sustainable use and management of natural resources such as wildlife, water, air, and Earth's resources

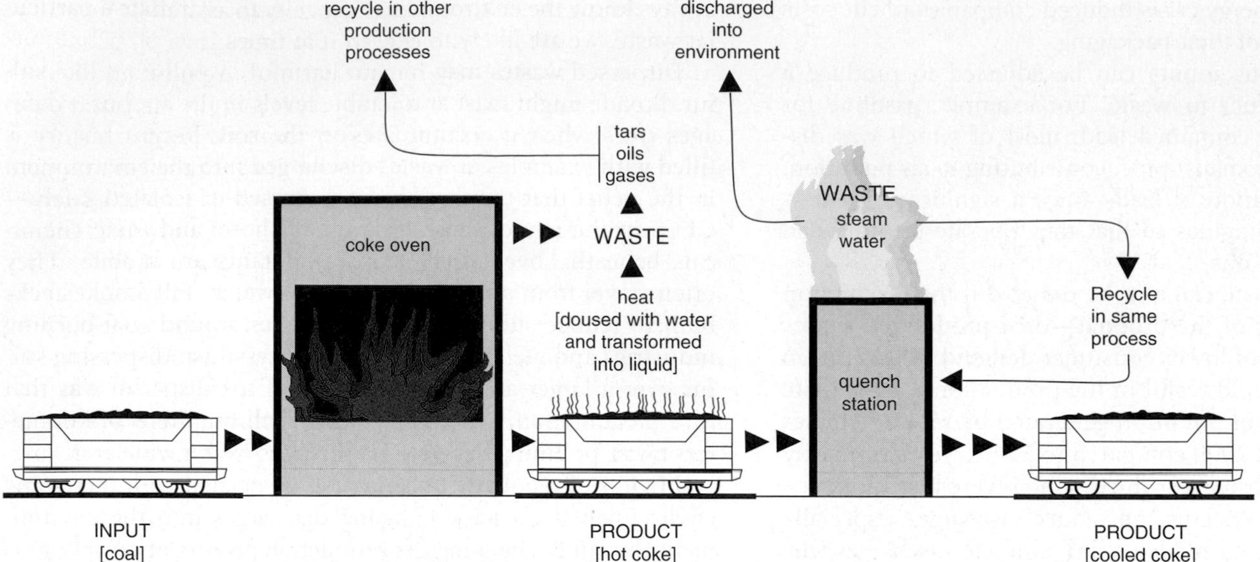

FIGURE 14-32 Comparing pollution reduction strategies. A coking plant, used for steelmaking, illustrates the application of principal alternatives for reducing pollution. The main input into a coking plant is a mixture of coal types, and the intended product is coke, which becomes an input in steel production. The coal is placed in a coke oven and cooked at very high temperatures to form coke. Four unwanted by-products result—gases, tars, oils, and heat.

Discharging the heat into the environment can cause air pollution. To reduce air pollution from the heat, a coking plant increases the capacity of the environment to accept discharges in two ways. First, the hot coke is taken to a quench station and doused with water to cool it. This process transforms the residual (hot gas) into a liquid (dirty water) as well as another gas (steam). In this way, the waste is transformed and discharged into different parts of the environment. Then the steam is discharged into the environment from a tall smokestack, an example of making more efficient use of whatever initially received the discharge (air).

The coking plant also minimizes pollution by reducing discharges. The dirty water produced at the quench station is reused to cool more hot coke, an example of recycling in the same production process. Meanwhile, the three unwanted by-products from the coke oven (other than heat)—gases, tars, and oils—are captured and sold to other companies for recycling in other processes. The other alternative for reducing discharges—changing the mix of coal used as inputs—is also employed, because the amount of gases emitted by the burning of coke varies depending on the mix of coal.

to meet human needs, including food, medicine, and recreation. Renewable resources such as trees are conserved if they are consumed at a less rapid rate than they can be replaced. Nonrenewable resources such as fossil fuels are conserved if remaining reserves are maintained for future generations. Conservation differs from **preservation**, which is the maintenance of resources in their present condition, with as little human impact as possible. Preservation takes the view that the value of nature does not derive from human needs and interests, but from the fact that every plant and animal living on Earth has a right to exist and should be preserved regardless of the cost.

Preservation does not regard nature as a resource for human use. In contrast, conservation is compatible with development but only if natural resources are utilized in a careful rather than a wasteful manner. An increasingly important approach to careful utilization of resources is sustainable development, based on promotion of biodiversity.

Sustainability and Economic Growth

The UN's "sustainable development" definition originated in the 1987 Brundtland Report, named for the World Commission on Environment and Development's chair, Gro Harlem Brundtland, former prime minister of Norway. Titled *Our Common Future*, the Brundtland Report was a landmark in recognizing sustainable development as a combination of environmental and economic elements.

The report argued that sustainable development had to recognize the importance of economic growth while conserving natural resources. Environmental protection, economic growth, and social equity are linked because economic development aimed at reducing poverty can at the same time threaten the environment (Figure 14-33). Plans to protect the environment will fail unless LDCs promote economic growth in a way that meets basic needs of employment, food, and energy, as well as water and sanitation. "Environment and development are not separate challenges: they are inexorably linked," concluded the Brundtland Report. "Development cannot exist on a deteriorating environmental base; the environment cannot be protected when growth leaves out of account the costs of environmental protection."

A rising level of economic development generates increased pollution, at least until a country reaches a GDP of about $5,000 per person, according to economists Gene Grossman and Alan Krueger (Figure 14-34). In the early stages of industrialization, pollution-control devices are an unpopular luxury that makes cars and other consumer goods more expensive. Consequently, twentieth-century environmental improvements in the MDCs of North America and Western Europe are likely to be offset by increased pollution in LDCs during the twenty-first century.

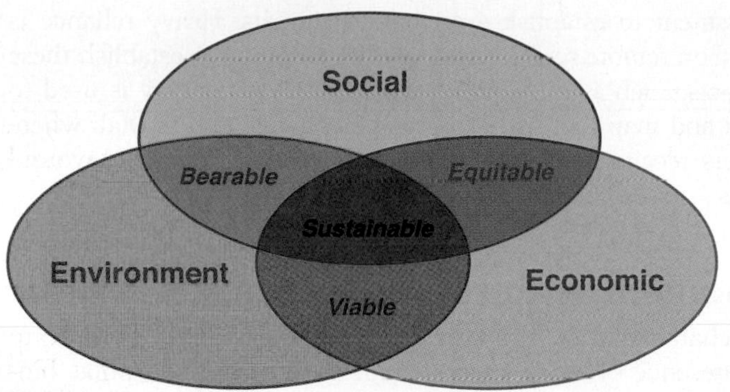

FIGURE 14-33 Sustainable development. Environmental improvement takes into consideration social and economic factors.

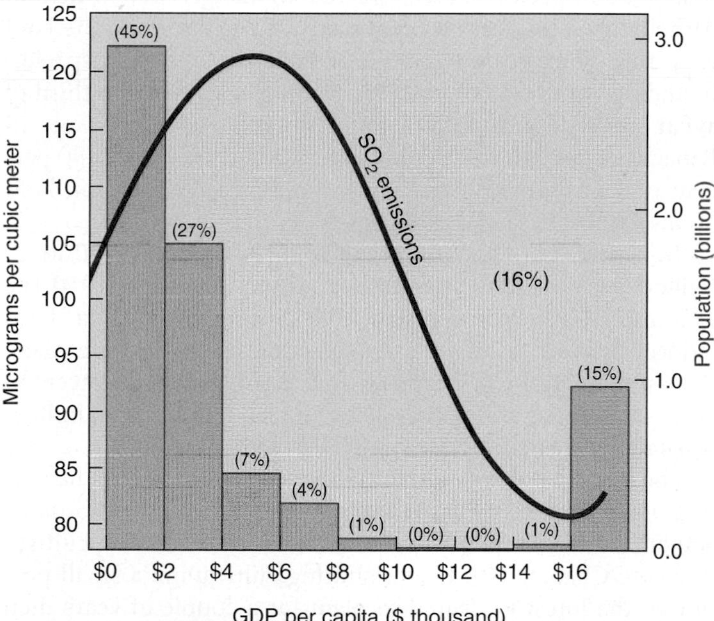

FIGURE 14-34 Pollution compared to a country's wealth. As a country's GDP per capita increases, discharge of sulfur dioxide increases, until GDP reaches about $5,000. Then, discharges tend to decrease as a country begins to spend money on pollution-control devices. The green bars show the percentage of people in each GDP per capita group.

Critical to world pollution in the twenty-first century is China. The rapid economic transformation of China has resulted in rapidly rising levels of pollution. The country has 16 of the 20 most polluted cities, according to the World Bank. Wastewater discharge is increasing 2 percent per year, and most of it is not being treated. Air pollution is especially severe, because China depends heavily on burning coal to produce electricity as well as home heat and cooking. Chinese cars have not been subject to effective emission controls. Sulfur dioxide emissions from China are even crossing the Pacific Ocean and being deposited in the western United States. The World Bank estimates that 10 percent of China's GDP is being lost to direct damage from pollution, including destruction of crops, medical bills, and sick-leave payments.

The Brundtland Report was optimistic that environmental protection could be promoted at the same time as economic growth and social equity. By gradually changing development practices, economic growth and social equity can be made compatible with protecting the environment and conserving resources. In recent years, the World Bank and other international development agencies have embraced the concept of sustainable development. Planning for development involves the consideration of many more environmental and social issues today than was the case in the past. Sustainable development is based on limiting the use of renewable resources to the level at which the environment can continue to supply them indefinitely. The amount of timber cut down in a forest, for example, or the number of fish removed from a body of water can be controlled at a level that does not reduce future supplies.

Sustainability's Critics

Some environmentally oriented critics have argued that it is too late to discuss sustainability. One critic, the World Wildlife Fund (WWF), claims that the world surpassed its sustainable level around 1980. The WWF Living Planet Report reaches its pessimistic conclusion by comparing the amount of land that humans are currently using with the amount of "biologically productive" land on Earth. "Biologically productive land" is defined as the amount of land required to produce the resources currently consumed and handle the wastes currently generated by the world's 7 billion people at current levels of technology. The WWF calculates that humans are currently using about 13 billion hectares of Earth's land area, including 3 billion hectares for cropland, 2 billion for forest, 7 billion for energy, and 1 billion for fishing, grazing, and built-up areas. However, according to the WWF, Earth has only 11.4 billion hectares of biologically productive land, so humans are already using all of the productive land and none is left for future growth.

Others criticize sustainability from the opposite perspective: Human activities have not exceeded Earth's capacity, they argue, because resource availability has no maximum, and Earth's resources have no absolute limit because the definition of resources changes drastically and unpredictably over time. Environmental improvements can be achieved through careful assessment of the outer limits of Earth's capacity.

Critics and defenders of sustainable development agree that one important recommendation of the UN report has not been implemented—increased international cooperation to reduce the gap between more developed and less developed countries. Only if resources are distributed in a more equitable manner can LDCs reduce the development gap with MDCs.

Biodiversity

Biological diversity, or **biodiversity** for short, refers to the variety of species across Earth as a whole or in a specific place. Biodiversity is an important development concept because it is a way of summing the total value of Earth's resources available for human use. Sustainable development is promoted when the biodiversity of a particular place or Earth as a whole is protected.

Biological and Geographic Biodiversity

Species variety can be understood from several perspectives. Geographers are especially concerned with biogeographic diversity, whereas biologists are especially concerned with genetic diversity. For biologists, biodiversity refers particularly to the maintenance of genetic diversity within populations of plants and animals. Genetic diversity embraces species variation in genetic material, such as genes and chromosomes.

Scientists have classified about 2.5 million species, including 900,000 insects, 41,000 vertebrates, and 250,000 plants, and more than a million invertebrates, fungi, algae, and microorganisms. About 1.4 million species have been given names. Estimates of Earth's total number of species range from 3 to 100 million, with 10 million as a median "guess," meaning that humans have not yet "discovered," classified, and named most of Earth's species. New species are constantly being identified—for example, three new bird species are found annually—but human actions are exterminating species more rapidly than they are "discovering" new ones. Human actions are responsible for the extinctions by destroying habitats, primarily through pollution of air and water, removal of existing plants and animals, and introduction of foreign or exotic species.

For geographers, biodiversity is measurement of the number of species within a specific region or habitat. A community containing a large number of species is said to be species-rich, whereas an area with few species is species-poor. Two communities may have the same number of species and the same total population of individuals, yet one may be more diverse than the other, depending on the distribution of the total population among the various species. A community with a large population of many species is regarded as more diverse than a community that contains a preponderance of one species and a very small number of all of the others.

Strategies to protect genetic diversity have been established on a global scale. Some endangered species have been protected by the Convention on International Trade in Endangered Species of Wild Fauna and Flora. Examples include the curtailing of logging, whaling, and taking of porpoises in tuna seines (nets). Strategies to protect biogeographic diversity vary among countries. Luxembourg protects 44 percent of its land and Ecuador 38 percent, whereas Cambodia, Iraq, and some former Soviet Union republics have no land under conservation.

Frustrated by the inability to precisely measure environmental impacts, the United Nations created the Millennium Ecosystem Assessment to establish systematic data sets. Heavy reliance is placed on remote sensing and satellite mapping to establish these data sets, such as in Namibia where satellite imagery is used to count and map the distribution of elephants, and in Mali where farmers receive satellite updates about storms on hand-wound radios.

Biodiversity in the Tropics

The characteristics of the tropical forest biome contribute to the presence of more species than in temperate or polar biomes. Higher temperatures, greater climate predictability, and longer growing seasons all create a more inviting habitat for a greater diversity of species. Thus reduction of biodiversity through species extinction is especially important in tropical forests, where six species per hour are extinguished and more than 5,000 species are considered in danger of extinction. Although tropical forests occupy only 6 percent of Earth's land area, they contain more than one-half of the world's species, including two-thirds of vascular plant species and one-third of avian species. At a small scale, a single stand of 19 trees in Panama examined in 1980 yielded 1,200 beetle species, 80 percent previously unknown. One gram of tropical soil can hold 90 million bacteria and other microbes.

Tropical rain forests are disappearing at the rate of 10 to 20 million hectares (25 to 50 million acres) per year. Since 1950, the area of Earth's surface allocated to tropical rain forests has been reduced by more than half, and unless drastic measures are taken, the last rain forest will disappear around 2050. Only 6 percent of Earth's rainforests are protected, leaving the remaining 33 million square kilometers (13 million square miles) vulnerable.

The principal cause of the high rate of extinction is the cutting down of forests, which is the result of changing economic activities in the tropics, especially a decline in shifting cultivation (see Chapter 10). Under shifting cultivation, a small portion of the forest is cleared to plant for a couple of years then left to regenerate over a couple of decades. Shifting cultivation is being replaced by logging, cattle ranching, and cultivation of cash crops, which require cutting down vast expanses of forest. Governments in LDCs support the destruction of rain forests because they view activities such as selling timber to builders or raising cattle for fast-food restaurants as more effective strategies for promoting economic development than shifting cultivation. Shifting cultivation is also regarded as a relatively inefficient approach to growing food in a hungry world; compared to other forms of agriculture it can support only a low level of population in an area without causing environmental damage. Until recently, the World Bank has provided loans to finance development proposals that required clearing forests.

SUMMARY

We have examined the problems of depletion and degradation of Earth's resources. The distribution of resources, as well as patterns of use and abuse, varies locally. But actions with regard to resources in one region can affect people everywhere.

Some scientists believe that further depletion and destruction of Earth's resources will lead to disaster in the near future. In 1968, a group of scientists known as the Club of Rome presented a particularly influential statement of this position in a report titled *The Limits to*

Growth. According to these scientists, many of whom were professors at the Massachusetts Institute of Technology, the combination of population growth, resource depletion, and unrestricted use of industrial technology will disrupt the world's ecology and economy and lead to mass starvation, widespread suffering, and destruction of the physical environment. In a recent update, the authors argued that environmental destruction is proceeding at a more rapid rate than they had originally thought. If new sets of attitudes and policies toward environmental protection are not in place within 20 years, the environment will be permanently damaged, and people's standards of living will fall.

The threat of irreparable global environmental damage is heightened by the confrontation between MDCs and LDCs. The MDCs have achieved wealth in part by using large percentages of the world's resources and discharging large percentages of the world's pollutants. Now LDCs are being asked to promote development with greater sensitivity to the environment than today's MDCs showed in the past. People in MDCs are increasingly willing to allocate some of their wealth to clean up the environment. Subsistence farmers in LDCs cannot afford to invest in environmental protection.

Most geographers recognize that unrestricted industrial and demographic growth will have negative consequences, but they do not believe that the dire predictions of *The Limits to Growth* are inevitable. Human actions have depleted some resources, but substitutes may be available. Although pollution degrades the physical environment, industrial growth can be compatible with environmental protection. Demand for food is increasing, but human actions are also expanding the capacity of Earth to provide food.

Here again are the key issues in Chapter 14:

1. **Why Are Resources Being Depleted?** As we consume resources, we are depleting Earth's supply. Fossil fuels and minerals are distributed unevenly across Earth, and supplies are not found in places where demand is highest.
2. **Why Are Resources Being Polluted?** Human beings are damaging and destroying Earth's resources through pollution. Pollution is the discharge of waste at a rate that exceeds the environment's capacity to absorb it. Pollutants are discharged into the atmosphere and water and onto land.
3. **Why Are Resources Being Reused?** Depletion and destruction of scarce resources can be minimized by converting from nonrenewable to renewable sources of energy and by recycling more unwanted waste.
4. **Why Should Resources Be Conserved?** Sustainable development promotes economic development while not reducing the world's current resource base. Especially important in conserving natural resources is maintaining biodiversity through minimizing species extinction in the development process.

CASE STUDY REVISITED / Future Directions

Geographers emphasize that each resource in the physical environment has a distinctive capacity for accommodating human activities. Just as a good farmer knows how many animals can be fed on a parcel of land, a scientist can pinpoint the constraints that resources place on population density or economic development in a particular region. With knowledge of these constraints, we will be able to maintain agricultural and industrial development in the future.

Rapid population growth means that rapidly growing cities in LDCs face pressure to expand economic opportunities and material benefits for the people, regardless of environmental impact. Stricter enforcement of pollution controls would require shutting down many businesses and eliminating jobs.

Mexico City has taken steps to reduce air pollution. Polluting factories have been closed. Cars have been banned from the CBD, and motorists are not allowed to use their cars one day each week, depending on the last digit of the license plate. Cars must now have catalytic converters and use unleaded fuel, and older buses and taxicabs have been removed from service. As a result, Mexico City is recording much lower levels of lead, sulfur dioxide, carbon monoxide, and ozone than in the 1990s.

A generation ago environmentalists coined the phrase "Think global, act local" so that we would recognize that our actions in our own communities—and even in our own backyards—could affect the entire planet (Figure 14-35). Now geographers urge us to "think global *and* think local." In an age of globalization, we cannot lose sight of the importance and pleasure of the diversity of local physical conditions and human behavior. Think both global and local, and act wherever you can to do some good. ■

FIGURE 14-35 Carbon footprint of orange juice. A ½-gallon carton of Tropicana orange juice generates the equivalent of 1.7 kg (3.75 lb) of carbon dioxide. Around one-third of the carbon dioxide is generated in growing the oranges, especially application of fertilizer. Processing, packaging, and distribution share most of the remainder.

KEY TERMS

Acid deposition (p. 452) Sulfur oxides and nitrogen oxides, emitted by burning fossil fuels, enter the atmosphere—where they combine with oxygen and water to form sulfuric acid and nitric acid—and return to Earth's surface.

Acid precipitation (p. 452) Conversion of sulfur oxides and nitrogen oxides to acids that return to Earth as rain, snow, or fog.

Active solar energy systems (p. 463) Solar energy systems that collect energy through the use of physical devices like photovoltaic cells or flat-plate collectors.

Air pollution (p. 451) Concentration of trace substances, such as carbon monoxide, sulfur dioxide, nitrogen oxides, hydrocarbons, and solid particulates, at a greater level than occurs in average air.

Animate power (p. 441) Power supplied by people or animals.

Biochemical oxygen demand (BOD) (p. 456) Amount of oxygen required by aquatic bacteria to decompose a given load of organic waste; a measure of water pollution.

Biodiversity (p. 469) The number of species within a specific habitat.

Biomass fuel (p. 441) Fuel that derives from plant material and animal waste.

Breeder reactor (p. 460) A nuclear power plant that creates its own fuel from plutonium.

Chlorofluorocarbon (CFC) (p. 452) A gas used as a solvent, a propellant in aerosols, a refrigerant, and in plastic foams and fire extinguishers.

Conservation (p. 467) The sustainable use and management of a natural resource through consuming it at a less rapid rate than it can be replaced.

Ferrous (p. 449) Metals, including iron, that are utilized in the production of iron and steel.

Fission (p. 459) The splitting of an atomic nucleus to release energy.

Fossil fuel (p. 441) Energy source formed from the residue of plants and animals buried millions of years ago.

Fusion (p. 461) Creation of energy by joining the nuclei of two hydrogen atoms to form helium.

Geothermal energy (p. 462) Energy from steam or hot water produced from hot or molten underground rocks.

Greenhouse effect (p. 451) Anticipated increase in Earth's temperature caused by carbon dioxide (emitted by burning fossil fuels) trapping some of the radiation emitted by the surface.

Hydroelectric power (p. 461) Power generated from moving water.

Inanimate power (p. 441) Power supplied by machines.

Nonferrous (p. 450) Metals utilized to make products other than iron and steel.

Nonrenewable energy (p. 441) A source of energy that is a finite supply capable of being exhausted.

Ozone (p. 452) A gas that absorbs ultraviolet solar radiation, found in the stratosphere, a zone between 15 and 50 kilometers (9 to 30 miles) above Earth's surface.

Passive solar energy systems (p. 463) Solar energy systems that collect energy without the use of mechanical devices.

Photochemical smog (p. 454) An atmospheric condition formed through a combination of weather conditions and pollution, especially from motor vehicle emissions.

Photovoltaic cell (p. 463) Solar energy cells, usually made from silicon, that collect solar rays to generate electricity.

Pollution (p. 451) Addition of more waste than a resource can accommodate.

Potential reserve (p. 442) The amount of a resource in deposits not yet identified but thought to exist.

Preservation (p. 468) Maintenance of a resource in its present condition with as little human impact as possible.

Proven reserve (p. 441) The amount of a resource remaining in discovered deposits.

Radioactive waste (p. 459) Materials from a nuclear reaction that emit radiation; contact with such particles may be harmful or lethal to people; therefore, the Materials must be safely stored for thousands of years.

Recycling (p. 465) The separation, collection, processing, marketing, and reuse of unwanted material.

Renewable energy (p. 441) A resource that has a theoretically unlimited supply and is not depleted when used by humans.

Resource (p. 440) A substance in the environment that is useful to people, is economically and technologically feasible to access, and is socially acceptable to use.

Sanitary landfill (p. 458 A place to deposit solid waste, where a layer of earth is bulldozed over garbage each day to reduce emissions of gases and odors from the decaying trash, to minimize fires, and to discourage vermin.

Sustainable development (p. 467) The level of development that can be maintained in a country without depleting resources to the extent that future generations will be unable to achieve a comparable level of development.

THINKING GEOGRAPHICALLY

1. What steps has your community taken to recycle solid waste and to conserve energy?
2. U.S. carmakers must meet a standard for Corporate Average Fuel Efficiency (CAFE). This means that the average miles per gallon achieved by all models of a company's American-made cars must meet a government-mandated level. If they do not, the company must pay a stiff fine. Should the United States raise the CAFE standard to conserve fuel and reduce air pollution, even if the result is a loss of American jobs? Explain.
3. A recent study compared paper and polystyrene foam drinking cups. Conventional wisdom is that foam cups are bad for the environment because they are made from petroleum and do not degrade in landfills. However, the manufacture of a paper cup consumes 36 times as much electricity and generates 580 times as much wastewater.

Further, as they degrade in landfills, paper cups release methane gas, a contributor to the greenhouse effect. Which types of cups should companies such as McDonald's be encouraged to use? Why?

4. Pollution is a by-product of producing almost anything. How can MDCs, which historically have been responsible for generating the most pollution, encourage LDCs to seek to minimize the adverse effects of pollution as they improve their levels of development?
5. Malthus argued 200 years ago that overpopulation was inevitable because population increased geometrically while food supply increased arithmetically. Was Malthus correct? Why or why not?

RESOURCES

Some recent and classic books and articles on environmental geography:

Adams, W. M. *The Future of Sustainability: Re-thinking Environment and Development in the Twenty-first Century*. Gland, Switzerland: World Conservation Union. 2006.

Aubrecht, Gordon J., II. *Energy: Physical, Environmental, and Social Impact,* 3rd ed. Upper Saddle River, NJ: Pearson Prentice Hall, 2006.

Balling, Robert C., Jr. "The Geographer's Niche in the Greenhouse Millenium." *Annals of the Association of American Geographers* 90 (2000): 114–22.

Barr, Stewart, "Strategies for Sustainability: Citizens and Responsible Environmental Behaviour." *Area* 35 (2003): 227–40.

Bickerstaff, K., and G. Walker. "The Place(s) of Matter: Matter Out of Place—Public Understandings of Air Pollution." *Progress in Human Geography* 27 (2003): 45–68.

Brown, Lester R., et al. *State of the World*. New York and London: W. W. Norton & Co., annually since 1984.

Buckingham, Hatfield S. "Gender Equality: A Pre-Requisite for Sustainable Development." *Geography* 87 (2002): 227–33.

Bulkeley, Harriet. "Governing Climate Change: The Politics of Risk Society?" *Transactions of the Institute of British Geographers*, New Series 26 (2001): 430–47.

Buttimer, Anne, ed. *Sustainable Landscapes and Lifeways: Scale and Appropriateness*. Cork, Ireland: Cork University Press, 2001.

Commoner, Barry. *Making Peace with the Planet*. New York: New Press, 1990.

Cutter, Susan L., and William H. Renwick. *Exploitation, Conservation, Preservation: A Geographic Perspective on Natural Resource Use*, 4th ed. Danvers, MA: John Wiley, 2004.

Ehrlich, Paul R., and Anne H. Ehrlich. *Betrayal of Science and Reason: How Anti-Environmental Rhetoric Threatens Our Future*. Washington: Island Press, 1998.

Forsyth, Tim. *Critical Political Ecology: The Politics of Environmental Science*. London: Routledge, 2003.

Gerdes, Louise I., ed. *Pollution: Opposing Viewpoints*. Detroit: Greenhaven Press, 2006.

Gonzalez, George A. *The Politics of Air Pollution: Urban Growth, Ecological Modernization, and Symbolic Inclusion*. Albany: State University of New York Press, 2005.

Kidd, J. S., and Renee A. Kidd. *Air Pollution: Problems and Solutions*. New York: Chelsea House, 2006.

Leone, Daniel A., ed. *Is the World Heading Toward an Energy Crisis?* Farmington Hills, MI: Greenhaven Press, 2006.

Livingston, James V., ed. *Focus on Water Pollution Research*. New York: Nova Science Publishers, 2006.

Mallon, Karl, ed. *Renewable Energy Policy and Politics: A Handbook for Decision-making*. London and Sterling, VA: Earthscan, 2006.

Meadows, Donnela H., Jorgen Randers, and Dennis L. Meadows. *Limits to Growth: The 30-Year Update*. White River Junction, VT: Chelsea Green Publishing, 2004.

Myers, Garth Andrew. "Local Communities and the New Environmental Planning: A Case Study from Zanzibar." *Area* 34 (2002): 149–59.

National Geographic. *Energy: Special Report*. Washington, DC: National Geographic Society, 1981.

Ristinen, Robert A., and Jack J. Kraushaar. *Energy and the Environment*, 2nd ed. Hoboken, NJ: John Wiley, 2006.

Turner, B. L., II, Robert W. Kates, and William C. Clark. *The Earth as Transformed by Human Action*. Cambridge: Cambridge University Press, 1990.

Journals featuring environmental geography:

Ecological Economics, Ecologist, Energy Journal, Energy Policy, Environment, Environmental Management, Environmental Pollution, Journal of Environmental Management, and *World Watch*

Key Internet sites:

www.eia.doe.gov Statistics on production, consumption, and reserves of different energy sources in the United States and worldwide can be founded at the web site of the U.S. Department of Energy's Energy Information Administration.

www.epa.gov The U.S. Environmental Protection Agency provides information organized by various types of pollution on its web site.

www.earth-policy.org The Earth Policy Institute tracks environmental indicators that relate to economic development, such as carbon dioxide emissions and fish production.

Log in to www.mygeoscienceplace.com for videos, interactive maps, RSS feeds, case studies, and self-study quizzes to enhance your study of Resource Issues.

Afterword

CAREERS IN GEOGRAPHY

An increasing number of students recognize that geographic education is practical as well as stimulating. Employment opportunities are expanding for students trained in geography, especially in teaching, government service, and business.

Teaching. A doctorate in geography was offered at 68 U.S. and 23 Canadian universities in 2009, and the master's was the highest available degree at 74 U.S. and 4 Canadian universities. Traditionally, most trained geographers became teachers in high schools, colleges, or universities.

A career as a geography teacher is promising because schools throughout North America are expanding their geography curriculum. Educators increasingly recognize geography's role in teaching students about global diversity (Figure A-1). AP Human Geography has grown rapidly in U.S. high schools.

Some university geography departments have emphasized good teaching over research; others are increasingly concerned with research. The Association of American Geographers includes several dozen specialty groups organized around research themes, including agricultural, industrial, medical, and transportation geography.

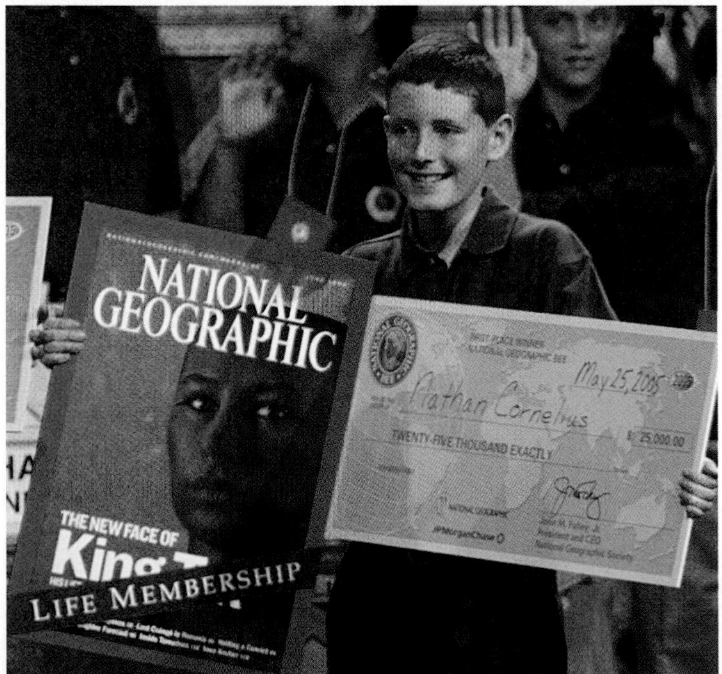

FIGURE A-1 Nathaniel Cornelius, from Cottonwood, Minnesota, won the 2005 National Geographic Bee at age 13.

Government. Geographers contribute their knowledge of the location of activities, the patterns underlying the distribution of various activities, and the interpretation of data from maps and satellite imagery to local, state, and national governments. Employment opportunities with cities, states, provinces, and other units of local government are typically found in departments of planning, transportation, parks and recreation, economic development, housing, zoning, or other similarly titled government agencies. Geographers may be hired to conduct studies of local economic, social, and physical patterns; to prepare information through maps and reports; and to help plan the community's future.

Many federal government agencies also employ geographers:

- The Department of Agriculture hires geographers for the Forest Service and Natural Resources Conservation Service to enhance environmental quality.
- The Department of Commerce hires geographers for the Bureau of the Census to study changing population trends and for the Economic Development Administration to promote rural development.
- The Department of Defense hires geographers for the Defense Intelligence Agency and the National Imagery and Mapping Agency to analyze satellite imagery.
- The Department of Energy hires geographers for the Office of Environmental Policy and Guidance to administer environmental protection programs.
- The Department of Housing and Urban Development hires geographers to help revitalize American cities.
- The Department of Interior hires geographers for the U.S. Geological Survey to study land use and create topographic maps and for the Office of Environmental Policy and Guidance to administer environmental protection programs.
- The Department of State hires geographers for foreign service.
- The Department of Transportation hires geographers to plan new transportation projects.

Business. An increasing number of American geographers are finding jobs with private companies. The list of possibilities is long, but here are some common examples:

- Developers hire geographers to find the best locations for new shopping centers.
- Real estate firms hire geographers to assess the value of properties.
- Supermarket chains, department stores, and other retailers hire geographers to determine the potential market for new stores.

- Banks hire geographers to assess the probability that a loan applicant has planned a successful development.
- Distributors and wholesalers hire geographers to find ways to minimize transportation costs.
- Transnational corporations hire geographers to predict the behavior of consumers and officials in other countries.
- Manufacturers hire geographers to identify new sources of raw materials and markets.
- Utility companies hire geographers to determine future demand at different locations for gas, electricity, and other services.

For more information on careers in geography, contact the Association of American Geographers in Washington, D.C. or the National Council for Geographic Education at Indiana University of Pennsylvania.

Woody Allen's Final Word to the Graduates. This book is supposed to end with a word about the future to graduates. Woody Allen is not a geographer, but his speech to graduates has captured effectively some of the interrelationships among various human actions and the physical environment, which form the core of human geography (Figure A-2).

> *More than any other time in history, mankind faces a crossroads. One path leads to despair and utter hopelessness. The other, to total extinction. Let us pray we have the wisdom to choose correctly. . . .*
>
> *Science is something we depend on all the time. If I develop a pain in the chest I must take an X ray. But what if the radiation from the X ray causes me deeper problems? Before I know it, I'm going in for surgery. Naturally, while they're giving me oxygen an intern decides to light up a cigarette. The next thing you know I'm rocketing over the World Trade Center in bed clothes. . . .*
>
> *At no other time in history has man been so afraid to cut into his veal chop for fear that it will explode. Violence breeds more violence and it is predicted that . . . kidnapping will be the dominant mode of social interaction. Overpopulation will exacerbate problems to the breaking point. Figures tell us that there are already more people on earth than we need to move even the heaviest piano. If we do not call a halt to breeding . . . there will be no room to serve dinner unless one is willing to set the table on the heads of strangers. Then they must not move for an hour while we eat. Of course energy will be in short supply and each car owner will be allowed only enough gasoline to back up a few inches. . . .*
>
> *Summing up, it is clear the future holds great opportunities. It also holds pitfalls. The trick will be to avoid the pitfalls, seize the opportunities, and get back home by six o'clock.*

FIGURE A-2 Woody Allen "poses" between Presidents Calvin Coolidge (left) and Herbert Hoover (right) for his 1983 film *Zelig*.

Human geographers do not know how to solve all of the world's problems of population growth, cultural and political conflict, economic development, and abuse of resources. This course has tried to expose you to the need to understand differences in human actions among different regions, the interdependencies between people and the environment, and other geographic perspectives on world problems. Above all, this book's aim is to heighten your sense of global awareness, that is, an understanding that our comfort—if not survival—requires greater knowledge of Earth's human and physical processes.

Appendix

MAP SCALE AND PROJECTIONS

Phillip C. Muercke

Unaided, our human senses provide a limited view of our surroundings. To overcome those limitations, humankind has developed powerful vehicles of thought and communication, such as language, mathematics, and graphics. Each of those tools is based on elaborate rules; each has an information bias, and each may distort its message, often in subtle ways. Consequently, to use those aids effectively, we must understand their rules, biases, and distortions. The same is true for the special form of graphics we call maps: we must master the logic behind the mapping process before we can use maps effectively.

A fundamental issue in cartography, the science and art of making maps, is the vast difference between the size and geometry of what is being mapped—the real world, we will call it—and that of the map itself. Scale and projection are the basic cartographic concepts that help us understand that difference and its effects.

Map Scale

Our senses are dwarfed by the immensity of our planet; we can sense directly only our local surroundings. Thus, we cannot possibly look at our whole state or country at one time, even though we may be able to see the entire street where we live. Cartography helps us expand what we can see at one time by letting us view the scene from some distant vantage point. The greater the imaginary distance between that position and the object of our observation, the larger the area the map can cover, but the smaller the features will appear on the map. That reduction is defined by the *map scale*, the ratio of the distance on the map to the distance on the earth. Map users need to know about map scale for two reasons: so that they can convert measurements on a map into meaningful real-world measures and so that they can know how abstract the cartographic representation is.

Real-World Measures. A map can provide a useful substitute for the real world for many analytical purposes. With the scale of a map, for instance, we can compute the actual size of its features (length, area, and volume). Such calculations are helped by three expressions of a map scale: a word statement, a graphic scale, and a representative fraction.

A *word statement* of a map scale compares X units on the map to Y units on the earth, often abbreviated "X unit to Y units." For example, the expression "1 inch to 10 miles" means that 1 inch

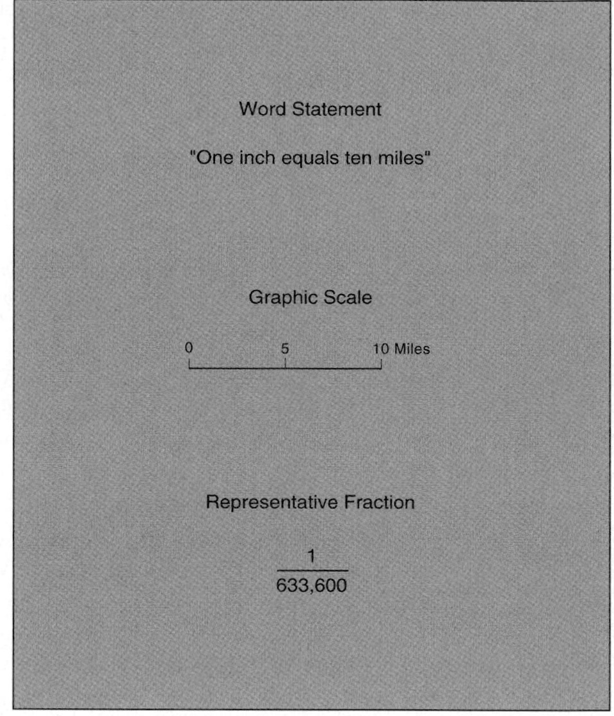

FIGURE A-1 Common expressions of map scale.

on the map represents 10 miles on the earth (Figure A-1). Because the map is always smaller than the area that has been mapped, the ground unit is always the larger number. Both units are expressed in meaningful terms, such as inches or centimeters and miles or kilometers. Word statements are not intended for precise calculations but give the map user a rough idea of size and distance.

A *graphic scale,* such as a bar graph, is concrete and therefore overcomes the need to visualize inches and miles that is associated with a word statement of scale (see Figure A-1). A graphic scale permits direct visual comparison of feature sizes and the distances between features. No ruler is required; any measuring aid will do. It needs only be compared with the scaled bar; if the length of 1 toothpick is equal to 2 miles on the ground and the map distance equals the length of 4 toothpicks, then the ground distance is 4 times 2, or 8 miles. Graphic scales are especially convenient in this age of copying machines, when we are more likely to be working with a copy than with the original map. If a map is reduced or enlarged as it is copied, the graphic scale will change in proportion to the change in the size of the map and thus will remain accurate.

The third form of a map scale is the *representative fraction* (RF). An RF defines the ratio between the distance on the map and the distance on the earth in fractional terms, such as 1/633,600 (also written 1:633,600). The numerator of the fraction always refers to the distance on the map, and the denominator always refers to the distance on the earth. No units of measurement are given, but both numbers must be expressed in the same units. Because map distances are extremely small relative to the size of the earth, it makes sense to use small units, such as inches or centimeters. Thus the RF 1:633,600 might be read as "1 inch on the map to 633,600 inches on the earth."

Herein lies a problem with the RF. Meaningful map-distance units imply a denominator so large that it is impossible to visualize. Thus, in practice, reading the map scale involves an additional step of converting the denominator to a meaningful ground measure, such as miles or kilometers. The unwieldy 633,600 becomes the more manageable 10 miles when divided by the number of inches in a mile (63,360).

On the plus side, the RF is good for calculations. In particular, the ground distance between points can be easily determined from a map with an RF. One simply multiplies the distance between the points on the map by the denominator of the RF. Thus a distance of 5 inches on a map with an RF of 1/126,720 would signify a ground distance of 5 X 126,720, which equals 633,600. Because all units are inches and there are 63,360 inches in a mile, the ground distance is 633,600/63,360, or 10 miles. Computation of area is equally straightforward with an RF. Computer manipulation and analysis of maps is based on the RF form of map scale.

Guides to Generalization. Scales also help map users visualize the nature of the symbolic relation between the map and the real world. It is convenient here to think of maps as falling into three broad scale categories (Figure A-2). (Do not be confused by the use of the words large AND small in this context; just remember that the larger the denominator, the smaller the scale ratio and the larger the area that is shown on the map.) Scale ratios greater than 1:100,000, such as the 1:24,000 scale of U.S. Geological Survey topographic quadrangles, are large-scale maps. Although those maps can cover only a local area, they can be drawn to rather rigid standards of accuracy. Thus they are useful for a wide range of applications that require detailed and accurate maps, including zoning, navigation, and construction.

At the other extreme are maps with scale ratios of less than 1:1,000,000, such as maps of the world that are found in atlases. Those are small-scale maps. Because they cover large areas, the symbols on them must be highly abstract. They are therefore best suited to general reference or planning, when detail is not important. Medium- or intermediate-scale maps have scales between 1:100,000 and 1:1,000,000. They are good for regional reference and planning purposes.

Another important aspect of map scale is to give us some notion of geometric accuracy; the greater the expanse of the real world shown on a map, the less accurate the geometry of

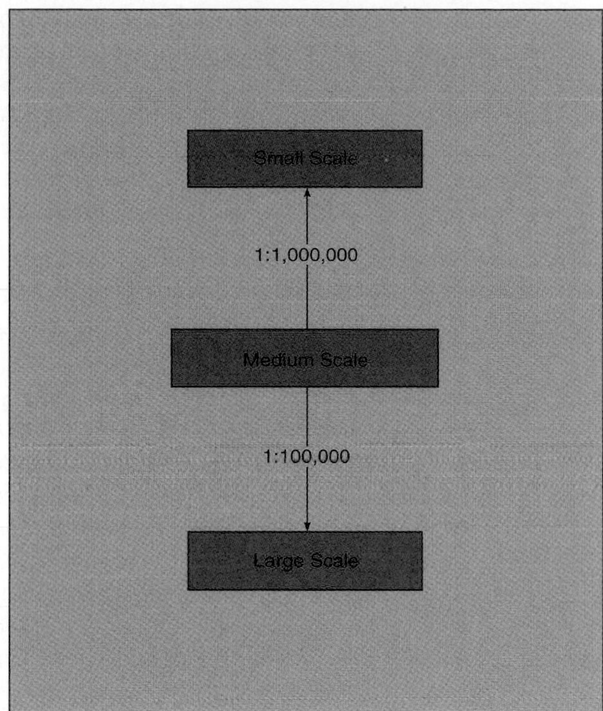

FIGURE A-2 The scale gradient can be divided into three broad categories.

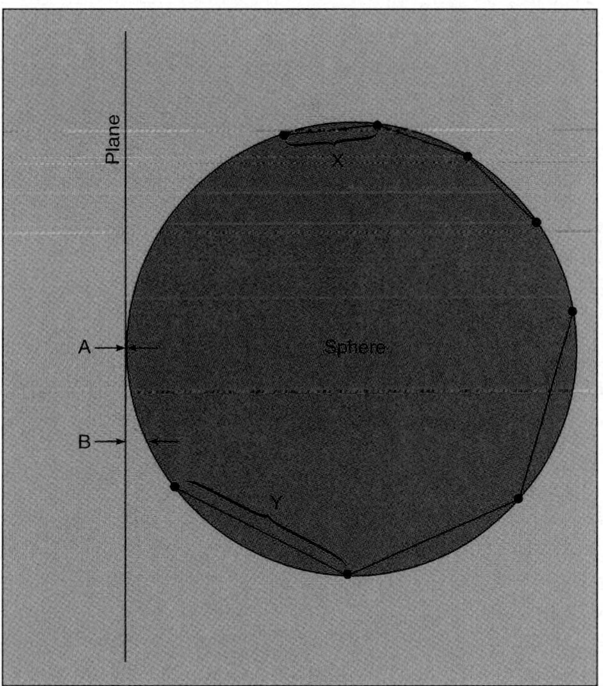

FIGURE A-3 Relationships between surfaces on the round earth and a flat map.

that map is. Figure A-3 shows why. If a curve is represented by straight line segments, short segments (*X*) are more similar to the curve than are long segments (*Y*). Similarly, if a plane is placed in contact with a sphere, the difference between the two surfaces is slight where they touch (*A*) but grows rapidly with increasing distance from the point of contact (*B*). In view of the large diameter and slight local curvature of the earth, distances will be well represented on large-scale maps (those with small denominators) but will be increasingly poorly represented at smaller scales. This close relationship between map scale and map geometry brings us to the topic of map projections.

Map Projections

The spherical surface of the earth is shown on flat maps by means of map projections. The process of "flattening" the earth is essentially a problem in geometry that has captured the attention of the best mathematical minds for centuries. Yet no one has ever found a perfect solution; there is no known way to avoid spatial distortion of one kind or another. Many map projections have been devised, but only a few have become standard. Because a single flat map cannot preserve all aspects of the earth's surface geometry, a mapmaker must be careful to match the projection with the task at hand. To map something that involves distance, for example, a projection should be used in which distance is not distorted. In addition, a map user should be able to recognize which aspects of a map's geometry are accurate and which are distortions caused by a particular projection process. Fortunately, that objective is not too difficult to achieve.

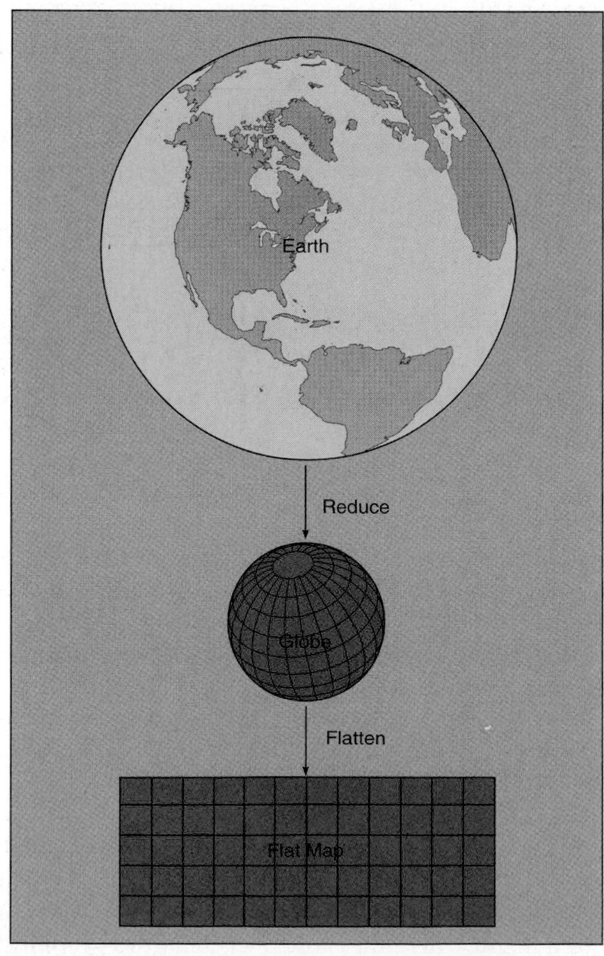

FIGURE A-4 The two-step process of creating a projection.

It is helpful to think of the creation of a projection as a two-step process (Figure A-4). First, the immense earth is reduced to a small globe with a scale equal to that of the desired flat map. All spatial properties on the globe are true to those on the earth. Second, the globe is flattened. Since that cannot be done without distortion, it is accomplished in such a way that the resulting map exhibits certain desirable spatial properties.

Perspective Models. Early map projections were sometimes created with the aid of perspective methods, but that has changed. In the modern electronic age, projections are normally developed by strictly mathematical means and are plotted out or displayed on computer-driven graphics devices. The concept of perspective is still useful in visualizing what map projections do, however. Thus projection methods are often illustrated by using strategically located light sources to cast shadows on a projection surface from a latitude/longitude net inscribed on a transparent globe.

The success of the perspective approach depends on finding a projection surface that is flat or that can be flattened without distortion. The cone, cylinder, and plane possess those attributes and serve as models for three general classes of map projections: *conic, cylindrical,* and *planar* (or azimuthal). Figure A-5 shows those three classes, as well as a fourth, a false cylindrical class with an oval shape. Although the oval class is not of perspective origin, it appears to combine properties of the cylindrical and planar classes (Figure A-6).

The relationship between the projection surface and the model at the point or line of contact is critical because distortion of spatial properties on the projection is symmetrical about, and increases with distance from, that point or line. That condition is illustrated for the cylindrical and planar classes of projections in Figure A-7. If the point or line of contact is changed to some other position on the globe, the distortion pattern will be recentered on the new position but will retain the same symmetrical form. Thus centering a projection on the area of interest on the earth's surface can minimize the effects of projection distortion. And recognizing the general projection shape, associating it with a perspective model, and recalling the characteristic distortion pattern will provide the information necessary to compensate for projection distortion.

Preserved Properties. For a map projection to truthfully depict the geometry of the earth's surface, it would have to

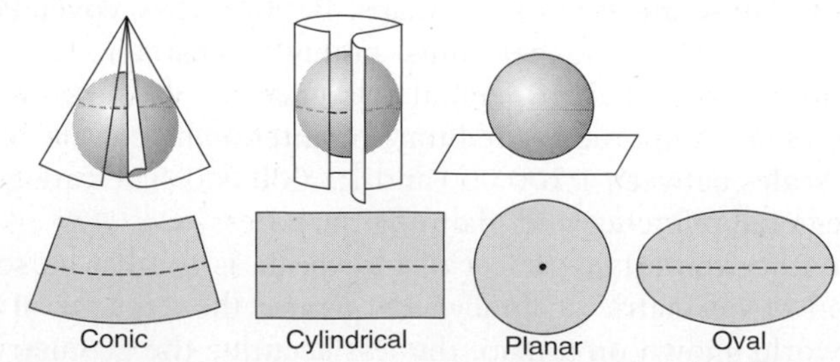

FIGURE A-5 General classes of map projections. *(Courtesy of ACSM)*

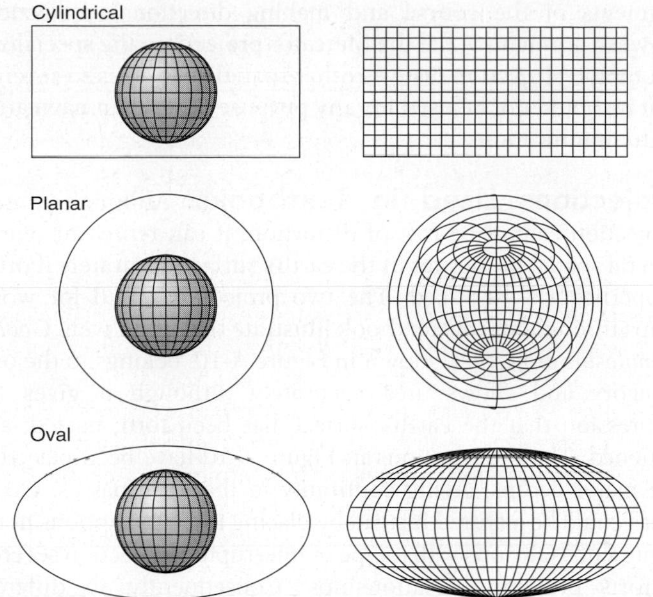

FIGURE A-6 The visual properties of cylindrical and planar projections combined in oval projections. *(Courtesy of ACSM)*

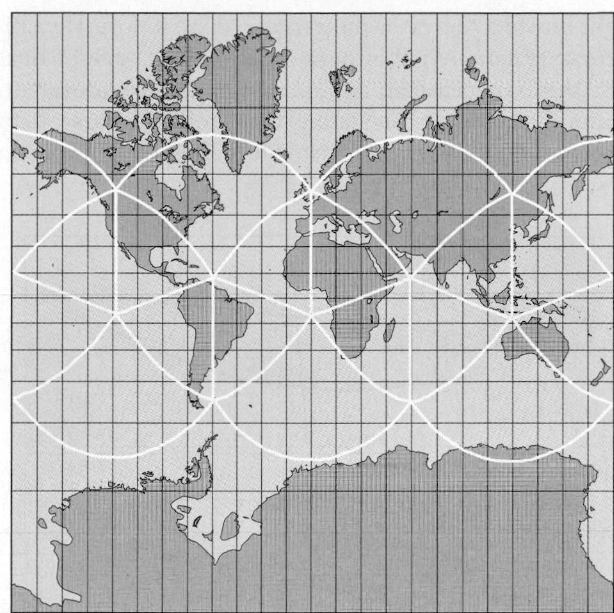

FIGURE A-8 The useful Mercator projection, showing extreme area distortion in the higher latitudes. *(Courtesy of ACSM)*

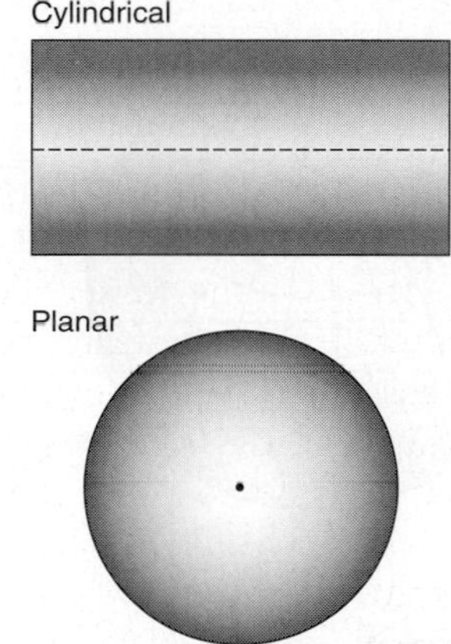

FIGURE A-7 Characteristic patterns of distortion for two projection classes. Here, darker shading implies greater distortion. *(Courtesy of ACSM)*

preserve the spatial attributes of *distance, direction, area, shape*, and *proximity*. That task can be readily accomplished on a globe, but it is not possible on a flat map. To preserve area, for example, a mapmaker must stretch or shear shapes; thus area and shape cannot be preserved on the same map. To depict both direction and distance from a point, area must be distorted. Similarly, to preserve area as well as direction from a point, distance has to be distorted. Because the earth's surface is continuous in all directions from every point, discontinuities that violate proximity relationships must occur on all map projections. The trick is to place those discontinuities where they will have the least impact on the spatial relationships in which the map user is interested.

We must be careful when we use spatial terms, because the properties they refer to can be confusing. The geometry of the familiar plane is very different from that of a sphere; yet when we refer to a flap map, we are in fact making reference to the spherical earth that was mapped. A shape-preserving projection, for example, is truthful to local shapes—such as the right-angle crossing of latitude and longitude lines—but does not preserve shapes at continental or global levels. A distance-preserving projection can preserve that property from one point on the map in all directions or from a number of points in several directions, but distance cannot be preserved in the general sense that area can be preserved. Direction can also be generally preserved from a single point or in several directions from a number of points but not from all points simultaneously. Thus a shape-, distance-, or direction-preserving projection is truthful to those properties only in part.

Partial truths are not the only consequence of transforming a sphere into a flat surface. Some projections exploit that transformation by expressing traits that are of considerable value for specific applications. One of those is the famous shape-preserving *Mercator projection* (Figure A-8). That cylindrical projection was derived mathematically in the 1500s so that a compass bearing (called rhumb lines) between any two points on the earth would plot as straight lines on the map. That trait let navigators plan, plot, and follow courses between origin and destination, but it was achieved at the expense of extreme areal distortion toward the margins of the projection (see Antarctica in Figure A-8). Although the Mercator projection is admirably suited for its intended purpose, its widespread but inappropriate use for nonnavigational purposes has drawn a great deal of criticism.

The *gnomonic projection* is also useful for navigation. It is a planar projection with the valuable characteristic of showing the shortest (or great circle) route between any two points on the earth as straight lines. Long-distance navigators first plot the great circle course between origin and destination on a gnomonic projection (Figure A-9, top). Next they transfer the straight line to a Mercator projection, where it normally appears as a curve (Figure A-9, bottom). Finally, using straight-line segments, they construct an approximation of that course on the Mercator projection. Navigating the shortest course between origin and destination then involves following the straight segments of the course and making directional corrections between segments. Like the Mercator projection, the specialized gnomonic projection distorts other spatial properties so severely that it should not be used for any purpose other than navigation or communications.

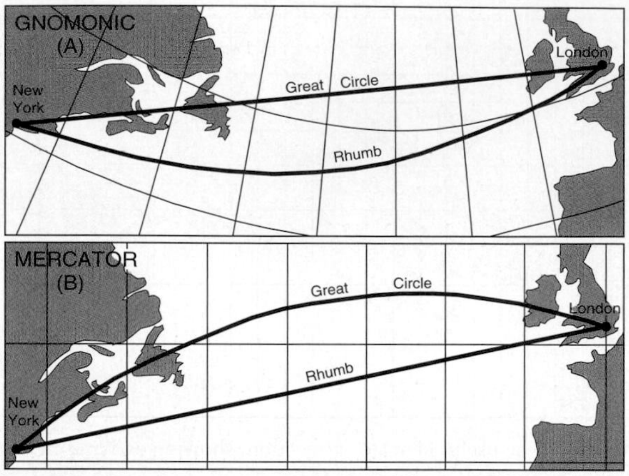

GNOMONIC (A)
New York
Great Circle
Rhumb
London
MERCATOR (B)

FIGURE A-9 A gnomonic projection (A) and a Mercator projection (B), both of value to long-distance navigators.

Projections Used in Textbooks. Although a map projection cannot be free of distortion, it can represent one or several spatial properties of the earth's surface accurately if other properties are sacrificed. The two projections used for world maps throughout this textbook illustrate that point well. *Goode's homolosine projection*, shown in Figure A-10, belongs to the oval category and shows area accurately, although it gives the impression that the earth's surface has been torn, peeled, and flattened. The interruptions in Figure A-10 have been placed in the major oceans, giving continuity to the land masses. Ocean areas could be featured instead by placing the interruptions in the continents. Obviously, that type of interrupted projection severely distorts proximity relationships. Consequently, in different locations the properties of distance, direction, and shape are also distorted to varying degrees. The distortion pattern mimics that of cylindrical projections, with the equatorial zone the most faithfully represented (Figure A-11).

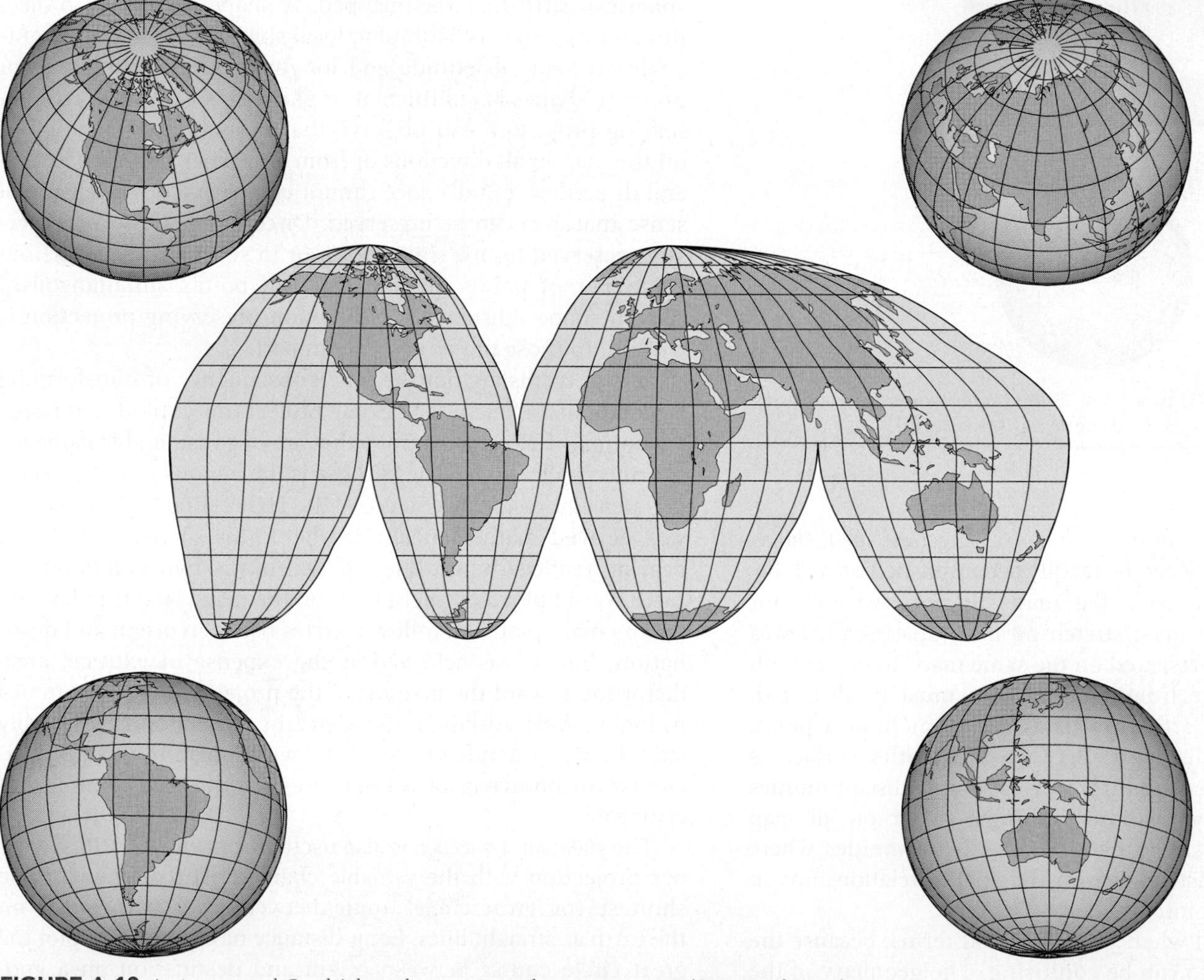

FIGURE A-10 An interrupted Goode's homolosine, an equal-area projection. *(Courtesy of ACSM)*

An alternative to special-property projections such as the equal-area Goode's homolosine is the compromise projection. In that case no special property is achieved at the expense of others, and distortion is rather evenly distributed among the various properties, instead of being focused on one or several properties. The *Robinson projection*, which is also used in this textbook, falls into that category (Figure A-12). Its oval projection has a global feel, somewhat like that of Goode's homolosine. But the Robinson projection shows the North Pole and the South Pole as lines that are slightly more than half the length of the equator, thus exaggerating distances and areas near the poles. Areas look larger than they really are in the high latitudes (near the poles) and smaller than they really are in the low latitudes (near the equator). In addition, not all latitude and longitude lines intersect at right angles, as they do on the earth, so we know that the Robinson projection does not preserve direction or shape either. However, it has fewer interruptions than the Goode's homolosine does, so it preserves proximity better. Overall, the Robinson projection does a good job of representing spatial relationships, especially in the low to middle latitudes and along the central meridian.

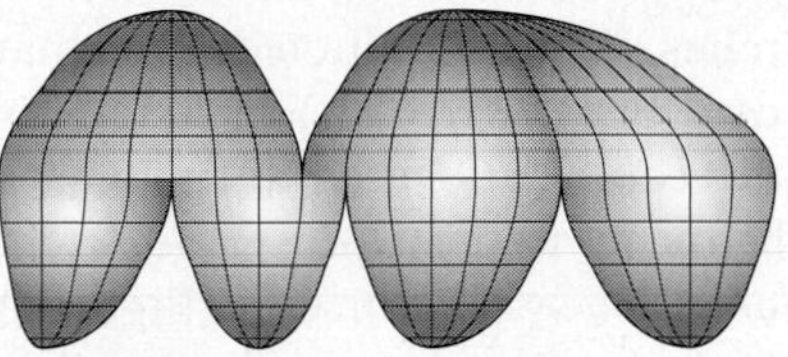

FIGURE A-11 The distortion pattern of the interrupted Goode's homolosine projection, which mimics that of cylindrical projections. *(Courtesy of ACSM)*

Scale and Projections in Modern Geography

Computers have drastically changed the way in which maps are made and used. In the preelectronic age, maps were so laborious, time-consuming, and expensive to make that

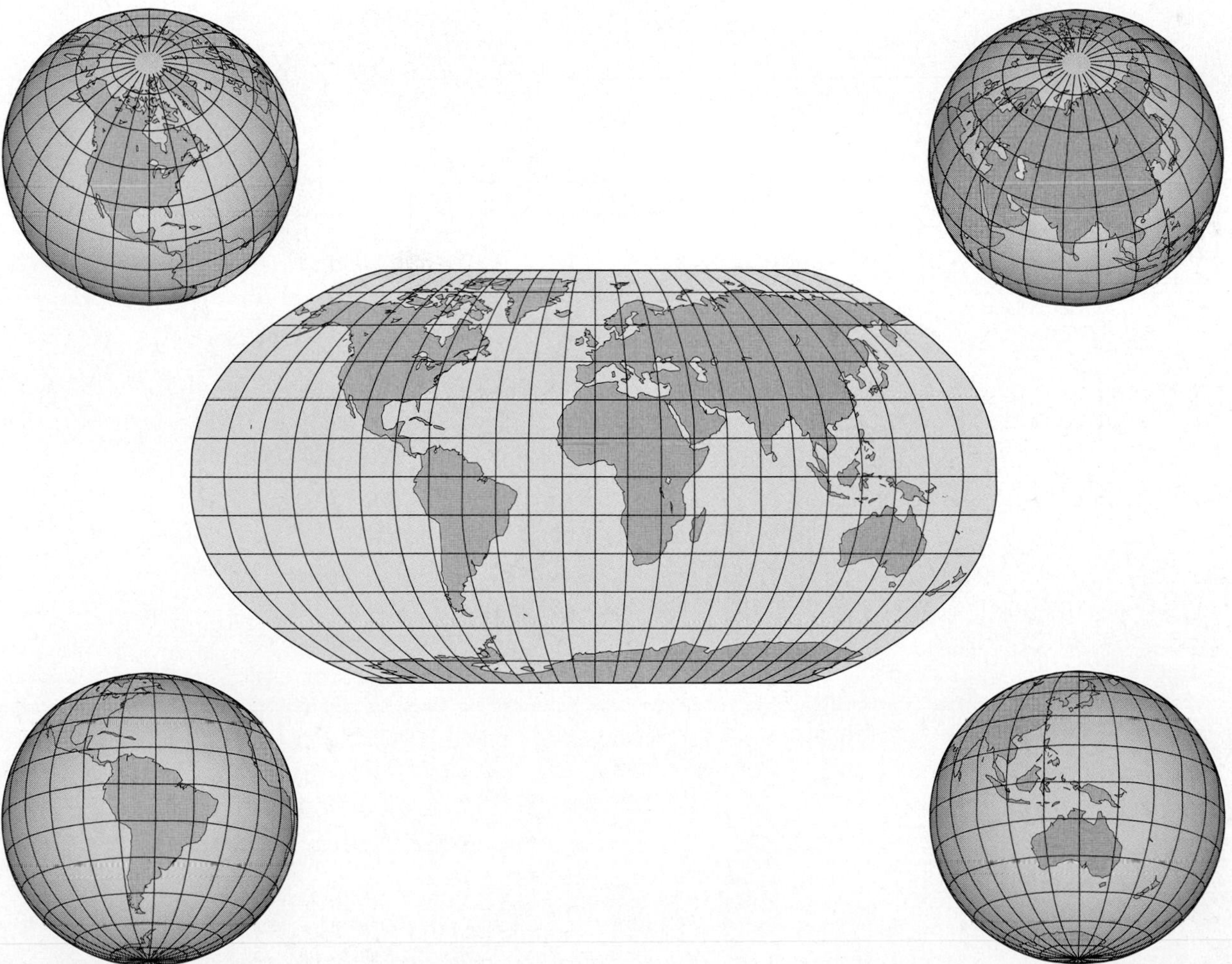

FIGURE A-12 The compromise Robinson projection, which avoids the interruptions of Goode's homolosine but preserves no special properties. *(Courtesy of ACSM)*

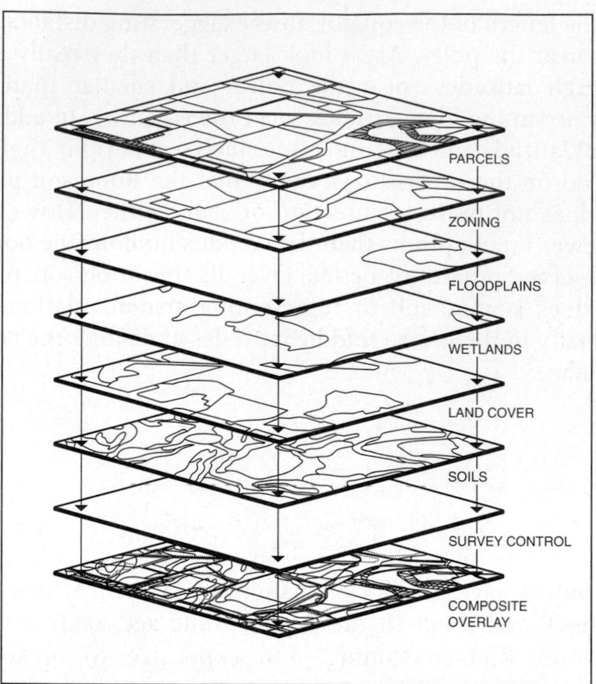

FIGURE A-13 Within a GIS, environmental data attached to a common terrestrial reference system, such as latitude/longitude, can be stacked in layers for spatial comparison and analysis.

relatively few were created. Frustrated, geographers and other scientists often found themselves trying to use maps for purposes not intended by the map designers. But today anyone with access to computer mapping facilities can create projections in a flash. Thus projections will be increasingly tailored to specific needs, and more and more scientists will do their own mapping rather than have someone else guess what they want in a map.

Computer mapping creates opportunities that go far beyond the construction of projections, of course. Once maps and related geographical data are entered into computers, many types of analyses can be carried out involving map scales and projections. Distances, areas, and volumes can be computed; searches can be conducted; information from different maps can be combined; optimal routes can be selected; facilities can be allocated to the most suitable sites; and so on. The term used to describe such processes is *geographical information system*, or GIS (Figure A-13). Within a GIS, projections provide the mechanism for linking data from different sources, and scale provides the basis for size calculations of all sorts. Mastery of both projection and scale becomes the user's responsibility because the map user is also the map maker. Now more than ever, effective geography depends on knowledge of the close association between scale and projection.

Key Terms

Acid deposition (p. 452) The combination of sulfur oxides and nitrogen oxides, emitted by burning fossil fuels, in the atmosphere with oxygen and water to form sulfuric acid and nitric acid and their return to Earth's surface.

Acid precipitation (p. 452) Conversion of sulfur oxides and nitrogen oxides to acids that return to Earth as rain, snow, or fog.

Active solar energy systems (p. 463) Solar energy systems that collect energy through the use of physical devices like photovoltaic cells or flat-plate collectors.

Agribusiness (p. 313) Commercial agriculture characterized by the integration of different steps in the food-processing industry, usually through ownership by large corporations.

Agricultural density (p. 33, 52) The ratio of the number of farmers to the total amount of land suitable for agriculture.

Agricultural revolution (p. 57) The time when human beings first domesticated plants and animals and no longer relied entirely on hunting and gathering.

Agriculture (p. 309) The deliberate effort to modify a portion of Earth's surface through the cultivation of crops and the raising of livestock for sustenance or economic gain.

Air pollution (p. 451) Concentration of trace substances, such as carbon monoxide, sulfur dioxide, nitrogen oxides, hydrocarbons, and solid particulates, at a greater level than occurs in average air.

Animate power (p. 441) Power supplied by people or animals.

Animism (p. 178) Belief that objects, such as plants and stones, or natural events, like thunderstorms and earthquakes, have a discrete spirit and conscious life.

Annexation (p. 424) Legally adding land area to a city in the United States.

Apartheid (p. 215) Laws (no longer in effect) in South Africa that physically separated different races into different geographic areas.

Arithmetic density (p. 32, 50) The total number of people divided by the total land area.

Autonomous religion (p. 195) A religion that does not have a central authority but shares ideas and cooperates informally.

Balance of power (p. 257) Condition of roughly equal strength between opposing countries or alliances of countries.

Balkanization (p. 232) Process by which a state breaks down through conflicts among its ethnicities.

Balkanized (p. 232) A small geographic area that could not successfully be organized into one or more stable states because it is inhabited by many ethnicities with complex, long-standing antagonisms toward each other.

Base line (p. 9) An east-west line designated under the Land Ordinance of 1785 to facilitate the surveying and numbering of townships in the United States.

Basic industries (p. 398) Industries that sell their products or services primarily to consumers outside the settlement.

Biochemical oxygen demand (BOD) (p. 456) Amount of oxygen required by aquatic bacteria to decompose a given load of organic waste; a measure of water pollution.

Biodiversity (p. 469) The number of species within a specific habitat.

Biomass fuel (p. 441) Fuel that derives from plant material and animal waste.

Blockbusting (p. 215) A process by which real estate agents convince white property owners to sell their houses at low prices out of fear that persons of color will soon move into the neighborhood.

Boundary (p. 247) Invisible line that marks the extent of a state's territory.

Brain drain (p. 92) Large-scale emigration by talented people.

Branch (p. 171) A large and fundamental division within a religion.

Break-of-bulk point (p. 355) A location where transfer is possible from one mode of transportation to another.

Breeder reactor (p. 460) A nuclear power plant that creates its own fuel from plutonium.

British Received Pronunciation (BRP) (p. 139) The dialect of English associated with upper-class Britons living in London and now considered standard in the United Kingdom.

Bulk-gaining industry (p. 352) An industry in which the final product weighs more or comprises a greater volume than the inputs.

Bulk-reducing industry (p. 350) An industry in which the final product weighs less or comprises a lower volume than the inputs.

Business services (p. 375) Services that primarily meet the needs of other businesses, including professional, financial, and transportation services.

Cartography (p. 5) The science of making maps.

Caste (p. 196) The class or distinct hereditary order into which a Hindu is assigned according to religious law.

Census (p. 62) A complete enumeration of a population.

Census tract (p. 412) An area delineated by the U.S. Bureau of the Census for which statistics are published; in urbanized areas, census tracts correspond roughly to neighborhoods.

Central business district (CBD) (p. 406) The area of a city where retail and office activities are clustered.

Central place (p. 388) A market center for the exchange of services by people attracted from the surrounding area.

Central place theory (p. 388) A theory that explains the distribution of services, based on the fact that settlements serve as centers of market areas for services; larger settlements are fewer and farther apart than smaller settlements and provide services for a larger number of people who are willing to travel farther.

Centripetal force (p. 219) An attitude that tends to unify people and enhance support for a state.

Cereal grain (p. 323) A grass yielding grain for food.

Chaff (p. 320) Husks of grain separated from the seed by threshing.

Chain migration (p. 90) Migration of people to a specific location because relatives or members of the same nationality previously migrated there.

Chlorofluorocarbon (CFC) (p. 452) A gas used as a solvent, a propellant in aerosols, a refrigerant, and in plastic foams and fire extinguishers.

Circulation (p. 80) Short-term, repetitive, or cyclical movements that recur on a regular basis.

City (p. 424) An Urban settlement that has been legally incorporated into an independent, self-governing unit.

City-state (p. 379, 243) A sovereign state comprising a city and its immediate hinterland.

Clustered rural settlement (p. 381) A rural settlement in which the houses and farm buildings of each family are situated close to each other and fields surround the settlement.

Colonialism (p. 244) An attempt by one country to establish settlements and to impose its political, economic, and cultural principles in another territory.

Colony (p. 243) A territory that is legally tied to a sovereign state rather than completely independent.

Combine (p. 326) A machine that reaps, threshes, and cleans grain while moving over a field.

Combined statistical area (CSA) (p. 425) In the United States, two or more contiguous core based statistical areas tied together by commuting patterns.

Commercial agriculture (p. 311) Agriculture undertaken primarily to generate products for sale off the farm.

Compact state (p. 247) A state in which the distance from the center to any boundary does not vary significantly.

Concentration (p. 33) The spread of something over a given area.

Concentric zone model (p. 410) A model of the internal structure of cities in which social groups are spatially arranged in a series of rings.

Connections (p. 5) Relationships among people and objects across the barrier of space.

Conservation (p. 467) The sustainable use and management of a natural resource, through consuming it at a less rapid rate than it can be replaced.

Consumer services (p. 375) Businesses that provide services primarily to individual consumers, including retail services and education, health, and leisure services.

Contagious diffusion (p. 39) The rapid, widespread diffusion of a feature or trend throughout a population.

Core based statistical area (CBSA) (p. 425) In the United States, the combination of all metropolitan statistical areas and micropolitan statistical areas.

Cosmogony (p. 188) A set of religious beliefs concerning the origin of the universe.

Cottage industry (p. 345) Manufacturing based in homes rather than in a factory; commonly found prior to the Industrial Revolution.

Council of government (p. 426) A cooperative agency consisting of representatives of local governments in a metropolitan area in the United States.

Counterurbanization (p. 99) Net migration from urban to rural areas in more developed countries.

Creole or creolized language (p. 149) A language that results from the mixing of a colonizer's language with the indigenous language of the people being dominated.

Crop (p. 309) Grain or fruit gathered from a field as a harvest during a particular season.

Crop rotation (p. 321) The practice of rotating use of different fields from crop to crop each year, to avoid exhausting the soil.

Crude birth rate (CBR) (p. 53) The total number of live births in a year for every 1,000 people alive in the society.

Crude death rate (CDR) (p. 53) The total number of deaths in a year for every 1,000 people alive in the society.

Cultural ecology (p. 24) Geographic approach that emphasizes human-environment relationships.

Cultural landscape (p. 17) The fashioning of a natural landscape by a cultural group.

Culture (p. 21) The body of customary beliefs, social forms, and material traits that together constitute the distinct tradition of a group of people.

Custom (p. 106) The frequent repetition of an act, to the extent that it becomes characteristic of the group of people performing the act.

Demographic transition (p. 56) The process of change in a society's population from a condition of high crude birth and death rates and low rate of natural increase to a condition of low crude birth and death rates, low rate of natural increase, and a higher total population.

Demography (p. 45) The scientific study of population characteristics.

Denglish (p. 164) Combination of German and English.

Denomination (p. 171) A division of a branch that unites a number of local religious congregations in a single legal and administrative body.

Density (p. 32) The frequency with which something exists within a given unit of area.

Density gradient (p. 427) The change in density in an urban area from the center to the periphery.

Dependency ratio (p. 59) The number of people under the age of 15 and over age 64, compared to the number of people active in the labor force.

Desertification (p. 335) The degradation of land, especially in semiarid areas, primarily because of human actions like excessive crop planting, animal grazing, and tree cutting.

Development (p. 274) A process of improvement in the material conditions of people through diffusion of knowledge and technology.

Dialect (p. 139) A regional variety of a language distinguished by vocabulary, spelling, and pronunciation.

Diffusion (p. 36) The spreading of a feature or trend from one place to another over time.

Dispersed rural settlement (p. 381) A rural settlement pattern characterized by isolated farms rather than clustered villages.

Distance decay (p. 36) The diminishing in importance and eventual disappearance of a phenomenon with increasing distance from its origin.

Distribution (p. 32) The arrangement of something across Earth's surface.

Double cropping (p. 321) Harvesting twice a year from the same field.

Doubling time (p. 53) The number of years needed to double a population, assuming a constant rate of natural increase.

Ebonics (p. 162) Dialect spoken by some African Americans.

Economic base (p. 398) A community's collection of basic industries.

Ecumene (p. 49) The portion of Earth's surface occupied by permanent human settlement.

Edge city (p. 427) A large node of office and retail activities on the edge of an urban area.

Elongated state (p. 247) A state with a long, narrow shape.

Emigration (p. 80) Migration *from* a location.

Enclosure movement (p. 384) The process of consolidating small landholdings into a smaller number of larger farms in England during the eighteenth century.

Environmental determinism (p. 24) A nineteenth- and early twentieth-century approach to the study of geography that argued that the general laws sought by human geographers could be found in the physical sciences. Geography was therefore the study of how the physical environment caused human activities.

Epidemiologic transition (p. 71) Distinctive causes of death in each stage of the demographic transition.

Epidemiology (p. 71) The branch of medical science concerned with the incidence, distribution, and control of diseases that affect large numbers of people.

Ethnic cleansing (p. 229) A process in which a more powerful ethnic group forcibly removes a less powerful one in order to create an ethnically homogeneous region.

Ethnic religion (p. 170) A religion with a relatively concentrated spatial distribution whose principles are likely to be based on the physical characteristics of the particular location in which its adherents are concentrated.

Ethnicity (p. 208) Identity with a group of people that share distinct physical and mental traits as a product of common heredity and cultural traditions.

Expansion diffusion (p. 38) The spread of a feature or trend among people from one area to another in a snowballing process.

Extinct language (p. 156) A language that was once used by people in daily activities but is no longer used.

Fair trade (p. 301) An alternative to international trade that emphasizes small businesses and worker-owned and democratically run cooperatives and requires employers to pay workers fair wages, permit union organizing, and comply with minimum environmental and safety standards.

Federal state (p. 254) An internal organization of a state that allocates most powers to units of local government.

Ferrous (p. 449) Metals, including iron that are utilized in the production of iron and steel.

Filtering (p. 419) A process of change in the use of a house, from single-family owner-occupancy to abandonment.

Fission (p. 459) The splitting of an atomic nucleus to release energy.

Floodplain (p. 82) The area subject to flooding during a given number of years according to historical trends.

Folk culture (p. 106) Culture traditionally practiced by a small, homogeneous, rural group living in relative isolation from other groups.

Forced migration (p. 84) Permanent movement compelled usually by cultural factors.

Fordist production (p. 368) Form of mass production in which each worker is assigned one specific task to perform repeatedly.

Foreign direct investment (p. 298) Investment made by a foreign company in the economy of another country.

Formal region (p. 17) (or uniform or homogeneous region) An area in which everyone shares in one or more distinctive characteristics.

Fossil fuel (p. 441) Energy source formed from the residue of plants and animals buried millions of years ago.

Fragmented state (p. 248) A state that includes several discontinuous pieces of territory.

Franglais (p. 164) A term used by the French for English words that have entered the French language; a combination of *français* and *anglais*, the French words for "French" and "English," respectively.

Frontier (p. 253) A zone separating two states in which neither state exercises political control.

Functional region (p. 19) (or nodal region) An area organized around a node or focal point.

Fundamentalism (p. 197) The literal interpretation and strict adherence to basic principles of a religion (or a religious branch, denomination, or sect).

Fusion (p. 461) Creation of energy by joining the nuclei of two hydrogen atoms to form helium.

Gender Empowerment Measure (GEM) (p. 289) Compares the ability of women and men to participate in economic and political decision making.

Gender-Related Development Index (GDI) (p. 289) Compares the level of development of women with that of both sexes.

Gentrification (p. 420) A process of converting an urban neighborhood from a predominantly low-income, renter-occupied area to a predominantly middle-class, owner-occupied area.

Geographic information system (GIS) (p. 12) A computer system that stores, organizes, analyzes, and displays geographic data.

Geothermal energy (p. 462) Energy from steam or hot water produced from hot or molten underground rocks.

Gerrymandering (p. 255) The process of redrawing legislative boundaries for the purpose of benefiting the party in power.

Ghetto (p. 185) During the Middle Ages, a neighborhood in a city set up by law to be inhabited only by Jews; now used to denote a section of a city in which members of any minority group live because of social, legal, or economic pressure.

Global Positioning System (GPS) (p. 9) A system that determines the precise position of something on Earth through a series of satellites, tracking stations, and receivers.

Globalization (p. 29) Actions or processes that involve the entire world and result in making something worldwide in scope.

Grain (p. 325) Seed of a cereal grass.

Gravity model (p. 391) A model that holds that the potential use of a service at a particular location is directly related to the number of people in a location and inversely related to the distance people must travel to reach the service.

Green revolution (p. 336) Rapid diffusion of new agricultural technology, especially new high-yield seeds and fertilizers.

Greenbelt (p. 429) A ring of land maintained as parks, agriculture, or other types of open space to limit the sprawl of an urban area.

Greenhouse effect (p. 451) The anticipated increase in Earth's temperature, caused by carbon dioxide (emitted by burning fossil fuels) trapping some of the radiation emitted by the surface.

Greenwich Mean Time (GMT) (p. 18) The time in that time zone encompassing the prime meridian, or 0° longitude.

Gross domestic product (GDP) (p. 275) The value of the total output of goods and services produced in a country in a given time period (normally 1 year).

Guest workers (p. 93) Workers who migrate to the more developed countries of Northern and Western Europe, usually from Southern and Eastern Europe or from North Africa, in search of higher-paying jobs.

Habit (p. 106) A repetitive act performed by a particular individual.

Hearth (p. 36) The region from which innovative ideas originate.

Hierarchical diffusion (p. 39) The spread of a feature or trend from one key person or node of authority or power to other persons or places.

Hierarchical religion (p. 194) A religion in which a central authority exercises a high degree of control.

Horticulture (p. 328) The growing of fruits, vegetables, and flowers.

Hull (p. 320) The outer covering of a seed.

Human Development Index (HDI) (p. 274) An indicator of the level of development for each country, constructed by the United Nations, combining income, literacy, education, and life expectancy.

Hydroelectric power (p. 461) Power generated from moving water.

Ideograms (p. 152) The system of writing used in China and other East Asian countries in which each symbol represents an idea or a concept rather than a specific sound, as is the case with letters in English.

Immigration (p. 80) Migration *to* a new location.

Imperialism (p. 245) Control of territory already occupied and organized by an indigenous group.

Inanimate power (p. 441) Power supplied by machines.

Industrial Revolution (p. 345, 57) A series of improvements in industrial technology that transformed the process of manufacturing goods.

Infant mortality rate (IMR) (p. 55) The total number of deaths in a year among infants under 1 year old for every 1,000 live births in a society.

Intensive subsistence agriculture (p. 319) A form of subsistence agriculture in which farmers must expend a relatively large amount of effort to produce the maximum feasible yield from a parcel of land.

Internal migration (p. 84) Permanent movement within a particular country.

International Date Line (p. 18) An arc that for the most part follows 180° longitude, although it deviates in several places to avoid dividing land areas. When you cross the International Date Line heading east (toward America), the clock moves back 24 hours, or one entire day. When you go west (toward Asia), the calendar moves ahead one day.

International migration (p. 84) Permanent movement from one country to another.

Interregional migration (p. 84) Permanent movement from one region of a country to another.

Intervening obstacle (p. 83) An environmental or cultural feature of the landscape that hinders migration.

Intraregional migration (p. 84) Permanent movement within one region of a country.

Isogloss (p. 139) A boundary that separates regions in which different language usages predominate.

Isolated language (p. 160) A language that is unrelated to any other languages and therefore not attached to any language family.

Labor-intensive industry (p. 356) An industry for which labor costs comprise a high percentage of total expenses.

Land Ordinance of 1785 (p. 9) A law that divided much of the United States into townships to facilitate the sale of land to settlers.

Landlocked state (p. 249) A state that does not have a direct outlet to the sea.

Language (p. 136) A system of communication through the use of speech, a collection of sounds understood by a group of people to have the same meaning.

Language branch (p. 143) A collection of languages related through a common ancestor that existed several thousand years ago. Differences are not as extensive or as old as with language families, and archaeological evidence can confirm that the branches derived from the same family.

Language family (p. 143) A collection of languages related to each other through a common ancestor that existed long before recorded history.

Language group (p. 144) A collection of languages within a branch that share a common origin in the relatively recent past and display relatively few differences in grammar and vocabulary.

Latitude (p. 15) The numbering system used to indicate the location of parallels drawn on a globe and measuring distance north and south of the equator (0°).

Less developed country (LDC) (p. 274) A country that is at a relatively early stage in the process of economic development.

Life expectancy (p. 55) The average number of years an individual can be expected to live, given current social, economic, and medical conditions. Life expectancy at birth is the average number of years a newborn infant can expect to live.

Lingua franca (p. 162) A language mutually understood and commonly used in trade by people who have different native languages.

Literacy rate (p. 278) The percentage of a country's people who can read and write.

Literary tradition (p. 136) A language that is written as well as spoken.

Location (p. 13) The position of anything on Earth's surface.

Longitude (p. 15) The numbering system used to indicate the location of meridians drawn on a globe and measuring distance east and west of the prime meridian (0°).

Map (p. 4) A two-dimensional, or flat, representation of Earth's surface or a portion of it.

Maquiladora (p. 344) Factories built by U.S. companies in Mexico near the U.S. border to take advantage of much lower labor costs in Mexico.

Market area (or hinterland) (p. 388) The area surrounding a central place from which people are attracted to use the place's goods and services.

Medical revolution (p. 58) Medical technology invented in Europe and North America that is diffused to the poorer countries of Latin America, Asia, and Africa. Improved medical practices have eliminated many of the traditional causes of death in poorer countries and enabled more people to live longer and healthier lives.

Mental map (p. 20) A representation of a portion of Earth's surface based on what an individual knows about a place, containing personal impressions of what is in a place and where places are located.

Meridian (p. 15) An arc drawn on a map between the North and South poles.

Metropolitan statistical area (MSA) (p. 425) In the United States, a central city of at least 50,000 population, the county within which the city is located, and adjacent counties meeting one of several tests indicating a functional connection to the central city.

Micropolitan statistical area (p. 425) An urbanized area of between 10,000 and 50,000 inhabitants, the county in which it is found, and adjacent counties tied to the city.

Microstate (p. 242) A state that encompasses a very small land area.

Migration (p. 80) A form of relocation diffusion involving a permanent move to a new location.

Migration transition (p. 84) A change in the migration pattern in a society that results from industrialization, population growth, and other social and economic changes that also produce the demographic transition.

Milkshed (p. 324) The area surrounding a city from which milk is supplied.

Millennium Development Goals (p. 302) Eight goals established by the United Nations to reduce disparities between more developed countries and less developed countries.

Missionary (p. 182) An individual who helps to diffuse a universalizing religion.

Mobility (p. 80) All types of movement from one location to another.
Monotheism (p. 178) The doctrine or belief of the existence of only one god.
More developed country (MDC) (p. 274) A country that has progressed relatively far along a continuum of development.
Multiethnic state (p. 219) A state that contains more than one ethnicity.
Multinational state (p. 219) A state containing two or more ethnic groups with traditions of self-determination that agree to coexist peacefully by recognizing each other as distinct nationalities.
Multiple nuclei model (p. 412) A model of the internal structure of cities in which social groups are arranged around a collection of nodes of activities.
Nationalism (p. 219) Loyalty and devotion to a particular nationality.
Nationality (p. 217) Identification with a group of people who share legal attachment and personal allegiance to a particular place as a result of being born there.
Nation-state (p. 217) A state whose territory corresponds to that occupied by a particular ethnicity that has been transformed into a nationality.
Natural increase rate (NIR) (p. 53) The percentage growth of a population in a year, computed as the crude birth rate minus the crude death rate.
Net migration (p. 80) The difference between the level of immigration and the level of emigration.
New international division of labor (p. 365) The transfer of some types of jobs, especially those requiring low-paid, less-skilled workers, from more developed to less developed countries.
Nonbasic industries (p. 398) Industries that sell their products primarily to consumers in the community.
Nonferrous (p. 450) Metals utilized to make products other than iron and steel.
Nonrenewable energy (p. 441) A source of energy that is in finite supply and thus capable of being exhausted.
Official language (p. 136) The language adopted for use by the government for the conduct of business and publication of documents.
Outsourcing (p. 365) A decision by a corporation to turn over much of the responsibility for production to independent suppliers.
Overpopulation (p. 46) The number of people in an area exceeding the capacity of the environment to support life at a decent standard of living.
Ozone (p. 452) A gas that absorbs ultraviolet solar radiation, found in the stratosphere, a zone between 15 and 50 kilometers (9 to 30 miles) above Earth's surface.
Paddy (p. 320) Malay word for wet rice, commonly but incorrectly used to describe a sawah.
Pagan (p. 182) A follower of a polytheistic religion in ancient times.
Pandemic (p. 71) Disease that occurs over a wide geographic area and affects a very high proportion of the population.
Parallel (p. 15) A circle drawn around the globe parallel to the equator and at right angles to the meridians.
Passive solar energy systems (p. 463) Solar energy systems that collect energy without the use of mechanical devices.
Pastoral nomadism (p. 318) A form of subsistence agriculture based on herding domesticated animals.
Pasture (p. 319) Grass or other plants grown for feeding grazing animals, as well as land used for grazing.
Pattern (p. 33) The geometric or regular arrangement of something in a study area.
Perforated state (p. 248) A state that completely surrounds another one.
Peripheral model (p. 427) A model of North American urban areas consisting of an inner city surrounded by large suburban residential and business areas tied together by a beltway or ring road.
Photochemical smog (p. 454) An atmospheric condition formed through a combination of weather conditions and pollution, especially from motor vehicle emissions.
Photovoltaic cell (p. 463) Solar energy cells, usually made from silicon, that collect solar rays to generate electricity.
Physiological density (p. 33, 51) The number of people per unit of area of arable land, which is land suitable for agriculture.
Pidgin language (p. 162) A form of speech that adopts a simplified grammar and limited vocabulary of a lingua franca, used for communications among speakers of two different languages.
Pilgrimage (p. 185) A journey for religious purposes to a place considered sacred.
Place (p. 5) A specific point on Earth distinguished by a particular character.
Plantation (p. 322) A large farm in tropical and subtropical climates that specializes in the production of one or two crops for sale, usually to a more developed country.
Polder (p. 27) A specific point on Earth distinguished by a particular character.
Pollution (p. 451) The addition of more waste than a resource can accommodate.
Polytheism (p. 178) Belief in or worship of more than one god.
Popular culture (p. 106) Culture found in a large, heterogeneous society that shares certain habits despite differences in other personal characteristics.
Population pyramid (p. 59) A bar graph representing the distribution of population by age and sex.
Possibilism (p. 24) The theory that the physical environment may set limits on human actions but that people have the ability to adjust to the physical environment and choose a course of action from many alternatives.
Post-Fordist production (p. 368) The adoption by companies of flexible work rules, such as the allocation of workers to teams that perform a variety of tasks.
Potential reserve (p. 442) The amount of a resource in deposits not yet identified but thought to exist.
Preservation (p. 468) The maintenance of a resource in its present condition, with as little human impact as possible.
Primary census statistical area (PCSA) (p. 425) In the United States, all of the combined statistical areas plus all of the remaining metropolitan statistical areas and micropolitan statistical areas.
Primary sector (p. 276) The portion of the economy concerned with the direct extraction of materials from Earth's surface, generally through agriculture, although sometimes by mining, fishing, and forestry.
Primate city (p. 393) The largest settlement in a country, if it has more than twice as many people as the second-ranking settlement.
Primate city rule (p. 393) A pattern of settlements in a country such that the largest settlement has more than twice as many people as the second-ranking settlement.
Prime agricultural land (p. 313) The most productive farmland.
Prime meridian (p. 15) The meridian, designated as 0° longitude, that passes through the Royal Observatory at Greenwich, England.
Principal meridian (p. 9) A north-south line designated in the Land Ordinance of 1785 to facilitate the surveying and numbering of townships in the United States.

Productivity (p. 276) The value of a particular product compared to the amount of labor needed to make it.

Projection (p. 8) The system used to transfer locations from Earth's surface to a flat map.

Prorupted state (p. 247) An otherwise compact state with a large projecting extension.

Proven reserve (p. 441) The amount of a resource remaining in discovered deposits.

Public housing (p. 420) Housing owned by the government; in the United States, it is rented to residents with low incomes and the rents are set at 30 percent of the families' incomes.

Public services (p. 376) Services offered by the government to provide security and protection for citizens and businesses.

Pull factor (p. 81) A factor that induces people to move to a new location.

Push factor (p. 81) A factor that induces people to leave old residences.

Quotas (p. 92) In reference to migration, laws that place maximum limits on the number of people who can immigrate to a country each year.

Race (p. 208) Identification with a group of people descended from a common ancestor.

Racism (p. 214) The belief that race is the primary determinant of human traits and capacities and that racial differences produce an inherent superiority of a particular race.

Racist (p. 214) A person who subscribes to the beliefs of racism.

Radioactive waste (p. 459) Materials from a nuclear reaction that emit radiation, contact with which may be harmful or lethal to people; therefore, the materials must be safely stored for thousands of years.

Ranching (p. 326) A form of commercial agriculture in which livestock graze over an extensive area.

Range (of a service) (p. 389) The maximum distance people are willing to travel to use a service.

Rank-size rule (p. 393) A pattern of settlements in a country, such that the *n*th largest settlement is 1/*n* the population of the largest settlement.

Reaper (p. 326) A machine that cuts cereal grain standing in the field.

Recycling (p. 465) The separation, collection, processing, marketing, and reuse of unwanted material.

Redlining (p. 420) A process by which banks draw lines on a map and refuse to lend money to purchase or improve property within the boundaries.

Refugees (p. 81) People who are forced to migrate from their home country and cannot return for fear of persecution because of their race, religion, nationality, membership in a social group, or political opinion.

Region (p. 5) The system used to transfer locations from Earth's surface to a flat map.

Regional (or cultural landscape) studies (p. 17) An approach to geography that emphasizes the relationships among social and physical phenomena in a particular study area.

Relocation diffusion (p. 37) The spread of a feature or trend through bodily movement of people from one place to another.

Remote sensing (p. 9) The acquisition of data about Earth's surface from a satellite orbiting the planet or other long-distance methods.

Renewable energy (p. 441) A resource that has a theoretically unlimited supply and is not depleted when used by humans.

Resource (p. 24, 440) A substance in the environment that is useful to people, is economically and technologically feasible to access, and is socially acceptable to use.

Ridge tillage (p. 331) The system of planting crops on ridge tops in order to reduce farm production costs and promote greater soil conservation.

Right-to-work state (p. 362) A U.S. state that has passed a law preventing a union and company from negotiating a contract that requires workers to join a union as a condition of employment.

Rush hour (p. 433) The four consecutive 15-minute periods in the morning and evening with the heaviest volumes of traffic.

Sanitary landfill (p. 458) A place to deposit solid waste, where a layer of earth is bulldozed over garbage each day to reduce emissions of gases and odors from the decaying trash, to minimize fires, and to discourage vermin.

Sawah (p. 320) A flooded field for growing rice.

Scale (p. 5) Generally, the relationship between the portion of Earth being studied and Earth as a whole; specifically, the relationship between the size of an object on a map and the size of the actual feature on Earth's surface.

Secondary sector (p. 276) The portion of the economy concerned with manufacturing useful products through processing, transforming, and assembling raw materials.

Sect (p. 171) A relatively small group that has broken away from an established denomination.

Section (p. 9) A square normally 1 mile on a side. The Land Ordinance of 1785 divided townships in the United States into 36 sections.

Sector model (p. 411) A model of the internal structure of cities in which social groups are arranged around a series of sectors, or wedges, radiating out from the central business district (CBD).

Self-determination (p. 217) The concept that ethnicities have the right to govern themselves.

Service (p. 374) Any activity that fulfills a human want or need and returns money to those who provide it.

Settlement (p. 374) A permanent collection of buildings and inhabitants.

Sex ratio (p. 60) The number of males per 100 females in the population.

Sharecropper (p. 212) A person who works fields rented from a landowner and pays the rent and repays loans by turning over to the landowner a share of the crops.

Shifting cultivation (p. 314) A form of subsistence agriculture in which people shift activity from one field to another; each field is used for crops for a relatively few years and left fallow for a relatively long period.

Site (p. 14) The physical character of a place.

Site factors (p. 356) Location factors related to the costs of factors of production inside the plant, such as land, labor, and capital.

Situation (p. 14) The location of a place relative to other places.

Situation factors (p. 350) Location factors related to the transportation of materials into and from a factory.

Slash-and-burn agriculture (p. 314) Another name for shifting cultivation, so named because fields are cleared by slashing the vegetation and burning the debris.

Smart growth (p. 429) Legislation and regulations to limit suburban sprawl and preserve farmland.

Social area analysis (p. 412) Statistical analysis used to identify where people of similar living standards, ethnic background, and life style live within an urban area.

Solstice (p. 189) Astronomical event that happens twice each year, when the tilt of Earth's axis is most inclined toward or away from the Sun, causing the Sun's apparent position in the sky to reach it most northernmost or southernmost extreme, and resulting in the shortest and longest days of the year.

Sovereignty (p. 241) The ability of a state to govern its territory free from control of its internal affairs by other states.

Space (p. 5) The physical gap or interval between two objects.

Space-time compression (p. 35) The reduction in the time it takes to diffuse something to a distant place as a result of improved communications and transportation systems.

Spanglish (p. 164) Combination of Spanish and English, spoken by Hispanic Americans.

Sprawl (p. 428) The development of new housing sites at relatively low density and at locations that are not contiguous to the existing built-up area.

Spring wheat (p. 325) Wheat planted in the spring and harvested in the late summer.

Squatter settlement (p. 417) An area within a city in a less developed country in which people illegally establish residences on land they do not own or rent and erect homemade structures.

Standard language (p. 139) The form of a language used for official government business, education, and mass communications.

State (p. 241) An area organized into a political unit and ruled by an established government with control over its internal and foreign affairs.

Stimulus diffusion (p. 39) The spread of an underlying principle, even though a specific characteristic is rejected.

Structural adjustment program (p. 300) Economic policies imposed on less developed countries by international agencies to create conditions encouraging international trade, such as raising taxes, reducing government spending, controlling inflation, selling publicly owned utilities to private corporations, and charging citizens more for services.

Subsistence agriculture (p. 310) Agriculture designed primarily to provide food for direct consumption by the farmer and the farmer's family.

Sustainable agriculture (p. 331) Farming methods that preserve long-term productivity of land and minimize pollution, typically by rotating soil-restoring crops with cash crops and reducing inputs of fertilizer and pesticides.

Sustainable development (p. 467) The level of development that can be maintained in a country without depleting resources to the extent that future generations will be unable to achieve a comparable level of development.

Swidden (p. 315) A patch of land cleared for planting through slashing and burning.

Taboo (p. 112) A restriction on behavior imposed by social custom.

Terroir (p. 114) The contribution of a location's distinctive physical features to the way food tastes.

Tertiary sector (p. 276) The portion of the economy concerned with transportation, communications, and utilities, sometimes extended to the provision of all goods and services to people in exchange for payment.

Textile (p. 356) A fabric made by weaving used in making clothing.

Thresh (p. 320) To beat out grain from stalks by trampling it.

Threshold (p. 389) The minimum number of people needed to support the service.

Toponym (p. 13) The name given to a portion of Earth's surface.

Total fertility rate (TFR) (p. 54) The average number of children a woman will have throughout her childbearing years.

Township (p. 9) A square normally 6 miles on a side. The Land Ordinance of 1785 divided much of the United States into a series of townships.

Transhumance (p. 319) The seasonal migration of livestock between mountains and lowland pastures.

Transnational corporation (p. 30, 299) A company that conducts research, operates factories, and sells products in many countries, not just where its headquarters or shareholders are located.

Triangular slave trade (p. 212) A practice, primarily during the eighteenth century, in which European ships transported slaves from Africa to Caribbean islands, molasses from the Caribbean to Europe, and trade goods from Europe to Africa.

Truck farming (p. 328) Commercial gardening and fruit farming, so named because *truck* was a Middle English word meaning *bartering* or the exchange of commodities.

Unauthorized immigrants (p. 90) People who enter a country without proper documents.

Underclass (p. 421) A group in society prevented from participating in the material benefits of a more developed society because of a variety of social and economic characteristics.

Uneven development (p. 39) The increasing gap in economic conditions between core and peripheral regions as a result of the globalization of the economy.

Unitary state (p. 254) An internal organization of a state that places most power in the hands of central government officials.

Universalizing religion (p. 170) A religion that attempts to appeal to all people, not just those living in a particular location.

Urban renewal (p. 420) A program in which cities identify blighted inner-city neighborhoods, acquire the properties from private owners, relocate the residents and businesses, clear the site, build new roads and utilities, and turn the land over to private developers.

Urbanization (p. 385) An increase in the percentage of the number of people living in urban settlements.

Urbanized area (p. 424) In the United States, a central city plus its contiguous built-up suburbs.

Value added (p. 276) The gross value of the product minus the costs of raw materials and energy.

Vernacular region (p. 19) (or perceptual region) An area that people believe exists as part of their cultural identity.

Voluntary migration (p. 84) Permanent movement undertaken by choice.

Vulgar Latin (p. 147) A form of Latin used in daily conversation by ancient Romans, as opposed to the standard dialect, which was used for official documents.

Wet rice (p. 319) Rice planted on dry land in a nursery and then moved to a deliberately flooded field to promote growth.

Winnow (p. 320) To remove chaff by allowing it to be blown away by the wind.

Winter wheat (p. 325) Wheat planted in the autumn and harvested in the early summer.

Zero population growth (ZPG) (p. 58) A decline of the total fertility rate to the point where the natural increase rate equals zero.

Zoning ordinance (p. 429) A law that limits the permitted uses of land and maximum density of development in a community.

Photo and Illustration Credits

CHAPTER 1 **Chapter opener**, Panoramic Images/Getty Images; Figure **1.9**, Adriao/Wikipedia; Figure **1.10a**, Google Inc; Figure **1.10b**, Nataq/Dreamstime; Figure **1.11a**, Used with permission-Google Maps; Figure **1.11c**, ROSLAN RAHMAN/ Newscom; Figure **1.20**, Kamal Kishore/Corbis; Figure **1.21b**, Ovak Arslanian/Getty Images; Figure **1.22b**, Mark Randall/ Newscom; Figure **1.24**, Jack Dabaghian/Corbis; Figure **1.27**, Comstock/JupiterImagesUnlimited; Figure **1.28**, David Paul Morris/Getty Images; Figure **1.32**, Photo Researchers; Figure **1.33**, David Butow/CORBIS SABA

CHAPTER 2 **Chapter opener**, Timothy Clary/Newscom; Figure **2.20**, TIZIANA FABI/Getty Images; Figure **2.26**, Magnum Photos; Figure **2.29**, RAVEENDRAN/Getty Images; Figure **2.34**, Newscom

CHAPTER 3 **Chapter opener**, Peter Parks/Newscom; Figure **3.1**, Corbis/Bettmann; Figure **3.3**, Alison Wright/Corbis; Figure **3.4**, Bettman/Corbis; Figure **3.6**, Omar TORRES/Newscom; Figure **3.10**, Google Inc.; Figure **3.11**, Melvin Levine/Getty Images; Figure **3.13**, David McNew/Getty Images; Figure **3.14**, Jonathan S. Blair/Getty Images; Figure **3.15**, Bettmann/Corbis; Figure **3.17**, Bettmann/CORBIS; Figure **3.19**, Dmitri Kessel/Getty Images

CHAPTER 4 **Chapter opener**, Gavin Hellier/Getty Images; Figure **4.1**, DAI KUROKAWA/Corbis; Figure **4.4b**, Joe Schneid/ Wikipedia; Figure **4.5**, Ully Bleil/Corbis; Figure **4.7**, David Silverman/Getty Images; Figure **4.20**, Petr Svarc/Alamy; Figure **4.22**, Paul Chesley; Figure **4.23**, Ann Johansson/Corbis; Figure **4.25**, Donovan Reese/Getty Images

CHAPTER 5 **Chapter opener**, Lo Mak/Corbis; Figure **5.1**, BAZUKI MUHAMMAD/Corbis; Figure **5.4**, Leonard Mccombe/Getty Images; Figure **5.6**, Pawel Libera/CORBIS; Figure **5.13**, Corbis/ Bettmann; Figure **5.21**, David Silverman/Getty Images; Figure **5.22**, David Lefranc/Corbis; Figure **5.23b**, Mark Renders/Getty Images; Figure **5.25**, Paul A. Souders/CORBIS; Figure **5.28**, Guillaume Lavallee/Getty Images; Figure **5.29**, Newscom

CHAPTER 6 **Chapter opener**, Akhtar Soomro/Corbis; Figure **6.4**, MADAREE TOHLALA/Getty Images; Figure **6.5**, DESHAKALYAN CHOWDHURY/Newscom; Figure **6.6**, ABIR SULTAN/Corbis; Figure **6.7**, Ahmad Faizal Yahya/Dreamstime; Figure **6.13**, Kaetana/Shutterstock; Figure **6.14**, Google Maps; Figure **6.15**, ayazad/Shutterstock; Figure **6.16**, Wojciech Zwierzynski/ istockphoto; Figure **6.18**, RONEN ZVULUN/ Corbis; Figure **6.19**, Danny Lehman/Corbis; Figure **6.20**, Magnum Photos; Figure **6.24b**, Getty Images

CHAPTER 7 **Chapter opener**, ALEXANDER JOE/Newscom; Figure **7.7**, The New-York Historical Society; Figure **7.9**, Courtesy of Library of Congress; Figure **7.12**, Corbis/Bettmann; Figure **7.13**, Corbis/Sygma/Corbis/Sygma; Figure **7.16**, Brykaylo Yuriy/ Shutterstock; Figure **7.20**, Zurab Kurtsikidze/Corbis; Figure **7.22**, Reuters/Corbis; Figure **7.24**, Patrick Robert/Corbis; Figure **7.32**, CORBIS- NY/CORBIS- NY/Live Forever/Wikipedia

CHAPTER 8 **Chapter opener**, Gian Luca Faccio/Corbis; Figure **8.3**, ABDELHAK SENNA/Getty Images; Figure **8.4**, Dave Pape/ Wikipedia; Figure **8.7**, Hulton Archive/Getty Images; Figure **8.12**, Winky/Wikipedia; Figure **8.13**, Damiano Luchetti/Wikipedia; Figure **8.15b**, ALEX MITA/AFP/Getty Images; Figure **8.17**, Feifei Liang/istockphoto; Figure **8.24**, Reza/Getty Images; Figure **8.30**, Jim Young/Newscom

CHAPTER 9 **Chapter opener**, Per-Anders Pettersson/Getty Images; Figure **9.13a**, Luis Costa/Newscom; Figure **9.13b**, Frederic J. BROWN/Newscom; Figure **9.14a**, DENIS SINYAKOV/AFP/ Newscom; Figure **9.14b**, Vitaly Belousov/Newscom; Figure **9.14c**, ADI WEDA/Corbis, Figure **9.15a**, NOAH SEELAM/Getty Images; Figure **9.15b**, Gideon Mendel/Corbis; Figure **9.29**, Bay ISMOYO/ Newscom; Figure **9.33**, Philippe Lissac/Corbis

CHAPTER 10 **Chapter opener**, Panoramic Images/Getty Images; Figure **10.1**, Louise Gubb/Corbis; Figure **10.6**, Cotton Coulson/Woodfin Camp&Associates; Figure **10.8**, AFP Pool/ Getty Images; Figure **10.9**, Frontpage/Shutterstock; Figure **10.10**, Dmytro Korolov/istockphoto; Figure **10.11**, Steve Estvanik/ dreamstime; Figure **10.13**, Hywit Dimyadi/Shutterstock; Figure **10.14**, Scott Sinklier/Corbis; Figure **10.16**, James p. Blair/ Corbis; Figure **10.18**, Jupiterimagesunlimited; Figure **10.20**, Patrick Dove/Newscom; Figure **10.22**, Samuel Aranda/Getty Images; Figure **10.23**, David H. Wells/CORBIS; Figure **10.25**, Macduff Everton/CORBIS; Figure **10.26**, YOSHIKAZU TSUNO/ Getty Images; Figure **10.30**, JES AZNAR/Newscom

CHAPTER 11 **Chapter opener**, PoodlesRock/CORBIS; Figure **11.10**, Hulton Archive/Getty Images; Figure **11.13**, Peter Endig/Corbis; Figure **11.14**, Vstock LK/Alamy; Figure **11.15**, donvictorio/shutterstock; Figure **11.16c**, Joerg Boethling/Alamy; Figure **11.17c**, Brownie Harris/CORBIS; Figure **11.18c**, Ted Soqui/Corbis; Figure **11.19**, Yonhap/Corbis; Figure **11.20**, James L. Amos/Corbis

CHAPTER 12 **Chapter opener**, MIKE CLARKE/Getty Images; Figure **12.5**, Nik Wheeler/CORBIS; Figure **12.6**, SIMELA PANTZARTZI/Corbis; Figure **12.7**, Alison Cornford-matheson/ Dreamstime; Figure **12.8**, Adam Woolfitt/Corbis; Figure **12.11**, Bettman/Corbis; Figure **12.12**, Jonathan Blair/Corbis; Figure **12.13**, Pegaz/Alamy; Figure **12.24**, Jack Sullivan/Alamy; Figure **12.27**, Belinda Lawley/Panos Pictures

CHAPTER 13 **Chapter opener**, David Brabyn/Corbis; Figure **13.2**, Michael Nagle/Corbis; Figure **13.3**, Atlantide Phototravel/Corbis; Figure **13.10b**, Jon Arnold/Alamy; Figure **13.10c**, Michel Setboun/Corbis; Figure **13.11a**, De Agostini Picture Library/ Getty Images; Figure **13.11b**, De Agostini Picture Library/Getty Images; Figure **13.12**, Aerial Archives/Alamy; Figure **13.13a**, Tom Brakefield/ Getty Images; Figure **13.15b**, luoman/istockphoto; Figure **13.27**, Gary Woods/Loop Images/Corbis; Figure **13.29a**, Spencer Platt/ Getty Images; Figure **13.29b**, Joseph Sohm-Visions of America/ Getty Images

CHAPTER 14 **Chapter opener**, Panoramic Images/Getty Images; Figure 14.9, Craig Aurness/Woodfin Camp & Associates; Figure 14.14, EIGHTFISH/Alamy; Figure 14.16, Peter Bennetts/Getty Images; Figure 14.17, Oxford Scientific/Getty Images; Figure 14.20, PIYAL ADHIKARY/Corbis; Figure 14.21d, NASA/Corbis; Figure 14.22, Penny Tweedie / Alamy; Figure 14.25, MAXIM KNIAZKOV/Getty Images; Figure 14.27, AFP Photo/Newscom; Figure 14.29, Pallava Bagla/Corbis; Figure 14.31, ALEX HOFFORD/Corbis; Figure 14.35, PRNewsFoto/Newscom

Afterword Figure A.1, KAREN BLEIER/Getty Images; Figure A.2, Bettman/Corbis

Map Index

Index